ISBN 978-0-331-37719-4
PIBN 11114235

nt.

A SYSTEMATIC CATALOGUE OF BRITISH INSECTS:

BEING

AN ATTEMPT TO ARRANGE ALL THE HITHERTO DISCOVERED INDIGENOUS INSECTS IN ACCORDANCE WITH THEIR NATURAL AFFINITIES.

CONTAINING ALSO

THE REFERENCES TO EVERY ENGLISH WRITER ON ENTOMOLOGY, AND TO THE PRINCIPAL FOREIGN AUTHORS.

WITH ALL THE PUBLISHED BRITISH GENERA TO THE PRESENT TIME.

BY

JAMES FRANCIS STEPHENS, F.L.& Z.S. &c.

"...... *Ten thousand* different tribes
People the blaze." *Thomson.*

LONDON:
PUBLISHED FOR THE AUTHOR,
BY BALDWIN AND CRADOCK, PATERNOSTER-ROW.
1829.

LONDON:
PRINTED BY RICHARD TAYLOR,
RED LION COURT, FLEET-STREET.

INTRODUCTION.

THE following Catalogue was originally designed to have appeared as the precursor of my "Illustrations of British Entomology," but the magnitude of the undertaking, as well as the necessity for previously commencing the latter work, so far frustrated my designs as to render it precursive in part only.

I have been induced to publish it in its present form (as the work above alluded tò will not only require several years for its elaboration and completion, but must necessarily be proportionally more expensive and bulky,) in order to meet the wishes of several entomological friends, who have repeatedly solicited me to furnish them with an arrangement of all our indigenous insects, from the circumstance that scarcely one-third have yet been commemorated as British; and the works hitherto published on the subject are either confined to particular orders, families, or genera, or are so limited when embracing all the orders, and are nearly all, moreover, either arranged in conformity with the Linnean system, or totally destitute of arrangement, that they render but little information to the student who is desirous of obtaining a more perfect knowledge of the classification, either of the genera or of the species.

Although fully aware of the objections which have been so frequently urged by the most learned entomologists against the publication of mere catalogues, I feel confident that the plan and object of the following will so far have their utility, as at least to disarm my compatriots of the opposition *they* have hitherto exhibited towards similar performances, as in fact it may be considered as an appendage to my "Illustrations;" and it cannot have escaped the notice of the most cursory inspector of entomological collections, that the nomenclature of each differs in numberless instances, and that *various names* are applied in different cabinets to the *same identical species.* Such confusion arises from several causes; primarily from the difficulty there is in many instances of correctly ascertaining the name given by the first describer, either from the description being so vague and indefinite as to preclude the possibility of accurately determining the species intended, or from the impracticability of obtaining the books in which the first description is to be found; secondarily, from the contemporaneous appearance of two or more publications upon the same group of insects, in which the new species are necessarily (unless by the merest accident) called by different names; and lastly from the dissimilar MS. *names* which have been applied by various practical entomologists to such species as they have not been able to identify in any printed works. An endeavour to assimilate such discrepancies, and also to render the labours of my countrymen more generally useful, prompted me to attempt the synonymical portion of the Catalogue, in which I have also introduced such MS. names as are in general use in the more extensive collections, so far as circumstances will permit; and also those which I have applied to the *apparently undescribed* species in my own cabinet*, by way of showing the present extent of the various groups.

Upon the propriety of introducing the MS. names either of genera or species, there is, and probably ever will be, considerable diversity of opinion. Latreille exclaims against the publication of mere generic names without characters, in the following terms: "Je remarque que plusieurs naturalistes s'empressent, comme par une anticipation titulaire, de donner des noms à quelques coupes, qui leur paroissent devoir former de nouveaux genres, sans le donner la peine d'en établir les caractères. Ces ne sont que de simple indications, et qui n'imposent aucune loi." But this celebrated entomologist appears to forget that naturalists generally, however zealous in the pursuit of knowledge, have other and more paramount avocations to follow than those of attending to the minutiæ of science, and that all are not equally fortunate with himself in being able to devote their exclusive attention to that branch thereof which their inclination prompts them to study. An individual thus necessarily occupied with the ordinary callings of life, and devotedly attached to the study of natural objects, discovers while arranging a group of animals, a specimen which differs so essentially from the rest in general characters, as to satisfy him that it belongs to a genus whose characters have not been described, and the elaboration of which his want of time *alone* prevents him from attempting: nevertheless he is desirous of giving it a name by which he may more readily remember it, or register any peculiarities he may have observed relative to its habits, locality, &c., than by attaching a mark or number thereto. Is he not therefore entitled to the merit (if such it be) of discovering the new group; and ought he not to be permitted to apply a name to it, and to publish that name even before his leisure allows him to elaborate its characters? M. Latreille †, I grant, does not deny such right; but merely infers that

* Excepting to the Hymenoptera Pupivora, for the reasons stated in the note at p. 343, to the last section of that order; and to the Hemiptera and Homoptera, from want of time to investigate their contents sufficiently: the number of species in the respective genera is however indicated.

† This author in his last work, "Familles Naturelles," has himself introduced names without distinguishing characters.

the name may be adopted or not, at the pleasure of the individual whose time permits him to describe its characters, and thereby stamp an additional value thereon, by rendering it more generally known.

With regard to the MS. names in the following pages, it is of little consequence whether those which I have proposed be adopted or not; as they will ultimately appear in my "Illustrations," unless the species to which they are applied should be described by others in the interim. It is sufficient for my purpose to have pointed out the new species, and the divisions (by whatever names caprice or *convenience* may please to term them), and to have retained *all* such as could satisfactorily be identified, which have been imposed by others; not only *in justice* to their authors, but as I do not choose *wilfully* to create additional inconvenience by applying new ones for the mere purpose of securing the paltry fame dependent upon mere priority of nomenclature.

Now with reference to such simple indications, as Latreille terms them, not imposing any law, as to the adoption of the generic or specific names, proposed by others without characters, I would inquire whether such names (at least those of genera), especially when applied to well known species, do not as satisfactorily point out the new groups, as when the characters are given in a language peculiar to one nation alone, and unknown to a foreign student; e. g. the genera of Lepidoptera by Ochsenheimer and Treitschke are invariably characterized in German; and many of Latreille's recent ones in French alone, and are consequently useless to the mere English student, as are those of the Englishman or Frenchman to the German, when given in their respective languages. I therefore conceive that if writers are to be governed by the Latreillian precept, the characters to be efficient should be given in that language which is more or less known to all men of science throughout the civilized world, notwithstanding it has been remarked that a work is *more generally* useful for being *entirely written* in English! But, as I before stated, I do not choose unnecessarily to incumber the science with additional names; I have considered the genera of Ochsenheimer, &c. here alluded to as sufficiently characterized, though my ignorance of German prevents me from correctly ascertaining their typical species; and I have also endeavoured, so far as is consistent with its nature, to show how my predecessors have cleared the way towards the ascertaining of the new indigenous species, by referring to *every* English writer, and to such foreign ones as lay within my reach, including even such names as have hitherto appeared in the various German and French catalogues when known to me by the inspection of original specimens transmitted to this country by the respective authors.

By the foregoing remarks it may be supposed that I am laying too much stress upon nomenclature; but as Linné says, "Nomina si periit, periit et cognitio rerum;" and without names, how can any one communicate to another the knowledge he has acquired relative to any particular fact that he may have observed, either of physiology, habit, utility, or locality*? Is he upon the discovery of an injurious or useful property belonging to any species of whose name he *chooses* to remain in ignorance, to give a detailed description thereof when he wishes to develop its qualities? Such an antiquated proceeding, at this period, would doubtless tend to disparage the science, and to justify the appellation of a zoological writer, who terms the study of insects "specious idleness†!" But as this subject has been fully discussed by others, I shall merely add that it appears to me very great advantage must accrue from having all natural objects distinguished, and as far as possible by appropriate names.

The arrangement I have adopted, though obviously defective, appears to me more consonant to Nature in its details than any that has hitherto been promulgated: but my very limited acquaintance with exotic groups will inevitably lead me into several errors, as every attempt to classify the productions of a limited spot must necessarily be extremely imperfect, especially the first.

With reference to one of the alterations from the Latreillian system which I have published in my "Illustrations," it has been truly remarked "that no *good system* ought to be disturbed without solid reasons:" it is therefore merely necessary to show the imper-

---

* The following anecdote admirably illustrates the value of a knowledge of nomenclature, especially of synonymy:—Two individuals, residing upwards of two hundred miles apart, in the course of their correspondence communicated to each other the intelligence that their respective collections had received the accession of two specimens of *Achatia piniperda* (part ii. p. 87.); one of them calling the insect by the name of *Noctua piniperda*, the other by that of *Bombyx spreta*. This was followed by a request on the one part to exchange one specimen of the former for one of the latter, whereby each cabinet might be mutually benefited by the addition of a new species! The result (as may be anticipated) was far from satisfactory: the two specimens, after the exchange was ratified, proving to be of the same species; one party having sacrificed a fine and beautiful example, reared from the larva, for an injured one captured by mothing, and both parties having been put to some expense without obtaining their object.

† In the Monthly Epitome for February 1825 are some remarks upon this point, contained in a notice of Professor Kidd's anatomy of the Gryllus gryllotalpa, Linné (Gryllotalpa vulgaris, p. 303.), which tend to show the feeling of literary persons regarding the study of zoology, in which, concernedly do I add, the writer is supported in his illiberal sentiments by the learned editor, who seems, in the warmth of his zeal for his own favourite pursuit, to imagine that all sciences should succumb to the hypercritical acumen of the elocutionist; and that the study of enunciation is of infinitely more importance than an insight into the structure of the wonderful and multifarious works of the divine Creator and disposer of all things.

fection of *the good systems* already promulgated, by the following succession of groups extracted from the last work of Latreille,—whose system, alluded to in the above quotation, has been termed, "par excellence," the "*Natural System*,"—the names of the typical genera being given for the sake of brevity:—Cicindela, Carabus, Dyticus, Gyrinus, Staphylinus, Buprestis, Elater, Lampyris, Clerus, Ptinus, Hister, Silpha, Scydmænus, Dermestes, Byrrhus, Heterocerus, Hydrophilus, Sphæridium, Scarabæus, Lucanus; thence to Blaps, &c.—which in the majority of instances have no affinity with each other, beyond the mere circumstance of possessing five joints to all the tarsi; a fact so notoriously manifest, that I shall not waste the reader's time by attempting to show it. Now as Latreille professes to arrange his subjects according to their organization and affinities, I am at a loss to conceive why the Buprestidæ, &c. are interposed between the Dyticidæ and the Hydrophilidæ in all his works; an arrangement that the merest tyro in entomology must perceive cannot be natural. Many other defects in his "good system" might be pointed out with facility without the aid of figures, by the *only correct* standard—that of Nature herself; but sufficient has been adduced to show that even the system of the "first of entomologists" may be advantageously revised. I do not however pretend to be capable of executing so arduous a task; nevertheless, as it is possible that others may benefit from a fortuitous hit in the position of some of the groups, I have ventured to remove several genera from situations that experience has convinced me were untenable: and notwithstanding the remark of Mr. Bicheno, that "he who is constantly dwelling upon particulars, and following them out to still minuter elements, is taking the opposite course to another, whose object is combination and enlargement,"—I contend that the "analysis of species, which after all is our chief business," must conduce towards the attainment of the knowledge of the true natural system.

It is almost needless to add, that I have gleaned much valuable information from the erudite labours of Mr. MacLeay, in his valuable Horæ Entomologicæ and Annulosa Javanica, and that I have endeavoured, though at a very humble distance, and without the necessary acquaintance with exotic groups, to follow his steps; as I am convinced that natural objects cannot be arranged agreeably to their affinities, otherwise than by a series of circles "returning," as the above author expresses it, "into themselves," although I continue sceptical as to the quinary arrangement being universal throughout Nature.—With such views I have attempted the following Catalogue, the grouping of which may be more readily comprehended by means of the subjoined tables.

MANDIBULATA.
I. Coleoptera.
II. Dermaptera.
III. Orthoptera.
IV. Neuroptera.
V. Trichoptera.
VI. Hymenoptera.
VII. Strepsiptera.

Diploptera et Fossores, Latr.

Anthophila, Latr. VI. Heterogyna, Latr.

VII. Parasitica : Pupophaga, Latr.
Terebrantia, Latr.

Adephaga, Latr. . Rypophaga mihi . . . Serricornes, Latr. Lamellicornes, Latr. . . . Helocera, Dum. Clavicornes, Latr. Imbricornes, Latr.

Brachelytra. I. Longicornes, Latr. . Rhinchophori, Latr.

Heteromera p, Latr. Trimeri, Dum. Cyclici, Latr. Eupodina, Latr.

II. Blattina. III. Gryllina. Locustina.

Megaloptera, MacL. . . . Termitina, MacL.
Panorpina, MacL. IV. Libellulina, MacL.
Anisoptera, Lea.

HAUSTELLATA.

Homoptera. Hemiptera.
V . . . Lepidoptera. Aptera.
Diptera. Aphaniptera.
Homaloptera.

In order to render the views I have adopted more distinct, some of the details of the Mandibulata are introduced; though I must remark that a great mass of the Coleoptera, comprising most of the Heteromera of Latreille, seems to require extensive removals; but whether the Œdemeridæ imitate the Telephoridæ and Melyridæ by any real affinity, or merely by the property which opposite points of the circle possess of resembling each other, I am not prepared to show: it is however supposed that these families are allied:—the situation of the Scydmænidæ is very doubtful.

Some of the remarkable affinities in the Order Coleoptera may be still further exemplified by the following more detailed tabular view of the first seven groups; which it is unnecessary to observe, might be rendered still more complete were it not that want of space necessarily obliges me to be concise.

Dyti-... 88 cus. Gyri- . nus. . . Haliplus.	110. Hetero- Sphæri- { cerus. dium. { Parnus. Hydro- . philus. 93 . Helophorus.	Buprestis. ——— Cetonia.	
Bembidium. . . Ela- 57 phrus. . Carabus. Cicindela. . 11 Dromius.	. 120 . . Silpha . . Dermes- . tes. 126 . . 164 . 149	Sino- dendron. Trox 180	Bostri- chus. Hylur- gus.
Lesteva. Oma- lium.	Nitidu- Micro- la. . . peplus:	Byrrhus 179 Abræus. Hister.	Rhin- colus.

By this view the affinities of Dromius to Lesteva, Elaphrus to Cicindela, Bembidium to Haliplus?, Gyrinus to Heterocerus and Parnus, Helophorus to Silpha, Micropeplus to Abræus, Byrrhus to Trox, Sinodendron to Bostrichus, Hylurgus to Rhincolus, Omalium to Nitidula, &c. are rendered evident, notwithstanding in a linear arrangement several hundreds or even thousands of species necessarily intervene, and *apparently* disturb the series, thereby exemplifying the utter impracticability of placing the whole in one continuous succession*.

* The figures in the table point out the position of other genera, which are introduced in order to render the circular distribution of each group more evident; the numbers corresponding with those of the respective genera in the Catalogue, as do those in the first and last of the following tables.

I shall now introduce a table of the distribution of the Lepidoptera.

Ægeriidæ : Sesiidæ. Papilionidæ : Nymphalidæ.
. II. Sphingidæ : Hesperiidæ. I. .
Zygænidæ. Lycænidæ.

Notodontidæ : Hepialidæ. Geometridæ : Platyptericidæ.
. III. . V.*
Bombycidæ : Arctiidæ. . Pyralidæ.

. . .
. Lithosiidæ : Tineidæ. Tortricidæ.
Noctuidæ. IV. . VI.* Yponomeutidæ.
Pterophoridæ.

My limits will not, however, permit me to dilate upon the above; I shall therefore merely remark, that it appears to me the Platyptericidæ are far more intimately allied to the Lep.-semidiurna than to the Lep.-pomeridiana; their affinity to the former in their perfect state being evident from some peculiar characters they possess in common with the conterminous genera of the Geometridæ and Pyralidæ, as I shall elsewhere have occasion to notice; while their resemblance in their larva state to the Lep.-pomeridiana I conceive to be analogical, and to result from their relative positions in the circle.

I have now to refer to the two remaining tables, in order to show how nearly the arrangement of the Lep.-pomeridiana which I have adopted, from a consideration of the perfect insect, accords with Dr. Horsfield's sketch deduced from the larva.

Pygæra. | Hepialus.
Notodonta. II. | Cossus. I.
Endromis. | Zeuzera : *Oiketicos.*
. Aglia. | Psyche.

Saturnia. III. Limacodes. Nudaria.

Lasiocampa. Eutricha. | Penthophera.
IV. Cnethocampa | : Dasychira. V. Arctia.
Eriogaster. | Hypercampa.

The above represents the five groups which Dr. Horsfield appears to recognise in his "Descriptive Catalogue;" and the subjoined, the four into which I have divided the same subjects in the following pages.

* Strictly, these two sections ought to have been reversed in the following arrangement as here given; but not being prepared to enter into details, I preferred a slight modification of the usual beaten track, rather than its total subversion.

		Pygæra.			Hepialus.
Notodonta.	II.		Cossus.	I.	
	Endromis.			Zeuzera :	*Oiketicos.*
....	....	....	....	....	....
		Aglia.		Psyche.	
Saturnia.			Limacodes.		Nudaria.
Lasiocampa.	III.	Eutricha.		IV.	Penthoptera.
		Cnethocampa :	Dasychira.		Arctia.
Eriogaster.				Hypercampa.	
				......┼......	
				Callimorpha.	

Upon an inspection of the above tables, it will be manifest that the principal difference between the two methods exists in the conterminous genera alone; though it must be added that Dr. Horsfield has not entered into details as to the succession of his groups, or their affinities to each other; neither does the first table exhibit the five groups in accordance with the schemes employed by Mr. MacLeay; my object being simply to show that the difference in our arrangements is in reality more apparent than real, and seems to depend upon the relative position of the lines of separation, as indicated by the dots;—a fact which is again observable at that point of the fourth circle which impinges upon the Lithosiidæ, and consequently renders the position of the genus Hypercampa debateable.

The number of genera herein given will doubtless alarm many persons: it however seems to me, that the extensive and multifarious nature of the subject demands the adoption of such as I have *indicated* in the following pages *, without being shackled by *numerical* convenience. Upon this point there will probably always exist considerable diversity of opinion; but it is allowed by all parties, that some addition should be made to the genera instituted, as adopted by Linnæus, though the followers of that great man rest contented by admitting a few new genera only, and totally disregard *every* attempt to increase the number; stating as reasons, that they merely tend to burthen the memory unnecessarily, and to introduce confusion into the science. Another reason assigned by them for not introducing new genera into their systems, is, that many of them contain one species only; not considering the very confined knowledge we have of the latter, and the vast accessions

* With regard to the number of genera, it has been well observed by Mr. MacLeay, that writers are frequently guided in their views by the extent of their collections; persons with small collections not seeing that necessity for subdivision which those with larger ones think expedient. It may be added, that a writer, who most strenuously opposes the formation of new genera, has adopted, without restriction, *all* the genera which I have proposed in my "Illustrations," to the Sphæridiidæ, and which are correspondent with those in the following pages; and that other writers who have succeeded me in the Haustellata have subdivided some of the genera therein given.

to their numbers that are daily arising from abroad, and even discovered in this country alone: for if they were, they would find that many of the genera, formerly supposed to contain one species only, have received such additions since Linné's time, as to have their numbers augmented to a considerable extent: they would also repeatedly find, that not a single species even is noticed by him in the "Systema Naturæ," of many of the modern genera which are now known to contain a great number, especially amongst the Carabidæ and Staphylinidæ; nay, the entire and singular family of Pselaphidæ is totally unnoticed by him. These are points, however, that few persons trouble themselves to investigate; the majority of entomologists taking all for granted that they see advanced in print, or meeting every argument in support of new genera by some sweeping conclusion against their adoption*: and I have frequently heard it asserted, that modern entomologists have run to such excessive lengths in their career, as to have separated the individuals belonging to *one* variable *species* into *three* genera! and that it is no uncommon occurrence to find the sexes placed in two!

Now as I never found an individual who could point out a single instance of the former allegation, nor have I ever noticed the fact myself, I conceive no refutation is necessary; but with regard to the second, there may be some plausibility in the objection, as instances of such a proceeding are upon record:—e. g. in the delightful "Horæ Entomologicæ," p. 98, the learned author has given the female of Pholidotus as a new genus, by the name of Casignetus, accompanied by a query whether it ought to be considered distinct: but in this, as in other examples, the mistake has arisen from a laudable desire to be accurate, and not with a view to attain reputation by the mere imposition of a generic name; and the author has promptly corrected his error in his second part. Other examples might be adduced, but the above is fully sufficient to justify the remarks of the objectors when applied to the modern innovators, as they are pleased to term them. Yet the moderns are not alone guilty of such misconceptions: will not therefore these objectors be inclined to palliate such venial errors as the one referred to, when the justly-celebrated Linné is shown to have committed similar ones? This "great master," in the last edition of his "Systema Naturæ," and "Fauna Suecica," has placed the sexes of a well known and common little beetle (the Cercyon quisquilium, p. 65.) in *two* different genera, calling the male *Scarabæus* quis-

* Amongst the absurdities that I have heard advanced in opposition to new genera, a naturalist of some eminence once asserted that a genus in question could not be a good one because the name applied to it was one that Linnæus would not have used! He might as well have asserted that there were no new objects in nature because Linnæus had not supplied names for them.

quilius (Syst. Nat. ii. 558.—*Faun.* Suec. No. 397.), and the female *Coccinella!* unipunctata (Syst. Nat. ii. 579.—*Faun.* Suec. No. 470.); which is the more to be wondered at, as, from the insect being far from uncommon in Sweden, he might have had frequent opportunities of detecting the identity of the sexes. But in the instance noticed from Mr. MacLeay, the insect in question is a native of South America, and consequently the writer had no means of ascertaining any portion of its history, but was subsequently indebted to the observations of Mr. Swainson, whose zeal for entomology led him to explore the distant region of Brazil, for correcting the mistake into which he had accidentally fallen*. But as I have elsewhere advocated the expediency of generic division, I shall proceed to give some cursory observations upon species, and upon the difficulty of discriminating them. It cannot be denied that much obscurity prevails upon these subjects: it therefore becomes the duty of every inquirer to endeavour to dispel the mist, and to contribute his mite towards their elucidation; though I do not assume that such should be the sole object of every one. I would rather that attention should be paid to the economy, &c. of the interesting objects of the entomologist's research; though I would ask, how can an individual communicate to others the knowledge he has acquired by his observations, unless he be able to define with accuracy the *very* animal which furnished it? Upon this consideration, therefore, it is manifest that an examination into the true distinctive characters of species should occupy at least *a portion* of his time, instead of devoting it almost exclusively to searching after the relationship of affinities,—which has been so much insisted on by late writers, almost to the total exclusion of specific investigation; or in generalizing upon structure before he is acquainted with particulars.

In pursuing this investigation, the first question which presents itself is, In what do the characteristics of a species consist? I need scarcely add, that its solution is one of the utmost difficulty, as the characters of many species are very obscure, and are not readily detected, owing to the innumerable variations to which they are subject from age, locality, or climate, or from the close approximation of those in kindred species. Of the latter description, the genus Cucullia (part ii. p. 102.) offers a remarkable example; most of the species of which it is composed resembling each other so closely in their final state, that were it not from a knowledge of them during their previous appearance (their larvæ being totally dissimilar, and their food and time of flight very different), they might

* A more recent example, however, occurs in a work which assumes superior accuracy in the *generic* distribution of insects; in which the Musca putris of Linné is converted into *three* species belonging to *two* genera. Vide part ii. p. 319. Nos. 9126 and 9140.

still continue, as they formerly were, to be considered as mere varieties of one inconstant species.

Again, Acronycta (another Lepidopterous genus) contains also similar examples; among which may be pointed out the two well-known insects A. Psi and A. tridens, which approximate so closely to each other in the winged state, and moreover are each subject to such considerable variations, that none but an experienced entomologist can reduce all the varieties to their proper species; yet their larvæ are singularly dissimilar.

Many similar examples might be cited amongst the Coleoptera, especially among the Protean genus Pœcilus, the species of which are so remarkably alike, and so prone to vary, that even the most acute entomologists err upon the point of their specific differences, from not attending sufficiently to their minutiæ.

It has been observed, that the chief points of specific distinction consist of "form, sculpture, and disposition of colour." But a very slight acquaintance with the works of Nature will show that each of these characters must be handled with the greatest circumspection, as they are all subject to much variation: however, the first is the most stable; sculpture the next so, though in many of the Coleoptera a considerable deviation from the usual standard takes place, as instanced in the Geodephaga generally*, and more particularly so in Carabus monilis: slight variations of this character also occur in some of the other Orders. Disposition of colour is, however, a most inconstant specific diagnostic, though its various shades and gradations admirably distinguish the larger groups, as I have elsewhere taken occasion to notice, and as Mr. Kirby has indicated amongst the Geodephaga. Do we not, for example, find the Dyticidæ characterized by prevalent tints of brown, fuscous, or olive, rarely with metallic shades? and is not the succeeding family (the Gyrinidæ) usually of a deep leaden hue, and mostly glossed with æneous? and are not the Chrysomelidæ eminently distinguished by the splendour of their brilliant golden or cupreous tints, most frequently upon a green or bluish ground? while in the neighbouring family of Coccinellidæ, the species of which *seemingly* differ greatly in colour, all metallic splendour vanishes, and the only tints of colour which occur are black and white, with various combinations of red and yellow. Numerous other examples might be pointed out, as may be readily seen by an inspection of several of the groups in this Catalogue. There are, however, many species, especially amongst the Geodephaga and the Crioceridæ, that partake of nearly every colour of the rainbow: Donacia Proteus,

---

* The monstrous variations in the sculpture of the elytra,—such as the interrupted confluent striæ, or the rugose crenulated appearance of the surface,—are clearly not indicative of specific difference, although such has been employed. Vide Nos. 190, 196, &c.

for instance, varying in different individuals by almost imperceptible changes, from coppery-red to gold-colour, bronze, green of various tints, bright blue, purple, violaceous and even intense metallic black. The same may be remarked of all the Pœcili, several of the Carabi, Harpali, &c.; and it may be added, that all metallic greens have a tendency in some specimens to become blue, and *vice versâ*. Numberless similar examples might be adduced; but enough has been said to show the slight dependence to be placed on colour alone in the detection of species. Some very extraordinary discrepancies in the Lepidoptera may, however, be pointed out. In this Order, the chief distinguishing specific marks consist in the position and number of the various coloured lines, fasciæ, ocellated or simple spots, &c. which adorn the wings; yet such remarkable changes often take place in their disposition, as to be tantamount to a variation in colour:—e. g. Abraxas grossulariata, in its usual appearance, has the wings white, with spots and streaks of black, and two orange fasciæ on the anterior ones,—one at the base, and the other towards the posterior margin; but specimens have occurred totally suffused with black, with the usual black markings of a darker hue, and the orange fasciæ barely perceptible; others have been found nearly of an immaculate white; while occasionally this latter colour is totally obliterated, and a yellow or orange tint adorns the wings. Of the latter description, I once saw an example, captured near Edinburgh, nearly of as brilliant an orange as Colias Edusa, and but sparingly maculated with black. Arctia Caja offers another conspicuous example: in this insect the usual sinuated white or cream-coloured strigæ on the anterior wings sometimes become changed to a deep fuscous, while the rest of the wing becomes of a paler colour than usual: and I possess a specimen in which the anterior wings are white speckled with brown, instead of being brown or black with sinuated white streaks.

Among the more conspicuous changes of colour in this Order may be mentioned that of Colias Edusa, which is usually of a bright fulvous orange; but specimens occur nearly of a pure white *. Again, Melitæa Euphrosyne also varies from deep tawny to pale luteous. Lycæna Phlæas is sometimes found of a clear glossy white, in lieu of its brilliant fiery copper-colour. The well known "illuminated females," as they are termed, of the Polyommati may also be cited as examples of the variation of colour; and P. Argus has once been found of the exact colour of Hipparchia Pamphilus,—pale fulvous! And in conclusion, Anthrocera Hippocrepidis and Callimorpha Jacobææ have each occurred with the brilliant sanguineous markings on their anterior wings, and similarly coloured posterior ones, converted into a clear pale luteous. Sufficient, therefore, has been adduced to show the very little de-

---

* A beautiful intermediate specimen is in the collection at the British Museum.

pendence that is to be placed on colour alone in discriminating species *. And that form and markings are supposed to be inconstant, the following observation on Cicindela hybrida of DeJean sufficiently testifies:—"I feel satisfied, from the variation in form and markings, that Ci. riparia and Ci. aprica are only varieties of Ci. hybrida †." In fine, it may be remarked that even the presence or absence of particular organs is not always indicative of specific difference: e. g. Clivinæ Fossor and collaris, the Dromii, Patrobi, &c. the species of which are sometimes winged, at others apterous.

From my experience I conceive that, where an insect differs from its nearest congener by some trifling variation of form ‡, combined with a diversity of sculpture, dissimilar bulk, or prevalent discrepancy of colour, either in the disposition of particular markings or in an universal change, we are perfectly justified in considering such examples as distinct species, unless we are enabled by experience to show incontestibly that such is contrary to nature, as in the examples of the Donaciæ before alluded to, and in several of the species of Coccinellæ and Cicadæ, as hereafter united in the synonymy. Such principles have guided me in my endeavours to discriminate the indigenous species, as given in the following pages: but to expect that I should invariably be correct in my ideas respecting their real distinction, I need not remark would be absurd; although it is observed in the "Horæ Entomologicæ," that "in describing species, each of which Nature has manifestly insulated, a failure in precision denotes an inability to seize their characteristic marks—it is in short a glaring fault." But I think if the learned writer of the above paragraph had been better acquainted with some of the indigenous groups, especially such genera as Agrotis, Cucullia, Peronea, &c., he would at least have altered the concluding sentence; for, as Mr. Kirby with propriety observes, "As we do not know the value and weight of the momenta by which climate, food, and other supposed fortuitous circumstances operate upon animal forms, we cannot point out any certain diagnostic by which in all cases a species may be distinguished from a variety; for these characters that in some are constant, in others vary."

* One consequence of assuming colour to be indicative of specific distinction has been to falsify physiological facts. Marsham, p. 169, says, "Ex coitu Coccinellarum inter omnes familias promiscuo, varietates plurimæ ortæ sunt." Whereas this apparent promiscuous intercourse of the families (sections) arises entirely from regarding the mere varieties, differing in trifling spots of colour, as distinct species, and distributing these varieties, as Marsham has done, into different sections, according to their colour.

† Curtis, No. I. second edition;—a writer who in other places considers form as conclusive of specific difference.

‡ Considerable diversity of form is rather indicative of sectional or generic difference than of specific.

If, therefore, it be so difficult for entomologists of experience occasionally to discriminate the distinction of the perfect productions of the Creator, how much more so must be the task of detecting them through the imperfect descriptions of the creature †?

It now only remains for me to add a few words upon the Catalogue itself. Although I have endeavoured to render it as perfect as its nature will admit, so far as relates to the number of species at present discovered, many are doubtless omitted; for if no single cabinet has ever yet contained a specimen of each of the indigenous Lepidoptera diurna, the most conspicuous and eagerly sought after of our insects (and such has been the case with those which I have inspected, at least two hundred in number), how much less chance is there of obtaining a knowledge of all the minute and hitherto neglected species? And as the following Catalogue is necessarily in several places entirely prepared from the contents of my own cabinet; from the extreme difficulty of recognizing kindred species in particular groups, added to the facts that *no* British collection has as yet been named throughout, and that the less favoured insects are usually kept in a confused and unnamed mass, a correct knowledge of such species is perfectly unattainable; nevertheless, as *less than* four thousand species have hitherto been recorded by name, by *all* preceding English writers, to inhabit Britain, and nearly ten thousand are named in the following pages, the fact of my having *more than* doubled the number sufficiently evinces the exertions I have made to attain perfection: and by way of increasing its utility, I have endeavoured to form this work into an Entomologia Londinensis, by placing a * at the end of every species which to my knowledge has been captured *within* twenty-five miles of St. Paul's. I have also given an authority for the insertion of *every* species, where practicable, by a set of characters which require explanation.—To such insects as I have never

† That this is an object of difficulty may be seen with facility by an inspection of the numerous synonyms, many doubtless arising from misconception, and may be illustrated by a reference to the genus Cicindela, as being the first in the series, and containing insects of considerable size. Marsham gives three indigenous species of this genus; since his time, other species have turned up, amongst which one termed Ci. hybrida by Sowerby has occurred in no little profusion. Upon this insect I hazarded the conjecture that it was the true Ci. hybrida of Linné, as it had been *invariably* recorded by English writers, and suspected to be by the best continental ones; and upon the re-examination of the series in my collection, I detected a single example of an insect which has since proved to appertain to a distinct species, and which has recently been given as the true Ci. hybrida, by a writer who previously gave the former species that appellation, notwithstanding he had repeated opportunities of examining both kinds in my collection, and at the time of his first recording Ci. hybrida, Linn. as an indigenous insect, that species had not been detected by English entomologists, but only the one which he now terms Ci. maritima!

seen, a . is placed after its number *in the genus* (e.g. Chrysomela Sparshallii, No. 2299.); to those which I have merely seen preserved in cabinets, a , is similarly attached (as to Cicindela Sylvicola, No. 4.); a : to such as have been seen alive by me (as Drypta emarginata, No. 7.); and a ; to those insects which have actually been captured by myself (as Cicindela sylvatica, No. 1.*). The uncertain species are placed beneath those of which they may be varieties, and are known by having their names within [ ] (as Dromius notatus, No. 20, which may prove to be a variety of No. 19.). Such insects of which I have obtained foreign specimens alone, for the sake of illustration are indicated by a ‡; and such as I do not possess, by a †; and *every collection* in which such species are respectively contained, is referred to so far as in my power; distinguishing the instances where *I have seen* the insects themselves, by printing the possessor's name in Italics; and where *I have not seen* them, in Roman characters (e. g. Anisoplia Donovani, No. 1196.). The specimens of uncertain British origin† are inserted in their proper

* What merit there can possibly be in being the discoverer of a *new species*, I am totally at a loss to comprehend, knowing from experience that such an occurrence is purely a matter of chance, and that as many valuable acquisitions have been made by the mere *ignorant mechanic* as by the most *assiduous entomologist;* hence the absurdity of attempting to indicate such species *as appear* to have been detected by myself, which would answer *no good purpose,* but probably tend *to mislead;* e. g. in a recent compilation, grossly infringing upon my copyright, the compiler assumes that he discovered, *inter alia*, the four following insects placed consecutively:—Chlænius sulcicollis, Licini Silphoides and depressus, and Badister cephalotes. Of these insects, the first, which is unique as British, was found by Mr. Charles Curtis: the second was taken by the late Mr. Swainson *before* he left London, some thirty years ago; two specimens, captured at Dover, were in the collection of the late Rev. J. Lyon, which was sold by auction in 1815, and two in that of Mr. Spratt, purchased by Dr. Leach in 1813; the last four specimens are now in the British Museum. Licinus depressus was taken near Hull, as it appears by Mr. Marsham's MSS., in 1807, by Mr. Watson; and Badister cephalotes was found by Mr. Haworth, I believe, in the same neighbourhood about the same time, and shortly after by the Rev. W. Kirby in Kent. The above dates contrasted with those given in the places cited in pages 14 and 15 of the following Catalogue, Nos. 97, 100, 102 and 105, fully justify the position I have above advanced.

† It may not be irrelevant in this place to add a few words upon Papilio Europome of Haworth, *not of* Esper. Of this species it has been observed, with a total disregard to facts, "It is not a British insect, those described and even figured as such being the P. Philodice of Godart, a North American species, of which there is no testimony of a single specimen having been taken in Britain, nor, it may be added, in Europe: the old examples in the cabinets of the late Mr. Francillon, and Mr. Swainson, and two in Mr. Plastead's, were no doubt placed there as representatives of C. Hyale." Upon this paragraph I must remark, that Mr. Marsham declared to me that *every* insect in his British cabinet had been captured in Britain; and a memorandum in a Catalogue of Mr. Francillon's collection, which I once saw in the possession of B. Clark, Esq., stated that "every insect in that collection, except those set in the German

situations, with their synonyms at the foot of the page, the authority for their insertion being shown by a (!) after the reference (as Scarites *subterraneus*, No. 42, which is printed in Italics to denote the species to be extra-european): while the same character (!) in the body of the page indicates that the author referred to gives England as the habitat of the species (e. g. Lebia turcica, No. 27; Carabus turcicus, *Fab.*). Finally, as in all entomological works, even in those treating upon genera, the *real synonyms* of the genera are not pointed out, I have endeavoured to remedy that inconvenience by attaching a *p* to such genera as are not strictly synonymous (as Lamprias of Bonelli includes a *portion only* of Latreille's Lebiæ, while it is really synonymous with Echimuthus of Leach*). And I may add, that although occasionally the oldest name may by accident occupy a secondary position, I have attempted in all instances to give it the priority.

I cannot conclude this Introduction, without returning my most sincere acknowledgements to the several individuals who have so kindly and handsomely assisted me by the presentation of rare insects, or the loan of scarce and valuable publications. To particularize might appear invidious; but as the List of Subscribers contains the names of those to whom I have been thus indebted, I

---

style, had been taken at large in England." I have recently inspected the late Mr. Swainson's cabinet (in which are fine specimens of the insect in question), which his son, W. Swainson, Esq., tells me is still arranged in the precise state it was left by his father, who never introduced a foreign specimen into his cabinet without attaching a memorandum thereto; and the insects in question are without. Now as all the above cabinets contained *not only* examples of Colias Europome, *but also* of Co. Hyale and Edusa (and I possess specimens *of each* from the two first cabinets, set in the English style), it is palpable that the former species could not be introduced as representatives of C. Hyale; and the concurrent testimony of the three respectable and unprejudiced individuals above referred to is at least as conclusive as the mere assertions of a writer who, "without note or comment," gives Cantharis longicornis and Noctua subterranea of Fabricius, both of which are natives of *South* America, as British insects. And as Mr. Plastead was in the constant habit of purchasing specimens of Mr. Latham the dealer (and I speak *advisedly* upon this point), it is most probable his insects were not British, any more than his specimen of Hipparchia Arcanius, which there is every reason to believe was also obtained from the same source. In fine, the mere silence of continental writers regarding Co. Europome is no proof of its not inhabiting Europe; as till within these few months the Hesperia Artaxerxes of Fabr. was likewise unknown to them, though comparatively common in Britain.

* And I have also endeavoured to show where a genus or a species *has not been described*, but only the name *published*, by including the references to its author or recorder in ( ), as Drypta, (Kirby,) and Polistichus fasciolatus, (Ing. Inst. 90.), page 2: the references in such cases being made to the genera, as indicating the authors who have adopted them, and to the species, as showing where their localities, habits, peculiarities of structure, &c., have been recorded by English writers.

beg to repeat my thanks to the persons therein named, as without their support the following work would not have appeared. There is, however, one absent name, that of my esteemed and regretted friend Dr. W. E. Leach, of whom it has been said, "*nihil non tetigit, et omnia quæ tetigit ornavit,*"—as well as that of his commender, my friend the Rev. W. Kirby, from whom I have received such signal marks of assistance, that it would be unpardonable were I to pass them over tacitly.

*** *Every Species is referred to, which are mentioned in the Works quoted herein, to which a * is attached in the subjoined List.*

AUTHORS QUOTED.

Act. Hall.—Abhandlungen der Hallischen Naturforschenden Gesell-schaft. Dessau und Liepzig. 8vo.

Act. H. N. Paris.—Annales de la Société Linnéene de Paris. 1827, &c. 8vo.

Act. H., or *Act. Holl.*—Kongl. Svenska Vetenskaps Academiens Hand-lingar. Stockholm. 1739—1829. 8vo.

Act. Taur.—Mem. della R. Acad. delle Scienze di Torino. Tom. xxiv. Taurino. 1819. 4to.

Act. U., or *Act. Ups.*—Nova Acta Regiæ Societatis Scientiarum Upsa-liensis. 1773, &c. 4to.

Ahr. F., or *Ahrens F.*—A. Ahrensii Fauna Insectorum Europæ. Halæ. 1812. 12mo.

**Albin.*—A Natural History of English Insects, illustrated with 100 cop-per plates, curiously engraven from the life, by E. Albin. London. 1720. 1 vol. 4to.

Albin Sp.—A Natural History of Spiders and other curious Insects, by E. Albin. London. 1736. 1 vol. 4to.

Annal. du Mus. N. H.—Annales du Muséum National d'Histoire Natu-relle à Paris. An xii. (1804), &c. 4to.

Ann. Sci. Nat.—Annalės des Sciences Naturelles. Paris. 8vo.

Anon. Rem. L. G.—Remarks on the Linnean Orders of Insects. Lon-don. 1828. 4to.

Baker M.—Of Microscopes and the Discoveries made thereby. By H. Baker. London. 1785. 2 vols. 8vo.

Barb. G.—The Genera Insectorum of Linnæus exemplified by various specimens of English Insects, drawn from Nature. By J. Barbut. London. 1781. 1 vol. 4to.

Beck. Beyt.—Beitrage zur Baierschen Insecten Fauna, &c. Von L. Beck. Ausburg. 1817.

Bergstraesser Nom.—Nomenclatur und Beschreibung der Insekten in der Graffschaft Hanau-Münzenberg. Von J. A. B. Bergstraesser. Ha-nau. 1778. 4to.

**Berk. S.*—Synopsis of the Natural History of Great Britain and Ireland. By J. Berkenhout, M.D.; being A second edition of the Outlines, &c. London. 1789. 2 vols. 8vo.

Biche. Add.—An Address delivered at the Anniversary Meeting of the Zoological Club of the Linnean Society. By J. E. Bicheno. 1826. 8vo.

Billberg.—G. J. Billberg Monographia Mylabridum. Holmiæ. 1813. 8vo.

Bingley.—Animal Biography, or Popular Zoology. By W. Bingley. 4th edition. London. 1813. 3 vols. 8vo.

Boisd.—Europæorum Lepidopterorum Index Methodicus: auctore J. A. Boisduval. Pars 1ma, sistens genera Papilio, Sphinx, Bombyx et Noctua Linne. Parisiis. 1829. 1 vol. 8vo.

Boisd. Zy.—Essai sur une Monographie des Zygénides, suivi du Tableau Methodique des Lépidoptères d'Europe: par J. A. Boisduval. Paris. 1829. 1 vol. 8vo.

Boit. M.—Manuel d'Entomologie, ou Histoire Naturelle des Insectes: par M. Boitard. 2 vols. 18mo. Paris. 1828.

Bonelli Obs.—Observations Entomologiques: par *F.* A. Bonelli. Turin. 1809. 4to.

Bonsd. Cu.—Historia Naturalis Curculionum Suecicæ. Pars prima et secunda. Auctore G. Bonsdorff. Upsaliæ. 1785. 4to.

Bork. E. S.—Naturgeschichte der Europaischen Schmetterlinge nach Systematischen Ordnung, von M. B. Borkhausen. Frankfort. 1788, &c. 4 tom. 8vo.

Brahm I. K.—Insekten-Kalender von N. J. Brahm. Mainz. 1790. 8vo.

**Brit. B.*—British Butterflies, their distinctions, generic and specific; with Lithographic Illustrations of each genus. Birmingham. 1828. 24mo. 1 vol.

Bull. Sci. Nat.—Bulletin des Sciences Naturelles. Paris. 8vo.

Cederh. F.—Faunæ Ingricæ Prodromus, exhibens methodicam descriptionem Insectorum agri Petropolensis, &c. Auctore J. Cederhielm. Lipsiæ. 1798. 1 vol. 8vo.

Charp. & Z. S.—Die Zünsler, Wickler, Schaben und Geistchen (Pyralis, Tortrix, Tinea et Alucita) des Systematischen Verzeichnisses der Schmetterlinge der Wiener Gegend, &c. Von T. von Charpentier, und J. L. Th. Fr. Zincken G. Sommer. Braunschweig. 1821. 1 vol. 8vo.

Childr. Address.—An Address delivered at the Anniversary Meeting of the Zoological Club of the Linnean Society. By J. G. Children. London. 1827. 8vo.

Christius B., or *Hy.*—Naturgeschichte, Klassification und Nomenclatur der Insekten vom Bienen, Wespen und Ameisengeschlect. Von J. C. Christius. Frankfurt am Main. 1791. 4to.

Clairv., or *Clairv. E. H.*—Entomologie Helvetique, ou Catalogue des Insectes de la Suisse rangés d'apres une nouvelle methode (par J. Clairville). Zuric. 1798—1806. 2 vols. 8vo.

Clark.—An Essay on the Bots of Horses and other Animals. By B. Clark. London. 1815. 4to.

Clerck Ic.—Icones Insectorum rariorum cum nominibus eorum trivialibus. Auctore C. Clerck. Holm. 1764. 4to.

Coqueb., *Coqueb. Ic.*, or *Coqueb. Ill.*—Illustratio iconographica Insectorum quæ in musæis Parisinis observavit et in lucem edidit J. C. Fabricius. Auctore J. A. Coquebert. Parisiis. 1779. 4to.

Cramer.—Papillons exotiques des trois parties du Monde, L'Asia, L'Afrique, et L'Amerique: par P. Cramer. Utrecht. 1779. 4to.

Creutz. E. V.—C. Creutzer Entomologische Versuche. Wien. 1799. 8vo.

**Curtis.*—British Entomology; being illustrations and descriptions of the genera of Insects found in Great Britain and Ireland, &c. By J. Curtis. London. 1824. 6 vols. 8vo.

**Curtis 2nd edit.*—Idem: 2nd edition (No. I. only).

Curtis G.—A Guide to an Arrangement of British Insects. By J. Curtis. June 1829. 8vo.

Curtis C.—Idem: Catalogue on No. 45 of the above.

Curtis W.—A short history of the brown-tailed Moth, &c. By W. Curtis. London. 1782. 4to.

Czempinski.—Totius regni animalis Genera. Auctore P. Czempinski. Vindob. 1778. 8vo.

Dalman A. E.—Analecta Entomologicæ. Auctore J. W. Dalman. Holmiæ. 1823. 1 vol. 4to. cum tabulis æneis.

DeGeer.—Mémoires pour servir à l'histoire des Insectes: par Charles DeGeer. Stockholm. 1752, &c. 7 tom. 4to.

DeGeer G.—Caroli Lib. Bar. DeGeer Genera et Species Insectorum, curante A. J. Retzio. Lipsiæ. 1783. 8vo.

DeJean C.—Catalogue de la Collection de Coléoptères de M. le Baron DeJean, &c. Paris. 1821. 8vo.

DeJean Col.—Histoire Naturelle et Iconographie des Insectes Coléoptères d'Europe, par M. Latreille et M. le Baron DeJean, &c. 1—3 livr. Paris. 1822. 8vo.

DeJean Sp.—Species Général des Coléoptères de la Collection de M. le Comte DeJean. Paris. 1824—1828. 3 tom. 8vo.

**Denny M.*—Monographia Pselaphidorum et Scydmænidarum Britanniæ. By H. Denny. Norwich. 1825. 8vo.

**Don.*—The Natural History of British Insects, explaining them in their several states; illustrated with coloured figures, &c. By J. E. Donovan. London. 1792—1816. 16 vols. 8vo.

Dufts. F.—Fauna Austriæ: Oder Beschreibung der Osterreichischen Insecten für angehende Freunde der Entomologie. Von C. Duftschmid. Linz. 1803—1825. 3 vols. 12mo.

Dumeril.—Traité élémentaire d'histoire naturelle. Par A. M. C. Dumeril. Paris. 1807. 8vo.

Ent. Hefte.—Entomologische Hefte, enthaltend Beiträge zur weitern Kenntniss und Aufklärung der Insectengeschichte, &c. Frankfurt am Main. 1803. 8vo.

**Ent. Trans.*—Transactions of the Entomological Society of London. London. 1807. 1 vol. 8vo.

Esper.—Die Schmetterlinge in abbildungen nach der natur mit Beschreibungen, E. J. C. Esper. Erlangen. 1777, &c. 4to.

Fabr.—Joh. Chr. Fabricii Entomologia Systematica emendata et aucta. Hafniæ. 1792. 4 vols. 8vo.

Fabr. S.—Idem: Supplementum. Hafniæ. 1794. 8vo.

Fabr. A.—J. C. Fabricii Systema Antliatorum. Brunsvigæ. 1805. 8vo.

Fabr. E.—J. C. Fabricii Systema Eleutheratorum. Kiliæ. 1801. 2 vols. 8vo.

Fabr. Gen.—J. C. Fabricii Genera Insectorum, &c. Chilonii. 8vo.

Fabr. M.—J. C. Fabricii Mantissa Insectorum, &c. Haffniæ. 1787. 8vo.

Fabr. P.—J. C. Fabricii Systema Piezatorum. Brunsvigæ. 1804. 8vo.

Fabr. R.—J. C. Fabricii Systema Rhyngotorum. Brunsvigæ. 1803. 8vo.

Fabr. S. E.—J. C. Fabricii Systema Entomologiæ, sistens Insectorum Classes, Ordines, Genera, Species, &c. Flensburgi et Lipsiæ. 1775. 8vo.

Fabr. S. I.—J. C. Fabricii Species Insectorum, &c. Hamburgi et Kilonii. 1781. 2 vols. 8vo.

Fallen C. S.—Monographia Cimicum Sueciæ. Auctore C. F. Fallen. Hafniæ. 1807. 8vo.

Fall. D. S.—Diptera Sueciæ descripta a C. F. Fallen. Lundæ. 1814. 4to.

Fall. H.—Specimen novam Hemiptera disponendi methodum exhibens: a C. F. Fallen. Lundæ. 1814. 4to.

Fallen Hy.—Specimen novam Hymenoptera disponendi methodum exhibens: a C. F. Fallen. Lundæ. 1813. 4to.

Fallen M. C.—Monographia Cantharidum et Malachiorum Sueciæ: a C. F. Fallen. Lundæ. 1807. 4to.

Fallen Obs. Ent.—Observationes Entomologicæ, &c.: a C. F. Fallen. Lundæ. 1802. 4to.

Faun. F.—Faune Française, ou Histoire Naturelle, générale et particulière, des Animaux qui se trouvent en France, &c. Par MM. Viellot, Desmarest, de Blainville, Audinet-Serville, St.-Fargeau et Walckenaer. Paris. 1824—1829. 8vo.

Faun. I.—Vide *Cederh. F.*

Fisch. E. R.—Entomographia Imperii Russici, &c. Vol. 1. par E. R. Fischer. Mosquæ. 1820. 4to.

**Forst. C.*—Novæ Species Insectorum. Centuria 1ª. Auct. J. R. Forstero. Londini. 1771. 8vo.

Fourc., or *Fourc. P.*—Entomologia Parisiensis, sive Catalogus Insectorum quæ in Agro Parisiensi reperiuntur, &c. edente A. F. Fourcroy. Paris. 1785. 2 vols. 12mo.

Frisch.—J. L. Frisch Beschreibung von allerley Insekten in Deutschland. Berlin. 1766. 4to.

Fuessl. Ar.—Archives de l'Histoire des Insectes publiées en Allemand par Jean Caspar Fuessly, traduits en François. Winterthur. 1794. 4to.

Fuessly M.—Neues Magazin für die Liebhaber der Entomologie. Herausgegeben von J. C. Fuessly. Zürich und Winterthur. 1782. 8vo. 2 St.

Geoff.—Histoire Abregée des Insectes, dans laquelle ces animaux sont rangés suivant un Ordre Methodique, par M. Geoffroy, M.D. Paris. 1764. 2 vols. 4to.

Germ. I. N.—Insectorum Species novæ aut minùs cognitæ, descriptionibus illustratæ. Auctore E. F. Germar. Halle. 1824. 8vo.

Germ. M.—Mágazin der Entomologie, von E. F. Germar. Halle. 1813, &c. 4 vols. 8vo.

Gmel.—Caroli a Linné Systema Naturæ, &c. Editio 13ª, aucta, reformata. Cura J. F. Gmelin. Liepsiæ. 1788. 3 tom. 8vo.

Godart E. M.—Encyclopédie Methodique. Art. Papillon. 4to.

Godart. Dup.—Histoire Naturelle des Lépidoptères de France. Par M. Godart, et M. Duponchel. Paris. 1818, &c. 8vo.

Goëz.—Entomologische Beitrage, &c. Von J. A. E. Goëze. Leipzic. 1777. 3 vols. 8vo.

**Gor. & Pr.*—The Natural History of several new, popular and diverting living Objects for the Microscope, &c. By C. R. Goring, M.D., and A. Pritchard. London. 1829. 8vo.

Gould A.—An Account of English Ants. By Rev. W. Gould. London. 1747. 1 vol. 18mo.

Grav. I. P.—Monographia Ichneumonum pedestrium, &c. Auctore J. L. C. Gravenhorst, &c. Lipsiæ. 1815. 1 vol. 8vo.

Grav. Mi.—Coleoptera Microptera Brunsvicensia, necnon exoticorum quotquot extant in collectionibus Brunsvicensium. Auct. J. L. C. Gravenhorst. Brunsvigæ. 1802. 8vo.

Grav. Mo.—Monographia Coleopterorum Micropterorum. Auct. J. L. C. Gravenhorst. Gottingæ. 1806. 1 vol. 8vo.

Gyll.—Insecta Succica descripta a Leonardo Gyllenhal. Classis I. Coleoptera sive Eleutherata. Scaris. 1808—1828. 4 vols. 8vo.

**Harr. A.*—The Aurelian; or Natural History of English Insects; namely, Moths and Butterflies. By M. Harris. London. 1778. Fol.

**Harr. Ex.*—An Exposition of English Insects, &c. By M. Harris. London. 1782. 4to.

**Harr. V. M.*—The English Lepidoptera, or the Aurelian's Pocket Companion, &c. By M. Harris. London. 1775. 8vo.

**Haw.*—Lepidoptera Britannica, sistens Digestionem novam insectorum Lepidopterorum quæ in Magna Britannia reperiuntur, &c. Auct. A. H. Haworth. Londini. 1803—1828. 8vo.

**Haw. Pr.*—Prodromus Lepidopterorum Britannicorum. A concise Catalogue of British Insects, with the times and places of appearance in the winged state. By a Fellow of the Linnean Society. Holt. 1802. 4to.

Hellwig.—Fauna Etrusca, &c. iterum edita et Annotatis perpetuis aucta, a D. J. C. Lud Hellwig. Helmstadii. 1795. 8vo.

Herbst Ar.—Archiv der Insectengeschichte herausgegeben von J. C. Fuessly. Zürich und Winterthur. 1781. 4to.

Herbst C.—Natursystem aller bekaunten in und ausländischen Insecten,

&c. von C. G. Jablonsky, und forgetgesetzt von J. F. W. Herbst. Berlin. 1789. 40 vols. 8vo.

Herbst Gem. Nat.—Herbst Gemeinnützige Naturgeschichte.

Herbst Nat.—Natursystem aller bekannen in und auslandischen Insecten, &c., von C. G. Jablonsky. Berlin. 1785, &c. 8vo.

Herman.—Mémoire Apterologique par J. F. Herman, M.D. Strasburgh. 1804. Folio.

Hist. Ins.—A short History of Insects (extracted from works of credit), designed as an Introduction to the Study of that branch of Natural History, &c. Norwich. 1 vol. 12mo.—No date.

Höppe.—Davidi Henrici Höppe, M.D. Enumeratio Insectorum Elytratorum circa Erlangam indigenarum. Erlangæ. 1795. 8vo.

Horsfield.—A Descriptive Catalogue of the Lepidopterous Insects contained in the Museum of the Honourable East India Company. By T. Horsfield, M.D. London. 1827—1829. 4to.

Hort. Trans.—Transactions of the Horticultural Society. London. 1814, &c. 4to.

Hübn. B.—Beitrage zur geschichte der Schmetterlinge von J. Hübner. Augsburg. 1786—1789. 2 tom. 8vo.

Hübn.—Der Samlung Europaischer Schmetterlinge von J. Hübner. Augsburg. 1776, &c. 4to.

Illig. K.—Verzeichniss der Käfer Preussens entsvorfen von J. Gottleib. Kugelann, ausgean beitet von J. K. W. Illiger, &c. &c. Halle. 1798. 8vo.

Illig. M.—Magazin für Insectenkunde herausgegeben von J. K. W. Illiger. Braunschweig. 1801—1806. 7 tom. 8vo.

**Ing. Inst.*—Instructions for collecting, rearing and preserving British Insects. By A. Ingpen. London. 1827. 18mo.

Ins. Suec.—Vide *Thun. I. S.*

Isis.—Isis von Oken. Jena. 4to.

**Jerm.*—The Butterfly Collector's Vade Mecum; or A Synoptical Table of English Butterflies. Ipswich. 1824. 12mo.

**Id. 2 edit.*—Idem, 2nd edition. Ipswich. 1827.

Jurine Hy.—Nouvelle Methode de Classes les Hyménoptères et les Diptères. Par L. Jurine. Tom. I. Hyménoptères. A Genève. 1807. 4to.

Kirby & Sp. I. E.—Introduction to Entomology. By the Rev. W. Kirby and W. Spence. London. 1818—1827. 4 vols. 8vo. (Vols. 1 and 2, 4th edit.; vols. 3 and 4, 1st edit., quoted in this work.)

**Kirby M. A.*—Monographia Apum Angliæ. By William Kirby, B.A. Ipswich. 1802. 2 vols. 8vo.

Klug Bl.—Die Blattwespen nach ihren Gattungen und Arten zusammengestellt von F. Klug. Berlin. 1818, 1819. 4to.

Klug Ci.—Versuch der Blattwespen gattung Cimbex, Fab. Von Dr. F. Klug. Berlin. 1819. 4to.

Klug Si.—Monographia Siricum Germaniæ atque Generum illis adnumeratorum. Auct. D. F. Klug. Berolini. 1803. 1 vol. 4to.

Knoch Beitr.—Beiträge zur Insecten geschichte. Von A. W. Knoch. Leipzig. 1781. 8vo.

Kunzé E. F.—Entomologische Fragmente. Von G. Kunze. Halle. 1818. 8vo.

Laich. T.—J. N. V. von Laicharting der Tyroler Insecten. Zurich. 1781. 8vo.

Lama.—Histoire Naturelle des Animaux sans Vertèbres, &c. Par M. de Lamarck. Tom. 3 et 4. Paris. 1816, 1817. 8vo.

Lasp.—Sesiæ Europæ Iconibus et Descriptionibus illustratæ. Auct. J. H. Laspeyres. Berolini. 1801. 4to.

Latr. Essai.—Essai sur l'Hist. des Fourmis de la France. Par P. A. Latreille. Paris. 8vo.

Latr. F.—Histoire Naturelle des *F*ourmis et recueil des Mémoires et d'observations sur les Abeilles, &c. Par P. A. Latreille. A Paris. 1802. 8vo.

Latr. G.—Genera Crustaceorum et Insectorum secundum ordinem naturalem in familias disposita, iconibus exemplisque plurimis explicata. Auct. P. A. Latreille. Parisiis et Argentorati. 1806—1810. 8vo.

Latr. H.—Histoire Naturelle générale et particulière des Crustacés et des Insectes. Par P. A. Latreille. Paris. 1802, &c. 14 tom. 8vo.

**Leach Ar.*, or *Leach E. I.*—On the genera and species of Eproboscideous Insects, and on the arrangement of Œstrideous Insects. By Dr. Leach (From the Memoirs of the Wernerian Nat. Hist. Society). Edinburgh. 1817. 8vo.

**Leach E. E.*—Leach in Edinburgh Encyclopædia: articles Entomology and Insecta. Edinburgh. 1810, &c. 4to.

**Leach E. B. Sup.*—Leach in Supplement to the Encyclopedia Britannica: articles Annulosa and Entomology. Edinburgh. 1818, &c.

**Leach Z. M.*—The Zoological Miscellany; or descriptions of new, rare, or highly interesting Animals. By W. E. Leach. London. 1814. 3 vols. 8vo.

Lehman D.—De Sensibus externis Animalium exsanguium Commentatio, &c. Goëttinb. 1798. 8vo.

**Lew.*—The Papilios of Great Britain. By W. Lewin. London. 1795. 4to.

Linn., or *Linn. S.*—Caroli a Linné Systema Naturæ per Regna tria Naturæ. Editio decima tertia. Vindobonæ. 1767. 3 tom. 8vo.

Linn. F.—Ejusdem *F*auna Suecica. Editio altera auctior. Stockholmiæ. 1761. 8vo.

Linn. M.—Ejusdem Mantissa Plantarum altera generum editionis sexta, &c. Holmiæ. 1771. 8vo.

Linn. Mus.—Ejusdem Systema Naturæ per Ulricæ Reginæ. Holmiæ. 1764. 8vo.

Linn. Trans.—Transactions of the Linnean Society of London. 1791—1829. 16 vols. 4to.

List. Goëd.—Johannes Goëdartius de Insectis in Methodum redactus, cum Notularum Additione, Operâ M. Lister. Londini. 1685. 8vo.

Id. (*Ang.*)—Johannes Goëdartius of Insects, done into English, and methodized, with the addition of Notes. By M. L. The *F*igures etched upon copper. By Mr. *F.* York. 1682. (*In my copy there is a MS. systematic arrangement and description of the indigenous Papiliones, dated* 1692.)

Loudon M.—The Magazine of Natural History: conducted by J. C. Loudon. 1828—1829. 8vo.

MacL. A.J.—Annulosa Javanica. By W. S. MacLeay. London. 1824. 4to.

MacL. H. E.—Horæ Entomologicæ, or Essays on the Annulose Animals. By W. S. MacLeay. 1819. 8vo.

Macq.—Insectes Dipterès du nord de la France, Tipulaires, &c. Par J. Macquart. Paris. 1825-8. 8vo.

Man. Euc.—Eucnemis, Insectorum genus, Monographiæ tractatum a C. G. de Mannerheim. Petropoli. 1823. 8vo.

**Marsh.*—Entomologia Britannica, sistens Insecta Britanniæ indigena secundum methodum Linnæanam disposita. Auctore T. Marsham. Tomus 1. Coleoptera. Londini. 1802. 8vo.

Mart. C.—The English Entomologist, exhibiting all the Coleopterous Insects found in England, &c. By T. Martyn. London. 1792. Large 4to. (*References on the authority of Marsham.*)

Mart. V. M.—The Aurelian's Vade Mecum; containing an English alphabetical and Linnean systematical Catalogue of Plants affording nourishment to Butterflies, &c. By M. Martyn. Exeter. 1785. 12mo.

Meig. Kl.—Klassifikazion und Beschreibung der Europäischen Zweiflügligen Insekten, von J. W. Meigen. Braunschweig. 1804. 2 vols. 4to.

Meig. Zw.—Systematische beschreibung der bekannten Europ. Zweiflügeligen Insecten, von J. W. Meigen. Aachen. 1818—1828. 5 vols. 8vo.

Merret Pinax.—Pinax Rerum Naturalium Britannicarum, &c. Auctore C. Merret. Londini. 1667. 12mo.

Mikan. B.—Monographia Bombyliorum Bohemiæ. Auct. J. C. Mikan. 1796. 8vo.

**Mill. B. E.*—Outlines of British Entomology, in Prose and Verse. By S. W. Millard. Bristol. 1821. 12mo.

Mouff.—Insectorum sive minimorum Animalium Theatrum, &c. Auct. T. Mouffetti. Londini. 1634. Folio.

Müll. F. F.—Otto Fr. Müller Fauna Insectorum Fridrichsdalina. Hafniæ et Lipsiæ. 1764. 8vo.

Müll. Z. D.—Zoologiæ Danicæ Prodromus, seu Animalium Daniæ et Norvegiæ indigenarum, &c. Auctore O. F. Müller. Hafniæ. 1786. 8vo.

Nat. Comp.—The Naturalist's and Traveller's Companion. By J. C. Lettsom. London. 1772. 8vo.

Naturf., or *Naturforscher.*—Der Naturforscher. Halle. 1774, &c.

Nicol. Col. Hal.—Diss. sistens Coleopterorum Species agri Halensis. Auct. E. A. Nicolai. Halæ. 1822. 8vo.

Nov. Act. Hal.—Neue Schriften der Naturforschenden G. Halle. Halæ. 1811—1818. 8vo.

Nova Act. U.—Nova Acta Regiæ Societatis scientiarum Upsaliensis. Upsal. 1773, &c. 4to.

Ochs.—Die Schmetterlinge von Europa. Von F. Ochsenheimer. 4 vols. 8vo. 1807—1816.

Ochs. Tr.—Die Schmetterlinge von Europa (Fortsekung des Ochsenheimer'schen Werks). Von F. Treitschke. 3 vols. 1825—1828.

O. Fab.—Fauna Groenlandica. Auct. O. Fabricio. Hafniæ et Lipsiæ. 1780. 8vo.

Olfers.—De Vegetativis et Animatis corporibus in corporibus animatis reperiundis commentarius. Auctore J. Fr. M. de Olfers. Pars 1. Berolini. 1816. 12mo.

Oliv. E.—Entomologie, ou Histoire Naturelle des Insectes, avec des caractères génériques et specifiques, &c. Par M. Olivier, M.D. a Paris. 1789. 8 Tom. 4to.

Oliv. E. M.—Encyclopédie Methodique. Histoire Naturelle des Insectes, &c. Par M. Olivier. 1782, &c. 4to.

Pallas Sp. Zo.—Spicilegia Zoologica quibus novæ imprimis atque obscuræ animalium species describuntur, &c. Auct. P. S. Pallas. Hagæ Com. 1766. 4to.

Panz. E.—Entomologia Germanica, exhibens Insecta per Germaniam Indigena, &c. Auct. G. W. F. Panzeri. Norimbergæ. 1795. 12mo.

Panz. F.—Fauna Insectorum Germanicæ Initia. Auct. Dr. G. W. F. Panzer. Nurnberg. 1789, &c. 12mo.

Panz. I.—Index Entomologicus sistens omnes Insectorum species, &c. Auct. G. W. F. Panzeri. Norimbergæ. 1813. 12mo.

Panz. Beit.—Dr. G. W. F. Panzer Beytrage zur Geschichte der Insecten. Erlangæ. 1793. 12mo.

Panz. Krit. Rev.—Ejusdem Kritische Revision der Insectenfaune Deutschlands nachdem System bearbeiter. Nurnberg. 1805. 12mo.

Panz. Schæf.—D. J. C. Schæfferi Iconum Insectorum circa Ratisbonam

Indigenorum Enumeratio Systematica; opera et studio Dr. G. W. F. Panzeri. Erlangæ. 1804. 4to.

Pap. d'Eur.—Papillons d'Europe peints d'après Nature, par M. Ernst, &c. décrits par le R. P. Engramelle. Paris. 1779, &c. 8 vols. 4to.

Payk. F.—G. Paykull Fauna Suecica. Insecta. Upsaliæ. 1798, &c. 8vo.

Payk. Ca.—Monographia Caraborum Sueciæ. 1790.

Payk. Cu.—Monographia Curculionum Sueciæ. 1792.

Payk. M.—Monographia Staphylinorum Sueciæ: a G. Paykull. Upsaliæ. 1789. 8vo.

Payk. M. S.—Vide *Payk. M.*

Payk. M. H.—Monographia Histeroidum. Auct. G. Paykull. Upsaliæ. 1811. 8vo.

Petagna I. C.—V. Petagna Specimen Insectorum Ulterioris Calabriæ. Francofurti et Moguntiæ. 1787. 4to.

Pet. Cent.—Musei Petiveriani centuria prima. Auct. J. Petiver, &c. London. 1695. 12mo.

Pet. Gaz.—Jacobi Petiveri Gazophylaceum Naturæ et Artis. London. sine anno. Folio.

**Pet. Pap.*—Papilionum Britanniæ Icones, Nomina, &c.; containing the Figures, Names, Places, Seasons, &c. of above Eighty English Butterflies, &c. By J. Petiver. London. 1717.

Preysler B.—Verzeichniss Böhmischer Insekten. Von J. D. Preysler. Prag. 1790. 4to.

Quensel Diss.—Dissertatio Historico-Naturalis, Ignotas Insectorum species continens. Auct. C. Quensel. Lundæ. 1790. 4to.

Raii or *Ray S.*—Historia Insectorum. Auctore Joanne Raio. Londini. 1710. 4to.

Reaum. Ins.—Mémoires pour servir à l'Histoire des Insectes. 6 Tom. Par R. A. S. Reaumur. A Paris. 1734. 4to.

Redi.—Experimenta circa Generationem Insectorum. Amstelodami. 1671. 24mo.

Rees's Cyclop.—The New Cyclopædia. By A. Rees. London. 1802, &c. &c. 4to.

Règne Animal.—Le Règne Animal distribue d'après son Organisation. Tom. 3. Par M. Latreille. A Paris. 1817. 8vo.

Reich M.—Mantissæ Insectorum Iconibus illustratæ, species novas aut nondum depictas exhibentis, Fasc. I. Auct. G. C. Reich, &c. Norimbergæ. 1797. 8vo.

Reich. Ps.—Monographia Pselaphorum. Auct. H. F. L. Reichenbach. Lipsiæ. 1816. 8vo.

Rhein. M.—Rheinisches Magazine zu erweiterung der Naturkunde. Von M. B. Borkhausen. Giesen. 1793. 8vo.

Rœm. G.—Genera Insectorum Linnæi et Fabricii Iconibus illustrata, a J. J. Rœmer. Vitoduri Helvetorum. 1789. 4to.

Rossi F.—Fauna Etrusca, sistens Insecta quæ in Provinciis Florentina et Pisana præsertim collegit P. Rossius, &c. Liburni. 1794. 2 vols. 4to.

Rossi M.—Mantissa Insectorum exhibens species nuper in Etruria collectas, a P. Rossio, adjectis Faunæ Etruscæ illustrationibus ac emendationibus. Pisis. 1792, 1794. 2 tom. 4to.

Sahlb. I. F.—Diss. Ent. Insecta Fennica enumerans. Præs. C. R. Sahlberg. Aboæ. 1817—1823. 8vo.

St. Farg. M.—Monographia Tenthredinetarum Synonymia Extricata. Auct. Am. Le P. de Saint-Fargeau. Parisiis. 1823.

**Samou.*—The Entomologist's Useful Compendium, or an Introduction to the Knowledge of British Insects, &c. By G. Samouelle. London. 1819. 8vo.

**Sam. I.*—A Nomenclature of British Entomology, alphabetically arranged. By G. Samouelle. London. 1819. 8vo.

Schæff. Ic.—D. Jacobi Christiani Schæfferi Icones Insectorum circa Ratisbonam indigenorum, &c. Regensburg. 1769. 3 tom. 4to.

Schell. G. M.—Genres des mouches Diptères representés en xlii planches, a J. R. Schellenberg. Zuric. 1803. 8vo.

Schell. H.—Cimicum in Helvetiæ aquis et terris degentium, &c. Auct. J. R. Schellenberg. Turici. 1800. 8vo.

Schneid. Mag.—Nuestes Magazin fur die Liebhaber der Entomologie, herausgegeben von D. H. Schneider. Stralsund. 1791—1794.

Schön., or *Schön. S.*—Synonymia Insectorum, &c. von C. J. Schönherr. Stockholm. 1806—1827. 4 tom. 8vo.

Schön. Cu.—Curculionidum dispositio methodica. Auct. C. J. Schönherr. Lipsiæ. 1826. 8vo.

Schrank, or *Schrank A.*—F. de P. Schrank, Enumeratio Insectorum Austriæ: cum figuris. Augustæ Vindeliorum. 1781. 1 vol. 8vo.

Schra. B., or *Schrank B.*—Ejusdem Fauna Boica: Durchgedachte Geschichte der in Bagern einheiwischen und zahmen Thiere. Nurnberg. 1798, &c. 8vo.

Schrank N.—Vide *Naturforscher.*

Schwatz N.—Nomenclator über die in den Roselschen Insecten belustigungen und Kleemanschen Beytragen zur Insectengeschichte, &c. Erste Abtheilung: Kafer. Nurnberg. 1793. 4to.

Scop. C.—Entomologia Carniolica, exhibens insecta Carnioliæ indigena et distributa in ordines, genera, species, varietates, methodo Linneana. Auct. J. A. Scopoli. Vindobonæ. 1763. 8vo.

Scriba B.—Beiträge zu der insekten geschichte. Von L. G. Scriba. Frankfurt. 1790. 4to.

Scriba Jo.—Journal für die Liebhaber der Entomologie, herausgegeben von L. G. Scriba. Frankfurt. 1790. 8vo.

Sepp.—Beschouwing der Wonderen Gods in de Minstgeachte Schepzelen of Nederlandsche Insecten. Door J. C. Sepp. Amsterdam. 1760—1829. 5 vols. 4to.

Shaw G. Z.—General Zoology, vol. 6. Insecta. By George Shaw, M.D. &c. 1806. 8vo.

Shaw N. M.—Vivarium Naturæ, or the Naturalist's Miscellany. By G. Shaw. The Figures by R. P. Nodder. London. 1790, &c. 8vo.

Smith and Abb.—The Natural History of the Rare Lepidopterous Insects of Georgia, collected from the observations of John Abbott. By J. E. Smith. Folio. 1797. 2 vols.

**Sower. B. M.*—British Miscellany, or Coloured Figures of new, rare, or little-known Animal Subjects, many not before ascertained to be natives of the British Isles, &c. By J. Sowerby. London. 1806. 8vo.

Spinola I. L.—Insectorum Liguriæ Species novæ aut rariores, quas in Agro Ligustico nuper detexit, descripsit et Iconibus illustravit. Auct. M. Spinola. Genuæ. 1806—1808. 4to.

Stark E.—Elements of Natural History, adapted to the present state of the science, containing the generic characters of nearly the whole animal kingdom, and descriptions of the principal species. By J. Stark. Edinburgh. 1828. 2 vols. 8vo.

Steph. Nom.—The Nomenclature of British Insects. By J. F. Stephens. London. 1829. 12mo.

**Steph. Ill. (H.)*—Illustrations of British Entomology (Haustellata). By J. F. Stephens. London. 1827—1829. 2 vols. 8vo.

**Steph. Ill. (M.)*—Idem: Mandibulata. London. 1827—1829. 2 vols. 8vo.

**Stew.*—Elements of Natural History; being an Introduction to the Systema Naturæ of Linnæus. London and Edinburgh. 1802. 2 vols. 8vo.

Stoll. Ci.—Représentation exactement colorée d'après nature des Punaises, &c. Par C. Stoll. Amsterdam. 1788. 4to.

Stoll. Cic.—Idem: Cigales. Amsterdam. 1788. 4to.

Sturm D. F.—Deutschland's Fauna in Abbildungen nach der Natur mit Beschreibungen von J. Sturm. 5e Abtheilung; die Insecten. Kafer. Nurnberg. 1805—1828. 7 Band. 18mo.

Sturm V.—Verzeichnis meiner Insecten-Samlung oder Entomologisches Handbuch für Leibhaber und Sammler: von J. Sturm. Nurnberg. 1800. 1 tom. 8vo.

Sulzer, or *Sulz. K.*—Dr. Sulzer's Abgekurzte Geschichte der Insecten, 2 theile. Winterthur. 1776. 4to.

Swam. B. N.—The Book of Nature, or the Book of Insects, &c. By John Swammerdam. Translated by John Hill. London. 1758. Fol.

Thunb. A. U.—Museum Naturalium Academie Upsaliensis, &c. Auct. P. C. P. Thunberg. Ups. 1800. 4to.

Thunb. Diss.—Ejusdem Dissertatio Entomologica Novas Insectorum Species sistens, &c. Upsaliæ. 1781—1791.

Thunb. Ins. S.—Dissertatio Entomologica sistens Insecta Sueciæ. Auct. P. C. P. Thunberg. Upsaliæ. 1784—1795.

Thunb. Mus.—Ejusdem Museum Naturalium Academiæ Upsaliensis, &c. Upsaliæ. 1787—1800.

**Turt.*—A General System of Nature through the three grand kingdoms of Animals, Vegetables, and Minerals, &c. Translated from Gmelin's last edition of Systema Naturæ, by W. Turton, M.D. London. 1806, &c. 8vo. Vols. 2 and 3.

Vieillot Anal.—Analyse d'une nouvelle Ornithologie élémentaire. Par L. P. Vieillot. Paris. 1816. 8vo.

Vieweg T. V.—Quoted on the authority of Treitschke. (I am ignorant of the title.)

Vill., *Vill. E.*, *Villers*, or *Villars.*—Car. Linnæi Entomologia &c. curante et augente Carolo de Villers. Lugduni. 1789. 4 tom. 8vo.

Voët C.—J. E. Voët. Descriptiones et Icones Coleopterorum. 4to.

Voët (Pz.)—J. E. Voët Icones Insectorum Coleopterorum, &c. Illustravit Dr. G. W. F. Panzer, &c. Erlangæ. 1794. 4to.

Walck. M.—Mémoires pour servir a l'Histoire Naturelle des Abeilles solitaires qui composent le genre Halicte. Par C. A. Walckenaer. Paris. 1817. 1 vol. 8vo.

Walck. F.—Faune Parisienne, Insectes, ou Histoire abrégée des Insectes des environs de Paris, &c. Par C. A. Walckenaer. 2 tom. 8vo. An xi. 1802.

Weber Obs.—Observationes Entomologicæ continentes Novorum quæ condidit Generum Characteres. Auct. F. Weberi. Killiæ. 1801.

Wied. Z. M.—Zoologische Magazine. Von J. Wiedeman. 1816, &c. 8vo.

Wien. V.—Systematisches Verseichniss der Schmetterlinge der Wienergegend, &c. Wien. 1776. 4to.

**Wilkes.*—One Hundred and Twenty Plates of English Moths and Butterflies, &c. By B. Wilkes. London. 1773. 4to.

**Wilkes D.*—Twelve New Designs of English Butterflies. By B. Wilkes. London. 1742. imp. 4to.

Wilkin Catal.—Catalogue of some rare Insects in the Collection of S. Wilkin. Norwich. 1816. 8vo.

Will. Ich.—Fr. Willoughby de Historia Piscium. Oxf. 1686. Fol.

Wolff C.—Icones Cimicum, descriptionibus illustratæ: fasc. 1—5. Erlangæ. 1800—1806. 4to.

**Wood.*—Illustrations of the Linnæan Genera of Insects. By W. Wood. London. 1821. 12mo.

Zool. Journ.—The Zoological Journal. Edited by N. A. Vigors, &c. London. 1825, &c. 8vo.

LIST OF SUBSCRIBERS.

Charles Anderson, Esq., Lea, Lincolnshire.
Mrs. Ashton, Bexley.
Charles C. Babington, Esq., St. John's College, Cambridge.
J. O. Backhouse, Esq., Darlington.
T. Backhouse, Esq., York.
Mr. Bainbridge, London.
Mr. Henry Baines, York.
Rev. F. Baker.
Mr. J. S. Barnes, St. Luke's Hospital, Old Street.
P. Barrow, Esq., Manchester.
Mr. T. Beck, Mile End.
J. J. Bedingfield, Esq., Ditchingham Hall, Norfolk.
E. T. Bennett, Esq. F.L.S. &c. Edward Street, Portman Square, London.
Mr. Bentley, City Road.
J. E. Bicheno, Esq. F.R.S. &c. Furnival's Inn, London.
Rev. C. S. Bird, M.A. F.L.S., Burghfield, near Reading.
Captain Blomer, Teignmouth.
Mr. Peter Bown, Scarborough.
Rev. W. T. Bree, M.A., Allersley, near Coventry.
W. J. Broderip, Esq. F.L.S. &c., Raymond's Buildings, Gray's Inn.
Mr. Brown, Brighton.
Rev. J. Bulwer, B.A. F.L.S. &c., Dublin.
Mr. Burrough, London.
Cambridge Philosophical Society.
R. Chambers, Esq. F.L.S., Castle Street, Leicester Square.
Miss Champernowne, Teignmouth.
Mr. Chant, City Road.
J. G. Children, Esq. F.R.S. &c., British Museum. 3 copies.
C. C. Clark, Esq., Twickenham.
William Barnard Clarke, Esq., Jun., Ipswich.
W. Clear, Esq., Cork.
T. Cocks, Esq., High Buckington, Devon.
A. Cooper, Esq. R.A., New Millman Street, London.
J. Curtis, Esq., F.L.S., Grove Place, Lisson Grove. 2 copies.
J. C. Dale, Esq. F.L.S., Glanvilles Wootton, Dorset. 2 copies.
J. Dallinger, Esq., Hertford.
C. Darwin, Esq., Cambridge.
A. H. Davis, Esq. F.L.S., Nelson Square, London.
Mr. C. J. Derbyshire.

Mr. B. Dillon, Wrexham.
L. W. Dillwyn, Esq. F.R.S., &c. Pentlegare, Glamorganshire.
J. Dixon, Esq., Worthing.
Mr. H. Doubleday, Epping.
Lady Emily Drummond.
W. H. Dikes, Esq. M.G.S., Lincolnshire.
Rev. — Dyson.
T. W. Edwards, Esq. F.L.S. &c., Chelsea.
E. C. Faithfull, Esq., Gray's Inn.
Rev. J. Fleming, D.D. F.R.S.E. &c., Flisk, Fifeshire.
W. D. Fox, Esq., Christ College, Cambridge. 3 copies.
Rev. J. Francis, M.A. F.L.S., Edgefield, Norfolk.
Right Hon. Lord Galway.
Charles Ganneys, Esq., Bungay.
Rev. W. L. P. Garnons, B.D. F.L.S. &c., Sidney Sussex College, Cambridge.
Mr. S. Gibson, Hebden Bridge, Yorkshire.
Mr. J. B. Giles, London.
Rev. E. Goodenough, D.D. F.R.S. &c., Little Cloisters, Westminster.
A. Griesbach, Esq., Brompton.
J. H. Griesbach, Esq., York Street, Baker Street.
J. E. Gray, Esq. F.G.S. &c., British Museum.
Right Hon. Baroness Grey de Ruthyn, Beaksbourne, near Canterbury.
W. S. Grey, Esq., St. John's College, Cambridge.
Right Hon. Countess Dowager of Guildford.
A. H. Haliday, Esq., Clifton, near Belfast, Ireland.
Mr. S. Hanson, London. 2 copies.
Major General Thomas Hardwicke, F.L.S. &c., Clapham.
Robert Harrison, Esq., Hull.
Mr. J. Hatchett, F.L.S., Kingsland Road.
Rev. T. T. Haverfield, M.A., Kew.
A. H. Haworth, Esq. F.L.S. &c., Little Chelsea.
G. S. Heales, Esq., Doctors' Commons.
Dr. Heinekin, Madeira.
H. Helsham, Esq., Norfolk.
Rev. J. S. Henslow, M.A. F.L.S. &c., Professor of Botany, Cambridge.
W. C. Hewitson, Esq., Newcastle.
Mrs. Hey.
T. C. Heysham, Esq., Carlisle.
Miss Hillman, Teignmouth.
Mr. Hoare, Queen's College, Cambridge.
Mr. Holdsworth, St. Paul's Churchyard. 2 copies.
H. Hole, Esq., Ebberley House, Devon.
Rev. F. W. Hope, M.A. F.L.S. &c., Upper Seymour Street, Portman Square. 5 copies.
M. J. Horner, Darlington.

T. Horsfield, Esq., M.D. F.R.S. &c., Raymond's Buildings.
— Howitt, Esq., M.D., Nottingham.
H. Hursley, Esq.
G. C. Hyndeman, Esq., Portsmouth.
Rev. Dr. Jacob, Plymouth.
Rev. L. Jenyns, M.A. F.L.S. &c., Bottisham Hall, Cambridge.
Mr. Thomas Ingall, Bank.
Mr. A. Ingpen, A.L.S., Chelsea.
Mr. E. Ingpen, Clarendon Square, Somer's Town.
George Johnson, Esq. M.P., Berwick.
J. Jones, Esq.
S. Irvine, Esq., Surgeon R.N.
Miss Kenrick, Oswalds, near Canterbury.
Mrs. Kenrick, Broom, near Dorking.
Rev. W. Kirby, M.A. F.R.S. &c., Barham, near Ipswich. 2 copies.
J. Lamb, Esq., Nottingham.
R. Latham, Esq. F.L.S. &c., Great Russell Street.
W. F. Latham, Esq., Umberleigh House, near Barnstaple.
Mr. Thomas Letts, Royal Exchange.
Robert Leyland, Esq., Halifax.
Mr. R. Lightbody, Liverpool.
J. Lindley, Esq. F.R.S. &c., Prof. of Botany to the London University.
Rev. Thomas Lloyd, M.A., Hertford.
J. C. Loudon, Esq. F.L.S. &c., Bayswater.
George Lyell, jun. Esq. F.R.S. &c., Temple.
T. Lupton, Esq., Leigh Street, Burton Crescent.
W. S. MacLeay, Esq. F.L.S. &c., Havannah.
Mr. J. Marlow, Nottingham.
T. Marshall, Esq., London.
Rev. George May, Monkton Vicarage, Kent.
Mr. T. Maynell, jun.
George Milne, Esq., F.L.S. &c., Forfar.
J. Milne, Esq., Sun Fire Office.
Mr. George Music, Peterhead.
The Norfolk and Norwich Institute.
F. Paget, Esq., Christchurch.
Hon. Capt. W. Percy, R.N., Portman Square.
H. Philips, Esq., York.
William Raddon, Esq., Sidmouth Street, Gray's Inn Lane.
Wm. Reader, Esq., Stoneham, Hants.
J. Rennie, Esq. M.A., Lewisham.
T. Richards, Esq.
M. J. Robinson, Commercial Road.
Rev. G. T. Rudd, M.A., Kimpton Rectory, near Andover.
F. Russell, Esq., Stockwood House, Brislington, near Bristol.
H. Salter, Esq., Plymouth.
Mr. G. Samouelle, A.L.S., British Museum.
J. Scales, Esq., Beachamwell, Norfolk.
E. W. Sergeant, Esq., Manchester.

Miss Simcoe, Exeter.
Lady Charlotte Shaw.
Rev. T. Skrimshire, M.A., South Creak, near Fakenham.
J. Smith, Esq., North Lodge, Rippon, York.
J. de C. Sowerby, Esq. F.L.S. &c., Mead Place, Lambeth.
Mr. J. Sparshall, Norwich.
Wm. Spence, Esq. F.R.S. &c., Brussels. 2 copies.
C. H. Spragg, Esq., Cambridge.
Rev. E. Stanley, M.A. F.L.S. &c., Alderley Rectory, Cheshire.
J. Stark, Esq. F.R.S.E., Edinburgh.
Mr. Stent.
J. Stephenson, Esq. M.D. F.L.S., Millbank Street, Westminster.
Charles Stokes, Esq. F.R.S. &c., Verulam Buildings. 2 copies.
Mr. Stone, Ratcliff.
Rev. J. Streathfield.
H. Strickland, Esq., Ripon.
W. Swainson, Esq. F.R.S. &c., Tittenhanger Green, St. Albans.
J. Tardy, Esq., Dublin.
Mr. William Tayleure, Teignmouth.
C. J. Thompson, Esq. F.L.S., Fulham.
Miss Trotter, Horton, Epsom.
J. R. Turner, Esq., Manchester.
N. A. Vigors, Esq. M.A. F.R.S. &c., Bruton Street, Berkeley Square.
George Wailes, Esq., Newcastle-on-Tyne.
Rev. — Walford, Long Stratton, Norfolk.
Sir P. Walker, Knt. F.R.S.E. &c., Drumseugh, near Edinburgh.
Mr. Ward, Wellclose Square.
Mr. G. Waterhouse, Brompton.
P. W. Watson, Esq., Cottingham, near Hull.
J. Watton, Esq., Upper Bedford Place, Russell Square.
Mr. R. Weaver, Birmingham.
William Wenman, Esq., Trinity Hall, Cambridge.
J. O. Westwood, Esq. F.L.S., Little Chelsea. 2 copies.
Mrs. Wilde, Guilford Street, Russell Square.
Rev. W. H. Wilder, F.L.S., Purley Hall, Reading.
H. Williams, Esq.
Thomas Williams, Esq. M.D., Guilford Street, Russell Square.
Mr. J. Williamson, Scarborough.
J. Wilmore, Esq., Birmingham.
Rev. Edward Wilson, M.A., Swinton, near Rotherham.
W. F. Witherington, Esq., Hadlow Street, Burton Crescent.
William Wood, Esq. F.R.S. &c., Strand. 2 copies.
Mr. G. Wood, Manchester.
William Wolley, Esq., Hull.
Rev. T. Yonge, M.A., Robertsbourne, near Salisbury.

W. Beadon, Esq., Stoneham, Hants.
Rev. James Browne, Norwich.

Works published by the Author.

THE

NOMENCLATURE

OF

BRITISH INSECTS;

Being a Compendious List of such Species as are contained in his *Systematic Catalogue of British Insects*, and forming a Guide to their Classification, &c. &c. Price 4*s.* 6*d.* in cloth boards; or, printed upon one side for labelling Cabinets, &c., price 5*s.*

AND

(In Monthly Numbers, price 5s. each,)

ILLUSTRATIONS

OF

BRITISH ENTOMOLOGY;

OR

A Synopsis of Indigenous Insects;

Containing their Generic and Specific distinctions, with an Account of their metamorphoses, times of appearance, localities, food, and œconomy, as far as practicable: illustrated with Coloured Figures of some of the rarer species.

(To be completed in 87 Numbers; 28 of which are published.)

A

SYSTEMATIC CATALOGUE

OF

BRITISH INSECTS.

INSECTA MANDIBULATA.

ORDO I. COLEOPTERA.

ELEUTHERATA, *Fabricius.*—ELYTROPTERA, *Clairville.*
Anglice "*Beetles, Chaffers, Dors, Clocks and Bobs.*"

SECTIO I. ADEPHAGA, *Clairville.*

ADEPHAGANA, *Kirby.*—ENTOMOPHAGA, *Latreille.*—CARNIVORA, *Cuvier.*
(PENTAMERA *p, Latr.*—CHILOPODOMORPHA *p, MacLeay.*—CHILOPODOMORPHITA *p, Kirby.*)

Subsectio 1. GEODEPHAGA, *MacLeay.*

(CICINDELETÆ *et* CARABICI, *Latr.*—GEODEPHAGENA, *Kirby.*)
"*Ground-beetles.*"

Familia I. CICINDELIDÆ, *Leach.*

(EUPTERINA, *Kirby.*)

Genus 1. CICINDELA, *Linné, &c.* *Scale-beetle, or Sparkler.*
BUPRESTIS *p, Geoffroy.*—ARENARIUS, *Voët.*

No. Sp.
1. 1; sylvatica*. *Linn.* ii. 658.—*Don.* x. 69. *pl.* 351. *f.* 1.—*Stew.* ii. 74.—*Marsh.* i. 391.—*Shaw* G. Z. vi. 86.—*Leach E. E.* ix. 78.—*Samou.* 144. *pl.* 3. *f.* 8.—(*Curtis l. c. infra.*)—*Steph. Ill.* (*M.*) i. 7.
Ar. fuscus. *Voët.* (*Pz.*) ii. 95.

2. 2; hybrida. *Linn.* ii. 657.—*Sower. B. M.* i. *pl.* 18.—*Turt.* ii. 395.—*Ent. Trans.* (*Burrell.*) i. 202.—*Leach E. E.* ix. 78.—

(*Sam. I.* 11.)—*Wood.* i. 61.—(*Curtis l. c. infra.*)—*Steph. Ill.* (*M.*) i. 8.
Ci. maculata. *DeGeer.* iv. 115?
Ci. maritima. *DeJean Col.* i. 52. *pl.* iv. *f.* 5.

3. 3, riparia*? *Megerle.*—*DeJean Col.* i. 50. *pl.* iv. *f.* 2.—*Steph. Ill.* (*M.*) i. 9. *pl.* 1. *f.* 1.

3. *4, aprica. *Steph. Ill.* (*M.*) i. 18.
Ci. hybrida. *DeJean Col.* i. 48. *pl.* iv. *f.* 1.—(*Curtis C. no.* 46.)

4. 5, Sylvicola*. *Megerle.*—*DeJean Sp.* i. 67.—*Curtis.* i. *pl.* 1.—*Steph. Ill.* (*M.*) i. 10.
Ci. hybrida. *Dufts. F.* ii. 225.

5. 6; campestris*. *Linn.* ii. 657.—*Barb. G. pl.* 6. *f.* 12.—*Berk. S.* i. 104.—*Don.* i. 29. *pl.* 12.—*Stew.* ii. 73.—*Marsh.* i. 389.—*Turt.* ii. 395.—*Shaw G. Z.* vi. *p.* 86.—*Leach E. E.* ix. 79.—*Samou.* 57.—(*Curtis l. c. supra.*)—*Steph. Ill.* (*M.*) i. 11.
Ci. austriaca. *Schra. B.* i. 69.
Ar. viridis. *Voët.* (*Pz.*) ii. 96.

6. 7; germanica*. *Linn.* ii. 657.—*Panz. F.* vi. *f.* 5.—*Marsh.* i. 391.—(*Sam. I.* 11.)—(*Curtis l. c. supra.*)—*Steph. Ill.* (*M.*) i. 11.
Ar. carniolicus. *Voët.* (*Pz.*) ii. 97.

Familia II. BRACHINIDÆ, *MacLeay.*

CA.-TRUNCATIPENNES, *Latreille.*

(EUPODINA-AMAURONA *p*, *Kirby.*—CARABIDÆ *p*, *Leach.*)

(CARABUS *p*, CICINDELA *p*, *Linné*, *Marsh. &c.*—BUPRESTIS *p*, *Geoff.*)

Genus 2. DRYPTA, *Fabricius*, *Leach*, *Samouelle*, (*Kirby*,) *Stephens.*

7. 1: emarginata. *Fabr. E.* i. 230.—*Leach E. E.* ix. 81.—*Samou.* 156.—*Steph. Ill.* (*M.*) i. 13. *pl.* 1. *f.* 2.
Ca. chrysostomos. *Marsh.* i. 469.—*Sowcr. B. M.* i. *pl.* 59.
Ca. dentatus. *Rossi F.* i. 222.

Genus 3. POLISTICHUS, *Bonelli*, *Steph.*

GALERITA *p*, *Fabricius.*—ZUPHIUM *p*, *Latreille.*—LEBIA *p*, *Duftschmid.*

8. 1: fasciolatus. *Bonelli.*—(*Ing. Inst.* 89.)—*Steph. Ill.* (*M.*) i. 13. *pl.* 1. *f.* 3.
Ga. fasciolata. *Fabr. E.* i. 216.

Genus 4. ODACANTHA, *Paykul*, *Leach*, *Samou.*, (*Kirby*,) *Steph.*—DRYPTA *p*, *Lamarck.*—ATTELABUS *p*, *Linné.*

9. 1, melanura. *Payk. F.* i. 169.—*Leach E. E.* ix. 82.—(*Samou.* 156.)—*Steph. Ill.* (*M.*) i. 14.
At. melanurus. *Linn.* ii. 620.—*Don.* xv. *pl.* 513.

Ci. angustata. *Fabr. E. S.* 169..
Ca. angustatus. *Sower. B. M.* i. *pl.* 36.—*Ent. Trans* (*Burrell.*) i. 225.

Genus 5. DEMETRIAS, *Bonelli, Samou., Curtis, Steph.*

LEBIA *p, Latr.*—DROMIUS, *Germar.*—RHYZOPHILUS, *Leach.*— (RISOPHILUS *in E. E.* ix. 81.)

10. 1; atricapillus *. *Samou.* 156.—(*Curtis l. c. infra.*)—*Steph. Ill.* (*M.*) i. 15.
Ca atricapillus. *Linn.* ii. 673.—*Mart. C. pl.* 38. *f.* 36?—*Oliv. E.* iii. 111. *pl.* 9. *f.* 106. *a, b*?—*Marsh.* i. 462.
Ri. atricapillus. *Leach E. E.* ix. 81.
Dr. elongatus. *DeJean Sp.* i. 232.

11. 2, monostigma. *Samou.* 156.—*Curtis.* iii. *pl.* 119.—*Steph. Ill.* (*M.*) i. 15.
Le. atricapilla c. *Gyll.* ii. 188?
Ri. monostigma. *Leach MSS.*
Dr. unipunctatus. *Creutzer.*—*Germ. I. N.* i. 1.

Genus 6. DROMIUS, *Bonelli, Samou., Steph.*

DEMETRIAS *p, Panzer.*—LEBIA *p, Latr., Leach, &c.*—DROMIA, (*Kirby.*)

A. Alati.

12. 1; meridionalis*. *DeJean Sp.* i. 242.—(*Ing. Inst.* 88.)—*Steph. Ill.* (*M.*) i. 16.
Ca. agilis. *Panz. F.* lxxv. *f.* 11.

13. 2; agilis*. *DeJean Sp* i. 240.—*Steph. Ill.* (*M.*) i. 21.
Ca. agilis. *Fabr. E.* i. 185.
Ca. 4-maculatus e. *Illig. K.* i. 102.
β; *Steph. Ill. l. c.*
Ca. atricapillus. *Panz. F.* xxx. *f.* 9.—*Turt.* ii. 447.
Ca. truncatus. *Fabr. E.* i. 209.
Dr. rufescens. (*Sam. I.* 15.)
Ca. planatus. *Brahm?*
γ; *Steph. Ill. l. c.*
Ca. fenestratus. *Fabr. E.* i. 209.
Ca. arcticus. *Oliv. E.* iii. 97. *pl.* 12. *f.* 145.
δ; *Steph. Ill. l. c.*
Dr. bimaculatus. (*DeJean C.* 3.)

14. 3; quadrimaculatus*. *DeJean Sp.* i. 239.—(*Samou.* 155.)—*Steph. Ill.* (*M.*) i. 21.
Ca. 4-maculatus. *Linn.* ii. 673.—*Mart. C. pl.* 38. *f.* 47.—*Stew.* ii. 82.—*Marsh.* i. 459.—*Turt.* ii. 460.—*Panz. F.* lxxv. *f.* 10.
Le. 4-maculata. *Leach E. E.* ix. 81.—(*Kirby & Sp. I. E.* ii. 443.)
Ca. crux-major. *Schrank.* 216.

15. 4; quadrinotatus*. *DeJean Sp.* i. 238.—*Steph. Ill.* (*M.*) i. 21.
Ca. 4-notatus. *Panz. F.* lxxiii. *f.* 5.
Le. 4-notata. (*Kirby & Sp. I. E.* ii. 443.)
Ca. punctomaculatus. *Marsh.* i. 460.
Dr. punctomaculatus. (*Sam. I.* 15.)
Le. fasciata b. *Gyll.* ii. 190.
β; *Steph. Ill.* (*M.*) i. 22. *pl.* 1. *f.* 4.

16. 5; melanocephalus*. (*DeJean C.* 3.)—*DeJean Sp.* i. 234. —*Steph. Ill.* (*M.*) i. 22. *pl.* 1. *f.* 5.
Le. pusilla. *Kirby MSS.*
Dr. pusillus. (*Sam. I.* 15.)
Le. linearis b. *Gyll.* ii. 187?
Dr. pallidus. *Sturm.*
β; *Steph. Ill. l. c.*
Le. venustula. *Spence MSS.*
γ; *Steph. Ill. l. c.*
Dr. scutellaris *mihi* (*olim*).

17. 6; impunctatus*. *Kirby MSS.*—*Steph. Ill.* (*M.*) i. 23.
Dr. spilotus var? *DeJean Sp.* i. 246?
Dr. æratus *mihi* (*olim*).

18. 7, humeralis. *Curtis MSS.*—*Steph. Ill.* (*M.*) i. 23.
Dr. quadrillum. *Creutzer?*—*DeJean Sp.* i. 249?

B. Apteri.

a. *Corpore oblongo.*

19. 8, fasciatus. *DeJean Sp.* i. 238.—*Steph. Ill.* (*M.*) i. 24.
Ca. fasciatus. *Fab. E* i. 186.
Ca. atricapillus var. f. *Illig. K.* i. 204.

20. [9, notatus.] (*Ing. Inst.* 88.)—*Steph. Ill.* (*M.*) i. 24.

21. 10; linearis*. (*Sam. I.* 15.)—*DeJean Sp.* i. 233.—*Steph. Ill.* (*M.*) i. 25.
Ca. linearis. *Oliv. E.* iii. 111. *pl.* 14. *f.* 167. *a*, *b*.—*Marsh.* i. 163.
Le. punctato-striata. *Dufts. F.* ii. 258.
Dr. parasiticus. *Leach MSS.*

22. 11; glabratus*. *DeJean Col.* i. 190. *pl.* xv. *f.* 3.—*Steph. Ill.* (*M.*) i. 25.
Le. glabrata. *Dufts. F.* ii. 248.
Le. lævissima. *Kirby MSS.*

23. 12; femoralis*. *Steph. Ill.* (*M.*) i. 25.
Ca. femoralis. *Marsh.* i. 463.

b. *Corpore brevi.*

24. 13; truncatellus*. (*Panz. I.* i. 73.)—*Steph. Ill.* (*M.*) i. 26.
Ca. truncatellus. *Linn.* ii. 673.—*Panz. F.* lxxv. *f.* 12.—*Marsh.* i. 464.—*Turt.* ii. 462.

52. 14; foveolus*. *Steph. Ill.* (*M.*) i. 26.

Le. foveola. *Gyll.* ii. 183.
Le. truncatella. *Latr.* G. i. 193.
Le. punctatella. *Dufts. F.* ii. 248.
Dr. punctatellus. *DeJean Col.* i. 194. *pl.* xv. *f.* 7.

Genus 7. LEBIA, *Latreille, Samou., Curtis, Steph.*

26. 1, crux-minor*. (*Samou.* 155.)—(*Curtis l. c. infra.*)—*Steph. Ill.* (*M.*) i. 24. *pl.* 1. *f.* 6.
Ca. crux-minor. *Linn.* ii. 673.—*Marsh.* 1. 465.—*Turt.* ii. 458.
Ca. crux-major. *Oliv. E.* iii. 96. *pl.* 4. *f.* 41. *a, b.*—*Stew.* ii. 82.
Ca. Andreæ. *Rossi F.* i. 221.
♀, Ca. erratus. *Rossi M.* i. 91?

27 † 2, turcica. *Curtis.* ii. *pl.* 87.—*Steph. Ill.* (*M.*) i. 27.
Ca. turcicus. *Fabr. E.* i. 203. (!)
Le. humeralis. *Sturm.* In Museo *Britannico.*

28 ‡ 3, hæmorrhoidalis. (*Panz. I.* 72.)—(*Ing. Inst.* 88.)—*Steph. Ill.* (*M.*) i. 28.
Ca. hæmorrhoidalis. *Fabr. E.* i. 203.—*Panz. F.* lxxv. *f.* 6.
In Mus. D. *Hope.*

Genus 8. LAMPRIAS, *Bonelli, Samou., Steph.*

LEBIA *p, Latr.*—BRACHINUS *p, Clairv.*—ECHIMUTHUS, *Leach.*

29. 1; cyanocephalus*. *Samou.* 155.—*Steph. Ill.* (*M.*) i. 29. *pl.* 2. *f.* 1.
Ca. cyanocephalus. *Linn.* ii. 671.—*Berk. S.* i. 107.
Ec. cyanocephalus. *Leach E. E.* ix. 81.—(*Sam. I.* 16.)

30. [2; nigritarsis*.] *Leach MSS.*—(*Ing. Inst.* 88.)—*Steph. Ill.* (*M.*) i. 29.

31. 3; chlorocephalus*. *Samou.* 155.—*Steph. Ill.* (*M.*) i. 30.
Ca. cyanocephalo affine. *Illig. K.* i. 207 nota.
Ca. chlorocephalus. *Ent. Hefte.* ii. 117.
Ca. cyanocephalus. *Mart. C. pl.* 36. *f.* 14.—*Don.* iii. 35. *pl.* 86.—*Stew.* ii. 81.—*Marsh.* i. 451.—*Turt.* ii. 457.
Le. cyanocephala. *Lama.* iv. 503.

32. 4, rufipes. (*Ing. Inst.* 88.)—*Steph. Ill.* (*M.*) i. 30.
Le. rufipes. *DeJean Sp.* i. 258?—*DeJean Col.* i. 161. *pl.* xiii *f.* 1.

Genus 9. TARUS, *Clairville, Steph.*

CYMINDIS, *Latr., Leach, Samou.,* (*Kirby.*)—CYMIDIS, *Gyllenhal ex errore.*—ANOMÆUS, *Fischer.*—LEBIA *p, Dufts.*

33. 1, angularis. *Steph Ill.* (*M.*) i. 31.
Cy. angularis. *Gyll.* ii. 173.—*DeJean Col.* i. 144. *pl.* xii. *f.* 1
Ca. humeralis β. *Payk. F.* i. 122.

34. 2: lævigatus*. *Steph. Ill.* (*M.*) i. 32. *pl.* 2. *f.* 2.

35. 3, macularis. *Steph. Ill.* (*M.*) i. 32. *pl.* 2. *f.* 3.
Cy. macularis. *Fisch. E. R.* (*Mannerheim.*) iii. 25. *pl.* 3. *f.* 4 β.

36 ‡ 4, humeralis. *Clairv.* ii. 96.—*Steph. Ill.* (*M.*) i. *pl.* 2. *f.* 4.
Ca. humeralis. *Marsh.* i. 452.
Cy. humeralis. *Leach E. E.* ix. 81.—(*Samou. I.* 14.)
Ca. Dianæ. *Fanz. F.* xxx. *f.* 8.
Ca. humerosus. *Schon. S.* i. 184. In Mus. *Brit. et D. Hope.*

Genus 10. BRACHINUS, *Weber, Leach, Samou.,* (*Kirby*) *Steph.*

37. 1; crepitans*. *Fabr. E.* i. 219.—*Leach E. E.* ix. 81.—*Samou.* 154. *pl.* 3. *f.* 19.—(*Kirby & Sp. I. E.* ii. 246. *Id.* iv. 129.) —*Steph. Ill.* (*M.*) i. 34. *pl.* 2. *f.* 6.
Ca. crepitans. *Linn.* ii. 671.—*Berk. S.* i. 107.—*Mart. C. pl.* 36. *f.* 17.—*Marsh.* i. 468.—*Stew.* ii. 80.—*Turt.* ii. 440. —*Don.* xiv. 49. *pl.* 486.—*Mill. B. E.* 221.
Bombardier. *Bingley.* iii. 147.
Bup. erythrocephalus anglus. *Voët.* (*Panz.*) ii. 78.

37 * 2; immaculicornis*. *DeJean Sp.* ii. 466.—*Steph. Ill.* (*M.*) i. app.
Br. crepitans var. A. *DeJean Sp.* i. 319.
Br. pectoralis. *Dahl.*

38. 3, explodens. *Dufts. F.* ii. 234.—*Steph. Ill.* (*M.*) i. 35.

39. 4; glabratus. *Bonelli.*—*DeJean Sp.* i. 320.—*DeJean Col.* i. *pl.* 8. *f.* 8.—(*Ing. Inst.* 87.)—*Steph. Ill.* (*M.*) i. 36.
Br. strepitans. *Dufts. F.* ii. 235?

40. 5, sclopeta. *Fabr. E.* i. 220.—(*Ing. Inst.* 87.)—*Steph. Ill.* (*M.*) i. 36. *pl.* 2. *f.* 5.

Familia III. SCARITIDÆ, *MacLeay.*

CA.-BIPARTITI, *Latreille.*

(EUPODINA-AMAURONA *p, Kirby.*—CARABIDÆ *p, Leach.*)

(TENEBRIO *p, Linné.*—CARABUS *p, Marsh.*—SCARITES *p, Fabr.*)

Genus 11. SCARITES, *Fabricius,* (*Kirby,*) *Steph.*

A. Tibiis intermediis bispinosis.

41 † 1, Beckwithii. *Steph. Ill.* (*M.*) i. 37. *pl.* iii. *f.* 1.
In Mus. *Brit. D. Swainson et Vigors.*

42 † 2. *subterraneus*[a].

B. Tibiis intermediis unispinosis.

43 ‡ 3. lævigatus[b].

Genus 12. OXYGNATHUS[c].

Genus 13. CLIVINA, *Latreille, Leach, Samou.*, (*Kirby*,) *Curtis, Steph.*

ATTELABUS *p*, *DeGeer.*

45. 1; Fossor*. *Leach E. E.* ix. 79.—(*Sam. I.* 11.)—*Steph. Ill.* (*M.*) i. 39.—(*Curtis l. c. infra.*)
Te. Fossor. *Linn.* ii. 675.—*Turt.* ii. 477.
Sc. arenarius. *Fabr. E.* i. 125.—*Panz. F.* xliii. *f.* 11.
Cl. arenaria. *Latr.*—(*Samou.* 153.)
Ca. distans. *Marsh.* i. 472.
Pseudocupis minor. *Voët.* (*Panz.*) ii. 63.
β, Cl. discicollis. *Megerle?*

46. 2; collaris*. *Steph. Ill.* (*M.*) i. 40. *pl.* iii. *f.* 3.—*Curtis.* iv. *pl.* 175.
Ca. collaris. *Herbst A.* 141. *pl.* 29. *f.* 15.
Te. collaris. *Gmel.* iv. 1994.—*Turt.* ii. 477.
Sc. Fossor β. *Illig. K.* i. 121.
β, Cl. discipennis. *Megerle?*

Genus 14. DYSCHIRIUS, *Panzer, Samou., Steph.*

CLIVINA *p*, *Gyll., Leach.*

47. 1, nitidus. *Steph. Ill.* (*M.*) i. 40.
Cl. nitida. *DeJean Sp.* i. 421.
Cl. elloughtonensis. *Spence MSS.?*

47 * † 2. æratus. *Zool. Journ.* (*Halliday.*) iv.—*Steph. Ill.* (*M.*) i. app. In Mus. D. Halliday.

48. 3, thoracicus. (*Panz. I.* 67.)—*Steph. Ill.* (*M.*) i. 41.

---

[a] 42 † 2. subterraneus. *Fabr. E.* i. 124.—*Oliv. E.* iii. 8. *pl.* i. *f.* 10.—*Steph. Ill.* (*M.*) i. 37, nota.
Te. subterraneus. *Gmel.* iv. 1993.—*Turt.* ii. 476. (!)
At. Fossor. *DeGeer.* iv. 350. *pl.* 13. *f.* 1. 2.
Ca. interruptus. *Herbst A.* 133. *pl.* 29. *f.* 4.
Ca. subterraneus. *Mart. C. pl.* 38. *f.* 46. (!)

[b] 43 ‡ 3. lævigatus. *Fabr. E.* i. 124.—*Wilkins' Catal.* (!)—*Steph. Ill.* (*M.*) i. 38, nota.
Sc. littoralis. *Creutzer.* i. 134.
Sc. sabulosus. *Oliv. E.* iii. 11. *pl.* 1. *f.* 8.

[c] Genus 13. OXYGNATHUS? *DeJean, Steph.*—ARPEPHORUS, *Hope MSS.*
44 † 1, anglicanus. *Steph. Ill.* (*M.*) i. 38. *pl.* iii. *f.* 2, nota.
Ar. anglicanus. *Hope MSS.* In Mus. *D. Hope.*

Sc. thoracicus. *Illig. K.* i. 125.—*Panz. F.* lxxxiii. *f.* 2.—*Ent. Trans.* (*Burrell.*) i. 250, 313.

49. 4, politus*. *Steph. Ill.* (*M.*) i. 41.
Cl. polita. *DeJean Sp.* i. 422.

50. 5, cylindricus. *Steph. Ill.* (*M.*) i. 41.
Cl. cylindrica. *DeJean Sp.* i. 423.
Dy. maritimus *mihi* (*olim*).

51. 6, arenosus. *Leach MSS.*—*Steph. Ill.* (*M.*) i. 42.

52 7, digitatus. *Steph. Ill.* (*M.*) i. 42.
Cl. digitata. *DeJean Sp.* i. 427?

53. 8, æneus*. *Steph. Ill.* (*M.*) i. 42.
Cl. ænea. *Ziegler.*—*DeJean Sp.* i. 423.

54. 9; tristis*——? *MSS.*—*Steph. Ill.* (*M.*) i. 43.
Cl. obscura. *Gyll.* iv. 456?

55. 10; gibbus*. (*Panz. I.* i. 67.)—(*Samou.* 153.)—(*Ing. Inst. pl. f.* 6.)—*Steph. Ill.* (*M.*) i. 43.
Sc. gibbus. *Fabr. E.* i. 126.—*Panz. F.* v. *f.* 1.
Ca. remotus. *Marsh.* i. 473.
Sc. globosus. *Herbst A.* 142. *pl.* 29. *f.* 17.
Cl. gibba. *Leach E. E.* ix. 79.

Genus 15. DISTOMUS, *Leach MSS.*—*Steph.*

DITOMUS *p*, *Bonelli.*—ARISTUS *p*, *Ziegler.*

56 † 1, Leachii. (*Ing. Inst.* 88.)—*Steph. Ill.* (*M.*) i. 44. *pl.* iii. *f.* 4.
Dit. fulvipes. *DeJean Sp.* i. 444?
Ar. sulcatus. (*Bich. Add.* 13. (!) In Mus. *Brit.*

Familia IV. CARABIDÆ, *MacLeay.*

CA.-ABDOMINALES, *Latreille.*

(EUPODINA-LAMPRONA, *Kirby.*)

(CARABUS *p*, *Linné*, *Marsham*, *&c.*—BUPRESTIS *p*, *Geoffroy.*)

Genus 16. CYCHRUS, *Fabricius*, *Leach*, *Samou.*, (*Kirby*,) *Steph.*—TENEBRIO *p*, *Linné.*

57. 1; rostratus*. *Fabr. E.* i. 165.—*Leach E. E.* ix. 82.—(*Samou.* 145.)—*Steph. Ill.* (*M.*) i. 45.
Te. rostratus. *Linn.* ii. 677.
Te. caraboides. *Linn.* ii. 677.
Ca. rostratus. *Mart. C. pl.* 37. *f.* 27.—*Marsh.* i. 470.—*Turt.* ii. 442.—*Don.* xiv. 90. *pl.* 504. *f.* 2.
Ca. coadunatus. *DeGeer.* iv. 92. *pl.* 3. *f.* 13.

58 † 2. elongatus [a].

59 † 3. attenuatus [b].

Genus 17. PROCRUSTES [c].

Genus 18. CARABUS *Auctorum.*

TACHYPUS, *Weber.*

A. Elytris haud sulcatis.

a. *Elytris elevato-striatis, striis alternis interruptis: corpore oblongo subconvexo.* Granulati. *Panz. I.* i. 34.

61· 1, intricatus. *Linn. F. no.* 780.—*Don.* xv. *pl.* 526. *fig. inf.* —*Leach E. E.* ix. 83.—*Samou.* 145.—*Steph. Ill.* (*M.*) i. 47. *pl.* iii. *f.* 5.
Ca. cyaneus. *Fabr. E.* i. 117.—*Turt.* ii. 439.

62 ‡ 2. purpurascens [d].

63. 3; catenulatus *. *Fabr. E.* i. 170.—*Panz. F.* iv. *f.* 6.—*Leach E. E.* ix. 82.—*Samou.* 145.—*DeJean Sp.* ii. 68 (!).—*Steph. Ill.* (*M.*) i. 48.
Ca. intricatus. *Marsh.* i. 432.
Ca. problematicus. *Herbst A.* 177. *pl.* 47. *f.* 5?
Ca. purpurascens. *Payk. Ca.* 13.

64 † 4, agrestis? *Creutz. E. V.* 110. *pl.* 2. *f.* 15. *A*?—*Steph. Ill.* (*M.*) i. 49.
Ca. Leachiellus *mihi* (*olim.*) In Mus. *Brit.*

65. 5; monilis *. *Fabr. E.* i. 171.—*Panz. F.* cviii. *f.* 1.—*Samou.* 146.—*Steph. Ill.* (*M.*) i. 49.
Ca. catenulatus. *Linn. F. no.* 781.—*Oliv. E.* iii. 30. *pl.* 3. *f.* 29.—*Marsh.* i. 434.—*Leach E. E.* ix. 83.
Ca. granulatus. *Berk. S.* i. 107.
β; *Steph. Ill.* (*M.*) i. 50.
Ca. interlineatus. *Wilk. MSS.*
Ca. inæqualis. *Kirby MSS.*
Ca. consitus. *Panz. F.* cviii. *f.* 3.
γ; *Steph. Ill. l. c.*
Ca. rugulosus. *Marsh.* i. 434.

66. 6; cancellatus *. *Fabr. E.* i. 176. *Steph. Ill.* (*M.*) i. 50.

---

[a] 58 † 2. elongatus. *DeJean Sp.* ii. 7.—*Literary Gazette,* 11*th March* 1827. (!)—*Steph. Ill.* (*M.*) i. 46, nota.

[b] 59 † 3. attenuatus. *Fabr. E.* i. 166.—*Steph. Ill.* (*M.*) i. 46, nota.
Ca. proboscideus. *Oliv. E.* iii. 45. *pl.* 11. *f.* 128.

[c] Genus 17. PROCRUSTES, *Bonelli, Stephens.*

60 ‡ 1. coriaceus. *Sturm D. F.* iii. *pl.* liv.—*Steph. Ill.* (*M.*) i. 46, nota.
Ca. coriaceus. *Linn.* ii. 668.—*Turt.* ii. 439 (!).

[d] 62 ‡ 2. purpurascens. *Fabr. E.* i. 170.—*Oliv. E.* iii. 20. *pl.* 4. *f.* 40.—*Stew.* ii. 80 (!).—*Steph. Ill.* (*M.*) i. 48, nota.

Ca. granulatus, b. c. *Linn.* ii. 668?—*Stew.* ii. 79.—*Mart. C. pl.* 37. *f.* 20.—*Don.* vii. 15. *pl.* 222. *f.* 3.—*Marsh.* i. 433.—*Turt.* ii. 442.

Ca. morbillosus. *Leach E. E.* ix. 83.—*Samou.* 146. *pl.* 3. *f.* 17.

67. 7, granulatus *. *Linn.* ii. 668.—*Steph. Ill.* (*M.*) i. 51. *pl.* iv. *f.* 1.

Ca. cancellatus. *Illig. K.* i. 154.

Ca. clathratus. *Scriba B.* i. 12. *pl.* 1. *f.* 6.

68. 8; arvensis *. *Fabr. E.* i. 174.—*Ent. Trans.* (*Sowerby, &c.*) i. 93. *et* 246. *pl.* 4. *fig. med.*—*Leach E. E.* ix. 83.—(*Sam. I.* 9.)—*Steph. Ill.* (*M.*) i. 51.

b. *Elytris lævibus aut punctulatis: corpore valdè elongato convexo.* Punctati. *Panz. I.* i. 37.

69. 9; violaceus *. *Linn.* ii. 669.—*Barb.* G. *pl.* 7. *f.* 2.—*Berk. S.* i. 107.—*Mart. E. pl.* 36. *f.* 16.—*Don.* vii. 13. *pl.* 222. *f.* 1.—*Marsh.* i. 432.—*Stew.* ii. 80.—*Turt.* ii. 439.—*Shaw* G. *Z.* vi. 99.—*Leach E. E.* ix. 82.—*Samou.* 145.—(*Kirby & Sp. I. E.* ii. 247.)—*Steph. Ill.* (*M.*) i. 52.

Ca. coriaceus. *Scop. C. no.* 265.

β; Ca. marginalis. *Fabr. E.* i. 169?

70. 10, glabratus *? *Fabr. E.* i. 170.—*Ent. Trans.* (*Haworth.*) i. 93. *pl.* 2. *fig. inf.*—*Don.* xv. *pl.* 506.—(*Sam. I.* 9.)—(*Kirby & Sp. I. E.* iv. 499.)—*Steph. Ill.* (*M.*) i. 52.

Ca. coriaceus. *DeGeer.* iv. 90.

Ca. convexus. *Herbst A.* 130. *pl.* 29. *f.* 2.

Ca. violaceus. *Müll. Z.* D. *no.* 810.

c. *Elytris confertissimè striatis, singulis seriebus tribus punctorum impressorum distinctissimorum: corpore abbreviato valdè convexo.* Convexi. *Panz. I.* i. 37.

71 ‡ 11, convexus. *Fabr. E.* i. 172.—(*Ing. Inst.* 88.)—*Steph. Ill.* (*M.*) i. 53. *pl.* iv. *f.* 2. In Mus. *Brit. D. Hope, Kirby & Vigors.*

72. 12; hortensis *. *Fabr. E.* i. 172.—*Marsh.* i. 433.—*Steph. Ill.* (*M.*) i. 53.

Ca. nemoralis. *Illig. K.* i. 152.—*Stew.* ii. 79.—*Turt.* ii. 440.—*Leach E. E.* ix. 83.—*Samou.* 145.

Ca. violaceus. *DeGeer.* iv. 89.

Ca. gemmatus. *Don.* vii. 14. *pl.* 222. *f.* 2.

Ca. catenulatus. (*Mill. B. E.* 221. *pl.* 3. *f.* 2.)

d. *Elytris confertissimè striatis, singulis seriebus tribus punctorum excavatorum dilatatorum: corpore oblongo subconvexo.* Striatopunctati. *Panz. I.* i. 38.

73 † 13. gemmatus [a].

---

[a] 73 † 13. gemmatus. *Fabr. E.* i. 172.—*Panz. F.* lxxiv. *f.* 2.—*Stew.* ii. 79 (!).—*Turt.* ii. 440 (!).—*Steph. Ill.* (*M.*) i. 53, nota.

Ca. hortensis. *Linn.* ii. 668.—*Shaw G. Z.* vi. 99 (!).

Ca. striatus. *DeGeer.* iv. 90. *pl.* 3. *f.* 1.

B. Elytris sulcatis.

a. *Elytris seriebus tribus punctorum excavatorum.*

74. 14, clathratus. *Linn.* ii. 669.—*Turt.* ii. 442.—*Ent. Trans.* (*Haworth.*) i. 338.—(*Don.* xv. *pl.* 526. *fig. sup.*)—(*Sam. I.* 9.)—*Wood.* i. 70. *pl.* 25.—*Steph. Ill.* (*M.*) i. 54.
Ca. æneopunctatus. *DeGeer.* iv. 87.
Ca. adspersus. *Gmel.* iv. 1968.

b. *Elytris tantum porcis costisve elevatis tribus.*

75 ‡ 15. auronitens. *Fabr. E.* i. 175.—*Panz. F.* iv. *f.* 7.—*Steph. Ill.* (*M.*) i. 54. In Mus. D. I. H. Griesbach.

76. 16, auratus. *Linn.* ii. 669.—*Stew.* ii. 80.—*Turt.* ii. 441.—(*Kirby & Sp. I. E.* i. 269.)—(*Ing. Inst.* 88.)—*Steph. Ill.* (*M.*) i. 55. *pl.* iii. *f.* 6.
Ca. sulcatus. *DeGeer.* iv. 104. *pl.* 17. *f.* 20.

77. 17 : nitens. *Linn.* ii. 669.—*Don.* ix. 53. *pl.* 313.—*Marsh.* i. 435.—*Sower. B. M.* i. *pl.* 27.—(*Sam. I.* 9.)—*Steph. Ill.* (*M.*) i. 56.
Ca. aureus. *DeGeer.* iv. 91.

Genus 19. CALOSOMA, *Weber, Leach, Samou.*, (*Kirby,*) *Steph.*

78. 1, Sycophanta* ? *Fabr. E.* i. 212.—*Ent. Trans.* (*Haworth.*) i. 252.—*Leach E. E.* ix. 82.—*Leach Z. M.* ii. 94.—(*Sam. I.* 8.)—(*Kirby & Sp. I. E.* i. 269—275. *pl.* 1. *f.* 1.)—*Steph. Ill.* (*M.*) i. 57.
Ca. Sycophanta. *Linn.* ii. 670.—*Turt.* ii. 452.—*Don.* xiv. 27. *pl.* 477. *f.* 1.
Ca. nitens. *Scop. C. no.* 262.

79 ‡ 2. *Scrutator* [a].

80. 3 ; Inquisitor*. *Fabr. E.* i. 212.—(*Leach Z. M.* ii. 93.)—(*Sam. I.* 8.)—*Steph. Ill.* (*M.*) i. 58.
Ca. Inquisitor. *Linn.* ii. 669.—*Mart. C. pl.* 38. *f.* 42.—*Turt.* ii. 452.—*Don.* xiv. 89. *pl.* 504. *f.* 1.

Genus 20. NEBRIA, *Latr., Leach, Samou., Curtis,* (*Kirby,*) *Steph.*

81. 1, complanata. *Leach E. E.* ix. 83.—*Samou.* 146. *pl.* 3. *f.* 18.—(*Curtis l. c. infra.*)—(*Kirby & Sp. I. E.* iii. 513.)—*Steph. Ill.* (*M.*) i. 59.
Ca. complanatus. *Linn.* ii. 671.—*Don.* xiv. 53. *pl.* 488.
Ca. arenarius. *Fabr. E.* i. 179.—*Turt.* ii. 443.
Ne. arenaria. *Latr. H.* viii. 276 (!).—*DeJean Sp.* ii. 223 (!)

---

[a] 79 ‡ 2. Scrutator. *Fabr. E.* i. 213.—*Leach Z. M.* ii. 94. *pl.* 93.—*Steph. Ill.* (*M.*) i. 57, nota.
Ca. Sycophanta. *Linn. Mus. Ulr. no.* 95.—*DeGeer.* iv. 105. *pl.* 17. *f.* 19.

82 ‡ 2. collaris[a].

83. 3, livida. *Steph. Ill.* (*M.*) i. 60.
Ca. lividus. *Linn.* ii. 670?
Ne livida a, *Gyll.* ii. 38?—(*Sam. I.* 28.)—*Curtis* i. *pl.* 6.
Ca. sabulosus. *Fabr. E.* i. 179?

84. 4, picicornis. *DeJean Sp.* ii. 227.—(*Ing. Inst.* 89.)—*Steph. Ill.* (*M.*) i. 60. *pl.* iv. *f.* 3.
Ca. picicornis. *Fabr. E.* i. 180.—*Panz. F.* xcii. *f.* 1.
Ca. erythrocephalus. *Fabr. M.* i. 199.

Genus 21. HELOBIA, *Leach MSS.*, *Curtis*, *Steph.*

NEBRIA *p*, *Gyllenhal.*

85. 1; brevicollis*. (*Curtis l. c. infra.*)—*Steph. Ill.* (*M.*) i. 61.
Ca. brevicollis. *Fabr. E.* i. 191.—*Sturm D. F.* iii. *pl.* lxvii.
Ca. infidus. *Rossi M.* i. 88.
Ca. rugimarginatus. *Marsh.* i. 444.
Ne. brevicollis. *Leach E. E.* ix. 83.—(*Sam. I.* 28.)
β, Ne. fuscata. *Bonelli Obs.* i. 44.

87. 3, Gyllenhalii. *Curtis* iii. *pl.* 103.—*Steph. Ill.* (*M.*) i. 62.
Ca. Gyllenhalii. *Schon. S.* i. 196.
Ne. Gyllenhalii. (*Sam. I.* 28.)
Ca. nivalis β. *Payk. F.* i. 119.

86. 2, Marshallana *mihi.*—*Steph. Ill.* (*M.*) i. 61. *pl.* iv. *f.* 4.—(*Curtis C. no.* 45.)
Ne. arctica. *DeJean Sp.* ii. 235?

Genus 22. ALPÆUS, *Bonelli*, *Steph.*

NEBRIA *p. DeJean.*

88 ‡ 1, castaneus. *Bonelli Obs.* i. 55?—*Steph. Ill.* (*M.*) i. 63.
Ne. castanea. *DeJean Sp.* ii. 250.
Ne. concolor. (*DeJean Cat.* 7.) In Mus. *Brit.*

Genus 23. LEISTUS, *Frölich*, *Samou.*, *Curtis*, *Steph.*

POGONOPHORUS, *Latr.*, *Leach.* (*Kirby.*)—MANTICORA, *Jurine.*

89. 1; spinibarbis*. (*DeJean C.* 7.)—*Steph. Ill.* (*M.*) i. 63.—(*Curtis l. c. infra.*)
Ca. spinibarbis. *Fabr. E.* i. 181.—*Stew.* ii. 80.—*Marsh.* i. 451. *Turt.* ii. 445.—*Panz. F.* xxx. *f.* 6.
Po. cæruleus. *Latr. G.* i. 223.—*Leach E. E.* ix. 83.
Le. cæruleus. (*Samou.* 147.)
β; Ma. pallipes. *Panz. F.* lxxxix. *f.* 2?

---

[a] 82 ‡ 2. collaris. *Steph. Ill.* (*M.*) i. 59, nota.
Ca. collaris. *Act. Not.* (*Thunb.*) iv. 50. *pl. f.* 12.
Ca. lateralis. *Fabr. E.* 180.—*Turt.* ii. 443 (!).

90. 2; fulvibarbis*. *Hoffmannsegg.*—*DeJean Sp.* ii. 215.(!)—*Steph. Ill.* (*M.*) i. 64.)—*Curtis.* iv. *pl.* 176.
Ca. rufibarbis. *Fabr. E.* i. 201?
Le. Rawlinsii. *Leach MSS.*—(*Sam. I.* 24.)
Ca. xantholoma. *Kirby MSS.*

91. 3, montanus. *Dale MSS.*—*Steph. Ill.* (*M.*) i. 64. *pl.* iv. *f.* 5. —(*Curtis C. no.* 45.)

92. 4; spinilabris*. *DeJean Sp.* ii. 216.—*Steph. Ill.* (*M.*) i. 65. —(*Curtis l. c. supra.*)
Ca. spinilabris. *Fabr. E.* i. 204.—*Panz. F.* xxxix. *f.* 11.
Le. testaceus. *Naturf.* (*Fröl.*) xxvii. 8.
Ca. brunneus. *Marsh.* i. 455.
Le. brunneus. (*Sam. I.* 24.)
Ca. rufescens α. *Payk. F.* i. 123.
Ma. fuscoænea. *Panz. F.* lxxxix. *f.* 3.

93. 5; rufescens*. (*Sam. I.* 24.)—*Steph. Ill.* (*M.*) i. 65.—(*Curtis l. c. supra.*)
Ca. rufescens. *Fabr. E.* i. 205. (!)—*Stew.* ii. 81.—*Marsh* i. 458. —*Turt.* ii. 459.
Ca. terminatus. *Panz. F.* vii. *f.* 2.
Po. spinilabris b. *Gyll.* ii. 47.
Le. melanocephalus. *Leach MSS.*

Familia V. HARPALIDÆ, *MacLeay.*

CA.-THORACICI, *Latreille.*
(EUPODINA-AMAURONA, *Kirby.*—CARABIDÆ *p*, *Leach.*)
(CARABUS *p*, *Linné*, *Marsham*, *&c.*—HARPALUS, *Latr.*, *&c.*)
Imps, Sunshiners.

Genus 24. LORICERA, *Latreille, Leach, Samou., Steph.*

94. 1; pilicornis*. *Steph. Ill.* (*M.*) i. 69.
Ca. pilicornis. *Fabr. E.* i. 193.—*Mart. C. pl.* 36. *f.* 18.—*Turt.* ii. 454.—*Marsh.* i. 446.—*Don.* xi. 21. *pl.* 367. *f.* 1.
Ca. seticornis. *Müll. Z. D. no.* 860.
Lo. ænea. *Latr. G.* i. 224.—*Leach E. E.* ix. 82.—(*Samou.* 150.)

Genus 25. PANAGÆUS, *Latreille, Leach, Samou., Steph.*

95. 1; quadripustulatus*. *Sturm D. F.* iii. 72. *pl.* lxxiii. *f. P. p.* (*Ing. Inst.* 89.)—*Steph. Ill.* (*M.*) i. 70. *pl.* iv. *f.* 6.

96. 2; crux-major*. *Latr. G.* i. 220.—*Leach E. E.* ix. 82.—(*Samou.* 147.)—*Steph. Ill.* (*M.*) i. 71.
Ca. crux-major. *Linn.* ii. 673.—*Mart. C. pl.* 38. *f.* 49.—*Marsh.* i. 470.—*Turt.* ii. 458.—*Don.* xiv. 28. *pl.* 477. *f.* 2.
Ca. bipustulatus. *Fabr. M.* i. 204.—*Stew.* ii. 81.
Ca. nobilis. *Gmel.* iv. 1986.

Bu. equestris. *Fourc.* i. 45.
Pa. Crux. *Gyll.* ii. 78.

Genus 26. BADISTER, *Clairville, Leach, Samou., Curtis, Steph.*

AMBLYCHUS, *Gyllenhal.*—LICINUS *p, Latreille.*

97. 1; cephalotes*. *DeJean Sp.* ii. 406.—*Curtis* iii. *pl.* 139.—*Steph. Ill.* (*M*) i. 72.
Ba. megacephalus. *Kirby MSS.*

98. 2; bipustulatus*. *Clairv.* ii. 92.—*Leach E. E.* ix. 80.—(*Samou.* 147.)—(*Curtis l. c. infra.*)—*Steph. Ill.* (*M.*) i. 72.
Ca. bipustulatus. *Fabr. E.* i. 203.—*Marsh.* i. 464.—*Turt.* ii. 459.—*Don.* xv. *pl.* 516.
Ca. crux-minor. *Oliv. E.* iii. 99. *pl.* 8. *f.* 9. *a, b.*—*Stew.* ii. 81.
Ca. balteatus. *Naturf.* (*Schra.*) xxiv. 88.

99. 3; suturalis*. *Steph. Ill.* (*M.*) i. 73. *pl.* v. *f.* 1.

Genus 27. LICINUS, *Latreille, Leach, Curtis, Steph.*

100. 1: depressus*. *Curtis.* i. *pl.* 75.—*Steph. Ill.* (*M.*) i. 73.
Ca. depressus. *Payk. F.* i. 110.
Ca. cassideus. *Illig. K.* i. 159.
Li. cossyphoides. *Sturm D. F.* iii. *pl.* lxxiv. *f.* O. o.
Ca. Watsoni. *Marsh? MSS.*
Ca. olens. *Brahm?*

101 ‡ 2, cassideus. *Clairv.* ii. 104. *pl.* 16. *f. a.*—(*Curtis l. c. supra.*)—*Steph. Ill.* (*M*) i. 74.
Ca. cassideus. *Fabr. E.* i. 190.
Ca. emarginatus. *Oliv. E.* iii. 55. *pl.* 13. *f.* 150.
Li. emarginatus. *Leach E. E.* ix. 80. In Mus. *Brit.*

102. 3: Silphoides*. *Clairv.* ii. 106.—*Sturm D. F.* lxxiv. *f. a—n.*—(*Curt. l. c. supra.*)—*Steph. Ill.* (*M.*) i. 74. *pl.* v. *f.* 2.
Ca. Silphoides. *Fabr. E.* i. 190.
Ca. Agricola. *Oliv. E.* iii. 55. *pl.* 5. *f.* 53?

Genus 28. REMBUS[a].

Genus 29. EPOMIS, *Bonelli, Samou., Steph.*

CHLÆNIUS *p, Sturm.*

104. 1, circumscriptus. *DeJean. Sp.* ii. 369.—*Steph. Ill.* (*M.*) i. 76. *pl.* v. *f.* 3.
Ch. circumscriptus. *Dufts. F.* ii. 166.

---

[a] Genus 28. REMBUS, *Latreille, Steph.*

103. 1, impressus. *Steph. Ill.* (*M.*) i. 75. nota.
Ca. impressus. *Fabr. E.* i. 188.
Ca. subæqualis. *Marsh.* i. 442. Et in Mus. *D. Kirby.*

Ca. cinctus. *Panz. I.* xxx. *f.* 7.—*Mart. C. pl.* 38. *f.* 39.—*Marsh.* i. 449.
Ca. crœsus. *Fabr. E.* i. 183.
β[a]?

Genus 30. CHLÆNIUS, *Bonelli, Samou., Curtis, Steph.*

POECILLUS *p, Samouelle.*—CHLÆNIA, (*Kirby.*)

A. Thorace subquadrato.

105 † 1, sulcicollis. *Curtis.* ii. *pl.* 83.—*Steph. Ill.* (*M.*) i. 77.
Ca. sulcicollis. *Payk. F.* i. 153. In Mus. D. *Curtis.*

106. 2: holosericeus. (*Curtis l. c. supra*). *Steph. Ill.* (*M.*) i. 77. *pl.* v. *f.* 4.
Ca. holosericeus. *Fabr. E.* i. 193.—*Panz. F.* xi. *f.* 9. *a.*
Ca. carbonarius. *Rossi F.* i. 216.

107. 3; nigricornis*. (*Curtis l. c. supra.*)—(*Kirby & Sp. I. E.* iv. 554.)—*Steph. Ill.* (*M.*) i. 77.
Ca. nigricornis. *Fabr. E.* i. 198.—*Marsh.* i. 441.
Ca. nitidulus. *Act. H.* (*Thunb.*) iv. 20.
Ca. holosericeus var. *Panz. F.* xi. *f.* 9. *b, c.*
Ca. Upsaliensis. *Gmel.* iv. 1980.
Po. nigricornis. (*Sam. I.* 34.)

108. 4, melanoceras. (*Ing. Inst.* 88.)—*Steph. Ill.* (*M.*) i. 78.
Ch. melanocornis. *Ziegler.*—*DeJean Sp.* ii. 350.
Ch. nigricornis. *Sturm D. F.* v. 135.
Ca. holosericeus var. δ. (*Schon. S.* i. 198.)

109. 5, fulgidus. *Steph. Ill.* (*M.*) i. 78. *pl.* v. *f.* 5.

110. 6, Agrorum. *DeJean Sp.* ii. 313.—*Steph. Ill.* (*M.*) i. 79. *pl.* v. *f.* 6.
Ca. Agrorum. *Oliv. E.* iii. 86. *pl.* 12. *f.* 144.
Epomis cincta. (*Samou.* 152.)

B. Thorace posticè attenuato.

111. 7; vestitus*. (*Curtis l. c. supra*).—(*Kirby & Sp. I. E.* iv. 554.) —*Steph. Ill.* (*M.*) i. 79.
Ca. vestitus. *Fabr. E.* i. 200.—*Panz. F.* xxxi. *f.* 5.—*Marsh.* i. 448.
Ca. marginatus. *Linn.* ii. 670.
Ca. dubius. *Hoppe.*
Ch. festivus. (*Samou.* 151.)

112 † 8, xanthopus. *Steph. Ill.* (*M.*) i. 80. In Mus. D. *Swainson.*

---

[a] β? Ep. nigricans. *DeJean Sp.* ii. 371.

Genus 31. CALLISTUS, *Bonelli, Samou., Curtis, Steph.*

ANCHOMENUS *p, Sturm.*

113. 1, lunatus. (*Panz. I.* i. 52.)—(*Samou.* 150.)—*Curtis.* iv. *pl.* 180.—*Steph. Ill.* (*M.*) i. 81.
Ca. lunatus. *Fabr. E.* i. 205.—*Marsh.* i. 466.—*Don.* xv. *pl.* 530. —*Turt.* ii. 459.
Ca. crux-minor. *Herbst C.* xxiv. 88.
Ca. Eques. *Schra. A.* 215.
Bu. plateosus. *Fourc.* i. 53.

Genus 32. ANCHOMENUS, *Bonelli, Samou., Steph.*

AGONUM *p, Samouelle.*

A. Alati.

114. 1; prasinus*. (*Panz. I.* i. 55.)—(*Samou.* 151.)—*Steph. Ill.* (*M.*) i. 82.
Ca. prasinus. *Fabr. E.* i. 195.—*Panz. F.* xvi. *f.* 6.
Ha. prasinus. (*Kirby & Sp. I. E.* ii. 246.)
Ca. viridanus. *Fabr. M.* i. 204.—*Mart. C. pl.* 37. *f.* 25.—*Marsh.* i. 467.
Ca. violaceus. *Act. H.* (*Thunb.*) iv. 20.
Ca. viridis. *Gmel.* iv. 1986.

115. 2; albipes*. *Steph. Ill.* (*M.*) i. 82.
Ca. albipes. *Illig. K.* i. 54.
Ca. pallipes: (albipes *Sup.*) *Fabr. E.* i. 187.
Ca. pavidus. *Panz. F.* lxxxiii. *f.* 7.
Ca. circulatus. *Marsh.* i. 450.
Ca. dilutus. *Marsh.* i. 469.
Ag. albipes. (*Sam. I.* 2.)

116. [3, sordidus.] *Steph. Ill.* (*M.*) i. 82.
Ca. sordidus. *Marsh.* i. 457.
Ag. sordidum. (*Sam. I.* 2.)

B. Apteri.

117. 4; oblongus*. (*DeJean C.* 10.)—*Steph. Ill.* (*M.*) i. 82.
Ca. oblongus. *Fabr. E.* i. 186.
Ca. tæniatus. *Panz. F.* xxxiv. *f.* 3.
Ca. elegantulus. *Marsh.* i. 470.
Ca. obscurus. *Herbst A.* 139. *pl.* 29. *f.* 12.

Genus 33. PLATYNUS, *Bonelli, Steph.*

ANCHOMENUS *p, Sturm.*—SPHODRUS *p, Samou.*

A. Alati.

118. 1; angusticollis*. (*DeJean C.* 10.)—*Steph. Ill.* (*M.*) i. 83.
Ca. angusticollis. *Fabr. E.* i. 182.
Ca. assimilis. *Payk. F.* i. 119.—*Panz. F.* lxxiii. *f.* 9.

Ca. collaris. *Marsh.* i. 443.
Sp. collaris. (*Sam. I.* 38.)
Ca. junceus. *Scop. C.* 89.
Sp. affinis. *Kirby MSS.*

B. Apteri.

119 ‡ 2, scrobiculatus. *Steph. Ill.* (*M.*) i. 83.
Ca. scrobiculatus. *Fabr. E.* i. 178.—*Panz. F.* cix. *f.* 7.
In Mus. *Brit.*

Genus 34. SPHODRUS, *Clairville, Leach, Samou.*, (*Kirby*,) *Steph.*

A. Alati.

120. 1; leucophthalmus*. (*Kirby & Sp. I. E.* iii. 332.)—*Steph. Ill.* (*M.*) i. 84.
Ca. leucophthalmus. *Linn.* ii. 668.—*Berk. S.* i. 107.—*Stew.* ii. 79.—*Marsh.* i. 431.
Ca. planus. *Fabr. E.* i. 179.—*Panz. F.* xi. *f.* 4.
Sp. planus. *Clairv.* ii. 86.—*Leach E. E.* ix. 80.—(*Samou.* 152.)
Ca. spiniger. *Payk. F.* i. 114.
Ca. obsoletus. *Rossi F.* i. 209.

B. Apteri.

121. 2; Terricola*. (*Sam. I.* 38.)—*Steph. Ill.* (*M.*) i. 85.
Ca. Terricola. *Payk. M.* 17.—*Panz. F.* xxx. *f.* 3.—*Marsh.* i. 443.
Ca. subcyaneus. *Illig. M.* i. 57.

Genus 35. AGONUM, *Bonelli, Samou., Steph., Curtis.*

A. Thorace suborbiculato.

122. 1; marginatum*. (*Panz. I.* i. 52.)—(*Curtis l. c. infra.*)—*Steph. Ill.* (*M.*) i. 85.
Ca. marginatus. *Linn. F. no.* 804.—*Mart. C. pl.* 36. *f.* 19.—*Marsh.* i. 449.—*Panz. F.* xxx. *f.* 14.—*Turt.* ii. 457.
Ca. Agrorum. *Oliv. E. M.* v. 346?

123. 2; sexpunctatum*. (*Panz. I.* i. 53.)—*Samou.* 150. *pl.* 3. *f.* 20.—(*Curtis l. c. infra.*)—*Steph. Ill.* (*M.*) i. 86.
Ca. sexpunctatus. *Linn.* ii. 672.—*Berk. S.* i. 108.—*Stew.* ii. 82.—*Marsh.* i. 447.—*Turt.* ii. 456.
β? Ca. impressus. *Illig. K.* i. 195?—*Panz. F.* xxxvii. *f.* 17?

124 ‡ 3: austriacum. (*Panz. I.* i. 53.)—*Curtis.* iv. *pl.* 183.—*Steph. Ill.* (*M.*) i. 87.
Ca. austriacus. *Fabr. E.* i. 198.
Ca. nigricornis. *Oliv. E.* iii. 83.
Ca. modestum. *Sturm D. F.* v. 205. In Mus. *Brit. et D. Hope.*

125. 4; parumpunctatum*. (*Panz. I.* i. 53.)—(*Curtis l. c. supra.*)—*Steph. Ill.* (*M.*) i. 87.

Ca. parumpunctatus. *Fabr. E.* i. 199.—*Panz. F.* xcii. *f.* 4.
Ca. cærulescens. *Mart. C. pl.* 35. *f.* 1.—*Marsh.* i. 446.
Ag. cærulescens. (*Sam. I.* 2.)
Ca. sexpunctatus. *Müll. Z. D. pr. no.* 78.
Ca. Mülleri. *Herbst A.* 139.

126. 5, plicicolle. *Nicol. Coll. Hal. p.* 19.—*Steph. Ill.* (*M.*) i. 87.

127. 6; viduum*. (*DeJean C.* 10.)—(*Curtis l. c. supra.*)—*Steph. Ill.* (*M.*) i. 88.
Ca. viduus. *Illig. K.* i. 196.—*Panz. F.* xxxvii. *f.* 8.
Ca. vernalis. *Payk. F.* i. 133.—*Ent. Trans.* (*Haworth.*) i. 94.—*Id.* (*Burrell.*) i. 218.

128. 7; versutum*. *Sturm D. F.* v. 191. *pl.* cxxxii. *f. a. A.*—(*Curtis l. c. supra.*)—*Steph. Ill.* (*M.*) i. 88.

129. 8; læve*. (*DeJean C.* 10.)—*Steph. Ill.* (*M.*) i. 88.
Ha. lævis. *Gyll.* iv. 451.
Ag. nitidulum. *Ziegler.*

130. 9; emarginatum*. *Gyll.* iv. 450.—*Steph. Ill.* (*M.*) i. 89.
Ag. æsopus. *Hoffmansegg MSS.*—(*Curtis l. c. supra.*)
Ag. humerosum. *Kirby MSS.*

131. 10; mœstum*. *Sturm D. F.* v. 187. *pl.* cxxxiv. *f. b. B.*—(*Curtis l. c. supra.*)—*Steph. Ill.* (*M.*) i. 89.
Ca. mœstus. *Ziegler.*—*Dufts. F.* ii. 138.

132. [11; afrum*.] (*Curtis l. c. supra.*)—*Steph. Ill.* (*M.*) i. 89.
Ca. afer. *Ziegler.*—*Dufts. F.* ii. 138.
Ag. afer. *Sturm D. F.* v. 188. *pl.* cxxxiv. *f. a. A.*
Ag. lugubre. (*DeJean C.* 10.)
Ag. complanatum. *Kirby MSS?*
Ag. triste *mihi* (*olim*).

133 † 12. Bogemanni. (*Curtis l. c. supra.*)—*Steph. Ill.* (*M.*) i. 89.
Ha. Bogemanni. *Gyll.* iii. 697. In Mus. D. Curtis.

B. Thorace oblongo-ovato, angulis posticis sæpè subacutis.

134. 13, quadripunctatum. *Sturm D. F.* v. 217.—*Steph. Ill.* (*M.*) i. 90. *pl.* vi. *f.* 1.
Ca. quadripunctatus. *DeGeer.* iv. 102.—*Oliv. E.* iii. 107. *pl.* 13. *f.* 158. *a, b.*
Ca. foveolatus. *Illig. M.* i. 61.
Ca. cærulescens. *Act. U.* (*Thunb.*) v. 117.
Ca. substriatus. *Gmel.* iv. 1987.
Ag. violaceum *mihi.*

135. 14; consimile*. *Steph. Ill.* (*M.*) i. 90.
Ha. consimile. *Gyll.* ii. 160.

136. 15, atratum* *Sturm D. F.* v. 189. *pl.* cxxxv. *f. a. A.*—*Steph. Ill.* (*M.*) i. 91.—(*Curtis l. c. supra.*)

Ca. atratus. *Dahl.*—*Dufts. F.* ii. 138.
Ag. nigrum. (*DeJean C.* 10.)

137. [16; cursitor*.] *Kirby MSS.*—*Steph. Ill.* (*M.*) i. 91.

138. 17; micans*. *Nicol. Col. Hal.* 19.—*Steph. Ill.* (*M.*) i. 91.
Ag. nitidum *mihi.* (*Curtis l. c. supra?*)

139. 18; piceum*. (*Curtis l. c. supra?*)—*Steph. Ill.* (*M.*) i. 91.
Ca. piceus. *Linn.* ii. 672?

140. 19; Simpsoni*. *Spence MSS.*—(*Sam. I.* 2.)—*Steph. Ill.* (*M.*) i. 92.—(*Curtis l. c. supra.*)
Ca. pilipes. *Fabricius?*

141. [20; pullum*.] *Hoffmansegg MSS.*—*Steph. Ill.* (*M.*) i. 92.—(*Curtis l. c. supra.*)

142. 21; striatum* *mihi.*—*Steph. Ill.* (*M.*) i. 92.

143. 22, fuscipenne. *Nicol. Col. Hal.* 20.—*Steph. Ill.* (*M.*) i. 93.

144. 23; fuliginosum*. (*Panz. I.* i. 54.)—*Steph. Ill.* (*M.*) i. 93.
Ca. fuliginosus. *Knoch.*—*Panz. F.* cviii. *f.* 5.

145. 24; picipes*. (*Sam. I.* 2.)—*Steph. Ill.* (*M.*) i. 93.—(*Curtis l. c. supra.*)
Ca. picipes. *Fabr. E.* i. 203.
Ca. lutescens. *Panz. F.* xxx. *f.* 20.
Ca. pelidnus. *Herbst A.* 139.
Ag. brunnipes. (*DeJean C.* 10.)

146 [25; gracile*.] *Sturm* D. *F.* v. 197. *pl.* cxxxvi. *f. a. A.*—*Steph. Ill.* (*M.*) i. 94.—(*Curtis l. c. supra.*)
Ha. picipes *b.* G*yll.* ii. 152.

147. 26; pelidnum*. *Steph. Ill.* (*M.*) i. 94.
Ca. pelidnus. *Payk. F.* i. 134.

148. 27; affine* *mihi.*—*Steph. Ill.* (*M.*) i. 94.
Ag. pelidnum. *Sturm* D. *F.* v. 194. *pl.* cxxxv. *f. b. B.*—(*Curtis l. c. supra?*)

149. 28, pusillum*. *Steph. Ill.* (*M.*) i. 95.

150. 29, livens. *Steph. Ill.* (*M.*) i. 95.—(*Curtis l. c. supra.*)
Ha. livens. *Gyll.* ii. 149.
Ag. bipunctatum. *Sturm* D. *F.* v. 184. *pl.* cxxxiii. *f. b. B.*

Genus 36. ODONTONYX *mihi.*

AGONUM *p, Bonelli, &c.*

151. 1; rotundatus*. *Steph. Ill.* (*M.*) i. 96. *pl.* vi. *f.* 2.
Ca. rotundatus a. *Payk. F.* i. 136.
Ca. ochropus. *Marsh.* i. 471.
Ca. flavipes. *Panz. F.* cviii. *f.* 9?
Ha. picicornis *Museorum Norfolciensium?*

152. [2, rotundicollis.]—*Steph. Ill.* (*M.*) i. 96.

Ca. rotundicollis. *Marsh.* i. 471.
Ca. Sturmii. *Dufts. F.* ii. 143?
Ag. Sturmii. *Sturm* D. *F.* v. 198. *pl.* cxxxvi. *f. b. B?*
Ha. hypoxanthus. *Kirby MSS.?*

Genus 37. SYNUCHUS, *Gyllenhal, Samou., Steph.*

TAPHRIA, *Bonelli.*—AGONUM *p, Panzer.*

153. 1; vivalis*. *Gyll.* ii. 77.—(*Sam. I.* 40.)—*Steph. Ill.* (*M.*) i. 97.
Ca. vivalis. *Illig. K.* i. 197.—*Ent. Trans.* (*Haworth.*) i. 96.
Ca. rotundatus, b. *Payk. F.* i. 136.
Ca. nivalis. *Panz. F.* xxxvii. *f.* 19.
Ca. impiger. *Panz. F.* cviii. *f.* 7.

Genus 38. CALATHUS, *Bonelli, Samou., Curtis, Steph.*

A. Apteri.

a. *Thorace posticè subrotundato.*

154. 1; piceus*. (*Curtis l. c. infra.*)—*Steph. Ill.* (*M.*) i. 98. *pl.* vi. *f.* 3.
Ca. piceus. *Linn.* ii. 672?—*Turt.* ii. 444?—*Marsh.* i. 444.
Ha. interritus. *Spence MSS.*
Cal. Stephensii. *Leach MSS.*

b. *Thorace posticè angulis acutis.*

155. 2; melanocephalus*. (*Panz. I.* i. 60.)—(*Sam. I.* 8.)—(*Curtis l. c. infra.*)—*Steph. Ill.* (*M.*) i. 98.
Ca. melanocephalus. *Linn.* ii. 671.—*Berk. S.* i. 107.—*Mart. C. pl.* 37. *f.* 26.—*Stew.* ii. 81.—*Marsh.* i. 438.—*Don.* xiv. *pl.* 480.—*Turt.* ii. 457.

156. 3; Cisteloides*. (*Panz. I.* 60.)—(*Sam. I.* 8.)—(*Curtis l. c. infra.*)—*Steph. Ill.* (*M.*) i. 99.
Ca. Cisteloides*. *Illig. K.* i. 163.—*Panz. F.* xi. *f.* 12.
Ca. flavipes. *Oliv. E.* iii. 76. *pl.* 8. *f.* 86.—*Marsh.* i. 437.
Ca. punctulatus. *Rossi M.* i. 76.
Cal. frigidus. *Sturm* D. *F.* v. 101. *pl.* cxxi.
Ca. fuscipes. *Gmel.* iv. 1987.
β; Ca. obscurus. *Marsh.* i. 437.

157 ‡ 4, latus. (*DeJean C.* 11.)—*Curtis.* iv. *pl.* 184.—*Steph. Ill.* (*M.*) i. 99. In Mus. *Brit.*

158. 5; crocopus*. *Hoffmansegg MSS.*—*Steph. Ill.* (*M.*) i. 99.
Ca. flavipes var. *Payk. M.* 87?
Cal. flavipes. *Sturm* D. *F.* v. 112. *pl.* cxxii. *f. a, A.*—(*Curtis l. c. supra.*)
Cal. fulvipes. (*DeJean. C.* 11?)

159. 6, fuscus. (*Curtis l. c. supra?*)—*Steph. Ill.* (*M.*) i. 100.
Ca. fuscus. *Fabr. E.* i. 191.
Car. ambiguus. *Oliv. E.* iii. 77. *pl.* 12. *f.* 147.

B. Alati.

160. 7; rufangulus*. *Steph. Ill.* (*M.*) i. 100. *pl.* vi. *f.* 4.
Ca. rufangulus. *Marsh.* i. 441.
Ca. fuscus b. *Payk. F.* i. 165.—(*Curtis l. c. infra?*)

161. 8; mollis*. *Steph. Ill.* (*M.*) i. 101.
Ca. mollis. *Marsh.* i. 456.
Cal. littoralis. *Hoffmansegg.*—(*Sam. I.* 8.)—(*Curtis l. c. infra.*)
Cal. micropterus. *Sturm D. F.* v. 113. *pl.* cxxii. *f. b. B?*
Ha. melanocephalus b. *Gyll.* ii. 129.

Genus 39. PLATYDERES *mihi.*

CALATHUS *p? DeJean.*

162. 1; ruficollis*. *Steph. Ill.* (*M.*) i. 101.
Ca. ruficollis. *Marsh.* i. 456.
Cal. rufus. (*DeJean C.* 11?)
β; Ca. rotundatus. *Marsh.* i. 457.
γ; Pl. ferrugineus *mihi* (*olim*).

Genus 40. ARGUTOR, (*Megerle,*) *Steph.*

PŒCILUS *p, Latreille.*—PLATYSMA *p, Sturm.*—OMASEUS *p, DeJean.*

A. Elytris striâ abbreviatâ suturali nullâ.

163. 1; inquinatus*.
Ca. inquinatus. *Megerle.*
Pl. inquinata. *Sturm D. F.* v. 79. *pl.* cxvi. *f.* c. *C.*

164. 2; vernalis*. (*DeJean C.* 11.)
Ca. vernalis. *Fabr. E.* i. 207.—*Panz. F.* xxx. *f.* 17.
Ca. tibialis. *Marsh.* i. 445.
Ha. tibialis. (*Sam. I.* 20.)

165. 3; inæqualis*.
Ca. inæqualis. *Marsh.* i. 456.
Ca. crenulatus. ——? *MSS.*
Ca. crenata. *Dufts. F.* ii. 92?
Pl. crenata. *Sturm D. F.* v. 73. *pl.* cxv. *f. b. B.*

166. [4; Scalesii*.]
Ca. Scalesii. *Marsh. MSS.*

167. 5, longicollis.
Ca. longicollis. *Dufts. F.* ii. 180.
Pl. longicollis. *Sturm D. F.* v. 80. *pl.* cxvi. *f. d.* D.

B. Elytris striâ suturali abbreviatâ.

168. 6; diligens*. (*DeJean C.* 11.)
Pl. diligens. *Sturm D. F.* v. 81. *pl.* cxvii. *f. A. a.*
Ar. elongatulus *mihi.*

169. 7; interstinctus*.

Pl. interstincta. *Sturm D. F.* v. 77. *pl.* cxvi. *f. b. B.*
Ar. affinis *mihi.*

170. 8; erythropus*.
Ca. erythropus. *Marsh.* i. 461.
Ha. erythropus. (*Sam. I.* 20.)

171. 9; strenuus*. (*DeJean C.* 11.)
Ca. strenuus. *Illig. K.* i. 185.—*Panz. F.* xxxviii. *f.* 6.—*Ent. Trans.* (*Burrell.*) i. 225.
Ca. exaratus. *Marsh. MSS.*

172. 10; pullus*.
Ha. pullus. *Gyll.* iv. 429.
Ar. eruditus. (*DeJean. C.* 11?)
Ar. politus *mihi.*

173. 11; Anthracinus*.
Ca. Anthracinus. *Gyll.* ii. 89.
Ca. Nigrita var. *Payk. F.* i. 157.
Om. minor. (*DeJean C.* 10.)
Ca. Nigrita. *Panz. F.* xi. *f.* 11.

Genus 41. POGONUS, (*Ziegler,*) *Curtis, Steph.*

RAPTOR, (*Megerle.*)—PLATYSMA *p, Sturm.*

174. 1, Burrellii. *Curtis* i. *pl.* 47.
Ca. Burrellii. *Haworth MSS.*
Po. pallidipennis. (*DeJean C.* 9?)
Po. iridipennis. *Nicol. Col. Hal.* 17?

175. 2; chalceus*. *Steph. Ill.* (*M.*) i. *pl.* vi. *f.* 5.
Ca. chalceus. *Marsh.* i. 460.
Ca. littoralis. *Dufts. F.* ii. 183.
Pl. littoralis. *Sturm D. F.* v. 67. *pl.* cxv. *f. a, A.*
Ra. pilipes. *Germar.*
β; Ca. parallelipipedus. *Marsh.* i. 469.
γ; Ca. orichalceus. *Kirby MSS.*

176 [3, brevis] *mihi.*
Po. halophilus. *Nicol. Col. Hal.* 16?

177. 4, æruginosus *mihi.* *Steph. Ill.* (*M.*) i. *pl.* vi. *f.* 6.
Po. oceanicus. (*DeJean C.* 9?

Genus 42. PŒCILUS, *Bonelli,* (*Kirby,*) *Curtis, Steph.*

PLATYSMA *p, Sturm.*—POECILLUS, *Samou.*

A. Apteri.

178. 1; lepidus*. (*Sam. I.* 34.)—(*Curtis l. c. infra.*)—*Steph. Ill.* (*M.*) i. *pl.* vii. *f.* 1.
Ca. lepidus. *Fabr. E.* i. 189.
Ca. vulgaris. *DeGeer.* iv. 97.

† β, Ca. cærulescens. *Herbst. A.* v. 133.
Pœ. lepidus var. f. *Curtis.* iv. *pl.* 107.
β in Mus. *Brit.*, D. *Jenyns, et Samouelle.*

B. Alati.

179. 2; dimidiatus*. (*Sam. I.* 34.)—(*Curtis l. c. supra.*)
Ca. dimidiatus. *Oliv. E.* iii. 72. *pl.* ii. *f.* 121.—*Marsh.* i. 445.
—*Don.* xvi. *pl.* 565.
Ca. Kugellani. *Illig. K.* i. 166.
Ca. tricolor. *Fabr. E.* i. 185?

180 [3, crenatostriatus*] *mihi.*

181. 4; cupreus*. (*Sam. I.* 34.)—(*Curtis l. c. supra.*)
Ca. cupreus. *Linn.* ii. 672.—*Mart. C. pl.* 35. *f.* 2. 3.—*Stew.* ii. 81.—*Marsh.* i. 439.—*Don.* xvi. *pl.* 554.—*Turt.* ii. 454.—*Shaw G. Z.* vi. 100.
β; Ca. cærulescens. *Linn.* ii. 672?—*Stew.* ii. 81.—*Turt.* ii. 454.

182 † 5. rufifemoratus *mihi.*
Pl. affinis. *Ziegler?*—*Sturm* D. *F.* v. *pl.* cxx. *f. a. A.?*
In Mus. D. Halliday.

183. 6; versicolor*. (*Curtis l. c. supra.*)
Pl. versicolor. *Ziegler.*—*Sturm* D. *F.* v. 99. *pl.* cxx. *f. b. B. c.*
Ca. assimilis. *Marsh? MSS.*
Poë. assimilis. (*Steph. Ill.* (*M.*) i. 69. *nota.*)
Poë. micans *mihi* (*olim*).

184 [7; punctatostriatus*] *mihi.*

Genus 43. SOGINES, *Leach MSS., Steph.*

POECILUS *p, DeJean.*

185 † 1, punctulatus.
Ca. punctulatus. *Illig. K.* i. 175.—*Panz. F.* xxx. *f.* 10.
In Mus. *Brit.*

Genus 44. OMASEUS, (*Ziegler,*) *Curtis, Steph.*

PLATYSMA *p, Samou.*—PTEROSTICHUS *p, Sturm.*—PŒCILUS *p, Bonelli.*

A. Alati, elytris punctis excavatis ornatis.

186. 1: aterrimus. (*Nicol. Col. Hal. p.* 22.)—*Curtis.* i. *pl.* 15.
Ca. aterrimus. *Fabr. E.* i. 198.

187. 2, Bulwerii *mihi.*—*Steph. Ill.* (*M.*) i. *pl.* vii. *f.* 2.

188. 3, orinomum. *Leach MSS.*—(*Curtis l. c. supra.*)—*Steph. Ill.* (*M.*) i. *pl.* vii. *f.* 3.

189. 4, Anthracinus. (*DeJean C.* 12.)
Ca. Anthracinus. *Illig. K.* i. 181.
Om. rotundicollis *mihi.*

190. 5; Nigrita*. (*Nicol. Col. Hal. p.* 22.)—(*Curtis l. c. supra.*)

Ca. Nigrita. *Fabr. E.* i. 200.—*Ent. Trans.* (*Burrell.*) i. 218.
Pl. nigritum. (*Sam. I.* 34.)
Ca. aterrimus. *Panz. F.* xxx. *f.* 12.—*Marsh.* i. 444.
β; Ca. confluens. *Panz. F.* xxx. *f.* 22.

191. 6, lævigatus *mihi.*
Ca. lævicollis. *Duft. F.* ii. 163?

192 † 7. rufifemoratus *mihi.* In Mus. D. Halliday.

B. Apteri, elytris haud excavato-punctatis. (Molops *p? Bonelli.*)

193. 8; melanarius*. (*Nicol. Col. Hal. p.* 22.)
Ca. melanarius. *Illig. K.* i. 163.
Ca. leucophthalmus. *Fabr. E.* i. 177.—*Panz. F.* xxx. *f.* 1.—*Turt.* ii. 442.
Ca. angustior. *Marsh.* i. 442.
Ab. angustior. (*Sam. I.* i.)

194. [9; rugicollis*] *mihi.*

195 [10; affinis*.]
Ca. affinis. *Marsh. MSS.*
Ca. intermedius. *Kirby MSS.?*

Genus 45. STEROPUS, (*Megerle,*) *Curtis, Steph.*

Pterostichus *p? Bonelli.*—Molops *p, Sturm.*

196. 1; madidus*. (*DeJean C.* 13.)—(*Curtis l.* c. *infra.*)
Ca. madidus. *Fabr. E.* i. 181.—*Stew.* ii. 81.—*Marsh.* i. 443.—*Turt.* ii. 444.
St. concinnus. (*DeJean C.* 13.)
Mo. concinnus. *Sturm* D. *F.* iv. 175. *pl.* civ. *f.* c.
β; Ha. crenulatus. *Leach MSS.*

197. 2, Æthiops. (*DeJean C.* 13?)
Ca. Æthiops. *Panz. F.* xxxvii. *f.* 22.
St. concinnus. *Curtis.* iv. *pl.* 171.

Genus 46. BROSCUS, *Panzer, Samou.,* (*Kirby,*) *Steph.*

Scarites *p, Olivier.*—Cephalotes[a], *Bonelli.*

198. 1; cephalotes. (*Panz. I.* i. 62.)—(*Samou.* 153.)
Ca. cephalotes. *Linn.* ii. 669.—*Mart. C. pl.* 38. *f.* 41.—*Stew.* ii. 80.—*Marsh.* i. 472.—*Turt.* ii. 449.—*Don.* xiv. *pl.* 484.
Ce. vulgaris. *Bonelli.*
Pseudocupis major. *Voët.* (*Panz.*) ii. 64. *pl.* 33. *f.* 2.

Genus 47. STOMIS, *Clairville, Leach, Samou.,* (*Kirby,*) *Steph.*

199. 1; pumicatus*. *Clairv.* ii. 48. *pl.* vi. *f. A. a.*—*Leach E. E.* ix. 80.—(*Samou.* 153.)

---

[a] Cephalotes: Genus Mammaliorum. Vide Cuvier *Règne Animal. v.* i. *p.* 127.

Ca. pumicatus. *Illig. K.* i. 186.
Ca. tenuis. *Marsh.* i. 468.
β, Ca. picicolor var. minor. *Marsh.* i. 458.

Genus 48. PATROBUS, (*Megerle,*) *Steph.*, *Curtis.*

AGONUM *p*, *Samou.*—BLETHISA *p*, *Panz.*

200. 1; rufipes*. (*DeJean C.* 10.)—(*Ing. Inst.* 89.)—(*Curtis l. c. infra.*)
Ca. rufipes. *Fabr. E.* i. 184.
Ca. excavatus. *Payk. F.* i. 123.—*Panz. F.* xxiv. *f.* 2.—*Ent. Trans.* (*Haworth*) i. 95.
Ca. picicolor var. major. *Marsh.* i. 458.
Ag. rufipes. (*Sam. I.* 2.)
β, Ha. ochropterus. *Kirby MSS.*

201 † 2. alpinus. *Curtis.* iv. *pl.* 192.
Ha. septentrionalis. *Schönher?*—*Gyll.* iv. 428?
In Mus. D. Curtis.

Genus 49. PTEROSTICHUS, *Bonelli*, (*Kirby,*) *Steph.*, *Curtis.*

PŒCILUS *p*, *Panzer.*

202 † 1, fasciatopunctatus. (*Panz. I.* 70.)—(*Curtis l. c. infra.*)
Ca. fasciatopunctatus. *Fabr. E.* i. 178.—*Panz. F.* lxvii. *f.* 9.
In Mus. *Brit.*

203 † 2, brunnipes. *Samou. MSS.*—(*Curtis l. c. infra.*) In Mus. *Brit.*

204 † 3, elongatus. *Samou. MSS.*—*Curtis.* v. *pl.* 196. In Mus. *Brit.*

205 † 4, Panzeri. (*DeJean C.* 12?)
Ca. Panzeri. *Megerle?*—*Panz. F.* lxxxix. *f.* 8?
Pt. Selmanni? (*Curtis l. c. supra?*) In Mus. *Brit.*

206. 5, oblongopunctatus*. (*DeJean C.* 12.)—(*Curtis l. c. supra.*)
Ca. oblongopunctatus. *Fabr. E.* i. 183.—*Panz. F.* lxxiii. *f.* 2.
Ca. hafniensis. *Gmel.* iv. 1977.

207 [6, octopunctatus*?]—*Steph. Ill.* (*M.*) i. *pl.* vii. *f.* 4.—(*Curtis l. c. supra.*)
Ca. octopunctatus. *Marsh.* i. 447.

208. 7; macer*. *Steph. Ill.* (*M.*) i. *pl.* vii. *f.* 5.—(*Curtis l. c. supra.*)
Ca. macer. *Marsh.* i. 466.
Ca. piceus. *Oliv. E.* iii. 58. *pl.* ii. *f.* 123?
Ca. picimanus. *Dufts. F. A.* 159?
Ca. striatus. *Rossi M.* i. 76?
β, Ca. corticalis. *Marsh? MSS.*

Genus 50. PLATYSMA, *Bonelli*, *Steph.*

SIMOTHELUS, (*Megerle.*)—PTEROSTICHUS *p*, *Sturm.*—ABAX *p*, *Samou.*

209. 1; niger*. (*DeJean C.* 12.)

Ca. niger. *Fabr. E.* i. 178.
Ca. nigrostriatus. *DeGeer.* iv. 96.
Ca. striatus. *Payk. F.* i. 115.—*Ent. Trans.* (*Haworth*) i. 93.
Ca. leucopthalmus. *Oliv. E.* iii. 48. *pl.* i. *f.* 4.
Ca. Frishii. *Gmel.* iv. 1981.
Ab. melanarius. (*Sam. I.* 1.)

Genus 51. ABAX, *Bonelli, Samou.*, (*Kirby,*) *Steph.*

PTEROSTICHUS *p, Sturm.*

210. 1; striola*. (*Sam. I.* 1.)
Ca. striola. *Fabr. E.* i. 188.—*Fanz. F.* xi. *f.* 6.
Ca. depressus. *Marsh.* i. 435.

211 ‡ 2, metallicus. (*DeJean C.* 12.)
Ca. metallicus. *Fabr. E.* i. 101.—*Fanz. F.* xi. *f.* 6.
In Mus. D. *I. H. Griesbach.*

Genus 52. OÖDES, *Bonelli, Samou., Steph.*

212. 1; Helopioides*. (*Sam. I.* 32.)—*DeJean Sp.* ii. 378.
Ca. Helopioides. *Fabr. E.* i. 196.—*Fanz. F.* xxx. *f.* 11.
Ca. ovatus *α*. *Payk. M.* 132.

Genus 53. AMARA, *Bonelli, Samou., Steph.*

213. 1; ærata*. *Kirby MSS.*—(*Sam. I.* 2.)—*Steph. Ill.* (*M.*) i. *pl.* vii. *f.* 6.
Am. lata. *Sturm D. F.* vi. 23. *pl.* cxl. *f. B. b?*
β, Am. viridipennis. *Kirby MSS.*

214. 2; acuminata*. *Sturm D. F.* vi. 42. *pl.* cxliii. *f.* c. *C.*
Ca. acuminatus. *Payk. F.* i. 166.
Ca. vulgaris. *Oliv. E.* iii. 75. *pl.* iv. *f.* 36.
Ca. æneus. *DeGeer.* iv. 98.

215 [3; eurynota*.]—(*Panz. I.* 58.)
Ca. eurynotus. *Illig. K.* i. 167.—*Panz. F.* xxxvii. *f.* 23.—*Ent. Trans.* (*Burrell.*) i. 218.

216. 4; similata*. *Sturm D. F.* vi. 40. *pl.* cxliv. *f. a, A.*
Ha. similatus. *Gyll.* ii. 138.

217. 5; vulgaris*. (*Sam. I.* 2.)
Ca. vulgaris. *Linn.* ii. 672.—*Berk. S.* i. 108.—*Mart. C. pl.* 37. *f.* 28?—*Panz. F.* xl. *f.* 1.—*Stew.* ii. 81.—*Marsh.* i. 438.—*Turt.* ii. 455?
Ca. dispar a. *Payk. F.* i. 167.
Ca. ovatus. *Rossi E.* i. 89.
β? Am. obsoleta. *Sturm.* vi. 52. *pl.* cxlv. *f. A. a.*
Ca. obsoleta. *Dufts. F.* ii. 116.

218. 6; ovata*. *Sturm D. F.* vi. 51. *pl.* cxliv. *f. B, b.*
Ca. ovatus. *Fabr. E.* i. 196?

219. 7; trivialis*. (*Nicol. Col. Hal. p.* 14.)
Ca. trivialis. *Dufts. F.* ii. 116.

220. 8; nitida*. *Sturm* D. *F.* vi. 35. *pl.* cxlii. *f. B. b.*
Am. lævis *mihi.*

221. 9; lævis*. *Sturm* D. *F.* vi. 34. *pl.* cxlii. *f. A. a.*
Am. nitens *mihi.*

222. 10; cursor*. *Sturm* D. *F.* vi. 57. *pl.* cxlvi. *f.* D. *d.*
Am. subpunctata. *Kirby MSS.*

223. 11; bifrons*. (*DeJean C.* 9.)
Ha. bifrons. *Gyll.* ii. 144.
Am? castanea. *Ziegler?*

224. 12, brunnea*. (*DeJean C.* 9.)—*Sturm* D. *F.* vi. 56. *pl.* cxlvi. *f.* C. c.
Ha. brunneus. *Gyll.* ii. 143.
Am. æneofusca. *Kirby MSS.*

225. 13, discrepans.
Ca. discrepans. *Marsh. MSS.*

226. 14, crassa *mihi.*

227. 15; convexior*.
Ha. convexior. *Wilk. MSS.*
Am. atrocærulea. *Sturm* D. *F.* vi. 54. *pl.* cxlvi. *f. A. a?*

228. 16; plebeia*. *Sturm* D. *F.* vi. 25.
Ha. plebeius. *Gyll.* ii. 141.

229. 17, obtusa* *mihi.*

230. 18; cognata* *mihi.*

231. 19; latescens* *mihi.*
Am. lata *mihi* (*olim*).

232 † 20, laticollis* *mihi.*
Ca. erraticus. *Dufts. F.* ii. 120?
Am. erratica. *Sturm* D. *F.* vi. 55. *pl.* cxlvi. *f. B. b?*
In Mus. D. *Hatchett.*

233. 21; communis*. (*Fanz. I.* 59.)
Ca. communis. *Fabr. E.* i. 195.—*Panz. F.* xl. *f.* 2.—*Marsh.* i. 439.
Ca. dispar b. *Payk. F.* i. 167.
Ca. ovatus γ. *Payk. M.* 81.
Ha. trivialis. *Gyll.* ii. 140.

234. 22; familiaris*. *Creutzer.*—(*Nicol. Col. Hal. p.* 15.)
Ca. dispar c. *Payk. F.* i. 168.
Ca. ovatus β. *Payk. M.* 81.
Ca. communis. *Illig. K.* i. 168.
Ca. viridis. *Dufts. F.* ii. 120.
Am. intacta. *Kirby MSS.*

235. 23; lucida*.
Ca. lucida. *Andersch.*—*Dufts. F.* ii. 121.
Ha. familiaris b. *Gyll.* iv. 445.
Am. limbata *mihi.*

236. 24; erythropa*.
Ca. erythropus. *Marsh. MSS.*
Am. viridis. *Sturm D. F.* vi. 60. *pl.* cxlvii. *f. B. b.?*

237. 25, atra* *mihi.*

238. 26; tibialis*. (*DeJean C.* 9.)
Ca. tibialis. *Payk. F.* i. 168.—*Ent. Trans.* (*Burrell.*) i. 218.

239. 27; infima*. (*DeJean C.* 9.)
Ca. infimus. *Dufts. F.* ii. 114.
Am. nitidula *mihi.*

Genus 54. BRADYTUS *mihi.*

AMARA *p, Bonelli?*—(*Samou.*)

240. 1; consularis*.
Ca. consularis. *Dufts. F.* ii. 112.
Am. consularis. *Sturm D. F.* vi. 26. *pl.* cxxxix. *f. a. A.*
Ha. latus. *Gyll.* ii. 133?

241. 2; apricarius*.
Ca. apricarius. *Fabr. E.* i. 205.—*Panz. F.* xl. *f.* 3.—*Ent. Trans.* (*Burrell.*) i. 222.
Ha. apricarius. (*Sam. I.* 20.)
Ca. par. *Marsh.* i. 460.
Ca. latus. *Fabr. E.* i. 196?

242. 3; fulvus*.
Ca. fulvus. *DeGeer.* iv. 101?—*Panz. F.* xxxix. *f.* 10.
Ca. apricarius a. *Payk. F.* i. 162.

243. 4; ferrugineus*.
Ca. ferrugineus. *Linn.* ii. 672.—*Panz. F.* xxxix. *f.* 9.—*Stew.* ii. 81.—*Marsh.* i. 440.—*Turt.* ii. 455.
Ha. ferrugineus. (*Sam. I.* 20.)
Ca. concolor. *Oliv. E.* iii. 80. *pl.* 12. *f.* 136?

Genus 55. CURTONOTUS *mihi.*

AMARA *p, Bonelli?*

244. 1; convexiusculus*. *Steph. Ill.* (*M.*) i. *pl.* viii. *f.* 1.
Ca. convexiusculus. *Marsh.* i. 462.

245. 2; aulicus*.
Ca. aulicus. *Illig. K.* i. 174.—*Panz. F.* xxxviii. *f.* 3.
Ha. aulicus. (*Sam. I.* 20.)
Ca. bicolor. *Payk. F.* i. 159.
Ca. ruficornis. *DeGeer.* iv. 95.

Ca. spinipes. *Linn.* ii. 671?
Ca. convexus. *Marsh.* i. 445.
Am. picea. *Sturm* D. *F.* vi. 10.

Genus 56. ZABRUS, *Clairville, Leach, Samou., Steph., Curtis.*

246. 1; gibbus*. (*Sam. I.* 44.)—*Sturm* D. *F.* iv. 128. *pl.* xcviii. —(*Curtis l. c. infra.*)
Ca. gibbus. *Fabr. E.* i. 189.
Ca. tenebroides. *Panz. F.* lxxiii. *f.* 8.
Ca. gibbosus. *Marsh.* i. 436.
Ca. spinipes. *Scop. C.* 267.
Ca. piger. *Fourc.* i. 42.
Ha. tardus. *Latr.* G. i. 205.
Za. tardus. *Leach E. E.* ix. 80.

247 ‡ 2, obesus. (*DeJean C.* 13.)—*Curtis.* iv. *pl.* 188.
Za. obesus. *Latreille?* In Mus. *Brit.*

Genus 57. HARPALUS, *Latreille, Leach, Samou., Steph.*

A. Thoracis angulis posticis obtusis.

a. *Elytris apice sinuato-truncatis.*

1. Thorace posticè haud attenuato.

* *Corpore brevi, convexo.*—Cyrtoma *mihi* (*olim.*)

248. 1, serripes*. *Sturm* D. *F.* iv. *pl.* lxxxi. *f. b. B.*
Ca. serripes. *Schon. S.* i. 199. *pl.* 3. *f.* 4.

249. 2, tardus*. *Sturm* D. *F.* iv. 34.
Ca. tardus. *Illig. K.* i. 168.—*Panz. F.* xxxvii. *f.* 24.
Ca. proteus f. *Payk. F.* i. 163?

250 [3; rufimanus*].
Ca. rufimanus. *Marsh.* i. 441.
Ca. tardus b. *Illig. K.* i. 51?

251. 4, stygius. *Wilkin MSS?*
Cy. obscura *mihi* (*olim*).

252. 5; fuliginosus*. *Sturm* D. *F.* iv. 91. *pl.* xcii. *f. d.* D.
Ca. fuliginosus. *Dufts. F.* ii. 83.

253. 6; latus*.
Ca. latus. *Linn.* ii. 672.—*Schæff. Ic.* 194. *f.* 7.—*Marsh.* i. 440.

254. 7; fuscipalpis*. *Sturm* D. *F.* iv. 66. *pl.* lxxxviii. *f. b. B.*

** *Corpore elongato, subdepresso.*

255. 8; nigripes*. *Sturm* D. *F.* iv. 69. *pl.* lxxxviii. *f. d.* D.

256. 9; piger*. *Sturm D. F.* iv. 31. *pl.* lxxxii. *f.* c. C.
Ca. piger. *Creutzer.*—*Dufts. F.* ii. 104.
Ha. ruficrus? *Kirby MSS.*

257. 10, anxius*. *Sturm* D. *F.* iv. 72. *pl.* lxxxix. *f. b. B.*—*Steph. Ill.* (*M.*) i. *pl.* viii. *f.* 2.

Ca. anxius. *Dufts. F.* ii. 101.
Ha. rufimarginatus. *Wilk. MSS.*
Ha. cognatus. *Kirby MSS.*

258. 11, femoralis *mihi.*

259. 12, maritimus. *Kirby MSS.*
Ha. coracinus. *Sturm* D. *F.* iv. 45. *pl.* lxxxiv. *f. a. A?*

260. 13, flaviventris*. *Sturm* D. *F.* iv. 47. *pl.* lxxxiv. *f. b. B?*
Ca. proteus c. *Payk. F.* i. 164.

*** *Corpore lato, valdè depresso.*

261. 15; thoracicus*. *Leach MSS.—Steph. Ill.* (*M.*) i. *pl.* viii. *f.* 3.
Ca. crassipes. *Dufts. F.* iii. 95?
Ha. crassipes. *Sturm* D. *F.* iv. 14. *pl.* lxxix. *f. b. B?*
Ha. crassus. (*Steph. Ill.* (*M.*) i. 68.)

262 † 14. depressus*. *Sturm* D. *F.* iv. 15. *pl.* lxxx. *f. a. A.*
Ca. depressus. *Dufts. F.* iii. 73. In Mus. D. Waterhouse.

2. Thorace posticè attenuato.

263. 16, cupreus. (*DeJean. C.* 14.)—*Steph. Ill.* (*M.*) *pl.* viii. *f.* 4.

264. 17; rubripes*. *Gyll.* ii. 118.—*Sturm* D. *F.* iv. 55. *pl.* lxxxvi. *f. a. A.*
Ca. azurescens. *Gyll.* iv. 432.

265. [18; azureus*.] *Sturm* D. *F.* iv. 42. *pl.* lxxxiii. *f. c. C.*
Ca. azureus. *Fabr. E.* i. 196.

266. [19; cyanopterus*.] *Kirby MSS.*

267. [20; chloropterus*.] *Kirby MSS.*

268. [21; subsinuatus*.] *Sturm* D. *F.* iv. 52. *pl.* lxxxv. *f. b. B.*
Ca. subsinuatus. *Dufts. F.* ii. 80.
Ha. xanthoceras. *Kirby MSS.*

269. 22; marginellus*. *Ziegler.*—(*DeJean C.* 14.)
Ca. pisseropus. *Kirby MSS.*

270. 23; punctiger* *mihi.*

271. 24; lentus*. *Sturm* D. *F.* iv. 28. *pl.* lxxxii. *f. a. A.*

272. 25; fulvipes*. *Sturm* D. *F.* iv. 58. *pl.* lxxxvi. *f.* c. *C.*

273. 26; limbatus*. *Sturm* D. *F.* iv. 50. *pl.* lxxxv. *f. a. A.*
Ca. limbatus. *Dufts. F.* ii. 84.
Ha. nitidus. *Ziegler?*
Ha. rubripes b, f. *Gyll.* ii. 118.
β; Ha. xanthopus. *Kirby MSS.*

274. 27; niger* *mihi.*

275. 28; annulicornis*. *Kirby MSS.*
Ha. cærulescens. *Sturm.*

276. [29; atrocærulescens*.] *Kirby MSS.*

277. 30; rufipalpis*. *Sturm D. F.* iv. 70. *pl.* lxxxix. *f. a. A.*

278. 31; obscuricornis. *Megerle.—Sturm* D. *F.* iv. 67. *pl.* lxxxviii. *f.* c. *C.*

279. 32; notatus*. *Kirby MSS.*

280. 33; gracilis*. *Wilkin MSS.*

281. 34; cuniculinus*. *Sturm D. F.* iv. 61. *pl.* lxxxvii. *f. b. B.*
Ca. cuniculinus. *Dufts. F.* ii. 87.

282. 35, servus. *Sturm D. F.* iv. 73. *pl.* lxxxix. *f.* c. *C.*
Ca. servus. *Creutzer.—Dufts. F.* ii. 101.

283. 36, luteicornis. *Sturm D. F.* iv. 60. *pl.* lxxxvii. *f. a. A.*
Ca. luteicornis. *Dufts. F.* ii. 86.

284. 37, nitidus. *Sturm* D. *F.* iv. 40. *pl.* lxxxiii. *f. b. B.*

285. 38, ignavus. *Sturm* D. *F.* iv. 44. *pl.* lxxxiii. *f. d.* D.
Ca. ignavus. *Creutzer.—Dufts. F.* ii. 85.
Ha. honestus. (*DeJean C.* 14.]

3. Thoracis lateribus angulisque posticis rotundatis.

* *Antennæ mediocres.*

286. 39, cyaneus *mihi.*

287. 40, pœciloides* *mihi. Steph. Ill.* (*M.*) i. *pl.* viii. *f.* 5.
Ha. virens. *DeJean.*

288. 41, scaritides*. *Sturm D. F.* iv. 81. *pl.* xci. *f.* c. *C?*

** *Antennæ breviores.*

289. 42, vernalis.
Ca. vernalis. *Dufts. F.* ii. 102.
Ha. picipennis. *Sturm* D. *F.* 14. 75. *pl.* xc. *f. a. A.*
Ha. Orfordensis. *Spence MSS.*
Ha. brevicornis. *Kirby MSS.*
Ha. nanus *mihi* (*olim.*)

290. 43, pumilus. *Sturm* D. *F.* iv. 77. *pl.* xc. *f. b, B.*

b. *Elytris apice eroso-dentatis.*

291. 44; æneus*. *Gyll.* ii. 116.
Ca. æneus. *Fabr. E.* i. 197.—*Marsh.* i. 439.—*Panz. F.* lxxxv. *f.* 4.
Ca. discolor. *Marsh.* i. 450.
Ca. azureus. *Marsh.* i. 450.—*Mart. C. pl.* 36. *f.* 15.—*Panz. F.* lxxxv. *f.* 3.
Ca. Proteus. *Payk. F.* i. 163.

292. [45, confinis*.] *Kirby MSS.*
Ha. suturalis *mihi.*

293. [46; æneopiceus*.] *Kirby MSS.*

Ha. æneus d. *Gyll.* ii. 117.
β? Ha. castanipes. *Kirby MSS.*
294. [47; dentatus *.] *Kirby MSS.*
β? Ha. oblongus *mihi.*
295. [48; subcæruleus *.] *Kirby MSS.*
Ha. æneus e. *Gyll.* ii. 117.
β; Ha. brevis *mihi.*

B. Thoracis angulis posticis acutis.

296. 49, caffer *? *Sturm. D. F.* iv. 33. *pl.* lxxxii. *f. d.* D.
Ca. caffer. *Dufts. F.* ii. 99.
Ca. tardus. *Fabr. E.* i. 194.
Ca. nitidulus. *Mus. Marsham.*
Ha. piciceps. *Kirby MSS.*
297. 50; binotatus *. *Gyll.* ii. 122.
Ca. binotatus. *Fabr. E.* i. 193.—*Panz. F.* xcii. *f.* 3.—*Ent. Trans.* (*Haworth.*) i. 94.
Ha. bizonatus. (*Sam. I.* 20.)
298 [51; spurcaticornis.] *Ziegler.*—(*DeJean C.* 15.)
Ca. binotatus var. *Dufts. F.* ii. 78.
299. 52; ruficornis *. *Gyll.* ii. 107.—*Leach E. E.* ix. 80.—*Sturm D. F.* iv. 8. *pl.* lxxvi. *f. a. A.*—(*Sam. I.* 20.)
Ca. ruficornis. *Fabr. E.* i. 180.—*Stew.* ii. 80.—*Mart. C. pl.* 36. *f.* 13.—*Marsh.* i. 436.—*Turt.* ii. 443.
Ca. rufipes. *DeGeer.* iv. 96.
Ca. fuscus. *Gmel.* iv. 1987.
Ca. pulverulentus. *Rossi F.* i. 213.
β; Ca. bicolor. *Marsh.* i. 436.
Ha. bicolor. (*Sam. I.* 20.)
Ca. pubescens. *Müll. Z. D. pr. no.* 77.
300. [53. griseus.] (*DeJean C.* 14.)
Ha. griseus. *Panz. F.* xxxviii *f.* 1.
Ca. ruficornis b. *Illig. K.* i. 171.

Genus 58. OPHONUS, (*Ziegler,*) *Steph., Curtis.*

301. 1, stictus *? *Steph. Ill.* (*M.*) i. *pl.* viii. *f.* 6.
Ca. stictus. *Marsh. MSS.*
Op. elongatulus. (*DeJean C.* 13?)
302. 2; obscurus *. (*Ing. Inst.* 89.)—(*Curtis l. c. infra.*)
Ca. obscurus. *Fabr. E.* i. 192.—*Turt.* ii. 453.
Ha. obscurus. (*Sam. I.* 20.)
Ca. purpuro-cæruleus. *Marsh.* i. 450.
Ha. obscurus. *Sturm D. F.* iv. 85. *pl.* xcii. *f. a. A.*
β, Ha. venustus. *Kirby MSS.*
303. [3; Sabulicola *.] (*DeJean C.* 13.)—*Curtis l. c. infra.*)
Ca. Sabulicola. *Fabr. E.* i. 190?

Ca. azureus. *Oliv. E.* iii. 76. *pl.* 12. *f.* 135.
Ca. obscurus β. (*Schon. S.* i. 197.)
Ha. Sabulicola. *Sturm* D. *F.* iv. 87. *pl.* xcii. *f. B.* (thorax.)

304. 4; punctatulus*. (*DeJean C.* 13.)
Ca. punctatulus. *Dufts. F.* ii. 89.
Ha. punctatulus. *Sturm* D. *F.* iv. 101. *pl.* xciii. *f. d.* D.

305. 5; nitidulus*.
Ca. nitidulus. *Schra. A.* 213.
Ca. holosericeus. *Marsh. MSS.*
Op. Millardii. *Leach MSS.*
Op. nitidulus. (*Curtis l. c. infra?*)

306. 6; azureus*. (*DeJean C.* 13.)—(*Curtis l. c. infra.*)
Ca. azureus. *Illig. M.* i. 51.—*Turt.* ii. 455.
Ca. chlorophanus. *Zenker.*—*Fanz. F.* lxxiii. *f.* 3.

307. 7, germanus. (*DeJean C.* 14.)—*Curtis.* iv. *pl.* 191.—*Steph. Ill.* (*M.*) i. *pl.* ix. *f.* 1.
Ca. germanus. *Linn.* ii. 672.—*Panz. F.* xvi. *f.* 4.
Ha. germanus. (*Sam. I.* 20.)

308. 8; puncticollis*. (*DeJean C.* 13.)—(*Curtis l. c. supra.*)
Ca. puncticollis. *Payk. F.* i. 120.
Ha. puncticollis. *Sturm* D. *F.* iv. 103. *pl.* xciii. *f. a. A?*

309 [9; foraminulosus*.]
Ca. foraminulosus. *Marsh.* i. 457.

310. 10; cribellum*. *Leach MSS.*—(*Curtis l. c. supra.*)
Op. brevicollis. (*DeJean C.* 13?)

311. 11; punctatissimus* *mihi.*
Ha. rupicola. *Sturm* D. *F.* iv. 105. *pl.* xciv. *f. b?*
Op. rupicola. (*Curtis l. c. supra?*)
Op. subcordatus. (*DeJean C.* 13?)

312. 12; puncticeps *mihi.*
Op. angustatus. *Curtis l. c. supra?*

313. 13; subpunctatus *mihi.*

Genus 59. STENOLOPHUS, (*Ziegler,*) *Steph.*

AGONUM *p, Samou.*—TRECHUS *p, Sturm.*

314. 1: vaporariorum*. (*DeJean C.* 15.)—(*Ing. Inst.* 89.)—*Steph. Ill.* (*M.*) i. *pl.* ix. *f.* 2.
Ca. vaporariorum. *Linn.* ii. 671.—*Fanz. F.* xvi. *f.* 7.—*Turt.* ii. 461.
Ca. teutonus. *Schra. A.* i. 214.
Ag. vaporariorum. (*Sam. I.* 2.)

315. [2, Skrimshiranus] *mihi.*

316. 3 : vespertinus*. (*DeJean C.* 15.)—(*Ing. Inst.* 89.)—*Steph. Ill.* (*M.*) i. *pl.* ix. *f.* 3.
Ca. vespertinus. *Illig. K.* i. 197.—*Panz. F.* xxxvii. *f.* 21.
Ca. mixtus. *Herbst A.* 143.
Ca. meridianus, e. *Payk. F.* i. 147.
Ca. lateralis. *Marsh.* i. 457.

317 † [4, Ziegleri.] (*DeJean C.* 15.)
Ca. Ziegleri. *Megerle.*—*Panz. F.* cviii. *f.* 8. In Mus. *Brit.*

318. 5 : dorsalis*. (*DeJean C.* 15.)—*Steph. Ill.* (*M.*) i. *pl.* ix. *f.* 4.
Ca. dorsalis. *Fabr. E.* i. 208.
Tr. dorsalis. *Sturm* D. *F.* vi. 72. *pl.* cxlix. *f.* *b. B.*
Ca. vaporariorum. *Linn. F. no.* 796?
Ca. meridianus, *α.* *Payk. F.* i. 147.
Ca. Kiloniensis. G*mel.* iv. 1978.

319. 6, parvulus*.
Tr. parvulus. *Sturm* D. *F.* vi. 79. *pl.* cl. *f.* *B. b*?
Tr. discoidalis. *Kirby MSS.*

320. 7, cognatus*.
Ca. cognatus. *Gyll.* iv. 455.

Genus 60. TRECHUS, *Clairville, Samou.*, (*Kirby*,) *Steph.*

BEMBIDIUM *p*, *Gyllenhal.*—ELAPHRUS *p*, *Illiger.*—OPHONUS *p*, *DeJean.*

321. 1; pubescens*.
Ca. pubescens. *Payk. F.* i. 124.
Ca. echinatus. *Marsh.* i. 455.
β; Ca. punctulatus. *Marsh.* i. 455.
γ; Ha. limbatus. *Kirby MSS.*
δ; Ha. ruficeps. *Kirby MSS.*

322. 2; pallipes* *mihi.*

323. 3; fulvus*. (*Sam. I.* 42.)
Ca. fulvus. *Marsh.* i. 456.
Ca. collaris. *Payk. F.* i. 146?
Tr. collaris. *Sturm* D. *F.* vi. 74. *pl.* cl. *f.* *a. A.*

324. [4; pallidus*] *mihi.*
Tr. testaceus *mihi* (*olim*).

325. 5; suturalis*. *Leach MSS.*
Tr. fulvescens. (*DeJean. C.* 16?)

326. [6; ruficollis*] *mihi.*

327. 7; flavicollis*. *Sturm* D. *F.* vi. 87. *pl.* cli. *f.* *c. C.*
Tr. fulvicollis. *Leach MSS.*

328. 8; meridianus*. *Clairv.* ii. 24. *pl.* 2. *f.* *A. a.*—(*Sam. I.* 42.)
Ca. meridianus. *Linn.* ii. 673.—*Panz. F.* lxxv. *f.* 9.—*Marsh.* i. 454.—*Turt.* ii. 460.

Ca. cruciger. *Fabr. E.* i. 209?
Ca. DeGeeri. *Gmel.* iv. 1983.

329. 9; minutus *.
Ca. minutus. *Fabr. E.* i. 210.
El. aquatilis. *Illig. K.* i. 232.
Ca. ustulatus, e. *Payk. F.* i. 142.
Ca. aquaticus. *Fanz. F.* xxxviii. *f.* 10.—*Marsh.* i. 461.
Tr. aquaticus. (*Sam. I.* 42.)
Be. Doris, b. *Gyll.* ii. 24.

330. [10; fuscipennis *.]
Ha. fuscipennis. *Wilk.? MSS.*

331. 11; tristis *.
Ca. tristis. *Payk. F.* i. 145.
Ca. nigriceps. *Marsh.* i. 461.
Be. quadristriatum, b. *Gyll.* ii. 31.

Genus 61. BLEMUS, (*Ziegler*,) *Steph.*

BEMBIDIUM *p*, *Gyllenhal.*—TRECHUS *p*, *Samou.*

332. 1, rubens.
Ca. rubens. *Fabr. E.* i. 187.
Tr. rubens. *Clairv.* ii. 26. *pl.* ii. *f. B. b.*
Be. paludosum. *Gyll.* ii. 34.
Ca. tristis var. (*Schon. S.* i. 220.)

333. 2; discus *. (*DeJean C.* 16.)—(*Ing. Inst.* 87.)
Ca. discus. *Fabr. E.* i. 207.
Tr. discus. (*Sam. I.* 42.)
Ca. micros. *Herbst. A.* 142.—*Fanz. F.* xl. *f.* 4.

334. 3, unifasciatus *. (*Ing. Inst.* 87.)
Ca. unifasciatus. *Fanz. F.* xxxviii. *f.* 7.—*Marsh.* i. 466.
Ca. elevatus. *Fabr. E.* i. 204.

335 † 4. longicornis. (*Curtis Cat. no.* 45.)
Tr. longicornis. *Sturm* D. *F.* vi. 83. *pl.* cli. *f. a. A.*
In Mus. D. Curtis.

336. 5; humeralis *. *Steph. Ill.* (*M.*) i. *pl.* ix. *f.* 5.
Tr. humeralis. *Leach MSS.*—(*Sam. I.* 42.)

337. 6, consputus *. *Steph. Ill.* (*M.*) i. *pl.* ix. *f.* 6.
Ca. consputus. *Dufts. F.* ii. 148.
Tr. consputus. *Sturm* D. *F.* vi. 71. *pl.* cxlix. *f. a. A.*
Ha. Ephippiger b. *Gyll.* iv. 454.

Genus 62. EPAPHIUS, *Leach, Samou., Steph.*

BEMBIDIUM *p*, *Gyllenhal.*—TRECHUS *p*, *Sturm.*

338. 1, Secalis. *Leach MSS.*—(*Sam. I.* 16.)
Ca. Secalis. *Payk. F.* i. 146.—*Ent. Trans.* (*Haworth.*) i. 96.
Tr. Secalis. *Sturm* D. *F.* vi. 96. *pl.* clii. *f. d.* D.

Ca. Bruntoni. *Kirby MSS.*
β, Ca. testaceus. *Fabr. E.* i. 209?—*Turt.* ii. 461?

Genus 63. AËPUS, *Leach, Samou., Steph.*
BLEMUS *p, DeJean.*

339. 1, fulvescens. *Leach MSS.—Samou.* 149.

Familia VI. BEMBIDIIDÆ *mihi.*

CA.-SUBULIPALPI, *Latreille.*
(EUPODINA-AMAURONA *p, Kirby.*—CARABIDÆ *p, Leach.*)
(CARABUS *p*, BEMBIDION *et* BEMBIDIUM *Auctorum.*—ELAPHRUS *p, Illiger.*—OCYS, *Kirby MSS.*—OCYDROMUS, *Clairv.*)

Genus 64. LYMNÆUM *mihi.*

340. 1, nigropiceum.
Ca. nigropiceus. *Marsh.* i. 468.

341. 2, depressum.
Tachys? depressus. (*Curtis Cat. no.* 45?)

Genus 65. CILLENUM, *Leach MSS., Samou., Curtis.*
BLEMUS, *p, DeJean.*

342. 1, laterale. *Leach MSS.—Samou.* 148.—*Curtis.* v. *pl.* 200.

Genus 66. TACHYS, (*Ziegler.*)

343. 1, scutellaris *mihi.*
Ci. minimum. (*Curtis. fo.* 200?)

344. 2; binotatus * *mihi.*

345. 3; vittatus * *mihi.*

346. 4; immunis *.
Oc. immunis. *Kirby MSS.*
Ta. violaceus *mihi.*

347. [5; obtusus *.]
Be. obtusum. *Sturm D. F.* vi. 165. *pl.* clxi. *f.* c. *C.*
Oc. fuscipennis. *Kirby MSS.?*
Ta. piceus *mihi.*

348. 6; pusillus *. (*DeJean C.* 16.)

349. 7, gracilis *mihi.*

350. 8, minutissimus. *Leach? MSS.*

Genus 67. PHILOCHTHES *mihi.*
LEJA [a], (*Megerle.*)

351. 1, ænea.
Be. æneum. *Spence MSS.—Germ. I. N.* i. 28.

---

[a] Leia: Genus Dipterorum:—vide *Meig. Zw.* i. 253.

352. 2; fuscipes*.
Le. fuscipes. (*DeJean C.* 17.)
Ca. Doris. *Marsh.* i. 453.

353. 3; subfenestrata*.
Le. subfenestrata. *Megerle.*
Be. apicalis. *Hoffmansegg.*
Ca. tristis. *Sturm: teste* DeJean *C.* 17.

354. 4; biguttatus*.
El. biguttatus. *Illig. K.* i. 230.
Be. biguttatum. *Sturm* D. *F.* vi. 162. *pl.* clxi. *f. b. B.*
Ca. ustulatus. *Berk. S.* i. 108?

355. 5; guttula*.
El. guttula. *Illig. K.* i. 229.
Be. guttula. (*Sam. I.* 6.)
Ca. riparius. *Oliv. E.* iii. 115. *pl.* xiv. *f.* 162. *A. b.*

356. 6; hæmorrhous*.
Oc. hæmorrhous. *Kirby MSS.*

Genus 68. OCYS.

357. 1; currens*. *Kirby MSS.*
Be. cyaneum. *Wilkin MSS.*

358. 2; tempestivus.
Ca. tempestivus. *Fanz. F.* lxxiii. *f.* 6.

359. [3; melanocephalus*.]
Be. melanocephalum. *Leach MSS.*
Be. ruficolle. *Gyll.* iv. 401?

Genus 69. PERYPHUS, (*Megerle.*)

A. Elytris maculatis.

360. 1; femoratus*. (*DeJean C.* 17.)
Be. femoratus. *Sturm* D. *F.* vi. 117. *pl.* clv. *f. B. a.*
Oc. cruciger. *Kirby MSS.*
Be. crucigerum. (*Sam. I.* 6.)

361. 2; concinnus*.
Oc. concinnus. *Kirby MSS.*

362. 3; saxatile*.
Be. saxatile. *Gyll.* iv. 407?
Pe. femoralis *mihi* (*olim.*)

363. 4; littoralis*.
Ca. littoralis. *Oliv. E.* iii. 110. *pl.* 9. *f.* 103.—*Marsh.* i. 452.—*Fanz. F.* xxxv. *f.* 6.
Ca. ustulatus β. *Payk. F.* i. 141.
El. rupestris. *Fabr. E.* i. 246.
Be. littorale. *Leach E. E.* ix. 79.—(*Sam. I.* 6.)

364. 5, lunatus. (*DeJean C.* 17.)
El. lunatus. *Dufts. F.* ii. 24.
Be. lunatum. *Sturm* D. *F.* vi. 119. *pl.* clv. *f. C.* c.

365. 6, ustus. (*DeJean C.* 17.)
Ca. ustus. *Schon. S.* i. 221.
Ca. Spencii. *Marsh. MSS.*
Be. Spencii. (*Sam. I.* 6.)

366. 7; decorus*. (*DeJean C.* 17.)
Be. decorum. *Fanz. F.* lxxiii. *f.* 4.
Ca. immunis. *Marsh.* i. 452.
Be. rufipes, b. *Gyll.* ii. 18.

B. Elytris haud maculatis.

367. 8 viridiæneus *mihi.*
Be. virens. *Gyll.* iv. 407?

368. 9; nitidulus*.
Ca. nitidulus. *Marsh.* i. 454.
Oc. celer. *Kirby MSS.*

369. 10, agilis.
Be. agilis. *Spence MSS.*—(*Sam. I.* 6.)
Oc. auratus. *Kirby MSS.*
Pe. fuscicornis. (*DeJean C.* 17?)
Be. luridum. *Sturm* D. *F.* vi. 125. *pl.* clvi. *f. b. B?*

370. 11, tibialis.
Be. tibiale. *Leach MSS.*

371. 12, Leachii.
Be. Leachii. *Spence MSS.*

372. 13, picipes.
Oc. picipes. *Kirby MSS.*

373. 14, atrocæruleus *mihi.*
El. prasinus. *Dufts. F.* ii. 201?
Be. olivaceum. *Gyll.* iv. 408?

Genus 70. NOTAPHUS, (*Megerle.*)

A. Elytris maculatis aut variegatis.

374. 1; articulatus*. (*DeJean C.* 16.)
Be. articulatus. *Sturm* D. *F.* vi. 156. *pl.* clx. *f.* D. *d.*
Be. undulatum. *Sturm?*
Be. majus. *Gyll.* iv. 411.

375. 2; ustulatus*. (*DeJean C.* 16.)
El. ustulatus. *Illig. K.* i. 231.
Ca. ustulatus. *Linn.* ii. 673.—*Stew.* ii. 82.—*Turt.* ii. 461.—
Ent. Trans. (*Burrell.*) i. 224.

376. 3, nebulosus *mihi.*
Ca. ustulatus. *Fanz. F.* xl. *f.* 7?
Ca. dentellus. *Thunb. Mus. Ups.* iv. 50?

377. 4; semipunctatus.
Ca. semipunctatus. *Don.* xi. 22. *pl.* 367. *f.* 2.
Be. varium. *Latr. G.* i. 185?
Be. bifasciatum. *Leach MSS.*

378. 5, obliquus.
Be. obliquus. *Sturm* D. *F.* vi. 160. *pl.* clxi. *f. A. a.*
No. confinis *mihi.*

379. 6; stictum* *mihi.*

380. 7, fumigatus. (*DeJean C.* 16.)
El. fumigatus. *Creutzer.*—*Dufts. F.* ii. 204.

B. Elytris immaculatis.

381. 8: ephippium.
Ca. ephippium. *Marsh.* i. 462.
Be. ephippium. (*Sam. I.* 6.)

382. 9; castanopterus* *mihi.*

Genus 71. LOPHA, (*Megerle.*)

LEJA *p*, *DeJean.*

383. 1; pœcila*. (*DeJean C.* 18.)
Be. pœcilum. *Hoffmansegg MSS.*—(*Sam. I.* 6.)
Ca. articulatus. *Fanz. F.* xxx. *f.* 21?
Ca. subglobosus β. *Payk. F.* i. 143.
El. quadrimaculatus β. *Illig. K.* i. 233.
Ca. nebulosus. *Marsh. MSS.*

384. 2; quadriguttata*. (*DeJean C.* 17.)
Ca. quadriguttatus. *Fabr. E.* i. 207.—*Mart. C. pl.* 38. *f.* 37. —*Panz. F.* xl. *f.* 5.—*Stew.* ii. 82.—*Marsh.* i. 459.—*Turt.* ii. 460.
Be. quadriguttatum. (*Sam. I.* 6.)
Bu. ustulatus. *Fourc. P.* i. 46.
Chlorocephalotes mas. *Voët.* (*Pz.*) ii. *pl.* 35. *f.* 20.

385. 3; quadrimaculata*. (*DeJean. C.* 17.)
Ci. 4-maculata. *Linn.* ii. 658.
Ca. subglobosus. *Payk. F.* i. 142.
Ca. pulchellus. *Panz. F.* xxxviii. *f.* 8.
Be. 4-notatum. *Mus. Marsham.*
Ca. 4-maculatus. *Berk. S.* i. 108.

386. 4, pulchella.
Ca. pulchellus. *Marsh.* i. 454.

387. 5, Doris.

El. Doris. *Illig. K.* i. 232.—*Fanz. F.* xxxviii. *f.* 9.
Be. Kirbii. *Spence? MSS.*

388. 6; Spencii *.
Oc. Spencii. *Kirby MSS.*

389. 7; assimilis *.
Be. assimile. *Gyll.* ii. 26.
Oc. hæmorrhoidalis. *Kirby MSS.*

390. 8, Kirbii * *mihi.*

391. 9; nigra *.
Be. nigra. *Wilkin MSS.*
Oc. pulchellus. *Kirby MSS.*

392. 10; pulicarius *.
Ca. pulicarius. *Marsh. MSS.*
Oc. nitidulus? *Kirby MSS.*

393. 11; minima *.
Ca. minimus. *Fabr. E.* i. 210?

Genus 72. TACHYPUS. (*Megerle?*)

LEJA *p, DeJean.*

A. Corpore brevi, ovato.

394. 1, Andreæ.
Ca. Andreæ. *Fabr. E.* i. 204.
El. pallidipennis. *Illig. M.* i. 489.

B. Corpore elongato, oblongo.

395. 2, striatus.
El. striatus. *Dufts. F.* ii. 198.
Be. striatus. *Sturm* D. *F.* iv. 186. *pl.* clxiii. *f. B. b.*
Be. puncticolle. *Leach MSS.*—(*Sam. I.* 6.)
Oc. chalconotus. *Kirby MSS.*

396. 3, chlorophanus. (*Curtis Cat. no.* 45.)
Be. chlorophanus. *Sturm* D. *F.* vi. 187. *pl.* clxiii.

397. 4; bipunctatus *.
Ca. bipunctatus. *Linn.* ii. 672.—*Oliv. E.* iii. 112. *pl.* xiv. *f.* 163. *a, b.*—*Marsh.* i. 453.
Be. bipunctatum. (*Sam. I.* 6.)

398. 5; properans *.
Be. properans. *Hoffmansegg MSS.*—(*Sam. I.* 6.)
Ca. rufipes. *Marsh.* i. 453.
Be. rufipes. (*Sam. I.* 6.)

399. 6; celer *.
Ca. celer. *Fabr. E.* i. 210.—*Turt.* ii. 461.
Ca. pygmæus. *Fayk. F.* i. 148.
Ca. rufipes. *Oliv. E.* iii. 112. *pl.* 14. *f.* 164. *a. b?*
Ca. Lampros. *Herbst A.* 143?

400 [7; acutus*.]
Ca. acutus. *Marsh.* i. 461.
Be. acutum. (*Sam. I.* 6.)

401. 8; orichalcicus*.
El. orichalcicus. *Illig. K.* i. 228.
Ca. orichalceus. *Fanz. F.* xxxviii. *f.* 11.

402. 9; chalceus* *mihi.*

Genus 73. BEMBIDIUM, *Illiger, Leach, Samou.*
CICINDELA *p, Linné.*

403 ‡ 1, paludosum. (*Fanz. I.* 47.)
El. paludosus. *Fanz. F.* xx. *f.* 4. In Mus. *Brit. et* D. *Haworth.*

404. 2, impressum. *Gyll.* ii. 13.
El. impressus. *Fabr. E.* i. 246.—*Fanz. F.* xl. *f.* 8.
Ci. striata. *Marsh.* i. 393.

405. 3; flavipes*. *Gyll.* ii. 12.—*Leach E. E.* ix. 79.—(*Sam. I.* 6.)
Ci. flavipes. *Linn.* ii. 658.—*Fanz. F.* xx. *f.* 2.—*Marsh.* i. 394.
—*Turt.* ii. 400.

406. 4, pallipes. *Sturm* D. *F.* vi. 111. *pl.* cliv. *f.* *B. b.*
Ca. pallipes. *Dufts. F.* ii. 197.
Be. pallipes. (*Sam. I.* 6.)
Be. cromerense. *Wilkin MSS.*

Familia VII. ELAPHRIDÆ *mihi.*

(CA.-ABDOMINALES *p, Latreille.*—EUPODINA-LAMPRONA *p, Kirby.*)

(CARABUS *p,* CICINDELA *p, Linné, &c.*—ELAPHRUS *p, Fabr. &c.*)

Genus 74 NOTIOPHILUS, *Dumeril, Leach, Samou.,* (*Kirby.*)

407. 1; aquaticus*. *Leach E. E.* ix. 79.—(*Sam. I.* 30.)
Ci. aquatica. *Linn.* ii. 658.—*Berk. S.* i. 105.—*Stew.* ii. 74.—
Don. viii. *pl.* 351. *f.* 2.—*Marsh.* i. 393.—*Turt.* ii. 400.
Ci. striata. *DeGeer* iv. 18.
β; El. semipunctatus. *Fabr. E.* i. 246.
Ci. semipunctata. *Marsh.* i. 394.
γ, No. purpuro-cæruleus *mihi.*

408. 2; biguttatus*. *Leach E. E.* ix. 79.—(*Sam. I.* 30.)
El. biguttatus. *Fabr. E.* i. 247.
Ci. biguttata. *Marsh.* i. 395.
El. aquaticus β. *Illig. K.* i. 224.

Genus 75. ELAPHRUS, *Fabricius, Leach, Samou.,* (*Kirby,*) *Curtis.*

409. 1; uliginosus*. *Fabr. E.* i. 245.—*Leach E. E.* ix. 79.—
(*Sam. I.* 16.)—*Curtis* iv. *pl.* 179.
Ci. uliginosa. *Marsh.* i. 392.
Ci. riparia. *Schrank. A.* 192.

410. 2; cupreus*. *Megerle.—Dufts. F.* ii.194.—(*Ing. Inst.* 88.)—(*Curtis l. c. supra.*)
El. riparius. *Oliv. E.* ii. 34. *pl.* i. *f.* 1. *a—e.*
El. uliginosus. *Ill. K.* i. 225.

411. 3; riparius*. *Fabr. E.* i. 245.—*Leach E. E.* ix. 79.—(*Sam. I.* 16.)—(*Curtis l. c. supra.*)
Ci. riparia. *Linn.* ii. 658.—*Berk. S.* i. 105.—*Stew.* ii. 74.—*Don.* ix. *pl.* 301.—*Marsh.* i. 392.—*Turt.* ii. 400.
El. paludosus. *Oliv. E.* ii. 5. *pl.* 1. *f.* 4. *a, b.*
Arenarius parvus. *Voet.* (*Pz.*) ii. 98. *pl.* 40. *f.* 7.
β, El. purpurascens. *Kirby? MSS.*

Genus 76. BLETHISA, *Bonelli, Samou.*

NEBRIA *p, Latreille.*—HELOBIUM. (*Leach E. E.*)—BLETHISUS, (*Kirby.*)

412. 1; multipunctata*. (*Sam. I.* 6.)
Ca. multipunctatus. *Linn.* ii. 672.—*Panz. F.* xi. *f.* 5.—*Marsh.* i. 467.—*Turt.* ii. 445.
He. multipunctatum. *Leach E. E.* ix. 83.

Subsectio 2. HYDRADEPHAGA, *MacLeay.*

(HYDROCANTHARI, *Latreille.*—HYDRADEPHAGENA, *Kirby.*)
Water-Beetles, Toe-biters.

Familia VIII. DYTICIDÆ, *Leach.*

EUNECHINA, *Kirby.*
(DYTISCUS *p, Linné, &c.*)

Genus 77. HALIPLUS, *Latreille, Leach, Samou.*, (*Kirby.*)
CNEMIDOTUS, *Illiger.*—HOPLITUS, *Clairville.*

A. "Elytris lineis elevatis." *Leach.*

413. 1; elevatus*. *Gyll.* i. 545.—(*Sam. I.* 20.)
Dy. elevatus. *Panz. F.* xiv. *f.* 9.

B. "Elytris striis nullis elevatis." *Leach.*

414. 2; impressus*. *Latr. G.* i. 234. *pl.* 6. *f.* 6.—*Leach E. E.* ix. 84.—(*Sam. I.* 20.)
Dy. impressus. *Fabr. E.* i. 271.—*Panz. F.* xiv. *f.* 7.—*Turt.* ii. 429.
Dy. laminatus. *Gmel.* iv. 1952.
Dy. thoracicus. *Fourc. F.* i. 69.
Dy. flavicollis. *Marsh.* i. 430.
Ha. flavicollis. (*Sam. I.* 20.)
Dy. minutus. *Don.* ii. *pl.* 68. *f.* 1, 2.—*Turt.* ii. 430.
β; Ha. excavatus. *Kirby MSS.*
γ; Ha. nigriceps. *Kirby MSS.*
δ; Ha. ruficeps. *Kirby MSS.*

415. 3; mucronatus. *Leach MSS.*
Ha. ophthalmicus. *Kirby MSS?*

416. 4; ferrugineus*. *Gyll.* i. 546.—(*Sam. I.* 20.)
Dy. ferrugineus. *Linn.* ii. 666.
β; Dy. fulvus. *Fabr. E.* i. 271.
Dy. assimilis. *Marsh.* i. 428.
Ha. assimilis. (*Sam. I.* 20.)
γ; Dy. interpunctatus. *Marsh.* i. 429.

417. 5; obliquus*. *Gyll.* i. 550.—(*Sam. I.* 20.)
Dy. obliquus. *Illig. K.* i. 268.—*Fanz. F.* xiv. *f.* 6.
Dy. amœnus. *Oliv. E.* iii. 40?
Dy. obliquus. *Ent. Trans.* (*Burrell.*) i. 215.

418. 6; confinis*. *Kirby MSS.*
β; Ha. nebulosus. *Kirby MSS.*
Ha. varius. *Nicol. Col. Hal. p.* 54?

419. 7; lineato-collis*. *Gyll.* i. 549.—(*Sam. I.* 20.)
Dy. lineato-collis. *Marsh.* i. 429.
Dy. bistriolatus. *Dufts. F.* i. 285.

420. 8; marginepunctatus. (*Schon. S.* ii. 27.)
Dy. marginepunctatus. *Fanz. F.* xiv. *f.* 10.

421. 9; ruficollis*. (*Sam. I.* 20.)
Dy. ruficollis. *DeGeer* iv. 404. *pl.* 16. *f.* 9, 10?—*Marsh.* i. 428.
Ha. impressus var. c. *Gyll.* i. 548.
Dy. cæsus. *Dufts. F.* i. 284.

422. 10; affinis* *mihi.*

423. 11, brevis. *Kirby MSS.*

424. 12, melanocephalus* *mihi.*

Genus 78. PÆLOBIUS, *Schönher, Leach, Samou.*, (*Kirby.*)
HYGROBIA, *Clairville.*—HYDRACHNA *p, Fabricius.*

425. 1; Hermanni*. *Leach E. E.* ix. 84.—*Samou.* 157. *pl.* 3. *f.* 14.
Dy. Hermanni. *Fabr. E.* i. 255.—*Mart. C. pl.* 33. *f.* 17.—*Marsh.* i. 418.—*Don.* xiv. *pl.* 501. *f.* 2.
Pæ. tardus. (*Schon. S.* ii. 27.)

Genus 79. HYPHIDRUS, *Illiger.*
HYDRACHNA *p, Fabricius.*—HYDRACHNE, (*Haworth.*)—HYPHYDRUS, *Leach Z. M.*—*Samou.*, (*Kirby.*)

426. 1; ovatus*. *Samou.* 157.
Dy. ovatus. *Linn.* ii. 667.—*Mart. C. pl.* 33. *f.* 12.—*Marsh.* i. 419.—*Turt.* ii. 423.
Dy. ferrugineus. *Don.* ii. *pl* 68. *f.* 3, 4.—*Stew.* ii. 78?
Hy. ferrugineus. *Latr. G.* i. 233.—*Leach E. E.* ix. 84.
Hy. gibbus. *Fabr. E.* i. 256.—*Ent. Trans.* (*Haworth.*) i. 88.
Dy. sphæricus. *DeGeer* iv. 402.

427 [2; ovalis*.] *Gyll.* i. 519.
Hy. ovalis. *Fabr. E.* i. 256.—*Ent. Trans.* (*Haworth.*) i. 88.
Dy. ovatus. *Oliv. E.* iii. 33. *pl.* 3. *f.* 28. *a*, *b*.
Dy. ovatus var. *Marsh.* i. 419.
β; Dy. brunneus. *Kirby MSS.*
γ; Dy. variatus. *Kirby MSS.*

428 [3, variegatus?] *Illiger?*—(*DeJean C.* 20?)

Genus 80. HYGROTUS *mihi.*

HYPHIDRUS *p*, *Illig.*—HYDRACHNA *p*, *Fabr.*—HYDROPORUS *p*, *Clairv.*—HYDROPORUS** *Samouelle.*—AQUARIUS ——? *MSS.*

429. 1, fluviatilis.
Hyd. fluviatilis. *Leach MSS.*—(*Sam. I.* 22.)
Hydro. assimilis β. *Kunzé. E. F.* 64.

430. 2, assimilis.
Dy. assimilis. *Payk. F.* i. 236.—*Ent. Trans.* (*Haworth.*) i. 91.
Hy. halensis. *Graven.* 105.
Hyp. rotundatus. *Illig. M.* (*Müller.*) iv. 211.

431 † 3, decoratus.
Hyp. decoratus. *Gyll.* ii. *add.* 16.
Hyp. cuspidatus. *Mus. Germar.*—*Ahr. F.* v. *pl.* 4.
Hydro. cuspidatus. *Kunzé. E. F.* 68. In Mus. *Brit.*

432. 4; confluens*.
Dy. confluens. *Fabr. E.* i. 270.—*Panz. F.* xiv. *f.* 5.—*Marsh.* i. 424.
Hyd. confluens. (*Sam. I.* 22.)

433. 5; collaris*.
Dy. collaris. *Panz. F.* xxvi. *f.* 4.
Dy. recurvus. *Marsh.* i. 424.
Dy. inæqualis γ. *Payk. F.* i. 238.
Hyp. reticulatus b. *Gyll.* ii. 521.

434 [6; affinis*] *mihi.*

435 [7; reticulatus*.]
Dy. reticulatus. *Fabr. E.* i. 273.
Hyd. variegatus. *Kirby MSS.*

436. 8; inæqualis*.
Dy. inæqualis. *Fabr. E.* i. 272.—*Oliv. E.* iii. 36. *pl.* 3. *f.* 29. *a*, *b*.
Dy. trifidus. *Marsh.* i. 423.
Hyd. trifidus. (*Sam. I.* 22.)
Dy. versicolor. *Gmel.* iv. 1952.

437. 9; scitulus*.
Dy. scitulus. *Spence MSS.*

438. 10; pictus*.
Dy. pictus. *Fabr. E.* i. 273.
Dy. arcuatus. *Panz. F.* xxvi. *f.* 1.

Dy. flexuosus. *Marsh.* i. 425.
Hyd. flexuosus. (*Sam. I.* 22.)
β, Dy. Crux. *Fabr. E.* i. 271?

Genus 81. HYDROPORUS, *Clairville, Leach, Samou.*, (*Kirby.*)
HYPHIDRUS *p*, *Illiger.*

A. "Thoracis lateribus rotundatis." *Kirby MSS.*

a. *Elytris apice dentatis.*

439. 1: frater*. *Spence MSS.*—*Kunzé. E. F.* 62.
Dy. depressus. *Dufts. F.* i. 272.

440. 2: areolatus. (*DeJean C.* 19.)
Dy. areolatus. *Dufts. F.* i. 274.
Hy. halensis. *Kunzé E. F.* 66.
Dy. picipes. *Thunb. I. S.* 76.
Dy. griseostriatus. *DeGeer* iv. 103.

441. 3; depressus*. *Kunzé E. F.* 63.—(*Sam. I.* 22.)
Dy. depressus. *Fabr. E.* i. 268.
Dy. elegans. *Illig. K.* i. 265.—*Panz. F.* xxiv. *f.* 5.—*Marsh.* i. 421.
Dy. 12-pustulatus var. minor. *Oliv. E.* iii. 31. *pl.* 5. *f.* 46.
Dy. Neuhoffi. *Cederh. F.* 32. *pl.* 2. *f.* 1.

b. *Elytris apice haud dentatis.*

442. 4; duodecimpustulatus*. (*Sam. I.* 22.)
Dy. 12-pustulatus. *Fabr. E.* i. 270. *Oliv. E.* iii. 21. *pl.* 5. *f.* 46. *a, b.*—*Marsh.* i. 422.—*Don.* xiv. 73. *pl.* 496.

443. 5; consobrinus. *Kunzé E. F.* 61.

444. 6, nigrolineatus. *Gyll.* iii. 688.
Hyp. nigrolineatus. *Schon. S.* ii. 33. *pl.* 4. *f.* 2. *a, b.*

445. 7, lineatus*.
Dy. lineatus. *Marsh.* i. 426.
Dy. paralellogrammus. *Ahr. F.* ii. *f.* 2.
Hy. nigrolineatus. *Kunzé E. F.* 60.

446. [8; picipes*.] *Kunzé E. F.* 61.
Dy. punctatus. *Marsh.* i. 426.

447 † 9, alternans. *Kunzé E. F.* 62.
Hyp. lineellus. *Gyll.* i. 529.
Hyd. lineellus. (*Sam. I.* 22.)
Hyp. lineatus. (*Schon. S.* ii. 10.)
Hy. stragula. *Illig. Misc.* In Mus. *Brit.*

448. 10; dorsalis*. (*Sam. I.* 22.)
Dy. dorsalis. *Fabr. E.* i. 269.—*Marsh.* i. 421.
Dy. rufifrons. *Fabr. E.* i. 198.
β; Hy. immaculatus. *Kirby MSS.*

γ; Dy. dorsalis. *Fanz. F.* xiv. *f.* 2.
Hy. triguttatus. *Kirby MSS.*

449. 11; alpinus. *Kunzé E. F.* 67.
Dy. alpinus. *Payk. F.* i. 226?
Hy. rivalis. *Leach MSS.—Gyll.* iv. 384. (!)
β? Hyp. borealis. *Gyll.* iv. 586?

450. 12; sexpustulatus*. (*DeJean C.* 20.)
Dy. sexpustulatus. *Fabr. E.* i. 269.—*Ent. Trans.* (*Haworth.*) i. 89.
Dy. palustris. *Illig. M.* i. 76.
Dy. lituratus. *Fanz. F.* xiv. *f.* 4.—*Marsh.* i. 423.
Hyd. lituratus. (*Sam. I.* 22.)
Dy. variegatus. *Fourc.* i. 68.
β; Dy. palustris. *Linn.* ii. 667?
γ; Hy. rufescens. *Kirby MSS.*
Dy. fimbriatus. *Gmel.* iv. 1957.

451 [13; proximus.]
Hy. proximus. *Wilk.? MSS.*

452. 14; tristis*. (*DeJean C.* 20.)
Dy. tristis. *Payk. F.* i. 232.

453. 15, umbrosus.
Hyp. umbrosus. *Gyll.* i. 538.

B. "Thoracis lateribus in elytrorum arcû." *Kirby MSS.*

a. *Elytris apice subtruncatis.*

454. 16; rufifrons*.
Dy. rufifrons. *Dufts. F.* i. 270.
Hyp. axillaris. *Schönher?*
Hy. punctulatus *mihi.*

455. 17; marginatus*. *Kunzé E. F.* 67.
Dy. marginatus. *Dufts. F.* i. 269.
Hy. Lapponum. *Gyll.* i. 532.
Hy. basalis. *Mus. Hoffmann.*

456. 18; deplanatus*.
Hyp. deplanatus. *Gyll.* iv. 391?

b. *Elytris apice haud truncatis.*

457. 19, Scalesianus *mihi.*

458. 20; unistriatus*. (*Sam. I.* 22.)
Dy. unistriatus. *Illig. K.* i. 266.
Dy. parvulus a. *Payk. F.* i. 232.—*Fanz. F.* xcix. *f.* 2.
Hy. impressicollis. *Kirby MSS.*
Dy. monostictus. *Fourc.* i. 69.
Dy. nanus. *Gmel.* iv. 1956.

459. 21; geminus*. (*DeJean C.* 20.)

Dy. geminus. *Fabr. E.* i. 272.—*Ent. Trans.* (*Haworth.*) i. 92.
Dy. trifidus. *Panz. F.* xxvi. *f.* 2.
Dy. parvulus, b. *Payk. F.* i. 233.
Dy. pygmæus. *Oliv. E.* iii. 39.
Dy. unistriatus. *Schra. A.* 205.

460. 22; minimus*.
Dy. minimus. *Scop. E.* 297.

461. 23; granularis*. (*Sam. I.* 22.)
Dy. granularis. *Linn.* ii. 667.—*Oliv. E.* iii. 33. *pl.* 2. *f.* 13. *a, b.*—*Marsh.* i. 426.—*Turt.* ii. 429.
Dy. unilineatus. *Gmel.* iv. 1953.

462. 24; ovalis*.
Dy. ovalis. *Marsh.* i. 425.
Dy. lineatus. *Fabr. E.* i. 272?
Hyd. lineatus. (*DeJean C.* 20.)
Hy. stagnicola. *Leach MSS.*
β; Hy. trilineatus. *Kirby MSS.*

463. 25; nigrita*. (*DeJean C.* 20.)
Dy. nigrita. *Fabr. E.* i. 273.—*Ent. Trans.* (*Haworth.*) i. 91.
Hy. lateralis. *Kirby MSS.*

464. 26; trivialis*. *Spence? MSS.*

465. 27; minutus* *mihi.*
Hy. memnonius. *Nicol. Col. Hal.* 33?

466. 28, planus*. (*Sam. I.* 22.)
Dy. planus. *Marsh.* i. 425.

467. 29; melanocephalus*. (*Sam. I.* 22.)
Dy. melanocephalus. *Marsh.* i. 423.

468. 30; erythrocephalus*. (*DeJean C.* 20.)
Dy. erythrocephalus. *Linn. F. no.* 774.—*Panz. F.* ci. *f.* 3.—*Ent. Trans.* (*Haworth.*) i. 90.

469. 31; flavipes*. (*Sam. I.* 22.)
Dy. flavipes. *Fabr. E.* i. 273.—*Ent. Trans.* (*Haworth.*) i. 90.
Dy. rufipes. *Oliv. E.* iii. 30. *pl.* 4. *f.* 39. *a, b.*
Dy. fusculus. *Illig. K.* i. 264.
Dy. planus. *Payk. F.* i. 223.
Dy. lividus. *Fourc.* i. 68.
Dy. humeralis. *Marsh.* i. 422.
Hy. humeralis. (*Sam. I.* 22.)
Hy. fusculus. *Leach E. E.* ix. 84.
β; Hy. picipennis. *Kirby MSS.*

470. 32, ater?
Dy. ater. *Forst. C.* i. 54.

471. 33; holosericeus*.

Dy. holosericeus. *Marsh.* i. 427.
Dy. planus. *Fabr. E.* i. 268.
Dy. erythrocephalus. *Linn.* ii. 666?—*Turt.* ii. 427.

472. 34, pubescens*.
Hyp. pubescens. *Gyll.* i. 536.
Hy. erythropus. *Kirby MSS.*

473. 35, fuscatus*. *Kirby MSS.*
Hy. scopularis. *Illiger?*

474. 36, piceus*. *Kirby MSS.*

Genus 82. NOTERUS, *Clairville, Leach, Samou.,* (*Kirby.*)

HALIPLUS *p, Schon.*

475. 1; crassicornis*. *Leach E. E.* ix. 84.
Dy. crassicornis. *Fabr. E.* i. 273.—*Marsh.* i. 420.—*Panz. F.* ci. *f.* 5.
Dy. clavicornis. *DeGeer.* iv. 402.
Dy. capricornis. *Herbst A.* 128.
No. Geerii. *Leach Z. M.* iii. 71.—*Samou.* 158.

476. 2; sparsus*. *Leach Z. M.* iii. 71.—*Samou.* 158.
Dy. sparsus. *Marsh.* i. 430.—*Ent. Trans.* (*Burrell.*) i. 211.

Genus 83. LACCOPHILUS, *Leach, Samou.*

477. 1; minutus*. *Leach E. E.* ix. 84.—*Samou.* 158.
Dy. minutus. *Linn.* ii. 667.—*Stew.* ii. 78.—*Mart: C. pl.* 33. *f.* 13.—*Marsh.* i. 419—*Samou.* 158.
Dy. hyalinus. *DeGeer.* iv. 406.
Dy. marmoreus. *Oliv. E.* iii. 27.
Dy. obscurus. *Panz. F.* xxvi. *f.* 3.

478. 2; interruptus*.
Dy. interruptus. *Panz. F.* xxvi. *f.* 5.
Dy. hyalinus. *Marsh.* i. 420.
La. hyalinus. (*Samou.* 158.)

Genus 84. COLYMBETES, *Clairville, Leach, Samou.,* (*Kirby.*)

479. 1: oblongus*. (*Sam. I.* 12.)
Dy. oblongus. *Illig. M.* i. 72.

480. 2; guttatus*. (*DeJean C.* 19.)
Dy. guttatus. *Payk. F.* i. 211.
Dy. fenestratus. *Panz. F.* xc. *f.* 1.
Dy. picinus. *Marsh.* i. 428.

481 [3; fontinalis.] *Leach MSS.*—(*Sam. I.* 12.)

482. 4; angustior*.
Dy. angustior. *Gyll.* i. 500.
Dy. nitidus. *Kirby MSS?*

483 † 5, vittiger.
Dy. vittiger. *Gyll.* iv. 379? In Mus. *Brit.*

484. 6, immunis *mihi.*

485. 7; fuliginosus*. (*Sam. I.* 12.)
Dy. fuliginosus. *Fabr. E.* i. 263.
Dy. lacustris. *Fabr. E.* i. 264.—*Panz. F.* xxxviii. *f.* 14.
Dy. uliginosus. *Mart. C. pl.* 32. *f.* 2–5.—*Marsh.* i. 416.
Dy. fœtidus. *Müll. Z. D. p.* 71.

486 † 8, lævigatus*. *Wilkin MSS.*
Dy. glaber. *Forst. C.* i. 55?—*Stew.* ii. 78?—*Turt.* ii. 434?
Co. elongatus. *Leach MSS.* In Mus. *Brit. et D. Vigors.*

487. 9, fuscus*?
Dy. fuscus. *Linn.* ii. 665?
Dy. striatus. *Panz. F.* lxxxvi. *f.* 5.

488 [10; striatus*.] (*Sam. I.* 12.)
Dy. striatus. *Linn.* ii. 665.—*Mart. C. pl.* 34. *f.* 27.—*Marsh.* i. 414.—*Turt.* ii. 424.
Dy. transverse-striatus. *DeGeer.* iv. 399. *pl.* 15. *f.* 16.

489. 11; pulverosus*.
Dy. pulverosus. *Knoch.*
Dy. conspersus. *Gyll.* i. 482.
Dy. Gyllenhalii *mihi* (*olim.*)

490. 12; notatus*. (*Sam. I.* 12.)
Dy. notatus. *Fabr. E.* i. 267.—*Oliv. E.* iii. 29. *f.* 5. *f.* 47.—*Ent. Trans.* (*Haworth.*) i. 89.—*Id.* (*Burrell.*) i. 209.
Dy. roridus. *Müll. Z. D. pr.* 771.
Dy. frontalis. *Marsh.* i. 425. (mas.)
Dy. virgulatus. *Illig. M.* v. 225. (fœm.)
Co. suturalis. (*DeJean C.* 19.)

491. 13; collaris*. (*Sam. I.* 12.)
Dy. collaris. *Payk. F.* i. 200.—*Marsh.* i. 415.
Dy. adspersus. *Illig. K.* i. 261.—*Panz. F.* xxxviii. *f.* 18.
Dy. oculatus. *Herbst A.* 125.
Dy. exsoletus. *Forst. C.* 57.—*Stew.* ii. 78.—*Turt.* ii. 434.

492. 14, agilis. (*Sam. I.* 12.)
Dy. agilis. *Fabr. E.* i. 266.—*Panz. F.* xc. *f.* 2.—*Ent. Trans.* (*Haworth.*) i. 251.
Dy. bilineatus var. β. *DeGeer.* iv. 400.

493. 15, adspersus*. (*DeJean C.* 19.)
Dy. adspersus. *Fabr. E.* i. 267.
Dy. collaris var. *Marsh.* i. 415.
Dy. adsperso affino. *Illig. M.* i. 71, nota.
Dy. funebris. *Schra. B.* i. 713.

494. 16; bipunctatus*. (*Sam. I.* 12.)

Dy. bipunctatus. *Illig. K.* i. 262.—*Mart. C. pl.* 33. *f.* 19.—*Marsh.* i. 418.—*Don.* ix. *pl.* 303.
Dy. nebulosus. *Forst. C.* 56.—*Stew.* ii. 78.—*Turt.* ii. 434.

495. 17, subnebulosus* *mihi.*

496. 18; conspersus*. (*Sam. I.* 12.)
Dy. conspersus. *Marsh.* i. 427.

497. 19; Sturmii*. (*DeJean C.* 19.)
Dy. Sturmii. *Gyll.* i. 493.
Dy. umbrinus. *Sturm.*

498. 20, congener. (*DeJean C.* 19.)
Dy. congener. *Payk. F.* i. 214.
Dy. brunnipennis. *Kirby MSS.*

499. 21; paludosus*. (*DeJean C.* 19.)
Dy. paludosus. *Fabr. E.* i. 266.
Dy. politus. *Marsh.* i. 419.
Dy. congener β. *Schon. S.* ii. 22.
Co. politus. (*Sam. I.* 12.)

500. 22; maculatus*. *Samou.* 158. *pl.* 3. *f.* 15.—(*Sam. I.* 12.)
Dy. maculatus. *Linn.* ii. 666.—*Oliv. E.* iii. 27. *pl.* 2. *f.* 16.—*Stew.* ii. 78.—*Marsh.* i. 418.—*Turt.* ii. 427.—*Don.* xiv. *pl.* 501. *f.* 1.
Dy. ornatus. *Gmel.* iv. 1953.
β; Dy. inæqualis. *Fanz. F.* xiv. *f.* 8.—*Marsh.* i. 417.

501. 23, abbreviatus. (*Sam. I.* 12.)
Dy. abbreviatus. *Fabr. E.* i. 265.—*Fanz. F.* xiv. *f.* 1.—*Ent. Trans.* (*Burrell.*) i. 210.
Dy. undulatus. *Schra. A.* 202.

502. 24, vitreus*. (*Sam. I.* 12.)
Dy. vitreus. *Payk. F.* i. 217.—*Ent. Trans.* (*Burrell.*) i. 210.—*Id.* (*Haworth.*) i. 251.
Dy. didymus. *Oliv. E.* iii. 26. *pl.* 4. *f.* 37.
Dy. abbreviatus var. *Illig. K.* i. 263.
Dy. fenestratus. *Rossi F.* i. 201.

503. 25; chalconotus*. (*Sam. I.* 12.)
Dy. chalconotus. *Illig. K.* i. 260.—*Fanz. F.* xxxviii. *f.* 17.—*Ent. Trans.* (*Burrell.*) i. 213.
Dy. concinnus. *Marsh.* i. 427.
Dy. bipustulatus minor. *Rossi.*

504 [26; montanus*.] *Leach MSS.*

505. 27; nigroæneus*.
Dy. nigroæneus. *Marsh.* i. 428.

506. 28; femoralis*. (*DeJean C.* 19.)
Dy. femoralis. *Payk. F.* i. 215.

507. 29; affinis. (*DeJean C.* 19)
Dy. affinis. *Payk. F.* i. 211.

508 † 30, æratus *mihi.* In Mus. *Brit.*

509. 31; niger. (*DeJean C.* 19?)
Dy. niger. *Illig. M.* i. 73?

510. 32; bipustulatus*. *Leach E. E.* ix. 84.—(*Sam. I.* 12.)
Dy. bipustulatus. *Linn.* ii. 667.—*Oliv. E.* iii. 21. *pl.* 3. *f.* 26.
—*Marsh.* i. 415.—*Turt.* ii. 426.
Dy. carbonarius. *Fabr. E.* i. 263.
Dy. luctuosus. *Fourc.* i. 67.
β; Dy. acuductus. *Marsh.* i. 416.
γ; Dy. phæopterus. *Kirby MSS.*

511. 33; subopacus* *mihi.*

512. 34, striolatus.
Dy. striolatus. *Gyll.* i. 508.

513. 35, uliginosus. (*Sam. I.* 12.)
Dy. uliginosus. *Linn.* ii. 667.—*Stew.* ii. 78.—*Turt.* ii. 427.
Dy. Hybneri. *Marsh.* i. 417.

514. 36; guttiger*. (*DeJean C.* 19.)
Dy. guttiger. *Gyll.* i. 499.

515. 37; fenestratus*. (*Sam. I.* 12.)
Dy. fenestratus. *Fabr. E.* i. 264.—*Marsh.* i. 416.
Dy. æneus. *Illig. K.* i. 259.—*Panz. F.* xxxviii. *f.* 16.

516. 38; ater*. (*DeJean C.* 19.)
Dy. ater. *DeGeer.* iv. 401.—*Panz. F.* xxxviii. *f.* 15.—*Ent. Trans.* (*Haworth.*) i. 88.
Dy. fenestratus. *Oliv. E.* iii. 23.

517. [39; obscurus*.] (*Sam. I.* 12.)
Dy. obscurus. *Marsh.* i. 414.

518 † 40, Grapii. *Ahn. F.* vi. *f.* 4.—(*Ing. Inst.* 88.)
Dy. Grapii. *Gyll.* i. 505.
In Mus. D. *Bentley, Chant, Curtis et Vigors.*

Genus 85. HYDATICUS, *Leach, Samou., Curtis.*

519. 1; transversalis*. (*Sam. I.* 22.)—(*Curtis l. c. infra.*)
Dy. transversalis. *Fabr. E.* i. 265.—*Panz. F.* lxxxvi. *f.* 6.—
Turt. ii. 426.

520. 2; Hybneri*. *Samou.* 159.—(*Curtis l. c. infra.*)
Dy. Hybneri. *Fabr. E.* i. 265.—*Oliv. E.* iii. 24. *pl.* 4. *f.* 33.
Dy. seminiger. *DeGeer.* iv. 401.
Dy. parapleurus. *Marsh.* i. 427.

521. 3, stagnalis. (*Sam. I.* 22.)—(*Curtis l. c. infra*).
Dy. stagnalis. *Fabr. E.* i. 265.—*Panz. F.* xci. *f.* 7.

522 † 4, cinereus*. *Curtis.* ii. *pl.* 95.
Dy. cinereus. *Linn.* ii. 666.—*Stew.* ii. 77.
In Mus. D. *Bentley et Chant.*

Genus 86. DYTICUS, *Geoffroy, Leach, Samou., Curtis.*

HYDROCANTHARUS, *Ray.*

A. " Elytra margine non dilatato." *Leach.*

523. 1, circumcinctus*. *Nov. Act. Hall.* (*Ahr.*) i. 55.
Dy. marginatis fœm. var. *Nov. Act. Hall.* (*Ahr.*) i. 51.
Dy. circumscriptus ♀. *Eschscholtes.*
♀ Dy. dubius. *Gyll.* iv. 372.

524. 2, conformis.
Dy. conformis. *Nov. Act. Hall.* (*Kunzé.*) ii. 58.
Dy. circulatus. *Kirby MSS.*—(*Vide Leach Z. M.* iii. 97.)
Dy. flavoscutellatus. *Latr. H.* viii. 162?
Dy. flavomaculatus. (*Curtis l. c. supra.*)
Dy. excrucians *mihi.*
Dy. marginalis b. *Gyll.* i. 467.

525. 3; circumflexus*. *Illig. M.* i. 370.—*Leach. E. E.* ix. 84.—*Samou.* 159. *pl.* 3. *f.* 13. *b.* (sternum)—(*Curtis l. c. supra.*)

526. 4, angustatus *mihi.*—(*Curtis l. c. supra.*)

527. 5; marginalis*. *Linn.* ii. 665.—*Berk. S.* i. 106.—*Mart. C. pl.* 34. *f.* 25, 26.—*Stew.* ii. 77.—*Don.* v. *pl.* 161.—*Turt.* ii. 423.—*Shaw G. Z.* vi. 91.—*Bingley.* iii. 145.—*Leach E. E.* ix. 84.—*Samou.* 159. *pl.* 3. *f.* 13.—(*Mill. B. E. pl.* 3.)—(*Kirby & Sp. I. E.* ii. 255.—*Id.* iv. 254.)—(*Curtis l. c. supra.*)
Dy. totomarginalis. *DeGeer.* iv. 391. *pl.* 16. *f.* 1. ♀, *f.* 2. ♂.
♀; Dy. semistriatus. *Linn.* ii. 665.—*Berk. S.* i. 106.—*Stew.* ii. 77.

528. [6; submarginatus*] *mihi.*

529. 7; punctulatus*. *Illig. M.* i. 67.—*Marsh.* i. 412.—*Mart. C. pl.* 34. *f.* 24.—*Don.* xv. *pl.* 540.—(*Sam. I.* 15.)—(*Curtis l. c. supra.*)
Dy. marginalis β. *Illig. K.* i. 253.
Dy. punctatus. *Oliv. E.* iii. 12.
Dy. laterali-marginalis. *DeGeer.* iv. 396.
♀; Dy. porcatus. *Thunb. I. S.* 74.
Dy. semisulcatus. *Müll. Z. D. pr.* 666.

530. 8, dimidiatus. *Curtis.* iii. *pl.* 99.
Dy. dimidiatus. *Bergst. Num. pl.* 7. *f.* 1.

B. " Elytra margine dilatato." *Leach.*

531 ‡ 9. latissimus[a].

---

[a] 531 ‡ 9. latissimus. *Linn.* ii. 665.—*Panz. F.* lxxxvi. *f.* 1. ♂. *f.* 2. ♀.—*Berk. S.* i. 106 (!)—*Stew.* ii. 77 (!)—*Turt.* ii. 423 (!)
♀. Dy. amplissimus. *Müll. Z. D. pr.* 662.

Genus 87. CYBISTER, *Curtis.*

TROGUS[a], *Leach.*

532 † 1, Rœselii. *Curtis* iv. *pl.* 151.
Dy. Rœselii. *Fabr. E.* i. 259.
Dy. dispar. *Rossi F.* i. 199.
Dy. dissimilis. *Rossi M.* i. 66.
Dy. glaber. *Bergs. Nom. pl.* 6. *f.* 4. 5.—*pl.* 8. *f.* 4. *pl.* 9. *f.* 2.
Dy. virens. *Müll. Z. D. pr.* 70.
Dy. virescens. *Gmel.* iv. 1958.
Dy. intricatus. *Act. Hal.* (*Schall.*) i. 311.
Dy. punctulatus. *Schwatz. Nom.* i. 34.
In Mus. D. *I. H. Griesbach.*

Genus 88. ACILIUS, *Leach, Samou., Curtis,* (*Kirby.*)

533. 1; sulcatus*. (*Samou.* 159.)—(*Curtis l. c. infra.*)
Dy. sulcatus. *Linn.* ii. 666.—*Berk. S.* i. 107.—*Mart. C. pl.* 32. *f.* 7. ♂. *f.* 8. ♀.—*Don.* ii. *pl.* 68. *f.* 5.—*Stew.* ii. 78.—*Turt.* ii. 424.
Dy. fasciatus. *DeGeer.* iv. 397.
Dy. cinereus. *Fabr. E. S.* i. 262.—*Berk. S.* i. 106.—*Marsh.* i. 413.—*Turt.* ii. 425?—*Shaw G. L.* vi. 93.—(*Kirby & Sp. I. E.* ii. 251.)
Dy. punctatus. *Scop. C.* 295.
Dy. Scopolii. *Gmel.* iv. 1954.

534 † 2, varipes *mihi.* In Mus. D. *Cooper.*

535 † 3. scoticus. *Curtis MSS.* In Mus. D. Curtis.

536. 4, canaliculatus.
Dy. canaliculatus. *Knock.*—*Nicol. Col. Hal. p.* 29.
Dy. sulcipennis. *Sahlb. I. F.* 157.
Dy. dispar. *Ziegler.*—(*DeJean C.* 18.)
Ac. caliginosus. *Curtis.* ii. *pl.* 63.
Ac. cinereus. (*Curtis l. c.* olim.)

Familia IX. GYRINIDÆ, *Leach.*

GYRONECHINA, *Kirby.*

(OTIOPHORI *p, Latreille.*—AMPHIBII *p, Gyllenhal.*)

Genus 89. GYRINUS *Auctorum. Waterflea* or *Whirlwig.*

DYTISCUS *p, Linn. F.*

A. Elytris glabris.

537. 1; æneus*. *Leach MSS.*—(*Sam. I.* 20.)—(*Curtis l. c. infra.*)

538. 2; marinus. *Gyll.* i. 143.—(*Sam. I.* 20.)—(*Curtis l. c. infra.*)

---

[a] TROGUS: Genus Hymenopterorum. Vide *Panz. Krit. Rev.* ii. 80.

β; Gy. dorsalis. *Gyll.* i. 142.
Gy. marginatus. *Kirby MSS.*

539. 3; minutus*. *Fabr. E.* i. 276.—(*Sam. I.* 20.)—(*Kirby & Sp. I. E.* ii. 241.)—(*Curtis l. c. infra.*)
Gy. bicolor. *Oliv. E.* iii. 14. *pl.* 1. *f.* 8. *a, b.*
Gy. natator b. *Payk. F.* i. 239.
Gy. Kirbii. *Marsh.* i. 100.

540. 4; natator*. *Linn.* ii. 567.—*Berk. S.* i. 91.—*Mart. C. pl.* 7. *f.* 1?—*Stew.* ii. 35.—*Marsh.* i. 99.—*Shaw G. Z.* vi. 39. *pl.* 11.—*Leach E. E.* ix. 84.—*Samou.* 159. *pl.* 2. *f.* 2.—(*Kirby & Sp. I. E.* ii. 241.)—(*Curtis l. c. infra.*)
Pulex aquaticus. *Merret Pinax,* 203.
Scarabæus aquaticus. *Ray I.* 87.
β; Gy. analis. *Kirby MSS.*

541. 5; substriatus* *mihi.*
Gy. mergus. *Scopoli?*—*Ahr. F.* ii. *f.* 6?

542. 6; bicolor*. *Fabr. E.* i. 274.—*Ent. Trans.* (*Burrell.*) i. 125.—*Curtis.* ii. *pl.* 79.
Gy. elongatus. *Marsh.* i. 100.—(*Sam. I.* 20.)

B. Elytris villosis.—(Potamobius, *Leach MSS.*)

543. 7, villosus. *Illig. K.* i. 271.—(*Sam. I.* 20.)—(*Kirby & Sp. I. E.* ii. 241. *Id.* iv. 56.)—(*Curtis l. c. supra.*)
Gy. Modeeri. *Marsh.* i. 100.

Subsectio 3. PHILHYDRIDA, *MacLeay.*

(Palpicornes et Clavicornes *p, Latreille.*—Helocera *p, Duméril.*)

Familia X. PARNIDÆ, *Leach.*

(Otiophori *p, Latreille.*—Amphibii *p, Gyllenhal.*)

Genus 90. PARNUS, *Fabricius, Leach, Samou., Millard,* (*Kirby,*) *Curtis.*

Dryops *p, Olivier.*—Elater *p, Linné.*

544. 1; prolefericornis*. *Fabr. E.* i. 332.—*Panz. F.* xiii. *f.* 1.—*Marsh.* i. 399.—*Curtis l. c. infra.*
El. dermestoides. *Gmel.* iv. 1910.—*Turt.* ii. 393.

545. [2; sericeus*.] *Leach MSS.*—*Samou.* 185. *pl.* 3. *f.* 10.

546. [3, impressus*.] *Curtis.* ii. *pl.* 80.

547. 4; auriculatus*. *Illig. K.* i. 351.—*Curtis l. c. supra.*
Dr. auriculatus. *Oliv. E.* iii. 4. *pl.* 1. *f.* 1. *a—e.*
Pa. villosus. *Bonelli?*

548. [† 5, bicolor*.] *Curtis l. c. supra.* In Mus. D. *Curtis.*

Familia XI. HETEROCERIDÆ, *MacLeay.*

(BYRRHII *p, Latreille.*—AMPHIBII *p, Gyllenhal.*)

Genus 91. HETEROCERUS, *Bosc, Leach, Samou., Millard, (Kirby.)*

DERMESTES *p, Thunberg.*—APATE *p, Fabr.*

549. 1; marginatus*. *Fabr. E.* i. 355.—*Panz. F.* xxiii. *f.* 12.—*Marsh.* i. 400.—*Leach E. E.* ix. 95.—*Samou.* 185. *pl.* 3. *f.* 11.
De. fenestratus. *Act. Ups.* (*Thunb.*) iv. 3.
De. villosus. *Gmel.* iv. 1600.
De. marginatus. *Turt.* ii. 84.
He. lævigatus. *Ent. Trans.* (*Burrell.*) i. 204.

550. 2, obsoletus. *Marsh. MSS.*

551. 3: lævigatus*. *Fabr. E.* i. 356.—*Panz. F.* xxiii. *f.* 11.—(*Ing. Inst.* 88.)
He. marginatus b. *Gyll.* i. 137.
He. marginatus. *Ent. Trans.* (*Burrell.*) i. 204.

552. 4, nebulosus *mihi.*

553. 5, flexuosus* *mihi.*

554 † 6, minutus? (*DeJean C.* 50?) In Mus. D. *Bentley et Chant.*

Familia XII. LIMNIIDÆ.

(BYRRHII *p, Latreille.*—HYDROCANTHARI *p, Gyllenhal.*

Genus 92. GEORYSSUS, *Latreille, Leach,* (*Kirby.*)

PIMELIA *p, Fabricius.*—TROX *p, Panzer.*—BYRRHUS *p, Rossi.*

555. 1; pygmæus*. (*DeJean C.* 49.)
Pi. pygmæa. *Fabr. E.* i. 133.
Tr. dubius. *Panz. F.* lxii. *f.* 5.—*Ent. Trans.* (*Haworth.*) i. 253.
By. crenulatus. *Rossi M.* ii. *App.* 81.

Genus 93. ELMIS, *Latreille, Millard,* (*Kirby,*) (*MacLeay.*)

LIMNIUS, *Müller, Leach, Samou.*—DYTISCUS *p, Panzer.*—CHRYSOMELA *p, Marsham.*

A. Thorace lineis elevatis.

a. Vide *Schon. S.* ii. 117. *A.*

556. 1; Volkmari*. *Latr.* G. ii. 51. (!)
Li. Volkmari. *Illig. M.* (*Müller.*) v. 195.—(*Sam. I.* 25.)
Dy. Volkmari. *Panz. F.* vii. *f.* 4.
Ch. buprestoides. *Marsh.* i. 192.
El. Maugetii. *Latr. H.* ix. 228. *pl.* 78. *f.* 3.

557. 2, tuberculatus.
Li. tuberculatus. *Illig. M.* (*Müller.*) v. 199.
El. lineatus. *Kirby MSS.*

558. 3, variabilis.
Li. variabilis. *Leach MSS.*
El. Dargelasi. *Latr. G.* ii. 51?

559. 4, lacustris. *Spence? MSS.*
El. stagnalis. *Kirby MSS?*

560. 5, fluviatilis *mihi.*
Li. Troglodytes. *Gyll.* iv. 395?

561. 6, parallelipipedus.
Li. parallelipipedus. *Illig. M.* (*Müller.*) v. 200.
b. Vide *Schon. S.* ii. 117. *B.*

562. 7, æneus*. (*Kirby & Sp. I. E.* ii. 258.)
Li. æneus. *Illig. M.* (*Müller.*) v. 202.
Lim. Megerlei. *Illiger?*

563. 8, Maugetii.
El. Maugetii. *Latr.* G. ii. 50?

B. Thorace lineis elevatis nullis.

564. 9: cupreus*.
Li. cupreus. *Illig. M.* (*Müller.*) v. 205.
β, Lim. Spencii. *Leach MSS.*

Familia XIII. HELOPHORIDÆ, *Leach.*

(Hydrophilii II. *Latreille.*—Sphæridiota *p, Gyllenhal.*)
(Hydrophilus *p, Marsham, &c.*)

Genus 94. HYDROCHUS, *Germar, Leach, Samou.*

Elophorus *p, Fabr.*—Nitidula *p, Gmel.*

565. 1; elongatus*. (*Sam. I.* 22.)
El. elongatus. *Fabr. E.* i. 277.—*Panz. F.* xxvi. *f.* 7.
He. elongatus. (*Leach E. E.* ix. 95.)
Hy. cicindeloides. *Marsh.* i. 411.
Hyd. cicindeloides. (*Sam. I.* 22.)

566. 2; crenatus*. (*Sam. I.* 22.)
El. crenatus. *Fabr. E.* i. 278.
El. elongatus. *Oliv. E.* iii. 8. *pl.* 1. *f.* 4. *a, b?*

567. 3: brevis. (*Sam. I.* 22.)
El. brevis. *Herbst C.* v. 141. *pl.* 49. *f.* 10. *K. K.*—*Ent. Trans.* (*Haworth.*) i. 252.

Genus 95. HELOPHORUS, *Leach, Samou.*

Elophorus, *Fabricius.*—Silpha *p, Linné.*—Peltis *p, Müller.*—Dermestes *p, Geoffroy.*—Buprestis *p, Forster.*—Nitidula *p, Gmelin.*

A. Elytris lineis vix elevatis ornatis.

568. 1; aquaticus*.

Si. aquatica. *Linn.* ii. 575.—*Berk. S.* i. 92.
El. aquaticus. *Panz. F.* xxvi. *f.* 6.—(*Kirby & Sp. I. E.* ii. 257.)
Ni. aquatica. *Stew.* ii. 39.—*Turt.* ii. 112.
El. grandis. *Illig. K.* i. 272.
El. flavipes. *Herbst C.* v. 138.
Hy. æneus. *DeGeer.* iv. 379.
Hy. stagnalis. *Marsh.* i. 409.
He. stagnalis. (*Sam. I.* 21.)

569. 2; granularis*.
Bu. granularis. *Linn.* ii. 663.—*Berk. S.* i. 105.—*Stew.* ii. 74. —*Turt.* ii. 420.
El. granularis. *Gyll.* i. 127.
El. aquaticus. *Illig. K.* i. 273.
El. flavipes. *Oliv. E.* iii. 7. *pl.* 1. *f.* 3. *a, b.*

570. 3; griseus*. (*Sam. I.* 21.)
El. griseus. *Illig. K.* i. 273.
El. minutus. *Oliv. E.* iii. 7. *pl.* 1. *f.* 6. *a, b.*
Ni. minuta. *Stew.* ii. 39.
El. flavipes. *Thunb.* v. 66.
Hy. affinis. *Marsh.* i. 409.

571. 4; dorsalis*.
Hy. dorsalis. *Marsh.* i. 410.

572. 5; viridicollis*. *Kirby MSS.*

573. 6; tuberculatus.
El. tuberculatus. *Gyll.* i. 129.

B. Elytris lineis elevatis. (OPATRUM *p, Fabr.*)

574. 7; Fennicus*. (*Sam. I.* 21.)
El. Fennicus. *Payk. F.* i. 243.
Hy. cinereus. *Marsh.* i. 410.

575. 8; nubilus*. (*Sam. I.* 21.)
El. nubilus. *Oliv. E.* iii. 6. *pl.* 1. *f.* 2. *a, b.*
Hy. nubilus. *Marsh.* i. 410.
Ni. nubila. *Turt.* ii. 112.
Op. minutum. *Fabr. E.* i. 120.

Genus 96. OCHTHEBIUS, *Leach, Samou.*

ELOPHORUS *p, Fabricius.*—HYDRÆNA *p, Illiger.*

A. Elytris punctato-striatis.

576. 1; marinus. (*Leach E. E.* ix. 95.)—(*Sam. I.* 31.)
El. marinus. *Payk. F.* i. 245.—*Ent. Trans.* (*Haworth.*) i. 87.
Hy. margipallens. *Marsh.* i. 408.

577. 2; dilatatus*. *Leach MSS.*

578. 3; pygmæus*. (*Sam. I.* 31.)—*Ahr. F.* viii. *f.* 7.

El. pygmæus. *Fabr. E.* i. 278.
Hy. riparia. *Illig. K.* i. 279.
Oc. riparius. (*Leach E. E.* ix. 95.)—(*Sam. I.* 31.)

579. [4; impressus *.]
Hy. impressus. *Marsh.* i. 408.

580. 5, bicolon. *Germ. I. N.* i. 92.
Hy. bicolon. *Kirby MSS.*

581. 6, rufimarginatus *mihi.*

582. 7, nanus *mihi.*

583. 8, æratus * *mihi.*

B. Elytris punctatis, haud striatis.

584. 9, punctatus *mihi.*

Genus 97. HYDRÆNA, *Kugellan, Leach, Samou.*, (*Kirby.*)

ELOPHORUS *p*, *Gyllenhal.*—NITIDULA *p*, *Gmel.*

585. 1; riparia *. *Schneid. Mag.* (*Kugellan.*) 578.
Hy. longipalpis. *Marsh.* i. 407.
El. minutus. *Fabr. E.* i. 278.
Ni. minima. *Turt.* ii. 113.
El. minimus. *Ahr. F.* viii. *f.* 6.
Hy. Kugellani. (*Leach Z. M.* iii. 91.)—(*Sam. I.* 22.)—(*Kirby & Sp. I. E.* iv. 522.)

586. 2; pusilla * *mihi.*

587. 3, minutissima. (*Schon. S.* ii. 42?)
El. minutissimus. *Gyll.* i. 136?

Familia XIV. HYDROPHILIDÆ, *Leach.*

(HYDROPHILII I. *Latreille.*—SPHÆRIDIOTA *p*, *Gyllenhal.*)

(HYDROPHILUS, *Marsham*, *&c.*)

Genus 98. LIMNEBIUS, *Leach, Samou.*

DERMESTES *p*, *Marsham.*

588. 1; ater * *mihi.*

589. 2; affinis * *mihi.*

590. 3; truncatellus *.
Hy. truncatellus. *Fabr. E.* i. 255.
Hy. parvulus. *Herbst C.* vii. 314. *pl.* 114. *f.* 10. *K.*

591. 4; marginalis * *mihi.*

592. 5; nigrinus *. (*Sam. I.* 25.)
De. nigrinus. *Marsh.* i. 77.

593. 6, lutosus.
Hy. lutosus. *Marsh.* i. 407.

594. 7; mollis*. (*Sam. I.* 25.)
Hy. mollis. *Marsh.* i. 407.
Hy. truncatellus b. *Gyll.* i. 124.

595. 8, minutus.
Hy. minutus. *Fabr. E.* i. 254.
Hy. picinus. *Marsh.* i. 407.

596. 9, nitidus. (*Sam. I.* 25.)
Hy. nitidus. *Marsh.* i. 407.

Genus 99. HYDRÖUS, *Linn. MSS.*, *Leach*, *Samou.*

DYTICUS *p*, *Linné.*

597. 1; piceus*. *Linn. MSS.*—(*Leach E. E.* ix. 96.)—(*Sam. I.* 22.)
Dy. piceus. *Linn.* ii. 664.—*Berk. S.* i. 105.—*Barb.* G. 74. *pl.* 6. *f.* 24.—*Mart. C. pl.* 32. *f.* 3, 4.
Hy. piceus. *Stew.* ii. 76.—*Marsh.* i. 401.—*Turt.* ii. 420.—*Shaw N. M.* ix. *pl.* 292.—*Shaw G. Z.* vi. 95. *pl.* 34.—*Sam.* 58.—(*Kirby & Sp. I.* iii. 33.)
Hy. ruficornis. *DeGeer.* iv. 371.

Genus 100. HYDROPHILUS *Auctorum.*

DYTISCUS *p*, *Linné.*

598. 1; caraboides*. *Fabr. E.* i. 250.—*Stew.* ii. 76.—*Marsh.* i. 402.—*Turt.* ii. 421.—(*Leach E. E.* ix. 96.)—(*Samou.* 187. *pl.* 3. *f.* 1 6.)—*Curtis.* iv. *pl.* 159.
Dy. caraboides. *Linn.* ii. 664.—*Berk. S.* i. 106.—*Mart. C. pl.* 34. *f.* 28.
Hy. nigricornis. *DeGeer.* iv. 376.

Genus 101. SPERCHEUS, *Fabricius*, *Leach*, *Samou.*

599. 1, emarginatus*. *Fabr. E.* i. 248.—*Panz. F.* xci. *f.* 4.
Hy. sordidus. *Marsh.* i. 403.
Sp. sordidus. (*Leach Z. M.* iii. 93.)—(*Sam. I.* 38.)
β, Hy. verrucosus. *Marsh.* i. 403.

Genus 102. BEROSUS, *Germar*, *Leach*, *Samou.*

DYTISCUS *p*, *Linné.*

600. 1; luridus*. (*Leach Z. M.* iii. 93.)—(*Sam. I.* 6.)
Dy. luridus. *Linn.* ii. 665.—*Mart. C. pl.* 33. *f.* 10.
Hy. luridus. *Fabr. E.* i. 253.—*Stew.* ii. 77.—*Panz. F.* vii. *f.* 3.—*Marsh.* i. 404.—*Turt.* ii. 422.—(*Leach E. E.* ix. 96.)
Hy. fuscus. *DeGeer.* iv. 378.

601. 2; æriceps*.
Hy. æriceps. *Spence? MSS.*

602. 3; obsoletus* *mihi.* (*Ing. Inst.* 87.)

Genus 103. HYDROBIUS, *Leach, Samou.*

DYTISCUS *et* SCARABÆUS *p, Linné.*

A. Elytris striatis.

603. 1; picipes *.
Hy. picipes. *Fabr. E.* i. 251.
Hy. oblongus. *Herbst C.* vii. 300. *pl.* 113. *f.* 10.
Hy. substriatus. *Marsh. MSS.*

604. 2; fuscipes *. (*Sam. I.* 22.)
Dy. fuscipes. *Linn.* ii. 664.—*Berk. S.* i. 106.—*Mart. C. pl.* 33. *f.* 15.
Hy. fuscipes. *Stew.* ii. 77.—*Marsh.* i. 403.—*Panz. F.* lxvii. *f.* 13.
Hy. niger. *Gmel.* iv. 1944.
Sca. aquaticus. *Linn. F. no.* 404.
Hy. scarabæoides. *Fabr. E.* i. 251.—*Turt.* ii. 421.
Dy. gyrinoides. *Schrank A.* 372.

605. 3; chalconotus *. *Leach MSS.*—(*Sam. I.* 22.)

606. 4; subrotundus * *mihi.*

B. Elytris haud striatis.

a. *Corpore elongato, convexo.*

607. 5; melanocephalus *. (*Sam. I.* 22.)
Hy. melanocephalus. *Fabr. E.* i. 253.—*Ent. Trans.* (*Haworth.*) i. 86.
Hy. minutus. *Payk. F.* i. 182.
Hy. 4-punctatus. *Herbst C.* vii. 307. *pl.* 114. *f.* 4. D.
β; Hy. torquatus. *Marsh.* i. 405.
Hy. torquatus. (*Sam. I.* 24.)
Hy. grisescens. *Gyll.* iv. 276.
γ; Dy. dermestoides. *Forst. C.* i. 53.
Hy. dermestoides. *Marsh.* i. 405.

608. 6; testaceus.
Hy. testaceus. *Fabr. E.* i. 252?

609. [7, óchropterus.]
Hy. ochropterus. *Marsh.* i. 409.
Hy. melanocephalus e. *Gyll.* i. 120.

610. 8; bicolor *.
Hy. bicolor. *Payk. F.* i. 184.
Hy. fulvus. *Marsh.* i. 408.
Hy. fulvus. (*Sam. I.* 22.)
Hy. lividus. *Latr. G.* ii. 66.

611. 9; griseus *. (*Sam. I.* 22.)
Hy. griseus. *Illig. K.* i. 246.—*Ent. Trans.* (*Burrell.*) i. 207.
Hy. bicolor. *Fabr. E.* i. 252.
Hy. chrysomelinus. *Panz. F.* lxvii. *f.* 15.

Dy. lividus. *Forst. C.* i. 52.
Hy. lividus. *Marsh.* i. 405.
Hy. obscurus. *Müll. Z. D. pr.* 69.
Hy. pallidus. *Rossi M.* i. 66.
β; Hy. variegatus. *Herbst C.* viii. 304. *pl.* 114. *f.* 3, *E.*

612. 10; marginellus*. (*Sam. I.* 22.)
Hy. marginellus. *Fabr. E.* i. 252.—*Herbst C.* viii. 303. *pl.* 114. *f.* 2, *B.*—*Ent. Trans.* (*Haworth.*) i. 85.

613. [11; affinis*]
Hy. affinis. *Payk. F.* i. 185.
Hy. marginellus β. *Illig. M.* i. 66.

b. *Corpore subgloboso.*

614. 12; atricapillus*. *Marsh.*? *MSS.*—(*Sam. I.* 22.)

615. 13; orbicularis*. (*Sam. I.* 22.)
Hy. orbicularis. *Fabr. E.* i. 252.—*Panz. F.* lxvii. *f.* 13.—*Marsh.* i. 403.
Hy. Pilula. *Müll. Z. D. p.* 69.

616. 14; bipunctatus*.
Hy. bipunctatus. *Fabr. E.* i. 254.—*Panz. F.* lxvii. *f.* 14.—*Marsh.* i. 406.—*Turt.* ii. 422.
Chry. minuta. *Linn.* ii. 593?
Hy. chrysomelinus. *Müll. Z. D. pr.* 69.
Hy. coccinelloides. *Rossi F.* i. 197.
Dy. coccinelloides. *Mart. C. pl.* 33. *f.* 9.
Dy. marginellus. *Herbst A.* v. 129.

617. [15; striatulus*.]
Hy. striatulus. *Fabr. E.* i. 254.

618. 16; Colon* *mihi.*

619. 17; bipustulatus*. (*Sam. I.* 22.)
Hy. bipustulatus. *Marsh.* i. 406.

620. 18; minutus*. (*Sam. I.* 22.)
Hy. minutus. *Fabr. E.* i. 254.—*Marsh.* i. 406.—*Turt.* ii. 422.
Dy. minutus. *Mart. C. pl.* 33. *f.* 16.

621. 19; ochraceus* *mihi.*

622. 20; lutescens* *mihi.*

623. 21; foveolatus*.
Hy. foveolatus. *Ent. Trans.* (*Haworth.*) i. 86.

624. 22, sordens. *Kirby MSS.*

625. 23; globulus*.
Hy. globulus. *Payk. F.* i. 188.
Hy. minutus. *Herbst C.* vii. 313. *pl.* 114. *f.* 8. *H.*

626. 24; seminulum*. (*Sam. I.* 22.)
Hy. seminulum. *Payk. F.* i. 190.—*Herbst C.* vii. 314. *pl.* 114. *f.* 11, *L.*

Familia XV. SPHÆRIDIIDÆ *Leach.*

SPHÆRIDIOTA, *Latreille.*—SPHÆRIDIIDÆ *p*, *MacLeay.*

(DERMESTES *p*, *Linné*, *&c.*)

Genus 104. CERCYON, *Leach*, *Samou.*, (*Kirby.*)

SPHÆRIDIUM *p*, *Fabr.*—HYDROPHILUS *p*, *Fabr.*—PHALACRUS *p*, *Illiger.*—COCCINELLA *et* SCARABÆUS *p*, *Linné.*—SILPHA *p*, *Gmel.*

627. 1; ruficorne *.
Sp. ruficorne. *Kirby MSS.*

628. 2; littorale *.
Sp. littorale. *Gyll.* i. 111.
Hy. obscurus. *Fabr. E.* i. 253.
De. bipustulatus. *Marsh. MSS.*

629. [3; binotatum] *mihi.*

630. 4; dilatatum *mihi.*

631. 5; depressum * *mihi.*

632. 6; aquaticum *.
Sp. aquaticum. *Kirby MSS.*

633. 7; flavipes *.
Sp. flavipes. *Fabr. E.* i. 97.—*Panz. F.* ciii. *f.* 2.—*Ent. Trans.* (*Haworth.*) i. 251.
Sp. melanocephalus c. *Gyll.* i. 103?
Sc. hæmorrhoidalis. *Schra. A.* 9.

634. 8; terminatum *. (*Sam. I.* 10.)
De. terminatus. *Marsh.* i. 70.
Hyd. analis. *Payk. F.* i. 187?
Sp. leucorrhœum. *Kirby MSS.*
Sp. aquaticum. (*DeJean C.* 274.)

635. 9; acutum *.
Sp. acutum. *Kirby MSS.*
β? Sp. nitidum. *Kirby MSS.*
γ? Sp. nigriclave. *Kirby MSS.*

636. 10; Calthæ *.
De. Calthæ. *Scop. C.* 18.—*Marsh.* i. 70.

637. 11, bolitophagum *.
De. bolitophagus. *Marsh.* i. 72.

638. 12; immune *.
Sp. immune. *Kirby MSS.*

639. 13; lævigatum *.
Sp. lævigatum. *Kirby MSS.*

640. 14; apicale * *mihi.*

641. 15; obsoletum *.

Sp. obsoletum. *Gyll.* i. 107.
Sp. atomarium. *Payk. F.* i. 58.
Sp. lugubre. *Oliv. E.* ii. 7. *pl.* 2. *f.* 12. *a, b.*
De. lugubris. *Marsh.* i. 67.

642. 16; piceum*.
De. piceus. *Marsh.* i. 69.
Sp. hæmorrhoidale b. *Payk. F.* i. 59.

643. 17; picinum*.
De. picinus. *Marsh.* i. 69.

644. 18; simile*. (*Sam. I.* 20.)
De. similis. *Marsh.* i. 68.
Sp. melanocephalum b. *Payk. F.* i. 62.

645. 19; suturale* *mihi.*

646. 20; laterale*. (*Sam. I.* 10.)
De. lateralis. *Marsh.* i. 69.

647. 21; hæmorrhoidale*. (*Kirby & Sp. I. E.* iv. 500.)
Sp. hæmorrhoidale. *Fabr. E.* i. 96.—*Panz. F.* lxi. *f.* 1.
Sp. melanocephalum b. *Illig. K.* i. 66.

648. 22; femorale*.
Sp. femorale. *Kirby MSS.*

649. 23; hæmorrhoum*.
Sp. hæmorrhoum. *Gyll.* i. 107.
Hy. hæmorrhoidalis. *Fabr. E.* i. 252?

650. 24, xanthorhœum*. *Leach MSS.*

651. 25, xanthocephalum*.
Sp. xanthocephalum. *Kirby MSS.*

652. 26; infuscatum* *mihi.*

653. 27, impressum.
Sp. impressum. *Sturm* D. *F.* ii. 9. *pl.* xxii. *f. a. A.*

654. 29; melanocephalum*. (*Sam. I.* 10.)
De. melanocephalus. *Linn.* ii. 563.—*Herbst* C. iv. 74. *pl.* 37. *f.* 10. *K.*—*Marsh.* i. 68.
Si. melanocephala. *Turt.* ii. 106.
β; Sp. nigripes. *Kirby MSS.*

655. 29, conspurcatum*.
Sp. conspurcatum. *Sturm* D. *F.* ii. 15. *pl.* xxii. *f. b. B.*

656. 30, atomarium*.
Sp. atomarium. *Fabr. E.* i. 96.
Si. atomaria. *Turt.* ii. 106.
Sp. minutum. *Payk. F.* i. 63.
Sp. crenatum. *Panz. F.* xxiii. *f.* 3.

657. 31, sordidum. (*Sam. I.* 10.)
De. sordidus. *Marsh.* i. 69.

658. 32, convexium *.
Sp. convexium. *Kirby MSS.*

659. 33, convexior *.
De. convexior. *Marsh. MSS.*

660. 34, convexiusculum *.
De. convexiusculum. *Marsh. MSS.*

661. 35, nigrinum.
De. nigrinum. *Mus. Marsham.*

662. 36, immaculatum *.
Sp. immaculatum. *Kirby MSS.*

663. 37; stercorator *.
Sp. stercorator. *Kirby MSS.*

664. 38; merdarium *.
Sp. merdarium. *Sturm* D. *F.* ii. 26. *pl.* xxii. *f. f.* *F.*

665. 39; pygmæum *.
Sp. pygmæum. *Illig. K.* i. 104.

666. 40; erythropum *.
Sp. erythropum. *Kirby MSS.*

667. 41, lugubre *.
Sp. lugubre. *Fabr. E.* i. 96.
Sp. triste. *Illig. M.* i. 39.

668. 42, minutum *. (*Sam. I.* 10.)
Sp. minutum. *Fabr. E.* i. 98. (!)
De. minutus. *Marsh.* i. 75.
Sp. flavipes. *Payk. F.* i. 60.
Si. minuta. *Turt.* ii. 107.

669. 43, obscurum.
De. obscurus. *Marsh.* i. 72.

670. 44, contaminatum *.
Sp. contaminatum. *Kirby MSS.*

671. 45; concinnum *.
De. concinnus. *Marsh.* i. 74.
Sp. ferrugineum. *Herbst C.* iv. 70. *pl.* 37. *f.* 5, *E?*
Sp. melanocephalum var. pygmæa. *Payk. F.* i. 62.

672. 46, ferrugineum *.
De. ferrugineus. *Marsh.* i. 74.

673. 47, stercorarium *.
De. stercorarius. *Marsh.* i. 76.

674. 48; fuscescens * *mihi.*

675. 49, immundum.
Sp. immundum. *Sturm* D. *F.* ii. 25. *pl.* xxii. *f.* c. C. D.

676. 50, nigriceps.
De. nigriceps. *Marsh.* i. 72.

677. 51, atricapillum.
De. atricapillus. *Marsh.* i. 72.

678. 52, atriceps.
Sp. atriceps. *Kirby MSS.*

679. 53, læve.
De. lævis. *Marsh.* i. 73.

680. 54, testaceum.
Sp. testaceum. *Kirby MSS.*

681. 55, inustum*.
De. inustus. *Marsh.* i. 76.
Sp. centrimaculatum. *Sturm* D. *F.* ii. 23. *pl.* xxii. *f.* e. *E.*

682. 56, ustulatum.
Sp. ustulatum. *Kirby MSS.*

683. [57, bimaculatum*] *mihi.*

684. 58; quisquilium* (*Sam. I.* 10.)
Sp. dispar. *Payk. F.* i. 62.
♂; Sca. quisquilius. *Linn.* ii. 558.—*Stew.* ii. 27.
De. quisquilius. *Marsh.* i. 71.
Sca. minimus. *Scop. C.* 29.
Sp. melanocephalum var. *Herbst C.* iv. 63. *pl.* 37. *f.* 10, *L.*
Sp. xanthopterum. *Laich. T.* i. 86.
♀; Coc. unipunctata. *Linn.* ii. 579.
De. unipunctatus. *Marsh.* i. 70.
Si. unipunctata. *Turt.* ii. 107.
Ce. unipunctatum. (*Sam. I.* 10.)
Hyd. cordiger. *Herbst Ar.* 122.

685. [59; flavum*.]
De. flavus. *Marsh.* i. 71.

686. 60; scutellare* *mihi.*

Genus 105. SPHÆRIDIUM, *Fabricius, Leach, Samou.*, (*Kirby.*)

SCARABÆUS *p, Linné.*—HISTER *p, DeGeer.*—SILPHA *p, Gmelin.*

687. 1; Scarabæoides*. *Fabr. E.* i. 92.—*Leach E. E.* ix. 96.—*Samou.* 187. *pl.* 3. *f.* 12.—*Gyll.* iv. 274. (!)
De. Scarabæoides. *Linn.* ii. 563.—*Marsh.* i. 65.—*Don.* vii. *pl.* 231. *f.* 4.
Si. Scarabæoides. *Stew.* ii. 39.—*Turt.* ii. 105.
Hi. testudinarius. *DeGeer.* iv. 345.
De. 4-maculatus. *Schra. A.* 43.
De. hortensis. *Fourc.* i. 21.

688. [2; bipustulatum*.] *Fabr. E.* i. 93.

Sp. Scarabæoides, ♂. *Illig. K.* i. 66.
Sp. marginatum b. *Payk. F.* i. 56.
De. bipustulatus. *Marsh.* i. 66.
De. hæmorrhous. *Schra. A.* 44.
De. testudinarius. *Fourc.* i. 24.
Si. stercorea. *Gmel.* iv. 1626.

689. 3; quadrimaculatum*.
De. 4-maculatus. *Marsh.* i. 66.
Sp. marginatum c. *Gyll.* i. 101.

690. 4; lunulatum. *Kirby MSS.*

691. 5; lunatum*. *Fabr. E.* i. 93.
Sp. bipustulatum. *Herbst C.* iv. 66. *pl.* 37. *f.* 2.
Sp. Scarabæoides β. *Illig. K.* i. 65.
De. lunatus. *Marsh.* i. 66.

692. 6; marginatum*. *Fabr. E.* i. 93.—*Herbst C.* iv. 67. *pl.* 37. *f.* 3.—(*Samou. I.* 38.)
Sp. Scarabæoides γ. *Illig. K.* i. 65.
De. marginatus. *Marsh.* i. 66.
De. gagatinus. *Fourc.* i. 21.

693. 7, Daltoni. *Kirby MSS.*

Familia XVI. ANISOTOMIDÆ.

DIAPERIALES *p. et* EROTYLENÆ *p, Latreille.*—SPHÆRIDIIDÆ *p, MacLeay.*

Genus 106. TRITOMA, *Fabricius, Leach, Samou., Millard.*

DERMESTES *p, Marsham.*

694. 1; bipustulatum*. *Fabr. E.* ii. 571.—*Stew.* ii. 40.—*Turt.* ii. 116.—(*Leach E. E.* ix. 115.)—*Samou.* 51. *pl.* 2. *f.* 9.
De. humeralis. *Marsh.* i. 67.

Genus 107. PHALACRUS, *Paykul, Leach, Samou.* (*Kirby.*)

SPHÆRIDIUM *p, Fabricius.*—DERMESTES *p, Marsham.*—ANISOTOMA *p, Illiger, Samou.*—VOLVOXIS *p, Kugellan.*—ANTHRIBUS *p, Geoffroy.*—SILPHA *p, Gmelin.*—TETRATOMA *p, Herbst.*

695. 1; maritimus *mihi.*

696. 2; æneus*. *Payk. F.* iii. 349.—(*Sam. I.* 33.)
Sp. æneus. *Panz. F.* ciii. *f.* 3.
De. æneus. *Marsh.* i. 79.

697. 3; ovatus*.
De. ovatus. *Marsh.* i. 76.

698. [4, cognatus*] *mihi.*

699. 5; corticalis*. *Illig. M.* i. 40.—*Sturm* D. *F.* ii. 74. *pl.* xxx. —(*Sam. I.* 33.)
De. politus. *Marsh.* i. 75.

700. 6; affinis*. *Sturm* D. *F.* ii. 76. *pl.* xxxi. *f. a?*

701. 7; corruscus*. *Payk. F.* iii. 438.—(*Sam. I.* 33.)—(*Kirby & Sp. I. E.* iv. 503.)
An. corruscum. *Panz. F.* xxvii. *f.* 10.
Sp. fimetarium. *Fabr. E.* i. 97.
Si. fimetaria. *Stew.* ii. 39.—*Turt.* ii. 107.
De. fimetarius. *Marsh.* i. 74.
Te. atra. *Herbst C.* iv. 86. *pl.* 38. *f.* 4.
An. lævis. *Kirby MSS.*

702. 8; picipes. *Kirby? MSS.*

703. 9; substriatus*. *Gyll.* iii. 428.
An. corruscum. *Illig. K.* i. 79.

704. 10, nigrinus.
De. nigrinus. *Marsh.* i. 77.

705. 11; Caricis*. *Sturm* D. *F.* ii. 80. *pl.* xxi. *f.* D. *d.*—(*Sam. I.* 33.)
An. striatula. *Kirby MSS.*
β, An. picicornis. *Kirby? MSS.*

706. 12; Millefolii*. *Payk. F.* iii. 439.—*Panz. F.* ciii. *f.* 5.—(*Sam. I.* 33.)

707. 13; Achillææ*. *Mus. Marsham.*
An. ovalis. *Kirby MSS.*

708. [14, æneopiceus.]
An. æneopicea. *Kirby MSS.*

709. 15; bicolor*. *Payk. F.* iii. 439.—(*Leach E. E.* ix. 116.)—(*Sam. I.* 33.)
Sp. bicolor. *Panz. F.* ciii. *f.* 4.
De. nitidus. *Marsh.* i. 75.

710. 16; Leachiellus* *mihi.*

711. 17; flavicornis*. *Sturm* D. *F.* ii. 78. *pl.* xxi. *f. b.* B.

712. 18; Stephensii*. *Leach MSS.*

713. 19; rufipes*. ——? *MSS.*

714. 20; geminus*. *Sturm D. F.* ii. 75.—(*Sam. I.* 33.)
An. geminus. *Illig. K.* i. 80.
An. testaceum. *Panz. F.* xxxvii. *f.* 12.

715. 21; consimilis*. (*Sam. I.* 33.)
De. consimilis. *Marsh.* i. 75.

716. 22, piceus*. *Knoch.*

717. 23; Ulicis. *Kirby MSS.*—*Gyll.* iii. 430.

718. 24; pygmæus*. *Sturm D. F.* ii. 84. *pl.* xxxii. *f.* c. *C.*
De. æneus var. *Mus. Marsham.*

719. 25, globosus*. *Sturm D. F.* ii. 82. *pl.* xxxii. *f. a. A?*

720. 26; piceorrhœus*.
De. piceorrhœus. *Marsh.* i. 78.
Ph. dimidiatus. *Sturm D. F.* ii. 85. *pl.* xxxii. *f. d. D.*

721. 27, pulchellus.
De. pulchellus. *Marsh.* i. 72.

722. 28, nitens.
De. nitens. *Marsh.* i. 79.

Genus 108. EPHISTEMUS. *Westwood MSS.*

DERMESTES *p, Marsham.*

723. 1, Gyrinoides.
De. Gyrinoides. *Marsh.* i. 77.

724. 2; nigriclavis* *mihi.*

725 † 3, confinis* *mihi.* In Mus. D. *Westwood.*

Genus 109. LEIODES, *Latreille, Leach, Samou.*

ANISOTOMA *p, Illiger,* (*Kirby,*)—VOLVOXIS *p, Kugellan.*—SILPHA *p, Marsh.*—DERMESTES *p, Marsh.*—TRITOMA *p, Panzer.*—SPHÆRIDIUM *p, Payk.*—AGATHIDIUM *p, Samou.*

726. 1; punctata*.
An. punctata. *Sturm D. F.* ii. 42. *pl.* xxv. *f. a. A.*
An. testaceum. *Kirby MSS.*

727. 2, aciculata*.
An. aciculatum. *Kunzé.*

728. 3, dentipes*.
An. dentipes. *Gyll.* ii. 567?

729. 4; brunnea*.
An. brunneum. *Gyll.* ii. 566.

730. [5, badia.]
An. badia. *Sturm D. F.* ii. 41. *pl.* xxiv. *f. e. E.*
Le. rufa. ——? *MSS.*

731. 6; pallens*.
An. pallens. *Sturm D. F.* ii. 39. *pl.* xxiv. *f. b. B.*

732. 7; multistriata*.
An. multistriatum. *Gyll.* ii. 708.

733. 8, castanea*.
An. castaneum. *Illig. K.* i. 77.—*Sturm D. F.* ii. 48. *pl.* xxv. *f.* c. *C.*

734. 9; Marshami * *mihi.*
De. castaneus. *Marsh.* i. 78.

735. 10; punctulata.
An. punctulatum. *Gyll.* ii. 566.

736. 11; thoracicus * *mihi.*

737. 12; polita *. (*Sam. I.* 24.)
Si. polita. *Marsh.* i. 124.

738. 13, nigriclavis * *mihi.*

739. 14, ferruginea *.
An. ferrugineum. *Fabr. E.* i. 99.
Te. ferruginea. *Herbst C.* iv. 90. *pl.* 33. *f.* 9. *I?*

740. 15; maxillosa *. *Kirby MSS.*

741. 16, testaceus *mihi.*

742. 17; rufipennis *.
Sp. rufipenne. *Payk. F.* i. 70.
Ag. rufipenne. (*Sam. I.* 2.)
An. dubium. *Illig. K.* i. 78.
De. ruber. *Act. Ups.* (*Thunb.*) iv. 3.

743. 18, Lycoperdi *mihi.*

744. 19, cinnamomea.
Te. cinnamomea. *Panz. F.* xii. *f.* 15.

745. 20, Tuberis *mihi.*

746. 21, picea *. *Latr.* G. ii. 181.—(*Leach E. E.* ix. 103.)—(*Sam. I.* 24.)
An. picea. *Illig. K.* i. 75.—*Panz. F.* xxxvii. *f.* 8.—(*Sam. I.* 3.)
An. piceum. (*Sam. I.* 3.)
Vo. armata. *Schneid. M.* (*Kugell.*) v. 536.
Sp. ferrugineum. *Payk. F.* i. 72.

747. 22, armata.
An. armata. *Sturm D. F.* ii. 34. *pl.* xxiv. *f. a. A.*

B. Corpore subhemisphærico.

748. 23; humeralis *. (*Sam. I.* 24.)
An. humeralis. *Fabr. E.* ii. 558.
Sp. humeralis. *Panz. F.* xxiii. *f.* 1.

Genus 110. AGATHIDIUM, *Illiger, Leach, Samou.* (*Kirby*).

ANISOTOMA *p, Fabricius, Samou.*—SILPHA *p, Linné.*—SPHÆRIDIUM *p, Payk.*—VOLVOXIS *p, Kugellan.*—DERMESTES *p, Marsh.*—TETRATOMA *p, Herbst.*—PELTIS *p, Thunb.*

749. 1; ruficolle *. *Sturm D. F.* ii. 68. *pl.* xxix. *f. d.* D.
De. ruficollis. *Marsh.* i. 68.

Leioides ruficollis. (*Sam. I.* 24.)
Ag. nigripenne. (*Leaoh E. E.* ix. 116.)—(*Sam. I.* 2.)
Ani. nigripenne. (*Sam. I.* 3.)

750. 2, ferrugineum*. *Sturm* D. *F.* ii. 66. *pl.* xxix. *f. b. B.*
An. staphylæum. *Gyll.* ii. 569?

751. 3; atrum*.
An. atrum. *Payk. F.* iii. 417.
Ag. affine. *Kirby MSS.*

752. 4; globus*. (*DeJean C.* 129?)
An. globus. *Payk. F.* iii. 437.
Ag. ruficolle. *Sturm* D. *F.* ii. 68?

753. 5; seminulum*. *Sturm* D. *F.* ii. 59. *pl.* xxvi.
Si. seminulum. *Linn.* ii. 570.
An. seminulum. *Fabr. E.* i. 100. (!)
Ag. globosum. *Illig. K.* i. 83.

754. 6; orbiculare*.
Te. orbicularis. *Herbst C.* iv. 91. *pl.* 38. *f.* 10. *K.*
Sp. seminulum b. *Payk. F.* i. 67.

755. 7; mandibulare*. *Sturm* D. *F.* ii. 58. *pl.* xxvii. *f. c. C.*

756. 8, nigrinum*. *Sturm* D. *F.* ii. 56. *pl.* xxvii. *f. a. A.*

757. 9, affine.
De. affinis. *Marsh. MSS.*
Ag. marginatum. *Sturm* D. *F.* ii. 62. *pl.* xxviii. *f. a. A?*

758. 10, carbonarium*. *Sturm* D. *F.* ii. 61. *pl.* xxvii. *f. d.* D.
Ag. aquaticum. *Kirby MSS.?*

759. 11, atratum*. *Sturm* D. *F.* ii. 63. *pl.* xxviii. *f. d.* D.

760. 12, minutum*. *Sturm* D. *F.* ii. 64. *pl.* xxviii. *f. e.* E.

761. 13; nanum*. (*Sam. I.* 1.)
Ag. atomarium. *Sturm* D. *F.* ii. 65. *pl.* xxix. *f. a. A?*
Ag. punctulum. (*DeJean C.* 129.)—*Gyll.* iv. 515.

Genus 111. CLAMBUS, *Fischer.*

MONOMERA, *Latr.*—DERMESTES *p*, *DeGeer.*, *Marsh.*—SCAPHIDIUM *p*, *Gyllenhal.*

762. 1, Armadillus*. *Zool. Jour.* (*Halliday et Steph.*) iv.
De. Armadillus. *DeGeer.* iv. 220. *pl.* 8. *f.* 21—23.
De. convexus. *Marsh.* i. 73.

763. 2, coccinelloides.
An. coccinelloides. *Kirby MSS.*

764. 3; enshamensis*. *Westwood MSS.*

Genus 112. CLYPEASTER, *Andersch.*

CORYLOPHUS, *Leach MSS.*—COSSYPHUS, *Gyll.*—DERMESTES *p, Marsh.*

765. 1; cassidoides*.
De. cassidoides. *Marsh.* i. 77.
Co. lateralis. *Megerle.*—*Gyll.* iv. 516.

Genus 113. ————

DERMESTES *p, Marsh.*—ANISOTOMA *p, Kirby MSS.*

766. 1, Punctum.
De. Punctum. *Marsh.* i. 80.

767. 2 [picea.]
An. picea. *Kirby? MSS.*

768. 3, picatus.
De. picatus. *Marsh.* i. 80.

769. 4, truncatus.
An. truncatus. *Kirby MSS.*

770. 5; nigrescens* *mihi.*

771. 6, Atomos.
An. Atomos. *Kirby MSS.*

Genus 114. ————

SCAPHIDIUM *p, Marsh.*

772. 1; thoracicum* *mihi.*
Sc. dubium. *Marsh.* i. 234.

Subsectio 4. NECROPHAGA, *MacLeay.*

(NECROPHAGI, *et* XYLOPHAGI *p, Latreille.*—HELOCERA *p, Dumeril.*—DERMESTIDEÆ *et* NITIDULARIÆ *p, Gyllenhal.*)

(SILPHA *et* DERMESTES *p, Linné, &c.*)

Familia XVII. SCAPHIDIDÆ, *MacLeay.*

SCAPHIDILIA, *Latreille.*

Genus 115. SCAPHIDIUM *Auctorum.*

773. 1; quadrimaculatum*. *Oliv. E.* ii. 20. *pl.* 1. *f.* 1. *a. c.*—*Marsh.* i. 233.—*Leach E. E.* ix. 89.—*Samou.* 168.

Genus 116. SCAPHISOMA, *Leach, Samou.*

SCAPHIDIUM *p, Fabr., Marsh.*—SPHÆRIDIUM *p, Rossi.*

774. 1; Agaricinum*. *Leach E. E.* ix. 89.—*Samou.* 168.
Si. Agaricina. *Linn.* ii. 570.—*Panz. F.* ii. *f.* 2.

Sp. pulicarium. *Rossi M.* 11. 46.
Sc. Boleti. *Marsh.* i. 234.

775. 2; Boleti*.
Sc. Boleti. *Panz. F.* xii. *f.* 16.
Sc. Agaricinum b. *Gyll.* i. 187.

Genus 117. MYLÆCHUS, *Latreille, Leach, Samou.*

CATOPS *p, Payk.*—HALLOMINUS *p, Panz.*—CHOLEVA *p, Spence.*

776 † 1, brunneus. *Latr. G.* ii. 30. *pl.* 8. *f.* 11.—*Leach E. E.* ix. 89. —*Samou.* 169.
Ca. brevicornis. *Payk. F.* i. 140.
Ha. testaceus. *Panz. F.* lvii. *f.* 23.
Ch. brunnea. *Linn. Trans.* (*Spence*) xi. 158.
In Mus. D. *Jenyns?* Spence et Watson.

Genus 118. PTOMAPHAGUS, *Illiger.*

CHOLEVA *p, Latr., Spence, Millard.*—CATOPS *p, Panzer.*—MORDELLA *p, Marsham.*—PELTIS *p, Geoffroy.*—MYCETOPHAGUS *p, Kugellan.*—HELOPS *p, Panzer.*—CISTELA *p, Olivier.*—PTOMOPHAGUS, *Samou.*—TENEBRIO *p, Gmel.*

777. 1; truncatus*. *Illig. M.* i. 42.—(*Sam. I.* 35.)
Ch. villosa. *Latr.* G. ii. 29.—*Linn. Trans.* (*Spence.*) xi. 152.
Pt. villosa. (*Sam. I.* 35.)
Te. villosa. *Turt.* ii. 480?
Mo. silphoides. *Marsh.* i. 493.
My. picipes. *Schneid M.* (*Kugellan.*) i. 558.
He. dermestoides. *Panz. F.* lvii. *f.* 2?
He. sericeus. *Panz. F.* lxxiii. *f.* 10?

778. 2; velox*.
Ch. velox. *Linn. Trans.* (*Spence.*) xi. 154.
Ca. agilis. *Panz. F.* xcv. *f.* 10?

779. 3; fumatus*. (*Sam. I.* 35.)
Ch. fumatus. *Linn. Trans.* (*Spence.*) xi. 155.
Ca. agilis. *Gyll.* i. 279.
Ci. fusca. *Oliv. E.* iii. 10. *pl.* 1. *f.* 14. *a. b.*?

780. 4, Watsoni.
Ch. Watsoni. *Linn. Trans.* (*Spence.*) xi. 156.

781. 5; anisotomoides*.
Ch. anisotomoides. *Linn. Trans.* (*Spence.*) xi. 157.

782. 6; Wilkinii*.
Ch. Wilkinii. *Linn. Trans.* (*Spence.*) xi. 158.

Genus 119. CATOPS, *Paykul, Samou.*

PTOMAPHAGUS *p, Illiger.*—LUPERUS *p, Frölich.*—MORDELLA *p,*

Forster; *Marsh.*—HELOPS *p*, *Fanz.*—CISTELA *p*, *Fabr.*—TRITOMA *p*, *Fabr.*—CHRYSOMELA *p*, *Thunb.*—CHOLEVA *p*, *Latr.*, *Spence*, *Millard.*

A. "Thorace margine basilari prope angulos exciso." *Spence.*

783. 1; fornicatus *.
De. fornicatus. *DeGeer.* iv. 216. *pl.* viii. *f.* 15?
Lu. niger. *Naturf.* (*Frölich.*) xxviii. *f.* 3. *pl.* 1. *f.* 17?
Ch. nigricans. *Linn. Trans.* (*Spence.*) xi. 141.
Ca. nigricans. (*Sam. I.* 9.)
β; Mo. cicatricata. *Marsh.* i. 495.

784. 2; sericeus *. *Payk. F.* i. 342.—*Samou.* 168.
Lu. fusca. *Naturf.* (*Frölich.*) xxviii. *f.* 24. *pl.* 1. *f.* 16?
Ch. sericea. *Linn. Trans.* (*Spence.*) xi. 142.

785. 3; tristis *.
He. tristis. *Panz. F.* viii. *f.* 1.
Ch. tristis. *Latr. G.* ii. 28.—*Linn. Trans.* (*Spence.*) xi. 144.
Pt. fornicatus. *Illig. K.* i. 89.
Ca. Morio. *Payk. F.* i. 344.
Mo. clavicornis. *Forst. C.* i. 166.—*Stew.* ii. 87.—*Marsh.* i. 494.
Ci. ovata. *Oliv. E.* iii. 10. *pl.* 1. *f.* 11. *a. b*?
Ch. gibbosa. *Nova. A. U.* (*Thunb.*) iv. 14.

786. 4; festinans *.
Ch. festinans. *Linn. Trans.* (*Spence.*) xi. 145.

787. 5; affinis *. *Mus. Marsham.*

788. 6; elongatus * *mihi.*

B. "Thorace margine basilari prope angulos recto." *Spence.*

a. "*Thoracis lateribus ad angulos posticos subrectis.*" Spence.

789. 7; chrysomeloides *. (*Sam. I.* 9.)
He. chrysomeloides. *Panz. F.* lvii. *f.* 1.
Ch. chrysomeloides. *Linn. Trans.* (*Spence.*) xi. 146.

790. 8; Leachii *.
Ch. Leachii. *Linn. Trans.* (*Spence.*) xi. 148.

791. 9; Kirbii.
Ch. Kirbii. *Linn. Trans.* (*Spence.*) xi. 148.

792. 10; Spencii * *mihi.*

793. 11; caliginosus * *mihi.*

b. "*Thoracis lateribus rotundatis.*" Spence.

794. 12, Marshami.
Ch. Marshami. *Linn. Trans.* (*Spence.*) xi. 149.

795. 13; dissimulator *.
Ch. dissimulator. *Linn. Trans.* (*Spence.*) xi. 150.

Genus 120. CHOLEVA, *Latreille, Spence, Leach, Samou.*

PTOMAPHAGUS *p, Illiger.*—CATOPS *p, Paykul.*—MORDELLA *p, Marsham.*—HELOPS *p, Panzer.*—LUPERUS *p, Frölich.*—CRYPTOCEPHALUS *p? Stewart, Turton.*—TRITOMA *p? Fabr.*

796. 1; angustata*.
Ca. angustata. *Fabr. E. S.* i. *b.* 46.
Ch. oblonga. *Latr. G.* ii. 27.—*Linn. Trans.* (*Spence.*) xi. 138. —*Leach E. E.* ix. 89.—*Samou.* 168. *pl.* 4. *f.* 9.
Ca. elongatus. *Payk. F.* i. 345.
Pt. rufescens. *Illig. K.* i. 87.
Mo. picea. *Marsh.* i. 494.
Lu. cisteloides. *Naturfors.* (*Frölich.*) xxviii. 25. *pl.* 1. *f.* 15.
Cr. angustata. *Stew.* ii. 50?—*Turt.* ii. 184?

797. 2; agilis*. *Linn. Trans.* (*Spence.*) xi. 140.—(*Sam. I.* 10.)
Pt. agilis. *Illig. K.* i. 88.
He. fuscus. *Panz. F.* xviii. *f.* 1?
Ca. fuscus. *Gyll.* i. 281.
Tr. dubia. *Fabr. E. S.* i. *b.* 506?
β, Ch. testacea. *Latr.* G. ii. 26.

798. 3, gomphosata*. *Spence MSS.*

Familia XVIII. SILPHIDÆ, *Leach.*

(SILPHALES, *Latreille.*)

Carrion Beetles.

Genus 121. NECROPHORUS, *Fabricius, Leach, Curtis.*

DERMESTES *p, Geoffroy.*—NECROPHAGUS, *Leach E. E.*—*Samou.*

Grave digger.

A. Tibiis posticis rectis: trochanteribus simplicibus.

a. *Thorace anticè valdè dilatato.*

799. 1, germanicus*? *Fabr. E.* i. 333.—(*Sam. I.* 28.)—*Curtis.* ii. *pl.* 71.
Si. germanica. *Linn.* ii. 569.—*Marsh.* i. 113.—*Turt.* ii. 105. —*Wood.* i. 24. *pl.* 8.
Si. nigra major. *DeGeer.* iv. 173. *pl.* 6. *f.* 4–6.
De. Listerianus. *Fourcroy.* i. 17.
Pollinictor niger. *Voët.* (*Pz.*) *pl.* 30. *f.* 4.

800. 2; anglicus*. *Leach MSS.*—(*Curtis l. c. supra.*)
Ne. anglicanus. (*Sam. I.* 28.)
Ne. britannicus. *Wilk.? MSS.*

b. *Thorace anticè vix dilatato.*

801. 3; humator*. *Fabr. E.* i. 333.—(*Sam. I.* 28.)—(*Curtis l. c. supra.*)
Ne. germanicus b. *Payk. F.* i. 323.
Si. germanica. (*Shaw G. Z.* vi. 51. *pl.* 14.)

Si. nigra major var. *DeGeer.* iv. 173.
Si. humator. *Oliv.* ii. 8. *pl.* 1. *f.* 2. c.—*Marsh.* i. 114.—*Don.* xv. *pl.* 537. *f.* 1.
β, Si. bimaculata. *Ent. Trans.* (*Haworth.*) i. 82. *pl.* 2. *fig. sup.*

802. 4; vestigator*. *Illig. Mag.* (*Herschel.*) vi. 274.—(*Curtis l. c. supra.*)
Si. Vespillo. *Panz. F.* ii. *f.* 2.
Ne. investigator. *Act. Holm.* (*Zettersted.*) 1824. *p.* 151.

803. 5; Sepultor*. (*DeJean C.* 42.)—*Gyll.* iv. 308.
Ne. Vespillo. (*Samou. pl.* 2. *f.* 6.)

804. 6; Mortuorum*. *Fabr. E.* i. 335.—*Panz. F.* xli. *f.* 3.—(*Sam. I.* 28.)—(*Curtis l. c. supra.*)
Ne. Vespillo g. *Payk. F.* i. 325.
Si. Mortuorum. *Marsh.* i. 115.—*Don.* xv. *pl.* 537. *f.* 2.
Ne. Vespilloides. *Herbst A.* 32.
Silpha. *Mill. B. E. pl.* 2. *f.* 5.

B. Tibiis posticis curvis: trochantibus spinosis.

805. 7; Vespillo*. *Fabr. E.* i. 335.—*Illig. Mag.* (*Herschel.*) vi. 274.—(*Kirby & Sp. I. E.* i. 350.)—*Samou.* 166.—(*Curtis l. c. infra.*)
Si. Vespillo. *Linn.* ii. 569.—*Barb. G.* 30. *pl.* 3. *f.* 8.—*Berk. S.* i. 91.—*Don.* i. *pl.* 23?—*Stew.* ii. 37.—*Marsh.* i. 114.—*Turt.* ii. 105.—*Shaw G. Z.* vi. 50.—*Bingley.* iii. 126.—*Samou.* 51.
Ne. spinipes. *Kirby.*—*Leach E. E.* ix. 88.—*Samou.* 166.
Ne. curvipes. *Megerle.*

Genus 122. NECRODES, *Wilk. MSS.*, *Leach*, *Samou.*
Shore Beetle.

CYCLOPHORUS, *Kirby.*—PELTIS *p*, *Müller.*

806. 1; littoralis*. *Leach Z. M.* ii. 88.—*Samou.* 166.
Si. littoralis. *Linn.* ii. 570.—*Panz. F.* xl. *f.* 15.—*Marsh.* i. 116.—*Turt.* ii. 99.—*Leach E. E.* ix. 88.
Si. rufo-clavata. *DeGeer.* iv. 176.
Pe. gibbosa. *Fourc.* i. 30.
♂; Pe. femorata. *Müll. Z. D. pr.* 64.
♀; Si. simplicipes. (*DeJean C.* 42.)
β; Ne. Curtisii. *Leach E. E.* ix. 89.

Genus 123. OICEOPTOMA, *Leach*, *Samou.*

PELTIS *p*, *Geoffroy.*

A. " Elytra in utroque sexû integra." *Leach.*

807 † 1, *marginalis*[a] * ?

---

[a] 807 † 1, marginalis * ? *Fabr. E.* i. 338? In Mus. *D. Vigors.*

808. 2; thoracica*. *Leach E. E.* ix. 89.—*Samou.* 167.
Si. thoracica. *Linn.* ii. 571.—*Don.* ii. *pl.* 63. *f.* 1.—*Stew.* ii. 38.—*Marsh.* i. 117.—*Turt.* ii. 100.

809. 3; rugosa*. *Leach.*—(*Sam. I.* 32.)
Si. rugosa. *Linn.* ii. 571.—*Panz. F.* xl. *f.* 17.—*Stew.* ii. 38. —*Marsh.* i. 120.—*Turt.* ii. 100.
Pe. complicata. *Fourc.* i. 30.
Si. scabra. *Scop. C.* 59.

B. "Elytra fœminæ apice emarginata." *Leach.* (THANATOPHILUS *Leach.*)

810. 4; sinuata*. (*Sam. I.* 32.)
Si. sinuata. *Fabr. E.* i. 341.—*Oliv. E.* ii. 18. *pl.* 2. *f.* 12.—*Stew.* ii. 38.—*Marsh.* i. 120.—*Turt.* ii. 101.—*Don.* xv. *pl.* 539.
♀; Si. opaca. *Marsh.* i. 119.
Pe. scabra. *Fourc.* i. 30.
♂ β, Si. rugicollis. *Kirby MSS.*

811. 5, dispar.
Si. dispar. *Illig. K.* i. 359.—*Herbst C.* v. 204. *pl.* 52. *f.* 1.
Si. sinuata b. *Payk. F.* i. 333.

Genus 124. SILPHA *Auctorum.*

A. "Thorace anticè emarginato." *Gyllenhal.*

812. 1; quadripunctata*. *Linn.* ii. 571.—*Berk. S.* i. 91.—*Don.* ii. *pl.* 86. *f.* 2.—*Stew.* ii. 38.—*Marsh.* i. 118.—*Turt.* ii. 101.—*Samou.* 51.
Si. quadrimaculata. (*Samou.* 167. *pl.* 2. *f.* 7.)

B. "Thorace anticè integro." *Gyllenhal.*

a. *Elytris lineis elevatis tribus.*

813. 2; obscura*. *Linn.* ii. 572.—*Don.* ii. *pl.* 63. *f.* 4.—*Marsh.* i. 118.—*Turt.* ii. 101.—*Leach E. E.* ix. 89.—*Samou.* 167.
Si. atrata. *Herbst C.* v. 183. *pl.* 51. *f.* 1.
β; Si. tenebrosa. *Schonher teste Gyll.* iv. 310.

814. 3, tristis. *Illig. K.* i. 366.—(*Sam. I.* 38.)
Si. recta. *Marsh.* i. 117.
Si. granulata. *Thunb. I. S.* 72.
Si. punctata ♀. *DeGeer.* iv. 177?

815 † 4, Griesbachiana *mihi.* In Mus. *Brit.*

816. 5, nigrita. *Creutz. E. V.* i. 116. *pl.* 2. *f.* 20.—(*Ing. Inst.* 89.)
Si. nitidiuscula. *Kirby MSS.*—(*Sam. I.* 38.)
β, Si. bicolor. *Ent. Trans.* (*Haworth.*) i. 82.

817. 6; reticulata*. *Fabr. E. E.* i. 341.—(*Sam. I.* 38.)
Si. rugosa. *Panz. F.* v. *f.* 9.
Si. granulata. *Marsh.* i. 119.

Si. cancellata. *Gmel.* iv. 1622.
Pe. undata. *Müll. Z. D. pr.* 64.

818. 7; opaca*. *Linn.* ii. 571.—*Turt.* ii. 101.—(*Sam. I.* 38.)
Si. tomentosa. *DeGeer.* iv. 183.—*Herbst C.* v. 203. *pl.* 51. *f.* 16.—*Marsh.* i. 120.
Si. hirta. *Gmel.* iv. 1622.
Si. villosa. *Act. Holm.* (*Næzen.*) 1792. 168. *pl.* 7.

b. *Elytra lævia simplicia.*

819. 8; lævigata*. *Fabr. E.* i. 340.—*Oliv. E.* ii. 14. *pl.* 1. *f.* 1. *a, b.*—*Marsh.* i. 119.—*Turt.* ii. 100.—(*Sam. I.* 38.)
Si. polita. *Sulzer.*

Genus 125. PHOSPHUGA, *Leach, Samou.*

PELTIS *p, Müller.*

820. 1; atrata*. *Leach Z. M.* iii. 75.—*Samou.* 167.
Si. atrata. *Linn.* ii. 571.—*Stew.* ii. 37.—*Marsh.* i. 116.—*Turt.* ii. 100.
Si. punctata. *Herbst C.* v. 199. *pl.* 51. *f.* 13.
β; Si. brunnea. *Herbst C.* v. 202. *pl.* 51. *f.* 15.
γ; Si. fusca. *Herbst C.* v. 200. *pl.* 51. *f.* 14.

821. 2, subrotundata. *Leach Z. M.* iii. 75.—*Samou.* 167.

Genus 126. AGYRTES[a].

Genus 127. PELTIS[b].

Familia XIX. NITIDULIDÆ *MacLeay.*

NITIDULARIÆ, *Latr.*—(NITIDULA, *Fabr., &c.*)

Genus 128. THYMALUS, *Latreille, Curtis.*

PELTIS *p, Paykul.*—CASSIDA *p, Olivier.*

824. 1: limbatus*. *Curtis.* i. *pl.* 39.
Ca. limbata. *Fabr. E. S.* i. *a.* 294.

---

[a] Genus 126. AGYRTES, *Frölich.*

MYCETOPHAGUS *p, Fabr.*

822 † 1. castaneus. *Naturf.* (*Frölich.*) 28. *pl.* 1. *f.* 11.—(*Leach E. E.* ix. 89.) (!)
My. spinipes. *Panz. F.* xxiv. *f.* 20.

[b] Genus 127. PELTIS, *Kugellan.*

OSTOMA *p, Laicharting.*—THYMALUS, *Samou.*

823 † 1. ferrugineus. *Schneid. Mag.* (*Kugellan.*) 509.
Si. ferruginea. *Linn.* ii. 572.—*Panz. F.* lxxv. *f.* 17.—*Stew.* ii. 38. (!)
Si. ferruginea. (*Samou.* 170. (!)
Si. cimicoides. *DeGeer.* iv. 183.

Pe. limbata. (*Kirby & Sp. I. E.* iv. 136.)
Pe. brunnea. *Payk. F.* i. 340.
Ca. rubiginosa. *Gmel.* iv. 1643.
Ca. ænea. *Wilkin MSS.*

Genus 129. NITIDULA *Auctorum.*

PELTIS *p, Müller.*—OSTOMA *p, Laicharting.*—STRONGYLUS *p, Herbst.*

825. 1; marginata*. *Fabr. E.* i. 348.—(*Sam. I.* 28.)
Ni. biloba. *Panz. F.* xxxv. *f.* 10.

826. 2; punctatissima*. *Schneid. Mag.* (*Illiger.*) v. 598.—*Panz. F.* xxv. *f.* 7.
Si. maculata. *DeGeer.* iv. 184.
Ni. elevata. *Kirby MSS.*

827. [3; grisea*.] *Marsh.* i. 134.—(*Sam. I.* 28.)
Si. grisea. *Linn.* ii. 574.—*Linn. Trans.* (*Curtis.*) ii. 86. *pl.* 5. *f.* 6—11.—*Turt.* ii. 102.
Ni. variegata. *Oliv. E.* ii. 4. *pl.* 1. *f.* 1. *a, c.*
Ni. varia. *Fabr. E.* i. 350.
Si. ferruginea. *Scop. C.* 21.

828. 4; depressa*. *Marsh.* i. 133.—(*Sam. I.* 28.)
Si. depressa. *Linn.* ii. 573.
Ni. sordida. *Fabr. E.* i. 351.
Ni. varia. *Oliv. E.* ii. 12. *pl.* 2. *f.* 10. *a, b.*
Ni. ferruginea. *Herbst A.* 159.
Ni. Colon. *Herbst C.* v. 234. *pl.* 53. *f.* 5. *a. E.*
Ni. immaculata. *Oliv. E.* ii. 11. *pl.* 2. *f.* 16. *a, b.*

829. 5; Colon*. *Fabr. E.* i. 351.—*Oliv. E.* ii. 13. *pl.* 1. *f.* 4. *a, b.* —*Marsh.* i. 132.
Si. Colon. *Linn.* ii. 573.
β; Ni. hæmorrhoidalis. *Fabr. E.* i. 352?

830. 6; discoidea*. *Fabr. E.* i. 252.—*Stew.* ii. 40.—*Panz. F.* lxxxiii. *f.* 5.—*Marsh.* i. 133.—*Turt.* ii. 111.—*Samou.* 51. *pl.* 2. *f.* 5.
Ni. hæmorrhoidalis b. *Payk. F.* i. 252.

831. 7; bipustulata*. *Fabr. E.* i. 347.—*Berk. S.* i. 91.—*Panz. F.* iii. *f.* 10.—*Stew.* ii. 40.—*Marsh.* i. 129.—*Turt.* ii. 110.—*Leach E. E.* ix. 90.—*Samou.* 170.
Si. bipustulata. *Linn.* ii. 570.
De. Scarabæoides. *Scop. C.* 15.
De. bipunctatus. *Fourc.* i. 18.

832. 8; obscura*. *Fabr. E.* i. 348.—*Oliv. E.* ii. 12. *pl.* 1. *f.* 3. *a, b.*—*Marsh.* i. 130.—(*Sam. I.* 28.)
Si. rufipes. *Linn.* ii. 573.
De. fulvipes. *Fourc.* i. 22.

β; Ni. immunis. *Kirby.*
Ni. ossium. *Marsh. MSS.*

833. 9, quadripustulata. *Fabr. E. S.* i. *a.* 255?
Ni. guttalis. *Herbst C.* v. 247. *pl.* 54. *f.* 7. *g.* G.
Si. carnaria. *Gmel.* iv. 1619.

834. [10, variata] *mihi.*
Ni. variegata. *Kirby MSS.*

835. 11; oblonga*. *Herbst C.* v. 245. *pl.* 54. *f.* 4. *d.* D.

836. 12, Silacea*. *Herbst C.* v. 232. *pl.* 53. *f.* 3. c.

837. 13; æstiva*. *Oliv. E.* ii. 16. *pl.* 3. *f.* 23. *a, b.*
Si. æstiva. *Linn.* ii. 574.—*Turt.* ii. 110.

838. 14; depressa*. *Illig. K.* i. 386.
Ni. æstiva. *Fabr. E.* i. 348.—*Panz. F.* lxxxiv. *f.* 10.—*Marsh.* i. 135.
Ni. villosa. *Thunb. I. S.* v. 70.

839. 15; obsoleta*. *Illig. K.* i. 384.—*Marsh.* i. 135.—(*Sam. I.* 28.)
Si. biguttata. *Act. Ups.* (*Thunb.*) iv. 9.
Ni. unicolor. *Oliv. E.* ii. 17. *pl.* 2. *f.* 9. *a, b*?

840. 16; variegata*. *Herbst C.* v. 245. *pl.* 54. *f.* 3. c?
Ni. obsoleta var. *Illig. K.* i. 384. (nota.)
β; Ni. thoracica *mihi.*

841. 17, impressa*. *Kirby MSS.*

842. 18; pusilla*. *Illig. K.* i. 386?
Ni. pubescens. *Schneid. M.* (*Kugell.*) iv. 512.

843. 19, pygmæa. *Gyll.* i. 226.

844. 20; truncata*. *Kirby MSS.*

845. 21, affinis*. *Marsh? MSS.*

846. 22, melanocephala. *Marsh.* i. 136.

847. 23; limbata*. *Fabr. E.* i. 352.—*Oliv. E.* ii. 20. *pl.* 3. *f.* 18. *a, b.*
Ni. Agaricina. *Kirby MSS.*
Ni. nebula ——?

848. 24, fervida*. *Oliv. E.* ii. 15. *pl.* 4. *f.* 32. *a, b.*

849. 25; decemguttata*. *Fabr. E.* i. 350.—*Oliv. E.* ii. 10. *pl.* 3. *f.* 24. *a, b.*—*Marsh.* i. 135.—(*Sam. I.* 28.)

Genus 130. STRONGYLUS, *Herbst.*

SPHÆRIDIUM *p*, *Panzer.*

A. Elytris haud striatis.

850. 1; strigatus*. *Herbst C.* iv. 187. *pl.* 43. *f.* 7. *g.* G.
Ni. strigata. *Fabr. E.* i. 350.—*Panz. F.* lxxxiii. *f.* 4.—(*Ing. Inst.* 89.)

Ni. undata. *Marsh.* i. 134.
Ni. Verbasci. *Thunb. I. S.* v. 71.
De. graphicus. *Schran. B.* 423.

851. 2; imperialis*. (*DeJean C.* 43.)
Ni. imperialis. *Fabr. E.* i. 350.
Ni. undulata. *Oliv. E.* ii. 14. *pl.* 3. *f.* 17. *a, b.*
Ni. nebulosa. *Marsh.* i. 134.

B. Elytris striatis.

852. 3; ferrugineus*. (*DeJean C.* 43.)
Ni. ferruginea. *Fabr. E.* i. 349. (!)—*Turt.* ii. 110.
Sp. ferrugineum. *Panz. F.* lxxxiv. *f.* 2.
Ni. fulva. *Marsh.* i. 136.—(*Sam. I.* 28.)
St. æstivus. *Herbst C.* iv. 186. *pl.* 43. *f.* 6. *e. E.*
Sp. pilosum. *Rossi M.* ii. 84.
Ni. striata. *Oliv. E.* ii. 14. *pl.* 1. *f.* 7. *a. b.*

Genus 131. CAMPTA, *Kirby MSS.*

STRONGYLUS *p, Herbst.*—SPHÆRIDIUM *p, Fabricius.*—ANTHRIBUS *p, Olivier.*—CYCHRAMUS *p, Kugellan.*

853. 1; lutea*.
St. luteus. *Herbst C.* iv. 183. *pl.* 43. *f.* 3. c. *C.*
De. luteus. *Marsh.* i. 73.
De. Rosæ. *Scop. C.* 15.
An. unicolor. *Oliv. E. M.* iv. 160.
Ni. Boleti. (*Sam. I.* 28.)
β?; De. ovalis. *Marsh.* i. 73.

Genus 132. MELIGETHES, *Kirby MSS.*

SCARABÆUS *p, Geoffroy.*—ANTHRIBUS *p, Olivier.*—STRONGYLUS *p, Herbst.*—LARIA, *Scopoli.*

854. 1, Dulcamaræ*.
La. Dulcamaræ. *Scop. C.* 22.
Ni. olivacea. *Gyll.* iii. 679?
Ni. Aceris. *Kirby MSS.*

855. 2; rufipes*.
Si. rufipes. *DeGeer.* iv. 188.
Ni. rufipes. *Marsh.* i. 130.—*Turt.* ii. 112.—(*Sam. I.* 28.)
Ni. atrata. *Oliv. E.* ii. 18. *pl.* 4. *f.* 31. *a, b.*

856. [3, flavimanus.] *Kirby MSS.*
Ni. ænea. *Marsh.* i. 131?

857. 4; viridescens*.
Ni. viridescens. *Fabr. E.* i. 353.—*Panz. F.* lxxxiii. *f.* 7.—*Marsh.* i. 131.
Ni. Pedicularia g. *Payk. F.* i. 353.
Ni. ænea c. *Gyll.* i. 238.

β; Ni. rufipes. *Fabr. E.* i. 355.—*Oliv. E.* ii. 21. *pl.* 5. *f.* 33. *a*, *b*.
De. Brassicæ. *Scop. C.* 45.
γ; Ni. ænea. *Fabr. E.* i. 353.—*Panz. F.* lxxxiii. *f.* 6.

858. [5; cæruleus*.]
Ni. cærulea. *Marsh.* i. 132.
Ni. ænea b. *Gyll.* i. 238.

859. [6; Urticæ*.]
Ni. Urticæ. (*Sam. I.* 28.)

860. 7, nigrescens *mihi*.

861. 8; Pedicularius*.
Si. Pedicularia. *Linn.* ii. 574?
Ni. Pedicularia. *Fabr. E.* i. 352.—*Oliv. E.* ii. 19. *pl.* 3. *f.* 21. *a*, *b*.
Ni. solida. *Illig. K.* i. 389.
Ni. latipes. *Marsh.* i. 131.

862. 9, xanthoceras *mihi*.
Ni. serripes. *Gyll.* iv. 301?

863. 10, nigricornis. *Kirby MSS.*

864. 11; subrugosa*.
Ni. subrugosa. *Gyll.* i. 236.

865. 12; nigrina*.
Ni. nigrina. *Marsh.* i. 138.—(*Sam. I.* 28.)
Ni. æthiops. *Latr. G.* ii. 14?

866. 13; erythropa*.
Ni. erythropa. *Marsh.* i. 132.—(*Sam. I.* 28.)

Genus 133. PRIA, *Kirby MSS.*

867. 1, truncatella.
Si. truncatella. *Marsh.* i. 123.

Genus 134. CARPOPHILUS, *Leach.*

COLOBUS, *Kirby?*

868. 1; flexuosus*.
Ni. flexuosa. *Fabr. E. S.* i. *a*. 258.—*Herbst C.* v. 246. *pl.* 34. *f. e. E.*—*Marsh.* i. 133.
De. hemipterus. *Linn.* ii. 565?
Ni. bimaculata. *Gyll.* i. 244.
Ni. quadriguttata. *Thunb. I. S.* v. 70.

869. 2; pusillus* *mihi*.

Genus 135. CATERETES, *Herbst*, *Leach*, *Samou.*, *Millard.*

SPHÆRIDIUM *p*, *Fabricius*.—SCAPHIDIUM *p*, *Panzer*.—STRONGYLUS *p*, *Herbst*.—BRACHYPTERUS, *Kugellan*.—CERCUS,

Latreille.—ANTHRIBUS *p, Fourcroy.*—GYMNURA, *Kirby MSS.* —CATHERETES (*Kirby.*)

870. 1; bipustulatus*. *Gyll.* i. 248.—(*Sam. I.* 9.)
De. bipustulatus. *Payk. F.* i. 286.
Si. histeroides. *Act. Hol.* (*Thunb.*) iv. 8.
Si. bimaculata. *Marsh.* i. 126.
Gy. biguttata. *Kirby MSS.*
β; Si. punctulata. *Marsh.* i. 126.
Ni. pusilla. *Thunb. I. S.* v. 69.
γ, Gy. fusciventris. *Kirby MSS.*

871. 2; pedicularius*. *Herbst C.* v. 12. *pl.* 45. *f.* 1. *a. A.*
De. pedicularius. *Linn.* ii. 564.—*Panz. F.* vii. *f.* 5.—*Turt.* ii. 82.
Si. unicolor. *Marsh.* i. 127.
β; Si. similis. *Marsh.* i. 127.
Ca. gravidus. *Illig. K.* i. 395.—(*Kirby & Sp. I. E.* i. 295.)

872. 3; Urticæ*. *Illig. K.* i. 395.—(*Kirby & Sp. I. E.* i. 295.)
De. Urticæ. *Fabr. E. S.* i. *a.* 235.
Sca. scutellatum. *Panz. F.* iv. 11.
St. abbreviatus. *Herbst C.* iv. 190. *pl.* 43. *f.* 10. *k. K.*
Si. pulicaria. *Linn.* ii. 574?—*Berk. S.* i. 93.—*Stew.* ii. 39.—*Marsh.* i. 128.—*Turt.* ii. 108.
De. pulicarius. *Linn.* ii. 564?—*Berk. S.* i. 89?
De. hemipterus. *Naturf.* (*Panzer.*) xxiv. 11. *pl.* 1. *f.* 14. *b.*

873. 4, Linariæ. *Kirby MSS.*

874. 5; nitidus*. *Kirby MSS.*
De. glaberrimus. *Payk. F.* i. 287?

875. 6; ruficornis*.
Ni. ruficornis. *Marsh.* i. 131.

876. 7; pyrrhopus*.
Ni. pyrrhopus. *Marsh.* i. 138.

877. 8, rufilabris. *Samou.* 170.
Ce. rufilabris. *Latr. G.* ii. 16.
Si. collaris. *Marsh. MSS.*

878. [9, Junci.] *Kirby? MSS.*

879. 10, Caricis. *Spence MSS.*

880. 11, Agarici. *Kirby MSS.*

Genus 136. MICROPEPLUS, *Latreille, Leach, Samou.,* (*Kirby.*)

STAPHYLINUS *p, Paykul.*—OMALIUM *p, Gyllenhal.*—CLATHRIDIUM, *Kirby MSS.*

881. 1; porcatus*. *Latr. G.* iv. 377.—*Leach. E. E.* ix. 90.—*Samou.* 171.
St. porcatus. *Payk. M. S.* 50.—*Oliv. E.* iii. 35. *pl.* 4. *f.* 33. *a, b.*

Ni. porcata. *Marsh.* i. 137.
Ni. sulcata. *Herbst Col.* v. 247. *pl.* 54. *f.* 6. *f.*

882. 2; staphylinoides*. (*Sam. I.* 27.)
Ni. staphylinoides. *Marsh.* i. 137.

Familia XX. ENGIDÆ, *MacLeay.*

NITIDULARIÆ *p,*—IPSIDES, *Latr.*—XYLOPHAGI *et* EROTYLENÆ *p, Latr.*—DERMESTIDEÆ *p, Gyllenhal.*—SILPHIDÆ *p,* CISSIDÆ *p,* DIAPERIDÆ *et* MYCETOPHAGIDÆ, *Leach.*

(SILPHA *et* DERMESTES *p, Linné, &c.*)

Genus 137. TRICHOPTERYX, (*Kirby.*)

LATRIDIUS *p, Herbst.*—OPATRUM *p, Brongniart.*—CATHERETES *p, Wolf.*—SCAPHIDIUM *p, Gyllenhal.*—PTYCHOPTERYX, *Kirby MSS.* (*olim.*)—PTILIUM, *Schüpp.*

883. 1; atomaria*. (*Kirby & Sp. I. E.* iii. 40. *nota*)
De. atomarius. *DeGeer.* iv. 218. *pl.* 8. *f.* 16–20.
Si. minutissima. *Marsh.* i. 125.
La. fasciculatus. *Herbst C.* v. 8. *pl.* 44. *f.* 7. *g. G.*?
De. Armadillo. *Payk. F.* i. 296?
Ca. atomos. *Wolf. Beit.*
Op. plumigerum. *Brongniart.*

884. 2; minima*.
De. minimus. *Marsh.* i. 80.

885. 3; pusilla*.
Sc. pusillum. *Gyll.* i. 189.

886. 4; nana* *mihi.*

887. 5; minuta* *mihi.*
Pt. punctatum. *Gyll.* iv. 293?

888. 6, trisulcata ——?

Genus 138. ATOMARIA, (*Kirby,*) *Millard.*

CORTICARIA *p, Marsh.*—LATRIDIUS *p, Kugellan.*—CATERETES *p, Herbst.*—CRYPTOPHAGUS** *p, Gyll., Samou.*—PTILIUM *p, Schüpp.*

A. Corpore lato, subovato.

889. 1; punctula* *mihi.*

890. 2; minutissima*. *Kirby MSS.*

891. 3, perpusilla.
De. perpusillus. *Marsh.* i. 80.

892. 4; Melas*.
De. Melas. *Marsh.* i. 78.

893. 5; nitida* *mihi.*

894. 6, pilosella.
De. pilosellus. *Marsh.* i. 78.

895. 7, nitidula.
De. nitidulus. *Marsh.* i. 79.

896. 8, brunnea.
De. brunneus. *Marsh.* i. 78.

B. Corpore sublineare.

a. *Antennis basi remotis.*

897. 9; fulvicollis *mihi.*

898. 10; thoracica* *mihi.*

899. 11; evanescens*.
Si. evanescens. *Marsh.* i. 126.

900. 12; phæogaster*.
Si. phæogaster. *Marsh.* i. 125.

901. 13, basella. *Kirby MSS.*

902. 14; atricapilla*. *Kirby MSS.*

903. 15; hirta*.
Si. hirta. *Marsh.* i. 124.
Cr. hirtus. (*Sam. I.* 13.)
Cr. villosus. *Wolf. Beit.*

904. 16, testacea. *Kirby MSS.*

905. 17; castanea*. *Kirby MSS.*

906. 18; nigripennis*.
De. nigripennis. *Payk. F.* i. 292.
Cr. ruficollis. *Panz. F.* xcix. *f.* 13.—(*Sam. I.* 13.)
La. ater var. *Schneid. M.* (*Kugellan.*) v. 578.

907. 19; mesomelas*.
De. mesomelus. *Herbst C.* iv. 143. *pl.* 41. *f.* 7. *g.* G.
Si. phæorrhœa. *Marsh.* i. 124.
Cr. phæorrhœus. (*Sam. I.* 13.)
Cr. globulus. *Gyll.* i. 184?

908. 20; dimidiata*.
Co. dimidiata. *Marsh.* i. 112.

909. 21; atra*.
Ca. ater. *Herbst C.* v. 15. *pl.* 45. *f.* 5. *e.* *E.*
De. fimetarius. *Payk. F.* i. 293.
Si. nitidula. *Marsh.* i. 123.

910. 22; carbonaria*.

911. 23; rufipes*.
Cr. umbrinus. *Schüppel.*—*Gyll.* iv. 291?
Cr. fimetarius c. *Gyll.* i. 182.

912. 24; ruficornis*.
Si. ruficornis. *Marsh.* i. 125.

913. 25; dorsalis*. *Kirby MSS.*

914. 26, fuscipes.
Cr. fuscipes. *Gyll.* i. 182.

b. *Antennis basi approximatis.*

915. 27; nigrirostris*. *Kirby MSS.*

916. 28, nigriventris*. *Kirby MSS.*

917. 29; linearis* *mihi.*

Genus 139. TYPHÆA, *Kirby MSS.*

MYCETOPHAGUS *p, Fabr., Curtis.*—BOLETARIA *p, Marsh MSS.*—CRYPTOPHAGUS *p, Leach MSS?*—PENTAPHYLLUS *p, DeJean.*

918. 1; Sparganii*.
Cr. Sparganii. *Leach? MSS?*
Ty. suturalis. *Kirby MSS?*

919. 2; ferruginea*.
Si. ferruginea. *Marsh.* i. 125.
My. ferrugineus. (*Curtis. fo.* 156.)

920. 3; testaceus*.
My. testaceus. *Fabr. E.* ii. 570.

921. [4, tomentosa*.]
Bo. tomentosa. *Marsh. MSS.*

Genus 140. CRYPTOPHAGUS, *Herbst, Leach, Samou., Millard,* (*Kirby,*) *Curtis.*

IPS *p, Oliv.*—CORTICARIA *p, Marsh.*—ONCINUM, *Kirby MSS.*

A. Thoracis lateribus denticulatis.

922. 1; Populi*. *Payk. F.* iii. 355.—(*Sam. I.* 13.)—*Curtis.* iv. *pl.* 160.

923. 2; bituberculatus*. *Kirby MSS.*—(*Curtis l. c. supra.*)

924. 3; humeralis* *mihi.*

925. 4; uncinnata*.
Co. uncinnata. *Marsh. MSS.*
Co. affinis. ——? *MSS.*—(*Curtis l. c. supra?*)

926. 5, Marshami *mihi.*
Co. acutangulus. *Gyll.* iv. 285?

927. 6, Lycoperdi*. *Herbst C.* iv. 176. *pl.* 42. *f.* 13. *n. N.*—(*Curtis l. c. supra.*)
De. Fungorum. *Panz. F.* xxxix. *f.* 14.
Co. rufa. *Marsh.* i. 111.
On. testaceum. *Kirby MSS.*

928. 7; fumatus*. *Gyll.* i. 167.—(*Curtis l. c. supra.*)
Co. fumata. *Marsh.* i. 110.

929. 8; Illicis*. *Kirby MSS.*—(*Curtis l. c. supra.*)

930. 9; Abietis*. *Payk. F.* iii. 356.—(*Curtis l. c. supra.*)
Co. obcordata. *Marsh.* i. 112.
De Vini. *Panz. F.* xl. *f.* 14?

931. 10; cellaris*. *Payk. F.* iii. 356.—*Leach E. E.* ix. 89.—*Samou.* 169.—(*Curtis l. c. supra.*)
Co. denticulata. *Marsh.* i. 111.
Cr. denticulatus. (*Sam. I.* 13.)
β; Cr. crenatus. *Herbst C.* iv. 177. *pl.* 42. *f.* 14. *o. O.*

932. 11; serratus*. *Gyll.* i. 171?—(*Sam. I.* 13.)—(*Curtis l. c. supra.*)
Co. serrata. *Marsh.* i. 109.

933. 12, ruficornis *mihi.*

B. Thoracis lateribus vix crenulatis.—BYTURUS *p, Latr.*

934. 13; Caricis. *Gyll.* ii. iv.
Ips Caricis. *Oliv. E.* ii. 15. *pl.* 5. *f.* 23. *a, b.*

935. 14; Typhæ*. *Gyll.* i. 174.—(*Sam. I.* 13.)
De. fulvipes. *Marsh. MSS.*
Cr. Typhæ. *Fallen. Obs. Ent.* i. 16.

Genus 141. ANTHEROPHAGUS, (*Megerle.*)

TENEBRIO *p, Linné, Marsh.*—MYCETOPHAGUS *p, Fabr.*—CRYPTOPHAGUS *p, Gyll.*—STENE *p, Kirby MSS.*

936. 1; pallens*. (*DeJean C.* 45.)
Te. pallens. *Linn.* ii. 675.—*Oliv. E.* iii. 19. *pl.* 2. *f.* 25.—*Mart. C. pl.* 39. *f.* 3.—*Marsh.* i. 477.
Cr. pallens. (*Sam. I.* 13.)

937. 2; Silaceus*.
Ips Silaceus. *Herbst C.* iv. 169. *pl.* 42. *f.* 7. *g. G.*
Te. pallens. *Payk. F.* i. 90.
My. nigricornis. *Fabr. E.* ii. 369.—
An. nigricornis. (*Ing. Inst.* 87.)
Cr.? ferrugineus. *Latr.*
St. ferruginea. *Kirby MSS.*

938. 3, glaber*.
Cr. glaber. *Gyll.* i. 178.
St. Bombi. *Kirby MSS?*

Genus 142. BYTURUS, *Latreille, Leach, Samou.,* (*Kirby.*)

IPS *p, Olivier.*—TRIXAGUS *p, Kugellan.*

939. 1; tomentosus*. *Latr.* G. ii. 18.—*Leach E. E.* ix. 90.—*Samou.* 170.—(*Kirby & Sp. I. E.* i. 194.)

De. tomentosus. *Fabr. E. S.* i. 316.—*Stew.* ii. 32.—*Marsh.* i. 64.—*Panz. F.* xcvii. *f.* 4.—xl. *f.* 12.—*Turt.* ii. 81.
De. Sambuci. *Scop. C.* 16.
β; De. fumatus. *Fabr. E. S.* i. 231.—*Panz. F.* xcvii. *f.* 3.
De. flavescens. *Marsh.* i. 64.
De. ochraceus. *Scriba. Jo.* 153.
γ. Tr. nigriceps. *Kirby MSS.*

Genus 143. MYCETÆA, *Kirby MSS.*

CRYPTOPHAGUS *p*, *Payk.*—MYCETOPHAGUS *p*, *Gyll.*, *Curtis.*—BITURUS *p*, *Samou.*—TRIPHYLLUS *p*, *DeJean.*

940. 1, fumata*.
De. fumatus. *Linn.* ii. 564.—*Berk. S.* i. 89.—*Stew.* ii. 32.—*Marsh.* i. 65.—*Turt.* ii. 81.
Bi. fumatus. (*Sam. I.* 6.)
Mycetop. fumatus. *Gyll.* iii. 599.—(*Curtis. fo.* 156.)
De. stercoreus. *Linn. F. no.* 432.
Cr. variabilis. *Payk. F.* iii. 354.
De. variabilis. *Herbst C.* iv. 141. *pl.* 41. *f.* 5?
De. Rosæ. *Scop. C.* 39.
My. fungorum. *Kirby MSS.*

Genus 144. TRIPHYLLUS, (*Megerle.*)

IPS *p*, *Fabr.*—MYCETOPHAGUS *p*, *Gyll.*, *Curtis.*—CRYPTOPHAGUS *p*, *Payk.*—MYCETÆA *p*, *Kirby MSS.*—BOLETARIA *p*, *Donovan.*

941. 1; punctatus*. (*DeJean C.* 102.)
Ips punctata. *Fabr. E.* ii. 579.
My. punctatus. *Panz. F.* xii. *f.* 12.—(*Curtis. fo.* 156.)
Bo. punctata. *Don.* xv. *pl.* 538. *f.* 3.
Cr. pilosus. *Herbst C.* iv. 177. *pl.* 42. *f.* 15.
Si. humeralis. *Marsh.* i. 120.

942. 2; bifasciatus*. (*DeJean C.* 102.)
Ips bifasciata. *Fabr. E.* ii. 579.
My. bifasciatus. (*Curtis. fo.* 156.)
De. fasciatus. *Act. Ups.* (*Thunb.*) iv. 57.
My. signatus. *Panz. F.* lvii. *f.* 20.
β, Ips marginalis. *Panz. F.* ii. *f.* 24.

Genus 145. MYCETOPHAGUS, *Fabr.*, *Don.*, *Leach*, *Samou.*—(*Kirby.*)—*Curtis.*

BOLETARIA, *Marsh.*, *Millard.*—CARABUS *p*, *Linné.*—CHRYSOMELA *p*, *Linné.*

943. 1; quadripustulatus*. (*Curtis l. c. infra.*)
Ch. quadripustulata. *Linn.* ii. 597.—*Turt.* ii. 161.
Bo. quadripustulata. *Marsh.* i. 138.—(*Sam. I.* 6.)
My. quadripustulatus. (*Leach E. E.* ix. 110.)—(*Sam. I.* 28.)

My. quadrimaculatus*. *Fabr. E.* ii. 565.—*Don.* vi. *pl.* 185. *f.* 5.
Si. quadrimaculata. *Turt.* ii. 102.
β; My. assimilis *mihi* (*olim*).

944. 2; atomarius*. *Fabr. E.* ii. 568.—*Panz. F.* xii. *f.* 10.—(*Sam. I.* 28.)—(*Curtis l. c. infra.*)
Bo. atomaria. *Marsh.* i. 141.—*Don.* xv. *pl.* 538. *f.* 2.

945. 3; multipunctatus*. *Fabr. E.* ii. 568.—*Panz. F.* xii. *f.* 11.—(*Sam. I.* 28.)—(*Curtis l. c. infra.*)
Bo. multipunctata. *Marsh.* i. 139.—*Don.* xv. *pl.* 538. *f.* 1.
β, Bo. varia. *Marsh.* i. 140.
My. varius. (*Sam. I.* 28.)

946. 4; variabilis*. *Schneid. Mag.* (*Hellwig.*) iv. 397.
α, My. piceus. *Fabr. E.* ii. 569.—*Panz. F.* i. *f.* 22.—*Curtis.* iv. *pl.* 156.
β, My. piceus. *Payk. F.* iii. 318.—*Panz. F.* ii. *f.* 5.
γ, My. lunaris. *Fabr. E.* ii. 568.
δ, My. brunneus. *Panz. F.* lvii. *f.* 21.
ε, My. atomarius, b. *Payk. F.* iii. 318.
ζ, Bo. undulata. *Marsh.* i. 140.
My. undulatus. (*Sam. I.* 28.)
η, Bo. similis. *Marsh.* i. 141.
Mi. similis. (*Sam. I.* 28.)

947. 5, Populi? *Fabr. E.* ii. 570.
Bo. rufa. *Marsh.* i. 140.
My. rufus. (*Sam. I.* 28.)

948. 6, pubescens *mihi.*

Genus 146. BIPHYLLUS, *DeJean.*

COLOBICUS *Museorum Londinensium.*—DERMESTES *p*, *Fabricius?* BITOMA *p*, *Gyllenhal.*—NITIDULA *p*, *Rossi?*—SPHÆRIOPHAGUS, *Kirby MSS.*

949. 1, Sphæriæ*.
Si. Sphæriæ. *Marsh.* i. 122.
De. lunatus. *Fabr. E.* i. 317?—*Panz. F.* lxxv. *f.* 14?
Ni. punctata. *Rossi F.* i. 58?

Genus 147. TRIPLAX, *Paykul, Fabricius, Leach, Samou.*

TRITOMA *p*, *Latr.*—IPS, *Panzer.*

950. 1; russica*. *Herbst C.* v. 147. *pl.* 49. *f.* 12. *m.*—(*Leach E. E.* ix. 115.)—(*Sam. I.* 42.)
Si. russica. *Linn.* ii. 570.—*Marsh.* i. 121.
Tr. nigripennis. *Fabr. E.* ii. 581.
Ips nigripennis. *Panz. F.* xxx. *f.* 7.
Si. nigripennis. *Turt.* ii. 104.

951. [2; castanea*.]
Si. castanea. *Marsh.* i. 122.

952. 3; ænea*. *Fabr. E.* ii. 582.
Cr. æneus. *Herbst C.* iv. 173. *pl.* 42. *f.* 9. *i.*
Si. ænea. *Turt.* ii. 104.

953. 4; bicolor*. *Gyll.* i. 206.—(*Sam. I.* 42.)
Si. bicolor. *Marsh.* i. 122.

954. 5, rufipes*. *Fabr. E.* ii. 582.—(*Sam. I.* 42.)
Ips rufipes. *Panz. F.* xiii. *f.* 17.
Si. collaris. *Act. Hall.* (*Schall.*) i. 256.
Si. rufipes. *Turt.* ii. 104.

955. 6, ruficollis. (*DeJean. C.* 129?)

Genus 148. TETRATOMA, *Herbst, Leach, Samou., Curtis.*

BOLETARIA *p, Marsh.*

956. 1; fungorum*. *Fabr. E.* ii. 574.—*Panz. F.* ix. *f.* 10.—(*Sam. I.* 40.)—(*Curtis l. c. infra.*)
Bo. bicolor. *Marsh.* i. 141.
Te. dermestoides. *Herbst C.* iv. 88.
De. collaris. *Quensel. Diss.* 8.

957 † 2, ancora*. *Fabr. E.* ii. 575.—*Curtis.* iii. *pl.* 123.
In Mus. D. *Beck, Bennett, Bentley, Curtis* et Walker.

Genus 149. ENGIS, *Latreille, Leach, Samou.*

IPS *p, Herbst.*—EROTYLUS *p, Olivier.*—DACNE, *Latreille.*

958. 1; rufifrons*. *Fabr. E.* ii. 585.—(*Sam. I.* 16.)
Ips rufifrons. *Fabr. M.* i. 46.—*Panz. F.* xxxvi. *f.* 19.
Ni. rufifrons. *Marsh.* i. 136.
Si. rufifrons. *Stew.* ii. 37.—*Turt.* ii. 105.
En. humeralis b. *Payk. F.* i. 350.
β; Ni. rufa. *Marsh.* i. 136.

959. 2; humeralis*. *Fabr. E.* ii. 583.—*Leach E. E.* ix. 90.—*Samou.* 169.
De. scanicus. *Linn.* ii. 564.—*Panz. F.* iv. *f.* 9.—*Ent. Trans.* (*Haworth.*) i. 251.
Ni. scanica. *Ent. Trans.* (*Burrell.*) i. 137.
De. bipustulatus. *Act. Ups.* (*Thunb.*) iv. 4.
β; Si. flava. *Marsh.* i. 122.
En. humeralis β. *Gyll.* i. 203.

960. [3, angustata.] *Kirby MSS.*

Genus 150. IPS, *Herbst, Leach, Samou.*

LYCTUS *p, Kugellan.*—ENGIS *p, Samou. I.*—MYCETARIA, *Kirby MSS.*

A. Corpore elongato-cylindrico.

961. 1, ferruginea. *Fabr. E.* ii. 580.—*Leach E. E.* ix. 90.—(*Sam. I.* 23.)
De. ferrugineus. *Linn.* ii. 564.
En. ferrugineus. (*Sam. I.* 16.)
Ly. dermestoides. *Panz. F.* viii. *f.* 15.

B. Corpore depresso.

962. 2, quadriguttata. *Fabr. E.* ii. 580.—*Panz. F.* iii. *f.* 18.
Ips 4-maculatus. (*Sam. I.* 23.)

963. 3, quadripustulata*. *Fabr. E.* ii. 579.—(*Sam. I.* 23.)
Si. pustulata. *Linn.* ii. 570.—*Berk. S.* i. 91.—*Turt.* ii. 104.
Ni. 4-pustulata. *Oliv. E.* ii. 8. *pl.* 3. *f.* 22. *a, b.*—*Marsh.* i. 130.

964. 4, quadripunctata. *Herbst C.* iv. 165. *pl.* 42. *f.* 2. *b.*
Ni. 4-punctata. *Oliv. E.* ii. 12. *pl.* 3. *f.* 19. *a. b.*

Genus 151. NEMOSOMA, *Latr., Leach.*

COLYDIUM *p, Herbst.*—NEMOZOMA, *Gyllenhal.*

965 † 1; elongatum*. *Latr.* G. iii. 13. *pl.* 11. *f.* 4.—(*Leach E. E.* ix. 110.)—(*Ing. Inst.* 89.)
De. elongatus. *Linn.* ii. 561.
Co. fasciatum. *Herbst C.* viii. 81.—*Panz. F.* xxxi. *f.* 22.
In Mus. *Brit.*

Genus 152. SYNCHITA, *Hellwig.*

LYCTUS *p, Fabr.*—MONOTOMA *p, Herbst.*—CERYLON *p, Gyll.*

966. 1; Juglandis*. *Schneid. M.* (*Hellwig.*) iv. 403.
Ly. Juglandis. *Fabr. E.* ii. 561.—(*Kirby & Sp. I. E.* ii. 231.)
Mo. striata. *Herbst C.* v. 23. *pl.* 46. *f.* 1. *a.*

Genus 153. CERYLON, *Latr., Leach, Samou.*

LYCTUS *p. Fabr.*—IPS *p, Olivier.*—SYNCHITA *p, Hellwig.*—CORTICARIA *p, Marsh.*—RHYZOPHAGUS, (*Sam. I.*)

967. 1; histeroides*. *Latr.* G. iii. 14.—(*Leach E. E.* ix. 110.)—(*Sam. I.* 10.)
Ly. hysteroides. *Panz. F.* v. *f.* 16.
Rh. histeroides. (*Sam. I.* 36.)
968. [2; pilicorne*.]
Co. pilicornis. *Marsh.* i. 112.

Genus 153 a. TRIBOLIUM, *MacLeay*:—*Vide* Appendix.

969. 3; picipes *.
Ips picipes. *Oliv. E.* ii. 18. *pl.* 2. *f.* 12. *a—d.*
Co. picipes. *Marsh.* i. 106.
Ly. minutus. *Turt.* ii. 483?

970. 4; ferrugineum * *mihi.*

971. 5; obsoletum.
Ly. obsoletus. *Spence MSS.*
β? Ni. Oryzæ. *Marsh. MSS.*

Genus 154. CICONES, *Curtis.*

972 † 1, Carpini *. *Curtis.* iv. *pl.* 149.
In Mus. D. *Bainbridge et Beck.*

Genus 155. RHYZOPHAGUS, *Herbst.*

LYCTUS *p, Fabr.*—TRITOMA *p, Thunb.*—SYNCHITA *p, Hellwig.* —IPS *p, Olivier.*—MONOTOMA, *Leach, Samou.*

973. 1; ferrugineus *. *Gyll.* iii. 421.
Ly. ferrugineus. *Payk. F.* iii. 326.
Tr. stercorea. *Thunb. M. U.* iii. 39.

974. 2; cylindricus *.
Ly. cylindricus. *Panz. F.* xxxv. *f.* 18.

975. 3; rufus *.
Co. rufa. *Mus. Marsh.*

976. 4; dispar *. *Gyll.* iii. 424.
Ly. dispar. *Payk.* iii. 328.
Ips elongata. *Oliv. E.* ii. 18. 8. *pl.* 2. *f.* 15. *a. b.*
Rh. bipunctatus. *Herbst C.* v. 19. *pl.* 45. *f.* 9. 1.
Mo. Juglandis. (*Leach E. E.* ix. 110.)—(*Sam. I.* 27.)
Co. taxicornis. *Marsh.* i. 106.—(*Sam. I.* 13.)
Si. taxicornis. *Turt.* ii. 108.

977. 5; bipustulatus. (*DeJean C.* 103.)
Ly. bipustulatus. *Fabr. E.* ii. 561.
Rh. clavicornis. *Herbst C.* v. 20. *pl.* 45. *f.* 10. *K.*
Rh. dispar c. *Gyll.* iii. 425.

978. 6, parvulus *. *Gyll.* iii. 424.
Ly. parvulus. *Payk. F.* iii. 329.

Genus 156. MONOTOMA, *Herbst.*

LYCTUS *p, Payk.*—LATRIDIUS *p, Kugellan.*—CORTICARIA *p, Marsh., Millard.*—CERYLON *p, Gyll.*

979. 1; picipes *. *Herbst C.* v. 24. *pl.* 46. *f.* 2.
La. monotomus. *Schneid M.* (*Kugellan.*) v. 576.
Co. contracta. *Marsh.* i. 110.

980. 2; angustata*.
Co. angustata. *Marsh. MSS.*

981. 3, pallida *mihi.*

Genus 157. CRYPTA, *Kirby MSS.*

NOTOXUS *p, Panzer.*—ANTHICUS *p, Fabr.*—CORTICARIA *p, Marsh.*—LATRIDIUS *p, DeJean.*—CERYLON *p, Samou. I.*

982. 1; bipunctata*.
An. bipunctatus. *Fabr. E.* i. 291. (!)
No. bipunctatus. *Panz. F.* xxvi. *f.* 9.
Co. bipunctata. *Marsh.* i. 108.
Ce. bipunctatum. (*Sam. I.* 11.)

Genus 158. SILVANUS, *Latreille, Leach, Samou.*

COLYDIUM *p, Payk.*—LYCTUS *p, Kugellan.*—IPS *p, Olivier.*

A. Thoracis lateribus multidentatis.

983. 1; *Surinamensis* *[a].

984. 2; *dentatus* *[b].

B. Thoracis lateribus unidentatis.

985. 3, unidentatus*. *Latr. G.* iii. 20. *pl.* 11. *f.* 2.—(*Leach E. E.* ix. 111.)
De. unidentatus. *Fabr. E.* i. 317.
Co. planum. *Herbst C.* vii. 235. *pl.* 113. *f.* 3.

Genus 159. BITOMA, *Herbst.*

LYCTUS *p, Fabr.*—IPS *p, Olivier.*—MONOTOMA *p, Panz.*—SYNCHITA *p, Hellwig.*—DITOMA, *Latr., Leach,* (*Kirby.*)

986. 1, crenata*. *Herbst C.* v. 30.
Mo. crenata. *Panz. F.* i. *f.* 24.
Ly. crenatus. *Ent. Trans.* (*Haworth.*) i. 339.
Di. crenata. *Leach E. E.* ix. 110.

Genus 160. CORTICARIA, *Marsh, Millard.*

LATRIDIUS *p, Herbst.*

987. 1; pubescens*.
La. pubescens. *Illiger.*—*Gyll.* iv. 123.

---

[a] 983. 1; Surinamensis*.
De. Surinamensis. *Linn.* ii. 565.
Ips frumentaria. *Oliv. E.* ii. 10. *pl.* 2. *f.* 13. *a. d.*
Cor. frumentaria. *Marsh.* i. 107. (!)
Si. frumentarius. (*Samou.* 208 (!)
De. sexdentatus. *Panz. F.* xiv. *f.* 11.—*Stew.* ii. 32 (!).

[b] 984. 2; dentatus*.
Cor. dentata. *Marsh.* i. 108 (!)

De. fenestralis. *Payk. F.* i. 197.
De. longicornis. *Herbst A.* iv. 23. *pl.* 20. *f.* 8.
Co. punctulata. *Marsh.* i. 109.
La. serratus. *Latr.* G. iii. 18 ?
β ? Co. testacea. *Kirby MSS.*

988. 2; crenulata *.
La. crenulatus. *Schüppel.—Gyll.* iv. 125.

989. 3; denticulata *.
La. denticulatus. *Schüppel.—Gyll.* iv. 126.

990. 4; longicornis *.
La. longicornis. *Herbst C.* v. 4. *pl.* 44. *f.* 1.
La. ruficornis. *Schneid. Mag.* (*Kugellan.*) v. 574.

991. 5; linearis *.
De. linearis. *Payk. F.* i. 302.

992. 6; ferruginea *. *Marsh.* i. 121.
De. fenestralis. *Linn.* ii. 565 ?

993. 7, elongatus.
Co. immunis. *Kirby MSS.?*

994. 8; gibbosa *.
La. gibbosus. *Herbst C.* v. 5. *pl.* 44. *f.* 2.
De. minutus. *Fabr. E.* i. 319.
Co. impressa. *Marsh.* i. 110.
La. impressus. (*Sam. I.* 24.)

995. 9; transversalis *.
La. transversalis. *Schüppel.—Gyll.* iv. 133.
Co. sulcicollis. *Kirby MSS.*

996. 10, pallida. *Marsh.* i. 112.
La. fusculus. *Megerle.*—*Gyll.* iv. 133 ?
La. angulatus. *Gyll.* (*olim.*)

997. 11, similata *.
La. similatus. *Schüppel.—Gyll.* iv. 134.

Genus 161. LATRIDIUS, *Herbst, Leach, Samou., Millard.*

TENEBRIO *p, DeGeer.*—IPS *p, Oliv.*—CORTICARIA *p, Marsh.*—
RISSA, *Kirby MSS.*

998. 1; quadratus *. *Herbst C.* v. 8. *pl.* 44. *f.* 6. *F.*
Te. lardarius. *DeGeer.* v. 45. *pl.* 2. *f.* 25–31.
Co. nigricollis. *Marsh.* i. 113.
La. nigricollis. (*Sam. I.* 24.)
La. nigricollis. (*Sam. I.* 24 ?)
De. acuminatus. *Payk. F.* i. 299.
Ri. acuta. *Kirby MSS.*

999. 2; porcatus*. *Herbst. C.* v. *pl.* 41. *f* 4. *d.*—(*Leach E. E.* ix. 111.)—(*Sam. I.* 24.)
Co. pulla. *Marsh.* i. 111.
De. marginatus. *Payk. F.* i. 300.
La. minutus. *Latr. G.* iii. 18.
Ips minuta. *Oliv. E.* ii. 18. *pl.* 3. *f.* 22. *a, b.*
Te. minuta. *Linn.* ii. 675?

1000. 3; ferrugineus*. *Kirby MSS.*
β; Ri. testacea. *Kirby MSS.*
γ; Ri. flava. *Kirby MSS.*

1001. 4; hirsutulus*.
Co. hirsutulus. *Marsh. MSS.*
La. hirtus. *Gyll.* iv. 139?

1002. 5; transversus*. (*Sam. I.* 24.)
Ips transversus. *Oliv. E.* ii. 14. *pl.* 3. *f.* 20. *a, b.*
Co. transversa. *Marsh.* i. 109.
La. transversalis. (*DeJean C.* 102.)
La. sculptilis. *Gyll.* iv. 141.
La. asperatus. (*Megerle.*)

1003. [6; crassicornis*.]
Ri. crassicornis. *Kirby MSS.*

1004. 7; ruficollis*. (*Sam. I.* 24.)
Co. ruficollis. *Marsh.* i. 111.
La. constrictus. *Gyll.* iv. 139.

1005. 8; angustatus* *mihi.*
La. filiformis. *Gyll.* iv. 143?

Genus 162. LYCTUS, *Fabr., Leach, Samou.*

IPS *p, Olivier.* BITOMA *p, Herbst.*—CORTICARIA *p, Marsh.*—SYNCHITA *p, Hellwig.*

1006. 1; oblongus*. *Latr.* G. iii. 116.—(*Leach E. E.* ix. 111.)—(*Sam. I.* 26.)
Ips oblonga. *Oliv. E.* ii. 18. 7. *pl.* 1. *f.* 5. *a. b.*
Co. oblongus. *Marsh.* i. 111.—(*Sam. I.* 13.)
Ly. canaliculatus. *Fabr. E.* ii. 562.
Bi. unipunctata. *Herbst C.* v. 26. *pl.* 46. *f.* 3, c.
Si. fusca. *Linn.* ii. 573?
De. linearis. *Thunb. A. U.* iv. 4.

1007. [2; pusillus*.] *Kirby MSS.*
Ly. pubescens. *Panz. F.* vi. *f.* 7?

Genus 163. ———

1008. 1; parasiticus* *mihi.*

Familia XXI. DERMESTIDÆ, *Leach.*

(DERMESTINI, *Latreille.*—DERMESTIDEÆ *p*, *Gyllenhal.*)
(DERMESTES, *Linné*, *&c.*)

Genus 164. THROSCUS, *Latreille*, *Leach*, *Samou.*, *Curtis.*

ELATER *p*, *Linné.*—TRIXAGUS *p*, *Kugellan.*—CISTELA *p*, *Millard.*

1009. 1; dermestoides*. *Leach E. E.* ix. 94.—*Samou.* 183.—(*Curtis l. c. infra.*)
El. dermestoides. *Linn.* ii. 656.
De. adstrictor. *Payk. F.* i. 284.—*Panz. F.* lxxv. *f.* 15.
El. clavicornis. *Oliv. E.* ii. 54. *pl.* 8. *f.* 85. *a*, *b.*

1010. 2; obtusus*. *Westwood MSS.*—(*Ing. Inst.* 89.)—*Curtis.* iv. *pl.* 163.

Genus 165. MEGATOMA, *Herbst*, *Leach.*

ATTAGENUS *p*, *Samou.*

1011. 1; Serra*. *Latr. G.* ii. 35. *pl.* 8. *f.* 10.
De. Serra. *Fabr. E.* i. 319.—*Marsh.* i. 63.
At. Viennensis. *Herbst C.* vii. 336. *pl.* 115. *f.* 10. K?
At. Serra. (*Sam. I.* 5.)

Genus 166. ATTAGENUS, *Latreille.*

MEGATOMA *p*, *Herbst*, *Leach*, *Samou.*

1012. 1; undatus*. *Latr. H.* ix. 143. *pl.* 78. *f.* 7.
De. undatus. *Linn.* ii. 562.—*Panz. F.* lxxv. *f.* 13.—*Marsh.* i. 62.—*Turt.* ii. 79.
Me. undatus. *Leach E. E.* ix. 84.—*Samou.* 182.
♂; Me. undulata. *Herbst C.* iv. 96. *pl.* 39. *f.* 4. *a*, *b.*

1013 † 2, trifasciatus*. *Leach E. E.* ix. 94.
De. trifasciatus. *Fabr. E.* i. 313.—*Oliv. E.* ii. 13. *pl.* 1. *f.* 7. *a*, *b.* In Mus. *Brit. et* D. *Haworth.*

1014. 3; pellio*. *Leach E. E.* ix. 94.—*Samou.* 182.
De. pellio. *Linn.* ii. 562.—*Berk. S.* i. 89.—*Don.* vii. *pl.* 231. *f.* 3.—*Stew.* ii. 31.—*Marsh.* i. 63.—*Turt.* ii. 79.—*Shaw G. Z.* vi. 32. *pl.* 7.—(*Kirby & Sp. I. E.* i. 231.)
De. bipunctatus. *DeGeer.* iv. 197.
♂; Me. atra. *Herbst C.* iv. 95. *pl.* 39. *f.* 2. *a*, *b.*
Me. Schrankii. *Schneid. M.* (*Kugellan.*) i. 480.
De. cylindricornis. *Naturf.* (*Schrank.*) xxiv. 65.

Genus 167. DERMESTES *Auctorum.* *Leather-eater.*

1015. 1; lardarius*. *Linn.* ii. 561.—*Barb. G. pl.* 3. *f.* 1.—*Berk. S.* i. 89.—*Stew.* ii. 31.—*Marsh.* i. 69.—*Turt.* ii. 78—*Shaw G. Z.* vi. 31. *pl.* 7.—*Bingley* iii. 122.—*Leach E. E.* ix. 94.—*Samou.* 181.

1016. 2; murinus*. *Linn.* ii. 563.—*Marsh.* i. 61.—*Turt.* ii. 80. *Shaw.* G. *Z.* vi. *pl.* 7.—*Don.* xv. *pl.* 515.—(*Leach E. E.* ix. 94.)—*Samou.* 49. *pl.* 1. *f.* 4.
De. Catta. *Herbst C.* iv. 123. *pl.* 40. *f.* 4.—*Panz. F.* xl. *f.* 11.
De. nebulosus. *DeGeer.* iv. 197.

1017. 3, liniarius? *Illig. M.* i. 85.
De. ater. *Oliv. E.* ii. 9. *pl.* 2. *f.* 12. *a, b?*
De. macellarius. *Herbst C.* iv. 126. *pl.* 40. *f.* 7. *e, E.*

1018. 4; vulpinus[a].

1019. 5, tessellatus. *Fabr. E.* i. 315.—*Turt.* ii. 80.—*Leach. E. E.* ix. 94.)—(*Sam. I.* 15.)
De. murinus. *Oliv. E.* ii. 8. *pl.* 1. *f.* 3. *a, b.*
De. vulpecula. *Illig. K.* i. 313, *nota.*

Sectio II.

PENTAMERA *et* HETEROMERA *p, Latreille.*—CHILOGNATHOMORPHA, *MacLeay.*

Subsectio 1.

(CLAVICORNES *p, Latreille.*)

Familia XXII. BYRRHIDÆ *Leach.*

(BYRRHII *p, Latreille.*—DERMESTIDEÆ *p, Gyllenhal.*)

(BYRRHUS, *Linné, &c.*)

Genus 168. ANTHRENUS, *Geoffroy, Leach, Samou.,* (*Kirby.*)

1020. 1; Verbasci*. *Oliv. E.* ii. 7. *pl.* 1. *f.* 2. *a—d.*
By. Verbasci. *Linn.* ii. 568.—*Berk. S.* i. 91.—*Stew.* ii. 37.—*Marsh.* i. 101.—*Turt.* ii. 98.—(*Sam. I.* 3.)
An. varius a. *Illig. K.* i. 399.—*Panz. F. c. f.* 3.
An. tricolor. *Herbst C.* vii. 333. *pl.* 115. *f.* 8. *H.*
An. Scrophulariæ. (*Samou. pl.* 2. *f.* 4.)

1021. 2; varius*. (*Fabr. E.* i. 108.)

1022. 3; Museorum*. *Fabr. E. S.* i. 61.—*Turt.* ii. 98.—*Shaw G. Z.* vi. 48. *pl.* 13.—(*Sam. I.* 3.)—(*Kirby & Sp. I. E.* ii. 226. *Id.* iii. 177.)
By. Museorum. *Linn.* ii. 568.

---

[a] 1018. 4; vulpinus*. *Fabr. E. S.* i. 229.—(*Leach E. E.* ix. 94.)
De. maculatus. *DeGeer.* iv. 223.
De. murinus. *Panz. F.* xl. *f.* 10.
De. marginatus. *Ins. Suec.* (*Thunb.*) 7. *f.* 6.
De. Frishii. *Schneid. M.* (*Kugell.*) i. 478.
De. cadaverinus. *Rossi M.* i. 33.
De. tessellatus. *Marsh.* i. 61 (!).

An. varius b—e. *Illig. K.* i. 399.
An. Verbasci. *Panz. F.* c. *f.* 2.
An. nebulosus. *Forst. C.* i. 10 ?

1023. 4, Scrophulariæ *. *Fabr. E.* i. 104.—*Turt.* ii. 98.—*Shaw G. Z.* vi. 47. *pl.* 13.—*Sturm D. F.* ii. 123. *pl.* xxxvi.—*Leach E. E.* ix. 94.—*Samou.* 182.
By. Scrophulariæ. *Linn.* ii. 568.—*Berk. S.* i. 91.—*Stew.* ii. 36.
De. variegatus. *Scop. C.* 41.

1024. 5; Pimpinellæ *. *Fabr. S. E.* i. 61.—*Panz. F.* c. *f.* 1.
By. Pimpinellæ. *Marsh.* i. 101.
An. Scrophulariæ. *Fourc.* i. 27.

Genus 169. TRINODES, (*Megerle.*)

ANTHRENUS *p*, *Fabr.*—NITIDULA *p*, *Gmel.*

1025. 1, hirtus. (*DeJean C.* 47.)
An. hirtus. *Fabr. E.* i. 108.—*Panz. F.* xi. *f.* 16.
β, An. pubescens. *Fabr. E.* i. 108 ?
An. pilosus. *Herbst A.* 39. *pl.* 21. *f.* G. *g.*
By. rufipes. *Schneid. M.* (*Kugell.*) i. 485.

Genus 170. ASPIDIPHORUS, (*Ziegler.*)

NITIDULA *p*, *Gyll.*—ARPIDIPHORUS, (*DeJean.*)

1026 † 1. orbiculatus. (*DeJean C.* 47.)—(*Ing. Inst.* 87.)
Ni. orbiculatus. *Gyll.* i. 242.
As. viennensis. *Ziegler.* In Mus. D. Spence.

Genus 171. NOSODENDRON, *Latreille, Leach.*

1027. 1, fasciculare. *Latr. G.* ii. 44.—(*Leach E. E.* ix. 95.)—(*Ing. Inst.* 89.)
Sp. fasciculare. *Fabr. E. S.* i. 81.
By. fascicularis. *Panz. F.* xxiv. *f.* 2.

1028 † 2, setigerum. (*Ing. Inst.* 89.)
By. setiger. *Illig. K.* i. 95.—*Sturm D. F.* ii. 116. *pl.* xxxv. *f. d*, D. In Mus. *Brit.*

Genus 172. BYRRHUS *Auctorum.* *Pill-Beetle.*

CISTELA, *Forst.*, *Marsh*, *Millard.*

1029. 1; pilula *. *Linn.* ii. 568.—*Stew.* ii. 36.—*Panz. F.* iv. *f.* 3. —*Turt.* ii. 97.—*Shaw G. Z.* vi. 48. *pl.* 13.—(*Leach E. E.* ix. 94.) —*Samou.* 50. *pl.* 2. *f.* 3.—*Wood.* i. 22. *pl.* 7.—(*Curtis l. c. infra.*)
Ci. pilula. *Marsh.* i. 102.
Ci. viridescens. *Fourc.* i. 28.
β; Ci. striata. *Forst. C.* i. 15.—*Marsh.* i. 103.
By. ater. *Illig. K.* i. 91.—*Panz. F.* xxxii. *f.* 2.
γ; Ci. ferruginea. *Marsh.* i. 104.

δ; Ci. concolor. *Kirby MSS.*
ε, By. albopunctatus. *Fabr. E.* i. 103?

1030. [2; Dennii.] *Kirby MSS.*—*Curtis.* iii. *pl.* 135.
Ci. elegans *mihi* (*olim.*)

1031. 3; fasciatus*. *Fabr. E. S.* i. 85.—(*Sam. I.* 8.)—(*Curtis l. c. supra.*)
By. pilula ♂. *Payk. F.* i. 75.
Ci. fasciata. *Forst. C.* i. 12.—*Ent. Trans.* (*Burrell.*) i. 127.
Ci. undulata. *Marsh.* i. 103.
β; By. Dianæ. *Fabr. E.* i. 103.
By. ornatus. *Panz. F.* xxiv. *f.* 1.
By. cinctus. *Illig. K.* i. 91.—(*Ing. Inst.* 87.)
γ; Ci. similis. *Kirby MSS.*
Ci. annulus. *Marsh. MSS.*

1032. 4, oblongus*. *Sturm* D. *F.* ii. 97. *pl.* xxxiv. *f. a. A.*—(*Curtis l. c. supra.*)
Ci. ater. *Mus. Marsham.*
By. ater. (*Curtis l. c. supra.*)

1033. 5; dorsalis*. *Fabr. E.* i. 85.—(*Sam. I.* 8.)—(*Curtis l. c. supra.*)
Ci. dorsalis. *Marsh.* i. 104.
By. fasciatus. *Illig. K.* i. 92.—*Panz. F.* xxxii. *f.* 1.
By. pilula ε. *Payk. F.* i. 75.
β; By. ater. *Fabr. E. S.* i. 85?
By. Morio. *Illig. K.* i. 93.—*Panz. F.* xxxvii. *f.* 15.
Ci. nigra. *Forst. C.* i. 14?
By. bimaculata. *Kirby MSS.*

1034. 6; sericeus*.
Ci. sericea. *Forst. C.* i. 16.—*Marsh.* i. 104.
De. pilula. *DeGeer.* iv. 213. *pl.* 7. *f.* 23, 26.
By varius. *Fabr. E. S.* i. 85.—*Stew.* ii. 36.—*Panz. F.* xxxii. *f.* 3.—*Turt.* ii. 97.—(*Sam. I.* 8.)—(*Curtis l. c. supra.*)
β; Ci. bicolor. *Marsh.* i. 105.
γ; Ci. stoica. *Müll. Z. D. pr.* 58.

1035. 7; fuscus. (*Curtis l. c. supra.*)
Ci. fusca. *Marsh.* i. 105.
By. varius β, *Gyll.* i. 197.

1036. 8: murinus. *Fabr. E.* i. 104.—*Panz. F.* xxv. *f.* 1.—(*Sam. I.* 8.)—(*Curtis l. c. supra.*)
By. undulatus. *Panz. F.* xxxvii. *f.* 14.
By. rubidus. *Schneid. M.* (*Kugell.*) i. 484
Ci. bilineata. *Marsh. MSS.*

Genus 173. SIMPLOCARIA, *Marsh. MSS.*

CISTELA *p*, *Marsh.*

1037. 1; semistriata*.
By. semistriatus. *Illig. K.* i. 97.—*Panz. F.* xxv. *f.* 2.—(*Sam. I.* 8.)
Ci. picipes. *Marsh.* i. 105.
β; Ci. picea. *Marsh.* i. 106.
By. rufipes. *Schneid. M.* (*Kugell.*) i. 485.
By. minutus. *Thunb. Ins. S.* 68?

1038. 2, picipes*?
By. picipes. *Oliv. E.* ii. 89. *pl.* 2. *f.* 9. *a. b.*

Genus 174. ————

CHÆTOPHORA[a], (*Kirby.*)—CISTELA *p*, *Marsh.*—GEORYSSUS**, *Leach.*

1039. 1, maritimus.
Ci. maritima. *Marsh.* i. 105.

1040. 2, arenarius.
By. arenarius. *Sturm* D. *F.* ii. 117. *pl.* xxxv. *f. e. E.*
Ci. cretaria. *Marsh. MSS.*
Ge. arenifera. *Kirby MSS.*—(*Kirby & Sp. I. E.* ii. 258.)

1041. 3; cretiferus.
Ch. cretifera. *Mus. Marsham.*

Familia XXIII. HISTERIDÆ, *Leach.*

(BYRRHII *p*, *Latreille.*—HISTEROIDES, *Gyll.*)

(HISTER, *Linné, &c.*)

Genus 175. ABRÆUS, *Leach, Samou.*

1042. 1, globosus. (*Leach Z. M.* iii. 79.)
Hi. globosus. *Ent. Hefte.* i. 100. *pl.* 2. *f.* 1.
Hi. minimus. *Rossi F.* i. 30.
Hi. perpusillus. *Marsh.* i. 99.
Ab. perpusillus. (*Sam. I.* 1.)

1043. 2; minutus*. (*Leach Z. M.* iii. 79.)
Hi. minutus. *Fabr. E.* i. 90.—*Herbst C.* iv. 41. *pl.* 36. *f.* iv. *a, b.*
Hi. minimus. *Marsh.* i. 99.
Hi. nigricornis. *Ent. Hefte.* i. 127.
Hi. atomos. *Rossi M.* i. 15.
Hi. atomarius. *Schneid. M.* (*Kugell.*) i. 305.
Hi. lævigatus. *Kirby MSS.*

---

[a] CHÆTOPHORA: Genus Fungorum. Vide *Greville's Cryptog. of Scotland.*

Genus 176. ONTHOPHILUS, *Leach, Samou.,* (*Kirby.*)

1044. 1; striatus *. (*Leach Z. M.* iii. 79.)—(*Sam. I.* 32.)
Hi. striatus. *Forst. C.* i. 11.—*Fabr. E.* i. 90.—*Stew.* ii. 35. —*Marsh.* i. 97.
Hi. sulcatus. *Oliv. E.* i. 17. *pl.* 1. *f.* 6.—*Turt.* ii. 94.

1045. 2, sulcatus. (*Leach Z. M.* iii. 79.)—(*Sam. I.* 32.)
Hi. sulcatus. *Fabr. E.* i. 89.—*Panz. F.* lxxx. *f.* 5.—*Ent. Trans.* (*Burrell.*) i. 124.
Hi. globulosus. *Oliv. E.* i. 16. *pl.* 2. *f.* 15. *a, b.*
Hi. striatus var. major. *Herbst C.* iv. 38. *pl.* 36. *f.* 1. *a, b.*

Genus 177. HISTER *Auctorum. Mimic-Beetle.*

ATTELABUS *p, Geoffroy.*

A. " Elytris striatis; striis externis integris." *Leach.*

a. " *Thorace lateribus longitudinalitèr bistriatis.*" Leach.

1. " Elytris striâ marginali." *Leach.*

1046. 1, Marshami *mihi.*

1047. 2, merdarius *. *Ent. Hefte.* i. 39. *pl.* 1. *f.* 3.—(*Ing. Inst.* 88.)

1048. 3; unicolor *. *Linn.* ii. 567.—*Barb.* G. 24. *pl.* 3. *f.* 5 ?—*Berk. S.* i. 90 ?—*Stew.* ii. 35 —*Ent. Hefte.* i. *pl.* 1. *f.* 1.—*Turt.* ii. 94.—*Shaw G. Z.* vi. 38. *pl.* 10.—(*Leach Z. M.* iii. 78.)—(*Sam. I.* 22.)—*Wood.* i. 15. *pl.* 5.
Hi. inæqualis. *Marsh.* i. 93.
Hi. ater. *DeGeer.* iv. 342. *pl.* 12. *f.* 12.
β, Hi. undulatus. *Kirby MSS.*

1049. 4; cadaverinus *. *Ent. Hefte.* i. 34. *pl.* 1. *f.* 2.—(*Leach Z. M.* iii. 78.)—(*Sam. I.* 22.)
Hi. unicolor. *Marsh.* i. 92.
Hi. duplicatus. *Kirby MSS.*
β, Hi. brunneus. *Fabr. E.* i. 86.

1050. 5, quadrimaculatus. *Linn.* ii. 567.—*Turt.* ii. 95. (!)—*Shaw G. Z.* vi. 38. (!)
Hi. sinuatus. *Herbst C.* iv. 43.
Hi. reniformis. *Oliv. E.* i. 10. *pl.* 1. *f.* 5. *a.*
Hi. 4-notatus. *Scriba. I.* i. 72.
Hi. lunatus. *Fabr. E.* i. 86 ?

2. " Elytris striâ marginali nullâ." *Leach.*

1051. 6; sinuatus *. *Illig. K.* i. 57.—*Panz. F.* lxxx. *f.* 1.—(*Leach Z. M.* iii. 78.)—(*Sam. I.* 22.)
Hi. 4-maculatus. *Marsh.* i. 94.—*Don.* xv. *pl.* 525.

1052. 7, quadrinotatus. *Illig. K.* i. 58.—(*Leach Z. M.* iii. 78.)—(*Sam. I.* 22.)
Hi. 4-maculatus. *DeGeer.* iv. 344.—*Panz. F.* lxxx. *f.* 2.

b. *Thorace lateribus unistriatis.*

1. "Elytris striâ marginali nullâ." *Leach.*

1053. 8; duodecimstriatus*. *Payk. F.* i. 39.—*Marsh.* i. 96.—*Ent. Hefte.* i. 53. *pl.* 1. *f.* 6, *a.*—*Leach Z. M.* iii. 78.—(*Sam. I.* 22.)
Hi. bis-sexstriatus. *Fabr. E.* i. 84.
β, Hi. picatus. *Kirby MSS.*

1054. 9; bimaculatus*. *Linn.* ii. 567.—*Berk. S.* i. 90.—*Stew.* ii. 35.—*Marsh.* i. 93.—*Turt.* ii. 95.—*Payk. M. H.* 34. *pl.* 3. *f.* 6.—(*Leach E. E.* ix. 78.)—(*Sam. I.* 22.)
Hi. erythropterus. *Fabr. E.* i. 88.
Hi. apicatus. *Schran.* ii. 452.

1055. 10; parvus*. *Marsh.* i. 93.—(*Leach Z. M.* iii. 78.)—(*Sam. I.* 22.)
Hi. bis-sexstriatus, a. *Payk. M. H.* 32. *pl.* 3. *f.* 3.

2. "Elytris striâ marginali." *Leach.*

1056. 11; stercorarius*. *Ent. Hefte.* i. 57. *pl.* i. *f.* 5.—(*Leach Z. M.* iii. 79.)—(*Sam. I.* 22.)
Hi. obscurus. *Panz. E. F.* 20?

1057. 12; neglectus*. *Megerle.*—(*Leach Z. M.* iii. 79.)—(*Sam. I.* 22.)

1058. 13; Leachii* *mihi.*

1059. 14; carbonarius*. *Ent. Hefte.* i. 54. *pl.* 1. *f.* 4.—(*Leach Z. M.* iii. 79.)—(*Sam. I.* 22.)
Hi. 12-striatus. *Fabr. E.* i. 85.

1060. 15; quisquilius* *mihi.*

1061. 16; purpurascens*. *Fabr. E.* i. 87.—(*Leach Z. M.* iii. 79.)—(*Sam. I.* 22.)—*Payk. M. H.* 38. *pl.* 3. *f.* 7.
Hi. bimaculatus. *Schra. A.* 39.
Hi. bipustulatus. *Marsh.* i. 94.
β, Hi. brunneus. *Marsh.* i. 97.

1062. 17; Kirbii* *mihi.*

1063. 18; castanipes*. *Kirby MSS.*

1064. 19; Nigrita* *mihi.*

1065. 20, caliginosus* *mihi*

B. "Elytris striatis: striis externis abbreviatis." *Leach.*

1066. 21; nitidulus*. *Fabr. E.* i. 85.—*Payk. M. H.* 53. *pl.* 5. *f.* 3.—(*Leach Z. M.* iii. 79.)—*Samou.* 184. *pl.* 2. *f.* 1.
Hi. semipunctatus. *Payk. F.* i. 45.—*Marsh.* i. 95.—*Samou.* 49.
Hi. semistriatus. *Herbst C.* iv. 30. *pl.* 35. *f.* 6.
Hi. acuminatus var. *Fabr. E.* i. 86.
Hi. æneus. *Rossi F.* i. 29.
Hi. unicolor. *Scop. C.* 30.

1067. 22; pulcherrimus*. *Weber Obs.* 37.—(*Schon. S.* i. 97.)
Hi. speculifer. *Latr. G.* ii. 48.—*Payk. M. H.* 71. *pl.* 6. *f.* 4. —(*Leach Z. M.* iii. 79.)—(*Sam. I.* 22.)
Hi. personatus. *Illig. M.* vi. 39.

1068. 23, maritimus. *Kirby MSS.*

1069. 24, variabilis. *Leach MSS.*
Hi. rugosus *mihi.*
Hi. immundus. *Schuppel?*—*Gyll.* iv. 266?

1070. 25, quadristriatus. *Payk. F.* i. 45.—*Ent. Hefte.* i. 85. *pl.* 1. *f.* 9.

1071. 26; æneus*. *Fabr. E.* i. 88.—*Stew.* ii. 35.—*Panz. F.* xciii. *f.* 2.—*Marsh.* i. 95.—*Turt.* ii. 95.—(*Leach Z. M.* iii. 79.)—(*Sam. I.* 22.)
At. æneus. *Fourc.* i. 68.

1072. 27, metallicus. *Fabr. E.* i. 89.—*Payk. M. H.* 67. *pl.* 6. *f.* 3.
Hi. rugifrons. *Payk. F.* i. 47.

1073. 28; smaragdulus* *mihi.*

1074. 29, violaceus. *Marsh.* i. 96.

1075. 30; virescens*. *Payk. F.* i. 48.—*Marsh.* i. 96.—(*Leach Z. M.* iii. 79.)—(*Sam. I.* 22.)—*Payk. M. H.* 69. *pl.* 6. *f.* 7.

Genus 178. DENDROPHILUS, *Leach, Samou., Curtis.*

PLATYSOMA *p, Leach, Samou.*

A. Elytris striatis: corpore subgloboso.

1076 † 1, Sheppardi. *Curtis.* iii. *pl.* 131.
Hi. Sheppardi. *Kirby MSS.* In Mus. *D. Kirby.*

B. Elytris striis nonnullis externis.

1077. 2; rotundatus*.
Hi. rotundatus. *Fabr. E.* i. 90.—*Ent. Hefte.* i. 87. *pl.* i. *f.* 11.
Hi. piceus. *Marsh.* i. 97.
Hi. pygmæus. *DeGeer.* iv. 344.
Hi. punctatus. *Payk. F.* i. 49.
Hi. nanus. *Herbst C.* iv. 56.

1078. 3; punctatus*. (*Sam. I.* 14.)—(*Curtis l. c. supra.*)
Hi. punctatus. *Illig. K.* i. 60.—*Ent. Hefte.* i. 92. *pl.* 1. *f.* 12.
Hi. pygmæus. *Fabr. E.* i. 89?
Hi. corticalis. *Payk. F.* i. 50.
Hi. abbreviatus. *Rossi F.* i. 30.

1079. [4, seminulum.] *Kirby MSS.*—(*Curtis l. c. supra.*)—(*Ing. Inst.* 88.)
Hi. pygmæus. *Fabr. E.* i. 89?
Dr. Curtisii. *Leach MSS.*—(*Curtis l. c. supra.*)

C. Elytris striis nullis: corpore sublineare.

1080. 5, flavicornis. (*Curtis l. c. supra.*)
Hi. flavicornis. *Herbst C.* iv. 40. *pl.* 36. *f.* 2. *a, b.*
Pl. flavicornis. (*Sam. I.* 34.)
Hi. parvulus. *Rossi M.* i. 14.
Hi. picipes. *Payk. F.* i. 52.
Hi. minutus. *Panz. F.* xciii. *f.* 3.

1081. 6, picipes. (*Curtis l. c. supra.*)
Hi. picipes. *Fabr. E.* i. 92.
Pl. picipes. (*Sam. I.* 34.)
Hi. pusillus. *Schneid. M.* (*Kugell.*) i. 305.
Hi. parallelipipedus. *Herbst C.* iv. 37. *pl.* 35. *f.* ii. *a, b.* *L.*
Hi. pygmæus. *Linn.* ii. 567?—*Marsh.* i. 98.

1082 † 7, Milleri. *Leach MSS.*—(*Curtis l. c. supra.*) In Mus. *Brit.*

Genus 179. PLATYSOMA, *Leach, Samou.*

HOLOLEPTA *p, DeJean.*

1083. 1; depressus*. (*Sam. I.* 34.)
Hi. depressus. *Fabr. E. S.* i. 10.—*Panz. F.* lxxx. *f.* 6.—*Marsh.* i. 98.—(*Sam. I.* 22.)
Hi. compressus. *Herbst A.* 20.

1084 ‡ 2, oblongus. (*Sam. I.* 34.)
Hi. oblongus. *Fabr. E. S.* i. 13.—*Panz. F.* xciii. *f.* 5.
Hi. elongatus. *Oliv. E.* i. 16. *pl.* 2. *f.* 14. *a, b.* In Mus. *Brit.*

Subsectio 2. LAMELLICORNES, *Latreille.*

(PETALOCERA, *Dumeril.*)

Familia XXIV. LUCANIDÆ, *Leach.*

(LUCANIDES, *Latreille.*—LUCANOIDES, *Gyll.*)

(LUCANUS, *Linné, &c.*)

Genus 180. PLATYCERUS. *Geoffroy, Leach.*

1085. 1, caraboides. *Fourc.* i. 3.—(*Leach E. E.* ix. 100.)
Lu. caraboides. *Linn.* ii. 561.—*Mart. C. pl.* 5. *f.* 5.—*Berk. S.* i. 89.—*Stew.* ii. 31.—*Marsh.* i. 50.—*Turt.* ii. 77.
Lu. Caprea. *DeGeer.* iv. 334. *pl.* 12. *f.* 11.

Genus 181. DORCUS, *MacLeay.*

PLATYCERUS *p, Latr.*

1086. 1; parallelipipedus*. *MacLeay H. E.* i. 111.
Lu. parallelipipedus. *Linn.* ii. 561.—*Barb. G.* 16. *pl.* 2. *f.* 3. *Mart. C. pl.* 5. *f.* 3, 4.—*Berk. S.* i. 88.—*Don.* viii. *pl.* 264. *f.* 1.—*Stew.* ii. 31.—*Marsh.* i. 48.—*Turt.* ii. 77.—(*Sam. I.* 26.)—(*Kirby & Sp. I. E.* ii. 227.)

Lu. infractus. *Bergs. N.* i. 8. *pl.* 8. *f.* 2.
♀; Lu. Capra. *Panz. F.* lviii. *f.* 12.
Lu. bipunctatus. *Schra. B.* i. 376.
Lu. Dama. *Müll. Z. D. pr.* 52.
Do. tuberculatus. *MacLeay H. E.* i. 112.

Genus 182. LUCANUS *Auctorum. Stag-beetle. Pinch-bob.*

PLATYCERUS *p, Geoffroy.*

1087. 1; Cervus*. *Linn.* ii. 559.—*Barb.* G. 15. *pl.* 2.—*Mart. C. pl.* 5. *f.* 1. ♂, *f.* 2. ♀.—*Berk. S.* i. 88.—*Don.* i. *pl.* 13.—*Stew.* ii. 30.—*Marsh.* i. 46.—*Turt.* ii. 76.—*Shaw* G. *Z.* vi. 27. *pl.* 6.—*Bingley.* iii. 120.—*Ent. Trans.* (*Haworth.*) i. 78.—(*Leach E. E.* ix. 100.)—(*Samou.* 48 & 192. *pl.* 1. *f.* 3.)—(*Kirby & Sp. I. E.* iii. 33. *et* 314.)
β; Lu. dorcas. *Panz. F.* lviii. *f.* 11.—*Ent. Trans.* (*Haworth.*) i. 77.
Lu. hircus. *Herbst C.* iii. 299. *pl.* 33. *f.* 4, 5.
Lu. Capreolus. *Fabr. E.* ii. 249.—*Mill. B. E.* 153. *pl.* 2. *f.* 4.
Lu. Capra. *Oliv. E.* i. 11. *pl.* 1. *f. e. pl.* 2. *f.* 1. *g.*
γ; Lu. grandis. *Ent. Trans.* (*Haworth.*) i. 77.
♀; Lu. inermis. *Marsh.* i. 48.—*Don.* xii. 13. *pl.* 400.

Genus 183. SINODENDRON, *Fabricius, Leach, Samou., Millard.*

SCARABÆUS *p, Linné.*

1088. 1; cylindricum*. *Fabr. E. S.* i, *b.* 358.—*Don.* xi. *pl.* 368. —*Turt.* ii. 89.—*Leach E. E.* ix. 98.—*Samou.* 190.
Sc. cylindricus. *Linn.* ii. 544.—*Mart. C. pl.* 3. *f.* 20, 21.
Lu. cylindricus. *Laich. Ty.* i. 3.—*Marsh.* i. 50.
Lu. Tenebroides. *Scop. C.* 5.

Familia XXV. SCARABÆIDÆ, *MacLeay.*

(COPROPHAGI *p, Latreille.*—SCARABÆIDES *p, Gyll.*)

(SCARABÆUS *p, Linné, &c.*)

Genus 184. COPRIS, *Geoffroy, Leach, Samou.,* (*Kirby.*)

1089. 1; lunaris*. *Fabr. E.* i. 36.—(*Leach E. E.* ix. 96.)—(*Sam. I.* 12.)—(*Kirby & Sp. I. E.* ii. 240.)
Sc. lunaris. *Linn.* ii. 543.—*Mart. C. pl.* 2. *f.* 15, 16.—*Berk. S.* i. 86.—*Don.* v. *pl.* 154. *f.* 4.—*Stew.* ii. 23.—*Marsh.* i. 31.—*Turt.* ii. 21.—*Wood.* i. 5. *pl.* 1.
Sc. 4-dentatus. *DeGeer.* vii. 638.
♀, Sc. emarginatus. *Fabr. E. S.* i. 46.—*Marsh.* i. 32.

Genus 185. ONTHOPHAGUS, *Latreille, Leach, Samou.,* (*Kirby,*) *Curtis.*

COPRIS *p, Fabr.*—ATEUCHUS *p, Fabr.*

A. Capite cornuto.

1090 ‡ 1, Taurus. *Latr. H.* x. 213.—*Curtis.* ii. *pl.* 52.

Sc. Taurus. *Linn.* ii. 547.
♂, Sc. illyricus. *Scop. C. no.* 25.
♀, Sc. rugosus. *Scop. C. no.* 23. In Mus. D. *Stone.*

1091. 2, Vacca*? *Latr. H.* x. 115.—(*Leach E. E.* ix. 97.)—(*Sam. I.* 32.)—(*Curtis l. c. supra.*)
Sc. Vacca. *Linn.* ii. 547.—*Panz. F.* xii. *f.* 4. ♂.—*Marsh.* i. 34.—*Don.* xvi. 59. *pl.* 561.
Sc. conspurcatus. *Fourc.* i. 14.
Sc. gibbulus. *Pall. I.* 7. *pl. A. f.* 6. ♀.

1092. 3; medius*. *Latr. H.* x. 114.—(*Ing. Inst.* 89.)
Co. media. *Illig. K.* i. 41.
Sc. media. *Panz. F.* xxxvii. *f.* 4.—*Ent. Trans.* (*Burrell.*) i. 110.

1093. 4; fracticornis*. (*DeJean C.* 53.)
Sc. fracticornis. *Preysler. B.* i. 99.—*Panz. F.* xlix. *f.* 9.
Sc. Xiphias. *Panz. F.* xlix. *f.* 8.—*Mart. C. pl.* 1. *f.* 6.—*Marsh.* i. 33.
On. Xiphias. (*Sam. I.* 32.)
Sc. nuchicornis. *Payk. F.* i. 31.
β; On. affinis. *Mus. Marsham.*

1094. 5; cænobita*. *Latr. H.* x. 112.—(*Sam. I.* 32.)—(*Curtis l. c. supra.*)
Sc. cænobita. *Fabr. E.* i. 49.—*Mart. C. pl.* 1. *f.* 5?—*Panz. F.* xlviii. *f.* 6.—*Marsh.* i. 33.
Sc. tenuicornis. *Preysler. B.* i. 44.
β; On. obsoletus. *Kirby MSS.*

1095 † 6. austriacus. (*Curtis l. c. supra.*) (!)
Sc. austriacus. *Schneid.*—*Panz. F.* xii. *f.* 6. In Mus. ——?

1096. 7, Dillwynii. *Leach MSS.*—(*Sam. I.* 32.)—(*Curtis l. c. supra.*)
On. reticulatus. *Kirby MSS.*
Co. similis. *Scriba. B.* i. 35. *pl.* 4. *f.* 5?
Sc. nuchicornis. *Mus. Linn.* (*teste D. Kirby.*)

1097. 8; nuchicornis*. *Latr. H.* x. 113.—(*Sam. I.* 32.)—(*Curtis l. c. supra.*)
Sc. nuchicornis. *Linn.* ii. 547.—*Berk. S.* i. 86.—*Don.* viii. *pl.* 255. *f.* 2.—*Stew.* ii. 24.—*Turt.* ii. 34.
Sc. Xiphias. *Fabr. E.* i. 50.
Sc. planicornis. *Herbst C.* ii. 210. *pl.* 14. *f.* 13.
β; On. muticus. *Kirby MSS.*
♀; Sc. acornis. *Fourc. P.* i. 14.

1098. 9; nutans*. *Latr. H.* x. 111.—(*Sam. I.* 32.)—(*Curtis l. c. supra.*)
Sc. nutans. *Fabr. E.* i. 50.—*Don.* viii. *pl.* 255. *f.* 1.—*Marsh.* i. 35.—*Turt.* ii. 34.
Sc. verticicornis. *Laich. T.* i. 22.

B. Capite haud cornuto. (ATEUCHUS *p*, *Fabr.*)

1099. 10; ovatus*. *Latr. H.* x. 110.—(*Sam. I.* 32.)—(*Curtis l. c. supra.*)
Sc. ovatus. *Linn.* ii. 551.—*Mart. C. pl.* 4. *f.* 36.—*Stew.* ii. 27. —*Panz. F.* xlviii. *f.* 11.—*Marsh.* i. 35.—*Turt.* ii. 45.

Genus 186. ONITICELLUS[a].

Familia XXVI. GEOTRUPIDÆ, *MacLeay.*

(GEOTRUPINI, *Latreille.*—SCARABÆIDES, *Gyllenhal.*)

(SCARABÆUS *p*, *Linné, &c.*)

Genus 187. BOLBOCERUS, *Kirby.*

ODONTÆUS, (*Köppe*), (*Samou.*)

1101. 1; mobilicornis*. (*Kirby & Sp. I. E.* iii. 329.)—(*Ing. Inst.* 87.)
Sc. mobilicornis. *Fabr. E.* i. 24.—*Mart. C. pl.* 4. *f.* 40, 41.—*Stew.* ii. 23.—*Marsh.* i. 8.—*Turt.* ii. 18.—*Sturm D. F.* i. 21. *pl.* vi. *f. s—v.*
Sc. bicolor. *Gmel.* iv. 1548.
Od. mobilicornis. (*Sam. I.* 31.)

1102. 2, testaceus.
Sc. testaceus. *Fabr. E.* i. 26.—*Panz. F.* xxviii. *f.* 5.—*Stew.* ii. 26.—*Marsh.* i. 16.—*Turt.* ii. 28.
Sc. mobilicornis β. *Herbst C.* ii. 135. *pl.* 6. *f.* 7. ♂.—*pl.* 12. *f.* 3. ♀.

1103 † 3. quadridens.
Sc. quadridens. *Fabr. E.* i. 23.—*Panz. F.* xii. *f.* 1.—*Ent. Trans.* (*Skrimshire.*) i. 316. (!)
Sc. quadridentatus. *Oliv. E.* i. 62.
Sc. Aeneas. *Panz. Beit. p.* 34.
Sc. unicornu. *Schrank. N.* 24. *pl.* 61.
In Mus. D. G. Skrimshire?

Genus 188. TYPHŒUS, *Leach, Samou.* *Bull-comber.*

GEOTRUPES *p*, *Latreille.*

1104. 1; vulgaris*. (*Leach E. E.* ix. 97.)—(*Samou.* 189. *pl.* 1. *f.* 1.)—(*Kirby & Sp. I. E.* ii. 475.)

---

[a] Genus 186. ONITICELLUS, (*DeJean.*)

ONITIS *p?* *Fabr.*—COPRIS *p*, *Fabr.*—ONTHOPHAGUS *p*, *Samou.*

1100 † 1. verticicornis.
Sc. verticicornis. *Fabr. E. S.* i. 61. (!)—*Stew.* ii. 24. (!)—*Marsh.* i. 34. (!)
Sc. verticornis. *Turt.* ii. 34. (!)
Ont. verticornis. (*Sam. I.* 32.) (!)
Oni. flavipes β. (*Schon. S.* i. 33.)

Sc. Typhœus. *Linn.* ii. 543.—*Barb.* G. *pl.* 1. *f.* 1.—*Mart. C. pl.* 1. *f.* 3. 4.—*Berk. S.* i. 86.—*Stew.* ii. 23.—*Panz. F.* xi. *f.* 23. —*Marsh.* i. 7.—*Turt.* ii. 17.—*Bingley.* iii. 111.—*Samou.* 47.
Sc. ovinus. *Rai. Ins.* 103.
β; Sc. pumilus. *Marsh.* i. 8.—*Sowerb. B. M.* i. *pl.* 57.

Genus 189. GEOTRUPES, *Latreille, Samou. Dor, Clock or Shard-borne-Beetle.*

1105. 1; vernalis*. *Latr. H.* x. 146.—(*Sam. I.* 19.) Spring Beetle.
Sc. vernalis. *Linn.* ii. 551.—*Barb.* G. i. *pl.* 1. *f.* 6.—*Berk. S.* i. 87.—*Mart. C. pl.* 2. *f.* 9.—*Stew.* ii. 86.—*Marsh.* i. 23.—*Turt.* ii. 36.—*Bingley.* iii. 111.—*Don.* xvi. *pl.* 547. *f.* 1.—(*Leach E. E.* ix. 97.)

1106. 2; lævis*.
Sc. lævis. *Ent. Trans.* (*Haworth.*) i. 79.

1107. 3; sylvaticus*. *Latr. H.* x. 146.—(*Sam. I.* 19.)
Sc. sylvaticus. *Payk. F.* i. 5.—*Panz. F.* xlix. *f.* 3.—*Marsh.* i. 23.—*Don.* xvi. *pl.* 547. *f.* 2.
Sc. stercorosus. *Scriba. I.* 250.

1108. 4; niger*. (*Sam. I.* 19.)
Sc. niger. *Marsh.* i. 22.

1109. 5; foveatus*.
Sc. foveatus. *Marsh.* i. 21.—*Sowerb. B. M.* i. *pl.* 55. *f.* 2.

1110. 6, punctatostriatus. *Kirby MSS.*

1111. 7; mutator*.
Sc. mutator. *Marsh.* i. 22.—*Mill. B. E.* 132. *pl.* 2. *f.* 1.
Sc. politus. *Molinousky.*
Ge. politus. (*Sam. I.* 19.)
β; Ge. splendidus. *Kirby MSS.*
Sc. caballinus. *Köhler?*

1112. 8; puncticollis*. (*Sam. I.* 19.)
Sc. puncticollis. *Molinousky.*
Sc. bulbilus. *Köhler?*

1113. 9; sublævigatus* *mihi.*

1114. 10; stercorarius*. *Latr. H.* x. 146.—(*Sam. I.* 19.)—*Kirby & Sp. I. E.* i. 253.—ii. 234.)
Sc. stercorarius. *Linn.* ii. 550.—*Mart. C. pl.* 3. *f.* 25.—*Berk. S.* i. 87.—*Don.* viii. *pl.* 264. *f.* 3.—*Stew.* ii. 26.—*Marsh.* i. 20. —*Turt.* ii. 36.—*Sturm D. F.* i. 22. *pl.* vi. *f. a—r.*—*Bingley.* iii. 111.
♂; Sc. spiniger. *Marsh.* i. 21.—*Sowerb. B. M.* i. *pl.* 35. *f.* 1.

Familia XXVII. APHODIIDÆ, *MacLeay*.

(COPROPHAGI *p*, *Latreille*.—SCARABÆIDES *p*, *Gyllenhal*.)
Dung-beetles.

(SCARABÆUS *p*, *Linné*, *&c.*)

Genus 190. APHODIUS, *Illiger*, *Leach*, *Samou.*, (*Kirby*,) *Curtis*.

A. Scutellum magnum subelongatum.

a. *Corpore convexo.*

1115. 1; Fossor*. *Fabr. E.* i. 67.—*Sturm* D. *F.* i. 81. *pl.* 12.—(*Sam. I.* 4.)—(*Curtis l. c. infra.*)
Sc. Fossor. *Linn.* ii. 548.—*Mart. C. pl.* 3. *f.* 30.—*Berk. S.* i. 87.—*Stew.* ii. 25.—*Marsh.* i. 16.—*Turt.* ii. 26.—*Don.* xii. *pl.* 417. *f.* 3.

1116. 2; subterraneus*. *Illig. K.* i. 20.—(*Sam. I.* 4.)
Sc. subterraneus. *Linn.* ii. 548.—*Mart. C. pl.* 4. *f.* 33.—*Fanz. F.* xxviii. *f.* 3.—*Marsh.* i. 18.

1117. 3; hæmorrhoidalis*. *Illig. K.* i. 23.—(*Sam. I.* 4.)
Sc. hæmorrhoidalis. *Linn.* ii. 548.—*Mart. C. pl.* 4. *f.* 31.—*Berk. S.* i. 87.—*Panz. F.* xxviii. *f.* 8.—*Stew.* ii. 25.—*Marsh.* i. 19.
β; Sc. sanguinolentus. *Herbst A.* 6. *pl.* 19. *f.* 4.—*Marsh.* i. 28.

b. *Corpore depresso.*

1118. 4; erraticus*. *Illig. K.* i. 34.—(*Sam. I.* 4.)
Sc. erraticus. *Linn.* ii. 548.—*Mart. C. pl.* 4. *f.* 37.—*Marsh.* i. 9.

B. Scutellum parvum, breve.

a. *Clypeo tuberculato, sæpiùs emarginato.*

1. Corpore convexo.

1119 ‡ 5. Scrutator. *Fabr. E.* i. 69.—(*Sam. I.* 4.)—(*Curtis Cat. No.* 45.)
Sc. Scrutator. *Panz. F.* xxxi. *f.* 1.—*Marsh.* i. 11.
Sc. rubidus. *Oliv. E.* i. 77. *pl.* 26. *f.* 224.
In Mus. D. Atkinson et MacLeay.

1120. 6; fimetarius*. *Illig. K.* i. 31.—(*Sam. I.* 4.)
Sc. fimetarius. *Linn.* ii. 548.—*Mart. C. pl.* 3. *f.* 22.—*Berk. S.* i. 87.—*Stew.* ii. 25.—*Marsh.* i. 10.—*Panz. F.* xxxi. *f.* 2.—*Turt.* ii. 28.—*Don.* xii. 29. *pl.* 404. *f.* 1.
Sc. bicolor. *Fourc. P.* i. 9.
Sc. pedellus. *DeGeer.* iv. 266. *pl.* 10. *f.* 8.

1121. 7; fœtens*. *Illig. K.* i. 31.—(*Sam. I.* 4.)
Sc. fœtens. *Fabr. E. S.* i. 24.—*Panz. F.* xlviii. *f.* 1.—*Marsh.* i. 17.
Sc. vaccinarius. *Herbst C.* ii. 138. *pl.* 12. *f.* 5.

1122. 8; scybalarius*. *Illig. K.* i. 33.
Sc. scybalarius. *Panz. F.* xlvii. *f.* 1.
Sc. coprinus. *Marsh.* i. 12.—*Don.* xii. *pl.* 404. *f.* 4.
Ap. coprinus. (*Sam. I.* 4.)
Sc. conflagrans. *Turt.* ii. 28?
β; Sc. conflagratus. *Fabr. E. S.* i, *a.* 27.—*Don.* ii. *pl.* 70.—*Stew.* 26.—*Marsh.* i. 11.—*Turt.* ii. 28.
Ap. conflagratus. (*Sam. I.* 4.)
Sc. conspurcatus. *Laich. T.* i. 12.

1123. 9; rufescens*. *Fabr. E.* i. 74.
Ap. sordidus β. *Illig. K.* i. 32.
Sc. fœtens. *Oliv. E.* i. 85. *pl.* 9. *f.* 71. *a, b.*
β; Sc. unicolor. *Marsh.* i. 11.
Ap. unicolor. (*Sam. I.* 4.)

1124. [10; castaneus*.]
Sc. castaneus. *Marsh.* i. 12.

1125. 11; ochraceus* *mihi.*

1126. 12; sordidus*. *Illig. K.* i. 32.—(*Sam. I.* 4.)
Sc. sordidus. *Fabr. E. S.* i. 29.—*Mart. C. pl.* 2. *f.* 13.—*Panz. F.* xlviii. *f.* 2.—*Marsh.* i. 10.—*Stew.* ii. 25.—*Turt.* ii. 29.—*Don.* xii. *pl.* 404. *f.* 2, 3.
Sc. fimetarius β. *Linn.* ii. 548.
Sc. conspurcatus. *DeGeer.* iv. 268.
Sc. quadripunctatus. *Naturf.* (*Panz.*) xxiv. *pl.* 1. *f.* 4.
β, Ap. marinus. *Kirby MSS.*

1127. 13; nitidulus*. *Fabr. E.* i. 75.
Sc. nitidulus. *Fabr. E. S.* i. 30.
Sc. ictericus. *Payk. F.* i. 17.
Ap. ictericus. *Creutz. E. V.* i. 52. *pl.* 1. *f.* 8. *a.*—*Ent. Trans.* (*Haworth.*) i. 80.—*Id.* (*Burrell.*) i. 105.—(*Sam. I.* 4.)
Sc. merdarius. *Panz. F.* xlviii. *f.* 3.

1128. 14; lividus*. *Creutz. E. V.* i. 44. *pl.* 1. *f.* 7. *a.*
Sc. lividus. *Oliv. E.* i. 86. *pl.* 26. *f.* 222. *a, b.*
Ap. vespertinus. *Panz. F.* lxvii. *f.* 3.
Ap. anachoreta. *Fabr. E.* i. 74.
Sc. bilituratus. *Marsh.* i. 15.
β, Sc. vespertinus var. *Panz. F.* lxvii. *f.* 4.
γ, Sc. Limicola. *Panz. F.* lviii. *f.* 6.

1129. 15, conspurcatus*. (*DeJean C.* 54.)
Sc. conspurcatus. *Linn.* ii. 549.—*Stew.* ii. 25.—*Marsh.* i. 12.

1130. 16; sticticus*. *Creutz. E. V.* i. 26.
Sc. sticticus. *Panz. F.* lviii. *f.* 4.
Ap. prodromus. *Fabr. E.* i. 70.
Sc. conspurcatus var. *Marsh.* i. 13.

Sc. lineolatus. *Marsh. MSS.*
β, Sc. nemoralis. *Panz. F.* lxvii. *f.* 1.

1131. 17; inquinatus*. *Creutz. E. V.* i. 24.—(*Sam. I.* 4.)
Sc. inquinatus. *Fabr. E. S.* i. 28.—*Panz. F.* xxviii. *f.* 7.—*Marsh.* i. 13.
Sc. vaginosus. *Voet. C.* i. *pl.* 21. *f.* 149.
Sc. distinctus. *Müll. Z. D. pr.* 53.
β; Sc. attaminatus. *Marsh.* i. 13.—*Don.* xii. *pl.* 417. *f.* 1.
Ap. attaminatus. (*Sam. I.* 4.)
γ; Sc. fœdatus. *Marsh.* i. 14.
Ap. fœdatus. (*Sam. I.* 4.)
δ; Sc. centrolineatus. *Panz. F.* lviii. *f.* 1.—*Marsh.* i. 14.
ε; Ap. Panzeri. *Kirby MSS.*
Ap. lunula. *Kirby MSS.*
Ap. nubilus. *Leach? MSS.*
ζ; Ap. putridus. *Kirby MSS.*

1132. 18, tessulatus*. *Sturm D. F.* i. 112.—(*Ing. Inst.* 87.)
Sc. tessulatus. *Payk. F.* i. 20.
Sc. inquinatus var. *Herbst C.* ii. 156.
Sc. contaminatus. *Panz. F.* xlvii. *f.* 7.

1133. 19; terrestris*. *Fabr. E.* i. 71.—(*Sam. I.* 4.)
Sc. terrestris. *Fabr. Sp.* i. 16.—*Mart. C. pl.* 2. *f.* 14?—*Stew.* ii. 25.—*Marsh.* i. 17.—*Turt.* ii. 27.
Sc. ater. *DeGeer.* iv. 270.—*Panz. F.* xliii. *f.* 1.

1134. [20; obscurus*]. (*Sam. I.* 4.)
Sc. obscurus. *Marsh.* i. 18.

1135. 21; terrenus*. *Kirby MSS.*
Sc. pusillus. *Marsh.* i. 18.
Ap. pusillus. (*Sam. I.* 4.)

1136. 22; ater*. *Fabr. E.* i. 71?

1137. 23; nitidus*. *Kirby MSS.*

1138. 24; lucens*. *Gyll.? MSS?*

1139. 25; hæmorrhous*. *Kirby? MSS.*

1140. 26; niger*. *Illig. K.* i. 24.
Sc. niger. *Panz. F.* xxxvii. *f.* 1.
Sc. terrestris. *Payk. F.* i. 22.

1141 ‡ 27, bimaculatus. *Fabr. E.* i. 71.—(*Ing. Inst.* 87.)
Sc. bimaculatus. *Panz. F.* xliii. *f.* 2.
Sc. terrestris β. *Illig. K.* i. 24.
Ap. humeralis. (*Sam. I.* 4.) In Mus. *Brit.*

1142. 28; granarius*. *Gyll.* i. 18.—(*Sam. I.* 4.)
Sc. granarius. *Linn.* ii. 547.—*Marsh.* i. 19.—*Stew.* ii. 26.
Ap. carbonarius. *Sturm D. F.* i. 128. *pl.* xiv. *f.* c, C.

Ap. niger. *Creutz. E. V.* 20.
Sc. hæmorrhoidalis. *DeGeer.* iv. 271.

1143. 29; emarginatus * *mihi.*

1144. 30; melanopus *. *Kirby.*

1145. 31; borealis *. *Gyll.* iv. 248?

2. Corpore depresso.

1146. 32, porcus *. *Illig. K.* i. 31.
Sc. porcus. *Fabr. E. S.* i. 26.
Sc. anachoreta. *Panz. F.* xxxv. *f.* 1.
Sc. turpis. *Marsh.* i. 15.
Ap. turpis. (*Sam. I.* 4.)

1147. [33, ruficrus.]
Sc. ruficrus. *Marsh.* i. 16.

1148. 34, fulvicrus *mihi.*

b. *Clypeo haud tuberculato.*

1. Corpore depresso; clypeo integro.

1149. 35; rufipes *. *Fabr. E.* i. 76.—(*Leach E. E.* ix. 97.)—(*Sam. I.* 4.)—(*Curtis l. c. infra.*)
Sc. rufipes. *Linn.* ii. 559.—*Mart. C. pl.* 3. *f.* 29.—*Marsh.* i. 26.—*Don.* xii. *pl.* 417. *f.* 4.
Sc. capitatus. *DeGeer.* iv. 263. *pl.* 10. *f.* 6.
Sc. oblongus. *Scop. C.* 19.

1150. 36; muticus * *mihi.*

1151. 37, depressus *. *Illig. K.* i. 19.—*Ent. Trans.* (*Sowerby.*) i. 246. *pl. fig. sup.*—(*Sam. I.* 4.)
Sc. depressus. *Panz. F.* xxxix. *f.* 1.
Ap. abdominalis. ——?

1152. 38; nigripes *. *Fabr. E.* i. 76.
Sc. nigripes. *Panz. F.* xlvii. *f.* 9.
Ap. rufipes a. *Illig. K.* i. 28.
Sc. luridus b. *Payk. F.* i. 14.
Sc. Arator. *Herbst A.* 9.
Sc. gagates. *Oliv. E.* i. 87.—*Mart. C. pl.* 4. *f.* 39.—*Marsh.* i. 26.
Sc. gagatinus. *Fourc.* i. 10.
Ap. rufitarsis. ——?

1153. 39; luridus *. *Fabr. E.* i. 76.—(*Sam. I.* 4.)—(*Kirby & Sp. I. E.* iv. 397.)
Sc. luridus. *Panz. F.* xlvii. *f.* 6.—*Mart. C. pl.* 3. *f.* 23.—*Marsh.* i. 27.—*Turt.* ii. 29.
Ap. rufipes. *Illig. K.* i. 28.
β; Sc. variegatus. *Herbst A.* 9. *pl.* 19. *f.* 12.—*Marsh.* i. 26.
Sc. varius. *Gmel.* i. 1553.
γ; Sc. interpunctatus. *Herbst A.* 8. *pl.* 19. *f.* 11.

Sc. luridus. *Don.* ix. 75. *pl.* 323.
δ; Sc. nigro-sulcatus. *Marsh.* i. 27.

2. Corpore subconvexo; clypeo subemarginato.

1154. 40; contaminatus*. *Fabr. E.* i. 77.—*Creutz. E. V.* i. 34. *pl.* 1. *f.* 5. *a.*
Sc. ciliaris. *Marsh.* i. 14.

1155. 41; prodromus*. *Creutz. E. V.* i. 37.
Sc. contaminatus. *Payk. F.* i. 21.
Ap. consputus. *Fabr. E.* i. 77.

1156. 42; sphacelatus*. *Gyll.* i. 37.
Sc. sphacelatus. *Panz. F.* lviii. *f.* 5.—*Marsh.* i. 15.—*Don.* xii. *pl.* 417. *f.* 2.
Ap. prodromus β. *Illig. M.* i. 27.

1157. 43; marginalis* *mihi.*
Ap. punctato-sulcatus. *Sturm D. F.* i. 113. *pl.* 13. *f. a. A. B*?

1158 ‡ 44, Pecari*. *Fabr. E.* i. 80.—(*Ing. Inst.* 87.)
Sc. Pecari. *Fanz. F.* xxxi. *f.* 3.
Sc. satellitius. *Herbst C.* ii. 281. *pl.* 19. *f.* 1. In Mus. *Brit.*

1159. 45; merdarius*. *Illig. K.* i. 34.—(*Sam. I.* 4.)
Sc. merdarius. *Oliv. E.* i. 94. *pl.* 19. *f.* 173. *a, b.*—*Mart. C. pl.* 4. *f.* 34.—*Marsh.* i. 30.
Sc. quisquilius. *Panz. F.* xlviii. *f.* 4.
Sc. ictericus. *Laich. T.* 1. 14.
β; Ap. affinis. *Kirby MSS.*

1160. 46, phæopterus. *Marsham MSS.*

1161. 47; pusillus*. *Sturm D. F.* i. 160.
Sc. pusillus. *Fanz. F.* xlix. *f.* 11.
Ap. granum c. *Gyll.* i. 19.

1162. 48; granum*. *Gyll.* i. 19.
Ap. granarius. *Fabr. E.* i. 75.
Sc. granarius. *Panz. F.* xliii. *f.* 3.
Sc. hæmorrhoidalis. *Herbst C.* ii. 152. *pl.* 12. *f.* 11.

1163. 49; cænosus*.
Sc. cænosus. *Panz. F.* lviii. *f.* 7.
Ap. granarius c. *Sturm D. F.* i. 131.

1164. 50; tristis*. *Illig. M.* ii. 193.—*Ent. Trans.* (*Haworth.*) i. 80.
Sc. tristis. *Zenker.*—*Panz. F.* lxxiii. *f.* 1.

1165. 51, quadrimaculatus*. *Illig. K.* i. 35.
Sc. quadrimaculatus. *Linn.* ii. 558.—*Don.* ii. *pl.* 70. *f.* 3.—*Stew.* ii. 27.—*Marsh.* i. 28.—*Turt.* ii. 38.
Ap. quadripustulatus. *Fabr. E.* i. 78.

1166. 52, plagiatus. *Fabr. E.* i. 79.
Sc. plagiatus. *Linn.* ii. 559.—*Fanz. F.* xliii. *f.* 6.—*Ent. Trans.* (*Skrimshire.*) i. 318.

1167. 53, arenarius. *Illig. K.* i. 22.—*Sturm V.* i. 50. *pl.* 2. *f. v. V.*
Sc. rhododactylus. *Marsh.* i. 29.
Sc. pusillus. *Panz. F.* lviii. *f.* 8.

1168. 54, Scrofa. *Fabr. E.* i. 80.
Sc. Scrofa. *Panz. F.* xlvii. *f.* 12.
Sc. tomentosus. *Schneid. M.* i. 269.
Sc. minutus. *Herbst C.* ii. 269. *pl.* 18. *f.* 7.
Sc. fuscus. *Rossi M.* i. 8.
Ips! platycephalus. *Marsh.* i. 56.

1169 † 55, villosus. *Gyll.* i. 40.—*Curtis.* i. *pl.* 27.
In Mus. D. *Vigors.*

1170. 56, Sus. *Illig. K.* i. 27.—(*Sam. I.* 4.)
Sc. Sus. *Fabr. E. S.* i. 78.—*Panz. F.* xxviii. *f.* 11.—*Marsh.* i. 29.
Sc. pubescens. *Oliv. E.* i. 91. *pl.* 24. *f.* 205. *a, b.*

1171. 57; testudinarius*. *Illig. K.* i. 35.—(*Sam. I.* 4.)
Sc. testudinarius. *Fabr. E. S.* i. 38.—*Panz. F.* xxviii. *f.* 12.—*Don.* ii. *pl.* 70. *f.* 1.—*Stew.* ii. 27.—*Marsh.* i. 29.—*Turt.* ii. 39.

Genus 191. PSAMMODIUS, *Gyllenhal, Leach, Samou.*, (*Kirby.*)

APHODIUS *p*, *Illiger*, *Curtis.* i. *fo.* 27.

A. Thorace haud sulcato.

1172. 1, elongatus *mihi.*

1173. 2, sabuleti. *Gyll.* i. 7.
Ap. Sabuleti. *Illig. K.* i. 21.—*Sturm* D. *F.* i. 169. *pl.* 15. *f. a. A. B.*
Sc. punctulatus. *Marsh.* i. 30.

1174. 3; porcatus*. *Gyll.* i. 8.
Sc. porcatus. *Fabr. E. S.* i. 38.—*Panz. F.* xxxviii. *f.* 13.—*Marsh.* i. 30.
Ap. porcatus. (*Curtis l. c. supra.*)

1175. 4; cæsus. (*DeJean C.* 55.)
Ap. cæsus. *Fabr. E.* i. 82.—(*Curtis l. c. supra.*)
Sc. cæsus. *Panz. F.* xxxv. *f.* 2.

B. Thorace sulcato.

1176. 5, asper. *Gyll.* i. 9.
Ap. asper. *Fabr. E.* i. 82.—(*Curtis l. c. supra.*)
Sc. asper. *Panz. F.* xlvii. *f.* 13.
Ptinus germanus. *Linn.* ii. 566.

1177. 6, sulcicollis. *Gyll.* i. 9.—(*Sam. I.* 35.)
Ap. sulcicollis. *Illig. M.* i. 20.
Sc. sulcicollis. *Panz. F.* xcix. *f.* 1.
Sc. asper. *Payk. F.* i. 29.

Familia XXVIII. TROGIDÆ, *MacLeay*.

(SCARABÆIDES-XYLOPHILI *p*, *Latr.*—SCARABÆUS *p*, *Marsh. &c.*)

Genus 192. ÆGIALIA, *Latreille*, *Leach*, *Samou.*

APHODIUS *p*, *Illiger.*—PSAMMODIUS *p*, *Gyll.*

1178. 1, globosa. *Latr. G.* ii. 97.—*Leach E. E.* ix. 98.—*Samou.* 190.
Ap. globosus. *Illig. K.* i. 20.
Sc. globosus. *Panz. F.* lvii. *f.* 2.—*Don.* xiv. 13. *pl.* 470.—*Ent. Trans.* (*Burrell.*) i. 109.
Sc. arenarius. *Payk. F.* i. 27.
Sc. ovalis. *Sower. B. M.* i. *pl.* 34.

Genus 193. TRACHYSCELIS, *Latreille*, *Leach.*

1179. 1, Aphodioides. *Latr.* G. iv. *app.*—(*Leach E. E.* ix. 103.) (*Ing. Inst.* 89.)

Genus 194. TROX, *Fabricius*, *Leach*, *Samou.*, (*Kirby.*)

SILPHA *p*, *Linné.*

1180. 1; sabulosus *. *Fabr. E.* i. 110.—(*Sam. I.* 42.)—(*Kirby & Sp. I. E.* ii. 248.)
Sc. sabulosus. *Linn.* ii. 551.—*Mart. C. pl.* 3. *f.* 26.—*Stew.* ii. 30.—*Marsh.* i. 24.—*Turt.* ii. 74.
Sc. femoralis. *DeGeer.* iv. 269. *pl.* 10. *f.* 12.
Sc. subterraneus. *Fourc.* i. 8.

1181. 2; arenosus *. *Gyll.* i. 11.
Tr. arenarius. *Payk. F.* i. 80.
Tr. niger. *Rossi M.* i. 9?
Tr. hispidus. (*DeJean C.* 56?)

1182. 3; arenarius *. *Fabr. E.* i. 111.—*Panz. F.* xcvii. *f.* 1.—(*Sam. I.* 42.)
Si. scabra. *Linn.* ii. 573.
Tr. hispidus. *Payk. F.* i. 81?
Sc. arenarius. *Marsh.* i. 25.
Sc. arenosus. *Gmel.* iv. 1586.
Tr. barbarus. *Laich. T.* i. 31.

1183 † 4, lutosus.
Sc. lutosus. *Marsh.* i. 25. In Mus. D. *Kirby.*

Familia XXIX. DYNASTIDÆ, *MacLeay*.

(SCARABÆIDES-XYLOPHILI *p*, *Latreille.*)

Genus 195. ORYCTES, *Illiger*, *Leach.*

GEOTRUPES *p*, *Fabricius.*

1184 ‡ 1, nasicornis *. *Illig. K.* i. 14.—*Sturm D. F.* i. 8. *pl.* 4 & 5.

Sc. nasicornis. *Linn.* ii. 544.
Ge. nasicornis. *Fabr. E.* i. 13.—*Ent. Trans.* (*Haworth.*) 1.76.
In Mus. D. *Haworth.*

1185 ‡ 2, inermis.
Sc. inermis. *Mart. C. pl.* 4. *f.* 35.—*Marsh.* i. 9.
In Mus. D. *Kirby et Vigors.*

Genus 196. DYNASTES?[a]

Familia XXX. MELOLONTHIDÆ, *MacLeay.*

(SCARABÆIDES-PHYLLOPHAGI, *Latreille.*)

(SCARABÆUS *p*, *Linné*, *Marsham.*—MELOLONTHA, *Fabricius.*)

Genus 197. SERICA, *MacLeay*, (*Kirby.*)

OMALOPLIA *p*, (*Megerle.*)—TROX *p*, *Schrank.*

1186. 1; brunnea*. *MacLeay H. E.* i. 147.
Sc. brunneus (brunnus). *Linn.* ii. 556.—*Mart. C. pl.* 3. *f.* 24.
—*Panz. F.* xcv. *f.* 7.—*Marsh.* i. 38.—*Turt.* ii. 51.
Me. brunneus. (*Sam. I.* 27.)
Sc. fulvus. *DeGeer.* iv. 277.
Sc. fulvescens. *Fourc.* i. 10.
Tenuicrusta rubens. *Voet. C.* i. 34. *pl.* 7. *f.* 53. ♂. *f.* 54. ♀.

Genus 198. OMALOPLIA, (*Köppe.*)

AMALOPLIA, (*Samou.*)

1187. 1; Ruricola*. (*DeJean C.* 59.)
Me. Ruricola. *Fabr. E. S.* i. 173.
Sc. Ruricola. *Mart. C. pl.* i. *f.* 7.—*Stew.* ii. 28.—*Marsh.* i. 39.
—*Don.* xi. *pl.* 378.—*Turt.* ii. 54.
An. ruricola. (*Samou.* 191.)
Me. nigromarginata. *Herbst A.* 150. *pl.* 43. *f.* 7.
Sc. marginatus. *Fourc.* i. 9.
Me. floricola. *Laich. T.* i. 41.
β; Sc. varius. *Marsh.* i. 39.
Me. humeralis. *Schon. S.* iii. 185.

Genus 199. ZANTHEUMIA, *Leach MSS. Fern Web.*

AMPHIMALLA, (*Latreille.*)

1188. 1; solstitialis*. *Leach MSS.*

---

[a] Genus 196. DYNASTES? *MacLeay.*

GEOTRUPES *p*, *Fabricius.*

1185 † 1. Juvencus. (*Wilkin Catalogue* (!).)
Ge. Juvencus. *Fabr. E.* i. 20.
Sc. Juvencus. *Oliv. E.* i. 45. *pl.* 16. *f.* 143. ♂.—*pl.* 8. *f.* 66. ♀.

Sc. solstitialis. *Linn.* ii. 554—*Mart. C. pl.* 2. *f.* 17.—*Berk. S.* i. 88.—*Stew.* ii. 29.—*Marsh.* i. 38.—*Turt.* ii. 47.
Me. solstitialis. *Leach E. E.* ix. 99.—(*Sam. I.* 27.)—(*Kirby & Sp. I. E.* i. 205.)
Sc. autumnalis. *Fourc.* i. 6.

Genus 200. MELOLONTHA *Auctorum. Cockchaffer.*

1189. 1; vulgaris*. *Fabr. E.* ii. 161.—(*Leach E. E.* ix. 99.)—(*Sam. I.* 27.)
Sc. Melolontha. *Linn.* ii. 554.—*Mart. C. pl.* 2. *f.* 12.—*Berk. S.* i. 87.—*Don.* viii. *pl.* 264. *f.* 2.—*Stew.* ii. 28.—*Marsh.* i. 36.—*Turt.* ii. 46.—*Shaw G. Z.* vi. 21. *pl.* 3.—(*Bingley.* iii. 112.)—*Mill. B. E.* 142. *pl.* 2. *f.* 2.—(*Kirby & Sp. I. E.* i. 177, 205.)
Brown-tree Beetle, Blind Beetle, Chaffer, Cockchaffer, Jack-horner, Jeffry-cock, Maybug, Tree-beetle, Brown Clock, Dor, Miller. *Bingley l. c.* Acre-bob, May-bob (*in Surry*), Oak Web (*Devon*).

1190. 2: Fullo. *Fabr. E.* ii. 160.—(*Sam. I.* 27.)
Sc. Fullo. *Linn.* ii. 553.—*Mart. C. pl.* i. *f.* 1, 2.—*Don.* iv. *pl.* 112.—*Stew.* ii. 27.—*Marsh.* i. 36.—*Turt.* ii. 45.—*Shaw G. Z.* vi. 26. *pl.* 5.

Genus 201. ANOMALA, (*Köppe,*) *Samou.*

A. Capite thoraceque valdè pubescentes.

1191. 1; Horticola*. (*Sam. I.* 3.) *Bracken-Clock.*
Sc. Horticola. *Linn.* ii. 554.—*Mart. C. pl.* 4. *f.* 43.—*Berk. S.* i. 87.—*Stew.* ii. 28.—*Panz. F.* xlvii. *f.* 15.—*Marsh.* i. 44.—*Turt.* ii. 53.
Me. Horticola. *Leach E. E.* ix. 99.—(*Kirby & Sp. I. E.* i. 205.)
Sc. viridicollis. *DeGeer.* iv. 278. *pl.* 10. *f.* 18.
Sc. adiaphorus. *Scop. C.* 5.
β; Sc. Arvicola. *Marsh.* i. 40.

1192. 2, errans*. (*DeJean C.* 58.)
Sc. errans. *Fabr. E.* ii. 173. (!)—*Oliv. E.* i. 5. *pl.* 8. *f.* 92.—*Turt.* ii. 52.

B. Capite thoraceque glabris.

1193. 3; Frishii*. (*Sam. I.* 3.)
Me. Frishii. *Fabr. E.* ii. 172.
Sc. Frishii. *Mart. C. pl.* 4. *f.* 42.—*Marsh.* i. 40.—*Don.* xi. *pl.* 390. *f.* 2.—*Turt.* ii. 52.
Me. Vitis. *Latr. G.* ii. 111.
β; Me. Julii. *Fabr. E.* ii. 171.
Me. dubia. *Herbst C.* iii. 128. *pl.* 25. *f.* 9.
Me. holosericea. *Fabr. E.* ii. 171.
γ, Sc. æneus. *DeGeer.* iv. 277. *pl.* 10. *f.* 16.

Me. oblonga. *Fabr. E.* ii. 165?
♂, Me. cyanocephala. *Fabr. E.* ii. 169?
Sc. Scopolii. *Fuesl.* v. 2.
Sc. teres. G*mel.* iv. 1568.

1194. 4, Vitis*? (*DeJean C.* 58.)
Sc. Vitis. *Vill. E.* i. 38.—*Mart. C. pl.* 1. *f.* 8.—*Marsh.* i. 41. —*Turt.* ii. 52.
Sc. dubius. *Frisch.* iv. 29. *pl.* 14.
Me. Vitis. (*Leach E. E.* ix. 99.)

Genus 202. ANISOPLIA, (*Köppe.*)

ANOMALA *p, Samou.*

1195. 1, Agricola. (*DeJean C.* 58.)
Sc. Agricola. *Linn.* ii. 553.—*Panz. F.* xlvii. *f.* 18.—*Marsh.* i. 43.—*Don.* xi. *pl.* 390. *f.* 1.
Me. Agricola. (*Leach E. E.* ix. 99.)
Ano. Agricola. (*Sam. I.* 3.)
Sc. cyathiger. *Scop. C.* 46.
Me. crucifer. *Herbst C.* iii. 100. *pl.* 24. *f.* 9?
Me. graminicola β. *Latr.* G. ii. 114.

1196 † 2, Donovani.
Sc. Donovani. *Marsh.* i. 44.
Ano. Donovani. (*Sam. I.* 3.)
Me. graminicola γ. *Latr. G.* ii. 114?
Me. Agricola. *Fabr. E.* ii. 176? In Mus. *Brit. et* D. Donovan.

Genus 203. HOPLIA, *Illiger, Leach, Samou.*

1197. 1; argentea*. *Dufts. F.* i. 180.
Me. argentea. *Oliv. E.* i. 67. *pl.* 3. *f.* 22. *a—d.*
Sc. argenteus. *Mart. C. pl.* 2. *f.* 11.—*Marsh.* i. 45.—*Turt.* ii. 55.
Ho. pulverulenta. (*Leach E. E.* ix. 99.)—(*Sam. I.* 23.)—(*Kirby & Sp. I. E.* i. 178.—*Id.* ii. 234.)
Ho. philanthus. *Latr.* G. ii. 116. (!)
♀; Sc. pulverulentus. *Fabr. E.* ii. 181.—*Marsh.* i. 46.

Familia XXXI. CETONIADÆ, *MacLeay.*

(SCARABÆIDES-ANTHOBII, *Latreille.*)

(SCARABÆUS *p, Linné, &c.*)

Genus 204. TRICHIUS, *Fabricius, Leach, Samou.*, (*Kirby.*)

MELOLONTHA *p, Herbst.*—CETONIA *p, Oliv.*

A. Corpore tomentoso.

1198. 1, fasciatus. *Fabr. E. S.* i. 119.—(*Leach E. E.* ix. 99.)—(*Sam. I.* 42.)
Sc. fasciatus. *Linn.* ii. 556.—*Mart. C. pl.* 2. *f.* 10.—*Berk. S.*

i. 88.—*Don.* iv. *pl.* 140.—*Stew.* ii. 29.—*Marsh.* i. 43.—
Turt. ii. 72.—(*Mill. B. E. pl.* i. *f.* 1.)

1199. [2, succinctus.] *Fabr. E. S.* ii. 132.

B. Corpore glabro.

1200. 3: variabilis*. *Ent. Trans.* (*Haworth.*) i. 81. *pl.* 1. *fig. inf.*
—(*Sam. I.* 42.)—(*Ing. Inst.* 89.)
Sc. variabilis. *Linn.* ii. 558.
Tr. octopunctatus. *Fabr. E.* ii. 131.
Sc. cordatus. *Fabr. M.* i. 27.
Tr. decempunctatus. *Schran. B.* i. 414.

1201. 4; nobilis*. Rose-beetle. *Fabr. E.* ii. 130.—(*Leach E. E.* ix. 99.)—(*Samou.* 191. *pl.* 1. *f.* 2.)
Sc. nobilis. *Linn.* ii. 558.—*Mart. C. pl.* 3. *f.* 28.—*Berk. S.* i. 88.—*Don.* v. *pl.* 154. *f.* 1, 2, 3.—*Stew.* ii. 29.—*Marsh.* i. 42. —*Turt.* ii. 71.
Sc. cuspidata. *Fabr. M.* i. 27.
Sc. viridulus. *DeGeer.* iv. 297.
Sc. auratus. *Schrank. A.* 9.

1202 ‡ 5. hemipterus[a].

Genus 205. CETONIA, *Fabricius, Leach, Samou.*, (*Kirby.*)

1203. 1; aurata*. *Fabr. E. S.* i. *b.* 127.—(*Leach E. E.* ix. 99.) —(*Sam. I.* 10.)—(*Kirby & Sp. I. E.* iii. 344.)
Sc. auratus. *Linn.* ii. 557.—*Barb. G.* 14. *pl.* 1. *f.* 2.—*Mart. C. pl.* 3. *f.* 27.—*Berk. S.* i. 88.—*Stew.* ii. 29.—*Marsh.* i. 41.—*Turt.* ii. 71.—*Bingley.* iii. 116.—*Millard.* 150. *pl.* 2. *f.* 3.
Sc. Smaragdulus. *DeGeer.* iv. 279. *pl.* 11. *f.* 1.
Sc. nobilis. *Schrank. A.* 10.
Rose May Chaffer, Green Beetle, Brass Beetle, *Bingley. l. c.* —June-bob (in Surry).

1204 ‡ 2, stictica. *Fabr. E.* ii. 155.—*Panz. F.* i. *f.* 4.
Sc. sticticus. *Linn.* ii. 552.
Ce. albopunctata. *DeGeer.* iv. 301. 29. *pl.* 10. *f.* 22.
Sc. funestus. *Schrank. A.* 13.
Ce. funeraria. *Fourc.* i. 8.
Sc. Greenii. *Don.* xii. *pl.* 418.
In Mus. *D. Curtis,* Donovan, *I. H. Griesbach et Haworth.*

---

[a] 1202 ‡ 5. hemipterus. *Fabr. E.* ii. 132.—*Herbst C.* iii. 187. *pl.* 27. *f.* 13, 14.
Sc. hemipterus. *Linn.* ii. 555.—*Berk. S.* i. 88. (!)—*Stew.* ii. 30. (!) —*Turt.* ii. 72. (!)
Sc. variegatus. *Scop. C.* 12.
Sc. squamulatus. *Müll. Z. D. pr.* 55.

Subsectio 3.

(SERRICORNES: STERNOXI, *Latreille.*)

Familia XXXII. BUPRESTIDÆ, *Leach.*

(BUPRESTIDES, *Latr.*)

(BUPRESTIS, *Linné, &c.*) *Burncows.*

Genus 206. BUPRESTIS *Auctorum.*

A. Elytris apice dentatis vel serratis.

a. "*Elytrorum margine apicis serrato.*" Gyllenhal.

1205 ‡ 1. chrysostigma[a].

1206 ‡ 2, rutilans *. *Fabr. E.* ii. 192.—*Fanz. F.* xxii. *f.* 8.
In Mus. D. ——?

b. "*Elytris apice bi- vel tri-dentatis.*" Gyllenhal.

1207 † 3, ænea. *Linn.* ii. 662.—(*Ing. Inst.* 87.)
Bu. carniolica. *Fabr. E.* ii. 189.
Bu. subrugosa. *Herbst C.* ix. 92. *pl.* 143. *f.* 2.
In Mus. D[r]. *Hole.*

1208. 4, rustica*? *Linn.* ii. 660.—*Berk. S.* i. 105.—*Panz. F.* lxviii. *f.* 19.—*Stew.* ii. 75.—*Marsh.* i. 395.—*Turt.* ii. 412.
Bu. violacea. *DeGeer.* iv. 130.

1209. 5, flavomaculata *. *Fabr. E.* ii. 193.—*Panz. F.* xxii. *f.* 9.
Bu. flavopunctata. *DeGeer.* iv. 129.
Elater tetrastichon. *Linn.* ii. 656?

1210 ‡ 6. octoguttata[b].

B. "Elytris integris, muticis." *Gyllenhal.*

a. *Corpore haud depresso.*

1211 † 7. splendens *[c].

b. *Corpore valdè depresso.*

1212. 8, Salicis *. *Fabr. E.* ii. 216.—*Don.* iv. *pl.* 127.—*Stew.* ii. 75. —*Marsh.* i. 397.—*Turt.* ii. 417.—(*Curtis l. c. infra.*)

1213 ‡ 9, nitidula. *Linn.* ii. 662.—*Turt.* ii. 417.—*Curtis.* i. *pl.* 31.

---

[a] 1205 ‡ 1. chrysostigma. *Linn.* ii. 660.—*Berk. S.* i. 105. (!)—*Panz. F.* lxviii. *f.* 18.—*Stew.* ii. 75. (!)—*Turt.* ii. 409. (!)

[b] 1210 ‡ 6. octoguttata. *Linn.* ii. 659.—*Turt.* ii. 411. (!)—*Herbst A.* v. 118. *pl.* 28. *f.* 7. *a, b.*
Bu. albopunctata. *DeGeer.* iv. 132.

[c] 1212 † 7. splendens *. *Fabr. E.* ii. 104.—(*Linn. Trans.* (*Marsham.*) x. 399. *pl.* xxxii. *f.* 1, 2. (!)
Bu. splendida. *Payk. F.* ii. 229.
Bu. pretiosa. *Herbst C.* ix. 127?

1214 ‡ 10. manca[a].

1215 † 11. quadripunctata[b].

Genus 207. AGRILUS, (*Megerle,*) *Curtis.*

A. Corpore lineari-elongato, acuminato.

1216. 1; biguttata*. (*Curtis l. c. infra.*)
Bu. biguttata. *Fabr. E.* ii. 212.—*Stew.* ii. 75.—*Marsh.* i. 396. —*Panz. F.* xc. *f.* 8.—*Turt.* ii. 416.—*Leach E. E.* ix. 85.— *Samou.* 58 & 160.

1217. 2; viridis*. (*Curtis l. c. infra.*)
Bu. viridis. *Linn.* ii. 663.—*Don.* v. *pl.* 174.—*Stew.* ii. 75.— *Marsh.* i. 397.—*Turt.* ii. 416.—(*Sam. I.* 8. *pl.* 3. *f.* 9.)

1218 † 3, chryseis*. *Ziegler.*—*Curtis.* ii. *pl.* 67.
In Mus. *D. Griesbach et Stone.*

B. Corpore cylindrico.

1219 ‡ 4. Rubi[c].

1220. 5, novemmaculata.
Bu. novemmaculata. *Linn.* ii. 662?—*Marsh.* i. 396.—*Panz. F.* lxviii. *f.* 17.—(*Linn. Trans.* (*Marsham.*) x. 402.)

Genus 208. APHANISTICUS, *Latr., Leach, Samou.*

1221. 1, emarginatus*. (*Leach E. E.* ix. 85.)—(*Sam. I.* 4.)— (*Ing. Inst.* 87.)
Bu. emarginatus. *Fabr. E.* ii. 213.—*Herbst C.* ix. 261. *pl.* 155. *f.* 7. *a. b.*
Bu. pusilla. *Gyll.* i. 460?

Genus 209. TRACHYS, *Fabr., Leach, Samou., Millard.*

1222. 1: pygmæa*. *Fabr. E.* ii. 219.
Bu. pygmæa. *Don.* viii. *pl.* 282.—*Marsh.* i. 398.—*Turt.* ii. 415.
Tr. viridis. (*Sam. I.* 42.)

1223. 2; minuta*. *Fabr. E.* ii. 219.—*Leach E. E.* ix. 85.— *Samou.* 160.
Bu. minuta. *Linn.* ii. 663.—*Don.* viii. *pl.* 256.—*Marsh.* i. 398. —*Turt.* ii. 415.

1224. 3; nana*. *Fabr. E.* ii. 220.—*Panz. F.* xcv. *f.* 9.—(*Ing. Inst.* 89.)

---

[a] 1214 ‡ 10. manca. *Fabr. E.* ii. 211.—*Turt.* ii. 415. (!)

[b] 1215 † 11. quadripunctata. *Linn.* ii. 662.—*Turt.* ii. 415. (!)—*Oliv. E.* ii. 80. *pl.* 10. *f.* 117. *a, b.*

[c] 1219 ‡ 4. Rubi.
Bu. Rubi. *Linn.* ii. 661.—*Turt.* ii. 413. (!)

Genus 210. MELASIS, *Olivier, Leach, Samou., Curtis.*

ELATER *p, Linné.*—HISPA *p, Gmelin.*—PTILINUS *p, Kugell.*

1225. 1, buprestoides*. *Curtis.* ii. *pl.* 55.
El. buprestoides. *Linn.* ii. 656.
Me. flabellicornis. *Fabr. E.* ii. 331.—*Leach E. E.* ix. 85.—*Samou.* 160.
Hi. flabellicornis. *Stew.* ii. 53.—*Turt.* ii. 201.

Familia XXXIII. ELATERIDÆ, *Leach.*

(ELATERIDES, *Latreille.*)

(ELATER, *Linné, &c.*)

Blacksmiths, Clickers, Snaps, Spring beetles and Skipjacks.

Genus 211. CERATOPHYTUM, *Leach, Samou.*

CEROPHYTUM, *Latreille.*—MELASIS *p, Latreille H.*

1226 † 1, Latreillii. *Leach MSS.*—(*Sam. I.* 10.)
Me. elateroides. *Latr. H.* ix. 76.
Ce. elateroides. (*Leach E. E.* ix. 85.) In Mus. *Brit.*

Genus 212. EUCNEMIS, *Mannerheim.*

1227 † 1, pygmæus. (*DeJean C.* 34?)—*Mann. Euc.* 30. *pl.* 2. *f.* 4, 5, 6.
El. pygmæus. *Fabr. E.* ii. 246? In Mus. *D. Vigors.*

Genus 213. HEMIRHIPUS, *Latreille.* Larvæ: *Wire worms.*

A. Thorace sublineari.

1228. 1; limbatus*.
El. limbatus. *Fabr. E.* ii. 242.
El. pusillus. *Herbst C.* x. 85. *pl.* 165. *f.* 11.
El. nitidulus. *Marsh.* i. 380.—(*Sam. I.* 16.)
β, El. pallens. *Fabr. E.* ii. 242.

1229. 2; acuminatus* *mihi.*

1230. 3; marginatus*.
El. marginatus. *Linn.* ii. 654.—*Turt.* ii. 381.—*Leach E. E.* ix. 85.—*Samou.* 162.
El. lateralis. *Oliv. E.* ii. 31. *pl.* 8. *f.* 80. *a, b.*
El. lineatus. *DeGeer.* iv. 158.
El. dorsalis. *Payk. F.* iii. 4.
El. suturalis. *Marsh.* i. 379.
β; El. fulvus. *Marsh.* i. 379.

B. Thorace valdè convexo.

1231. 4; sputator*.
El. sputator. *Linn.* ii. 654.—*Stew.* ii. 72.—*Turt.* ii. 382?—(*Sam. I.* 16.)

El. variabilis. *Herbst C.* x. 75. *pl.* 164. *f.* 11.
El. fusculus. *Illig. M.* iv. 101.

1232. 5; obscurus*.
El. obscurus. *Linn.* ii. 655.—*Mart. C. pl.* 30. *f.* 1.—*Stew.* ii. 72.—*Marsh.* i. 377.—*Turt.* ii. 381.—(*Sam. I.* 16.)—(*Kirby & Sp. I. E.* i. 179.)
El. obtusus. *GeGeer.* iv. 147.
El. variabilis. *Fabr. E.* ii. 241.
El. hirtellus. *Herbst C.* x. 94. *pl.* 166. *f.* 11.
β; El. similis. *Kirby MSS.*

1233. 6; lineatus*.
El. lineatus. *Linn.* ii. 653.—*Marsh.* i. 377.—(*Sam. I.* 16.)—(*Kirby & Sp. I. E.* i. 179.)
El. segetis. *Act. Holm.* (*Bierkander.*) 1779. 284. *pl.* 10. *f.* 1–3.
El. striatus. *Fabr. E.* ii. 241.—*Panz. F.* xciii. *f.* 13.

Genus 214. ELATER *Auctorum.*

A. Corpore subcylindrico.

a. *Thorace brevi, anticè convexo.*

1234. 1: fugax. *Fabr. E.* ii. 237.—(*Ing. Inst.* 88.)
El. vulgaris. *Herbst A.* i. 114.
El. brunneus β. *Payk. F.* iii. 29.

1235. 2: brunneus. *Linn.* ii. 653.—*Turt.* ii. 381.—*Herbst A.* v. 112. *pl.* 27. *f.* 6.—(*Ing. Inst.* 88.)—(*Curtis Cat. no.* 45.)

1236 † [3. maritimus.] (*Curtis Cat. no.* 45.)? In Mus. D. Curtis.

1237. 4, fulvicollis *mihi.*

b. *Thorace elongato, medio convexo.*

1238. 5; cylindricus*. *Payk. F.* iii. 24.—(*Sam. I.* 16.)
El. obsoletus. *Marsh.* i. 387.

1239. 6, nigripes. *Gyll.* i. 395.
El. obtusicornis. *Kirby MSS.*

c. *Thorace oblongo, posticè convexo.*

1240. 7; serraticornis*. *Payk. F.* iii. 21.—(*Ing. Inst.* 88.)
El. serraticornis. *Herbst C.* x. 100. *pl.* 167. *f.* 7.
El. quadricollis. *Kirby MSS.*

1241. 8; minutus*. *Linn.* ii. 656.—*Marsh.* i. 381.—*Turt.* ii. 383.—(*Sam. I.* 16.)
El. angustus. *Herbst C.* x. 98. *pl.* 167. *f.* 4.
β El. lævicollis. *Kirby MSS.*

1242. 9, nigroæneus. *Marsh.* i. 384.

1243 † 10. Bructeri. *Fabr. E.* ii. 243.—*Panz. F.* xxxiv. *f.* 13.—(*Wilk. Catal.*)

El. minutus. *Payk. F.* iii. 40.
El. æneo-niger. *DeGeer.* iv. 159. In Mus. D. Vigors?

1244 † 11, subrugosus *mihi.* In Mus. D. C. *Griesbach.*

B. Corpore plus minusve depresso.

a. *Thorace æquali, lateribus haud dilatatis.*

1245. 12, ephippium*. *Fabr. E.* ii. 238.—*Panz. F.* v. *f.* 14.—*Marsh.* i. 383.—(*Sam. I.* 16.)
El. sanguinolentus. *Schra. A.* 341.
El. sanguineus var c. *Payk. F.* iii. 33.

1246. 13; sanguineus*. *Linn.* ii. 654.—*Berk. S.* i. 104.—*Mart. C. pl.* 40. *f.* 6?—*Stew.* ii. 72?—*Marsh.* i. 382.—*Turt.* ii. 382.—*Shaw G. Z.* vi. 85.—*Don.* xv. *pl.* 508. *f.* 2.—*Samou.* 57. *pl.* 3. *f.* 6?

1247. 14, rufipennis. *Hoffmansegg. MSS.*—(*Sam. I.* 16.)

1248. [15, semiruber.] *Hoffmansegg. MSS.*—(*Sam. I.* 16.)

1249. 16, Pomonæ*. *Hoffmansegg. MSS.*—(*Sam. I.* 16.)

1250. 17; præustus*. *Fabr. E.* ii. 238.—*Panz. F.* xciii. *f.* 8.—(*Sam. I.* 16.)
El. sanguineus b. *Payk. F.* iii. 33.

1251. 18; balteatus*. *Linn.* ii. 654.—*Berk. S.* i. 104.—*Mart. C. pl.* 30. *f.* 5.—*Stew.* ii. 72.—*Marsh.* i. 384.—*Panz. F.* xciii. *f.* 9.—*Turt.* ii. 382.—(*Sam. I.* 16.)—*Wood.* i. 58. *pl.* 21.

b. *Thorace inæquali, lateribus dilatatis.*

1. Corpore subpubescenti.

1252. 19; ustulatus. *Payk. F.* iii. 32.—(*Ing. Inst.* 88.)

1253. 20; bipustulatus*. *Linn.* ii. 652.—*Berk. S.* i. 104.—*Stew.* ii. 71.—*Panz. F.* lxxvi. *f.* 10.—*Marsh.* i. 375.—(*Sam. I.* 16.)

1254. 21, aterrimus. *Linn.* ii. 653?—*Mart. C. pl.* 31. *f.* 19.—*Stew.* ii. 72.—*Marsh.* i. 380.—*Turt.* ii. 378?

2. Corpore densè pubescenti.

* *Antennæ articulo secundo haud dilatato.*

1255. 22; holosericeus*. *Fabr. E.* ii. 228.—*Marsh.* i. 386.—(*Sam. I.* 16.)
El. undulatus. *Herbst C.* x. 41. *pl.* 161. *f.* 9.

** *Antennæ articulo secundo dilatato; primo magno.*

1256. 23; murinus*. *Linn.* ii. 655.—*Mart. C. pl.* 31. *f.* 17.—*Herbst C.* x. 39. *pl.* 161. *f.* 8.—*Marsh.* i. 385.—*Turt.* ii. 378.—*Leach E. E.* ix. 85.—*Samou.* 162.

1257 ‡ 24. fasciatus [a].

---

[a] 1257 ‡ 24. fasciatus. *Linn.* ii. 655.—*Panz. F.* lxvi. *f.* 2.—*Turt.* ii. 379(!).
El. inæqualis. *DeGeer.* iv. 148.

Genus 215. ———

A. Antennis articulis 2, 3, et 4 brevibus.

1258. 1, riparius*.
El. riparius. *Fabr. E.* ii. 243.—(*Sam. I.* 16.)
El. littoreus. *Herbst C.* x. 86. *pl.* 165. *f.* 12.
El. æneus. *Marsh.* i. 388.

1259. 2, rivularis.
El. rivularis. *Gyll.* i. 403.
El. riparius. *Panz. F.* xxxiv. *f.* 12.

B. Antennis articulis 2, et 3 brevibus.

1260. 3; quadripustulatus*.
El. quadripustulatus. *Fabr. E.* ii. 248.—*Ent. Trans.* (*Burrell.*) i. 202.—*Don.* xvi. *pl.* 545.—(*Sam. I.* 16.)—(*Curtis Cat. no.* 45.)
β; El. Dermestoides. *Herbst C.* x. 85. *pl.* 165. *f.* 10?

1261. 4, pulchellus?
El. pulchellus. *Linn.* ii. 656.—*Panz. F.* lxxvi. *f.* 8.—*Turt.* ii. 384.
β, El. trimaculatus. *Fabr. E.* ii. 245.
γ, El. nanus. *Vigors MSS.?*

Genus 216. ———

1262. 1; fulvipes*.
El. fulvipes. *Herbst C.* x. 46. *pl.* 162. *f.* 2.
El. castanipes. *Payk. F.* iii. 23.—*Marsh.* i. 381.—(*Sam. I.* 16.)
El. obscurus. *Fabr. E.* ii. 233.
El. fuscus-major. *DeGeer.* iv. 146.
β, El. rugosus. *Marsh.* i. 381.

Genus 217. LUDIUS, *Latreille.*

1263. 1, ferrugineus*.
El. ferrugineus. *Linn.* ii. 654.—*Mart. C. pl.* 31. *f.* 12.—*Don.* x. *pl.* 356. *f.* 1.—*Marsh.* i. 382.—*Turt.* ii. 378.—*Shaw* G. Z. vi. 85. *pl.* 30.—*Leach E. E.* ix. 85.—*Samou.* 161.

Genus 218. CLENIOCERUS, *Leach MSS.*

A. Antennis pectinatis.

1264 † 1, aulicus.
El. aulicus. *Panz. F.* lxxvii. *f.* 6.—(*Ing. Inst.* 88.)
In Mus. *Brit.*

1265. 2, pectinicornis.
El. pectinicornis. *Linn.* ii. 655.—*Berk. S.* i. 104.—*Mart. C. pl.* 3. *f.* 10, 10.—*Don.* x. *pl.* 356. *f.* 2.—*Stew.* ii. 73.—*Marsh.* i. 387.—*Turt.* ii. 379.—*Sam. I.* 16.—(*Curtis Cat. no.* 45.)
El. æneo-pectinicornis. *DeGeer.* iv. 145.

El. flabellicornis. *Voët.* (*Pz.*) ii. 120. *pl.* 45. *f.* 31.
♀ β? El. picipes. *Kirby? MSS.*

1266. 3 : cupreus*.
El. cupreus. *Fabr. E.* ii. 231.—*Mart. C. pl.* 31. *f.* 16.—*Stew.* ii. 73.—*Panz. F.* lxxvii. *f.* 22.—*Marsh.* i. 384.—*Turt.* ii. 380. *Don.* xv. *pl.* 508.—(*Sam. I.* 16.)—(*Curtis Cat. no.* 45.)
β, El. chalybeus. *Sowerb. B. M.* i. *pl.* 72.
El. æruginosus. *Fabr. E.* ii. 231?

1267 ‡ 4, castaneus.
El. castaneus. *Linn.* ii. 654.—*Berk. S.* i. 104.—*Panz. F.* lxxvii. *f.* 4.—*Stew.* ii. 72.—*Turt.* ii. 380.—*Samou.* 161.
El. flavo-pectinicornis. *DeGeer.* iv. 153. In Mus. D. *Vigors.*

B. Antennis haud pectinatis.

1268. 5 : sanguinicollis*.
El. sanguinicollis. *Hellwig.*—*Panz. F.* vi. *f.* 13.—(*Ing. Inst.* 88.)
El. ruficollis. (*Don.* xv. *pl.* 518. *f.* 1.)

1269. 6; tessellatus*.
El. tessellatus. *Linn.* ii. 655.—*Herbst C.* x. 32. *pl.* 161. *f.* 1. —*Marsh.* i. 386.—*Turt.* ii. 379.—*Shaw G. Z.* vi. 85.—(*Sam. I.* 16.)
El. rufounguiculatus. *DeGeer.* iv. 48.
β; El. assimilis. *Gyll.* i. 394.
El. nigripes. *Kirby MSS.*

1270. 7; metallicus*.
El. metallicus. *Gyll.* i. 392.—(*Sam. I.* 16.)
El. nigricornis. *Panz. F.* lxi. *f.* 5.

Genus 219. ———

1271. 1, æneus*.
El. æneus. *Linn.* ii. 655.—*Turt.* ii. 379.—(*Sam. I.* 16.)
El. impressus. *Marsh.* i. 387.—*Don.* xv. *pl.* 535. *f.* 2.
El. æneus rufipes. *DeGeer.* iv. 149.
β, El. germanus. *Linn.* ii. 655.—*Turt.* ii. 379.
El. cyaneus. *Marsh.* i. 388.—*Sowerb. B. M.* i. *pl.* 26.—*Don.* xv. *pl.* 535. *f.* 1.—*Samou.* 57. *pl.* 3. *f.* 7.

1272 ‡ 2, cruciatus*? *Linn.* ii. 653.—*Stew.* ii. 72.—*Turt.* ii. 380. —*Panz. F.* lxxvi. *f.* 6. In Mus. *Brit.*

1273 † 3, bimaculatus. (*Ing. Inst.* 88.)
El. bimaculatus. *Fabr. E.* ii. 121.—*Panz. F.* lxxvi. *f.* 9. In Mus. *Brit.*

Genus 220. ———

1274. 1, thoracicus*.
El. thoracicus. *Fabr. E.* ii. 236.—*Panz. F.* vi. *f.* 12.—*Mart. C. pl.* 31. *f.* 12.—*Stew.* ii. 72.—*Marsh.* i. 376.—*Turt.* ii. 381. —(*Sam. I.* 16.)

1275. 2, ruficollis*.

El. ruficollis. *Linn.* ii. 653.—*Berk. S.* i. 104.—*Mart. C. pl.* 30. *f.* 3.—*Stew.* ii. 72.—*Marsh.* i. 376.—*Turt.* ii. 381.—*Don.* xv. *pl.* 7.—(*Sam. I.* 16.)

1276. 3; Equiseti *.
El. Equiseti. *Herbst C.* x. 67. *pl.* 163. *f.* 12.—(*Ing. Inst.* 88.)
El. pilosus. *Payk. F.* iii. 25.

1277. [4; cordiger *.]
El. cordiger. *Kirby MSS.*

1278 † 5, cylindrus. *Leach? MSS.* In Mus. *Brit.*

Genus 221. ———

1279. 1: pubescens.
El. pubescens. *Marsh. MSS.*—(*Ing. Inst.* 88.)
El. Bydderi. *Wilk. MSS.*
El. rufus. *Fabr. E.* ii. 225?—*Panz. F.* x. *f.* 11?
El. melanophthalum. *Gmel.* iv. 1914.

1280. 2; niger *.
El. niger. *Linn.* ii. 656.—*Berk. S.* i. 104.—*Stew.* ii. 73.—*Panz. F.* ci. *f.* 16?—*Turt.* ii. 375.—(*Sam. I.* 16.)
El. aterrimus. *Fabr. E.* ii. 227.
El. hirtus. *Herbst A.* v. 114.

1281. [3; nigrinus *.]
El. nigrinus. *Marsh.* i. 389.

1282 ‡ 4, varius *.
El. varius. *Fabr. E.* ii. 229.—*Panz. F.* lxxvi. *f.* 3.—(*Ing. Inst.* 88.)
El. Quercus. *Herbst A.* iv. *pl.* 27. *f.* 11. *C?*
In Mus. D. *H. Griesbach.*

1283. 5, hirsutus *mihi.*

1284. 6; ruficaudis *.
El. ruficaudis. *Gyll.* i. 409.
El. fuscus minor. *DeGeer.* iv. 147.
El. analis. *Herbst C.* x. 66. *pl.* 163. *f.* 11.
El. sputator. *Oliv. E.* ii. 30.—*Don.* iii. *pl.* 96. *f.* 4.—*Mart. C. pl.* 31. *f.* 20.—*Marsh.* i. 384.
El. obscurus. *Payk. F.* iii. 2.
El. hæmorrhoidalis. *Fabr. E.* ii. 235.

1285. [7, elongatus.]
El. elongatus. *Marsh.* i. 385.

1286. 8; subfuscus *.
El. subfuscus. *Gyll.* i. 411.
El. linearis. *Payk. F.* iii. 3.
El. varians. ———? *MSS.*

1287. 9; vittatus *.

El. vittatus. *Fabr. E.* ii. 231.—*Panz. F.* xcviii. *f.* 6.—*Ent. Trans.* (*Burrell.*) i. 202.—*Don.* xv. *pl.* 518. *f.* 2.—(*Sam. I.* 16.)
El. marginatus. *Oliv. E.* ii. 31. *pl.* 3. *f.* 29.

1288. 10, angularis *mihi.*
El. erythrogonos. *Germ.* (*M.*) *Müller.*) iv. 186?—*Ahr. F.* v. *f.* 7?
El. auritus. *Schon. S.* iii. *App.* 139?

1289. 11, testaceus.
El. testaceus. *Fabr. E.* ii. 238.—(*Ing. Inst.* 88.)

1290. [12, ochropterus] *mihi.*

1291. 13; rufipes *.
El. rufipes. *Fabr.* ii. 242.—*Marsh.* i. 389.—*Panz. F.* xciii. *f.* 14.—(*Sam. I.* 16.)

1292. [14, tibialis *.]
El. tibialis. *Kirby MSS.*

1293. 15; longicollis *.
El. longicollis. *Fabr. E.* ii. 241.—(*Sam. I.* 16.)
El. marginatus. *Payk. F.* iii. 15.—(*Mart. C. pl.* 31. *f.* 15. —*Marsh.* i. 379.
El. marginellus. *Herbst C.* x. 75. *pl.* 164. *f.* 10.
β; El. unicolor. *Marsh.* i. 379.—(*Sam. I.* 16.)
El. longicollis. *Panz. F.* xciii. *f.* 12.
El. impressus. *Kirby MSS.*
♀; El. lateralis. *Marsh.* i. 380.

Genus 222. CAMPYLIS, *Fisher.*
EXOPTHALMUS, (*Latreille.*)—CAMPYLUS (*Gyll.*)

1294. 1; dispar *.
El. linearis. *Linn.* ii. 653.—*Turt.* ii. 380.
El. variabilis. *DeGeer.* iv. 154.
El. dispar. *Payk. F.* iii. 37.
El. bicolor. *Panz. F.* viii. *f.* 11.—*Marsh.* i. 378.
β, El. livens. *Fabr. E.* ii. 232.—*Turt.* ii. 380.
γ, El. mesomelus. *Linn.* ii. 653.—*Panz. F.* vii. *f.* 6.—*Marsh.* i. 378.—*Turt.* ii. 380.—(*Sam. I.* 16.)

Subsectio 4.

(SERRICORNES: MALACODERMI *p, et* XYLOPHAGI *p, Latreille.*)

Familia XXXIV. CEBRIONIDÆ?

(CEBRIONITES, *Latreille.*)

Genus 223. ATOPA. *Paykul, Leach.*
CHRYSOMELA *p, Linn., Don.*—PTINUS *p, DeGeer.*—CISTELA *p, Olivier.*—CRIOCERIS *p, Marsh.*—DASCILLUS[a], *Latr., Samou.,* (*Kirby.*)—CRYPTOCEPHALUS *p, Gmel.*

---

[a] DASCILLUS: Piscis antiquorum.

1295. 1; cervina*. *Payk. F.* ii. 117.—*Leach E. E.* ix. 85.
Ch. cervina. *Linn.* ii. 602.—*Don.* iii. *pl.* 78, *f.* 3, 4.
Cryp. cervinus. *Stew.* ii. 50.—*Turt.* ii. 182.
Da. cervinina. *Samou.* 162.
♂; Cri. cervina. *Marsh.* i. 220.
♀; Cri. cinerea. *Marsh.* i. 220.
Cry. cinereus. *Turt.* ii. 182.
Pti. testaceo-villosus. *DeGeer.* iv. 235. *pl.* 9. *f.* 8.

Familia XXXV. CYPHONIDÆ.

Genus 224. SCIRTES, *Illiger, Leach, Samou.*

CYPHON *p, Fabr.*—ELODES *p, Latr.*—CHRYSOMELA *p, Linné, Marsh.*—ALTICA *p, Panz.*

1296. 1; hemisphæricus*. *Leach E. E.* ix. 86.—*Samou.* 163.
Ch. hemisphærica. *Linn.* ii. 595.—*Panz. F.* xcvi. *f.* 7.—*Marsh.* i. 199.
Cy. hemisphæricus. (*Sam. I.* 14.)
Ch. fusca. *DeGeer.* iv. 348.
Al. latiuscula. *Müll. Z. D. pr.* 926.

1297. 2, pallescens* *mihi.*
Sc. testaceus. *Leach MSS.*

Genus 225. CYPHON, *Payk., Leach.*

CHRYSOMELA *p, Linn., Marsh.*—CRIOCERIS *p, Marsh, Millard.*—CISTELA *p, Fanz.*—LAMPYRIS *p? Linné.*—ELODES, *Latr., Samou.*

A. Capite thorace abscondito.

a. *Corpore hemisphærico.*

1298. 1; chrysomeloides*. *Wilkin MSS.*
Ch. pubescens. *Marsh.* i. 183.
Cy. serraticornis. *Germ. M.* (*Müller.*) iv. 221.

b. *Corpore oblongo.*

1299. 2; melanurus*. *Fabr. E.* i. 502.
El. melanura. (*Sam. I.* 16.)
Cy. pallidus. *Fabr. E.* i. 501.—*Leach E. E.* ix. 85.—(*Sam. I.* 14.)
Ci. pallida. *Panz. F.* viii. *f.* 7.—*Turt.* ii. 184.
Ci. pallida. *Marsh.* i. 227.
El. pallida. *Samou.* 162.
Cry. pallidus. *Stew.* ii. 51.
La.? minuta. *Linn.* ii. 645?
♂; Cy. villosus. *Kirby MSS.*

1300. [3: læta.]
Ci. læta. *Panz. F.* viii. *f.* 8.
Cy. melanurus β. *Gyll.* i. 366.

1301. 4; assimilis* *mihi*.

1302. 5; lividus*. *Fabr. E.* i. 501.
Cy. pallidus a. *Payk. F.* ii. 118.
Cr. mollis. *Marsh.* i. 225.
El. mollis. (*Sam. I.* 16.)

1303. 6; testaceus* *mihi*.
Cy. flavicornis. *Kirby MSS.* ?

1304. 7; obscurus* *mihi*.

1305. 8, marginatus. *Fabr. E.* i. 502.
Ci. nimbata. *Panz. F.* xxiv. *f.* 15.
Cy. pallidus c. *Payk. F.* ii. 119.

1306. 9; pubescens*. *Fabr. E.* i. 502.
Can. variabilis. *Mus. Acad.* (*Thunb.*) iv. 54.

1307. [10, dorsalis.]
Cryptocephalus! dorsalis. *Marsh.* i. 210.—(*Curtis l. c.* i. *folio* 35.)

1308. 11; griseus*. *Fabr. E.* i. 502.
Cy. coarctatus. *Payk. F.* ii. 120.
Can. variabilis var. *Mus. Acad.* (*Thunb.*) iv. 54.
El. fuscescens. *Latr. G.* ii. 253.
β; Cr. nigricans. *Marsh.* i. 226.
El. nigricans. (*Sam. I.* 16.)
Cy. niger. *Kirby MSS.*
γ, Cr. concolor. *Marsh.* i. 226.
Cy. latus. *Kirby MSS.*

1309. 12; Padi*. *Gyll.* i. 371.
Ch. Padi. *Linn.* ii. 588.
Cr. Padi. *Marsh.* i. 226.
Cy. coarctatus var. pygmæa. *Payk. F* ii. 121.
Cy. discolor. *Panz. F.* xcix. *f.* 8.

1310. 13; ater* *mihi*.

B. Capite exserto.

1311. 14; angulosus*.
Cr. angulosa. *Marsh.* i. 228.
Cy. bicolor. ——?

1312. 15, dubius* *mihi*.

Familia XXXVI. LAMPYRIDÆ, *Kirby*.

(LAMPYRIDES, *Latreille*.)

Genus 226. LAMPYRIS *Auctorum*. *Glow-worm*.

1313. 1; noctiluca*. *Linn.* ii. 643.—*Berk. S.* i. 102.—*Stew.* ii. 68.—*Marsh.* i. 361.—*Turt.* ii. 358.—*Shaw. G. Z.* vi. 77. *pl.* 28.—*Bingley* iii. 140.—*Leach E. E.* ix. 86.—*Samou.* 55. *pl.* 3. *f.* 1. ♂.

f. 2, ♀.—*Millard.* 217. *pl.* 3. *f.* 1.—*Wood.* i. 53. *pl.* 20.—(*Kirby & Sp. I. E.* ii. 410.)

1314 ‡ 2. splendidula [a].

Genus 227. LYCUS, *Fabr.*, *Leach*, *Samou.*, *Millard*, (*Kirby.*)

HOMALYSUS *p*, *Illig.*

1315 ‡ 1. sanguineus [b].

1316 † 2. festivus [c].

1317. 3; minutus*. *Fabr. E.* ii. 117.—*Fanz. F.* xli. *f.* 11.—*Leach E. E.* ix. 86.—*Samou.* 163.
La. nigrorubra. *DeGeer.* iv. 46.
La. pusilla. *Marsh.* i. 363.

Familia XXXVII. TELEPHORIDÆ, *Leach.*

(CANTHARIS, *Linné*, *Marsh*, *&c.*)

Soft-wings, *Soldiers*, *Sailors.*

Genus 228. TELEPHORUS, *DeGeer*, *Leach*, *Samou.*, (*Kirby.*)

A. Thoracis lateribus vix rotundatis.

a. *Thorace transverso, lateribus postice excisis*: (*Antennæ serratæ, articulo secundo minuto.*)

1318. 1, ruficollis*. (*Sam. I.* 40.)
Ca. ruficollis. *Fabr. E.* ii. 299.—*Stew.* ii. 70.—*Marsh.* i. 366. —*Mart. C. pl.* 29. *f.* 11?—*Turt.* ii. 367.
Ca. torquata. *Gyll.* iv. 340?

b. *Thorace transverso, angulis posticis integris.*

1. Antennæ articulo secundo tertio multo breviore.

1319. 2; thoracicus*. *Oliv. E.* ii. 12. *pl.* 1. *f.* 2. *a*, *b.*
Ca. fulvicollis. *Illig. K.* i. 302.—(*Sam. I.* 40.)—*Ent. Trans.* (*Burrell.*) i. 196.
Ca. bicolor. *Herbst A.* v. 108.
Ca. fulva. *Gmel.* iv. 1896.

1320. 3; fulvicollis*. *Sahl. I. F.* viii. 113.—*Fabr. E.* i. 300?
Te. sanguinicollis *mihi* (*olim.*)

1321. [4; affinis*.] *mihi.*

1322. 5; ater*. *Oliv. E.* ii. 13. *pl.* 1. *f.* 3. *a*, *b.*

---

[a] 1314 ‡ 2. splendidula. *Linn.* ii. 644.—*Stew.* ii. 68. (!)—*Marsh.* i. 362. (!) —*Leach E. E.* ix. 86. (!)—(*Sam. I.* 24. (!)—*Panz. F.* xli. *f.* 8.

[b] 1315 ‡ 1. sanguineus. *Fabr. E.* ii. 116.—*Panz. F.* xli. *f.* 9.
La. sanguinea. *Linn.* ii. 646.—*Stew.* ii. 68. (!)
La. villosa. *DeGeer.* iv. 37.

[c] 1316 † 2. festivus.
La. festiva. *Don.* xvi. *pl.* 544. (!)

Ca. atra. *Linn.* ii. 649.—*Turt.* ii. 366.
Ca. Iridis. *Marsh? MSS.*

1323. 6; flavilabris*.
Ca. flavilabris. *Fallen. M. C.* i. 12.—*Gyll.* i. 337.
Ca. picea. *Kirby MSS.*
β, Ca. nigrina. *Kirby MSS.*
γ, Ca. dissimilis. *Marsh. MSS.*

1324. 7, pulicarius. *Oliv. E.* ii. 26. *pl.* 3. *f.* 20. *a, b.*
Te. æthiops. (*Curtis Cat. no.* 45?)

2. Antennæ articulis secundo et tertio ferè æqualibus.

1325. 8; lateralis*. (*Sam. I.* 40.)
Ca. lateralis. *Linn.* ii. 648.—*Turt.* ii. 366.—*Ent. Trans.* (*Burrell.*) i. 196.

1326. 9; marginatus* *mihi.*

1327. 10; testaceus*. *DeGeer.* iv. 71.—(*Sam. I.* 40.)
Ca. testacea. *Linn.* ii. 649.—*Stew.* ii. 69.—*Mart. C. pl.* 29. *f.* 6. —*Marsh.* i. 367.—*Turt.* ii. 369.

1328. [11; pallipes*.] *mihi.*
Ca. testacea. *Fanz. F.* lvii. *f.* 4.
Ca. flavipes. *Kirby MSS.?*

c. *Thorace elongato.* (*Antennæ elongatæ.*)

1329. 12; pallidus*. *Oliv. E.* ii. 14. *pl.* 2. *f.* 9. *a, b.*—(*Sam. I.* 40.)
Ca. pallida. *Fabr. E. S.* i. *a.* 217.—*Marsh.* i. 368.
β; Te. femoralis. *Ziegler.*

1330. 13; fuscicornis*. *Oliv. E.* ii. 11. *pl.* 1. *f.* 4. *a, b.*
Ca. melanocephala. *Panz. F.* xxxix. *f.* 12.
Ca. flavicollis. *Marsh.* i. 375.

1331. 14; melanurus*. *Oliv. E.* ii. 8. *pl.* 3. *f.* 21.—(*Sam. I.* 40.)
Ca. melanura. *Fabr. E.* i. 302.—*Stew.* ii. 70.—*Marsh.* i. 368. —*Turt.* i. 368.
Te. bimaculata. *DeGeer.* iv. 71.

1332. 15; pilosus*. (*DeJean C.* 37.)
Ca. pilosa. *Payk. F.* i. 264.
Te. longicornis *mihi.*

B. Thoracis lateribus manifestè rotundatis.

a. *Thorace anticè rotundato.*

1. Antennæ articulo secundo tertio vix longiore.

1333. 16; clypeatus.
Ca. clypeata. *Illig. K.* i. 299.
Ca. nivea. *Panz. F.* lvii. *f.* 5.
Ca. testacea. *Scop. C.* 123.
Ca. discoidalis. *Kirby MSS.*

1334. 17; dispar*.
Ca. dispar. *Fabr. E.* i. 295.
Ca. livida var. a. *Illig. K.* i. 296.
Ca. rufipes. *Herbst A.* v. 107.
Te. opacus. *Müll. Z. D. pr.* 61.

1335. 18; nigricans*. *Müll. Z. D. pr.* 61.
Ca. nigricans. *Fabr. E.* i. 296.
Ca. obscura. *Linn. F.* 706.—*Marsh.* i. 365. *desc. Faun. Suec.*
Ca. obscura var. *Payk. F.* i. 262.

1336. [19; discoideus*.] *mihi.*
Ca. nigricans var. *Gyll.* i. 334.

1337. 20; obscurus*? *Oliv. E.* ii. 8. *pl.* 2. *f.* 10. *a, b.*—(*Sam. I.* 40.)—(*Curtis Cat. no.* 45.)
Ca. obscura. *Linn.* ii. 648.—*Marsh.* i. 365. *desc. Syst. Nat.*—*Turt.* ii. 366.

2. Antennæ articulo secundo tertio breviore.

1338. 21, tricolor. *Marsh. MSS.*
Ca. abdominalis. *Panz. F.* lxxxiv. *f.* 5.—(*Curtis Cat. no.* 45.)

1339 † 22. violaceus[a].

1340. 23; pellucidus*.
Ca. pellucida. *Fabr. E.* i. 296.

1341. [24; cantianus*] *mihi.*

1342. 25, fuscus*. *DeGeer.* iv. 60. *pl.* 2. *f.* 5—15.—*Leach E. E.* ix. 86.—*Samou.* 164.
Ca. fuscus. *Linn.* ii. 647.—*Barb. G. pl.* 6. *f.* 3.—*Berk. S.* i. 103. —*Stew.* ii. 69.—*Marsh.* i. 365.—*Turt.* ii. 365.—*Samou.* 56.
Ca. antica. *Illiger.*

1343. 26; rusticus*.
Ca. rustica. *Gyll.* i. 330.—*Fallen. M. C.* i. 9.
Te. fuscus. *Oliv. E.* ii. 6. *pl.* 1. *f.* 1. *a*—*c.*—*Mart. C. pl.* 29. *f.* 13. 14?
Ca. femoralis. *Kirby MSS.*

1344. 27; lividus*. *Oliv. E.* ii. 7. *pl.* 2. *f.* 8.—(*Sam. I.* 40. *pl.* 3. *f.* 4.)—(*Kirby & Sp. I. E.* ii. 312.)
Ca. livida. *Linn.* ii. 647.—*Barb. G. pl.* 6. *f.* 8.—*Berk. S.* ii. 103. —*Mart. C. pl.* 29. *f.* 7?—*Stew.* ii. 69.—*Marsh.* i. 366.—*Turt.* ii. 365.
Te. flavus. *DeGeer.* iv. 70.
Ca. dispar ♂. *Payk. F.* i. 259.

1345. 28; confinis* *mihi.*
Te. nigricornis. *Kirby MSS.?*

---

[a] 1339 † 22. violaceus. *Wilk. Catal.* (!)
Ca. violacea. *Payk. F.* i. 260.

1346. 29; lituratus*.
Ca. liturata. *Gyll.* i. 348.
Ca. assimilis var. b. *Payk. F.* i. 262.
Ca. nigriventris. *Kirby MSS.*

1347. 30; analis*.
Ca. analis. *Fabr. E.* i. 295.

1348. 31; rufus*. *Müll. Z. D. pr.* 62.—(*Sam. I.* 40.)
Ca. rufa. *Linn.* ii. 647.—*Ent. Trans.* (*Burrell.*) i. 197.
Ca. livida. *Herbst A.* v. 107.—*Panz. F.* lvii. *f.* 3.

1349. 32; bicolor*.
Ca. bicolor. *Fabr. E.* i. 303.—*Panz. F.* xxxix. *f.* 13.

b. *Thorace anticè emarginato.* (*Antennæ articulis 2 et 3 brevibus.*)

1350. 33; Alpinus*.
Ca. alpinus. *Payk. F.* i. 259.
Te. angulatus. (*Wilk. Catal.?*)

Genus 229. MALTHINUS, *Latreille, Leach, Samou.*

NECYDALIS *p, Geoff.*—TELEPHORUS *p, Oliv.*—APOTOMA, *Kirby MSS.*

A. Capite rhomboidali: oculis prominulis.

1351. 1; flavus*. *Latr.* G. i. 262.—*Leach. E. E.* ix. 86.—*Samou.* 164.
Ca. flaveola. *Payk. F.* iii. *app.* 446.
Te. minimus. *Oliv. E.* ii. *pl.* 1. *f.* 6. *a.*

1352. [2; immunis*.] (*Sam. I.* 26.)
Ca. immunis. *Marsh.* i. 374.

1353. 3; humeralis*. (*Schon. S.* ii. 74.)—(*Sam. I.* 26.)
Ca. humeralis. *Marsh.* i. 374.

1354. [4; luteola.] *Kirby MSS.*

1355. 5; fasciatus*. (*Schon. S.* ii. 74.)
Te. fasciatus. *Oliv. E.* ii. 18. *pl.* 4. *f.* 14. *a, b.*

1356. 6; collaris. *Latr.* G. i. 262.—*Leach E. E.* ix. 87.

1357. 7; biguttulus*. (*Schon. S.* ii. 74.)
Ma. longicornis. *Kirby MSS.*
Ca. biguttula. *Payk. F.* iii. *App.* 445.
Te. biguttatus. *Oliv. E.* ii. 26. *pl.* 2. *f.* 12. *a, b.*

1358. 8; frontalis*.
Ca. frontalis. *Marsh.* i. 363.

1359. [9; immaculatus*.] *Kirby MSS.*
Ca. biguttulus b. *Gyll.* i. 342.

1360. 10; Pinicola* *mihi.*

B. Capite rotundato.

a. *Thorace elongato.*

1361. 11; biguttatus*. (*Schon. S.* ii. 74.)
Ca. biguttata. *Linn.* ii. 648.—*Berk. S.* i. 103.—*Mart. C. pl.* 29. *f.* 5.—*Stew.* ii. 69.—*Marsh.* i. 372.—*Turt.* ii. 369.

1362. 12; nigricollis*. *Kirby MSS.*

1363. 13; sanguinicollis*. (*Schon. S.* ii. 75.)
Ca. sanguinicollis. *Fallen. M. C.* 15.
Ca. minima var. *Payk. F.* i. 268.
Ma. ruficollis. *Latr. G.* ii. 261?—*Leach E. E.* ix. 86.

1364. 14, fulvicollis*. *Kirby MSS.*

1365. 15; minimus*.
Ca. minima. *Linn.* ii. 649.—*Berk. S.* i. 103.—*Stew.* ii. 69.—*Turt.* ii. 369.
Ma. marginatus. *Latr. G.* i. 261.—*Leach E. E.* ix. 86.

1366. 16; cognatus* *mihi.*
Ca. minimus. *Marsh.* i. 373.

1367. 17, melanocephalus*.
Ca. melanocephalus. *Marsh.* i. 374.

1368. 18; concolor*. *Kirby MSS.*
Ca. minima b. *Gyll.* i. 344?
Ma. marginatus β. (*Schon. S.* ii. 75?)

b. *Thorace transverso.*

1369. 19, brevicollis*. (*Schon. S.* ii. 75.)
Ca. brevicollis. *Payk. F.* i. 269.

Familia XXXVIII. MELYRIDÆ.

(MELYRIDES *p*, *Leach.*)

Genus 230. MALACHIUS, *Fabr.*, *Leach*, *Samou.*, *Millard*, *Curtis.*

CANTHARIS *p*, *Linn.*, *Marsh.*—CICINDELA *p*, *Geoff.*—TELEPHORUS *p*, *DeGeer.*

1370. 1; æneus*. *Fabr. E.* i. 306.—*Leach E. E.* ix. 87.—*Samou.* 165.—(*Curtis l. c. infra.*)
Ca. ænea. *Linn.* ii. 648.—*Mart. C. pl.* 29. *f.* 4.—*Berk. S.* i. 103.—*Stew.* ii. 70.—*Don.* iii. *pl.* 96. *f.* 2.—*Marsh.* i. 369.—*Turt.* ii. 372.—*Wood.* i. 55. *pl.* 20.
Donacia Ranunculorum. *Voet.* (*Panz.*) ii. 129. *pl.* 46. *f.* 7.

1371. 2; bipustulatus*. *Fabr. E.* i. 306.—(*Sam. I.* 26. *pl.* 3. *f.* 5.)—(*Curtis l. c. infra.*)—(*Kirby & Sp. I. E.* ii. 323.)
Ca. bipustulatus. *Linn.* ii. 648.—*Mart. C. pl.* 29. *f.* 1.—*Stew.* ii. 70.—*Marsh.* i. 369.—*Turt.* ii. 372.—*Don.* xv. *pl.* 528. *f.* 2.

Ma. biguttatus. (*Sam. I.* 26.)
Ca. biguttata. *Samou.* 56.
Donacia Asparagorum. *Voet.* (*Panz.*) iii. 128. *pl.* 46. *f.* 6.

1372. 3; viridis*. *Fabr. E.* i. 307.—*Oliv. E.* ii. 7. *pl.* 3. *f.* 11. *a, b.* —(*Curtis l. c. infra.*)
Ma. bipustulatus, β. *Illig. K.* i. 303.
Ca. cyanea. *Thunb. M. U.* iv. 55.
β, Ma. elegans. *Fabr. E.* i. 307.—*Oliv. E.* ii. 6. *pl.* 3. *f.* 12. *a, b.*

1373. 4, marginellus. *Fabr. E.* i. 307.—*Oliv. E.* ii. 6. *pl.* 3. *f.* 18. *a, b.*—(*Curtis l. c. infra.*)
Ma. bipustulatus γ. *Illig. K.* i. 303.
♂, Ma. bispinosus. *Curtis MSS.*—*Curtis.* iv. *pl.* 167.

1374. 5: sanguinolentus*. *Fabr. E.* i. 307.—*Oliv. E.* ii. 7. *pl.* 3. *f.* 13. *a, b.*—(*Sam. I.* 26.)—(*Curtis l. c. supra.*)
Ca. sanguinolenta. *Marsh.* i. 370.
Ca. coccinea. *Gmel.* iv. 1894.
Ma. erythromelas. *Gmel.* iv. 1899.
Ma. rufus. *Herbst A.* 108.

1375 † 6, ruficollis*. *Fabr. E.* i. 307?—*Panz. F.* ii. *pl.* 8.—(*Ing. Inst.* 88.)—(*Curtis l. c. supra.*) In Mus. D. *Chant*, *Dale*, *et Stone.*

1376 † 7, thoracicus*. *Fabr. E.* i. 308.—*Oliv. E.* ii. 9. *pl.* 2. *f.* 10. *a, b.*—(*Curtis l. c. supra.*) In Mus. *Brit.*

1377. 8; rubricollis*. *Gyll.* i. 362.—(*Curtis l. c. supra.*)
Ca. rubricollis. *Marsh.* i. 371.
Ma. ruficollis. *Oliv. E.* ii. 9. *pl.* 2. *f.* 9. *a, b.*—(*Sam. I.* 26.)
Ma. pulicarius β. *Illig. K.* i. 306.
♂, Ma. obliqua. *Kirby MSS.*

1378. 9, pulicarius*. *Fabr. E.* i. 308—*Panz. F.* x. *f.* 4. ♂.—(*Curtis l. c. supra.*)
Ca. nemoralis. *Gmel.* iv. 1898.

1379 † 10. pedicularius[a].

1380. 11; productus*. *Oliv. E.* ii. 13. *pl.* 3. *f.* 17. *a, b.* ♂.
Ma. præustus. *Fabr. E.* i. 308.
♀; Ma. flavipes var. *Fabr. E. S.* i. 225.
Ma. apicalis. *Leach MSS.*—(*Curtis l. c. supra.*)

1381. 12; fasciatus*. *Fabr. E.* i. 309.—*Panz. F.* x. *f.* 5.—(*Sam. I.* 26.)—(*Curtis l. c. supra.*)
Ca. fasciata. *Linn.* ii. 648.—*Berk. S.* i. 103.—*Stew.* ii. 70.—*Marsh.* i. 371.—*Turt.* ii. 374.—*Don.* xv. *pl.* 528. *f.* 1.

1382. [13; bituberculatus*] *mihi.*—(*Curtis l. c. supra.*)
Ma. fasciatus b. *Gyll.* i. 361.

---

[a] 1379 † 10. pedicularius. *Oliv. E.* ii. 8. *pl.* 1. *f.* 3. *a, b.*—*Turt.* ii. 373. (!)
♂, Ca. Cardiacæ. *Linn.* ii. 649?

1383 † 14, equestris. *Fabr. E.* i. 309.—*Panz. F.* x. *f.* 6.
Ma. bipunctatus. *Herbst A.* 108.
Ca. Herbstii. *Gmel.* iv. 1899.
Melöe Govani. *Gmel.* iv. 2021.
Ma. fasciatus γ. (*Schon. S.* ii. 82.) In Mus. *Brit.*

1384 † 15, humeralis. *Leach MSS.*—(*Curtis l. c. supra.*) In Mus. *Brit.*

Genus 231. ELICOPIS, *Besser.*

DASYTES *p, Fabr.*—LAGRIA *p, Fabr.?*—MELYRIS *p, Illiger.*

1385. 1; impressus *. (*Ing. Inst.* 88.)
Cr. impressus. *Marsh.* i. 226.
Da. nigricornis. *Payk. F.* ii. 158?

1386 † 2, femoralis? *Illig. M.* vi. 302.
Me. nigricornis var. *Illig. M.* i. 82. *nota.* In Mus. *Brit.*

1387 ‡ 3, quadripustulatus? (*Schon. S.* iii. 12.)
Hi. quadripustulatus. *Fabr. E.* ii. 59.
Me. quadrimaculatus. *Oliv. E.* ii. 10. *pl.* 1. *f.* 2. *a, b.* In Mus. *Brit.*

1388. [‡ 4, apicalis *mihi.*] In Mus. *Brit.*

Genus 232. DASYTES, *Paykul, Leach, Samou.*

DERMESTES *p, Linné.*—CICINDELA *p, Geoff.*—MELYRIS *p, Oliv.*—LAGRIA *p, Rossi.*—TILLUS *p, Marsh.*—CANTHARIS *p, Scopoli.*—TELEPHORUS *et* CLERUS *p, DeGeer.*—ANOBIUM *p, Fabr.*

1389 † 1, niger *. *Fabr. E.* ii. 72.—*Panz. F.* xcvi. *f.* 9.
De. niger. *Linn.* ii. 564.
Ca. nigra. *Fabr. E.* i. 303.—*Turt.* ii. 368.
Me. villosus. *Oliv. E.* ii. 9. *pl.* 2. *f.* 10. *a, b.* In Mus. *Brit.*

1390 † 2, ater *. *Fabr. E.* ii. 71.—(*Leach E. E.* ix. 87.—*Samou.* 164.—(*Kirby & Sp. I. E.* iii. 690.)
De. hirtus. *Linn.* ii. 563?
Me. ater. *Oliv. E.* ii. 9. *pl.* 2. *f.* 8. *a—e.*
Ca. pilosa. *Scop. C.* 131. In Mus. *Brit.*

1391. 3, cæruleus. *Fabr. E.* ii. 73.—*Panz. F.* xcvi. *f.* 10.—(*Sam. I.* 14.)
Pt. cyaneus. *Gmel.* iv. 1605.

1392. 4, viridis. (*Sam. I.* 14.)
La. viridis. *Rossi M.* i. 35.
Me. cyaneus. *Oliv. E.* ii. 8. *pl.* 2. *f.* 9. *a—d.*
Da. nobilis. *Illig. K.* i. 309. *nota ad Mel. cæruleum, teste Schonh.*

1393. 5; æratus * *mihi.*

Ti. æneus. *Marsh.* i. 230.
Da. æneus. (*Sam. I.* 14.)

1394. [6; serricornis*.] *Kirby MSS.*
Da. fusculus. (*Schon. S.* iii. 14)?—*Gyll.* iv. 336?

1395. 7; flavipes*. *Fabr. E.* ii. 73.—(*Sam. I.* 14.)
Me. flavipes. *Oliv. E.* ii. 12. *pl.* 3. *f.* 16. *a, b.*
Me. plumbea. *Illig. K.* i. 310.
Ti. virens. *Marsh.* i. 230.

1396 † 8, linearis. *Fabr. E.* ii. 73.
Ti. filiformis. *Creutz. E. V.* 121. *pl.* 3. *f.* 25. *a.*
In Mus. D. *Kirby.*

Genus 233. DRILUS, *Olivier, Leach, Samou.*

PTILINUS *p, Geoffroy.*—CANTHARIS *p, Marsh.*—HISPA *p, Rossi.* —CRIOCERIS *p, Thunb.*—COCHLEOCTONUS, *Mielzinsky.*

1397. 1; flavescens*. *Oliv. E.* ii. 4. *pl.* 1. *f.* 1. *a. e.*—*Leach E. E.* ix. 86.—*Samou.* 163.—(*Kirby & Sp. I. E.* iv. 479 *note.*)
Ca. serraticornis. *Marsh.* i. 374.
Ca. parisina. *Act. Ups.* (*Thunb.*) v. 93.
† ♀. Co. vorax. *Bull. Sci. Nat.* (*Meilz.*) 1824. iii. 297.

Familia XXXIX. TILLIDÆ, *Leach.*

(CLERII, *Latreille.*)—(TILLUS *et* CLERUS, *Marsh.*)

Genus 234. TILLUS, *Olivier, Leach, Samou., Millard.*

CHRYSOMELA *p, Linn.*—CLERUS *p, Fabr.*—LAGRIA *p, Panz.*—CRYPTOCEPHALUS *p, Gmelin.*

A. Thorace cylindrico.

1398. 1; elongatus*. *Oliv. E.* ii. 4. *pl.* 1. *f.* 1. *a–e.*—*Marsh.* i. 229.—*Samou.* 165.—*Leach E. E.* ix. 88.
Chr. elongata. *Linn.* ii. 603.
Cry. elongatus. *Stew.* ii. 52.—*Turt.* ii. 198.
La. ruficollis. *Herbst A.* 68. *pl.* 23. *f.* 35.

1399. 2, ambulans*. *Fabr. E.* i. 282.—*Marsh.* i. 230.—(*Ing. Inst.* 89.)
Ti. elongatus var. *Payk. F.* i. 154.—*Leach E. E.* ix. 88.
La. atra. *Panz. F.* viii. *f.* 9.
Cryp. Marchiæ. *Gmel.* iv. 1731.
β, Ti. bimaculatus. *Don.* xii. *pl.* 411. *f.* 2.

B. Thorace subcordato. (ATTELABUS *p, Gmelin.*)

1400. 3; unifasciatus*. *Latr. H.* ix. 145.—*Marsh.* i. 231.—*Leach E. E.* ix. 88.—*Samou.* 165.
Cl. unifasciatus. *Oliv. E.* iv. 17. *pl.* 2. *f.* 21. *a, b, c.*

Genus 235. OPILUS, *Latreille, Leach, Samou.*

ATTELABUS *p, Linné.*—NOTOXUS *p, Fabricius.*—EUPOCUS, *Illiger.*—OPILO? (*Kirby.*)

1401. 1; mollis*. *Latr. H.* ix. 149. *pl.* 77. *f.* 2, 3.—*Leach E. E.* ix. 88.—*Samou.* 166. *pl.* 12. *f.* 1.
At. mollis. *Linn.* ii. 621.—*Mart. C. pl.* 23. *f.* 7.
Cl. mollis. *Illig. K.* i. 285.—*Marsh.* i. 322.—*Don.* xii. *pl.* 411. *f.* 1.
Cl. fuscofasciatus. *DeGeer.* v. 159. *pl.* 5. *f.* 6.
No. mollis. *Fabr. E.* i. 287.—*Stew.* ii. 62.—*Turt.* ii. 286.

1402. 2, fasciatus*. *Wilk.? MSS.*

Genus 236. THANASIMUS, *Latreille, Leach, Samou.,* (*Kirby.*)

ATTELABUS *p, Linn.*—CLERIOIDES, *Schæffer.*

1403. 1; formicarius*. *Latr. G.* i. 270.—*Leach E. E.* ix. 88.—*Samou.* 165.
Cl. formicarius. *Fabr. E.* i. 280.—*Marsh.* i. 321.
At. formicarius. *Linn.* ii. 620.—*Mart. C. pl.* 23. *f.* 8.—*Berk. S.* i. 100.—*Don.* vii. *pl.* 231. *f.* 2.—*Stew.* ii. 62.—*Turt.* ii. 285.

Genus 237. CLERUS, *Geoffroy, Leach, Curtis,* (*Kirby.*)

TRICHODES, *Herbst.*

1404. 1: apiarius. *Fabr. E. S.* i. 208.—*Leach E. E.* ix. 88.—(*Curtis l. c. infra.*)
At. apiarius. *Linn.* ii. 620.—*Berk. S.* i. 101.—*Don.* vii. *pl.* 231. *f.* 1.—*Stew.* ii. 62.—*Turt.* ii. 286.

1405 ‡ 2, alvearius*. *Fabr. E. S.* i. 209.—*Curtis.* i. *pl.* 44.—(*Ing. Inst.* 88.)
Tr. apiarius var. *Herbst C.* iv. 156.

Genus 238. NECROBIA, *Olivier, Leach, Samou.,* (*Kirby.*)

DERMESTES *p, Linné.*—CORYNETES *p, Paykul, Samou.,* (*Kirby.*) —TILLUS *p, Samou.*

1406. 1; violaçea*. *Oliv. E.* iv. 5. *pl.* 1. *f.* 1. *a, b, c.*—(*Sam. I.* 28.)
Cl. quadra. *Marsh.* i. 323.
Ti. quadra. (*Sam. I.* 41.)

1407. 2; ruficollis*. *Oliv. E.* iv. 6. *pl.* 1. *f.* 3. *a, b.*—*Leach E. E.* ix. 88.—*Samou.* 166.
Cl. ruficollis. *Marsh.* i. 324.
Co. ruficollis. (*Sam. I.* 13.)—(*Kirby & Sp. I. E.* i. 255.)

1408. 3; rufipes*. *Oliv. E.* iv. 5. *pl.* 1. *f.* 2. *a, b.*—(*Sam. I.* 28.)
De. rufipes. *Fabr. E. S.* i. 230.

Genus 239. CORYNETES, *Paykul,* (*Samou.*)—(*Kirby.*)

DERMESTES *p, Linné.*—(Fam. XYLOPHAGI *p, Gyllenhal.*)

1409. 1; violaceus. *Payk. F.* i. 275.

De. violaceus. *Linn.* ii. 563.—*Mart. C. pl.* 6. *f.* 7?—*Stew.* ii. 32.—*Turt.* ii. 80.—*Wood.* i. 12. *pl.* 3.
Cl. violacea. *Marsh.* i. 323.
Cl. cæruleus. *DeGeer.* v. 163. *pl.* 5. *f.* 13.
At. Geoffroyanus. *Laich. T.* i. 247.

Familia XL. PTINIDÆ, *Leach.*

(PTINIORES *et* XYLOPHAGI *p*, *Latreille.*—CISSIDÆ *p*, *Leach.*)

(PTINUS, *Linné*, *&c.*)

Genus 240. XILETINUS, *Latreille*, *Leach.*

SERROCERUS, *Kugellan.*—PTILINUS *p*, *Fabr.*—ANOBIUM *p*, *Illiger.*—DERMESTES *p*, *Marsh.*—BRUCHUS *p*, *Quensel.*—ORION, (*Megerle.*)

1410. 1, striatus?
Le. striatus. *Schneid. M.* (*Kugell.*) i. 486?
Pt. pectinatus. *Fabr. E.* i. 329.—*Panz. F.* vi. *f.* 9.
Bru. rufipes. *Quensel. Diss.* 11.
Pti. nigricornis. *Kirby MSS.*

1411. [2; ater*.] (*DeJean C.* 40.)
Pt. ater. *Panz. F.* xxxv. *f.* 9.
An. serraticorne. *Rossi F.* i. 42.
Pt. pectinatus b. *Gyll.* i. 302.
De. rufipes. *Marsh.* i. 62.
Pt. serratus. *Fabr. E.* i. 330.
Or. laticollis. *Megerle.*

Genus 241. PTILINUS, *Fabr.*, *Leach*, *Samou.*, (*Kirby.*)

DERMESTES *p*, *Linné.*—ANOBIUM *p*, *Illig.*—LIGNIPERDA *p*, *Herbst.*—HISPA *p*, *Gmel.*, *Turt.*—BOSTRICHUS *p*, *Laich.*

1412. 1; pectinicornis*. *Fabr. E.* i. 329.—*Leach E. E.* ix. 93. —*Samou.* 181.
Pt. cylindricus. *Müll. Z. D. pr.* 81.
Ptin. pectinicornis. *Linn.* ii. 565.—*Berk. S.* i. 90.—*Don.* x. *pl.* 326.—*Marsh.* i. 81.—*Shaw G. Z.* vi. 37. *pl.* 9.
Hi. pectinicornis. *Stew.* ii. 52.—*Turt.* ii. 200.
♀; Pt. serraticornis. *Marsh.* i. 81.
Pt. fuscus. *Fourc.* i. 4.
Bo. pectinatus. *Laich. Ty.* ii. 68.
β? Pt. costatus. *Gyll.* iv. 329.

Genus 242. PTINUS *Auctorum.* *Chair-eater.*

BRUCHUS *p*, *Geoffroy.*—CERAMBYX *p*, *Linn. F.*

1413. 1; imperialis*. *Linn.* ii. 565.—*Panz. F.* v. *f.* 7.—*Marsh.* i. 88.—*Turt.* ii. 92.—*Samou.* 49.
Br. cruciatus. *Fourc.* i. 58.

1414. 2; Germanus*. *Oliv. E.* ii. 7. *pl.* 1. *f.* 6. *a, b.*—*Marsh.* i. 89.—(*Sam. I.* 35.)
Pt. imperialis. *Samou. pl.* 1. *f.* 6?
1415. 3; rufipes*. *Fabr. E.* i. 325.—*Oliv. E.* ii. 8. *pl.* 2. *f.* 8. *a. b.*—(*Sam. I.* 35.)
Pt. germanus. *Payk. F.* i. 312.—*Leach E. E.* ix. 93.
♀; Pt. elegans. *Fabr. E.* i. 325.
Pt. bimaculatus. *Kirby MSS.*

1416. 4; Museorum*. *Leach MSS.*—(*Sam. I.* 35.)
Pt. sexpunctatus. *Panz. F.* i. *f.* 20?

1417. 5; Fur*. *Linn.* ii. 566.—*Berk. S.* i. 190.—*Stew.* ii. 34.—*Marsh.* i. 89.—*Turt.* ii. 92.—*Shaw* G. *Z.* vi. 36. *pl.* 9.—*Don.* xii. 79. *pl.* 422.—*Leach E. E.* ix. 93.—*Samou.* 180.—*Wood.* i. 14. *pl.* 4.—(*Kirby & Sp. I. E.* i. 238.)
Br. furunculus. *Müll. Z. D. pr.* 57.
Pt. rapax. *DeGeer.* iv. 231. *pl.* 9. *f.* 1—7.
♂; Pt. testaceus. *Marsh.* i. 91.
Pt. latro. *Fabr. E.* i. 326.—*Stew.* ii. 34.
Pt. clavipes. *Panz. F.* xcix. *f.* 4.
Pt. longipes. *Rossi M.* i. 20.

1418. 6; crenatus*. *Fabr. E.* i. 326.—(*Sam. I.* 35.)
Pt. ovatus. *Marsh.* i. 90.

1419. [7; Cerevisiæ*.] *Marsh.* i. 90.—(*Sam. I.* 35.)
Pt. minutus. *Illig. K.* i. 347.
Pt. crenatus alter sexus?

1420. 8; Lichenum*. *Marsh.* i. 89.—(*Sam. I.* 35.)

1421. [9; similis*.] *Marsh.* i. 90.
Pt. bidens. *Oliv. E.* ii. 8. *pl.* 2. *f.* 10. *a, b*?
Pt. Lichenum ♂?

Genus 243. MEZIUM, *Leach*, (*Samou.*)

GIBBIUM *p*, (*Samou.*)

1422. 1; sulcatum*. *Leach MSS.*—(*Samou.* 180.)
Pt. sulcatus. *Fabr. E.* i. 327.—*Marsh.* i. 91.
Gi. sulcatus. (*Sam. I.* 19.)

Genus 244. GIBBIUM, *Kugellan, Leach, Samou.*, (*Kirby.*)

SCOTIAS, *Czemspinski.*

1423 ‡ 1, Scotias*. *Schneid. M.* (*Kugellan.*) iv. 502.—*Leach E. E.* ix. 93.—(*Sam. I.* 19.)
Pt. Scotias. *Fabr. E.* i. 327.—*Panz. F.* v. *f.* 8.—*Turt.* ii. 93.
Pt. seminulum. *Schra. A.* 36.
Pt. apterus. *Gmel.* iv. 1608.
Sco. psylloides. *Czemspinski.*
In Mus. *Brit. D. Haworth et* Hope.

Genus 245. DORCATOMA, *Herbst, Leach*, (*Kirby.*)

ANOBIUM *p*, *Illig.*—DERMESTES *p*, *Panzer.*—SERROCERUS *p*, *Kugellan.*

1424. 1, Dresdensis. *Herbst C.* iv. 104. *pl.* 39. *f.* 8. *a*, *b.*—(*Leach E. E.* ix. 93.)
De. bistriatus. *Payk. F.* ii. 318.
An. Dorcatoma. *Illig. K.* i. 334.
De. Serra. *Panz. F.* xxvi. *f.* 10.
Se. Serra. *Marsh. MSS.*
Se. glaber. *Schneid. M.* (*Kugellan.*) iv. 486.

1425. 2, Bovistæ. *Ent. Hefte.* ii. 100. *pl.* 3. *f.* 11.

Genus 246. ANOBIUM, *Fabricius, Leach, Samou.*, (*Kirby.*)

DERMESTES *p*, *Thunb.*—BYRRHUS *p*, *Geoff.*

A. Elytris striatis.

1426. 1; castaneum*. *Fabr. E.* i. 322.—*Oliv. E.* ii. 7. *pl.* 1. *f.* 2. *a*, *b.*—(*Sam. I.* 3.)
An. excavatum. *Schneid. M.* (*Kugell.*) iv. 488.
Pt. castaneus. *Marsh.* i. 83.
By. mollis. *Fourc.* i. 26.
Pt. ferrugineus. *Gmel.* iv. 1606.

1427. 2; rufipes*. *Fabr. E.* i. 322.—(*Sam. I.* 5.)
An. elongatum. *Payk. F.* i. 303.
Pt. rufipes. *Marsh.* i. 83.
β; Pt. cylindricus. *Marsh.* i. 83.
An. Juglandis. *Herbst C.* v. 61. *pl.* 47. *f.* 8. 9?

1428. 3; pertinax*. *Oliv. E.* ii. 6. *pl.* 1. *f.* 4. *a*, *b.*
Pt. pertinax. *Linn.* ii. 565.—*Berk. S.* i. 90.—*Marsh.* i. 82.
An. striatum. *Fabr. E.* i. 321.
Pt. rugosus. *Gmel.* iv. 1605.
Pt. rufus. *Gmel.* iv. 1606.

1429. [4; Fagi*.] *Herbst C.* v. 57. *pl.* 47. *f.* 4. c. C.

1430. 5; striatum*. *Oliv. E.* ii. 9. *pl.* 2. *f.* 7. *a*, *b.*—*Leach E. E.* ix. 93.—*Samou.* 181.
Pt. punctatus. *DeGeer.* iv. 230.
An. pertinax. *Fabr. E.* i. 322.—*Stew.* ii. 34.—*Turt.* ii. 90.
By. domesticus. *Fourc.* i. 26?

1431. 6; paniceum*. *Fabr. E.* i. 323.—*Panz. F.* lxvi. *f.* 6.
De. paniceus. *Linn.* ii. 564.
An. paniceum. (*Sam. I.* 3.)
An. ferrugineum. *Herbst A.* 27. *pl.* 20. *f.* 13.
Pt. testaceus. *Act. Ups.* (*Thunb.*) iv. 6.
Pt. Upsaliensis. *Gmel.* iv. 1608.

♂; Pt. tenuicornis. *Marsh.* i. 84.
♀; Pt. rubellus. *Marsh.* i. 85.—(*Kirby & Sp. I. E.* i. 386.)

B. Elytris vagè punctatis.

1432. 7; tessellatum*. *Fabr. E.* i. 321.—*Panz. F.* xlvi. *f.* 3.—*Leach E. E.* ix. 93.—*Samou.* 181.—(*Kirby & Sp. I. E.* i. 36.—*Id.* ii. 387.)
Pt. tessellatus. *Marsh.* i. 85.—*Turt.* ii. 90.—*Bingley.* iii. 124.
By. nigrofuscus. *Geoffr.* i. 112.
An. pulsatorium. *Scriba. Jo.* 156.
Pt. rufovillosus. *DeGeer.* iv. 230.
Pt. fatedicus. *Blumenb.*—*Shaw N. M.* iii. *pl.* 104.—*Shaw* G. Z. vi. 34. *pl.* 8.
Pt. pulsator. *Gmel.* iv. 1605.—*Turt.* ii. 91.
Pt. faber. *Gmel.* iv. 1608?
Pt. fuscus. *Gmel.* iv. 1606?
By. pertinax. *Fourc.* i. 26.
By. rubiginosus. *Müll. Z. D. pr.* 57.
Death-watch. *Shaw et Bingley l. c. supra.*

1433. 8; Abietis*. *Fabr. E.* i. 323.—*Panz. F.* lxvi. *f.* 7.—(*Sam. I.* 3.)
An. molle b. *Payk. F.* i. 307.
Pt. lævis. *Marsh.* i. 84.

1434. 9; molle*. *Fabr. E.* i. 323.—*Herbst C.* v. 61. *pl.* 47. *f.* 9. *h. H.*—*Leach E. E.* ix. 93.—(*Sam. I.* 3.)
Pt. mollis. *Linn.* ii. 565.—*Berk. S.* i. 90.—*Stew.* ii. 34.—*Marsh.* i. 84.—*Turt.* ii. 90.
Pt. testaceus. *Gmel.* iv. 1606.

1435 † 10, erythropum. *Leach MSS.* In Mus. *Brit.*

Genus 247. OCHINA? (*Ziegler?*)

ANOBIUM *p*, *Schonher.*—CRIOCERIS *p*, *Marsh.*

1436. 1; ptinoides*.
Cr. ptinoides. *Marsh.* i. 228.
An. ptinoides. (*Sam. I.* 3.)
β, Pt. bifasciatus. *Marsh. MSS.*

Genus 248. CHORAGUS, *Kirby.*—(Fam. CHORAGIDÆ, *Kirby.*)

1437 † 1; Sheppardi*. *Linn. Trans.* (*Kirby.*) xii. 448. *pl.* xxii. *f.* 14.—(*Kirby & Sp. I. E.* ii. 315.) In Mus. D. *Kirby et* Sheppard.

1438 † 2, niger. *Kirby MSS.* In Mus. D. *Kirby.*

Genus 249. CISS, *Latreille, Leach, Samou.*

ANOBIUM *p*, *Fabr.*—PTINUS *p*, *Marsh.*—DERMESTES *p*, *Scop.*—HYLESINUS *p*, *Fabr.*—BOSTRICHUS *p*, *Kugell.*—BOLITESTES, *Kirby MSS.* (*olim.*)

1439. 1; Boleti*. *Latr. G.* iii. 12.—(*Leach E. E.* ix. 110.—(*Sam. I.* 11.)

An. Boleti. *Fabr.* i. 323.
De. picipes. *Herbst C.* iv. 137.
An. bidentatum. *Oliv. E.* ii. 16. 11. *pl.* 2. *f.* 5. *a, b, c.*
De. variabilis. *Fabr. E.* i. 319.
Pt. Boletorum. *Marsh.* i. 85.
β? Ci. picipes. *Kirby MSS.*

1440. 2; concinnus*. (*Sam. I.* 11.)
Pt. concinnus. *Marsh.* i. 87.

1441. 3, flavus. *Kirby MSS.*

1442. 4; micans*. *Gyll.* iii. 379.
An. micans. *Fabr. E.* i. 324.—*Herbst C.* v. 64. *pl.* 47. *f.* 11. *K.*
Pt. villosulus. *Marsh.* i. 86.

1443. 5; hispidus*. *Gyll.* iii. 380.
An. hispidum. *Payk. F.* i. 310.
An. micans. *Panz. F.* x. *f.* 8.

1444. 6, pyrrhocephalus.
Pt. pyrrhocephalus. *Marsh.* i. 86.

1445. 7, pygmæus.
Pt. pygmæus. *Marsh.* i. 86.
An. festivum. *Panz. F.* vi. *f.* 7?

1446. 8, rhododactylus.
Pt. rhododactylus. *Marsh.* i. 87.

1447. 9; nigricornis*.
Pt. nigricornis. *Marsh.* i. 87.

1448. 10, ruficornis*.
Pt. ruficornis. *Marsh.* i. 87.
Ci. perforatus. *Gyll.* iii. 385.

1449. 11; nitidus*. *Gyll.* iii. 382.
An. nitidum. *Fabr. E.* i. 324.
β; Ci. nitidulus. *Kirby MSS.*
An. nitidum. *Panz. F.* x. *f.* 9.

1450. 12; fronticornis*. (*DeJean C.* 102.)
An.? fronticornis. *Frölich.*
Apate fronticornis. *Panz. F.* xcviii. *f.* 7?
Ci. emarginatus. *Kirby MSS.*?

1451. 13; bidentatus*. *Gyll.* iii. 383.—(*Sam. I.* 11.)
♂; Pt. bidentatus. *Marsh.* i. 86.
♀; Pt. inermis. *Marsh.* i. 87.

Familia XLI. BOSTRICIDÆ, *Leach.*

(Xylophagi *p*, *et* Curculionites-scolitarii, *Latreille.*)

Genus 250. APATE, *Fabricius.*

Dermestes *p*, *Linné.*—Bostrichus, *Leach.*

A. Elytris posticè dentatis.

1452. 1, muricata *. (*Ing. Inst.* 87.)
De. muricatus. *Linn.* ii. 562.
Sy. muricatum. *Fabr. E.* ii. 377.—*Panz. F.* xxxv. *f.* 7.
Bo. bispinosus. *Oliv. E.* iv. 77. *pl.* 1. *f.* 1.

B. Elytris posticè muticis. Ligniperda *p*, *Herbst.*

1453 ‡ 2, capuzina *. *Fabr. E.* ii. 381.—*Fanz. F.* xliii. *f.* 18.—*Ent. Trans.* (*Haworth.*) i. 81.
De. capucinus. *Linn.* ii. 562.
Bo. capucinus. *Childr. Address*, 7.
In Mus. D. *Haworth* et Sparshall.

Genus 251. BOSTRICHUS, *Olivier.*

Ips *p*, *DeGeer.*—Dermestes *p*, *Linné.*—Apate *p*, *Fabr.*

1454. 1, dispar. *Illig. M.* iv. 129.
♂, Bo. brevis. *Panz. F.* xxxiv. *f.* 20.
♀, Bo. thoracicus. *Panz. F.* xxxiv. *f.* 18.

1455. 2, domesticus. *Gyll.* iii. 365.
De. domesticus. *Linn.* ii. 563.—*Berk. S.* i. 89.—*Stew.* ii. 32.—*Turt.* ii. 85.
Ap. limbatus. *Fabr. E.* ii. 382.—*Panz. F.* xliii. *f.* 19.

Genus 252. TOMICUS, *Latreille, Leach, Samou.*, (*Kirby.*)

Dermestes *p*, *Linné.*—Ips *p*, *DeGeer.*—Scolytus *p*, *Olivier.*—Bostrichus *p*, *Fabr.*—Eccoptogaster *p*, *Gyllenhal.*

A. "Elytris posticè integris, rotundatis." *Gyllenhal.*

a. *Thorace anticè angusto.*

1456. 1; polygraphus *.
De. polygraphus. *Linn.* ii. 562.
Bo. polygraphus. *Panz. F.* xv. *f.* 5.
Ips polygraphus. *Marsh.* i. 52.

1457. 2; affinis * *mihi.*

b. *Thorace anticè capite latiore.*

1458. 3; villosus *.
Bo. villosus. *Payk. F.* iii. 154.—*Panz. F.* xv. *f.* 8.

β, Ips villosus. *Marsh.* i. 53.
De. micrographus. *Linn. F. no.* 419?

1459. 4; micrographus *.
Bo. micrographus. *Payk. F.* iii. 155.—*Panz. F.* lxvi. *f.* 11.—*Turt.* ii. 87.

B. "Elytris posticè retusis, plerumque dentatis." *Gyllenhal.*

a. *Elytris posticè muticis.*

1460. 5; fuscus *. (*Sam. I.* 41.)
Ips fuscus. *Marsh.* i. 53.
Bo. retusus. (*DeJean C.* 101.)

1461. 6; pubescens *.
Ips pubescens. *Marsh.* i. 58.
Bo. cinereus. *Herbst C.* v. 116. *pl.* 48. *f.* 15. O.—*Gyll.* iii. 370?
Bo. Pini. *Kirby MSS.*

b. *Elytris posticè dentatis.*

1462. 7; bidens *.
Bo. bidens. *Fabr. E.* ii. 389.—*Panz. F.* xxxix. *f.* 21.
Sc. bidentatus. *Oliv. E.* iv. 78. *pl.* 2. *f.* 13. *a, b.*
Ips micrographus. *Marsh.* i. 52.

1463. 8; Laricis *.
Bo. Laricis. *Fabr. E.* ii. 386.—*Panz. F.* xv. *f.* 3.
Sc. chalcographus. *Oliv. E.* iv. 78. *pl.* 1. *f.* 8. *a*—c.

1464. 9, monographus.
Bo. monographus. *Fabr. E.* ii. 387.
Bu. tuberculatus. *Herbst C.* v. 113. *pl.* 48. *f.* 12.

1465 † 10, flavus. *Wilk.? MSS.?*
Bo. octodentatus. *Payk. F.* iii. 146?
In Mus. D. *Curtis et Vigors.*

1466. 11, Typographus *? *Latr. G.* ii. 276.—(*Leach E. E.* ix. 109.)—(*Sam. I.* 41.)
Ips Typographus. *DeGeer.* v. 193. *pl.* 6. *f.* 1—7.—*Marsh.* i. 51.
Bo. Typographus. *Fabr. E.* ii. 385.—*Panz. F.* xv. *f.* 2.—*Stew.* ii. 33.—*Turt.* ii. 86.—(*Sam. I.* 7.)—(*Kirby & Sp. I. E.* i. 210.)

Genus 253. PLATYPUS, *Herbst, Leach, Samou., Curtis,* (*Kirby.*)
BOSTRICHUS *p, Fabr.*—SCOLYTUS *p, Oliv.*—CYLINDRA, *Dufts.*

1467. 1, cylindrus. *Herbst C.* v. 120. *pl.* 49. *f.* 3.—*Curtis* ii. *pl.* 51.
Sc. cylindricus. *Oliv. E.* iv. 78. *pl.* 1. *f.* 2. *a, b.*
Pl. cylindricus. (*Sam. I.* 34.)
Bo. cylindricus. (*Sam. I.* 7.)
Cy. platypus. *Dufts. F. A.* iii. 87.

Genus 254. HYLESINUS, *Fabr., Leach, Samou.,* (*Kirby.*)
SCOLYTUS *p, Olivier.*—BOSTRICHUS *p, Panz.*—IPS *p, Marsh.*

A. Corpore pubescente.

a. *Corpus subelongatum.*

1468. 1; crenatus*. *Fabr. E.* ii. 390.—(*Sam. I.* 22.)
Bo. crenatus. *Panz. F.* xv. *f.* 7.
Ips sulcatus. *Marsh.* i. 59.
Hylur. sulcatus. (*Ing. Inst.* 88.)

1469. 2; hæmorrhoidalis*.
Ips hæmorrhoidalis. *Marsh.* i. 56.
Bo. minutus. *Panz. F.* xv. *f.* 11?

1470. 3; picipennis* *mihi.*

b. *Corpus breve, latum.*

1471. 4; scaber*.
Ips scaber. *Marsh.* i. 56.

B. Corpore squamulis tecto.

1472. 5; Fraxini*. *Fabr. E.* ii. 390.
Bo. Fraxini. *Fanz. F.* lxvi. *f.* 15.
Ips varius. *Marsh.* i. 54.
Hy. varius. (*Sam. I.* 22.)
β; Sc. varius. *Oliv. E.* iv. 78. 11. *pl.* 2. *f.* 17. *a, b.*
Ips griseus. *Marsh.* i. 55.
γ; Ips rufescens. *Marsh.* i. 55.

1473. 6, furcatus.
Ips furcatus. *Marsh.* i. 55.

1474. 7; coadunatus*.
Ips coadunatus. *Marsh.* i. 58.
Ips maculosus. *Kirby MSS.*

1475. 8; sericeus*.
Ips sericeus. *Marsh.* i. 55.

Genus 255. SCOLYTUS, *Geoffroy, Leach, Samou.*, (*Kirby,*) *Curtis.*

HYLESINUS *p*, *Fabr.*, *MacLeay.*—ECKOPTOGASTER, *Herbst.*—COPTOGASTER, *Illig.*—IPS *p*, *Marsh.*—BOSTRICHUS *p*, *Payk.*

1476. 1; Destructor*. *Oliv. E.* iv. 78. 5. *pl.* 1. *f.* 4. *a, b, c.*—*Leach E. E.* ix. 109.—(*Sam. I.* 37. *pl.* 1. *f.* 5.)—*Curtis.* i. *pl.* 43.
Hy. scolytus. *Fabr. E.* ii. 390.
Bo. scolytus. *Stew.* ii. 33.—*Turt.* ii. 87.
De. scolytus. *Samou.* 48.
♂; Ips scolytus. *Marsh.* i. 53.
♀; Ips inermis. *Marsh.* i. 53.

1477. [2; picicolor*.] *Kirby MSS.*

1478. 3; pygmæus*. *Oliv. E.* iv. 78. 6. *pl.* 1. *f.* 5. *a. b.*
Ips multistriatus. *Marsh.* i. 54.
Sc. multistriatus. (*Sam. I.* 37.)

Genus 256. HYLURGUS, *Latr., Leach, Samou., (Kirby,) Curtis.*

HYLESINUS *p, Fabr.*—IPS *p, Marsh.*—BOSTRICHUS *p, Paykul.*—
DERMESTES *p, Linné.*

1479. 1; piniperda*. *Latr. G.* ii. 275. (!)—(*Leach E. E.* ix. 109.)
—(*Sam. I.* 23.)—*Curtis* iii. *pl.* 104.
De. piniperda. *Linn.* ii. 563.
Bo. piniperda. *Stew.* ii. 33.
Ips piniperda. *Marsh.* i. 57.
β; Hyle. testaceus. *Fabr. E.* ii. 393.
Ips fumatus? *DeGeer.* v. 195.

1480. 2; ater*. (*Sam. I.* 22.)—(*Curtis l. c. supra.*)
Hyle. ater. *Fabr. E.* ii. 394.
Bo. angustatus. *Schneid. M.* (*Kugell.*) v. 524.
Ips Boleti. *Marsh.* i. 59.
Hylu. Boleti. (*Curtis l. c. supra.*)
β; Ips niger. *Marsh.* i. 59.
Hylu. niger. (*Sam. I.* 23.)

1481. 3; palliatus*. (*DeJean C.* 100.)
Hyle. palliatus. *Gyll.* iii. 340.
Bo. ater, b. *Payk. F.* iii. 154.
Bo. angustata *var.* *Herbst C.* v. 111.

1482. 4; angustatus*. (*Curtis l. c. supra.*)
Bo. angustatus *var.* nigra. *Herbst C.* v. 111. *pl.* 48. *f.* 9. *h.*
Ips ater. *Marsh.* i. 59.
Hy. opacus. (*DeJean C.* 100.)

1483. 5; obscurus*. (*Sam. I.* 23.)—(*Curtis l. c. supra.*)
Ips obscurus. *Marsh.* i. 57.
β; Ips hispidus. *Kirby MSS.*

1484. 6; rufescens* *mihi.*

1485. 7; rufus*. (*Curtis l. c. supra.*)
Ips rufus. *Marsh.* i. 57.

1486. 8; piceus*. (*Curtis l. c. supra.*)
Ips piceus. *Marsh.* i. 58.

1487. 9, fuscescens *mihi.*

1488. 10; rhododactylus*. (*Curtis l. c. supra.*)—*Gyll.* iv. 619. (!)
Ips rhododactylus. *Marsh.* i. 58.
Bo. Ulicis. *Kirby MSS.*

SECTIO III.

(TETRAMERA *p*, *et* HETEROMERA *p*, *Latreille.*—HELMINTHOMORPHA? (larvæ vermiformes) *MacLeay.*)

Subsectio 1.

(RHINCHOPHORI, *Clairville.*—RHINCHOPHORA, *Latreille.*)

Familia XLII. CURCULIONIDÆ, *Leach.*

(CURCULIONITES *proprii*, *Latreille.*)

(CURCULIO, *Linné*, *&c.*—CURCULIO, RHYNCHÆNUS, *Fabr.*, *&c.*)

Weevils, Longsnouts.

Genus 257. RHYNCOLUS [a].

Genus 258. COSSONUS, *Clairville*, *Leach*, *Samou.*, *Curtis.*

1490. 1, Tardii. *Vigors' MSS.*—*Curtis.* ii. *pl.* 59.

1491. 2; linearis*. *Fabr. E.* ii. 496.—*Clairv. E. H.* i. 60. *pl.* 1. *f.* 1. 2.—(*Leach E. E.* ix. 109.)—(*Sam. I.* 13.)—(*Curtis l. c. supra.*)
Cu. linearis. *Panz. F.* xviii. *f.* 7.—*Marsh.* i. 271.
Cu. parallelipipedus. *Herbst C.* vi. 275.

1492. [3, elongatus*] *mihi.* (*Curtis l. c. supra.*)

Genus 259. CALANDRA, *Clairville*, *Leach*, *Samou.*, (*Kirby*).
RHYNCOPHORUS, *Herbst.*

1493. 1; granaria*. *Clairv. E. H.* i. 62. *pl.* 2. *f.* 1. 2.—(*Leach E. E.* ix. 109.)—(*Sam. I.* 8.)—(*Kirby & Sp. I. E.* i. 171. et iii. 84.)
Cu. granarius. *Linn.* ii. 608.—*Berk. S.* i. 97.—*Stew.* ii. 55.—*Marsh.* i. 274.—*Turt.* ii. 218.—*Shaw. G. Z.* vi. 64.—*Bingley.* iii. 132.

1494. [2, unicolor.]
Cu. unicolor. *Marsh.* i. 275.

1495. 3; *Oryzæ** [b].

1496. 4; *Frumentarius** [c].

---

[a] Genus 257. RHYNCOLUS, (*Germar.*), *Schön. Cu.* 332.
HYLESINUS *p*, *Fabricius.*—COSSONUS *p*, *Gyll.*

1489 † 1, ater.
Cu. ater. *Linn.* ii. 617.
Hyl. chloropus. *Fabr. E.* ii. 593.
Cu. chloropus. *Panz. F.* xix. *f.* 14.

[b] 1495. 3; *Oryzæ**. *Fabr. E.* ii. 438.
Cu. Oryzæ. *Marsh.* i. 275.
Cu. frugilegus. *DeGeer.* v. 273.
Rh. Oryzæ. *Herbst C.* vi. 18. *pl.* 60. *f.* 9.

[c] 1496. 4; *Frumentarius** *mihi.*

Genus 260. BARIS? (*Germar.*)

COSSONUS *p*, *Gyllenhal.*—CALANDRA *p*, *Samou.*—RHYNCOLUS *p*, *Schön.?*

1497. 1; lignarius*.
Cu. lignarius. *Marsh.* i. 275.
Co. lignaria. (*Sam. I.* 8.)

1498. [2; piceus*.]
Co. piceus. *Wilk. MSS.*
Cu. picinus. *Marsh. MSS.*
Co. æneopiceus. *Kirby MSS.*

1499. 3, truncorum.
Co. truncorum. *Schüppel.*—*Germ. I. N.* i. 308.

Genus 261. BARIDIUS, *Schön. Cu.*

BARIS *p*, (*Germar.*)—COSSONUS *p*, *Samou.*—CALANDRA *p*, *Fabr.* —MICHORHYNCHUS *et* STENORHYNCHUS, (*Megerle.*)—STENOSOMA, *Kirby MSS.*

1500. 1; Atriplicis*.
Cu. Atriplicis. *Payk. F.* iii. 243.
Cu. funereus. *Herbst C.* vi. 164. *pl.* 71. *f.* 3.
Cu. hypoleucos. *Marsh.* i. 274.
Co. hypoleucos. (*Sam. I.* 13.)
Cu. T. album. *Gyll.* iii. 79.

1501. 2; pilistriatus*. *Kirby MSS.*

1502. 3; Artemisiæ*. *Germ. I. N.* i. 199.—(*Schön. Cu.* 276.)
Rh. Artemisiæ. *Fabr. E.* ii. 456.
Cu. Artemisiæ. *Panz. F.* xviii. *f.* 10.
Cu. laticollis. *Marsh.* i. 276.

1503. 4; picicornis*.
Cu. picicornis. *Marsh.* i. 276.
Cu. cærulescens. *Sturm?*

1504. 5, impunctatus. *Kirby MSS.*

Genus 262. MECINUS, (*Germar.*), *Schön. Cu.* 321.

BALANINUS *p*, *Samou.*

1505. 1; semicylindricus*. (*Schön. Cu.* 322.)
Cu. semicylindricus. *Marsh.* i. 294.
Ba. semicylindricus. (*Sam. I.* 5.)
Cu. Pyrastri. *Herbst C.* vi. 252. *pl.* 78. *f.* 6.

1506. 2, circulatus*.
Cu. circulatus. *Marsh.* i. 274.
Me. marginatus. (*Germar?*)

1507. 3, hæmorrhoidalis*.
Cu. hæmorrhoidalis. *Herbst C.* vi. 226. *pl.* 80. *f.* 4.

Genus 263. GYMNAËTRON, *Schön. Cu.* 319.

CIONUS *p*, *Germ.*—CLEOPUS *p*, *DeJean.*

1508. 1, Beccabungæ. (*Schönher.*)
Cu. Beccabungæ. *Linn.* ii. 611.—*Berk. S.* i. 99.—*Stew.* ii. 58. —*Turt.* ii. 243.
Cu. grisescens. *Kirby MSS.*
β; Ci. Veronicæ. *Germ. Mag.* iv. 305.

Genus 264. RHINUSA, *Kirby MSS.*

CLEOPUS *p*, *DeJean.*—CIONUS *p*, *Germar.*—GYMNAËTRON 3 *p*, *Schön. Cu.*

1509. 1; Antirrhini*.
Cu. Antirrhini. *Payk. F.* iii. 257.—*Panz. F.* xxvi. *f.* 18.—*Marsh.* i. 264.
Cu. ellipticus. *Herbst C.* vi. 171. *pl.* 71. *f.* 13?

1510. 2; acephalus*.
Cu. acephalus. *Marsh.* i. 271.

1511. 3; intaminata*. *Kirby MSS.*

1512. 4; tricolor.
Cu. tricolor. *Marsh.* i. 259.
Cu. labilis. *Herbst C.* vi. 244. *pl.* 77. *f.* 12.
β; Naso hæmorrhous. *Kirby MSS.*

Genus 265. MIARUS, *Schön. Cu. olim.*

CLEOPUS *p*, *DeJean.*—CEUTORHYNCHUS *p*, *Germar.*—NASO, *Kirby MSS.*—GYMNAËTRON 2, *Schön.*

A. Femoribus dentatis.

1513. 1; Campanulæ*. (*Schön. Cu.* 320.)
Cu. Campanulæ. *Linn.* ii. 607.—*Turt.* ii. 215.—*Herbst C.* vi. 161. *pl.* 70. *f.* 15.
Cu. ellipticus. *Herbst C.* vi. 171. *pl.* 71. *f.* 13?

1514. 2, Nasturtii.
Cr. Nasturtii. *Spence MSS.*
Ceu. Nasturtii. *Germ. I. N.* i. 233.

B. Femoribus posticis solè dentatis.

1515. 3, Graminis. (*Schön. Cu.* 320.)
Rh. Graminis. *Gyll.* iii. 210.
Cu. Campanulæ ♀. *Payk. F.* iii. 212.

1516. 4, Linariæ.
Cu. Linariæ. *Panz. F.* xxvi. *f.* 18.

Genus 266. CIONUS, *Clairville, Leach, Samou.*, (*Kirby.*)

1517. 1; Scrophulariæ*. (*Leach. E. E.* ix.108.)—(*Sam. I.* ii. *pl.* 2. *f.* 21.)—(*Kirby & Sp. I. E.* ii. 274.)

Cu. Scrophulariæ. *Linn.* ii. 614.—*Berk. S.* i. 99.—*Mart. C. pl.* 19. *f.* 25.—*Don.* ii. *pl.* 60.—*Stew.* ii. 59.—*Marsh.* i. 276. —*Turt.* ii. 233.—*Samou.* 54.

1518. [2; Verbasci*.] (*DeJean C.* 83.)
Rh. Verbasci. *Fabr. E.* ii. 479.
Cu. Scrophulariæ var. *Linn.* ii. 614.

1519. 3, Thapsi. (*Sam. I.* 11.)
Cu. Thapsi. *Fabr. E.* ii. 479.—*Herbst C.* vi. 187. *pl.* 73. *f.* 2. —*Marsh.* i. 277.

1520. 4; Hortulanus*. (*Sam. I.* 11.)
Cu. Hortulanus. *Vill.* i. 202.—*Stew.* ii. 59.—*Marsh.* i. 278.

1521. 5; Blattariæ*. *Clairv. E. H.* i. 66. *pl.* 3. *f.* 1. 2.
Cu. bipustulatus. *Marsh.* i. 278.
Ci. bipustulatus. (*Sam. I.* 11.)
Cu. Fraxini. *DeGeer.* v. 212?

Genus 267. CLEOPUS, (*Megerle.*)
CIONUS *p, Schön., Samou.*

1522. 1; Solani*. (*DeJean C.* 83.)
Cu. Solani. *Payk. F.* iii. 457.
Cu. pulchellus. *Herbst C.* vi. 356. *pl.* 88. *f.* 1.
Cu. immunis. *Marsh.* i. 278.
Ci. immunis. (*Sam. I.* 11.)
Cu. Hortulanus. *Don.* vi. 62. *pl.* 205. *f.* 2.

1523. [2; rigidus*.] —— *MSS.?*

1524. [3, flavus.] —— *MSS.?*

Genus 268. SPHÆRULA, (*Megerle.*)
CIONUS *p, Clairv.*—OROBITIS *p, DeJean.*—CRYPTORHYNCHUS *p, Samou.*—NANODES [a] *Schön.*

1525. 1; Lythri*.
Cu. Lythri. *Payk. F.* iii. 263.—*Panz. F.* xvii. *f.* 8.—*Marsh.* i. 252.
Cr. Lythri. (*Sam. I.* 13.)
Cu. pygmæus. *Herbst C.* vi. 142. *pl.* 69. *f.* 7.
β; Rh. Salicariæ. *Fabr. E.* ii. 449.
Cu. Salicariæ. *Stew.* ii. 55.—*Panz. F.* xvii. *f.* 4.—*Marsh.* i. 251. —*Turt.* ii. 215.
γ; Or. varius. *Kirby MSS.*

Genus 269. OROBITIS, *Germar, Schön.*
SPHÆRULA *p, Megerle.*—ATTELABUS *p, Fabricius.*—CRYPTORHYNCHUS *p, Samou.*

1526. 1; cyaneus*.

---

[a] NANODES: Genus Avium. Vide *Shaw Gen. Zool. v.* xiv. *p.* 118.

Cu. cyaneus. *Linn.* ii. 606.—*Berk. S.* i. 97.—*Stew.* ii. 54.
At. globosus. *Fabr. E.* ii. 426.—*Fanz. F.* lvii. *f.* 10.
Cu. hypoleucos. *Quensel. D.* 16.
Sp. gigantea. *Megerle.*
Cr. globosus. (*Sam. I.* 13.)
Or. globosus. (*Ing. Inst.* 89.)

Genus 270. CEUTORHYNCHUS, (*Schüpp.*), *Schön.*

ENSIFER? (*Megerle.*)—CRYPTORHYNCHUS *p*, *Samou.*—NEDYUS *p*, *Schön.*

A. Thorax non tuberculatus, nec anticè reflexus.

1527. 1; Quercus*.
Cu. Quercus. *Herbst C.* vi. 412. *pl.* 92. *f.* 7.
Cu. pallens. *Marsh.* i. 251.

1528. [2; melanorhynchus*.]
Cu. melanorhynchus. *Marsh.* i. 250.
Cr. melanorhynchus. (*Sam. I.* 13.)
Rh. Quercus, c. *Gyll.* iii. 138.

1529. [3; ruber*.]
Cu. ruber. *Marsh.* i. 251.—*Don.* xi. *pl.* 389. *f.* 1.
Cr. ruber. (*Sam. I.* 13.)
Rh. Quercus, b. *Gyll.* iii. 138.

1530. 4; inermis*.
Cu. inermis. ——? *MSS.*

1531. 5; rufirostris*.
Cu. rufirostris. ——? *MSS.*

1532. 6; rubicundus*.
Cu. rubicundus. *Payk. F.* iii. 263.—*Panz. F.* xcix. *f.* 12.—(*Rh. dumetorum.*)

1533. [7; melanocephalus*.]
Cu. melanocephalus. *Marsh.* i. 253.

1534. 8; hæmorrhoidalis*.
Cu. hæmorrhoidalis. *Panz. F.* xcix. *f.* 11.
Cr. analis. (*Panz. I.* 193.)

B. "Thorax tuberculatus, aut margo anticus elevatus, reflexus." *Schön.*

a. *Femora mutica.*

1535. 9; Geranii*. (*Schön. Cu.* 298.)
Cu. Geranii. *Payk. F.* iii. 256.
En.? Lima. *Megerle.*

b. *Femora dentata.*

1536. 10; didymus*. (*Schön. Cu.* 298.)
Cu. didymus. *Payk. F.* iii. 213.—*Herbst C.* vi. 397. *pl.* 91. *f.* 6.—*Ent. Trans.* (*Burrell.*) i. 176.

Cu. arquata. *Herbst C.* vi. 396. *pl.* 91. *f.* 5?
Cu. viduus. *Panz. F.* xlii. *f.* 17.—*Ent. Trans.* (*Burrell.*) i. 176.
Cr. viduus. (*Sam. I.* 13.)
Cu. urticarius. *Clairv. E. H.* i. 98. *pl.* xi. *f.* 1. 2.

1537. [11; Urticæ*.]
Cu. Urticæ. *Marsh.* i. 281.
Cr. Urticæ. (*Sam. I.* 13.)

1538. 12; guttula*. (*Schön. Cu.* 298.)
Rh. guttula. *Fabr. E.* ii. 482.
Cu. Cardui. *Herbst A. pl.* 24. *f.* 22?

1539. [13; fuliginosus*.]
Cu. fuliginosus. *Marsh.* i. 280.

1540. 14; ruficornis* *mihi.*
Rh. Lamii. *Fabr. E.* ii. 483?

1541. 15; subrufus*.
Cu. subrufus. *Marsh.* i. 282.

1542. 16; sulcicollis*.
Cu. sulcicollis. *Payk. F.* iii. 217.
Cu. pleurostigma. *Marsh.* i. 282.—(*Kirby & Sp. I. E.* i. 453.)
Cr. pleurostigma. (*Sam. I.* 13.)

1543. [17; affinis*.]
Cu. affinis. *Panz. F.* lvii. *f.* 15.
Rh. Alauda. *Fabr. E.* ii. 454.

Genus 271. NEDYUS *mihi.*

FALCIGER, (*Megerle.*)—CEUTORHYNCHUS *p, Schüp.*—CRYPTORHYNCHUS *p, Samou.*—NEDYUS *p, Schön.*

A. Thorax nec tuberculatus, nec anticè reflexus.

1544. 1; Sisymbrii*.
Cu. Sisymbrii. *Herbst C.* vi. 159.—*Panz. F.* xvii. *f.* 6.—*Marsh.* i. 253.
Rh. Sisymbrii. (*Sam. I.* 36.)

B. Thorax tuberculatus, aut margo anticus elevatus, reflexus.

a. *Femoribus muticis.*

1545. 2; assimilis*.
Cu. assimilis. *Payk. F.* iii. 257.—*Marsh.* i. 257.
Cr. assimilis. (*Sam. I.* 13.)
Cu. Alauda. *Herbst C.* vi. 410. *pl.* 92. *f.* 5.

1546. 3; obstrictus.
Cu. obstrictus. *Marsh.* i. 255.
Cr. obstrictus. (*Sam. I.* 13.)
Ce. Syrites. *Germ. Ins. Nov.* i. 232.

1547. 4; Erysimi*.
Cu. Erysimi. *Payk. F.* iii. 265.—*Panz. F.* xvii. *f.* 7.—*Marsh.* i. 257.—*Turt.* ii. 217.
Cr. Erysimi. (*Leach E. E.* ix. 108.)—(*Sam. I.* 13.)

1548. 5; chloropterus*.
Cr. chloropterus. *Kirby MSS.*

1549. 6; contractus*.
Cu. contractus. *Marsh.* i. 250.—(*Kirby & Sp. I. E.* i. 186 & 453.)
Cr. contractus. (*Sam. I.* 13.)

1550. 7; Cochleariæ*.
Rh. Cochleariæ. *Gyll.* iii. 144.
Cr. subglobosus. *Kirby MSS.*

1551. 8; depressicollis*.
Rh. depressicollis. *Gyll.* iii. 147.
Cu. nigrinus. *Marsh.* i. 250.

1552. 9; constrictus*.
Cu. constrictus. *Marsh.* i. 258.
Rh. floralis, c. *Gyll.* iii. 145.

1553. 10; floralis*.
Cu. floralis. *Payk. F.* iii. 266.

1554. 11, monostigma.
Cu. monostigma. *Marsh.* i. *MSS.*

1555. 12; pyrrhorhynchus*.
Cu. pyrrhorhynchus. *Marsh.* i. 257.

1556. 13; melanarius*.
Cr. melanarius. *Kirby MSS.*

1557. 14; sulculus*.
Cr. sulculus. *Kirby MSS.*—(*Sam. I.* 13.)

1558. 15, phæorhynchus.
Cu. phæorhynchus. *Marsh.* i. 258.
Cr. phæorhynchus. (*Sam. I.* 13.)
† β? Cr. phæopus. *Kirby MSS.* β, in Mus. D. *Kirby.*

1559. 16; ruficrus*.
Cu. ruficrus. *Marsh.* i. 258.

1560. 17, Ericæ.
Rh. Ericæ. *Gyll.* iii. 147.
Fa. rufirostris. (*DeJean C.* 84.)

b. *Femoribus dentatis.*

1561. 18; ovalis*.
Cu. ovalis. *Linn.* ii. 612?—*Marsh.* i. 279.
Cr. ovalis. (*Sam. I.* 13.)

1562. 19; leucomelanus*.

Cr. leucomelanus. *Kirby MSS.*
Cu. quadrimaculatus. *Linn.* ii. 609?—*Turt.* ii. 217?

1563. [20, detritus*] *mihi.*

1564. 21; pollinarius*.
Cu. pollinarius. *Forst. C.* i. 33.—*Stew.* ii. 60.—*Turt.* ii. 240.
Cu. dentatus. *Marsh.* i. 280.
Cr. dentatus. (*Sam. I.* 13.)
Rh. Raphani. *Fabr. E.* ii. 485?

1565. 22; nigrinus*.
Cr. nigrinus. *Kirby MSS.*

1566. 23; Quercicola*.
Cu. Quercicola. *Payk. F.* iii. 215.—*Herbst C.* vi. 403. *pl.* 91. *f.* 13.

1567. 24; Borraginis*.
Cu. Borraginis. *Payk. F.* iii. *Add.* 457.
Cu. quadridens. *Panz. F.* xxxvi. *f.* 13.
Cu. Quercicola. *Marsh.* i. 280.
Cr. Quercicola. (*Sam. I.* 13.)
Cr. rufitarsis. *Kirby MSS.*

1568. [25, pallidactylus*.]
Cu. pallidactylus. *Marsh.* i. 259.

1569. 26; rugulosus*.
Cu. rugulosus. *Herbst C.* vi. 406. *pl.* 91. *f.* 16?
Cu. Quercicola, var. *Payk. F.* iii. 216.
Cu. melanostictus. *Marsh.* i. 282.
Cr. melanostictus. (*Sam. I.* 13.)

1570. 27; Chrysanthemi*.
Ceu. Chrysanthemi. *Germ. I. N.* i. 221.
Rh. rugulosus c. *Gyll.* iii. 231.

1571. 28, melanostigma.
Cu. melanostigma. *Marsh.* i. 256.

1572. 29, cinereus.
Cu. cinereus. *Marsh.* i. 283.

1573. 30; scutellatus* *mihi.*

1574. 31; litura*.
Cu. litura. *Payk. F.* iii. 210.
Cu. cruciger, var. *Herbst C.* vi. 395.

1575 † 32, Crux.
Cr. Crux. *Leach MSS.*
Rh. Crux. *Fabr. E.* ii. 455? In Mus. *Brit.*

1576. 33; Asperifoliarum*.
Cr. Asperifoliarum. *Kirby MSS.*

Rh. Asperifoliarum. *Gyll.* iii. 221.
Ceu. intersectus. *Germ. I. N.* i. 227.

1577. 34; quadrimaculatus*.
Cu. quadrimaculatus. *Linn.* ii. 609?—*Marsh.* i. 252.

1578. 35; Echii*.
Rh. Echii. *Fabr. E.* ii. 482.
Cu. Echii. *Marsh.* i. 279.—*Panz. F.* xvii. *f.* 12.
Cu. geographicus. *Vill.* i. 201.
Rh. glyphicus. *Schall.*

C. Thorace mutico, anticè reflexo.

a. *Elytris squamulis setiformibus armatis.*

1579. 36; horridus*.
Cu. horridus. *Panz. F.* lxxxiv. *f.* 9.
Cr. horridus. (*Sam. I.* 13.)

1580. 37; troglodytes*.
Cu. troglodytes. *Fabr. E.* ii. 485.—*Marsh.* i. 281.

1581. [38; pusio*.]
Cu. pusio. *Panz. F.* xxxvi. *f.* 15.
Rh. troglodytes, b. *Gyll.* iii. 233.

1582. [39; spiniger.]
Cu. spiniger. *Herbst C.* vi. 410. *pl.* 92. *f.* 4.
Rh. troglodytes, c. *Gyll.* iii. 233.

b. *Elytris haud setosis.*

1583. 40; marginatus*.
Cu. marginatus. *Payk. F.* iii. 211.
Cu. Resedæ. *Marsh.* i. 256.

Genus 272. MONONYCHUS, (*Schüpp.*)—*Schön. Cu.* 299.

FALCIGER *p*, *DeJean.*

1584 † 1, Pseudacori. (*Schön. Cu.* 300.)
Rh. Pseudacori. *Fabr. E.* ii. 450.
Cu. Pseudacori. *Panz. F.* xvii. *f.* 5. In Mus. D. *Curtis.*

Genus 273. RHINONCUS.

CAMPYLIRHYNCHUS, (*Megerle.*)—RHINONCUS, *Schön. olim.*—CEUTORHYNCHUS 2, *Schön.*—PACHYRHINUS *p*, *Kirby MSS.*—PACHYLORHYNCHUS *p*, (*Megerle.*)

A. Thorace bituberculato.

a. *Tuberculis thoracis obtusis.*

1585. 1; Pericarpius*.
Cu. Pericarpius. *Linn.* ii. 609.—*Berk. S.* i. 98.—*Stew.* ii. 56.—*Herbst C.* vi. 401. *pl.* 91. *f.* 12.—*Turt.* ii. 216.

1586. 2; Spartii*. *Kirby.*
Ceu. Spartii. *Kirby MSS.*

1587. 3; tibialis* *mihi.*

b. *Tuberculis thoracis acutis.*

1588. 4, flavipes *mihi.*

1589. 5; Castor*.
Rh. Castor. *Fabr. E.* ii. 451.
Cu. fruticulosus. *Herbst C.* vi. 400. *pl.* 91. *f.* 10.

1590. [6, leucostigma.]
Cu. leucostigma. *Marsh.* i. 255.
Rh. scabratus. *Fabr. E.* ii. 454?

1591. [7; interstitialis*.]
Cu. interstitialis. *Reich. M.* i. 6. *pl.* 1. *f.* 2.
Cu. seniculus. *Grav.*
Rh. Castor, b. *Gyll.* ii. 160.

1592. 8; rufipes* *mihi.*

1593. 9; crassus*.
Cu. crassus. *Marsh.* i. 264.

1594. 10; canaliculatus*.
Cu. canaliculatus. *Mus. Marsham.*

B. Thorace quadrituberculato.

1595. 11; quadrituberculatus*.
Rh. quadrituberculatus. *Fabr. E.* ii. 448.
Cu. quadrituberculatus. *Herbst C.* vi. 409. *pl.* 92. *f.* 3.

1596. 12, inconspectus*.
Cu. inconspectus. *Herbst C.* vi. 405. *pl.* 91. *f.* 15.
Cu. accipitrinus. *Reich. M.* i. 5. *pl.* 1. *f.* 1.
Rh. suturalis. *Oliv. E.*

1597. 13, quadricornis.
Rh. quadricornis. *Gyll.* iii. 154.
Cu. quadrituberculatus var. *Herbst C.* vi. 409.

Genus 274. ACALLES, *Schön. Cu.*

CRYPTORHYNCHUS *p, Samou.*—COMASINUS *p, Megerle?*

1598. 1; Ptinoides*. (*Schön. Cu.* 296.)
Cu. Ptinoides. *Marsh.* i. 258.
Cr. Ptinoides. (*Sam. I.* 13.)

1599. 2; variegatus* *mihi.*

Genus 275. ———

CEUTORHYNCHUS 1, *p, Schön.*

1600. 1; globulus*.
Cu. globulus. *Herbst C.* vi. 398. *pl.* 91. *f.* 7.

Genus 276. CRYPTORHYNCHUS *Illiger.*

1601. 1; Lapathi*. (*Sam. I.* 13.)
Cu. Lapathi. *Linn.* ii. 608.—*Berk. S.* i. 98.—*Linn. Trans.* (*Curtis.*) i. 86. *pl.* 5. *f.* 1–5.—*Don.* vi. *pl.* 205. *f.* 1.—*Stew.* ii. 85.—*Marsh.* i. 254.—*Turt.* ii. 230.
Cu. albicaudis. *DeGeer.* v. 223.

Genus 277. BROTHEUS *mihi.*

1602. 1, porcatus*.
Cu. porcatus. *Marsh.* i. 255.

Genus 278. BAGOÜS, (*Germar.*), (*Samou.*), *Schön. Cu.*, (*Kirby.*)

1603. 1, binodulus*. (*Schön. Cu.* 290.)
Cu. binodulus. *Herbst C.* vi. 247. *pl.* 77. *f.* 15.
Cu. atrirostris. *Payk. F.* iii. 227.
Rh. atrirostris. (*Sam. I.* 36.)
Ba. hydröus. *Kirby MSS.*

1604. 2, lutosus. (*Schön. Cu.* 290.)
Rh. lutosus. *Gyll.* iii. 85.
Ba. tubercularis. *Kirby MSS.*

1605. 3; lutulentus*. (*Schön. C.* 290.)
Rh. lutulentus. *Gyll.* iii. 86.
Cu. productus. *Illiger.*
β, Cu. collignensis. *Herbst C.* vii. 50. *pl.* 98. *f.* 7. *C.*

1606. 4; binotatus*.
Rh. binotatus. ——? *MSS.*

1607. 5; tibialis* *mihi.*

Genus 279. LYPRUS, *Schön. Cu.*

LIXUS *p*, *Ahrens.*—BAGOÜS *p*, *DeJean.*

1608. 1, cylindrus*. (*Schön. Cu.* 289.)
Rh. cylindrus. *Gyll.* iii. 78.
Ba. attenuatus. (*DeJean C.* 89.)
Li. attenuatus. *Act. Hal.* (*Ahrens.*) ii. 18. *pl.* 1. *f.* 9.

Genus 280. PACHYRHINUS.

CEUTORHYNCHUS 3, *Germar.*—PACHYLORHYNCHUS *p*, *Megerle.*—PACHYRHINUS *p*, *Kirby.*—HYDATICUS[a], *Schön.*—CRYPTORHYNCHUS *p*, *Samou.*

A. Thorace bituberculato.

1609. 1; leucogaster*.
Cu. leucogaster. *Marsh.* i. 253.

---

[a] HYDATICUS: Genus Coleoptratorum. Vide *Leach Z. M.* iii. 69.

Cr. leucogaster. (*Sam. I.* 13.)
Rh. asperatus. (*Gyll.* iv. 583.)

1610. 2; rufescens* *mihi.*

1611. 3; Comari*.
Cu. Comari. *Herbst C.* vi. 411. *pl.* 92. *f.* 6.

B. Thorace quadrituberculato.

1612. 4; quadridentatus* *mihi.*

1613. 5, quadrinodosus.
Rh. quadrinodosus. *Gyll.* iii. 155.
Ceu. mucronulatus. *Germ. I. N.* i. 259.

1614. 6, Hydrolapathi.
Cu. Hydrolapathi. *Mus. Marsham.*

Genus 281. ANOPLUS, (*Schüpp.*) *Schön.*

1615. 1; plantaris*. (*DeJean C.* 85.)
Rh. plantaris. *Act. Holm.* (*Næzin.*) 1794. 270.—*Gyll.* iii. 252.
Cu. brevis. *Marsh.* i. 265.
Rh. brevis. (*Sam. I.* 36.)
Rh. Alniarius. *Kirby MSS.*

1616. [2; nitidulus*.]
Rh. nitidulus. *Wilk.? MSS.*

1617. [3; atratus*.]
Rh. atratus. *Wilk.? MSS.*

Genus 282. AMALUS, *Schön. Cu.*

CRYPTORHYNCHUS *p*, *Samou.*

1618. 1; scortillum*. (*Schön. Cu.* 241.)
Cu. scortillum. *Herbst C.* vi. 418. *pl.* 92. *f.* 13.
Cu. Agricola. *Payk. F.* iii. 260.
Cu. inflexus. *Marsh.* i. 253.
Cr. inflexus. (*Sam. I.* 13.)

1619. 2; castaneus*.
Cr. castaneus. *Wilk.? MSS.*

Genus 283. TYCHIUS, (*Germar.*) *Schön. Cu.*

SIBINIA *p*, *DeJean.*—CRYPTORHYNCHUS *p*, *Samou.*—MICCOTROGUS, (*Schönher.*)

A. Femora dentata.

1620. 1, quinquepunctatus*. (*Schön. Cu.* 246.)
Cu. quinquepunctatus. *Linn.* ii. 614.—*Panz. F.* lxxxiv. *f.* 8.

1621. 2; venustus*. (*Schön. Cu.* 246.)
Cu. venustus. *Fabr. E. S.* ii. 456. (!)—*Herbst C.* vi. 163. *pl.* 71. *f.* 2.—*Stew.* ii. 56.

Cu. vernalis. *Reich. M.* i. 8. *pl.* 1. *f.* 4.—*Marsh.* i. 272.
β; Cu. parallelus. *Panz. F.* xviii. *f.* 5.
Cu. bivittatus. *Marsh.* i. 281.

1622. [3; nervosus*.]
Cu. nervosus. *Marsh.* i. 282.
β, Rh. Beckwithii. *Kirby MSS.*

B. Femora mutica.

1623. 4; Meliloti*.
Si. Meliloti. *Kirby MSS.*

1624. 5; flavicollis*.
Si. flavicollis. *Kirby MSS.*

1625. 6; picirostris*. (*Schön. Cu.* 246.)
Cu. picirostris. *Payk. F.* iii. 253.

1626. [7; tomentosus*.]
Cu. tomentosus. *Herbst C.* vi. 278. *pl.* 81. *f.* 7.
Cu. villosus. *Marsh.* i. 260.

1627. 8; cinerascens*.
Cu. cinerascens. *Marsh.* i. 248.
Mi. cinerascens. *Schönher.*

1628. 9; parvulus*.
Si. parvula. *Kirby MSS.*

1629. 10; canescens*.
Cu. canescens. *Marsh.* i. 259.
Cr. canescens. (*Sam. I.* 13.)

1630 † 11, lineatulus.
Si. lineatula. *Kirby MSS.* In Mus. D. *Kirby.*

Genus 284. SIBINIA, *Germar.*

SIBYNES, *Schön. Cu.*

1631. 1, Arenariæ*. *Kirby MSS.*

1632. 2, primita*. (*DeJean C.* 83.)
Cu. primitus. *Herbst C.* vi. 104. *pl.* 66. *f.* 8.
Cu. signatus. *Panz. F.* xcix. *f.* 10.

1633 † 3. Viscariæ[a].

Genus 285. ORCHESTES, *Illiger, Leach, Samou.*, (*Kirby.*)

SALIUS *p, Germar.*

A. Femora postica haud dentata.

1634. 1; Quercus*.
Cu. Quercus. *Linn.* ii. 609.—*Berk. S.* i. 98.—*Stew.* ii. 57.

---

[a] 1633 † 3. Viscariæ. (*Schön. Cu.* 249.)
Cu. Viscariæ. *Linn.* ii. 609.—*Berk. S.* i. 98. (!)—*Stew.* ii. 56. (!)—*Turt.* ii. 212. (!)

Cu. saltator Ulmi. *DeGeer.* v. 260. *pl.* 8. *f.* 5-11.
Rh. Viminalis. *Fabr. E.* ii. 494.
Cu. rufus. *Don.* xi. 81. *pl.* 389. *f.* 3.

1635. 2; scutellaris*. (*DeJean C.* 82.)
Rh. scutellaris. *Fabr. E.* ii. 495.
Cu. rufus. *Schra. A.* 116.—*Stew.* ii. 58.—*Don.* vii. *pl.* 249. *f.* 1.—*Marsh.* i. 261.—*Turt.* ii. 242.
Or. rufus. (*Sam. I.* 32.)
Cu. Alni, c. *Payk. F.* iii. 221.

1636. 3; ferrugineus*. (*Sam. I.* 32.)
Cu. ferrugineus. *Marsh.* i. 260.

1637. 4; atricapillus*. (*Sam. I.* 32.)
Cu. atricapillus. *Marsh.* i. 261.

1638. 5; nigricollis*. (*Sam. I.* 32.)
Cu. nigricollis. *Marsh.* i. 261.

1639. 6; Alni*. (*Leach E. E.* ix. 108.)—(*Sam. I.* 32.)
Cu. Alni. *Linn.* ii. 611.—*Berk. S.* i. 99.—*Don.* vii. *pl.* 249. *f.* 2.—*Stew.* ii. 58.—*Marsh.* i. 260.—*Turt.* ii. 242.

1640. 7; depressus*. (*Sam. I.* 32.)
Cu. depressus. *Marsh.* i. 262.

1641. 8; Ilicis*. (*DeJean C.* 82.)
Cu. Ilicis. *Payk. F.* iii. 218.
Cu. pilosus. *Herbst C.* vi. 426. *pl.* 93. *f.* 8.—*Marsh.* i. 262.
Cu. saltator Segetis. *DeGeer.* v. 264.

1642. 9; pilosus*. (*Sam. I.* 32.)
Cu. pilosus. *Payk. F.* iii. 218.—*Stew.* ii. 58.—*Turt.* ii. 242.

1643. 10; Calcar*.
Rh. Calcar. *Fabr. E.* ii. 493.
Cu. rhododactylus. *Marsh.* i. 262.
Or. rhododactylus. (*Sam. I.* 32.)
Rh. Fagi. *Gyll.* iii. 243.

1644. [11; Fragariæ*.]
Cu. Fragariæ. *Payk. F.* iii. 217.—*Herbst C.* vi. 423. *pl.* 93. *f.* 3.—*Marsh.* i. 263.

1645. 12, Populi*. (*DeJean C.* 83.)
Rh. Populi. *Fabr. E.* ii. 495?
Cu. Populi. *Panz. F.* xviii. *f.* 17.
Cu. rhodopus. *Marsh.* i. 263.

B. *Femora postica haud dentata.*

1646. 13; signifer*. *Creutzer.* 125. *pl.* 3. *f.* 29.
Cu. bifasciatus β. *Payk. F.* iii. 270.
Cu. Avellanæ. *Don.* vi. *pl.* 205. *f.* 3.—*Stew.* ii. 61.—*Marsh.* i. 263.—*Turt.* ii. 251.
Or. Avellanæ. (*Sam. I.* 32.)

1647. 14; x-album*. *Kirby MSS.*

1648. 15; bifasciatus*. (*DeJean C.* 83.)
Cu. bifasciatus. *Payk. F.* iii. 270.
Sa. Rusci. *Germ. M.* iv. 332.

1649. 16, confinis*. *Kirby MSS.*

1650. 17; affinis* *mihi.*

Genus 286. TACHYERGES, *Schönher.*

ORCHESTES *p, Illiger.*—SALIUS *p, Germar.*

1651. 1; Salicis*. (*Schön. Cu.* 256.)
Cu. Salicis. *Linn.* ii. 611.—*Stew.* ii. 58.—*Marsh.* i. 264.
Or. Salicis. (*Sam. I.* 32.)
Cu. scrutator. *Herbst Col.* vi. 191.
Cu. Capreæ. *Fabr. E. S.* ii. 409. (!)—*Don.* iv. *pl.* 121. *f.* 5–7.
—*Stew.* ii. 56.—*Turt.* ii. 216.
Sa. bifasciatus. *Germ. M.* iv. 332.

1652. 2; Saliceti*. (*Schön. Cu.* 256.)
Rh. Saliceti. *Fabr. E.* ii. 493.

1653. 3, scutellatus*.
Or. scutellatus. *Wilk.? MSS.*

1654. 4; nitidulus*.
Or. nitidulus. *Kirby MSS.*

1655. 5; rufitarsis.
Or. rufitarsis. (*DeJean C.* 82.)

1656. 6, Stigma*.
Rh. Iota. *Payk. F.* iii. 271.
Or. impressus. *Kirby MSS.*
Sa. Stigma. *Germ. M.* iv. 334.
Or. alboscutellatus. (*DeJean C.* 82.)

Genus 287. BALANINUS, *Germar, Samou.,* (*Kirby.*)

ARCHARIAS, (*Megerle.*)

A. Pygidium ab elytris non tectum.

1657. 1; Nucum*. (*Sam. I.* 5. *pl.* 2. *f.* 20.)—(*Kirby & Sp. I. E.* iii. 84.)
Cu. Nucum. *Linn.* ii. 613.—*Mart. C. pl.* 19. *f.* 26.—*Berk. S.* i. 99.—*Panz. F.* xlii. *f.* 21.—*Stew.* ii. 59.—*Marsh.* i. 283.—*Turt.* ii. 236.—*Shaw G. Z.* vi. 63. *pl.* 21.—*Samou.* 54.—*Bingley.* iii. 134.—(*Kirby & Sp. I. E.* i. 201.—*Id.* ii. 274.)
β, Ba. admotus. *Kirby MSS.*
γ, Ba. rufogriseus. *Kirby MSS.*

1658. 2; glandium*. (*Kirby & Sp. I. E.* iii. 84.)
Cu. glandium. *Marsh.* i. 284.

Ba. gladius. (*Sam. I.* 5.)
♂, Ba. longirostris. *Kirby MSS.*

1659. 3; villosus*. (*DeJean C.* 86.)
Cu. villosus. *Herbst C.* vi. 195. *pl.* 73. *f.* 8.
Cu. Cerasorum. *Panz. F.* xlii. *f.* 22.—*Stew.* ii. 59.—*Marsh.* i. 284.—*Turt.* ii. 237.
Ba. Cerasorum. (*Sam. I.* 5.)

1660. 4; tenuirostris*. (*Sam. I.* 5.)
Cu. tenuirostris. *Herbst C.* vi. 210.—*Mart. C. pl.* 20. *f.* 40.—*Don.* vii. *pl.* 249. *f.* 3.—*Stew.* ii. 59.—*Marsh.* i. 284.—*Turt.* ii. 237.
β, Ba. rostratus. *Kirby MSS.*

1661. 5; Betulæ* *mihi.*
Rh. Cerasorum. *Gyll.* iii. 204?—*Herbst C.* vi. 196. *pl.* 73. *f.* 9?

B. Pygidium ab elytris ferè tectum.

1662. 6, Salicivorus. (*DeJean C.* 86.)
Cu. Salicivorus. *Payk. F.* iii. 214.—*Herbst C.* vi. 200. *pl.* 73. *f.* 12.
Rh. Brassicæ. *Fabr. E.* ii. 483?
Cu. arcuatus. *Marsh.* i. 288.

1663. 7; curvatus*.
Cu. curvatus. *Marsh.* i. 287.
Rh. Salicivorus, b. *Gyll.* iii. 206.

1664. 8; scutellaris*.
Rh. scutellaris. —— *MSS.*

1665. 9; intermedius*.
Cu. intermedius. *Marsh.* i. 288.

1666. 10; pyrrhoceras*. (*Schön. Cu.* 240.)
Cu. pyrrhoceras. *Marsh.* i. 288.

1667. 11, brunneus*.
Cu. brunneus. *Marsh.* i. 248.

Genus 288. ANTHONOMUS, *Germar.*

PALLENE, (*Megerle.*)—BALANINUS *p*, *Samoü.*

A. Femoribus acutè dentatis.

1668. 1; incurvus*. *Germ. M.* iv. 323.
Cu. incurvus. *Panz. F.* xxxvi. *f.* 17.
Rh. Pomorum b. *Gyll.* iii. 188.
Cu. fasciatus. *Don.* xii. 58. *pl.* 414. *f.* 3.

1669. 2; Pomorum*. (*DeJean C.* 87.)
Cu. Pomorum. *Linn.* ii. 612.—*Herbst C.* vi. 157. *pl.* 70. *f.* 11.—*Berk. S.* i. 99.—*Stew.* ii. 58.—*Marsh.* i. 285.—*Turt.* ii. 238.
Ba. Pomorum. (*Sam. I.* 5.)

1670. 3; Ulmi*. (*Schön. Cu.* 237.)
Cu. Ulmi. *DeGeer.* v. 215. *pl.* 6. *f.* 26—30.—*Marsh.* i. 287.
Cu. Pomorum var. *Herbst C.* vi. 138.
An. clavatus. (*DeJean C.* 87.)

1671. [4; fasciatus*.]
Cu. fasciatus. *Marsh.* i. 286.
Ba. fasciatus. (*Sam. I.* 5.)
Rh. Ulmi, b. *Gyll.* iii. 189.
β, An. maculosus. *Kirby MSS.*
γ, An. rubescens. *Kirby MSS.*

1672. 5; pedicularius*.
Cu. pedicularius. *Linn.* ii. 615.—*Marsh.* i. 286.
Rh. Avarus. *Fabr. E.* ii. 488.
Cu. Druparum, var. *Herbst C.* vi. 157. *pl.* 70. *f.* 10.
Rh. Ulmi. c. *Gyll.* iii. 190.

1673. 6; Druparum*. (*DeJean C.* 87.)
Cu. Druparum. *Linn.* ii. 614.—*Herbst C.* vi. 156. *pl.* 70. *f.* 9.
Rh. Druparum. *Fabr. E.* ii. 489. (!)

B. Femoribus obsoletè dentatis.

1674. 7; obscurus* *mihi.*

1675. 8; ater*.
Cu. ater. *Marsh.* i. 285.

1676. 9; Rubi*. (*DeJean C.* 87.)
Cu. Rubi. *Herbst C.* vi. 167. *pl.* 71. *f.* 8.
Cu. melanopterus. *Marsh.* i. 289.

1677. 10, clavatus.
Cu. clavatus. *Marsh.* i. 285.

Genus 289. HYDRONOMUS, *Schön.*

BAGOÜS *p*, *Germar.*

1678. 1; Alismatis*. (*Schön. C.* 232.)
Cu. Alismatis. *Marsh.* i. 273.
Rh. Alismatis. (*Sam. I.* 36.)
β, Cu. cnemerythrus. *Marsh.* i. 268.

Genus 290. GRYPIDIUS, *Schön. Cu.*

GRYPUS[a], (*Germar.*), (*Samou.*)—ARACHNIPUS, (*Megerle.*)

1679. 1; Equiseti*. (*Schön. Cu.* 231.)
Rh. Equiseti. *Fabr. E.* ii. 443. (!)—(*Sam. I.* 36.)
Cu. Equiseti. *Panz. F.* xlii. *f.* 4.—*Stew.* ii. 56.—*Marsh.* i. 254. —*Turt.* ii. 213.
Cu. nigro-gibbosus. *DeGeer.* v. 224.

---

[a] GRYPUS: Genus Avium. Vide *Gen. Zoology* xiv. *p.* 256.

Genus 291. ERIRHINUS.

NOTARIS *p*, *Germar.*—ERIRHINUS 2, *Schön.*

1680. 1, Festucæ *. (*Schön. Cu.* 230.)
Cu. Festucæ. *Herbst C.* vi. 327. *pl.* 85. *f.* 13.
Cu. Caricis. *Thunb. Mus. Ups. App.* vi. 111.
Rh. lunula. *Kirby MSS.*

1681. 2; Nereis *. (*Schön. Cu.* 230.)
Cu. Nereis. *Payk. F.* iii. 240.
Rh. Nereis. (*Sam. I.* 36.)

1682. 3; inquisitor *.
Cu. inquisitor. *Herbst C.* vii. 42. *pl.* 97. *f.* 12.
Rh. Nereis. *Gyll.* iii. 77?
Rh. Typhæ. *Ahr. F.* iv. *f.* 69.

1683. 4; Arundineti *.
Rh. Arundineti. *Kirby MSS.*

Genus 292. NOTARIS, *Germar.*

ERIRHINUS 1, *p. Schön.*

1684. 1; acridulus *.
Cu. acridulus. *Linn.* ii. 607.—*Berk. S.* i. 97.—*Panz. F.* xlii. *f.* 10.—*Stew.* ii. 55.—*Turt.* ii. 217.
Cu. resinosus. *Marsh.* i. 268.
Rh. resinosus. (*Sam. I.* 36.)

1685. [2; punctum *.]
Rh. punctum. *Fabr. E.* ii. 442.
Cu. rigidus. *Marsh.* i. 270.

1686. 3; bimaculatus *.
Rh. bimaculatus. *Fabr. E.* ii. 442.—*Herbst C.* vi. 291. *pl.* 82. *f.* 8.
Rh. Sparganii. *Kirby MSS.*

Genus 293. DORYTOMUS, *Germar.*

BALANINUS *p*, *Samou.*—ERIRHINUS 1. *p*, *Schön.*

A. Pedibus anticis elongatis.

1687. 1; vorax *. (*DeJean C.* 86.)
Cu. vorax. *Herbst C.* vi. 165.—*Panz. F.* xviii. *f.* 13.
Cu. longimanus. *Forst. C.* i. 32.—*Marsh.* i. 293.
Ba. longimanus. (*Sam. I.* 5.)
Cu. Forsteri. *Stew.* ii. 60.

1688. 2, ventralis. (*DeJean C.* 86?)

B. Pedibus anticis haud elongatis.

1689. 3; Tortrix *. (*DeJean C.* 86.)
Cu. Tortrix. *Linn.* ii. 615.—*Mart. C. pl.* 13. *f.* 2.—*Berk. S.* i. 100.—*Panz. F.* xviii. *f.* 14.—*Marsh.* i. 291.—*Turt.* ii. 238.

Rh. Tortrix. *Leach E. E.* ix. 108.
Ba. Tortrix. (*Sam. I.* 5.)
Cu. fulvus. *DeGeer.* v. 214.

1690. 4; arcuatus*.
Cu. arcuatus. *Panz. F.* xviii. *f.* 6.
Rh. pectoralis, c. *Gyll.* iii. 179.

1691. 5; pectoralis*. (*DeJean C.* 86.)
Cu. pectoralis. *Panz. F.* xxxvi. *f.* 16.
Cu. rubellus. *Marsh.* i. 293.
Ba. rubellus. (*Sam. I.* 5.)

1692. 6; melanopthalmus*.
Cu. melanopthalmus. *Payk. F.* iii. 196.
Cu. fructuum. *Marsh.* i. 292.
Ba. fructuum. (*Sam. I.* 5.)
Rh. pectoralis, b. *Gyll.* iii. 179.

1693. 7, affinis*.
Cu. affinis. *Payk. F.* iii. 190.
Do. flavipes. (*DeJean C.* 86.)

1694. 8; Tremulæ*. (*DeJean C.* 86.)
Cu. Tremulæ. *Payk. F.* iii. 189.—*Marsh.* i. 291.
Ba. Tremulæ. (*Sam. I.* 5.)

1695. 9; maculatus*.
Cu. maculatus. *Marsh.* i. 292.
Ba. maculatus. (*Sam. I.* 5.)
Rh. maculatus. (*Sam. I.* 36.)
Rh. tæniatus. *Fabr. E.* ii. 492. (!)
Cu. tæniatus. *Herbst C.* vi. 271. *pl.* 80. *f.* 12.
Cu. fumosus. *Rossi M.* i. 124.
Rh. Tremulæ, b. *Gyll.* iii. 172.

1696. 10; tæniatus*. (*DeJean C.* 86.)
Rh. tæniatus. *Fabr. E.* ii. 492?—*Stew.* ii. 60.—*Turt.* ii. 238.

1697. 11, majalis*.
Cu. majalis. *Payk. F.* iii. 177?

1698 † 12. dorsalis[a].

Genus 294. PISSODES, *Germar.*

PISSOCLES, (*DeJean.*)—PINIPHILUS, (*Megerle.*)

1699. 1, Pini. (*Schön. Cu.* 226.)

---

[a] 1698 † 12. dorsalis. (*DeJean C.* 86.)
Cu. dorsalis. *Linn.* ii. 608.—*Berk. S.* i. 98. (!)—*Panz. F.* xvii. *f.* 9. —*Stew.* ii. 55 (!)—*Turt.* ii. 217. (!)

Cu. Pini. *Linn.* ii. 608.—*Panz. F.* xlii. *f.* 1.—*Stew.* ii. 56.—
—*Turt.* ii. 211.
Cu. numero 9. *Berk. S.* i. 98.
Rh. Abietis. *Leach E. E.* ix. 108.—(*Sam. I.* 36.)

1700 † 2, notatus*. (*Schön. Cu.* 226.
Lixus notatus. *Fabr. E.* ii. 501.
Cu. notatus. *Herbst C.* vi. 65. *pl.* 97. *f.* 3.
In Mus. D. *Sparshall.*

1701 † 3, Fabricii. *Leach MSS.*
Cu. Pineti. *Herbst C.* vi. 293. *pl.* 82. *f.* 10? In Mus. *Brit.*

Genus 295. GRONOPS, *Schön.*

BAGOÜS *p*, *Germ.*—ANLACUS, (*Megerle.*)

1702. 1, lunatus*. (*Schön. Cu.* 158.)
Cu. lunatus. *Fabr. E.* ii. 524. (!)—*Stew.* ii. 61.—*Turt.* iii. 252.
Rh. Lathburii. *Kirby MSS.*—(*Sam. I.* 36.)
Cu. percussor. *Herbst C.* vi. 250. *pl.* 78. *f.* 4.
Rh. rubricus. *Act. Hal. Nov.* (*Ahrens*) ii. 16. *pl.* 1. *f.* 7.

1703. [2, costatus.]
Rh. costatus. *Gyll.* iii. 89.
Cu. Sheppardi. *Marsh. MSS.*

Genus 296. ORTHOCHÆTES, *Müller.*

STYPHLUS, *Schön.?*

1704. 1; setiger*. *Germ. I. N. v.* i. 304.
St. penicillus. *Schon. Cu.* 259?

Genus 297. ———

HYPERA *p*, *Curtis l. c. fo.* 116.

1705. 1; picipes*.
Cu. picipes. *Marsh.* i. 272.
Hy. picipes. (*Curtis l. c. supra.*)
Cu. subnebulosus. *Wilkin MSS.*
Rh. subnebulosus. (*Sam. I.* 36.)

1706. 2, pyrrhodactylus.
Cu. pyrrhodactylus. *Marsh.* i. 259.

Genus 298. HYPERA, *Germar*, (*Samou.*) *Curtis*, (*Kirby.*)

DONUS, (*Megerle.*)—PHYTONOMUS, *Schön.*

A. Corpus breve; ovatum.

1707. 1; punctata*. (*Curtis l. c. infra.*)
Cu. punctatus. *Fabr. E.* ii. 529.
Cu. medius. *Marsh.* i. 302.
Cu. austriacus. *Herbst C.* vi. 243. *pl.* 77. *f.* 11.—*Marsh.* i. 302.
Rh. austriacus. (*Sam. I.* 36.)

1708. 2, fasciculosa*. *Curtis.* iii. *pl.* 116.
Rh. fasciculosa. *Gyll.* iii. 107.
Cu. fasciculatus. *Herbst C.* vi. 289. *pl.* 82. *f.* 6.
Rh. sticticus. *Kirby MSS.*
Rh. Dauci. *Oliv. E.* v. 124. *pl.* 35. *f.* 542.

B. Corpus oblongum.

a. *Alati.*

1709. 3; Polygoni*. (*Curtis l. c. supra.*)
Cu. Polygoni. *Fabr. E.* ii. 520.
Cu. fasciatus. *DeGeer.* v. 234.
Cu. lineatus. *Herbst C.* vi. 497. *pl.* 95. *f.* 5. *a, b.*

1710. 4; arator*. (*Curtis l. c. supra.*)
Cu. arator. *Linn. M.* 531.—*Marsh.* i. 266.
Cu. Polygoni. *Fanz. F.* xix. *f.* 10.
Cu. Polygoni b. *Gyll.* iii. 110.
Rh. ærator. (*Sam. I.* 36.)

1711. 5; canescens* *mihi.*

1712. 6; Viciæ*. (*Curtis l. c. supra.*)
Rh. Viciæ. *Gyll.* iii. 101.

1713. 7; lineatus* *mihi.*

1714. 8, picicornis *mihi.*

1715. 9; Rumicis*. (*Curtis l. c. supra.*)
Cu. Rumicis. *Linn.* ii. 614.—*Stew.* ii. 57.—*Marsh.* i. 266.—
Turt. ii. 218.—*Bingley* iii. 136.
Rh. Rumicis. (*Sam. I.* 36.)
Cu. Acetosæ. *Panz. F.* xlii. *f.* 9.
β, Hy. albicans. *Kirby MSS.*

1716. 10; palustris*. (*Curtis l. c. supra.*)
Rh. palustris. *Leach MSS.*—(*Sam. I.* 36.)

1717. 11; sublineata*. *Kirby MSS.*—(*Curtis l. c. supra.*)

1718. 12; nebulosus* *mihi.*

1719. 13; Pollux*. (*Curtis l. c. supra.*)
Rh. Pollux. *Fabr. E.* ii. 457.
Cu. commaculatus. *Herbst C.* vi. 230. *pl.* 76. *f.* 8.
Cu. Rumicis, d. *Payk. F.* iii. 229.
Cu. interruptus. *Marsh.* i. 269.
Rh. interruptus. (*Sam. I.* 36.)
Hy. dorsiger. *Kirby MSS.*

1720. 14; villosula*. ——? *MSS.*—(*Curtis l. c. supra.*)

1721. 15; picipes* *mihi.*

1722. 16; murina*. (*Curtis l. c. supra.*)
Cu. murinus. *Fabr. E.* ii. 520.

Cu. Plantaginis. *Marsh.* i. 265.
Rh. Plantaginis. (*Sam. I.* 36.)

1723. 17, nigrirostris *. (*Curtis l. c. supra.*)
Cu. nigrirostris. *Fabr. E.* ii. 448.—*Panz. F.* xxxvi. *f.* 14.—*Stew.* ii. 54.—*Marsh.* i. 267.—*Turt.* ii. 215.
Rh. nigrirostris. (*Sam. I.* 36.)
Cu. virescens. *Quensel. Diss.* 16.
β; Cu. variabilis. *Fabr. E.* ii. 449.

1724. 18; fulvipes * *mihi.*

b. *Apteri.*

1725. 19; Plantaginis *. (*Curtis l. c. supra.*)
Cu. Plantaginis. *DeGeer.* v. 237. *pl.* 7. *f.* 17—21.
Hy. interstinctus. *Kirby MSS.*

1726. 20; Trifolii *. (*Curtis l. c. supra.*)
Cu. Trifolii. *Herbst C.* vi. 266. *pl.* 80. *f.* 5.
Cu. trilineatus. *Marsh.* i. 268.
Hy. meles. *Germ. M.* iv. 340.

1727. [21, straminea *.] (*Curtis l. c. supra.*)
Cu. stramineus. *Marsh.* i. 267.
Rh. stramineus. (*Sam. I.* 36.)
Rh. Trifolii, d. *Gyll.* iii. 111.
Hy. pallida. (*DeJean C.* 88.)

1728. 22; postica *. (*Curtis l. c. supra.*)
Rh. postica. *Gyll.* iii. 113.
Cu. hæmorrhoidalis. *Herbst C.* vi. 266. *pl.* 80. *f.* 4.
Cu. bimaculatus. *Marsh.* i. 266.
Hy. cordicollis. *Kirby MSS.*

1729. 23; phæopa * *mihi.*

1730. 24; rufipes * *mihi.*

1731. 25; variabilis *. (*Curtis l. c. supra.*)
Cu. variabilis. *Herbst C.* vi. 263. *pl.* 80. *f.* 1.

1732. 26, Arundinis. (*Curtis l. c. supra.*)
Cu. Arundinis. *Fabr. E.* ii. 521.—*Panz. F.* xix. *f.* 11.
Rh. Sii. *Leach MSS.*

1733. 27, elongata.
Cu. elongata. *Payk. F.* iii. 236.
Cu. fusco-cinerea. *Marsh.* i. 271.
Hy. fusco-cinerea. (*Curtis l. c. supra.*)

1734. 28; pedestris *?
Cu. pedestris. *Payk. F.* iii. 233?
Hy. senex. *Kirby MSS.?*

1735. 29; miles *. (*Curtis l. c. supra.*)
Cu. miles. *Payk. F.* iii. 233.

Cu. suspiciosus. *Herbst C.* vi. 265. *pl.* 80. *f.* 3.
Cu. bitæniatus. *Marsh.* i. 268.
♂; Cu. octolineatus. *Marsh. MSS.*

Genus 299. ELLESCUS, (*Megerle.*)

PHYTONOMUS *p, Schön.*—HYPERA *p, Germ.*

1736. 1, bipunctatus *. (*DeJean C.* 87.)
Cu. bipunctatus. *Linn.* ii. 609.—*Panz. F.* xlii. *f.* 7.

Genus 300. PLINTHUS, *Germar.*

MELEUS, (*Megerle.*)—LIXUS *p, Fabricius.*—LIPARUS *p, Olivier.*

1737. 1, caliginosus *. *Germ. I. N.* i. 330.—(*Ing. Inst.* 89.)
Cu. caliginosus. *Fabr. E. S.* ii. 427. (!)—*Stew.* ii. 59.—*Marsh.* i. 292.—*Turt.* ii. 230.
Cu. didymus. *Don.* xvi. *pl.* 570.
Cu. reticulatus. *Marsh. MSS.*

Genus 301. LIPARUS, *Olivier, Leach, Samou.*

MOLYTES, *Schön.*

1738. 1; germanus. (*Leach E. E.* ix. 108.)—(*Sam. I.* 25.)
Cu. germanus. *Linn.* ii. 613.—*Mart. C. pl.* 22. *f.* 68.—*Stew.* ii. 59.—*Marsh.* i. 290.
Cu. fuscomaculatus. *Fabr. E.* ii. 537.—*Herbst C.* vi. 327. *pl.* 86. *f.* 2.

1739. 2; anglicanus *. (*Sam. I.* 25.)
Cu. anglicanus. *Mart. C. pl.* 18. *f.* 7.—*Don.* i. *pl.* 34. *f.* 2.—*Marsh.* i. 290.
Cu. germanus. *Fabr. E.* ii. 475.—*Turt.* ii. 233.

Genus 302. LEIOSOMA, *Kirby MSS.*

LIPARUS *p, Samou.*—MOLYTES *p, Schön.*

1740. 1; punctata *.
Cu. punctatus. *Marsh.* i. 291.
Li. punctatus. (*Sam. I.* 25.)

Genus 303. HYLOBIUS, *Germar.*

LIPARUS *p, Olivier.*

1741. 1, Abietis. (*DeJean C.* 88.)—*Zool. Journ.* (*MacLeay.*) i. 445.
Cu. Abietis. *Linn.* ii. 613.—*Mart. C. pl.* 21. *f.* 60.—*Panz. F.* xliii. *f.* 14.—*Stew.* ii. 59.—*Turt.* ii. 230.
Cu. Pini. *Marsh.* i. 289.—*Don.* xv. 529.
Rh. Pini. (*Leach E. E.* ix. 108.)—(*Sam. I.* 36.)
Curculio. (*Millard B. E.* 179. *pl.* 2. *f.* 8.)

1742 † 2, Pinastri? (*DeJean C.* 88?)
Rh. Pinastri. *Gyll.* iii. 168? In Mus. *Brit.*?

Genus 304. TANYSPHYRUS, *Germar*, (*Kirby*.)

1743. 1; Lemnæ*.
Rh. Lemnæ. *Fabr. E.* ii. 455.—*Panz. F.* xvii. *f.* 10.
Cu. inconspectus. *Herbst C.* vi. 301. *pl.* 83. *f.* 9.
Rh. nudus. *Kirby MSS.*
β? Ta. nanus *mihi.*

Genus 305. ALOPHUS, *Schön. Cu.*

LEPYRUS, *Germar*, (*Samou.*)—LIPARUS *p*, *Leach*, (*Samou.*)—GRAPTUS, *Schön. olim.*

1744. 1; triguttatus*. (*Schön. Cu.* 167.)
Cu. triguttatus. *Fabr. E.* ii. 521.—*Stew.* ii. 60.—*Marsh.* i. 300.—*Turt.* ii. 251.
Cu. cordiger. *Clairv. E. H.* i. 86. *pl.* 7. *f.* 3, 4.
Li. triguttatus. (*Leach E. E.* ix. 108.)
♂; Cu. striatirostris. *Marsh.* i. 269.
β, Cu. rufimanus. *Marsh.* i. 270.

1745. [2, Vau*.]
Cu. Vau. *Schrank. A.* 227.—*Marsh.* i. 299.—*Don.* xii. *pl.* 414. *f.* 1.
Li. Vau. (*Sam. I.* 25.)

1746. [3, trinotatus.]
Cu. bipunctatus. *Marsh. MSS.*

Genus 306. BARYNOTUS, *Germar.*

LIPARUS *p*, *Samou.*—BRIUS *p*, *Megerle.*

1747. 1; Mercurialis*. (*Schön. Cu.* 166.)
Cu. Mercurialis. *Fabr. E.* ii. 530.
Cu. elevatus. *Herbst C.* vi. 235. *pl.* 77. *f.* 3.
Cu. Lapidarius. *Payk. F.* iii. 292.
Cu. Æscidii. *Marsh.* i. 307.
Li. Æscidii. (*Sam. I.* 25.)
♀; Cu. tomentosus. *Marsh.* i. 270.

Genus 307. MERIONUS, (*Megerle.*)

LIPARUS *p*, *Samou.*—BARYNOTUS *p*, *Schön.*

1748. 1; obscurus*. (*DeJean C.* 90.)
Cu. obscurus. *Fabr. E.* ii. 530.
Cu. honorus. *Herbst C.* vii. 59. *pl.* 100. *f.* 3.
Cu. murinus. *Bonsd. Cu.* ii. 37. *f.* 31.
Cu. pilosulus. *Marsh.* i. 299.
Li. pilosulus. (*Sam. I.* 25.)

1749. 2; elevatus*.
Cu. elevatus. *Marsh.* i. 306.
Li. elevatus. (*Sam. I.* 25.)
Cu. Bohemanni. *Gyll.* iv. 611.
β; Cu. carinatus. *Marsh.* i. 299.

Genus 308. LIOPHLÆUS, *Germar*.

GASTRODUS, (*Megerle*,)—LIPARUS *p*, *Samou*.

1750. 1; nubilus*. (*Schön. Cu.* 160.
Cu. nubilus. *Fabr. E.* ii. 538.—*Panz. F.* cvi. *f.* 5.
Cu. tessellatus. *Bonsd. Cu.* ii. 38. *f.* 32.
♂; Cu. subclavatus. *Marsh.* i. 314.
Cu. chrysopterus. *Herbst C.* vi. 339. *pl.* 86. *f.* 10.
♀; Cu. floccosus. *Marsh.* i. 314.—*Millard. B. E. pl.* 2. *f.* 9.
Cu. subglobosus. *Marsh.* i. 313.
Li. subglobosus. (*Sam. I.* 25.)

1751. 2; maurus*.
Cu. maurus. *Marsh.* i. 316.
Li. maurus. (*Sam. I.* 25.)

Genus 309. OTIORHYNCHUS, *Germar*.

LOBORHYNCHUS, (*Megerle*.)—BRACHYRHYNCHUS, (*Megerle*.) —BRACHYRHINUS *p*, *Latreille*.—PACHYGASTER[a], *Germar*, (*Samou*.)—MICOCERUS, *Billb*.—LIPARUS *p*, *Oliv*., (*Samou*.)

A. Femora dentata.

a. *Corpus ovatum.*

1752. 1; Ligustici*. (*Schön. Cu.* 206.)
Cu. Ligustici. *Linn.* ii. 615.—*Bonsd. Cu.* ii. 38. *pl. f.* 33.—*Marsh.* i. 313.—*Turt.* ii. 273.
Li. Ligustici. (*Sam. I.* 25.)
Cu. rugosus. *Schrank.?*

b. *Corpus oblongum.*

1753. 2; sulcatus*. (*Schön. Cu.* 206.)
Cu. sulcatus. *Fabr. E.* ii. 539.—*Herbst C.* vi. 347. *pl.* 87. *f.* 5. —*Marsh.* i. 315.
Li. sulcatus. (*Sam. I.* 25.)
Cu. griseopunctatus. *DeGeer.* v. 217.

1754. 3; notatus*.
Cu. notatus. *Bonsd. Cu.* ii. 39. *f.* 34.
Cu. picipes. *Fabr. E.* ii. 346?
Cu. granulatus. *Fuesl. Ar. pl.* 24. *f.* 33?
Cu. vastator. *Marsh.* i. 300.
Li. vastator. (*Sam. I.* 25.)
Cu. Corruptor. *Host?*

1755. 4; singularis*.
Cu. singularis. *Linn.* ii. 1066?

---

[a] PACHYGASTER: Genus Dipterorum. Vide *Meig. Klass.* i. 146.

Cu. asper. *Marsh.* i. 301.
Li. asper. (*Sam. I.* 25.)

1756. [5; squamiger *.]
Cu. squamiger. *Marsh.* i. 301.
Li. squamiger. (*Sam. I.* 25.)

1757. 6; scaber.
Cu. scaber. *Linn.* ii. 609.—*Berk. S.* i. 98.—*Stew.* ii. 56.

1758. 7; rugifrons. (*Schön. Cu.* 206.)
Cu. rugifrons. *Gyll.* iii. 319.

1759. 8; ovatus*. (*Schön. Cu.* 206.)
Cu. ovatus. *Linn.* ii. 615.—*Herbst C.* vi. 357. *pl.* 88. *f.* 2.—
Marsh. i. 315.—*Turt.* ii. 275.
Li. ovatus. (*Sam. I.* 25.)
Cu. Rosæ. *DeGeer.* v. 219.
Cu. rufipes. *Scop. C.* 32.

1760. [9; pabulinus *.]
Cu. pabulinus. *Panz. F.* lvii. *f.* 19.
Cu. ovatus b. *Gyll.* iii. 320.
Pa. confinis. *Kirby MSS.*

1761. 10; Dilwynii.
Pa. Dilwynii. *Kirby? MSS.*

B. Femora mutica.

a. *Elytris haud striatis.*

1762. 11, Lima.
Cu. Lima. *Marsh.* i. 298.

1763. 12; tenebricosus*. (*Schön. Cu.* 205.)
Cu. tenebricosus. *Herbst C.* vi. 333. *pl.* 86. *f.* 5.
Cu. morio. *Payk. F.* iii. 294.
Cu. clavipes. *Bons. Cu.* ii. 40. *f.* 36.
Cu. niger. *Marsh.* i. 297.
Li. niger. (*Sam. I.* 25.)
Cu. maritimus, var. *Don.* xv. 63. *pl.* 533. *f.* 2.

1764. 13, atroapterus.
Cu. atroapterus. *DeGeer.* v. 243. *pl.* 7. *f.* 22—24.
Cu. nigerrimus. *Marsh. MSS.*

1765. 14, ater. (*Schön. C.* 205.)
Cu. ater. *Herbst C.* vi. 332. *pl.* 86. *f.* 4.
Cu. niger. *Payk. F.* iii. 295.
Cu. maritimus. *Don.* xv. 63. *pl.* 533. *f.* 1.
Cu. hypolaus. *Marsh. MSS.*
Cu. arenosus. *Leach MSS.*

1766. 15, lævigatus.
Cu. lævigatus. *Fabr. E.* ii. 531.—*Herbst C.* vi. 347. *pl.* 87. *f.* 6.

b. *Elytris striatis; glabris.*

1767. 16; piceus *.
Cu. piceus. *Marsh.* i. 305.
Li. piceus. (*Sam. I.* 25.)

c. *Elytris striatis, plus minusve scabrosis.*

1768. 17; scabrosus *.
Cu. scabrosus. *Marsh.* i. 298.
Li. scabrosus. (*Sam. I.* 25.)

1769. 18; scabridus *. *Kirby MSS.*

1770. 19; rugicollis *mihi.*
Larinus? rugicollis. (*Curtis. Catal. no.* 45.?)

1771. 20, maurus. (*Schön. C.* 205.)
Cu. maurus. *Gyll.* iii. 293.
Cu. morio. *Bons. Cu.* ii. 36. *f.* 29.
Cu. sulcatus, b. *Payk. F.* iii. 276.

1772. 21; raucus *. (*Schön. C.* 205.)
Cu. raucus. *Fabr. E.* ii. 529.—*Marsh.* i. 300.
Li. raucus. (*Sam. I.* 25.)
Cu. tristis. *Bons. Cu.* ii. 36. *f.* 28.—*Turt.* ii. 255?
Cu. arenarius. *Fuesly. A.* 127.
β; Cu. tristis. *Fabr. E.* ii. 529. (!)—*Turt.* ii. 255?

Genus 310. TRACHYPHLÆUS, *Germar.*

LIPARUS *p*, *Samou.*

1773. 1; tessellatus *.
Cu. tessellatus. *Marsh.* i. 307.

1774. 2; confinis * *mihi.*
Tr. grisescens. *Kirby MSS.* ?

1775. 3; ventricosus *. *Germ. I. N.* i. 405.

1776. 4; scabriculus *. (*DeJean C.* 96.)
Cu. scabriculus. *Linn. M.* ii. 531.—*Herbst C.* vi. 351. *pl.* 87. *f.* 10?—*Marsh.* i. 304.
Li. scabriculus. (*Sam. I.* 25.)
Br. scabriculus. (*Kirby & Sp. I. E.* ii. 219.)

1777. 5; hispidulus *.
Cu. hispidulus. *Herbst C.* vi. 354. *pl.* 87. *f.* 14.
Cu. scabriculus, b. *Gyll.* iii. 309.
Cu. bifoveolatus. *Beck.*

1778. 6; spinimanus *. *Germ. I. N.* i. 405.
Tr. nanus *mihi.*

Genus 311. PHILOPÈDON, *Schön. olim.*

THYLACITES *p*, *Germ.*, (*Samou.*)—LIPARUS *p*, *Samou.*—CNEORHINUS *p*, *Schön.*

1779. 1, geminatus. (*Schön. C.* 97.)

Cu. geminatus. *Fabr. E.* ii. 323.—*Herbst C.* vi. 348. *pl.* 87. *f.* 7.
Cu. plagiatus. *Act. Hal.* (*Schall.*) i. 284.
Cu. maritimus. *Marsh.* i. 307.
Li. maritimus. (*Sam. I.* 25.)
β, Cu. lineatocollis. *Marsh. MSS.*
γ, Cu. squamulatus. *Marsh. MSS.*

1780. 2, parapleurus*.
Cu. parapleurus. *Marsh.* i. 306.

1781. 3; exaratus*. (*Schön. C.* 97.)
Cu. exaratus. *Marsh.* i. 303.—*Don.* xii. 58. *pl.* 414. *f.* 2.
β, Cu. sexstriatus. *Marsh.* i. 305.
Li. sexstriatus. (*Sam. I.* 25.)

1782. 4, plumbeus.
Cu. plumbeus. *Marsh.* i. 302.

Genus 312. STROPHOSOMUS, *Billberg.*

THYLACITES *p*, *Germ.*—BRYSSUS, (*Megerle.*)—LIPARUS *p*, *Samou.*

A. Thorax basi truncatus.

a. "*Rostrum basi per strigam transversam a capite distinctum.*" Schön.

1783. 1; Coryli*. (*Schön. C.* 97.)
Cu. Coryli. *Fabr. E.* ii. 524.—*Mart. C. pl.* 19. *f.* 20.—*Panz. F.* xix. *f.* 12.—*Stew.* ii. 60.—*Marsh.* i. 303.—*Turt.* ii. 252.
Li. Coryli. (*Sam. I.* 25.)
Th. Coryli. (*Samou.* 205.)
Cu. capitatus. *Bons. Cu.* ii. 23. *f.* 24.

1784. 2; rufipes* *mihi.*

1785. 3; Asperifoliarum*.
Th. Asperifoliarum. *Kirby MSS.*

1786. 4; obesus*.
Cu. obesus. *Marsh.* i. 304.
Li. obesus. (*Sam. I.* 25.)
Cu. Coryli, c. *Gyll.* iii. 305.

1787. 5; cognatus* *mihi.*

1788. 6; atomarius*.
Cu. atomarius. *Marsh.* i. 312.

1789. 7; subrotundus*.
Cu. subrotundus. *Marsh.* i. 304.
Li. subrotundus. (*Sam. I.* 25.)

b. "*Rostrum capiti contiguum, nec per strigam transversam impressam a fronte distinctum.*" Schön.

1790. 8; nigricans*.
Th. nigricans. *Kirby MSS.*

1791. 9, scrobiculatus.
Cu. scrobiculatus. *Marsh.* i. 307.

1792. 10; squamulatus*. (*Schön. Cu.* 98.)
Cu. squamulatus. *Fabr. E.* ii. 527.—*Herbst C.* vi. *pl.* 87. *f.* 12.

1793. 11; retusus*.
Cu. retusus. *Marsh.* i. 306.
Cu. Coryli, b. *Gyll.* iii. 304.
Cu. cervinus. *Fabr. E.* ii. 528?
Th. affinis. (*DeJean C.* 95.)

1794. 12; nebulosus* *mihi.*

B. "Thorax basi sub-bisinuatus." *Schön.*

1795. 13; chætophorus* *mihi.*

1796. 14; Sus*.
Th. Sus. *Kirby MSS.*
Cu. limbatus. *Marsh.* i. 301.—*Fabr. E.* ii. 527?

1797. 15; pilosellus*. (*Schön. Cu.* 97.)
Cu. pilosellus. *Gyll.* iii. 300.
Cu. pilosus. *Herbst C.* vi. 237. *pl.* 77. *f.* 6.

1798. 16; septentrionis*.
Cu. septentrionis. *Herbst C.* vi. 360. *pl.* 88. *f.* 6.
Cu. setosus. *Fabr. E.* ii. 527.
Cu. scaber. *Bons. Cu.* ii. 35.
Cu. griseo-punctatus. *DeGeer.* v. 244?

Genus 313. SCIAPHILUS, *Schön.*

THYLACITES *p*, *Germar.*—LIPARUS *p*, *Samou.*

1799. 1; muricatus*. (*Schön. Cu.* 99.)
Cu. muricatus. *Fabr. E.* ii. 544.—*Herbst C.* vi. 323. *pl.* 87. *f.* 9.—*Ent. Trans.* (*Burrell.*) i. 183.
Cu. setosus. *Marsh.* i. 304.
Li. setosus. (*Sam. I.* 25.)

1800. [2; asperatus*.]
Cu. asperatus. *Fabr. E.* ii. 541.—*Bons. Cu.* ii. 34. *f.* 25.

1801. 3; pusillus* *mihi.*

Genus 314. BRACHYSOMUS, *Schön. Cu.*

TRACHYPHLÆUS *p*, *Germ.*?

1802. 1; hirsutulus*. (*Schön. Cu.* 99.)
Cu. hirsutulus. *Fabr. E.* ii. 526.—*Panz. F.* vii. *f.* 7.—*Marsh.* i. 305.
Tr. villosulus. *Germ. I. N.* i. 406?
Cu. echinatus. *Bons. Cu.* ii. 21. *f.* 22?

Genus 315. SITONA, *Germar.*, (*Samou.*)

A. " Oculi parvi, subimmersi." *Kirby MSS.*

a. " *Elytra hispida.*" Kirby.

1. Apteri.

1803. 1; Ulicis*. *Kirby MSS.*

1804. 2; Spartii*. *Kirby MSS.*

1805. 3; femoralis* *mihi.*

1806. 4; pleuritica*. *Kirby MSS.*

2. Alati.

1807. 5; hispidula*. (*Samou.* 204.)
Cu. hispidulus. *Fabr. E.* ii. 526.—*Marsh.* i. 310.
Cu. crinitus. *Herbst C.* vi. 245. *pl.* 77. *f.* 13.

1808. 6, pallipes *mihi.*

b. " *Elytra pubescentia, haud pilosa.*" Kirby.

1809. 7; lineata*. (*Samou.* 204.)
Cu. lineatus. *Linn.* ii. 616.—*Marsh.* i. 309.—*Don.* xi. *pl.* 389. *f.* 2.—*Turt.* ii. 252.
Cu. cinerascens. *Herbst C.* vi. 214. *pl.* 75. *f.* 2.
Cu. Mus. *Fabr. E.* ii. 524.

1810. [8; pisivora*] *mihi.*

1811. 9; grisea*.
Cu. griseus. *Fabr. E.* ii. 520. (!)—*Stew.* ii. 60?—*Marsh.* i. 313.—*Turt.* ii. 251.
Cu. lineatus a. *Gyll.* iii. 279.

1812. 10; ruficlavis*.
Cu. ruficlavis. *Marsh.* i. 312.
Cu. lineatus d. *Gyll.* iii. 280.
Cu. chloropus. *Bons. C.* ii. 31. *f.* 19?

1813. 11; nigriclavis*.
Cu. nigriclavis. *Marsh.* i. 312.

1814. 12; canina*. (*DeJean C.* 94.)
Cu. caninus. *Fabr. E.* ii. 524.
Cu. lineatus b. *Payk. F.* iii. 308.

1815. 13; flavescens*.
Cu. flavescens. *Marsh.* i. 311.
Cu. lineatus b. *Bons. C.* ii. 30. *f.* 16.
Cu. canina b. *Gyll.* iii. 278.
Si. octopunctatus. *Germ. I. N.* i. 416.

1816. 14; puncticollis*.
Cu. puncticollis. *Herbst C?*

1817. 15; longiclavis*.
Cu. longiclavis. *Marsh?* *MSS.*

1818. 16, suturalis *mihi.*
Cu. rufipes. *Marsh.* i. 310.

1819. 17; subaurata*. *Kirby MSS.*

1820. 18; tibialis*. (*DeJean C.* 94.)
Cu. tibialis. *Herbst C.* vi. 217. *pl.* 75. *f.* 5.
Cu. sulcifrons. *Thunb. Mus. Up. App.* 6. 113.
Cu. chloropus. *Marsh.* i. 310.
β; Si. affinis. *Kirby MSS.*

1821. 19; humeralis*. *Kirby MSS.*

1822. 20; Pisi* *mihi.*

B. "Oculi magni, prominuli." *Kirby MSS.*

a. *Elytra hispida.*

1823. 21; crinita*. (*DeJean C.* 94.)—*Gyll.* iv. 610. *nota.* (!)
Cu. crinitus. *Oliv. E.* v. 382. *pl.* 3. *f.* 550.
Cu. macularius. *Marsh.* i. 312.
Cu. Occator. *Herbst C.* vi. 219. *pl.* 75. *f.* 8?
Cu. lineellus c. *Gyll.* iii. 281.

1824. 22; lineella*.
Cu. lineellus. *Linn. F. No.* 546.—*Bons. C.* ii. 30. *f.* 18.
Cu. lineatus d. *Payk. F.* iii. 308.

1825. 23; albescens*. *Kirby MSS.*

b. *Elytra subpubescentia.*

1. Corpore brevi, ovato.

1826. 24; cambrica*. *Kirby MSS.*
β, Si. rugulosa. *Kirby MSS.*

2. Corpore elongato, sublineari.

1827. 25; fusca*.
Cu. fuscus. *Marsh.* i. 313.
β, Cu. trisulcus. *Kirby MSS.*

Genus 316. POLYDRUSUS, *Germar.*

POLYDROSUS *p, Schön.*

A. Femoribus dentatis.

1828. 1; amaurus*.
Cu. amaurus. *Marsh.* i. 319.
Cu. Rubi. *Gyll.* iii. 329.

1829. 2, confluens. *Kirby MSS.*

1830. 3; marginatus* *mihi.*

1831. 4; pulchellus* *mihi.*

1832. 5; cervinus*.
Cu. cervinus. *Linn.* ii. 615.—*Marsh.* i. 317.—*Turt.* ii. 254.
Cu. Iris. *Fabr. E.* ii. 546.
Cu. Messor. *Herbst C.* vi. 222. *pl.* 75. *f.* 11.
Cu. griseo-æneus. *DeGeer.* v. 220.

1833. 6; melanotus*. *Kirby MSS.*

1834. 7; sericeus*.
Cu. sericeus. *Act. Hall.* (*Schall.*) i. 286.—(*Sam. I.* 14.)
Cu. splendidus. *Herbst A.* 5. 82. *pl.* 24. *f.* 30.
Po. squamosus. (*Germ. I. N.* i. 452.)

B. Femoribus muticis.

1835. 8; micans*. (*DeJean C.* 93.)
Cu. micans. *Fabr. E.* ii. 519.
Cu. Pyri. *Linn. F. no.* 623.—*Marsh.* i. 317.—*Don.* iv. *pl.* 121. *f.* 314.

1836. 9; flavipes*. (*DeJean C.* 93.)
Cu. flavipes. *DeGeer.* v. 245.—*Marsh.* i. 311.
Cu. sericeus. *Herbst C.* vii. 37. *pl.* 97. *f.* 10. *F.*

Genus 317. ——

POLYDROSUS *p*, *Schönher.*—DICHROA, *Kirby MSS.* (*olim.*)

A. Femoribus muticis.

1837. 1; undatus*.
Cu. undatus. *Fabr. E.* ii. 525.
Cu. tereticollis. *DeGeer.* v. 246.
Cu. albofasciatus. *Herbst C.* vi. 220. *pl.* 75. *f.* 9.
Cu. seleneus. *Marsh.* i. 269.—*Millard. B. E.* 201.

1838. 2; fulvicornis*.
Cu. fulvicornis. *Fabr. E.* ii. 525.
Cu. ruficornis. *Bons. Cu.* ii. 27. *f.* 13.

B. Femoribus dentatis.

1839. 3; oblongus*.
Cu. oblongus. *Linn.* ii. 615.—*Panz. F.* xix. *f.* 15.—*Mart. C.* *pl.* 20. *f.* 33.—*Marsh.* i. 316.—*Turt.* ii. 275.—(*Sam. I.* 14.)
Cu. floricola. *Herbst C.* vi. 220. *pl.* 75. *f.* 10.
β; Cu. rufescens. *Marsh.* i. 316.
γ; Di. testacea. *Kirby MSS.*

Genus 318. PHYLLOBIUS, *Schön.*

POLYDRUSUS *p*, *Germ.*—CHRYUS, (*Megerle.*)—DASCIRUS, (*Megerle.*)—MURANUS *p*, (*Megerle.*)—EUSOMUS *p?* *Germ.*—CHLOROLEPIS *p?* (*Megerle?*)

A. Femoribus dentatis.

1840. 1; Pyri*. (*Schön. Cu.* 182.)

Cu. Pyri. *Linn.* ii. 615.—*Berk. S.* i. 100.—*Panz. F.* cvii. *f.* 4. —*Stew.* ii. 60.—*Turt.* ii. 274.—(*Samou.* 54. *pl.* 2. *f.* 19.)
Cu. æruginosus. *Bons. Cu.* ii. 23. *f.* 8.

1841. [2; cæsius*.]
Cu. cæsius. *Marsh.* i. 318.
β, Po. angustatus. *Kirby MSS.*

1842. 3; Alneti*.
Cu. Alneti. *Fabr. E.* ii. 542.
Cu. cnides. *Marsh.* i. 318.—(*Sam. I.* 14.)
Cu. Pyri b. *Gyll.* iii. 323.

1843. 4; maculicornis*. *Germ. I. N.* i. 449.
Cu. Urticæ. *DeGeer.* v. 219?

1844. 5; argentatus*. (*Schön. Cu.* 182.)
Cu. argentatus. *Linn.* ii. 615.—*Berk. S.* 100.—*Don.* iii. *pl.* 107.—*Stew.* ii. 60.—*Marsh.* i. 318.—*Turt.* ii. 274.—*Shaw.* G. Z. vi. 66.—(*Leach E. E.* ix. 108.)—(*Sam. I.* 14.)
β; Ph. flavidus, *Kirby MSS.*
γ; Ph. femoralis. *Kirby MSS.*
δ; Ph. nigripes. *Kirby MSS.*
ε; Ph. angustior. *Kirby MSS.*

1845. 6; Mali*. (*Schön. Cu.* 182.)
Cu. Mali. *Fabr. E.* ii. 542.—*Mart. C. pl.* 19. *f.* 21, 22.—*Herbst C.* vi. 261. *pl.* 79. *f.* 5. *a, b.*—*Marsh.* i. 317.—(*Sam. I.* 14.)
β; Cu. Padi. *Bons. Cu.* ii. 26. *f.* 11.
γ; Cu. fulvipes. *Fabr. E.* ii. 525.
Cu. vespertinus. *Fabr. E.* ii. 542?

B. Femoribus muticis.

1846. 7; Pomonæ*. *Germ. I. N.* i. 450.
Cu. Pomonæ. *Oliv. E.* v. 455. *pl.* 35. *f.* 548.

1847. 8; uniformis*.
Cu. uniformis. *Marsh.* i. 311.
β; Po. obscurior. *Kirby MSS.*

1848. 9; albidus* *mihi.*
Po. canescens. *Leach MSS.*—*nec* Po. canescens. *Germ. l. c.*

1849. 10; parvulus*. (*Schön. Cu.* 182.)
Cu. parvulus. *Fabr. E.* ii. 528?
Cu. fulvipes. *Payk. F.* iii. 310.
Cu. argentatus b. *Bons. Cu.* ii. 27.

1850. 11; minutus* *mihi.*

1851. 12; viridicollis*. (*Schön. Cu.* 182.)
Cu. viridicollis. *Fabr. E.* ii. 528.—*Panz. F.* xix. *f.* 13.

Genus 319. TANYMECUS, *Germar.*

1852. 1; palliatus*. (*DeJean C.* 94.)
Cu. palliatus. *Fabr. E.* ii. 513.—*Panz. F.* xix. *f.* 5.
Cu. canescens. *Herbst C.* vi. 240.
Cu. diffinis. *Marsh.* i. 309.
Cu. graminicola. *Oliv. E?*

1853. [2; affinis* *mihi.*]

Genus 320. BRACHYDERES[a].

Genus 321. CLEONUS, *Schön.*

LIXUS *p, Illig., Leach.*—EPIMECES *p, Billb.*—CLEONIS *p,* (*Megerle.*)—GEOMORUS, *Schön. olim.*

A. Rostro carinato.

1855. 1, distinctus*.
Cu. distinctus. *Fabr. E.* ii. 516.
Cu. ophthalmicus. *Panz. F.* lvii. *f.* 7.

1856. 2: nebulosus*. (*Schön. Cu.* 146.)
Cu. nebulosus. *Linn.* ii. 617.—*Berk. S.* i. 100.—*Herbst C.* vi. 76. *pl.* 64. *f.* 8.—*Stew.* ii. 61.—*Marsh.* i. 308.—*Turt.* ii. 248.—(*Kirby & Sp. I. E.* ii. 219.)
Cu. carinatus. *DeGeer.* v. 241.

1857. 3, glaucus*. (*Ing. Inst.* 88.)
Cu. glaucus. *Fabr. E.* ii. 516.—*Panz. F.* xix. *f.* 6.
Cu. nebulosus. *Payk. F.* iii. 298.

B. Rostro canaliculato.

1858. 4; sulcirostris*. (*Schön. Cu.* 146.)
Cu. sulcirostris. *Linn.* ii. 617.—*Mart. C. pl.* 19. *f.* 23.—*Marsh.* i. 308.—*Don.* xv. *pl.* 509. *f.* 1.
Li. sulcirostris. *Latr.* G. ii. 260.—(*Leach E. E.* ix. 108.)

Genus 322. BOTHYNODERES, *Schön.*

LIXUS *p, Oliv.*—EPIMECES *p, Billb.*—CLEONIS *p, DeJean.*

1859. 1, albidus. (*Schön. Cu.* 148.)
Cu. albidus. *Fabr. E.* ii. 517.—*Panz. F.* xix *f.* 7.

---

[a] Genus 320. BRACHYDERES, *Schön.*

NAUPACTUS, (*Megerle.*)—THYLACITES *p, Germ., Samou.*

1854 † 1. incanus. (*Schön. Cu.* 103.)
Cu. incanus. *Linn.* ii. 616.—*Berk. S.* i. 100. (!)—*Stew.* ii. 61. (!)—*Turt.* ii. 250. (!)
Th. incanus. (*Samou.* 205. (!)

Cl. albidus. (*Ing. Inst.* 88.)
Cu. niveus. *Bons. Cu.* ii. 21. *f.* 6.

Genus 323. LIXUS, *Fabr., Leach, Samou.*

A. Elytris apice acuminatis:—(LEPTOSOMA, *Leach.*)

1860 ‡ 1, Ascanii. *Fabr. E.* ii. 503.
Cu. Ascanii. *Linn.* ii. 610.—*Panz. F.* xlii. *f.* 13.
In Mus. D. *Sparshall?*

1861. 2, paraplecticus. *Fabr. E.* ii. 498.—(*Leach E. E.* ix. 108.) —(*Sam. I.* 25.)
Cu. paraplecticus. *Linn.* ii. 610.—*Don.* x. *pl.* 348. *f.* 2.—*Stew.* ii. 57.—*Marsh.* i. 272.—*Turt.* ii. 219.—*Bingley.* iii. 135.

1862. 3; productus*. (*Sam. I.* 25.)
Cu. productus. *Marsh. MSS.*

B. Elytris apicè muticis.

1863. 4; angustatus. *Fabr. E.* ii. 502. (!)
Cu. angustatus. *Stew.* ii. 57.—*Marsh.* i. 273.—*Turt.* ii. 220. —*Mart. C. pl.* 20. *f.* 43.—*Panz. F.* xlii. *f.* 2.

Genus 324. LARINUS, (*Schüppel.*), *Germar.*

RHINOBATUS *p*, (*DeJean.*)

1864. 1, Sturnus*. *Germ. I. N.* i. 384.
Cu. Sturnus. *Herbst C.* vi. 126. *pl.* 68. *f.* 5.
Cu. Jaceæ mas. *Herbst C.* vi. 122. *pl.* 68. *f.* 2.
Rh. Fringilla. (*DeJean C.* 98.)

Genus 325. RHINOBATUS, (*Megerle.*)

LARINUS *p*, *Germar.*—RHYNCHÆNUS *p*, *Samou.*

1865. 1; planus. *Germ. I. N.* i. 389.
Cu. planus. *Fabr. E.* ii. 441.
Cu. odontalgicus. *Oliv. E?*
Cu. ebeneus. *Marsh.* i. 270.—(*Don.* xv. *pl.* 509. *f.* 2.)
Rh. ebeneus. (*Sam. I.* 36.)

Genus 326. RHINOCYLLUS, (*Germar.*)

LIXUS *p*, *Oliv.*—RHINOBATUS *p*, *DeJean.*—RHINOMACER, *Leach.*

1866. 1, thaumaturgus. (*Schön. Cu.* 59.)
Cu. thaumaturgus. *Rossi.*
Li. antiodontalgicus. *Illiger?*
Cu. Cardui. *Don.* xv. *pl.* 512.
Rh. Cardui. *Leach MSS.*
Cu. hydrorhynchus. *Marsh. MSS.*

Genus 327. CHLOROPHANUS[a].

Genus 328. MAGDALIS, *Germar*, (*Samou.*)

RHINODES *p*, *DeJean.*—BALANINUS *p*, *Samou. I.*—THAMNOPHILUS[b] 1. *Schönh.*

1868. 1; carbonaria*. (*Germ. I. N.* i. 193.)
Cu. carbonarius. *Linn.* ii. 612.—*Herbst C.* vi. 70. *pl.* 64. *f.* 3. —*Turt.* ii. 236?

1869. 2; atramentaria*. (*Germ. I. N.* i. 193.)
Cu. atramentarius. *Marsh.* i. 293.
Ba. atramentarius. (*Sam. I.* 5.)
Cu. aterrimus. *Herbst C.* vi. 72. *pl.* 64. *f.* 5.
♂; Rh. atratus. *Gyll.* iii. 187.
Cu. Cerasi ♂. *Payk. F.* iii. 193.
♀; Rh. carbonarius. *Gyll.* iii. 185.

1870. 3; aterrima*. (*Germ. I. N.* i. 193.)
Rh. aterrimus. *Fabr. E.* ii. 486.
Cu. Cerasi ♂. *Oliv. E.* v. 229. *pl.* 22. *f.* 309.
Cu. stygius. *Marsh.* i. 294.
Ba. stygius. (*Sam. I.* 5.)

1871. 4; Cerasi*.
Rh. Cerasi. *Fabr. E.* ii. 486.
Cu. Cerasi. *Turt.* ii. 236.—*Panz. F.* xlii. *f.* 19.
Ma. asphaltina. (*Germ. I. N.* i. 193.)

Genus 329. RHINODES.

RHINA *p*, *Olivier.*—MAGDALIS *p*, *Germar.*—EDO? (*Germar.*)—THAMNOPHILUS 2. *Schön.*—RHINODES *p*, *DeJean.*

1872. 1; Pruni*. (*DeJean C.* 98.)
Cu. Pruni. *Linn.* ii. 607.—*Berk. S.* i. 97.—*Turt.* ii. 214.
Cu. erythroceras. *Herbst C.* vi. 73. *pl.* 64. *f.* 6.
Rh. ruficornis. *Schönher.*

1873. [2, caliginosus] *mihi.*
Cu. Pruni. *Marsh.* i. 247.

1874. 3; Cerasi*. (*DeJean C.* 98?)
Cu. Cerasi. *Linn.* ii. 607.—*Berk. S.* i. 97.—*Stew.* ii. 55.—*Marsh.* i. 265.—*Herbst C.* vi. 68. *pl.* 64. *f.* 1. ♀.
Rh. Armeniacæ. *Fabr. E.* ii. 447.

---

[a] Genus 327. CHLOROPHANUS, (*Dalman.*)

CHLORIMA, (*DeJean.*)—PLATYRHINCHUS *p*, *Megerle.*—BRACHYRHINUS *p*, *Latr.*

1867 ‡ 1. viridis. (*Schön. Cu.* 54.)
Cu. viridis. *Linn.* ii. 616.—*Panz. F.* cvii. *f.* 3.—*Stew.* ii. 61. (!)
Cu. flavocinctus. *DeGeer.* v. 256.

[b] THAMNOPHILUS: Genus Avium. Vide *Viellot Anal.* 40.

1875. [4; Rhina*.]
Rh. Rhina. *Gyll.* iii. 83.
Rhina Cerasi. *Illig. M.* (*Latr.*) iii. 104?

Genus 330. PANUS, *Schönh.*

RHINA *p, Latr.*—MAGDALIS *p, Germ.*

1876. 1; barbicornis*.
Rh. barbicornis. *Latr. G.* ii. 264.
Rh. Rhina b. *Gyll.* iii. 83.
Ma. barbicornis. *Germ. I. N.* i. 192.

Genus 331. APION, *Herbst, Kirby, Leach, Samou.*

ATTELABUS *p, Fabr.*—RHINOMACER *p, Clairv.*—OXYSTOMA *p, Dumeril.*—APIUS *p, Billberg.*

A. Rostrum apice subulatum.

a. *Rostro brevi.*

1877. 1; Craccæ*. *Herbst C.* vii. 102. *pl.* 102. *f.* 2. *B.*—*Linn. Trans.* (*Kirby.*) ix. 29.—(*Sam. I.* 3.)
Cu. Craccæ. *Linn.* ii. 606.—*Marsh.* i. 245.
Cu. Viciæ. *DeGeer.* v. 253.

1878. [2; ruficorne*.] *Herbst C.* vii. 110. *pl.* 102. *f.* 8. *U.*—*Linn. Trans.* (*Kirby.*) ix. 30.—(*Sam. I.* 3.)
Ap. Craccæ, ♂. *Gyll.* iii. 38.

1879. 3; Pomonæ*. *Gyll.* iii. 39.
Cu. Pomonæ. *Fabr. E.* ii. 425.
Cu. cærulescens. *Marsh.* i. 245.
Ap. cærulescens. *Linn. Trans.* (*Kirby.*) ix. 27. *pl.* 1. *f.* 4.—(*Sam. I.* 3.)
β; Cu. glaber. *Marsh.* i. 245.

b. *Rostro elongato.*

1880. 4; subulatum*. *Linn. Trans.* (*Kirby.*) ix. 28. *pl.* 1. *f.* 5.—(*Sam. I.* 3.)

1881. 5; Marshami* *mihi.*

B. Rostrum filiforme, aut cylindricum.

a. *Rostro brevi.*

1. Antennis in medio rostri positis.

1882. 6, Limonii. *Linn. Trans.* (*Kirby.*) ix. 78. *pl.* 1. *f.* 20.

1883. 7; Rumicis*. *Linn. Trans.* (*Kirby.*) ix. 67.—(*Sam. I.* 3.)

1884. 8; affine*. *Linn. Trans.* (*Kirby.*) ix. 68.

1885. 9; Curtisii*. *Kirby MSS.*

1886. 10; Spartii*. *Linn. Trans.* (*Kirby.*) ix. 56.—(*Sam. I.* 3.)

1887. 11; curtirostre*. *Germ. M.* ii. 230.
Ap. brevirostre. *Linn. Trans.* (*Kirby.*) ix. 68.

1888. [12; humile*.] *Germ. M.* ii. 232. *pl.* 3. *f.* 1.

1889. 13; velox*. *Linn. Trans.* (*Kirby.*) x. 349.
β; Ap. minimum. *Germ. M.* ii. 234. *pl.* 3. *f.* 9?

1890. 14; simile*. *Linn. Trans.* (*Kirby.*) x. 351.

1891. 15; tenue*. *Linn. Trans.* (*Kirby.*) ix. 61.

1892. 16, seniculus. *Linn. Trans.* (*Kirby.*) ix. 61.
Ap. tenuius. *Gyll.* iii. 57.

1893. [17; plebeium*.] *Germ. M.* iii. 43?

1894. 18; violaceum*. *Linn. Trans.* (*Kirby.*) ix. 65. *pl.* 1. *f.* 16. —(*Sam. I.* 3.)
Cu. Fagi var. *Linn. Mus.*

1895. 19; Hydrolapathi*. *Linn. Trans.* (*Kirby.*) ix. 66. *pl.* 1. *f.* 17.—(*Sam. I.* 3.)
Cu. Hydrolapathi. *Marsh.* i. 249.
Ap. cyaneum. *Herbst C.* vii. 108. *pl.* 102. *f.* 7?
At. cyaneus. *Turt.* ii. 284?

1896. 20; cæruleopenne* *mihi.*

1897. 21; Malvæ*. *Linn. Trans.* (*Kirby.*) ix. 20. *pl.* 1. *f.* 2.—(*Sam. I.* 3.)
At. Malvæ. *Fabr. E.* ii. 426.
Cu. Malvæ. *Marsh.* i. 246.—*Stew.* ii. 54.—*Turt.* ii. 284.

1898. 22; hæmatodes*. *Linn. Trans.* (*Kirby.*) ix. 76.—*Germ. M.* ii. 251. *pl.* 4. *f.* 25.—(*Sam. I.* 3.)
Ap. frumentarium. *Gyll.* iii. 32.

1899. 23; frumentarium*. *Linn. Trans.* (*Kirby.*) ix. 77.—(*Sam. I.* 3.)
Cu. frumentarius. *Linn.* ii. 608.—*Panz. F.* xx. *f.* 14.—*Marsh.* i. 242.—*Turt.* ii. 284.—*Shaw* G. *Z.* vi. 65.
Cu. sanguineus. *DeGeer.* v. 251.

2. Antennis ad basin rostri positis.

1900. 24; vernale*. *Herbst C.* vii. 123.—*Linn. Trans.* (*Kirby.*) ix. 21.—*Germ. M.* ii. 131. *pl.* 2. *f.* 7. *a*, *b*.—(*Sam. I.* 3.)
Cu. vernalis. *Payk. F.* iii. 183.
Cu. concinnus. *Marsh.* i. 248.
Cu. Lythri. *Panz. F.* xvii. *f.* 8.
Cu. urticarius. *Fuesly. A.* 74.

1901. 25; Onopordi*. *Linn. Trans.* (*Kirby.*) ix. 71.—*Germ. M.* ii. 240. *pl.* 2. *f.* 14.

1902. 26; scabricolle* *mihi.*

1903. 27; caliginosum* *mihi.*

1904. 28; elongatum * *mihi.*

1905. 29; Radiolus*. *Linn. Trans.* (*Kirby.*) ix. 56.—*Germ. M.* ii. 246. *pl.* 2. *f.* 13.
Cu. Radiolus. *Marsh.* i. 247.
Cu. aterrimus. *Marsh.* i. 244.
Ap. oxurum. *Linn. Trans.* (*Kirby.*) ix. 57.—(*Sam. I.* 3.)

1906. 30; nigrescens * *mihi.*

1907. 31; lævigatum *. *Linn. Trans.* (*Kirby.*) ix. 70.—(*Sam. I.* 3.)

1908. 32; æneum *. *Herbst C.* vii. 101. *pl.* 102. *f.* 1. *A.*—*Linn. Trans.* (*Kirby.*) ix. 74.—(*Sam. I.* 3.)
Cu. æneus. *Fabr. E.* ii. 423.—*Stew.* ii. 56.—*Marsh.* i. 243.—*Turt.* ii. 214.
Ap. Craccæ. *Panz. F.* xx. *f.* 10.
β; Cu. chalceus. *Marsh.* i. 243.
At. cyaneus. *Panz. F.* xx. *f.* 12.

1909. 33; Carduorum *. *Linn. Trans.* (*Kirby.*) ix. 72. *pl.* 1. *f.* 19. —(*Sam. I.* 3.)
Cu. Sorbi. *Marsh.* i. 244.
Cu. cyaneus. *DeGeer.* v. 252.
At. æneus β. *Payk. F.* iii. 180.
Ap. gibbirostre. *Gyll.* iii. 52.

b. *Rostro longo vel mediocri.*

1. Antennæ basales.

* *Thorace subcylindrico.*

1910. 34; rufirostre *. *Herbst C.* vii. 111. *pl.* 102. *f.* 10. *K.*—*Linn. Trans.* (*Kirby.*) ix. 37.
Cu. rufirostris. *Fabr. E. S.* ii. 390. (!)—*Stew.* ii. 55.—*Marsh.* i. 246.—*Turt.* ii. 283.

1911. [35; Malvarum *.] *Linn. Trans.* (*Kirby.*) ix. 33.—(*Sam. I.* 3.)
Cu. Trifolii. *Marsh.* i. 246.—*Linn. Trans.* (*Markwick.*) vi.

1912. 36, pallipes. *Linn. Trans.* (*Kirby.*) ix. 38. *pl.* 1. *f.* 7.
† β? Ap. geniculatum. *Germ. M.* ii. 175. *pl.* 3. *f.* 25?

1913. 37; confluens *. *Linn. Trans.* (*Kirby.*) ix. 62. *pl.* 1. *f.* 15.

1914. 38; pusillum *. *Germ. M.* ii. 209. *pl.* 2. *f.* 4.
β, Ap. atomarium. *Germ. M.* ii. 209.

1915. 39; pubescens *. *Linn. Trans.* (*Kirby.*) x. 350.

1916. 40, vicinum. *Linn. Trans.* (*Kirby.*) ix. 25. *pl.* 1. *f.* 3.—(*Sam. I.* 3.)
♀? Ap. incrassatum. *Germ. M.* ii. 140. *pl.* 2. *f.* 3. *a, b?*

** *Thorace subgloboso.*

1917. 41, Hookeri. *Linn. Trans.* (*Kirby.*) ix. 69. *pl.* 1. *f.* 18.

2. Antennis submediis.

* *Coleoptra subglobosa.*

1918. 42; Pisi*. *Germ. M.* ii. 90.
At. Pisi. *Fabr. E.* ii. 425.
Cu. striatus. *Marsh.* i. 249.
Ap. striatum. *Linn. Trans.* (*Kirby.*) ix. 52. *pl.* 1. *f.* 10.

1919. 43; atratulum*. *Germ. M.* ii. 192. *pl.* 3. *f.* 16.

1920. 44; immune*. *Linn. Trans.* (*Kirby.*) ix. 52.—(*Sam. I.* 3.)

1921. [45; carbonarium*.] *Germ. M.* ii. 176. *pl.* 3. *f.* 17.

1922. 46, Sorbi. *Herbst C.* vii. 111. *pl.* 102. *f.* 9. *I.*—*Linn. Trans.* (*Kirby.*) ix. 46.—(*Sam. I.* 3.)
At. Sorbi. *Fabr. E.* ii. 426.
Cu. lævigatus. *Payk. M. Cu.* 133.
Cu. viridescens. *Marsh.* i. 249.

1923. 47; Ervi*. *Linn. Trans.* (*Kirby.*) ix. 23.—*Germ. M.* ii. 133. *pl.* 3. *f.* 13. *a, b.*—(*Sam. I.* 3.)

1924. 48; sulcifrons*. *Herbst C.* vii. 132. *pl.* 103. *f.* 12. *M.*

1925. 49; punctigerum*. *Germ. M.* ii. 188.
At. punctiger. *Panz. F.* iii. 179.
Ap. sulcifrons. *Linn. Trans.* (*Kirby.*) ix. 50.—(*Sam. I.* 3.)

1926. 50, Spencii. *Linn. Trans.* (*Kirby.*) ix. 57. *pl.* 1. *f.* 13.

1927. 51; subcæruleum* *mihi.*

1928. 52, unicolor. *Linn. Trans.* (*Kirby.*) ix. 58.
Ap. Æthiops. *Gyll.* iii. 54?

** *Coleoptra ovata.*

1929. 53; virens*. *Linn. Trans.* (*Kirby.*) ix. 53.—(*Sam. I.* 3.)—*Germ. M.* ii. 193. *pl.* 4. *f.* 2.
Ap. æneocephalum. *Gyll.* iii. 49.

1930. 54; Marchicum*. *Linn. Trans.* (*Kirby.*) ix. 54.—(*Sam. I.* 3.)—*Germ. M.* ii. 196. *pl.* 3. *f.* 15.
Ap. violaceum. *Gyll.* iii. 50.

1931. 55, Astragali. *Linn. Trans.* (*Kirby.*) ix. 55. *pl.* 1. *f.* 12.—*Ent. Trans.* (*Haworth.*) i. 339.—(*Sam. I.* 3.)
At. Astragali. *Payk. F.* iii. 180.

1932. 56, Loti. *Linn. Trans.* (*Kirby.*) ix. 53.—(*Sam. I.* 3.)
♀? Ap. incrassatum. *Germ. M.* ii. 140. *pl.* 2. *f.* 3?

1933. 57; civicum*. *Germ. M.* ii. 234. *pl.* 3. *f.* 12.

1934. 58; auratum* *mihi.*

1935. 59; pavidum*. *Germ. M.* ii. 203. *pl.* 4. *f.* 4.

1936. 60; scutellare*. *Linn. Trans.* (*Kirby.*) x. 355.

1937. 61, obscurum. *Linn. Trans.* (*Kirby.*) ix. 33.
Cu. obscurus. *Marsh.* i. 244.

1938. 62; flavipes*. *Herbst C.* vii. 106. *pl.* 102. *f.* 5. *C.*—*Linn. Trans.* (*Kirby.*) ix. 37.—(*Sam. I.* 3.)

1939 † 63. Fagi[a].

3. Antennis mediis.

* "Coleoptris ovalibus." *Germar.*

1940. 64; nigritarse*. *Linn. Trans.* (*Kirby.*) ix. 36. *pl.* 1. *f.* 6. —(*Sam. I.* 3.)

1941. 65; assimile*. *Linn. Trans.* (*Kirby.*) ix. 42.—(*Sam. I.* 3.)

1942. 66; flavifemoratum*. *Linn. Trans.* (*Kirby.*) ix. 42.—(*Sam. I.* 3.)—(*Kirby & Sp. I. E.* i. 176.)
Ap. apricans. *Herbst C.* vii. 127. *pl.* 103. *f.* 5.
Ap. flavipes b. *Gyll.* iii. 36.
Cu. ochropus. *Müll. Z. D. pr.* 1018.
Cu. Trifolii var. *Marsh.* i. 246.

1943 † 67, lævicolle. *Linn. Trans.* (*Kirby.*) x. 348.
In Mus. D. *Kirby.*

1944. 68; æstivum*. *Germ. M.* ii. 169. *pl.* 4. *f.* 16.
Ap. flavifemoratum β. *Linn. Trans.* (*Kirby.*) ix. 43.
Cu. flavipes. *Laich. T.* i. 132.
At. flavipes. *Turt.* ii. 284?

1945. [69; ruficrus*.] *Germ. M.* ii. 171. *pl.* 4. *f.* 17.

1946. 70, Leachii* *mihi.*
Ap. Ononidis. *Gyll.* iv. 539?

1947. 71; Gyllenhalii*. *Linn. Trans.* (*Kirby.*) ix. 63.—(*Sam. I.* 3.)
Ap. æthiops. *Gyll.* iii. 54?
At. flavipes β. *Payk. F.* iii. 182.

1948. 72; varipes*. *Germ. M.* ii. 173. *pl.* 4. *f.* 19.
Ap. flavipes c. *Gyll.* iii. 37.
Ap. flavifemoratum γ. *Linn. Trans.* (*Kirby.*) ix. 43.

1949 † 73, difforme. *Germ. M.* iii. 46. In Mus. D. *Haworth.*

1950. 74; filirostre. *Linn. Trans.* (*Kirby.*) ix. 46.

1951 † 75. glabratum. *Germ. M.* iii. 47. (!) In Mus. D. Spence.

1952. 76; ebeninum*. *Linn. Trans.* (*Kirby.*) ix. 55. *pl.* 1. *f.* 11.

1953 † 77, Kirbii. *Leach MSS.*—*Germ. M.* iii. 50.
In Mus. *Brit. et* D. *Kirby.*

** "Coleoptris obovalibus." *Germar.*

1954. 78, Viciæ. *Linn. Trans.* (*Kirby.*) ix. 31.—*Germ. M.* ii. 150. *pl.* 4. *f.* 15. *a, b.*—(*Sam. I.* 3.)
At. Viciæ. *Payk. F.* iii. 181.

---

[a] 1939 † 63. Fagi. *Linn. Trans.* (*Kirby.*) ix. 40. *pl.* 1. *f.* 8.—(*Sam. I.* 3) (!)
Cu. Fagi. *Linn.* ii. 611.—*Berk. S.* i. 99. (!)—*Stew.* ii. 58. (!)—*Turt.* ii. 243. (!)

1955. 79, Ononis. *Linn. Trans.* (*Kirby.*) ix. 25.—*Germ. M.* ii. 137. *pl.* 3. *f.* 24. *a, b.*—(*Sam. I.* 3.)

1956. 80; Lathyri*. *Linn. Trans.* (*Kirby.*) ix. 24.—(*Sam. I.* 3.)

1957. 81; vorax*. *Herbst C.* vii. 129. *pl.* 103. *f.* 8. *H.*—*Linn. Trans.* (*Kirby.*) ix. 26.—(*Sam. I.* 3.)
♂; Cu. fuscicornis. *Marsh.* i. 244.
♀; Cu. villosulus. *Marsh.* i. 250.

*** "Coleoptris subglobosis." *Germar.*

1958. 82; punctifrons*. *Linn. Trans.* (*Kirby.*) ix. 50. *pl.* 1. *f.* 9.

1959. 83, subsulcatum. *Linn. Trans.* (*Kirby.*) ix. 46.—(*Sam. I.* 3.)
Cu. subsulcatum. *Marsh.* i. 249.
Ap. cæruleum. *Herbst C.* vii. 123. *pl.* 102. *f.* 11. *L?*

1960 † 84, foveolatum. *Linn. Trans.* (*Kirby.*) ix. 48.
Ap. cyaneum. *Gyll.* iii. 45. In Mus. D. *Kirby.*

1961. 85; Meliloti*. *Linn. Trans.* (*Kirby.*) ix. 64.—(*Sam. I.* 3.)

Genus 332. OXYSTOMA.

Apion *p, Herbst, Kirby, Samou.*—Attelabus *p, Fabricius.*—Oxystoma *p, Dumeril.*—Apius *p, Billb.*—Eurhynchus? *Kirby? MSS.*

1962. 1; fuscirostris*.
At. fuscirostris. *Fabr. E.* ii. 424.
Ap. albovittatum. *Herbst C.* vii. 126. *pl.* 103. *f.* 4. D?
Cu. melanopus. *Marsh.* i. 248.
Ap. melanopum. *Linn. Trans.* (*Kirby.*) ix. 19. *pl.* 1. *f.* 1.—(*Sam. I.* 3.)

1963. 2; Ulicis*.
Cu. Ulicis. *Forst. C.* i. 31.—*Stew.* ii. 57.—*Marsh.* i. 256.—*Turt.* ii. 222.
Ap. Ulicis. *Linn. Trans.* (*Kirby.*) ix. 18.—(*Sam. I.* 3.)—(*Kirby & Sp. I. E.* iv. 487.)

1964. 3; Genistæ*.
Ap. Genistæ. *Linn. Trans.* (*Kirby.*) x. 347.

Genus 333. RAMPHUS, *Clairville, Leach,* (*Kirby.*)

Rhynchænus *p, Gyll.*

1965. 1; flavicornis*. *Clairv. E. H.* i. 104. *pl.* xii. *f.* 1–4.—(*Leach. E. E.* ix. 105.)
Cu. Oxyacanthæ. *Marsh.* i. 263.

1966. [2; pulicarius*.]
Cu. pulicarius. *Herbst C.* vi. 429. *pl.* 93. *f.* 12.

Genus 334. DEPORAÜS, *Leach MSS.*, *Samou.*

ATTELABUS *p*, *Linn.*, *Marsh.*—RHYNCHITES *p*, *Herbst.*—APODERUS *p*, (*Kirby.*)

1967. 1; Betulæ*. (*Sam. I.* 14.)
At. Betulæ. *Linn.* ii. 620.—*Panz. F.* xx. *f.* 15.—*Marsh.* i. 321.
—*Wood.* i. 42. *pl.* 15.
Cu. Populi b. *Scop. no.* 74.
Cu. excoriato-niger. *DeGeer.* v. 259.
At. Betuleti. *Fabr. E.* ii. 421.—*Turt.* ii. 285.

Genus 335. RHYNCHITES, *Herbst*, *Leach*, *Samou.*, (*Kirby.*)

ATTELABUS *p*, *Linné.*—RHINOMACER *p*, *Geoffroy.*—MECHORIS, *Billb.*

A. Rostro brevi.

a. *Corpore subcylindrico, glabro.*

1968. 1; cylindricus*. *Kirby MSS.*—(*Sam. I.* 36.)

b. *Corpore pubescente.*

1969. 2; pubescens*. *Herbst C.* vii. 138. *pl.* 105. *f.* 3.—(*Sam. I.* 36.)
Cu. pubescens. *Marsh.* i. 240.
At. pubescens. *Turt.* ii. 282.

1970. 3; ophthalmicus* *mihi.*

B. Rostro elongato.

a. *Corpore angustato, nudo.*

1971. 4; angustatus*. *Leach MSS.*—(*Sam. I.* 36.)

b. *Corpore ovato.*

1. Elytris distinctè striatis. (Thorace in utroque sexû mutico.)

1972. 5; Alliariæ*. *Gyll.* iii. 26.—(*Sam. I.* 36.)
Cu. Alliariæ. *Linn.* ii. 606.—*Berk. S.* i. 97.—*Stew.* ii. 54.—
Marsh. ii. 238.
At. Alliariæ. *Turt.* ii. 283.
Cu. cæruleus. *DeGeer.* v. 251.

1973. 6; interpunctatus*. *Wilkin? MSS.*

1974. 7; nanus*. *Gyll.* iii. 28.—(*Sam. I.* 36.)
At. nanus. *Payk. F.* iii. 776.
Cu. nanus. *Marsh.* i. 238.

1975. 8; atrocæruleus* *mihi.*

1976. 9; æquatus*. *Herbst C.* vii. 132.—(*Sam. I.* 36.)
Cu. æquatus. *Linn.* ii. 607.—*Don.* iv. *pl.* 121. *f.* 1, 2.—
Stew. ii. 55.—*Marsh.* i. 238.
At. æquatus. *Turt.* ii. 283.
β; At. nigripes. *Kirby MSS.*

1977 † 10, cæruleocephalus. (*Schön. Cu.* 45.)
At. cæruleocephalus. *Fabr. E.* ii. 423.
At. cyanocephalus. *Herbst A. pl.* 24. *f.* 11? In Mus. *Brit.*

1978. 11; minutus*. *Herbst C.* vii. 135.
Cu. æneovirens. *Marsh.* i. 239.
Rh. æneovirens. (*Sam. I.* 36.)

1979. 12, cupreus*. *Herbst C.* vii. 138. *pl.* 105. *f.* 21?—(*Sam. I.* 36.)
Cu. cupreus. *Linn.* ii. 608.—*Marsh.* i. 239.
At. cupreus. *Turt.* ii. 283.

2. Elytris punctatis vix striatis. (Thorace in uno sexû spinoso.)

1980. 13; Betulæ*. (*Sam. I.* 36.)
Cu. Betulæ. *Linn.* ii. 611.—*Berk. S.* i. 99.—*Don.* iii. *pl.* 74. —*Stew.* ii. 57.—*Marsh.* i. 241.
At. Betulæ. *Turt.* iii. 282.
Rh. Betuleti. *Herbst C.* vii. 246.
At. Populi. *Payk. F.* iii. 170.
β; Cu. nitens. *Marsh.* i. 242.
Rh. nitens. *Samou.* 54. *pl.* 2. *f.* 18.

1981. 14; Populi*. *Herbst C.* vii. 128.—(*Sam. I.* 36.)
Cu. Populi. *Linn.* ii. 611.—*Panz. F.* xx. *f.* 7.—*Marsh.* i. 241.
At. Populi. *Turt.* ii. 283.

1982. 15, Bacchus*. *Herbst C.* vii. 124.—(*Leach E. E.* ix. 107.) —(*Sam. I.* 36.)—(*Kirby & Sp. I. E.* i. 196 & 202.)
Cu. Bacchus. *Linn.* ii. 611.—*Don.* i. *pl.* 34. *f.* 1.—*Stew.* ii. 57. —*Marsh.* i. 240.—*Wood.* i. 41. *pl.* 14.
At. Bacchus. *Turt.* ii. 282.
β, Cu. purpureus. *Linn.* ii. 607?—*Berk. S.* i. 98?—*Stew.* ii. 55?

Genus 336. ATTELABUS, *Linné, Marsh, Leach, Samou.*

RHINOMACER *p*, *Laich.*—RHYNCHITES *p*, *Illig.*—CHYPHUS, *Thunb.*

1983. 1; curculionoides*. *Linn.* ii. 619.—*Mart. C. pl.* 23. *f.* 6. —*Don.* v. *pl.* 149.—*Stew.* ii. 62.—*Marsh.* i. 320.—*Turt.* ii. 282. —(*Leach E. E.* ix. 107.)—(*Sam. I.* 5.)
Cu. nitens. *Payk. Cu.* 122.
At. Coryli. *Berk. S.* i. 100.

Genus 337. APODERUS, *Oliv.*, *Leach*, *Samou.*, (*Kirby.*)

ATTELABUS *p*, *Fabr.*, *Marsh.*—APODERES, *Schön.*

1984. 1; Avellanæ*. *Olivier.*—(*Leach E. E.* ix. 107.)
At. Avellanæ. *Linn.* ii. 619.—*Marsh.* i. 319.
At. Coryli β. *Fabr. E.* ii. 416.—*Clairv. E. H.* i. 118. *pl.* 15. *f.* 1, 2.
At. Coryli. *Stew.* ii. 62.

β. Cu. collaris. *Scop. C.* 25.
Cu. excoriato-ruber. *DeGeer.* v. 257. *pl.* 8. *f.* 3. 4.

1985 † 2. Coryli[a].

Familia XLIII. BRUCHIDÆ, *Leach.*

(BRUCHELÆ, *Latreille.*)

(CURCULIO *p*, *et* BRUCHUS, *Linné*, *&c.*)

Genus 338. RHINOMACER[b].

Genus 339. ANTHRIBUS, *Fabr.*

MACROCEPHALUS *p*, *Oliv.*—PLATYRHINUS *p*, *Clairv.*, *Leach*, *Samou.*—ANTHRODUS, (*Megerle.*)—AMBLYCERUS *p*, *Thunb.*

1987. 1: albinus*. *Fabr. E.* ii. 408.—(*Kirby & Sp. I. E.* iii. 319.)
Cu. albinus. *Linn.* ii. 616.—*Don.* x. *pl.* 348. *f.* 3.—*Marsh.* i. 295.—*Turt.* ii. 278.
Rh. albinus. (*Sam. I.* 34.)
β, Cu. leucopsis. *Marsh. MSS.*

Genus 340. PLATYRHINUS, *Clairv.*, *Leach*, *Samou.*

MACROCEPHALUS *p*, *Oliv.*—ANTHRIBUS *p*, *Fabr.*

1988. 1; latirostris*. (*Leach E. E.* ix. 107.)—(*Sam. I.* 34.)
Cu. latirostris. *Bons. Cu.* ii. *f.* 3.—*Don.* x. *pl.* 348. *f.* 1.—*Marsh.* i. 295.
Pl. costirostris. *Clairv. E. H.* i. 114. *pl.* 14. *f.* 1, 2.

Genus 341. TROPIDERES, *Schön.*

MACROCEPHALUS *p*, *Oliv.*—ANTHRIBUS *p*, *Fabr.*—AMBLYCERUS *p*, *Thunb.*—PLATYRHINUS *p*, *Billb.*, *Samou.*

1989. 1, albirostris. (*Schön. Cu.* 35.)
An. albirostris. *Fabr. E.* ii. 408.—*Panz. F.* xv. *f.* 13.
Pl. fascirostris. *Clairv. E. H.* 116. *pl.* 14. *f.* 3. 4?

1990. 2; niveirostris*. (*Schön. Cu.* 35.)
Ma. niveirostris. *Oliv. E.* iv. 80. *pl.* 1. *f.* 8. *a*, *b*.
An. brevirostris. *Panz. F.* lvii. *f.* 9.
Cu. brevirostris. *Marsh.* i. 296.
Pl. brevirostris. (*Sam. I.* 34.)

---

[a] 1985 † 2. Coryli.
At. Coryli. *Fabr. E.* ii. 416.—*Turt.* ii. 281. (!)—(*Leach E. E.* ix. 107.)—*Samou.* 54. (!)

[b] Genus 338. RHINOMACER, *Fabricius*, *Samou.*

RHYNCHITES *p*, *Gyll.*—ANTHRIBUS *p*, *Oliv.*

1986 † 1. Attelaboides. *Fabr. E.* ii. 428.—*Sam.* 200. (!)
An. Rhinomacer β. *Payk. F.* iii. 166.

Genus 342. PHLOËOBIUS? *Schön.*

ANTHRIBUS *p, Fabr.*—SCAPHIDIUM *p, Marsh.*

1991. 1; griseus*. (*Schön. Cu.* 37.)
An. griseus. *Fabr. E.* ii. 410?
Sca. griseum. *Marsh.* i. 233.

Genus 343. BRACHYTARSUS, *Schön.*

ANTHRIBUS *p, Fabr., Leach, Samou.*—MACROCEPHALUS *p, Oliv.* —PAROPES, (*Megerle.*)

1992. 1; scabrosus*. *Fabr. E.* ii. 411.—*Panz. F.* xv. *f.* 15.—(*Leach E. E.* ix. 107.)—(*Sam. I.* 3.)
Br. scabrosus. *Stew.* ii. 53.—*Marsh.* i. 235.
An. fasciatus. *Forst. C.* 9.
Cu. scabrosus. *Turt.* ii. 278.

1993. 2, varius. *Fabr. E.* ii. 411.—*Panz. F.* xv. *f.* 16.—(*Sam. I.* 3.)

Genus 344. BRUCHUS, *Linn., Don., Marsh., Leach, Samou., Millard*, (*Kirby.*)

MYLABRIS *p, Geoffroy.*—LARIA *p, Scop.*

A. Femora dentata.

1994. 1; Pisi*. *Linn.* ii. 604.—*Panz. F.* lxvi. *f.* 11.—*Turt.* ii. 201.—*Samou.* 53.—*Wood.* i. 38. *pl.* 13.—(*Kirby & Sp. I. E.* i. 175.)

1995. 2; granarius*. *Linn.* ii. 605.—*Panz. F.* lxi. *f.* 8.—*Marsh.* i. 235.—*Turt.* ii. 203.—*Shaw G. Z.* vi. 61.—(*Kirby & Sp. l. c. supra.*)
Cu. atomarius. *Linn. F. no.* 628.
Br. Pisi. (*Samou. pl.* 2. *f.* 17.)

1996. 3; affinis* *mihi.*
Br. Viciæ. *Kirby MSS.?*

1997. 4; Loti. *Payk. F.* iii. 158.

1998. 5; Lathyri*. *Kirby MSS.*

B. Femora mutica.

1999. 6; seminarius*. *Linn.* ii. 605.—*Fabr. E.* ii. 400. (!)—*Oliv. E.* iv. 79. *pl.* 2. *f.* 12. *a.*—*Marsh.* i. 236.—*Turt.* ii. 203.—*Shaw* G. *Z.* vi. 61.—(*Sam. I.* 7.)
β, Br. immaculatus. *Kirby MSS.*

2000. 7; Cisti*. *Fabr. E.* ii. 400.—*Panz. F.* lxvi. *f.* 15.
Bru. ater. *Marsh.* i. 236.—(*Sam. I.* 7.)

Familia XLIV. SALPINGIDÆ, *Leach.*

(CISTELENIÆ II. *Latreille.*)

Genus 345. MYCTERUS, *Clairville, Leach, Samou., Millard.*

RHINOMACER *p, Fabr.*—MYLABRIS *p, Schæffer.*—ANTHRIBUS *p, Payk.*

2001. 1, griseus. *Clairv. E. H.* i. 124. *pl.* 16. *f.* 1, 2, 3.—(*Leach E. E.* ix. 106.)—(*Samou.* 199.)
Rh. curculionoides. *Fabr. E.* ii. 428.
An. Rhinomacer. *Payk. F.* iii. 166.

Genus 346. SALPINGUS, *Illiger, Leach, Samou.*

CURCULIO *p, Linn., Marsh.*—RHINOSIMUS, *Latreille.*—ANTHRIBUS *p, Fabr.*

2002. 1; ruficollis*. *Gyll.* ii. 640.
Cu. ruficollis. *Linn.* ii. 609.—*Marsh.* i. 296.
An. ruficollis. *Panz. F.* xxiv. *f.* 19?
Cu. rostratus. *DeGeer.* v. 252. *pl.* 7. *f.* 27, 28.
Cu. Roboris. *Payk. M. Cu.* 118.
Sa. Roboris. (*Leach E. E.* ix. 106.)—(*Sam. I.* 36.)

2003. 2; viridipennis*.
Rh. viridipennis. *Ziegler?*
Sa. ruficollis b. *Gyll.* ii. 640.
Sp. dimidiatus. *Kirby MSS.*

2004. 3; bicolor*.
Cu. bicolor. *Mus. Marsham.*

2005. 4; planirostris. *Gyll.* ii. 641.
An. planirostris. *Fabr. E.* ii. 410.—*Panz. F.* xv. *f.* 14.
Cu. planirostris. *Marsh.* i. 297.
An. fulvirostris. *Payk. M. Cu.* 117.
Sa. rufirostris. (*Sam. I.* 37.)

Genus 347. SPHÆRIESTES, *Kirby MSS.*

TENEBRIO *p, Linn.?*—CURCULIO *p, Marsh.*—SALPINGUS *p, Gyll.*—DERMESTES *p, Payk.*—RHINOSIMUS *p, Latreille.*

A. Thoracis lateribus integris.

2006. 1, ater.
Sa. ater. *Gyll.* ii. 642.
De. ater. *Payk. F.* i. 298.

2007. 2; quadripustulatus*.
Cu. quadripustulatus. *Marsh.* i. 297.
Sa. quadripustulatus. (*Sam. I.* 36.)
Rh. quadriguttatus. *Latr.?*
β; Sp. humeralis. *Kirby MSS.*

2008. [3; immaculatus *] *mihi.*
Sa. rufescens. (*DeJean C.* 77?)

B. Thoracis lateribus crenulatis.

2009. 4; denticollis *.
Sa. denticollis. *Gyll.* iii. 715.
Te. Cursor. *Linn.* ii. 675?—*Turt.* ii. 477.

Subsectio 2.

(Xylophagi *p*, Platysoma *et* Longicornes, *Latreille.*)

Familia XLV. CUCUJIDÆ.

(Trogositariæ *et* Cucujipes, *Latreille.*—Lucanoides *p*, *Gyllenhal.*—Mycetophagidæ *p*, *Leach.*)

Genus 348. TROGOSITA, *Fabricius, Leach, Samou.*, (*Kirby.*)

Tenebrio *p*, *Linn.*, *Marsh.*—Platycerus *p*, *Geoffroy.*

2010. 1; mauritanica *. *Oliv. E.* ii. 6. *pl.* 1. *f.* 2. *a, b.*—(*Leach E. E.* ix. 111.)—(*Sam. I.* 42.)
Te. mauritanicus. *Linn.* ii. 674.—*Marsh.* i. 478.—*Ent. Trans.* (*Kirkup.*) i. 329.
Tr. caraboides. *Fabr. E.* i. 151.—*Panz. F.* iii. *f.* 4.—(*Kirby & Sp. I. E.* i. 171.)
Te. caraboides. *Stew.* ii. 83.—*Turt.* ii. 482.
Te. piceus. *Gmel.* iv. 1998.
Pl. striatus. *Fourc. P.* i. 3.

Genus 349. CUCUJUS, *Fabricius, Leach.*

Cantharis *p*, *Linné.*—Corticaria *p*, *Marsh.*—Colydium *p*, *Herbst.*—Brontes *p*, *Fabricius.*

2011 † 1. depressus [a].

2012. 2; dermestoides. *Fabr. E.* ii. 94.—*Panz. F.* iii. *f.* 13.
Co. depressum. *Herbst C.* vii. 286. *pl.* 113. *f.* 4. *D.*
Co. dermestoides. *Marsh.* i. 107.

2013. 3; testaceus *. *Payk. F.* iii. 168.
Br. testaceus. *Fabr. E.* ii. 98.

Genus 350. ULEIOTA, *Latreille, Leach.*

Brontes *p*, *Fabr.*—Cerambyx *p*, *Linné.*—Cucujus *p*, *Gmel.*

2014. 1, flavipes. *Latr. G.* iii. 26.
Ce. planatus. *Linn.* ii. 628.

---

[a] 2011 † 1. depressus. *Fabr. E.* ii. 93.—*Ræm. G. I.* 48. *pl.* 34. *f.* 28.—(*Leach E. E.* ix. 111.) (!)
Ca. sanguinolenta. *Linn.* ii. 647.

Br. flavipes. *Fabr. E.* ii. 97.—*Panz. F.* xcv. *f.* 4.
Cu. flavipes. *Turt.* ii. 364. (!)
Ce. compressus. *Fourc. p.* i. 76.

2015. 2; monilicornis*.
Co. monilicornis. *Marsh. MSS.*
Br. pallens. *Fabr. E.* ii. 98?
Cu. pallens. *Turt.* ii. 364?
Cu. filicornis. *Kirby MSS.*
Cu. complanatus. *Wilk. MSS.*—(*Ing. Inst.* 88.)

Familia XLVI. PRIONIDÆ, *Leach.*

(Prionii, *Latreille.*)

Genus 351. PRIONUS, *Geoffroy, Leach, Samou.*, (*Kirby.*)

Cerambyx *p, Linné, Marsh.*

2016. 1; coriarius*. *Fabr. E.* ii. 260.—(*Leach E. E.* ix. 112.) —(*Sam. I.* 34.)—(*Kirby & Sp. I. E.* iii. 34.)
Ce. coriarius. *Linn.* ii. 622.—*Mart. C. pl.* 24. *f.* 4.—*Berk. S.* i. 101.—*Stew.* ii. 63.—*Marsh.* i. 325.—*Turt.* ii. 291.—*Shaw* G. Z. vi. 73. *pl.* 25.—*Don.* xiv. 61. *pl.* 491.
Ce. Prionus. *DeGeer.* v. 59. *pl.* 3. *f.* 5.

Familia XLVII. CERAMBYCIDÆ, *Kirby.*

(Cerambycini, *Latreille.*)

(Cerambyx *p, Linné, &c.*)

Bucks, Capricorns, Goatchafers, Timber-Beetles.

Genus 352. HAMATICHERUS, (*Megerle.*)

2017 ‡ 1, Heros. (*DeJean C.* 105.)
Ce. Heros. *Fabr. E.* ii. 270.—*Panz. F.* lxxxii. *f.* 1.
Ce. Cerdo, *var. major.* *Linn.* ii. 629. In Mus. *Brit.*

2018 ‡ 2. Cerdo[a].

Genus 353. CERAMBYX *Auctorum.* *Musk-beetle.*

2019. 1; Moschatus*. *Linn.* ii. 627.—*Mart. C. pl.* 24. *f.* 7.—*Berk. S.* i. 101.—*Don.* iii. *pl.* 94. *f.* 2.—*Stew.* ii. 64.—*Marsh.* i. 327.—*Turt.* ii. 293.—*Shaw* G. Z. vi. 72.—(*Leach E. E.* ix. 112.) —(*Sam. I.* 9.)—*Rem.* L. G. (*Anon*). *pl.* 1.
Ce. odoratus. *DeGeer.* v. 63.

---

[a] 2018 ‡ 2. Cerdo. (*DeJean C.* 105.)
Ce. Cerdo. *Fabr. E.* ii. 270.—*Panz. F.* lxxxii. *f.* 2.
Ce. Heros ♂. *Bergst. N.* i. 67. *pl.* 11. *f.* 6.
Ce. Scopoli. *Laich. T.* ii. 8.
Ce. piceus. *Fourc. P.* i. 74.

Genus 354. MONOCHAMUS, (*Megerle.*)

LAMIA *p, Fabricius, Samou.*

2020. 1: Sutor*. (*DeJean C.* 106.)
Ce. Sutor. *Linn.* ii. 628.—*Oliv. E.* iv. 111. *pl.* 3. *f.* 20. *a*—*c.* ♂.
—*Marsh.* i. 329.—*Don.* xiii. 5. *pl.* 435. *f.* 1.
La. Sutor. *Fabr. E.* ii. 294.—(*Sam I.* 24)
Ce. atomarius. *DeGeer.* v. 65.
Ce. anglicus. *Voët. C.* (*Panz.*) iii. 14. *pl.* 5. *f.* 6.

2021 † 2: Sartor*. (*Ing. Inst.* 89.)
La. Sartor. *Fabr. E.* ii. 294.—*Panz. F.* xix. *f.* 3. ♂.—*Ent. Trans.* (*Haworth.*) i. 339.
La. Sutor. *Panz. F.* xix. *f.* 2. ♀.
In Mus. *Brit*, D. *Beck, Cooper, Samouelle et Vigors.*

Genus 355. ACANTHOCINUS, (*Megerle.*)

LAMIA *p, Fabricius, Leach, Samou.*

2022. 1: ædilis*. (*DeJean C.* 106.)
Ce. ædilis. *Linn.* ii. 628.—*Don.* ii. *pl.* 72.—*Mart. C. pl.* 26. *f.* 4.—*Marsh.* i. 328.—*Turt.* ii. 303.—*Shaw* G. *Z.* vi. 73.
La. ædilis. *Payk. F.* iii. 62.—(*Leach E. E.* ix. 112.)—(*Sam. I.* 24.)

Genus 356. POGONOCERUS, (*Megerle.*)

LAMIA *p, Latreille, Leach, Samou.*

2023. 1; nebulosus*. (*DeJean C.* 107.)
Ce. nebulosus. *Linn.* ii. 627.—*Mart. C. pl.* 24. *f.* 6.—*Marsh.* i. 326.—*Turt.* ii. 298.—*Don.* xi. 95. *pl.* 394.
La. nebulosa. *Latr.* G. iii. 37.—(*Leach E. E.* ix. 112.)—(*Sam. I.* 24.)

2024. 2: fasciculatus. (*DeJean C.* 107.)
Ce. fasciculatus. *Fabr. E.* ii. 277.
Ce. fascicularis. *Panz. F.* xiv. *f.* 15.
Ce. hispidus α. *Linn.* ii. 627.—*Marsh.* i. 327.
Ce. setifer. *Müll. Z.* D. *pr.* 92.

2025. 3; hispidus*. (*DeJean C.* 107.)
Ce. hispidus. *Linn.* ii. 627.—*Don.* ii. *pl.* 64. *f.* 2, 3.—*Mart. C. pl.* 24. *f.* 10.—*Stew.* ii. 63.—*Marsh.* i. 326.—*Turt.* ii. 298.—*Wood.* i. 46. *pl.* 16.
La. hispida. *Latr.* G. iii. 37.—(*Sam. I.* 24.)

2026. 4; pilosus*. (*DeJean C.* 107.)
Ce. pilosus. *Fabr. E.* ii. 278.—*Marsh.* i. 327.
La. pilosa. *Latr.* G. iii. 37.—(*Sam. I.* 24.)
Ce. hispidus. *Panz. F.* xiv. *f.* 16.
Ce. dentatus. *Fourc. P.* i. 76.

Genus 357. LAMIA *Auctorum.*

A. Thoracis lateribus muticis.

2027. 1; nubila*. (*Schön. S.* iii. 384.)—(*Sam. I.* 24.)—*Curtis.* iv. *pl.* 172.
Ce. nubilus. *Oliv. E.* iv. 109. *pl.* 3. *f.* 15.—*Marsh.* i. 332.
La. nebulosa. *Fabr. E.* ii. 293.
Ce. nebulata. *Turt.* ii. 306.

B. Thoracis lateribus spinosis.

2028. 2, textor. *Fabr. E.* ii. 285.—(*Leach E. E.* ix. 112.)—(*Samou.* 209. *pl.* 2. *f.* 24.)—(*Curtis l. c. supra.*)
Ce. textor. *Linn.* ii. 629.—*Turt.* ii. 302. (*tesstor.*)
Ce. nigro-rugosus. *DeGeer.* v. 64.
Ce. cephalotes. *Voët. C.* (*Panz.*) iii. 28. *pl.* 9. *f.* 36.

Genus 358. SAPERDA, *Fabricius, Leach, Samou.*

LAMIA *p, Latr., Leach, Samou.*—LEPTURA *p, Geoffroy.*

2029 ‡ 1, carcharias. *Fabr. E.* ii. 317.—*Panz. F.* lxix. *f.* 1.—(*Ing. Inst.* 89.)
Ce. carcharias. *Linn.* ii. 631.
Ce. punctatus. *DeGeer.* v. 73. *pl.* 3. *f.* 19.
Sa. similis. *Laich. T.* ii. 3.
In Mus. *Brit.*, D. *I. H. Griesbach, Jenyns et Ingpen.*

2030. 2, Scalaris*. *Fabr. E.* ii. 318.
Ce. Scalaris. *Linn.* ii. 632.—*Marsh.* i. 329.—*Don.* xi. 93. *pl.* 393.
La. Scalaris. (*Sam. I.* 24.)

2031. 3, oculata. *Fabr. E.* ii. 319.
Ce. oculatus. *Linn.* ii. 633.—*Don.* v. *pl.* 305.—*Mart. C. pl.* 26. *f.* 26.—*Marsh.* i. 332.—*Turt.* ii. 315.
La. oculata. (*Leach E. E.* ix. 112.)—(*Sam. I.* 24.)
Ce. melanocephalus. *Voët. C.* (*Panz.*) ii. 49. *pl.* 18. *f.* 81.

2032. 4; cylindrica*. *Fabr. E.* ii. 320.—*Panz. F.* lxix. *f.* 4.
Ce. cylindricus. *Linn.* ii. 633.—*Marsh.* i. 331.—*Turt.* ii. 316.
Ce. cinereus. *DeGeer.* v. 75.
Ce. Silphoides. *Gmel.* iv. 1843.

2033 ‡ 5. linearis [a].

2034 † 6, Tremulæ. *Fabr. E.* ii. 327.—*Panz. F.* i. *f.* 7.
Ce. octopunctatus. *Schra. A. no.* 267.
Sa. punctata. *Laich. T.* ii. 32. In Mus. D. *Curtis.*

2035. 7; populnea*. *Fabr. E.* ii. 327.—*Panz. F.* lxix. *f.* 7.

---

[a] 2033 ‡ 5. linearis. *Fabr. E.* ii. 320.—*Panz. F.* vi. *f.* 14.
Ce. linearis. *Linn.* ii. 632.—*Turt.* ii. 315. (!)
Le. fulvipes. *Fourc. P.* ii. 79.

Ce. populneus. *Linn.* ii. 632.—*Mart. C. pl.* 24. *f.* 8. 9.—*Marsh.* i. 330.—*Turt.* ii. 318.
Ce. decempunctatus. *DeGeer.* v. 78.
La. populnea. (*Sam. I.* 24.)

2036. 8: Cardui. *Fabr. E.* ii. 325.
Ce. lineatocollis. *Don.* vi. *pl.* 209.—*Marsh.* i. 331.—*Turt.* ii. 300.
Sa. lineatocollis. (*Leach Z. M.* i. 13.)—(*Leach E. E.* ix. 112.) (*Sam. I.* 37.)
Ce. villoso-viridescens. *DeGeer.* v. 76.
Lep. virens. *Voët. C.* (*Panz.*) iii. 58. *pl.* 20. *f.* 98.
Ce. viridescens. *Gmel.* iv. 1864.

Genus 359. TETROPS, *Kirby MSS.*

SAPERDA *p, Fabricius.*—LEPTURA *p, Linné.*—LAMIA *p, Samou.*

2037. 1; præusta*.
Le. præusta. *Linn.* ii. 641.—*Mart. C. pl.* 28. *f.* 12.
Ce. præustus. *Berk. S.* i. 102.—*Stew.* ii. 65.—*Marsh.* i. 333.—*Turt.* ii. 320.
La. præusta. (*Sam. I.* 24.)
Le. pilosa. *Fourc. P.* ii. 78.

Genus 360. OBRIUM, (*Megerle,*) *Curtis.*

CALLIDIUM *p, Fabricius.*—SAPERDA *p, Oliv.*—STENOCHORUS *p, Schönher.*—LAMIA *p, Samou.*

2038. 1; minutum*.
Sa. minuta. *Fabr. M.* i. 150.
Ce. minutus. *Gmel.* iv. 1842.—*Stew.* ii. 64.—*Turt.* ii. 325.—*Don.* xvi. *pl.* 553. *f.* 2.
La. minuta. (*Sam. I.* 24.)
Ca. Vini. *Panz. F.* lxvi. *f.* 10.
Ca. pygmæum. *Fabr. E.* ii. 339. (!)—(*Kirby & Sp. I. E.* i. 231.)

2039 † 2, cantharinum*. *Curtis.* ii. *pl.* 91. ♂. ♀.
Ca. cantharinus. *Linn.* ii. 637.
Ce. fuscicornis. *Gmel.* iv. 1837.
♂, Sa. brunnea. *Fabr. E.* ii. 331.
♀, Sa. ferruginea. *Fabr. E.* ii. 330.
In Mus. D. *Curtis et Sparshall.*

Genus 361. CALLIDIUM, *Fabr., Leach, Samou.,* (*Kirby.*)

2040. 1; bajulum*. *Fabr. E.* ii. 333.—(*Sam. I.* 8. *pl.* 12. *f.* 2.)
Ce. bajulus. *Linn.* ii. 636.—*Stew.* ii. 65.—*Mart. C. pl.* 24. *f.* 1–5.—*Marsh.* i. 334.—*Turt.* ii. 323.
Ce. caudatus. *DeGeer.* v. 86.
β; Ca. Linneanum. *Laich. T.* ii. 69.
Ce. similis. *Marsh.* i. 335.

2041. 2: sanguineum*. *Fabr. E.* ii. 340.—*Panz. F.* lxx. *f.* 9.
Ce. sanguineus. *Linn.* ii. 636.—*Berk. S.* i. 101.—*Stew.* ii. 65.—*Marsh.* i. 336.—*Turt.* ii. 327.—*Don.* xvi. *pl.* 553. *f.* 1.

2042. 3; violaceum*. *Fabr. E.* ii. 335.—(*Sam. I.* 8.)
Ce. violaceus. *Linn.* ii. 635.—*Don.* ii. *pl.* 64. *f.* 1.—*Stew.* ii. 64.—*Mart. C. pl.* 25. *f.* 15.—*Linn. Trans.* (*Kirby.*) v. 257. *pl.* 12.—*Marsh.* i. 333.—*Turt.* ii. 324.—*Bingley.* iii. 138.

2043. 4; variabile*. (*DeJean C.* 110.)
Ce. variabilis. *Linn. F. no.* 669.
Ca. fennicum. *Fabr. E.* ii. 334.—*Panz. F.* lxx. *f.* 2.—(*Ing. Inst.* 87.)
Ce. fennicus. *Gmel.* iv. 1851.—*Mart. C. pl.* 25. *f.* 16.—*Marsh.* i. 336.—*Turt.* ii. 323.
Le. rubricollis. *Voët. C.* (*Panz.*) iii. 57. *pl.* 20. *f.* 97.

2044. [5; testaceum*.] *Fabr. E.* ii. 341.—(*Ing. Inst.* 88.)
Ce. testaceus. *Linn.* ii. 635.—*Mart. C. pl.* 26. *f.* 3.—*Marsh.* i. 334.—*Turt.* ii. 327.
Ce. variabilis β. *Linn. F. no.* 669.

2045 † 6. dilatatum [a].

2046 ‡ 7. rusticum. *Fabr. E.* ii. 338.
Ce. rusticus. *Linn.* ii. 634.—*Berk. S.* i. 101.—*Panz. F.* lxx. *f.* 8.—*Stew.* ii. 64.—*Turt.* ii. 325.
Ca. triste. *Fabr. E.* ii. 342.
Ce. lugubris. *Gmel.* iv. 1847. In Mus. D. Curtis.

2047 † 8. striatum [b].

2048 † 9. luridum [c].

2049 † 10. undatum [d].

2050. 11, russicum. *Fabr. E.* ii. 336.—*Oliv. E.* iv. 54. *pl.* 4. *f.* 49.

2051. 12; Alni*. *Fabr. E.* ii. 338.—*Panz. F.* lxx. *f.* 10.
Le. Alni. *Linn.* ii. 639.—*Mart. C. pl.* 28. *f.* 14.
Ce. Alni. *Gmel.* iv. 1855.—*Marsh.* i. 338.—*Turt.* ii. 333.

---

[a] 2045 † 6. dilatatum. *Payk. F.* iii. 91.
Ca. variabile. *Fabr. E.* ii. 337.—*Panz. F.* lxx. *f.* 6.
Ce. variabilis. *Gmel.* iv 1850.—*Turt.* ii. 325. (!)
Ca. æneum. *Herbst A.* 96. *pl.* 26. *f.* 12.
Ce. aurichalceus. *Gmel.* iv. 1857.

[b] 2047 † 8. striatum. *Fabr. E.* ii. 343.—*Panz. F.* lxx. *f.* 13.
Ce. striatus. *Linn.* ii. 635.—*Turt.* ii. 328. (!)

[c] 2048 † 9. luridum. *Fabr. E.* ii. 342.—*Panz. F.* lxx. *f.* 10.
Ce. luridus. *Gmel.* iv. 1846.—*Turt.* ii. 327. (!)

[d] 2049 † 10. undatum. *Fabr. E.* ii. 344.—*Panz. F.* lxx. *f.* 15.
Ce. undatus. *Linn.* ii. 636.—*Turt.* ii. 329. (!)
Ce. sulphuratus. *Voët. C.* (*Panz.*) iii. *pl.* 19. *f.* 87.

Clytus Alni. (*Sam. I.* 11.)
Ce. globifer. *Voët. C.* (*Panz.*) iii. *pl.* 18. *f.* 80.

2052. 13, luteum.
Ce. luteus. *Marsh. MSS.*

Genus 362. CLYTUS, *Fabricius, Leach, Samou.*, (*Kirby*,) *Curtis.*
LEPTURA *p, Linné.*—CALLIDIUM *p, Latr.*

2053. 1; arcuatus*. *Fabr. E.* ii. 347.—*Samou.* 55. *pl.* 2. *f.* 25.—(*Curtis l. c. infra.*)
Le. arcuata. *Linn.* ii. 640.—*Barb.* G. 57. *pl.* 5. *f.* 20.—*Berk. S.* i. 102.—*Mart. C. pl.* 27. *f.* 2, 3.—*Don.* iii. *pl.* 84. *f.* 1.
Ce. arcuatus. *Stew.* ii. 65.—*Marsh.* i. 338.—*Turt.* ii. 330.
Ce. lunatus. *Gmel.* iv. 1852.
Ce. detritus. *Voët. C.* (*Panz.*) iii. 55. *pl.* 19. *f.* 99.

2054 ‡ 2. detritus[a].

2055 † 3. plebeius[b].

2056. 4, Upsilon*.
Ce. Upsilon. *Marsh. MSS.*

2057. 5; mysticus*. *Fabr. E.* ii. 352.—(*Sam. I.* 11.)—(*Curtis l. c. infra.*)
Le. mystica. *Linn.* ii. 639.—*Don.* iii. 29. *pl.* 84. *f.* 2.—*Wood.* i. 47. *pl.* 17.
Ce. mysticus. *Gmel.* iv. 1855.—*Stew.* ii. 65.—*Marsh.* i. 337.—*Turt.* ii. 332.
Ce. albofasciatus var. *DeGeer.* v. 82.
Ce. quadricolor. *Scop. C.* 177.
Ce. litteratus. *Gmel.* iv. 1857.

2058. 6; Arietis*. *Laich. T.* ii. 92.—(*Leach E. E.* ix. 113.)—(*Sam. I.* 11.)—(*Curtis l. c. infra.*)
Le. Arietis. *Linn.* ii. 640.—*Berk. S.* i. 102.—*Mart. C. pl.* 27. *f.* 4.—*Don.* i. *pl.* 27.—*Shaw* G. *Z.* vi. 74. *pl.* 26?
Ce. Arietis. *Stew.* ii. 65.—*Marsh.* i. 339.—*Turt.* ii. 330.
Ce. quadrifasciatus. *DeGeer.* v. 81.
β? Cl. gazella. *Fabr. E.* ii. 348?

2059 ‡ 7. quadripunctatus. *Fabr. E.* ii. 352.—*Child. Add.* 7. (!)—*Curtis.* v. *pl.* 199.

---

[a] 2054 ‡ 2. detritus. *Fabr. E.* ii. 350.
Le. detrita. *Linn.* ii. 640.
Ce. detritus. *Gmel.* iv. 1854.—*Turt.* ii. 331. (!)
Ca. detritum. *Panz. F.* xciv. *f.* 5.

[b] 2055 † 3. plebeius. *Fabr. E.* ii. 349.
Ce. plebeius. *Gmel.* iv. 1853.—*Stew.* ii. 65. (!)—*Turt.* ii. 331. (!)
Ca. plebeium. *Panz. F.* lxxxiii. *f.* 7.
Le. rustica. *Müll. Z. D. pr. no.* 1041.

Le. villosa. *Vill. E.* i. 272. *pl.* i. *f.* 31.
Le. nævia. *Gmel.* iv. 1877. In Mus. D. Sparshall.

Genus 363. MOLORCHUS, *Fabricius, Leach, Samou., Millard, Curtis,* (*Kirby.*)

NECYDALIS *p, Linn.*—GYMNOPTERION, *Schrank.*

2060 † 1. major[a].

2061. 2; minor*. *Curtis.* i. *pl.* 11.
Ne. minor. *Linn.* ii. 641.—*Mart. C. pl.* 23. *f.* 1.—*Stew.* ii. 67.—*Marsh.* i. 358.—*Turt.* ii. 354.—*Shaw G. Z.* vi. 76.—*Wood.* i. 49. *pl.* 18.
Ne. ceramboides. *DeGeer.* v. 151.
Mo. dimidiatus. *Fabr. E.* ii. 375.—(*Sam. I.* 27.)

2062. 3; Umbellatarum*. *Fabr. E.* ii. 375.—(*Sam. I.* 27.)—(*Curtis l. c. supra.*)
Ne. Umbellatarum. *Linn.* ii. 641.—*Mart. C. pl.* 23. *f.* 2.—*Marsh.* i. 358.—*Turt.* ii. 354.
Ne. minima. *Scop. C.* 180.

Genus 364. STENOPTERUS[b].

Familia XLVIII. LEPTURIDÆ, *Leach.*

(LEPTURETES, *Latreille.*—CERAMBYCINI *p, Gyllenhal.*)
(CERAMBYX *p, et* LEPTURA, *Linné, &c.*)

Genus 365. RHAGIUM, *Fabr., Leach, Samou.,* (*Kirby.*)

CERAMBYX *p, Linné.*—LEPTURA *p, Marsh.*—STENOCHORUS *p, Olivier.*

A. Antennis medio incrassatis. (HARGIUM, *Leach, Samou.*)

2064. 1, Indagator. *Fabr. E.* ii. 313.—*Panz. F.* lxxxii. *f.* 5.
Le. inquisitor α. *Linn.* ii. 630.
Ce. nubeculus. *Bergs. N.* i. 25. *pl.* 4. *f.* 4.
Ha. inquisitor. (*Sam. I.* 20.)

---

[a] 2030 † 1. major. (*Sam. I.* 27. (!)
Ne. major. *Linn.* ii. 641.—*Turt.* ii. 354.
Gy. magus. *Schr. F. B.* i. 688.
Ne. ichneumonea. *DeGeer.* v. 148. *pl.* 5. *f.* 1.
Mo. abbreviatus. *Fabr. E.* ii. 374.—*Panz. F.* xli. *f.* 20.

[b] Genus 364. STENOPTERUS, *Illiger.*

NECYDALIS *p, Linné.*—MOLORCHUS *p, Fabr.*

2063 ‡ 1. rufus. (*DeJean C.* 111.)
Ne. rufa. *Linn.* ii. 642.—*Oliv. E.* iv. 74. *pl.* 1. *f.* 6. *a, b.*—*Turt.* ii. 356. (!)
Le. attenuata. *Fourc. P.* i. 84.

B. Antennis setaceis. (RHAGIUM, *Leach, Samou.*)

2065. 2; inquisitor*. *Fabr. E.* ii. 313.—*Panz. F.* lxxxii. *f.* 4.—*Mart. C. pl.* 25. *f.* 19.—(*Leach E. E.* ix. 113.)
Le. inquisitor β. *Linn.* ii. 630.
Le. inquisitor. *Marsh.* i. 341.—*Turt.* ii. 321.
Le. mordax. *DeGeer.* v. 124. *pl.* 4. *f.* 6.
Ce. bifasciatus. *Schra. A.* 259.
Rh. vulgare. (*Sam. I.* 36.)

2066. 3; bifasciatum*. *Fabr. E.* ii. 314.—*Don.* iii. *pl.* 94. *f.* 1.—*Mart. C. pl.* 25. *f.* 18.—(*Sam. I.* 36.)
Le. bifasciata. *Marsh.* i. 342.
Ce. bifasciatus. *Turt.* ii. 322.
Ce. anglicus. *Gmel.* iv. 1844.—*Stew.* ii. 64.
β, Le. nigrolineata. *Don.* x. *pl.* 253. *f.* 1.—*Marsh.* i. 343.
γ, Le. bimaculata. *Marsh.* i. 343.
δ, Le. dorsalis. *Marsh.* i. 343.—*Don.* xi. 97. *pl.* 395. *f.* 1.
Ce. elegans. *Gmel.* iv. 1845.

Genus 366. TOXOTUS, (*Megerle*).

LEPTURA *p*, *Linné.*—STENOCHORUS *p*, *Olivier.*—RHAGIUM *p*, *Fabricius.*—CERAMBYX *p*, *Gmelin.*

2067 ‡ 1. Cursor[a].

2068. 2; meridianus*. (*DeJean C.* 112.)
Ce. meridianus. *Linn.* ii. 630.—*Mart. C. pl.* 25. *f.* 11. 12.—*Stew.* ii. 65.—*Don.* xiii. 6. *pl.* 435. *f.* 2.
Le. meridiana. *Panz. F.* xlv. *f.* 20.—*Marsh.* i. 340.—*Turt.* ii. 346.—(*Sam. I.* 24.)
St. sericeus. *Oliv. E.* iv. 20. *pl.* 1. *f.* 8. ♀.
β; Le. humeralis. *Kirby MSS.*
γ; Le. femoralis. *Kirby MSS.*

2069. [3; chrysogaster*.] (*DeJean C.* 112.)
St. chrysogaster. *Oliv. E.* iv. 19. *pl.* 3. *f.* 23.
Le. splendens. *Laich. T.* ii. 136.
Le. meridiana, var. δ. (*Schön. S.* i. 479.)
Le. rufiventris. *Marsh.* i. 341.—(*Sam. I.* 24.)

Genus 367. LEPTURA *Auctorum.* *Wood-Beetles.*

A. Elytris apice plus minusve excisis.

2070. 1; elongata*. *DeGeer.* v. 134.—*Don.* iii. *pl.* 84. *f.* 4.—*Mart. C. pl.* 27. *f.* 10.—*Marsh.* i. 355.—(*Leach E. E.* ix. 113.)—(*Sam. I.* 24.)

---

[a] 2067 ‡ 1. Cursor. (*DeJean C.* 112.)
Ce. Cursor. *Linn.* ii. 630.—*Turt.* ii. 322. (!)
St. Cursor. *Oliv. E.* iv. 14. *pl.* 1. *f.* 9.
Ce. vittatus. *Gmel.* iv. 1865.
♂, Ce. noctis. *Linn.* ii. 630.—*Turt.* ii. 322. (!)
St. noctis. *Oliv. E.* iv. 17. *pl.* 1. *f.* 10.

Ce. elongatus. *Turt.* ii. 330.
Le. armata. *Gmel.* iv. 1872.
Le. fasciata. *Hoppe. E.* 59.
♂; Le. calcarata. *Fabr. E.* ii. 363.
♀; Le. subspinosa. *Fabr. E.* ii. 363.
Le. quinquemaculata. *Gmel.* iv. 1868.
β; Le. sinuata. *Fabr. E.* ii. 363.

2071. 2, attenuata. *Linn.* ii. 639.—*Stew.* ii. 67.—*Oliv. E.* iv. 17. *pl.* 1. *f.* 8.—*Marsh.* i. 492.—*Turt.* ii. 349.—(*Sam. I.* 24.)
Ce. fasciatus. *Scop. C.* 172.

2072. 3; revestita*. *Linn.* ii. 638.—*Marsh.* i. 350.—(*Sam. I.* 24.)
Le. villica. *Fabr. E.* ii. 357.—*Mart. C. pl.* 27. *f.* 1.—*Panz. F.* xxii. *f.* 13.
Le. villa. *Turt.* ii. 346.
♀, Le. fuscicornis. *Marsh.* i. 357.

2073 † 4. rubrotestacea [a].

2074. 5: aurulenta. *Fabr. E.* ii. 364.—*Marsh.* i. 356.—*Panz. F.* xc. *f.* 5.—(*Sam. I.* 24.)
Le. quadrifasciata. *Rossi F.* i. 161.

2075. 6; quadrifasciata*. *Linn.* ii. 639.—*Mart. C. pl.* 27. *f.* 9.—*Marsh.* i. 354.—*Turt.* ii. 349.—*Samou.* 55. *pl.* 2. *f.* 26.—(*Ing. Inst.* 89.)
Le. octomaculata. *DeGeer.* v. 131. *pl.* 4. *f.* 11.

2076. 7; apicalis*. *Haworth. MSS.*—(*Sam. I.* 24.)

2077. 8, virens. *Linn.* ii. 638.—*Berk. S.* i. 102.—*Panz. F.* lxix. *f.* 13.—*Stew.* ii. 66.—*Turt.* ii. 347.

2078. 9: scutellata*. *Fabr. E.* ii. 359.—*Panz. F.* lxix. *f.* 15.—(*Ing. Inst.* 89.)
Ce. niger. *Gmel.* iv. 1845.
St. funereus. *Fourc. P.* i. 89.

2079. 10: tomentosa. *Fabr. E.* ii. 355.—*Oliv. E.* iv. 12. *pl.* 2. *f.* 13. c.
Le. fulva. *DeGeer.* v. 136.
Le. lutescens. *Vill.* i. 274.
Le. affinis. *Marsh.* i. 353.—(*Sam. I.* 24.)
St. testaceus. *Fourc. P.* i. 87.

2080. 11, sanguinolenta*. *Linn.* ii. 638.—*Panz. F.* lxix. *f.* 8.—*Turt.* ii. 346.—*Don.* xvi. *pl.* 557.—(*Sam. I.* 24.)
Le. variabilis. *DeGeer.* v. 137.

---

[a] 2073 † 4. rubrotestacea. *Illig. M.* iv. 122.
Le. dispar. *Payk. F.* iii. 107.
Le. Umbellatarum. *Laich. T.* ii. 161.
♂. Le. testacea. *Linn.* ii. 638.—*Panz. F.* lxix. *f.* 12.—*Turt.* ii. 347. (!)
♀. Le. rubra. *Linn.* ii. 638.—*Panz. F.* lxix. *f.* 11.—*Turt.* ii. 347. (!)

St. ignitus. *Fourc. P.* i. 89.
Le. melanura var. *Linn.* ii. 637.

2081. 12; melanura*. *Linn.* ii. 637.—*Berk. S.* i. 102.—*Mart. C. pl.* 27. *f.* 7.—*Stew.* ii. 66.—*Panz. F.* lxix. *f.* 19. ♀.—*Marsh.* i. 350.—*Turt.* ii. 346.—(*Sam. I.* 24.)
Le. sutura-nigra. *DeGeer.* v. 138.
Le. similis. *Herbst A.* 101. *pl.* 26. *f.* 22.

2082. 13; nigra*. *Linn.* ii. 639.—*Mart. C. pl.* 27. *f.* 8.—*Marsh.* i. 351.—*Turt.* ii. 348.—(*Sam. I.* 24.)
St. piceus. *Fourc. P.* i. 87.

B. Elytris apice vix aut non excisis.

2083. 14; sexguttata*. *Fabr. E.* ii. 364.—*Panz. F.* lxix. *f.* 22.—(*Sam. I.* 24.)
♂; Le. punctomaculata. *Marsh.* i. 357.
β, Le. exclamationis. *Fabr. E.* ii. 359.—*Oliv. E.* iv. 29. *pl.* 2. *f.* 19.—*Marsh.* i. 356.

2084. 15; abdominalis* *mihi.* (*Ing. Inst.* 89.)
Le. abdominalis. ——? *MSS.?*

2085. 16, lævis*. *Fabr. E.* ii. 355.—*Panz. F.* xxxiv. *f.* 16.—*Marsh.* i. 351.—(*Sam. I.* 24.)
Le. tabacicolor. *DeGeer.* v. 139.
Le. chrysomeloides. *Schra. A.* 158.
Le. melanura var. b. *Müll. Z. D. pr.* 94.
Le. solstitialis. *Gmel.* iv. 1873.
Le. genii. *Gmel.* iv. 1877.

2086. 17; ruficornis*. *Fabr. E.* ii. 360.—*Sturm V.* 1796. 51. *pl.* 2. *f.* 9.
Le. Parisina. *Nov. Act. Ups.* (*Thunb.*) iv. 16.
Le. pumila. *Gmel.* iv. 1874.
Le. lævis. *Gmel.* iv. 1873.
Le. femorata. *Marsh.* i. 352.—(*Sam. I.* 24.)

2087. 18; pallipes *mihi.*

2088. 19: præusta. *Fabr. E.* ii. 360.—*Panz. F.* xxxiv. *f.* 17.—*Turt.* ii. 348.—(*Ing. Inst.* 89.)
Le. ustulata. *Gmel.* iv. 1874.
Le. adusta. *Gmel.* iv. 1872.
Le. splendida. *Herbst A.* 103.

2089 † 20. atra*.

Genus 368. PACHYTA, (*Megerle.*)

LEPTURA *p, Linné.*—CERAMBYX *p, Linné.*—STENOCHORUS *p, Fabr.*

2090. 1; livida*.

2089 † 20. atra. *Fabr. E.* ii. 359.—*Panz. F.* lxix. *f.* 14.—*Turt.* ii. 347. (!)
Le. melanaria. *Herbst A.* 101.

Le. livida. *Fabr. E.* ii. 355.—*Mart. C. pl.* 28. *f.* 13.—*Marsh.* i. 352.—(*Sam. I.* 24.)
Le. nigripes. *DeGeer.* v. 136.
Le. Pastinacæ. *Panz. E. G.* i. 275.

2091. 2; collaris*. (*DeJean C.* 112.)
Le. collaris. *Linn.* ii. 639.—*Mart. C. pl.* 27. *f.* 5.—*Marsh.* i. 349.—*Turt.* ii. 350.—(*Sam. I.* 24.)
Le. ruficollis. *DeGeer.* v. 143.
Le. thalassina. *Schrank. A.* 160.

2092 † 3. virginea[a].

2093. 4; octomaculata*. (*DeJean C.* 112.)
Le. octomaculata. *Fabr. E.* ii. 361.—*Turt.* ii. 348.
Le. decempunctata. *Oliv. E.* iv. 26. *pl.* 4. *f.* 42.
Le. sexmaculata. *Don.* x. *pl.* 353. *f.* 2.—*Mart. C. pl.* 28. *f.* 22.—*Stew.* ii. 67.—*Marsh.* i. 353.—(*Sam. I.* 24.)
Le. quadrimaculata. *Scop. C.* 47.
Le. cerambyciformis. *Gmel.* iv. 1873.

2094 † 5. sexmaculata[b].

2095 † 6. Lamed[c].

SECTIO IV.

(TETRAMERA *p, et* TRIMERA, *Latreille.*—ANOPLURIMORPHA? *MacLeay.*)

Subsectio 1. EUPODA, *Latreille.*

Familia XLIX. CRIOCERIDÆ, *Leach.*

(CRIOCERIDES, *Latreille.*—LEMOIDEÆ, *Gyllenhal.*)

Genus 369. DONACIA, *Fabricius, Leach, Samou., Millard,* (*Kirby.*)

LEPTURA *p, Linné.*—STENOCHORUS *p, Geoffroy.*

A. Femoribus posticis bi- vel tri-dentatis, magnis, clavatis.

2096. 1; cincta*.—*Germar.*—(*Kunzé E. F.* 5.)
Do. clavipes. *Fabr. E.* ii. 128.

---

[a] 2092 † 3. virginea. (*DeJean C.* 112.)
Le. virginea. *Linn.* ii. 639.—*Oliv. E.* iv. 28. *pl.* 2. *f.* 24. *a, b.*—*Turt.* ii. 350. (!)
Le. violacea. *DeGeer.* v. 144.

[b] 2094 † 5. sexmaculata. (*DeJean C.* 112.)
Le. sexmaculata. *Linn.* ii. 638.—*Oliv. E.* iv. 26. *pl.* 4. *f.* 43.—*Turt.* ii. 348. (!)
Le. testaceo-fasciata. *DeGeer.* v. 133.

[c] 2095 † 6. Lamed. (*DeJean C.* 112.)
Ce. Lamed. *Linn.* ii. 630.
Le. Lamed. *Don.* xi. 98. *pl.* 395. *f.* 2. (!)
Le. pedella. *DeGeer.* v. 129. *pl.* 4. *f.* 10.

Do. bidens. *Gyll.* iii. 648.
Le. micans. *Marsh.* i. 344.—*Turt.* ii. 343.
Do. micans. (*Leach E. E.* ix. 113.)—(*Sam. I.* 15.)
Le. aquatica. *Mart. C. pl.* 28. *f.* 16. 17.
Le. versicolorea. *Brahm?*

2097. 2; crassipes*. *Fabr. E.* ii. 126.
Le. aquatica β. *Linn.* ii. 637.—*Berk. S.* i. 101.—*Don.* iii. 31.
—*Stew.* ii. 66.—*Turt.* ii. 342?—*Shaw* G. *Z.* vi. 74. *pl.* 26?
Do. striata. *Panz. F.* xxix. *f.* 1.

2098. 3; dentata*. *Kunzé E. F.* 7.—*Hoppe.* 40. *f.* 2.—(*Ing. Inst.* 88.)
Do. Potamogetonis. *Kirby MSS.*—(*Ing. Inst.* 88.)

2099. 4; angustata*. *Kunzé E. F.* 8.
Do. bidens. *Oliv. E.* iv. 11. *pl.* 2. *f.* 12. *a, b?*

2100. 5; melanocephala*. (*Sam. I.* 15.)
Le. melanocephala. *Marsh.* i. 348.
Do. Sparganii. (*Kunzé E. F.* 12?)

B. Femoribus posticis mediocribus, unidentatis.

a. *Pedibus corpore concoloribus.*

2101. 6; Lemnæ*. *Fabr. E.* ii. 128.
Le. vittata. *Marsh.* i. 345.
Do. vittata. (*Sam. I.* 15.)
Do. marginata. *Hoppe.* 42. *f.* 4.

2102. 7; dentipes*. *Fabr. E.* ii. 127.
Do. vittata. *Oliv. E.* iv. 7. *pl.* 1. *f.* 5. *a, b.*
Do. fasciata. *Hoppe.* 42. *f.* 3.—(*Sam. I.* 15.)
Le. aquatica fasciata. *DeGeer.* v. 142.
Le. fasciata. *Marsh.* i. 344.
Le. aquatica α. *Linn.* ii. 637.
Do. nitida. *Gmel.* iv. 1066.

2103. 8; Sagittariæ*. *Fabr. E.* ii. 128.—(*Sam. I.* 15.)
Le. Sagittariæ. *Marsh.* i. 345.
Do. aurea. *Hoppe.* 43. *f.* 5.
Do. aquatica. *Act. Ups.* (*Thunb.*) v. 118.
Le. bicolor. *Gmel.* iv. 1867.
β, Do. collaris. *Panz. F.* xxix. *f.* 8.

2104. 9; brevicornis. *Kunzé E. F.* 16.

2105. 10; obscura. *Gyll.* iii. 654.
Do. impressa. *Nov. Act. Hall.* (*Ahrens.*) i. 3.
Do. lacunosa. *Kirby MSS.*
Le. nigripes. *Marsh. MSS.*
Do. hybrida. *Schüppel.*
Do. Scirpiola. *Eschscholtz.*

2106. 11, thalassina. *Germar.—Kunzé E. F.* 18.
Do. impressa b. *Gyll.* iii. 656.

2107. 12; impressa*. *Gyll.* iii. 655.
Do. hybrida. *Schüppel.*
β? Do. similis. *Kirby MSS.?*

2108. 13; Proteus*. *Kunzé E. F.* 23.
α; Do. crassipes b. *Oliv. E.* iv. 4. *pl.* 1. *f.* 1.
Do. armata. *Payk. F.* ii. 194.
Do. discolor. ♀. *Panz. F.* xxix. *f.* 6.
Le. Nymphææ. *Marsh.* i. 347.—*Turt.* ii. 343.
Do. Nymphææ. (*Sam. I.* 15. *pl.* 2. *f.* 27.)
Do. sericea. *Gyll.* iii. 658.
Le. sericea. *Turt.* ii. 350?
Do. cærulea. *Kirby MSS.*
β; Do. violacea. *Hoppe.* 44. *f.* 7.—*Gyll.* iii. 660.
Do. Festucæ. *Panz. F.* xxix. *f.* 2.
Le. Festucæ. *Marsh.* i. 346.
Do. discolor ♂. *Panz. F.* xxix. *f.* 3.
γ; Do. viridis. *Kirby MSS.*

2109. 14; micans*. *Ahrens.—Kunzé E. F.* 25.
Do. crassipes a, c. *Oliv. E.* iv. 75. *f.* 1.
Do. sericea a, d. *Gyll.* iii. 657.
Le. aquatica-ænea. *DeGeer.* v. 143.
Do. ænea. *Hoppe. E.* 44. *f.* 6.
Le. ænea. *Marsh.* i. 346.

b. *Pedibus pallidis.*

2110. 15; rustica*. *Schüppel.—Kunzé E. F.* 31.
Do. discolor. *Hoppe. E.* 45. *pl.* 1. *f.* 8, 9. ♂. ♀.
♂; Le. fusca. *Marsh.* i. 349.
Do. fusca. (*Sam. I.* 15.)
Do. flavipes. *Kirby MSS.*
♀; Le. discolor. *Marsh.* i. 346.
Do. discolor b. *Gyll.* iii. 661.

2111. 16; nigra*. *Fabr. E.* ii. 128.
Do. abdominalis. *Oliv. E.* iv. 75. 9.
Do. palustris. *Herbst C.* v. 100.—*Panz. F.* xxix. *f.* 10.—(*Sam. I.* 15.)
Le. palustris. *Marsh.* i. 349.

B. Femoribus posticis simplicibus.

2112. 17; Menyanthedis*. *Fabr. E.* ii. 129.
Do. clavipes. *Oliv. E.* iv. 75. 8. *pl.* 1. *f.* 6. *a*, *b*.
Le. aquatica mutica. *DeGeer.* v. 142.
Do. simplex. *Payk. F.* ii. 189.
Le. simplex. *Mart. C. pl.* 28. *f.* 25.—*Marsh.* i. 348.—*Turt.* ii. 345.

2113. 18; simplex*. *Fabr. E.* ii. 129.—*Stew.* ii. 66.—(*Leach E. E.* ix. 113.)—(*Sam. I.* 15.)
Do. semicuprea. *Panz. F.* xxix. *f.* 14.—(*Ing. Inst.* 88.)
Le. vulgaris. *Gmel.* iv. 1867.

2114. 19; linearis*. *Hoppe.* 46. *f.* 10.—(*Sam. I.* 15.)
Le. linearis. *Marsh.* i. 347.
Do. simplex. *Oliv. E.* iv. 75. 11.
Le. aquatica ænea, var. *DeGeer.* v. 143.
Le. aquatica. *Don.* iii. *pl.* 84. *f.* 3.

2115. 20; Typhæ*. *Brahm.*—*Kunzé E. F.* 47.
Do. linearis b. *Gyll.* iii. 665.

2116. 21; Hydrochæridis*. *Fabr. E.* ii. 129.—*Panz. F.* xxix. *f.* 17.
Do. Hydrocharis. *Oliv. E.* iv. 75. 12. *pl.* 2. *f. a, b.*—(*Sam. I.* 15.)
Le. Hydrocharis. *Marsh.* i. 347.
Do. tersata. *Panz. F.* xxix. *f.* 16.
Do. cinerea. *Herbst A.* v. 100.
β? Do. tomentosa. *Illiger.*—*Nov. Act. Hall.* (*Ahrens.*) i. 3?

Genus 370. MACROPLEÄ, *Hoffmansegg*, (*Samou.*), (*Kirby.*)

HÆMONIA, *Megerle.*—RHAGIUM *p*, *Paykul.*—DONACIA *p*, *Fabr.*

2117. 1, Zosteræ*? (*Kirby & Sp. I. E.* iv. 522.)
Do. Zosteræ. *Fabr. E.* ii. 127.—*Ent. Trans.* (*Haworth.*) i. 85.
Rh. muticum. *Payk. F.* iii. 69.

2118 † 2, Equiseti*.
Do. Equiseti. *Fabr. E.* ii. 127.
Do. mucronata. *Hoppe. E.* 47. *f.* 12.
Do. appendiculata. *Panz. F.* xxiv. *f.* 17.
In Mus. *Brit.* et D. Jenyns?

Genus 371. ORSODACNA, *Latreille*, *Leach.*

CRIOCERIS *p*, *Fabricius.*—LEMA *p*, *Illiger.*—GALERUCA *p*, *Fabricius.*—CRYPTOCEPHALUS *p*, *Stew.*

2119. 1; chlorotica*. *Latr. G.* iii. 44.
Cr. Cerasi. *Fabr. E.* i. 456.
Cr. fulvicollis. *Payk. F.* ii. 77.—*Panz. F.* lxxxiii. *f.* 8.
β, Cr. glabrata. *Fabr. E.* i. 455.—*Panz. F.* xxxiv. *f.* 6.

2120. 2: nigriceps*. *Latr.* G. iii. 44.
Cr. lineola. *Fabr. E.* i. 462.—*Panz. F.* xxxiv. *f.* 5.

2121. 3: humeralis*. *Latr.* G. iii. 45.—(*Ing. Inst.* 89.)
β? Or. cærulea *mihi.*

2122 † 4. cantharoides[a].

---

[a] 2122 † 4. cantharoides. *Dufts. F. A.* iii. 249.
Cr. cantharoides. *Fabr. S. E.* i. 462. (!)—*Marsh.* i. 229. (!)
Cry. cantharoides. *Stew.* ii. 51. (!)—*Turt.* ii. 187. (!)

Genus 372. CRIOCERIS, *Geoffroy, Leach, Samou.*, (*Kirby.*)

CHRYSOMELA *p*, *Linné.*—LEMA *p*, *Fabricius.*—AUCHENIA *p*, *Marsh.*, *Millard.*—CRYPTOCEPHALUS *p*, *Gmelin.*

A. Corpore brevi, thoracis lateribus utrinque incisis.

2123. 1: merdigera*. *Latr.* G. iii. 47.—*Panz.* F. xlv. *f.* 2.—(*Leach E. E.* ix. 113.)—(*Samou.* 211. *pl.* 2. *f.* 14.)
Ch. merdigera var. capite &c. nigro. *Linn.* ii. 599.—*Mart. C. pl.* 17. *f.* 47.—*Shaw. G. Z.* vi. 58. *pl.* 18.
Ch. rubra Liliorum. *DeGeer.* v. 338.
Att. Lilii. *Scop. C.* 112.
Au. merdigera. *Marsh.* i. 213.
Cry. merdiger. *Turt.* ii. 188.

B. Corpore brevi, thoracis lateribus haud incisis.

2124. 2, duodecimpunctata*. *Panz. F.* xlv. *f.* 3.—(*Sam. I.* 13.)
Ch. 12-punctata. *Linn.* ii. 601.—*Berk. S.* i. 96.—*Mart. C. pl.* 16. *f.* 36.
Cry. 12-punctatus. *Stew.* ii. 51.—*Turt.* ii. 188.
Au. 12-punctata. *Marsh.* i. 214.

2125. 3; puncticollis*. *Spence MSS.*—(*Sam. I.* 13.)
Le. cyanella ♀. *Gyll.* iii. 640.

2126. 4; cyanella*. *Panz. F.* lxxi. *f.* 1.—(*Sam. I.* 13.)
Ch. cyanella. *Linn.* ii. 600.—*Berk. S.* i. 96.—*Mart. C. pl.* 15. *f.* 25.
Cry. cyanellus. *Stew.* ii. 51.—*Turt.* ii. 190.
Au. cyanella. *Marsh.* i. 215.

2127. 5, obscura *mihi.*
Le. cyanella b. *Gyll.* iii. 639?

C. Corpore elongato.

2128. 6; melanopa*. (*Sam. I.* 13.)
Ch. melanopa. *Linn.* ii. 601.—*Mart. C. pl.* 17. *f.* 53.
Cry. melanopus. *Turt.* ii. 190.
Au. melanopa. *Marsh.* i. 215.

2129. 7; Asparagi*. (*Sam. I.* 13.)
Ch. Asparagi. *Linn.* ii. 601.—*Berk. S.* i. 96.—*Don.* i. *pl.* 28.—*Mart. C. pl.* 17. *f.* 49.—*Shaw G. Z.* vi. 58.
Cry. Asparagi. *Stew.* ii. 51.—*Turt.* ii. 190.
Au. Asparagi. *Marsh.* i. 214.

Genus 373. ZEUGOPHORA, *Kunzé.*

CRIOCERIS *p*, *Fabr.*—AUCHENIA *p*, *Marsham.*—LEMA *p*, *Gyll.*—CRYPTOCEPHALUS *p*, *Gmel.*—CREVIA, ——? *MSS.*

2130. 1; subspinosa*. *Kunzé E. F.* 75.
Le. subspinosa. *Panz. F.* lxxxiii. *f.* 10.
Cr. subspinosa. *Fabr. E.* i. 461.—*Oliv. E.* vi. 74. *pl.* 2. *f.* 5, 6.—(*Sam. I.* 13.)

Cry. subspinosus. *Stew.* ii. 51.—*Turt.* ii. 190.
Au. subspinosa. *Marsh.* i. 216.

2131. 2, flavicollis *. *Kunzé E. F.* 75.
Au. flavicollis. *Marsh.* i. 217.
Cr. flavicollis. (*Sam. I.* 13.)
Le. subspinosa, ♀. *Gyll.* iii. 640.

Subsectio 2. CYCLICA, *Latreille.*

Familia L. GALERUCIDÆ.

(GALERUCOIDEÆ, *Gyllenhal.*—GALERUCITES, *Latr.*)
(CHRYSOMELA *p*, *Linné*, *&c.*—GALERUCA, *Fabr.*)

Genus 374. AUCHENIA *mihi.*

AUCHENIA *p*, *Marsh.*—CRIOCERIS *p*, *Panz.*—LUPERUS *p*, *Müll.*

2132. 1: quadrimaculata*. *Marsh.* i. 218.
Ch. quadrimaculata. *Linn.* ii. 600.
Cr. bimaculata. *Panz. F.* xlviii. *f.* 16.
Cr. melanogaster. *Gmel.* iv. 1720.

Genus 375. ADIMONIA, *Schrank*, *Leach*, *Samou.*
LUPERUS *p*, *Müller.*

2133. 1: Alni*. (*Leach E. E.* ix. 114.)—(*Sam. I.* 1.)
Ch. Alni. *Linn.* ii. 587.—*DeGeer.* v. 314. *pl.* 9. *f.* 18.—*Berk. S.* i. 94.—*Stew.* ii. 46.—*Marsh.* i. 172.—*Turt.* ii. 159.
Ad. violacea. *Laich. T.* i. 193.

2134. 2; halensis*. (*Leach E. E.* ix. 114.)—(*Sam. I.* 1.)
Ch. halensis. *Linn.* ii. 589.—*Mart. C. pl.* 15. *f.* 20.—*Marsh.* i. 177.
Cr. nigricornis. *Fabr. E.* i. 453.—*Panz. F.* xci. *f.* 9.
Ga. viridis. *Fourc. P.* i. 104.

Genus 376. GALERUCA, *Geoffroy*, *Leach*, *Samou.*, *Millard*, (*Kirby.*)

CRIOCERIS *p*, *Marsh.*—AUCHENIA *p*, *Marsh.*—TENEBRIO *p*, *Scop.*—ADIMONIA *p*, *Laich.*

A. Corpore ovato.

2135. 1; Tanaceti*. *Fabr. E.* i. 481.—(*Leach E. E.* ix. 114.)—(*Samou.* 212. *pl.* 2. *f.* 13.)
Ch. Tanaceti. *Linn.* ii. 587.—*Berk. S.* i. 94.—*Mart. C. pl.* 16. *f.* 31.—*Stew.* ii. 45.—*Marsh.* i. 177.—*Turt.* ii. 159.—*Samou.* 53.
Te. tristis. *Scop. C.* 256.

2136. 2, rustica. *Fabr. E.* i. 481.—*Panz. F.* cii. *f.* 1.—(*Ing. Inst.* 88.)
Ad. Tanaceti γ. *Laich. T.* i. 192.
Silpha ferruginea. *Petagna I. C.* i. 7. *f.* 16.

2137. 3; Capreæ*. *Fabr. E.* i. 487.—(*Sam. I.* 17.)
Ch. Capreæ. *Linn.* ii. 600.—*Mart. C. pl.* 17. *f.* 45.—*Turt.* ii. 161.
Cri. Capreæ. *Fabr. S. E.* 118.—*Marsh.* i. 225.
Cry. longicornis. *Gmel.* iv. 1688.
Cry. pallescens. *Gmel.* iv. 1724.
Ad. polygonata. *Laich. T.* i. 193.

2138. 4; Viburni*. *Payk. F.* ii. 89.—(*Sam. I.* 17.)
Cr. Viburni. *Marsh.* i. 224.

2139. 5; Cratægi*.
Ch. Cratægi. *Forst. C.* i. 28.—*Mart. C. pl.* 14. *f.* 4.
Cri. Cratægi. *Marsh.* i. 228.
Cry. Cratægi. *Stew.* ii. 52.—*Turt.* ii. 191.
Ga. Cratægi. (*Sam. I.* 17.)
Ga. sanguinea. *Fabr. M.* i. 87.

2140. 6, saturata *mihi.*

B. Corpore elongato-ovato.

2141. 7; Nymphææ*. *Fabr. E.* i. 486.—*Panz. F.* cii. *f.* 6.
Ch. Nympheæ. *Linn.* ii. 600.—*Mart. C. pl.* 17. *f.* 46.—*Turt.* ii. 161.
Cri. Nympheæ. *Marsh.* i. 224.

2142. 8; Sagittariæ*. *Gyll.* iii. 511.

2143. 9; Calmariensis*.
Ch. calmariensis. *Linn.* ii. 600.—*Stew.* ii. 51.—*Turt.* ii. 162.
Cr. calmariensis. *Marsh.* i. 227.
Ga. Lythri b. *Gyll.* iii. 513.

2144. 10, Lythri*.
Ga. Lythri a. *Gyll.* iii. 513.—(*Schön. S.* ii. 296.)
Ch. grisea Alni ♂. *DeGeer.* v. 325.

2145. 11, lineola*. *Fabr. E.* i. 486.—*Panz. F.* cii. *f.* 5.
Au. lineola. *Ent. Trans.* (*Burrell.*) i. 162.
Ch. grisea Alni ♀. *DeGeer.* v. 325. *pl.* 9. *f.* 36.

2146. 12, xanthomelæna.
Ga. Calmariensis. *Fabr. E.* i. 488.—*Latr. G.* iii. 62.
Ch. xanthomelæna. *Schrank. E.* 78.
Ga. Ulmi. *Oliv. E. M.* vi. 589?

2147. 13; tenella*. *Fabr. E.* i. 490.—*Panz. F.* cii. *f.* 9.
Ch. tenella. *Linn.* ii. 600.—*Mart. C. pl.* 16. *f.* 34.—*Turt.* ii. 162.
Au. tenella. *Marsh.* i. 219.—*Ent. Trans.* (*Burrell.*) i. 162.
Cri. parva. *Herbst A.* 66.

Genus 377. ————

ALTICA *p, Panzer.*—LUPERUS *p, Schön.*

2148. 14; circumfusa*.
Cr. circumfusa. *Marsh.* i. 227.

At. Brassicæ. *Panz. F.* xxi. *f.* 18.
Ha. Spartii. *Ent. Hefte.* ii. 76.
Lu. suturella. (*Schön. S.* ii. 276.)

Genus 378. LUPERUS, *Geoffroy, Leach, Samou.*

CRIOCERIS *p, Fabr.*—AUCHENIA *p, Marsh.*

2149. 1; flavipes*. *Leach E. E.* ix. 114.—(*Sam. I.* 26.)
Ch. flavipes. *Linn.* ii. 601.
Cr. flavipes. *Panz. F.* xxxii. *f.* 4.
Au. flavipes. *Marsh.* i. 216.
Cry. ochropus. *Gmel.* iv. 1723.
Cry. nigricans. *Gmel.* iv. 1725.
♀, Lu. brevicornis. *Kirby MSS.*

2150. 2; rufipes*. *Leach E. E.* ix. 114.—(*Sam. I.* 26.)
Cr. rufipes. *Fabr. E.* i. 461.—*Panz. F.* xxxii. *f.* 5.
Cr. flavipes, β. *Payk. F.* ii. 79.
Au. rufipes. *Marsh.* i. 217.
Ch. xanthopoda. *Schr.* 177.
Cry. saxonicus. *Gmel.* iv. 1723.
Cry. erythromelas. *Gmel.* iv. 1725.
♂; Pti. longicornis. *Fabr. E.* i. 325.

Genus 379. HALTICA, *Illiger, Leach, Samou., Millard,* (*Kirby.*)

ALTICA, *Geoffroy.*

A. Tibiis posticis haud dentatis.

a. *Antennis* (*in masculis*) *nodosis.*

1. Corpore oblongo: elytris apice obtusis.

2151. 1; antennata*. *Óliv. E.* vi. 82. *pl.* 5. *f.* 82.—*Ent. Heft* ii. 67. *pl.* 3. *f.* 4.
Ha. nodicornis. (*Sam. I.* 20.)
Ch. nodicornis. *Marsh.* i. 204.

2. Corpore ovato: elytris apice rotundatis.

2152. 2; Brassicæ*. *Oliv. E. M.* iv. 111.—(*Sam. I.* 20.)
Ha. quadripustulata. *Ent. Heft.* ii. 73. *pl.* 3. *f.* 5.
Ch. exclamationis. *Thunb. Act. Up.* iv. 14.
β; Ha. quadriguttata. *Kirby MSS.*

b. *Antennis simplicibus.*

1. Corpore oblongo, aut ovato.

* *Thorace posticè haud transversè impresso.*

† Elytris punctulatis, non striatis: tarsi postici in apice tibiæ inserti.

+ *Tarsi postici breves.*

2153. 3; Nemorum*. Turnip Fly: Black-jack.—*Ent. Heft.* ii. 70.—*Panz. F.* xxi. *f.* 19.—(*Sam. I.* 20.)—(*Kirby & Sp. I. E.* i. 186.)

Ch. Nemorum. *Linn.* ii. 595.—*Berk. S.* i. 96.—*Stew.* ii. 50. —*Marsh.* i. 197.—*Turt.* ii. 168.—*Don.* xvi. *pl.* 569. *f.* 1.
Ch. fasciata. *DeGeer.* v. 347.

2154. 4; flexuosa*. *Ent. Heft.* ii. 71.—(*Sam. I.* 20.)
Ch. flexuosa. *Marsh.* i. 198.
Al. Nasturtii. *Fanz. F.* xxi. *f.* 9.
Ch. Nemorum, β. *Payk. F.* ii. 98.

2155. [5, sinuata*] *mihi.*
Al. flexuosa. *Panz. F.* xxv. *f.* 12.

2156. 6, vittata* *mihi.*
Cr. Brassicæ. *Fabr. E.* i. 468?
Ch. Brassicæ. *Ent. Trans.* (*Burrell.*) i. 157.

2157. 7; Lepidii*. *Ent. Heft.* ii. 64.
Ch. nigripes. *Fabr. E.* ii. 447. (!)—*Stew.* ii. 49? (!)—*Turt.* ii. 166.? (!)
Al. nigripes. *Panz. F.* xxi. *f.* 5.
Al. hortensis. *Oliv. E. M.* iv. 108.

2158. [8; lens*.]
Ch. lens. *Thunb. Act. Up.* iv. 13.
Ha. Lepidii β. *Gyll.* iii. 528.

2159. 9; obscurella*. *Illig. M.* vi. 154.
Ha. atra. *Ent. Heft.* ii. 63.

2160. [10; punctulata*.]
Ch. punctulata. *Marsh.* i. 200.

2161. 11; melæna*. *Illig. M.* vi. 154.
Ga. atra. *Payk. F.* ii. 100?

2162. 12, nigroænea. (*Sam. I.* 20.)
Ch. nigroænea. *Marsh.* i. 197.

2163. 13; quadriguttata*.
Ch. quadripustulata. *Stew.* ii. 49.—*Marsh.* i. 198.—*Turt.* ii. 167.
Ha. quadripustulata. (*Sam. I.* 20.)

2164. 14, dorsalis*. *Ent. Heft.* ii. 79. *pl.* 3. *f.* 7.
Cr. dorsalis. *Fabr. E.* i. 465.—*Stew.* ii. 49.
Ch. dorsata. *Turt.* ii. 167.

2165. 15, Cyparissiæ. *Ent. Heft.* ii. 80. *pl.* 3. *f.* 8.

2166. 16; Euphorbiæ*. *Ent. Heft.* ii. 58.
Cr. Euphorbiæ. *Fabr. E.* i. 467.
Ch. Euphorbiæ. *Marsh.* i. 204.

2167. 17; atrocærulea* *mihi.*

2168. 18; cærulea*. *Ent. Heft.* ii. 55.
Ch. flavipes. *Herbst A.* iv. 61.
β; Al. Hyoscyami. *Panz. F.* xxi. *f.* 4.

2169. 19; Pseudacori*. (*Sam. I.* 20.)
Ch. Pseudacori. *Marsh.* i. 196.
Ha. violacea. *Ent. Heft.* ii. 56.

2170. 20; fuscicornis*. *Ent. Heft.* ii. 51.
Ch. fuscicornis. *Linn.* ii. 595.
Cr. fulvipes. *Fabr. E.* i. 463.
Ch. rufipes. *DeGeer.* v. 343.—*Marsh.* i. 198.—*Don.* xi. *pl.* 365. *f.* 3.
Al. rufipes. *Panz. F.* xxi. *f.* 10.
Ha. rufipes. (*Sam. I.* 20.)

+ + *Tarsi postici elongati.*

2171. 21; quadripustulata. *Illig. M.* vi. 168.—*Panz. F.* lxxxviii. *f.* 2.
Al. quatuorpustulata. *Oliv. E. M.* iv. 109?
Ch. Cynoglossi. *Marsh.* i. 205.
Ha. quadrimaculata. *Ent. Heft.* ii. 128.
Ha. quadrinotata. *Dufts. F. A.* iii. 259.

2172. 22; Verbasci. *Ent. Heft.* ii. 84. *pl.* 3. *f.* 8.—(*Sam. I.* 20.)
Ch. Verbasci. *Marsh.* i. 202.

2173. 23, Thapsi.
Ch. Thapsi. *Marsh.* i. 202.
Cr. Sisymbrii. *Fabr. E.* i. 465?

2174. 24; tabida*. *Gyll.* iii. 542.—*Panz. F.* xxi. *f.* 15.—(*Sam. I.* 20.)
Cr. tabida. *Fabr. E.* i. 467.—*Stew.* ii. 50.
Ch. tabida. *Marsh.* i. 203.—*Turt.* ii. 168.

2175. 25; flavicornis*. *Kirby MSS.*

2176. 26; atricilla*. *Ent. Heft.* ii. 86?—(*Sam. I.* 20.)
Ch. atricilla. *Linn.* ii. 594.—*Marsh.* i. 200.—*Turt.* ii. 167.
β? Ha. atricapilla. *Dufts. F. A.* iii. 257.

2177. 27; piciceps* *mihi.*
Ha. oblongiuscula. *Dufts. F. A.* iii. 261?

2178. 28; femoralis*. (*Sam. I.* 20.)—*Gyll.* iv. 657. (!)
Ch. femoralis. *Marsh.* i. 201.
Ha. melanocephala. *Gyll.* iii. 545.

2179. 29; confinis*. *Kirby MSS.*

2180. 30; atriceps* *mihi.*
Ch. exoleta. *Marsh.* i. 201.

2181. 31, lutescens. *Gyll.* iii. 546.

2182. 32; ochroleuca*. (*Sam. I.* 20.)
Ch. ochroleuca. *Marsh.* i. 202.

2183. 33; suturalis*. (*Sam. I.* 20.)
Ch. suturalis. *Marsh.* i. 201.

2184. 34; Nasturtii*.
Cr. Nasturtii. *Fabr. E.* i. 463.
Ha. pratensis b. *Ent. Heft.* ii. 88. *pl.* 3. *f.* 9.
Ga. atricilla γ. *Payk. F.* ii. 103.

2185. 35; thoracica*. *Kirby MSS.*

2186. 36; fuscicollis*. *Kirby MSS.*

2187. 37; Ballotæ*.
Ch. Ballotæ. *Marsh.* i. 205.

2188. 38; pratensis*. *Ent. Heft.* ii. 88.—*Panz. F.* xxi. *f.* 16.
Ga. atricilla α. *Payk. F.* ii. 102.

2189. 39; pallens*. *Kirby MSS.*

2190. 40; abdominalis*. *Megerle.*—*Dufts. F. A.* iii. 262.

2191. 41; pusilla*. *Gyll.* iii. 549.
Ha. Anglica. *Marsh. MSS.*

2192. 42; collaris* *mihi.*

2193. 43; lurida*. *Gyll.* iii. 537.
Ga. atricilla, δ. *Payk. F.* ii. 104.

2194. 44; castanea*. *Megerle.*—*Dufts. F. A.* iii. 260.

2195. 45; brunnea*. *Dufts. F. A.* iii. 260?

2196. 46; lævis*. *Dufts. F. A.* iii. 261.

2197. 47; fuscescens*. *Kirby MSS.*
Al. pratensis. *Ent. Heft.* ii. 88?

2198. 48; nigricans*. *Kirby MSS.*

2199. 49; læta*. *Kirby MSS.*

2200. 50, Anchusæ. *Ent. Heft.* ii. 62.
Ga. Anchusæ. *Payk. F.* ii. 101.
Ch. Anchusæ. *Ent. Trans.* (*Burrell.*) i. 157.

2201. 51; parvula*. *Ent. Heft.* ii. 59.
Ha. pumila. *Illig. M.* vi. 170.

2202. 52, pulex.
Ch. Pulex. *Marsh.* i. 204.

2203. 53; pilaris*. *Kirby MSS.*

2204. 54; Holsatica*. *Ent. Heft.* ii. 60.—*Oliv. E.* vi. 721. *pl.* 5. *f.* 74.
Ch. Holsatica. *Linn.* ii. 595.—*Ent. Trans.* (*Burrell.*) i. 156.

†† Elytris punctato-striatis.

+ *Tarsi postici in apice tibiæ inserti.*

2205. 55; rustica*. *Illig. M.* vi. 159.
Ch. rustica. *Linn.* ii. 595.

2206. 56; semiænea*. *Ent. Heft.* ii. 43.—(*Sam. I.* 20.)
Ch. semiænea. *Fabr. E.* i. 448.—*Marsh.* i. 194.

2207. 57, Chrysanthemi. *Ent. Heft.* ii. 45.

2208. 58, ænea.
Ch. ænea. *Mus. Marsham.*

2209. 59, obtusata. *Gyll.* iii. 379.

2210. 60; ærata. (*Sam. I.* 20.)
Ch. ærata. *Marsh.* i. 204.

2211. 61; striatula*.
Ch. striatula. *Marsh.* i. 205.
Ha. striata. (*Sam. I.* 20?)

2212. 62; Rubi*. *Ent. Heft.* ii. 43.
Cr. Rubi. *Fabr. E.* i. 468.

2213. 63; brunnicornis* *mihi.*

2214. 64; fuscipes*. *Ent. Heft.* ii. 10.—*Panz. F.* xxi. *f.* 11.—(*Sam. I.* 20.)
Ch. fuscipes. *Marsh.* i. 199.—*Turt.* ii. 167.

2215. 65; tripudiens*. *Kirby MSS.*

++ *Tarsi postici elongati, ab apice tibiæ remoti.*
MACRONEMA, (*Megerle.*)

2216. 66; chalcomera*. *Illig. M.* vi. 75.

2217. 67, Hyoscyami. *Ent. Heft.* ii. 27. *pl.* 2. *f.* 6.—(*Sam. I.* 20.)
Ch. Hyoscyami. *Linn.* ii. 594.—*Berk. S.* i. 95.—*Stew.* ii. 49.—*Mart. C. pl.* 14. *f.* 15.—*Marsh.* i. 193.—*Turt.* ii. 166.

2218. 68; Napi*. *Gyll.* iii. 567.
Ch. Hyoscyami, b. *Payk. F.* ii. 105.

2219. 69; Rapæ*. *Illig. M.* vi. 174.
Ha. violacea. *Kirby? MSS.*

2220. 70; chrysocephala*. *Ent. Heft.* ii. 31. *pl.* 2. *f.* 8.
Ch. chrysocephala. *Linn.* ii. 594.—*Marsh.* i. 193.
Al. Napi. *Panz. F.* xxi. *f.* 3.

2221. 71; rufilabris*. *Ent. Heft.* ii. 33. *pl.* 2. *f.* 9.

2222. 72; brunnipes*? *Megerle.*—*Dufts. F. A.* iii. 280.
Ha. Sinapis *mihi.*

2223. 73; nigricollis*. (*Sam. I.* 20.)
Ch. nigricollis. *Marsh.* i. 205.
Ch. anglicana. *Stew.* ii. 49?
Ch. anglica. *Gmel.* iv. 1693.—*Turt.* ii. 167.

2224. [74; sordida*.] *Kirby.*

2225. 75; picicornis*. *Kirby MSS.*

2226. 76; exoleta*. *Illig. M.* vi. 176.—(*Sam. I.* 20.)
Ch. exoleta. *Linn.* ii. 594?—*Berk. S.* i. 95?—*Stew.* ii. 49?—*Turt.* ii. 167.
Ha. nigriceps. *Kirby MSS.*

*** Elytris lævigatis.

2227 † 77, Kirbii *mihi.*
Ha. cyanoptera. *Kirby MSS.*—non Ha. cyanoptera. *Illig. M.* vi. 174. In Mus. *D. Kirby.*

** *Thorace posticè transversè impresso.*

† Elytris punctato-striatis.

2228. 78; ferruginea*. *Schr.*—*Illig. M.* vi. 109.
Ch. transversa. *Marsh.* i. 203:
Ha. transversa. (*Sam. I.* 20.)

2229. 79; flava*.
Mordella flava. *Linn. F. ed.* i. *no.* 535.
Ch. exoleta. *Linn. Faun. ed.* ii. 541. *descr.*
Ha. ferruginea, var. minor. *Illig. M.* vi. 109.

2230. 80; similis*. *Kirby MSS.*

2231. 81; affinis*. *Ent. Heft.* ii. 35.—(*Sam. I.* 20.)
Al. atricilla. *Panz. F.* xxi. *f.* 8.
Ch. atricilla. *Don.* xvi. 69. *pl.* 566.
Ch. affinis. *Ent. Trans.* (*Burrell.*) i. 157.

2232. 82; Salicariæ*. (*Schön. S.* ii. 311.)
Ga. Saliçariæ. *Payk. F.* iii. *App.* 453.
Ga. affinis capite pallido. *Payk. F.* ii. 109.

2233. 83; Modeeri*. *Ent. Heft.* ii. 47.—*Panz. F.* xxi. *f.* 7.—(*Sam. I.* 20.)
Ch. Modeeri. *Linn.* ii. 594.—*Marsh.* i. 194.—*Don.* xvi. *pl.* 569. *f.* 2.
β; Ha. immaculata. *Kirby MSS.*

2234. 84; rufipes*. *Ent. Heft.* ii. 11:
Ch. rufipes. *Linn.* ii. 595.
Al. ruficornis. *Panz. F.* xxi. *f.* 12.
Ch. ruficornis. *Marsh.* i. 199.—*Don.* xi. *pl.* 365. *f.* 2.
Ha. ruficornis. (*Sam. I.* 20.)
Ch. cæruleo-striata. *DeGeer.* v. 343.

2235 † 85, femorata. *Gyll.* iii. 559?
Ha. rufipes, b. *Schön.* ii. 307.
Ha. lineata. *Kirby? MSS.* In Mus. *D. Kirby.*

2236. 86; Helxines*. *Ent. Heft.* ii. 15.—*Panz. F.* xxi. *f.* 6.—(*Sam. I.* 20.)
Ch. Helxines. *Linn.* ii. 594.—*Marsh.* i. 194.
Ch. viridi-aurata. *DeGeer.* v. 345.
β; Ch. aurata. *Marsh.* i. 195.
Ha. aurata. (*Sam. I.* 20.)
γ; Ch. violaceo-punctata. *DeGeer.* v. 344.

2237. [87; fulvicornis*.]

Ch. fulvicornis. *Fabr. E.* i. 447.
Ha. Helxines, b. *Ent. Heft.* ii. 15.

2238. 88, nitidula. *Ent. Heft.* ii. 13.—(*Sam. I.* 20.)
Ch. nitidula. *Linn.* ii. 594.—*Berk. S.* i. 95.—*Mart. C. pl.* 14. *f.* 14.—*Stew.* ii. 49.—*Marsh.* i. 195.—*Don.* viii. *pl.* 273.—*Turt.* ii. 166.

2239. 89, cyanea. (*Sam. I.* 20.)
Ch. cyanea. *Marsh.* i. 196.

2240 † 90, gaudens. *Kirby MSS.* In Mus. D. *Kirby.*

2241. 91; apicalis* *mihi.*

†† Elytris temerè punctulatis.

2242. 92; Oleracea*. *Ent. Heft.* ii. 54.—*Panz. F.* xxi. *f.* 1.—(*Leach E. E.* ix. 114.)—(*Sam. I.* 2—20.)
Ch. Oleracea. *Linn.* ii. 593.—*Berk. S.* i. 95.—*Stew.* ii. 49.—*Marsh.* i. 192.—*Turt.* ii. 165.

2243. [93; indigacea*.] *Illig. M.* vi. 114?

2244. 94; Erucæ*. *Panz. F.* xxi. *f.* 2.—(*Sam. I.* 20.)
Ga. Erucæ. *Fabr. E.* i. 497.
Ch. Erucæ. *Marsh.* i. 193.
Ha. oleracea, c. *Gyll.* iii. 322.

2. Corpore hemisphærico.

* *Thorace haud transversè impresso.*

† Elytris non striatis.

2245. 95; testacea*. *Ent. Heft.* ii. 50.—*Panz. F.* xxi. *f.* 13.—(*Leach. E. E.* ix. 114.)—(*Sam. I.* 2—20.)
Ch. testacea. *Fabr. E.* i. 448.—*Marsh.* i. 202.—*Turt.* ii. 167.

2246. 96; Centaureæ*. *Kirby MSS.*—(*Sam. I.* 20.)

2247. 97; Cardui*. *Kirby MSS.*—*Gyll.* iv. 659. (!)

†† Elytris striatis.

2248. 98; orbiculata*. (*Sam. I.* 20.)
Ch. orbiculata. *Marsh.* i. 200.
Ha. Graminis. *Ent. Heft.* ii. 47.
Ha. Hederæ. *Illig. M.* vi. 164.

2249. 99; globosa*.
Al. globosa. *Panz. F.* xxv. *f.* 13.
Ha. conglomerata. *Illig. M.* vi. 164.

** *Thorace lineâ transversâ impressâ: elytris striatis.*

2250. 100; Mercurialis*. *Hellwig.*—*Illig. M.* vi. 117.—(*Sam. I.* 20.)
Ch. pernigra. *Marsh. MSS.*

b. Tibiis posticis dentatis.

a. *Capite prominulo: elytris punctato-striatis.*

2251. 101; aridella*. *Ent. Heft.* ii. 41. *pl.* 3. *f.* 2.

2252. 102; æneofusca*. *Kirby MSS.*—(*Sam. I.* 20.)

2253. 103; picina*. (*Sam. I.* 20.)
Ch. picina. *Marsh.* i. 206.

2254. 104; concinna*. (*Sam. I.* 20.)—(*Kirby & Sp. I. E.* i. 182.)
Ch. concinna. *Marsh.* i. 196.
Ha. dentipes. *Ent. Hefte.* ii. 38. *pl.* 3. *f.* 1.

2255. 105; picipes*. *Kirby MSS.*

2256. 106; Mannerheimii. *Gyll.* iv. 664. (!)

2257 † 107, saltitans. *Kirby MSS.* In Mus. D. *Kirby.*

b. *Capite infrà thoracem retracto: elytris obsoletè striatis.*

2258. 108; exaltans*. *Kirby MSS.*

2259. 109, Cynoglossi. *Ent. Hefte.* ii. 20. *pl.* 1. *f.* 2.
Ch. pallipes. *Marsh. MSS.*

Familia LI. CHRYSOMELIDÆ, *Leach.*

(Chrysomelini, *Latr.*—Chrysomelinæ *p, et* Cryptocephaloideæ, *Gyllenhal.*)

(Chrysomela *p. Linné, Fabr., &c.*—Cryptocephalus, *Fabr., Marsh., &c.*)

Genus 380. CHRYSOMELA *Auctorum.* Golden-*beetles.*

A. Palpi filiformes.

a. *Antennæ subclavatæ, breves.*

2260. 1; Armoraciæ* *Linn.* ii. 588. *Panz. F.* xliv. *f.* 14.—*Marsh.* i. 179.—*Turt.* ii. 152.
Ch. cæruleo-Salicis. *DeGeer* v. 318. *pl.* 9. *f.* 24.
Ga. Salicis. *Panz. E.* i. 172.

2261. [2; clavicornis*.] *Kirby MSS.*—(*Sam. I.* 10.)

b. *Antennæ extrorsum crassiores, elongatæ.*

1. Elytris striatis.

* *Corpore hemisphærico.*

2262. 3, Betulæ*. *Linn.* ii. 587.—*Berk. S.* i. 94.—*Stew.* ii. 46. —*Marsh.* i. 178.—*Turt.* ii. 159.—(*Sam. I.* 10.)
Ch. Plantaginis. *DeGeer.* v. 322.
Ch. Cochleariæ, b. *Gyll.* iii. 480.
Ch. Cochleariæ. *Panz. F.* xliv. *f.* 15.

2263. 4; Cochleariæ. *Fabr. E.* i. 445.
Alt. Erucæ. *Panz. F.* xxi. *f.* 2.

Ch. atroviolascens. *Marsh.* i. 184.
β; Ch. violascens. *Kirby MSS.*

2264. [5, Gomphoceros *.] *Leach MSS.*

2265. 6: concinna *mihi.*

2266. 7; tumidula *. *Kirby.*—*Germ. I. N.* i. 588.
β; Ch. confinis. *Kirby MSS.*

** *Corpore oblongo.*

2267. 8; aucta *. *Fabr. E.* i. 442.—*Mart. C. pl.* 15. *f.* 27.—*Marsh.* i. 181.—*Don.* xi. *pl.* 373. *f.* 2.—*Turt.* ii. 150.—(*Sam. I.* 10.)
Ch. analis. *Schr. A.* 80.

2268. 9; Vitellinæ *. *Linn.* ii. 589.—*Mart. C. pl.* 14. *f.* 1.—*Stew.* ii. 47.—*Marsh.* i. 180.—*Turt.* ii. 160.—(*Sam. I.* 11.)
β; Ch. vulgatissima. *Linn.* ii. 589.—*Mart. C. pl.* 14. *f.* 8.—*Marsh.* i. 179.
Ch. cæruleo Betulæ. *DeGeer.* v. 317.
Ga. Betulæ. *Panz. F.* cii. *f.* 4.
γ; Ch. nigrina. *Kirby MSS.*

2269. 10; unicolor *. *Marsh.* i. 185.—(*Sam. I.* 10.)
β; Ch. intermedia. *Kirby MSS.?*

2. Elytris punctatis.

* *Corpore ovato-hemisphærico.*

2270. 11; Polygoni *. *Linn.* ii. 589.—*Berk. S.* i. 95.—*Mart. C. pl.* 14. *f.* 2.—*Don.* iii. *pl.* 96. *f.* 1.—*Stew.* ii. 47.—*Marsh.* i. 178.—*Turt.* ii. 148.—(*Sam. I.* 10.)
Ga. ruficollis. *Fabr. E.* i. 479. (!)—*Turt.* ii. 158.
Bu. salicina. *Scop. C.* 199.

2271. 12; Raphani *. *Fabr. E.* i. 430.—*Herbst A.* 59. *pl.* 23. *f.* 42.
Ch. viridula. *DeGeer.* v. 311.
Ch. Polygoni β. *Linn. F.* 520.
Ch. Hypochæridis. *Marsh.* i. 184.—*Don.* xi. *pl.* 373. *f.* 3.—(*Sam. I.* 10.)
Ch. holosericea. *Marsh MSS.*
Bu. Syngenesiæ. *Scop. C.* 193.
Ch. Rumicum. *Kirby MSS.*

** *Corpore oblongo.*

2272. 13; fastuosa *. *Linn.* ii. 470.—*Mart. C. pl.* 14. *f.* 11.—*Don.* vi. *pl.* 194.—*Stew.* ii. 46.—*Marsh.* i. 174.—*Turt.* ii. 149.—(*Sam. I.* 10.)—(*Curtis l. c. infra folio* 111.)

B. Palpi securiformes.

a. *Elytris punctato-striatis.*

1. Thoracis margine haud incrassato.

2273 † 14, Adonidis. *Fabr. E.* i. 431.—*Curtis* iii. *pl.* 111.
In Mus. *Brit.*

2274. 15; pallida*. *Linn.* ii. 589.—*Mart. C. pl.* 16. *f.* 35—*Marsh.* i. 174.—*Turt.* ii. 147.—(*Sam. I.* 10.)—*Wood.* i. 34. *pl.* 11.
Ch. Padi. *DeGeer.* v. 301.
Ch. dispar α. *Payk. F.* ii. 66.

2275. 16; decempunctata*. *Linn.* ii. 590.—*Marsh.* i. 175.—*Turt.* ii. 147.—(*Sam. I.* 10.)
Ch. nigripes. *DeGeer.* v. 296.
Ch. viminalis. *Panz. F.* lxxviii. *f.* 3.—(*Curtis l. c. supra.*)

2276. 17; rufipes*. *DeGeer. v.* 295. *pl.* 8. *f.* 25.—(*Curtis l. c. supra.*)
Ch. decempunctata β. *Linn.* ii. 590.—*Mart. C. pl.* 16. *f.* 40.
Ch. decemnotata. *Marsh.* i. 175.—*Don.* xi. *pl.* 373. *f.* 1.—(*Sam. I.* 10.)

2277. 18, sexpunctata*? *Fabr. E.* i. 436.—*Panz. F.* xxvi. *f.* 11.

2278. 19; Litura*. *Fabr. E.* i. 429.—*Mart. C. pl.* 16. *f.* 37.—*Stew.* ii. 48.—*Marsh.* i. 182.—*Turt.* ii. 152.—(*Leach. E. E.* ix. 115.)—(*Sam. I.* 10.)
Ch. vernalis. *Forst. C.* i. 23.—*Turt.* ii. 157.—*Stew.* ii. 48.
β; Ch. olivacea. *Forst. C.* i. 22.—*Marsh.* i. 181.
Ch. flavicans. *Fabr. E.* i. 429.
γ; Ch. rufitarsis. *Marsh.* i. 182.
δ; Ch. stygia. *Marsh.* i. 183.

2279 † 20, festa. *Kirby MSS.* In Mus. *D. Vigors.*

2. Thoracis margine incrassato.

* *Corpore oblongo.*

2280. 21, marginata*. *Linn.* ii. 591.—*Panz. F.* xvi. *f.* 11.—*Marsh.* i. 190.—(*Sam. I.* 10.)—(*Curtis l. c. supra.*)

2281 † 22, lepida. *Leach. MSS.*—(*Curtis l. c. supra.*)
In Mus. *Brit.*

2282. 23; geminata*. *Payk. F.* ii. 65.—(*Curtis l. c. supra.*)
Ch. quinquejugis. *Marsh.* i. 173.—(*Sam. I.* 10.)

2283. 24; Hyperici*. *Forst. C.* i. 20.—*Marsh.* i. 173.—(*Sam.* i. 10.)
Ch. anglica. *Gmel.* ii. 1689.—*Stew.* ii. 48.—*Turt.* ii. 156.
Ch. gemellata. *Rossi.* i. 30.—*Panz. F.* xliv. *f.* 6.
Ch. fucata. *Fabr. E.* i. 444.
Ch. bicolor. *Act. Hol.* (*Zetterstedt.*) 1818. 256.

** *Corpore ovato.*

2284 † 25, lurida*. *Linn.* ii. 590. *Panz. F.* lxxviii. *f.* 1.—(*Sam. I.* 10.—(*Curtis l. c. supra.*) In Mus. *Brit.*

2285. 26; lamina*. *Fabr. E.* i. 430.—*Panz. F.* xliv. *f.* 5.
Ch. olivacea. *Act. Hol.* (*Schaller.*) i. 272.
Ch. orichalcea. *Müll. Z. D. pr.* 82.
Ch. incrassata. *Marsh.* i. 186.
Ch. bulgarensis. *Schrn. A.* 70.

2286. 27; Banksii*. *Fabr. E.* i. 430.—*Mart. C. pl.* 16. *f.* 42.—*Don.* iv. *pl.* 138. *f.* 4.—*Stew.* ii. 45.—*Marsh.* i. 187.—*Turt.* ii. 142.—(*Leach. E. E.* ix. 115.)—(*Sam. I.* 10.)

b. *Elytris punctatis.*

1. Thoracis margine incrassato.

* *Corpore oblongo.*

2287. 28: limbata*. *Fabr. E.* i. 441.—*Panz. F.* xvi. *f.* 8.—*Marsh.* i. 191.—*Turt.* ii. 150.—(*Sam. I.* 10.)—*Curtis l. c. supra.*)

** *Corpore ovato.*

2288. 29; sanguinolenta*. *Linn.* ii. 591.—*Berk. S.* i. 95.—*Mart. C. pl.* 15. *f.* 23.—*Don.* iv. *pl.* 111. *f.* 3, 4.—*Stew.* ii. 48.—*Marsh.* i. 190.—*Turt.* ii. 150.—(*Sam. I.* 10.)
Ch. rubromarginata. *DeGeer.* v. 298.
Ch. marginata. *Mus. Lond.* (*sæpè.*)

2289 † 30, Carnifex. *Payk. F.* xvi. *f.* 9.
Ch. marginata. *Ræm. Ins.* 7. *pl.* 3. *f.* 10.
Ch. sanguinolenta. *Rossi F.* i. 77.
Ch. Rossia. *Illiger.* In Mus. D. *Curtis.*

2290. 31; Staphylæa*. *Linn.* ii. 590.—*Mart. C. pl.* 14. *f.* 3.—*Stew.* ii. 47.—*Marsh.* i. 186.—*Turt.* ii. 145.—(*Sam. I.* 10.)
Ch. cuprea. *DeGeer.* v. 294. *pl.* 8. *f.* 24.

2291. 32; polita*. *Linn.* ii. 590.—*Mart. C. pl.* 14. *f.* 5.—*Berk. S.* i. 95.—*Stew.* ii. 47.—*Marsh.* i. 188.—*Turt.* ii. 145.—(*Sam. I.* 10.)

2. Thoracis margine haud incrassato.

* *Corpore oblongo.*

2292. 33: fulgida*. *Fabr. E.* i. 432.—*Ent. Trans.* (*Haworth.*) i. 250.—(*Sam. I.* 10.)—(*Curtis l. c. supra.*)
Ch. graminis, c. *Gyll.* iii. 468.

2293. 34: graminis. *Linn.* ii. 587.—*Barb. G.* 40. *pl.* 4. *f.* 3.—*Berk. S.* i. 94.—*Mart. C. pl.* 14. *f.* 10.—*Stew.* ii. 46.—*Marsh.* i. 172.—*Don.* xi. *pl.* 365. *f.* 1. (*var. viridi.*)—*Turt.* ii. 144.—*Shaw. G. Z.* vi. 58. *pl.* 17.—(*Sam. I.* 10.)

2294 † 35, violacea. *Fabr. E.* i. 433.—*Panz. F.* xliv. *f.* 8.
Ch. graminis (var. violacea.) *Don.* xi. *pl.* 365. *f.* 2?
In Mus. *Brit. et* D. *I. H.* *Griesbach.*

2295. 36, cerealis[a] *.

** *Corpore ovato.*

2296. 37; varians *. *Fabr. E.* i. 433.—(*Sam. I.* 11.)
Ch. Marshami. *Don.* viii. *pl.* 286. *f.* 1.—*Marsh.* i. 173.—*Turt.* ii. 149.
Coccinella Betulæ. *Scop. C.* 221.
β; Ch. Centaurei. *Fabr. E.* i. 428.—*Panz. F.* xliv. *f.* 10.
γ; Ch. Hyperici. *DeGeer.* v. 312. *pl.* 9. *f.* 13.
δ; Ch. æthiops. *Fabr. E.* i. 429?
ε; Ch. viridiænea. *Marsh.* i. 184.

2297. 38, hæmoptera. *Linn.* ii. 587.—*Berk. S.* i. 94.—*Mart. C. pl.* 15. *f.* 22.—*Stew.* ii. 46.—*Marsh.* i. 171.—*Turt.* ii. 144.—(*Sam. I.* 10.)—*Kirby & Sp. I. E.* iv. 201.
Ch. Hottentotta. *Fab. E.* i. 429.
β, Ch. picicornis. *Marsh. MSS.*

2298. 39; goettingensis *. *Linn.* ii. 586.—*Mart. C. pl.* 16. *f.* 41. —*Don.* viii. *pl.* 286. *f.* 2.—*Marsh.* i. 171.—*Turt.* ii. 142.—(*Sam. I.* 10.)
Ch. violacea-nigra. *DeGeer.* v. 298.
Ch. hæmoptera. *Payk. F.* ii. 52.

2299 † 40. Sparshalli[b]. *Curtis MSS.*—*Childr. Address.* 7. (!)
In Mus. D. Sparshall.?

Genus 381. TIMARCHA, (*Megerle*,) (*Samou.*,) (*Kirby.*)

TENEBRIO *p*, *Linn.*, *Wood.*

2300. 1; Tenebricosa *. (*DeJean C.* 122.) Bloody-nose-Beetle.
Ch. tenebricosa. *Fabr. E.* i. 423.—*Mart. C. pl.* 16. *f.* 43.—*Don.* viii. *pl.* 276.—*Marsh.* i. 169.—*Turt.* 141.—(*Sam. I.* 10.)—(*Kirby & Sp. I. E.* ii. 321. & iii. 99.)
Ch. tenebrioides. *Gmel.* ii. 1667.
Te. lævigatus. *Linn.* ii. 678.—*Wood.* i. 72. *pl.* 26.
Te. cæruleus. *Berk. S.* i. 109.
Ch. grossa. *Müll. Z. D. pr.* 81.

2301. 2; coriaria *. (*DeJean C.* 122.)—(*Samou. pl.* 2. *f.* 12.)
Ch. coriaria. *Fabr. E.* i. 424.—*Panz. F.* xliv. *f.* 2.—*Marsh.* i. 170.—*Samou.* 53.
Ch. goettingensis. *Payk. F.* ii. 51.
β; Ch. violacea-nigra. *DeGeer.* v. 298.
γ; Ch. iopa. *Marsh.* i. 170.

---

[a] 2295. 36, cerealis *. *Linn.* ii. 558.—*Berk. S.* i. 94. (!)—*Don.* iv. *pl.* 115. (!)—*Stew.* ii. 46. (!)—*Marsh.* i. 185. (!)—*Turt.* ii. 148. (!)
[b] An hujus divisio?

Genus 382. ————

A. Corpore ovato aut subelongato.

a. *Elytris punctatis.*

2302. 1; Populi*.
Ch. Populi. *Linn.* ii. 390.—*Mart. C. pl.* 15. *f.* 18, 19.—*Berk. S.* i. 95.—*Stew.* ii. 47.—*Marsh.* i. 188.—*Turt.* ii. 145.—*Shaw G.Z.* vi. 57. *pl.* 17.—(*Sam. I.* 10.)—(*Kirby & Sp. I. E.* ii. 245.)

2303. 2; Tremulæ*. *Fabr. E.* i. 434.—*Marsh.* i. 189.—(*Sam. I.* 10.)
Ch. Populi β. *Laich. T.* i. 150.

b. *Elytris striatis.*

2304. 3, vigintipunctata. *Fabr. E.* i. 442.—*Panz. F.* vi. *f.* 10.—*Stew.* ii. 48.—*Marsh.* i. 191.—*Turt.* ii. 161.
Ch. vicies maculata. *Bergs. N.* i. 87. *pl.* 13. *f.* 10.

2305 † 4. Lapponica[a].

2306. 5, ænea. *Linn.* ii. 587.—*Panz. F.* xxv. *f.* 9.—*Marsh.* i. 189. *Turt.* ii. 151.
Ch. viridis Alni. *DeGeer.* v. 305.

B. Corpore elongato.

2307. 6; marginella*. *Linn.* ii. 591.—*Mart. C. pl.* 15. *f.* 26.—*Stew.* ii. 48.—*Marsh.* i. 181.—*Turt.* ii. 150.—(*Leach. E. E.* ix. 115.)—(*Sam. I.* 10.)
Ch. marginella Ranunculi. *DeGeer.* v. 304.

Genus 383. HELODES, *Paykul, Leach, Samou., Millard,* (*Kirby.*)

PRASOCURIS, *Latr.*—CRYPTOCEPHALUS *p, Gmel.*—CRIOCERIS *p, Panzer.*

2308. 1; Phellandrii*. *Fabr. E.* i. 469.—(*Leach. E. E.* ix. 115.)—(*Sam. I.* 21.)
Ch. Phellandrii. *Linn.* ii. 601.—*Berk. S.* i. 96.—*Marsh.* i. 185.
Cry. Phellandrii. *Gmel.* iv. 1723.—*Stew.* ii. 51.—*Turt.* ii. 190.
Cri. Phellandrii. *Panz. F.* lxxxiii. *f.* 9.
Ch. calmariensis. *Don.* vi. 13. *pl.* 185. *f.* 1.

2309. 2; Beccabungæ*. *Payk. F.* iii. 451.
Ch. Beccabungæ. *Panz. F.* xxv. *f.* 11.—*Marsh.* i. 186.
He. violacea. *Fabr. E.* i. 470.—(*Leach. E. E.* ix. 115.)—(*Sam. I.* 21.)

Genus 384. CLYTHRA, *Laicharting, Leach, Samou., Millard.*

MELOLONTHA *p, Müller.*

2310. 1; quadripunctata*. *Fabr. E.* ii. 31.—(*Leach. E. E.* ix. 115.)—(*Sam. I.* 11.)

---

[a] 2305 † 4. Lapponica. *Linn.* ii. 591.—*Panz. F.* xxiii. *f.* 13.—*Turt.* ii. 148. (!)
Ch. curvilinea. *DeGeer.* v. 302. *pl.* 9. *f.* 3.

Cr. quadripunctatus. *Payk.* ii. 123.—*Stew.* ii. 50.—*Marsh.* i. 207.—*Turt.* ii. 172.
Ch. quadripunctata. *Linn.* ii. 596.—*Berk. S.* i. 96.—*Don.* iv. *pl.* 111. *f.* 1, 2.—*Mart. C. pl.* 17. *f.* 50.

2311. 2; tridentata*. *Gyll.* iii. 587.—(*Sam. I.* 11.)
Ch. tridentata. *Linn. S.* ii. 596.—*Panz. F.* xlviii. *f.* 12.—*Marsh.* i. 206.
Cr. tridentatus. *Panz. F.* xlviii. *f.* 12.
♂; Cl. longimana. *Fabr. E.* ii. 37.—*Panz. F.* xlviii. *f.* 14.

2312. 3, taxicornis. *Fabr. E.* ii. 34.
Ch. similis. *Illig. M.* (*Schneid.*) ii. 191. ♀.
Cr. tridentatus. *Petag. I. C.* 11. *pl. f.* 8.

2313 ‡ 4, longipes?
Ch. longipes. *Fabr. E.* ii. 28?—*Ent. Trans.* (*Leach.*) i. 248. *pl.* 8. *fig. sup.* ♂.—*fig. media*, ♀.—*Don.* xv. *pl.* 520.
In Mus. *Brit.*

2314 ‡ 5, Hordei. *Fabr. E.* ii. 41. In Mus. *Brit.*

Genus 385. CRYPTOCEPHALUS, *Geoffroy*, *Marsh.*, *Leach*, *Samou.*, *Millard*, *Curtis.*

BUPRESTIS *p*, *Scop.*

A. Elytris vagè punctatis.

2315. 1; sexpunctatus*. *Fabr. E.* ii. 46.—*Panz. F.* lxviii. *f.* 7. —*Marsh.* i. 208.—(*Sam. I.* 13.)—(*Curtis l. c. infra.*)
Ch. sexpunctata. *Linn.* ii. 599.

2316. 2; Coryli*. *Fabr. E.* ii. 45.—*Marsh.* i. 208.—(*Sam. I.* 13.) —(*Curtis l. c. infra.*)
Ch. Coryli. *Don.* ix. *pl.* 321. *f.* 1.
♂; Cr. Vitis. *Panz. F.* lxviii. *f.* 5.

2317. 3, Histrio. *Fabr. E.* ii. 55.—*Oliv. E.* v. 96. *pl.* 3. *f.* 31. *a*, *b*.

2318. 4; sericeus*. *Fabr. E.* ii. 49.—*Stew.* ii. 50.—*Marsh.* i. 209.—*Turt.* ii. 176.—(*Leach E. E.* ix. 115.)—(*Sam. I.* 13.)—(*Curtis l. c. infra.*)
Ch. sericea. *Linn.* ii. 598.—*Berk. S.* i. 96.—*Don.* ix. *pl.* 321. *f.* 2.—*Mart. C. pl.* 14. *f.* 6, 7.
Bu. Syngenesiæ. *Scop. C.* 63.

2319. 5; similis*. (*Sam. I.* 13.)—(*Curtis l. c. infra.*)
Cr. violaceus. *Fabr. E.* ii. 47?

2320. 6, nitens*. *Fabr. E.* ii. 49.—*Panz. F.* lxviii. *f.* 8.—*Marsh.* i. 209.—(*Sam. I.* 13.)—(*Curtis l. c. infra.*)
Ch. nitens. *Linn.* ii. 598.
† ♂, Cr. flavifrons. *Fabr. E.* ii. 51.

B. Elytris punctato-striatis.

2321. 7: bipustulatus. *Fabr. E.* ii. 54.—*Turt.* ii. 178.—*Curtis.* i. *pl.* 35.
Cr. biguttatus. *Herbst A.* vii. 163.
Cr. dispar, e. *Gyll.* iii. 615.

2322. 8; Lineola*. *Fabr. E.* ii. 44.—*Marsh.* i. 207.—*Turt.* ii. 178.—*Samou.* 53. *pl.* 2. *f.* 15.—(*Curtis l. c. supra.*)
Cr. dispar, b. *Gyll.* iii. 615.
Ch. biliturata. *Don.* iii. *pl.* 99. *f.* 1-3.—*Stew.* ii. 48.
Ch. bipunctata. *Berk. S.* i. 96?
Bu. sanguinolenta. *Scop. C.* 66.

2323. 9, nigripennis *mihi.*

2324. 10; Moræi*. *Fabr. E.* ii. 52.—*Panz. F.* lxviii. *f.* 11. ♀.—*Marsh.* i. 212.—(*Sam. I.* 13.)
Ch. Moræi. *Linn.* ii. 597.

2325. 11, vittatus. *Fabr. E.* ii. 50.—*Herbst A.* 62. *pl.* 23. *f.* 23.

2326. 12; flavilabris*. *Fabr. E.* ii. 51.—(*Sam. I.* 13.)—(*Curtis l. c. supra.*)

2327 † 13, Barbareæ. *Marsh.* i. 212.—(*Curtis l. c. supra.*)
Ch. Barbareæ. *Linn.* ii. 598.
Cr. flavilabris, b. *Gyll.* iii. 625. In Mus. *Brit. et* D. *Kirby.*

2328. 14; punctiger*. *Payk.* ii. 146.—(*Curtis l. c. supra.*)
Cr. cyaneus. *Wilk. MSS.*
Ch. chrysocephala. *DeGeer.* v. 337.

2329 † 15, frontalis. *Marsh.* i. 211.—(*Curtis l. c. supra.*)
Cr. labiatus β. *Schneid. M.* ii. 204.
In Mus. *Brit. et* D. *Kirby.*

2330. 16; labiatus*. *Fabr. E.* ii. 51.—*Panz. F.* lxviii. *f.* 9.—*Marsh.* i. 211.—(*Sam. I.* 13.)—(*Curtis l. c. supra.*)
Ch. labiata. *Linn.* ii. 598.

2331. 17; exilis*. *Schüppel.*—(*Curtis l. c. supra.*)

2332. 18, bilineatus. *Payk. F.* ii. 170.—*Turt.* ii. 177.—*Ahr. F.* vii. *f.* 8.—(*Curtis l. c. supra.*)
Ch. bilineata. *Linn.* ii. 597.
Cr. bilituratus. (*Sam. I.* 13.)

2333. 19; minutus*. *Fabr. E.* ii. 57.—*Panz. F.* xxxix. *f.* 18.
Ch. exoleta. *DeGeer.* v. 338.
Cr. fulvus. *Gmel.* iv. 1711.
♂; Cr. pusillus. *Fabr. E.* ii. 56.—*Marsh.* i. 210.—(*Sam. I.* 13.)—(*Curtis l. c. supra.*)
Cr. exilis. *Gmel.* ii. 1686.
♀; Chry. marginella. *Don.* x. 27. *pl.* 335.
Cr. marginellus. *Marsh.* i. 211.—(*Sam. I.* 13.)—(*Curtis l. c. supra.*)

2334. [20; ochraceus*.] (*Curtis l. c. supra.*)
2335. [21; gracilis*.] *Fabr. E.* ii. 57.—*Panz. F.* xcviii. *f.* 5.
Cr. sanguinicollis. *Naturf.* (*Frölich.*) xxvi. 130.
Cr. rufipes. *Gmel.* ii. 1711.
Cr. minutus ♂. (*Schön. S.* ii. 370.)

Genus 386. EUMOLPUS, *Fabr.*, *Leach.*

2336 ‡ 1, obscurus. *Fabr. E.* i. 423.
Ch. obscura. *Linn.* ii. 599.
Cr. obscurus. *Panz. F.* v. *f.* 12.
Ch. nigro-quadrata. *DeGeer.* v. 336. In Mus. *Brit.*
2337 ‡ 2. Vitis[a].
2338. 3, pygmæus. *Oliv. E.* vi. 916. *pl.* 2. *f.* 32.—(*Ing. Inst.* 88.)

Familia LII. CASSIDIADÆ, *Leach.*

(CASSIDARIÆ, *Latreille.*—CHRYSOMELINÆ *p*, *Gyllenhal.*)

Genus 387. CASSIDA *Auctorum.* *Tortoise-beetles.*

A. Apteri.

2339. 1: vittata*. *Fabr. E.* i. 391.—(*Curtis l. c. infra.*)—(*Ing. Inst.* 88.)
Ca. ocellata. *Herbst C.* viii. 244. *pl.* 131. *f.* 4. *b.*

B. Alati.

a. *Thoracis angulis posticis haud rotundatis.*

2340. 2; Murræa*. *Linn.* ii. 575.—*Herbst A.* 50. *pl.* 22. *f.* 28. —*Marsh.* i. 147.—*Samou.* 52. *pl.* 2. *f.* 10.
β; Ca. maculata. *Linn.* ii. 575.—*Berk. S.* i. 92.—*Don.* viii. *pl.* 285.—*Stew.* ii. 41.—*Linn. Trans.* (*Kirby.*) iii. 10.—*Marsh.* i. 147.—*Turt.* ii. 118.—(*Curtis l. c. infra.*)
2341. 3; rubiginosa*. *Illig. K.* i. 479.
Ca. viridis. *Latr. G.* iii. 52.—*Panz. F.* xcvi. *f.* 4.—*Wood.* i. 28. *pl.* 9.—(*Curtis l. c. infra.*)
Ca. prasina. *Fabr. E.* i. 388.
Ca. similis. *Marsh.* i. 144.—(*Sam. I.* 9.)
2342. 4. thoracica? *Fabr. E.* i. 388?—*Panz. F.* xxxviii. *f.* 24?
Ca. viridis var. *Act. Holm.* (*Paykul.*) 1801. 120. *b*?

b. *Thoracis angulis posticis rotundatis.*

1. Thorace brevissimo, transverso.

2343. 5; equestris*. *Fabr. E.* i. 388.—*Panz. F.* xcvi. *f.* 5.—(*Leach E. E.* ix. 114.)—(*Sam. I.* 9.)—(*Curtis l. c. infra.*)

---

[a] 2337 ‡ 2. Vitis. *Fabr. E.* i. 422.—*Panz. F.* lxxxix. *f.* 12.
Cr. Vitis. *Turt.* ii. 175. (!)
Ch. villosula. *Schra. E.* 95.

Ca. viridis. *Illig. K.* i. 480.—*Berk. S.* i. 92?—*Stew.* ii. 41? —*Barb.* G. *pl.* 34. *f.* 3.—*Linn. Trans.* (*Kirby.*) iii. 9.—*Shaw* G. Z. vi. 53. *pl.* 15.—*Marsh.* i. 143.—*Turt.* ii. 117.
Ca. Cardui. *DeGeer.* v. 174.

2. Thorace mediocri, haud transverso.

1. *Corpore ovato.*

2344. 6; Vibex*. *Linn.* ii. 575.—*Panz. F.* xcvi. *f.* 6.
Ca. liriophora. *Marsh.* i. 145.—*Linn. Trans.* (*Kirby.*)iii. 8. —*Turt.* ii. 117.—(*Curtis l. c. infra.*)
Ca. viridis β. *Payk. F.* ii. 46.

2345. 7; nebulosa*. *Linn.* ii. 575.—*Berk. S.* i. 92.—*Stew.* ii. 41. —*Marsh.* i. 145.—*Turt.* ii. 118.—(*Sam. I.* 9.)—(*Curtis l. c. infra.*)
Ca. affinis. *Fabr. E.* i. 388.
β; Ca. tigrina. *DeGeer.* v. 168. *pl.* 5. *f.* 15, 16.

2346. 8; ferruginea*. *Fabr. E.* i. 391.—*Herbst C.* viii. 245. *pl.* 130. *f.* 14.—*Marsh.* i. 148.
Ca. fusca. *Laich. T.* i. 2. 112.
Ca. obsoleta. (*Curtis l. c. infra.*)

2347. 9; obsoleta*. *Illig. K.* i. 484.
Ca. pallida. *Payk. F.* ii. 50.
Ca. flaveola. *Thunb. I. G.* viii. 108.
Ca. marcida? *Hoff.*?—(*Sam. I.* 9.)—(*Curtis l. c. infra.*)

2348. 10; sanguinolenta*. *Fabr. E.* i. 389.—(*Curtis l. c. infra.*)
Ca. cruentata. *Don.* ii. *pl.* 63. *f.* 2, 3.—*Stew.* ii. 41.—*Marsh.* i. 145.—*Turt.* ii. 117.—(*Sam. I.* 9.)

2349. 11, singularis*. *Sturm.*

2. *Corpore elongato.*

2350. 12; viridula*. *Payk. F.* ii. 49.
Ca. margaritacea var. *Herbst C.* viii. 225. *pl.* 130. *f.* 16.

2351. [13, Spergulæ*.] *Marsh.* i. 144.—(*Sam. I.* 9.)—*Curtis l. c. infra.*)

2352. 14; Salicorniæ*. *Curtis.* iii. *pl.* 127.

3. *Corpore suborbiculato.*

2353. 15; nobilis*. *Linn.* ii. 575.—*Berk. S.* i. 92.—*Don.* iv. *pl.* 138. *f.* 1, 2, 3.—*Stew.* ii. 41.—*Marsh.* i. 146.—*Turt.* ii. 120. —(*Sam. I.* 9.)—(*Curtis l. c. supra.*)

2354. 16; splendidula*. *Marsh.* i. 147.—(*Sam. I.* 9.)
Ca. pulchella. *Panz. F.* xxxix. *f.* 15.
Ca. lævis. *Herbst C.* viii. 250.
Ca. nobilis β. *Illig.* i. 485.

2355. 17; margaritacea*. *Fabr. E.* i. 397.—(*Curtis l. c. supra.*)
Ca. mutabilis. *Vill. E.* i. 93. *pl.* 1. *f.* 11.

2356. 18, reticulata*. Kirby? MSS.
Ca. reticularis. Wilk. MSS.
Ca. anglica. (Curtis l. c. supra.)

2357. 19, hemisphærica*. Herbst C. viii. 220. pl. 129. f. 9. G.
Ca. hæmorrhoa. Leach? MSS.

Subsectio 3. TRIMERI, *Dumeril.*

(TRIMERA *et* TETRAMERA *p*, *Latreille.*)

Familia LIII. COCCINELLIDÆ, *Latreille.*

(TRIMERA APHIDIPHAGA, *Latreille.*)

(COCCINELLA, *Linné, &c.*) *Lady-birds, Lady-cows.*

Genus 388. CHILOCORUS, *Leach, Samou.*, (*Kirby.*)

PSILA, *Kirby MSS.*

2358. 1; bipustulatus*. (*Sam. I.* 10.)
Co. bipustulata. *Linn.* ii. 585.—*Berk. S.* i. 94.—*Mart. C. pl.* 13. *f.* 21.—*Stew.* ii. 44.—*Marsh.* i. 164.—*Turt.* ii. 138. —*Ent. Trans.* (*Haworth.*) i. 285.
Co. fasciata. *Payk. F.* ii. 25.
Co. frontalis. *Act. Ups.* (*Thunb.*) v. 105.

2359. 2; renipustulatus*.
Co. renipustulata. *Illig. K.* i. 474.
Co. bipustulata. *Payk. F.* ii. 24.—*Herbst C.* v. 383. *pl.* 59. *f.* 11.
Co. mediopustulata. *Schra. B.* i. 465.
Co. similis. *Rossi* i. 68.
Co. Cacti. *Marsh.* i. 163.—*Ent. Trans.* (*Haworth.*) i. 285.
Co. Cacti. (*Leach E. E.* ix. 116.)—(*Sam. I.* 10.)
Co. abdominalis. *Thunb. I. S.* ix. 111.
Coccinella. (*Millard. B. E.* 173. *pl.* 2. *f.* 7.)

2360. 3, hæmorrhoidalis*.
Co. hæmorrhoidalis. *Thunb. I. S.* ix. 111.
Co. atra. *Gmel.* iv. 1664?

2361. 4; quadriverrucatus*. (*Sam. I.* 10.)
Co. quadriverrucata. *Fabr. E.* i. 381.—*Marsh.* i. 163.—*Ent. Trans.* (*Haworth.*) i. 286.
Co. quadripustulata. *Linn.* ii. 585?
Co. sexpustulata ♀. *Fuesly. M.* (*Schr.*) ii. 148.
Co. lunulata. *Gmel.* iv. 1662.
Co. Cassidoides. *Don.* vii. *pl.* 243. *f.* 3.

Genus 389. COCCINELLA *Auctorum.*

A. Glabræ.

2362. 1, lateralis*. *Illig. K.* i. 472.—*Panz. F.* xxiv. *f.* 9.—(*Sam. I.* 12.)—(*Curtis l. c. infra.*)

Co. campestris. *Herbst C.* v. 386.
Co. frontalis α, β. *Payk. F.* ii. 28.

2363. 2; duodecimpunctata*. *Fabr. E.* i. 367.—*Herbst A.* 45. *pl.* 58. *f.* 16.—*Ent. Trans.* (*Haworth.*) i. 282.—(*Sam. I.* 12.)—(*Curtis l. c. infra.*)
Co. undecimpunctata. *Gmel.* iv. 1652.
β; Co. duodecimpunctata. *Linn.* ii. 581.—*Marsh.* i. 155.
γ; Co. sedecimpunctata. *Linn.* ii. 582.—*Berk. S.* i. 93.—*Stew.* ii. 43.

2364. 3; quatuordecimguttata*. *Linn.* ii. 583.—*Berk. S.* i. 94. —*Mart. C. pl.* 13. *f.* 28.—*Don.* vii. *pl.* 243. *f.* 1.—*Stew.* ii. 44.—*Marsh.* i. 161.—*Turt.* ii. 136.—*Ent. Trans.* (*Haworth.*) i. 294.—*Samou.* 52.—(*Curtis l. c. infra.*)

2365. 4, bis-sexguttata*. *Illig. K.* i. 432.—(*Sam. I.* 11.)—(*urtis l. c. infra.*)
Co. duodecimguttata. *Herbst A.* 47. *pl.* 22. *f.* 21.

2366. 5, bis-septemguttata. *Illig. K.* i. 433.—*Herbst C.* v. 331. *pl.* 59. *f.* 9.
† β. Co. quindecimguttata. *Fabr. E.* i. 375.
Co. duodecimgeminata. *Herbst C.* v. 376. *pl.* 59. *f.* 2.

2367. 6; sedecimguttata*. *Linn.* ii. 584.—*Mart. C. pl.* 13. *f.* 27. —*Marsh.* i. 161.—*Turt.* ii. 136.—*Shaw* G. Z. vi. 59. *pl.* 16.—*Ent. Trans.* (*Haworth.*) i. 295.—(*Sam. I.* 12.)—(*Curtis l. c. infra.*)
β, Co. hyalinata. *Kirby MSS.*

2368 ‡ 7. decemguttata[a].

2369. 8; oblongo-guttata*. *Linn.* ii. 584.—*Mart. C. pl.* 13. *f.* 25. —*Stew.* ii. 44.—*Marsh.* i. 162.—*Turt.* ii. 136.—*Don.* xi. *pl.* 362. *f.* 1.—*Ent. Trans.* (*Haworth.*) i. 295.—(*Sam. I.* 12.)—(*Curtis l. c. infra.*)

2370. 9; ocellata*. *Linn.* ii. 582.—*Panz. F.* lxxix. *f.* 6.—(*Sam. I.* 12.)—*Curtis.* v. *pl.* 208.

2371 † 10. marginepunctata[b].

2372. 11; septempunctata*. *Linn.* ii. 581.—*Berk. S.* i. 93.—*Don.* ii. *pl.* 39. *f.* 5. *Id. pl.* 40. *f.* 1. *larva.*—*Stew.* ii. 43.—*Marsh.* i. 152.—*Turt.* ii. 128.—*Shaw* G. Z. vi. 58. *pl.* 16.—*Bingley.* iii. 130.

---

[a] 2368 ‡ 7. decemguttata. *Fabr. E.* i. 374.—*Herbst A.* 47. *pl.* 22. *f.* 16. g —*Turt.* ii. 136. (!)

[b] 2371 † 10. marginepunctata. *Fabr. E.* i. 358.—*Panz. F.* lxxix. *f.* 1.
Co. marginella. *Müll. Z. D. pr.* 66.
Co. notata. *Oliv. E.* vi. 51.
Co. albida. *Gmel.* iv. 1646.
β. Co. sedecimpunctata. *Fabr. E.* i. 370.—*Panz. F.* lxxix. *f.* 2.—*Turt.* ii. 134. (!)

—*Ent. Trans.* (*Haworth.*) i. 270.—(*Leach E. E.* ix. 116.)—(*Sam. I.* 12.)—*Wood.* i. 31. *pl.* 10.—(*Kirby & Sp. I. E.* ii. 9.)—(*Curtis l. c. supra.*)

Coccinella. *Harr. A. pl.* 22. *f. n—t.*

δ; Co. septempunctata β. *Ent. Trans.* (*Haworth.*) i. 270.

θ, Co. circularis. *Marsh. MSS.*

† ι, Co. æthiops. *Haworth. MSS.* ι in Mus. D. *Haworth.*

2373. 12; quinquepunctata*. *Linn.* ii. 580.—*Berk. S.* i. 93.—*Mart. C. pl.* 13. *f.* 26.—*Stew.* ii. 42.—*Marsh.* i. 151.—*Turt.* ii. 127.—*Ent. Trans.* (*Haworth.*) i. 271.—*Don.* xvi. *pl.* 572. *f.* 1.—(*Sam. I.* 12.)—(*Curtis l. c. supra.*)

β; Co. tripunctata. *Rossi* i. 61.

γ; Co. tripunctata. *Linn.* ii. 580?

2374 † 13. trifasciata [a].

2375. 14; vigintiduopunctata*. *Linn.* ii. 582.—*Mart. C. pl.* 13. *f.* 14.—*Berk. S.* i. 93.—*Don.* ii. *pl.* 39. *f.* 1. 4.—*Marsh.* i. 158.—*Stew.* ii. 43.—*Ent. Trans.* (*Haworth.*) i. 281.—(*Sam. I.* 12.)—(*Kirby & Sp. I. E.* ii. 230.)—(*Curtis l. c. supra.*)

Co. vigintiduonotata. *Turt.* ii. 134.

Co. octodecimpunctata. *Shaw G. Z.* vi. 56?

β; Co. vigintipunctata. *Fabr. E.* i. 371. (!)—*Marsh.* i. 158.—*Turt.* ii. 134.

2376. 15; variabilis*. *Illig. K.* i. 447.—(*Sam. I.* 12.)—(*Curtis l. c. supra.*)

Co. tredecimmaculata. *Ent. Trans.* (*Haworth.*) i. 277.

α; Co. flava. *Marsh.* i. 160.

Co. immaculata. *Oliv. E.* vi. 50.

Co. tredecimmaculata μ: (innotata). *Ent. Trans.* (*Haworth.*) i. 279.

β; Co. tredecimmaculata γ: (bina). *Ent. Trans.* (*Haworth.*) i. 279.

γ; Co. marginepunctata. *Marsh.* i. 150.

Co. tredecimmaculata. *Ent. Trans.* (*Haworth.*) i. 279.

δ; Co. subpunctata. *Schr.* 57.

Co. tredecimmaculata ξ: (terna). *Ent. Trans.* (*Haworth.*) i. 279.

ε; Co. tredecimmaculata ο: (quaterna). *Ent. Trans.* (*Haworth.*) i. 279.

Co. quadripunctata. *Linn.* ii. 580.—*Marsh.* i. 151.

ζ; Co. tredecimmaculata κ. *Ent. Trans.* (*Haworth.*) i. 279.

η; Co. tredecimmaculata π: (quina). *Ent. Trans.* (*Haworth.*) i. 280.

θ; Co. tredecimmaculata ρ: (sena). *Ent. Trans.* (*Haworth.*) i. 280.

ι; Co. sexpunctata. *Müll. Z. D. pr.* 622.—*Act. Ups.* (*Thunb.*) iv. *pl.* 1. *f.* 3.

---

[a] 2374 † 13. trifasciata. *Linn.* ii. 580.—*Herbst C.* v. 330. *pl.* 57. *f.* 13.—*Stew.* ii. 43. (!)

Co. obliterata. *DeGeer.* v. 382.
κ; Co. sexpunctata. *Linn.* ii. 580.—*Mart. C. pl.* 13. *f.* 15.—*Marsh.* i. 152.—*Turt.* ii. 127.
Co. sexmaculata. *Thunb. I. S.* ix. 106.
Co. tredecimmaculata ϑ. *Ent. Trans.* (*Haworth.*) i. 279.
λ; Co. octopunctata. *Fabr. E.* i. 365.—*Mart. C. pl.* 13. *f.* 19.—*Marsh.* i. 153.—*Turt.* ii. 128.
Co. tredecimmaculata ζ, η. *Ent. Trans.* (*Haworth.*) i. 278.
μ; Co. tredecimmaculata σ: (septena). *Ent. Trans.* (*Haworth.*) i. 280.
ν; Co. decempunctata. *Linn.* ii. 581.—*Marsh.* i. 154.
Co. tredecimmaculata δ, ε. *Ent. Trans.* (*Haworth.*) i. 278.
Co. tredecimmaculata. *Don.* xii. 93. *pl.* 428. *f.* 1 & 3.
ξ; Co. undecimnotata. *Marsh.* i. 155.
Co. tredecimmaculata γ. *Ent. Trans.* (*Haworth.*) i. 278.
ο; Co. duodecimpunctata. *Müll. Z. D. pr.* 66.
π; Co. tredecimmaculata. *Fabr. E.* i. 369.—*Mart. C. pl.* 13. *f.* 18.—*Forst. C.* i. 18.—*Marsh.* i. 157.—*Turt.* ii. 133.—*Don.* xii. *pl.* 428. *f.* 4.
Co. tredecimmaculata α. *Ent. Trans.* (*Haworth.*) i. 277.
Co. quatuordecimpunctata. *Müll. Z. D. pr.* 67.
ρ; Co. tredecimmaculata τ: (suturalis). *Ent. Trans.* (*Haworth.*) i. 280.
σ; Co. conglomerata. *Linn.* ii. 583?
Co. tredecimmaculata β. *Ent. Trans.* (*Haworth.*) i. 278.
† τ. Co. tredecimmaculata. *Don.* xii. 93. *pl.* 428. *f.* 2.

2377. [16; instabilis*] *mihi.* (*Sam. I.* 12.)—(*Curtis l. c. supra.*)
Co. variabilis ρ ad ω. *Illig. K.* i. 447, *&c.*
Co. decempustulata. *Ent. Trans.* (*Haworth.*) i. 292.
α; Co. decemguttata minor. *Act. Hal.* (*Schall.*) i. 263.
Co. decempustulata α: (rubripustulata). *Ent. Trans.* (*Haworth.*) i. 292.
β; Co. decempustulata β: (flavipustulata). *Ent. Trans.* (*Haworth.*) i. 292.
γ; Co. variabilis σ. *Illig. K.* i. 449.
Co. decempustulata γ: (albipustulata). *Ent. Trans.* (*Haworth.*) i. 292.
δ; Co. decemmaculata. *Scop. C.* 251.
Co. decempustulata δ: (parvipustulata). *Ent. Trans.* (*Haworth.*) i. 292.
ε; Co. humeralis. *Gmel.* iv. 1665.
Co. decempustulata ε: (fulvipustulata). *Ent. Trans.* (*Haworth.*) i. 292.
ζ; Co. decemguttata. *Mart. C. pl.* 13. *f.* 23.—*Don.* vii. *pl.* 243. *f.* 1.—*Marsh.* i. 161.
η; Co. decempustulata. *Linn.* ii. 585.—*Marsh.* i. 165.—*Turt.* i. 139.

ϑ; Co. pantherina. *DeGeer.* v. 392.
Co. similata. *Thunb. I. S.* 112.
ι; Co. decempustulata η: (sinuosa). *Ent. Trans.* (*Haworth.*) i. 293.
κ; Co. decempustulata ϑ: (rubripunctata). *Ent. Trans.* (*Haworth.*) i. 293.
λ; Co. guttato-punctata. *Linn.* ii. 583.
Co. punctato-guttata. *Gmel.* iv. 1658.
μ; Co. decempustulata ι: (obliterata). *Ent. Trans.* (*Haworth.*) i. 293.
ν; Co. decempustulata κ: (octoguttata). *Ent. Trans.* (*Haworth.*) i. 293.

2378. [17; humeralis*.] (*Schön. S.* ii. 163.)—(*Sam. I.* 12.)—(*Curtis l. c. supra.*)
Co. lunæpustulata. *Ent. Trans.* (*Haworth.*) i. 288.
α; Co. variabilis αα. *Illig. K.* i. 449.
Co. bipustulata. *Herbst A.* 48. *pl.* 22. *f.* 22. *w.*
Co. marginata. *Act. Ups.* (*Thunb.*) iv. 10.
Co. marginella. *Thunb. I. S.* 110.
Co. Thunbergii. *Gmel.* iv. 1666.
Co. lunæpustulata α: (nigra). *Ent. Trans.* (*Haworth.*) i. 288.
β; Co. variabilis. *Fabr. E.* i. 380.
Co. unifasciata. *Scrib. B.* ii. 107. *pl.* 8. *f.* 14.
Co. bipunctata var. 11. *Herbst C.* v. 345. *pl.* 58. *f.* 10.
Co. dispar 1. *Payk. F.* ii. 19.
Co. austriaca. *Schra.* 63.
Co. lunigera. *Brahm. I. K.* i. 119.
Co. limbata. *Gmel.* iv. 1662.
Co. lunularis. *Marsh.* i. 168.
γ; Co. biguttata. *Fabr. E.* i. 374.
Co. bimaculosa. *Herbst A.* 160. *pl.* 43. *f.* 13. *e. f.*
Co. lunæpustulata β: (biguttata). *Ent. Trans.* (*Haworth.*) i. 289.
δ; Co. lunæpustulata γ: (fulva). *Ent. Trans.* (*Haworth.*) i. 289.
ε; Co. lunæpustulata δ: (flava). *Ent. Trans.* (*Haworth.*) i. 289.

2379. 18; conglomerata*. *Fabr. E.* i. 372.—*Stew.* ii. 44.—*Turt.* ii. 135.—(*Sam. I.* 12.)—(*Curtis l. c. supra.*)
Co. conglobata. *Illig. K.* i. 462.
Co. tessellata. *Schneid. M.* 158.
α; Co. quatuordecimmaculata. *Fabr. E.* i. 370.—*Marsh.* i. 157.—*Ent. Trans.* (*Haworth.*) i. 281.
Co. quatuordecimpunctata. *Herbst A.* 44. *pl.* 22. *f.* 5.
Co. tetragona. *Laich. T.* ii. 125.
β; Co. quatuordecimpunctata. *Linn.* ii. 552.—*Berk. S.* i. 93.—*Stew.* ii. 43.—*Marsh.* i. 157.—*Turt.* ii. 130 & 134.—*Ent. Trans.* (*Haworth.*) i. 280.
γ; Co. conglobata. *Linn.* ii. 583.

Co. conglomerata. *Fabr. E.* i. 372.—*Herbst A.* 46. *pl.* 22. *f.* 15. ♂.
Co. tessulata. *Scop. C.* 78.
δ; Co. bis-sexpustulata. *Fabr. E.* i. 384.
Co. duodecimpustulata. *Fabr. E.* i. 385.—*Marsh.* i. 168.—*Ent. Trans.* (*Haworth.*) i. 286.
Co. dentata. *Thunb. I. S.* ix. 113.
Co. quatuordecimpunctata var. 3. *Herbst C.* v. 336. *pl.* 57. *f.* 18.

2380. 19; quatuordecimpustulata*. *Linn.* ii. 585.—*Stew.* ii. 45.—*Herbst C.* v. 388. *pl.* 59. *f.* 17.—(*Sam. I.* 12.)—(*Curtis l. c. supra.*)
Co. quatuordecimmaculata. *Laich. T.* ii. 139.
Co. quatuordecimpunctata. *Goeze.* viii. 340.

2381. 20; dispar*. *Illig. K.* i. 455.—(*Sam. I.* 12.)—(*Curtis l. c. supra.*)
α; Co. bipunctata. *Linn.* ii. 580.—*Mart. C. pl.* 13. *f.* 17.—*Berk. S.* i. 93.—*Stew.* ii. 42.—*Marsh.* i. 150.—*Turt.* ii. 127.—*Bingley.* iii. 131.—*Ent. Trans.* (*Haworth.*) i. 272.
β; Co. bipunctata β: (flava). *Ent. Trans.* (*Haworth.*) i. 272.
γ; Co. perforata. *Marsh.* i. 151.—*Ent. Trans.* (*Haworth.*) i. 271.
δ; Co. bipunctata γ: (sesquipunctata). *Ent. Trans.* (*Haworth.*) i. 272.
ε; Co. bipunctata δ: (interpunctata). *Ent. Trans.* (*Haworth.*) i. 272.
Co. quadripunctata. *Don.* xvi. 3. *pl.* 542.
ζ; Co. bipunctata ε: (rhombipunctata). *Ent. Trans.* (*Haworth.*) i. 272.
η; Co. unifasciata. *Fabr. E.* i. 268.—*Mart. C. pl.* 13. *f.* 24.—*Marsh.* i. 149.—*Ent. Trans.* (*Haworth.*) i. 272.
ϑ; Co. hastata. *Oliv. E. M.* vi. 77.
ι; Co. annulata. *Fabr. E.* i. 359.—*Don.* vii. 7. *pl.* 243. *f.* 2.—*Stew.* ii. 42.—*Marsh.* i. 149.—*Turt.* ii. 126.
κ; Co. bipunctata var. 7. *Herbst C.* v. 342. *pl.* 58. *f.* 6.
Co. cincta. *Schæff. pl.* 30. *f.* 14.
Co. quadrimaculata. *Scop. C. no.* 250.
Co. bis-quadripustulata. *Ent. Trans.* (*Haworth.*) i. 290.
λ; Co. septempustulata. *Marsh.* i. 165.—*Ent. Trans.* (*Haworth.*) i. 290.
Co. tripustulata. *Gmel.* iv. 1662.
Co. pantherina. *Fabr. E.* i. 385.
μ; Co. sexpustulata. *Linn.* ii. 585.—*Mart. C. pl.* 13. *f.* 22.—*Berk. S.* i. 94.—*Don.* ii. *pl.* 39. *f.* 3.—*Stew.* ii. 44.—*Marsh.* i. 165.—*Turt.* ii. 138.—*Ent. Trans.* (*Haworth.*) i. 290.
ν; Co. quadripustulata. *Fabr. E.* i. 381.—*Marsh.* i. 164.—*Turt.* ii. 138.—*Ent. Trans.* (*Haworth.*) i. 289.
Co. cincta. *Müll. Z. D. pr.* 68.

2382. 21; impustulata*. *Linn.* ii. 584.—*Ent. Trans.* (*Haworth.*) i. 288.—(*Sam. I.* 12.)—(*Curtis l. c. supra?*)

α; Co. conglobata var. 3. *Herbst C.* v. 356. *pl.* 58. *f.* 15.
Co. marginata. *Marsh. MSS.?*
β; Co. flavipes. *Thunb. I. S.* ix. 110.
γ; Co. impustulata β: (rufo-cincta). *Ent. Trans.* (*Haworth.*) i. 288.
δ; Co. curvipustulata. *Ent. Trans.* (*Haworth.*) i. 287.
ε; Co. bis-triverrucata. *Ent. Trans.* (*Haworth.*) i. 287.

2383. [22; conglobata*.] *Fabr. E.* i. 373.—*Berk. S.* i. 93.—*Stew.* ii. 44.—*Turt.* ii. 135.—(*Sam. I.* 12.)—(*Curtis l. c. supra.*)
Co. impustulata α—ε. *Illig. K.* i. 459—460.
α; Co. sedecimmaculata. *Schneid. M.* 156.
β; Co. gemella. *Herbst A.* 44. *pl.* 22. *f.* 7. *g.*
γ; Co. sedecimmaculata. *Fabr. E.* i. 370.
Co. sedecimpunctata. *Schra.* 57.
Co. octodecimpunctata. *Scop. C.* 77.
Co. rosea. *DeGeer.* v. 378.

2384. 23; undecimpunctata*. *Linn.* ii. 581.—*Herbst C.* v. 352. *pl.* 58. *f.* 13.—*Marsh.* i. 155.—*Turt.* ii. 129.—*Ent. Trans.* (*Haworth.*) i. 274.—*Don.* xvi. *pl.* 572. *f.* 2.—(*Sam. I.* 12.)—(*Curtis l. c. supra.*)
α; Co. collaris γ. *Payk. F.* ii. 38.
Co. undecimmaculata. *Thunb. I. S.* ix. 108.
Co. triangularis. *Thunb. I. S.* ix. 108.
β; Co. nigrofasciata. *Rossi* i. 62.
Co. undecimpunctata β: (confluens). *Ent. Trans.* (*Haworth.*) i. 274.
δ; Co. novempunctata. *Linn.* ii. 581.—*Berk. S.* i. 93.—*Stew.* ii. 43.—*Marsh.* i. 154.—*Turt.* ii. 128.—*Ent. Trans.* (*Haworth.*) i. 274.
ε; Co. decempunctata. *Fabr. E.* i. 81.

2385. 24; hieroglyphica*. *Linn.* ii. 580.—(*Sam. I.* 12.)—(*Curtis l. c. supra.*)
Co. lineolata. *Marsh.* i. 153.—*Ent. Trans.* (*Haworth.*) i. 273.
β; Co. flexuosa. *Fabr. E.* i. 362.
Co. trilineata. *Herbst. A.* 46. *pl.* 22. *f.* 12.
Co. quadrilineata. *Gmel.* iv. 1658.
Co. octopustulata. *Thunb. I. S.* 112.
Co. sinuata. *Marsh.* i. 160.
γ; Co. sinuosa. *Marsh.* i. 160.

2386. 25; octodecimguttata*. *Linn.* ii. 585.—*Marsh.* i. 162.—*Ent. Trans.* (*Haworth.*) i. 294.—(*Samou. pl.* 2. *f.* 11.)—(*Curtis l. c. supra.*)
Co. ornata. *Herbst A.* 47. *pl.* 22. *f.* 19.

2387. 26; mutabilis*. *Illig. K.* i. 427.—(*Sam. I.* 12.)—(*Curtis l. c. supra.*)
Co. septemnotata. *Ent. Trans.* (*Haworth.*) i. 275.

β; Co. læta. *Fabr. E.* i. 369.
Co. similis. *Schra.* 56.
Co. sexpunctata var. 3. *Herbst C.* vi. 328.
ε; Co. undecimpunctata. *Schra. E.* 54.
θ; Co. novempunctata. *Schra. E.* 54.
Co. sexpunctata var. 2. *Herbst C.* vi. 328.
Co. septemnotata β: (9-punctata). *Ent. Trans.* (*Haworth.*) i. 275.
ι; Co. constellata. *Laich. T.* ii. 121.
Co. septemnotata γ: (9-notata). *Ent. Trans.* (*Haworth.*) i. 275.
κ; Co. septemnotata. *Fabr. E.* i. 365.—*Marsh.* i. 153.—*Panz. F.* lxxix. *f.* 5.—*Don.* xi. *pl.* 362. *f.* 3.—*Ent. Trans.* (*Haworth.*) i. 275. *a.*
Co. fennica. *Thunb. I. S.* ix. 106.
Co. novempunctata var. *DeGeer.* v. 374.
μ; Co. observe-punctata. *Schr.* 53.
ν; Co. quinquemaculata. *Fabr. E.* i. 364.
Co. septemnotata δ: (quinquemaculata). *Ent. Trans.* (*Haworth.*) i. 276.
ς; Co. limbata. *Fabr. E.* i. 356.
τ; Co. septemnotata ε: (3-punctata). *Ent. Trans.* (*Haworth.*) i. 278.

2388. 27; tredecimpunctata*. *Linn.* ii. 582.—*Stew.* ii. 43.—*Marsh.* i. 156.—*Turt.* ii. 134.—*Don.* xi. *pl.* 362. *f.* 2.—*Ent. Trans.* (*Haworth.*) i. 276.—(*Sam. I.* 12.)—(*Curtis l. c. supra.*)
Co. quatuordecimpunctata. *Don.* ii. 2. *pl.* 39. *f.* 2.

2389. 28; novemdecimpunctata*. *Linn.* ii. 582.—*Marsh.* i. 158.—*Herbst A.* 45. *pl.* 22. *f.* 9.—*Ent. Trans.* (*Haworth.*) i. 277.—(*Sam. I.* 12.)—(*Curtis l. c. supra.*)

B. Pubescentes.

2390. 29; globosa*. *Illig. K.* i. 469.—(*Sam. I.* 12.)—(*Curtis l. c. supra.*)
α; Co. impunctata var. *Marsh.* i. 148.—*Herbst C.* v. 358. *pl.* 58. *f.* 17.
β; Co. impunctata. *DeGeer.* v. 369.
Co. livida. *Turt.* ii. 126.—*Herbst A.* 42. *pl.* 22. *f.* 1. *a.*
δ; Co. Colon. *Herbst A.* 42. *pl.* 22. *f.* 2. *b.*
ε; Co. quadrinotata. *Fabr. E.* i. 363.
ζ; Co. rufa γ. *Ent. Trans.* (*Haworth.*) i. 284.
η; Co. rufa α. *Ent. Trans.* (*Haworth.*) i. 284.
ϑ; Co. vigintiduopunctata. *Fabr. E.* i. 371.—*Turt.* ii. 130.
Co. vigintiquatuorpunctata β. *Ent. Trans.* (*Haworth.*) i. 283.
ι; Co. rufa β. *Ent. Trans.* (*Haworth.*) i. 284.
κ; Co. vigintiquatuorpunctata. *Linn.* ii. 583.—*Marsh.* i. 159. β.—*Don.* xi. *pl.* 362. *f.* 4.
λ; Co. vigintiquinquepunctata. *Linn.* ii. 583.—*Turt.* ii. 130.

μ; Co. vigintiquatuorpunctata. *Marsh.* i. 159.—*Don.* xi. *pl.* 362. *f.* 5.—*Ent. Trans.* (*Haworth.*) i. 283.
Co. vigintitrespunctata. *Linn.* ii. 582.
ν; Co. hæmorrhoidalis. *Fabr. E.* i. 379.

2391 † 30. aptera. *Payk. F.* ii. 11.
Co. impunctata. *Linn.* ii. 579.—*Panz. F.* xxxvi. *f.* 4.—*Mart. C. pl.* 13. *f.* 20. (!)—*Marsh.* i. 148. (!)—*Turt.* ii. 126. (!)—*Ent. Trans.* (*Haworth.*) i. 284. (!) In Mus. —— ?

Genus 390. SPHÆROSOMA, *Leach MSS.*—(*Samou.*)

LEPTIUM, *Kirby MSS.*

2392. 1; Quercus*. *Leach MSS.*—(*Sam. I.* 38.)
† β? Le. Boleti. *Kirby MSS.* β In Mus. D. *Kirby.*

Genus 391. SCYMNUS, *Herbst,* (*Samou.*)

NITIDULA *p, Fabricius.*—SPHÆRIDIUM *p, Thunberg.*—TRITOMA *p, Panzer.*—BYRRHUS *p, Marsham.*—CHRYSOMELA *p, Schrank.*

2393. 1; ater*. *Schneid. M.* (*Kugellan.*) i. 548.
Co. minima. *Payk. F.* ii. 8.

2394. 2; nigrinus*. *Herbst C.* vii. 344.—*Panz. F.* xxiv. *f.* 12.—(*Sam. I.* 37.)
Sc. ater. *Thunb. I. S.* ix. 105.
Co. Morio. *Payk. F.* ii. 8.
Co. minima. *Müll. Z. D. pr.* 65.

2395. 3; fulvifrons*. (*Sam. I.* 37.)
Co. fulvifrons. *Marsh.* i. 168.
Co. flavipes. *Fabr. E.* i. 377.
Co. parvula β. *Payk. F.* ii. 9.
Co. capitata. *Fabr. Sup.* 79?

2396. [4; parvulus*.] (*Sam. I.* 37.)
Co. parvula. *Fabr. E.* i. 377.—*Panz. F.* xiii. *f.* 2.—*Marsh.* i. 167.
Co. pygmæa. *Rossi M.* ii. 88.
Sc. collaris. *Herbst C.* vii. 345. *pl.* 116. *f.* 9.
Sc. auritus. *Thunb. I. S.* ix. 105.

2397. 5; quadripustulatus*. *Schneid. M.* (*Kugellan.*) i. 547.—(*Sam. I.* 37.)
Co. quadrimaculata. *Rossi F.* i. 71.—*Marsh.* i. 167.
Co. quadriguttata. *Brahm. I. K.* i. 221.
Co. bis-biverrucata. *Panz. F.* xxiv. *f.* 8. (*ed.* 2-*da.*)
Co. villosa. *Fourc.* i. 149.
β; Co. bis-bipustulata γ. *Illig. K.* i. 415.

2398. 6, bis-bipustulatus*. (*Sam. I.* 37.)
Co. bis-bipustulata. *Fabr. E.* i. 382.—*Panz. F.* xiii. *f.* 5?—*Marsh.* i. 167.—*Ent. Trans.* (*Haworth.*) i. 284.
Sc. quadripustulatus. *Herbst C.* vii. 344. *pl.* 116. *f.* 7. *G.*

2399. 7; bipustulatus*. *Thunb. I. S.* ix. 105.—(*Sam. I.* 37.)
Sc. bipunctatus. *Panz. F.* xxiv. *f.* 11.
Co. biverrucata. *Illig. K.* i. 415.
Co. affinis. *Payk. F.* ii. 10.
Co. bimaculata. *Marsh.* i. 166.

2400. 8, flavilabris*. (*DeJean C.* 132.)
Co. flavilabris. *Payk. F.* ii. 6.
Co. humeralis. *Panz. F.* (*ed.* 2-*da*). xxiv. *f.* 10.—*Marsh.* i. 166.
Co. frontalis. *Rossi M. App.* 86.
Sc. bimaculatus. *Thunb. I. S.* ix. 105.
Sp. bipunctatum. *Thunb. I. S.* viii. 102.
Sp. bipustulatum. *Thunb. I. S.* viii. 102.

2401. 9; analis*. (*Sam. I.* 37.)
Co. analis. *Fabr. E.* i. 378.—*Panz. F.* xiii. *f.* 3.
Co. parvula α. *Illig. K.* i. 414.
Co. ruficollis. *Oliv. E. M.* vi. 81.
β; Co. hæmorrhoidalis. *Herbst C.* v. 342. *pl.* 116. *f.* 4. *D.*—*Marsh.* i. 168.
γ; Co. biliturata. *Marsh.* i. 166.

2402. 10; discoideus*. (*Sam. I.* 37.)
Co. discoidea. *Fabr. E.* i. 377.
Sc. pilosus. *Herbst C.* vii. 343. *pl.* 116. *f.* 6. *F.*
Sc. suturalis. *Thunb. I. S.* ix. 106.
Sc. plagiatum. *Thunb. I. S.* viii. 101.
By. Pini. *Marsh.* i. 102.

2403. 11, atriceps* *mihi.*

2404. 12; limbatus*. *Kirby MSS.*

Genus 392. RHYZOBIUS *mihi.*

NITIDULA *p, Fabricius.*—DERMESTES *p, Marsh.*—STRONGYLUS *p, Herbst.*—ANTHRIBUS *p, Oliv.*—SCYMNUS *p, Samou.*—CACIDULA *p, DeJean.*—RYZOBIUS *p, Leach. MSS.*

2405. 1; Litura*.
Ni. litura. *Fabr. E.* i. 353.
Co. Aurora. *Panz. F.* xxxv. *f.* 5.
De. coadunatus. *Marsh.* i. 76.
Sc. litura. (*Sam. I.* 37.)
β; De. hypomelanus. *Marsh.* i. 77.
γ; Ni. fasciata. *Fabr. E.* i. 353.
St. chrysomeloides. *Herbst C.* iv. 180.
An. lividus. *Oliv. E. M.* iv. 161.
Ni. testacea. *Panz. E.* i. 130.
δ; De. pallidus. *Marsh.* i. 79.
ε; De. Absinthii. *Marsh.* i. 77.

Genus 393. CACICULA, (*Megerle.*)

CHRYSOMELA *p*, *Fabr.*—NITIDULA *p*, *Fabr.*—DERMESTES *p*, *Herbst.*—ANTHRIBUS *p*, *Oliv.*—STRONGYLUS *p*, *Herbst.*—SILPHA *p*, *Marsh.*—LASIUS, *Kirby MSS.*—RYZOBIUS *p*, *Leach MSS.*—CACIDULA, (*DeJean,*) *Curtis.*

2406. 1; pectoralis *. (*DeJean C.* 132.)—(*Curtis l. c. infra.*)
Ch. pectoralis. *Fabr. E.* i. 443.—*Panz. F.* lxxviii. *f.* 5.
De. rufus. *Herbst A.* 22. *pl.* 20. *f.* 7.
De. testaceus. *Thunb. I. S.* vi. 80.
Si. rosea. *Marsh.* i. 123.

2407. 2, scutellata*. (*DeJean C.* 132.)—*Curtis.* iii. *pl.* 144.
Ci. scutellata. *Fabr. E.* i. 443.
St. quinquepunctatus. *Herbst C.* iv. 181. *pl.* 43. *f.* 2. *b.*
Ni. bipunctata. *Gmel.* iv. 1630.
Si. melanophthalma. *Gmel.* iv. 1627.
La. bifasciatus. *Kirby MSS.*

Familia LIV. ENDOMYCHIDÆ, *Leach.*

(TRIMERA FUNGICOLA, *Latreille.*)

Genus 394. ENDOMYCHUS, *Paykul*, *Leach*, *Samou.*, *Millard.*

CHRYSOMELA *p*, *Linné.*—TENEBRIO *p*, *Marsh.*

2408. 1; coccineus*. *Payk. F.* ii. 112.—(*Leach E. E.* ix. 116.)—(*Samou.* 215. *pl.* 4. *f.* 3.)
Te. coccineus. *Marsh.* i. 477.
Ch. coccinea. *Linn.* ii. 592.—*Mart. C. pl.* 16. *f.* 38.—*Don.* iv. *pl.* 111. *f.* 5, 6.—*Stew.* ii. 48.—*Turt.* ii. 161.
Ch. quadrimaculata. *DeGeer.* v. 301. *pl.* 9. *f.* 1.
Ch. quadripunctata. *Turt.* ii. 157.

Genus 395. LYCOPERDINA, *Latr.*, *Leach*, *Samou.*

ENDOMYCHUS *p*, *Fabr.*—TENEBRIO *p*, *Marsh.*

2409. 1; Bovistæ*. *Payk. F.* ii. 115.—(*Leach E. E.* ix. 116.)—(*Sam. I.* 26.)
En. Bovistæ. *Panz. F.* viii. *f.* 4.
Te. Bovistæ. *Marsh.* i. 478.
Ly. immaculata. *Latr. G.* iii. 73.
En. Lycoperdi. *Latr. H.* xii. 78.

Familia LV. HISPIDÆ, *Kirby*.

(CHRYSOMELINI *p*, *Latreille*.—HISPOIDEÆ, *Gyllenhal*.)

Genus 396. HISPA, *Linné*, *Marsh.*, *Leach*.

CRIOCERIS *p*, *Geoffroy*.

2410. 1, atra*? *Linn.* ii. 603.—*Barb.* G. 42. *pl.* 4. *f.* 16.—*Turt.* ii. 198.—*Marsh.* i. 231.—*Shaw G. Z.* vi. 59. *pl.* 19.
Hi. spinosa. *Fabr. E.* ii. 53.
Cr. spinosissima. *Fourc.* i. 96.

SECTIO V.

(HETEROMERA *p et* PENTAMERA *p*, *Latreille*.—THYSANURIMORPHA? *p*, *MacLeay*.)

Familia LVI. TENEBRIONIDÆ.

(TENEBRIONITES *et* DIAPERIALES *p*, *Latreille*.—TENEBRIONIDÆ *et* DIAPERIDÆ *p*, *Leach*.)

(TENEBRIO *p*, *Linné*, *&c.*)

Genus 397. SARROTRIUM, *Illiger*, *Leach*, *Samou.*, *Millard*.

HISPA *p*, *Linné*, *Marsh*.—PTILINUS *p*, *Paykul*.—DERMESTES *p*, *Linn. F.*—ORTHOCERUS, *Latreille*.

2411. 1; muticum*. *Illig. K.* i. 344.—(*Leach E. E.* ix. 102.)—(*Samou.* 193. *pl.* 2. *f.* 16.)—(*Kirby & Sp. I. E.* iii. 522.)
Hi. mutica. *Linn.* ii. 604.—*Marsh.* i. 232.—*Don.* xvi. *pl.* 575.—*Millard. B. E.* 167. *pl.* 2. *f.* 6.—*Wood.* i. 36. *pl.* 12.
Pt. muticus. *Turt.* ii. 200.
De. clavicornis. *Linn. F.* 413.
Te. hirticornis. *DeGeer.* v. 47. *pl.* 3. *f.* 1.

Genus 398. HYPOPHLÆUS, *Fabricius*, *Leach*.

IPS *p*, *Olivier*.

2412 † 1, castaneus. *Fabr. E.* ii. 558.—*Panz. F.* xii. *f.* 13.
Ips taxicornis. *Oliv. E.* ii. 11. *pl.* 1. *f.* 2. *a, b.*
In Mus. *Brit.* D. *Kirby et Vigors.*

2413. 2; bicolor*. *Fabr. E.* ii. 559.—*Panz. F.* xii. *f.* 14.—(*Leach E. E.* ix. 102.)—(*Ing. Inst.* 88.)

2414. 3; depressus*. *Fabr. E.* ii. 559.—*Panz. F.* i. *f.* 23.
Si. depressa. *Turt.* ii. 108.
Ips unicolor. *Oliv. E.* ii. 12. *pl.* 2. *f.* 8. *a, b.*
Melinus. *Herbst A.* iv. 37. *pl.* 21. *f. B. b.*

Genus 399. TENEBRIO *Auctorum*. Larvæ: *Meal-worms*.

2415. 1; Molitor*. *Linn.* ii. 674.—*Mart. C. pl.* 39. *f.* 2.—

Berk. S. i. 108.—*Stew.* ii. 83.—*Marsh.* i. 474.—*Turt.* ii. 478.—*Panz. F.* xliii. *f.* 12.—*Shaw* G. Z. vi. 102. *pl.* 36.—(*Leach E. E.* ix. 102.)—*Samou.* 59. *pl.* 4. *f.* 1.

2416. 2, laticollis* *mihi.*

2417. 3; obscurus*. *Fabr. E.* i. 146.—*Panz. F.* xliii. *f.* 13.—(*Leach E. E.* ix. 102.)—(*Sam. I.* 40.)
Te. Morio. *Herbst C.* vii. 247. *pl.* 111. *f.* 7.

Genus 400. STENE.

COLYDIUM *p, Herbst.*—TROGOSITA *p, Fabr.*—PHALERIA *p, Gyll.*—ULOMA *p, DeJean?*—STENE *p, Kirby MSS.*

2418. 1; ferrugineus*. *Kirby MSS.*
Te. ferrugineus. *Oliv. E.* iii. 18. *pl.* 2. *f.* 24. *a, b.*—(*Ing. Inst.* 89.)
Co. testaceum. *Herbst C.* vii. 282. *pl.* 112. *f.* 3. E.
Te. testaceus. (*Schön. S.* i. 153.)
Ly. navalis. *Fabr. E. S.* i. *b.* 504?
♀, St. culinaris. *Kirby MSS.*

Genus 401. USOMA, (*Megerle.*)

PHALERIA *p, Latr.*—TROGOSITA *p, Fabr.*—ULOMA, *DeJean.*

2419. 1; cornuta*.
Ph. cornuta. *Latr.* G. ii. 175. *pl.* 10. *f.* 4. ♂. *f.* 5. ♀. (*head*).—*Millard B. E.* 118. (!)—(*Ing. Inst.* 89.)
Tr. cornuta. *Fabr. E.* i. 155.

Genus 402. DIAPERIS, *Geoffroy, Leach, Samou., Millard,* (*Kirby.*)

CHRYSOMELA *p, Linn.*—TENEBRIO *p, DeGeer.*—SCOLYTUS *p, Panzer.*—COCCINELLA *p, Scop.*—MYCETOPHAGUS *p, Fabr.*

2420. 1, Boleti*. *Fabr. E.* ii. 585.—(*Leach E. E.* ix. 103.)—(*Sam. I.* 15.)
Ch. Boleti. *Linn.* ii. 591.—*Mart. C. pl.* 16. *f.* 30.—*Don.* iii. *pl.* 78. *f.* 1. 2.—*Stew.* ii. 47.—*Marsh.* i. 176.—*Turt.* ii. 146.—*Shaw G. Z.* vi. 59. *pl.* 18.
Di. fasciata. *Fourc.* i. 153.

2421. 2: violacea. *Fabr. E.* ii. 586.
Ch. dytiscoides. *Rossi.* i. 86. *pl.* 2. *f.* 6.—*pl.* 4. *f.* 13.

2422. 3; ænea*. *Payk. F.* iii. 359.—(*Ing. Inst.* 88.)
Di. bicolor. *Panz. F.* xciv. *f.* 9.
My. metallicus. *Fabr.* i. *b.* 570.

2423. [4, ahenea.] (*Sam. I.* 15.)
Ch. ahenea. *Marsh.* i. 176.

Genus 403. PHALERIA, *Latr., Leach, Samou., Millard,* (*Kirby*).

2424. 1, cadaverina. (*Leach E. E.* ix. 102.)—(*Sam. I.* 33.)—(*Kirby & Sp. I. E.* i. 173.)

Te. cadaverina. *Fabr. E.* i. 149.—*Sturm D. F.* ii. 230. *pl.* xlvii. *f. a. A. B.*—*Ent. Trans.* (*Haworth.*) i. 339.

Genus 404. ALPHITOBIUS *mihi.*

2425. 1; picipes*.
Te. Fagi. *Panz. F.* lxi. *f.* 3?

Genus 405. BOLITOPHAGUS, *Fabr., Samou.*

ELEDONA, *Latr., Leach, Millard.*—OPATRUM *p, Oliv., Marsh.*

2426. 1; Agricola*. *Fabr. E.* i. 114.
Op. Agricola. *Oliv. E.* iii. 11. *pl.* 1. *f.* 11. *a, b.*—*Marsh.* i. 143.
Op. sulcatum. *Thunb. I. S.* v. 67.
El. Agaricicola. *Latr.* G. ii. 178.—(*Leach E. E.* ix. 103.)
Bo. Agaricola. (*Sam. I.* 6.)

2427. [2, brunneus.] *Wilkin. MSS.*

Genus 406. OPATRUM, *Fabr., Marsh., Leach, Samou.*, (*Kirby.*)

SILPHA *p, Linné.*

2428. 1; sabulosum*. *Fabr. E.* i. 116.—*Stew.* ii. 40.—*Marsh.* i. 142.—*Turt.* ii. 113.—(*Leach E. E.* ix. 102.)—*Samou.* 51. *pl.* 2. *f.* 8.
Si. sabulosa. *Linn.* ii. 572.—*Berk. S.* i. 92.
Te. rugosus. *DeGeer.* v. 43. *pl.* 2. *f.* 21.

2429. 2, tibiale. *Fabr. E.* i. 119.—*Panz. F.* xliii. *f.* 10.—*Marsh.* i. 142.—*Ent. Trans.* (*Burrell.*) i. 312.—(*Sam. I.* 32.)

Genus 407. ——

OPATRUM *p, Mus. Marsh.*

2430. 1, Marshami *mihi.*

Genus 408. ——

2431. 1, obsoletus.
Te. obsoletus. *Marsh.* i. 475.

Genus 409. PEDINUS, *Latr., Leach, Samou.*, (*Kirby.*)

BLAPS *p, Fabricius.*—HELOPS *p, Oliv.*—OPATRUM *p, Illiger.*

2432. 1, femoralis. *Latr.* G. ii. 165.
Pe. maritimus. (*Leach E. E.* ix. 102.)—(*Sam. I.* 33. *pl.* 4. *f.* 2.)
♂, Bl. femoralis. *Fabr. E.* i. 143.—*Panz. F.* xxxix. *f.* 5.
Te. gemellatus. *Marsh.* i. 475.
♀, Bl. laticollis. *Herbst C.* viii. 191. *pl.* 129. *f.* 3.
Te. arenosus. *Marsh.* i. 475.
Te. femoralis. *Linn.* ii. 679.—*Panz. F.* xxxix. *f.* 6.

Genus 410. CRYPTICUS, *Latreille.*

BLAPS *p, Fabr.*—HELOPS *p, Oliv.*

2433. 1, glaber. *Latr.* G. ii. 164.

Bl. glabra. *Fabr. E.* i. 143.
Te. quisquilius. *Linn.* iii. 676?—*Marsh.* i. 475.
Te. humeralis. *Marsh.* i. 476.
Te. nigrinus. *Marsh.* i. 476.
♂, He. laticollis. *Panz. F.* xxxvi. *f.* 1.
Bl. pusilla. *Herbst C.* viii. 199. *pl.* 129. *f.* 4.
Pi. lævis. *Gmel.* iv. 2011.

2434. [2, stygius.]
Te. stygius. *Marsh.* i. 476.

Familia LVII. BLAPSIDÆ, *Leach.*

(BLAPSIDES, *Latreille.*—TENEBRIONITES *p*, *Gyllenhal.*)

Genus 411. BLAPS, *Fabr.*, *Marsh.*, *Leach*, *Samou.*, *Millard*, (*Kirby*,) *Curtis.*

Black-beetle. Slow-legged-beetle.

TENEBRIO *p*, *Linné.*—PIMELIA *p*, *Gmel.*

2435. 1; mortisaga*. *Fabr. E.* i. 141.—*Panz. F.* iii. *f.* 3.—*Marsh.* i. 479.—*Leach E. E.* ix. 101.—*Samou.* 59. *pl.* 4. *f.* 4.—(*Kirby & Sp. I. E.* ii. 242. *et* iii. 344.)—(*Curtis l. c. infra.*)
Te. mortisagus. *Linn.* ii. 676.—*Berk. S.* i. 108.
Te. acuminatus. *DeGeer.* v. 31.
Pi. mortisaga. *Gmel.* iv. 2001.—*Stew.* ii. 83.—*Turt.* ii. 490.

2436. 2; obtusa*. *Fabr. E.* i. 141?—*Curtis.* iv. *pl.* 148.
Te. lethifer. *Schæff. Ic. pl.* 37. *f.* 6.
Bl. lethifera. *Marsh.* i. 479.—(*Sam. I.* 6.)
Bl. similis. *Latr. H.* x. *pl.* 279.

2437 ‡ 3, Gigas. *Herbst C.* viii. 181. *pl.* 128. *f.* 1.—(*Curtis l. c. supra.*)
Te. Gigas. *Linn.* ii. 676.
Bl. Gages. *Fabr. E.* i. 141. In Mus. *D. I. H. Griesbach.*

Familia LVIII. HELOPIDÆ.

(HELOPII *et* CISTELINÆ *p*, *Latreille.*—DIAPERIDÆ *p*, *Leach.*)

Genus 412. HELOPS, *Fabricius*, *Leach*, *Samou.*, *Millard.*

TENEBRIO *p*, *Linné.*—BLAPS *p*, *Marsh.*—PIMELIA *p*, *Gmel.*

2438. 1; cæruleus*. *Fabr. E.* i. 156.
He. chalybeus. *Oliv. E.* iii. 14. *pl.* 2. *f.* 9.
Te. chalybeus. *Linn.* ii. 674?—*Herbst C.* vii. 254. *pl.* 112. *f.* 1.
Bl. violacea. *Marsh.* i. 480.
He. violacea. (*Sam. I.* 21.)
Pi. cærulea. *Gmel.* iv. 2009.—*Stew.* ii. 83.—*Turt.* ii. 494.

2439. 2, lanipes. *Fabr. E.* i. 157.—*Panz. F.* l. *f.* 2.—(*Leach E. E.* ix. 103.)—(*Sam. I.* 21.)

Te. lanipes. *Linn. M.* 533.
Te. æneus. *Scop. C.* 255.
Te. arboreus. *Schra. A.* 319.
2440. 3; caraboides*. *Panz. F.* xxiv. *f.* 3.
He. dermestoides. *Illig. K.* i. 120.
He. striatus. *Oliv. E.* iii. 6. *pl.* 1. *f.* 4.
Bl. Spartii*. *Marsh.* i. 481.
β; He. ruficollis. *Fabr. E.* i. 163.

2441 † 4. quisquilius[a].

Genus 413. ——

HELOPS *p, Fabr.*—CRIOCERIS *p, Marsh.*—BLAPS *p, Marsh.*—PYROCHROA *p, DeGeer.*

2442. 1; ater*.
He. ater. *Fabr. E.* i. 161.—*Panz. F.* l. *f.* 3.
Py. nigra. *DeGeer.* iv. 25. *pl.* 1. *f.* 23, 24.
Cr. nigra. *Marsh.* i. 221.—(*Sam. I.* 11.)
Bl. atrata. *Marsh.* i. 480.

Genus 414. MYCETOCHARUS, *Latreille.*

MYCETOPHILA[b], *Gyll.*—CISTELA *p, Panzer.*

2443. 1; scapularis*.
My. scapularis. *Gyll.* ii. 549.
Ci. humeralis. *Panz. F.* xxv. *f.* 14.—(*Sam. I.* 11.)
Cr. bipustulatus. *Marsh. MSS.*

Genus 415. CISTELA, *Fabr., Leach, Samou.,* (*Kirby.*)

CHRYSOMELA *p, Linné.*—PYROCHROA *p, DeGeer.*—CRIOCERIS *p, Marsh., Millard.*

A. Antennis elongatis.

a. *Antennis serratis.*

2444. 1; Ceramboides*. *Fabr. E.* ii. 16.—(*Leach E. E.* ix. 104.)—(*Sam. I.* 11.)
Ch. Ceramboides*. *Linn.* ii. 602.
Cr. Ceramboides. *Marsh.* i. 222.
Py. rufa. *DeGeer.* v. 23. *pl.* 1. *f.* 20—22.

b. *Antennis haud serratis.*

2445. 2; castanea*. (*Schön. S.* ii. 338.)—(*Sam. I.* 11.)
Cr. castanea. *Marsh.* i. 223.
Ci. ferruginea. *Fabr. E.* ii. 22?

---

[a] 2441 † 4. quisquilius. *Fabr. E.* i. 163.—*Panz. F.* l. *f.* 5.
Pi. quisquilia. *Gmel.* iv. 2011.—*Stew.* ii. 84. (!)—*Turt.* ii. 496. (!)

[b] MYCETOPHILA: Genus Dipterorum. Vide *Meig. K.* i. 89.

2446. 3; fulvipes*. *Fabr. E.* ii. 19.—(*Sam. I.* 11.)
Ci. Luperus. *Herbst A.* 65. *pl.* 23. *f.* 30.
Ci. erythropa. *Marsh.* i. 223.

B. Antennis brevibus.

2447. 4; murina*. *Fabr. E.* ii. 19.—(*Sam. I.* 11.)
Ch. murina. *Linn.* ii. 602.—*Mart. C. pl.* 16. *f.* 32.
Cr. murina. *Marsh.* i. 222.
Cry. murinus. *Turt.* ii. 183.
Ci. reppensis. *Gmel.* iv. 1716.—*Herbst A.* 65. *pl.* 23. *f.* 32.

2448. [5; fusca*.] *Panz. F.* xxv. *f.* 19?—(*Sam. I.* 11.)
Cr. fusca. *Marsh.* i. 223.
Ci. maura. *Fabr. E.* ii. 20.
Ci. murina β. *Gyll.* ii. 626.

Genus 416. ALLECULA?

CHRYSOMELA *p*, *Linn.*—CISTELA *p*, *Fabr.*—CRIOCERIS *p*, *Marsh.*, *Millard.*—CRYPTOCEPHALUS *p*, *Gmel.*

2449. 1; sulphurea.
Ch. sulphurea. *Linn.* ii. 602.—*Mart. C. pl.* 17. *f.* 44.
Cr. sulphurea. *Marsh.* i. 219.
Cry. sulphureus. *Turt.* ii. 183.
Ci. sulphurea. (*Sam. I.* 11.)
Te. flavus. *Scop. C.* 260.

2450. [2, bicolor.]
Ci. bicolor. *Fabr. E.* ii. 18.—*Panz. F.* xxxiv. *f.* 8.
Cry. sulphuratus. *Gmel.* iv. 1717.

Familia LIX. MELANDRYIDÆ, *Leach.*

(HELOPII *p*, *Latreille.*)

Genus 417. LAGRIA, *Fabricius*, *Leach*, *Samou.*, (*Kirby.*)

CHRYSOMELA *p*, *Linné.*—CANTHARIS *p*, *Geoffroy.*—TENEBRIO *p*, *DeGeer.*—AUCHENIA *p*, *Marsh.*, *Millard.*

2451. 1; hirta*. *Fabr. E.* ii. 70.—(*Leach E. E.* ix. 104.)—(*Sam. I.* 24.)
Ch. hirta. *Linn.* ii. 602.—*Mart. C. pl.* 17. *f.* 55.
Au. hirta. *Marsh.* i. 218.
Cry. hirtus. *Turt.* ii. 196.
Te. villosus. *DeGeer.* v. 44. *pl.* 2. *f.* 23, 24.
♀; Ch. pubescens. *Linn.* ii. 600.
La. pubescens. *Fabr. E.* ii. 70.—*Panz. F.* cvii. *f.* 1.

Genus 418. MELANDRYA, *Fabricius, Leach, Samou., Curtis.*

CHRYSOMELA *p, Linné.*—SERROPALPUS *p, Illig.*—HELOPS *p, Payk.*—CRIOCERIS *p, Marsh.*—TENEBRIO *p, Müll.*

2452. 1; caraboides*. *Latr.* G. 191.—(*Leach E. E.* ix. 104.)—(*Sam. I.* 26.)—(*Curtis l. c. infra.*)
Ch. caraboides. *Linn.* ii. 602.—*Mart. C. pl.* 17. *f.* 58.
Cr. caraboides. *Marsh.* i. 221.
Me. serrata. *Fabr. E.* i. 163.
Pi. serrata. *Gmel.* iv. 2009.—*Stew.* ii. 84.—*Turt.* ii. 494.
Te. rufibarbis. *Act. Hall.* (*Schall.*) i. 324.

2453 † 2, canaliculata. *Fabr. E.* i. 164.—*Curtis.* iv. *pl.* 155.—(*Ing. Inst.* 89.)
Te. dubius. *Act. Hall.* (*Schall.*) i. 326. In Mus. D. *Bentley.*

Genus 419. SCRAPTIA, *Latreille, Leach, Samou.*

MELANDRYA *p, Latr. H.*—DIRCÆA *p, Schön.*

2454 † 1, fusca*. *Latr.* G. ii. 200.—(*Leach. E. E.* ix. 104.)—(*Sam. I.* 37.)
Di. sericea. *Schön. S.* iii. *App.* 19. In Mus. *Brit. et* D. *Curtis.*

2455. 2; nigricans* *mihi.*

Genus 420. SERROPALPUS, *Hellenius, Samou.* 195, *Millard.*

DIRCÆA *p, Gyll.*

2456. 1, rufipes*. *Gyll.* ii. 519.
Xi. rufipes. (*Ing. Inst.* 89.)
Se. Vaudoueri. *Latr. MSS.*?—(*DeJean C.* 70.)

Genus 421. XILITA[a].

Genus 422. DIRCÆA, *Fabricius.*

SERROPALPUS *p, Latr.*?

2458 † 1, variegata*. *Fabr. E.* ii. 90.
Se. variegatus. *Act. H. N. Paris.* 1. *pl.* 10. *f.* 2. In Mus. *Brit.*

Genus 423. HYPULUS, *Paykul.*

DIRCÆA *p, Fabr.*—SERROPALPUS *p, Illiger.*—ELATER *p, Quensel.*—NOTOXUS *p, Panz.*

2459. 1; Quercinus*. *Payk. F.* i. 252.—(*Ing. Inst.* 88.)
Di. dubia. *Fabr. E.* ii. 90.

---

[a] Genus 421. XILITA, *Paykul.*

DIRCÆA *p, Fabr.*—SERROPALPUS *p, Illig.*—XYLETA, *Millard.* (!)

2457 † 1. buprestoides. *Payk. F.* i. 249.—*MacLeay H. E.* i. 464. (!)
Di. discolor. *Fabr. E.* ii. 89.
Se. lævigatus. *Illig. K.* i. 131.
Ly. lævigatum. *Panz. F.* xxiv. *f.* 16.

No. dubius. *Panz. F.* xi. *f.* 13.
El. Blekingensis. *Schneid. M.* (*Rhèn.*) ii. 141.

Genus 424. ——

MORDELLA *p*, *Marsh.*—ANASPIS *p*, *Samou.*—HALLOMENUS *p?* *Payk.*

2460. 1; bifasciata*.
Mo. bifasciata. *Marsh.* i. 493.
An. bifasciatus. (*Sam. I.* 2.)
Ha. fasciatus. *Payk. F.* ii. 182?

2461 † 2, trifasciata *mihi.* In Mus. D. Curtis? *Ingpen, Stone, et Vigors.*

Genus 425. ORCHESIA, *Latreille, Samou., Millard, Curtis.*

DIRCÆA *p*, *Fabr.*, *Leach.*—SERROPALPUS *p*, *Illig.*—HALLOMENUS *p*, *Paykul.*—MEGATOMA *p*, *Herbst.*—MORDELLA *p*, *Marsh.*—ANASPIS *p*, *Latreille.*

2462. 1, micans*. *Latr. G.* ii. 195.—(*Sam. I.* 32.)—(*Curtis l. c. infra.*)
Me. picea. *Herbst C.* iv. 97. *pl.* 39. *f.* 5. *a*, *b*, c.
Mo. Boleti. *Marsh.* i. 494.
An. clavicornis. *Latr. H.* x. 417.
Di. micans. (*Leach. E. E.* ix. 104.)

2463 † 2, fasciata. (*DeJean C.* 69.)—(*Ing. Inst.* 89.)—*Curtis.* v. 197. In Mus. D. Dale *et Vigors.*

Familia LX. MORDELLIDÆ, *Leach.*

(MORDELLONÆ, *Latreille.*)

(MORDELLA, *Linné, &c.*)

Genus 426. ANASPIS, *Geoffroy, Leach, Samou., Millard.*

2464. 1; frontalis*. *Latr. G.* ii. 210.—(*Leach E. E.* ix. 105.)—(*Sam. I.* 2.)
Mo. frontalis. *Linn.* ii. 682.—*Panz. F.* xiii. *f.* 13.—*Marsh.* i. 491.—*Turt.* ii. 506.—*Wood.* i. 75. *pl.* 28.
β; An. similis. *Kirby MSS.*
Mo. atra. *Fabr. E.* ii. 126?
γ; An. affinis. *Kirby MSS.*
δ?; An. rufilabris. *Sturm?*—*Gyll.* iv. 521.

2465. 2; lateralis*. (*Schön. S.* iii. 88.)
Mo. lateralis. *Fabr. E.* ii. 125.—*Gyll.* ii. 616.

2466. 3; ruficollis*. (*Schön. S.* iii. 89.)—(*Sam. I.* 2.)
Mo. ruficollis. *Oliv. E.* iii. 9. *pl.* 1. *f.* 9. *a*, *b.*—*Marsh.* i. 491.

2467. 4; fuscescens* *mihi.*

2468. 5; obscura*. (*Schön. S.* iii. 89.)—(*Sam. I.* 2.)
Mo. obscura. *Marsh.* i. 492.
Mo. melanopa. *Forst. C.* i. 64.—*Stew.* ii. 87.—*Turt.* ii. 507.
An. maculata. (*DeJean C.* 73.)

2469. 6; pallida*.
Mo. pallida. *Marsh.* i. 492.

2470. 7, lurida. *Kirby MSS.*

2471. 8; flava*. (*Schön. S.* iii. 89.)
Mo. flava. *Linn.* ii. 682.—*Panz. F.* xiii. *f.* 14.—*Marsh.* i. 495.
—*Turt.* ii. 506.
Mo. thoracica β. *Payk. F.* ii. 185.

2472. 9; fusca*. (*Schön. S.* iii. 90.)
Mo. fusca. *Marsh.* i. 493.
β; Mo. testacea. *Marsh.* i. 493.

2473. 10, testacea. (*Schön. S.* iii. 90.)
Mo. testacea. *Forst. C.* i. 63.—*Stew.* ii. 86?
Mo. Oxyacanthæ. *Turt.* ii. 506.
An. melanocephala. *Kirby MSS.*?

2474. 11; subfasciata* *mihi.*

2475. 12; quadripustulata*.
Mo. quadripustulata. *Marsh. MSS.*

2476. 13; nigricollis*.
Mo. nigricollis. *Marsh.* i. 492.
Mo. fasciata. *Forst. C.* i. 65.
Mo. bicolor. *Stew.* ii. 87.—*Turt.* ii. 507.

2477. 14; quadrinotata* *mihi.*

2478. 15; biguttata*. (*Sam. I.* 2.)
Mo. biguttata. *Marsh.* i. 492.

2479. 16, scapularis* *mihi.*

2480 † 17. thoracica[a].

Genus 427. MORDELLA, *Linné, Marsh., Samou., Millard.*
OXURA, *Kirby MSS.* (*olim.*)

2481. 1; abdominalis*. *Fabr. E.* ii. 125.—*Oliv. E.* iii. 7. *pl.* 1. *f.* 5. *a, b.*—*Marsh.* i. 489.—*Turt.* ii. 506.—(*Sam. I.* 27.)

2482. 2; pumila*. *Gyll.* ii. 605.

2483. 3; aculeata*. *Linn.* ii. 682.—*Berk. S.* i. 109.—*Oliv. E.* iii. 4. *pl.* 1. *f.* 1. *a, b, c.*—*Stew.* ii. 86.—*Marsh.* i. 488.—*Turt.* ii. 505.—*Shaw* G. *Z.* vi. 107. *pl.* 38.—(*Sam. I.* 27.)

2484. 4; ventralis*. *Fabr. E.* ii. 125.
Mo. abdominalis var. b. *Payk. F.* ii. 187.
Mo. nigra. *Marsh.* i. 490.

---

[a] 2480 † 17. thoracica.
Mo. thoracica. *Linn.* ii. 682.—*Oliv. E.* iii. 9. *pl.* 1. *f.* 10. *a, b.*—*Turt.* ii. 506. (!)

2485. 5; humeralis*. *Linn.* ii. 682.—*Panz. F.* lxii. *f.* 3.—*Marsh.* i. 489.—*Turt.* ii. 506.

2486. 6; variegata*. *Fabr. E.* ii. 122.
Mo. lateralis. *Oliv. E.* iii. 8. *pl.* 1. *f.* 8. *a, b.*
Mo. dorsalis. *Panz. F.* xiii. *f.* 15.
Mo. bicolor. *Marsh.* i. 490.—(*Sam. I.* 27.)
Mo. huméralis, c. *Payk. F.* ii. 188.

2487. 7; ferruginea*. *Marsh.* i. 490.—(*Sam. I.* 27.)
β; Mo. flavescens. *Marsh.* i. 490.

2488. 8: fasciata*. *Fabr. E.* ii. 122.—*Oliv. E.* iii. 4. *pl.* 1. *f.* 2. *a, b.*—*Marsh.* i. 488.—(*Samou.* 60. *pl.* 4. *f.* 8.)
Mo. aculeata, var. *Linn.* ii. 684.—*Shaw* G. *Z.* vi. 107.

Genus 428. RIPIPHORUS, *Fabr., Leach.*

RHIPIPHORUS, *Samou., Curtis,* (*Kirby.*)

2489. 1: paradoxus*. *Fabr. E.* ii. 119.—*Panz. F.* xxvi. *f.* 14.—(*Leach E. E.* ix. 105.)
Mo. paradoxa. *Linn.* ii. 682.—*Marsh.* i. 491.
Rh. paradoxus. (*Sam. I.* 36.)—*Curtis.* i. *pl.* 19.
♀, Ri. angulatus. *Panz. F.* xc. *f.* 3.

Familia LXI. ŒDEMERIDÆ, *Leach.*

(ŒDEMERITES, MALACODERMI *p*, CANTHARIDIÆ *p*, *et* CISTELINIÆ *p*, *Latreille.*)

Genus 429. SITARIS, *Latr.*

NECYDALIS *p*, *Fabr., Forst., Don., Marsh.*—CANTHARIS *p*, *Geoff.*—MENYS, *Kirby MSS.*

2490. 1: humeralis. *Latr.* G. ii. 222.
Ne. humeralis. *Fabr. E.* ii. 371.—*Mart. C. pl.* 23. *f.* 7.—*Don.* x. *pl.* 358. *f.* 1.—*Marsh.* i. 359.
Ne. muralis. *Forst. C.* i. 48.—*Stew.* ii. 67.

Genus 430. ——

NECYDALIS *p*, *Linné, Marsh., Millard.*—ŒDEMERA *p*, *Oliv.*—CANTHARIS *p*, *Geoff.*

2491. 1; cærulea*.
Ne. cærulea. *Linn.* ii. 642.—*Mart. C. pl.* 23. *f.* 6.—*Marsh.* i. 359.—*Turt.* ii. 356.—*Shaw* G. *Z.* vi. 76. *pl.* 27.—*Don.* xvi. *pl.* 558.—*Samou.* 55.
Œd. cærulea. (*Samou.* 198. *pl.* 2. *f.* 28.)
♀; Ne. ceramboides. *Forst. C.* i. 47.—*Mart. C. pl.* 23. *f.* 5.—*Marsh.* i. 360.

2492 † 2, marginata?
Ne. marginata. *Fabr. S.* 155?

♂, Ne. femorata. *Panz. F.* xxxvi. *f.* 12?
In Mus. D. *Haworth.* et ——?

Genus 431. ONCOMERA *mihi.*

NECYDALIS *p, Linné.*—ŒDEMÈRA *p, Sam.*

2493. 1; Podagrariæ*.
Ne. Podagrariæ. *Linn.* ii. 642.—*Oliv. E.* iii. 10. *pl.* 1. *f.* 10. *a, b.*—*Marsh.* i. 360.
Œd. Podagrariæ. (*Sam. I.* 31.)
♀: Ne. testacea. *Fabr. E.* ii. 373.
Ne. simplex. *Don.* x. *pl.* 358. *f.* 2.—*Marsh.* i. 361.
Ne. melanocephala. *Panz. F.* xxxvi. *f.* 9.

Genus 432. ŒDEMERA, *Olivier, Leach, Samou.*

CANTHARIS *p, Linné, Marsh.*—NECYDALIS *p, Fabr., Millard.*

2494. 1; lurida*. (*Sam. I.* 31.)
Ne. lurida. *Marsh.* i. 360.

2495 [† 2, Leontodontis.] *Leach MSS.* In Mus. *Brit.*

2496 † 3. viridissima[a].

2497. 4; cærulescens*. *Oliv. E.* iii. 12. *pl.* 2. *f.* 17. *a—d.*
Ne. cærulescens. *Turt.* ii. 355.
Ca. cærulea. *Linn.* ii. 650.
Ne. cyanea. *Fabr. E.* ii. 369.
Ne. viridissima. *Stew.* ii. 67.
Ca. viridissima. *Marsh.* i. 372.
Œd. viridissima. (*Sam. I.* 36.)

2498. 5, fulvicollis?
Ne. fulvicollis. *Fabr. E.* ii. 372?
Œd. ruficollis. (*Sam. I.* 31.)

2499. 6; melanura*. *Oliv. E.* iii. 9. *pl.* 1. *f.* 8. *a, b.*
Ca. melanura. *Linn.* ii. 651.
β; Ne. notata. *Payk. F.* iii. 132.
Ca. nigripes. *Fabr. E.* i. 302.
Œd. nigripes. (*Sam. I.* 31.)
Ca. acuta. *Marsh.* i. 372.
♂; Ne. Lepturoides. *Act. Ups.* (*Thunb.*) iv. 18.

Genus 433. LYMEXYLON[b].

---

[a] 2496 † 3. viridissima. *Oliv. E.* iii. 13. *pl.* 2. *f.* 15. *a—d.*
Ne. viridissima. *Fabr. E.* ii. 369.—*Turt.* ii. 354. (!)

[b] Genus 433. LYMEXYLON, *Fabr.*

CANTHARIS *p, Linné.*—PTEROPHORUS, *Herbst.*

2500 † 1. navale. *Fabr. E.* ii. 88.—*Panz. F.* xxii. *f.* 5.
Ca. navalis. *Linn.* ii. 650.—*Stew.* ii. 70. (!)—*Turt.* ii. 375. (!)
♂. Ly. flavipes. *Fabr. E.* ii. 88.—*Panz. F.* xxii. *f.* 6.
Ca. saxonica. *Gmel.* iv. 1900.
Pterophorus. *Herbst A.* 105. *pl.* 27. *f.* 1. *a, b.*

Genus 434. HYLECÆTUS[a].

Genus 435. NOTHUS? *Ziegler.*

2502 † 1, bimaculatus?* *Oliv. E. M.* ——? In Mus. *Brit.*

Genus 436. CONOPALPUS, *Gyll.*

NOTHUS *p*, *Mus. Lond.*—MELYRIS *p*, *Olivier.*—DASYTES *p*, *Schön.*—HELLENIA, *MacLeay.*—ZONITIS, *Curtis.*

2503. 1; testaceus*.
Me. testacea. *Oliv. E.* ii. 21. *pl.* 3. *f.* 15. *a*, *b*.
Zo. testacea. *Curtis.* iii. *pl.* 112.
No. pallida. *Mus. Lond.*

2504. 2, Vigorsii.
He. Vigorsii. *MacLeay MSS.*
Co. flavicollis. *Gyll.* ii. 547?
Co. ruficollis *mihi.*

Genus 437. CALOPUS[b].

Familia LXII. PYROCHROIDÆ, *Leach.*

(PYROCHROÏDES, *Latreille.*)

Genus 438. PYROCHROA, *Fabr.*, *Don.*, *Marsh.*, *Leach*, *Samou.*, *Millard.*

CANTHARIS *p*, *Linné.*—LAMPYRIS *p*, *Schall.*

Cardinal-beetle.

2506. 1; rubens*. *Fabr. E.* ii. 109.—*Marsh.* i. 363.—(*Leach E. E.* ix. 104.)—*Samou.* 56.
Py. purpurata. *Müll. Z. D. pr.* 80.
Py. Satrapa. *Schran. A.* 174.
Py. coccinea. *Don.* ii. *pl.* 56. *f.* 1.
Ca. serraticornis. *Scop. C.* 42.

---

[a] Genus 434. HYLECÆTUS, *Latreille.*

CANTHARIS *p*, *Linné.*—LYMEXYLON *p*, *Fabr.*—LYTTA *p*, *Herbst.*—HORIA, *Stewart.*

2501 † 1. dermestoides. *Latr. G.* i. 266.—*Latr. R. A.* iii. 252. (!)
Ca. dermestoides. *Linn.* ii. 650.
Lym. dermestoides. *Fabr. E.* ii. 87.—*Panz. F.* xxii. *f.* 2.
Lyt. francofurtana. *Herbst A.* 145. *pl.* 30. *f.* 4.
Ho. dermestoides. *Stew.* ii. 68. (!)
♂. Lym. proboscideum. *Fabr. E.* ii. 87.—*Panz. F.* xxii. *f.* 3.

[b] Genus 437. CALOPUS, *Paykul*, *Leach.*

CERAMBYX *p*, *Linné.*

2505 † 1. serraticornis. *Payk. F.* iii. 65.—*Panz. F.* iii. *f.* 15.—*Turt.* ii. 342. (!)
Ce. serraticornis. *Linn.* ii. 634.

2507. 2; coccinea*. *Fabr. E.* ii. 108.—*Marsh.* i. 364.—(*Leach E. E.* ix. 104.)—*Samou.* 56. *pl.* 3. *f.* 3.
Ca. coccinea. *Linn. F.* 705.
La. coccinea. *Turt.* ii. 361.
Py. rubra. *DeGeer.* v. 20. *pl.* 1. *f.* 14—17.
Py. rubens. *Don.* xi. 65. *pl.* 383.
Py. purpurata. *Schr. A.* 173.

Familia LXIII. CANTHARIDÆ, *Leach.*

(Cantharidiæ, *Latreille.*—Melooides *p, Gyllenhal.*)

Genus 439. PROSCARABÆUS, *Leach MSS. Oil-beetle.*

2508. 1; violaceus*.
Me. violaceus. *Marsh.* i. 482.—*Linn. Trans.* (*Leach.*) xi. 45. *pl.* vii. *f.* 3, 4, 5.—*Samou.* 369. *pl.* 4. *f.* 7. ♀.
Me. similis. *Marsh.* i. 482.
Me. tecta. *Don.* vii. *pl.* 240.—*Stew.* ii. 86.

2509. 2; vulgaris*.
Me. proscarabæus. *Linn.* ii. 679.—*Berk. S.* i. 109.—*Don.* ii. *pl.* 43. *fig. inf.*—*Mart. C. pl.* 39. *f.* 6.—*Stew.* ii. 88.—*Marsh.* i. 481.—*Turt.* ii. 501.—*Shaw* G. Z. vi. 103. *pl.* 37. ♀. —*Linn. Trans.* (*Leach.*) xi. 46. *pl.* vii. *f.* 6, 7.—(*Sam. I.* 27.) —(*Kirby & Sp. I. E.* ii. 250.)

2510. 3; tectus*.
Me. tectus. *Panz. F.* x. *f.* 14.—*Turt.* ii. 502.—*Linn. Trans.* (*Leach.*) xi. 47. *pl.* vii. *f.* 8, 9.—(*Sam. I.* 27.)

2511. 4, rugicollis* *mihi.*

2512. 5, autumnalis*.
Me. autumnalis. *Oliv. E.* vi. 45. *pl.* 1. *f.* 2.
Me. punctatus. *Marsh.* i. 483.
Me. glabratus. *Linn. Trans.* (*Leach.*) xi. 43. *pl.* vii. *f.* 1, 2.—(*Sam. I.* 27.)

Genus 440. MELOË, *Leach MSS.*

2513. 1, brevicollis*. *Fabr. E.* ii. 588.—*Linn. Trans.* (*Leach.*) xi. 41. *pl.* vi. *f.* 9.—(*Sam. I.* 27.)
Me. scabriuscula. *Besser.*

2514. 2, punctatus.
Me. punctata. *Fabr. E.* ii. 588. (!)—*Turt.* ii. 502.
Me. rugosus. *Marsh.* i. 483.
Me. autumnalis. *Linn. Trans.* (*Leach.*) xi. 40. *pl.* vi. *f.* 7, 8. —(*Sam. I.* 27.)

2515. 3, cicatricosus.
Me. cicatricosus. *Linn. Trans.* (*Leach.*) xi. 39. *pl.* vi. *f.* 5, 6. —(*Sam. I.* 27.)
Me. punctato-radiatus. *Latr. H.* x. 391.

2516. 4; scabrosus. *Marsh.* i. 483.—*Illig. M.* iv. 168.—*Shaw G. Z.* vi. 105.
Me. variegatus. *Don.* ii. *pl.* 67.—*Mart. C. pl.* 39. *f.* 1.—*Turt.* ii. 501.—*Linn. Trans.* (*Leach.*) xi. 37. *pl.* vi. *f.* 1, 2. —(*Sam. I.* 27.)
Me. maialis. *Schæff. Pl.* 3.

Genus 441. CANTHARIS, *Geoffroy, Leach, Samou.*, (*Kirby.*)

LYTTA *p, Marsh., Millard. Blister Fly.*

2517. 1, vesicatoria*. *Geoffroy.* i. 341.—(*Leach E. E.* ix. 106.) —(*Samou.* 198. *pl.* 4. *f.* 5.)—(*Kirby & Sp. I. E.* i. 312.—ii. 227.)
Me. vesicatoria. *Linn.* ii. 679.—*Mart. C. pl.* 39. *f.* 2.—*Shaw* G. *Z.* vi. 106.
Ly. vesicatoria. *Fabr. E.* ii. 76.—*Stew.* ii. 85.—*Marsh.* i. 484. —*Bingley.* iii. 148.—*Don.* xv. *pl.* 534.—*Samou.* 59.

Genus 442. ——

LYTTA *p, Fabr.?*—NECYDALIS *p, Marsh. MSS.*

2518. 1, immunis*?
Ne. immunis. *Marsh. MSS.*

Familia LXIV. NOTOXIDÆ.

(ANTHICITES, *Latreille.*—PYROCHROIDÆ 2, *Leach.*—MELOOIDES *p, Gyllenhal.*)

Genus 443. NOTOXUS, *Illiger, Leach, Samou., Millard.*

MELOE *p, Linn., Don.*—ANTHICUS *p, Fabr.*—LYTTA *p, Marsh.*

2519. 1; monoceros*. *Illig. K.* i. 287.—*Stew.* ii. 63.—*Turt.* ii. 288.—(*Leach E. E.* ix. 104.)—*Samou.* 54. *pl.* 2. *f.* 23.
Me. monoceros. *Linn.* ii. 651.—*Don.* vi. *pl.* 182.
Ly. monoceros. *Marsh.* i. 487.
β; No. nigricans. *Kirby MSS.*

2520. 2, Rhinoceros.
An. Rhinoceros. *Fabr. E.* i. 289.
No. serricornis. *Panz. F.* xxxi. *f.* 17.

Genus 444. ANTHICUS, *Fabr., Leach, Samou., Millard.*

MELOE *p, Linn.*—NOTOXUS *p, Illiger.*—LYTTA *p, Marsh.*—CRYPTOCEPHALUS *p, Gmel.*

2521. 1; Antherinus*. *Fabr. E.* i. 291.—(*Leach E. E.* ix. 105.) —(*Sam. I.* 3.)
Me. Antherinus. *Linn.* ii. 681.—*Mart. C. pl.* 39. *f.* 3.
Ly. Antherinus. *Marsh.* i. 486.
No. Antherinus. *Turt.* ii. 288.—*Panz. F.* xi. *f.* 14.
β? An. confinis *mihi* (*olim.*)

2522. 2, quadrinotatus. *Gyll.* ii. 498.

2523. 3, ater. *Act. Holm.* (*Payk.*) 1801. 117.
An. Antherinus var. c. *Payk. F.* i. 256.
No. ater. *Illig. K.* i. 290.—*Panz. F.* xxxi. *f.* 15.

2524. 4; fuscus*. (*Leach E. E.* ix. 105.)—(*Sam. I.* 3.)
Ly. fusca. *Marsh.* i. 486.
No. floralis. *Panz. F.* xxiii. *f.* 4.

2525. 5; floralis*. *Fabr. E.* i. 291.—(*Sam. I.* 3.)
Me. floralis. *Linn.* ii. 681.—*Mart. C. pl.* 39. *f.* 4.
Ly. floralis. *Marsh.* i. 485.
No. floralis. *Turt.* ii. 288.

2526. 6; equestris*.
No. equestris. *Panz. F.* lxxiv. *f.* 6.
No. pedestris β. (*Panz. I.* 88.)

2527 † 7, pedestris. *Fabr. E.* i. 291? In Mus. D. Curtis?

2528. 8; gracilis*.
No. gracilis. *Kugellan.*—*Panz. F.* xxxviii. *f.* 21.

Genus 445. ADERUS, *Westwood MSS.*

Lytta *p*, *Marsh.*—Anthicus *p*, *Paykul.*—Notoxus *p*, *Panzer.*

2529. 1; Boleti*.
Ly. Boleti. *Marsh.* i. 486.
No. melanocephalus. *Panz. F.* xxxv. *f.* 5.
An. ferrugineus. *Payk. F.* i. 257.
An. pygmæus. *Gyll.* ii. 502?
♂. Ce. pygmæus. *DeGeer.* v. 80. *pl.* 4. *f.* 5?

2530 † 2, nigricollis.
Ly. nigricollis. *Marsh.* i. 487.
An. oculatus. *Gyll.* ii. 501? In Mus. D. *Kirby.*

Familia LXV. SCYDMÆNIDÆ, *Leach.*

(Palpatores, *Latreille.*)

Genus 446. MASTIGUS[a].

Genus 447. SCYDMÆNUS, *Latreille, Leach, Samou., Millard, Denny.*

Anthicus *p*, *Fabr.*—Notoxus *p*, *Panzer.*—Lytta *p*, *Marsh.*—Pselaphus *p*, *Illiger.*

A. Thorace angusto, sæpiùs hirsuto.

2532. 1; tarsatus*. *Kunzé M.* 11.—*Denny M.* 57. *pl.* 11. *f.* 11.

---

[a] Genus 446. MASTIGUS, *Latreille, Leach.*

2531 † 1, palpalis. *Latr. G.* i. 281. *pl.* 8. *f.* 5.—(*Leach E. E.* ix. 92.) In Mus. *Brit.*

No. minutus. *Panz. F.* xxiii. *f.* 5.
Ly. picea. *Marsh.* i. 486.
Sc. Hellwigii. *Latr. H.* ix. 156.—*Samou.* 180.
Sc. Illigeri. *Schüppel.*

2533. 2, ruficornis*. *Kirby MSS.—Denny M.* 59. *pl.* 11. *f.* 2.

2534. 3; hirticollis*. *Gyll.* i. 286.—*Denny M.* 62. *pl.* 12. *f.* 1.
Ps. hirticollis. *Ent. Trans.* (*Haworth.*) i. 250.
Sc. pilosicollis. *Dahl.*

2535 † 4, rutilipennis. *Kunzé M.* 17.—*Denny M.* 63. *pl.* 12. *f.* 2. In Mus. *Brit.*

2536 † 5, denticornis. *Kunzé M.* 20.—*Denny M.* 64. *pl.* 13. *f.* 1. In Mus. *Brit.*

2537 † 6. elongatulus. *Kunzé M.* 19.—*Denny M.* 65. *pl.* 13. *f.* 2. In Mus. D. Burrell.

2538. 7; Sparshallii*. *Denny M.* 66. *pl.* 13. *f.* 3.

2539. 8; scutellaris*. *Kunzé M.* 23.—*Denny M.* 67. *pl.* 12. *f.* 3.
Sc. Dahlii. *Hoffmansegg.*

2540 † 9, bicolor. *Curtis MSS.—Denny M.* 68. *pl.* 13. *f.* 4. In Mus. D. *Curtis et Hope.*

2541. 10, collaris. *Kunzé M.* 26.—*Denny M.* 69. *pl.* 14. *f.* 2.
Sc. minutus. *Gyll.* i. 286.
Ly. nigripes. *Marsh. MSS.*

2542. 11; pusillus*. *Kunzé M.* 25.—*Denny M.* 70. *pl.* 14. *f.* 1.

2543 † 12. Wighami. *Denny M.* 71. *pl.* 14. *f.* 3. In Mus. D. Wigham.

2544. 13; Dennii* *mihi.*

B. Thorace latissimo.

2545. 14, thoracicus. *Kunzé M.* 12.—*Denny M.* 61. *pl.* 11. *f.* 3.
Sc. auripes. *Wilk. MSS.*

SECTIO VI. BRACHELYTRA, *MacLeay.*

(DIMERA *et* PENTAMERA *p*, *Latr.*—CHILOPODOMORPHA *p*, *MacLeay.*)

(STAPHYLINII, *Latreille.*—MICROPTERA, *Gravenhorst.*)

Rove-beetles, Cock-tails.

Familia LXVI. PSELAPHIDÆ, *Leach.*

(COLEOPTERA-DIMERA, *Latreille.*)

Genus 448. PSELAPHUS, *Herbst, Leach, Samou., Denny.*

ANTHICUS *p*, *Panz.*

2546. 1; Herbstii*. *Reich. Ps.* 25.—*Leach Z. M.* iii. 87.—(*Samou.* 179. *pl.* 4. *f.* 15.)—*Denny M.* 43. *pl.* 9. *f.* 1.
Ps. Heisii ♂. *Herbst C.* iv. 110. *pl.*
Ps. brevipalpis β. *Schr. B.* i. 438.

2547. 2; Heisii*. *Herbst C.* iv. 109.—*Leach Z. M.* iii. 87.—(*Sam. I.* 35.)—*Denny M.* 45. *pl.* 9. *f.* 2.
Ps. gracilicollis. *Dahl.*
Ps. eurygaster. *Bech. Beyt.* ii. *pl.* 2. *f.* 8.

2548. 3; longicollis*. *Reich. Ps.* 30.—*Leach Z. M.* iii. 87.—(*Sam. I.* 35.)—*Denny M.* 46. *pl.* 10. *f.* 1.
An. Dresdensis. *Panz. F.* xcviii. *f.* 1.

2549. 4, Dresdensis. *Herbst C.* iv. 110.—*Leach Z. M.* 87.—(*Sam. I.* 35.)—*Denny M.* 47. *pl.* 10. *f.* 2.
Ps. Heisii. *Payk. F.* iii. 364.

Genus 449. BRYAXIS, *Knoch., Leach, Samou., Denny,* (*Kirby.*)
STAPHYLINUS *p, Linné.*—ANTHICUS *p, Fabr.*

A. Antennarum articulo ultimo acuto.

2550. 1; longicornis*. *Leach Z. M.* iii. 85.—(*Sam. I.* 7.)—*Denny M.* 32. *pl.* 7. *f.* 2.—*Zool. J.* (*Leach.*) ii. 451.
Ps. foveolatus. *Kirby MSS.*

2551. 2; sanguinea*. *Leach Z. M.* iii. 85.—(*Sam. I.* 7.)—*Denny M.* 34. *pl.* 7. *f.* 3.—*Zool. J.* (*Leach.*) ii. 451.
St. sanguineus. *Linn.* ii. 685.
Ps. mucronatus. *Panz. F.* lxxxix. *f.* 10.
Ps. Paykullii. *Gyllenhal MSS.*

B. Antennarum articulo ultimo obtuso. (REICHENBACHIA, *Leach.*)

2552. 3; fossulata*. *Leach Z. M.* iii. 86.—(*Sam. I.* 7.)—*Denny M.* 37. *pl.* 8. *f.* 1.
Br. tripunctatus. *Knoch.*
Ps. cruentatus. *Knoch.*
Ps. pratensis. *Schüppel.*
Ps. ruficornis. *Kirby MSS.*

2553. 4; impressa*. *Leach Z. M.* iii. 86.—(*Sam. I.* 7.)—*Denny M.* 36. *pl.* 7. *f.* 4.
Ps. impressus. *Panz. F.* lxxxix. *f.* 10.—*Ent. Trans.* (*Haworth.*) i. 256.—*Leach E. E.* ix. 117.
Ps. tripunctatus. *Kirby MSS.*

2554 † 5, nigriventris*. *Denny M.* 41. *pl.* 7. *f.* 1.
Ps. nigriventris. *Kirby MSS.*
In Mus. D. Curtis, *Hope et Kirby.*

2555. 6, hæmatica. *Leach Z. M.* iii. 86.—(*Sam. I.* 7.)—*Denny M.* 38. *pl.* 8. *f.* 2.
Ps. hæmatica. *Reich. Ps.* 52.
Ps. rubellus. *Schüppel.*
Ps. sanguineus. *Germar.*

2556. 7; Juncorum*. *Leach Z. M.* iii. 86.—(*Sam. I.* 7.)—*Denny M.* 40. *pl.* 8. *f.* 3.
Rei. Juncorum. *Zool. J.* (*Leach.*) ii. 452.

Genus 450. TYCHUS, *Leach, Samou., Denny.*

2557. 1; niger*. *Leach Z. M.* iii. 84.—(*Sam. I.* 42.)—*Denny M.* 30. *pl.* 6. *f.* 1.—*Zool. J.* (*Leach.*) ii. 450.
♂; Ps. niger. *Payk.* iii. 365.
Ps. nodicornis. *Bech. Beyt.* 12. *pl.* 2. *f.* 10.
♀; Ps. ruficornis. *Dahl.*
Ps. detritus. *Essenbeck.*
Ps. congener. *Essenbeck.*
β; castaneus* *mihi* (*olim.*)

Genus 451. ARCOPAGUS, *Leach, Samou., Denny.*

BOSTRICHUS *p, Schrank.*

2558. 1, glabricollis*. *Leach Z. M.* iii. 83.—(*Sam. I.* 5.)—*Denny M.* 28. *pl.* 5. *f.* 4.
Ps. glabricollis. *Reich. Ps.* 43.

2559 † 2, clavicornis. *Leach Z. M.* iii. 84.—(*Sam. I.* 5.)—*Denny M.* 25. *pl.* 5. *f.* 2.
Ps. clavicornis. *Panz. F.* xcix. 3.
Ps. antennarius. *Hoffmansegg.* In Mus. *Brit.*

2560. 3; puncticollis*. *Denny M.* 26. *pl.* 5. *f.* 3.

2561. 4; bulbifer*. *Leach Z. M.* iii. 84.—(*Sam. I.* 5.)—*Denny M.* 24. *pl.* 5. *f.* 1.
Ps. bulbifer. *Reich. Ps.* 37.
Ps. Klugii. *Hellwig.*
Bo. anomalus. *Schrank.*

Genus 452. BYTHINUS, *Leach, Samou., Denny.*

2562. 1; Curtisii*. *Leach Z. M.* iii. 83.—(*Sam. I.* 8.)—*Denny M.* 20. *pl.* 3. *f.* 1.
Ps. binodis. *Kirby MSS.*
By. Curtisianus. *Zool. J.* (*Leach.*) ii. 446.

2563. 2, securiger*. *Leach Z. M.* iii. 83.—(*Sam. I.* 8.)—*Denny M.* 21. *pl.* 3. *f.* 2.
Ps. securiger. *Reich. Ps.* 45.
Ps. grandipalpus. *Kirby MSS.*

2564 † 3. Burrellii. *Denny M.* 22. *pl.* 4. *f.* 1. In Mus. D. Burrell.

Genus 453. EUPLECTUS, *Kirby, Leach, Samou., Denny.*

STAPHYLINUS *p, Marsh.*

2565 † 1, Kirbii. *Leach MSS.*—*Denny M.* 14. *pl.* 2. *f.* 1. In Mus. *Brit.*

2566. 2; Karstenii. *Denny M.* 12. *pl.* 1. *f.* 3.
Ps. Karstenii. *Reich. Ps.* 71.
St. sanguineus. *Panz. F.* xi. *f.* 9.
Eu. castaneus. *Leach MSS.*

2567. 3, minutus.
St. minutus. *Marsh.* i. 511.

2568. [4; sanguineus.] *Denny M.* 10. *pl.* 1. *f.* 2.

2569 † 5, signatus. *Denny M.* 13. *pl.* 1. *f.* 4.
Ps. signatus. *Reich. Ps.* 73.
Eu. unicolor. *Kirby.* In Mus. D. *Kirby.*

2570 † 6, Reichenbachii. *Leach Z. M.* iii. 82.—(*Sam. I.* 17.)—*Denny M.* 9. *pl.* 1. *f.* 1.
Ps. nanus. *Reich. Ps.* 69.
Eu. staphylinoides. *Kirby.*
In Mus. D. Curtis, *Kirby et* Skrimshire.

2571. 7; pusillus. *Leach MSS.*—*Denny M.* 15. *pl.* 2. *f.* 2.
Eu. tenuicornis. *Kirby MSS.*

2572. 8, bicolor. *Kirby.*—*Denny M.* 17. *pl.* 2. *f.* 3.

2573 † 9, brevicornis. *Denny M.* 18. *pl.* 2. *f.* 4.
Ps. brevicornis. *Reich. Ps.* 47.
Eu. Kunzii. *Leach MSS.* In Mus. *Brit.* D. *Vigors, &c.*

2574 † 10. Easterbrookianus. *Zool. Journ.* (*Leach.*) ii. 445.
In Mus. D. Easterbrook.

Familia LXVII. TACHYPORIDÆ, *MacLeay.*

(ST.-MICROCEPHALI *et* -DEPLANATI *p, Latreille.*)

(STAPHYLINUS *p, Linné, &c.*)

Genus 454. AUTALIA, *Leach MSS.*—(*Samou.*)

ALEOCHARA *p, Gravenhorst.*

2575. 1; impressa*.
St. impressus. *Oliv. E.* iii. *p.* 23. *pl.* 5. *f.* 41. *a, b.*
Al. impressa. *Grav. Mi.* 72.—(*Leach E. E.* ix. 92.)—(*Sam. I.* 2.)

2576. [2, plicata*.]
Al. plicata. *Kirby MSS.*

2577. 3; ruficornis* *mihi.*

2578. 4; aterrima* *mihi.*

2579. 5; rivularis*.
Al. rivularis. *Grav. Mi.* 73.—(*Sam. I.* 2.)
Al. pselaphoides. *Kirby MSS.*

2580. 6; angusticollis* *mihi.*

Genus 455. FALAGRIA, *Leach MSS.*—(*Samou.*)

ALEOCHARA *p, Grav.*

2581. 1; sulcata*.
St. sulcatus. *Payk. F.* iii. 385.—*Oliv. E.* iii. 23. *pl.* 6. *f.* 52. *a, b.*—*Marsh.* i. 507.
Al. sulcata. (*Sam. I.* 2.)

2582. 2, sulcatula.
Al. sulcatula. *Grav. Mo.* 151.

2583. 3, confinis.
Al. confinis. *Kirby MSS.*

2584. 4; obscura*.
Al. obscura. *Grav. Mi.* 74.—(*Sam. I.* 2.)
Al. immunis. *Kirby MSS.*

2585. 5, nigra.
Al. nigra. *Grav. Mi.* 75.

2586. 6, picea.
Al. picea. *Grav. Mi.* 75.

2587. 7, flavipes.
St. flavipes. *Linn.* ii. 685?
St. crassicornis. *Fabr. E.* ii. 601?
Al. flavipes. *Kirby MSS.*

2588. 8, nitens.
Al. nitens. *Kirby MSS.*

2589. 9, floralis. ——? *MSS.*

2590 † 10, thoracica.
Al. thoracica. *Kirby MSS.* In Mus. D. *Kirby.*

Genus 456. ——

DRUSILLA[a], *Leach MSS.*—(*Samou.*)—ALEOCHARA *p*, *Graven.*

2591. 1; canaliculata*.
St. canaliculatus. *Fabr. E.* ii. 599.—*Panz. F.* xxvii. *f.* 13.—*Marsh.* i. 507.
Al. canaliculata. *Leach E. E.* ix. 92.—*Samou.* 176.
Dr. canaliculata. *Leach MSS.*

Genus 457. DINARDA, *Leach MSS.*—(*Samou.*)

LOMECHUSA *p*, *Grav.*

2592 † 1, dentata.
Lo. dentata. *Grav. Mo.* 181.—(*Sam. I.* 25.)
St. strumosus. *Payk. F.* iii. 402. In Mus. *Brit.*

Genus 458. ——

GONIODES[b], *Kirby MSS.*—LOMECHUSA *p*, *Graven.*

2593. 1; acuminata*.
Go. acuminata. *Kirby MSS.*
Lo. paradoxa. *Gyll.* ii. 438?—*Ahr. F.* v. *f.* 12?

---

[a] DRUSILLA: Genus Lepidopterorum. Vide *Swain. Z. Ill.*

[b] GONIODES: Genus Pediculorum. Vide *Germ. M.* (*Nitzsch.*) iii. 293.

2594 † 2, strumosa*.
St. strumosus. *Fabr. E.* ii. 597.
Lo. strumosa. *Grav. Mo.* 179. In Mus. *Brit.*

Genus 459. LOMECHUSA, *Gravenhorst, Leach, Samou.*, (*Kirby.*)

2595 † 1, emarginata. *Grav. Mo.* 179.—*Samou.* 177.
St. emarginatus. *Fabr. E.* ii. 600.—*Oliv. E.* iii. 31. *pl.* 2. *f.* 12. *a, b.* In Mus. *D. Vigors.*

Genus 460. ALEOCHARA, *Knoch, Leach, Samou.*

A. "Thorace latitudine longiore." *Kirby MSS.*

2596. 1; longitarsis*. *Kirby MSS.*

2597. 2; aterrima*. *Grav. Mo.* 83.
Al. longicornis, b. *Gyll.* ii. 405.

2598. [3; leucopus*.]
St. leucopus. *Marsh.* i. 506.
β, Al. euceras. *Kirby MSS.*

2599 † 4. frontalis. *Kirby MSS.* In Mus. D. Sheppard.

2600. 5, ruficrus. *Kirby MSS.*

2601. 6; foveata*. *Kirby MSS.*

2602. 7, attenuata.
St. attenuatus. *Mus. Marsham.*

2603 † 8, basella. *Kirby MSS.* In Mus. *Brit. D. Kirby et* Spence.

B. "Thorace longitudine vix latior." *Kirby MSS.*

a. "*Antennis articulis intermediis campanulatis.*" Kirby MSS.

1. "Corpore plerumque nigro." *Kirby MSS.*

* "*Sericantes.*" Kirby.

2604. 9, subsericea. *Kirby MSS.*
Al. dubia. *Grav. Mo.* 173.

2605. 10, micans. *Kirby MSS.*

2606. 11; Kirbii* *mihi.*

2607. 12, crassicornis. *Kirby MSS.*

** *Vix sericantes.* † Elytris discoloribus.

2608. 13, xanthopus. *Kirby MSS.*
Al. Boleti, var. 2. 1. a. *Grav. Mi.* 80.
Al. Fungi. *Gyll.* ii. 410.?

2609. 14, tricolor. *Kirby MSS.*

2610. 15, longicornis. *Grav. Mi.* 87.

2611. 16, antennata*. *Kirby MSS.*

2612. 17, pusilla. *Grav. Mi.* 78.

2613 † 18, angustata. *Kirby MSS.* In Mus. *D. Kirby.*

2614 † 19, contigua. *Kirby MSS.* In Mus. *D. Kirby.*

2615. 20, xanthoptera. *Kirby MSS.*
Al. Boleti var. 1. 1 a. *Grav. Mi.* 80.
St. sordidus. *Marsh.* i. 514.
β, St. socialis. *Oliv. E.* iii. 37. *pl.* 3. *f.* 25. *a, b?*

2616. 21, vicina. *Kirby MSS.*

2617. 22, consimilis*. *Kirby MSS.*

2618. 23; assimilis*. *Kirby MSS.*

2619. 24, teres. *Grav. Mi.* 79.

2620. [25, elongata.] *Kirby MSS.*

2621 [† 26, nitidula.] *Kirby MSS.* In Mus. *D. Kirby.*

2622. 27, longiuscula. *Grav. Mi.* 80?
β, Al. pallipes. *Kirby MSS.*

2623. 28, sericoptera. *Kirby MSS.*

2624 † 29, striola.
Al. linearis. *Grav. Mi.* 69. In Mus. *D. Kirby.*

2625 † 30. pilosella.
Al. angustula. *Gyll.* ii. 393? In Mus. *D.* Spence.

2626. 31, foveolaris. *Kirby MSS.*
Al. quisquiliarum. *Gyll.* ii. 398?

2627. 32; nigricornis*. *Kirby MSS.*

2628 [† 33, migripalpis.] *Kirby MSS.* In Mus. *D. Kirby.*

2629 † 34, Brassicæ. *Kirby MSS.* In Mus. *D. Kirby.*

2630. 35; sulcifrons*. *Kirby MSS.*

2631 † 36, terminalis. *Grav. Mo.* 160. In Mus. *D. Kirby.*

2632 † 37, nitidiuscula *mihi.*
Al. nitidula. *Leach? MSS.* In Mus. *Brit.*

2633. 38, femorata*.
St. femoratus. *Mus. Marsh.*

2634. 39; consobrina* *mihi.*

2635. 40; ochropa* *mihi.*

2636. 41; caliginosa* *mihi.*

2637. 42; confinis* *mihi.*

†† Elytris concoloribus.

2638 † 43, carbonaria. *Kirby MSS.*
Al. cursitans. *Kirby? MSS.* In Mus. *D. Kirby et Vigors.*

2639. 44; subpubescens*. *Kirby MSS.*

2640. 45; rufitarsis*. *Kirby MSS.*
Al. æthiops. *Grav. Mi.* 77?

2641 † 46, foveola. *Leach MSS.* In Mus. *Brit. et D. Kirby.*

2. Corpore fusco.

2642. 47; lunulata*. *Gyll.* ii. 386.
St. lunulatus. *Payk. F.* iii. 415.
Al. cincta. *Grav. Mo.* 166.

2643. 48; nigricollis*. *Grav. Mi.* 84.
St. nigricollis. *Payk. F.* iii. 400.

2644. 49; picata*. *Kirby MSS.*

3. Corpore basi rufo. * *Thorace rufo.*

2645. 50, collaris. *Grav. Mi.* 72.
St. collaris. *Payk. F.* iii. 401.—*Oliv. E.* iii. 37. *pl.* 11. *f.* 13. *a, b.*

** *Thorace nigro.*

2646. 51, Haworthi. *Kirby MSS.*

b. "*Antennis articulis intermediis transversis.*" Kirby MSS.

1. Antennis rufis.

2647 † 52, cingulata. *Kirby MSS.*
Al. alternans. *Grav. Mi.* 85? In Mus. D. *Kirby.*

2648. 53; corticalis*. *Grav. Mi.* 76.
Al. obtusicornis. *Kirby MSS.*

2649 † 54, ochropus. *Kirby MSS.* In Mus. D. *Kirby.*

2650. 55; elongatula*. *Grav. Mi.* 79.

2651. [56, rufescens.] *Kirby MSS.*
β; Al. nigricans. *Kirby MSS.*

2652. 57, brachyptera. *Kirby MSS.*

2653 † 58, inquinula. *Grav. Mi.* 78.
In Mus. *Brit.* D. *Kirby et* Spence.

2654. [59, foveolata.] *Kirby MSS.*

2655. 60, nigrofusca. *Kirby MSS.*
Al. Boleti var.?

2656. 61, Boleti. *Kirby MSS.*—*Gyll.* ii. 416.
St. Boleti. *Linn.* ii. 686.—*DeGeer.* iv. 26. *pl.* 1. *f.* 15—17.

2657 † 62. immunis. *Kirby MSS.* In Mus. D. G. I. Hooker.

2658. 63; analis* *mihi.*

2659. 64; contigua* *mihi.*

2. Antennis nigris.

2660. 65; atramentaria*. *Gyll.* ii. 408.

2661 † 66, admota. *Kirby MSS.* In Mus. *D. Kirby.*

2662. 67; pumila. *Kirby MSS.*
β? Al. flavipes. *Kirby? MSS.*

2663. 68, livipes*. *Kirby MSS.*
β, Al. dissimilis. *Kirby MSS.*

2664. 69; erythropus*. *Kirby MSS.*

2665 † 70. obscuriuscula. *Kirby MSS.* In Mus. D. Sheppard.

2666 † 71, auricula. *Kirby MSS.* In Mus. D. *Kirby.*

2667 † 72, picipennis. *Kirby MSS.* In Mus. D. *Kirby et* Spence.

2668. 73; fuscula* *mihi.*

2669. 74; brunnipes* *mihi.*

2670. 75, atricornis* *mihi.*

2671. 76, fimetaria* *mihi.*

C. "Thorace longitudine latiore." *Kirby MSS.*

a. "*Antennis articulis intermediis campanulatis.*" Kirby MSS.

1. Corpore nigro.

2672. 77; socialis*. *Gyll.* ii. 406.
St. socialis. *Payk. F.* iii. 407.
Al. Boleti var. *Grav. Mi.* 80.
Al pallipennis. *Kirby MSS.*

2673 † 78. foveatocollis. *Kirby MSS.* In Mus. D. *Kirby?*

2674. 79, euryptera.
Al. Boleti var. 2. 1. b. *Grav. Mi.* 80?

2675. 80; obfuscata*. *Grav. Mi.* 87.

2676. 81, laticollis*. *Kirby MSS.*

2677. 82; infuscata*. *Kirby MSS.*

2678. 83, rufangula. *Kirby MSS.*

2679 † 84, luripes. *Kirby MSS.* In Mus. D. *Kirby.*

2680. [85, similis.] *Kirby MSS.*

2681 † 86, obscurata. *Kirby MSS.*
Al. pumilio. *Grav. Mi.* 98?
In Mus. D. *Haworth, Kirby* et Spence.

2682. 87, fulvipes. *Kirby MSS.*

2683. [88, pyrrhopus.] *Kirby MSS.*

2684 [† 89, Agarici.] *Kirby MSS.* In Mus. D. *Haworth et Kirby.*

2685. 90; atricollis* *mihi.*

2686. 91; brunniceps* *mihi*

2687. 92; microcephala* *mihi.*

2688. 93, littoralis ——?
St. littoralis. *Mus. Marsham.*

2689. 94; angusta *mihi*.

2. Corpore vario.

2690. 95; cinnamomea*. *Grav. Mi.* 88.—(*Sam. I.* 2.)
β; Al. dimidiata. *Kirby MSS.*

2691. 96; zonalis*. *Kirby MSS.*

2692. 97; atriceps*. *Kirby MSS.*
Al. alternans. *Grav. Mi.* 85?
St. melanocephalus. *Payk. F.* iii. 416?

2693. 98; nigriceps*.
St. nigriceps. *Marsh.* i. 515.

2694. 99; lucida*. *Grav. Mi.* 70.
Al. lunulata b. *Gyll.* ii. 386.

2695. 100; thoracica* *mihi*.

2696. 101; concinna*. *Leach? MSS.*

b. "*Antennis articulis intermediis transversis.*" Kirby MSS.

1. Antennæ versus apicem sensim crassiores.

2697. 102, nigritula*. *Grav. Mi.* 85.
Al. lineato-collis. *Kirby MSS.*

2698 [† 103, pyrrhoceras.] *Kirby MSS.*
Al. ochropus. *Kirby? MSS.* In Mus. D. *Kirby*.

2699 † 104, liturata. *Kirby MSS.*
Al. Boleti var. 2, 2, a. *Grav. Mi.* 80? In Mus. D. *Kirby*.

2700. 105, fusca.
St. fuscus. *Marsh.* i. 514.

2701 † 106, tenuior. *Kirby MSS.* In Mus. *D. Kirby*.

2702. 107, castanipes. *Kirby MSS.*

2703 † 108, parvula. *Kirby MSS.* In Mus. *Brit.* D. *Kirby* et Spence.

2704 † 109, pilosula. *Kirby MSS.* In Mus. D. *Kirby*.

2705 † 110, punctulata. *Kirby MSS.*
Al. opaca. *Grav. Mi.* 89? In Mus. D. *Kirby*.

2706. 111; lunicollis* *mihi*.

2707. 112; opaca* *mihi*.

2708. 113, fuscicornis* *mihi*.

2709. 114; glauca* *mihi*.

2. Antennis clavatis.

2710. 115; picipes*. *Kirby MSS.*

2711. 116; Pusio*. *Kirby MSS.*

2712. 117; pumilio. *Grav. Mi.* 98?
Al. casei. *Kirby MSS.*

2713. 118; fuscipennis*. *Kirby MSS.*

2714. 119, minutissima. *Kirby MSS.*
Al. brevicornis. *Kirby? MSS.*

D. "Thorace postice utrinque angulato." *Kirby MSS.*

2715. 120; sordida*. *Kirby MSS.*

2716. 121, curvipes*. *Kirby MSS.*

2717 † 122, livida. *Kirby MSS.* In Mus. D. *Haworth.*

2718. 123; umbrata*. *Grav. Mi.* 90.

2719. [124, emarginata*.] *Kirby MSS.*

2720 † 125. Sheppardi. *Kirby MSS.* In Mus. D. Sheppard.

2721. 126, erosa*. *Kirby MSS.*

2722 † 127, melanocephala. *Kirby MSS.*
In Mus. D. *Kirby* et Sheppard.

2723 † 128, brevicornis. *Kirby MSS.* In Mus. D. *Kirby.*

2724 † 129, nigrofusca. *Kirby MSS.* In Mus. D. *Kirby.*

2725. 130, ruficornis. *Grav. Mi.* 91.

2726. 131; lata* *mihi.*

2727. 132; pallipes* *mihi.*

2728. 133; acuminata* *mihi.*

E. "Thorace brevi, vix capite longiore: capite rhombeo: elytris transversis." *Kirby MSS.*

2729. 134; fasciata*.
St. fasciatus. *Marsh.* i. 514.

2730. 135; nana*. *Grav. Mi.* 98.
St. nanus. *Payk. F.* iii. 408.

2731. [136, angustata.] *Kirby MSS.*

2732. 137; polita*. *Grav. Mi.* 99.

2733. 138; pallicornis* *mihi.*

2734. 139; latissima* *mihi.*

2735. 140; marginata *mihi.*

F. "Thorace lato: elytris transversis: antennis crassis." *Kirby MSS.*

a. *Elytris nigris.*

2736. 141; concolor*. *Grav. Mi.* 96?

2737 † 142, nigrina. *Kirby MSS.* In Mus. D. *Kirby* et Sheppard.

2738. 143; immaculata*. *Kirby MSS.*

2739 [† 144, bilineata.] *Gyll.* ii. 436. In Mus. D. *Kirby.*

2740. 145, sericea. *Leach? MSS.*

2741 † 146, agilis. *Kirby MSS.* In Mus. *D. Kirby.*

2742 † 147, Cursitor. *Kirby MSS.* In Mus. *Brit.*

2743. 148; minutula*. *Kirby MSS.*
Al. Morion. *Grav. Mi.* 97?
β, Al. pusilla. *Kirby MSS.*

b. *Elytris piceis.*

2744 † 149. fumata. *Grav. Mi.* 96. In Mus. D. Sheppard.

2745 † 150, puncticollis. *Kirby MSS.* In Mus. D. *Kirby?*

2746 † 151, villosula. *Kirby MSS.* In Mus. D. *Kirby.*

2747 † 152, fulvicornis. *Kirby MSS.* In Mus. D. *Kirby.*

2748 † 153, erythroceras *mihi.*
Al. ruficornis. *Kirby MSS.* In Mus. D. *Kirby.*

2749 † 154, rufipes. *Kirby MSS.* In Mus. *Brit. et* D. *Kirby.*

c. *Elytris maculatis.*

2750. 155; lanuginosa*. *Grav. Mi.* 94.—(*Sam. I.* 2.)

2751. 156, bimaculata. *Kirby MSS.*

2752. 157; bipunctata*. *Grav. Mi.* 93.
St. fuscipes var. minor. *Payk. F.* iii. 397.
Al. oxura. *Kirby? MSS.*

2753. 158; terminata*. *Kirby MSS.*
St. bipunctatus. *Oliv. E.* iii. 31. *pl.* 5. *f.* 44.
Al. biguttata *mihi.*

2754. 159, nitida*. *Grav. Mi.* 97.
Al. biguttulus. *Kirby? MSS.*

2755. 160; velox*. *Kirby MSS.*

2756. 161; Cursor*. *Kirby MSS.*

2757. 162; dorsalis* *mihi.*

d. *Elytris rufis.*

2758. 163; fuscipes*. *Grav. Mi.* 92.—(*Sam. I.* 2.)
St. fuscipes. *Payk. F.* iii. 397.—*Marsh.* i. 525.—*Turt.* ii. 513.
Al. recurva. *Kirby MSS.*

2759 † 164, sanguinea.
St. sanguineus. *Linn.* ii. 685? In Mus. D. *Kirby.*

2760 † 165, rufipennis. *Kirby MSS.* In Mus. D. *Kirby.*

2761 † 166, celer. *Kirby MSS.* In Mus. D. *Kirby.*

2762 † 167, cursitans. *Kirby MSS.* In Mus. D. *Kirby.*

2763. 168, phæopus. *Kirby MSS.*
St. lugens. *Grav. Mi.* 95?
St. fuscipes, α, β. *Payk. F.* iii. 397?

2764 † 169, Simsii. *Wilkin MSS.* In Mus. *D. Vigors.*
G. "Thorace lato, posticè utrinque subangulato: antennis crassis." *Kirby MSS.*

2765. 170; limbata*. *Grav. Mi.* 69.
St. limbatus. *Payk. F.* iii. 399.—*Marsh.* i. 509.

2766. [171; divisa*.]
St. divisus. *Marsh.* i. 510.

2767. 172; humeralis*. *Grav. Mi.* 90.
Al. limbata major. *Grav. Mo.* 167.

Genus 461. ——
ALEOCHARA H, *Kirby MSS.*

2768. 1, Daltoni*.
Al. Daltoni. *Kirby MSS.*
β, Al. castanea. *Kirby MSS.*

Genus 462. ENCEPHALUS, *Kirby MSS.*
ALEOCHARA *p*, (*Kirby.*)

2769. 1; complicans*. *Kirby MSS.*
Al. complicans. (*Kirby & Sp. I. E.* ii. 232.)
β, En. lata. *Kirby MSS.*

Genus 463. CALLICERUS, *Gravenhorst.*

2770 † 1. Spencii. *Kirby MSS.* In Mus. D. Spence.

Genus 464. MEGACRONUS *mihi.*
TACHINUS *p*, *Gravenhorst.*

2771. 1; analis*.
St. analis. *Oliv. E.* iii. 28. *pl.* 3. *f.* 24. *a, b.*—*Fabr. E.* ii. 598. —*Marsh.* i. 522.
Ta. analis. (*Sam. I.* 40.)

2772. 2; rufipennis* *mihi.*

2773. 3; merdarius*.
St. merdarius. *Oliv. E.* iii. 29. *pl.* 5. *f.* 45. *a, b.*—*Fabr. E.* ii. 598.
St. analis, b. *Payk. F.* iii. 394.

2774. 4, cernuus*.
Ta. cernuus. *Grav. Mo.* 31.
Ta. piceus. *Kirby MSS.*

2775. 5; castaneus* *mihi.*

2776. 6, nitidus* *mihi.*

2777. 7, atriceps* *mihi.*

Genus 465. ISCHNOSOMA *mihi.*
TACHINUS *p*, *Gravenhorst.*—TACHYPORUS *p*, *Gyllenhal.*

2778. 1; splendens*.
St. splendens. *Marsh.* i. 524.

2779. 2; tenuis*.
St. tenuis. *Fabr. E.* ii. 599.
Ta. tenuis. *Kirby MSS.*

2780. 3; lepidus*.
Ta. lepidus. *Grav. Mo.* 26.
St. punctato-striatus, b. *Payk. F.* iii. 417.

2781. [4; punctato-striatus*.]
St. punctato-striatus, α. *Payk. F.* iii. 417.
St. lepidus, γ. *Gyll.* ii. 248.

2782. 5; clavicornis*.
Ta. clavicornis. *Kirby MSS.*

2783. 6, splendidus*.
Ta. splendidus. *Grav. Mo.* 24.
St. punctato-striatus, d. *Payk. F.* iii. 417.

2784. 7; melanurus*.
Ta. melanurus. *Kirby MSS.*

2785. 8, brunneus.
St. brunneus. *Marsh.* i. 524.

2786. 9, rufescens.
Ta. rufescens. *Kirby MSS.*

2787. 10; punctus*.
Ta. punctus. *Grav. Mo.* 27.

Genus 466. BOLITOBIUS, *Leach MSS.*, (*Samou.*)

TACHINUS *p*, *Grav.*—OXYPORUS *p*, *Panzer.*

2788. 1; lunulatus*.
St. lunulatus. *Linn.* ii. 684.—*Mart. C. pl.* 22. *f.* 15.—*Marsh.* i. 523.—*Turt.* ii. 520.—*Don.* xv. *pl.* 532. *f.* 1.
Ta. lunulatus. (*Sam. I.* 40.)

2789. 2; atricapillus*.
St. atricapillus. *Fabr. E.* ii. 599. (!)—*Stew.* ii. 89.—*Marsh.* i. 522.—*Turt.* ii. 513.
Ox. lunulatus. *Panz. F.* xxii. *f.* 15.

2790. 3; trimaculatus*.
St. trimaculatus. *Payk. F.* iii. 422.
Ta. trimaculatus. (*Sam. I.* 40.)
Ta. melanocephalus, var. 2. *Grav. Mi.* 144.

2791. 4; apicalis* *mihi.*

2792. 5; melanocephalus.
St. melanocephalus. *Marsh.* i. 523.

2793. 6; angularis* *mihi.*

2794. 7; ochraceus* *mihi.*

2795. 8; bimaculatus* *mihi.*

2796. 9, pygmæus.
Ox. pygmæus. *Panz. F.* xxvii. *f.* 19.
Ta. melanocephalus var. *Grav. Mi.* 144.

2797. 10; marginalis* *mihi.*

2798. 11; thoracicus*.
Ox. thoracicus. *Fabr. E.* ii. 606?

2799. 12; nitidulus*. *Leach MSS.*

2800. 13; brunnipennis* *mihi.*

2801. 14; discoideus* *mihi.*

2802. 15; biguttatus*.
Ta. biguttatus. *Kirby MSS.*

2803. 16; ruficollis* *mihi.*

2804. 17; piciceps* *mihi.*

2805 † 18. Simpsoni.
Ta. Simpsoni. *Spence? MSS.* In Mus. D. Simpson.

Genus 467. TACHYPORUS, *Gravenhorst, Leach, Samou.,* (*Kirby.*)

OXYPORUS *p, Fabr.*

A. Corpore elongato.

2806. 1, diffinis*. *Kirby MSS.*

2807. 2; testaceus*.
Ox. testaceus. *Fabr. E.* ii. 607.
Ta. abdominalis, b. *Gyll.* ii. 239.

2808. 3, flavicornis. *Kirby MSS.*
Ta. nitidulus. *Grav. Mi.* 126?

2809. [4, nitidus.]
St. nitidulus. *Oliv. E.* iii. 34. *pl.* 3. *f.* 28. *a, b.*

2810. 5, livens*. *Kirby MSS.*

2811. 6, flavescens. *Kirby MSS.?*

2812. 7; angustatus*. *Kirby MSS.*

2813. 8, minimus. *Leach MSS.*

2814. 9; basalis* *mihi.*

2815. 10; angustatus* *mihi.*

2816. 11; libens*. *Kirby MSS.*

2817. 12; brunneus* *mihi.*

2818. 13; thoracicus* *mihi.*

2819. 14; pyrrhoceras*. *Kirby MSS.*

2820. 15; ruficollis*.
St. dispar. ♂. *Payk. F.* iii. 424.
Ta. flavicollis. *Kirby MSS.*

2821. 16; nigripennis* *mihi.*

2822. 17, pusillus.
Ta. pusillus. *Grav. Mo.* 9.

2823. 18, atriceps* *mihi.*

2824. 19; pyrrhopterus*. *Kirby MSS.*

B. Corpore brevi. a. *Antennis simplicibus.*

2825. 20; chrysomelinus*. *Grav. Mi.* 128.—*Leach E. E.* ix. 92. —*Samou.* 176.
St. chrysomelinus. *Linn.* ii. 685.—*Mart. C. pl.* 42. *f.* 37.—*Berk. S.* i. 110.—*Stew.* ii. 88.—*Marsh.* i. 519.—*Turt.* ii. 521.
Ox. chrysomelinus. *Panz. F.* ix. *f.* 14.
St. dispar α. *Payk. F.* iii. 423.

2826. [21; merdarius*.]
Ox. merdarius. *Panz. F.* xxvii. *f.* 18.
St. merdarius. *Marsh.* i. 521.—*Turt.* ii. 521.
St. chrysomelinus. *Oliv. E.* iii. 33. *pl.* 3. *f.* 22. *a, b.*

2827. 22, marginellus*.
St. marginellus. *Mus. Marsham.*
Ta. nigricornis. *Gyll.* iv. 469?

2828. 23; obtusus*.
St. obtusus. *Linn.* ii. 684.—*Stew.* ii. 88.—*Marsh.* i. 520.
St. dispar ε. *Payk. F.* iii. 424.

2829. [24; analis*.] *Grav. Mi.* 129.—(*Sam. I.* 40.)
Ox. analis. *Fabr. E.* ii. 606.—*Panz. F.* xxii. *f.* 16.

2830. [25, melanurus.]
St. melanurus. *Marsh.* i. 525.

2831. 26, collaris* *mihi.*

2832. 27; testaceus* *mihi.*

2833. 28, marginatus. *Grav. Mi.* 127.—(*Sam. I.* 40.)
St. dispar β. *Payk. F.* iii. 424.

2834. [29; nitidulus*.] (*Sam. I.* 40.)
St. nitidulus. *Oliv. E.* iii. 34. *pl.* 3. *f.* 28. *a, b.*—*Stew.* ii. 88. —*Marsh.* i. 520.

2835. [30, hypnorum.]
St. hypnorum. *Fabr. E.* ii. 607. (!)—*Stew.* ii. 88.—*Marsh.* i. 525.—*Turt.* ii. 522.

2836. 31; nitidicollis* *mihi.*

2837. 32, erythropterus* *mihi.*

2838. 33, lateralis. *Kirby MSS.*

2839. 34, putridus* *mihi.*

2840. 35, abdominalis. *Grav. Mi.* 127.
Ox. abdominalis. *Fabr. E.* ii. 607?

2841. 36, apicalis* *mihi.*

2842. 37, dimidiatus*. *Kirby MSS.*
Ox. dimidiatus. *Fabr. E.* ii. 607.

2843. 38; macropterus* *mihi.*

b. *Antennis nodosis, verticillato-pilosis.*

2844. 39, nodicornis. *Kirby MSS.*

Genus 468. CYPHA, *Kirby MSS,* (*Samou.*)

TACHYPORUS I. *Gravenhorst.*—SCAPHIDIUM *p, et* SILPHA *p, Marsham.*—HYPOCYPHTUS *p, Schüppel?*

2845. 1, Agaricina.
Si. Agaricina. *Linn.* ii. 570?—*Marsh.* i. 129.

2846. 2; rufipes*. *Kirby MSS.*
Ta. granulum. *Grav. Mo.* 3.
Ta. granum. (*Sam. I.* 40.)
Sc. acuminatum. *Marsh.* i. 234.

2847. 3; nigripes* *mihi.*

2848. 4; parvula* *mihi.*

2849. 5; Anisotomoides* *mihi.*

Genus 469. CONURUS *mihi.*

TACHYPORUS *p*, *Grav.*—OXYPORUS *p*, *Fabr.*

2850. 1; pubescens*.
Ta. pubescens. *Grav. Mi.* 130.—(*Sam. I.* 40.)
St. cellaris β. *Payk. F.* iii. 421.

2851. [2, Marshami] *mihi.*
St. conicus. *Marsh.* i. 522.

2852. 3; cellaris*.
Ox. cellaris. *Fabr. E.* ii. 605.
St. littoreus. *Linn.* ii. 685?

2853. 4; bipustulatus*.
Ox. bipustulatus. *Fabr. E.* ii. 606.—*Panz. F.* xvi. *f.* 21.
Ta. bimaculatus. *Grav. Mo.* 5.

2854. 5; immaculatus*.
Ta. bipunctatus. *Grav. Mi.* 133.
Ta. immaculatus. *Kirby MSS.*

2855. 6; pusillus*.
Ta. pusillus. *Kirby MSS.*

2856. 7, obscuratus.
Ta. obscuratus. *Kirby MSS.*

2857. 8, obscuripennis.
Ta. obscuripennis. *Kirby MSS.*

2858. 9; flavus*.
Ta. flavus. *Kirby MSS.*
Ta. truncatellus. *Grav. Mo.* 5?

2859. 10, flavipes* *mihi.*

2860. 11, melanopterus* *mihi.*

Genus 470. TACHINUS, *Grav. Mi., Leach,* (*Kirby.*)
TACHYPORUS *p, Grav. Mo.*—OXYPORUS *p, Fabr.*

A. Pubescentes.

2861. 1, fimetarius. *Grav. Mi.* 141.

2862. 2, castaneus*. *Grav. Mi.* 140.
St. fuscipes. *Panz. F.* xxvii. *f.* 12?
Ta. fimetarius b. *Gyll.* ii. 263.

2863. 3, angularis* *mihi.*

2864. 4; Sowerbii*. *Kirby MSS.*

B. Glabri. a. *Corpore lato, vix elongate.*

2865. 5; collaris*. *Grav. Mi.* 143.

2866. [6; corticinus*.] *Grav. Mi.* 141.

2867. 7; Silphoides*.
Sta. silphoides. *Linn.* ii. 684.—*Marsh.* i. 528.
Sta. dispar ε. *Payk. F.* iii. 423.
β; Ox. suturalis. *Panz. F.* xviii. *f.* 20.

2868. 8; lævigatus*.
St. lævigatus. *Marsh.* i. 519.

2869. [9; marginellus*.] *Grav. Mi.* 143.—(*Sam. I.* 40.)
St. rufipes ε. *Payk. F.* iii. 419.—*Turt.* ii. 522.
St. marginatus. *Don.* xv. 60. *pl.* 532. *f.* 2.

2870. 10, laticollis*. *Grav. Mi.* 141.

2871. 11, brunnipennis* *mihi.*

2872. 12, nigripes. *Kirby MSS.*
St. rufipes. *Oliv. E.* iii. 32. *pl.* 4. *f.* 35. *a, b?*

2873. 13, apicalis* *mihi.*

2874. 14, fulvipes* *mihi.*

2875. 15; pullus*. *Grav. Mi.* 140.
St. rufipes. *DeGeer.* iv. 24.—*Oliv. E.* iii. 32. *pl.* 4. *f.* 35. *c, d.*

2876. 16; rufipes*. *Grav. Mi.* 137.—*Leach E. E.* ix. 92.—*Samou.* 176.
St. rufipes. *Linn.* ii. 685.—*Panz. F.* xxvii. *f.* 20.—*Marsh.* i. 518.—*Turt.* ii. 522.

2877. 17; limbosus*. *Kirby MSS.*

2878. 18; intermedius*. *Kirby MSS.*
St. rufipes var. 4, 5. *Grav. Mi.* 137?

2879. 19; latus*.
St. latus. *Marsh.* i. 524.
Ox. marginatus. *Fabr. E.* ii. 605?

2880. 20; cinctus*.
St. cinctus. *Marsh.* i. 519.
St. rufipes β. *Payk. F.* iii. 418.
Ta. humeralis var. 2, 3. *Grav. Mi.* 130.

2881. 21, scapularis* *mihi.*

2882. 22; bipustulatus*. *Grav. Mi.* 135.
St. subterraneus var. *Payk. F.* iii. 420.
St. bipustulatus. *Don.* xv. 61. *pl.* 532. *f.* 3.

2883. 23; subterraneus*. *Grav. Mi.* 135.—(*Sam. I.* 40.)
St. subterraneus. *Linn.* ii. 684.—*Marsh.* i. 518.

2884. 24, pallens*. *Gyll.* ii. 259?
St. rufipes var. d. *Payk. F.* iii. 419?
b. *Corpore valdè elongato.*

2885. 25, aterrimus *mihi.*

2886. 26, elongatus*. *Gyll.* ii. 251.

Familia LXVIII. STAPHYLINIDÆ, *Leach.*

(ST.-FISSILABRES, *Latreille.*)

(STAPHYLINUS *p*, *Linné*, *Marsh.*, *&c.*)

Genus 471. VELLEIUS, *Leach*, (*Samou.*)

2887. 1, dilatatus*? *Leach MSS.*—(*Samou.* 172.)
St. dilatatus. *Payk. F.* iii. 389.
St. concolor. *Marsh.* i. 498.—*Sower. B. M.* i. *pl.* 54.

Genus 472. CREOPHILUS, (*Kirby.*)

2888. 1; maxillosus*. *Kirby MSS.*—(*Samou.* 172.)—(*Kirby & Sp. I. E.* iii. 433.)
St. maxillosus. *Linn.* ii. 683.—*Berk. S.* i. 109.—*Don.* iii. *pl.* 96. *f.* 3.—*Stew.* ii. 88.—*Marsh.* i. 498.—*Turt.* ii. 510.—(*Sam. I.* 39.)
St. balteatus. *DeGeer.* iv. 18. *pl.* 1. *f.* 7—10.

2889 † 2, ciliaris. *Leach MSS.* In Mus. *Brit.*

Genus 473. EMUS, *Leach*, (*Samou.*)

(CREOPHILUS *p*, *Kirby.*)

2890. 1, hirtus*. (*Samou.* 172.)—(*Ing. Inst.* 88.)
St. hirtus. *Linn.* ii. 683.—*Panz. F.* iv. *f.* 19.—*Marsh.* i. 496.—*Turt.* ii. 509.—*Don.* xvi. *pl.* 552.
St. Bombylius. *DeGeer.* iv. 20.

Genus 474. STAPHYLINUS *Auctorum*.

A. Corpore toto lanugine tecto.

2891. 1; nebulosus*.
St. nebulosus. *Fabr. E.* ii. 590.
St. murinus. *Mus. Linn.*
St. hybridus. *Marsh.* i. 500.—(*Sam. I.* 39.)—*Don.* xvi. *pl.* 563.
St. murinus. *Oliv.* iii. 15. *pl.* 6. *f.* 51. *b.*—*Panz. F.* lxvi. *f.* 16.

2892. 2; murinus*. *Linn.* ii. 683?—*Mart. C. pl.* 41. *f.* 22.—*Berk. S.* i. 109.—*Stew.* ii. 87.—*Oliv. E.* iii. 15. *pl.* 6. *f.* 51. *a.*—*Marsh.* i. 499.—*Turt.* ii. 510.—(*Sam. I.* 39.)

2893. 3; pubescens*. *Fabr. E.* ii. 590.—*Marsh.* i. 500.—(*Sam. I.* 39.)
β; St. fulviceps. *Kirby MSS.*

B. Corpore glabro. a. *Elytris rufis.*

2894. 4; erythropterus*. *Linn.* ii. 683.—*Don.* ix. *pl.* 308.—*Stew.* ii. 88.—*Marsh.* i. 499.—*Turt.* ii. 510.—*Leach E. E.* ix. 91.—*Samou.* 171. *pl.* 4. *f.* 10.—*Wood.* i. 78. *pl.* 29.

2895. 5, castanopterus. *Grav. Mi.* 10.—(*Sam. I.* 39.)
St. erythropterus var. minor. *Oliv. E.* iii. 12. *pl.* 11. *f.* 14.

2896. 6; stercorarius*. *Oliv. E.* iii. 18. *pl.* 3. *f.* 23.—*Marsh.* i. 499.—(*Sam. I.* 39.)
St. brunnicornis. *Kirby MSS.*

2897. [7; æriceps*.] *Kirby MSS.*

2898. [† 8, chalcocephalus.] *Fabr. E.* ii. 593. In Mus. Brit.?

b. *Elytris concoloribus.*

2899. 9; æneocephalus*. *Payk. F.* iii. 374.—(*Sam. I.* 39.)
St. cupreus. *Oliv. E.* iii. 16. *pl.* 2 *f.* 16.
St. sericeus. *Marsh.* i. 508.
β; St. leucophthalmus. *Marsh.* i. 511.

2900. 10; compressus*. *Marsh.* i. 503.

2901. 11; cantianus*.
Oc. cantianus. *Kirby MSS.*

Genus 475. GOËRIUS, *Leach.*

Ocypus *p*, *Kirby.*

2902. 1; olens*. *Devil's Coach-horse.*
St. olens. *Fabr. E.* ii. 591.—*Mart. C. pl.* 41. *f.* 21.—*Stew.* ii. 87.—*Panz. F.* xxvii. *f.* 1.—*Marsh.* i. 497.—*Turt.* ii. 510.—(*Sam. I.* 39.)—(*Kirby & Sp. I. E.* ii. 236. & iii. 33.)
St. major. *DeGeer.* iv. 17.—*Shaw G. Z.* vi. 108. *pl.* 39.

2903 † 2, macrocephalus. *Leach MSS.* In Mus. *Brit.*

2904. 3; cyaneus*.
St. cyaneus. *Fabr. E.* ii. 592.—*Panz. F.* xxvii. *f.* 3.—*Marsh.* i. 501.—(*Kirby & Sp. I. E.* i. *pl.* 1. *f.* 2.)

2905. 4; punctulatus*.
St. punctulatus. *Mart. C. pl.* 42. *f.* 27?—*Marsh.* i. 501.—(*Sam. I.* 39.)
St. subpunctatus. *Gyll.* iv. 474.

2906. [5; Morio*.]
St. Morio. *Grav. Mi.* 6.—(*Sam. I.* 39.)
St. politus. *DeGeer.* iv. 22. In Mus. *Brit.*

2907 † 6, Kirbii. *Leach MSS.*

2908. 7; fuscatus*.
St. fuscatus. *Grav. Mi.* 164?—*Gyll.* iv. 474. (!)
St. obscurus. *Marsh.* i. 514.
St. similis. *Payk. F.* iii. 371?
Oc. obscuratus. *Kirby MSS.*

2909 † 8; erythropus*.
St. erythropus. *Payk. F.* iii. 372.
St. tricolor. *Grav. Mi.* 7? In Mus. *Brit. et* D. *Ingpen.*

Genus 476. OCYPUS, (*Kirby.*)

2910. 1; similis*.
St. similis. *Oliv.* iii. 13. *pl.* 5. *f.* 42.—(*Sam. I.* 39.)

2911. 2; confinis*.
St. confinis. *Kirby MSS.*

2912. 3; picipes* *mihi.*

2913. 4; angustatus*. *Kirby MSS.*

2914. 5; brunnipes*.
St. brunnipes. *Oliv. E.* iii. 13. *pl.* 1. *f.* 7. *a, b.*—*Mart. C. pl.* 41. *f.* 15.—*Stew.* ii. 88.—*Marsh.* i. 502.—*Turt.* ii. 511.—(*Sam. I.* 39.)

2915. 6; phæopus* *mihi.*

Genus 477. TASGIUS, *Leach.*

ASTRAPÆUS *p, Latr.*

2916. 1; rufipes*.
As. rufipes. *Latr. G.* i. 285.

Genus 478. ASTRAPÆUS, *Gravenhorst, Leach.*

2917 † 1; rufipennis*. *Leach MSS.*
St. Ulmi. *Oliv. E.* iii. 42. *pl.* 4. *f.* 37?
St. Ulmineus. *Fabr. E.* ii. 595? In Mus. *Brit.*

Genus 479. QUEDIUS, *Leach.*

A. Oculis magnis.

2918. 1; tristis*.
St. tristis. *Grav. Mi.* 34.—(*Sam. I.* 39.)
St. picipennis, var. *Payk. F.* iii. 373.
St. dilatatus. *Marsh.* i. 504.
St. fulvicornis. *Kirby MSS.*

2919. [2; picicornis*]
St. picicornis. *Kirby MSS.*

2920. 3, gracilis *mihi.*

2921. 4; pyrrhopus.
St. pyrrhopus. *Kirby MSS.*

2922. 5; picipennis*.
St. picipennis. *Payk. F.* iii. 373.—(*Sam. I.* 39.)
St. molochinus. *Grav. Mo.* 46.
β; St. ruficornis. *Grav. Mo.* 50.

2923. [6, denudatus*.]
St. denudatus. *Kirby.*

B. Oculis mediocribus: thoracis lateribus haud sinuatis.

a. *Thorace brevi, tripunctato.*

2924. 7; hæmorrhous.
St. fulgidus. *Marsh.* i. 503.
St. hæmorrhous. *Kirby MSS.*—(*Sam. I.* 39.)

2925. 8; hæmopterus.
St. hæmopterus. *Kirby MSS.*

2926. 9; variabilis*.
St. nitidus. *Grav. Mi.* 31.
St. mesomelinus. *Marsh.* i. 510.
St. variabilis. *Gyll.* ii. 303.

2927 † 10, atriceps.
St. atriceps. *Kirby MSS.*
β; St. nitidus var. 8. *Grav. Mi.* 32. In Mus. *D. Kirby.*

2928. 11; Lathburii*.
St. Lathburii. *Kirby MSS.*
St. flavopterus. *Oliv.* iii. 27. *pl.* 4. *f.* 40. *a, b?*

2929. 12; hæmorrhoidalis*. *Leach MSS.*
St. fulgidus. *Mus. Marsham.*

2930. 13; fulvipes*.
St. fulvipes. *Fabr. E.* ii. 597.
St. nitidus β. *Payk.* iii. 390.
St. lævigatus. *Gyll.* ii. 306.
St. impressus var. *Grav. Mo.* 40.—*Panz. F.* xxxvii. *f.* 9?

2931. 14; impressus*.
St. impressus. *Grav. Mi.* 35.—*Panz. F.* xxxvi. *f.* 21.
St. cinctus. *Payk. F.* iii. 395.
St. marginellus. *Stew.* ii. 89.—*Marsh.* i. 505.—(*Sam. I.* 39.)

2932. 15; sericopterus* *mihi.*

2933. 16; rufitarsis.
St. rufitarsis. *Marsh.* i. 512.
St. variabilis e. *Gyll.* ii. 304.

2934 † 17, seminitidus.
St. seminitidus. *Kirby MSS.* In Mus. D. *Kirby.*

2935. 18; humeralis*.
St. humeralis. *Kirby MSS.*

b. *Thorace seriebus dorsalibus quatuor punctatis.*

2936. 19; picatus*.
St. picatus. *Kirby MSS.*
St. bimaculatus. *Grav. Mi.* 38?

2937. 20; picipes*.
St. picipes. *Kirby MSS.*

2938. 21; unicolor*.
St. unicolor. *Kirby MSS.*

2939. 22, hirtipennis.
St. hirtipennis. *Kirby MSS.*

2940 † 23, sericans.
St. sericans. *Kirby MSS.* In Mus. D. *Kirby.*

2941 † 24, fuscipennis.
St. fuscipennis. *Kirby MSS.* In Mus. *Brit.*? D. *Kirby et* Spence.

2942. 25; fulvipes* *mihi.*

2943. 26; suturalis*.
St. suturalis. *Marsh.* i. 509.
St. discoideus. *Grav. Mi.* 38?
St. flavescens. *Fabr. E.* ii. 597?

2944. 27, nigricornis* *mihi.*

2945 † 28, nitescens *mihi.* In Mus. *Brit.*

2946. 29, caliginosus* *mihi.*

2947. 30, picicornis* *mihi.*

c. *Thoracis seriebus dorsalibus quinque punctatis.*

2948. 31; lepidulus*.
St. lepidulus. *Kirby MSS.*
St. lepidus. *Grav. Mi.* 31?

2949 † 32, inquinatus.
St. inquinatus. *Kirby MSS.*
In Mus. *Brit.* D. *Kirby, T. Skrimshire, et* Spence.

2950. 33; picicollis*.
St. picicollis. *Kirby MSS.*

2951. 34, castanopterus* *mihi.*

d. *Thorace haud punctato.*

2952. 35, Skrimshiranus *mihi.*

C. Oculis mediis: thoracis lateribus sinuatis.

2953. 36; nitidus*.
St. nitidus. *Fabr. E.* ii. 596.
St. cænosus. *Grav. Mo.* 51.
St. ferrugineus. *Rossi F.* i. 248.

2954. 37, erythropterus* *mihi.*

D. (An. PHILONTHUS?)

2955 † 38, cyanipennis.
St. cyanipennis. *Fabr. E. S.* i. 597.
St. amœnus. *Oliv. E.* iii. 42. *pl.* 4. *f.* 36? In Mus. *Brit.*

Genus 480. PHILONTHUS, *Leach MSS.*

A. Thoracis disco impunctato.

a. *Thoracis lateribus rotundatis.*

2956. 1; laminatus*.
St. laminatus. *Creutz. E. V.* i. 128. *pl.* 3. *f.* 31. *a.*—(*Sam. I.* 39.)

2957. [2; æneus*.]
St. æneus. *DeGeer.* iv. 23.—*Marsh.* i. 511.

2958. [3; chalceus*] *mihi.*

b. *Thoracis lateribus subsinuatis.*

2959. 4; splendens*.
St. splendens. *Fabr. E.* i. 594.—*Ent. Trans.* (*Burrell.*) i. 236.
—(*Sam. I.* 39.)

2960. 5, æratus*.
St. æratus. *Kirby MSS.*

B. Thoracis seriebus dorsalibus 4-punctatis.

a. *Chalcopteri.*

2961. 6; puncticollis*.
St. puncticollis. *Kirby MSS.*
St. æneus. *Grav. Mi.* 17.
St. similis. *Marsh.* i. 497.
St. politus. *Mus. Linné.*

2962. 7; politus*.
St. politus. *Linn.* ii. 683.—*Oliv. E.* iii. 25. *pl.* 2. *f.* 10. ♂.—
Stew. ii. 88.—*Marsh.* i. 501.—*Turt.* ii. 511.—*Leach E. E.*
ix. 91.—(*Sam. I.* 39.)

2963. 8; cognatus* *mihi*.

2964. 9; microcephalus* *mihi*.

2965. 10; melanopterus*.
St. melanopterus. *Wilk.? MSS.*

2966. 11; nigripennis*.
St. nigripennis. *Kirby MSS.*

2967. [12; maculicornis*.]
St. maculicornis. *Kirby MSS.*—(*Sam. I.* 39.)

2968. 13; decorus*.
St. decorus. *Grav. Mi.* 19.—(*Sam. I.* 39.)

2969. 14; sericeus* *mihi*.

2970. 15; pilipes*.
St. pilipes. *Kirby MSS.*—(*Sam. I.* 39.)

2971. [16; chalcopterus.]
St. chalcopterus. *Mus. Marsham.*
St. varians. *Grav. Mi.* 20?
St. varius. *Gyll.* ii. 321?
St. æratus. *Kirby MSS.*

2972. 17; nigro-æneus *mihi*.

2973. 18; atratus*.
St. atratus. *Grav. Mi.* 21.

2974. 19; fimetarius*.
St. fimetarius. *Grav. Mi.* 175.—(*Sam. I* 39.)

2975. 20; æripennis*.
St. æripennis. *Kirby MSS.*
St. lucidus. *Grav. Mi.* 21?

2976. 21, nigripes* *mihi*.

2977. 22, fuscipes* *mihi*.

b. *Melanopteri*.

2978. 23; marginatus*.
St. marginatus. *Fabr. E.* ii. 597.—*Oliv. E.* iii. 26. *pl.* 3. *f.* 29. *a, b.*—*Marsh.* i. 512.—*Turt.* ii. 512.—(*Sam. I.* 39.)

C. Thoracis seriebus dorsalibus quinque punctatis.

a. *Chalcopteri*.

2979. 24; concinnus*.
St. concinnus. *Grav. Mi.* 21.—(*Sam. I.* 39.)

2980. 25; obscurus*.
St. obscurus. *Grav. Mi.* 174.
St. cognatus. *Kirby MSS.*

2981. 26; varians*.
St. varians. *Oliv. E.* iii. 27. *pl.* 5. *f.* 46.—*Grav. Mo.* 82.—(*Sam. I.* 39.)

2982. 27; simplex*.
St. simplex. *Marsh.* i. 505.

b. *Melanopteri.* 1. Elytris immaculatis.

2983 † 28, punctiventris.
St. punctiventris. *Kirby MSS.* In Mus. D. *Kirby.*

2984. 29; intaminatus*.
St. intaminatus. *Kirby MSS.*
St. varians a. *Payk.* iii. 393?

2985. 30, aterrimus.
St. aterrimus. *Marsh.* i. 513.
St. opacus. *Grav. Mi.* 26?

2986. [31, opacus.]
St. opacus. *Grav. Mo.* 64.

2987. 32, phæopus.
St. phæopus. *Kirby MSS.*

2988 † 33, nitens.
St. nitens. *Grav. Mi.* 26? In Mus. D. Sheppard.

2989. 34; obscuripénnis*.
St. obscuripennis. *Kirby MSS.*—(*Sam. I.* 39.)

2990. 35; longicornis*.
St. longicornis. *Kirby MSS.*

2991. 36, agilis.
St. agilis. *Grav. Mo.* 77.—*Gyll.* ii. 341?
St. debilis. *Dahl.*—(*DeJean C.* 23.)

2. Elytris maculatis.

2992. 37; lituratus*.
St. lituratus. *Kirby MSS.*—(*Sam. I.* 39.)
St. opacus a. *Gyll.* ii. 340?

2993. 38; bipustulatus*.
St. bipustulatus. *Fabr. E.* ii. 598.—*Panz. F.* xxvii. *f.* 10.—*Marsh.* i. 525.—*Turt.* ii. 512.—(*Sam. I.* 39.)
St. varians b. *Payk.* iii. 393. var.

2994. 39, aciculatus* *mihi.*

2995. 40; sanguinolentus*.
St. sanguinolentus. *Grav. Mi.* 36.—(*Sam. I.* 39.)
St. trilineatus. *Kirby MSS.*

2996. 41; bimaculatus*.
St. bimaculatus. *Grav. Mi.* 38.

c. *Erythropteri.*

2997. 42; rubripennis.
St. rubripennis. *Kirby MSS.*
St. fulvipes. *Grav. Mi.* 24?

2998. 43, corruscus.
St. corruscus. *Grav. Mi.* 33.
St. nitidus. *Marsh.* i. 511.

D. Thoracis seriebus dorsalibus sex-punctatis.

2999. 44; micans*.
St. micans. *Grav. Mi.* 25.
St. varians, c. *Payk. F.* iii. 393?

E. Thoracis seriebus dorsalibus octo-punctatis.

3000 † 45, Watsoni.
St. Watsoni. *Kirby MSS.* In Mus. *D. Kirby et* Watson.

F. Thoracis seriebus dorsalibus multi-punctatis.

3001 † 46, punctus.
St. punctus. *Grav. Mi.* 20.
St. politus. *Panz. F.* xxvii. *f.* 7. In Mus. *Brit.*? *et* D. Spence.

3002 [† 47, minax.]
St. minax. *Kirby MSS.*
St. punctatus. *Latr. M.* ix. 309? In Mus. *D. Kirby et* Spence.

Genus 481. RAPHIRUS, *Leach MSS.*

3003. 1; semiobscurus*.
St. semiobscurus. *Marsh.* i. 512.—(*Sam. I.* 39.)
St. rufipes. *Grav. Mi.* 171?
St. mauro-rufus. *Gyll.* ii. 309.

3004. 2; attenuatus*.
St. attenuatus. *Grav. Mi.* 27.—(*Sam. I.* 39.)
St. oculatus. *Kirby MSS.*
St. Boops. *Grav. Mi.* 21?

3005. 3; nitipennis*. *Leach MSS.*
St. nitipennis. (*Sam. I.* 39.)

3006. 4, rufipennis.
St. rufipennis. *Kirby MSS.*

3007. 5; fuscipes*.
St. fuscipes. *Kirby MSS.*—(*Kirby & Sp. I. E.* iv. 141.)

3008. 6, semiæneus.
St. semiæneus. *Kirby MSS.*

3009 † 7, ruficollis.
St. ruficollis. *Kirby MSS.* In Mus. *Brit. D. Kirby et* Spence.

3010 † 8, fumatus.
St. fumatus. *Kirby MSS.*
St. nitidus var. 7. *Grav. Mi.* 32? In Mus. *D. Kirby.*

3011. 9; præcox*.
St. præcox. *Grav. Mi.* 172.

Genus 482. CAFIUS, *Leach MSS.*

3012. 1, fucicola. *Leach MSS.*
St. fucicola. (*Sam. I.* 39.)

3013. 2, xantholoma.
St. xantholoma. *Grav. Mo.* 41.

3014. 3, lateralis.
St. lateralis. *Kirby MSS.*—(*Sam. I.* 39.)

3015. [4, littoralis] *mihi.*

3016. [5, tessellatus] *mihi.*

Genus 483. BISNIUS, *Leach MSS.*

3017. 1, cephalotes.
St. cephalotes. *Grav. Mi.* 22.

3018. 2, rotundiceps.
St. rotundiceps. *Kirby MSS.*

3019 † 3, simplex. *Leach? MSS.* In Mus. *Brit.*

3020. 4, fuscicornis.
St. fuscicornis. *Kirby MSS.*
St. debilis. *Grav. Mi.* 35?

3021. 5; fulvipes* *mihi.*

Genus 484. GABRIUS, *Leach MSS.*

3022. 1; suaveolens*.
St. suaveolens. *Kirby MSS.*—(*Kirby & Sp. I. E.* iv. 141.)
St. varians, γ. *Payk. F.* iii. 394?

3023. 2, pygmæus.
St. pygmæus. *Kirby MSS.*

3024. 3, phæopus* *mihi.*

3025. 4, aterrimus.
St. aterrimus. *Grav. Mi.* 41.
St. diffinis. *Kirby MSS.*

3026. 5, pallipes.
St. pallipes. *Kirby MSS.*
St. nigritulus. *Grav. Mi.* 41?

3027. 6, basalis* *mihi.*

3028. 7, nanus*.
St. nanus. *Grav. Mo.* 96.

3029. 8; attenuatus*. *Leach MSS.*

3030. 9, albipes* *mihi.*
St. splendidulus. *Grav. Mi.* 41?

3031 † 10, ventralis.
St. ventralis. *Kirby MSS.* In Mus. D. *Kirby.*

3032 † 11, cinerascens.
St. cinerascens. *Grav. Mi.* 49. In Mus. D. *Kirby.*

3033 † 12, semipunctatus.
St. semipunctatus. *Kirby MSS.* In Mus. D. *Kirby.*

3034 † 13, villosulus.
St. villosulus. *Kirby MSS.* In Mus. D. *Kirby.*

Genus 485. OTHIUS, *Leach MSS.*

GYROHYPNUS *p*, *Kirby.*—XANTHOLINUS *p*, *Dahl.*—PÆDERUS *p*, *Panz.*

A. Capite ovato.

3035. 1; fulgidus*.
St. fulgidus. *Payk. F.* iii. 377.
St. fulminans. *Grav. Mi.* 47.
Pæ. fulvipennis. *Panz. F.* xxvii. *f.* 23?

3036. 2, alternans.
St. alternans. *Grav. Mi.* 48.
St. pilicornis β, γ. *Payk. F.* iii. 379.

3037. 3, pilicornis.
St. pilicornis α. *Payk. F.* iii. 379.
St. palmula. *Grav. Mi.* 48.
Gy. Lathburii. *Kirby MSS.*

3038. 4, glabricornis.
St. glabricornis. *Kirby MSS.*

3039. 5, angustus.
Gy. angustus. *Kirby MSS.*
St. melanocephalus. *Grav. Mo.* 107?

3040. 6; ustulatus*.
St. ustulatus. *Grav. Mi.* 46.

3041. 7, læviusculus*.
Gy. læviusculus. *Kirby MSS.*

3042 † 8. obscurus. *Leach MSS.* In Mus. Brit.?

B. Capite orbiculato.

3043. 9, subiliformis*.
St. subiliformis. *Grav. Mi.* 29.
St. picicolor. *Kirby MSS.*

3044 † 10. Scoticus.
St. Scoticus. *Kirby MSS.* In Mus. D. MacLeay.

Genus 486. HETEROTHOPS, *Kirby MSS.*

3045. 1, binotatus. *Kirby MSS.*
St. binotatus. *Grav. Mi.* 28.
St. subiliformis, β. *Gyll.* ii. 312.

3046. 2, Holmensis. *Kirby MSS.*

3047. 3, Kirbiellus *mihi.*

Genus 487. GYROHYPNUS, (*Kirby.*)

XANTHOLINUS *p, Dahl.*—PÆDERUS *p, Fabr.*

A. Elytris lævioribus.

3048. 1; pyropterus *.
St. pyropterus. *Grav. Mo.* 102.
Gy. puncticeps. *Kirby MSS.*
Pæ. fulgidus. *Fabr. E. S.* 537?

3049. 2, rotundicollis. *Kirby MSS.*

3050 † 3, longicollis. *Kirby MSS.* In Mus. *D. Kirby.*

3051. 4; diaphanus *. *Kirby MSS.*
St. diaphanus. *Marsh.* i. 514.

3052. 5; sulcifrons *. *Kirby MSS.*

3053. 6, apicalis *mihi.*

B. Elytris punctatis.

3054. 7; cruentatus *. *Kirby MSS.*
St. cruentatus. *Marsh.* i. 516.
St. glabratus. *Grav. Mi.* 178.

3055 † 8, quadratus. *Kirby MSS.* In Mus. *D. Kirby.*

3056. 9; tricolor *.
St. tricolor. *Payk. F.* iii. 378.
St. elegans. *Oliv. E.* iii. 19. *pl.* 5. *f.* 50. *a, b.*

3057. [10, affinis.]
St. affinis. *Marsh.* i. 517.
St. tricolor β. *Payk. F.* iii. 378.
St. lentus. *Grav. Mo.* 101.

3058. 11; linearis *.
St. linearis. *Marsh.* i. 516.
St. punctulatus. *Grav. Mi.* 177.

3059. 12; longiceps *.
St. longiceps. *Grav. Mi.* 177.
St. bicolor. *Payk. F.* iii. 381.
St. linearis. *Oliv. E.* iii. 19. *pl.* 4. *f.* 38. *a, b.*
St. varians δ. *Payk. M. S.* 46.

3060. 13; semistriatus *. *Kirby MSS.*

3061. 14; punctulatus *.
St. punctulatus. *Payk. F.* iii. 380.
St. elongatus. *Grav. Mi.* 45.

3062. 15; angustatus *. *Kirby MSS.*
St. elongatus β. *Grav. Mi.* 45.

3063. 16; ochraceus*.
St. ochraceus. *Grav. Mi.* 43.
St. punctulatus β. *Payk. F.* iii. 381.
Xa. Batychrus? *Knoch?*
Gy. proximus. *Kirby MSS.*

3064 † 17, quadrisulcus. *Kirby MSS.*
St. ochraceus var. 1. *Grav. Mi.* 43? In Mus. D. *Kirby.*

3065. 18; pusillus* *mihi.*

Genus 488. ACHENIUM, *Leach MSS.*, (*Samou.*), *Curtis.*

LATHROBIUM *p*, *Grav.*

3066. 1: depressum*.
La. depressum. *Grav. Mi.* 182.
La. depressior. *Kirby MSS.*
β, Ac. trinotatum *mihi* (*olim.*)

3067. [2, angustatum*] *mihi.*
Ac. depressum. *Curtis.* iii. *pl.* 115.

Genus 489. LATHROBIUM, *Gravenhorst*, *Leach*, *Samou.*

PÆDERUS *p*, *Fabr.*

3068. 1; quadratum*. *Gyll.* ii. 367.—(*Sam. I.* 24.)
St. quadratus. *Payk. M. S.* 29.—*Marsh.* i. 527.
Pæ. filiformis a. *Payk. F.* iii. 429.
La. pilosum. *Grav. Mi.* 131.

3069. 2; brunnipes*. *Grav. Mi.* 56.
Pæ. brunnipes. *Fabr. E. S.* ii. 537.
Pæ. elongatus c. *Payk. F.* iii. 429.
St. dentatus. *Marsh.* i. 515.
La. dentatum. (*Sam. I.* 24.)

3070. 3; atriceps*. *Kirby MSS.*

3071. 4; elongatum*. *Grav. Mi.* 55.
St. elongatus. *Linn.* ii. 685.—*Mart. C. pl.* 40. *f.* 1.—*Marsh.* i. 515.—*Turt.* ii. 522.—*Don.* xvi. *pl.* 573. *f.* 3.—*Leach E. E.* ix. 91.—*Samou.* 172.
Pæ. elongatus. *Panz. F.* ix. *f.* 12.

3072. 5; fulvipenne*. *Gyll.* ii. 365.
Pæ. fulvipennis. *Fabr. E.* ii. 609.—*Panz. F.* xvii. *f.* 23.
La. elongatum γ. *Grav. Mo.* 133.

3073. 6; rufipenne*. *Gyll.* iii. 704.

3074. 7, ochraceum.
St. ochraceus. *Mus. Marsham.*

3075. 8, punctato-striatum*. *Kirby MSS.*
La. striatum. *Latr. H.* ix. 341?
Pæ. testaceus var. *Oliv. E.* iii. 5. *pl.* 1. *f.* 6. *a, b.*

3076. 9, multipunctatum. *Grav. Mi.* 52.

3077. 10; lineare*. *Grav. Mi.* 54.

3078. 11; longulum*. *Grav. Mi.* 53.
La. dentipes. *Kirby MSS.*
La. minutum. (*DeJean C.* 24.)

3079. 12, fovulum* *mihi.*

3080. 13; nanum* *mihi.*

3081. 14; erythrocephalum* *mihi.*

Genus 490. OCHTHEPHILUM *mihi.*

LATHROBIUM *p, Gravenhorst.*—PÆDERUS *p, Fabricius.*

3082. 1; fracticorne*.
Pæ. fracticornis. *Payk. F.* iii. 430.
Pæ. filiformis. *Fabr. E.* ii. 609.

Familia LXIX. STENIDÆ, *MacLeay?*

(ST.-LONGIPALPATI *et* -DEPLANATI *p, Latreille.*)

(PÆDERUS *p, Fabricius.*)

Genus 491. SUNIUS, *Leach MSS.*

A. Thorace subquadrato.

3083. 1; melanocephalus*.
Pæ. melanocephalus. *Fabr. E.* ii. 610.—(*Sam. I.* 32.)
Pæ. bicolor. *Oliv. E.* iii. 77. *pl.* 1. *f.* 4. *a, b.*

3084. [2; tricolor*.]
St. tricolor. *Marsh.* i. 516.

3085. 3; ochraceus*.
Pæ. ochraceus. *Grav. Mi.* 59?
Pæ. nigriceps. *Kirby MSS.*

3086. 4; rubricollis*.
Pæ. rubricollis. *Grav. Mo.* 138.—*Gyll.* ii. 376.
Pæ. obscurus. *Kirby MSS.*

B. Thorace suborbiculato.

3087. 5; brunneus*.
Pæ. brunneus. *Kirby MSS.*

3088. 6; angustatus*.
Pæ. angustatus. *Payk. F.* iii. 431.—(*Sam. I.* 32.)
St. angustatus. *Fabr. E.* ii. 599.—*Panz. F.* xi. *f.* 18.—*Marsh.* i. 527.—*Don.* xvi. *pl.* 573. *f.* 1.
β; St. gracilis. *Fabr. E.* ii. 599.

3089. 7; immaculatus.
Pæ. immaculatus. *Kirby MSS.*

3090. 8, dimidiatus.
Pæ. dimidiatus. *Kirby MSS.*

3091 † 9, sulcicollis.
Pæ. sulcicollis. *Kirby MSS.* In Mus. D. *Kirby.*

Genus 492. RUGILUS, *Leach*, (*Samou.*), *Curtis.*

3092. 1; orbiculatus*. (*Curtis l. c. infra.*)
Pæ. orbiculatus. *Fabr. E.* ii. 609.—*Panz. F.* xliii. *f.* 21.—(*Sam. I.* 24.)
St. orbiculatus. *Marsh.* i. 528.

3093. 2; immunis*. (*Curtis l. c. infra.*)
Pæ. immunis. *Kirby MSS.*—(*Sam. I.* 32.)

3094. 3, punctipennis. (*Curtis l. c. infra.*)
Pæ. punctipennis. *Kirby MSS.*

3095 † 4, fragilis*. *Curtis.* iv. *pl.* 168.
Pæ. fragilis. *Grav. Mi.* 140. In Mus. D. *Cooper.*

Genus 493. PÆDERUS *Auctorum.*

3096. 1, littoralis. *Grav. Mi.* 61.—(*Curtis l. c. infra.*)

3097. 2; riparius*. *Fabr. E.* ii. 608.—*Don.* v. *pl.* 167.—(*Leach E. E.* ix. 91.—*Samou.* 173. *pl.* 4. *f.* 12.—(*Curtis l. c. infra.*)
St. riparius. *Linn.* ii. 684.—*Berk. S.* i. 110.—*Stew.* ii. 89.—*Marsh.* i. 503.—*Turt.* ii. 522.

3098. 3; fulvipes*. *Curtis.* iii. *pl.* 108.
Pæ. thoracicus *mihi.*

3099. 4, ruficollis. *Fabr. E. S.* ii. 537.—*Panz. F.* xxvii. *f.* 22.—(*Curtis l. c. supra.*)

3100 † 5, sanguinicollis *mihi.* (*Curtis l. c. supra.*)
Pæ. cyanurus. *Kirby? MSS.* In Mus. *Brit.*

Genus 494. STENUS, *Latreille, Leach, Samou., Curtis.*

A. Abdomine immarginato.

a. *Pedibus pallidis.*

3101. 1; oculatus*. *Grav. Mi.* 155.—(*Sam. I.* 39.)
Sta. clavicornis. *Marsh.* i. 527.
Ste. clavicornis. (*Curtis l. c. infra.*)
Sta. similis. *Herbst A.* v. 151.
Pæ. 2-guttatus var. *Oliv.* iii. 44. *pl.* 1. *f.* 3. *e, d.*

3102. 2; scabrior*. *Kirby MSS.*

3103. 3, cicindeloides*. *Grav. Mi.* 155.—(*Sam. I.* 39.)—(*Curtis l. c. infra.*)
Sta. rugosus ——?
Ste. similis, b. *Liungh.* i. 65.

3104. 4, cognatus *mihi.*

3105. 5; similis*. *Kirby MSS.*

3106. 6; fulvicornis*. *Kirby MSS.*
Ste. flavicornis. (*Sam. I.* 39.)

3107. 7, pallipes. *Grav. Mi.* 157.—(*Curtis l. c. infra.*)

3108. 8, Marshami* *mihi.*
Sta. immunis. *Mus. Marsham.*

3109. 9; brunnipes*. *Kirby MSS.*—(*Sam. I.* 39.)

3110 † 10, curvipes. *Kirby MSS.* In Mus. *D. Kirby et Vigors.*

b. *Pedibus nigris.*

3111. 11, nigriclavis*. *Kirby MSS.*
Ste. clavicornis. *Grav. Mi.* 156.
Ste. buphthalmus. *Liungh.* i. 67.
Ste. tarsalis c. *Gyll.* ii. 472.

3112. 12, tarsalis. *Gyll.* ii. 472.—(*Curtis l. c. infra.*)
Ste. clavicornis. (*DeJean C.* 25.)

3113. 13; rufitarsis*. *Kirby MSS.*—(*Sam. I.* 39.)

3114. 14; unicolor*. *Kirby MSS.*

3115. 15; assimilis*. *Kirby MSS.*

3116. 16, fornicatus. *Kirby MSS.*

3117. [17, crassus.] *Kirby MSS.*

3118. 18, gracilis* *mihi.*

B. Abdomine marginato: ano simplici.

a. *Elytris immaculatis.* 1. Pedibus pallidis.

3119. 19; phæopus*. *Kirby MSS.*

3120. 20; picipes*. *Kirby MSS.*

3121. 21; argyrostoma* *mihi.*

3122. 22; fuscipes*. *Grav. Mi.* 157.—(*Curtis l. c. infra.*)

3123. 23; flavipes*. *Kirby MSS.*

3124. 24; circularis*. *Grav. Mi.* 157.—(*Curtis l. c. infra.*)
Ste. brevis. *Kirby MSS.*

3125. 25, immunis. *Kirby MSS.*
Sta. immunis. *Marsh.* i. 528.
Ste. circularis b. *Gyll.* ii. 479.

3126. 26; subrugosus* *mihi.*

3127. 27; ossium*. *Kirby MSS.*

3128. 28; tenuicornis* *mihi.*

3129. 29; geniculatus* *mihi.*

3130. 30; gonymelas*. *Kirby MSS.*

3131. 31; Aceris*. *Kirby MSS.*—(*Sam. I.* 39.)

3132. 32; Juncorum. *Leach MSS.*—(*Sam. I.* 39.)

3133. 33; nitidiusculus*. *Kirby MSS.*

3134. 34; punctatissimus *mihi.*

3135. 35; nigricornis*. *Kirby MSS.*—(*Sam. I.* 39.)
Sta. clavicornis. *Fabr. E.* ii. 603?
Ste. speculator. *Dahl?*

3136. 36; boops*. *Grav. Mo.* 226. *var.*—(*Curtis l. c. infra.*)
Sta. Cicindeloides. *Liungh.* i. 61.

3137. 37, clavicornis.
Sta. clavicornis. *Fabr. E. S.* ii. 527?

3138. 38; atricornis* *mihi.*

3139. 39; canaliculatus*. *Kirby MSS.*
Pæ. proboscideus. *Oliv. E.* iii. 6. *pl.* 1. *f.* 5. *a, b.*
Ste. buphthalmus var.? *Grav. Mo.* 230.

3140. 40; submarginatus*. *Kirby MSS.*
Ste. binotatus. *Web. A.* (*Liungh.*) i. 66?

3141. 41, sulcicollis*. *Kirby MSS.*

3142. 42; longicollis* *mihi.*

2. Pedibus nigris.

3143. 43; lineatulus*. *Kirby MSS.*

3144. 44; buphthalmus*. *Grav. Mi.* 156.
Sta. boops. *Liungh.* ii. 12.
Ste. clavicornis. *Panz. F.* xxvii. *f.* 11.

3145. 45; pubescens*. *Kirby MSS.*—(*Sam. I.* 39.)

3146. [46; lævior*.] *Latr. H.* ix. 353?

3147. 47, lævis.
Sta. lævis. *Mus. Marsham.*

3148. 48; pallitarsis*. *Kirby MSS.*

3149. 49, pilosulus. *Kirby MSS.*

3150. 50; affinis*. *Kirby MSS.*

3151. 51; angustatus*. *Kirby MSS.*—(*Sam. I.* 39.)

3152. 52; melanopus*. *Kirby MSS.*
Sta. melanopus. *Marsh.* i. 528.

3153. 53, melanarius. *Kirby MSS.*

3154. 54; nitidulus*. *Leach MSS.*

3155. 55; nitidus*. *Kirby MSS.*

3156. 56; nitens*. *Kirby MSS.*

3157. 57; pusillus*. *Kirby MSS.*—(*Sam. I.* 39.)

3158. 58, nanus *mihi.*

b. *Elytris maculatis.*

3159. 59; bimaculatus*. *Gyll.* ii. 466.
Sta. biguttatus. *Marsh.* i. 526.
Ste. Juno. *Grav. Mi.* 154.

3160. 60; biguttatus*. *Grav. Mi.* 154.—*Panz. F.* xi. *f.* 17.—*Don.* xvi. *pl.* 573. *f.* 2.—*Leach E. E.* ix. 91.—*Samou.* 173. *pl.* 4. *f.* 18.—(*Curtis l. c. infra.*)
Ste. Juno β. *Payk. F.* iii. 433.

3161. 61; bipunctatus*. *Kirby MSS.*—(*Curtis l. c. infra.*)

3162. 62; bipustulatus*. (*Curtis l. c. infra.*)
Sta. bipustulatus. *Linn.* ii. 685.—*Panz. F.* xxvii. *f.* 10.—*Marsh.* i. 527.
Sta. biguttatus, var. *Oliv. E.* iii. 5. *pl.* 1. *f.* 3. *a, b.*

3163. 63; Kirbii*. *Leach MSS.*—(*DeJean C.* 25.)—*Curtis.* iv. *pl.* 164.
Ste. biguttatus b. *Gyll.* ii. 464.
St. guttula. *Germ. Mag.* (*Müll.*) iv. 223.

3164. 64; binotatus*. *Kirby? MSS.*—(*Curtis l. c. supra?*)

C. Abdomine marginato: ano processû utrinque. ZOLMÆNUS, *Leach MSS.*

3165. 65, Juno. *Gyll.* ii. 467.—(*Curtis l. c. supra?*)
Sta. Juno. *Payk. F.* iii. 433.
Ste. boops. *Grav. Mo.* 226.

Genus 495. DIANOÜS, *Leach MSS.*, (*Samou.*), *Curtis.*

STENUS *p*, *Gyll.*

3166. 1, rugulosus. *Leach MSS.*
Ste. rugulosus. *Hoffmansegg MSS.*—(*Sam. I.* 39.)
Ste. cærulescens. *Gyll.* ii. 463?—(*Sam. I.* 39.)
Di. cærulescens. *Curtis.* iii. *pl.* 107.
Ste. biguttatus. *Liungh.* i. 62?

Genus 496. OXYPORUS, *Fabricius, Leach, Samou.*, (*Kirby.*)

3167. 1; rufus*. *Fabr. E.* ii. 604.—*Leach E. E.* ix. 91.—*Samou.* 174. *pl.* 4. *f.* 11.
St. rufus. *Linn.* ii. 684.—*Berk. S.* i. 110.—*Stew.* ii. 89.—*Mart. C. pl.* 41. *f.* 25.—*Marsh.* i. 502.—*Turt.* ii. 520.

3168 ‡ 2, maxillosus. *Fabr. E.* ii. 605.—*Panz. F.* xvi. *f.* 20.
In Mus. *Brit. et D. Kirby.*

Genus 497. SIAGONIUM, *Kirby, Westwood.*

SIAGONUM, *Curtis.*—PROGNATHUS, *Latr.*—LEPTOCHEIRUS, (*Germar.*)?

3169. 1: quadricorne*. (*Kirby & Sp. I. E.* i. *pl.* 1. *f.* 3. 315.)—*Curtis.* i. *pl.* 23.—(*Zool. Journ.* (*Westwood.*) iii. 61.)
Si. corticalis. (*Curtis l. c. supra.*)

Genus 498. BLEDIUS, *Leach MSS.*, (*Samou.*), *Curtis, Westwood.*

OXYTELUS *p, Grav.*

3170. 1, tricornis. (*Curtis l. c. infra.*)
St. tricornis. *Payk. F.* iii. 396.—*Ent. Trans.* (*Haworth.*) i. 97. *pl.*
St. armatus. *Panz. F.* lxxvi. *f.* 17.—*Ent. Trans.* (*Burrell.*) i. 236 & 310.
Ox. armatus. (*Samou.* 174.)
♀, Ox. subcornutus. *Kirby MSS.*

3171. 2, Skrimshirii. *Curtis.* iii. *pl.* 143.
Bl. Stephensii. *Zool. Journ.* (*Westwood.*) iii. *p.* 61, 301 & 509. *pl.* 11. *f.* 4. (caput et thorax.)
Ox. Taurus. (*DeJean C.* 25?)

Genus 499. HESPEROPHILUS *mihi.*

OXYTELUS *p, Gyll.*

3172. 1, fracticornis*.
St. fracticornis. *Payk. F.* iii. 382.
Ox. pallipes. *Grav. Mo.* 197?

3173. 2, hæmopterus*.
St. fracticornis β. *Payk. F.* iii. 382.
Ox. hæmopterus. *Kirby MSS.*
Ox. castaneipennis. *Ziegler.*

3174. 3, divisus. *Mus. Marsham.*
Ox. pallipes. *Gyll.* ii. 447?
Ox. lævicollis. (*DeJean C.* 25.)

3175 † 4, arenarius.
St. arenarius. *Payk. F.* iii. 382.
In Mus. *D. Kirby et* Sheppard.

3176 † 5, Talpa.
Ox. Talpa. *Gyll.* ii. 448. In Mus. *Brit.*

Genus 500. OXYTELUS, *Gravenhorst, Leach, Samou.*, (*Kirby.*)

A. Thorace convexo. a. *Thorace insculpto.*

3177. 1; brachypterus*. *Kirby MSS.*
St. brachypterus. *Marsh.* i. 510.
Ox. cælatus*. *Grav. Mi.* 103?

b. *Thorace canaliculato.*

3178. 2; cornutus*. *Grav. Mi.* 109.

3179. 3; morsitans*. *Grav. Mi.* 108.—(*Kirby & Sp. I. E.* iv. 141.)
St. morsitans. *Payk. F.* iii. 383.—*Marsh.* i. 508.
Ox. striolatus. *Ziegler.*

3180. [4; trilobus*.]
St. trilobus. *Oliv. E.* iii. 20. *pl.* 5. *f.* 48. *a, b.*—*Marsh.* i. 508.
Ox. morsitans, var. b. *Gyll.* ii. 451.

3181. 5, sulcatus*.
St. sulcatus. *Mus. Marsham.*

3182. 6, brunnipennis* *mihi.*

3183. 7, immunis. *Kirby MSS.*

3184. 8; foveatus*. *Kirby MSS.*

3185. 9, caliginosus* *mihi.*

3186. 10, pallidipennis. *Kirby MSS.*
St. pallidipennis. *Panz. F.* xxvii. *f.* 16.

B. Thorace depresso.

3187. 11; pulcher*. *Grav. Mi.* 107.
St. laqueatus. *Marsh.* i. 513.
Ox. carinatus var. c. *Gyll.* ii. 453.

3188. 12, carinatus*. *Grav. Mi.* 106.—*Panz. F.* lvii. *f.* 24.—*Leach E. E.* ix. 91.—*Samou.* 174.

3189. 13; rugosus*. (*Kirby & Sp. I. E.* iv. 141.)
St. rugosus. *Fabr. E.* ii. 601.—*Stew.* ii. 89.—*Marsh.* i. 506.—*Turt.* ii. 514.—(*Sam. I.* 32.)

3190. 14, laqueatus.
St. piceus. *Panz. F.* xxvii. *f.* 14?

3191. 15, nitens.
St. nitens. *Mus. Marsham.*

3192. 16, piceus. *Grav. Mi.* 105.
St. piceus. *Linn.* ii. 686.—*Oliv. E.* iii. 20. *pl.* 3. *f.* 38. *a, b.*—*Marsh.* i. 506.—*Turt.* ii. 515.

3193. 17, sculpturatus. *Grav. Mo.* 187.
St. sulculum. *Mus. Marsham.*

3194. 18, antennatus. *Kirby MSS.*

3195. 19, nitidus. *Kirby MSS.*

3196. 20, flavipes* *mihi.*

3197. 21; depressus*. *Grav. Mi.* 103.—*Gyll.* ii. 457.

3198. 22, fuscipennis* *mihi.*

3199. 23, pallipes*. *Kirby MSS.*

3200. 24; nitidulus*. *Grav. Mi.* 107.
St. piceus. *Payk. F.* iii. 384. var. minuta.
3201. [25, ruficrus.] *Kirby MSS.*
3202. 26, consobrinus* *mihi.*
3203. 27, angustatus*. *Kirby MSS.*—(*Sam. I.* 32.)
3204. 28; corticinus*. *Grav. Mo.* 192.
3205. 29; opacus*. *Kirby MSS.*—(*Sam. I.* 32.)
3206. 30; pusillus*. *Leach MSS.*

Genus 501. CARPALIMUS, (*Kirby.*)

CARPELIMUS, (*Samou.*)—OXYTELUS *p*, *Grav.*

A. Tarsi triarticulati.

3207. 1, bilineatus. *Kirby MSS.*
3208. 2, arcuatus. *Kirby MSS.*
3209 † 3. bicolon. *Kirby MSS.* In Mus. D. Spence.
3210. 4, fuliginosus.
Ox. fuliginosus. *Grav. Mi.* 102?
Ca. nitidulus. *Kirby MSS.*
β; Ca. crassicornis. *Kirby MSS.*
3211. 5, picipennis. *Kirby MSS.*
3212. 6, rufipennis. *Kirby MSS.*
3213. 7, affinis. *Kirby MSS.*
3214 † 8, obscurus. *Kirby MSS.*
In Mus. D. Sheppard, Spence *et Vigors.*
3215. 9; atratus* *mihi.*

B. Tarsis quinque articulatis.

3216 † 10. brunniceps. *Kirby MSS.* In Mus. D. Spence.

Familia LXX. OMALIDÆ, *MacLeay.*

(ST.-DEPLANATI *p*, *Latreille.*)

(OMALIUM, *Gravenhorst*, *&c.*)

Genus 502. EVÆSTHETUS, *Gravenhorst*, (*Kirby.*)

ERISTHETUS, (*Samou.*)—STENUS *p*, *Liungh.*—OXYTELUS *p*, *Samou.*

3217. 1, scaber. *Grav. Mo.* 202.—*Ahr. F.* vii. *f.* 13.
Ste. bipunctatus. *Liungh.* 68.
Ox. scaber. (*Sam. I.* 32.)

Genus 503. ——

EVÆSTHETUS, (*Kirby.*)

3218. 2; nigroæneus*. *Kirby MSS.*
Om. æneum. *Gyll.* iv. 466?

Genus 504. MEGARTHRUS, *Kirby MSS.*

DERMESTES *p*, *Panz.*—SILPHA *p*, *Marsh.*

3219. 1; rufescens*. *Kirby MSS.*

3220. 2; retusus*. *Kirby MSS.*

3221. [3; flavus*.] *Kirby MSS.*

3222. 4; depressus*.
St. depressus. *Payk. F.* iii. 412.—*Oliv. E.* iii. 36. *pl.* 3. *f.* 26. *a, b.*
Om. depressum. (*Sam. I.* 32.)

3223. [5; emarginatus*.] *Kirby? MSS.*

3224. 6; macropterus*.
Om. macropterum. *Grav. Mo.* 215.
Si. abbreviata. *Marsh.* i. 128.

3225. 7, pusillus *mihi.*

3226. 8, marginatus *mihi.*

3227. 9, affinis *mihi.*

Genus 505. PROTEINUS, *Latreille, Leach, Samou.*

DERMESTES *p*, *Fabr.*—CATERETES *p*, *Schönher.*

3228. 1; brachypterus*. *Latr.* G. i. 298.—*Leach E. E.* ix. 92. —*Samou.* 175.
De. brachypterus. *Payk. F.* i. 288.—*Panz. F.* iv. *f.* 10.
Om. ovatum. *Grav. Mo.* 215.
Pr. consimilis. *Kirby MSS.*

3229. 2, clavicornis*. *Kirby MSS.*

3230. 3; ovalis*. *Kirby MSS.*

Genus 506. ANTHOBIUM, *Leach MSS.*

SILPHA *p*, *Marsh.*

3231. 1; nigricorne* *mihi.*

3232. 2; subsulcatum* *mihi.*

3233. 3, sulculum.
Om. sulculum. *Kirby MSS.*

3234. 4: nitidum *mihi.*

3235. 5, Ranunculi*.
Om. Ranunculi. *Grav. Mi.* 118.
Si. minuta. *Fabr. E.* i. 342.—*Stew.* ii. 39.

3236. [6, picipenne.]
Om. picipenne. *Kirby MSS.*

3237. 7, ruficorne.
Om. ruficorne. *Mus. Marsham.*

3238. 8; Sorbi*.
Om. Sorbi. *Gyll.* ii. 206.
Om. ophthalmicum. *Grav. Mo.* 216.

3239. [9; Ulmariæ*.]
Om. Ulmariæ. *Kirby MSS.*
Si. lutea. *Marsh.* i. 128.

3240. 10; ophthalmicum*.
St. ophthalmicus. *Payk. F.* iii. 409.
Om. pallidum. *Grav. Mo.* 217.

3241. 11; torquatum*.
Si. torquata. *Marsh.* i. 127.
Si. semicoleoptrata. *Fanz. F.* xxiv. *f.* 6.

3242. [12; mucronatum*.]
Om. mucronatum. *Kirby MSS.*

3243. 13; brunneum* *mihi.*

3244. 14; canaliculatum. *Kirby MSS.*
Si. ruficollis. *Panz. F.* 25. *f.* 5.—*Marsh.* i. 127.

3245. 15, tectum.
St. tectus. *Payk. F.* iii. 411.—*Oliv. E.* iii. 36. *pl.* 3. *f.* 21. *a, b.*

3246. 16; grossum*.
Om. grossum. *Kirby MSS.*—(*Sam. I.* 32.)

3247. 17; melanocephalum*. (*Samou.* 175.)
Si. melanocephala. *Marsh.* i. 127.
St. melanocephalus. *Oliv. E.* iii. 38. *pl.* 4. *f.* 32. *a, b.*

3248. 18; assimile*.
St. assimilis. *Payk. F.* iii. 409.

3249. 19, atrocephalum.
Om. atrocephalum. *Gyll.* iv. 463.

3250. 20, longipenne.
Om. longipenne. *Kirby MSS.*
Om. cryptura. *Mus. Marsham.*
β, Om. ventrale. *Kirby MSS.*

3251. 21, unicolor.
St. unicolor. *Mus. Marsham.*

3252. 22, nigriventre* *mihi.*

Genus 507. CORYPHIUM, *Kirby MSS.*

3253 † 1, angusticolle. *Kirby MSS.*
β. Co. villosulum. *Kirby MSS.* In Mus. D. *Kirby et* Sheppard.

Genus 508. ELONIUM, *Leach,* (*Samou.*)

3254. 1; striatulum*.
Sta. striatulus. *Fabr. E.* ii. 596.

Om. rugosum. *Grav. Mi.* 115.
Om. striatum. (*Sam. I.* 32.)

Genus 509. OMALIUM, *Gravenhorst, Leach, Samou.,* (*Kirby.*)

A. Thorace obcordato.

3255. 1; planum*. *Grav. Mi.* 112.—(*Sam. I.* 32.)
St. planus. *Payk. F.* iii. 405.

3256. 2; deplanatum*. *Grav. Mi.* 113.

3257. 3; concinnum*.
St. concinnus. *Marsh.* i. 510.

3258. 4; brunnipes *mihi.*

3259. 5; monilicorne*. *Gyll.* ii. 219.
Om. monilicorne. *Kirby MSS.*

3260. 6; brunnipenne*.
St. brunnipennis. *Mus. Marsham.*

3261. 7; picinum*. *Kirby MSS.*

B. Thorace subquadrato.

3262. 8; brunneum. *Grav. Mi.* 113.
St. brunneus. *Payk. F.* iii. 404.

3263. 9; iopterum*. *Kirby MSS.*
St. brunneus β. *Payk. F.* iii. 405.

3264. 10, sordidum. *Kirby MSS.*

3265. 11, striatum. *Grav. Mi.* 119.
St. minutus. *Oliv. E.* iii. 38. *pl.* 6. *f.* 53. *a, b*?

3266. 12; subpubescens*. *Kirby MSS.*

C. Thorace transverso.

a. *Thorace obsoletè foveolato.*

3267. 13; florale*.
St. floralis. *Payk. F.* iii. 406.
Om. Viburni. *Grav. Mi.* 117.

3268. 14; Salicis*. *Gyll.* ii. 226.

3269. 15, ruficornis*. *Kirby MSS.*

3270. 16, pilosulum. *Kirby MSS.*

3271. 17, læve*. *Grav. Mo.* 211.
Om. læviusculum. *Gyll.* iv. 464.

b. *Thorace profundè foveolato.*

3272. 18, foveolatum. *Kirby MSS.*

3273. 19; piceum*. *Kirby MSS.*
Om. cæsum, b. *Gyll.* ii. 216?
Om. incisum. *Knoch.*
Om. rivulare var. b. *Grav. Mo.* 209.

3274. 20, cæsum. *Knoch.*—*Grav. Mo.* 209.

3275. 21; rivulare*. *Grav. Mi.* 116.—*Leach E. E.* ix. 92.—*Samou.* 174.
St. rivularis. *Payk. F.* iii. 407.—*Oliv. E.* iii. 35. *pl.* 3. *f.* 27. *a, b.*—*Marsh.* i. 513.

3276. 22; Oxyacanthæ*. *Knoch.*—*Gyll.* ii. 217.
Om. rivulare var. c. *Grav. Mo.* 210.

3277. 23, fuscum *mihi.*

3278. 24, excavatum. *Kirby MSS.*

3279. 25; subrugosum* *mihi.*

3280. 26, pygmæum. *Grav. Mo.* 206.
St. pygmæus. *Payk. F.* iii. 410.
Om. brunneum var. 3. *Grav. Mi.* 113.

3281 † 27, Primulæ. *Kirby MSS.*—(*Kirby & Sp. I. E.* ii. 242.) In Mus. D. *Kirby.*

3282 [† 28, nigricolle.] *Kirby MSS.* In Mus. D. *Kirby.*

3283. 29, latum* *mihi.*

Genus 510. ACIDOTA, *Kirby? MSS.*

3284. 1, crenata.
St. crenatus. *Fabr. E.* ii. 596.

3285 † 2. rufa.
Om. rufum. *Grav. Mi.* 115. In Mus. *Brit.* et D. *Kirby.*

Genus 511. LESTEVA, *Latr.*

CARABUS *p*, *Fabr.*, *Marsh.*—ANTHOPHAGUS, *Gravenhorst, Leach.* (*Kirby.*)—LESTIVA, *Samou.*

3286 † 1, Leachii. *Kirby MSS.* In Mus. *Brit.*

3287. 2, Caraboides*. (*Sam. I.* 25.)—(*Kirby & Sp. I. E.* iii. 506.)
St. caraboides. *Linn.* ii. 685.—*Oliv. E.* iii. 22. *pl.* 2. *f.* 17. *a, b.* —*Marsh.* i. 521.—*Turt.* ii. 514.
Ca. abbreviatus. *Fabr. E.* i. 209.
St. fulvus. *DeGeer.* iv. 25.

3288. 3, canaliculata. *Kirby MSS.*
Le. Scotica. *Leach? MSS.*

3289. 4, Hookeri. *Kirby MSS.*

3290. 5, alpina. *Latr. G.* i. 297.
St. alpinus. *Fabr. E.* ii. 598.—*Oliv. E.* iii. 32. *pl.* 6. *f.* 55. *a, b.*

3291. 6; testacea*.
An. testacea. *Grav. Mi.* 121?

3292. 7; obscura*. (*Sam. I.* 25.)
St. obscurus. *Payk. F.* iii. 388.

St. bicolor. *Fabr. E.* ii. 600?
Car. staphylinoides. *Marsh.* i. 464.
Le. punctulata. *Latr. H.* ix. 369.—*Samou.* 175.
An. punctulatus. *Leach E. E.* ix. 92.
Car. dimidiatus. *Panz. F.* xxxvi. *f.* 3.

3293. 8, plagiata.
St. plagiatus. *Payk. F.* iii. 387.

3294. 9, rufitarsis*. *Kirby MSS.*

3295. 10; nigripes* *mihi.*

3296 † 11, impressa. *Kirby MSS.* In Mus. D. *Kirby.*

3297. 12, planipennis. *Kirby MSS.*

3298. 13, Kirbii *mihi.*

Ordo II. DERMAPTERA, *DeGeer.*

(Coleoptera *p*, *Linné*, *&c.*—Ulonata *p*, *Fabr.*—Orthoptera *p*, *Latreille.*—Deratoptera, *Clairville.*)

Familia I: (71). FORFICULIDÆ.

Genus 1: (512). FORFICULA *Auctorum.* *Earwig.*

3299. 1; auricularia*. *Linn.* ii. 686.—*Berk. S.* i. 110.—*Panz. F.* lxxxvii. *f.* 8.—*Marsh.* i. 529.—*Stew.* ii. 89.—*Turt.* ii. 523.—*Shaw G. Z.* vi. 110. *pl.* 40.—*Bingley.* iii. 149.—*Leach E. E.* ix. 118.—*Samou.* 216.—(*Leach Z. M.* iii. 99.)—*Wood.* i. 80. *pl.* 30.
Fo. major. *Harrer.* 408.

3300. [2; media*.] *Marsh.* i. 530.

3301. 3; borealis*. *Leach MSS.*—(*Sam. I.* 17.)

3302. 4; forcipata* *mihi.*

Genus 2: (513). LABIA, *Leach, Samou.*, (*Kirby.*)

3303. 1; minor*. (*Leach E. E.* ix. 118.)—(*Leach Z. M.* iii. 99.)—*Samou.* 216. *pl.* 4. *f.* 16.
Fo. minor. *Linn.* ii. 686.—*Berk. S.* i. 110.—*Marsh.* i. 530.—*Stew.* ii. 89.—*Turt.* ii. 524.—(*Kirby & Sp. I. E.* iv. 514.)

Genus 3: (514). LABIDURA, *Leach, Samou.*, (*Kirby.*)

3304. 1, gigantea. *Leach E. E.* ix. 118.—*Samou.* 217.—*Leach E. B. Sup.* i. *pl.* 24.—*Leach Z. M.* iii. 99.
Fo. gigantea. *Fabr. E. S.* ii. 1.—*Ent. Trans.* (*Haworth.*) i. 255.—*Don.* xiv. *pl.* 500.—*Linn. Trans.* (*Bingley.*) x. 404.—(*Kirby & Sp. I. E.* ii. 237. *pl.* 1. *f.* 7.—*Id.* iii. 34.)
♀, Fo. erythrocephala. *Ent. Trans.* (*Haworth.*) i. 256?

Ordo III. ORTHOPTERA, *Olivier.*

(Hemiptera *p, Linné.*—Ulonata *p, Fabricius.*—Dermaptera *p, DeGeer.*—Deratoptera *p, Clairville.*)

Familia I: (72). GRYLLIDÆ, *Leach.*

(Gryllus *p, Linné, &c.*—Locustariæ, *Latreille.*—Conocephalidæ, *Kirby & Sp.*—Acridina, *MacLeay.*)

Genus 1: (515). ACRIDA, *Kirby, Curtis.* *Locusts.*

Conocephalus, *Thunberg, Samou.*—Locusta *p, Geoffroy, Leach.*

A. Alis incompletis.

a. *Elytris (in masculis) haud ocellatis.*

3305. 1; aptera*.
Lo. aptera. *Fabr. E. S.* ii. 45?
Gr. clypeatus. *Panz. F.* xxxiii. *f.* 4.
Ac. clypeata. (*Curtis l. c. infra.*)

b. *Elytris ocellatis, longioribus.*

3306. 2, brachyptera. (*Curtis l. c. infra.*)
Gr. brachypterus. *Linn. F.* 868.—*DeGeer.* iii. *pl.* 22. *f.* 2, 3?
—*Turt.* ii. 552.

3307. 3; Kirbii*. *Dale MSS.*—(*Curtis l. c. infra.*)
Gr. marginatus. *Schæff. Ic. pl.* 236. *f.* 4?

3308. 4, fusca. (*Curtis l. c. infra.*)
Lo. fusca. *Fabr. E. S.* ii. 43.—*Panz. F.* xxxiii. *f.* 2.

c. *Elytris subocellatis brevissimis.*

3309. 5; virescens* *mihi.*
Ac. aptera. (*Curtis l. c. infra.*)

B. Alis perfectis.

a. *Elytris (in masculis) ocellatis.*

1. Ovipositore incurvo.

3310. 6; grisea. (*Curtis l. c. infra.*)
Lo. grisea. *Fabr. E. S.* ii. 41.—*Sowerb. B. M.* i. *pl.* 64.
Co. griseus. (*Sam. I.* 12.)

2. Ovipositore ferè recto.

3311 † 7, Bingleii. *Dale MSS.*—*Curtis.* ii. *pl.* 82.
In Mus. D. *Dale et Haworth.*

3312. 8: verrucivora*? (*Curtis l. c. supra.*)
Gr. verrucivorus. *Linn.* ii. 698.—*Berk. S.* i. 112.—*Stew.* ii. 94.
—*Turt.* ii. 551.—*Panz. F.* lxxxix. *f.* 20. ♂. *f.* 21. ♀.—
Shaw G. Z. vi. 140.

Lo. verrucævora. *Ent. Trans.* (*Haworth.*) i. 256.
Co. verrucivorus. (*Sam. I.* 12.)

3313. 9; viridissima*. (*Curtis l. c. supra.*)—(*Kirby & Sp. I. E.* ii. 399: iii. 34.)
Gr. viridissimus. *Linn.* ii. 698.—*Don.* iv. *pl.* 130.—*Stew.* ii. 94.—*Turt.* ii. 551.
Lo. viridissima. *Leach E. E.* ix. 120.
Co. viridissimus. *Samou.* 218.

b. *Elytris* (*in utroque sexû*) *inocellatis.*

3314. 10; varia*. (*Curtis l, c. supra.*)
Lo. varia. *Fabr. E. S.* ii. 42.
Lo. thalassina. *DeGeer.* iii. 433.
Gr. varius. *Don.* iii. *pl.* 79. *f.* 1.—*Stew.* ii. 94.—*Turt.* ii. 552.
Co. varia. (*Sam. I.* 12.)

Familia II: (73). LOCUSTIDÆ, *Leach.*

(GRYLLUS, *Linné, &c.*—ACRIDII, *Latreille.*—LOCUSTINA, *MacLeay.*) *Locusts, Grasshoppers.*

Genus 2: (516). LOCUSTA, *Leach, Samou.*, (*Kirby.*)

3315. 1, migratoria*. (*Sam. I.* 25.)
Gr. migratorius. *Linn.* ii. 700.—*Don.* viii. *pl.* 270.—*Stew.* ii. 95.—*Turt.* ii. 560.—*Shaw G. Z.* vi. 129. *pl.* 48.—*Bingley.* iii. 166.—*Leach E. E.* ix. 120.

3316. 2; flavipes*. (*Sam. I.* 25.)
Gr. flavipes. *Gmel.* iv. 2088.—*Don.* xi. *pl.* 391.

3317 ‡ 3. cærulescens[a].

3318 † 4. stridula[b].

3319 † 5. grossa[c].

3320. 6; viridula*.
Gr. viridulus. *Linn.* ii. 702.—*Turt.* ii. 564.—*Sowerb. B. M.* i. *pl.* 63.

3321. 7; aprica* *mihi.*

3322. 8, ochropa* *mihi.*

---

[a] 3317 ‡ 3. cærulescens.
Gr. cærulescens. *Linn.* ii. 700.—*Panz. F.* lxxxvii. *f.* 11.—*Stew.* ii. 95. (!)

[b] 3318 † 4. stridula.
Gr. stridulus. *Linn.* ii. 701.—*Panz. F.* lxxxvii. *f.* 12.—*Stew.* ii. 95. (!) —*Turt.* ii. 561. (!)

[c] 3319 † 5. grossa.
Gr. grossus. *Linn.* ii. 702.—*Berk. S.* i. 112. (!)—*Panz. F.* xxxiii. *f.* 7.—*Stew.* ii. 95. (!)—*Turt.* ii. 564. (!)

3323. 9; lineata*.
Gr. lineatus. *Panz. F.* xxxiii. *f.* 9.

3324. 10; rhomboidea*.
Gr. rhomboidea. *Schæff. pl.* 288. *f.* 6, 7.

3325. 11; rubroviridata*. *Haworth? MSS.*

3326. 12, varipes* *mihi.*

3327. 13; vittata* *mihi.*

3328. 14; rosea* *mihi.*

3329. 15; crucigera* *mihi.*

3330. 16; consobrina* *mihi.*

3331. 17, varipes *mihi.*

3332. 18; rubicunda*.
Gr. rubicundus. *Schæff. pl.* 241. *f.* 5. 6.

3333. 19; venosa*.
Gr. venosus. *Mus. Lesk.* 49.

3334. 20, obscura *mihi.*

3335. 21, tricarinata* *mihi.*

3336. 22; pedestris*.
Gr. pedestris. *Linn.* ii. 703.—*Panz. F.* xxx. *f.* 8.

Genus 3: (517). GOMPHOCERUS, *Leach MSS.*, *Samou.*
GOMPHOCEROS, *Thunb.*—ACRYDIUM *p*, *Geoff.*

3337. 1; Sowerbii*. *Leach MSS.*

3338. 2; elegans* *mihi.*

3339. 3; biguttulus*. *Leach MSS.*
Gr. biguttulus. *Linn.* ii. 702?—*Panz. F.* xxxiii. *f.* 6.—*Don.* iii. *pl.* 79. *f.* 2.—*Stew.* ii. 95.—*Turt.* ii. 564.

3340. 4; rufus*. (*Sam. I.* 19.)—*Leach MSS.*
Gr. rufus. *Linn.* ii. 702.—*DeGeer.* iii. *pl.* 23. *f.* 13.—*Don.* xiv. *pl.* 482.—(*Sam. I.* 20.)—*Turt.* ii. 564.

3341. 5, calidoniensis. *Leach MSS.*

3342. 6; Ericetarius*. *Leach MSS.*

3343 † 7, sibiricus?
Gr. sibiricus. *Linn.* ii. 701?—*Panz. F.* xxiii. *f.* 20?
Gr. clavimanus. *Pallas. Sp. Zo.* ix. 21. *pl.* 1. *f.* 11?
In Mus. *D. Hope.*

Genus 4: (518). ACRYDIUM, *Fabr.*, *Leach*, *Samou.*, (*Kirby.*)
TETRIX, *Latr.*—ACHETA *p*, *Lamarck.*

3344. 1; subulatum*. *Fabr. E. S.* ii. 26.—*Sower. B. M.* i. *pl.* 74. —*Don.* xv. *pl.* 520.—*Leach E. E.* ix. 120.—*Samou.* 219. *pl.* 4. *f.* 18.

Gr. subulatus. *Linn.* ii. 693.—*Berk. S.* i. 111.—*Turt.* ii. 543.
β; Ac. undulatum. *Sower. B. M.* i. 74.
γ; Ac. nigricans. *Sower. B. M.* i. *pl.* 74.

3345. 2; bipunctatum*. *Fabr. E. S.* ii. 26.—*Panz. F.* v. *f.* 18.—(*Sam. I.* 1.)
Gr. bipunctatus. *Linn.* ii. 693.—*Berk. S.* i. 111.—*Stew.* ii. 93.—*Turt.* ii. 543.

3346. 3; brevipenne* *mihi.*

Familia IV: (74). ACHETIDÆ, *Leach.*

(Gryllides, *Latr.*—Gryllina, *MacLeay.*)

Genus 5: (519). GRYLLOTALPA, *Ray, Leach, Samou.,* (*Kirby.*)
Acheta *p, Fabr.* *Molecricket.*

3347. 1; vulgaris*. *Latr.* G. iii. 95.—(*Leach E. E.* ix. 119.)—*Samou.* 217.—(*Kirby & Sp. I. E.* ii. 366—398; iii. 35.)
Gryllus Gryllotalpa. *Linn.* ii. 693.—*Berk. S.* i. 111.—*Don.* v. *pl.* 147.—*Stew.* ii. 93.—*Turt.* ii. 544.—*Shaw. G. Z.* vi. 140. *pl.* 50.—*Bingley.* iii. 159.—*Wood.* i. 91. *pl.* 33.
Ac. Gryllotalpa. (*Kirby & Sp. I. E.* i. *pl.* 2. *f.* 2.)

Genus 6: (520). ACHETA, *Fabr., Leach, Samou.,* (*Kirby.*) *Cricket.*

3348. 1, campestris*. *Fabr. E. S.* ii. 31.—*Leach E. E.* ix. 119.—*Samou.* 218.—(*Kirby & Sp. I. E.* ii. 366.)
Gr. campestris. *Linn.* ii. 695.—*Berk. S.* i. 112.—*Stew.* ii. 94.—*Turt.* ii. 545.—*Sowerb. B. M.* i. *pl.* 64.—*Don.* xii. *pl.* 432.—*Bingley.* iii. 164.

3349. 2, sylvestris. *Fabr. E. S.* ii. 33.

3350. 3; domesticus*. *Fabr. E. S.* ii. 29.—*Panz. F.* lxxxviii. *f.* 6. ♂. *f.* 7. ♀.—(*Sam. I.* 1.)—(*Kirby & Sp. I. E.* i. 239.)
Gr. domesticus. *Linn.* ii. 694.—*Berk. S.* i. 111.—*Stew.* i. 93.—*Turt.* ii. 544.—*Don.* xi. *pl.* 409.—*Bingley.* iii. 161.

3351 † 4, italica. *Fabr. E. S.* ii. 32?
Gr. pellucens. *Panz. F.* xxii. *f.* 18? In Mus. D. *Haworth.*

Familia V: (75). BLATTIDÆ.

(Ordo Dictyoptera, *Leach.*)

(Blattariæ, *Latr.*—Blattina, *MacLeay.*)

Genus 7: (521). BLATTA, *Auctorum.* *Cockroach.*

3352 † 1, *gigantea** [a].

---

[a] 3352 † 1, gigantea*. *Pantologia* ii. (!)—*Shaw G. Z.* vi. *pl.* 41. (!)

3353. 2; *orientalis** [a].

3354. 3, *Maderæ** [b].

3355. 4; *Americana** [c].

3356. 5; germanica*. *Linn.* ii. 688.—*Don.* x. 39. *pl.* 341.—(*Kirby & Sp. I. E.* i. 239.)

3357. 6, pallens* *mihi.*

3358. 7; Lapponica*. *Linn.* ii. 688.—*Panz. F.* xcvi. *f.* 13.

3359. 8, perspicillaris. *Turt.* ii. 531.
Bl. Lapponica. *Fuesly A. pl.* 49. *f.* 11.

3360. 9, Panzeri *mihi.*
Bl. germanica. *Panz. F.* ii. *f.* 16.

3361. 10, nigripes *mihi.*

3362. 11; livida*. *Fabr. E. S.* ii. 10.—*Ent. Trans.* (*Haworth.*) i. 255.—(*Sam. I.* 7. *pl.* 4. *f.* 17.)
Bl. livens. *Turt.* ii. 529.
Bl. lapponica. *Don.* x. *pl.* 332.

3363. 12: pallida. *Olivier.*—*Latr. H.* xii. 97.

Ordo IV. NEUROPTERA.

(Synistata *p*, *et* Odonata, *Fabr.*—Dictyoptera, *Clairville.*)

Sectio I. PANORPINA.

(Planipennes *p*, *Latreille.*)

Familia I: (76). BOREIDÆ.

(Panorpatæ *p*, *Latr.*—Panorpidæ *p*, *Leach.*—Ordo Raphioptera, *MacLeay.*)

Genus 1: (522). BOREUS, *Latr.*, *Curtis*, (*Kirby.*)

Panorpa *p*, *Linn.*—Gryllus *p*, *Panz.*

3364. 1, hyemalis*. *Curtis.* iii. *pl.* 118.
Pa. hyemalis. *Linn.* ii. 914.—*Turt.* iii. 406.—(*Sam. I.* 33.)
Gr. proboscideus. *Panz. F.* xxii. *f.* 18.

---

[a] 3353. 2; orientalis*. *Linn.* ii. 688.—*Berk. S.* i. 110. (!)—*Panz. F.* xcvi. *f.* 12.—*Turt.* ii. 529. (!)—*Shaw G. Z.* vi. 116. *pl.* 41. (!)—*Bingley.* iii. 153. (!)—(*Sam. I.* 7. (!))—*Wood.* i. 85. *pl.* 31.

[b] 3354. 3, Maderæ*. *Fabr. E. S.* ii. 6.—*Don.* xiii. *pl.* 457. (!)

[c] 3355. 4; Americana*. *Linn.* ii. 687.—*Fuesly A. pl.* 49. *f.* 5, 6.—*Turt.* ii. 528. (!)—*Shaw G. Z.* vi. 116. *pl.* 41. (!)—*Bingley.* iii. 153. (!)

Familia II: (77). PANORPIDÆ, *Leach.*

(Panorpatæ *p*, *Latreille.*—Panorpina, *MacLeay.*)

Genus 2: (523). PANORPA *Auctorum.* *Scorpion-Fly.*

3365. 1; communis*. *Linn.* ii. 915.—*Barbut* G. 226. *pl.* 12.—*Berk. S.* i. 152.—*Panz. F.* 1. *f.* 10.—*Stew.* ii. 218.—*Don.* vi. *pl.* 201. *fig. inf.*—*Turt.* iii. 406.—*Shaw* G. Z. vi. 264. *pl.* 86. *fig. med.*—*Leach Z. M.* ii. 98. *pl.* 94. *f.* 1.—(*Samou.* 260.)—*Wood.* ii. 35. *pl.* 51.

3366. 2; affinis*. *Leach Z. M.* ii. 98. *pl.* 94. *f.* 2.
Pa. communis. *Don.* vi. *pl.* 201. *fig. sup.*—(*Shaw l. c. supra. fig. sup.*)—*Sam. I.* 33. *pl.* 7. *f.* 5.

3367. 3; apicalis* *mihi.*

3368. 4, germanica*. *Linn.* ii. 915.—*Turt.* iii. 406.—(*Sam. I.* 33.)

Sectio II. ANISOPTERA, *Leach MSS.*

(Subulicornes *p*, *Latreille.*)

Familia III: (78). EPHEMERIDÆ, *Leach.*

(Ephemerinæ, *Latreille.*—Ephemera, *Linné, &c.*)

Genus 3: (524). EPHEMERA *Auctorum.* *May- or Day-Flies.*

3369. 1; vulgata*. *Linn.* ii. 916.—*Barbut* G. 213. *pl.* 11. *fig. inf. dext.?*—*Berk. S.* i. 150.—*Panz. F.* xciv. *f.* 16.—*Stew.* ii. 211.—*Shaw* G. Z. vi. 249. *pl.* 81. *fig. sup.*—*Turt.* iii. 393.—(*Samou.* 260. *pl.* 7. *f.* 2. *immatura.*)—*Wood.* ii. 23. *pl.* 47. *pupa.*

3370. 2; cognata* *mihi.*
Ep. vulgata. *Don.* iv. 53. *pl.* 128.

3371. 3; Stigma *mihi.*

3372. 4, talcosa. *Mus. Marsham.*

3373. 5; lutea*. *Linn.* ii. 609.—*Turt.* iii. 393.

3374. 6; marginata*. *Linn.* ii. 906?—*Stew.* ii. 211?—*Shaw G. Z.* vi. *pl.* 81. *fig. inf.?*

3375. 7; submarginata* *mihi.*

3376. 8; dispar* *mihi.*

3377. 9; nigricans* *mihi.*

3378. 10; diluta* *mihi.*

3379. 11; apicalis* *mihi.*
Ep. halterata. *Fabr. E. S.* ii. 69?—*Shaw* G. Z. vi. 253. *pl.* 81. *fig. med.?*

3380. 12; rufescens* *mihi.*

3381. 13; dubia* *mihi.*

3382. 14; helvipes* *mihi.*

3383. 15; rosea* *mihi.*

3384 † 16. vespertina[a].

Genus 4: (525). ——

3385. 1; macrura* *mihi.*

3386. 2; brevicauda.
Ep. brevicauda. *Fabr. E. S.* ii. 69.
Ephemeron. *Harris Ex. pl.* vi. *f.* 3.

3387. 3; pennata* *mihi.*

Genus 5: (526). BAËTIS, *Leach, Samou.*

3388. 1, caudata* *mihi.*

3389. 2; venosa*. *Fabr. E. S.* ii. 70.—*DeGeer.* ii. *pl.* xviii. *f.* 1—4.

3390. 3; subfusca* *mihi.*

3391. 4; flavescens* *mihi.*

3392. 5; basalis* *mihi.*

3393. 6; striata*.
Eph. striata. *Linn.* ii. 907.—*Berk. S.* i. 150.—*Stew.* ii. 212.—*Turt.* iii. 394.
Eph. ciliaris. *Kirby? MSS.* (*pupa.*)

3394. 7; phæopa* *mihi.*

3395. 8; obscura* *mihi.*

3396. 9; horaria*.
Ep. horaria. *Linn.* ii. 907.—*Berk. S.* i. 150.—*Stew.* ii. 211.—*Turt.* iii. 394.

3397. 10; culiciformis*.
Ep. culiciformis. *Linn.* ii. 907.—*Berk. S.* i. 150.—*Stew.* ii. 211.—*Turt.* iii. 394.

3398. 11; fuscata*. *Linn.* ii. 907.—*Turt.* iii. 394.

3399. 12; bioculata*. (*Sam. I.* 5.)
Ep. bioculata. *Linn.* ii. 906.—*Panz. F.* xciv. *f.* 17.—*Barbut G. pl.* 11. *fig. inf. sinist.?*—*Stew.* ii. 211.—*Turt.* iii. 393.

3400 † 13. nigra[b].

3401. 14, cingulata* *mihi.*

Genus 6: (527). CLOËON, *Leach, Samou.*

3402. 1; diptera*.

---

[a] 3384 † 16. vespertina. *Linn.* ii. 906.—*Berk. S.* i. 150. (!)—*Stew.* ii. 211. (!)—*Turt.* iii. 393. (!)

[b] 3400 † 13. nigra.
Ep. nigra. *Linn.* ii. 907, (*pupa.*)—*Stew.* ii. 211.—*Turt.* iii. 393.

Ep. diptera. *Linn.* ii. 907.—*Stew.* ii. 212.—*Shaw* G. Z. vi. 253.
Cl. pallida. (*Leach E. E.* ix. 137.)—(*Sam. I.* 11.)

3403. 2; ochracea* *mihi.*

3404. 3; obscura* *mihi.*

3405. 4; hyalinata* *mihi.*

3406. 5; dorsalis* *mihi.*

3407. 6; cognata* *mihi.*

3408. 7; consobrinus* *mihi.*

3409. 8. Virgo*.
Ep. Virgo. *Oliv.*—*Latr. H.* xiii. 98?

Sectio III. LIBELLULINA, *MacLeay.*

(Subulicornes *p, Latreille.*)

Familia IV: (79). AGRIONIDÆ, *Leach.*

(Libellula *p, Linné, &c.*)

Genus 7: (528). AGRION, *Fabr., Leach, Samou.,* (*Kirby.*)

Libellula Puella. *Stew.* ii. 210.

3410. 1; rufipes* *mihi.*

3411. 2; zonata*. *Leach MSS.*
Ag. zonatus. (*Sam. I.* 2.)

3412. 3; ezonata* *mihi.*

3413. 4; Puella*. *Leach MSS.*—(*Sam. I.* 2.)
Li. Puella ♂. *Linn.* ii. 905.
Li. Puella. *Barbut* G. 208. *pl.* 11. *cum fig.*—*Berk. S.* i. 149. —*Shaw* G. Z. vi. 246.
Li. lucifugus. *Harris Ex. pl.* xxix. *f.* 5, 6.
Li. Puella c. *Turt.* iii. 392.

3414. 5; cingulata* *mihi.*

3415. 6; annularis*. *Leach MSS.*
Li. æreus. *Harris Ex. pl.* xxix. *f.* 3, 4?
Ag. annulare. (*Sam. I.* 2.)

3416. 7, rufescens*. *Leach MSS.*—(*Sam. I.* 2.)

3417. 8; sanguinea*. *Leach MSS.*
Li. Puella β. *Linn.* ii. 905.—*Turt.* iii. 192.
Li. nimius. *Harris Ex. pl.* xxix. *f.* 1, 2.
Li. Puella. *Barbut* G. 208. *pl.* 11. *cum fig.*—*Don.* i. *pl.* 36. *f.* 2.
Ag. sanguineus. (*Sam. I.* 2.)

3418. 9, Lincolniensis*. *Dale MSS.*

3419. 10; corea*. *Leach MSS.*—(*Sam. I.* 2.)
♂. Ag. albicans. *Leach MSS.*—(*Sam. I.* 2.)

Genus 8: (529). LESTES, *Leach, Samou.*

AGRION *p, Fabr.*

3420. 1; Sponsa*. (*Sam. I.* 25.)
Li. Puella α. *Linn.* ii. 905.
Ag. Sponsa. *Kirby? MSS.*

3421. [2; Nympha]*. *Leach? MSS.*

3422. 3; autumnalis*. *Leach MSS.*—(*Sam. I.* 25.)
Li. Puella γ. *Linn.* ii. 905.

Genus 9: (530). CALEPTERYX, *Leach, Samou.*, (*Kirby.*)

AGRION *p, Fabr.*—LIBELLULA Virgo. *Stew.* ii. 209.

3423. 1; Virgo*. (*Sam. I.* 8.)
Li. Virgo α, β, γ. *Linn.* ii. 904, 905.—*Anon. Rem. L. G.* 24. *pl.* 4.
Li. splendeo. *Harris Ex. pl.* xxx. *f.* 4, 5.
Li. Virgo a, b, c. *Turt.* iii. 391.

3424. 2; Ludoviciana*. *Leach MSS.*
Li. Virgo δ. *Linn.* ii. 905.—*Turt.* iii. 391.
Li. Virgo. *Barbut G.* 207. *pl.* 11. *cum fig.*—*Berk. S.* i. 149.—*Don.* i. *pl.* 36. *f.* 1.—*Shaw G. Z.* vi. 246. *pl. in front.*
Li. splendens. *Harris Ex. pl.* xxx. *f.* 1, 2, 3.
Ca. Ludovicia. (*Sam. I.* 8.)

Familia V: (80). LIBELLULIDÆ, *Leach.*

(LIBELLULA *p, Linné, &c.*) *Dragon-Flies, Adder-bolts, Horse-stingers.*

Genus 10: (531). ANAX, *Leach, Samou.*, (*Kirby.*)

3425. 1; Imperator*. (*Leach E. E.* ix. 137.)—(*Sam. I.* 2.)—(*Kirby & Sp. I. E.* iii. 35.)

Genus 11: (532). ÆSHNA, *Fabr., Leach, Samou.*, (*Kirby.*)

ÆSCHNA, (*Curtis.*)

3426 † 1, Dalii. *Leach MSS.*—(*Ing. Inst.* 90.)—(*Curtis Cat. No.* 45.)
In Mus. *Brit. D. Dale et Haworth.*

3427. 2; grandis*. *Fabr. E. S.* ii. 384.—(*Sam. I.* 1.)
Li. grandis. *Linn.* ii. 903.—*Berk. S.* i. 149.—*Stew.* ii. 209.—*Turt.* iii. 391.—*Don.* x. 30. *pl.* 337. *f.* 2.
Libellulæ (Large brown). *Harris Ex. pl.* xii. *f.* 1, 2.

3428. 3; varia*. *Shaw G. Z.* vi. 242. *pl.* 80.
Li. grandis. *Barbut G.* 206. *pl.* 11. *fig. sup. dext.*—*Don.* v. 77. *pl.* 166.
Li. viatica. *Leach MSS.*—(*Sam. I.* 1.)
Libellulæ (Large green). *Harris Ex. pl.* xvi.
β, Li. Anguis. *Harris Ex. pl.* xxiii. *f.* 4.

3429. 4; juncea*. *Leach MSS.*—(*Sam. I.* 1.)
Li. juncea. *Linn.* ii. 903.—*Turt.* iii. 390?

3430. 5; anglicana*. *Leach MSS.*—(*Sam. I.* 1.)
Li. Coluberculus. *Harris Ex. pl.* xxvii. *f.* 1.

3431. 6; teretiuscula*. *Leach MSS.*—(*Sam. I.* 1.)
♀; Li. Aspis. *Harris Ex. pl.* xxvii. *f.* 3.

Genus 12: (533). CORDULEGASTER, *Leach, Samou.*
ÆSHNA *p, Latr.*

3432. 1; annulatus*. *Leach E. E.* i. 136.—(*Samou,* 258.)
Æsh. annulata. *Latr. H.* xiii. 6.
Li. forcipata. *Linn.* ii. 903.—*Harris Ex. pl.* xxiii. *f.* 3.—*Berk. S.* i. 149.—*Stew.* ii. 209.—*Turt.* iii. 391.
Li. Boltoni. *Don.* xii. *pl.* 430.
Cordulia annulata. (*Sam. I.* 12.)

Genus 13: (534). GOMPHUS, *Leach, Samou.*

3433. 1; vulgatissimus*. *Leach E. E.* ix. 137.—(*Sam. I.* 19.)
Li. vulgatissimus. *Linn.* ii. 902.—*Berk. S.* i. 149.—*Stew.* ii. 208.
Li. forcipata. *Don.* xii. *pl.* 423.—*Turt.* iii. 388.

3434. 2; pulchellus *mihi.*—(*Ing. Inst.* 90.)

Genus 14: (535). CORDULIA, *Leach, Samou.*, (*Kirby.*)

3435. 1; ænea*. (*Sam. I.* 12.)
Li. ænea. *Linn.* ii. 902.—*Harris Ex. pl.* 27. *f.* 2.—*Berk. S.* i. 149.—*Don.* xii. *pl.* 415.—*Stew.* ii. 209.—*Turt.* iii. 390.—*Sower. B. M.* i. *pl.* 47.

3436. 2, compressa *mihi.*—(*Ing. Inst.* 90.)

Genus 15: (536). LIBELLULA *Auctorum.*

3437. 1; vulgata*. *Linn.* ii. 901.—*Harris Ex. pl.* xlvi. *f.* 3.—*Don.* x. *pl.* 337.—*Stew.* ii. 208.—*Turt.* iii. 388.—(*Sam. I.* 25.)

3438. 2; flaveola. *Linn.* ii. 901.—*Harris Ex. pl.* xlvi. *f.* 4.—*Berk. S.* i. 148.—*Stew.* ii. 208.—*Turt.* iii. 387.

3439. 3; basalis* *mihi.*

3440. 4; Scotica*. (*Leach E. E.* ix. 136.)—*Don.* xv. *pl.* 523.—(*Sam. I.* 25.)—(*Millard. pl.* 1. *f.* 4.)

3441. 5; Donovani*. (*Leach E. E.* ix. 136.)—(*Sam. I.* 25.)
Li. biguttata. *Don.* xiii. *pl.* 449.

3442. 6; cancellata*. *Linn.* ii. 902.—*Don.* xiv. *pl.* 472.—*Stew.* ii. 209.—*Turt.* iii. 388.—(*Sam. I.* 25.)—*Kirby & Sp. I. E.* i. *pl.* iii. *f.* 5.

3443. 7; conspurcata*. *Fabr. S.* 283.—*Sower. B. M.* i. *pl.* 46.—(*Sam. I.* 25.)
Li. quadrifasciata. *Don.* xii. *pl.* 425.
Li. rubicunda. *Stew.* ii. 208?—*Turt.* iii. 387?
Li. fugax. *Harris Ex. pl.* 46. *f.* 2.

3444. 8; quadrimaculata*. *Linn.* ii. 901.—*Barbut* G. 207. *pl.* 11. *cum fig.*—*Berk. S.* i. 148.—*Don.* xi. *pl.* 407.—*Stew.* ii. 207.—*Turt.* iii. 386.—(*Sam.* 65. *pl.* 7. *f.* 1.)—*Wood.* ii. 19. *pl.* 46.
Li. maculata. *Harris Ex. pl.* xlvi. *f.* 1.

3445. 9; depressa*. *Linn.* ii. 902.—*Harr. A. pl.* 26. *f. l, m.*—*Berk. S.* i. 148.—*Don.* i. *pl.* xxiv. ♀.—iii. *pl.* 81. ♂.—ii. *pl.* 44. *larva.*—*Stew.* ii. 208.—*Shaw* G. Z. vi. 245.—*Turt.* iii. 386.—*Samou.* 257.

Sectio IV. TERMITINA, *MacLeay.*

(Planipennes *p*, *Latreille.*)

Familia VI: (81). MYRMELEONIDÆ, *Leach.*

(Myrmeleonides, *Latreille.*)

Genus 16: (537). MYRMELEON[a].

Familia VII: (82). HEMEROBIDÆ, *Leach.*

(Hemerobius, *Linné, &c.*—Hemerobini, *Latreille.*)

Genus 17: (538). OSMYLUS, *Latr., Leach, Samou.,* (*Kirby.*)

3447. 1; maculatus*. *Leach E. E.* ix. 138.—*Samou.* 260. *pl.* 7. *f.* 4.
He. maculatus. *Fabr.* ii. 83.—*Turt.* iii. 401.
He. chrysops. *Ræm.* G. *pl.* 25. *f.* 1.—*Berk. S.* i. 151?—*Don.* vi. 21. *pl.* 188.—*Shaw* G. Z. vi. 259. *pl.* 83. *fig. med.*
He. fulvicephalus. *Vill.* iii. *pl.* 7. *f.* 7.

Genus 18: (539). DREPANEPTERYX, (*Leach.*)

3448 † 1. Phalænoides. (*Leach E. E.* ix. 138.)
He. Phalænoides. *Linn.* ii. 908.—*Panz. F.* lxxxvii. *f.* 15.—*Stew.* ii. 216.—*Turt.* iii. 401.—(*Curtis l. c. folio* 202.)
In Mus. D. Walker.

Genus 19: (540). CHRYSOPA, *Leach, Samou.* *Golden-eye.*

3449. 1; fulviceps *mihi.*
Ch. fulvocephala. (*Sam. I.* 11.)

3450. 2; capitata*. (*Sam. I.* 11.)
He. capitata. *Fabr.* ii. 82.—*Turt.* iii. 401.

3451. 3; reticulata*. (*Leach E. E.* ix. 138.)—(*Sam. I.* 11.)

---

[a] Genus 16: (537). MYRMELEON *Auctorum.* *Lion-Ant.*
Formicaleo, *Leach.*—Hemerobius *p*, *Berk.*

3446 ‡ 1. formicarium. *Linn.* ii. 914.—*Barbut G.* 221. *pl.* 12. (!)—*Stew.* ii. 217. *pl.* vii. *f.* 12—15. (!)—*Turt.* iii. 404. (!)—*Wood.* ii. 32. *pl.* 50.
Fo. formicarius. *Leach E. E.* ix. 138.
He. formicarium. *Berk. S.* i. 152. (!)

He. chrysops. *Linn.* ii. 912.—*Stew.* ii. 216.—*Turt.* iii. 401? —*Wood.* ii. 29. *pl.* 49.
He. perla. *Panz. F.* lxxxvii. *f.* 13?

3452. 4; maculata* *mihi.*

3453. 5; immaculata* *mihi.*

3454. 6; punctifrons* *mihi.*

3455. 7; alba*. (*Sam. I.* 11.)
He. albus. *Linn.* ii. 911.—*Panz. F.* lxxxvii. *f.* 14.—*Turt.* iii. 401.

3456. 8; affinis*. *Leach MSS?*

3457. 9, angusta* *mihi.*

3458. 10; Perla*. (*Leach E. E.* ix. 138.)—(*Sam. I.* 11.)
He. Perla. *Linn.* ii. 911.—*Barbut G.* 220. *pl.* 12.—*Stew.* ii. 216.—*Don.* viii. *pl.* 277. *f.* 2.—*Turt.* iii. 401.—*Shaw* G. Z. vi. 258. *pl.* 83. *fig. sup.*
He. pectinicornis. *Berk.* i. 151.
Albin. pl. 64.

Genus 20: (541). HEMEROBIUS *Auctorum.*

A. Alæ valdè reticulatæ.

3459. 1; hirtus*. *Linn.* ii. 912.—*Stew.* ii. 216.—*Don.* iv. *pl.* 113. *f.* 42.—*Turt.* iii. 401.—(*Curtis l. c. infra.*)
He. decussatus. *Leach MSS.*—(*Sam. I.* 21.)—(*Curtis l. c. infra?*)

3460. 2; concinnus* *mihi.*

3461. 3; fuscus*. *Leach MSS.*

3462. 4; nebulosus* *mihi.*

3463. 5; subnebulosus* *mihi.*

3464. 6; nervosus*. *Fabr.* ii. 85.—*Turt.* iii. 402?—(*Sam. I.* 21.) —(*Curtis l. c. infra.*)

3465. 7; Humuli*. *Linn.* ii. 912.—*Turt.* iii. 402.

3466. 8; lutescens*. *Fabr.* ii. 84?—*DeGeer.* ii. *pl.* 22. *f.* 8?—*Turt.* iii. 402.—(*Sam. I.* 21.)—(*Curtis l. c. infra?*)

3467. 9; affinis*. *Leach MSS.*—(*Sam. I.* 21.)—(*Curtis l. c. infra.*)

3468. 10; nemoralis*. *Leach MSS.*—(*Sam. I.* 21.)—(*Curtis l. c. infra.*)

3469. [11; obsoletus*] *mihi.*

3470. 12; paganus*. *Linn.* ii. 912.—*Turt.* iii. 403.

3471. 13; apicalis* *mihi.*

3472. 14; punctatus*. *Leach MSS.*—(*Sam. I.* 21.)—(*Curtis l. c. infra?*)

3473. 15; obscurus*. *Leach MSS.*—(*Sam. I.* 21.)—(*Curtis l. c. infra.*)

3474. 16; subfasciatus* *mihi.*

3475. 17; irroratus*. *Leach MSS.*—(*Sam. I.* 21.)—(*Curtis l. c. infra.*)

3476. 18; Pini*. *Leach MSS.*—(*Sam. I.* 21.)—(*Curtis l. c. infra.*)

3477. 19; Stigma* *mihi.*

3478. 20, fasciatus. *Fabr.* ii. 85.—(*Curtis l. c. infra.*)

3479. 21, angulatus* *mihi.*

3480. 22, pallidus* *mihi.*

3481. 23; variegatus*. *Fabr.* ii. 85.—*Turt.* iii. 402.—(*Sam. I.* 21.)—(*Curtis l. c. infra.*)

3482. 24; crispus*. (*Curtis l. c. infra.*)—*Schæff. Ic.* 122. *f.* 2, 3?

3483. 25; elegans* *mihi.*

3484. 26, Marshami *mihi.*

3485 † 27, fimbriatus. *Curtis.* v. *pl.* 202. In Mus. *Brit.* et D. Dale.

B. Alæ vix reticulatæ.

3486. 28; fuscatus*. *Fabr.* ii. 84.—*Turt.* iii. 402.
He. Beckwithii. *Leach MSS.*—(*Sam. I.* 21.)—(*Curtis l. c. supra.*)

3487. 29, nitidulus*. *Fabr.* ii. 83.—*Turt.* iii. 401.

3488. [30, confinis*] *mihi.*

Familia VIII: (83). PSOCIDÆ, *Leach.*

(Psoquilii, *Latreille.*—Termes *p*, Hemerobius *p*, *Linné.*)

Genus 21: (542). PSOCUS, *Latr.*, *Leach*, *Samou.*, (*Kirby.*)

A. Antennis valdè pilosis, aut pectinatis.

3489. 1; pilicornis*. *Kirby? MSS.*

3490. 2; picicornis*. *Fabr. S.* 204.
He. picicornis. *Turt.* iii. 402.

B. Antennis vix pilosis.

3491. 3; fasciatus*. *Fabr. S.* 203.—*Panz. F.* xciv. *f.* 20.
He. fasciatus. *Turt.* iii. 402.

3492. 4; variegatus*. *Kirby? MSS.*

3493. 5; atomarius* *mihi.*

3494. 6; maculatus* *mihi.*

3495. [7; obsoletus*] *mihi.*

3496. 8; lineatus*. *Kirby MSS.*

3497. 9; nebulosus*. *Kirby? MSS.*

3498. 10; similis*. *Kirby? MSS.*

3499. 11; bifasciatus*. *Kirby? MSS.*

3500. 12; contaminatus*. *Kirby MSS.*

3501. 13; longicornis*. *Fabr. S.* 203.—*Panz. F.* xciv. *f.* 19.
He. longicornis. *Turt.* iii. 402.

3502. 14; immunis*. *Kirby MSS.*

3503. 15; vittatus*. *Kirby MSS.*

3504. 16; ochropterus*. *Kirby MSS.*

3505. 17; flavidum*. *Kirby MSS.*

3506. 18; flavicans*. *Fabr. S.* 203.
He. flavicans. *Linn.* ii. 913.—*Turt.* iii. 403.

3507. 19; obsoletus* *mihi.*

3508. 20; hyalinus* *mihi.*

3509. 21; bipunctatus*. *Fabr. S.* 204.—*Panz. F.* xciv. *f.* 21.—*Leach E. E.* ix. 139.—(*Sam. I.* 35.)
He. bipunctatus. *Linn. F. no.* 1514.—*Turt.* iii. 403.

3510. 22; sexpunctatus*. *Fabr. S.* 203.
He. sexpunctatus. *Linn.* ii. 913.—*Berk. S.* i. 151.—*Coqueb. pl.* 2. *f.* 10.—*Stew.* ii. 216.—*Turt.* iii. 402.

3511. 23; striatulus*. *Fabr. S.* 203.
He. striatulus. *Turt.* iii. 402.

3512. 24; quadrimaculatus*. *Kirby? MSS.*

3513. 25; flaviceps* *mihi.*

3514. 26; immaculatus*. *Kirby MSS.*

3515. 27; rufescens* *mihi.*

3516. 28; quadripunctatus*. *Fabr. S.* 204.—*Panz. F.* xciv. *f.* 22.
He. 4-punctatus. *Turt.* iii. 403.

3517. [29; subpunctatus*] *mihi.*

3518. 30; costalis* *mihi.*

3519. 31; nervosus* *mihi.*

3520. 32; nigricornis* *mihi.*

3521. 33; phæopterus*. *Kirby MSS.*

3522. 34; nigricans*. *Kirby MSS.*

3523. 35; abdominalis*. *Fabr. S.* 204. (!)—*Stew.* ii. 217.
He. abdominalis. *Fabr. E. S.* ii. 86. (!)—*Turt.* iii. 403.
Ps. fuscescens. *Kirby MSS.*

3524. 36; dubius* *mihi.*

Genus 22: (543). ATROPOS, *Leach, Samou.* *Death-watch.*

PSOCUS *p, Fabr.*—PEDICULUS *p, Geoff.*

3525. 1; pulsatorius*.

Te. pulsatorium. *Linn.* ii. 1015.—*Wood.* ii. 120. *pl.* 75.
Te. lignarius. *DeGeer.* vii. 41. *pl.* 4. *f.* 1.
At. lignarius. (*Leach E. E.* ix. 139.)—(*Sam. I.* 5.)

3526. 2; fatedicus*.
Te. fatedicum. *Linn.* ii. 1016.
Ps. fatedicus. *Fabr. S.* 204.

Familia IX: (84). RAPHIDIIDÆ, *Leach.*

(RAPHIDINI, *Latreille.*)

Genus 23: (544). RAPHIDIA *Auctorum.* *Crane-Fly.*

3527. 1; Ophiopsis*. *Linn.* ii. 916.—*Shaw G. Z.* vi. 265. *pl.* 87. *fig. inf.*—(*Samou.* 261. *pl.* 7. *f.* 6.)—*Curtis.* i. *pl.* 37.
Ra. notata. *Fabr. M.* i. 251.—*Stew.* ii. 218.—(*Kirby & Sp. I. E.* i. *pl.* 3. *f.* 6.)

3528. [2; megacephala*.] *Leach MSS.*—(*Sam. I.* 36.)—(*Curtis l. c. supra.*)

3529. 3; Londinensis*. *Leach MSS.*—(*Sam. I.* 36.)—(*Curtis l. c. supra.*)
Ra. Ophiopsis. *Barbut* G. 229. *pl.* 12?—*Stew.* ii. 218?—*Wood.* ii. 37. *pl.* 52?

3530. 4, affinis*. *Leach MSS.*—(*Sam. I.* 36.)—(*Curtis l. c. supra.*)

3531. 5, maculicollis*. *Leach MSS.*—(*Sam. I.* 36.)—(*Curtis l. c. supra.*)

3532. 6. confinis* *mihi.*

SECTIO V. MEGALOPTERA, *MacLeay.*

(PLANIPENNES *p*, *Latreille.*)

Familia X: (85). SIALIDÆ, *Leach.*

(PERLARIÆ *p*, *Latreille.*)

Genus 24: (545). SIALIS, *Latr.*, *Leach*, *Samou.*, (*Kirby.*)

HEMEROBIUS *p*, *Linné*, *Geoff.*—SEMBLIS *p*, *Fabricius.*

3533. 1; lutarius*.
He. lutarius. *Linn.* ii. 913.—*Stew.* ii. 216.—*Turt.* iii. 400.—*Shaw* G. *Z.* vi. *pl.* 83. *fig. inf.*
Si. niger*. *Latr. H.* xiii. 44.—(*Leach E. E.* ix. 139.)—(*Sam. I.* 38.)

Familia XI: (86). PERLIDÆ, *Leach.*

(PERLARIÆ *p*, *Latr.*—PHRYGANEA *p*, *Linné*, &c.)

Genus 25: (546). PERLA, *Geoffroy*, (*Kirby*,) *Curtis.*

SEMBLIS *p*, *Fabr.*

A. Abdomen validum.

3534 † 1. grandis*. (*Curtis l. c. infra.*) In Mus. —— ?

3535. 2, marginata. *Fabr.* ii. 73.—(*Curtis l. c. infra.*)
Se. marginata. *Panz. F.* lxxi. *f.* 3.
Ph. marginata. *Turt.* iii. 395.

3536. 3, cephalotes. *Curtis.* iv. *pl.* 190.

3537. 4; bicaudata*. *Latr. H.* xiii. 49.—(*Curtis l. c. supra.*)
Ph. bicaudata. *Linn.* ii. 980.—*Barbut G.* 216.—*Berk. S.* i. 150. —*Stew.* ii. 213.—*Turt.* iii. 395.
Se. bicaudata. *Panz. F.* lxxi. *f.* 4.
Se. fusca. *Geoffroy.* ii. 231. *pl.* 13. *f.* 2.

3538. 5, nervosa *mihi.*

B. Abdomen subgracile. a. *Nervis alarum distinctis.*

3539. 6; fuscipennis*. *Curtis l. c. supra?*

3540. 7; lutea*. *Latr. H.* xiii. 49.

3541. 8; media*. *Curtis l. c. supra.*

3542. 9; venosa* *mihi.*

b. *Nervis alarum vix distinctis, concoloribus.*

3543. 10; viridis*.
Se. viridis. *Fabr.* ii. 74.—*Curtis l. c. supra?*
Ph. viridis. *Turt.* iii. 395.

3544. 11; pallida*.
Se. minor. *Curtis l. c. supra?*

Genus 26: (547). NEMOURA, *Latr., Leach.*

SEMBLIS *p, Fabr.*—PERLA *p, Geoffroy.*

A. Oculis prominulis; capite lato.

3545. 1; nebulosa*. *Latr.* G. iii. 210.—*Leach E. E.* ix. 139.
Ph. nebulosa. *Linn.* ii. 980.—*Berk. S.* i. 150.—*Stew.* ii. 213. —*Turt.* iii. 395.

3546. [2; fuliginosa*] *mihi.*

3547. 3; pallida* *mihi.*

3548. 4; cruciata* *mihi.*

3549. [5; affinis*] *mihi.*

3550. 6; pusilla* *mihi.*

3551. 7; annulata* *mihi.*

3552. 8; pallipes* *mihi.*

3553. 9; luteicornis* *mihi.*

3554. 10; pallicornis* *mihi.*

3555. 11, nitidus*? *mihi.*

3556. 12; sulcicollis* *mihi.*

3557. 13; fumosus* *mihi.*

3558. 14, variegatus *mihi.*

B. Oculis vix prominulis; capite subangusto.

3559. 15; fusciventris* *mihi.*

3560. 16; abdominalis* *mihi.*

Genus 27: (548). ——

3561. 1; geniculata* *mihi.*

Genus 28: (549). ACENTRIA *mihi.*

PHRYGANEA *p, Olivier?*

3562. 1: nivosa*.
Ph. nivea. *Olivier?—Latr. H.* xiii. 93?

ORDO V. TRICHOPTERA, *Kirby.*

(NEUROPTERA *p, Linné.*—SYNISTATA *p, Fabr.*—NEU.-PLICIPENNES, *Latr.*)

(Genus PHRYGANEA, *Linné, &c.*)

Water-moths, Cadew-flies, Spring-flies. Larvæ; *Caddis-worms.*

Familia I: (87). PHILOPOTAMIDÆ *mihi.*

Genus 1: (550). TINODES, *Leach.*

A. Nervis alarum distinctis (in masculis, alis anticis basi membranâ elevatâ.)

3563. 1, fimbriata.
Ph. fimbriata. *Walker's MSS.*

3564. [2, simplex] *mihi.*

3565. 3, pallescens *mihi.*

3566. 4; flaviceps* *mihi.*

3567. 5; xanthocera* *mihi.*

3568. 6; picicornis* *mihi.*

3569. 7, annulicornis *mihi.*

3570. 8, albipunctata* *mihi.*

3571. 9; unipunctata* *mihi.*

B. Nervis alarum vix distinctis: (in masculis, alis simplicibus.)

3572. 10, pallipes *mihi.*

3573. 11; lutescens* *mihi.*

3574. 12, obscura *mihi.*

3575. 13; phæopus* *mihi.*

3576. 14; subochracea* *mihi.*

3577. 15; ciliaris* *mihi.*

3578. 16; subaurata* *mihi.*

Genus 2: (551). HYDROPTILA, *Dahlman.*

3579. 1; tineoides*. *Dahlman. A. E.* i. *p.* 47.

3580. [2; marginata*] *mihi.*

Genus 3: (552). ——

TINODES B, *Leach MSS.*

3581. 1; albipes* *mihi.*

3582. 2; pusillus*.
Ph. pusilla. *Fabr. E. S.* ii. 81?

3583. 3; funereus*.
Ph. funereus. *Olivier.—Latr. H.* xiii. 93.

3584. 4; opacus* *mihi.*

3585. 5; pygmæus* *mihi.*

3586. 6, fuliginosus* *mihi.*

3587. 7, Marshamellus *mihi.*

3588. 8, flavipes* *mihi.*

Genus 4: (553). ——

TINODES A *p? Leach MSS.*

3589. 1, pulchellus.
Ti. pulchellus. *Leach MSS.*

3590. 2, cognatus *mihi.*

Genus 5: (554). ——

3591. 1, nigripunctatus *mihi.*

Genus 6: (555). ——

TINODES *p? Leach MSS.*

3592. 1; subpunctatus* *mihi.*

3593. 2, picicornis *mihi.*

3594. 3; fuliginosus* *mihi.*

3595. [4; concinnus*] *mihi.*

3596. 5; pyrrhocerus* *mihi.*

3597. 6; flavo-maculatus* *mihi.*

3598. 7; subnebulosus* *mihi.*

Genus 7: (556). PHILOPOTAMUS, *Leach MSS.*

A. Margine postico alarum subrotundato.

3599. 1, Scopulorum. *Leach MSS.*

3600. 2, maculatus.
Phr. maculata. *Olivier.*—*Latr. H.* xiii. 89.—*Don.* xvi. *pl.* 548. *f.* 2.

3601. 3; variegatus*.
Phr. variegata. *Fabr. E. S.* ii. 79.—*Stew.* ii. 214.

3602 † 4. reticulatus[a].

3603. 5, Stigma* *mihi.*

3604. 6, obsoletus* *mihi.*

3605. 7, opacus* *mihi.*

B. Margine postico alarum subtruncato, aut recto.

3606. 8, nebulosus *mihi.*

3607. 9; obscurus*. *Leach MSS?*

3608. 10; affinis* *mihi.*

3609. 11; montanus.
Phr. montana. *Don.* xvi. *pl.* 548. *f.* 1.

3610. 12; obliquus* *mihi.*

3611. 13, bimaculatus *mihi.*

3612. 14, antennatus* *mihi.*

3613. 15, ochroleucus *mihi.*

3614. 16; dorsalis* *mihi.*

3615. 17; marginepunctatus *mihi.*

Genus 8: (557). ——

3616. 1; nubilus* *mihi.*

3617. 2; concolor* *mihi.*

3618. 3, costalis *mihi.*

Genus 9: (558). AMBLYPTERYX *mihi.*

3619. 1; nigripalpis* *mihi.*

3620. 2; rufipalpis *mihi.*

Genus 10: (559). CHIMARRA, *Leach MSS.*

3621. 1; marginata*. *Leach MSS.*
Ph. marginata. *Linn.* ii. 910.

Genus 11: (560). POTOMARIA, *Leach MSS.*

3622. 1; analis*.
Ph. analis. *Fabr. E. S.* ii. 75.

---

[a] 3602 † 4. reticulatus.
Ph. reticulata. *Linn.* ii. 980.—*Panz. F.* lxxi. *f.* 5.—*Turt.* iii. 395. (!)

3623. 2; assimilis* *mihi.*

3624. 3; hyalinus* *mihi.*

Genus 12: (561). NOTIDOBIA *mihi.*

3625. 1, pallipes.
Ph. pallipes. *Fabr. E. S.* ii. 76.

3626. 2; atrata*.
Ph. atratus. *Fabr. E. S.* ii. 78.—*Coqueb. Ic.* i. *f. A. B?*
Ph. atra. (*Kirby & Sp. I. E.* ii. 220?)

Genus 13: (562). PROSOPONIA, *Leach MSS.*

3627. 1; Leachii* *mihi.*
Ph. personatus. *Spence MSS.*

Genus 14: (563). GOËRA, *Hoffmansegg MSS., Leach.*

3628. 1; pilosa*.
Ph. pilosa. *Fabr. E. S.* ii. 76.

3629. 2; vulgata*.
Ph. vulgata. *Olivier.*—*Latr. H.* xiii, 91?

3630. 3; fuscata* *mihi.*

3631. 4; hirta*.
Ph. hirta. *Fabr. E. S.* ii. 80.—*Stew.* ii. 215.

3632. 5; immaculata* *mihi.*

3633. [6, nigromaculata*] *mihi.*

Familia II: (88). LEPTOCERIDÆ, *Leach.*

Genus 15: (564). ODONTOCERUS, *Leach, Samou.*

3634. 1; griseus*. *Leach E. E.* ix. 136.—(*Sam. I.* 31.)

Genus 16: (565). CERACLEA, *Leach MSS.*

3635. 1; nervosa*.
Ph. nervosa. *Latr. H.* xiii. 91.

Genus 17: (566). LEPTOCERUS, *Leach MSS., Samou., Curtis.*

3636. 1, ochraceus*. *Curtis.* ii. *pl.* 57.

3637. 2; dissimilis* *mihi.*

3638. 3; assimilis* *mihi.*

3639. 4; annulicornis* *mihi.*

3640. 5; perfuscus* *mihi.*

3641. 6, tarsalis *mihi.*

3642. 7, subannulatus *mihi.*

3643. 8; seminiger* *mihi.*

3644. 9, grossus* *mihi.*

3645. 10, bimaculatus*.
Ph. bimaculata. *Linn.* ii. 909.—*DeGeer.* ii. 416. *pl.* 15. *f.* 5. —*Turt.* iii. 397.

3646. 11, filosus*.
Ph. filosa. *Linn.* ii. 910.—*Fabr. E. S.* ii. 80. (!)—*Stew. E.* ii. 215.—*Turt.* iii. 397.

3647. 12, ater* *mihi.*

3648. 13; nigricans *mihi.*

3649. 14; aterrimus* *mihi.*

3650. 15; interruptus*. (*Leach E. E.* ix. 136.)—(*Sam. I.* 24.)—(*Curtis l. c. supra.*)
Ph. interrupta. *Fabr. E. S.* ii. 79.—*Stew.* ii. 215.—*Turt.* iii. 397.—*Don.* xvi. *pl.* 551.

3651. 16, bilineatus.
Ph. bilineata. *Linn.* ii. 910?—*Turt.* iii. 397.

3652. 17; rufogriseus* *mihi.*

3653. 18; rufus* *mihi.*

3654. 19; attenuatus* *mihi.*

3655. 20, longicornis*.
Ph. longicornis. *Linn.* ii. 910.—*Berk. S.* i. 151.—*Stew.* ii. 214. —*Turt.* iii. 397.

3656. 21; quadrifasciatus*.
Ph. quadrifasciata. *Fabr. E. S.* ii. 80.—*Turt.* iii. 397.

3657. 22, affinis. *Leach MSS.*

3658. 23, elongatus* *mihi.*

Genus 18: (567). MYSTAX? (*Latreille?*)

3659. 1; nigra.
Ph. nigra. *Linn.* ii. 909.—*DeGeer.* ii. 424. *pl.* 15. *f.* 21.—*Berk.* i. 151.—*Stew.* ii. 214.—*Turt.* iii. 397.

3660. 2; azurea.
Ph. azurea. *Linn.* ii. 909.—*Turt.* iii. 397.

3661. 3, phæa* *mihi.*

Familia III: (89). PHRYGANIDÆ.

Genus 19: (568). ANABOLIA *mihi.*

LINMEPHILUS *p*, *Leach*, *Samou.*

3662. 1; nervosa*.
Li. nervosus. *Leach MSS.*—(*Sam. I.* 25. *pl.* 7. *f.* 3.)

3663. 2; lurida* *mihi.*

Genus 20: (569). ——

LIMNEPHILUS *p*, *Leach*.

3664. 1; dubius* *mihi*.

3665. 2; planus* *mihi*.

3666. 3, annulatus *mihi*.

3667. 4, lutescens *mihi*.

3668. 5, picicornis *mihi*.

Genus 21: (570). CHÆTOPTERYX *mihi*.

LIMNEPHILUS *p*, *Leach*.

3669. 1; villosa*.
Ph. villosa. *Fabr. E. S. Sup.* 200.
Ph. granularis. *Kirby MSS.*
Li. echinatus. *Leach MSS.*—(*Sam. I.* 25.)

Genus 22: (571). ——

LIMNEPHILUS *p*, *Leach MSS.*

3670. 1; radiatus*. *Leach MSS.*—(*Sam. I.* 25.)

3671. 2, angulatus *mihi*.

3672. 3, angustatus *mihi*.

3673. 4; confinis* *mihi*.

3674. 5; flavus* *mihi*.

3675. 6; lateralis* *mihi*.

3676. 7; obscurus* *mihi*.

Genus 23: (572). PHRYGANEA *Auctorum*.

3677. 1; striata*. *Linn.* ii. 908.—*Berk. S.* i. 151.—*Stew.* ii. 213. —*Turt.* iii. 395.—*Geoff.* ii. 246. *pl.* 13. *f.* 5.

3678. 2; grandis*. *Linn.* ii. 909.—*DeGeer.* ii. 388. *pl.* 13. *f.* 1. —*Stew.* ii. 213.—*Panz. F.* xciv. *f.* 18.—*Shaw* G. Z. vi. 255. *pl.* 82*. *fig. inf.*—(*Sam. I.* 33.)—*Wood.* ii. 25. *pl.* 48.—(*Kirby & Sp. I. E.* iii. 68. *pl.* iii. *f.* 4.)—*Turt.* iii. 396.—*Leach E. E.* ix. 136.

3679. 3, Beckwithii. *Leach MSS.*

3680. 4, flexuosa* *mihi*.

3681. 5, atomaria*. *Fabr. E. S.* ii. 78.

3682. 6; varia*. *Fabr. E. S.* ii. 77.—*Don.* viii. *pl.* 277. *f.* 1.—*Stew.* ii. 214.—*Turt.* iii. 396.

3683 † 7. Phalænoides[a].

3684 † 8. flavilatera[b].

---

[a] 3683 † 7. Phalænoides. *Linn.* ii. 908.—*Turt.* iii. 396. (!)
[b] 3684 † 8. flavilatera. *Linn.* ii. 909.—*Berk. S.* i. 151. (!)—*Stew.* ii. 214. (!) An hujus generis?

Genus 24: (573). LIMNEPHILUS, *Leach.*

A. Alis integris.

3685. 1; striola*. *Leach MSS.*—(*Sam. I.* 25.)

3686. 2; bimaculatus* *mihi.*

3687. 3; dorsalis* *mihi.*

3688. 4; griseus*. (*Leach MSS.*)—(*Sam. I.* 25.)
Ph. grisea. *Linn.* ii. 909.—*DeGeer.* ii. 399. *pl.* 13. *f.* 21.—*Stew.* ii. 213.—*Turt.* iii. 396.

3689. 5; flavicornis*.
Ph. flavicornis. *Fabr. E. S.* ii. 77.—*Stew.* ii. 213.

3690. 6; nebulosus* *mihi.*

3691. 7; glaucopterus* *mihi.*

3692. 8; rhombicus*. (*Leach E. E.* ix. 136.)—(*Sam. I.* 25.)
Ph. rhombica. *Linn.* ii. 909.—*Don.* vii. *pl.* 220.—*Stew.* ii. 214.—*Turt.* iii. 396.—*Shaw G. Z.* vi. 255. *pl.* 82*. *fig. med.*
Ph. rhomboidica. *Berk. S.* i. 151.

3693. 9; affinis* *mihi.*

3694. 10, versicolor *mihi.*

3695. 11: tessellatus *mihi.*

3696. 12: signatus *mihi.*

3697. 13, obliquus *mihi.*

3698. 14; marginalis* *mihi.*

3699. 15; costalis* *mihi.*

3700. 16; punctulatus* *mihi.*

3701. 17; stigma* *mihi.*

3702. 18, lunatus *mihi.*

3703. 19; fenestralis* *mihi.*

3704. 20, geminus *mihi.*

3705. 21; assimilis* *mihi.*

3706. 22; fuscus* *mihi.*

3707. 23; punctatissimus* *mihi.*

3708. 24; quadrimaculatus* *mihi.*

3709. 25; confluens* *mihi.*

3710. 26; fuscatus* *mihi.*

3711. 27; nubilus* *mihi.*

3712. 28, irroratus *mihi.*

3713. 29, flavescens *mihi.*

3714. 30; testaceus* *mihi.*

3715. 31; subnebulosus* *mihi.*

3716. 32, notatus *mihi.*

3717. 33; interruptus* *mihi.*

3718. 34, vittatus *mihi.*

3719. 35, lineola *mihi.*

3720. 36, præustus *mihi.*

3721. 37, villosus *mihi.*

3722. 38; variabilis* *mihi.*

3723. 39, punctatus *mihi.*

3724. 40, ustulatus *mihi.*

3725. 41, apicalis *mihi.*

3726. 42, elongatus *mihi.*

3727. 43, fuliginosus *mihi.*

B. Alis anticis angulato-dentatis.

3728. 44; angulatus* *mihi.*

3729. 45; cognatus* *mihi.*

3730. 46, diaphanus *mihi.*

3731. 47; ornatus* *mihi.*

3732. 48; pictus* *mihi.*

Genus 25: (574). NEURONIA, *Leach MSS.*

3733. 1; fusca*.
Ph. fusca. *Linn.* ii. 910.—*Stew.* ii. 215.—*Turt.* iii. 395.

Ordo VI. HYMENOPTERA.

(Piezata, *Fabricius.*—Phleboptera, *Clairville.*)

Familia I: (90). TENTHREDINIDÆ, *Leach.*

(Tenthredo, *Linné, &c.* Ordo Trichoptera *p, MacLeay.*)
Saw-Flies.

Genus 1: (575). CIMBEX, *Olivier, Leach, Samou.,* (*Kirby,*) *Curtis.*
Crabro, *Geoffroy.*—Clavellaria *p, Lamarck.*

3734. 1. femorata*. *Fabr. P.* 15.—*St. Farg. M.* 31.
Ci. Europæa. *Leach Z. M.* iii. 104.—*Samou.* 262.
Ci. variabilis femorata. *Klug Ci. p.* 7.
♂, Te. femorata. *Linn.* ii. 920.—*Berk. S.* i. 153.—*Panz. F.* xxvi. *f.* 20.—*Stew.* ii. 221.—*Turt.* iii. 412.
Cr. lunulatus. *Fourc.* ii. 362.

Ci. tristis. *Fabr. P.* 17?
† ♀. Te. lutea. *Linn.* ii. 921. (*teste Klug.*)
Ci. sylvarum. *Fabr. P.* 16.—*Panz. F.* lxxxviii. *f.* 4.

3735. 2; varians*. *Leach Z. M.* iii. 105.—(*Sam. I.* 11.)—*St. Farg. M.* 32. (!)
Ci. variabilis femorata var. *Klug Ci. p.* 9.
Albin. pl. lix. *f. a—c?*

3736. [3, pallidus*] *mihi.*

3737 † 4, decem-maculata. *Leach Z. M.* iii. 106.—(*Sam. I.* 11.)—*Curtis.* i. *pl.* 41.—*St. Farg. M.* 32. (!)
Ci. variabilis femorata var. *Klug Ci. p.* 9. In Mus. *Brit.*

3738. 5, Griffinii*. *Leach Z. M.* iii. 107.—(*Sam. I.* 11.)—*St. Farg. M.* 30. (!)
♂, Te. lutea. *Don.* vi. 49. *pl.* 234.—*Stew.* ii. 222.
Ci. variabilis femorata var. *Klug Ci. p.* 10.

3739. 6, annulata. *Leach Z. M.* iii. 107.—(*Sam. I.* 11.)
Ci. variabilis lutea. *Klug Ci. p.* 11.
Ci. lutea. *Fabr. P.* 16.—*Berk. S.* i. 153.—*Turt.* iii. 412.

3740 † 7, maculata. *Leach Z. M.* iii. 106.—(*Sam. I.* 11.)
Cr. maculatus. *Fourc. P.* ii. 2.
Ci. variabilis montana. *Klug Ci. p.* 13.
Te. montana. *Panz. F.* lxxxiv. *f.* 12.
Te. connata. *Schrank B.* ii. 322. In Mus. *Brit.*

3741 † 8, humeralis. *Leach Z. M.* iii. 108.—(*Sam. I.* 11.)
Cr. humeralis. *Fourc. P.* ii. 361.
Te. axillaris. *Panz. F.* lxxxiv. *f.* 11.
Te. connata. *Villars E.* iii. 84. In Mus. *Brit.*

Genus 2: (576). TRICHIOSOMA, *Leach, Samou., Curtis.*
CIMBEX *p, Oliv.*

3742. 1; sylvaticum*. *Leach Z. M.* iii. 108.—*Samou.* 263.—(*Curtis l. c. infra.*)

3743. 2; laterale*. *Leach Z. M.* iii. 109.—(*Sam. I.* 42.)—*Curtis.* i. *pl.* 49.
Te. Vitellinæ. *Stew.* ii. 222.—*Turt.* iii. 412.

3744. 3; Lucorum*. *Leach Z. M.* iii. 109.—(*Curtis l. c. supra.*)
Te. Lucorum. *Linn.* ii. 921.
Te. Vitellinæ. *Linn.* ii. 921.—*Don.* iii. *pl.* 88. *f.* 3.
Ci. Amerinæ. *Latreille.*

3745. [4; tibiale*] *mihi.* (*Curtis l. c. supra.*)

3746. 5; Scalesii*. *Leach Z. M.* iii. 111.—(*Sam. I.* 42.)—(*Curtis l. c. supra.*)

3747. 6; unidentatum*. *Leach Z. M.* iii. 111.—(*Sam. I.* 42.)—(*Curtis l. c. supra.*)

3748. 7, pusillum *mihi.*

Genus 3: (577). CLAVELLARIA, *Lam.*, *Leach*, *Samou.*, *Curtis.*

CIMBEX *p*, *Oliv.*

3749. 1, Amerinæ*. *Leach Z. M.* iii. 112.—(*Sam. I.* 11.)—(*Curtis l. c. infra.*)
Te. Amerinæ. *Linn.* ii. 921.—*Panz. F.* lxv. *f.* 1.—*Turt.* iii. 412.

3750. [2, marginata*.] *Leach Z. M.* iii. 112.—*Samou.* 263.
Ci. marginata. *Linn.* ii. 920.—*Fabr. E. S.* ii. 106. (!)—*Turt.* iii. 412.
Ci. quadrifasciata. *Oliv. En.* v. 771.

Genus 4: (578). ZARÆA, *Leach*, *Samou.*, *Curtis.*

CIMBEX *p*, *Oliv.*

3751. 1, fasciata*. *Leach Z. M.* iii. 113.—*Samou.* 263.—*Curtis.* ii. *pl.* 97.
Te. fasciata. *Linn.* ii. 921.—*Don.* xii. 7. *pl.* 398.—*Turt.* iii. 412.

Genus 5: (579). ABIA, *Leach*, *Samou.*, *Curtis.*

CIMBEX *p*, *Fabr.*

3752. 1; nigricornis*. *Leach Z. M.* iii. 113.—*Samou.* 263.—*Curtis.* ii. *pl.* 89.
Te. nitens, ♀. *Linn. F.* 1539.
Ce. sericea var. *Fabr. P.* 18.
Ci. ænea. *Klug Ci. p.* 21.
Ci. nitens. (*Kirby & Sp. I. E.* iii. 189.)

3753. 2; sericea*. *Leach Z. M.* iii. 113.—(*Sam. I.* 1.)
♀; Te. sericea. *Linn.* ii. 921.—*Don.* xii. 23. *pl.* 402.—*Turt.* iii. 413.
Te. nitens. *Schrank B.* ii. *No.* 1997.

Genus 6: (580). AMASIS, *Leach*, *Samou.*

CIMBEX *p*, *Fabr.*

3754 † 1, obscura. *Leach Z. M.* iii. 114.—*Panz. F.* lxxxiv. *f.* 13?
Ci. obscura. *Fabr. P.* 18. In Mus. *Brit.*?

3755 † 2, læta. *Leach Z. M.* iii. 114.—*Samou.* 263.
Te. læta. *Fabr. P.* 18?—*Panz. F.* lxii. *f.* 6.
Ci. sylvatica. *Oliv. En.* v. 772.
Te. crassicornis. *Rossi.* ii. 21.
Ci. Jurinæ. *St. Farg. M.* 38. In Mus. *Brit.*

Genus 7: (581). HYLOTOMA, *Fabr.*, *Leach*, *Samou.*, *Curtis.*

ARGE, *Schrank.*—CRYPTUS *p*, *Panz.*—PTILIA *p*, *St. Farg.*

3756. 1; pilicornis*. *Leach Z. M.* iii. 121.—*Samou.* 264.—(*Curtis l. c. infra.*)
Pt. pilicornis. *St. Farg. M.* 50. (!)

3757 † 2, Berberidis*. *Klug B.* 55.—(*Sam. I.* 22.)—(*Curtis l. c. infra.*)
Arge Berberidis. *Schrank B.* ii. 229. In Mus. *Brit.*

3758. 3; Anglica*. *Leach Z. M.* iii. 122.—(*Curtis l. c. infra.*)
Hy. Angelica. *St. Farg. M.* 45?

3759. 4; enodis*. *Klug Bl.* 53.—*Fabr. P.* 23.—(*Sam. I.* 22.)—(*Curtis l. c. infra.*)
Te. enodis. *Linn.* ii. 922.—*Panz. F.* xlix. *f.* 12, 13.—*Turt.* iii. 413.
Te. cæruleipennis. *Retzius.* 72?
Te. violacea. *DeGeer.* ii. 282. *pl.* 40. *f.* 6?
♂. Arge ciliaris. *Schrank B.* ii. 227.

3760. 5; violacea*. *Klug Bl.* 57.—(*Sam. I.* 22.)—(*Curtis l. c. infra.*)

3761. 6; cærulea*. *Klug Bl.* 57.—(*Sam. I.* 22.)—(*Curtis l. c. infra.*)

3762. 7; ustulata*. *Fabr. P.* 23.—*Klug Bl.* 58.—(*Sam. I.* 22.)—(*Curtis l. c. infra.*)
♀; Te. ustulata. *Linn.* ii. 922.—*Panz. F.* lxxxi. *f.* 10.—*Turt.* iii. 413.
Te. ochroptera. *Fourc. P.* ii. 365.
♂. Te. ciliaris. *Linn.* ii. 922.

3763. 8; Leachii* *mihi.*
Te. atrata. *Forst. C.* i. 80?—*Stew.* ii. 222?—*Turt.* iii. 414?

3764. 9; Klugii*. *Leach Z. M.* iii. 122.—(*Sam. I.* 22.)—(*Curtis l. c. infra.*)

3765. [10; segmentaria*.] *Klug Bl.* 59.—(*Sam. I.* 22.)—(*Curtis l. c. infra.*)
Cry. segmentarius. *Panz. F.* lxxxviii. *pl.* 17.
Arge ustulata. *Schrank B.* ii. 226.

3766. 11; cyaneo-crocea*.
Te. cyaneo-crocea. *Forst. C.* i. 78.—*Stew.* ii. 222.
Hy. cærulescens. *Fabr. P.* 24. (!)—(*Sam. I.* 22.)—(*Curtis l. c. infra.*)
Te. cærulescens. *Fabr.* ii. 108.—*Panz. F.* xlix. *f.* 14.—*Stew.* ii. 223.—*Turt.* iii. 413.
Arge bicolor. *Schrank. B.* ii. 229.

3767. 12; femoralis*. *Klug Bl.* 63.—(*Sam. I.* 22.)—(*Curtis l. c. infra.*)
Hy. fasciata. *St. Farg. M.* 43.—*Faun. F. pl.* 2. *f.* 1.
Te. melanochra. *Gmel.* v. 2657?

3768. 13; Rosæ*. *Fabr. P.* 25.—(*Sam. I.* 22.)—(*Curtis l. c. infra.*)—(*Kirby & Sp. I. E.* iv. 85.)
Hy. Rosarum. *Klug Bl.* 60.

Te. Rosæ. *Panz. F.* xlix. *f.* 15.—*Don.* v. 71. *pl.* 164.
Te. ochropus. *Gmel.* v. 2657.

3769. 14; Stephensii*. *Leach Z. M.* iii. 123.—(*Sam. I.* 22.)—*Curt.* ii. *pl.* 65.—*St. Farg. M.* 43. (!)

3770. 15; pagana*. *Klug Bl.* 61.—(*Sam. I.* 22.)—(*Curtis l. c. supra.*)
Te. pagana. *Panz. F.* xlix. *pl.* 16.
Te. tricolor. *Gmel.* v. 2657.
Te. flaviventris. *Act. Holm.* (*Fallen.*) 1807. 202.

Genus 8: (582). SCHIZOCERUS, *Latreille, Curtis.*

CRYPTUS, *Jurine, Leach, Samou.,* (*Curtis, olim.*)—HYLOTOMA *p, Fabr., Samou.*

3771. 1, furcatus. (*Curtis l. c. infra.*)
Te. furcata. *Villars E.* iii. 86.—*Panz. F.* xlvi. *f.* 1.
Hy. furcata. (*Sam. I.* 22.)
Cr. Villersii. *Leach Z. M.* iii. 124.—*Sam.* 264.
Te. Rubi Idei. *Rossi.* ii. 31.

3772. 2; pallipes*. *Curtis.* ii. *pl.* 58.
Cr. pallipes. *Leach Z. M.* iii. 125.—(*Sam. I.* 14.)—*St. Farg. M.* 53. (!)
Hy. geminata. *Klug Bl.* 73?

Genus 9: (583). LOPHYRUS, *Latr., Leach,* (*Samou.,*) *Curtis.*

PTERONUS, *Jurine.*—HYLOTOMA, *Fabr.*—DIPRION, *Schrank.*

3773. 1, Pini. *Leach Z. M.* iii.—(*Sam. I.* 25.)—*Curtis.* ii. *pl.* 54. ♂ & ♀.
Ten. pectinata major. *Retzius. No.* 317.—*DeGeer. pl.* 36. *f.* 15–18.
† ♂, Ten. Pini. *Linn.* ii. 922.—*Stew.* ii. 222.—*Turt.* iii. 415.
♀, Hyl. dorsata. *Fabr. P.* 21.
Ten. dorsata. *Fabr.* ii. 111. (!)—*Stew.* ii. 222.—*Turt.* iii. 415.

3774. 2; rufus*. *Klug Bl.* 30.—(*Sam. I.* 25.)—(*Curtis l. c. supra.*)
Ten. Pini rufa. *Villars. E.* iii. 88.
Ten. Juniperi. *Christius B. p.* 432. *pl.* 49. *f.* 4, 5.
Ten. sertifera. *Fourc. P.* ii. 378.
Lo. Piceæ. *St. Farg. M.* 56.

3775 † 3, pallidus*. *Klug Bl.* 34.—(*Curtis l. c. supra.*)
Ten. pectinata minor. *Retzius. No.* 317.—*DeGeer. pl.* 35. *f.* 26.
Ten. Pini minor. *Villars E.* iii. 87.
Lo. minor. *St. Farg. M.* 54.
♀, Ten. dorsata. *Fallen.*—*Panz. F.* lxii. *f.* 9.
In Mus. *Brit. et D. Cooper.*

3776 † 4. Juniperi[a].

Genus 10: (584). CLADIUS, *Leach, Samou.*

NEMATUS *p, Leach.*—HYLOTOMA *p, Fabr.*—PTERONUS *p, Jurine.* —LOPHYRUS *p, Latreille.*

A. Cladius. *Leach Z. M. l. c.*

3777. 1; difformis*. (*Leach Z. M.* iii. 130.)—(*Sam. I.* 11.)
Ten. difformis. *Panz. F.* lxii. *f.* 10. ♂.

B. Nematus**. *Leach Z. M.* iii. 129.

3778. 2; rufipes*. *St. Farg. M.* 58.—*Faun. F. pl.* 12. *f.* 5.

3779. 3; pallipes*. *St. Farg. M.* 59.—*Faun. F. pl.* 12. *f.* 6.

3780. 4; Morio*. *St. Farg. M.* 58.

3781. 5, immunis* *mihi.*

3782. 6; luteicornis* *mihi.*

Genus 11: (585). PRISTIPHORA, *Latreille.*

PTERONUS *p, Jurine.*

A. Vide *St. Farg. M.* 59.

3783. 1; Myosotidis*. *St. Farg. M.* 59.
Te. Myosotidis. *Fabr. P.* 41.—*Panz. F.* xcviii. *f.* 13.

3784. 2; testacea*. *Latr.*—*St. Farg. M.* 59.
Pt. testaceus. (*Jurine. H.* 64. *pl.* 13. *fig. inf. med.*)

3785. 3; pallipes*. *St. Farg. M.* 60.

3786. 4; rufipes*. *St. Farg. M.* 60.—*Faun. F. pl.* 12. *f.* 2.

3787. 5; testaceicornis*. *St. Farg. M.* 60.—*Faun. F. pl.* 12. *f.* 1.

B. Vide *St. Farg. M.* 60.

3788. 6; duplex*. *St. Farg. M.* 61.—*Faun. F. pl.* 12. *f.* 3.

C. Vide *St. Farg. M.* 61.

3789. 7; varipes*. *St. Farg. M.* 61.

Genus 12: (586). NEMATUS, *Leach, Samou.*

3790. 1; pallipes*. *St. Farg. M.* 62.

3791. 2; crassicornis* *mihi.*

3792. 3; grandis*. *St. Farg. M.* 61.—*Faun. F. pl.* 10. *f.* 1.

3793. 4; pectoralis*. *St. Farg. M.* 62.—*Faun. F. pl.* 10. *f.* 2.

3794. 5, clitellatus*. *St. Farg. M.* 62.—*Faun. F. pl.* 10. *f.* 3.

---

[a] 3776 † 4. Juniperi. *St. Farg. M.* 55.
Te. Juniperi. *Linn.* ii. 923.—*Panz. F.* 76. *f.* 11. ♀.—*Turt.* iii. 415. (!)
Te. pterophorus. *Sulz. K.* 46. *pl.* 18. *f.* 110.

3795 † 6. Salicis[a].

3796. 7; dimidiatus*. *St. Farg. M.* 68.

3797. 8, dorsalis*. *St. Farg. M.* 70.

3798. 9; luteus*. (*Leach Z. M.* iii. 129.)—(*Sam. I.* 28.)—*St. Farg. M.* 70.—*Faun. F. pl.* 11. *f.* 7.
Te. lutea. *Fabr. P.* 41.

3799. 10; flavescens* *mihi.*

3800. 11; miliaris*. *St. Farg. M.* 70.
Te. miliaris. *Panz. F.* lxv. *f.* 13.

3801. 12; Capreæ*. *St. Farg. M.* 64.
Te. Capreæ. *Fabr. P.* 35. (*des. larvæ.*)—*Panz. F.* lxv. *f.* 8?—*Stew.* ii. 224.—*Turt.* iii. 418.—*Wood.* ii. 47. *pl.* 54? *Albin. pl.* 59. *f. g. h.*—(*Vide Kirby & Sp. I. E. l. c. infra.*)

3802. 13, viridis* *mihi.*

3803. 14; vittatus*. *St. Farg. M.* 64?—*Faun. F. pl.* 10. *f.* 5?

3804. 15; fallax*. *St. Farg. M.* 64?—*Faun. F. pl.* 10. *f.* 4?

3805. 16; affinis*. *St. Farg. M.* 69.

3806. 17; oblitus*. *St. Farg. M.* 69.

3807. 18; bicolor* *mihi.*

3808. 19; bipartitus*. *St. Farg. M.* 69.

3809. 20; Ribesii* *mihi.* (*Vide Kirby & Sp. I. E.* i. 195. *nota.*)

3810. 21; interruptus*. *St. Farg. M.* 65.—*Faun. F. pl.* 11. *f.* 1.

3811. 22; melanostigma* *mihi.*

3812. 23; gonymelas* *mihi.*

3813. 24; analis* *mihi.*

3814. 25; ruficornis*. *St. Farg. M.* 71.

3815. 26; proximus*. *St. Farg. M.* 67.

3816. 27; intercus*. *St. Farg. M.* 67.
Te. intercus. *Panz. F.* xc. *f.* 11?
Te. nigra. *DeGeer? pl.* 39. *f.* 9. 10.
Te. Salicis pentandræ. *Villars E.* iii. 8.

3817. 28; niger*. (*Leach Z. M.* iii. 129.)—(*Sam. I.* 28.)—*St. Farg. M.* 68.
Pt. niger. (*Jurine Hy.* 60. *pl.* 6. *gen.* 5.)

3818. 29, nigricornis. *St. Farg. M.* 63.

3819. 30; lucidus*. (*Leach Z. M.* iii. 129.)—(*Sam. I.* 28.)—*St. Farg. M.* 66.
Te. lucida. *Panz. F.* lxxxii. *f.* 10.

3820. 31, consobrinus* *mihi.*

---

[a] 3795 † 6. Salicis. *St. Farg. M.* 68.—*Faun. F. pl.* 11. *f.* 3.
Te. Salicis. *Fabr. P.* 40.—*Stew.* ii. 223. (!)—*Turt.* iii. 416. (!)

Genus 13: (587). CRÆSUS, *Leach*, *Samou.*, *Curtis.*

NEMATUS *p*, *Jurine.*

3821. 1; septentrionalis*. (*Leach Z. M.* iii. 129.)—(*Samou.* 266.) —*Curtis.* i. *pl.* 17.
Te. septentrionalis. *Linn.* ii. 926.—*Turt.* iii. 418.
Te. largipes. *DeGeer.* ii. 262. *pl.* 37. *f.* 26.
β? Cr. Stephensii. *Heysham. MSS.*
Cr. niger *mihi* (*olim*).

Genus 14: (588). MESSA, *Leach*, *Samou.*

TE.-EMPHYTUS *p*, *Klug.*

3822. 1; hortulana*. (*Leach Z. M.* iii. 126.)—(*Samou.* 264.)
Te. (Em.) hortulana. *Klug Bl.* 214.

Genus 15: (589). FENUSA, *Leach*, *Samou.*

TE.-ALLANTUS *p*, *Klug.*

3823. 1; pumila*. (*Leach Z. M.* iii. 126.)—(*Sam. I.* 17.)
Te. (All.) pumila. *Klug Bl.* 120.

3824. 2; pygmæa*.
Te. (All.) pygmæa. *Klug Bl.* 121.

Genus 16: (590). ATHALIA, *Leach*, *Samou.*

ALLANTUS *p*, *Jurine.*—TE.-ALLANTUS *p*, *Klug.*—HYLOTOMA *** *Fabr.*—NEMATUS *p*, *Spinola.*

3825. 1; spinarum*. (*Leach Z. M.* iii. 126.—(*Sam. I.* 5.)
Te. spinarum. *Fabr.* ii. 110.
Te. centifoliæ. *Panz. F.* xlix. *f.* 18.
At. centifolia. (*Sam. I.* 5.)
Te. Colibri. *Christius. B.* 434. *pl.* 50. *f.* 1.

3826. 2; Rosæ*. (*Leach Z. M.* iii. 126.)—(*Sam. I.* 5.)
Te. Rosæ. *Linn.* ii. 925.—*Berk. S.* i. 154.—*Stew.* ii. 224.—*Turt.* iii. 414.—(*Kirby & Sp. I. E.* i. 192.)

3827. 3; annulata*. (*Leach Z. M.* iii. 126.)—(*Sam. I.* 5.)
Hy. annulata. *Fabr. P.* 26.

3828. 4; lugens*.
Te. (All.) lugens. *Klug Bl.* 88.
At. abdominalis. *St. Farg. M.* 23.—*Faun. F. pl.* 13. *f.* 2.

3829. 5; Richardi*. *St. Farg. M.* 23.

Genus 17: (591). SELANDRIA, *Leach*, *Samou.*

HYLOTOMA *p*, *Fabr.*—TE.-ALLANTUS *p*, *Klug.*—NEMATUS *p*, *Spinola.*

3830. 1; serva*. (*Leach Z. M.* iii. 126.)—(*Sam. I.* 37.)
Hy. serva. *Fabr. P.* 26.

3831. [2; dorsalis* *mihi.*]

3832. 3; luteiventris*. (*Sam. I.* 37.)
Te. (All.) luteiventris. *Klug Bl.* 104.

3833. 4; Spinolæ*.
Te. (All.) Spinolæ. *Klug Bl.* 105.
Hy. ventralis. *Spinola I. L.* i. 1.

3834. 5; hyalina*.
Te. (All.) hyalina. *Klug Bl.* 106.

3835. 6, scapularis *mihi.*
Te. lepida. *St. Farg. M.* 104?

3836. 7, ferruginea*.
Hy. ferruginea. *Fabr. P.* 26.
Al. ferrugineus. *Panz. F.* xc. *f.* 9.
Te. (All.) brunnea. *Klug Bl.* 101.

3837. 8; testudinea*.
Te. (All.) testudinea. *Klug Bl.* 108.

3838. 9, verna?
Te. (All.) verna. *Klug Bl.* 103?
Te. punctigera. *St. Farg. M.* 110.—*Faun. F. pl.* 7. *f.* 6.

3839. 10; ephippium*.
Te. ephippium. *Panz. F.* lii. *f.* 5.
Te. dubia. *Gmel.* v. 2668?

3840. 11, adumbrata.
Te. (All.) adumbrata. *Klug Bl.* 112.

3841. 12; geniculata* *mihi.*

3842. 13, tibialis* *mihi.*

3843. 14; fuliginosa*. (*Sam. I.* 37.)
Te. fuliginosa. *Schr. A. No.* 334.
Te. trichocera. *St. Farg. M.* 81?

3844. 15; brevicornis*.
Te. (All.) brevicornis. *Klug Bl.* 113.
Te. nigrita. *St. Farg. M.* 81?
Ne. nigritus. *Spinola I. L.* ii. 155?

3845. 16; albipes*.
Te. albipes. *Gmel.* v. 2667.
Te. Morio. *St. Farg.* 105?

3846. 17; stramineipes.
Te. (All.) stramineipes. *Klug Bl.* 123.
Te. albipes. *St. Farg.* 105?

3847. 18; Betuleti*.
Te. (All.) Betuleti. *Klug Bl.* 121.

3848. 19, Morio*.
Te. Morio. *Fabr.* ii. 119.—*Panz. F.* lxix. *f.* 17.

3849. 20; fuscula*.
Te. (All.) fuscula. *Klug Bl.* 118.

3850. 21; Æthiops*.
Te. Æthiops. *Fabr.* ii. 121. (!)—*Stew.* ii. 224.—*Turt.* iii. 419.

3851. 22; cinereipes*. (*Leach Z. M.* iii. 126.)—(*Sam. I.* 37.)
Te. (All.) cinereipes. *Klug Bl.* 115.

3852. 23; annulipes*.
Te. (All.) annulipes. *Klug Bl.* 118.

3853. 24; varipes*.
Te. (All.) varipes. *Klug Bl.* 117.

3854. 25; fulvicornis*.
Te. fulvicornis. *Fabr. P.* 58.—*Panz. F.* lxxxii. *f.* 13.

3855. 26; pusilla*.
Te. (All.) pusilla. *Klug Bl.* 119.

3856. 27; nana*.
Te. (All.) nana. *Klug Bl.* 120.

3857. 28; tenuicornis*.
Te. (All.) tenuicornis. *Klug Bl.* 116.

3858. 29, alternipes*.
Te. (All.) alternipes. *Klug Bl.* 115.
Te. Maura. *Schrank B.* ii. 248.

3859. 30; ovata*. (*Leach Z. M.* iii. 126.)—(*Sam. I.* 37.)
Te. ovata. *Linn.* ii. 924.—*Panz. F.* lii. *f.* 3.—*Turt.* iii. 416.
Te. leucozona. *Schrank B.* ii. 251.
Te. gossypina. *Retzius. No.* 303.

3860. 31; lineolata*.
Te. (All.) lineolata. *Klug Bl.* 124?

3861. 32; Alni*.
Te. Alni. *Linn.* ii. 925.—*Schæff. Ic.* cx. *f.* 6. 7.—*Turt.* iii. 417.
Te. australis. *St. Farg. M.* 71.—*Faun. F. pl.* 3. *f.* 1.

3862. 33, rufa*.
Te. rufa. *Panz. F.* lxxii. *f.* 2.

3863. 34, uncta*.
Te. (All.) uncta. *Klug Bl.* 125.

3864. 35, pulchella* *mihi.*

3865. 36; costalis*.
Te. costalis. *Fabr.* ii. 109.—*Faun. F. pl.* 7. *f.* 5.
Te. fulvivenia. *Schra. A.* 338.

Genus 18: (592). ALLANTUS, *Panzer, Leach, Samou.*

TENTHREDO (ALLANTUS), *Klug.*

A. Antennæ extrorsum crassiores. a. *Antennis flavis.*

3866. 1; Scrophulariæ*.

Te. Scrophulariæ. *Linn.* ii. 923.—*Berk. S.* i. 154.—*Panz. F. C. f.* 10.—*Stew.* ii. 223.—*Turt.* iii. 415.—(*Samou.* 67. *pl.* 8. *f.* 2.)
Te. Rustica. *Schrank B.* ii. 235.

3867. 2; annulatus*.
Te. (All.) annulata. *Klug Bl.* 155.
Te. meridiana. *St. Farg. M.* 88.—*Faun. F. pl.* 5. *f.* 1.

b. *Antennarum scapo solè flavo.*

3868. 3; tricinctus*?
Te. tricincta. *Fabr. P.* 30?—*DeGeer.* ii. 951. *pl.* 34. *f.* 9–19.
Te. vespiformis. *Latr.*—*St. Farg. M.* 89.
Te. rustica. *Fourc. P.* ii. 367?
All. affinis. *Leach MSS.*

3869. 4; Viennensis*.
Te. Viennensis. *Panz. F.* lxv. *f.* 5.
Te. marginella. *Fabr. P.* 29?—(*Kirby & Sp. I. E.* i. 386.)
Te. sexannulata. *Schrank B.* ii. 236?

3870. 5; Zonula*.
Te. (All.) Zonula. *Klug Bl.* 161.
Te. bicincta. *Schæff. Ic.* vii. *f.* 8?
Te. luteiventris. *St. Farg. M.* 89.

3871. 6, Zona*.
Te. (All.) Zona. *Klug Bl.* 160.
Te. succincta. *St. Farg. M.* 93.—*Faun. F. pl.* 5. *f.* 5.

3872. 7; arcuatus*.
Te. arcuata. *Forst. C.* i. 79.—*Stew.* ii. 223.—*Turt.* iii. 415.
Te. (All.) Notha. *Klug Bl.* 164.
♀, Te. marginella. *Panz. F.* lxiv. *f.* 7.
Al. Notha. (*Leach Z. M.* iii. 128.)—(*Sam. I.* 2.)

c. *Antennis nigris.*

3873. 8, tenulus.
Te. tenula. *Scop. C. No.* 725.
Te. (All.) bifasciata. *Klug Bl.* 165.
All. Rossii. *Panz. F.* xci. *f.* 15.

3874. 9; rusticus*.
Te. rustica. *Linn.* ii. 923.—*Berk. S.* i. 153.—*Stew.* ii. 223.—*Turt.* iii. 418.—*Faun. F. pl.* 7. *f.* 2. ♂.
Te. sulphurata. *Gmel.* v. 2665.
♀; Te. notata. *Panz. F.* lxiv. *f.* 10.
♂. Te. carbonaria. *Fabr. P.* 30.

3875. 10; duodecimpunctatus*. (*Sam. I.* 2.)
Te. 12-punctata. *Linn.* ii. 926.—*Panz. F.* lii. *f.* 8.
Te. fera. *Fabr. P.* 37?

3876. 11; albicinctus*. (*Sam. I.* 2.)
Te. (All.) albicincta. *Klug Bl.* 149.
♀; Te. luctuosa. *St. Farg. M.* 103?

3877. 12; Ribis*.
Te. Ribis. *Schra. A.* 332.—*Panz. F.* lii. *f.* 12.
Te. leucopus. *Gmel.* v. 2666.

3878. 13; Punctum*. (*Sam. I.* 2.)
Te. punctum album. *Linn.* ii. 924.
Te. erythropus. *Schrank Beyt.* 86.
Te. punctum. *Fabr. P.* 36.—*Panz. F.* xxvi. *f.* 21.

3879. 14; hæmatopus*. (*Sam. I.* 2.)
Te. hæmatopus. *Panz. F.* lxxxi. *f.* 11.
Te. ocreata. *Schæff. Ic.* 232. *f.* 4, 5.
♂. Te. diversipes. *Schrank B.* ii. 236.

3880. 15; blandus*. (*Sam. I.* 2.)
♀, Te. blanda. *Fabr. P.* 36. (!)—*Panz. F.* lii. 9; lxv. 9.—*Stew.* ii. 223.—*Turt.* iii. 419.
Te. lacrymosa. *St. Farg. M.* 101.—*Faun. F. pl.* 6. *f.* 7.
♂, Te. cylindrica. *Fabr. P.* 32.—*Panz. F.* lxxi. *f.* 7.

3881. 16; neglectus*. (*Sam. I.* 2.)
Te. (All.) neglecta. *Klug Bl.* 136.
Te. blanda. *Schæff. Ic.* vii. *f.* 5.
Te. Schæfferi. *St. Farg. M.* 98.—*Faun. F. pl.* 6. *f.* 4.

3882. 17; cinctus*.
Te. cincta. *Panz. F.* lxiv. *f.* 2.
Te. bicincta. *Linn.* ii. 925?
Al. bicinctus. (*Sam. I.* 2.)
Te. vaga. *Fabr.* ii. 120. (!)—*Turt.* iii. 419.
♂; Te. semicincta. *Schrank A.* 665.
Al. semicincta. (*Leach Z. M.* iii. 128.)—(*Sam. I.* 2.)
Te. mesomelas. *Gmel.* v. 2660.

3883. 18; zonatus*. (*Leach Z. M.* iii. 128.)—(*Sam. I.* 2.)
Te. zonata. *Panz. F.* lxiv. *f.* 9.
Te. cincta. *Schæff. Ic.* lvi. *f.* 2.
Te. succincta. *Don.* xiii. 17. *pl.* 441. *f.* 2.
Te. latizona. *St. Farg. M.* 74.—*Faun. F. pl.* 3. *f.* 4.
♂; Te. equestris. *Panz. F.* cvii. *f.* 6.

3884 † 19, flavicornis.
Te. flavicornis. *Fabr. P.* 31.—*Panz. F.* lii. *f.* 2.
Te. flava. *Scop. C.* 731.
Te. poecilechroa. *Schrank A.* 324.
Te. luteicornis. *Fabr. P.* 21.—*Panz. F.* lxiv. *f.* 1.
Te. Umbellatarum. *Panz. E. V.* 30. In Mus. *Brit.*

B. Antennæ haud extrorsum crassiores.

a. *Antennarum apice albo.*

3885. 20; conspicuus*. (*Sam. I.* 2.)
Te. (All.) conspicua. *Klug Bl.* 170.
Te. rufiventris. *Panz. F.* lxv. *f.* 6.

3886. 21; rufiventris*. (*Sam. I.* 2.)—*Faun. F. pl.* 4.*f.* 3.
Te. rufiventris. *Fabr. P.* 33.
Te. rufipennis. *Fabr.* ii. 116.

3887. 22; solitaria*.
Te. solitaria. *Scop. C. No.* 738.
Te. Fagi. *Panz. F.* lii.*f.* 14.
Te. Maura. *Panz. Schæff.* 186.*f.* 2.

3888. 23, aterrimus* *mihi.*

3889. 24; Colon*.
Te. (All.) Colon. *Klug Bl.* 172.

3890. 25; Coryli*.
Te. Coryli. *Panz. F.* lxxi.*f.* 8.

3891. 26; bipunctula.
Te. (All.) bipunctula. *Klug Bl.* 175.

3892. 27; lividus*. (*Jurine H.* 54.)—(*Sam. I.* 2.)
Te. livida. *Linn.* ii. 925.—*Panz. F.* lii.*f.* 6.—*Turt.* iii. 417.
♂, Te. Carpini. *Panz. F.* lxxi.*f.* 9.
Te. albicornis. *Fourc. P.* ii. 371.
Te. Maura. *Fabr. P.* 33.
Te. annularis. *Schrank A.* 325.

b. *Antennis omninò nigris.*

3893. 28; ater*. (*Sam. I.* 2.)
Te. atra. *Linn.* ii. 924.—*Panz. F.* lii.*f.* 7.
Te. fuscipes. *Gmel.* v. 2667.

3894. 29; mandibularis*.
Te. mandibularis. *Panz. F.* xcviii.*f.* 9.

3895. 30; Tiliæ*. *Panz. F.* xci.*f.* 13.

3896. 31; viridis*. (*Jurine H.* 55.)—(*Sam. I.* 2.)
Te. viridis. *Linn.* ii. 925.—*Schæff. Ic.* clxxxi.*f.* 5, 6.—*Turt.* iii. 416.
Te. hebraica. *Fourc. P.* ii. 363.
Te. Rosæ. *Scop. C.* 275.
Te. annularis. *Villars E.* iii. 117.
♂; Te. mesomela. *Linn.* ii. 924.
Te. marginata. *Christius B.* 438. *pl.* 51.*f.* 1.

3897. 32; scalaris*.
Te. (All.) scalaris. *Klug Bl.* 184.
Te. viridis. *Panz. F.* lxiv.*f.* 2.—*Don.* xiii. 23. *pl.* 444.
Te. interruptus. *Fabr. P.* 40?
Te. Pini. *Villars E.* iii. 112.

3898. 33; punctulatus*.
Te. (All.) punctulata. *Klug Bl.* 185.

3899. 34; pictus.
Te. (All.) picta. *Klug Bl.* 185.

3900. 35; agilis*.
Te. (All.) agilis. *Klug Bl.* 198.

3901. 36; nitidus*.
Te. (All.) nitida. *Klug Bl.* 208.

3902. 37; bicolor*.
Te. (All.) bicolor. *Klug Bl.* 208?

3903. 38; Aucupariæ*.
Te. (All.) Aucupariæ. *Klug Bl.* 202.
Te. juvenilis. *St. Farg. M.* 99.—*Faun. F. pl.* 6. *f.* 5.

3904. 39; lateralis*. (*Sam. I.* 2.)
Te. lateralis. *Fabr.* ii. 118.—*Panz. F.* lxxxviii. *f.* 18.

3905. 40; melanocephalus*.
Te. melanocephala. *Fabr. S.* 216.—*Coqueb. Ill. pl.* 3. *f.* 6?

Genus 19: (593). TENTHREDO, *Leach, Samou.*

A. Antennæ graciles, longiores.

3906. 1; Rapæ*. *Linn.* ii. 926.—*Schæff. Ic.* clxxix. *f.* 1.—(*Leach Z. M.* iii. 128.)—(*Sam. I.* 40.)

3907. 2; simulans*. *Klug Bl.* 152.
♀; Te. scripta. *Gmel.* v. 2667?—*St. Farg. M.* 86.—*Faun. F. pl.* 4. *f.* 6.

3908. 3; antennata*. *Klug Bl.* 153.
Te. duplex. *St. Farg. M.* 87.

3909. 4; variegata*. *Klug Bl.* 153.

B. Antennæ haud graciles, mediocres.

a. *Corpore subdepresso.*

3910. 5; nassata*. *Linn.* ii. 924.—*Panz. F.* lxv. *f.* 2. ♂.—(*Leach Z. M.* iii. 128.)—(*Sam. I.* 40.)
Te. (All.) instabilis a, nassata. *Klug Bl.* 187.

3911. [6; melanorhæa*.] *Gmel.* v. 2668.—*St. Farg. M.* 85.—*Faun. F. pl.* 4. *f.* 4.

3912. 7; ambigua*. *Klug Bl.* 192.
Te. neglecta. *St. Farg. M.* 77.

3913. 8; scutellaris*. *Fabr. P.* 39?—*Panz. F.* xcviii. *f.* 12?
Te. (All.) instabilis b, scutellaris. *Klug Bl.* 188?
Te. interrupta. *Mus. Marsham.*

3914. 9; Pavida*. *Fabr. P.* 31.
Te. rufocincta. *Retzius. No.* 305.

3915. 10; ornata*. *St. Farg. M.* 77.—*Faun. F. pl.* 3. *f.* 5.

3916. 11; subinterruptus* *mihi.*
Te. dorsalis. *Spin. I. L.* ii. 17. *pl.* 4. *f.* 14?

3917. 12; tristis*. *Mus. Marsham.*
Te. spreta. *St. Farg. M.* 78?

3918. 13; caliginosa* *mihi.*

3919. 14; Coquebertii*. *Klug. Bl.* 192.
Te. Stigma. *Coqueb. Ill.* i. 15. *pl.* 3. *f.* 5?

3920. 15; fulvipes* *mihi.*

3921. 16; ignobilis*. *Klug. Bl.* 195.
Te. Stigma. *St. Farg. M.* 76.

3922. 17; analis* *mihi.*

3923. 18; dimidiata*. *Fabr. P.* 42.—(*Leach Z. M.* iii. 128.)—(*Sam. I.* 40.)—*Faun. F. pl.* 4. *f.* 2.
Te. (All.) instabilis c, dimidiata. *Klug. Bl.* 188?
Te. cordata. *Fourc. P.* ii. 368.
Te. varia. *Gmel.* iv. 2665.

b. *Corpore subcylindrico.*

3924. 19; socia*.
Te. (All.) socia. *Klug. Bl.* 97.

3925. 20; Geeri*. *Klug. Bl.* 203.
DeGeer. ii. 1002. *pl.* 38. *f.* 8–10.

3926. 21, xanthocera* *mihi.*

3927. 22; cingulata*. *Fabr.* ii. 113.

3928. 23; atricornis* *mihi.*

Genus 20: (594). DOSYTHEUS, *Leach, Samou.*

HYLOTOMA *p*, *Fabr.*—TE.-DOLERUS *p*, *Klug.*—DOLERUS *p*, *St. Fargeau.*

3929. 1; lateritius*.
Te. (Dol.) lateritia. *Klug. Bl.* 233.

3930. 2; Eglanteriæ*. (*Leach Z. M.* iii. 128.)—(*Sam. I.* 15.)
Te. (Dol.) Eglanteriæ. *Klug. Bl.* 229.
♂; Te. Abietis. *Linn.* ii. 923?—*Turt.* iii. 417.
♀; Te. Eglanteriæ. *Fabr.* ii. 196.
Te. pedestris. *Panz. F.* lxxxii. *f.* 11.
Te. Germanica. *Panz. F.* lii. *f.* 4?

3931. 3; trimaculatus*.
Dol. trimaculatus. *St. Farg. M.* 121.

3932. 4; fulviventris*.
Te. fulviventris. *Scop. C. No.* 736.
Dol. germanicus. *St. Farg. M.* 121.

3933. 5; xanthopus* *mihi.*

3934. 6, Junci*. (*Leach Z. M.* iii. 128.)—(*Sam. I.* 15.)
Te. (Dol.) palustris. *Klug. Bl.* 234?

3935. 7; timidus*.
Te. (Dol.) timidus. *Klug. Bl.* 238?
Te. Abietis. *Panz. F.* lxiv. *f.* 3.

3936. 8; tristis*.
Te. tristis. *Panz. F.* xcviii. *f.* 11?

Genus 21: (595). DOLERUS, *Leach, Samou.*, (*Kirby.*)
TE.-DOLERUS *p, Klug.*

3937. 1; hæmatodes*.
Te. hæmatodes. *Schrank A. No.* 678.
Te. opaca. *Panz. F.* lii. *f.* 10.
Do. opacus. (*Leach Z. M.* iii. 128.)—(*Sam. I.* 15.)
Te. collaris. *Don.* xiii. 17. *pl.* 441. *f.* 1.

3938. 2, anthracinus*.
Te. (Dol.) anthracinus. *Klug. Bl.* 204.

3939. 3; niger*. *St. Farg. M.* 125.
Te. nigra. *Linn.* ii. 925.—*Panz. F.* lii. *f.* 11.

3940. 4; fumosus* *mihi.*

3941. 5; coracinus*.
Te. (Dol.) coracinus. *Klug. Bl.* 204.

3942. 6; palmatus*.
Te. (Dol.) palmatus. *Klug. Bl.* 205.

3943. 7; picipes*.
Te. (Dol.) picipes. *Klug. Bl.* 205.

3944. 8; gonager*. (*Leach E. E.* ix. 140.)—(*Leach Z. M.* iii. 128.)—(*Sam. I.* 15.)
Te. gonagra. *Fabr.* ii. 117.
Te. crassa. *Panz. F.* lxv. *f.* 4.
Te. erythrogona. *Schrank A. No.* 681.
Te. geniculata. *Fourc. P.* ii. 374.

3945. 9; vestigialis*.
Te. (Dol.) vestigialis. *Klug. Bl.* 243.
Dol. rufipes. *St. Farg. M.* 124.

Genus 22: (596). EMPHYTUS, *Leach, Samou.*
TE.-EMPHYTUS, *Klug.*

3946. 1; succinctus*. (*Sam. I.* 16.)
Te. (Em.) succincta. *Klug. Bl.* 217.
Te. togata. *Panz. F.* lxxxii. *f.* 12.

3947. 2; cinctus*. (*Leach Z. M.* iii. 128.)—(*Sam. I.* 16.)
Te. cincta. *Linn.* ii. 925.—*Turt.* iii. 418.

3948. 3; togatus*.
Te. togata. *Fabr. P.* 32.

3949. 4; melanarius*.
Te. (Em.) melanaria. *Klug. Bl.* 220.

3950. 5; rufocinctus*.
Te. (Em.) rufocinctus. *Klug. Bl.* 224.
Te. cingulum. *Spinola*: teste *Klug*.

3951. 6; calceatus*.
Te. (Em.) calceata. *Klug. Bl.* 226.

3952. 7; nigricans*.
Te. (Em.) nigricans. *Klug. Bl.* 214.
Do. varipes. *St. Farg. M.* 119.—*Faun. F. pl.* 8. *f.* 4.

3953. 8; coronatus*.
Te. (Em.) coronatus. *Klug. Bl.* 214.

3954. 9; gilvipes*.
Te. (Em.) gilvipes. *Klug. Bl.* 244.

3955. 10; serotinus*.
Te. (Em.) serotina. *Klug. Bl.* 226.
Do. abdominalis. *St. Farg. M.* 118.—*Faun. F. pl.* 8. *f.* 3.

3956 † 11, cereus. (*Leach Z. M.* iii. 128.)
Te. (Em.) cerea. *Klug. Bl.* 227.
Em. ceria. (*Sam. I.* 16.) In Mus. *Brit.*

3957. 12; filiformis*.
Te. (Em.) filiformis. *Klug. Bl.* 223.

3958. 13; tibialis*. (*Leach Z. M.* iii. 128.)—(*Sam. I.* 16.)
Te. tibialis. *Panz. F.* lxii. *f.* 1.

3959. 14; patellatus*.
Te. (Em.) patellata. *Klug. Bl.* 221.

3960. 15; immersus*.
Te. (Em.) immersa. *Klug. Bl.* 222.
Do. pallimacula. *St. Farg. M.* 117.—*Faun. F. pl.* 8. *f.* 2.

3961. 16, impressus.
Te. (All.) impressa. *Klug. Bl.* 116.

Genus 23: (597). ——

3962. 1, ochroleucus *mihi*.

Genus 24: (598). TARPA, *Fabr.*, *Leach*, *Samou.*
MEGALODONTES, *Latr.*—DIPRION, *Schrank.*

3963 † 1, cephalotes. *Fabr. P.* 19.
Ta. plagiocephala. *Klug. Bl.* 7. *pl.* vii. *f.* 1.
Ta. Klugii. *Leach Z. M.* iii. 131.—*Samou.* 266.

3964. 2, Panzeri. *Leach Z. M.* iii. 132.—(*Sam. I.* 40.)
Te. cephalotes. *Panz. F.* lxii. *f.* 7. ♂.—*f.* 8. ♀.
Me. cephalotes. (*Leach E. E.* ix. 141.)

3965 † 3. *Fabricii*[a].

---

[a] 3965 † 3. Fabricii. *Leach Z. M.* iii. 130.—*Samou.* 266. (!)

Genus 25: (599). LYDA, *Fabr.*, *Leach*, *Samou.*, (*Kirby.*)
PAMPHILIUS, *Latr.*, *Leach E. E.*—CEPHALEIA, *Jurine.*—PSEN *p*, *Schrank.*

3966. 1; Sylvatica*. *Fabr. P.* 43.
Te. Sylvatica. *Linn.* ii. 926.—*Panz. F.* lxv. *f.* 10.—*Turt.* iii. 424.
♂, Ly. Nemorum. *Fabr. P.* 45.—(*Sam. I.* 26.)
Te. Nemorum. *Panz. F.* lxxxvi. *f.* 8.—*Stew.* ii. 224.
Te. fulvipes. *Retzius. No.* 323.

3967. 2; Stigma* *mihi.*

3968. 3; aurita*. *Klug. Bl.* 15. *pl.* vii. *f.* 3.

3969. 4, pratensis*. *Fabr. P.* 45.—*Schæff. Ic. pl.* xlii. *f.* 8, 9.
Te. vafra. *Linn.* ii. 927?
Te. stellata. *Christius B.* 458. *pl.* li. *f.* 4.

3970. 5, depressa*. *Klug. Bl.* 14.—*Panz. F.* lxv. *f.* 11.
Te. depressa. *Schrank A. No.* 691.

3971. 6, Sylvarum. ——? *MSS?*

3972 † 7. flava[a].

3973. 8, hortorum. *Klug. Bl.* 18.
Ly. fallax. *St. Farg. M.* 13?

3974. 9, inanita. *St. Farg. M.* 12.—*Faun. F. pl.* 14. *f.* 6.
Te. inanita. *Villers E.* iii. 125. *pl.* 7. *f.* 21.
Ly. inanis. *Klug. Bl.* 18.

3975. 10; Arbustorum*. *Fabr. P.* 46. (!)
Te. Arbustorum. *Fabr.* ii. 123. (!)—*Turt.* iii. 425.
Te. lucorum. *Fabr. M.* i. 256.

3976 † 11. Populi[b].

3977. 12; lutescens*. *St. Farg. M.* 8.

3978. 13. Cynosbati[c].

3979. 14; cingulata*. *Faun. F. pl.* 14. *f.* 3.
Pa. cingulatus. *Latr. E. No.* 15.

3980. 15; Betulæ*. *Fabr. P.* 44.—(*Sam. I.* 26.)
Te. Betulæ. *Panz. F.* lxxxvii. *f.* 18.—*Turt.* iii. 425.
Te. fulva. *Retzius. No.* 321.—*DeGeer. pl.* 40. *f.* 21.

3981. 16, erythrocephala*. *Fabr. P.* 43.—(*Sam. I.* 26.)

---

[a] 3972 † 7. flava. *Fabr. P.* 46.
Te. flava. *Linn.* ii. 927.—*Turt.* iii. 425. (!)

[b] 3976 † 11. Populi. *Fabr. P.* 44.
Te. Populi. *Linn.* ii. 927.—*Turt.* iii. 424. (!)

[c] 3978. 13. Cynosbati. *Fabr. P.* 44.
Te. Cynosbati. *Linn.* ii. 927.—*Berk. S.* i. 154. (!)—*Stew.* ii. 224. (!)—*Turt.* iii. 424. (!)

Te. erythrocephala. *Linn.* ii. 926.—*Panz. F.* vii. *f.* 9.—*Stew.* ii. 224.—*Turt.* iii. 424.
Pa. erythrocephalus. (*Leach E. E.* ix. 141.)

Familia II: (91). XIPHYDRIIDÆ, *Leach.*

Genus 26: (600). CEPHUS, *Latreille, Leach, Samou.*, (*Kirby.*)

ASTATUS, *Klug.*—TRACHELUS, *Jurine.*—SIREX *p, Fabr.*—SYREX, *Harr.*

3982. 1; troglodyta*. *Latr. H.* xiii. 143.—(*Leach E. E.* ix. 141.)
Si. troglodyta. *Fabr. P.* 250.
As. troglodyta. *Klug. Si.* 49. *pl.* vi. *f.* 1, 2.
Sy. niger. *Harr. Ex.* 94. *pl.* xxviii. *fig.* 2.

3983. [2, quinquefasciatus*] *mihi.*

3984. 3; Satyrus*.
As. Satyrus. *Panz. F.* lxxxv. *f.* 12.

3985. 4; pygmæus*. *Fabr. P.* 251.—(*Leach E. E.* ix. 141.)—(*Samou.* 267.)
Si. pygmæus. *Linn.* ii. 929.—*Turt.* iii. 429.
As. pygmæus. *Klug. Si.* 50. *pl.* vi. *f.* 3.
♀, Ce. spinipes. *Klug. Si.* 51. *pl.* vi. *f.* 4. *a–b.*
Banchus spinipes. *Panz. F.* lxxiii. *f.* 17.
Ba. viridator. *Fabr. P.* 127.

3986. 5; pallipes*. *Klug. Si.* 53. *pl.* vi. *f.* 6.

3987. 6; floralis*. *Klug. Si.* 53. *pl.* vi. *f.* 5. *a–b.*

3988. 7, analis*. *Klug. Si.* 54. *pl.* vii. *f.* 1.
Tr. hæmorrhoidalis. *Jur. Hy. pl.* 7. G. 9.

3989. 8, tabidus*. *Fabr. P.* 252. (!)
Si. tabidus. *Fabr.* ii. 131. (!)—*Stew.* ii. 225.—*Turt.* iii. 429.
As. tabidus. *Klug. Si.* 56. *pl.* vii. *f.* 3. *a–b.*

3990. 9, pusillus* *mihi.*
As. punctatus. *Klug. Si.* 55. *pl.* vii. *f.* 2. *a–b?*

3991. 10; immaculata* *mihi.*
Ce. phthiscus. *Fabr. P.* 251?

Genus 27: (601). JANUS *mihi.*

3992. 1; connectens* *mihi.*
(Vide *Kirby & Sp. I. E.* iv. 374. *nota.*)

Genus 28: (602). XYELA, *Dalman, Curtis,* (*Kirby.*)

PINICOLA, *Latr.*—MASTIGOCERUS, *Klug.*

3993. 1; pusilla*. *Dalman A. E.* 28. *pl.* iii. *f.* 1. *a.* ♂.—*f.* 2. *a.* ♀.—*Curtis.* i. *pl.* 30.
Pi. Julii. *Latreille Dict. Hist. Nat.* (*teste St. Farg.*)

Genus 29: (603). XIPHYDRIA, *Latreille, Leach, Samou.*, (*Kirby.*)
HYBONOTUS, *Klug.*—UROCERUS *p, Jurine.*—ASTATUS *p, Panzer.*
3994. 1; camelus*. *Latr. H.* xiii. 145.—(*Leach E. E.* ix. 141.) —(*Sam. I.* 44.)
Si. camelus. *Linn.* ii. 929.—*Turt.* iii. 428.
Hy. camelus. *Klug. Si.* 14. *pl.* 1. *f.* 4, 5.
3995 † 2, dromedarius. *Latr. H.* xiii. 146.—(*Leach E. E.* ix. 141.) —(*Sam. I.* 44.)
Si. dromedarius. *Fabr. M.* i. 258.—*Don.* xiv. 43. *pl.* 483.
Hy. dromedarius. *Klug. Si.* 15. *pl.* 1. *f.* 6, 7. In Mus. *Brit.*

Familia III: (92). UROCERIDÆ, *Leach.*

Genus 30: (604). ORYSSUS, *Latreille, Leach, Samou.*
SPHEX *p, Scopoli.*
3996 † 1: coronatus*. *Latr. H.* xiii. 160.—(*Leach E. E.* ix. 141.) —(*Sam. I.* 32.)
Or. vespertilio. *Klug. Si.* 7. *pl.* 1. *f.* 1–3.—(*Sam. I.* 32.)
Sp. abietina. *Scop. C.* 296. In Mus. *Brit.*

Genus 31: (605). SIREX, *Linn., Kirby, Curtis.*
UROCERUS, *Geoffroy, Leach, Samou*—ICHNEUMON *p, DeGeer.*
Tailed-Wasp.
3997. 1, gigas*. *Linn.* ii. 928.—*Berk. S.* i. 154.—*Don.* vi. 39. *pl.* 197. ♀.—*Stew.* ii. 225. *pl.* vii. *f.* 16–20.—*Turt.* iii. 426.—(*Linn. Trans.* (*Marsham.*) x. 403.)—(*Kirby & Sp. I. E.* i. 209. *pl.* iv. *f.* 1. ♀.)—*Wood.* ii. 50. *pl.* 55.—(*Curtis l. c. infra.*)
Ur. gigas. (*Fourc. P.* 363.)—(*Leach E. E.* ix. 141.)—*Samou. pl.* 8. *f.* 3.
† ♂, Si. mariscus. *Linn.* ii. 929.
β. Si. hungaricus. *Christi.* 414. *pl.* 47. *f.* 1.
Si. psyllius. *Fabr. E. S.* ii. 104.
Si. gigas var. 1. *Klug. Si.* 33. *pl.* ii. *f.* 2.
Ur. psyllius. (*Leach E. E.* ix. 141.)—(*Sam. I.* 43.)
3998. 2, bizonatus* *mihi.*
3999 † 3. augur[a].
4000. 4; Juvencus*. *Linn.* ii. 929.—*Turt.* iii. 427.—*Don.* xi. 99. *pl.* 396. ♀.—(*Linn. Trans.* (*Marsham.*) x. 403.)—(*Kirby & Sp. I. E.* i. 209.)—*Curtis.* vi. *pl.* 253. ♂ & ♀.
Ur. Juvencus. (*Ing. Inst.* 90.)
Sy. torvus. *Harr. Ex.* 96. *pl.* xxviii. *f.* 1.
♂, Si. Noctilio. *Fabr. E. S.* ii. 130.
β, Si. Juvencus var. 1. *Klug. Si.* 38. *pl.* iv. *f.* 1, 2.

---

[a] 3999 † 3. augur. *Klug. Si.* 34. *pl.* iv. *f.* 4?
Si. bimaculatus. *Don.* xiii. 15. *pl.* 440. (!)

4001 † 5. Spectrum [a].

4002. 6, Magus*. *Fabr.* ii. 125.—*Klug. Si.* 42. *pl.* v. *f.* 2–4.
Ur. Magus. (*Ing. Inst.* 90.)
† ♂, Si. nigrita. *Fabr.* ii. 127.

Sectio II. PUPOPHAGA, *Latreille.*

Familia IV: (93). EVANIIDÆ, *Leach.*

Evaniales, *Latr.*

Genus 32: (606). EVANIA, *Fabricius, Leach, Samou.*, (*Kirby.*)

Sphex *p, Linné.*—Ichneumon *p, DeGeer.*

4003. 1, appendigaster*. *Fabr. P.* 178?—(*Sam. I.* 17.)—(*Kirby & Sp. I. E. pl.* 4. *f.* 2.)—*Curtis fo.* 257.
Sp. appendigaster. *Linn.* ii. 943.—*Don.* x. 13. *pl.* 329.
Ev. lævigata. *Latr.* G. iii. 252?—(*Leach E. E.* ix. 142?)

Genus 33: (607). BRACHYGASTER, *Leach.*

Evania *p, Curtis.*

4004. 1, minutus. *Leach E. E.* ix. 142.
Ev. minuta. *Olivier.*—*Fabr. P.* 179.—*Curtis l.* c. *infra.*

4005 † 2. fulvipes.
Ev. fulvipes. *Curtis.* vi. *pl.* 257. In Mus. D. Curtis et Dale.

Genus 34: (608). FŒNUS, *Fabricius, Leach, Samou.*, (*Kirby.*)

Ichneumon *p, Linné.*—Gasteruption, *Latr.* (*olim.*)

4006. 1; jaculator*. *Fabr. P.* 141.—*Panz. F.* xcvi. *f.* 16. ♀.—(*Leach E. E.* ix. 142.)—(*Sam. I.* 17.)
Ic. jaculator. *Linn.* ii. 937.—*Berk. S.* i. 156.—*Stew.* ii. 230.—*Turt.* iii. 458.

4007. 2, assectator*. *Fabr. P.* 142.
Ic. assectator. *Linn.* ii. 937.—*Turt.* iii. 458.

Familia V: (94). ICHNEUMONIDÆ [b], *Leach.*

(Ichneumon, *Linné, &c.*—Ichneumonides, *Latr.*)

Genus 35: (609). XORIDES? *Latreille?*

Cryptus *p, Fabricius.*—Anomalon *p, Jurine.*

4008. 1; rufus* *mihi.*

---

[a] 4001 † 5. Spectrum. *Linn.* ii. 929.—*Stew.* ii. 225. (!)—*Turt.* iii. 427. (!)—*Klug. Si.* 39. *pl.* iv. *f.* 5, 6.—*pl.* v. *f.* 1.
♂. Si. emarginata. *Fabr.* ii. 128.

[b] Gravenhorst's proposed Monograph upon this and the two following families being about to be speedily published, I shall merely introduce such species as have been noticed by *English* writers, with the *few* I have already ascertained to be described, and the indication of some of the more conspicuous genera. I possess above 800 indigenous species, the numbers of which in the respective genera I have endeavoured to show as a guide to the student; but I must observe that time has not yet permitted me to examine their nomenclature satisfactorily throughout: this latter observation is applicable also to the three last families of Hymenoptera, and to the Hemiptera and Homoptera.

Genus 36: (610). ——

PIMPLA *p, Panz.*?

4009. 1; tarsalis* *mihi.*
Pi. excitator. *Panz. F.* xcii. *f.* 5?

Genus 37: (611). ——

PIMPLA *p, Fabr.*

4010. 1, dentator.
Pi. dentator. *Fabr. P.* 114.
Ic. ruspator. *Linn. F. No.* 1625?—*Berk. S.* i. 156?—*Stew.* ii. 230?—*Turt.* iii. 458?

Genus 38: (612). PIMPLA, *Fabricius, Leach, Samou., Curtis,* (*Kirby.*)

CRYPTUS *p, Fabr.*

4011. 1; Cossivora*. (*Curtis l. c. infra.*)

4012. 2, persuasoria*? *Fabr. P.* 112.—(*Curtis l. c. infra.*)
Ich. persuasorius. *Linn.* ii. 932.—*Berk. S.* i. 155.—*Stew.* ii. 228.—*Turt.* iii. 439.—*Don.* xv. *pl.* 522.—*Sowerb. B. M.* i. *pl.* 52.

4013. 3; manifestator*. *Fabr. P.* 113.—(*Kirby & Sp. I. E.* i. 121.—*Id.* iv. 210 & 211.)
Ich. manifestator. *Linn.* ii. 934.—*Berk. S.* i. 155.—*Linn. Trans.* (*Marsham.*) iii. 23. *pl.* 4. *f.* 1–5.—*Stew.* ii. 228.—*Turt.* iii. 451.—*Bingley.* iii. 251.—(*Kirby & Sp. I. E.* i. 354.)—(*Sam. I.* 23. *pl.* 8. *f.* 4?)—*Wood.* ii. 54. *pl.* 56.

4014. 4, abbreviator.
Ich. abbreviator. *Mus. Marsham.*

4015. 5; enervator*.
Cr. enervator. *Fabr. P.* 85?

4016. 6; segmentator*. *Fabr. P.* 114.
Ic. speculator. *Mus. Marsham.*

4017. 7; mediator*. *Fabr. P.* 117.—(*Curtis l. c. infra.*)
Ich. Scurra. *Panz. F.* xcii. *f.* 6.

4018. 8; extensor*. *Fabr. P.* 115.—(*Curtis l. c. infra.*)
Ic. extensor. *Linn.* ii. 935.—*Panz. F.* cix. *f.* 11.

4019. 9, Histrio*. *Panz. K. R.* ii. 82.
Ic. Histrio. *Fabr. P.* 85.—*Panz. F.* xcii. *f.* 7.

4020. 10, varicornis. *Fabr. P.* 119.—*Panz. F.* cix. *f.* 13.

4021. 11: Æthiops. *Curtis.* v. *pl.* 214.

4022. 12; Spectrum*. (*Curtis l. c. supra.*)
Si. Spectrum. *Don.* vii. 25. *pl.* 225.

4023. 13; instigator*. (*Curtis l. c. supra.*)
Cr. instigator. *Fabr. P.* 85.

4024. 14; flavicans*. *Fabr. P.* 119.—*Panz. F.* cix. *f.* 14.

4025. 15; pennata*. *Fabr. P.* 116.—(*Curtis l. c. supra.*)

4026. 16; examinator*. (*Curtis l. c. supra.*)
Cr. examinator. *Fabr. P.* 85.

4027. 17; accusator*. *Fabr. P.* 117.—*Panz. F.* cix. *f.* 12.—(*Curtis l. c. supra.*)

4028. 18, Turionellæ.
Ic. Turionellæ. *Linn.* ii. 938.—*Turt.* iii. 454.

4029. 19, Strobilellæ. *Fabr. P.* 115.
Ic. Strobilellæ. *Linn.* ii. 935.—*Turt.* iii. 455.

4030. 20, arundinator. *Fabr. P.* 116.

4031 ad 4110—(89 sp. adhuc examinandæ.)—109 species.

Genus 39: (613). CRYPTUS, *Fabricius, Leach, Samou.*, (*Kirby.*)

4111. 1, incubitor. *Fabr. P.* 83.
Ic. incubitor. *Linn.* ii. 933.—*Berk. S.* i. 155.—*Stew.* ii. 228.
Ic. incubator. *Turt.* iii. 446.

4112. 2. profligator. *Fabr. P.* 83.

4113. 3. festinatorius. *Mus. Marsham.*

4114. 4, irrigator.
Ic. irrigator. *Fabr. P.* 97.—*Panz. F.* lxxi. *f.* 16.

4115. 5, 4-dentatus. *Mus. Marsham.*

4116. 6, miniator. *Mus. Marsham.*

4117. 7; sponsor. *Fabr. P.* 83.

4118 ad 4309—(192 sp. adhuc examinandæ.)—199 species.

Ichneumonidæ insertæ sedes: Crypto affines?—TRYPHON *p*, *Fallen?*

4310. 1, punctorius. *Fabr. P.* 78.

4311 † 2, femorata. *Kirby M. A.* ii. 253? In Mus. D. Kirby.

4312. 3, corruscator.
Ic. corruscator. *Linn.* ii. 934.—*Turt.* iii. 451.

4313. 4; titillator*. *Fabr. P.* 86.
Ic. titillator. *Linn.* ii. 934.—*Turt.* iii. 454.

4314. 5; atrator*.
Ic. atrator. *Forst. C.* i. 84.
Ic. obscurator. *Gmel.*—*Stew.* ii. 228.—*Turt.* iii. 447.

4315 † 6. compunctor. *Fabr. P.* 84.—(*Kirby & Sp. I. E.* iv. 223.)
Ic. compunctor. *Linn.* ii. 934.—*Stew.* ii. 229.—*Turt.* iii. 452.

4316 † 7. peregrinator.
Ic. peregrinator. *Linn. F. No.* 1601.—*Berk. S.* i. 155.—*Stew.* ii. 228.—*Turt.* iii. 446.

4317. 8, indigator. *Mus. Marsham.*

4318. 9, reluctator. *Fabr. P.* 79.
Ic. reluctator. *Panz. F.* lxxi. *f.* 13.

4319. 10, cantellum. *Mus. Marsham.*

4320. 11, dubitator. *Fabr. P.* 85.
Ic. dubitator. *Panz. F.* lxxviii. *f.* 14.

4321. 12, minutorius. *Fabr. P.* 72.
Ic. rubricator. *Panz. F.* lxxxiv. *f.* 14.

Genus 40: (614). ICHNEUMON *Auctorum.*

4322. 1; comitator*. *Linn.* ii. 933.—*Berk. S.* i. 155.—*Stew.* ii. 228.—*Panz. F.* lxxi. *f.* 14.—*Kirby M. A.* ii. 253.—*Turt.* iii. 444.—(*Curtis l. c. infra.*)
Ich. auspex. *Müll. Pr. No.* 1792.
Ich. molitorius θ. *Schra. B. No.* 2072?

4323. 2; albiguttatus*. *Act. Taur.* (*Gravenhorst.*) xxiv. 280.

4324. 3; nigritarius*. *Act. Taur.* (*Gravenhorst.*) xxiv. 281.
Ich. molitorius δ. *Schran. B. No.* 2072?

4325. 4; lineator*. *Fabr.* ii. 168. (!)—*Stew.* ii. 229.—*Turt.* iii. 454.

4326. 5; nigrator*. *Fabr.* ii. 150.—(*Curtis l. c. infra.*)

4327. [6; narrator*.] *Fabr. P.* 67.—(*Curtis l. c. infra.*)
Ich. nigrator var.? *Isis* (*Trentepohl.*) 1826. 69.

4329. 8; bilineatus*. *Gmel.*
Ich. comitator. *Sulzer. Gesch. pl.* 26. *f.* 14.

4330 † 9. *nigratorius*[a].

4331. 10; fossorius. *Linn.* ii. 933.
Ich. subsericans. *Act. Taur.* (*Gravenhorst.*) xxiv. 285.

4332. 11; cognatus* *mihi.*
Ich. fossorius. *Fabr.* ii. 149?—*Turt.* iii. 441?

4333. 12; multicolor*. *Act. Taur.* (*Gravenhorst.*) xxiv. 286.

4334. 13, compunctor. *Mus. Marsham.*

4335. 14; monostagon*. *Act. Taur.* (*Gravenhorst.*) xxiv. 287.

4336. 15; albicillus*. *Act. Taur.* (*Gravenhorst.*) xxiv. 287.

4337. 16, annulator*. *Fabr.* ii. 151?—(*Curtis l. c. infra?*)
Ich. pedatorius. *Mus. Marsham.*

4338. 17; semiorbitalis*. *Act. Taur.* (*Gravenhorst.*) xxiv. 290.

4339. 18; leucocerus*. *Act. Taur.* (*Gravenhorst.*) xxiv. 289.

4340. 19; deliratorius*. *Linn.* ii. 932.—*Turt.* iii. 441.—(*Curtis l. c. infra.*)
Ich. palmarius. *Fourc. P.* ii. 411.
Ich. molitorius b. *Schra. B. No.* 2072.

4341. [20; molitorius*.] *Linn.* ii. 931.—*Panz. F.* xix. *f.* 16.—*Turt.* iii. 430.—(*Curtis l. c. infra.*)

4342. 21; edictorius*. *Linn.* ii. 932?—*Turt.* iii. 440.

---

[a] 4330 † 9. nigratorius. *Fabr. P.* 55.—(*Curtis l. c. infra.*) (!)

4343. 22; biannulatus*. *Act. Taur.* (*Gravenhorst.*) xxiv. 293.

4344. 23, quæsitorius. *Linn.* ii. 930?—*Turt.* iii. 432.

4345. 24; bimaculatorius*. *Panz. F.* lxxx. *f.* 8.

4346. 25; Panzeri* *mihi.*
Ich. nigratorius. *Panzer?*

4347. 26; albilarvatus*. *Act. Taur.* (*Gravenhorst.*) xxiv. 352.

4348. 27; luctuosus*. *Act. Taur.* (*Gravenhorst.*) xxiv. 289.

4349. 28, oratorius*. *Fabr.* ii. 138.—*Panz. F.* lxxx. *f.* 10.—(*Curtis l. c. infra.*)

4350. 29, pallipes*. *Act. Taur.* (*Gravenhorst.*) xxiv. 293.

4351. 30, fabricator*. *Fabr.* ii. 166?

4352. 31, Proteus*. *Isis.* (*Nees et Gravenhorst.*)

4353. 32; saturatorius*. *Act. Taur.* (*Gravenhorst.*) xxiv. 294.

4354. 33, calceatorius. *Panz. F.* lxxx. *f.* 15?
Ic. fossorius. *Fabr. P.* 65?

4357. 36, pedestrinus*. *Act. Taur..*(*Gravenhorst.*) xxiv. 296.

4358. 37, sepiatorius. *Mus. Marsham.*

4359. 38, castigator*. *Fabr. P.* 68.—(*Curtis l. c. infra.*)

4360. [39, custodiator.] *Fabr. P.* 68.

4366. 45; hæmorrhoidalis*. *Act. Taur.* (*Gravenhorst.*) xxiv. 348.

4368. 47, melanocastanus*. *Act. Taur.* (*Gravenhorst.*) xxiv. 351.

4371. 50; sputator*. *Fabr. P.* 66.—*Panz. F.* xix. *f.* 20.—(*Curtis l. c. infra.*)
Ic. ferruginosus. *Gmel. No.* 254.
Ic. molitorius, d. *Schrank B. No.* 2072.
β? Ic. culpatorius: (calpatorius). *Fabr. P.* 55.

4374. 53; culpator*. *Schrank B. No.* 2077.

4375. 54, fumigator*. *Act. Taur.* (*Gravenhorst.*) xxiv. 347.

4383. 62; equitatorius*. *Gmel.* v. 2679.—*Act. Taur.* (*Gravenhorst.*) xxiv. 321.
Ic. subdentatorius. *Mus. Marsham.*

4385. 64, luctatorius*. *Linn.* ii. 931.—*Schæff. Ic.* 244. *f.* 6.—*Berk. S.* i. 155.—*Stew.* ii. 227.—*Turt.* iii. 437.—(*Curtis l. c. infra.*)
Ic. auratus. *Gmel. No.* 238.

4387. 66; flaviniger*. *Act. Taur.* (*Gravenhorst.*) xxiv. 320.

4389. 68, armatorius*. *Forst. C.* i. 82.

4390. [69, fasciatorius*.] *Fabr.* ii. 143. (!)—*Stew.* ii. 227.—*Turt.* iii. 438.
Ic. bidentatus. *Oliv.* 172.
Ic. dimicatorius. *Gmel.* v. 2680.—*Stew.* ii. 227.—*Turt.* iii. 437.
Ic. armatorius. *Rossi F. No.* 752.
Ic. fasciatorius: (nugatorius.) *Fabr. P.* 61. (!)

4392. 71, bidentorius*. *Fabr. P.* 63.—(*Curtis l. c. infra.*)
Ic. desertorius. *Panz. F.* xlv. *f.* 15?

4393. 72, nugatorius: (fasciatorius). *Fabr. S.* 220.—*Panz. F.* lxxx. *f.* 12.
Ic. fasciatorius. (*Curtis l. c. infra.*)

4394. 73, vaginatorius. *Linn.* ii. 932.—*Panz. F.* lxxix. *f.* 8.—(*Curtis l. c. infra.*)
Ic. flavatus. *Gmel. No.* 237.
Ic. curvatorius. *Müll. Z. D. Pr. No.* 1782.
Ic. mercatorius: (vaginatorius). *Fabr. P.* 61.

4397. 76, designatorius. *Linn.* ii. 932?—*Turt.* iii. 440.
Ic. notatorius. *Panz. F.* lxxx. *f.* 9.
Ic. punctatorius. *Mus. Marsham.*

4399. 78, natatorius*. *Fabr. S.* 219.
Ic. constellatus. *Fourc. P.* ii. 412.
Ic. bipunctatus. *Schrank B. No.* 2080.
Ic. 4-punctorius. *Müll. Z. D. Pr. No.* 1773.
β, Ic. sugillatorius. *Fabr.* ii. 132.—*Turt.* iii. 430.
Ic. mediatorius. *Panz. F.* lxxx. *f.* 7.

4400. 79, sugillatorius*. *Linn.* ii. 930.—*Schæff. Ic.* 84. *f.* 9.—(*Curtis l. c. infra.*)
Ic. moratorius. *Fabr. P.* 54.—(*Curtis l. c. infra.*)

4402. 81; quadrimaculatus*. *Schrank B. No.* 2090.
Ic. bidentorius. *Rossi F. No.* 751.

4404. 83, infractorius*. *Linn.* ii. 931.—*Panz. F.* lxxviii. *f.* 9.—(*Curtis l. c. infra.*)
β, Ic. interruptorius. *Mus. Marsham.*

4406. 84, diversorius*. *Mus. Marsham.*

4407. 85; xanthorius*. *Forst. C.* i. 83.—*Stew.* ii. 227.—*Turt.* iii. 435.—*Act. Taur.* (*Gravenhorst.*) xxiv. 314.

4408. 86; mercatorius*. *Fabr.* ii. 143.—*Panz. F.* lxxviii. *f.* 11.
Ic. micratorius. *Mus. Marsham.*
Ic. vaginatorius: (mercatorius). *Fabr. P.* 61.

4409. 87; negatorius. *Fabr.* ii. 141. (!)—*Turt.* iii. 437.—(*Curtis l. c. infra.*)
Ic. ornatorius. *Panz. F.* lxxiii. *f.* 15.

4410. 88; ambulatorius*. *Fabr.* ii. 139. (!)—*Stew.* ii. 227.—*Turt.* iii. 433.—(*Curtis l. c. infra.*)

4411. 89, concinnus* *mihi.*
Ic. ambulatorius. *Panz. F.* lxxviii. *f.* 10.

4412. 90, occisorius*. *Fabr. P.* 61.—(*Curtis l. c. infra?*)
Ic. marginatorius. *Panz. F.* lxxiii. *f.* 14.

4413. 91; laminatorius*. *Fabr. S.* 220.—*Isis.* (*Trentepohl.*) 1826. 62.
Albin. pl. ix. *fig. sup.* (*sinist.*)

4414. [92, bilineator*.] *Don.* xiv. 31. *pl.* 478.

4417. 95, fusorius*: (fuscus). *Linn.* ii. 933.—*Turt.* iii. 440.—(*Curtis l.* c. *infra.*)

4418. 96, pisorius*. *Linn.* ii. 931.—*Turt.* iii. 434.—(*Curtis l. c. infra.*)
Ic. lentorius. *Panz. F.* lxxi. *f.* 11.

4419. 97, flavatorius. *Fabr. P.* 63?—*Panz. F.* lxviii. *f.* 12?
Ic. ictericus. *Christ.* 34.
Var.? Ic. flavatorius. *Panz. F.* c. *f.* 12.

4420. 98, falcatorius. *Fabr.* ii. 148?

4421. 99, nigricaudatus. *Mus. Marsham.*

4422. 100; Atropos*. *Curtis.* v. *pl.* 234.
Albin. pl. vii. *fig. dext. sup.*

4423. 101, lutorius. *Fabr.* ii. 147.

4424 ad 4426. 103 ad 105. (35 nova? species?)

Genus 41: (615). ——

CRYPTUS *p, et* ICHNEUMON *p, Fabr.*

4427. 1, primatorius*.
Ic. primatorius. *Forst. C.* i. 81.—*Stew.* ii. 227.—*Turt.* iii. 435.
Ic. grossorius. *Fabr. P.* 57.—*Panz. F.* lxxviii. *f.* 8.
Ic. bicinctus. *Christ. pl.* 35. *f.* 1.
Ic. extensorius. *Panz. F.* xix. *f.* 17.
Ic. gloriatorius. *Mus. Marsham.*

4429. 3; vadatorius. *Rossi F.* (*Illig.*) ii. 59.—*Act. Taur.* (*Gravenhorst.*) xxiv. 306.
Ic. sarcitorius. *Fabr. P.* 56.
Ic. pictus. *Vill. E.* iii. 141.
Ic. sanguineus. *Christ. pl.* 35. *f.* 7.

4430. 4; sarcitorius*.
Ic. sarcitorius. *Linn.* ii. 930.—*Stew.* ii. 226.—*Turt.* iii. 431.
Ic. bipartitus. *Vill. E.* iii. 142.

4432. 6, atripes*. *Act. Taur.* (*Gravenhorst.*) xxiv. 307.
Ic. armatorius. *Mus. Marsham.*

4433. 7; raptorius*. *Linn.* ii. 930.—*Stew.* ii. 226?—*Turt.* iii. 431?—*Don.* ii. 13. *pl.* xlii. *f.* 2. (4)?

4434. 8; extensorius*. *Linn.* ii. 930.—*Stew.* ii. 226.
Var. Ich. vexatorius. *Grav. V. No.* 3733.
Ich. lusorius. *Grav. V. No.* 3734.

4438. 12; confusor*. *Act. Taur.* (*Gravenhorst.*) xxiv. 300.

4439. 13, ammonius. *Act. Taur.* (*Gravenhorst.*) xxiv. 301?

4440. 14; stramentarius*. *Act. Taur.* (*Gravenhorst.*) xxiv. 302.

4441. 15, terminatorius*. *Act. Taur.* (*Gravenhorst.*) xxiv. 302.

4442. 16, cerinthium. *Act. Taur.* (*Gravenhorst.*) xxiv. 303.

4443. 17; deceptor*. *Act. Taur.* (*Gravenhorst.*) xxiv. 308.
Ic. deceptorius. *Vill. E.* iii. 142.

4444. 18, crispatorius.
Ic. crispatorius. *Linn.* ii. 931.—*Turt.* iii. 431.

4446. 20; hostilis*. *Act. Taur.* (*Gravenhorst.*) xxiv. 309.

4449. 23, saturatorius. *Linn.* ii. 931?—*Turt.* iii. 431?

4450. 24, anator. *Fabr.* ii. 169.
Ic. biscutatus. *Gmel.* v. *No.* 233.

4451. 25, annulator. *Fabr. P.* 65?

4453. 27, cinctorius*.
Ic. cinctorius. *Fabr.* ii. 149. (!)?—*Turt.* iii. 442.
Ic. tinctorius. *Stew.* ii. 228?

Genus 42: (616). ALOMYA, *Panzer, Leach, Samou., Curtis,* (*Kirby.*)
CRYPTUS *p, et* PIMPLA *p, Fabr.*

4455. 2; debellator*. (*Panz. K. R.* ii. 84.)—(*Curtis l. c. infra.*)
Cr. debellator. *Fabr. P.* 82.—*Panz. F.* lxxviii. *f.* 13.
♂? Ic. ovator. *Fabr.* ii. 163.

4457. 4; stercorator*.
Ich. stercorator. *Kirby MSS.*—(*Kirby & Sp. I. E.* i. 266.)

4460. 7, Victor. *Curtis.* iii. *pl.* 120.

4462. 9, mensurator.
Pi. mensurator. *Fabr.* ii. 116.

Genus 43: (617). PELTASTES, *Illiger, Curtis.*
METOPIUS, *Panzer.*—PELASTES, *Leach, Samou.*—PIMPLA *p, Fabr.*

4463. 1, polyzonius*.
Ich. polyzonius. *Forst. C.* i. 85.—*Stew.* ii. 230.—*Turt.* iii. 473.
Pe. dentatus. *Curtis 2 edit.* 1. *pl.* 4.
β, Ich. dentatus. *Fabr.* ii. 180.
Ich. micratorius. *Fabr. P.* 62.
Ich. chrysopus. *Linn. Trans.* (*Lewin.*) iii. *p.* 4. *pl.*—*Stew.* ii. 230.—*Turt.* iii. 469.
Pe. Pini. *Curtis.* i. *pl.* 4.

4464. 2; necatorius*. (*Curtis l. c. supra.*)—*Id.* 2 *edit. l. c. supra.* —(*Ing. Inst.* 90.)
Ich. necatorius. *Fabr.* ii. 144.
Ich. Vespoides. *Panz. F.* xlvii. *f.* 19.

4465. 3, dissectorius. (*Curtis l. c. supra.*)—*Id.* 2 *edit. l. c. supra.*
Ich. dissectorius. *Panz. F.* xcviii. *f.* 14.

Ichneumonidæ insertæ sedes: Ichneumone affinitates?—TRYPHON *p, et* PORIZON *p, Fallen?*

4466 † 1. annulatorius. *Fabr. P.* 62. (!)—*Stew.* ii. 228.—*Turt.* iii. 438. In Mus. Soc. Linn.?

4467. 2; similatorius*. *Fabr. P.* 64.
4468. 3; elongator*. *Fabr. P.* 67.—*Stew.* ii. 229.—*Turt.* iii. 452.
4469. 4; rutilator*. *Linn.* ii. 934.—*Turt.* iii. 451.
4470. 5; sulphurator*. *Mus. Marsham.*
4471. 6; triumphator*. *Mus. Marsham.*
4472. 7; prærogator*. *Linn.* ii. 936.—*Turt.* iii. 455.
4473. 8, amictorius. *Panz. F.* lxxx. *f.* 14.
4474. 9; maculatorius*.
Ba. maculatorius. *Fabr. P.* 96.
4475. 10; lætatorius*. *Fabr. P.* 63.—*Panz. F.* xix. *f.* 19. ♂.—*C. f.* 14. ♀.
4476. 11; leucorrhœus. *Don.* xiv. 25. *pl.* 476. *f.* 1.
4477. 12, costator. *Don.* xiv. 26. *pl.* 476. *f.* 2.
4478. 13, hospitator. *Mus. Marsham.*
4479. 14; tarsatorius*. *Panz. F.* cii. *f.* 19.
4480. 15, pedatorius*. *Panz. F.* cii. *f.* 20.
4481 ad 4629: (148 nova? species, 22? nova genera?)

Genus 65: (639). BANCHUS, *Fabricius, Leach, Samou.*
4630. 1; pictus*. *Fabr. P.* 129.—*Don.* xii. 55. *pl.* 413.
4631. 2; compressus. *Fabr. P.* 129.
4632. 3, varius. *Fabr. P.* 129.
4633. 4, falcator. *Fabr. P.* 128.
4634. 5; venator*. *Fabr. P.* 126.

Genus 66: (640). OPHION, *Fabricius, Leach, Samou.*, (*Kirby.*)
4635. 1; luteum*. *Fabr. P.* 130.—*Stew.* ii. 230.—*Turt.* iii. 468.—(*Kirby & Sp. I. E.* iv. 213.)
Ic. luteus. *Linn.* ii. 937.—(*Kirby & Sp. I. E.* i. 266.)
Ic. Lucus. *Berk. S.* i. 156?
4636. 2; Vinulæ* *mihi.*
Albin. pl. xi. *fig. g–h.*
4637. 3; ramidulum*. *Fabr. P.* 131.
Ic. ramidulum. *Linn.* ii. 178.—*Berk. S.* i. 156.—*Don.* ii. 13. *pl.* 42. *f.* 1?—*Stew.* ii. 230.—*Turt.* iii. 468.
4638. 4; delusor*. *Fabr. P.* 67?
4639. 5; latrator*. *Fabr. P.* 135. (!)
4640. 6; saltator. *Fabr. P.* 137. (!)
Ic. saltator. *Stew.* ii. 230.—*Turt.* iii. 457.
4641. 7; inculcator*. *Fabr. P.* 135.
Ic. inculcator. *Linn.* ii. 936.—*Berk. S.* i. 155.—*Turt.* iii. 457.
4642. 8; pugillator*.
Ic. pugillator. *Linn.* ii. 936.—*Berk. S.* i. 155.—*Turt.* iii. 457.

4643 ad 4659: (17 sp. adhuc examinandæ.) 25 sp.

Genus 67: (641). ANOMALON, *Jurine.*

4660. 1; amictum*.
Ic. amictus. *Fabr. P.* 133. (!)—*Stew.* ii. 230.—*Turt.* iii. 469.

4662. 2; circumflexum*.
Ic. circumflexus. *Linn.* ii. 181.—*Don.* iii. 56. *pl.* 93. *f.* 2?—*Turt.* iii. 470.

4663. 3; glaucopterum*.
Ic. glaucopterus. *Linn.* ii. 181.—*Turt.* iii. 469.

4664 ad 4679: (17 sp. adhuc examinandæ.) 20 sp.

Genus 68: (642). ——

ANOMALON, *Curtis.*

4680. 1, vesparum.
An. vesparum. *Curtis.* v. *pl.* 198.

Genus 69: (643). ENICOSPILUS *mihi.*

4681. 1; simulator* *mihi.*

Genus 70: (644). ACÆNITUS, *Latreille.*

CRYPTUS *p, Fabr.*—ANOMALON *p, Jurine.*

4682. 1; dubitator*. (*Latr.* G. iv. 9.)—(*Leach E. E.* ix. 143.)
Ic. dubitatorius. *Panz. F.* lxxviii. *f.* 14.

4683 ad 4685: (3 sp. adhuc examinandæ.) 4 sp.

Familia VI: (95). BRACONIDÆ.

(ICHNEUMONIDES *p, Latr.*—BRACONES, *Essenbeck.*)
(ICHNEUMON *p, Linné.*—BRACON, *Fabr.*)

Genus 71: (645.) SPATHIUS, *Essenbeck.*

CRYPTUS *p, Panz.*

4686. 1; clavatus*.
Cr. clavatus. *Panz. F.* cii. *f.* 15. ♂.—*f.* 16. ♀.

4687 & 4691: (5 sp. adhuc examinandæ.) 6 sp.

Genus 72: (646). BRACON, *Jurine, Leach, Samou.*, (*Kirby.*).

VIPIO, *Latr. olim.*

4692. 1; desertor*. *Fabr. P.* 155.—(*Leach E. E.* ix. 143.)—(*Sam. I.* 7.)
Ic. desertor. *Turt.* iii. 448.
Ic. sextus. *Schæff. Ic.* 20. *f.* 2, 3.
Vi. nominator. *Latr. H.* xiii. 176.

4693. 2, nominator. *Fabr. P.* 104.
Ic. terrefactor. *Vill. E.* iii. 195.

4694. 3, deflagrator. *Spin. I. L.* ii. 101.
Ic. desertor. *Linn. F. No.* 1605?

4695. 4, denigrator*. *Fabr. P.* ii. 109.—*Curtis.* ii. *pl.* 69.
Ic. denigrator. *Turt.* iii. 451.
Ic. impostor. *Scop. C. No.* 287.
Ic. incertus. *Christ. pl.* 39. *f.* 9.

4696. 5, flavator. *Fabr. P.* ii. 110.
Ic. denigrator. *Linn.* ii. 934?

4697. 6; minutator. *Fabr. P.* 110.

4698. 7; cinctus.
Ic. cinctus. *Linn.* ii. 938.—*Turt.* iii. 470.

4699. 8; areator.
Ic. areator. *Mus. Marsham.*

4700 ad 4758: 59 species adhuc examinandæ: 9 genera?

Genus 81: (655). AGATHIS, *Latr., Leach.*

4759. 1; Panzeri*. (*Leach E. E.* ix. 143.)
Ic. Panzeri. *Jurine Hy. pl.* 8. *g.* 1. *f.* 2. ♀.
Ag. Malvacearum. *Latr. G.* i. *pl.* 13. *f.* 175.—*Id.* iv. 9.

Genus 82: (656). MICRODUS, *Essenbeck?*

4760. 1, gloriatorius*.
Ic. gloriatorius. *Panz. F.* cii. *f.* 17.

Genus 83: (657). BASSUS? *Fabricius, Curtis.*
DIPLOZON? *Essenbeck.*

4761. 1, calculator. *Fabr. S.* 225.—*Curtis.* ii. *pl.* 73. ♂.
Ic. calculatorius. *Panz. F.* lxxxiii. *f.* 13. ♀.

4762 ad 4776: 15 sp. adhuc examinandæ. (16 sp.)

Genus 84: (658). MICROGASTER, *Latr., Leach,* (*Kirby.*)
CRYPTUS *p, et* CEROPALES *p, Fabr.*—BASSUS *p, Panz.*

4777. 1; globatus*. *Spin. I. L.* ii. 147.
Ic. globatus. *Linn.* ii. 940.—*Berk. S.* i. 156.—*Stew.* ii. 232.—*Turt.* iii. 475.—(*Kirby & Sp. I. E.* i. 266.)

4778. 2, auriculatus*. *Spin. I. L.* ii. 147?
Ic. auriculatus. *Fabr. P.* 69.

4779. 3; deprimator*. *Spin. I. L.* ii. 148.—(*Leach E. E.* ix. 143.)
Ic. deprimator. *Panz. F.* lxxix. *f.* 11.

4780. 4, sessilis*. *Spin. I. L.* ii. 148.
Ce. sessilis. *Fabr. P.* 187.

4781. 5, glomeratus*. *Spin. I. L.* ii. 149.
Ic. glomeratus. *Linn.* ii. 940.—*Berk. S.* i. 157.—*Stew.* ii. 232.—*Turt.* iii. 475.—(*Kirby & Sp. I. E.* iii. 269.)

4782. 6, Aphidum*. *Spin. I. L.* ii. 150.
Ic. Aphidum. *Linn.* ii. 940.—*Stew.* ii. 231.—*Turt.* iii. 476.

4783 ad 4786: 4 sp. adhuc examinandæ. (10 sp.)

Genus 85: (659). GELIS, *Thunberg.*

ICHNEUMONES-PEDESTRES, *Grav.*—ICHNEUMON *aut* CRYPTUS *p*, *Fabr.*—MUTILLA *p*, *Linné.*

4787. 1; hemipterus*.
Ic. hemipterus. *Fabr.* ii. 190.

4788. 2; micropterus*.
Ic. micropterus. *Grav. I. P.* 26.

4789. 3; brachypterus*.
Ic. brachypterus. *Grav. I. P.* 29.
Ic. abbreviator. *Panz. F.* lxxi. *pl.* 17.

4790. 4; agilis*.
Ic. agilis. *Fabr.* ii. 190.
Var. Ic. vagans. *Walk. P.* ii. 67.
Ic. fuscicornis. *Vill. E.* iii. 215.
Ic. apterus. *Fourc. P.* ii. 425.
Ic. Cursor. *Schrank B.* iii. 2121.
Ic. hortensis. *Christ.* 375.
Ic. bicolor. *Vill. E.* iii. 275.

4791. 5; abbreviator.
Ic. abbreviator. *Fabr. S.* 222.

4792. 6; atricapillus*.
Ic. atricapillus. *Grav. I. P.* 41.

4793. 7; pedestris*.
Ic. pedestris. *Fabr.* ii. 192.

4794. 8; cursitans*.
Ic. cursitans. *Fabr.* ii. 191.
Ic. ruficornis. *Vill. E.* iii. 215.

4795. 9; pulicarius*.
Ic. pulicarius. *Fabr.* ii. 191.

4796. 10; festinans*.
Ic. festinans. *Fabr. S.* 232.

4797. 11; clavipes*. *Thunb.?—Bull. S. N.* 1829. *p.* 152.

4798. 12; fasciatus*.
Ic. fasciatus. *Fabr.* ii. 191.—*Panz. F.* lxxix. *f.* 14.
Ic. melanocephalus. *Vill. E.* iii. 216.

4799. 13, Acarorum*.
Mu. Acarorum. *Linn.* ii. 968.
Ic. Acarorum. *Fabr.* ii. 191.—*Turt.* iii. 482.

4800. 14, rufogaster*. *Thunb.?—Bull. S. N.* 1829. *p.* 152.

4801. 15, frontalis*. *Thunb.?—Bull. S. N.* 1829. *p.* 152.

4802. 16, ruficornis*. *Thunb.?—Bull. S. N.* 1829. *p.* 152.

4803. 17, pedicularius*.
Ic. pedicularius. *Fabr.* ii. 192.

4804. 18; nigrocinctus *.
Ic. nigrocinctus. *Grav. I. P.* 35.
Ic. pedicularius. *Panz. F.* lxxxi. *f.* 13.

4805 ad 4811—(7 sp. adhuc examinandæ.)

Familia VII: (96). ALYSIIDÆ.

Genus 86: (660). ALYSIA, *Latreille, Leach, Curtis,* (*Kirby.*)

BRACON *p, Jurine.*—CECHENUS, *Illiger.*—BASSUS *p, Panzer.*—CRYPTUS *p, Fabr.*

4812. 1; manducator *. (*Leach E. E.* ix. 143.)—(*Kirby & Sp. I. E.* iv. 215.)—*Curtis l. c. infra.*
Cr. manducator. *Fabr. P.* 87.
Ic. manducator. *Panz. F.* lxxii. *f.* 4.

4813. 2; apicalis *. *Curtis l. c. infra.*

4814. 3; similis *. *Curtis l. c. infra.*

4815. 4, Pratellæ. *Curtis l. c. infra.*

4816. 5; gracilis *. *Curtis l. c. infra.*

4817. 6, pallida. *Curtis l. c. infra.*

4818. 7, Apii *. *Curtis.* iii. *pl.* 141.

4819. 8, pubescens *. *Curtis l. c. supra.*

4820. 9. succincta. *Curtis l. c. supra.*

Genus 87: (661). ROGAS, *Nees.*

BASSUS *p, Fabr.*

4821. 1, testaceus *.
Ba. testaceus. *Fabr. P.* 101.

Genus 88: (662). SIGALPHUS, *Latreille, Leach, Samou.*

SPHÆROPYX, *Hoffm.*—CRYPTUS *p, Fabr.*—BRACON *p, Jurine.*

4822. 1; irrorator *. *Latr. H.* xiii. 139.—(*Leach E. E.* ix. 143.)—(*Sam. I.* 38.)
Cr. irrorator. *Fabr. P.* 88.

Genus 89: (663). CHELONUS, *Jurine,* (*Kirby.*)

SIGALPHUS *p, Latreille.*

4825. 1; oculator *. (*Jurine Hy.* 291.)
Ic. oculator. *Fabr. P.* 68. (!)
♂, Ic. scabrator. *Fabr.* ii. 174.

4826. 2; sulcatus *. *Jurine Hy.* 291. *pl.* 12. *g.* 41?

4827. 3, dentator *. *Panz. F.* lxxxviii. *f.* 14.

4828. 4; rufescens *. *Latr. G.* iv. 13.

4829. 5, rufipes *. *Latr.* G. iv. 14.

4830. 6, affinis * *mihi.*

Sectio II. ACULEATA, *Latreille.*

Subsectio 1. HETEROGYNA, *Latreille.*

Familia VIII: (97). FORMICIDÆ, *Leach.*

(Formicariæ, *Latreille.*)

Genus 90: (664). MYRMICA, *Latr., Leach,* (*Kirby.*)

Manica, *Jurine.*—Myrmecia, *Fabr.*

4831. 1; rubra*. (*Latr.* G. iv. 131.)—(*Kirby & Sp. I. E.* ii. 48.)
Fo. rubra. *Linn.* ii. 963.—*Berk. S.* i. 160.—*Stew.* ii. 246.—*Turt.* iii. 570.—*Don.* xiv. 87. *pl.* 503.—*Shaw G. Z.* vi. *pl.* 100.
Fo. minima rubra. *Ray Ins.* 69.
The red Ant. *Gould Ants.* 2, 3.—*Bingley.* iii. 295.
Fo. vagans. *Fabr. P.* 407?

4832. 2; cæspitum*.
Fo. cæspitum. *Scop. C. No.* 837.—*Turt.* iii. 572.—*Latr. F.* 251. *pl.* x. *f.* 63.
Fo. binodis. *Linn. F. No.* 1726?

4833. 3, graminicola*.
Fo. graminicola. *Latr. F.* 255.
Fo. Acervorum. *Fabr. P.* 407.

4834. 4, unifasciata.
Fo. unifasciata. *Latr. F.* 257.

4835. 5, tuberum*.
Fo. tuberum. *Fabr. P.* 407.
Fo. tuberosa. *Latr. F.* 259.

4836. 6, Kirbii *mihi.*
Fo. pallida. *Mus. Marsham.*

Genus 91: (665). ——

4837 † 1, brevicornis* *mihi.* In Mus. D. *Westwood.*

Genus 92: (666). ——

4838 † 1, Westwoodii *mihi.* In Mus. D. *Westwood.*

Genus 93: (667). *FORMICA Auctorum. Ant, Emmet.*

Lasius *p, Fabricius.*

4839 † 1. ligniperda[a].

4840. 2; pubescens*. *Latr. F.* 96. *pl.* 1. *f.* 2?
Fo. vaga. *Scop. C. No.* 833?
Fo. atra. *Fabr.* ii. 352?
♀; Fo. fuscoptera. *Fourc.* ii. 452?

---

[a] 4839 † 1. ligniperda. *Latr. F.* 88. *pl.* 1. *f.* 1.
Fo. herculanea. *Linn.* ii. 962?—*Berk. S.* i. 159. (!)—*Stew.* ii. 245. (!)—*Turt.* iii. 569. (!)—(*Samou.* 273.) (!)

4841. 3; fuliginosa*. *Latr. F.* 140. *pl.* 5. *f.* 27.—(*Kirby & Sp. I. E.* ii. 48.)
The Jet Ant. *Gould Ants.* 2, 3.

4842. 4; rufa*: (The Pismire.) *Linn.* ii. 962.—*Berk. S.* i. 159. —*Stew.* ii. 246.—*Don.* xiv. 75. *pl.* 496. ♀.—*Turt.* iii. 570.— (*Sam. I.* 17.)—(*Kirby & Sp. I. E.* i. 231. ii. 48. 94 *nota.*)— *Wood.* ii. 72. *pl.* 61.
Fo. herculanea. (*Samou.* 69. *pl.* 8. *f.* 10.)
Fo. maxima. *Ray I.* 69.
The Hill or Horse Ant. *Gould Ants.* 2, 3.—*Bingley.* iii. 293.
♀; Fo. dorsata. *Panz. F.* liv. *f.* 1.

4843. 5; sanguinea*. *Latr. F.* 150. *pl.* v. *f.* 29.

4844. 6; cunicularia*. *Latr. F.* 151. *pl.* v. *f.* 30.
Fo. rufibarbis. *Fabr. P.* 402.
Fo. pratensis. *Oliv. E. M.* vi. 504.
Fo. obsoleta. *Latr. Essai.* 38.
Fo. media. *Ray. Ins.* 69?
♂, Fo. microcephala. *Panz. F.* liv. *f.* 2?

4845. 7; nigra*. *Linn.* ii. 963.—*Berk. S.* i. 159.—*Latr. F.* 156. *pl.* v. *f.* 31.—*Turt.* iii. 570.—*Shaw* G. Z. vi. 351.—(*Sam. I.* 17.)
Fo. atra. *Linn.* ii. 96.
Fo. minor. *Ray. Ins.* 69.

4846. 8; fusca*. *Linn.* ii. 963.—*Berk. S.* i. 159.—*Latr. F.* 159. *pl.* v. *f.* 32.—*Stew.* ii. 246.—*Turt.* iii. 570.—(*Sam. I.* 17.)— (*Kirby & Sp. I. E.* ii. 48.)
Fo. libera. *Scop. C. No.* 835?
Fo. media. *Ray. Ins.* 69?
The small black Ant. *Gould Ants.* 2, 3.
♂, Fo. flavipes. *Fourc.* ii. 52.

4847. 9; emarginata*. *Oliv. E. M.* vi. 494.—*Latr. F.* 163. *pl.* vi. *f.* 33. (!)
Fo. minor rubescens. *Ray. Ins.* 69?

4848. 10; flava*. *Latr. F.* 166. *pl.* vi. *f.* 36. (!)—(*Kirby & Sp. I. E.* ii. 48.)
Fo. minor rubescens. *Ray. Ins.* 69?
The common yellow Ant. *Gould Ants.* 2, 3.

4849. 11, brunnea. *Latr. F.* 168. *pl.* vi. *f.* 35.
α. Fo. pallida. *Latr. Essai.* 41.

4850. 12, cognata *mihi.*
Fo. lutaria. *Mus. Marsham.*

Familia IX: (98). MUTILLIDÆ, *Leach.*

(MUTILLARIÆ, *Latr.*—MUTILLA, *Linné.*)

Genus 94: (668). MUTILLA *Auctorum.*

SPHEX *p, DeGeer.*—APIS *p, Christius.*

4851. 1; Europæa*. *Linn.* ii. 966.—*Don.* vi. 77. *pl.* 212.—

Stew. ii. 246.—*Turt.* iii. 578.—*Shaw* G. Z. vi. 355. *pl.* 101. *fig. sup.*—(*Leach E. E.* ix. 148.)—(*Samou.* 70. *pl.* 8. *f.* 11.)—(*Kirby & Sp. I. E.* ii. 392.)
Ap. Europa. *Harr. Ex.* 166. *pl.* l. *f.* 18.
♀; Ap. aptera. *Christius. pl.* 12. *f.* 1, 2.
Ap. simile. *Harr. Ex.* 166. *pl.* l. *f.* 19.

4852. 2; calva*. (*Latr. G.* iv. 121.)—*Fabr. P.* 438.—(*Curtis l. c. infra.*)—*Coqueb. Ic.* 16. *f.* 10.
Mu. cincta. *Mus. Marsham.*
† ♂? Mu. bimaculata. *Jur. Hy. pl.* 12. *f.* 38?

4853. 3; Ephippium*. *Fabr. P.* 434.—*Curtis.* ii. *pl.* 77.
♀, Mu. rufipes. *Fabr. P.* 439.

Genus 95: (669). MYRMOSA, *Latr.*, *Leach*, *Samou.*

HYLÆUS *p*, *Fabricius.*

4854. 1; melanocephala*. (*Latr.* G. 120. *pl.* xiii. *f.* 6. ♀.)—(*Leach E. E.* ix. 198.)—(*Sam. I.* 28.)
Mu. melanocephala. *Fabr. P.* 439.—*Ent. Trans.* i. 254.
My. nigra. (*Latr.* G. 120. *pl.* xiii. *f.* 6. ♂.)
My. atra. *Panz. F.* lxxxv. *f.* 14.

Subsectio 2. FOSSORES, *Latreille.*

(RAPACIA, *Lamarck.*—PRÆDONES, *Latr. G.*)

Familia X: (99). SCOLIADÆ, *Leach.*

(SCOLIETÆ *p*, *Latr.*)

Genus 96: (670). TIPHIA, *Fabr.*, *Leach*, *Samou.*

SPHEX *p*, *Scop.*—BETHYLLUS, *Panz.*

4855. 1; femorata*. *Fabr.* ii. 223. (!)—*Panz. F.* lxxvii. *f.* 14.—*Stew.* ii. 235.—(*Sam. I.* 41.)—*Turt.* iii. 505.

4856. 2, Morio. *Fabr. E. S.* ii. 227?—*Panz. F.* lv. *f.* 1?—(*Sam. I.* 41.)

4857. 3; villosa*. *Fabr.* ii. 227.
Be. villosus. *Panz. F.* xcviii. *f.* 16.

4858 † 4. 5-cincta[a].

Familia XI: (100). SAPYGIDÆ, *Leach.*

(SCOLIETÆ *p*, *Latreille.*)

Genus 97: (671). SAPYGA, *Latr.*, *Leach*, *Samou.*, (*Kirby.*)

APIS *p*, *Linné.*—HELLUS *p*, *Fabr.*—VESPA *p*, *Geoff.*—SPHEX *p*, *Villers.*—MASARIS *p*, *Panzer.*—SIREX *et* SCOLIA *p*, *Fabr. olim.*

4859. 1; punctata*. *Klug. M. S.* 61. *pl.* vii. *f.* 4, 5, 6.

---

[a] 4858 † 4. 5-cincta. *Fabr.* ii. 233. (!)—*Stew.* ii. 235. (!)—*Turt.* iii. 505. (!) An hujus generis?

Sa. 5-punctata. *Latr. H.* iii. 346.
Ap. 5-cincta. *Don.* xiii. 11. *pl.* 438.
Sa. 6-punctata. (*Leach E. E.* ix. 149.)—(*Sam. I.* 37.)
♂; He. 6-guttatus. *Fabr. P.* 247.
Ap. clavicornis. *Linn.* ii. 953?
♀; He. 6-punctatus. *Fabr. P.* 246.
β; He. 4-guttatus. *Fabr. P.* 247.
γ; He. pacca. *Fabr. P.* 247.
4860 † 2. prisma. *Klug. M. S.* 63. *pl.* vii. *f.* 7, 8.
Sc. prisma. *Fabr.* ii. 236.
Ma. crabroniformis. *Panz. F.* xlvii. *f.* 22. In Mus. Brit.?

Familia XII: (101). POMPILIDÆ, *Leach.*

(SPHEGIMÆ *p*, *Latreille.*—SPHEX, *Linné.*)

Genus 98: (672). APORUS, *Spinola, Leach, Samou.*

4861. 1, unicolor. *Spinola I. L.* ii. 34.—(*Leach E. E.* ix. 149.) —(*Sam. I.* 4.) In Mus. *Brit.*

4862. 2, bicolor*. *Spinola I. L.* ii. 35.
Ap. abdominalis. ——?

Genus 99: (673). CRYPTOCHEILUS[a].

Genus 100: (674). POMPILUS, *Latr.*, *Leach, Samou.*, (*Kirby.*)
ICHNEUMON *p*, *Geoff.*—PSAMMOCHARES, *Latr.*?

4864. 1; viaticus*. *Fabr. S.* 246.—*Panz. F.* lxv. *f.* 16.—*Stew.* ii. 233.—*Turt.* iii. 489.—(*Sam. I.* 34.)—(*Kirby & Sp. I. E.* i. 121.) —(*Curtis l. c. infra.*)
Sp. viatica. *Linn.* ii. 944?—*Berk. S.* i. 157.
Sp. vagus. *Harr. Ex.* 95. *pl.* xxviii. *f.* 2?

4865. 2; nervosus* *mihi.*

4866. 3; cognatus* *mihi.*

4867. 4; basalis* *mihi.*

4868. 5; phæopterus* *mihi.*

4869. 6; apicalis *mihi.*

4870. 7; aterrima *mihi.*

4871. 8; bifasciatus*. *Fabr. S.* 248.
Sp. bifasciata. *Panz. F.* lxxxvi. *f.* 11.

4872. 9; hircanus*. *Fabr. S.* 251?—*Panz. F.* lxxxvii. *f.* 21.—(*Sam. I.* 34.)

---

[a] Genus 99: (673). CRYPTOCHEILUS, *Panzer.*
POMPILUS *p*, *Fabr.*, *Samou.*, *Curtis.*

4863 † 1. annulatus. *Panz. K. R.* ii. 121.
Po. annulatus. *Fabr. P.* 197.—*Panz. F.* lxxvi. *f.* 16.—(*Samou.* 274. (!)—(*Curtis l. c. fo.* 238.)

4873. 10, argyrostoma *mihi.*

4874. 11; pulcher*. *Fabr. P.* 193.—(*Curtis l. c. infra.*)

4875. 12, nigerrimus* *mihi.*

4876. 13; niger*. *Fabr. S.* 247.—*Panz. F.* lxxi. *f.* 19.—(*Curtis l. c. infra.*)

4877. 14, subfasciatus *mihi.*

4878. 15; exaltatus*. *Fabr. S.* 251.—(*Sam. I.* 34.)—(*Curtis l. c. infra.*)
Sp. exaltata. *Panz. F.* lxxxvi. *f.* 10.

4879. 16; nebulosus* *mihi.*

4880. 17; zonatus* *mihi.*

4881. 18, subnebulosus* *mihi.*

4882. 19; gibbus*. *Fabr. S.* 193.—(*Curtis l. c. infra.*)
Sp. gibba. *Linn.* ii. 946.—*Panz. F.* lxxvii. *f.* 13.—(*Sam. I.* 34.)
Sp. revo. *Harr. Ex.* 95. *pl.* xxviii. *f.* 3.

4883. 20; fuscus*. *Fabr.* ii. 210.—*Panz. F.* lxv. *f.* 15.—*Stew.* ii. 234.—*Turt.* iii. 489.—(*Sam. I.* 34.)
Po. fuscus. *Latr.* G. iv. 64.—(*Curtis l. c. infra.*)
Sp. perturbator. *Harr. Ex.* 95. *pl.* xxviii. *f.* 1?

4884. 21; formosus* *mihi.*

4885. 22; binotatus* *mihi.*

4886 † 23, bipunctatus. *Fabr. P.* 195.—*Panz. F.* lxxii. *f.* 8.—(*Curtis l. c. infra.*) In Mus. *Brit.*

4887 † 24, rufipes*. *Fabr. P.* 195.—*Curtis.* v. *pl.* 238.
Sp. rufipes. *Linn.* ii. 214.—*Turt.* iii. 492. (!)
Sp. variabilis. *Rossi F.* ii. 99?
In Mus. D. Curtis, et *Westwood.*

§ B. Ceropales *p, Fabricius.*

4888. 25; calcaratus*.
Sp. calcarata. *Mus. Marsham.*

4889. 26, punctum*. *Panz. F.* lxxxvi. *f.* 12.—(*Curtis l. c. supra.*)
Ce. punctum. *Fabr. P.* 187.

Genus 101: (675). CEROPALES, *Latr., Leach, Samou.*

Ichneumon *p, Geoff.*—Pompilus *p, Panz.*—Evania *p, Olivier.*

4890. 1; maculatus*. *Fabr. P.* 185.—(*Leach E. E.* ix. 149.)—(*Sam. I.* 10.)
Sp. maculata. *Turt.* iii. 483.
Ev. maculata. *Fabr.* ii. 193. (!)
♀; Po. frontalis. *Panz. F.* lxxii. *f.* 9.

4891. 2; nigripes* *mihi.*

4892. 3, rufipes *mihi.*

4893 † 4, variegatus. *Fabr. P.* 186. In Mus. *Brit.*

Familia XIII: (102). SPHECIDÆ, *Leach.*

(SPHEGIMÆ PROPRIÆ, *Latr.*—SPHEX *p*, *Linné.*) *Sand Wasps.*

Genus 102: (676). DOLICHURUS, *Latr.*, *Leach*, *Samou.*

PISON, *Jurine.*—POMPILUS *p*, *Spinola.*

4894 † 1, ater. *Latr. G.* iv. 58.—(*Leach E. E.* ix. 150.)—(*Sam. I.* 15.)
Po. corniculus. *Spin. I. L.* ii. 52. In Mus. *Brit.*

Genus 103: (677). PELOPÆUS[a].

Genus 104: (678). SPHEX *Auctorum.*

PEPSIS *p*, *Fabr.*—ICHNEUMON *p*, *Geoff.*

4896. 1, flavipennis. *Jurine Hy.* 129. *pl.* 8. *g.* 5. *f.* 2.—(*Leach E. E.* ix. 150.)—(*Samou.* 275.)
Pe. flavipennis. *Fabr. P.* 210.

Genus 105: (679). AMMOPHILA, *Kirby*, *Turt.*, *Sowerb.*, *Leach*, *Samou.*

PEPSIS *p*, *Fabr.*—MISCUS, *Jurine.*

4897. 1; hirsuta*. *Linn. Trans.* (*Kirby.*) iv. *p.* 206.—*Turt.* iii. 497.—*Sowerb. B. M.* i. *pl.* xxxiii. *f.* 1.

4898. 2; argentea*. *Linn. Trans.* (*Kirby.*) iv. 208.—*Turt.* iii. 498.

4899. 3, affinis*. *Linn. Trans.* (*Kirby.*) iv. 205.—*Turt.* iii. 497.
Am. arenaria. *Don.* xiii. 74. *pl.* 468. *f.* 2?

4900. 4; vulgaris*. *Linn. Trans.* (*Kirby.*) iv. 204.—*Turt.* iii. 497.
Sp. sabulosa. *Linn.* ii. 941.—*Don.* iii. 55. *pl.* 93. *f.* 1.—*Stew.* ii. 232.—*Shaw G. Z.* vi. 282. *pl.* 93. *fig. inf.*
Am. sabulosa. (*Sam. I.* 2.)
♂; Pe. lutaria. *Fabr. P.* 228.—*Panz. F.* lxv. *f.* 14.
The common Sand Wasp. *Bingley.* iii. 256.
β; Am. cyanura. *Kirby MSS.*

4901. [5; pulvillata*.] *Sowerb. B. M.* i. *pl.* xxxiii. *f.* 2.

Familia XIV: (103). LARRIDÆ, *Leach.*

(CRABRONITES *p*, *Latr.*)—(SPHEX *p*, *et* VESPA *p*, *Linné.*)

Genus 106: (680). PSEN, *Latr.*, *Leach*, *Samou.*, *Curtis.*

TRYPOXYLON *p*, *Fabr.*—PELOPÆUS *p*, *Fabr.*—SPHEX *p*, *Panzer.*—PSENIA, *Kirby MSS.*

4902. 1, compressicornis*. *Latr. H.* iii. 338.
Pe. compressicornis. *Fabr. P.* 204.

---

[a] Genus 103: (677). PELOPÆUS, *Latr.*, *Leach*, (*Kirby.*)

PEPSIS *p*, *Illiger*,—SCELIPHRONS, *Klug.*

4895 ‡ 1. spirifex. *Latr. G.* iv. 60.
Sp. spirifex. *Linn.* ii. 942?—*Barbut G.* *pl.* 30. (!)—*Stew.* ii. 233. (!)—*Turt.* iii. 486. (!)—*Don.* xv. *pl.* 531. (!)
The turner savage. *Bingley.* iii. 255.

Ps. serraticornis. *Jurine Hy. pl.* 8. *g*. 6.
Sp. atra. *Fabr. S.* 244.—*Panz. F.* lxxii. *f.* 7.
† ♀. Pe. unicolor. *Fabr. P.* 204.

4903. 2; ater*. *Latr. H.* xiii. 310.—(*Leach E. E.* ix. 151.)—(*Sam. I.* 35.)—(*Curtis l. c. infra.*)
Tr. atratum. *Fabr. P.* 182.—*Panz. F.* xcviii. *f.* 15.

4904. 3; basalis* *mihi.*
Ps. equestris. *Curtis.* i. *pl.* 25.

4905. 4; bicolor*. *Jurine Hy. pl.* 13. *g.* 9.

4906. 5; equestris*. (*Latr.* G. iv. 92.)—(*Ing. Inst.* 90.)
Tr. equestris. *Fabr. P.* 182.
Ps. rufa. *Panz. F.* xcvi. *f.* 17?

4907. 6; aterrimus* *mihi.*

4908. 7; caliginosus* *mihi.*

4909. 8, assimilis* *mihi.*

4910. 9; phæopterus* *mihi.*

Genus 107: (681). ALYSON, *Jurine, Leach.*

Pompilus *p, Fabr.*—Mellinus *p, Latr.*—Alysson, *Panzer.*

4911 † 1, spinosus. *Jurine Hy.* 196. *pl.* 10.
♂. Sp. fuscata. *Fabr. P.* 192?—*Panz. F.* li. *f.* 3.
β? Sp. bimaculata. *Panz. F.* li. *f.* 4. In Mus. D. *Curtis,* et Kirby?

Genus 108: (682). NYSSON, *Latr., Leach.*

Sphex *p, Schr.*—Mellinus *p, Fabr.*—Crabro *p, Panzer.*—Oxybelus *p, Fabr.*—Pompilus *p, Fabr.*

4912. 1; spinosus*.
Sp. spinosa. *Forst. C.* i. 87.—*Stew.* ii. 234.—*Turt.* iii. 496.

4913. 2; interruptus*. (*Panz. K. R.* ii. 189.)
Me. interruptus. *Panz. F.* lxxii. *f.* 13.
Ny. spinosus. *Latr. H.* xiii. 305?—(*Leach E. E.* ix. 151.)

4914. 3; tricinctus*. (*Latr.* G. iv. 91.)
Me. tricinctus. *Fabr. P.* 299.
Cr. spinosus. *Panz. F.* lxii. *f.* 15?

Genus 109: (683). ——

Nysson *p, Latr.*—Pompilus *p, Fabr.*—Crabro *p, Panzer.*

4915. 1; maculatus*.
Po. maculatus. *Fabr. P.* 196.
Cr. trimaculatus. *Panz. F.* lxxviii. *f.* 17.
♂. Ny. decemmaculatus. *Spin. I. L.* ii. 41.

Genus 110: (684). ARPACTUS.

Arpactus *p, Jurine.*—Pompilus *p, Panz.*—Mellinus *p, Panz. I.*

4916. 1; tumidus*.
Po. tumidus. *Panz. F.* lxxxi. *f.* 15.

4917. [2; consobrinus*] *mihi.*

Genus 111: (685). GORYTES, *Latr.*, *Leach*, *Samou.*

SPHEX *p*, *Linné.*—MELLINUS *p*, *et* OXYBELUS *p*, *Fabr.*—SPHEX *p*, *Rossi.*—ARPACTUS *p*, *Jurine.*—EUZONIA, *Kirby MSS.*

4918. 1; mystaceus*.
Ve. mystacea. *Linn.* ii. 944.—*Panz.* *F.* liii. *f.* 11.—*Stew.* ii. 239.—*Turt.* iii. 527.
Ve. inimicus. *Harr. Ex.* 128. *pl.* xxxvii. § 1. *f.* 6?
Me. mystaceus. (*Sam. I.* 27?)
Ve. flavicincta. *Don.* xiii. 73. *pl.* 468. *f.* 1?

4919. 2; arenarius*.
Ve. campestris. *Linn.* ii. 950?—*Turt.* iii. 528.—*Panz. F.* xlvi. *f.* 11, 12.
Me. arenarius. *Panz. F.* liii. *f.* 12. var.

4910*. 3, quadrifasciatus*.
Me. quadrifasciatus. *Fabr. P.* 298.—*Panz. F.* xcviii. *f.* 17.

4911*. 4, quinquefasciatus.
Me. quinquefasciatus. *Panz. F.* liii. *f.* 3.

4912*. 5, quinquecinctus.
Me. quinquecinctus. *Fabr. P.* 299.—*Panz. F.* lxxii. *f.* 14.

Genus 112: (686). ASTATA, *Latr.*, *Leach*, *Samou.*, *Curtis.*

SPHEX *p*, *Schr.*—DIMORPHA, *Jurine.*—TIPHIA *p*, *Panz.*

4913* † 1. abdominalis. *Latr. H.* xiii. 297.
Sp. boops. *Schr. A. No.* 777.
Di. oculata. *Jurine Hy. pl.* 9. *g.* 10. ♂?
Ti. abdominalis. *Panz. F.* liii. *f.* 5. ♀? In Mus. Brit.?

4914*. 2, Victor. *Curtis.* vi. *pl.* 261.
La. pompiliformis. *Don.* xii. 73. *pl.* 420.

Genus 113: (687). LYROPS, *Illiger*, *Leach*, *Samou.*, (*Kirby.*)

TACHYTES *p*, *Panz.*—LARRA *p*, *Fabr.*—APIS *p*, *Rossi?*—SPHEX *p*, *Linné?*

4915*. 1, pompiliformis.
La. pompiliformis. *Panz. F.* lxxxix. *f.* 13.

4916*. 2; tricolor*. (*Leach E. E.* ix. 151.)—(*Sam. I.* 26.)
La. tricolor. *Fabr. P.* 221.—*Panz. F.* lxxxiv. *f.* 19.

4917*. 3, bicolor *mihi.*

Genus 114: (688). LARRA, *Fabr.*, *Leach*, *Samou.*, (*Kirby.*)

SPHEX *p*, *Villers.*—LIRIS *p*, *Fabr.*

4918*. 1, ichneumoniformis. *Fabr. P.* 220.—*Panz. F.* lxxvi. *f.* 18.—(*Leach E. E.* ix. 151.)—(*Sam. I.* 24.)

Genus 115: (689). MISCOPHUS, *Jurine.*

4919*. 1, bicolor*. *Jurine Hy. pl.* 11. *g.* 25.—(*Leach E. E.* ix. 151.)

Genus 116: (690). DINETUS, *Jurine, Leach, Samou.*, (*Kirby.*)
SPHEX *p, Schæff.*—POMPILUS *p, Fabr.*—CRABRO *p, Rossi.*

4920. 1, pictus*. *Jurine Hy. pl.* 11.*f.* 26.—(*Leach E. E.* ix. 151.) —(*Sam. I.* 15.)
Cr. ceraunius. *Rossi: teste Latr.*

Genus 117: (691). TRYPOXYLON, *Latr., Leach, Samou.*, (*Kirby.*)
SPHEX *p, Linné.*—APIUS, *Jurine.*

4921. 1; figulus*. *Latr. H.* xiii. 310.—*Panz. F.* lxxx.*f.* 16.—(*Sam. I.* 42.)
Sp. figulus. *Wood.* ii. 58. *pl.* 57.—(*Shaw* G. *Z.* vi. 281.)

4922. [2; pygmæum* *mihi.*]

Genus 118: (692). OXYBELUS, *Latr., Leach, Samou.*, (*Kirby.*)
SPHEX *p, Schæff.*

4923. 1; uniglumis*. *Latr. H.* xiii. 307.—(*Leach E. E.* ix. 152.) —(*Sam. I.* 32.)
Ve. uniglumis. *Linn.* ii. 951.—*Turt.* iii. 532.
Cr. uniglumis. *Panz. F.* lxiv.*f.* 14?
Ve. decem-maculata. *Don.* xi. 43. *pl.* 376.*f.* 1?

4924. 2; mucronatus*. *Fabr. P.* 318.

4925. 3, concinnus *mihi.*
Ox. mucronatus. *Panz. F.* ci.*f.* 19?

4926. 4; tridens*. *Fabr. P.* 318?

4927. 5; trispinosus*. *Fabr. P.* 318.—*Latr.* G. i. *pl.* 13.*f.* 13.

Familia XV: (104). BEMBICIDÆ, *Leach.*
(BEMBECIDES, *Latreille.*)

Genus 119: (693). BEMBEX[a].

Familia XVI: (105). CRABRONIDÆ, *Leach.*
(CRABRONITES *p, Latreille.*—VESPA, *Linné.*)

Genus 120: (694). CRABRO, *Fabr., Leach, Samou.*, (*Kirby.*)
SPHEX *p, Linné.*—PEMPHREDON *p, Fabr.*

4929 † 1, clypeatus. *Fabr. P.* 312.
Cr. vexillatus. *Panz. F.* xlvi.*f.* 5. In Mus. Brit.?

4930. 2; scutatus*. *Fabr. P.* 312.—*Panz. F.* xv. *f.* 22. ♂.—*f.* 23. ♀.
Sp. palmaria. *Naturf.* (*Schreber.*) xx. 100. *pl.* 2.*f.* 9.

---

[a] Genus 119: (693). BEMBEX, *Fabr., Don.*
APIS *p, Linné.*—VESPA *p, Sulz.*
4928 ‡ 1. octopunctata. *Don.* xiv. 21. *pl.* 474. (!)

4931 † 3, pterotus. *Panz. F.* lxxxiii. *f.* 16. ♂. *f.* 17. ♀.
Var. Cr. dentipes. *Panz. F.* xlvi. *f.* 9. ♀. In Mus. *Brit.*
4932. 4, cribrarius *. *Panz. F.* xv. *f.* 18. ♂. *f.* 19. ♀.—(*Sam. I.* 13.)
Ve. cribraria. *Stew.* ii. 239.—*Don.* xii. 63. *pl.* 416?—*Turt.* iii. 531.
Sp. cribraria. *Linn.* ii. 945.—*Berk. S.* i. 157.
4933. [5. interruptus * *mihi.*]
Cr. clypeatus. *Panz. F.* xv. *f.* 20? ♂.—*f.* 21. ♀.
4934. [6; palmatus *.] *Panz. F.* xlvi. *f.* 3.
4935. 7; patellatus *. *Panz. F.* xlvi. *f.* 4.
4936. 8; cephalotes *. *Panz. F.* lxii. *f.* 16.
4937. 9; subinterruptus * *mihi.*
Cr. lituratus. *Panz. F.* xc. *f.* 13?
4938. 10, quadricinctus. *Fabr. P.* 310.
4939. 11, vagabundus. *Panz. F.* liii. *f.* 16.
4940. 12, agrestis * *mihi.*
4941. 13, divisus * *mihi.*
4942. 14; vespiformis *. *Panz. F.* liii. *f.* 14.
4943. 15, sexcinctus. *Fabr. P.* 309.—*Panz. F.* lxiv. *f.* 14.
4944. 16, serripes *. *Panz. F.* xlvi. *f.* 8.
4945. 17, subterraneus. *Fabr. P.* 309.—*Panz. F.* iii. *f.* 21.
4946. 18; fossorius *. *Panz. F.* lxxii. *f.* 11.
Sp. fossorius. *Linn.* ii. 946.
Ve. fossaria. *Stew.* ii. 238.
4947. 19; philanthoides *. *Panz. F.* lxxxiii. *f.* 2?
4948. 20, lapidarius *. *Fabr. P.* 309?—*Panz. F.* xc. *f.* 12.
4949. 21, duodecimguttatus * *mihi.*
4950. 22, pictipes * *mihi.*
4951. 23; tarsalis * *mihi.*
4952. 24; signatus *. *Panz. F.* liii. *f.* 15.
4953. 25; analis * *mihi.*
4954. 26, consobrinus * *mihi.*
4955. 27, quadrimaculatus *. *Fabr. P.* 308.
4956. 28, vagus *. *Fabr. P.* 313.—*Panz. F.* xlvi. *f.* 10.
Sp. vaga. *Linn.* ii. 946.—*Stew.* ii. 239.
4957. 29; aterrimus * *mihi.*
4958. 30, leucostoma. *Fabr.* ii. 301.—*Panz. F.* xv. *f.* 24?
4959. 31; spiniceps *. *Kirby MSS.*
4960. 32; geniculatus * *mihi.*
4961. 33; hyalinus * *mihi.*
4962. 34; Stigma * *mihi.*

4963. 35; tibialis* *mihi.*

4964. 36; scutellatus* *mihi.*

4965. 37; phæopterus* *mihi.*

4966. 38; nanus* *mihi.*

Genus 121: (695). ——

CRABRO *p, Fabr.?*

4967. 1; dimidiatus*.
Cr. dimidiatus. *Fabr. P.* 313?

4968. 2, rufifemoratus* *mihi.*

Genus 122: (696). RHOPALUM, (*Kirby.*)

CRABRO *p, Latr.*—PEMPHREDON *p, Panz.*

4969. 1; tibiale*.
Cr. tibialis. *Fabr. S.* 271.—*Panz. F.* lxxxiii. *f.* 24.
Pe. varicornis. *Fabr. P.* 315.—*Panz. F.*

4970. 2; zonatum* *mihi.*

4971. 3; rufiventre*. (*Ing. Inst.* 90.)
Cr. rufiventris. *Panz. F.* lxxii. *f.* 12.

Genus 123: (697). STIGMUS, *Jurine, Leach, Samou.*, (*Kirby.*)

4972. 1; ater*. *Jurine Hy. pl.* 9. *g.* 7.—(*Leach E. E.* ix. 152.)—(*Sam. I.* 39.)
St. pendulus. *Panz. F.* xiv. *f.* 7.

Genus 124: (698). CEMONUS, *Leach.*

CEMONUS *p, Jurine.*—PEMPHREDON *p, Fabr.*—PSEN *p, Panzer.* —STIGMUS *p, Latr.*

4973. 1; minutus*. (*Leach E. E.* ix. 152.)
Pe. minutus. *Fabr. P.* 316.
Ps. pallipes. *Panz. F.* lii. *f.* 22.

Genus 125: (699). PEMPHREDON, *Latr., Leach, Samou.*

CEMONUS *p, Jurine.*—PELOPÆUS *p, Fabr.*

A. Abdomine sessili.

4974. 1; geniculatus* *mihi.*

4975. 2; albilabris*. *Fabr. P.* 316.

4976. 3; caliginosa* *mihi.*

B. Abdomine petiolato.

4977. 4, lugubre*. *Fabr. P.* 315?

4978. 5, nitida* *mihi.*

4979. 6; Jurinii* *mihi.*
Ce. unicolor. *Jurine Hy. pl.* 11. *g.* 28.

4980. 7; unicolor*. (*Latr. G.* iv. 84.)
Pel. unicolor. *Fabr. P.* 204?

4981. 8; fuscipennis* *mihi.*

4982. 9, assimilis* *mihi.*

4983. 10, hyalipennis *mihi.*

Genus 126: (700). MELLINUS, *Fabr., Leach, Samou.,* (*Kirby.*)
SPHEX *p, DeGeer.*

4984. 1, ruficornis*. *Fabr. P.* 298.—*Panz. F.* lxxvii. *f.* 17.—
(*Leach E. E.* ix. 152.)
♂? Me. sabulosus. *Fabr. P.* 297.

4985. 2; frontalis*.
Cr. frontalis. *Panz. F.* xlvi. *f.* 11.

4986. [3; petiolatus*.]
Cr. petiolatus. *Panz. F.* xlvi. *f.* 12.

4987. 4; bifasciatus *mihi.*

4988. 5, interruptus *mihi.*

4989. 6; arvensis*. *Fabr.* ii. 287.
Ve. arvensis. *Linn.* ii. 950.—*Stew.* ii. 238.—*Turt.* iii. 528.
Ve. superbus. *Harr. Ex.* 129. *pl.* xxxvii. § 2. *f.* 3.
♂; Me. bipunctatus. *Fabr. P.* 298.
♀, Cr. H. flavum. *Panz. F.* xvii. *f.* 20.

4990. 7; pratensis*. (*Jurine Hy.* 191. *pl.* 10. *g.* 9.)

Genus 127: (701). CERCERIS, *Latr., Leach, Samou.,* (*Kirby.*)
SPHEX *p, Schæff.*—PHILANTHUS *p, Fabr.,* (*Kirby.*)—CRABRO *p,*
et BEMBEX *p, Rossi.*

4991. 1, aurita. *Latr. H.* xiii. 315.
Ph. aurita. *Fabr. S.* 263.

4992. 2, læta*.
Ph. lætus. *Fabr. P.* 305.—*Panz. F.* lxiii. *f.* 11. ♀.
Ve. exultus. *Harr. Ex.* 129. *pl.* xxxvii. § 11. *f.* 1.

4993. [3, assimilis] *mihi.*

4994. 4, quinquecincta*.
Ph. quinquecinctus. *Fabr. P.* 304.—*Panz. F.* lxiii. *f.* 12. ♂.
Ph. flavipes. *Don.* xiii. 61. *pl.* 462?
Ve. petulans. *Harr. Ex.* 129. *pl.* xxxvii. § 2. *f.* 2?
♀? Ph. quadricinctus. *Panz. F.* lxiii. *f.* 15.—(*Samou.* 179?)

4995. 5, arenaria*.
Sp. arenaria. *Linn.* ii. 946?
Ve. arenaria. *Turt.* iii. 529.
Ph. arenarius. *Panz. F.* xlvi. *f.* 12.
Ce. Sabulicola. *Kirby MSS.*

4996. [6, Arenicola*.] *Kirby MSS.*
Ph. trifidus. *Fabr. P.* 305?

4997. 7, atra *mihi*.
Sp. xanthocephala. *Forst. C.* i. 86?—*Stew.* ii. 234?—*Turt.* iii. 495?

4998. 8, labiata*. (*Latr.* G. iv. 94.)
Ph. labiata ♀. *Fabr. P.* 303.
Ph. interruptus. *Panz. F.* lxiii. *f.* 17.

4999. 9, nasuta*. *Latr.* G. iv. 94.
Ph. labiatus. *Panz. F.* lxiii. *f.* 16.

5000. 10, fulvipes *mihi*.

5001. 11; ornata*.
Ph. ornata. *Fabr. P.* 304.—*Panz. F.* lxv. *f.* 10.
Var.? Ph. semicinctus. *Panz. F.* xlvii. *f.* 24.

5002. 12, Ferox*. *Kirby? MSS.*
Ph. sabulosus. *Panz. F.* lxiii. *f.* 13?

Subsectio 3. DIPLOPTERA, *Latreille*.

Familia XVII: (106). VESPIDÆ, *Leach*.

(VESPARIÆ, *Latreille*.—VESPA *p*, *Linné*.)

Genus 128: (702). EUMENES, *Latr.*, *Leach*, *Curtis*, (*Kirby*.)

RYGCHIUM, *Spin.?*—PTEROCHEILUS, *Klug.?*

5003. 1, atricornis. *Fabr. P.* 289.—*Curtis*. i. *pl.* 13.

5004 † 2. coarctata[a].

Genus 129: (703). ODYNERUS, *Latr.*, *Leach*, *Samou.*, *Curtis*.

5005. 1; bidens*. (*Curtis l. c. infra.*)
Ve. bidens. *Linn.* ii. 951.

5006. 2; connexus*. *Curtis l. c. infra.*

5007. 3; Eumenoides* *mihi*.

5008. 4; quadricinctus*.
Ve. quadricincta. *Fabr. P.* 262?

5009. 5; emarginatus. (*Curtis l. c. infra.*)
Ve. emarginata. *Fabr. P.* 263.

5010. 6, simplex.
Ve. simplex. *Fabr. P.* 263.

5011 † 7. bifasciatus[b].

5012. 8; quadrifasciatus*. (*Curtis l. c. infra?*)
Ve. quadrifasciata. *Fabr. P.* 262?

---

[a] 5004 † 2. coarctata.
Ve. coarctata. *Linn.* ii. 950.—*Stew.* ii. 238. (!)—*Turt.* iii. 522. (!)
Ve. coronata. *Panz. F.* lxiv. *f.* 12.

[b] 5011 † 7. bifasciatus.
Ve. bifasciata. *Linn.* ii. 950.—*Stew.* ii. 238. (!)

5013. 9, quinquefasciatus*.
Ve. quinquefasciata. *Fabr. P.* 262?

5014. 10, sexfasciatus*. (*Curtis l. c. infra?*)
Ve. sexfasciata. *Fabr. P.* 262?

5015. 11, Scoticus. *Curtis l. c. infra.*

5016. 12; murarius. (*Kirby & Sp. I. E.* i. 447.)—(*Curtis l. c. infra.*)
Ve. muraria. *Linn.* ii. 950.—*Turt.* iii. 518.
Ve. parietina. *Panz. F.* xlix. *f.* 24.
Ve. coarctata. *Berk. S.* i. 158?
Ve. insolens. *Harr. Ex.* 128. *pl.* xxxvii. *f.* 7?

5017. 13; Antilope. (*Curtis l. c. infra.*)
Ve. Antilope. *Panz. F.* liii. *f.* 9.

5018. 14; parietum*. (*Curtis l. c. infra.*)
Ve. parietum. *Linn.* ii. 949.—*Stew.* ii. 238.—*Turt.* iii. 517.
Ve. quadrata. *Don.* xiv. 72. *pl.* 495. *f.* 2.
Od. similis. (*Curtis l. c. infra.*)

5019. 15; nigricornis*. *Curtis l. c. infra.*

5020. 16; parietinus*. (*Samou.* 279.)—*Curtis.* iii. *pl.* 137
Ve. parietina. *Linn. F. No.* 1679.

5021. 17; quadratus*. (*Curtis l. c. supra.*)
Ve. quadrata. *Panz. F.* lxiii. *f.* 3.
Ve. flavipes. (*Curtis l. c. supra?*)

5022. 18; angulatus*. (*Curtis l. c. supra.*)
Ve. angulata. *Don.* xiv. 71. *pl.* 495. *f.* 1.

5023. 19; rotundatus* *mihi.*

5024. 20; pictus*. (*Curtis l. c. supra.*)

5025. 21; auctus*.
Ve. aucta. *Fabr. P.* 267.—*Panz. F.* lxxxi. *f.* 17.

5026. 22, triangulus* *mihi.*

Genus 130: (704). EPIPONE, (*Kirby.*)
ODYNERUS *p, Curtis.*

5027. 1; spinipes*. (*Kirby & Sp. I. E.* i. 346.—*Id.* ii. 319.—*Id.* iv. 218.)
Ve. spinipes. *Linn.* ii. 950.—*Panz. F.* xvii. *f.* 18.—*Ent. Trans.* (*Haworth.*) i. 253.
Od. spinipes. (*Curtis fo.* 137.)
♀. Ve. Gazella. *Panz. F.* liii. *f.* 10?

Genus 131: (705). VESPA *Auctorum.* *Wasp.*

5028. 1; Crabro*: Hornet. *Linn.* ii. 948.—*Berk. S.* i. 157.—*Barbut. Gen. pl.* 14.—*Harr. Ex.* 127. *pl.* xxxvii. *f.* 1. ♀.—*Stew.* ii. 237.—*Turt.* iii. 513.—*Don.* xiv. 85. *pl.* 502.—*Shaw* G. Z. vi. 286.

pl. 95. *fig. inf. med.*—(*Kirby & Sp. I. E.* iii. 633 *nota.*—*Bingley.* iii. 261.—(*Samou.* 280. *pl.* 8. *f.* 8.)
♂. Ve. vexator. *Harr. Ex.* 127. *pl.* xxxvii. *f.* 2, 3.
5029. 2; vulgaris*: Common Wasp. *Linn.* ii. 949.—*Berk. S.* i. 158.—*Harr. Ex.* 128. *pl.* xxxvii. *f.* 5.—*Stew.* ii. 238.—*Don.* vii. *pl.* 226.—*Turt.* iii. 513.—*Bingley.* iii. 263.—(*Shaw* G. *Z.* vi. 285. *pl.* 95. *fig. inf. dext.*)—(*Kirby & Sp. I. E.* ii. 331.)—(*Sam. I.* 43.)
♂. Ve. parietum. *Harr. Ex.* 128. *pl.* xxxvii. *f.* 4.
5030. 3; rufa*. *Linn.* ii. 949.
5031. 4; saxonica*. *Fabr.* ii. 256.—*Panz. F.* xlix. *f.* 21.
5032. 5; holsatica*. *Fabr.* ii. 257.—*Shaw G. Z.* vi. 287. *pl.* 97.
Ve. anglica. *Leach MSS.*
Ve. sexcincta. *Don.* xiii. 46. *pl.* 455?
The campanular Wasp. *Bingley.* iii. 270.
5033. 6, Britannica. *Leach Z. M.* i. 112. *pl.* 50.—(*Sam. I.* 43.)

Subsectio 4. MELLIFERA, *Latreille.*
(ANTHOPHILA, *Latreille* G.)

Familia XVIII: (107). APIDÆ, *Leach.*
(APIARIÆ, *Latreille.*)—(APIS *p, Linné.*—APIS, *Kirby.*)

Genus 132: (706). SYSTROPHA, *Illiger, Leach,* (*Kirby.*)
HYLÆUS *p, Fabr.*—ANDRENA *p, Olivier.*—CERATINA *p, Jurine.* —ANTHIDIUM *p, Panzer.*—EUCERA *p, Scopoli.*

5034 ‡ 1, spiralis*. *Illiger.*—(*Leach E. E.* ix. 156.)
Hy. spiralis. *Fabr. P.* 320.
An. spirale. *Panz. F.* xxxv. *f.* 22. ♂. In Mus. *Brit.*

Genus 133: (707). PANURGUS, *Panzer, Leach, Samou., Curtis,* (*Kirby.*)
DASYPODA, *Illiger.*—APIS, (* *a.*) *Kirby.*—ERIOPS, *Klug.*—ANDRENA *p, Panzer.*—TRACHUSA *p, Jurine.*—PHILANTHUS *p, Fabr.*

5035. 1; ursinus*. *Curtis.* iii. *pl.* 101.
Ap. ursina. *Gmel.* iv. 2790.—*Kirby M. A.* ii. 178. *pl.* 16. *f.* 1.
Pa. ursina. (*Sam. I.* 33.)
♂; Ap. Banksiana. *Kirby M. A.* ii. 179.—*Don.* xii. 26. *pl.* 403. *f.* 2.
Tr. atra. *Panz. F.* xcvi. *f.* 19.
Pa. Banksianus. (*Sam. I.* 33.)
5036. 2; calcaratus*.
Ap. calcarata. *Scop. C.* 301.
Ph. ater. *Fabr. E. S.* ii. 292?
An. lobata. *Panz. F.* xcvi. *f.* 18. ♂.—lxxii. *f.* 16. ♀.
Pa. lobata. (*Curtis l. c. supra.*)
Ap. Linnæella. *Kirby M. A.* ii. 179. *pl.* 16. *f.* 2.
Pa. Linneella. (*Sam. I.* 33.)

Genus 134: (708). CERATINA, *Latr.*, *Leach*, *Samou.*, (*Kirby.*)

MEGILLA *p*, *Fabr.*—PROSOPIS *p*, *Fabr.*—APIS, (** *d*.2.*a*.) *Kirby.* —PITHITIS, *Klug.*—CLAVICERA, *Walcka.*

5037. 1; cærulea*. (*Leach E. E.* ix. 156.)—(*Sam. I.* 10.)
Ap. cærulea. *Vill. E.* iii. 319. *pl.* 8. *f.* 25.
Ap. cyanea. *Kirby M. A.* ii. 308. *pl.* 17. *f.* 8. ♂.—*f.* 7. ♀.
Ce. albilabris. (*Latr.* G. i. *pl.* 14. *f.* 11. ♂.)
Ce. callosa. (*Latr.* G. iv. 160.)

Genus 135: (709). CHELOSTOMA, *Latr.*, *Leach*, *Samou.*, (*Kirby.*)

HYLÆUS *p*, *Fabr.*—ANTHOPHORA *p*, *Illiger.*—ANTHIDIUM *p*, *Panzer.*—TRACHUSA *p*. *Jurine.*—APIS, (** c. 2. γ. p.) *Kirby.*—MEGACHILE *p*, *Latr. olim.*

5038. 1; florisomne*. (*Leach E. E.* ix. 157.)—(*Sam. I.* 10.)
Ap. florisomnis. *Linn.* ii. 954.——*Turt.* iii. 540.—(*Kirby & Sp. I. E.* iii. 320.)
Hy. florisomnis. *Panz. F.* xlvi. *f.* 13.—(*Sam. I.* 22.)
Ap. tumidus. *Harr. Ex.* 164. *pl.* 50. *f.* 12.
Ch. maxillosa. (*Kirby & Sp. I. E.* iii. 318, 320 & 341.)
♀, Ap. maxillosa. *Linn.* ii. 954.—*Turt.* ii. 540.
Hy. maxillosus. *Panz. F.* liii. *f.* 17.

Genus 136: (710). HERIADES, *Spinola*, *Leach*, *Samou.*, (*Kirby.*)

ANTHOPHORA *p*, *Fabr.*—ANTHIDIUM *p*, *Panzer.*—TRACHUSA *p*, *Jurine.*—APIS (** c. 2. γ. *p*.) *Kirby.*—MEGACHILE *p*, *Latr. olim.*

5039. 1; Campanularum*. (*Sam. I.* 21.)
Ap. Campanularum. *Kirby M. A.* ii. 256. *pl.* 16. *f.* 15. ♂.—*f.* 14. ♀.
Ap. florisomnis minima. *Christi. Hy.* 197. *pl.* 17. *f.* 8.

5040. 2, truncorum*. (*Leach E. E.* ix. 157.)—(*Sam. I.* 21.)
Ap. truncorum. *Linn.* ii. 954.—*Kirby M. A.* ii. 258.
Hy. truncorum. *Fanz. F.* lxiv. *f.* 15.

Genus 137: (711). STELIS, *Panzer*, *Leach*, *Samou.*, (*Kirby.*)

ANTHOPHORA *p*, *Illig.*—MEGACHILE *p*, *Latr.*—APIS, (** c. 1. β.) *Kirby.*—TRACHUSA *p*, *Jurine.*—GYRODROMA, *Klug.*—MEGILLA *p*, *Fabr.*

5041. 1; aterrima*. *Panz. E. V.* 247.—*Panz. F.* lvi. *f.* 7. ♂.
Ap. punctulatissima. *Kirby M. A.* ii. 231. *pl.* 16. *f.* 9.
St. punctulatissima. (*Sam. I.* 39.)

5042. 2; phæoptera*. (*Latr. G.* iv. 164.)
Ap. phæoptera. *Kirby M. A.* ii. 232.
St. phæoptera. (*Sam. I.* 39.)

Genus 138: (712). ANTHIDIUM, *Fabr., Leach, Samou., Curtis,* (*Kirby.*)

ANTHOPHORA *p*, *Illiger.*—MEGACHILE *p*, *Spinola.*—TRACHUSA *p*, *Jurine.*—APIS, (** c. 2. β.) *Kirby.*

5043. 1; manicatum*. *Fabr. P.* 364.—*Curtis.* ii. *pl.* 61.—(*Kirby & Sp. I. E.* iii. 316. 332 & 341.)
Ap. manicata. *Linn.* ii. 958.—*Berk. S.* i. 159.—*Kirby M. A.* ii. 248. *pl.* 16. *f.* 13. ♂.—*f.* 12. ♀.—*Stew.* ii. 242.—*Turt.* iii. 555.—*Don.* xiv. 57. *pl.* 489.
An. manicum. (*Sam. I.* 3.)
The Garden Bee. *Bingley.* iii. 275.
Ap. pervigil. *Harr. Ex.* 162. *pl.* xlix. *f.* 3.
♂; Ap. dentala. *Harr. Ex.* 161. *pl.* xlix. *f.* 1.

Genus 139: (713). OSMIA, *Panzer, Leach,* (*Kirby.*)

ANTHOPHORA *p*, *Fabr.*—MEGACHILE *p*, *Walck.*—TRACHUSA *p*, *Jurine.*—HOPLITIS? *Klug.*—AMBLYS, *Klug.*—APIS, (** c. 2. δ.) *Kirby.*

5044. 1, spinulosa*. (*Sam. I.* 32.)—(*Curtis l. c. infra.*)
Ap. spinulosa. *Kirby M. A.* ii. 261. *pl.* 17. *f.* 2. ♂.—*f.* 1. ♀.

5045. 2; leucomelana. (*Sam. I.* 32.)—(*Curtis l. c. infra.*)
Ap. leucomelana. *Kirby M. A.* ii. 260.

5046. 3; hirta*. (*Curtis l. c. infra.*)
Ap. hirta. *Fourc.* ii. 444.
Ap. Leaiana. *Kirby M. A.* ii. 263.
Os. Leaiana. (*Sam. I.* 32.)

5047. 4, parietina. *Curtis.* v. *pl.* 222.

5048. 5; cærulescens*. (*Leach E. E.* ix. 158.)—(*Sam. I.* 32.)—(*Curtis l. c. supra.*)
Ap. cærulescens. *Linn.* ii. 955.—*Kirby M. A.* ii. 264.—*Turt.* iii. 541.—(*Kirby & Sp. I. E.* i. 442.)
An. cærulescens. *Panz. F.* lxv. *f.* 18.
Ap. superbus. *Harr. Ex.* 164. *pl.* 49. *f.* 10.
♂; Ap. ænea. *Linn.* ii. 955.—*Turt.* ii. 542.
An. ænea. *Panz. F.* lvi. *f.* 3.

5049. 6; Tunensis*. (*Sam. I.* 32.)—(*Curtis l. c. supra.*)
Ap. Tunensis. *Fabr.* ii. 334.—*Kirby M. A.* ii. 269.
Ap. Tunetana. *Gmel.* iv. 2773.
Ap. aurulenta. *Panz. F.* lxiii. *f.* 22?

5050. 7; bicolor*. (*Sam. I.* 32.)—(*Curtis l. c. supra.*)
Ap. bicolor. *Schr. A.*—*Kirby M. A.* ii. 277.
Ap. rufescens. *Gmel.* iv. 2790.
Ap. rustica. *Fourc.* ii. 451.
Ap. fusca. *Panz. F.* lvi. *f.* 11.

5051. 8, atricapilla. *Curtis l. c. supra.*

5052. 9; bicornis*. *Panz. F.* lvi. *f.* 10.—lv. *f.* 15.—(*Curtis l.* c. *supra.*)
Ap. bicornis. *Kirby M. A.* ii. 271.—(*Kirby & Sp. I. E.* i. 442.)
Os. cornuta. (*Sam. I.* 32.)
♀; Ap. bicornis. *Linn.* ii. 954.—*Harr. Ex.* 162. *pl.* 49. *f.* 4.
Ap. fronticornis. *Panz. F.* lxiii. *f.* 20.
♂; Ap. rufa. *Linn.* ii. 954.—*Berk. S.* i. 158.—*Panz. F.* lvi. *f.* 10. —*Stew.* ii. 241.—*Turt.* iii. 557.
Ap. vernalis. (*Forst. Cat.* 720.)
Ap. frontalis. *Vill. E.* iii. 127. *pl.* 8. *f.* 28.
Ap. agino. *Harr. Ex.* 163. *pl.* 49. *f.* 7.

5053. 10, cornuta. *Latr.* G. iv. 164.—(*Leach E. E.* ix. 158.)
Me. cornuta. *Latr. H.* xiv. 59.
♂. Ap. rufa. *Rossi F.* ii. 169.
Ap. bicolor. *Vill. E.* iii. *pl.* 8. *f.* 27.
♀, Ap. bicornis. *Rossi M.* 310.

Genus 140: (714). MEGACHILE, *Latr.*, *Leach*, *Samou.*, *Curtis*, (*Kirby.*)

ANTHOPHORA *p*, *Fabr.*—TRACHUSA *p*, *Jurine.*—XYLOCOPA *p*, *Fabr.*—CENTRIS *p*, *Fabr.*—APIS, (** c. 2. *a.*) *Kirby.*—PHYLLOTOMA, *Dumeril.*

A. Tarsis anticis maris dilatato-ciliatis.

5054. 1; Willughbiella*. (*Sam. I.* 26.)—*Curtis.* v. *pl.* 218.
Ap. Willughbiella. *Kirby M. A.* ii. 233.—*Schæff. Ic. pl.* 20. *f.* 1.
Ap. lagopoda var. *Don.* xiii. 19. *pl.* 442.

B. Tarsis anticis maris haud dilatatis.

5055. 2; ligniseca*: Carpenter Bee.—*Shaw l.* c. (*Sam. I.* 26.)—(*Kirby & Sp. I. E.* iii. 341.)—(*Curtis l.* c. *supra.*)
Ap. ligniseca. *Kirby M. A.* ii. 243. *pl.* 16. *f.* 11.
Ap. centuncularis. *Panz. F.* lv. *f.* 12.—*Don.* iv. 31. *pl.* 120. —*Harr. Ex.* 162. *pl.* 49. *f.* 2.—*Shaw* G. *Z.* vi. 344.
An. rufiventris. *Fabr. P.* 378.

5056. 3, circumcincta*. (*Sam. I.* 26.)—(*Curtis l.* c. *supra.*)
Ap. circumcincta. *Kirby M. A.* ii. 246. *pl.* 16. *f.* 10.

5057 † 4. disjuncta[a].

5058. 5; xanthomelana*. (*Curtis l.* c. *supra.*)
Ap. xanthomelana. *Kirby M. A.* ii. 246.
Ap. parietina. *Fourc.* ii. 443?

5059. 6; maritima. (*Sam. I.* 26.)—(*Curtis l.* c. *supra.*)
Ap. maritima. *Kirby M. A.* ii. 242.

5060. 7; centuncularis*. (*Leach E. E.* ix. 159.)—(*Sam. I.* 26.) —(*Kirby & Sp. I. E.* iv. 553.)—(*Curtis l.* c. *supra.*)

---

[a] 5057 † 4, disjuncta.
Ap. disjuncta. *Don.* xii. 46. *pl.* 410. *f.* 2. In Mus. D. Donovan.

Ap. centuncularis. *Linn.* ii. 953.—*Berk. S.* i. 158.—*Barb. G. pl.* 15. *f.* 9.—*Kirby M. A.* ii. 237.—*Stew.* ii. 240.—*Turt.* iii. 558.—(*Kirby & Sp. I. E.* i. 192 & 444.)
The leaf-cutting Bee. *Bingley.* iii. 272.

5061 † 8, Leachella. (*Curtis l. c. supra.*)
Ap. Leachella. *Kirby MSS.* In Mus. *Brit. D.* Kirby? *et Vigors.*

Genus 141: (715). CŒLIOXYS, *Latr.*, *Leach*, *Samou.*, (*Kirby.*)

ANTHOPHORA *p*, *Fabr.*—MEGACHILE *p*, *Walck.*—TRACHUSA *p*, *Jurine.*—ANTHIDIUM *p*, *Panzer.*—HERIADES *p*, *Spinola.*—APIS, (** c. 1. *a.*) *Kirby.*

5062. 1; conica*. *Latr.*—(*Leach E. E.* ix. 159.)—(*Sam. I.* 8.) —(*Kirby & Sp. I. E.* iii. 341.)
Ap. conica. *Kirby M. A.* ii. 224.
♂; Ap. quadridentata. *Linn.* ii. 958.—*Panz. F.* lv. *f.* 13.
♀; Ap. conica. *Linn.* ii. 958.—*Berk. S.* i. 159.—*Stew.* ii. 243.—*Turt.* iii. 560.
Ap. bidentata. *Panz. F.* lvi. *f.* 7.
Ap. agilis. *Harr. Ex.* 163. *pl.* 49. *f.* 8.

5063 [† 2, inermis*.]
Ap. inermis. *Kirby M. A.* ii. 229. *pl.* 16. *f.* 8.
♂, Ap. quadridentata. *Miller MSS.*
♀, Ap. conica. *Fabr. E. S.* ii. 341. In Mus. *D. Kirby.*

Genus 142: (716). NOMADA, *Scopoli*, *Leach*, *Samou.*, (*Kirby.*)
APIS, (* *b.*) *Kirby.*

5064. 1; Goodeniana*. (*Sam. I.* 30.)
Ap. Goodeniana. *Kirby M. A.* ii. 180.—(*Kirby & Sp. I. E.* ii. 262.)

5065. 2; alternata*. (*Sam. I.* 30.)
Ap. alternata. *Kirby M. A.* ii. 182.

5066. 3, Lathburiana*. (*Sam. I.* 30.)
Ap. Lathburiana. *Kirby M. A.* ii. 184.

5067 † 4, varia. *Panz. F.* lv. *f.* 20.—(*Sam. I.* 30.)
Ap. varia. *Kirby M. A.* ii. 185. In Mus. *D. Kirby.*

5068. 5; flava*. *Panz. F.* liii. *f.* 21.—(*Sam. I.* 30.)
Ap. flava. *Kirby M. A.* ii. 186.

5069. 6, rufiventris*. (*Sam. I.* 30.)
Ap. rufiventris. *Kirby M. A.* ii. 187.

5070. 7; Kirbiella* *mihi.*

5071. 8; Marshamella*. (*Sam. I.* 30.)
Ap. Marshamella. *Kirby M. A.* ii. 188.

5072. 9; cornigera*. (*Sam. I.* 30.)
Ap. cornigera. *Kirby M. A.* ii. 190.—*Don.* xii. 41. *pl.* 408. *f.* 1?

5073. 10, subcornuta.
Ap. subcornuta. *Kirby M. A.* ii. 192.

5074 † 11, Capreæ. (*Sam. I.* 30.)
Ap. Capreæ. *Kirby M. A.* ii. 193. In Mus. D. *Kirby.*

5075. 12, lineola*. *Panz. F.* liii. *f.* 23.—(*Sam. I.* 30.)
Ap. lineola. *Kirby M. A.* ii. 194.

5076. 13, fucata*. *Panz. F.* liii. *f.* 19.
Ap. fucata. *Kirby M. A.* ii. 195.

5077 † 14, leucophthalma. (*Sam. I.* 30.)
Ap. leucophthalma. *Kirby M. A.* ii. 197. In Mus. D. *Kirby.*

5078 † 15, sexcincta. (*Sam. I.* 30.)
Ap. sexcincta. *Kirby M. A.* ii. 198. In Mus. D. *Kirby.*

5079. 16, Schæfferella. (*Sam. I.* 30.)
Ap. Schæfferella. *Kirby M. A.* ii. 199.
Schæff. Ic. pl. 81. *f.* 7.

5080. 17; connexa*. (*Sam. I.* 30.)
Ap. connexa. *Kirby M. A.* ii. 199.
No. sexfasciata. *Panz. F.* lxii. *f.* 18.

5081. 18; Jacobææ*. *Panz. F.* lxxii. *f.* 20.—(*Sam. I.* 30.)
Ap. Jacobææ. *Kirby M. A.* ii. 201.—*Don.* xii. 42. *pl.* 408. *f.* 2?

5082 † 19, flavopicta. (*Sam. I.* 30.)
Ap. flavopicta. *Kirby M. A.* ii. 202. In Mus. *D. Kirby.*

5083. 20; Solidaginis*. *Panz. F.* lxxii. *f.* 21.—(*Sam. I.* 30.)
Ap. Solidaginis. *Kirby M. A.* ii. 204.

5084. 21; picta*. (*Sam. I.* 30.)
Ap. picta. *Kirby M. A.* ii. 206.

5085. 22; rufopicta*. (*Sam. I.* 30.)
Ap. rufopicta. *Kirby M. A.* ii. 207.

5086 † 23. Hillana. (*Sam. I.* 30.)
Ap. Hillana. *Kirby M. A.* ii. 208. In Mus. D. Hill.

5087 † 24, ochrostoma.
Ap. ochrostoma. *Kirby M. A.* ii. 209.—*Don.* xii. 77. *pl.* 421. *f.* 3.
No. ochrostoma. (*Sam. I.* 30.) In Mus. D. Donovan? *et Kirby.*

5088. 25; ruficornis*. *Fabr. P.* 390.—*Panz. F.* lv. *f.* 18.—(*Sam. I.* 30.)—(*Kirby & Sp. I. E.* iv. 553.)
Ap. ruficornis. *Linn.* ii. 958.—*Kirby M. A.* ii. 210.
Ve. rubra. *Fourc. P.* ii. 438.
Ap. infractus. *Harr. Ex.* 134. *pl.* xxxix. *f.* 13.

5089. 26; apicalis* *mihi.*

5090. 27; xanthosticta*. (*Sam. I.* 30.)
Ap. xanthosticta. *Kirby M. A.* ii. 212.

5091. 28; Fabriciella*. (*Sam. I.* 30.)
Ap. Fabriciana. *Linn.* ii. 955.—*Kirby M. A.* ii. 213. *pl.* 16. *f.* 3.

5092. 29; quadrinotata*. (*Sam. I.* 30.)
Ap. quadrinotata. *Kirby M. A.* ii. 214.

5093. 30; flavoguttata*. (*Sam. I.* 30.)
Ap. flavoguttata. *Kirby M. A.* ii. 215.

5094. 31; rufocincta*. (*Sam. I.* 30.)
Ap. rufocincta. *Kirby M. A.* ii. 216.

5095. 32; Sheppardana*. (*Sam. I.* 30.)
Ap. Sheppardana. *Kirby M. A.* ii. 217.

5096. 33; ferruginata*. (*Sam. I.* 30.)
Ap. ferruginata. *Linn.* ii. 958.—*Kirby M. A.* ii. 218.
No. germanica. *Panz. F.* lxxii. *f.* 17.

5097. 34; nana* *mihi.*

Genus 143: (717). EPEOLUS, *Latreille, Leach, Samou.*, (*Kirby.*)
APIS, (** *b.*) *Kirby.*—NOMADA *p, Fabr.*

5098. 1; variegatus*. (*Leach E. E.* ix. 159.)—(*Sam. I.* 16.)
Ap. variegata. *Linn.* ii. 957.—*Kirby M. A.* ii. 222. *pl.* 16. *f.* 6. —*Don.* xii. 10. *pl.* 399. *f.* 2.—(*Kirby & Sp. I. E.* ii. 262.)
No. crucifera. *Panz. F.* lxi. *f.* 20.
Ap. muscaria. *Christi. Hy.* 195. *pl.* 17. *f.* 5.
Ap. agitabilis. *Harr. Ex.* 163. *pl.* xlix. *f.* 6?

Genus 144: (718). MELECTA, *Latr., Leach, Samou., Curtis,* (*Kirby.*)
APIS, (** *a.*) *Kirby.*—SYMMORPHA, *Klug.*—CROCISA *p, Jurine.*

5099. 1; punctata*. *Fabr. P.* 387. (!)—(*Leach E. E.* ix. 160.)—(*Sam. I.* 27.)—*Curtis.* iii. *pl.* 125.
Ap. punctata. *Vill. E.* iii. 312. (!)—*Kirby M. A.* ii. 219. *pl.* 16. *f.* 5. ♂.—*Stew.* ii. 240.—*Turt.* iii. 558.—*Don.* xi. 47. *pl.* 376. *f.* 4.
Ap. albifrons. *Forst. C.* i. 94.—*Stew.* ii. 244.—*Turt.* iii. 564.
Cr. atra. *Jurine Hy. pl.* 12. *g.* 34?
Ap. dubito. *Harr. Ex.* 137. *pl.* xl. *f.* 11.

Genus 145: (719). EUCERA, *Scopoli, Leach, Samou.*, (*Kirby.*)
APIS, (** *d.* 1.) *Kirby.*—ANDRENA *p, Panzer.*

5100. 1; longicornis*. *Scop. An.* iv. 8.—(*Sam. I.* 17.)—(*Kirby & Sp. I. E.* iv. 553.)
Ap. longicornis. *Kirby M. A.* ii. 278.
♂. Ap. longicornis. *Linn.* ii. 953.—*Harr. Ex.* 163. *pl.* xlix. *f.* 5. —*Barb. G. pl.* xv. *f.* 7.—*Stew.* ii. 240.—*Turt.* iii. 567.
Eu. longicornis. *Panz. F.* lxiv. *f.* 21.
♀, Ap. Bryorum. *Schra. C. No.* 812.
An. strigosa. *Panz. F.* lxiv. *f.* 16.

5101 † 2, linguaria*. *Fabr.* ii. 344.—*Panz. F.* liv. *f.* 22.
Ap. linguaria. *Kirby M. A.* ii. 282.

Ap. tumulorum. *Roem. G.* 61. *pl.* 27. *f.* 14.—*Don.* v. 19. *pl.* 151. *f.* 2.—*Stew.* ii. 240.—*Turt.* iii. 566.
In Mus. D. Donovan, *et Kirby.*

5102 † 3. pollinaris.
Ap. pollinaris. *Kirby M. A.* ii. 284. *pl.* 17. *f.* 3.
In Mus. *Soc. Linn.*

5103 † 4, Druriella.
Ap. Druriella. *Kirby M. A.* ii. 285. *pl.* 17. *f.* 4.—*Don.* xii. 9. *pl.* 399. *f.* 1. In Mus. *Brit.* D. Donovan? *et Kirby.*

Genus 146: (720). MELITTURGA, *Latreille, Leach.*

5104 † 1, clavicornis. *Latr. G.* iv. 177. *pl.* 14. *f.* 14? In Mus. *Brit.*

Genus 147: (721). SAROPODA, *Latreille, Leach, Samou.*, (*Kirby.*)
MEGILLA *p, Illiger.*—HELIOPHILA, *Klug.*—APIS (** *d.* 2. *a, p.*) *Kirby.*

5105. 1, bimaculata*.
Ap. bimaculata. *Panz. F.* lv. *f.* 17.—*Kirby M. A.* ii. 286.

5106. 2, furcata*.
Ap. furcata. *Panz. F.* lvi. *f.* 8. ♂.—*Kirby M. A.* ii. 288. *pl.* 17. *f.* 6. ♂.—*f.* 5. ♀.

5107. 3, vulpina*.
Ap. vulpina. *Panz. F.* lvi. *f.* 6.—*Kirby M. A.* ii. 290.

5108. 4, rotundata*. (*Sam. I.* 37.)
Ap. rotundata. *Panz. F.* lvi. *f.* 9.—*Kirby M. A.* ii. 291.
Me. rotundata. (*Sam. I.* 26.)
Ap. terrestris. *Fourc.* ii. 445.

5109. 5, subglobosa*.
Ap. subglobosa. *Kirby M. A.* ii. 295.

Genus 148: (722). ANTHOPHORA, *Latr., Leach, Samou.*, (*Kirby.*)
LASIUS, *Jurine.*—PODALIRIUS, *Walck.*—CENTRIS *p, Fabr.*—MEGILLA *p, Illiger.*—APIS, (** *d.* 2. *a. p.*) *Kirby.*

5110. 1; retusa*. (*Leach E. E.* ix. 160.)—(*Sam. I.* 3. *pl.* 8. *f.* 9.)—(*Kirby & Sp. I. E.* iii. 302. 306. & 332.)
Ap. retusa. *Kirby M. A.* ii. 296.
♂; Ap. pennipes. *Linn. MSS.*—*Don.* xiii. 3. *pl.* 434.
Ap. pilipes. *Fabr. P.* 329. (!)—*Panz. F.* lv. *f.* 8.—*Stew.* ii. 242.—*Turt.* iii. 552.
Ap. plumipes. *Schra. A.* 804.
Me. pilipes. (*Sam. I.* 26.)
Me. Hispanica. *Fabr. P.* 328.—*Panz. F.* lv. *f.* 6.
Ap. audax. *Harr. Ex.* 137. *pl.* xl. *f.* 14.
♀; Ap. retusa. *Linn.* ii. 954.
Me. Acervorum. *Fabr. P.* 328.—*Panz. F.* lxxviii. *f.* 18.—*Don.* iii. 98. *pl.* 108. *f.* 2.—*Stew.* ii. 244.—*Turt.* iii. 550.

Lasis pilipes. (*Sam. I.* 24.)
Ap. rufipes. *Miller MSS.—Christi. Hy.* 132. *pl.* 9. *f.* 1.
Ap. intrepidus. *Harr. Ex.* 131. *pl.* xxxviii. *f.* 7.

5111 † 2, Haworthana.
Ap. Haworthana. *Kirby M. A.* ii. 307.
In Mus. *Brit.*, *et* D. *Haworth.*

Genus 149: (723). XYLOCOPA[a].

Genus 150: (724). BOMBUS, *Latreille, Leach, Samou.*, (*Kirby.*)
BREMUS, *Jurine.*—APIS, (** *e.* 2.) *Kirby.*
Humble-bee. Bumble-bee. Bumble-dore.

5113. 1; rupestris*. *Fabr. P.* 348.—(*Sam. I.* 6.)
Ap. rupestris. *Fabr.* ii. 320.—*Kirby M. A.* ii. 369.
Ap. lapidaria var. *Brun. Prod.* 19. *nota a. a.*
Ap. subterranea. *Christi. Hy.* 125. *pl.* 6. *f.* 5?

5114. 2; campestris*. *Fabr. P.* 344.—(*Sam. I.* 6.)
Ap. campestris. *Panz. F.* lxxiv. *f.* 11.—*Kirby M. A.* ii. 335. *pl.* 18. *f.* 2.

5115. 3; Barbutellus*. (*Sam. I.* 6.)
Ap. Barbutella. *Kirby M. A.* ii. 343.—*Don.* xi. 71. *pl.* 385. *f.* 3.
β, Ap. saltuum. *Panz. F.* lxxv. *f.* 21.
Ap. monacha. *Christi. Hy.* 131. *pl.* 8. *f.* 7.
Ap. autumnalis. *Fabr.* ii. 324.

5116. 4; vestalis*. (*Sam. I.* 6.)
Ap. vestalis. *Fourc.* ii. 450.—*Kirby M. A.* ii. 347. *pl.* 18. *f.* 4. ♂.—*f.* 3. ♀.—*Don.* xiii. 65. *pl.* 464.

5117. 5; Muscorum*. *Fabr. P.* 349.—(*Sam. I.* 6.)
Ap. Muscorum. *Linn.* ii. 960.—*Christi. Hy.* 130. *pl.* 8. *f.* 3. ♀. —142. *pl.* 11. *f.* 8. ☿.—*Kirby M. A.* ii. 317.—*Stew.* ii. 243. —*Don.* xi. 70. *pl.* 382. *f.* 2.—*Turt.* iii. 550.—(*Kirby & Sp. I. E.* ii. 476.)
Ap. senilis. *Fabr.* ii. 324.
Ap. impavidus. *Harr. Ex.* 131. *pl.* xxxviii. *f.* 6.
Ap. melleus. *Harr. Ex.* 138. *pl.* xl. *f.* 17.
Ap. melinus. *Harr. Ex.* 138. *pl.* xl. *f.* 18.
The Cording Bee. *Bingley.* iii. 288.

5118 † 6. Hypnorum[b].

---

[a] Genus 149: (723). XYLOCOPA? *Latreille*, (*Kirby.*)
CENTRIS *p*, *Fabr.*—BOMBUS *p*, *Fabr.*—APIS, (** *d.* 2. β.) *Kirby.*
5112 ‡ 1, iricolor. *Illiger.*
Ap. iricolor. *Kirby M. A.* ii. 310. *pl.* 17. *f.* 9. (!)—*Don.* xii. 25. *pl.* 403. *f.* 1. (!)
Ap. virens. *Christi. Hy.* 123. *pl.* 6. *f.* 2?

[b] 5118 † 6. Hypnorum. *Fabr. P.* 349.
Ap. Hypnorum. *Linn.* ii. 960.—*Christi. Hy. pl.* 9. *f.* 7.—*Stew.* ii. 243. (!)—*Turt.* iii. 550. (!)

5119. 7; floralis*. (*Sam. I.* 6.)
Ap. floralis. *Gmel.* iv. 2785.—*Kirby M. A.* ii. 321. *pl.* 17. *f.* 14. ♂.—*Don.* xi. 47. *pl.* 376. *f.* 5.
Ap. fasciata. *Scop. A.* iv. *p.* 12.
Ap. vulgo. *Harr. Ex.* 137. *pl.* xl. *f.* 13. ☿.

5120. 8; Francillonellus*.
Ap. Francillonella. *Kirby M. A.* ii. 319. *pl.* 17. *f.* 13.
Po. Francillonana. (*Sam. I.* 6.)

5121. 9; Beckwithellus*. (*Sam. I.* 6.)
Ap. Beckwithella. *Kirby M. A.* ii. 323.

5122. 10; Sowerbianus*.
Ap. Sowerbiana. *Kirby M. A.* ii. 322.

5123 † 11, Lapponicus. *Fabr. P.* 345.
Ap. Lapponica. *Fabr. E. S.* ii. 318. In Mus. D. Kirby?

5124. 12; Forsterellus*. (*Sam. I.* 6.)
Ap. Forsterella. *Kirby M. A.* ii. 325.

5125. 13; Curtisellus*. (*Sam. I.* 6.)
Ap. Curtisella. *Kirby M. A.* ii. 324.

5126 † 14. Agrorum[a].

5127. 15; Sylvarum*. *Fabr. P.* 348.—(*Sam. I.* 6.)
Ap. Sylvarum. *Linn.* ii. 960.—*Barb.* G. *pl.* 15. *f.* 4, 5?—*Kirby M. A.* ii. 326. *pl.* 17. *f.* 16. ♂.—*f.* 15. ♀.—*Turt.* iii. 550.
Ap. scylla. *Christi. Hy. p.* 129. *pl.* 8. *f.* 1?

5128. 16; fragrans*. (*Sam. I.* 6.)
Ap. fragrans. *Pallas. Sp.* i. 474.—*Kirby M. A.* ii. 329.
Ap. pratorum. *Fabr.* ii. 222.

5129. 17; Latreillellus*. (*Sam. I.* 6.)
Ap. Latreillella. *Kirby M. A.* ii. 330.
Ap. Maura. *Christi. Hy.* 131. *pl.* 8. *f.* 8?

5130. 18; Rossiellus*. (*Sam. I.* 6.)
Ap. Rossiella. *Kirby M. A.* ii. 331. *pl.* 18. *f.* 1.
Ap. sylvarum. *Schra. A.* 817.

5131. 19, Leeanus*. (*Sam. I.* 6.)
Ap. Leeana. *Kirby M. A.* ii. 333.

5132. 20, Leachiellus *mihi.*

5133. 21; Tunstallanus*. (*Sam. I.* 6.)
Ap. Tunstallana. *Kirby M. A.* ii. 346.
Ap. veterana. *Fabr.* ii. 324?

---

[a] 5126 † 14. Agrorum. (*Sam. I.* 6.)
Ap. Agrorum. *Fabr.* ii. 321. (!)—*Kirby M. A.* ii. 326.—*Turt.* iii. 549. (!)—*Panz. F.* lxxxv. *f.* 20? In Mus. Soc. Linn.

5134. 22; Soroensis*. *Fabr. P.* 345.
Ap. Soroensis. *Fabr.* ii. 318.—*Kirby M. A.* ii. 354.—*Panz. F.* vii. *f.* 11.
Ap. Cardui. *Müll. Z. D. Pr.* 1929.
Bo. sorensis. (*Sam. I.* 6.)

5135. 23; Scrimshiranus*.
Ap. Scrimpshirana. *Kirby M. A.* ii. 342.
Bo. Scrimshirana. (*Sam. I.* 6.)

5136. 24; cognata* *mihi.*

5137. 25, Francisanus*. (*Sam. I.* 6.)
Ap. Francisana. *Kirby M. A.* ii. 334.

5138. 26; Hortorum*. *Fabr. P.* 347.—(*Sam. I.* 6.)
Ap. Hortorum. *Linn.* ii. 960.—*Kirby M. A.* ii. 339.—*Shaw G. Z.* vi. 348. *fig. inf. med.*
Ap. ruderata. *Fabr.* ii. 317.—*Christi. Hy.* 128. *pl.* 7. *f.* 4.
Ap. paludosa. *Müll. Z. D. Pr.* 1919.
Ap. Hypnorum. *Fourc.* ii. 450.
Ap. fidens. *Harr. Ex.* 130. *pl.* xxxviii. *f.* 3.

5139. 27; lucorum*. *Fabr. P.* 350.—(*Sam. I.* 6.)
Ap. lucorum. *Linn.* ii. 961.—*Kirby M. A.* ii. 336.—*Turt.* iii. 550.
Ap. Hypnorum. *Rossi F.* ii. 905.
Ap. cæspitum. *Panz. F.* lxxxi. *f.* 19.

5140. 28, Jonellus*. (*Sam. I.* 6.)
Ap. Jonella. *Kirby M. A.* ii. 338.

5141. 29, virginalis*. (*Sam. I.* 6.)
Ap. virginalis. *Fourc.* ii. 450.—*Kirby M. A.* ii. 349.
Ap. Hortorum. *Fabr.* ii. 320.—*Turt.* iii. 549.

5142. 30; terrestris*. *Fabr. P.* 343.—(*Sam. I.* 6.)
Ap. terrestris. *Linn.* ii. 960.—*Berk. S.* i. 159.—*Barb. G. pl.* 15. *f.* 2.—*Panz. F.* i. *f.* 16.—*Kirby M. A.* ii. 350.—*Stew.* ii. 243.—*Turt.* iii. 548.—*Shaw G. Z.* vi. 348. *pl.* 98. *fig. inf. sinist.*—*pl.* 99. *fig. inf.* nidus.—*Wood.* ii. 68. *pl.* 60.
Ap. hortorum. *Scop. C.* 817.
Ap. audax. *Harr. Ex.* 130. *pl.* xxxviii. *f.* 1.
♀, Ap. Graminum. *Marsh. MSS.*
Ap. terrestris. *Barb. G. pl.* 15. *f.* 1.—*Don.* iii. 41. *pl.* 88. *f.* 1.

5143. 31; subinterruptus*. (*Sam. I.* 6.)
Ap. subinterrupta. *Marsh. MSS.*—*Kirby M. A.* ii. 356. *pl.* 18. *f.* 5. ♀.
Ap. regalis. *Fourc.* ii. 449.
Ap. bistriata. *Christi. Hy.* 128. *pl.* 7. *f.* 3.

5144 † 32. Nemorum[a].

---

[a] 5144 † 32. Nemorum. *Fabr. P.* 345.
Ap. Nemorum. *Fabr.* ii. 317.—*Stew.* ii. 243. (!)

5145. 33; Burrellanus*. (*Sam. I.* 6.)
Ap. Burrellana. *Kirby M. A.* ii. 358.
Ap. regalis β. *Fourc.* ii. 449.
Schæff. Ic. pl. 261. *f.* 5, 6.

5146. 34; Donovanellus*. (*Sam. I.* 6.)
Ap. Donovanella. *Kirby M. A.* ii. 357. *pl.* 18. *f.* 6. ♂.

5147. 35; Cullumanus*. (*Sam. I.* 6.)
Ap. Cullumana. *Kirby M. A.* ii. 359.

5148. 36, pratorum*. (*Sam. I.* 6.)
Ap. pratorum. *Linn.* ii. 960.—*Kirby M. A.* ii. 360.
Ap. vercor. *Harr. Ex.* 136. *pl.* xl. *f.* 9.

5149. 37, Albinellus*. (*Sam. I.* 6.)
Ap. Albinella. *Kirby M. A.* ii. 361.
Ap. frutetorum. *Panz. F.* lxxv. *f.* 20.

5150. 38; Derhamellus*. (*Sam. I.* 6.)
Ap. Derhamella. *Kirby M. A.* ii. 363.
Ap. fidus. *Harr. Ex.* 131. *pl.* xxxviii. *f.* 4?

5151. 39; subterraneus*. *Fabr. P.* 350.—(*Sam. I.* 6.)
Ap. subterranea. *Linn.* ii. 961.—*Berk. S.* i. 159.—*Kirby M. A.* ii. 371.—*Stew.* ii. 244.—*Turt.* iii. 550.
Ap. perniger. *Harr. Ex.* 131. *pl.* xxxviii. *f.* 8. ♂.

5152. 40; Harrisellus*. (*Sam. I.* 6.)
Ap. Harrisella. *Kirby M. A.* ii. 373. *pl.* 18. *f.* 8. ♂.—*f.* 7. ♀.
Ap. retusa. *Christi. Hy.* 133. *pl.* 9. *f.* 4. ♂.

5153. 41; Raiellus*. (*Sam. I.* 6.)
Ap. Raiella. *Kirby M. A.* ii. 367.

5154. 42; lapidarius*. *Fabr. P.* 347.—(*Sam. I.* 6.)
Ap. lapidaria. *Linn.* ii. 960.—*Barb. G. pl.* 15. *f.* 3.—*Don.* iii. 97. *pl.* 108. *f.* 1.—*Id. pl.* 88. *f.* 2. β.—*Shaw N. M. pl.* 454. —*Kirby M. A.* ii. 364.—*Stew.* ii. 243.—*Turt.* iii. 549.—*Shaw G. Z.* vi. 347. *pl.* 98. *fig. inf. dext.*—*Don.* xi. 69. *pl.* 385. *f.* 1. ♂.
The orange-tailed Bee. *Bingley.* iii. 290.
Ap. pratorum. *Christi. Hy.* 141. *pl.* 11. *f.* 5. β.
Ap. audens. *Harr. Ex.* 130. *pl.* xxxviii. *f.* 2.
Ap. opis. *Harr. Ex.* 137. *pl.* xl. *f.* 12.
Ap. pertristis. *Harr. Ex.* 137. *pl.* xl. *f.* 15.
♂; Ap. arbustorum. *Fabr. E. S.* ii. 347.
Ap. coronata. *Fourc.* ii. 449.
Ap. strenuus. *Harr. Ex.* 131. *pl.* xxxviii. *f.* 5?
Ap. hæmorrhoidalis. *Christi. Hy.* 132. *pl.* 9. *f.* 2.
β? Ap. flavicollis. *Sowerb. B. M.* i. *pl.* 19.
Bo. flavicollis. (*Sam. I.* 6.)

Genus 151: (725). APIS *Auctorum.* *Honey-Bee.*

5155. 1; mellifica*. *Linn.* ii. 955.—*Harr. Ex.* 132. *pl.* xxxix.

f. 9. ♂.—*f.* 10. ☿.—(*Berk. S.* i. 158.)—*Kirby M. A.* ii. 312. *pl.* 17. *f.* 11. ♂.—*f.* 10. ♀.—*f.* 12. ☿.—*Stew.* ii. 241.—*Turt.* iii. 552.—(*Sam. I.* 4.)—*Shaw G. Z.* vi. 289. *pl.* 98. *fig. sup.*—*Don.* xiv. 63. *pl.* 492.—*Bingley.* iii. 282.

Ap. cerifera. *Scop. A.* iv. 16.
Ap. gregaria. *Geoff. H.* ii. 407.
Ap. mellifera. *Fourc.* ii. 442.
Ap. domestica. *Latr. H.* iii. 386.

Familia XIX: (108). ANDRENIDÆ, *Leach.*

(ANDRENETÆ, *Latr.*)—(APIS *p*, *Linné.*—MELITTA, *Kirby.*)

Genus 152: (726). ANDRENA, *Fabricius, Leach, Samou., Curtis,* (*Kirby.*)

MELITTA (** c.), *Kirby.*—NOMADA *p*, *Fabr.*

5156 † 1, Lathamana*?
Me. Lathamana. *Kirby M. A.* ii. 83. In Mus. *Brit.*

5157. 2; Rosæ*. *Panz. F.* lxxiv. *f.* 10.—(*Sam. I.* 2.)
Me. Rosæ. *Kirby M. A.* ii. 83.
Me. zonalis. *Kirby M. A.* ii. 87?

5158. 3; cingulata*. (*Sam. I.* 2.)
♀, No. cingulata. *Fabr. P.* 394.
Ap. Suecica. *Gmel.* ii. 2794.
Ap. Sphegoides. *Panz. F.* lvi. *f.* 24.
♂. An. labiata. *Fabr. P.* 324.
Ap. albilabris. *Panz. F.* lvi. *f.* 23.
Me. cingulata. *Kirby M. A.* ii. 88.

5159. 4; Schrankella*. (*Sam. I.* 2.)
Me. Schrankella. *Kirby M. A.* ii. 90.
Ap. Cetii. *Schr. A. No.* 818.
An. marginata. *Fabr. P.* 326.—*Panz. F.* lxxii. *f.* 15.
Ap. marginella. *Gmel.* ii. 2793.

5160. [5, affinis.] (*Sam. I.* 2.)
Me. affinis. *Kirby M. A.* ii. 92.

5161. 6, Hallana. (*Curtis l. c. infra.*)
Me. Hallana. *Kirby MSS.*

5162. 7; fulvago*. (*Sam. I.* 2.)
Ap. fulvago. *Christi. Hy.* 189. *pl.* 16. *f.* 7.
Me. fulvago. *Kirby M. A.* ii. 93.

5163. 8; albicans*. (*Sam. I.* 2.)
Ap. albicans. *Müll. Z. D. Pr.* 1930.
Me. albicans. *Kirby M. A.* ii. 94.
Ap. hæmorrhoidalis. *Christi. Hy.* 189. *pl.* 16. *f.* 8.
Ap. hiberus. *Harr. Ex.* 135. *pl.* xxxix. *f.* 18?

5164. 9; pilipes*. *Fabr. P.* 322. (!)—(*Sam. I.* 2.)
Ap. pilipes. *Kirby M. A.* ii. 96.

An. ciliata. *Fabr. S. I.* i. 474.
An. atra. *Schra. A. No.* 814.
An. aterrima. *Panz. F.* lxiv. *f.* 19.

5165. 10; hirtipes* *mihi.*
An. Yeatella. *Kirby MSS?*—(*Curtis l. c. infra?*)

5166. 11; cineraria*. *Fabr. P.* 323.
Ap. cineraria. *Linn.* ii. 953.—*Turt.* iii. 554.
Me. cineraria. *Kirby M. A.* ii. 98.
Ap. atra. *Scop. C.* 797.—*Panz. F.* lvi. *f.* 14.
♂, Ap. cinerea. *Fourc. P.* ii. 444.

5167. 12, aprica* *mihi.*
Ap. albescens. *Kirby MSS?*—(*Curtis l. c. infra?*)

5168. 13; pratensis*. (*Sam. I.* 2.)
Ap. pratensis. *Müll. Z. D. Pr. No.* 1912.
Me. pratensis. *Kirby M. A.* ii. 100.
Ap. carbonaria. *Fabr.* ii. 312.
Ap. cineraria. *Christi. Hy.* 201. *pl.* 17. *f.* 14.

5169. 14; thoracica*. *Fabr. P.* 322.—(*Sam. I.* 2.)
Ap. thoracica. *Fabr.* ii. 328.
An. bicolor. *Christi. Hy.* 177. *pl.* 14. *f.* 5.
Ap. assiduus. *Harr. Ex.* 138. *pl.* xl. *f.* 19.
Me. thoracica. *Kirby M. A.* ii. 101.
♂; An. bicolor. *Fabr.* ii. 310.—*Panz. F.* lxv. *f.* 19.

5170. 15; melanocephala*.
Me. melanocephala. *Kirby M. A.* ii. 103.

5171 † 16. cunicularia[a].

5172. 17; nitida*. (*Sam. I.* 2.)
Ap. nitida. *Fourc.* ii. 443.
Me. nitida. *Kirby M. A.* ii. 104.
Ap. cunicularia. *Miller MSS.*
Ap. fortis. *Harr. Ex.* 134. *pl.* xxxix. *f.* 11. ♂?

5173. 18; tibialis*. (*Sam. I.* 2.)
Me. tibialis. *Kirby M. A.* ii. 107.

5174. 19; Moufetella*. (*Sam. I.* 2.)
Me. Moufetella. *Kirby M. A.* ii. 108.

5175. 20; nigroænea*. (*Sam. I.* 2.)
Me. nigroænea. *Kirby M. A.* ii. 109.—*Sowerb. B. M.* i. *pl.* 38.

5176. 21; atriceps*. (*Sam. I.* 2.)
Me. atriceps. *Kirby M. A.* ii. 114.

5177 † 22, bimaculata*.
Me. bimaculata. *Kirby M. A.* ii. 115.

---

[a] 5171 † 16. cunicularia. *Fabr. P.* 323.
Ap. cunicularia. *Linn.* ii. 327.—*Turt.* iii. 552. (!)

5178. 23; Trimmerana*. (*Sam. I. 2.*)
Me. Trimmerana. *Kirby M. A.* ii. 116.

5179. 24; varians*. (*Sam. I. 2.*)
Ap. varians. *Rossi M.* 317.
An. varians. *Panz. F.* lvi. *f.* 12.
Me. varians. *Kirby M. A.* ii. 117.

5180. 25; helvola*. *Fabr. P.* 326.—(*Sam. I. 2.*)
Ap. helvola. *Linn.* ii. 955.—*Turt.* ii. 542.
Me. helvola. *Kirby M. A.* ii. 119. *pl.* 15. *f.* 9.

5181. 26; Gwynana*. (*Sam. I. 2.*)
Ap. helvola. *Rossi F. No.* 895.
Me. Gwynana. *Kirby M. A.* ii. 120.

5182 † 27, angustior.
Me. angustior. *Kirby M. A.* ii. 124.
In Mus. Soc. Linn. *et D. Kirby.*

5183. 28, picicornis.
Me. picicornis. *Kirby M. A.* ii. 123.

5184. 29; fulva*. (*Sam. I. 2.*)
Ap. fulva. *Schr. A. No.* 805.
Me. fulva. *Kirby M. A.* ii. 128.
Ap. vestita. *Fabr. P.* 323.—*Panz. F.* lv. *f.* 9.
Ap. vulpina. *Christi. Hy.* 161. *pl.* 12. *f.* 13.
Ap. aurea. *Marsham. MSS.*

5185. 30. Clarkella. (*Sam. I. 2.*)
Me. Clarkella. *Kirby M. A.* ii. 130.
Ap. icterica. *Christi. Hy.* 190. *pl.* 16. *f.* 9?

5186. 31, Smithella*. (*Sam. I. 2.*)
Me. Smithella. *Kirby M. A.* ii. 131.
Ap. pilipes. *Panz. F.* vii. *f.* 13.

5187 † 32, nigriceps. (*Sam. I. 2.*)
Me. nigriceps. *Kirby M. A.* ii. 134. In Mus. D. *Kirby.*

5188. 33, Spencella*. *Kirby MSS.*

5189 † 34, rufitarsis.
Me. rufitarsis. *Kirby M. A.* ii. 135. In Mus. *D. Kirby.*

5190. 35, fuscipes.
Me. fuscipes. *Kirby M. A.* ii. 136.

5191 † 36, lanifrons *?
Me. lanifrons. *Kirby M. A.* ii. 139. In Mus. D. *Haworth.*

5192. 37; Listerella. (*Sam. I. 2.*)
Me. Listerella. *Kirby M. A.* ii. 137.

5193. 38; fulvicrus*. (*Sam. I. 2.*)
Me. fulvicrus. *Kirby M. A.* ii. 138.
Ap. invictus. *Harr. Ex.* 134. *pl.* xxxix. *f.* 12?

5194 † 39. pubescens.

Ap. pubescens. *Fabr.* ii. 336.
Me. pubescens. *Kirby M. A.* ii. 141. In Mus. *D. Kirby.*

5195. 40; contigua*.
Me. contigua. *Kirby M. A.* ii. 140.

5196 † 41, dorsata.
Me. dorsata. *Kirby M. A.* ii. 144. In Mus. *D. Kirby.*

5197. 42; hæmorrhoidalis*. *Fabr. P.* 327.—*Panz. F.* lxv. *f.* 20. —(*Sam. I.* 2.)
Ap. dichroa. *Gmel.* vi. 2792.
Me. hæmorrhoidalis. *Kirby M. A.* ii. 141.

5198. 43; chrysosceles*. (*Sam. I.* 2.)
Me. chrysosceles. *Kirby M. A.* ii. 143.

5199. 44; proxima*.
Me. proxima. *Kirby M. A.* ii. 146.

5200. 45; Wilkella*.
Me. Wilkella. *Kirby M A.* ii. 145.

5201. 46; Coitana*.
Me. Coitana. *Kirby M. A.* ii. 147.

5202 † 47, barbata.
Me. barbata. *Kirby M. A.* ii. 150. In Mus. *Soc. Linn.*

5203. 48, labialis*.
Me. labialis. *Kirby M. A.* ii. 148.

5204. 49; Lewinella*. (*Sam. I.* 2.)
Me. Lewinella. *Kirby M. A.* ii. 149.

5205. 50, ovatula*. (*Sam. I.* 2.)
Me. ovatula. *Kirby M. A.* ii. 149.

5206 † 51, Collinsoniana. (*Sam. I.* 2.)
Me. Collinsoniana. *Kirby M. A.* ii. 153. In Mus. D. *Kirby.*

5207. 52; barbilabris*. (*Sam. I.* 2.)
Me. barbilabris. *Kirby M. A.* ii. 151.

5208. 53; barbatula*.
Me. barbatula. *Kirby M. A.* ii. 152.

5209. 54; nitidiuscula*.
Me. nitidiuscula. *Kirby M. A.* ii. 155.

5210. 55; nitidula* *mihi.*

5211. 56; combinata*.
Ap. combinata. *Christi. Hy.* 187. *pl.* 15. *f.* 9.
Me. combinata. *Kirby M. A.* ii. 153.

5212. 57. connectens*.
Me. connectens. *Kirby M. A.* ii. 157.

5213. 58; albicrus*. (*Sam. I.* 2.)
Me. albicrus. *Kirby M. A.* ii. 156.

5214. 59, subincana.
Me. subincana. *Kirby M. A.* ii. 158.

5215. 60, digitalis.
Me. digitalis. *Kirby M. A.* ii. 159.

5216. 61; Shawella*. (*Sam. I.* 2.)
Me. Shawella. *Kirby M. A.* ii. 160.

5217 † 62, pilosula.
Me. pilosula. *Kirby M. A.* ii. 164. In Mus. *D. Kirby.*

5218. 63; minutula*. (*Sam. I.* 2.)
Me. minutula. *Kirby M. A.* ii. 161.

5219. 64; nana*.
Me. nana. *Kirby M. A.* ii. 161.

5220. 65; parvula*. (*Sam. I.* 2.)
Me. parvula. *Kirby M. A.* ii. 162.

5221. 66, convexiuscula.
Me. convexiuscula. *Kirby M. A.* ii. 166.

5222. 67; xanthura*.
Me. xanthura. *Kirby M. A.* ii. 164.

5223. 68; fuscata. (*Sam. I.* 2.)
Me. fuscata. *Kirby M. A.* ii. 167.

5224. 69; Afzeliella*. (*Sam. I.* 2.)
Me. Afzeliella. *Kirby M. A.* ii. 169.

5225. 70, fulvescens. *Kirby MSS.*—(*Curtis l. c. infra.*)

5226. 71, Kirbiella *mihi.*—*Curtis.* iii. *pl.* 129.

5227. 72, denticulata.
Me. denticulata. *Kirby M. A.* ii. 133.

5228. 73; spinigera*. (*Sam. I.* 2.)
Me. spinigera. *Kirby M. A.* ii. 123.—(*Kirby & Sp. I. E.* ii. 262.)

5229. 74, armata*. (*Sam. I.* 2.)
Ap. armata. *Lin. Mus. Lesk.* 80.
Me. armata. *Kirby M. A.* ii. 124.

5230. 75, subdentata.
Me. subdentata. *Kirby M. A.* ii. 126.

5231. 76, angulosa*.
Me. angulosa. *Kirby M. A.* ii. 127.

5232. 77, tridentata. (*Sam. I.* 2.)
Me. tridentata. *Kirby M. A.* ii. 132.

5233 † 78. picipes.
Me. picipes. *Kirby M. A.* ii. 127. In Mus. D. Donovan
Ap. picipes. *Don.* xi. 45. *pl.* 410. *f.* 1?

Genus 153: (727). CILISSA, *Leach, Samou.*
MELITTA, (** c. *p*,) *Kirby.*—ANDRENA *p*, *Latr.*

5234. 1, tricincta*. (*Leach E. E.* ix. 155.)—(*Sam. I.* 11.)
Me. tricincta. *Kirby M. A.* ii. 171.—(*Sam. I.* 27.)

5235. 2, hæmorrhoidalis. (*Leach F. E.* ix. 155.)

An. hæmorrhoidalis. *Panz. F.* lxv. *f.* 20.
Me. chrysura. *Kirby M. A.* ii. 172.
An. chrysura. (*Sam. I.* 2.)

5236. 3: affinis* *mihi.*

Genus 154: (728). DASYPODA, *Latreille, Leach, Samou.*, (*Kirby.*)

TRACHUSA *p, Jurine.*—ANDRENA *p, Rossi.*—MELITTA (** c. *p*,) *Kirby.*

5237. 1; Swammerdamella*.
Me. Swammerdamella. *Kirby M. A.* ii. 174.—(*Sam. I.* 27.)
Da. plumipes. (*Leach E. E.* ix. 155.)—(*Sam. I.* 14.)
Ap. altercator. *Harr. Ex.* 164. *pl.* xlix. *f.* 9?
♂; Ap. farfarisequa. *Panz. F.* lv. *f.* 14.
♀; Ap. plumipes. *Panz. F.* xlvi. *f.* 16.

Genus 155: (729). COLLETES, *Latreille, Leach, Samou., Curtis,* (*Kirby.*)

ANDRENA *p, Fabr.*—HYLÆUS *p, Cuvier.*—EVODIÁ, *Panzer.*—MELITTA, (* *a.*) *Kirby.*

5238. 1; succincta*. (*Latr.* G. iv. 149.)—(*Sam. I.* 12.)—(*Curtis l. c. infra.*)
Ap. succincta. *Linn.* ii. 955.—*Christi. Hy.* 185. *pl.* 15. *f.* 7. —*Turt.* ii. 544.
Me. succincta. *Kirby M. A.* ii. 32.—(*Sam. I.* 27.)
Ev. calendarum. *Panz. F.* lxxxiii. *f.* 19.

5239. 2; fodiens*. (*Latr.* G. iv. 149.)—(*Sam. I.* 12.)—*Curtis.* ii. *pl.* 85.
Ap. fodiens. *Fourc. P.* ii. 444.
Me. fodiens. *Kirby M. A.* ii. 34. (26.) *pl.* 1. *f.* 2. ♂. *f.* 1. ♀.

5240. 3, Daviesana*. (*Curtis l. c. supra.*)
Me. Daviesana. *Kirby MSS.*

Genus 156: (730). SPHECODES, *Latreille, Leach, Samou.*, (*Kirby.*)

SPHEX *p, Linné.*—PROAPIS *p, DeGeer.*—NOMADA *p, Fabr.*—ANDRENA *p, Oliv.*—MELITTA, (** *a.*) *Kirby.*—DICHROA, *Illig.* —TIPHIA *p, Panzer.*

5241. 1; Geoffroyellus*.
Sp. gibbus. (*Latr. G.* iv. 153.)—(*Leach E. E.* ix. 155.)—(*Sam. I.* 38.)
No. gibba. *Fabr. P.* 393. (!)
Ap. rufescens. *Fourc.* ii. 447.
No. succincta? *Scop. A.* iv. 45.
Ap. gibba. *Christi. Hy.* 183. *pl.* 15. *f.* 3.
Me. gibba. *Kirby M. A.* ii. 42.
Di. analis. (*Illiger M.* v. 48.)
♀; Me. Geoffroyella. *Kirby M. A.* ii. 45. *pl.* 1. *f.* 5.

Ap. Geoffroyella. *Don.* xi. 45. *pl.* 376. *f.* 3.
Ti. rufiventris. *Panz. F.* liv. *f.* 4.

5242. 2; gibbus*.
Sp. gibba. *Linn.* ii. 946.—*Stew.* ii. 244.—*Turt.* iii. 546.
Sp. gibbus. (*Sam. I.* 38.)
Ap. rufa. *Christ. Hy.* 201. *pl.* 17. *f.* 12.
Me. Sphecoides. *Kirby M. A.* ii. 46.
Ap. Sphecoides. *Don.* xi. 44. *pl.* 376. *f.* 2.
♂. Me. monilicornis. *Kirby M. A.* ii. 47. *pl.* 1. *f.* 6. *β.*
Sp. monilicornis. (*Sam. I.* 38.)

5243. 3; piceus*.
Me. picea. *Kirby M. A.* ii. 48.
Sp. picea. (*Sam. I.* 38.)

5244. 4; divisus*.
Me. divisa. *Kirby M. A.* ii. 49.
Sp. divisa. (*Sam. I.* 38.)
Ap. minimus. *Harr. Ex.* 136. *pl.* xxxix. *f.* 21?

Genus 157: (731). HALICTUS, *Latreille,* (*Kirby.*)

ANDRENA *p, Oliv.*—HYLÆUS *p, Fabr.*—MELITTA, (** *b.*) *Kirby.*—MEGILLA *p, Fabr.*—PROSOPIS *p, Fabr.*—HYLÆUS, *Leach, Samou.*

5245 † 1, sexcinctus. *Latr. H.* xiii. 366.
♂, Hy. sexcinctus. *Fabr. P.* 320.
Hy. sexcinctus. (*Leach E. E.* ix. 155.)
Hy. arbustorum. *Panz. F.* xlvi. *f.* 14.
An. quadricincta. *Oliv. E. M.* iv. 138.
Hy. cinctus. *Spin. I. L.* i. 123.
♀, An. rufipes. *Spin. I. L.* i. 123.
Hy. xanthodon. *Illiger: teste Latr.* In Mus. *Brit.*

5246. 2, nidulans*. *Walck. M.* 69.
Hy. quadricinctus. *Fabr. P.* 319?—(*Sam. I.* 22.)
Me. quadricincta. *Kirby M. A.* ii. 51.
Ap. flavipes. *Panz. F.* lvi. *f.* 17.

5247. 3; xanthopus*.
Me. xanthopus. *Kirby M. A.* ii. 78.
Ap. maxillosa. *Christ. Hy.* 179. *pl.* 14. *f.* 7?
Ha. fodiens. *Walck. M.* 71.

5248. 4; sexnotatus*. *Walck. M.* 72.
Me. sexnotata. *Kirby M. A.* ii. 82. *pl.* 15. *f.* 8. ♂.—*f.* 7. ♀.

5249. 5; rubicundus*.
Ap. rubicunda. *Christ. Hy.* 190. *pl.* 16. *f.* 10.
Me. rubicunda. *Kirby M. A.* ii. 53.

5250. 6; flavipes*.
Hy. flavipes. *Fabr. P.* 321.

Me. flavipes. *Kirby M. A.* ii. 55.
Ap. crocipes. *Fourc.* ii. 446?

5251. 7; seladonius*.
Meg. seladonia. *Fabr. P.* 334.
Ap. subaurata. *Rossi F.* ii. 144.—*Panz. F.* lvi. *f.* 4.
Me. seladonia. *Kirby M. A.* ii. 57.

5252. 8; æratus*.
Me. ærata. *Kirby M. A.* ii. 58.

5253. 9; Smeathmanellus*.
Me. Smeathmanella. *Kirby M. A.* ii. 375.

5254. 10, leucopus.
Me. leucopus. *Kirby M. A.* ii. 59.

5255. 11; Morio*.
Hy. Morio. *Fabr. P.* 321.
Me. Morio. *Kirby M. A.* ii. 60.
Coqueb. Ill. i. 25. *pl.* 6. *f.* 5. *A. E.*

5256. 12; minutus*.
Ap. minuta. *Schr. A. No.* 829.
Me. minuta. *Kirby M. A.* ii. 61.
Reaum. vi. 95. *pl.* 9. *f.* 1?

5257. 13; villosulus*.
Me. villosula. *Kirby M. A.* ii. 64.

5258. 14; minutissimus*.
Me. minutissima. *Kirby M. A.* ii. 63.

5259. 15; nitidiusculus*.
Me. nitidiuscula. *Kirby M. A.* ii. 64.

5260. 16, Milneanus. *Leach MSS.*

5261. 17, lævis.
Me. lævis. *Kirby M. A.* ii. 65.

5262 † 18, punctulatus.
Me. punctulata. *Kirby M. A.* ii. 66. In Mus. D. *Kirby.*

5263. 19, malachurus*.
Me. melachura. *Kirby M. A.* ii. 67.

5264 † 20, fulvicornis.
Me. fulvicornis. *Kirby M. A.* ii. 67. In Mus. D. *Kirby.*

5265. 21; fulvocinctus*.
Me. fulvocincta. *Kirby M. A.* ii. 68.
Ap. generosus. *Harr. Ex.* 134. *pl.* xxxix. *f.* 14?
♂; Ap. bicincta. *Schra. A. No.* 826.
Hy. annulatus. *Panz. F.* xv. *f.* 3.
♀; Ap. nitida. *Müll. Z. D. Pr. No.* 1619.
Ap. fodiens. *Fourc. P.* ii. 444.
Hy. terebrator. *Walck. M.* 73.

5266. 22; albipes*.

Hy. albipes. *Fabr. P.* 294?
Ap. albipes. *Panz. F.* vii. *f.* 5.
Me. albipes. *Kirby M. A.* ii. 71.
Ap. desertus. *Harr. Ex.* 135. *pl.* 59. *f.* 19.

5267. 23; abdominalis*.
Hy. abdominalis. *Panz. F.* liii. *f.* 18.
Me. abdominalis. *Kirby M. A.* ii. 73.

5268 † 24, obovatus.
Me. obovata. *Kirby M. A.* ii. 75. In Mus. D. *Kirby.*

5269. 25, lævigatus*.
Me. lævigata. *Kirby M. A.* ii. 75.
Ap. lævigata. *Don.* xii. 76. *pl.* 421. *f.* 2?

5270. 26; leucozonius*.
Ap. leucozonia. *Schr. A. No.* 819.
Me. leucozonia. *Kirby M. A.* ii. 76.

5271. 27; quadrinotatus*.
Me. quadrinotata. *Kirby M. A.* ii. 79.

5272. 28; lugubris.
Me. lugubris. *Kirby M. A.* ii. 81.

Genus 158: (732). HYLÆUS, *Latreille.*

ANDRENA *p, Oliv.*—VESPA *p, Rossi.*—PROSOPIS *p, Jurine,* (*Kirby.*) —SPHEX *p, Panz.*—MELITTA, (* *b.*) *Kirby.*

5273. 1; annulatus*. (*Sam. I.* 22.)
Ap. annulata. *Linn.* ii. 958.—*Stew.* ii. 243.—*Turt.* iii. 540.
Me. annulata. *Kirby M. A.* ii. 36. *pl.* 15. *f.* 3.

5274. 2; annularis*.
Me. annularis. *Kirby M. A.* ii. 38.
Sp. annulata. *Panz. F.* liii. *f.* 1.
Pr. labiata. *Fabr. P.* 295.

5275. 3; signatus*. (*Sam. I.* 22.)
Sp. signata. *Panz. F.* liii. *f.* 2.
Ap. signata. *Don.* xii. 75. *pl.* 421. *f.* 1.
Ve. pratensis. *Fourc.* ii. 437.
Me. signata. *Kirby M. A.* ii. 41.
Pr. nigrita. *Fabr. P.* 296.

5276. 4; dilatatus*. (*Sam. I.* 22.)
Me. dilatata. *Kirby M. A.* ii. 39. *pl.* 15. *f.* 4.

5277 † 5, geniculatus. *Leach? MSS.*
Pr. bifasciatus. *Jurine Hy. pl.* 11. *g.* 30? In Mus. *Brit.*

Sectio III. TUBULIFERA.

(Terebrantia-pupivora *p*, *et* Tubulifera, *Latr.* G.)

Familia XX: (109). CHRYSIDIDÆ, *Leach.*

(Chrysidides, *Latreille.*—Chrysis, *Linné*, *&c.*) *Ruby-tails.*

Genus 159: (733). HEDYCHRUM, *Latr.*, *Leach*, *Samou.*, *Curtis.*

Vespa *p*, *Geoffroy.*—Sphex *p*, *Scop.*—Omalus *p*, *Panz.*

A. "Segmento ultimo abdominis integro." *Panzer.*

5278. 1; regium*. (*Sam. I.* 21.)—(*Curtis l. c. infra?*)
Ch. regia. *Fabr.* ii. 243.—*Panz. F.* li. *f.* 9?—*Ent. Trans.* i. 254.
He. punctatum. *Leach MSS.*—(*Curtis l. c. infra.*)

5279. 2; lucidulum*. *Latr. H.* xiv. 239.—(*Curtis l. c. infra.*)
Ch. lucidula. *Fabr.* ii. 242.—*Panz. F.* li. *f.* 5.
Sp. nobilis. *Scop. C. No.* 792.

5280. 3, ardens. *Latr.?*—*Curtis.* i. *pl.* 38.
Ch. ænea. *Fabr.* ii. 242?

B. "Segmento ultimo abdominis emarginato." *Panzer.*

5281. 4; auratum*. *Latr. H.* xiii. 239.—(*Sam. I.* 21.)
Ch. aurata. *Linn.* ii. 948?—*Stew.* ii. 237.—*Panz. F.* li. *f.* 8. —*Turt.* iii. 510.
Ch. lucidula var. *Hellwig.*

5282. 5; eximium* *mihi.*

5283. 6; imperiale*. *Leach MSS.*
He. violaceum. (*Curtis l. c. supra.*)—nec C. violacea *Rossii.*
Ch. politus. *Harr. Ex.* 70. *pl.* xix. *Ord.* III.
Ch. atrocærulea. *Kirby MSS.*
Om. nitidum. *Panz. F.* xcvii. *f.* 17?

5284. [7, cæruleum*.] *Leach MSS.*—(*Curtis l. c. supra.*)
Ch. nitidus. *Mus. Marsham.*

5285. 8, politum*. *Leach MSS.*

5286. 9, pusillum*.
Ch. pusilla. *Fabr. P.* 176?

Genus 160: (734). ELAMPUS, *Spinola*, *Leach*, *Samou.*

Hedychrum *p*, *Panzer.*

5287. 1; Panzeri*. *Spin. I. L.* i. 10.—(*Leach E. E.* ix. 146.)—(*Sam. I.* 16.)
Ch. Panzeri. *Fabr. P.* 172.
Ch. scutellaris. *Panz. F.* li. *f.* 11.
He. spina. *Ann. Mus.* (*Lepelletier.*) xxxviii. 121. *f.* 2, 3.

Genus 161: (735). CHRYSIS *Auctorum.*

SPHEX *p, Scop.*

A. Ano quadridentato.

5288. 1; fulgida*. *Linn.* ii. 948.—*Turt.* ii. 509.—*Shaw* G. Z. vi. 283.—*Curtis.* i. *pl.* 8.—(*Ing. Inst.* 90.)

Ch. ignita. *Samou.* 448. *pl.* 3. *f.* 7.

5289. 2; ignita*. *Linn.* ii. 947.—*Harr. Ex.* 68. *pl.* xix. *Ord.* I. *f.* 1?—*Berk. S.* i. 157?—*Don.* i. 19. *pl.* 7.—*Stew.* ii. 236.—*Turt.* iii. 510.—*Shaw* G. Z. vi. 283. *pl.* 94. *fig. sup.*—(*Leach E. E.* ix. 146.)—(*Sam. I.* 10.)

5290. 3; affinis*. *Leach MSS.*—(*Sam. I.* 10.)

5291. 4; nitens* *mihi.*

5292. 5; micans* *mihi.*

5293. 6; fulminans* *mihi.*

5294. 7; confinis* *mihi.*

5295. 8; effulgens*. *Leach MSS.*—(*Sam. I.* 10.)

Ch. fulgida. *Harr. Ex.* 69. *pl.* xix. *Ord.* I. *f.* 2.

5296. 9; pulchra* *mihi.*

5297. 10; ephippium* *mihi.*

5298. [11, obscura*] *mihi.*

5299. 12; Stroudera*. *Jurine H. pl.* 12. *f.* 42.—(*Sam. I.* 10.)

5300. 13; aurulenta* *mihi.*

5301. 14; dimidiata*. *Fabr. S.* 258?

5302. 15; bidentata*. *Linn.* ii. 947.—*Don.* i. 43. *pl.* 19.—*Stew.* ii. 236.—*Turt.* iii. 510.—(*Sam. I.* 12.)

Ch. hephæstites. *Harr. Ex.* 69. *pl.* xix. *Ord.* I. *f.* 5.

5303. 16; inermis* *mihi.*

B. Ano tridentato.

5304. 17, succincta. *Linn.* ii. 947.—*Panz. F.* lxxvii. *f.* 16.—*Turt.* ii. 510.

5305. 18; cyanea*. *Linn.* ii. 948.—*Don.* vii. 51. *pl.* 235.—*Stew.* ii. 237.—*Turt.* iii. 510.—(*Sam. I.* 10.)

Ch. veridans. *Harr. Ex.* 69. *pl.* xix. *Ord.* I. *f.* 4.

Var.? *Harr. Ex.* 69. *pl.* xix. *Ord.* II.?

C. Ano integro.

5306. 19; rufa*. *Panz. F.* lxxix. *f.* 16.

5307. 20; austriaca*. *Fabr. P.* 173?

Ch. radians. *Harr. Ex.* 69. *pl.* xix. *Ord.* I. *f.* 3?

5308 † 21, Leachii *mihi.* In Mus. *Brit.*

Genus 162: (736). EUCHREUS, *Latreille, Leach.*

5309 † 1, quadratus. *Leach MSS?*

Ch. sexdentata. *Panz. F.* li. *f.* 12?
Ch. violacea. *Panz. K. R.* ii. 103? In Mus. *Brit.*

Genus 163: (737). CLEPTES, *Latr., Leach, Samou.*, (*Kirby.*)
SPHEX *p*, *Linné.*—VESPA *p*, *Geoff.*—ICHNEUMON *p*, *Rossi.*
5310. 1; semiaurata*. *Latr. H.* xiii. 206.—(*Leach E. E.* ix. 146.) —(*Sam. I.* 11.)
Sp. semiaurata. *Linn.* ii. 946.
Ic. semiauratus. *Panz. F.* li. *f.* 2.
♀. Ic. auratus. *Panz. F.* lii. *f.* 1.
Cl. aurata. (*Sam. I.* 11.)
Ch. aurata. *Wood.* ii. 61. *pl.* 58.
5311. [2; fervida*.] *Mus. Marsham.*
5312. 3, nitidula*. *Fabr.* ii. 184.—*Coqueb. Ic.* i. *pl.* 4. *f.* 5.
Cl. zonata. *Leach? MSS?*

Familia XXI: (110). CHALCIDIDÆ, *Westwood.*

(CYNIPSERA, *Latr.*—CYNIPSIDÆ, *Leach.*)
(DIPLOLEPIS & CHALCIS, *Fabricius.*—CLEPTES *p*, *Fabr.*)

Genus 164: (738). CHALCIS *Auctorum.*
SMIERA, *Spinola Ann.*—SPHEX *p*, *et* VESPA *p*, *Linné.*
5313. 1; clavipes*. *Fabr. P.* 159.—*Panz. F.* lxxviii. *f.* 15.—*Don.* xi. 57. *pl.* 379.—(*Leach E. E.* ix. 144.)—(*Samou.* 271. *pl.* 8. *f.* 6.)
5314. 2; sispes*. *Fabr. P.* 159.—*Panz. F.* lxxvii. *f.* 11.—*Stew.* ii. 236.

Genus 165: (739). BRACHYMERIA, *Westwood MSS.*
CHALCIS, *Spinola Ann.*
5315. 1; flavipes*.
Ch. flavipes. *Panz. F.* lxxviii. *f.* 16?
5316. 2; minuta*.
Ch. minuta. *Fabr. P.* 165.—*Panz. F.* xxxii. *f.* 6.—(*Leach E. E.* ix. 144.)

Genus 166: (740). HALTICELLA, *Spinola Ann.*
5317. 1; pusilla*.
Ch. pusilla. *Fabr. P.* 167.
5318. 2; armata*.
Ch. armata. *Panz. F.* lxxiv. *f.* 9.
5319. 3; bispinosa*.
Ch. bispinosa. *Fabr. P.* 168.

Genus 167: (741). PERILAMPUS, *Latreille, Leach, Curtis.*
5320. 1; pallipes*. *Curtis.* iv. *pl.* 158.
5321. 2; auriceps* *mihi.*

5322. 3; aureo-viridis* *mihi.*

Genus 168: (742). CRATOMUS, *Dalman.*

5323. 1; megacephalus*.
Di. megacephalus. *Fabr. P.* 149.—*Ent. Trans.* (*Haworth.*) i. 252.

Genus 169: (743). EUCHARIS, *Latreille, Leach.*

ICHNEUMON *p, Rossi.*—CHALCIS *p, Jurine.*—CYNIPS *p, Olivier.*

5324. 1, ascendens. *Latr. H.* xiii. 210.—*Fanz. F.* lxxxviii. *f.* 18. —(*Leach E. E.* ix. 144.)

Genus 170: (744). EURYTOMA, *Illiger, Leach.*

ICHNEUMON *p, DeGeer.*—EUCHARIS *p, Fabr.*—FIGITES *p, Spinola.*

5325. 1; appendigaster. *Swederus.*

5326. 2, Abrotani*.
Euc. Abrotani. *Panz. I.* 97.
Ch. Abrotani. *Panz. F.* lxxvi. *f.* 14.

5327. 3, Stigma.
Di. Stigma. *Fabr. P.* 152.
Eur. biguttata. *Swederus.*

5328. 4; Serratulæ*. (*Leach E. E.* ix. 144.)
Cy. Serratulæ. *Fabr. P.* 147.
Op. abbreviator. *Panz. F.* lxxxi. *f.* 16. ♀.
Euc. compressa. *Fabr. P.* 157. ♀?

5329. 5; longula. *Dalm.*

5330. 6; aterrima*.
Cy. aterrimus. *Schrank A.* 320.

5331. 7; verticillata*.
Ich. verticillata. *Fabr. P.* 153?

5332 † 8. penetrans.
Ich. penetrans. *Linn. Trans.* (*Kirby.*) v. *pl.* 4. *f.* 10.—*Stew.* ii. 231. In Mus. D. Kirby?

5333 ad 5349: 17 sp. adhuc examinandæ. (25 sp.)

Genus 171: (745). SPALANGIA, *Latreille, Leach.*

5350. 1; nigra*. *Latr. H.* xiii. 228.—*Id. Gen.* iv. 29. *pl.* xiii. *f.* 7, 8. ♂.—(*Leach E. E.* ix. 144.)

Genus 172: (746). ——

5351. 1; cornigera* *mihi.*

Genus 173: (747). CALLIMONE, *Spinola?*

TORYMUS *p, Dalman.*—CYNIPS, *Geoff.*, *Latr.*, *Leach, Samou.*—ICHNEUMON *p, Linné.*—CLEPTES *p, Fabr.*—CHALCIS *p, Cuvier.*

A. MEGASTIGMUS, *Dalman.*

5352. 1, bipunctatus. *Swederus.*

B. TORYMUS, *Dalman.*

5353. 2; Cynipidis*.
Ic. Cynipidis. *Linn.* ii. 939.

5354. 3; Bedeguaris*.
Ic. Bedeguaris. *Linn.* ii. 939.—*Berk. S.* i. 156.—*Stew.* ii. 231.—*Turt.* iii. 474.

5355. 4; Capræa*.
Cy. Capræa. *Fabr. P.* 146.—*Turt.* iii. 409.—(*Leach E. E.* ix. 144.)

5356. 5, purpurascens.
Ic. purpurascens. *Fabr. P.* 151.

5357. 6' nigricornis*.
Ic. nigricornis. *Fabr. P.* 150.

5358. 7, cyaneus*.
Ic. cyaneus. *Fabr. P.* 151.—*Coqueb. Ic.* i. *pl.* 5. *f.* 4?

5359. 8, Juniperi.
Ic. Juniperi. *Linn.* ii. 939?

5360 ad 5387: 28 sp. adhuc examinandæ. (36 sp.)

Genus 174: (748). EUPELMUS, *Dalman.*

5388. 1; apicalis* *mihi.*

Genus 175: (749). COLAX[a], (COLAS) *Curtis, Westwood.*
PTEROMALUS *p, Dalman?*—MISOCAMPUS, *Latreille,* (*Kirby.*)

5389. 1, dispar. *Curtis.* iv. *pl.* 166.—(*Zool. Journ.* (*Westwood.*) iv. *pl.* 2. *f.* 3. *p.* 9.)

5390. 2; Puparum*.
Ic. Puparum. *Linn.* ii. 939?—*Berk. S.* i. 156.—*Stew.* ii. 231.—*Turt.* iii. 475?—(*Kirby & Sp. I. E.* i. 264?)

5391. 3, Muscarum.
Ic. Muscarum. *Linn.* ii. 938.—*Turt.* iii. 474.

5392. 4; fuscicornis* *mihi.*

5393. 5; nigricornis* *mihi.*

5394 ad 5400: 7 sp. adhuc examinandæ. (12 sp.)

Genus 176: (750). CHEIROPACHUS, *Westwood.*

DIPLOLEPIS *p, Fabr.*—CLEONYMUS, *Curtis.*—PTEROMALUS *p, Dalm.*

5401. 1; quadrum*. *Zool. Journ.* (*Westwood.*) iv. *pl.* ii. *f.* 2. ♀.
Di. quadrum. *Fabr. P.* 152.
Cl. maculipennis. *Curtis.* iv. *pl.* 194. ♂.

---

[a] COLAX. Genus Dipterorum. Vide *Meig. Zw.* iv. 165.

Genus 177: (751). ——

PTEROMALUS *p, Dalm.?*—CLEONYMUS *p, Curtis?*

5402. 1; cyaneus* *mihi.*

5403. 2; aureus* *mihi.*

5404. 3, pulcherrimus* *mihi.*

5405. 4; basalis* *mihi.*

5406. 5; apicalis* *mihi.*

5407. 6; obscurus* *mihi.*

5408. 7, splendidus* *mihi.*

5409. 8; pallipes* *mihi.*

5410. 9. contiguus *mihi.*

Genus 178: (752). CLEONYMUS, *Latreille, Leach, Westwood.*

DIPLOLEPIS *p, Fabr.*—PTEROMALUS *p, Dalm.?*

5411. 1; depressus*. (*Leach E. E.* ix. 144.)—(*Curtis fo.* 194.)—*Zool. Journ.* (*Westwood.*) iv. *p.* 16. *pl.* ii. *f.* 1. ♀.
Di. depressus. *Fabr. P.* 151.

Genus 179: (753). TRIGONODERUS, *Westwood MSS.*

5412. 1, obscurus. *West. MSS.*

5413 ad 5424: 13 sp. adhuc examinandæ. (14 sp.)

Genus 180: (754). ——

5425. 1; bifasciatus* *mihi.*

Genus 181: (755). PTEROMALUS, *Dalman, Leach.*

5426. 1, Tortricis. (*Latr.* G. iv. 31.)—(*Leach E. E.* ix. 145.)
Ich. Tortricis. *Schrank A.* 376.

5427. 2, intercus. (*Latr.* G. iv. 31.)
Ich. intercus. *Schrank A.* 377.

5428. 3, Gallarum*. (*Latr. G.* iv. 31.)
Ich. Gallarum. *Linn.* ii. 939.—*Turt.* iii. 474.

5429 ad 5481: 53 sp. adhuc examinandæ. (56 sp.)

Genus 182: (756). HALTICOPTERA, *Spinola.*

5482. 1; aterrima* *mihi.*

5483 ad 5489: 7 sp. adhuc examinandæ. (8 sp.)

Genus 183: (757). EULOPHUS, *Geoffroy, Leach,* (*Kirby,*) *Curtis.*

ICHNEUMON *p, Linné.*—ENTEDON, *Dalm.*

5490 † 1, ramicornis*. (*Latr.* G. iv. 28.)—(*Leach E. E.* ix. 144.)—*Curtis l. c. infra.*
Di. ramicornis. *Fabr. P.* 153.—*DeGeer.* ii. 313. *pl.* 15. *f.* 3.
In Mus. D. Haworth, *Kirby et Westwood.*

5491. 2; Apatelæ* *mihi.*

5492. 3; damicornis *. *Linn. Trans.* (*Kirby.*) xiv. 112.—*Curtis.* iii. *pl.* 133.

5493. 4; assimilis *. *Westwood MSS.*

5494. 5; Stephensii *. *Westwood MSS.*

5495 † 6, Kirbii *. *Westwood MSS.*—*Curtis l. c. supra.*
In Mus. D. Curtis *et Westwood.*

5496 † 7, minutissimus *. *Westwood MSS.* In Mus. D. *Westwood.*

5497 † 8, pectinicornis *. *Curtis l. c. supra.*
Ic. pectinicornis. *Linn.* ii. 941.—*DeGeer.* i. 589. *pl.* 35. *f.* 1–7. —*Berk. S.* i. 157.—*Stew.* ii. 232.—*Turt.* iii. 476.
In Mus. D. *Kirby et Westwood.*

5498. 9, Nemati *. *Westwood MSS.*

5499 † 10. Latreillii. *Curtis l. c. supra.* In Mus. D. Curtis.

5500 † 11. Larvarum [a].

Genus 184: (758). ——

EULOPHUS *p?* (*olim.*)

5501 † 1, Westwoodii * *mihi.* In Mus. D. *Westwood.*

5502 † 2, . In Mus. *D. Westwood.*

Genus 185: (759). ——

EULOPHUS *p? olim.*

5503. 1; fasciatus * *mihi.*

5504. 2; semitestaceus * *mihi.*

5505 ad 5516: 12 sp. adhuc examinandæ. (14 sp.)

Genus 186: (760). ENCYRTUS, *Latreille, Leach.*

ICHNEUMON *p, Rossi.*—MIRA, *Schellenberg.*

5517. 1; scutellaris *. *Dalman?*

5518 † 2. inserens.
Ic. inserens. *Linn. Trans.* (*Kirby.*) v. *pl.* 4.—*Stew.* ii. 231.
In Mus. D. Kirby?

5519 ad 5536: 18 sp. adhuc examinandæ. (20 sp.)

Familia XXII: (111). PROCTOTRUPIDÆ.

(PROCTOTRUPII *et* OXYURI, *Latreille.*—PROCTOTRUPIDES, *Leach.*)

Genus 187: (761). PLATYGASTER, *Latreille, Leach.*

SCELIO *p, Latreille olim.*

5537. 1, ruficornis. (*Latr.* G. iv. 32.)—(*Leach E. E.* ix. 146.)
Sc. ruficornis. *Latr. H.* xiii. 227?

---

[a] 5500 † 11. Larvarum. (*Latr. G.* iv. 28.)
Ic. Larvarum. *Linn.* ii. 187.—*Turt.* iii. 475. (!)

5538. 2, Tipulæ.
Ich. Tipulæ. *Linn. Trans.* (*Kirby.*) iv. 230.—*Stew.* ii. 231.—*Turt.* iii. 476.

5539. 3, ovulorum*.
Ic. ovulorum. *Linn.* ii. 940?—*Turt.* iii. 476?—(*Kirby & Sp. I. E.* i. 264?)

5540 † 4. secalis[a].

5541 ad 5586: 46 sp. adhuc examinandæ. (50 sp.)

Genus 188: (762). ——
PLATYGASTER *p*? *Latreille.*—PSILUS *p*, *Jurine.*

5587. 1; Boscii*.
Ps. Boscii. *Jurine Hy.* 518.

Genus 189: (763). ——
ICHNEUMON *p*, *Shaw.*

5588. 1; Punctum*.
Ic. Punctum. *Linn. Trans.* (*Shaw.*) iv. 189. *pl.* 18. *f.* 1.—*Stew.* ii. 232.—*Turt.* iii. 476.

5589. 2; Atomos* *mihi.*

5590. 3, Monas* *mihi.*

5591. 4, Termo* *mihi.*

5592 ad 5599: 8 sp. adhuc examinandæ. (12 sp.)

Genus 190: (764). DIAPRIA, *Latreille,* (*Kirby.*)
PSILUS, *Jurine, Leach,* (*Kirby*?)—ICHNEUMON *p*, *Villers.*—CHRYSIS *p*? *Rossi.*

5600. 1; elegans*.
Ps. elegans. *Jurine Hy. pl.* 13. *g.* 48. ♂.
Di. verticillata. *Latr. H.* xiii. 231.

5601. 2, conica. *Latr. H.* xiii. 231.
Ch. conica. *Fabr. P.* 167.

5602 ad 5621: 20 sp. adhuc examinandæ. (22 sp.)

Genus 191: (765). CINETUS, *Jurine.*

5622. 1; bicornis* *mihi.*

5623 ad 5631: 9 sp. adhuc examinandæ. (10 sp.)

Genus 192: (766). ——

5632 † 1. cornutus.
Psilus cornutus. *Panz. F.* lxxxiii. *f.* 11.—(*Leach E. E.* ix. 145.) In Mus. D. Haworth? et Kirby.

---

[a] 5540 † 4. secalis.
Ic. secalis. *Linn.* ii. 939.—*Stew.* ii. 231. (!)

Genus 193: (767). BELYTA, *Latreille, Leach.*
CINETUS *p, Jurine.*
5633. 1; bicolor*. *Jurine Hy. pl.* 14. *g.* 8.
5634. 2; basalis* *mihi.*
5635, 3, ater* *mihi.*
5636. 4; pusilla* *mihi.*

Genus 194: (768). ——
5637. 1, dilatatus *mihi.*

Genus 195: (769). PROCTOTRUPES, *Latreille, Leach.*
CODRUS, *Jurine.*—BANCHUS *p? et* BASSUS *p? Fabricius.*—
ERODORUS, *Walckenaer,* (*Kirby.*)
5638. 1; brevipennis*. (*Latr.* G. i. *pl.* xiii. *f.* 1.—iv. 38.)—(*Leach E. E.* ix. 145.)
Ero. bimaculatus. *Walck. F.* ii. 47.
5639. 2, niger*. (*Latr.* G. iv. 38.)
Co. niger. *Panz. F.* lxxxv. *f.* 9.
5640. 3, pallipes*.
Co. pallipes. *Jurine Hy. pl.* 10. *g.* 46.
5641 † 4. gravidator[a].
5642 ad 5663: 22 sp. adhuc examinandæ. (26 sp.)

Genus 196: (770). HELORUS, *Latreille, Leach.*
SPHEX *p, et* PSEN *p, Panzer.*—PACHYPODIUM, *Kirby MSS.*
5664. 1; anomalipes*. *Latr. H.* xiii. 230.—(*Leach E. E.* ix. 146.)
Sp. anomalipes. *Panz. F.* lii. *f.* 23.—*Id.* c. *f.* 18.
Pa. nigripes. *Kirby MSS.*

Genus 197: (771). BETHYLUS, *Latreille, Leach.*
OMALUS, *Jurine,* (*Kirby.*)—CERAPHRON, *Panzer.*
5665. 1, cenopterus*. *Latr. H.* xiii. 229.—(*Leach E. E.* ix. 146.)
Ti. cenoptera. *Panz. F.* lxxxi. *f.* 14.
5666. 2, fuscicornis*.
Om. fuscicornis. *Jurine Hy. pl.* 13. *g.* 43.
5667. 3, formicarius.
Ce. formicarius. *Panz. F.* xcvii. *f.* 16.
5668 ad 5670: 3 sp. adhuc examinandæ. (6 sp.)

Genus 198: (772). ——
DRYINUS[b] *p, Latreille? Leach? Curtis.*
5671 † 1: formicarius?*

---

[a] 5641 † 4. gravidator.
Ic. gravidator. *Linn.* ii. 936.—*Turt.* iii. 457. ([1])

[b] DRYINUS: Genus Hymenopterorum. Vide *Fabr. P.* 200.

Dr. formicarius. *Latr. H.* xiii. 228?—*Latr. G.* iv. 40. *pl.* 13. *f.* 6?—(*Leach E. E.* ix. 146?)—*Curtis l. c. infra. fo.* 206.
In Mus. D. *Westwood.*

5672 † 2. bicolor.
Dr. bicolor. *Haliday MSS.*—*Curtis l. c. infra fo.* 206.
In Mus. D. Haliday.

Genus 199: (773). GONATOPUS, *Klug.*
DRYINUS, *Curtis.*

5673. 1; flavicornis*. *Dalman A. E.*

5674. 2; basalis*. *Dalman A. E.*

5675. 3; gracilis*. *Westwood MSS.*

5676. 4; fuscicornis*. *Dalman A. E.*

5677. 5; tenuicornis*. *Dalman A. E.*

5678. 6; similis*. *Westwood MSS.*

5679 † 7. Cursor.
Dr. Cursor. *Haliday MSS.*—*Curtis.* v. *pl.* 206.
In Mus. D. Curtis et Haliday.

5680 † 8. rapax.
Dr. rapax. *Haliday MSS.*—*Curtis l. c. supra.*
In Mus. D. Haliday?

5681 † 9. lucidus.
Dr. lucidus. *Haliday MSS.*—*Curtis l. c. supra.*
In Mus. D. Haliday?

5682 † 10. crassimanus.
Dr. crassimanus. *Haliday MSS.*—*Curtis l. c. supra.*
In Mus. D. Haliday?

5683 † 11. fulviventris.
Dr. fulviventris. *Haliday MSS.*—*Curtis l. c. supra.*
In Mus. D. Haliday?

Genus 200: (774). NOMIOPUS, *Westwood MSS.*

5684. 1; apicalis*. *Westwood MSS.*

Genus 201: (775). APHELOPUS, *Dalman.*

5685. 1; atratus*. *Dalman A. E.*

5686 et 5687: 2 sp. adhuc examinandæ. (3 sp.)

Genus 202: (776). MEGASPILUS, *Westwood MSS.*
CERAPHRON, *Curtis.*

5688 † 1. Dux.
Ce. Dux. *Curtis l. c. infra.* In Mus. D. Curtis et Haliday?

5689 † 2. opacus.
Ce. opacus. *Hal. MSS.*—(*Curtis l. c. infra.*)
In Mus. D. Haliday?

5690 † 3. cimicoides. *Hal. MSS.*
Ce. cimicoides. (*Curtis l. c. infra.*) In Mus. D. Haliday?

5691. 4; rufiscapus*.
Ce. rufiscapus. (*Curtis l. c. infra?*)

5692. 5; rufipes*.
Ce. rufipes. (*Curtis l. c. infra?*)

5693. 6; nitidus*.
Ce. nitidus. (*Curtis l. c. infra?*)

5694 † 7. puliciformis.
Ce. puliciformis. (*Curtis l. c. infra.*) In Mus. D. Curtis?

5695 † 8. crispus.
Ce. crispus. *Hal. MSS.*—(*Curtis l. c. infra.*)
In Mus. D. Haliday?

5696. 9, Carpenteri*.
Ce. Carpenteri. *Curtis l. c. infra.*

5697 † 10. elegans.
Ce. elegans. (*Curtis l. c. infra.*) In Mus. D. Walker?

5698 † 11. gracilis.
Ce. gracilis. (*Curtis l. c. infra.*) In Mus. D. Haliday?

5699 ad 5712: 14 sp. adhuc examinandæ.

5713. 26, brachypterus* *mihi.*

5714. 27, ruficollis*.
Ce. ruficollis. *Hal. MSS.*—(*Curtis l. c. infra?*)

5715 † 28. Rubi.
Ce. Rubi. (*Curtis l. c. infra.*) In Mus. D. Curtis?

5716. 29, melanocephalus.
Ce. melanocephalus. *Hal. MSS.*—(*Curtis l. c. infra.*)

5717 † 30. Halidayi. *Curtis.* vi. *pl.* 249. In Mus. D. Haliday.

Genus 203: (777). ——

CERAPHRON *p*, *Curtis fo.* 249.

5718 † 1. ferrugineus.
Ce. ferrugineus. *Hal. MSS.*—(*Curtis l. c. supra.*)
In Mus. D. Haliday?

5719 † 2. discolor.
Ce. discolor. *Hal. MSS.*—(*Curtis l. c. supra.*)
In Mus. D. Haliday?

5720. 3, nubilipennis*. (*Curtis l. c. supra.*)

Genus 204: (778). CERAPHRON, *Jurine*, *Leach.*

5721 † 1, sulcatus*. *Jurine Hy. pl.* 14. *g.* 9.—*Leach E. E.* ix. 145.
(*Curtis fo.* 249?) In Mus. D. Curtis?

5722 † 2. longipennis. (*Curtis l. c. supra.*)
In Mus. D. Haliday? et Walker?

5723. 3, nigripes* *mihi.*

5724. 4, varipes* *mihi.*

Genus 205: (779). SPARASIÓN, *Latreille, Leach.*
CERAPHRON *p, Jurine.*
5725 † 1. frontale. *Latr. H.* xiii. 230.—(*Leach E. E.* ix. 145.)
Ce. cornutus. *Jurine H. pl.* 13. *g.* 44. ♀. In Mus. D. Curtis?

Familia XXIII: (112). CYNIPIDÆ, *Westwood.*

(DIPLOLEPARIÆ, *Latr.*—DIPLOLEPIDÆ, *Leach.*)
(CYNIPS, *Linné.*) *Gall-flies.*

Genus 206: (780). FIGITES, *Latreille, Leach, Samou.*
EUCHARIS *p? Panzer.*
5726. 1; scutellaris*. *Latr. H.* xiii. 210.—(*Leach E. E.* ix. 144.) —(*Sam. I.* 17.)
Cy. scutellaris. *Rossi M.* ii. *App.* 106.
5727 ad 5735: 9 sp. adhuc examinandæ. (10 sp.)

Genus 207: (781). CYNIPS, *Linné, Westwood,* (*Kirby.*)
DIPLOLEPIS, *Geoff., Latr., Leach, Samou.*
5736. 1; Fagi. *Linn.* ii. 919.—*Stew.* ii. 221.—*Turt.* iii. 409.
5737. 2, pilicornis. *Kirby MSS.*
5738. 3; Quercus gemmæ. *Linn.* ii. 919.—*Berk. S.* i. 153.—*Stew.* ii. 220.—*Turt.* iii. 409.—*Bingley.* iii. 247.
5739. 4; Quercus-petioli. *Linn.* ii. 918.—*Berk. S.* i. 153.—*Stew.* ii. 220.—*Turt.* iii. 409.—*Shaw* G. *Z.* vi. 269.
5740. 5, Rubi. *Kirby? MSS?*
5741. 6, Amerinæ. *Linn.* ii. 919?
5742. 7, Quercus-ramuli. *Linn.* ii. 918.—*Turt.* iii. 409.
5743. 8; Glechomæ*. *Linn.* ii. 917.—*Berk. S.* i. 152.—*Turt.* iii. 408.
Cy. Glechomatis. *Stew.* ii. 220.—*Bingley.* iii. 247.
5744. 9, Quercus-inferus. *Linn.* ii. 918.—*Turt.* iii. 409.
5745. 10, Urticæ. *Kirby MSS.*
5746. 11; Rosæ*. *Linn.* ii. 917.—*Stew.* ii. 220.—*Shaw G. Z.* vi. 269. *pl.* 89.
5747. 12, Quercus-radius. *Kirby? MSS.*
5748. 13, Quercus-pedunculi. *Linn.* ii. 918.—*Turt.* iii. 409.
5749. 14, Quercus baccarum. *Linn.* ii. 917.—*Berk. S.* i. 152.—*Stew.* ii. 220.—*Turt.* iii. 408.
5750. 15, Quercus folii. *Linn.* ii. 918.—*Berk. S.* i. 153.—*Stew.* ii. 220.—*Turt.* iii. 408.—*Bingley.* iii. 247.—*Shaw* G. *Z.* vi. 268. *pl.* 88.—(*Leach E. E.* ix. 144.)—(*Sam. I.* 14.)—*Wood.* ii. 42. *pl.* 53.
5751. 16, viminalis. *Linn.* ii. 919.—*Stew.* ii. 221.—*Turt.* iii. 409.—*Shaw* G. *Z.* vi. 270.
5752 ad 5786: 35 sp. adhuc examinandæ. (51 sp.)

Genus 208: (782). ——

5787. 1; aptera*. *Kirby? MSS.*

Genus 209: (783). IBALIA, *Latreille, Leach, Curtis.*

BANCHUS *p, Fabricius.*—SAGARIS, *Panzer.*—ICHNEUMON *p, Illiger.*

5788 † 1, cultellator. *Latr. H.* xiii. 205. *pl.* 100. *f.* 5.—*Curtis.* i. *pl.* 22.

Ba. cultellator. *Fabr. P.* 127.

Ich. leucospoides. *Berl. Mag.* (*Illiger.*)? vi. 345. *pl.* 8. *f.* 5, 6. In Mus. D. *Curtis.*

ORDO VII. STREPSIPTERA, *Kirby.*

(RHIPIPTERA, *Latr.*—HYMENOPTERA *p, Rossi.*)

Familia I: (113). STYLOPIDÆ.

Genus 1: (783). STYLOPS *Auctorum.*

5789 † 1, Melittæ. *Kirby M. A.* ii. 113. *pl.* 14. *No.* 11. *f.* 1–9.—*Sowerb. B. M.* i. *pl.* 45.—(*Sam. I.* 39.) In Mus. D. *Kirby.*

5790 † 2, Kirbii*. *Leach Z. M.* iii. 135. *pl.* 149. In Mus. *Brit.*

5791 † 3, tenuicornis*? *Linn. Trans.* (*Kirby.*) xi. 233.—(*Sam. I.* 39.) In Mus. D. *Kirby?*

5792 † 4, Haworthi* *mihi.* In Mus. D. *Haworth.*

5793. 5, Dalii*. *Curtis.* v. *pl.* 226.

END OF PART I.

APPENDIX

AD PARTEM PRIMAM[a].

COLEOPTERA.

Page. No.

1. 1; Cicindela sylvatica*. (*Kirby & Sp. I. E.* ii.378: iv. 502.) —*Curtis l. c. infra.*

2: Ci. —— hybrida. (*Kirby & Sp. I. E.* iv. 506.)—Ci. maritima. *Curtis l. c. infra.*

2. 3, Ci. —— riparia: Ci. hybrida, var. *Curtis l. c. infra.*

3*, Ci. —— aprica: Ci. hybrida, var. *Curtis l. c. infra.*

4, Ci. —— Sylvicola*. *Curtis 2 edit. pl.* I.

5; Ci. —— campestris*. (*Kirby & Sp. I. E.* iv. 503.)— *Curtis l. c. supra.*

6; Ci. —— Germanica*. *Curtis l. c. supra.*

8: Polistichus fasciolatus. *Curtis.* v. *pl.* 223.

9: Odacantha melanura. *Curtis.* v. *pl.* 227.

3. † 11*, Demetrias imperialis. *Megerle.—Sturm* D. *F.* vii. 63. *pl.* clxxiii. *f. b, B.—Steph. Ill.* (*M.*) i. 176.
In Mus. D. *Hope.*

—† 14*, Dromius Sigma*. *DeJean Col.* i. *pl.* 14. *f.* 7.—*Steph. Ill.* (*M.*) i. 176.

4. 16*, Dr. —— bifasciatus. *DeJean Sp.* i. 237.—De Sigma. —(*Curtis l. c. infra?*)

— 17; Dr. —— spilotus*. *DeJean Sp.* i. 246.—*Curtis.* v. *pl.* 231.—Dr. humeralis. *Curtis MSS.*

— 18: Dr. —— quadrillum. *DeJean Sp.* i. 249.—*Sturm* D. *F.* 45. *pl.* clxx. *f. a. A.—b. B. var.*—Dr. humeralis. *Steph. Ill.* (*M.*) i. 23.

— 19; Dr. —— fasciatus. *Sturm* D. *F.* vii. 43. *pl.* clxix. *f. C.* c.

23*; Dr. —— maurus*. *Ziegler.—Sturm* D. *F.* vii. 55. *pl.* clxxi. *f. d. D.—Steph. Ill.* (*M.*) i. 176.—(*Curtis l. c. supra.*)

6. ‡ 34*. Tarus homagricus. (*Curtis l. c. infra.*)—Le. homagrica.

---

[a] Containing the names of such species only as have been described, figured or discovered during the printing of the foregoing pages, or of such as were accidentally omitted in their proper places; those which have been introduced in mere catalogues (unless for the *first* time) being purposely omitted.

Page. No.

Duft. F. ii. 240.—Cy. homagrica. *Sturm D. F.* vii. 10. *pl.* clxv. *f. b. B.* In Mus. D. Curtis.

6. 36*, Tarus basalis. *Gyll.* ii. 174.—*Sturm D. F.* vii. 15. *pl.* clxvi. *f. b. B.*—*Steph. Ill.* (*M.*) i. 177.—*Curtis.* v. *pl.* 235.—Cy. punctata. *Bonelli.*—Cy. scapularis. *And.*

7. †48*. Dyschirius pusillus. *DeJean.*—(*Curtis* G. 3.)

8. —— Genus 14, b. ONCODERUS *mihi.*—SPHÆRODERUS *mihi* (*olim.*)

— 55, b, On. —— chalconotus. (*Steph. Nom.* 1.)—*Vide Steph. Ill.* (*M.*) ii. 137 *nota.*

— —— DISTOMUS: Genus Vermium. Vide *Lamarck.*

10. 68; Carabus arvensis*. (*Kirby & Sp. I. E.* iv. 554.)

11. 77: Ca. —— nitens*. (*Kirby & Sp. I. E.* iv. 502.)

12. 85*, Helobia Æthiops *mihi.* *Steph. Ill.* (*M.*) i. 179.

87, He. —— arctica. *DeJean Sp.* ii. 235.—(*Steph. Nom.* 2.)

14. —— Genus 26, b. TRIMORPHUS *mihi.*—BADISTER *p, Sturm?*

— 99. b; Tri. —scapularis*. *Steph. Ill.* (*M.*) i. 180.—Ca. dorsiger. *Dufts F. A.* ii. 151?—Ba. humeralis. *Bonelli?* —*DeJean Sp.* ii. 408?—Ca. sodalis. *Dufts F. A.* ii. 152?

— 99. c; Tri. —— confinis*. *Steph. Ill.* (*M.*) i. 181. *pl.* ix. *f.* 5.—Trechus humeralis. *Leach MSS.*—(*Sam. I.* 42.) —Blemus humeralis. *Supra No.* 336.

17. —— SPHODRUS. § B.—PRISTONYCHUS, *DeJean.*

19. —— ODONTONYX *mihi.*—OLISTOPHUS, *DeJean.*—(*Loudon M.* i. 55. *fig.* 23.)

20. 154; Calathus piceus*. Cal. rotundicollis. *DeJean Sp.* iii. 75.

—‡157, Cal. —— punctipennis. *Germ. I. N.* i. 13.—Cal. latus. *DeJean, &c.*—Ca. angusticollis. *Sturm Catal.*

21. —— PLATYDERUS *nec* PLATYDERES.—FERONIA *p, DeJean.*

23. 181; Pœcilus cupreus*. (*Kirby & Sp. I. E.* iv. 554.)

— 182, Dele †.

24. 192, Dele †.

25. —— PTEROSTICHUS, *Bon., &c.*—COPHOSIS *p, DeJean.*

—†204, Pt. —— elongatus.—Co. filiformis. *DeJean Sp.* iii. 337?

— 207, Pt. —— 8-punctatus. (*Loudon M.* i. 161. *fig.* 61. *b.*)

— 208; Pt. —— macer*. (*Loudon M.* i. 161. *fig.* 61. *a.*)

26. 213; Dele Am. lata. *Sturm F.*—Adde Am. acuminata*. *Sturm &c.,* cum synonymis ut supra.

— 214; Am. eurynota*. *Panz. &c.*—*Steph. Ill.* (*M.*) i. 126.

— 215; Am. lata*. *Sturm D. F.* vi. 23. *pl.* cxl. *f. B. b?*—*Steph. Ill.* (*M.*) i. 127.

27. 232. Dele †, et Am. erratica. *Sturm?*

32. †296*. Harpalus calceatus. *Creutzer?*—(*Curtis G.* 5.)

34. 316; Stenolophus lateralis*. *Steph. Ill.* (*M.*) ii. 71 *nota.*—St. vespertinus. *Supra cum synonymis.*—(*Steph. Nom.*

Page. No.
3.)—Ha. placidus. *Gyll.* iv. 453.—St. placidus. (*Steph. Ill.* (*M.*) i. 183.)—Var. β. *Steph. Ill.* (*M.*) i. 166.—St. Ziegleri. *Supra. No.* 317.
34. —— Genus 59 b, MASOREUS, (*Ziegler*), *DeJean.*—BADISTER *p*, *Creutzer.*—TRECHUS *p*, *Sturm.*
— 1, luxatus. (*DeJ. C.* 15.)—*DeJean Sp.* iii. 537.—(*Steph. Ill.* (*M.*) ii. 137 *nota.*)—(*Steph. Nom.* 3.)—Ba. luxatus. *Creutzer: teste DeJean l. c.*—Tr. laticollis. *Sturm D. F.* vi. 103. *pl.* 150. *f. d. D*:—haud Ha. Orfordensis. *Sp. MSS.*
— 318; Trechus dorsalis*: *nec* Stenolophus dorsalis.
— 319; Tr. —— parvulus*: *nec* Stenolophus parvulus.
— 320; Tr. —— cognatus*: *nec* Stenolophus cognatus.
— 321; Ophonus? pubescens*: *nec* Trechus pubescens.
35. †331*, Blemus pallidus. *Sturm?*—(*Curtis G.* 8.)—Bl. paludosus β. *Steph. Ill.* (*M.*) ii. 171?
36. 339, Aëpus fulvescens. *Curtis.* v. *pl.* 203.
— —— BEMBIDIIDÆ:—(TACHYS, *Knoch.*)
— 340, Lymnæum nigropiceum. *Steph. Ill.* (*M.*) ii. 3. *pl.* x. *f.* 1.
— —— PHILOCHTHUS *nec* PHILOCHTHES.
— 351; Ph. —— æneus* *nec* ænea.
37. 353; Ph. —— subfenestratus* *nec* subfenestrata.
— 359; Ocys melanocephalus*. *Steph. Ill.* (*M.*) ii. 10. *pl.* x. *f.* 2.
— 362; Peryphus saxatilis* *nec* saxatile.
— 362*, Pe. —— albipes. *Steph. Ill.* (*M.*) ii. 189.—Pe. —— assimilis. *Heysham MSS.*—(*Steph. Nom.* 4.)
38. 364, Pe. —— lunatus. *Steph. Ill.* (*M.*) ii. 13. *pl.* x. *f.* 3.
— 374; Notaphus undulatus*. *Steph. Ill.* (*M.*) ii. 17. *pl.* x. *f.* 4.
40. 389; Lopha assimilis. *Steph. Ill.* (*M.*) ii. 23. *pl.* x. *f.* 5.
— 389*, Lo. —— viridis. *Heysham MSS.*—(*Steph. Nom.* 4.)—*Steph. Ill.* (*M.*) ii. 189.
— 392; Lo. —— pulicaria* *nec* pulicarius.
— 394: Tachypus Andreæ. *Steph. Ill.* (*M.*) ii. 29. *pl.* x. *f.* 6.
41. 403. Dele †.
— 407; Notiophilus aquaticus*. *Curtis l. c. infra.*
—†407*, No. —— rufipes. *Heysham MSS.*—*Curtis.* vi. *pl.* 190. —No. fulvipes. (*Steph. Ill.* (*M.*) ii. 137 *nota.*)
In Mus. D. *Heysham.*
— 408; No. —— biguttatus*. *Curtis l. c. supra.*
—†408*, No. —— 4-punctatus. *DeJean Sp.* ii. 190.—(*Steph. Ill.* (*M.*) ii. 137 *nota.*)—*Curtis l. c. supra.*
In Mus. D. *Heysham.*
42. 412; Blethisa multipunctata*. (*Kirby & Sp. I. E.* iv. 502.)
— 414; Adde Dy. cæsus. *Dufts F.* i. 284.
43. 415; Haliplus mucronatus*. *Steph. Ill.* (*M.*) ii. 40. *pl.* xi. *f.* 1.
— 421; Dele Dy. cæsus. *Dufts F.* i. 284.
44. 429, Hygrotus fluviatilis. *Steph. Ill.* (*M.*) ii. 46. *pl.* xi. *f.* 2.

Page. No.
44. 437; Hy. —— scitulus*. *Steph. Ill. (M.)* ii. 49. *pl.* xi. *f.* 3.
45. 439; Hydroporus frater*. *Steph. Ill. (M.)* ii. 50. *pl.* xi. *f.* 4.
—†—, Hyd. —— 9-lineatus. *Rudd MSS.—Steph. Ill. (M.)* ii. 192. In Mus. D. *Rudd.*
46. †448*, Hyd. —— opatrinus. (*DeJean C.* 19?)—(*Steph. Nom.* 4.)—*Steph. Ill. (M.)* ii. 192. In Mus. D. *Hope.*
—†448**, Hyd. —— latus. *Rudd? MSS.—Steph. Ill. (M.)* ii. 192.—(*Curtis G.* 11?) In Mus. D. *Rudd.*
— 449, Hyd. —— alpinus. *Steph. Ill. (M.)* ii. 54. *pl.* xi. *f.* 5.
— 456*; Hyd. —— ferrugineus*. *Rudd MSS.*—(*Steph. Nom.* 4.)—*Steph. Ill. M.* ii. 193.
48. 475; Noterus crassicornis*. *Curtis l. c. infra.*
— 476; No. —— sparsus*. *Steph. Ill. (M.)* ii. 63. *pl.* xi. *f.* 6. —*Curtis.* v. *pl.* 236.
— 479; Colymbetes oblongus*. *Steph. Ill. (M.)* ii. 65. *pl.* xii. *f.* 1.
—† —— Co. arcticus. (*Steph. Nom.* 5.)—*Steph. Ill. (M.)* ii. 194. Dy. arcticus. *Panz. F.* i. 201? In Mus. *D. Hope.*
49. 483, Dele †.
—†486, Co. —— *cicur*[a].
— 489; Co. —— pulverosus*. *Steph. Ill. (M.)* ii. 69. *pl.* xii. *f.* 2.
— 491; Co. —— exoletus* *nec* Co. collaris.
— 493; Co. —— adspersus. *Steph. Ill. (M.)* ii. 71. *pl.* xii. *f.* 3.
— 494; Co. —— nebulosus* *nec* Co. bipunctatus.
50. 496; Co. —— conspersus*. *Steph. Ill. (M.)* ii. 72. *pl.* xii. *f.* 4.
51. †507*, Co. —— ferrugineus. *Steph. Ill. (M.)* ii. 79.—Dy. castaneus. *Schön. S.* ii. 21?
— 508, Co. —— uliginosus var. β. *Steph. Ill. (M.)* ii. 79.
— 509; Co. —— aterrimus*. *Steph. Ill. (M.)* ii. 79.
— 511; Co. —— confinis*. *Gyll.* i. 511.
—†517*, Co. —— 4-notatus. *Steph. Ill. (M.)* ii. 83.
— —— Genus 84 b. AGABUS, *Leach, Steph.*
—†518*, serricornis. *Ahr. F.* v. *f.* 3.—*Steph. Ill. (M.)* ii. 84.—Dy. serricornis. *Payk. F.* iii. *App.* 443.—Co. serricornis. (*Kirby & Sp. I. E.* iii. 324.)—Agabus Paykullii. (*Leach Z. M.* iii. 72.) In Mus. D. *Hope.*
— 521, Hydaticus stagnalis. *Steph. Ill. (M.)* ii. 85. *pl.* xii. *f.* 5.
52. 523. Genus LEIONOTUS, *Kirby.—Steph. Ill. (M.)* ii. 87. *pl.* xii.
— 524. Do. Do. *f.* 6.
— 527; Dyticus marginalis*. (*Kirby & Sp. I. E.* ii. 348.—iii. 305 *nota*, 344, 560, 588, 635.—iv. 24, 38, 40.)
53. 533; Acilius sulcatus*. (*Kirby & Sp. I. E.* iii. 305.)—Dy. cinereus. *Wood.* i. 67. *pl.* 24.

---

[a] 49. †486, Colymbetes cicur. *Steph. Ill. (M.)* ii. 67. *nota.* Dy. cicur. *Fabr. E.* i. 262.—Co. lævigatus, *supra, No.* 486, *cum synonymis.*—Co. consobrinus. *Curtis.* v. *pl.* 207. (!)

Page. No.
53. 534. Var. δ prioris. *Vide Steph. Ill.* (*M.*) ii. 93.
— 535. Var. ε prioris. *Vide Steph. Ill.* (*M.*) ii. 93.
54. 539; Gyrinus minutus*. *Steph. Ill.* (*M.*) ii. 96. *pl.* xiii. *f.* 2.
— 540; Gy. —— natator*. (*Kirby & Sp. I. E.* iii. 80.)—*Wood.* i. 19. *pl.* 6.
— 540*, Gy. —— lineatus. *Hoffm.*—*Steph. Ill.* (*M.*) ii. 97. *pl.* xiii. *f.* 1.

— Divisio 2. RYPOPHAGA.

55. 549; Heterocerus marginatus*: dele *Marsh.* i. 400.
— 550, He. —— Marshami *mihi**. *Steph. Ill.* (*M.*) ii. 101.
— 552; He. —— obsoletus*. *Curtis.* v. *pl.* 224.
— 554, He. —— lævigatus β. *Steph. Ill.* (*M.*) ii. 102.
— 555; Georyssus pygmæus*. *Steph. Ill.* (*M.*) ii. 105. *pl.* xiii. *f.* 3.
56. 558; Elmis variabilis*. *Steph. Ill.* (*M.*) ii. 107. *pl.* xiii. *f.* 4.
— 561, El. —— parallelipipedus. *Steph. Ill.* (*M.*) ii. 108. *pl.* xiii. *f.* 5.
— 564, El. —— cupreus*. *Steph. Ill.* (*M.*) ii. 109. *pl.* xiii. *f.* 6.
— —— Genus 93 b. ENICOCERUS *mihi.*
— 564*, En. —— viridi-æneus *mihi. Steph. Ill.* (*M.*) ii. 196. *pl.* xv. *f.* 6.—(*Steph. Nom.* 5.)
57. 571; Helophorus dorsalis*. *Steph. Ill.* (*M.*) ii. 112. *pl.* xiv. *f.* 1.
— 576. Dele Hy. margipallens. *Marsh.* i. 408.
58. 584, Ochthebius punctatus. *Steph. Ill.* (*M.*) ii. 117. *pl.* xiv. *f.* 2.
—†584*. Oc. —— hibernicus*. *Curtis.* vi. *pl.* 250.—*Steph. Ill.* (*M.*) ii. 196.
In Mus. D. Curtis, Haliday, et *Waterhouse.*
— 586; Hydræna pusilla*. *Steph. Ill.* (*M.*) ii. 118. *pl.* xiv. *f.* 3.
— 588; Limnebius ater*. *Steph. Ill.* (*M.*) ii. 119. *pl.* xiv. *f.* 4.
59. 597; Hydroüs piceus*. *Curtis.* v. *pl.* 239.—Hydroph. piceus. (*Kirby & Sp. I. E.* iii. 72, 336.—iv. 10, 24, 123.)
— 598; Hydrophilus caraboides*. (*Kirby & Sp. I. E.* iv. 39.)—Black oval Water-Beetle. *Harr. A. pl.* 26. *f. e–i.*
— 600; Berosus luridus*. *Curtis l. c. infra.*
— 601; Be. —— æriceps*. *Curtis.* v. *pl.* 240.
— 602; Be. —— obsoletus*. *Curtis l. c. supra.*—*Steph. Ill.* (*M.*) ii. 126. *pl.* xiv. *f.* 5.
60. 603, Hydrobius picipes*.—β. Hydrobius chalconotus. *Curtis.* vi. *pl.* 243.
— 607*; Hyd. —— torquatus*. *Steph. Ill.* (*M.*) ii. 129.—Hyd. melanocephalus β.
— 610; Hyd. —— fulvus*. *Steph. Ill.* (*M.*) ii. 130.—Dele bicolor *et* Hy. bicolor. *Payk. F.* i. 184.
— 611; Hyd. —— lividus* *nec* griseus.—Hy. lividus. (*Kirby & Sp. I. E.* iii. 72.)
61. 613; Hyd. —— affinis*.—Hyd. margipallens. *Marsh.* i. 408.

Page. No.
61. 614; Hydrobius atricapillus*. *Steph. Ill.* (*M.*) ii. 131.— Hy. bicolor. *Payk. F.* i. 184.
— 620; Hyd. —— minutus*. Dele *Fabr. E.* i. 254.
— 624; Hyd. —— æneus*. *Steven.*—*Germ. I. N.* i. 96.—(*Curtis l. c. supra.*)—*Steph. Ill.* (*M.*) ii. 135.
63. 643; Lege. De. picinus. *Mus. Marsh.* nec *Marsh.* i. 60.
66. 695; Phalacrus? maritimus. *Steph. Ill.* (*M.*) ii. 159. *pl.* xv. *f.* 1.
67. 704; Dele *Marsh.* i. 77.—Adde *Mus. Marsh.*
68. 725, Dele †: adde *Steph. Ill.* (*M.*) ii. 169. *pl.* xv. *f.* 2.
— —— LEIODES. § A. Corpore subovato, haud hemisphærico.
— 729; Le. —— Gyllenhalii*. (*Steph. Nom.* 6.)—*Steph. Ill.* (*M.*) ii. 197.
— 732*, Le. —— punctatissima. *Steph. Ill.* (*M.*) ii. 172. *pl.* xv. *f.* 3.
69. 742*; Le. —— brunnea*. (*Steph. Nom.* 6.)—*Steph. Ill.* (*M.*) ii. 197.—Anisotoma brunnea. *Sturm* D. *F.* ii. 40. *pl.* xxiv. *f. d. D.?*
— 744; Le. —— cinnamomea. *Steph. Ill.* (*M.*) ii. 176.— *Curtis.* vi. *pl.* 251.—β? Le. Tuberis. *Supra. No.* 745. —(*Steph. Nom.* 6.)
— 747*; Le. —— rugosa*. *Steph. Ill.* (*M.*) ii. 178.
— 748*; Le. —— abdominalis. *Steph. Ill.* (*M.*) ii. 179.—Sphæridium abdominale. *Payk. F.* i. 71.—Anisotoma glabra. *Illig. K.* i. 76.—*Panz. F.* xxxvii. *f.* 9.
— 749; Agathidium globus*. *Steph. Ill.* (*M.*) ii. 180. *pl.* xv. *f.* 4.
70. 762; Clambus Armadillus*. *Steph. Ill.* (*M.*) ii. 184. *pl.* xv. *f.* 5.
71. —— Genus 113. ORTHOPERUS, *Steph.*
— —— Genus 114. SERICODERUS, *Steph.*
72. †776. Dele in Mus. *D. Jenyns?*
74. 798; Choleva gausapata* *nec* gomphosata.
— —— Genus NECROPHORUS. *Burying Beetle.*
— 800; Ne. —— anglicus*. *Steph. Ill.* (*M.*) iii. *pl.* xvi. *f.* 1.
75. 803; Ne. —— Sepultor*. *Steph. Ill.* (*M.*) iii. *pl.* xvi. *f.* 2.
— 804; Ne. —— Mortuorum*. (*Kirby & Sp. I. E.* i. 387.)
— 805; Ne. —— Vespillo*. (*Kirby & Sp. I. E.* i. 255.—ii. 377. —iii. 89.—iv. 178.)
76. 809; Oiceoptoma rugosa*. (*Kirby & Sp. I. E.* iii. 120 *note.*)
77. 821, Phosphuga subrotundata. *Steph. Ill.* (*M.*) ii. *pl.* xvi. *f.* 3.
79. 833; Nitidula 4-pustulata*. *Steph. Ill.* (*M.*) iii. *pl.* xvi. *f.* 4.
80. 851; Strongylus imperialis*. *Steph. Ill.* (*M.*) ii. *pl.* xvi. *f.* 5.
83. †882*. Micropeplus tesserula. *Curtis.* v. *pl.* 204. In Mus. D. Haliday.
— 882; Mi. —— Staphylinoides*. *Steph. Ill.* (*M.*) iii. *pl.* xvi. *f.* 6.
90. 963, Ips 4-pustulata*.—Si. 4-pustulata. *Linn.* ii. 570.
— —— Genus 153 a. TRIBOLIUM[a].

---

[a] 90. —— Genus 153 a. TRIBOLIUM, *MacLeay.*
— 1. Tr. —— castaneum. *MacLeay A. J.* 47. (!)

Page. No.
93. 998; Latridius quadratus*.—Tenebrio lardarius. (*Kirby & Sp. I. E.* i. 225.)
95. 1011; Megatoma Serra*. *Curtis.* vi. *pl.* 244.
— 1012; Attagenus undatus*. Meg. undatus. *Curtis folio* 244.
—†1013, Att. —— trifasciatus*. *Curtis.* vi. *pl.* 247.
— 1014; Att. —— pellio*. (*Kirby & Sp. I. E.* iii. 325.)—*Curtis l. c. supra.*
— 1015; Dermestes lardarius*. (*Kirby & Sp. I. E.* i. 225.)
97. 1027; Nosodendron fasciculare. *Curtis.* vi. *pl.* 246.
99. 1041; (Chætophora) cretifera. (*Kirby & Sp. I. E.* iv. 503?)
100. 1045; Onthophilus sulcatus*. *Curtis.* v. *pl.* 220.
— 1049; Hister cadaverinus*. (*Kirby & Sp. I. E.* iii. 154.)
104. 1087; Lucanus Cervus*. (*Kirby & Sp. I. E.* ii. 224; iii. 103; 298 *nota*, 400 *et* 426.)
— 1089; Copris lunaris*. (*Kirby & Sp. I. E.* ii. 257; iii. 313.)
105. 1097; Onthophagus nuchicornis*. (*Kirby & Sp. I. E.* iii. 312.)
106. —— BOLBOCERUS, *Kirby.*—Bolboceras, *Curtis.*
— 1102; Bo. —— testaceus.—Bo. mobilicornis. *Curtis.* vi. *pl.* 259.
—†1103; Bo. —— quadridens. *Curtis l. c. supra.*
107. 1109; Geotrupes foveatus*. (*Kirby & Sp. I. E.* iv. 199.)
— 1114; Geo. —— stercorarius*. (*Kirby & Sp. I. E.* i. 349; ii. 349; iii. 442, 545; iv. 149, 176.)—Scarabæus stercorarius. (*Kirby & Sp. I. E.* i. 392 *nota.*)—(*Loudon M.* i. 423. *fig.* 180.)
108. 1118; Aphodius erraticus*. (*Kirby & Sp. I. E.* iv. 554.)
111. 1152; Ap. —— nigripes*. (*Kirby & Sp. I. E.* iv. 397.)
— 1153; Ap. —— luridus*. (*Kirby & Sp. I. E.* iv. 397.)
112. 1166; Ap. —— plagiatus*. (*Kirby & Sp. I. E.* iv. 503.)
113. 1171; Ap. —— testudinarius*. (*Kirby & Sp. I. E.* iv. 509.)
— 1177: Psammodius sulcicollis. *Curtis.* vi. *pl.* 258.
114. 1178, Ægialia globosa. (*Kirby & Sp. I. E.* iv. 506.)
—‡1184, Oryctes nasicornis*. (*Kirby & Sp. I. E.* iii. 90, 203, 312; iv. 8, 10, 23, 39 & 62.)
116. 1189; Melolontha vulgaris*. (*Kirby & Sp. I. E.* i. 392 *nota;* ii. 5, 349; iii. 203, 345, 398, 591; iv. 179, 513.)—Cockchafer. *Albin. pl.* li.
— 1190: Me. —— Fullo. (*Kirby & Sp. I. E.* iii. 423.)—An. Sc. Fullo. *Linn.* ii. 553?
— —— ANOMALA § A.—Phyllopertha, *Kirby.*
— 1191; An. —— Horticola*. Me. Horticola. (*Kirby & Sp. I. E.* i. 194; iii. 691; iv. 40.)
— 1192; An. —— Frischii*. (*Kirby & Sp. I. E.* iv. 506.)
117. 1197; Hoplia argentea*. (*Kirby & Sp. I. E.* ii. 5.)
118. ‡1202. Trichius hemipterus: Genus AÇANTHURUS, *Kirby.* —Acanthurus hemipterus. (*Steph. Nom.* 10.)

Page. No.
118. 1203; Cetonia aurata*. (*Kirby & Sp. I. E.* iii. 298 *nota;* iv. 507.)—Rose May Beetle. *Harr. A. pl.* 17. *f. m–q.*
120. 1216; Agrilus biguttatus* *nec* biguttata.
— 1220: Ag. —— 9-maculatus *nec* 9-maculata.
— 1221: Aphanisticus emarginatus*.—Aph. pusillus. *Curtis.* vi. *pl.* 262.
122. 1233; Elater lineatus*.—El. segetum. (*Kirby & Sp. I. E.* iii. 203.)
128. 1295; Atopa cervina*.—Dascillus cervinus. *Curtis.* v. *pl.* 216. —(*Loudon M.* i. 272. *fig.* 135.)
— 1296; Scirtes hemisphæricus*. (*Kirby & Sp. I. E.* i. 87.)
130. 1317; Lycus minutus*. *Curtis.* vi. *pl.* 263.
— 1318. Ly. festivus[a].
— 1324. Telephorus Æthiops. *Curtis folio* 215.
131. 1326; Telephorus marginatus* *mihi.* (*Curtis fo.* 215, excluso synonymo. *Fabr.?*)
132. 1327; Te. —— discoideus* *mihi.* (*Curtis fo.* 215, excl. syn. *Kirby's MSS.*
—†1337*. Te. —— *longicornis*[b].
— 1338, Te. —— tricolor.—Te. cyaneus. *Curtis.* vi. *pl.* 215.
— 1341; Te. —— cantianus* *mihi.* (*Curtis fo.* 215. excl. syn. *Leach?*)
135. 1375, Dele †.
137. 1396, Dele †.
138. 1403; Thanasimus formicarius*. (*Kirby & Sp. I. E.* i. 271.)
— 1404: Clerus Apiarius. (*Kirby & Sp. I. E.* i. 402.)
—‡1405, Adde in Mus. D. *Chant*, *Curtis* et Sparshall.
— 1409; Corynetes violaceus*.—Co. cæruleus. (*Kirby & Sp. I. E.* i. 255.)
139. 1413; Ptinus imperialis*. (*Kirby & Sp. I. E.* iv. 513.)
140. 1414; Pt. —— Germanus*. (*Kirby & Sp. I. E.* iv. 513.)
— 1417; Pt. —— Fur*. (*Kirby & Sp. I. E.* i. 231, 387.)
— 1422; Mezium sulcatum*. *Curtis.* v. *pl.* 232.
141. 1428; Anobium pertinax*. (*Kirby & Sp. I. E.* i. 237; ii. 235.)
— 1431; An. —— paniceum*. (*Kirby & Sp. I. E.* i. 229.)
142. —— CIS nec CISS.
144. —— APATE § A. SINOXYLON, *Dufts.*—(*Steph. Nom.* 12.)
— 1453*. N. G.[c].

---

[a] 130 ‡ 1318. Lycus festivus. *Curtis l. c. folio* 263. (!)
[b] 132 † 1337. Telephorus *longicornis. Curtis l. c. fol.* 215. (!)—Ca. longicornis. *Fab.—Curtis l. c.*
[c] 144. 1453*; N.G.—*pusilla**.
Sinodendron pusillum. *Fabr. E.* ii. 378.—(*Kirby & Sp. I. E.* i. 229. 231.)—Ptinus piceus. *Marsh.* i. 88. (!)—Ptinus fissicornis. *Marsh.* i. 82. (!)

Page. No.
150. 1517; Cionus Scrophulariæ*. Rh. Scrophulariæ.—(*Kirby & Sp. I. E.* iv. 552.)
153. 1542; Ceutorhynchus sulcicollis*: (*Kirby & Sp. I. E.* i. 187 *nota.*)
156. 1579; Nedyus horridus*.—Rh. horridus. (*Kirby & Sp. I. E.* iv. 506.)
158. 1601; Cryptorhynchus Lapathi*. (*Kirby & Sp. I. E.* iv. 107.) —Cu. Lapathi. (*Kirby & Sp. I. E.* i. 194 *nota.*)
— 1603; Bagoüs binodulus*. Ba. atrirostris.—(*Kirby & Sp. I. E.* iv. 500.)
161. 1644; Orchestes Fragariæ*.—Cu. Fragariæ. (*Kirby & Sp. I. E.* i. 193.)
165. 1687; Dorytomus Vorax*.—Cu. longimanus. (*Kirby & Sp. I. E.* iii. 330.)—Cu. Forsteri. *Turt.* ii. 239.
167. 1705; Adde (*Kirby & Sp. I. E.* i. 198.)
168. 1715; Hypera Rumicis*.—Curculio bitæniatus. *Don.* xv. *pl.* 524?
171. 1743; Tanysphyrus Lemnæ*. (*Kirby & Sp. I. E.* iv. 500.)
172. 1754; Otiorhynchus notatus*. (*Kirby & Sp. I. E.* i. 198.)
175. 1783; Strophosomus Coryli*: Nut beetle. *Harr. A. pl.* 14. *f. p.–m.*—Cu. melanogrammus. *Forst. C.* i. 36.—*Stew.* ii. 61. *Turt.* ii. 264.
177. 1809; Sitona lineata*.—Cu. lineatus. (*Kirby & Sp. I. E.* i. 189 *nota.*)
179. 1836*, Polydrusus speciosus. (*Steph. Nom.* 15.)
— 1839. Adde Cu. oblongus. (*Kirby & Sp. I. E.* i. 197 *nota.*)
182. 1861, Lixus paraplecticus. (*Kirby & Sp. I. E.* iii. 123, 500.)
— 1862*, Li. —— lateralis. (*Steph. Nom.* 15.)
183. 1868; Magdalis carbonarius*. *Curtis.* vi. *pl.* 212.
184.‡1881*. Apion Platalea. *Germ. M.* iii. 143. *pl.* 3. *f.* 23.—*Curtis fo.* 211.
188.†1949, Ap. —— difforme*. *Curtis.* v. *pl.* 211.—(*Loudon M.* i. 160. *fig.* 59.)—Adde in Mus. D. Curtis, et Haliday.
— 1952; Ap. —— ebeninum*. (*Kirby & Sp. I. E.* iv. 503.)
190. 1967; Deporaüs Betulæ*. Ah. Betulæ. (*Kirby & Sp. I. E.* iv. 107.)
1972; Rhynchites Alliariæ*. (*Kirby & Sp. I. E.* iii. 76.)—Cu. Alliariæ. (*Kirby & Sp. I. E.* i. 194 *nota.*)
191. 1981; Rh. —— Populi*. (*Kirby & Sp. I. E.* i. 209.)
196. —— Genus 351 b. ACROCINUS[a].
— 2019; Cerambyx Moschatus*. Callichroma Moschatum. (*Kirby & Sp. I. E.* iii. 573, 576; iv. 141.)

---

[a] 196. —— Genus 351 b. ACROCINUS, *Illiger.* (*Kirby.*)
‡1. *Ac. Accentifer**. *Schön. S.* iii. 348.—Prionus Accentifer. *Oliv. E.* iv. 66. 8. *pl.* 4. *f.* 16.

Page. No.
197. 2020: Monochamus Sutor*. La. Sutor. (*Kirby & Sp. I. E.* iii. 319.)
—†2021: Mo. —— Sartor*. *Curtis.* v. *pl.* 219.
— 2021*: Mo. —— *dentator* [a].
—†2021**. Mo. ——? *literatus* [b].
— —— Genus 354 b. STENOCORUS [c].
198. 2029. Dele ‡.
199. —— Genus 358 b. TETRAOPES [d].
— 2040; Callidium bajulum*. (*Kirby & Sp. I. E.* i. 232 *nota.*)
200. 2042; Ca. —— violaceum*. (*Kirby & Sp. I. E.* iii. 120 *nota;* iv. 151.)
201. 2055*, Clytus *fulminans* [e].
—†2059*. Cl. *erythrocephalus* [f].
210. 2129; Crioceris Asparagi*. Le. Asparagi.—(*Kirby & Sp. I. E.* i. 31; iv. 104.)
213. 2151; Haltica antennata*. Ch. nodicornis.—(*Kirby & Sp. I. E.* iii. 324.)
218. —— Pro *** Elytris lævigatis *lege* ††† Elytris lævigatis.
221. 2268; Chrysomela Vitellinæ*. Gal. Vitellinæ.—(*Kirby & Sp. I. E.* iv. 105.)
— 2270; Ch. —— Polygoni*. (*Kirby & Sp. I. E.* iv. 163.)
222.†2279*. Ch. —— *festiva* [g].
223. 2285*, Ch. —— Hobsoni. *Hope MSS.*

---

[a] 197. 2021*: Mo. *dentator**. (*DeJean C.* 106.)—*Curtis l.* c. *infra.* (!) —La. dentator. *Fabr. E.* ii. 294.—*Ent. Trans.* (*Haworth.*) i. 84. *pl.* 1. *fig. sup.* (!)—Cer. Caroliniensis. *Oliv. E.* iv. 67. 85. *pl.* 12. *f.* 88.

[b] ——† 2021**. Mo.? litteratus.—Cer. litteratus. *Don.* xvi. *pl.* 546.(!)

[c] —— —— Genus 354 b. STENOCORUS, *Fabr.* (*Kirby.*)

‡ 1: *spinicornis**. *Fabr. E.* ii. 306.—Cer. spinicornis. *Oliv. E.* iv. 67. 41. *pl.* 17. *f.* 130.—Cer. insularis. *Gmel.* iv. 1859.

2: *quadrimaculatus**. *Fabr. E.* ii. 308.—*Ent. Trans.* (*Haworth.*) i. 83. (!)—Cer. 4-maculatus. *Linn.* ii. 627.—*Oliv. E.* iv. 67. *pl.* 21. *f.* 164.—Cer. ramphygeus. *Linn.* ii. 633. —Cer. bimaculatus. *Panz.* (*Voet.*) iii. 41. *pl.* 15. *f.* 65.

[d] 199. —— Genus 358 b. TETRAOPES, *Dalman.*

—— 2036*. 1: *Tornator**.—Lamia Tornator. *Fabr. E.* ii. 301. (*Wilk. Catal.*) (!)—Cer. Tornator. *Oliv. E.* iv. 67. 103. *pl.* 8. *f.* 52. —Cer. tetrophthalmus. *Forst. C.* i. 41.

[e] 201. 2055*, Clytus *fulminans**. *Fabr. E.* ii. 346.—(*Curtis folio* 199.) (!) —Cer. fulminans. *Oliv. E.* iv. 32. *pl.* 5. *f.* 63.—*Sowerb. B. M.* i. *pl.* 58. (!)

[f] ——† 2059*. Clytus *erythrocephalus**. *Fabr. E.* ii. 350.—*Curtis fol.* 199. (!)—Cer. erythrocephalus. *Oliv.* iv. 47. *pl.* 5. *f.* 60. —Call. acuminatum. *Fabr. Sp.* 1. 243.—Cer. americanus. *Gmel.* iv. 1854.

[g] 222† 2279. *Ch. festiva.* *Fabr. E.* i. 440.—(*Wilk. Catal.*) (!)

Page. No.
223.†2294, Chrysomela violacea. (*Kirby & Sp. I. E.* iv. 108.)
224. 2300; Timarcha tenebricosa*. (*Kirby & Sp. I. E.* iv. 215.) —Ch. Tenebricosa. (*Id.* ii. 247, 309.)
225. 2302; Ch. Populi*. (*Kirby & Sp. I. E.* iii. 141; iv. 134.) *Albin. pl.* lxiii. *f.* 6.
— 2306*, Ch. cuprea. *Fabr. E.* i. 432?—*Panz. F.* xxv. *f.* 8?—(*Steph. Nom.* 18.)—Chr. ruficaudis. *DeGeer.* v. 305? —Coccinella Vitellinæ. *Scop. C. No.* 224?
226. 2311; Clythra tridentata*. ♂. Cly. longimana. (*Kirby & Sp. I. E.* iii. 330.)
228. —— Genus 386 b. COLASPIS[a].
230. 2361*. Chilocorus *Cacti*[b].
231. 2372; Coccinella 7-punctata*: The common Lady-bird. *Harr. A. pl.* 22. *f. n–t.*
235. 2381; Coccinella dispar*.—*α.* Co. bipunctata. (*Kirby & Sp. I. E.* iii. 102; iv. 133.)
240. 2406: Cacicula pectoralis*.—Silpha rosea. (*Kirby & Sp. I. E.* i. 225.)
242. 2415; Tenebrio Molitor*. (*Kirby & Sp. I. E.* iii. 330, 635.)
— 2419; Uloma cornuta*. (*Kirby & Sp. I. E.* i. 173.)
—†2420*. Diaperis *Hydni*[c].
243. —— MISOLAMPUS[d].
244. 2435; Blaps mortisaga*. (*Kirby & Sp. I. E.* i. 34 & 399; iv. 200.)
247. —— Genus 423. HYPULUS, *Payk.*—ULODES, *Millard.*
2459; Quercinus*. *Curtis folio* 255.—Ul. pilosulus. *Mill. B. E.* 113.
248. —— Genus 424. HYPULUS *p! Curtis.*
— 2460; bifasciata*. Hy. biflexuosus. *Curtis.* vi. *pl.* 255.
—†2461, trifasciata.—Hy. 4-fasciatus. *Curtis folio* 255.
250. 2490, Sitaris humeralis adde *.
251. 2499; Œdemera melanura*.—Œd. notata. (*Kirby & Sp. I. E.* iii. 320.)
252. 2502, Dele †.

---

[a] 228. —— Genus 386 b. COLASPIS, *Fabr.*—EROTYLUS, *Fabr., Gmel.*
—— † 1. *flavicornis. Fabr. E.* i. 412.—(*Wilk. Catal.* (!)—Er. flavipes. *Fabr. M.* i. 92.—Chr. occidentalis. *Linn.* ii. 388? *DeGeer.* v. 353. *pl.* 16. *f.* 14.—Erot. jamaicenis. *Gmel.* iv. 1729.

[b] 230 ‡ 2361*. Ch. *Cacti.*—Co. Cacti. *Linn.* ii. 584.

[c] 242 † 2420*. Diaperis *Hydni. Fabr. E.* ii. 585.—(*Wilk. Catal.*) (!)

[d] 243. —— Genus 410 b. MISOLAMPUS, *Latr.*
244 † 2434*. Mi. Pimelia. *Regne Anim.* (*Latr.*) iii. 297.—He. Pimelia. *Fabr. E.* i. 162. (!)—Pimelia Morio. *Fabr. Sp.* i. 318.—Helops anglica. *Gmel.* iv. 2011. (!)—Pi. anglica. *Stew.* ii. 84. (!)—*Turt.* ii. 496. (!)—Scaurus Vienensis. *Sturm. D. F.* ii. 180. *pl.* xli. ?

Page. No.
253. 2509; Proscarabæus vulgaris*. Meloë proscarabæus. (*Kirby & Sp. I. E.* ii. 309; iv. 133, 225.)
254. 2517: Cantharis vesicatoria*. (*Kirby & Sp. I. E.* iii. 395; iv. 77.)
274. 2888; Creophilus maxillosus*. (*Kirby & Sp. I. E.* i. 226.)
275. 2902; Goërius olens*. (*Kirby & Sp. I. E.* ii. 255, 434; iv. 41.)
279. 2959; Philonthus splendens*.—St. splendens. (*Kirby & Sp. I. E.* iii. 340.)
— 2962; Ph. —— politus*.—St. politus. (*Kirby & Sp. I. E.* iv. 105.)
291. 3166, Dianoüs rugulosus adde*.
292. 3169; Siagonium quadricorne*. *Stark E.* ii. 260. (!)
— 3170, Bledius tricornis. Oxytelus tricornis. (*Kirby & Sp. I. E.* iv. 506.)
294. —— Genus 503. SYNTOMIUM, *Curtis.*
— 3218; Sy. —— nigroæneum*. *Curtis.* v. *pl.* 228.

DERMAPTERA.

299. 3299; Forficula auricularia*. (*Kirby & Sp. I. E.* iii. 341.)
— 3304, Labidura gigantea. Fo. gigantea. (*Kirby & Sp. I. E.* iii. 374; iv. 493 *nota.*)

ORTHOPTERA.

300. 3312; Acrida verrucivora. (*Ing. Inst.* 89.)
301. 3313; Ac. —— viridissima*. (*Kirby & Sp. I. E.* iii. 424, 545, 569, 579, 587; iv. 8, 35, 40, 73, 150, 152, 236.) —Gryllus viridissimus. (*Kirby & Sp. I. E.* i. 386, 302 *nota.*)
— 3315, Locusta migratoria*.—Gryllus migratorius. (*Kirby & Sp. I. E.* ii. 327.)
302. 3326; Lo. —— variegata* *nec* varipes*.
303. 3347; Gryllotalpa vulgaris*. (*Kirby & Sp. I. E.* i. 191; ii. 422; iii. 411.)
— 3348; Acheta campestris*. (*Kirby & Sp. I. E.* i. 276; ii. 309, 351, 397.)
— 3349: Ac. —— sylvestris. (*Ing. Inst.* 89.)
— 3350; Ac. —— domestica*. (*Kirby & Sp. I. E.* ii. 351, 396.)
—†3351, Ac. —— italica? (*Ing. Inst.* 89.)
—†3352, Blatta *gigantea**. *Linn.* ii. 687.
304. 3356; Bl. —— germanica*. (*Kirby & Sp. I. E.* i. 230. *pl.* ii. *f.* 3.)
— 3358; Bl. —— Lapponica*. (*Kirby & Sp. I. E.* i. 226.)—(*Ing. Inst.* 89.)
— 3359; Bl. —— perspicillaris*. (*Ing. Inst.* 90.)
— 3360, Bl. —— Panzeri. (*Ing. Inst.* 90.)
— 3363; Bl. —— pallida. (*Ing. Inst.* 89.)

Page. No. NEUROPTERA.

305. 3369; Ephemera vulgata*. (*Kirby & Sp. I. E.* iii. 342; iv. 57.)
308. —— Genus ANAX. LINDENIA?
— 3425; Anax imperator*. (*Kirby & Sp. I. E.* i. 273; ii. 355.)
— 3427; Æshna grandis*. (*Kirby & Sp. I. E.* iv. 179.)
— 3428; Æs. —— varia*. Æshna viatica. (*Kirby & Sp. I. E.* iii. 455; iv. 254.)—Æshna maculatissima. *Latr. H.* xiii. 7. (!)
309. 3430; Æshna mixta*. *Latr. H.* xiii. 8. (!)
310. 3445; Libellula depressa*. (*Kirby & Sp. I. E.* iii. 303; iv. 39.)
312. 3489; Psocus pilicornis*. *Latr. H.* xiii. 73.—*Coqueb. Ic.* 2. *f.* 12.
— 3492; Ps. —— variegatus*. *Latr. H.* xiii. 73.—*Coqueb. Ic.* 2. *f.* 13.
— 3496; Ps. —— lineatus*. *Latr. H.* xiii. 72.—*Coqueb. Ic.* 2. *f.* 8. Chermes Buxi. *Roem.* G. ii. *pl.* 11. *f.* 7.
313. 3499; Ps. —— bifasciatus*. *Latr. H.* xiii. 72.—*Coqueb. Ic.* 2. *f.* 4.
— 3512; Ps. —— 4-maculatus*. *Latr. H.* xiii. 72.—*Coqueb. Ic.* 2. *f.* 6.
— 3521; Ps. —— phæopterus*. Psocus fuscopterus. *Latr. H.* xiii. 71? *Coqueb. Ic.* 2. *f.* 2?
— 3523; Ps. —— abdominalis*. Psocus pedicularius.—*Latr. H.* xiii. 71. *Coqueb. Ic.* 2. *f.* 1.
— 3525; Atropos pulsatorius*. Psocus pulsatorius.—(*Kirby & Sp. I. E.* ii. 383; iv. 546.)—Termes pulsatorium. *Berk. S.* i. 169.—*Barb.* G. *pl.* 18. *gen.* 3.—*Stew.* ii. 278. —*Turt.* iii. 694.
314. 3526, At. —— fatedicus*. Termes fatedicum. *Stew.* ii. 278.
— 3527; Raphidia Ophiopis*. *Don.* ix. 57. *pl.* 315?
— 3533; Sialus lutarius*. Semblis lutaria. (*Kirby & Sp. I. E.* iv. 58.)

HYMENOPTERA.

325. 3750: Clavellaria marginata*. *Curtis.* ii. *pl.* 93.
—†3754, Amasis obscura. (*Ing. Inst.* 90.)
344. 4021, Pimpla Æthiops. (*Loudon M.* i. 161. *fig.* 60.)
356. —— Genus 666. MYRMECINA, *Curtis.*
—†4837, Myr. —— Latreillii. *Curtis.* vi. *pl.* 265. Dele *.
—†4838: N. G. —— Westwoodii *adde* *.
359. 4870; Pompilus aterrimus* *nec* aterrima.
366. 4974; Pemphredon geniculatum* *nec* geniculatus.
— 4976; Pe. —— caliginosum* *nec* caliginosus.
— 4978; Pe. —— nitidum* *nec* nitida.
373. 5054; Megachile Willughbiella*. (*Loudon M.* i. 273. *fig.* 137.)

END OF APPENDIX I.

INSECTA HAUSTELLATA.

Ordo I: (8). LEPIDOPTERA.

(Glossata, *Fabricius.*—Lepidioptera, *Clairville.*)

Sectio I. LEPIDOPTERA-DIURNA, *Latreille.*

(Papilio, *Linné, &c.*) *Butterflies.*

Familia I: (114). PAPILIONIDÆ, *Leach.*

(Equites, Heliconii *et* Danai Candidi, *Linné.*—Danai Flavi, *Wien V.*—Parnassii, *Fabr.*)

(Chilognathiform, *or* Juliform Stirps. *Horsfield.*)

Genus 1: (784). PAPILIO *Auctorum.* *Swallow-tailed B.*

Pterourus *p*, *Scopoli.*—Pieris *p*, *Schrank.*—Principes, *Hübner.*

5794. 1: Machaon*. *Linn.* ii. 750.—*Berk. S.* i. 123.—*Barbut G.* 169. *pl.* 10. *f.* 33.—*Don.* vi. 75. *pl.* 211.—*Lew. pl.* 34.—(*Haw. Pr.* 1.)—*Shaw N. M.* xii. *pl.* 398.—*Haw.* 5.—*Turt.* iii. 19.—*Stew.* ii. 119.—*Shaw G. Z.* vi. 209. *pl.* 64.—*Leach E. E.* ix. 127.—(*Kirby & Sp. I. E.* ii. 224; iii. 35, 148, 186, 213 & 263.)—*Samou.* 235. *pl.* 5. *f.* 1.—*Jerm.* (24), 41. *pl.* 1.—*Steph. Ill.* (*H.*) i. 16.—*Id.* i. *App.* 145.—*Jerm.* 2 *edit.* (62), 84. *pl.* 1: *pl.* 3. *f.* 1. neuræ. *Brit. B.* i. *pl.* 1.

Pa. diurnus maximus, maximam partem flavescens, &c. *Merr. P.* 198.

Pa. diurnæ, &c. *Mouff.* 99. 2. *cum fig.* larva et imago.—*Raii I.* 110. 1–111. larva.

Pa. major caudata, &c. Royal William B. *Pet. Pap. pl.* 2. *f.* 5.

Swallow-tail B: Pa. Machaon. (*Wilkes.* 47. *pl.* 93.)—*Harr. A. pl.* 36.—*Harr. V. M.* 7.

Pa. Reginæ. *DeGeer.* G. 30. 5.

5795 ‡ 2. Podalirius[a].

---

[a] 5795 ‡ 2. Podalirius: Scarce Swallow-tail B.—*Don. l. c.* *Linn.* ii. 751. —*Berk. S.* i. 123. (!)—(*Forst. Cat.* 404.) (!)—(*Martyn V. M.* 36.) (!) —*Don.* iv. 1. *pl.* 109. (!)—*Lew. pl.* 35. (!)—(*Haw. Pr.* 1. (!)—*Haw.* 5. (!) —*Id.* 588. (!)—*Turt.* iii. 16. (!)—*Stew.* ii. 119. (!)—*Rees' Cyclop.* (*Donovan.*) *Article* Papilio. (!)—(*Leach E. E.* ix. 127.)—(*Samou.* 35.)—*Jerm.* (24), 42. (!)—*Ing. Inst.* 91. (!)—*Steph. Ill.* (*H.*) i. 6. nota.—*Id.* i. *App.* 145.—*Jerm.* 2 *edit.* (62), 86. (!)

Pa. diurnæ, &c. *Mouff.* 99. 3. *cum fig.* imago.—*Raii Ins.* 111. 3. (!)

Genus 2: (785). GONEPTERYX, *Leach, Samou., Jermyn, Steph., Curtis, Children.*

COLIAS *p, Fabr.*—PIERIS *p, Schrank.*

5796. 1; Rhamni*. *Leach E. E.* ix. 128.—*Samou.* 236.—(*Kirby & Sp. I. E.* iii. 252.)—*Steph. Ill.* (*H.*) i. 8.—*Jerm. 2 edit.* (62), 87. *pl.* 3. *f.* 2. neuræ.—*Brit. B.* 2. *pl.* 2.—*Phil. M. & A.* (*Children.*) iv. 351.

Pa. Rhamni. *Linn.* ii. 765.—*Berk. S.* i. 125.—*Barbut* G. 170. *pl.* 10. *f.* 106.—*Don.* v. 1. *pl.* 145.—*Lew. pl.* 31.—(*Haw. Pr.* 2.)—*Haw.* 14.—*Turt.* iii. 76.—*Stew.* ii. 122.—*Jerm.* (26), 46.
Pa. corpore et antennis livescentibus, &c. *Merr. P.* 198. 5.
Pa. diurnæ med. &c. *Mouff.* 103. 1. *cum fig.*
Pa. sulphureus. *Pet. Mus. Cent. I. No.* 1.
Pa. præcox sulphurea, &c. *Raii I.* 112. 4.
The Brimstone Butterfly. *Albin. pl.* 3. *f.* 3. *e–h.*—(*Wilkes.* 48. *pl.* 94.)—*Harr. V. M.* 1.
Pa. canicularis. *DeGeer.* G. 30. 1.
♂; Pa. sulphurea: the Brimstone B. *Pet. Pap. pl.* 2. *f.* 1.
♀; Pa. sulphureus pallidus: the male Straw B. *Pet. Pap. pl.* 2. *f.* 2.
† *β*, *Steph. Ill.* (*H.*) i. 9.—*Curtis.* iv. *pl.* 173.

β in Mus. D. *Haworth.*

Genus 3: (786). COLIAS, *Fabricius, Leach, Samou.,* (*Kirby,*) *Jerm., Steph.*

ARGYREUS *p, Scopoli.*—PIERIS *p, Schrank.*

5797. 1, Europome: Clouded Sulphur B.—*Haw. l. c. Steph. Ill.* (*H.*) i. 10. *pl.* I*. *f.* 1. ♂. *f.* 2. 3. ♀.—(*Ing. Inst.* 90.)—*Jerm. 2 edit.* (78), 156.—*Brit. B.* 2.

Pa. Europome. *Vill.* iv. 408?—*Esper.* i. *pl.* 42. *f.* 2?—*Haw.* 13.—*Jerm.* (26), 45.

5798 ‡ 2. Palæno [a].

5799. 3, Chrysotheme. (*Och.* iv. 32.)—*Steph. Ill.* (*H.*) i. 11. *pl.* II*. *f.* 1. ♂. *f.* 2. ♀.—(*Ing. Inst.* 90.)—*Brit. B.* 3.

Pa. Chrysotheme. *Esper.* i. *pl.* 65. *f.* 3. 4.
Pa. Chrysothome. (*Jerm.* 40.)
Co. Chrysothome. *Jerm. 2 edit.* (62), (78), 91.

5800. 4; Edusa*. (*Sam. I.* 12.)—*Steph. Ill.* (*H.*) i. 12. & 145. —*Jerm. 2 edit.* (62), 89. *pl.* 3. *f.* 3. neuræ.—*Brit. B.* 3. *pl.* 3.

Pa. Edusa. *Fabr.* iii. *a.* 206.—(*Haw. Pr.* 2.)—*Haw.* 11.—*Turt.* iii. 73.—*Stew.* ii. 121.—*Jerm.* (24), 45.—*Don.* vii. 60. *pl.* 238. *f.* 2. ♀.

---

[a] 5798 ‡ 2. Palæno. *Steph. Ill.* (*M.*) i. 11. nota.—*Jerm. 2 edit.* (80), 161.
Pa. Palæno. *Linn.* ii. 764.—(*Martyn. V. M.* 35.) (!)—*Hüb. Pa. pl.* 86. *f.* 434, 435.—(*Haw. p.* xxix. (!)
Pa. Europome var. *Bork. E. S.* ii. 214.

Pa. Hyale. (*Wien. V.* 165.)—*Esper.* i. *pl.* 4. *f.* 3.—*Don.* ii. 17. *pl.* 43. *fig. sup.* ♂.
Pa. Electra. *Lewin. pl.* 32.
Pa. croceus α. *Fourc. P.* ii. 250.
Pa. diurnæ, &c. *Mouff.* 100. 5. *cum fig.* imago ♂.
Pa. med. mag., alis exterioribus summa parte flavis, &c. *Raii I.* 112. 6.
Clouded yellow, or Saffron B. *Harris A. pl.* 29. *f. n.* ♂, *f. m.* ♀.—*Harr. V. M.* 7.
♂; Pa. crocea, &c.: the Saffron B. *Pet. Pap. pl.* 2. *f.* 3.
♀; Pa. crocea, &c.: the Spotted Saffron B. *Pet. Pap. pl.* 2. *f.* 4.
β, * White clouded Yellow B.—*Haw. l. c. Steph. Ill. l.* c. i. 13. *pl.* II*. *f.* 3. ♀.
Pa. Helice. *Hüb. Pa. pl.* 87. *f.* 440. 441.—*Haw.* 12.—*Jerm.* (24), 45.
Co. Helice. *Jerm. 2 edit.* (62), 90.
Pa. Edusa-alba. (*Haw. Pr.* 2.)
Pa. alis exterioribus, &c. fœmina. *Raii I.* 112. 6. ♀'.

5801. 5, Hyale*. (*Och.* iv. 32.)—(*Leach E. E.* ix. 127.)—(*Sam. I.* 12.)—*Steph. Ill.* (*H.*) i. 13.—*Jerm. 2 edit.* (62), 91.—*Brit. B.* 4.
Pa. Hyale. *Linn.* ii. 764.—*Berk. S.* i. 124.—*Lewin. pl.* 33.—(*Haw. Pr.* 2.)—*Haw.* 12.—*Don.* vii. 57. *pl.* 238. *f.* 1.—*Turt.* iii. 74.—*Stew.* ii. 121.—*Jerm.* (26), 45.
Pa. Palæno. *Esper.* i. *pl.* 4. *f.* 2.
Pa. croceus C. *Fourc. P.* ii. 250.
Pale clouded Yellow B. *Harr. V. M.* 7.

Genus 4: (787). PONTIA, *Fabricius, Leach, Samou.,* (*Kirby,*) *Jerm., Curtis, Steph.*

PIERIS *p, Latr.*—BATTUS *p, Scop.*—ASCIA *p, Scop.*

A. Pontiæ veræ. *Steph. Ill.* (*H.*) i. 15.

5802. 1; Brassicæ*: Large White B.—*Haw. l. c. Och.* iv. 30.—(*Leach E. E.* ix. 127.)—(*Sam. I.* 34.)—(*Curtis l. c. infra.*)—*Steph. Ill.* (*H.*) i. 15.—*Jerm. 2 edit.* (62), 94. *pl.* 3. *f.* 5. neuræ.—*Brit. B.* 4. *pl.* 4. *fig. sup.*
Pa. Brassicæ. *Linn.* ii. 759.—*Berk. S.* i. 124.—*Don.* xiii. 29. *pl.* 446.—*Lew. pl.* 25.—(*Haw. Pr.* 1.)—*Haw.* 7.—*Turt.* iii. 64.—*Stew.* ii. 120.—*Shaw G. Z.* vi. 211. *pl.* 69. *fig. inf. dext.* —(*Kirby & Sp. I. E.* ii. 11; iii. 98, 212, 214, 345, 352, 557 & 571; iv. 8, 10 & 221.)—*Jerm.* (24), 43.
Pa. diurnus maximus cum antennis porrect. &c. *Merret. P.* 198. 2?
Pa. No. 7. *List. Goed.* (*Angl.*) 16. *f.* 7.—*Id.* (*Lat.*) 9. *f.* 7.
Pa. Brassicaria alba major, &c. *Raii I.* 113. 1.
Large White garden B: Pa. Brassicæ. (*Wilkes.* 49. *pl.* 96.)
Large garden White B. *Harr. V. M.* 7.—*Bing.* iii. 206.

♂; Pa. alba major, &c. Great White Cabbage B. *Pet. Pap. pl.* 1. *f.* 3. 4.

♀; Pa. alba major, bimaculata. *Pet. Pap. pl.* 1. *f.* 5. 6.

5803. 2; Chariclea*. *Steph. Ill.* (*H.*) i. 17. *pl.* III*. *f.* 1. ♂. *f.* 2. ♀.—*Id.* i. *App.* 146.—*Brit. B.* 5.

The Great White Butterfly. *Albin. pl.* 1.

β, *Steph. Ill.* (*H.*) *l. c.*—Po. Brassicæ præcox. *Haworth MSS.*

5804. 3; Rapæ*: Small White B.—*Haw. l. c.* (*Och.* iv. 30.)—(*Leach E. E.* ix. 127.)—(*Kirby & Sp. I. E.* i. 188; ii. 11; iv. 24. 281.)—(*Sam. I.* 34.)—(*Curtis l. c. infra.*)—*Steph. Ill.* (*H.*) i. 18. —*Jerm.* 2 *edit.* (62), 96.—*Brit. B.* 5.

Pa. Rapæ. *Linn.* ii. 759.—*Berk. S.* i. 124.—*Lew. pl.* 26.—(*Haw. Pr.* 1.)—*Haw.* 8.—*Turt.* iii. 64.—*Stew.* ii. 120.—*Jerm.* (24), 44.

Pa. diurnus medius, &c. *Merret. P.* 198. 1.

Pa. No. 8. *List. Goed.* (*Angl.*) 12. *f.* 8.—*Id.* (*Lat.*) 22. *f.* 8.

Pa. alba media, &c. *Raii I.* 114. 2. ♀.—3. ♂.—*Albin. pl.* 51.

Small White garden B: Pa. Rapæ. (*Wilkes.* 50. *pl.* 97.)

Small garden White B. *Harr. V. M.* 7.

♂; Pa. alba media, maculata: Lesser White Cabbage B. *Pet. Pap. pl.* 1. *f.* 7. 8.

♀; Pa. alba media, bimaculata: Lesser White double-spotted B. *Pet. Pap. pl.* 1. *f.* 9. 10.

Var. Pa. Nelo. *Bork. E. S.* i. 127?

5805. 4; Metra*: Mr. Howard's white B. *Steph. Ill.* (*H.*) i. 19. —*Id.* i. *App.* 146.—*Brit. B.* 6.

♂; Pa. alba, media, immaculata: Lesser White unspotted B. *Pet. Pap. pl.* 1. *f.* 13. 14.

♀; Pa. alba media, trimaculata: Lesser White treble-spotted B. *Pet. Pap. pl.* 1. *f.* 11. 12.

5806. 5; Napi*. (*Och.* iv. 31.)—(*Leach E. E.* ix. 127.)—(*Sam. I.* 34.)—(*Curtis l. c. infra.*)—*Steph. Ill.* (*H.*) i. 20.—*Jerm.* 2 *edit.* (78), 96.—*Brit. B.* 6. *pl.* 4. *fig. med.*

Pa. Napi. *Linn.* ii. 760.—*Berk. S.* i. 124.—*Lew. pl.* 27.—*Don.* viii. 23. *pl.* 280. *f.* 1.—(*Haw. Pr.* 1.)—*Haw.* 9.—*Turt.* iii. 6.—*Stew.* ii. 121.—*Jerm.* (24), 44.

Pa. alis æreis, &c. *Merr. P.* 198. 7.

Pa. Brassicaria media, &c. *Raii I.* 114. 4.—*Albin. pl.* 52. *f. d–g.*

Pa. alba, media, maculata, &c.: Lesser spotted, White, veined B. *Pet. Pap. pl.* 1. *f.* 19. 20.

White B. with green veins: Pa. Napi. (*Wilkes.* 50. *pl.* 98.)

Green-veined White B. *Harr. V. M.* 7.

5807. [6; Napææ*.] *Steph. Ill.* (*H.*) i. 21.—*Jerm.* 2 *edit.* (78), 157. —*Brit. B.* 7.

Pa. Napææ. *Esper. Pa.* i. *pl.* 116. *f.* 5.

Pa. Napi var. *Hüb. Pa. pl.* 81. *f.* 407*.

5808. 7; Sabellicæ*. *Steph. Ill.* (*H.*) i. 22. *pl.* III*. *f.* 3. ♂; *f.* 4. ♀.
Pa. Sabellicæ, &c. *Pet. Pap. pl.* 1. *f.* 17. 18. ♂; *f.* 15. 16. ♀.
Pa. Bryoniæ. *Godart. E. M.* ix. 162.
Pa. Napi *β*. *Haw.* 9.

B. Vide *Steph. Ill.* (*H.*) i. 22.—MANCIPIUM, *Hübner.*

5809. 8; Daplidice*?: Bath White B.—*Lew. l. c.* Green-chequered white B.—*Haw. l. c.* (*Och.* iv. 31.)—(*Leach E. E.* ix. 127.)—(*Sam. I.* 34.)—*Curtis.* i. *pl.* 48.—*Steph. Ill.* (*H.*) i. 22.—*Jerm.* 2 *edit.* (64), 97.—*Brit. B.* 7.

Pa. Daplidice. *Linn.* ii. 760.—*Lew. pl.* 28.—*Don.* vi. 47. *pl.* 200.—(*Haw. Pr.* 1.)—*Haw.* 10.—*Turt.* iii. 67.—*Stew.* ii. 121.—*Jerm.* (24), 44.
Pa. Cardamines var. *Scop. C. No.* 454. 2, 3 ♂, 4. ♀.
Pa. Edusa. *Fabr. Gen. I. Mant.* 255.
Pa. leucomelanos Cantabrigiensis: the greenish-marbled Half-mourner. *Pet. Gaz. pl.* 1. *f.* 7.
Pa. med. mag. alis supina parte albis, &c. *Raii I.* 116. 10.
♂, Pa. alba minor, subtus viridescens, &c.: the Slight greenish Half-mourner, &c. *Pet. Pap. pl.* 2. *f.* 8.
♀; Pa. leucomelanos, &c.: Vernoun's greenish Half-mourner. *Pet. Pap. pl.* 2. *f.* 9.

5810. 9; Cardamines*. (*Och.* iv. 31.)—(*Leach E. E.* ix. 127.)—(*Kirby & Sp. I. E.* iii. 254; iv. 510.)—(*Sam. I.* 34.)—(*Curtis l. c. supra.*)—*Steph. Ill.* (*H.*) i. 23.—*Id.* i. *App.* 147.—*Jerm.* 2 *edit.* (64), 98.—*Brit. B.* 8. *pl.* 4. *fig. inf.*

Pa. Cardamines. *Linn.* ii. 760.—*Berk. S.* i. 124.—*Lew. pl.* 30.—*Don.* v. 83. *pl.* 169.—(*Haw. Pr.* 1.)—*Haw.* 11.—*Turt.* iii. 67.—*Stew.* ii. 121.—*Jerm.* (24), 45.
Pa. minor, &c.: The White-marbled B. *Raii I.* 115. 6. ♂. 7. ♀.
Orange tip, or Lady of the Woods: Pa. Cardamines. (*Wilkes.* 51. *pl.* 99.)—*Harr. A. pl.* 32. *f. f.* i.—*Harr. V. M.* 5.
♂; Pa. albus, subtus viridi, &c.: White-marbled male B. *Pet. Pap. pl.* 2. *f.* 6.
♀; Pa. alba, subtus viridi, &c.: White-marbled female B. *Pet. Pap. pl.* 2. *f.* 7.

Genus 5: (788). LEUCOPHASIA *mihi*, (*Horsfield,*) *Children.*

PIERIS *p*, *Latr.*—PONTIA *p*, *Leach*, *Samou.*—BATTUS *p*? *Scop.*

5811. 1; Sinapis*. *Steph. Ill.* (*H.*) i. 24.—*Id.* i. *App.* 147.—*Brit. B.* 8. *pl.* 5.—*Phil. M. & A.* (*Children.*) iv. 350.

Pa. Sinapis. *Linn.* ii. 760.—*Lew. pl.* 29.—*Don.* viii. 74. *pl.* 280. *f.* 2.—(*Haw. Pr.* 1.)—*Haw.* 9.—*Turt.* iii. 65.—*Stew.* ii. 121.—*Jerm.* (24), 44.
Po. Sinapis. (*Leach E. E.* ix. 127.)—(*Sam. I.* 34.)—*Jerm.* 2 *edit.* (64), 99.
Pa. corpore cinereo, &c. *Merr. P.* 199. 12.

Pa. alba minor, &c.: Small White B. *Raii I.* 116. 8.
Wood White. *Harr. A. pl.* 29. *f. t.–u.*—*Harr. V. M.* 7.
Pa. candidus. *DeGeer.* G. 30. 6.
♂; Pa. alba minor apice nigricante: White small tipt B. *Pet. Pap. pl.* 1. *f.* 21. 22.
♀; Pa. alba minor, mas.: Small White Wood B.- *Pet. Pap. pl.* 1. *f.* 23. 24.
β; *Steph. Ill.* (*H.*) *l. c. supra.*—*Pap. d'Eur.* i. 215. *pl.* 1. *f.* 105. *a.*
γ; *Steph. Ill.* (*H.*) *l. c. supra.*
Pa. Erysimi. *Bork. E. S.* i. 132.

Genus 6: (789). PIERIS *mihi*, (*Kirby*,) *Jermyn.*

PIERIS *p*, *Schrank.*—PONTIA *p*, *Och.*, *Leach*, *Samou.*

5812. 1; Cratægi*: Hawthorn B.—*Kirby l. c.* *Godart. E. M.* ix. 154.—(*Kirby & Sp. I. E.* iii. 98; iv. 200.)—*Steph. Ill.* (*H.*) i. 27. —*Jerm.* 2 *edit.* (62), 93. *pl.* 4. *f.* 4. neuræ.—*Brit. B.* 9. *pl.* 6.
Pa. Cratægi. *Linn.* ii. 758.—*Berk. S.* i. 124.—*Barbut G.* 169. *pl.* 10. *f.* 72.—*Lew. pl.* 24.—(*Haw. Pr.* 1.)—*Haw.* 6.—*Turt.* iii. 53.—*Stew.* ii. 120.—*Shaw* G. Z. vi. 211. *pl.* 69. *fig. inf.* sin.—*Don.* xiii. 45. *pl.* 454.—*Jerm.* (24), 43.
Po. Cratægi. *Leach E. E.* ix. 127.—*Samou.* 236.
Pa. alba, nervis alarum nigris, &c. *Raii I.* 115. 5.
Pa. alba venis nigris. *Pet. Pap. pl.* 1. *f.* 1. 2.
White Butterfly with black veins. *Albin. pl.* 2. *f.* 2. *a–d.*—Pa. Cratægi. (*Wilkes.* 49. *pl.* 95.)
Black-veined White B. *Harr. A. pl.* 9. *f. g–k.*—*Harr. V. M.* 7.
Pa. nigro venosus. *DeGeer G.* 30.

Genus 7: (790). DORITIS[a].

Familia II: (115). NYMPHALIDÆ, *Swainson.*

(DANAI FESTIVI, NYMPHALES PHALERATI *et* GEMMATI, *Linné.*—HELICONII, *Fabr.*)

(NYMPHALIS, *Latreille.*)

(CHILOPODIFORM *or* SCOLOPENDRIFORM, *and* THYSANURIFORM Stirpes, *Horsfield.*)

---

[a] Genus 7. DORITIS, *Fabricius*, *Leach*, *Steph.*, *Jermyn.*

PARNASSIUS, *Latreille*, (*Kirby.*)—PIERIS *p*, *Schrank.*

5813 ‡ 1. Apollo: Crimson-ringed B.—*Haw. E. T. l.* c. *Leach E. E.* ix. 127.—*Steph. Ill.* (*H.*) i. 25. nota.—*Jerm.* 2 *edit.* (80), 163.
Pap. Apollo. *Linn.* ii. 754.—*Haw. p.* xxix. (!)—*Ent. Trans.* (*Haworth.*) i. *p.* 332. (!)—*Don.* xiii. 1. *pl.* 433. (!)

5814 ‡ 2. Mnemosyne. *Steph. Ill.* (*H.*) i. 26. nota.—*Jerm.* 2 *edit.* (80), 163.
Pap. Mnemosyne. *Linn.* ii. 754.—*Turt.* iii. 53. (!)

Genus 8: (791). NEMEOBIUS *mihi*, *Horsfield.*

MELITÆA *p*, *Fabr.*, *Leach*, *Samou.*, *Jerm.*—GRAPHIUM *p*, *Scopoli.* —LEMONIADES *p*, *Hübner.*

5815. 1; Lucina*. *Steph. Ill.* (*H.*) i. 29.—*Id.* i. *App.* 147.—*Brit. B.* 9. *pl.* 7.

Pa. Lucina. *Linn.* ii. 784.—*Berk. S.* i. 127.—*Lew. pl.* 15.—*Don.* vii. 70. *pl.* 242. *f.* 2.—(*Haw. Pr.* 3.)—*Haw.* 37.—*Turt.* iii. 40.—*Stew.* ii. 128.—*Jerm.* (32), 56.

Me. Lucina. (*Och.* iv. 14.)—*Leach E. E.* ix. 128.—(*Sam. I.* 27.)—*Jerm.* 2 *edit.* (64), 104.

Pap. Fritillarius minor: Mr. Vernon's small Fritillary. *Pet. Gaz. pl.* 16. *f.* 10.—*Raii I.* 122. 12.—*Pet. Pap. pl.* III. *f.* 15. 16.

The Duke of Burgundy *F. Harr. A. pl.* 27. *f. n, o.*—*Harr. A.* 3.

Genus 9: (792). MELITÆA, *Fabr.*, *Leach*, *Samou.*, (*Kirby*,) *Jermyn*, *Steph.*

GRAPHIUM *p*, *Scopoli.*—LEMONIADES *p*, *Hübn.*—ARGYNNIS *p*, *Ochs.*

A. Vide *Steph. Ill.* (*H.*) i. 30.

5816. 1; Athalia*. (*Och.* iv. 14.)—*Steph. Ill.* (*H.*) i. 30.—*Brit. B.* 10.

Pa. Athalia. *Esper. Pa.* i. *pl.* 47. *f.* 1.

Pa. Dictynna. *Lewin. pl.* 14. *f.* 5. 6.—(*Haw. Pr.* 3.)—*Haw.* 34. —*Stew.* ii. 128.—*Turt.* iii. 41.—*Jerm.* (32), 55.

Pa. Maturna. *Fabr.* iii. *a.* 154.

Me. Dictynna. (*Leach E. E.* ix. 128.)—(*Sam. I.* 27.)—*Jerm.* 2 *edit.* (64), 102.

Pa. media, alis supinis rufis, &c: the May Fritillary. *Raii I.* 120. 8.

Pa. Fritillaria tessellata serotina, subtus albida, &c.: White May *F. Pet. Pap. pl.* III. *f.* 9. 10.

The Heath Fritillary B: Pa. Maturna. (*Wilkes.* 58. *pl.* 112.)

The Pearl-bordered Likeness. *Harr. A. pl.* 38. *f. f–g.*—*Harr. V. M.* 3.

5817. [†2, Pyronia*.] *Steph. Ill.* (*H.*) i. 31. *pl.* IV*. *f.* 1. 2.—*Jerm.* 2 *edit.* (78), 157.—*Brit. B.* 10.

Pa. Pyronia. *Hüb. Pa. pl.* 114. *f.* 585–588.

Pa. Eos: Dark underwing *F. Haw.* 35.—(*Ent. Trans.* (*Haworth.*) i. 333.)—*Jerm.* (32), 55.

Me. Eos. *Jerm.* 2 *edit.* (64), 102.

Me. Dictynna var. (*Och.* iv. 14.) In Mus. D. *Haworth.*

5818. [†3. tessellata*.] *Steph. Ill.* (*H.*) i. 31. *pl.* 5. *f.* 1. 2.—*Jerm.* 2 *edit.* (64), 101.—*Brit. B.* 11.

Pa. tessellata serotina, subtus straminea: Straw May *F. Pet. Pap. pl.* III. *f.* 11. 12.—*Haw.* 53.

Pa. tessellata. *Jerm.* (32), 54.

B. *α*. Vide *Steph. Ill.* (*H.*) i. 32.

5819. 4; Artemis*. (*Och.* iv. 13.)—(*Leach E. E.* ix. 128.)—(*Sam. I.* 27.)—*Steph. Ill.* (*H.*) i. 32.—*Jerm.* 2 *edit.* (64), 103. *pl.* 3. *f.* 6. neuræ.—*Brit. B.* 11. *pl.* 8. *fig. sup.*

Pa. Artemis. *Fabr. E. S.* iii. *a.* 255.—*Lew. pl.* 15.—(*Haw. Pr.* 3.)—*Haw.* 36.—*Turt.* iii. 42.—*Jerm.* (32), 56.

Pa. Maturna. *Esper. Pap.* 1. *pl.* 16. *f.* 2. *pl.* 61. *f.* 3: *pl.* 97. *f.* 4.

Pa. Lye. *Bork. E. S.* i. 225.; ii. 198.

Pa. Matutina. *Thunb. D.* iii. 45.

Pa. Fritillaria, &c. *Raii I.* 121. 11. ♂.—12. ♀.

Pa. Fritillaria minor meandris nigris: Small black F. *Pet. Pap. pl.* III. *f.* 21. 22. ♂.

Pa. Fritillaria media meandris nigris: Dandridge's Middling black F. *Pet. Pap. pl.* III. *f.* 19. 20. ♀.

The Small Fritillary B: Pa. Lucina. (*Wilkes.* 59. *pl.* 114.)

The Greasy F. *Harr. A. pl.* 28. *f.* e–i.—*Harr. V. M.* 3.

The Marsh F. *Bing.* iii. 210.

5820. 5: Cinxia*. (*Och.* iv. 13.)—(*Leach E. E.* ix. 128.)—(*Sam. I.* 27.)—(*Kirby & Sp. I. E.* iii. 114.)—*Jerm.* 2 *edit.* (64), 103.—*Steph. Ill.* (*H.*) i. 33.—*Brit. B.* 12.

Pa. Cinxia. *Linn.* ii. 784.—*Berk. S.* i. 128.—*Lew. pl.* 14.—*Don.* vii. 69. *pl.* 242. *f.* 1.—(*Haw. Pr.* 3.)—*Haw.* 36.—*Stew.* ii. 128.—*Turt.* iii. 40.—*Jerm.* (32), 56.

Pa. Delia. *Hüb. Pa. pl.* 2. *f.* 7. 8.

Pa. Pilosellæ. *Esper.* 1. *pl.* 47. *f.* 3. *pl.* c. *f.* 4.

Pa. Trivia. *Schra. B.* ii. 203.

Pa. Fritillarius Lincolniensis, &c. *Pet. Gaz. pl.* 18. *f.* 10.—*Raii I.* 121. 9.

Pa. Fritillaria minor, undulata, &c.: White Dullidge F. *Pet. Pap. pl.* III. *f.* 25. 26.

The Plantain Fritillary B.: Pa. Cinxia. *Wilkes.* 58. *pl.* 111.

The Glanville F. *Harr. A. pl.* 16. *f.* a–f.—*Harr. V. M.* 3.

Pa. abbacus. *DeGeer. G.* 32.

5821 † 6. Maturna[a].

b. *Alis posticis subtùs argenteo maculatis.* (ARGYNNIS *p*, *Och.*)

5822 ‡ 7, Dia[b].

5823. 8; Selene*. *Steph. Ill.* (*H.*) i. 34.—*Brit. B.* 12. *pl.* 8. *fig. inf.*

---

[a] 5821 † 6. Maturna. *Jerm.* 2 *edit.* (80), 164.—*Steph. Ill.* (*H.*) i. 33. nota.
Pa. Maturna. *Linn.* ii. 784.—*Stew.* ii. 128. (!)—*Turt.* iii. 41. (!)—*Rees' Cycl.* (*Donovan.*) *Art.* PAPILIO. (!)—*Jerm.* 62. (!)
Pa. Cynthia. *Esper. Pa.* i. *pl.* 37. *f.* 2.
Pa. Agrotera. *Bork. E. S.* i. 226.

[b] 5822 ‡ 7, Dia. *Jerm.* 2 *edit.* (80), 163.—*Steph. Ill.* (*H.*) i. 34. nota.
Pa. Dia. *Linn.* ii. 785.—*Hüb. Pa. pl.* 6. *f.* 31.—*Stew.* ii. 129. (!)—*Turt.* iii. 42. (!)—*Rees' Cycl.* (*Donovan.*) *Art.* PAPILIO.—*Jerm.* 62. (!)

Pa. Selene. *Fabr. E. S.* iii. *a.* 147.—(*Haw. Pr.* 3.)
Pa. Silene. *Haw.* 34.—*Jerm.* (32), 54.
Me. Silene. (*Leach*. *E. E.* ix. 128.)—(*Sam. I.* 27.)—*Jerm.* 2 *edit.* (64), 100.
Pa. Euphrasia. *Lew. pl.* 13.
Pa. Maturna. *Berk. S.* i. 128?
Pa. Euphrosyne var. *Esper. Pap. pl.* 30. *f.* 1.
Pa. Summe nigra, &c. *Merr. P.* 199. 7?
Pa. Fritillarius major, &c.: The April Fritillary. *Raii I.* 120. 7.
Pa. Fritillaria, maculata, præcox: April F. *Pet. Pap. pl.* III. *f.* 13. 14.
The Small Pearl Border F. *Harr. A. pl.* 31. *f.* i, k.—*Harr. V. M.* 3.
β; *Steph. Ill.* (*H.*) *l. c.*—*Pap. d'Eur.* i. *pl.* III. iii[e] *Sup. f.* 23. *f.*
† γ, *Steph. Ill.* (*H.*) i. *App.* 147. *pl.* iv*. *f.* 3.
Me. Euphrosyne ε. *Steph. Ill.* (*H.*) i. 35.
Pa. Thalia: Wedge Pearl Border.—*Haw. l. c. Hüb. Pa. pl.* 11. *f.* 57. 58.—*Ent. Trans.* (*Haworth.*) i. 333.
γ in Mus. D. *Haworth.*

5824. 9; Euphrosyne*. (*Leach E. E.* ix. 128.)—*Samou.* 237.—*Jerm.* 2 *edit.* (64), 100.—*Steph. Ill.* (*H.*) i. 35.—*Id.* i. *App.* 147.—*Brit. B.* 13.
Pa. Euphrosyne. *Linn.* ii. 786.—*Berk. S.* i. 128.—*Lew. pl.* 13.—(*Haw. Pr.* 3.)—*Haw.* 33.—*Stew.* ii. 130.—*Turt.* iii. 116.—*Don.* xi. 51. *pl.* 312.—*Jerm.* (32), 54.
Pa. Fritillaria maculata, &c.: April F. with few spots. *Pet. Pap. pl.* III. *f.* 17. 18.
Pa. princeps. *Linn. F.* 1 *edit.* 327.
The Pearl-bordered F. *Harr. A. pl.* 40. *f.* e, *f.*—*Harr. V. M.* 3.
Pa. argenticollis. *DeGeer. G.* 31. 13.
β; *Steph. Ill.* (*H.*) *l. c. var.* γ.—*Pap. d'Eur.* i. 249. *pl.* lxi. *Sup.* vii. *f.* 22. c. d.

Genus 10: (793). ARGYNNIS, *Fabr.*, *Leach*, *Samou.*, (*Kirby*,) *Jermyn*, *Steph.*

ARGYREUS *p*, *Scopoli.*—DRYADES, *Hübner.*

5825. 1; Lathonia*. (*Leach E. E.* ix. 128.)—(*Sam. I.* 5.)—*Jerm.* 2 *edit.* (66), 108.—*Steph. Ill.* (*H.*) i. 36.—*Brit. B.* 13.
Pa. Lathonia. *Linn.* ii. 786.—*Berk. S.* i. 128.—*Lew. pl.* 12.—*Don.* iii. *pl.* 73.—(*Haw. Pr.* 3.)—*Haw.* 33.—*Stew.* ii. 130.—*Turt.* iii. 116.—*Jerm.* (32), 54.
Pa. Principissa. *Linn. F.* 1 *edit. No.* 781.
Pa. Latonia. (*Wien. V.* 177.)
Pa. Lathona. *Hüb. Pa. pl.* 11. *f.* 59. 60.
Pa. Rigensis aureus, &c.: the lesser silver-spotted Fritillary. *Raii I.* 120. 6.

Pa. Fritillaria minor, &c.: Lesser silver-spotted, or Riga *F.* *Pet. Pap. pl.* III. *f.* 23. 24.
The Queen of Spain F. *Harr. V. M.* 3.

5826 ‡ 2. Niobe[a].

5827. 3; Adippe*. (*Och.* iv. 15.)—(*Leach E.E.* ix. 128.)—(*Sam. I.* 5.)—*Jerm. 2 edit.* (66), 108.—*Steph. Ill.* (*H.*) i. 38.—*Brit. B.* 14.
Pa. Adippe. *Linn.* ii. 786.—*Lew. pl.* 10.—(*Haw. Pr.* 2.)—*Haw.* 32.—*Turt.* iii. 115.—*Stew.* ii. 129.—*Don.* xiii. 32. *pl.* 448.—*Jerm.* (30), 54.
Pa. diurnæ, &c. *Mouff.* 101. 11. *cum fig.*
Pa. major, alis fulvis, &c.: the greater silver-spotted *F. Raii I.* 119. 5.
Pa. major, maculis subtus argenteis, &c. *Pet. Mus.* 320.—*Pet. Pap. pl.* III. *f.* 5. 6.
The High Brown F. *Harr. A. pl.* 28. *f. a, d.*—*Harr. V. M.* 3.
Pa. margaritaceus medius. *DeGeer. G.* 31. 12.
β, *Steph. Ill.* (*H.*) *l. c.*—*Pap. d'Eur.* i. 52. *pl.* xiii. *f.* 16. *i.*
† γ, *Steph. Ill.* (*H.*) *l. c.*—*Pap. d'Eur.* i. 238. *pl.* lviii. *Sup.* iv. *f.* 16. *l.*
† δ, *Steph. Ill.* (*H.*) i. *App.* 147.—*Pap. d'Eur.* i. 238. *pl.* lviii. *Sup.* iv. *f.* 16. *k.*
γ in Mus. D. *MacLean et Stone*: δ in Mus. D. *Seaman et Weaver.*

5828. 4; Aglaia*. (*Och.* iv. 15.)—(*Leach E. E.* ix. 128.)—(*Sam. I.* 5.)—*Jerm. 2 edit.* (66), 106. *pl.* 4. *f.* 7. neuræ.—*Steph. Ill.* (*H.*) i. 39.—*Brit. B.* 15. *pl.* 9. *fig. sup.*
Pa. Aglaia. *Linn.* ii. 785.—*Berk. S.* i. 128.—*Lew. pl.* 11.—*Don.* ix. 31. *pl.* 302. ♂.—(*Haw. Pr.* 2.)—*Haw.* 31.—*Stew.* ii. 129.—*Turt.* iii. 114.—*Jerm.* (30), 53.
Pa. Fritillaria major, subtus viridior, &c. *Pet. Pap. pl.* III. *f.* 7. 8.
The great Fritillary B. with silver spots: Pa. Aglaja. (*Wilkes.* 59. *pl.* 115.)
The dark-green *F. Harr. A. pl.* 26. *f. o–p.*—*Harr. V. M.* 3.
† β, *Steph. Ill.* (*H.*) *l. c.*
Pa. Charlotta: The Queen of England F. (*Haw. Pr.* 3.)—*Haw.* 32.—*Sowerb. B. M.* i. *pl.* 11.—*Jerm.* (30), 53.
Ar. Caroletta. *Jerm. 2 edit.* (66), 107.
† γ, *Steph. Ill.* (*H.*) *l. c.*—*Id.* i. *App.* 148.
β in Mus. D. *Dale, Haworth et Sowerby*: γ in Mus. D. *Weaver.*

5829. 5; Paphia*. (*Och.* iv. 15.)—(*Leach E. E.* ix. 128.)—(*Sam. I.* 5.)—(*Kirby & Sp. I. E.* iii. 203 & 253.)—*Jerm. 2 edit.* (66), 105.—*Steph. Ill.* (*H.*) i. 40.—*Brit. B.* 15.
Pa. Paphia. *Linn.* ii. 785.—*Berk. S.* i. 128.—*Lew. pl.* 9.—

---

[a] 5826. 2. Niobe. (*Och.* iv. 15.)—*Jerm. 2 edit.* (80), 164.—*Steph. Ill.* (*H.*) i. 37. nota.
Pa. Niobe. *Linn.* ii. 786.—*Hüb. Pa. pl.* 12. *f.* 61. 62.—*Stew.* ii. 130. (!)

Don. vii. 83. *pl.* 247. ♂.—(*Haw. Pr.* 2.)—*Haw.* 30.—*Stew.* ii. 129.—*Turt.* iii. 114.—*Jerm.* (30), 53.

Pa. major, alis fulvis, &c.: the greater silver-stroaked *F. Raii I.* 119. 4.

Pa. *Fritillaria* major crocea, &c.: Great silver-streakt Orange *F. Pet. Pap. pl.* III. *f.* 1 & 2. ♂.

Pa. eadem aurea: Great silver-streakt Golden *F. Pet. Pap. pl.* III. *f.* 2. 3. ♀.

Pa. Imperator. *Linn. F.* 1 *edit. No.* 779.

The great *Fritillary* B: Pa. Paphia. (*Wilkes.* 57. *pl.* 110.)

The Silver-washed *F. Harr. A. pl.* 34. *f. k–u.*—*Harr. V. M.* 3.

† β, *Steph. Ill.* (*H.*) *l. c.*—*Pap. d'Eur.* i. 316. *pl.* II. *Sup.* 111. *f.* 15. *a. b. tert.* β in Mus. *Brit.*

Genus 11: (794). VANESSA, *Fabr.*, *Leach*, *Samou.*, (*Kirby*,) *Jermyn*, *Curtis*, *Steph.*

HAMADRYADES, *Hübner.*

A. Vide *Steph. Ill.* (*H.*) i. 41.

5830. 1; C. album*. (*Och.* iv. 17.)—(*Leach E. E.* ix. 128.)—(*Sam. I.* 43.)—(*Curtis l. c. infra.*)—*Jerm.* 2 *edit.* (68), 116.—*Steph. Ill.* (*H.*) i. 42.—*Brit. B.* 16. *pl.* 10. *fig. sup.* mala.

Pa. C. album. *Linn.* ii. 778.—*Berk. S.* i. 127.—*Lew. pl.* 5.—*Don.* vi. 45. *pl.* 199.—(*Haw. Pr.* 2.)—*Haw.* 25.—*Stew.* ii. 126.—*Turt.* iii. 105.—*Jerm.* (30), 51.

Pa. Comma-alba. (*Millard. pl.* i. *f.* 3.)

Pa. G. album. *Fourc. P.* ii. 235.

Pa. cum spinosis crenis fuscis. *Merr. P.* 198. 3.

Pa. Ulmariæ similis, &c. *Raii I.* 115. 3.—119. larva.—*Albin. pl.* 54.

Pa. Testudinaria Comma dicta: the silver Comma. *Pet. Pap. pl.* IV. *f.* 5. 6. ♀.

Pa. eadem, alis magis laceratis: jagged winged Comma. *Pet. Pap. pl.* IV. *f.* 9. 10. ♂.

The Comma B: Pa. C. album. (*Wilkes.* 57. *pl.* 109.)—*Harr. A. pl.* 1. *f. a–d.*—*Harr. V. M.* 1.

β; *Steph. Ill.* (*H.*) *l. c.*

Pa. Testudinarius Comma minor: Small Comma. *Pet. Pap. pl.* IV. *f.* 11. 12.

γ; *Steph. Ill.* (*H.*) *l. c.*

Pa. Testudinarius Comma pallidior: the pale Comma. *Pet. Pap. pl.* IV. *f.* 7. 8.

B. Vide *Steph. Ill.* (*H.*) i. 42.

5831. 2; Polychloros*. (*Och.* iv. 17.)—(*Leach E. E.* ix. 128.)—(*Sam. I.* 43.)—(*Curtis l. c. infra.*)—(*Kirby & Sp. I. E.* iii. 110.)—*Jerm.* 2 *edit.* (68), 114.—*Steph. Ill.* (*H.*) i. 42.—*Brit. B.* 16.

Pa. Polychloros. *Linn.* ii. 777.—*Berk. S.* i. 127.—*Lew. pl.* 2.

—-*Don*. viii. 69. *pl.* 278.—(*Haw. Pr.* 2.)—*Haw.* 26.—*Stew.* ii. 126.—*Turt.* iii. 104.—*Jerm.* (30), 51.

Pa. No. 3. *List. Goëd.* (*Angl.*) 3. *pl.* 3.—*Id.* (*Lat.*) 5. *f.* 3.

Pa. Urticariam referens major, &c.: the greater Tortoise-shell B.* *Raii I.* 118. 2.—*Albin. pl.* 55.

Pa. testudinarius major: Great Tortoise-shell B. *Pet. Pap. pl.* IV. *f.* 1. 2.

The great Tortoise-shell B: Pa. Polychloros. (*Wilkes*. 56. *pl.* 108.)

The large Tortoise-shell B. *Harr. V. M.* 5.

γ; *Steph. Ill.* (*H.*) *l. c.*—*Jerm.* 2 *edit. l.* c. 114.

δ; *Steph. Ill.* (*H.*) *l. c.*—*Pap. d'Eur.* i. 8. *pl.* 111. *f.* 3. *d.*

5832. 3; Urticæ*. (*Och.* iv. 17.)—(*Leach E. E.* ix. 128.)—(*Sam. I.* 43.)—(*Curtis l. c. infra.*)—(*Kirby & Sp. I. E.* iii. 101, 110; iv. 223.)—*Jerm.* 2 *edit.* (68), 114.—*Steph. Ill.* (*H.*) i. 43.—*Id.* i. *App.* 148.—*Brit. B.* 17. *pl.* 10. *fig. med.* mala.

Pa. Urticæ. *Linn.* ii. 777.—*Berk. S.* i. 127.—*Barbut* G. 173. *pl.* 10. *f.* 167.—*Don.* ii. 49. *pl.* 55.—*Lew. pl.* 3.—(*Haw. Pr.* 2.)—*Haw.* 26.—*Stew.* ii. 126.—*Wood.* ii. 6. *pl.* 43.—*Jerm.* (30), 51.

Pa. Antennis ex luteo, &c. *Merr. P.* 198. 8.

Pa. diurnus major 11. *Mouff.* 101. *fig. inf.*

Pa. No. 2. *List. Goëd.* (*Angl.*) 2. *f.* 2.—*Id.* (*Lat.*) 3. *f.* 2.

Pa. Urticaria vulgatissima, &c.: the lesser Tortoise-shell B. *Raii I.* 117. 1.

Pa. Testudinarius minor: lesser or common Tortoise-shell B. *Pet. Pap. pl.* IV. *f.* 3. 4.

The lesser Tortoise-shell B. *Albin. pl.* iv. *f.* 6.

The small Tortoise-shell B: Pa. Urticæ. (*Wilkes*. 56. *pl.* 107. *f.* 4.)—*Harr. A. pl.* 2. *f.* *i–n.*—*Harr. V. M.* 5.

The Nettle Tortoise-shell B. *Bing.* iii. 211.

5833. 4; Io*. (*Och.* iv. 17.)—(*Leach E. E.* ix. 128.)—(*Kirby & Sp. I. E.* i. 9.)—(*Sam. I.* 43.)—(*Curtis l. c. infra.*)—*Jerm.* 2 *edit.* (66), 112. *pl.* 4. *f.* 8. neuræ.—*Steph. Ill.* (*H.*) i. 44.—*Id.* i. *App.* 148.—*Brit. B.* 17.

Pa Io. *Linn.* ii. 769.—*Berk. S.* i. 125.—*Barbut G.* 172. *pl.* 10. *f.* 131.—*Lew. pl.* 4.—*Don.* vi. 67. *pl.* 206.—(*Haw. Pr.* 3.)—*Haw.* 17.—*Stew.* ii. 123.—*Turt.* iii. 90.—*Jerm.* (26), 47.

Pa. diurnæ omnium Regina, &c. *Mouff.* 99. 4. *cum fig.* imago.

Pa. No. 1. *List. Goëd.* (*Angl.*) 1. *f.* 1.—*Id.* (*Lat.*) 1. *f.* 1.

Pa. oculus Pavonis dictus. *Pet. Cent.* i. 1.—*Pet. Mus.* 314.

Pa. elegantissima, &c.: the Peacock's Eye. *Raii I.* 122. 14–123. larva.

Pa. oculus Pavonus dictus: the Peacock's Eye. *Pet. Pap. pl.* v. *f.* 1.

The Peacock's Eye. *Albin. pl.* iv. *f.* 5.

The Peacock B: Pa. Io. (*Wilkes*. 55. *pl.* 106.)—*Harr. A. pl.* 8. *f.* *f–k.*—*Harr. V. M.* 5.—*Bing.* iii. 209.

5834. 5: Antiopa*: the White-bordered B. *Haw. l. c.*—(*Och.* iv. 17.)—(*Leach E. E.* ix. 128.)—*Samou.* 238.—*Jerm.* 2 *edit.* (68), 113.—*Steph. Ill.* (*H.*) i. 45.—*Brit. B.* 18.

Pa. Antiopa. *Linn.* ii. 776.—(*Nat. Comp.* (*Lettsom.*) *pl. in front.*)—*Berk. S.* i. 126.—*Don.* iii. 45. *pl.* 89.—*Lew. pl.* 1.—(*Haw. Pr.* 2.)—*Haw.* 27.—*Stew.* ii. 126.—*Turt.* iii. 102.—*Jerm.* (30), 52.

Pa. Morio. *Linn. F. edit.* 1. *No.* 232.

Va. Antiope. *Curtis.* ii. *pl.* 96.

The Willow B: Pa. Antiopa. (*Wilkes.* 58. *pl.* 113.)

Pa. maxime nigra, &c. *Raii I.* 135.

The grand Surprise, or Camberwell Beauty. *Harr. A. pl.* 12. *f. a–e.*—*Harr. V. M.* 5.

β, *Steph. Ill.* (*H.*) *l. c.*—*Pap. d'Eur.* i. 1. *pl.* 1. *f.* 1. *d.*

C. Vide *Steph. Ill.* (*H.*) i. 45.

5835. 6; Atalanta*. (*Och.* iv. 16.)—(*Leach E. E.* ix. 128.)—*Samou.* 238.—(*Curtis l. c. supra.*)—(*Kirby & Sp. I. E.* iii. 84, 114.)—*Jerm.* 2 *edit.* (66), 111.—*Steph. Ill.* (*H.*) i. 46.—*Brit. B.* 18. *pl.* 10. *fig. inf.*

Pa. Atalanta. *Linn.* ii. 779.—*Berk. S.* i. 127.—*Lew. pl.* 7.—*Don.* viii. 19. *pl.* 260.—(*Haw. Pr.* 2.)—*Haw.* 28.—*Stew.* ii. 127.—*Turt.* iii. 103.—*Jerm.* (30), 52.

Pa. diurnæ major, &c. *Mouff.* 100. 6. *cum fig.* imago.

Pa. No. 4, 5. *List. Goëd.* (*Angl.*) 6. *f.* 4–7. *f.* 5.—*Id.* (*Lat.*) 10. *f.* 4–11. *f.* 5.

Pa. major nigrescens, &c.: the Admiral. *Pet. Mus.* 327.—*Pet. Pap. pl.* II. *f.* 11.

Pa. major nigricans, &c.: the Admiral. *Raii I.* 126. 1.

Pa. Ammiralis. *Linn. F.* 1 *edit. No.* 777.

The Admirable. *Albin. pl.* 3. Pa. Atalanta. (*Wilkes.* 55. *pl.* 105.)—*Harr. A. pl.* 6. *f. a–h.*—*Harr. V. M.* 1.

Genus 12: (795.) CYNTHIA, *Fabricius, Jermyn, Steph.*

VANESSA *p, Och., Leach, Samou., Curtis.*—HIPPARCHIA *p, Jerm.*

5836. 1; Cardui*. *Jerm.* 2 *edit.* (66), 109.—*Steph. Ill.* (*H.*) i. 47.—*Id.* i. *App.* 148.—*Brit. B.* 19. *pl.* 11.

Pa. Cardui. *Linn.* ii. 774.—*Berk. S.* i. 126.—*Lew. pl.* 6.—*Don.* ix. 9. *pl.* 292.—(*Haw. Pr.* 3.)—*Shaw N. M.* xi. *pl.* 430.—*Haw.* 20.—*Stew.* ii. 125.—*Turt.* iii. 96.

Va. Cardui. (*Leach E. E.* ix. 128.)—*Samou.* 238.—(*Kirby & Sp. I. E.* iii. 261.)—*Jerm.* (28), 48.—(*Curtis l. c. fo.* 96.)

Pa. diurnæ, &c. *Mouff.* 101. *cum fig.* imago.

Pa. No. 6. *List. Goëd.* (*Angl.*) 8. *f.* 6.—*Id.* (*Lat.*) 14. *f.* 6.

Pa. eleganter, &c. *Bella donna* dicta: Painted Lady. *Pet. Mus.* 326.—*Pet. Pap. pl.* IV. *f.* 21. 22.—*Albin. pl.* 56.

Pa. major pulchra, &c.: the Painted Lady. *Raii I.* 122. 13.

The Painted Lady B: Pa. Cardui. (*Wilkes*. 56. *pl*. 107. *f*. 1.) *Harr. A. pl.* 11. *f. a–f.*—*Harr. V. M.* 5.

5837 † 2. Hampstediensis*. *Steph. Ill.* (*H.*) i. 48. *pl*. 5. *f*. 3. 4.—*Brit. B.* 20.

Pa. oculatus Hampstediensis ex aureo fuscus: Albins' Hampstead Eye. *Pet. Pap. pl.* v. *f*. 2.—*Haw.* 54.

Hi. Hampstediensis. *Jerm.* (28), 50.—*Jerm.* 2 *edit.* (70), 129.

5838 ‡ 3. Levana[a].

Genus 13: (796). APATURA, *Fabricius, Leach, Samou.,* (*Kirby,*) *Jerm., Steph.*

Argus *p*, *Scop.*—Maniola *p*, *Schra.*—Potamides, *Hübn.*

5839. 1; Iris*. (*Och.* iv. 19.)—(*Leach E. E.* ix. 129.)—*Samou.* 239. —(*Kirby & Sp. I. E.* iii. 114; iv. 500 & 517.)—*Jerm.* 2 *edit.* (68), 117. *pl*. 4. *f*. 9. neuræ.—*Steph. Ill.* (*H*). i. 50.—*Brit. B.* 20. *pl*. 12.

Pa. Iris. *Linn.* ii. 775.—*Berk. S.* i. 126.—*Don.* ii. 1. *pl*. 37. ♂. —*Lew. pl.* 16.—(*Haw. Pr.* 3.)—*Haw.* 18.—*Stew.* ii. 125.—*Turt.* iii. 99.—*Jerm.* (28), 48.

Pa. major nigra, &c. *Raii I.* 126. 2.

The Purple Highflyer, or Emperor of the Woods: Pa. Iris. (*Wilkes*. 63. *pl*. 120.)

The Purple Emperor B. *Harr. A. pl.* 3. *fig. sup.*—*Harr. V. M.* 1. —*Bing.* iii. 207.

Genus 14: (797). LIMENITIS, *Fabricius, Leach, Samou.,* (*Kirby,*) *Jermyn, Curtis, Steph.*

Najades, *Hübner.*

5840 ‡ 1. Populi[b].

5841. 2; Camilla*. (*Leach E. E.* ix. 129.)—(*Sam. I.* 25.)—*Curtis.* iii. *pl*. 124.—*Jerm.* 2 *edit.* (68), 120. *pl*. 5. *f*. 10. neuræ.—*Steph. Ill.* (*H.*) i. 52.—*Id.* i. *App.* 148.—*Brit. B.* 21. *pl*. 13.

Pa. Camilla. *Linn.* ii. 781.—*Lew. pl.* 8.—*Don.* viii. 75. *pl*. 244. —(*Haw. Pr.* 2.)—*Haw.* 29.—*Turt.* iii. 38.—*Jerm.* (30, 52.)

Pa. Sibilla. *Fabr.* iii. *a*. 246.—*Stew.* ii. 127.

Pa. mediæ mag. elegantissima, &c. *Raii I.* 127. 3.

Pa. fusca: White Admiral. *Pet. Pap. pl.* II. *f*. 12.

The White Admirable. *Harr. A. pl.* 30. *f. m, n.*—*Harr. V. M.* 1.

† β. *Steph. Ill.* (*H.*) i. *App. l. c.*—*Pap. d'Eur.* i. 30. *pl*. xi. *f*. 13. *e, f*. (!)

---

[a] 5838 ‡ 3. Levana. *Steph. Ill.* (*H.*) i. 49. nota.
Pa. Levana. *Linn.* ii. 783.—*Hüb. Pa. pl.* 20. *f*. 97. 98.—*Turt.* iii. 42. (!)—*Rees' Cycl.* (*Donovan.*) *article* Papilio. (!)—*Jerm.* 62. (!)
Va. Levana. (*Och.* iv. 17.)—*Jerm.* 2 *edit.* (80), 165.

[b] 5840 ‡ 1. Populi. (*Och.* iv. 17.)—*Jerm.* 2 *edit.* (80), 165.—*Steph. Ill.* (*H.*) i. 51. nota.
Pa. Populi. *Linn.* ii. 776.—*Hüb. Pa. pl.* 103. *f*. 108–10.—*Stew.* ii. 125. (!)—(*Haw. p.* xxix.) (!)
Pa. Tremulæ. *Esper. Pap. pl.* 114. *f*. 4.

5842 ‡ 3. Sibilla[a].

Genus 15: (798). HIPPARCHIA, *Fabr.*, *Leach*, *Samou.*, (*Kirby*,) *Jermyn*, *Steph.*, *Curtis*.

ARGUS *p*, *Scop.*—MANIOLA *p*, *Schrank.*—OREADES, *Hübner.*—SATYRUS *et* NYMPHALIS *p*, *Latreille.*

A. Oculi pubescentes.

5843. 1; Ægeria*. (*Leach E. E.* ix. 129.)—(*Sam. I.* 21.)—(*Kirby & Sp. I. E.* iii. 97.)—*Jerm.* 2 *edit.* (70), 128.—*Steph. Ill.* (*H.*) i. 54.—(*Curtis l. c. infra.*)—*Brit. B.* 22.
Pa. Ægeria. *Linn.* ii. 771.—*Berk. S.* i. 125.—*Lew. pl.* 19.—*Don.* xiv. 77. *pl.* 498.—(*Haw. Pr.* 3.)—*Haw.* 23.—*Turt.* iii. 92.—*Stew.* ii. 123.—*Jerm.* (28), 49.
Hi. Egeria. (*Och.* iv. 21.)
Pa. cum corpore, &c. *Merr. P.* 198. 2.
Pa. media pulla, &c. *Raii I.* 128. 5.
Pa. oculatus e fusco, &c.: the Enfield Eye. *Pet. Gaz. pl.* 24. *f.* 3.—*Pet. Pap. pl.* v. *f.* 5. ♀.
Pa. oculatus, obscurior, &c.: Brown Enfield Eye. *Pet. Pap. pl.* v. *f.* 6. ♂.
Wood Argus: Pa. Ægeria. (*Wilkes.* 53. *pl.* 103.)
The Speckled Wood B. *Harr. A. pl.* 41. *f. f*, *K.*—*Harr. V. M.* 7.

5844 ‡ 2. Mæra[b].

5845. 3; Megæra*. (*Och.* iv. 21.)—(*Leach E. E.* ix. 129.)—(*Sam. I.* 21.)—*Jerm.* 2 *edit.* (70), 127.—*Steph. Ill.* (*H.*) i. 55.—(*Curtis l. c. infra.*)—*Brit. B.* 22. *f.* 14. *fig.* 2.
Pa. Megæra. *Linn.* ii. 771.—*Lew. pl.* 21.—*Don.* viii. 71. *pl.* 279. ♂.—(*Haw. Pr.* 3.)—*Haw.* 22.—*Turt.* iii. 92.—*Stew.* ii. 123.—*Jerm.* (28), 49.
Pa. Mæra. *Berk. S.* i. 125.
Pa. oculo nigro, &c. *Merr. P.* 198. 6.
Pa. oculatus ex aureo et fusco marmoreatus: the London Eye. *Pet. Mus.* 312.—*Pet. Pap. pl.* v. *f.* 7. ♀.
Pa. idem area fusca: London Eye with a brown List. *Pet. Pap. pl.* v. *f.* 8. ♂.
Pa. media, &c.: the golden marbled B. with black eyes. *Raii I.* 123. 15.

---

[a] 5842 ‡ 3. Sibilla. *Jerm.* 2 *edit.* (80), 165.—*Steph. Ill.* (*H.*) i. 52. nota.
Pa. Sibilla. *Linn.* ii. 781.
Pa. Camilla. *Fabr.* iii. *a.* 246.—*Hüb. Pa. pl.* 22. *f.* 106, 107.—*Stew.* ii. 127. (!)

[b] 5844 ‡ 2. Mæra. (*Och.* iv. 21.)—*Jerm.* 2 *edit.* (80), 166.—*Steph. Ill.* (*H.*) i. 51. nota.
Pa. Mæra. *Linn.* ii. 771.—*Hüb. Pa. pl.* 39. *f.* 174, 175.—*Stew.* ii. 123.—*Turt.* iii. 30.—*Rees' Cycl.* (*Donovan.*) *Art.* PAPILIO.—*Jerm.* (63.)

The Great Argus: Pa. Mæra. (*Wilkes*. 53. *pl*. 102.)
The Wall B. *Harr. A. pl*. 27. *f. a–g*.—*Harr. V. M*. 7.
β, *Steph. Ill*. (*H*.) *l. c*.—*Pap. d'Eur*. i. 261. *pl*. LXV. *Sup*. xi. *f*. 50. *g*, *h*.

B—E. Oculi nudi, &c. Vide *Steph. Ill*. (*H*.) i. 56. 57. 58. 60. 61. 63. 64.

5846 ‡ 4. Phædra[a].

5847 † 5. Alcyone[b].

5848. 6; Semele*. (*Och*. iv. 20.)—(*Leach E. E*. ix. 129.)—(*Sam. I*. 21.)—*Jerm*. 2 *edit*. (70), 126.—*Steph. Ill*. (*H*.) i. 56.—(*Curtis l. c. infra*.)—*Brit. B*. 3.
Pa. Semele. *Linn*. ii. 777.—*Berk. S*. i. 126.—*Lew. pl*. 17.—*Don*. viii. 17. *pl*. 259. ♀.—(*Haw. Pr*. 3.)—*Haw*. 21.—*Turt*. iii. 32.—*Stew*. ii. 124.—*Jerm*. (28), 49.
Pa. oculis nigris, subtus marmoreus: the Tunbridge Grayling. *Pet. Mus*. 307.—*Pet. Pap. pl*. v. *f*. 3. ♀.
Pa. idem obscurior: Brown Tunbridge Grayling. *Pet. Pap. pl*. v. *f*. 4. ♂.
Pa. majuscula, alis pullis, &c. *Raii I*. 128. 6.
The Grayling B. *Harr. A. pl*. 44. *f. d, e*.—*Harr. V. M*. 3.

5849. 7; Galathea*. (*Leach E. E*. ix. 128.)—(*Sam. I*. 21.)—*Jerm*. 2 *edit*. (70), 127. *pl*. 4. *f*. 11. neuræ.—*Steph. Ill*. (*H*.) i. 57.—(*Anon. Rem. L. G*. 21. *pl*. 3.)—(*Curtis l. c. infra*.)—*Brit. B*. 23. *pl*. 14. *fig. sup*.
Pa. Galathea (Galatea). *Linn*. ii. 772.—*Berk. S*. i. 125.—*Lew. pl*. 28.—*Don*. viii. 15. *pl*. 258. ♂.—(*Haw. Pr*. 3.)—*Haw*. 22.—*Turt*. iii. 35.—*Stew*. ii. 124.—*Jerm*. (28), 49.
Pa. capite alisque lacteis, &c. *Merr. P*. 198. 9.
Pa. leucomelanos, &c.: our Half-mourner. *Pet. Mus. C*. 1. 3.—*Pet. Pap. pl*. 2. *f*. 10.
Pa. med. mag. alis albo et nigro, &c.: the Half-mourner. *Raii I*. 116. 9.
The Marmoris, or Marbled B: Pa. Galathæa. (*Wilkes*. 52. *pl*. 100.)
The Marbled White or Marmoress. *Harr. A. pl*. 11. *f. g–k*.—*Harr. V. M*. 7.
Var. Le demi-deuil. *Pap. d'Eur*. i. 135. *pl*. xxx. *f*. 60. (!)

5850. 8; Tithonus*: large Heath.—*Haw. l. c*. (*Och*. iv. 21.)—*Jerm*. 2 *edit*. (70), 131.—*Steph. Ill*. (*H*.) i. 58.—(*Curtis l. c. infra*.)—*Brit. B*. 24.

---

[a] 5846 ‡ 4. Phædra. *Jerm*. 2 *edit*. (80), 165.—*Steph. Ill*. (*H*.) i. 56. nota. Pa. Phædra. *Linn*. ii. 773.—*Hüb. Pa. pl*. 28. *f*. 127–9.—*Turt*. iii. 33. (!)—*Rees' Cycl*. (*Donovan*.) *Art*. Papilio. (!)—*Jerm*. 62.

[b] 5847 † 5. Alcyone. *Jerm*. 2 *edit*. (70), 132.—*Steph. Ill*. (*H*.) i. 56. nota. Pa. Alcyone. *Esper. pl*. 84. *f*. 4.—*Stew*. 2 *edit*. ii. (!)—*Jerm*. (40), 61.

Pa. Tithonus. *Linn. Mant.* i. 537.—*Fabr. Sp.* 80.—*Lew. pl.* 22.
Pa. Tithonius. *Vill. E.* ii. 26.
Pa. Herse. *Hüb. Pa. pl.* 35. *f.* 156, 157.—*pl.* 119. *f.* 612.
Pa. Phædra. *Esper. Pa.* i. *pl.* 9. *f.* 1.—*pl.* 28. *f.* 3.
Pa. Pilosellæ. *Fabr.* iii. *a.* 240.—(*Haw. Pr.* 3.)—*Haw.* 24.—*Don.* xii. 33. *pl.* 405.—*Jerm.* (28), 51.
Hi. Pilosellæ. (*Leach E. E.* ix. 129.)—(*Sam. I.* 21.)
Pa. Amaryllis. *Bork. E. S.* i. 238.
Pa. corpore gruino, &c. *Merr. P.* 199. 5.
Pa. aureo-fuscus minor oculatus &c.: Hedge Eye with double specks. *Pet. Mus.* 310.—*Pet. Pap. pl.* v. *f.* 12. ♀.—*f.* 11. ♂.
Pa. media ex fulvo seu rufo &c. *Raii I.* 124. 18.
The Gate Keeper. *Harr. A. pl.* 44. *f. f, g.*—*Harr. V. M.* 3.

5851. 9; Janira*. (*Och.* iv. 21.)—*Jerm.* 2 *edit.* (70), 129.—*Steph. Ill.* (*H.*) i. 59.—(*Curtis l. c. infra.*)—*Brit. B.* 24.
Pa. Jurtina. *Lew. pl.* 18.—*Don.* ix. 69. *pl.* 320.—(*Haw. Pr.* 3.)—*Haw.* 24.—*Jerm.* (28), 50.—(*Kirby & Sp. I. E.* iii. 98.)
Hi. Janira. (*Leach E. E.* ix. 129.)—(*Sam. I.* 21.)
The Meadow Brown B.: Pa. Hyperanthus. (*Wilkes.* 53. *pl.* 101.) *Albin. pl.* 53. *f. a–e.*
The Meadow Brown. *Harr. A. pl.* 32. *f. a–e.*—*Harr. V. M.* 1.
♂; Pa. Janira. *Linn.* ii. 1053.—*Turt.* iii. 36.—*Stew.* ii. 124.
Pa. ultima parte alæ &c. *Merr. P.* 198. 10.
Pa. pratensis oculatus fuscus: Brown Meadow Eye. *Pet. Mus.* 309.—*Pet. Pap. pl.* v. *f.* 10.
Pa. media alis supina facie pullis &c. *Raii I.* 124. 16.
♀; Pa. Jurtina. *Linn.* ii. 1052.—*Berk. S.* i. 126.—*Stew.* ii. 124.
Pa. pratensis oculatus aureus: Golden Meadow Eye. *Pet. Mus.* 308.—*Pet. Pap. pl.* v. *f.* 9.
Pa. media alis superioribus supernè media parte rufis &c. *Raii I.* 124. 17.
β; *Steph. Ill.* (*H.*) *l. c.*—*Pap. d'Eur.* i. 263. *pl.* lxvi. *Sup.* xii. *f.* 54. *l, m.*

5852. 10, Ligea. (*Och.* iv. 23.)—(*Ing. Inst.* 91.)—*Jerm.* 2 *edit.* (70), 132.—*Steph. Ill.* (*H.*) i. 61. *pl.* 6. *f.* 1. ♂. 2. 3. ♀.—(*Curtis l. c. infra.*)—*Brit. B.* 25.
Pa. Ligea. *Linn.* ii. 772.—(*Haw. p.* xxix.)—*Sower. B. M.* i. *pl.* 2.—*Jerm.* (40), 61.
Pa. Alexis. *Esper. Pa. pl.* 44. *f.* 1.
Pa. Philomela. *Hüb. Pa. pl.* 47. *f.* 218, 219.

5853. 11, Blandina: Scotch Argus B.—(*Leach E. E.* ix. 129.)—(*Sam. I.* 21.)—*Jerm.* 2 *edit.* (70), 133.—*Steph. Ill.* (*H.*) i. 62.—(*Curtis l. c. infra.*)—*Brit. B.* 25.
Pa. Blandina. *Fabr.* iii. *a.* 236.—*Sower. B. M.* i. *pl.* 7?—*Don.* xii. 87. *pl.* 426.—*Jerm.* (40), 61.
Pa. Æthiops. *Esper. Pa.* 1. *pl.* 25. *f.* 3.
Pa. Medusa. *Bork. E. S.* i. 235.

Pa. Medea. *Hüb. Pa. pl.* 48. *f.* 220–222.
♀, Pa. Alcyone (non descriptio). *Stew.* 2 *edit.* ii.
Pa. Amaranthus. *Walker's MSS.*

5854. 12, Cassiope. (*Och.* iv. 22.)—*Steph. Ill.* (*H.*) i. 63. *pl.* 8. *f.* 1. 3. ♂. 2. ♀.—(*Curtis l. c. infra.*)—*Brit. B.* 26. *pl.* 14. *f.* 3.
Pa. Cassiope. *Fabr.* iii. *a.* 238.
Pa. Mnemon: Small Ringlet. *Ent. Trans.* (*Haworth.*) i. 332. —*Jerm.* (40), 61.
Hi. Mnemon. *Jerm.* 2 *edit.* (70), (132).—(*Ing. Inst.* 91.)
Pa. Æthiops minor. *Vill. E.* ii. 37.
Pa. Melampus var. *Esper. Pa. pl.* 78. *f.* 2.

5855 † 13, Mnestra[a].

5856. 14; Hyperanthus*. (*Och.* iv. 21.)—(*Leach E. E.* ix. 128.) —(*Sam. I.* 21.)—(*Kirby & Sp. I. E.* iii. 98.)—*Jerm.* 2 *edit.* (68), 122.—*Steph. Ill.* (*H.*) i. 60.—(*Curtis l. c. infra.*)—*Brit. B.* 24.
Pa. Hyperanthus. *Linn.* ii. 768.—*Barbut G.* 171. *pl.* 10. *f.* 127. —*Berk. S.* i. 125.—*Lew. pl.* 20.—*Don.* viii. 53. *pl.* 271.— (*Haw. Pr.* 2.)—*Haw.* 14.—*Stew.* ii. 122.—*Turt.* iii. 25.— *Jerm.* (26), 46.
Pa. Polymeda. *Scop. C. No.* 434.—*Hüb. Pa. pl.* 38. *f.* 172, 173.
Pa. Arete. *Rhein. M.* (*Bork.*) i. 145.
Pa. dorso nigro &c. *Merr. P.* 198. 4.
Pa. medius omnino fuscus &c. *Pet. Mus.* 313.—*Pet. Pap. pl.* v. *f.* 13, 14.
Pa. media tota pulla &c. *Raii I.* 129. 7.
Pa. Tristan. *DeGeer.* G. 33. 21.
The Ringlet. *Harr. A. pl.* 35. *f. f, h.*—*Harr. V. M.* 5.
Var. Pa. Polymeda. *Hüb. Pa. l. c. f.* 173*.

5857. 15, Polydama. *Steph. Ill.* (*H.*) i. *App.* 149.
Pa. Iphis. (*Wien. V.* 321?)—*Rhein. Mag.* (*Borkhausen.*) i. 241.
Hi. Iphis. *Jerm.* 2 *edit.* (78), 158.—*Steph. Ill.* (*H.*) i. 64. *pl.* vii. *f.* 1. 2.—*Brit. B.* 26.
Pa. Tiphon. *Esper. Pa.* i. *pl.* 35. *f.* 4.
Pa. Typhon: Scarce Heath. (*Haw. Pr.* 2.)—*Haw.* 16.— *Jerm.* (26), 47.
Hi. Typhon. (*Sam. I.* 22.)—*Jerm.* 2 *edit.* (68), 124.—(*Curtis l. c. infra.*)
Pa. Hero. *Fabr.* iii. *a.* 222?—*Turt.* iii. 27.—*Stew.* ii. 122.
Var. Pa. Laidion. *Bork. E. S.* i. 91.
Var. Pa. Polydama: Marsh Ringlet. *Haw.* 16.—*Jerm.* (26).— *Brit. B.* 27.
Pa. Polymeda. *Jerm.* 47.
Hi. Polymeda. *Jerm.* 2 *edit.* (68), 124.

---

[a] *5855* † 13. Mnestra. (*Och.* iv. 22.)—*Jerm.* 2 *edit.* (78), 158.—*Steph. Ill.* (*H.*) i. 63. nota.
Pa. Mnestra. *Hüb. Pa. pl.* 106. *f.* 540–543.

5858. 16, Davus: Small Ringlet.—*Haw. l. c.* (*Och.* iv. 23.)—(*Sam. I.* 21.)—*Jerm. 2 edit.* (68), 123.—*Steph. Ill.* (*H.*) i. 67.—*Id.* i. *App.* 149.—(*Curtis l. c. infra.*)—*Brit. B.* 27.
Pa. Davus. *Fabr.* iii. *a.* 221.—*Haw.* 15.—*Jerm.* (26), 47.
Pa. Tullia. *Hüb. Pa. pl.* 52. *f.* 243, 244.
Pa. Philoxenus. *Esper. Pa.* i. *pl.* 54. *f.* 3.
Pa. Musarion. *Bork. E. S.* i. 244.
Ma. Tiphon. *Schra. B.* ii. 179.
Pa. Hero. *Don.* vi. 17. *pl.* 186.—*Lew. pl.* 23. *f.* 5. 6.—(*Haw. Pr.* 2.)

5859. 17, Hero*? Silver-bordered Ringlet.—*Haw. l. c.* Scarce Meadow brown B.—*Jerm. l. c.* (*Och.* iv. 23.)—*Jerm. 2 edit.* (70), 131.—(*Ing. Inst.* 91.)—*Steph. Ill.* (*H.*) i. 68.—*Id.* i. *App.* 149.—*Curtis.* v. *pl.* 205.—*Brit. B.* 28.
Pa. Hero. *Linn.* ii. 793.—(*Ent. Trans.* (*Haworth.*) i. 332.)—*Rees' Cycl.* (*Donovan.*) *Art.* PAPILIO.—*Jerm.* 61.
Pa. Sabæus. *Fabr.* iii. *a.* 222.
Le. Mœlibee. *Pap. d'Eur.* i. 132. *pl.* xxix. *f.* 59. *a. b.* (!)

5860 ‡ 18, Arcanius. *Curtis. pl.* 205*.—*Brit. B.* 28.
Pa. Arcanius. *Linn.* ii. 791.—*Turt.* iii. 27.—*Rees' Cycl.* (*Donovan.*) *Art.* PAPILIO.—*Jerm.* 61.
Hi. Arcanius. *Jerm. 2 edit.* (78), 158.—*Steph. Ill.* (*H.*) i. 69.
Pa. Arcania. *Hüb. Pa. pl.* 51. *f.* 240–242. In Mus. D. *Curtis.*

5861. 19; Pamphilus*. (*Och.* iv. 23.)—(*Leach E. E.* ix. 128.)—(*Sam. I.* 21.)—*Jerm.* (68), 124.—*Steph. Ill.* (*H.*) i. 69.—(*Curtis l. c. supra.*)—*Brit. B.* 28. *pl.* 14. *fig. inf.*
Pa. Pamphilus. *Linn.* ii. 791.—*Berk. S.* i. 129.—*Lew. pl.* 23. *f.* 3, 4.—(*Haw. Pr.* 2.)—*Haw.* 17.—*Turt.* iii. 27.—*Stew.* ii. 622.—*Jerm.* (26), 47.
Pa. Nephele. *Hüb. Pa. pl.* 51. *f.* 237–239.
Pa. alis omnibus pallescente flavo &c. *Merr. P.* 199. 8.
Pa. parva &c.: the small heath B. *Raii I.* 125. 19.
Papiliunculus aureus oculatus: Golden heath Eye. *Pet. Pap. pl.* v. *f.* 15. ♀.
Pa. idem margine fusco: Selvedged heath Eye. *Pet. Pap. pl.* v. *f.* 16. ♂.
The small Heath, or Gate Keeper. *Harr. A. pl.* 21. *f. e–h.*—*Harr. V. M.* 3.

Familia III: (116). LYCÆNIDÆ, *Leach.*

(PLEBEII RURALES, *Linné.*—POLYOMMATIDÆ, *Swainson.*)

(VERMIFORM Stirps, *Horsfield.*)

Genus 16: (799). THECLA, *Fabr., Leach, Samou., Jermyn, Steph.*

POLYOMMATUS *p*, *Latr.*—LYCÆNA *p*, *Och.*—HESPERIA *p*, *Fabr.*—RUSTICI *p*, *Hübn.*—CUPIDO *p*, *Schrank.*

5862. 1; Betulæ*. (*Leach E. E.* ix. 129.)—(*Sam. I.* 40.)—

Jerm. 2 *edit.* (72), 133.—*Steph. Ill.* (*H.*) i. 75.—*Id.* i. *App.* 149. *Brit. B.* 29. *pl.* 15. *fig. sup.*

Pa. Betulæ. *Linn.* ii. 787.—*Barbut* G. 173. *pl.* 10. *f.* 220.—*Berk. S.* i. 129.—*Lew. pl.* 42.—*Don.* viii. 89. *pl.* 250. ♂.—(*Haw. Pr.* 4.)—*Haw.* 37.—*Turt.* iii. 131.—*Stew.* ii. 130.—*Jerm.* (34), 56.

Pa. minor fuscus campo aureo &c. *Pet. Gaz. pl.* 11. *f.* 11. ♀.

Pa. minor alis exterioribus nigricantibus &c. *Raii I.* 130. 10.

Pa. minor fuscus &c: the Brown Hair Streak. *Pet. Pap. pl.* IIII. *f.* 23. 25. ♂.

Pa. minor fuscus agro aureo: Golden Hair Streak. *Pet. Pap. pl.* IIII. *f.* 24. ♀.

The Hair Streak B. *Albin. pl.* 5. *f.* 7.

The Brown Hair Streak B: Pa. Betulæ. (*Wilkes.* 61. *pl.* 117.) —*Harr. A. pl.* 42. *f. b, c, f, g.*—*Harr. V. M.* 5.

5862*. 2, Pruni. (*Steph. Ill.* (*H.*) ii. 69. *nota.*) Vide *Appendix.*

Pa. Pruni. *Linn.* ii. 788.—*Hüb. Pa. pl.* 76. *f.* 386, 387.

Pa. Prorsas. *Berl. Mag.* (*Hufnagle.*) ii. 68.

Th. Spini. *Brit. B.* 30.

5863. 3; W. album*. (*Steph. Ill.* (*H.*) ii. 69. *nota.*)

Pa. W. album. *Villers E.* ii. 83. *pl.* iv. *f.* 12.

Pa. Pruni. *Bergstraesser Nom. pl.* 75. *f.* 1. 2.—*Lewin. pl.* 44. —(*Haw. Pr.* 41.)—*Haw.* 38.—*Turt.* iii. 131.—*Don.* xiii. 9. *pl.* 437.—*Jerm.* (34), 56.

Th. Pruni. (*Leach E. E.* ix. 129.)—(*Sam. I.* 40.)—*Jerm.* 2 *edit.* (72), 134.—*Steph. Ill.* (*H.*) i. 77.—*Brit. B.* 30.

The Dark or Black Hair Streak B. *Harr. V. M.* 5.

5864 † 4, Spini. *Jerm.* 2 *edit.* (78), 159.—*Steph. Ill.* (*H.*) i. 78.—*Brit. B.* 30.

He. Spini. *Fabr.* iii. *a.* 278.—*Hüb. Pa. pl.* 75. *f.* 376–377.

Pa. Spini: Pale Brown Hair-Streak. *Ent. Trans.* (*Haworth.*) i. 334.

Pa. Lynceus. *Esper. Pa.* i. *pl.* xxxix. *f.* 3.

In Mus. D. *Haworth* et Sparshall.

5865. 5; Quercus*. (*Leach E. E.* ix. 129.)—(*Sam. I.* 41.)—*Jerm.* 2 *edit.* (72), 135. *pl.* 5. *f.* 12. neuræ.—*Steph. Ill.* (*H.*) i. 76. —*Id.* i. *App.* 149.—*Brit. B.* 29. *pl.* 15. *fig. inf.*

Pa. Quercus. *Linn.* ii. 788.—*Berk. S.* i. 129.—*Lew. pl.* 43.—(*Haw. Pr.* 4.)—*Haw.* 38.—*Turt.* iii. 131.—*Stew.* ii. 130.—*Don.* xiii. 57. *pl.* 460.—*Jerm.* (34), 57.

The Purple Hair Streak B: Pa. Quercus. (*Wilkes.* 61. *pl.* 116.) —*Harr. A. pl.* 10. *f. a–g.*—*Harr. V. M.* 5. *Albin. pl.* 52. *f. a–c.*

♂; Pa. minor fuscus agro cæruleo: our Blew Hair Streak. *Pet Gaz. pl.* 11. *f.* 9.—*Pet. Pap. pl.* IIII. *f.* 27.

Pa. minor, supina facie nigricante &c. *Raii I.* 129. 7.

♀; Pa. minor cærulescens &c.: Ray's Blew Hair Streak. *Pet. Mus.* 319.—*Pet. Pap. pl.* IIII.*f.* 26, 28. ♀.
Pa. è mediis miniuscula &c. *Raii I.* 130. 8.

5866. 6; Rubi*: Green Hair Streak.—*Haw. l. c.* (*Leach E. E.* ix. 129.)—*Steph. Ill.* (*H.*) i. 78.—*Id.* i. *App.* 149.—*Brit. B.* 30.
Pa. Rubi. *Linn.* ii. 791.—*Berk. S.* i. 129.—*Lew. pl.* 44.—(*Haw. Pr.* 4.)—*Haw.* 39.—*Turt.* iii. 135.—*Stew.* ii. 132.—*Don.* xiii. 21. *pl.* 443.—*Jerm.* (34), 57.
Th. Rubi. (*Sam. I.* 41.)—*Jerm. 2 edit.* (72), 135.
Pa. alis cæsiis, &c. *Merr. P.* 199. 2.
Pa. parva alis supinis pullis &c. *Raii I.* 133. 22.
Pa. minor supernè fuscus &c: Holly B. *Pet. Pap. pl.* vi. *f.* 13.
The green B: Pa. Rubi. (*Wilkes.* 62. *pl.* 118.)
The green Fly, or Bramble F. *Harr. A. pl.* 26. *f. a, b, d, g.*—*Harr. V. M.* 1.
Albin. pl. 5. *f.* 8.

Genus 17: (800). LYCÆNA, *Fabr.*, *Leach*, *Samou.*, (*Kirby*,) *Jermyn*, *Curtis*, *Steph.*

POLYOMMATUS *p*, *Latr.*—HESPERIA *p*, *Fabr.*—ARGUS *p*, *Scop.* —CUPIDO *p*, *Schrank.*

5867. 1; Phlæas*. (*Leach E. E.* ix. 129.)—(*Sam. I.* 26.)—*Jerm. 2 edit.* (72), 140.—*Steph. Ill.* (*H.*) i. 79.—*Brit. B.* 31. *pl.* 16. *fig. sup.*
Pa. Phlæas. *Linn.* ii. 793.—*Berk. S.* i. 130.—*Lew. pl.* 41.—(*Haw. Pr.* 4.)—*Haw.* 42.—*Turt.* iii. 146.—*Stew.* ii. 132.—*Don.* xiii. 69. *pl.* 466.—*Jerm.* (34), 57.
Pa. minor aureus, ex nigro permaculatus: Small Tortoise Shell B. *Pet. Mus.* 317.—*Pet. Pap. pl.* IIII. *f.* 13. 14.
Pa. parva alis exterioribus circa margines nigricantibus. *Raii I.* 125. 10.
The Copper B. *Harr. A. pl.* 34. *f. p.*—*Harr. V. M.* 1.
β, *Steph. Ill.* (*H.*) *l. c.*—*Pap. d'Eur.* i. 277. *pl.* lxxii. *Sup.* xviii. *f.* 91. *e.*
γ, *Steph. Ill.* (*H.*) *l. c.*—*Pap. d'Eur.* i. 186. *pl.* xliii. *f.* 91. *a.*
† η, *Steph. Ill.* (*H.*) *l. c.*—*Pap. d'Eur.* i. 278. *pl.* lxxii. *Sup.* xviii. *f.* 91. *g. h.* η in Mus. D. Hatchett? *et Kirby.*

5868. 2, Chryseis*. (*Leach E. E.* ix. 129.)—(*Sam. I.* 26.)—*Jerm. 2 edit.* (72), 139.—*Steph. Ill.* (*H.*) i. 80.—*Brit. B.* 31.
He. Chryseis. *Fabr.* iii. *a.* 309.
Pa. Chryseis: Purple-edged Copper. *Haw.* 41.—*Sowerb. B. M.* i. *pl.* 13.—*Jerm.* (34), 57.
Pa. Hippothoë var. *Esper. pl.* 62. *f.* 1.
Pa. Eurydice. *Bork. E. S.* i. 143.
Pa. cum alis externis coccineis &c. *Merr. P.* 199. 1.

5869. 3: dispar: Large Copper.—*Haw. l. c.* (*Leach E. E.* ix. 129.) —(*Sam. I.* 26.)—*Curtis.* i. *pl.* 12.—*Jerm.* 2 *edit.* (78).—*Steph. Ill.* (*H.*) i. 81.—*Brit.-B.* 32. *pl.* 16. *fig. inf.*

Pa. dispar. (*Haw. Pr.* 3.)—*Haw.* 40.—(*Kirby & Sp. I. E.* i. *pl.* 3. *f.* 1. ♂.)—*Jerm.* (34), 57.

Pa. Hippothoë var. *Esper. Pa. pl.* xiv. *f.* 1, 2.

Pa. Hippothoë. *Lewin. pl.* 40.—*Don.* vii. 1. *pl.* 217.—*Turt.* iii. 145.—*Stew.* ii. 133.

Ly. Hippothoë. *Jerm.* 2 *edit.* (72), 136. *pl.* 5. *f.* 13. neuræ.

5870. 4, Hippothoë: Dark underwinged Copper.—*Haw. l. c. Steph. Ill.* (*H.*) i. 82.—*Brit. B.* 32.

Pa. Hippothoë. *Linn.* ii. 793.—(*Ent. Trans.* (*Haworth.*) i. 333.) —*Rees' Cycl.* (*Donovan.*) *Art.* PAPILIO.—*Jerm.* 62.

Roës. I. B. ii. *pl.* 37. *f.* 6, 7. ♂.

5871. 5, Virgaureæ: Scarce Copper.—*Haw. l. c.*—(*Leach E. E.* ix. 129.)—(*Sam. I.* 26.)—(*Kirby & Sp. I. E.* iv. 503.)—*Jerm.* 2 *edit.* (72), 138.—*Steph. Ill.* (*H.*) i. 83. *pl.* ix. *f.* 1, 2. ♂. *f.* 3. ♀. —*Brit. B.* 32.

Pa. Virgaureæ. *Linn.* ii. 793.—*Don.* v. 93. *pl.* 173. ♂.—*Lew. pl.* 41. *f.* 1, 2.—(*Haw. Pr.* 4.)—*Haw.* 41.—*Turt.* iii. 145.—*Stew.* ii. 132.—*Jerm.* (34), 57.

Genus 18: (801). POLYOMMATUS, *Latr.*, *Jermyn*, *Steph.*, *Horsfield.*

ARGUS *p*, *Scopoli.*—HESPERIA *p*, *Fabr.*—CUPIDO *p*, *Schrank.*

A. Vide *Steph. Ill.* (*H.*) i. 84. PITHECOPS, *Horsfield.*

5872. 1; Argiolus*. *Jerm.* 2 *edit.* (74), 147.—*Steph. Ill.* (*H.*) i. 85.—*Id.* i. *App.* 149.—*Brit. B.* 33.

Pa. Argiolus. *Linn.* ii. 790.—*Berk. S.* i. 129.—*Lew. pl.* 36. *f.* 4, 5, 6.—(*Haw. Pr.* 4.)—*Haw.* 47.—*Turt.* iii. 138.—*Stew.* ii. 132.—*Don.* xiv. 39. *pl.* 481.—*Jerm.* (36), 59.

Ly. Argiolus. (*Leach E. E.* ix. 130.)—(*Sam. I.* 26.)

Pa. Acis. *Hüb. Pa. pl.* 57. *f.* 272–4.

Pa. Cleobis. *Esper. Pa.* i. *pl.* xl. *Supp.* xvi. *f.* 3.

Pa. minor, alis cæruleis &c. *Raii I.* 132. 16.

Papiliu. cærulescens subtus punctatus: Blue speckt B. *Pet. Pap. pl.* vi. *f.* 11. ♂.

Papiliu. cæruleus, apicibus nigris &c.: Blue speckt B. with black tipps. *Pet. Pap. pl.* vi. *f.* 12. ♀.

The Azure blue B. *Harr. V. M.* 1.

5873. 2; Alsus*: Small blue B.—*Haw. l. c. Jerm.* 2 *edit.* (74), 149.—*Steph. Ill.* (*H.*) i. 86.—*Id.* i. *App.* 149.—*Brit. B.* 33.

He. Alsus. *Fabr.* iii. *a.* 295.

Pa. Alsus. *Lewin. pl.* 39. *f.* 3, 4.—*Don.* ix. 73. *pl.* 322. *f.* 1.—(*Haw. Pr.* 4.)—*Haw.* 48.—*Turt.* iii. 139.—*Stew.* ii. 132.—*Jerm.* (36,) 59.

Ly. Alsus: Bedford Blue. (*Leach E. E.* ix. 130.)—(*Sam. I.* 26.)
Pa. minimus. *Esper. Pa. pl.* 34. *f.* 3.
Pa. Pseudolus. *Bork. E. S.* i. 177.

5874. 3, Acis: Mazarine Blue.—*Haw. l. c. Jerm.* 2 *edit.* (78), 159. —*Steph. Ill.* (*H.*) i. 86.—*Id.* i. *App.* 149.—*Brit. B.* 32. *pl.* 17. *f.* 2.
Pa. Acis. (*Wien. V.* 182.)
Pa. Argiolus. *Hüb. Pa. pl.* 56. *f.* 267–9.
Pa. semiargus. *Bork. E. S.* i. 172.
Pa. Cymon. *Lew. pl.* 38. *f.* 6. 7.—(*Haw. Pr.* 4.)—*Haw.* 48. —*Jerm.* (36), 59.
Ly. Cymon. (*Leach E. E.* ix. 130.)—(*Sam. I.* 26.)
Po. Cymon. *Jerm.* 2 *edit.* (74), 148.
Pa. minor, alis supinis purpuro-cæruleis &c. *Raii I.* 132. 17.

B. Vide *Steph. Ill.* (*H.*) i. 84.—POLYOMMATI veri, *Horsfield.*

5875. 4: Arion: Large Blue.—*Lewin. l.c. Jerm.* 2 *edit.* (74), 141. —*Steph. Ill.* (*H.*) i. 87.—*Brit. B.* 34. *pl.* 17. *fig. sup.*
Pa. Arion. *Linn.* ii. 789.—*Lew. pl.* 37.—*Don.* vi. 11. *pl.* 184. ♀. —(*Haw. Pr.* 4.)—*Haw.* 43.—*Turt.* iii. 138.—*Stew.* ii. 131. —*Jerm.* (34), 57.
Ly. Arion. (*Sam. I.* 26.)
The Argus blue B. *Harr. V. M.* 1?

5876. [† 5, Alcon?] *Jerm.* 2 *edit.* (78), 160.—*Steph. Ill.* (*H.*) i. 88. —*Brit. B.* 35.
Pa. Alcon. *Hüb. Pa. pl.* 55. *f.* 263, 264. ♂. *f.* 265. ♀?
Pa. Arcas. *Esper. Pa. pl.* 1. *f.* xxxiv. *Sup.* x. *f.* 4?
Pa. Diomedes. *Bork. E. S.* i. 169?
He. Argiades. *Fabr.* iii. *a.* 300? In Mus. D. *Haworth.*

5877. 6; Corydon*. *Jerm.* 2 *edit.* (74), 142. *pl.* 5. *f.* 14. neuræ. —*Steph. Ill.* (*H.*) i. 88.—*Brit. B.* 35.
He. Corydon. *Fabr.* iii. *a.* 298.
Pa. Corydon. *Lew. pl.* 36.—*Don.* vii. *pl.* 231. *f.* 1. ♂.— (*Haw. Pr.* 4.)—*Haw.* 43.—*Turt.* iii. 140.—*Stew.* ii. 131.— *Jerm.* (34), 58.
Ly. Corydon. (*Leach E. E.* ix. 129.)—(*Sam. I.* 26.)
Pa. minor cærulescens, limbis nigris: Pale blue Argus. *Pet. Gaz. pl.* 35. *f.* 2.—*Raii I.* 131. 14. ♂.—15. ♀.—*Pet. Pap. pl.* vi. *f.* 1. 2.
The Chalk hill Blue B. *Harr. V. M.* 1.
♀, Pa. Tiphys. *Esper. Pa. pl.* 21. *cont.* 1. *f.* 4.
β; *Steph. Ill.* (*H.*) i. 89.
Po. Calæthis. *Jerm.* 2 *edit.* 169.

5878. 7; Adonis*. *Jerm.* 2 *edit.* (74), 143.—*Steph. Ill.* (*H.*) i. 89.—*Brit. B.* 36.
He. Adonis. *Fabr.* iii. *a.* 299.

Pa. Adonis. *Lew. pl.* 38. *f.* 1–3.—(*Haw. Pr.* 4.)—*Haw.* 44. —*Turt.* iii. 140.—*Stew.* ii. 132.—*Jerm.* (36), 58.
Ly. Adonis. (*Leach E. E.* ix. 129.)—(*Sam. I.* 26.)—(*Kirby & Sp. I. E.* i. 41; iii. 361.)
Pa. Bellargus. *Esper. Pa. pl.* 32. *f.* 3.
Pa. Ceronus. *Hüb. Pa. pl.* 61. *f.* 295–297.
Pa. Argus ♀. *Don.* iv. 93. *pl.* 143. *f.* 1.
The Clifden blue B. *Harr. V. M.* 1.

5879. 8; Dorylas?* *Jerm.* 2 *edit.* (74), 144.—*Steph. Ill.* (*H.*) i. 90.—*Brit. B.* 36.
Pa. Dorylas. (*Wien. V.* 322.)
Pa. Dorylas. *Hüb. Pa. pl.* 60. *f.* 289. ♂.—*f.* 290, 291. ♀?
Pa. Hylas. *Esper. Pa. pl.* xlv. *Supp.* xxi. *f.* 3. ♂?
Pa. Thetis. *Esper. Pa. pl.* xxxiii. *Supp.* ix. *f.* 3. ♀?

5880. † 9, Icarius? Black-bordered Blue.—*Haw. l. c.* *Jerm.* 2 *edit.* (78), 161.—*Steph. Ill.* (*H.*) i. 91.—*Brit. B.* 16.
Pa. Icarius. *Esper. Pa.* i. *pl.* xcix. *cont.* liv. *f.* 4?
Pa. Amandus. *Hüb. Pa. pl.* 59. *f.* 283–4?—*Ent. Trans.* (*Haworth.*) i. 334. In Mus. D. *Haworth.*

5881. 10, Alexis: Common Blue.—*Haw. l. c.* *Jerm.* 2 *edit.* (78), 160.—*Steph. Ill.* (*H.*) i. 91.—*Brit. B.* 37.
Pa. Alexis. (*Wien. V.* 184.)—*Hüb. Pa. pl.* 60. *f.* 292.
Pa. Icarus. *Vill.* ii. 74.—*Lew. pl.* 38.—(*Haw. Pr.* 4.)—*Haw.* 45.—*Jerm.* (36), 58.
Pa. Argus. *Berk. S.* i. 129.—*Don.* iv. 93. *pl.* 143. ♂.
Ly. Dorylas. (*Leach E. E.* ix. 129.)—(*Sam. I.* 26.)
Pa. alis oculatis cyanum &c. *Merr. P.* 199. 4.
Pa. diurnæ minimarum. *Mouff.* 105. 4.
Papiliunculus cæruleus &c.: the little blue Argus. *Pet. Mus.* 318.—*Pet. Pap. pl.* vi. *f.* 3. ♂.
Pa. parva, alis supernè purpureo-cæruleis &c.: the most common small blue B. *Raii I.* 131. 11 & 13. ♂.—12. ♀.
Papiliu. è cæruleo fuscus: Mixt Argus. *Pet. Pap. pl.* vi. *f.* 5. ♀.
Papiliu. cæruleus limbis nigris: Selvedged blew Argus. *Pet. Pap. pl.* vi. *f.* 4. ♀ var.
The blue Argus B: Pa. Argus. (*Wilkes.* 63. *pl.* 119.)
The blue B. *Harr. A. pl.* 39. *f. g. i.*—*Harr. V. M.* 1.

β; *Steph. Ill.* (*H.*) i. 92. Light blue.—*Haw. l. c. infra.*
Pa. Hyacinthus. *Lew. pl.* 37. *f.* 4–6.—*Turt.* iii. 139.—(*Haw. Pr.* 4.)—*Haw.* 45.—*Jerm.* (36), 59.

γ; *Steph. Ill.* (*H.*) *l. c.*
Pa. Labienus. *Jerm.* (36), 58.

δ; *Steph. Ill.* (*H.*) *l. c.*
Po. Thestylis. *Jerm.* 2 *edit.* 167.

ε; *Steph. Ill.* (*H.*) *l. c.*

Po. Lacon. *Jerm.* 2 *edit.* 168.
Var. Po. dubius. *Kirby MSS.*

5882. 11; Eros*: the pale Blue.—*Haw. l. c. Jerm.* 2 *edit.* (78), 160.—*Steph. Ill.* (*H.*) i. 93.—*Brit. B.* 37.
Ly. Eros. (*Och.* iv. 26.)
Pa. Tithonus. *Hüb. Pa. pl.* 108. *f.* 555–6?—*Ent. Trans.* (*Haworth.*) i. 334.

5883. 12; Argus*. *Jerm.* 2 *edit.* (74), 145.—*Steph. Ill.* (*H.*) i. 93.—*Brit. B.* 37. *pl.* 17.
Pa. Argus. *Linn.* ii. 789.—*Lew. pl.* 39. *f.* 5–7.—*Berk. S.* i. 129.—(*Haw. Pr.* 4.)—*Haw.* 46.—*Turt.* iii. 139.—*Stew.* ii. 131.—*Jerm.* (36), 59.
Ly. Argus. (*Leach E. E.* ix. 129.)—(*Sam. I.* 26.)
Pap. cærulescens minor. *Pet. Pap. pl.* vi. *f.* 6. ♂.
Pap. è fusco cæruleus. *Pet. Pap. pl.* vi. *f.* 7. ♀.
Pap. plumbeus: Lead Argus. *Pet. Pap. pl.* vi. *f.* 8. ♀. var.
Pap. plumbeus parvus: Small Lead Argus. *Pet. Pap. pl.* vi. *f.* 9. ♀ var.?
Silver-studded Blue. *Harr. V. M.* 1.
♀; Pa. Idas. *Linn. F. No.* 1075.
β; *Steph. Ill.* (*H.*) i. 94.
He. Acreon. *Fabr.* iii. *a.* 301?
Pa. Argyrognomon. *Bork. E. S.* i. 152?
Pa. Argiades. *Esper. Pa.* i. *pl.* ci. *cont.* lvi. *f.* 6?
Po. Alcippe. *Kirby MSS.*
† γ, *Steph. Ill.* (*H.*) *l. c.*
Po. maritimus. *Haworth MSS.*
† δ, *Steph. Ill.* (*H.*) *l. c.*
Var. Po. Leodorus. *Esper. Pa.* i. *pl.* lxxx. *cont.* xxx. *f.* 1. 2.
γ in Mus. D. *Haworth*: δ in Mus. D. *Hatchett.*

5884. 13; Agestis*: Brown Argus.—*Haw. l. c. Jerm.* 2 *edit.* 159.—*Steph. Ill.* (*H.*) i. 94.—*Brit. B.* 38. *pl.* 17. *fig.* 3.
Pa. Agestis. (*Wien. V.* 184)—*Hüb. Pa. pl.* 62. *f.* 303–4.
Pa. Medon. *Esper. Pa.* i. *pl.* xxxii. *Supp.* viii. *f.* 1.
Pa. Idas. *Lew. pl.* 39. *f.* 1, 2.—*Don.* x. 74. *pl.* 322. *f.* 2.—(*Haw. Pr.* 4.)—*Haw.* 46.—*Jerm.* (36), 59.
Ly. Idas: Black spot brown. (*Leach E. E.* ix. 130.)—(*Sam. I.* 26.)
Po. Idas. *Jerm.* 2 *edit.* (74), 146.
Papiliunculus fuscus, marginibus aureis: Edged brown Argus. *Pet. Gaz. pl.* 35. *f.* 4.—*Pet. Pap. pl.* vi. *f.* 10.

5885. 14, Artaxerxes: Scotch Argus.—*Lewin. l. c. Jerm.* 2 *edit.* (74), 147.—*Steph. Ill.* (*H.*) i. 95.—*Brit. B.* 38.
He. Artaxerxes. *Fabr.* iii. *a.* 297. (!)
Pa. Artaxerxes. *Lew. pl.* 39. *f.* 8, 9.—*Turt.* iii. 139.—

(*Haw. Pr.* 4.)—*Haw.* 47.—*Stew.* ii. 131.—*Don.* xvi. 1. *pl.* 541.—*Jerm.* (36), 59.

Ly. Artaxerxes: White spot brown. (*Leach E. E.* ix. 130.) —.(*Sam. I.* 26.)

5886 † 15. Titus[a].

Familia IV: (117). HESPERIIDÆ.

(Plebeii Urbicoli, *Linné.*—Hesperides, *Latr.*—Hesperidæ, *Leach, Swainson.*—Hesperia *p, Fabricius.*)

(Anopluriform Stirps, *Horsfield.*)

Genus 19: (802). THYMELE, *Fabricius, Steph.*

Erinnys *p, Schra.*—Battus *p, Scop.*—Urbani, *Hübn.*—Hesperia *p, Latr., Leach, Samou., Jermyn,* (*Kirby.*)

5887. 1; Alveolus*. *Steph. Ill.* (*H.*) i. 97.—*Brit. B.* 39. *pl.* 18. *fig. sup.*

Pa. Alveolus. *Hüb. Pa. pl.* 92. *f.* 466–7.

Pa. Malvæ minor. *Esper. Pa. pl.* xxxvi. *f.* 5?

Pa. Malvæ. *Lew. pl.* 46. *f.* 8. 9.—*Berk. S.* i. 130.—(*Haw. Pr.* 5.)—*Haw.* 52.—*Turt.* iii. 165.—*Jerm.* (38), 60.

He. Malvæ. (*Leach E. E.* ix. 130.)—(*Sam. I.* 21.)—(*Kirby & Sp. I. E.* iii. 115.)—*Jerm. 2 edit.* (76), 154.

He. Sao. *Faun. F. pl.* 26. *f.* 7. 8.

Papiliunculus fuscus &c.: Small spotted brown Marsh Fritillary. *Pet. Gaz. pl.* 36. *f.* 6.—*Pet. Pap. pl.* IIII. *f.* 17. 18.

Pa. minima pulla &c. *Raii I.* 133. 21.

The grizzled B: Pa. Malvæ. (*Wilkes.* 54. *pl.* 104?)

The Grizzle, or Gristle B. *Harr. A. pl.* 32. *f. k–n.*—*Harr. V. M.* 3.

β; *Steph. Ill.* (*H.*) *l. c.* Pa. Fritillum. *Lew. pl.* 46. *f.* 4, 5.—*Turt.* iii. 166.—*Stew.* ii. 134.—*Rees' Cycl.* (*Donovan.*) *Art.* Papilio.—*Jerm.* 63.

Pa. Lavateræ. (*Haw. Pr.* 5.)—*Haw.* 52.—*Jerm.* (38), 60.

He. Lavateræ. *Jerm. 2 edit.* (76), 155.

Pa. fuscus &c.: Brown Marsh Fritillary. *Pet. Gaz. pl.* 36. *f.* 9.—*Raii I.* 132. 20.—*Pet. Pap. pl.* IIII. *f.* 15, 16.

5888 ‡ 2. Malvæ[b].

5889. 3; Tages*. *Steph. Ill.* (*H.*) i. 98.—*Brit. B.* 39.

---

[a] 5886 † 15. Titus. *Jerm. 2 edit.* (80), 166. (!)—*Steph. Ill.* (*H.*) i. 95. nota.
He. Titus. *Fabr.* iii. *a.* 297. (!)
Pa. Titus. *Turt.* iii. 140. (!)—*Rees' Cycl.* (*Donovan.*) *Art.* Papilio. (!)—*Jerm.* 62. (!)

[b] 5888 ‡ 2, Malvæ. *Steph. Ill.* (*H.*) i. 98. nota.
Pa. Malvæ. *Linn.* ii. 795.—*Stew.* ii. 133. (!)—*Don.* xvi. *pl.* 567. (!)
Pa. Malvarum. *Och.* i. *b.* 195.

Pa. Tages. *Linn.* ii. 795.—*Lew. pl.* 45. *f.* 3, 4.—(*Haw. Pr.* 5.) —*Haw.* 51.—*Turt.* iii. 167.—*Stew.* ii. 134.—*Jerm.* (38), 60.
He. Tages. (*Leach E. E.* ix. 130.)—(*Sam. I.* 21.)—*Jerm.* 2 *edit.* (76), 154.
Pa. corpore pedibus &c. *Merr. P.* 199. 6.
Pa. minor fuscus Hampstediensis, &c. *Pet. Gaz. pl.* 36. *f.* 3. ♀.—*Raii I.* 132. 18.
Pa. minor tota pulla &c. *Raii I.* 132. 19. ♂.
Pa. niger fuscus &c.: Handley's Small brown Fritillary. *Pet. Pap. pl.* IIII. *f.* 19. 20.
The dingy Skipper. *Harr. A. pl.* 34. *f.* o.—*Harr. V. M.* 5.

5890 † 4. *Oileus* [a].

Genus 20: (803). PAMPHILA, *Fabricius, Steph.*

ERINNYS *p, Schrank.*—BATTUS *p, Scop.*—URBANI *p, Hübn.*—HESPERIA *p, Latr., Leach, Samou., Jerm.*

A. a. Vide *Steph. Ill.* (*H.*) i. 160.

5891. 1, Paniscus. *Steph. Ill.* (*H.*) i. 160.—*Brit. B.* 40.
He. Paniscus. *Fabr.* iii. *a.* 328.—(*Leach E. E.* ix. 130.)—(*Sam. I.* 21.)—*Jerm.* 2 *edit.* (76), 152.
Pa. Paniscus: Checquered Skipper. *Don.* viii. 7. *pl.* 254. *f.* 1. —(*Haw. Pr.* 5.)—*Haw.* 49.—*Turt.* iii. 154.—*Jerm.* (38), 60.
Pa. Brontes. *Hüb. Pa. pl.* 94. *f.* 475, 476.

5892 ‡ 2. Sylvius [b].

5893. 3; Linea *. *Steph. Ill.* (*H.*) i. 101.
He. Linea. *Fabr.* iii. *a.* 326.—(*Leach E. E.* ix. 130.)—(*Sam. I.* 21.)—*Jerm.* 2 *edit.* (76), 153.—*Brit. B.* 40.
Pa. Linea. *Don.* vii. *pl.* 236. *f.* 2. ♂.—(*Haw. Pr.* 5.)—*Haw.* 50.—*Stew.* ii. 133.—*Jerm.* (38), 60.
Pa. Thaumas. *Esper. Pa. pl.* 36. *f.* 2, 3.—*Lew. pl.* 45. *f.* 5–7.—*Turt.* iii. 153.—*Stew.* ii. 133.—*Rees' Cycl.* (*Donovan.*) *Art.* PAPILIO.—*Jerm.* 63.
Pa. flavus. *Müll. Z. D. pr.* 1333.
Pa. Comma. *Barbut* G. 173. *pl.* 10. *f.* 256.
Pa. minima, &c. *Raii I.* 125. 21. ♂.—22. ♀.
Pa. minor aureus &c.: Streakt Golden Hog. *Pet. Gaz. pl.* 34. *f.* 9.—*Pet. Pap. pl.* vi. *f.* 15. ♂.
Pa. minor aureus immaculatus: Spotless Hog. *Pet. Gaz. pl.* 34. *f.* 9. desc.—*Pet. Pap. pl.* vi. *f.* 14. ♀.
The Small Skipper. *Harr. A. pl.* 42. *f.* i.—*Harr. V. M.* 5.

---

[a] 5890 † 4. Oileus. *Steph. Ill.* (*H.*) i. 99. nota.
Pa. Oileus. *Gmel.* iv. 2370?—Georgian Grizzle. *Ent. Trans.* (*Haworth.*) i. 334. (!)

[b] 5892 ‡ 2. Sylvius. *Steph. Ill.* (*H.*) i. 100. nota.
Pap. Sylvius. *Knoch.*—*Hüb. Pa. pl.* 94. *f.* 477, 478.

B. Vide *Steph. Ill.* (*H.*) i. 101.

5894. 4; Sylvanus*. *Steph. Ill.* (*H.*) i. 101.—*Brit. B.* 41. *pl.* 18. *fig. inf.*
He. Sylvanus. *Fabr.* iii. *a.* 326.—(*Leach E. E.* ix. 130.)—(*Kirby & Sp. I. E.* ii. 305.)—(*Sam. I.* 21.)—*Jerm.* 2 *edit.* (76). *pl.* 5. *f.* 15. neuræ.
Pa. Sylvanus. *Lew. pl.* 46. *f.* 1–3.—*Don.* viii. 8. *pl.* 254. *f.* 2. ♂.—(*Haw. Pr.* 4.)—*Haw.* 50—*Stew.* ii. 133.—*Turt.* iii. 153.—*Jerm.* (38), 60.
Pa. Comma. *Scop. C. No.* 463?
Pa. minor ex aureo, &c.: Streakt Cloudy Hog. *Pet. Gaz. pl.* 34. *f.* 8.—*Pet. Pap. pl.* VI. *f.* 16. ♂.
Pa. minor &c.: Cloudy Hog. *Pet. Gaz. pl.* 34. *f.* 7.—*Pet. Pap. pl.* VI. *f.* 17. ♀.
The large Skipper. *Harr. A. pl.* 42. *f. h.*—*Harr. V. M.* 5.

5895. 5; Comma*. *Steph. Ill.* (*H.*) i. 102.—*Brit. B.* 41.
Pa. Comma. *Linn.* ii. 793.—*Berk. S.* i. 130.—*Lew. pl.* 45. *f.* 1, 2.—*Don.* ix. 17. *pl.* 295.—(*Haw. Pr.* 4.)—*Haw.* 50.—*Stew.* ii. 133.—*Jerm.* (38), 60.—*Id.* 2 *edit.* 151. ♂.
Hi. Comma. *Fabr.* iii. *a.* 325.—(*Leach E. E.* ix. 130.)—(*Sam. I.* 21.)—*Jerm.* 2 *edit.* (76).
Pa. alis conchatis, &c. *Merr. P.* 199. 9.
The Pearl Skipper. *Harr. V. M.* 5.
Pa. Virgula. *DeGeer.* G. 31. 7.
♀; He. Sylvanus. *Jerm.* 2 *edit.* 151.

B. Vide *Steph. Ill.* (*H.*) i. 102.

5896 † 6, *Bucephalus*[a].

5897 † 7, *Vitellius*[b].

Sectio II. LEP.-CREPUSCULARIA, *Latreille.*

(Sphinx, *Linné, &c.*)

Familia V: (118). ZYGÆNIDÆ, *Leach.*

(Zygænides *p*, *Latreille.*)

Genus 21: (804). INO, *Leach, Samou., Steph.* *Forester M.*

Procris, *Fabr.*—Zygæna *p*, *Panzer, Haw.*—Adscita *p*, *DeGeer.*—Atychia, *Och.*—Chrysaöres, *Hübn.*

5898. 1; Statices*. (*Leach E. E.* ix. 131.)—(*Sam. I.* 23.)—*Steph. Ill.* (*H.*) i. 105.

---

[a] 5896 † 6, Bucephalus. *Steph. Ill.* (*H.*) i. 102. *pl.* 10. *f.* 2. ♂. nota. In Mus. *D. Raddon.*

[b] 5897 † 7, Vitellius. *Steph. Ill.* (*H.*) i. 103. nota.
He. Vitellius. *Fabr.* iii. *a.* 327.
Pap. Vitellius: the bordered Skipper. *Ent. Trans.* (*Haworth.*) i. 334. (!) In Mus. *D. Haworth.*

Sp. Statices. *Linn.* ii. 808.—*Don.* vi. 60. *pl.* 204. *f.* 2.—*Stew.* ii. 140.—(*Haw. Pr.* 6.)—*Turt.* iii. 191.
Zy. Statices. *Haw.* 73.
Pap. parva alis pendulis &c. *Raii I.* 134. 3.
Ad. Turcosa. *DeGeer.* G. 35.
Green Forester. *Harr. A. pl.* 34. *f. a–f.*—*Harr. V. M.* 27.

5899 † 2. Globulariæ[a].

Genus 22: (805). ANTHROCERA, *Scopoli, Steph.* *Burnet-Moth.*

ZYGÆNA[b], *Fabricius, Haw., Sam.*—ADSCITA *p, DeGeer.*

5900. 1; Meliloti*. *Steph. Ill.* (*H.*) i. 107.
Sp. Meliloti. *Esper. Sp.* ii. *pl.* xxxix. *cont.* xiv. *f.* 1–8.
Sp. Loniceræ var. *Esper. Sp.* ii. *pl.* xxv. *Sup.* vii. *f.* 3.
Sp. Loti. *Hüb. Sp. pl.* 17. *f.* 82. ♂.
Sp. Viciæ. *Rhein. Magaz.* (*Borkhausen.*) i. 638.

5901 † 2. Scabiosæ[c].

5902. 3; Trifolii*. *Steph. Ill.* (*H.*) i. 108.
Sp. Trifolii. *Esper. Sp.* ii. *pl.* xxxiv. *cont.* ix. *f.* 4. 5.
Ph. 4 minor gracilis cum quatuor &c. *Merr. P.* 198?
Pap. pratensis alis pendulis &c. *Raii I.* 134. 2.

5903. 4; Loti*: 5-spot B.—*Haw. l. c.* *Steph. Ill.* (*H.*) i. 109.
Zy. Loti. *Fabr.* iii. *a.* 387.—*Don.* ix. 67. *pl.* 319?—*Haw.* 74.
Zy. fulvia. *Fabr. M.* ii. 101.
Sp. Loniceræ. *Esper. Sp.* ii. *pl.* xxiv. *Supp.* vi. *f.* 1. *a, b.*—(*Haw. Pr.* 6.)
Sp. Graminis. *Villers E.* ii. 115.

5904. 5; Hippocrepidis*. *Steph. Ill.* (*H.*) i. 110.
Sph. Hippocrepidis. *Hüb. Sp. pl.* 17. *f.* 83. ♂.
Sph. Loti. *Hüb. Sp. pl.* 5. *f.* 32. ♀.
β; *Steph. Ill.* (*H.*) *l. c.*
Sph. Loti. *Esper. Sp.* ii. *pl.* xxxv. *cont.* x. *f.* 1?

5905. 6; Filipendulæ*: 6-spotted B.—*Berk. l. c.* *Steph. Ill.* (*H.*) i. 111.

---

[a] 5899 † 2. Globulariæ. *Steph. Ill.* (*H.*) i. 106. nota.
Sp. Globulariæ. *Hüb. Sp. pl.* 1. *f.* 2. ♂. *f.* 3. ♀.
Sp. Statices minor. *Naturf.* (——) xxviii. 83.

[b] ZYGÆNA: Genus Piscium antiquorum.

[c] 5901 † 2. Scabiosæ. *Steph. Ill.* (*H.*) i. 108. nota.
Zy. Scabiosæ: Triple-spotted B.—*Haw. l. c.* *Fabr.* iii. *a.* 386.—*Haw.* 74. (!)
Sph. Scabiosæ. *Hüb. Sp. pl.* 18. *f.* 86. ♂.
Zy. minor. *Schrank B.* ii. 240.
Zy. Pythia. *Rossi. F.* ii. 166.

Sp. Filipendulæ. *Linn.* ii. 805.—*Barbut* G. 181. *pl.* 10. *f.* 34. —*Berk.* *S.* i. 132.—*Don.* i. 17. *pl.* 6.—(*Haw.* *Pr.* 6.)—*Stew.* ii. 140.—*Turt.* iii. 182.
Zy. Filipendulæ. *Fabr.* iii. *a.* 386.—*Haw.* 73.—(*Sam.* *I.* 44.)
Burnet M.: Sp. Filipendulæ. (*Wilkes.* 45. *pl.* 91. *fig. sup.*)—*Harr.* *A.* *pl.* 1. *f.* e–h.—*Harr.* *V.* *M.* 15.
Ph. minor cum alis externis nigris &c. *Merr.* *P.* 198.
Ph. tertius &c. *Mouff.* 97. *fig. ult.*
Ph. No. 37. *List. Goëd.* (*Angl.*) 47. *f.* 37.—*Id.* (*Lat.*) 100. *f.* 37.
Ad. Aries. *DeGeer.* G. 35.

5906 † 7. Peucedani[a].

Familia VI: (119). SPHINGIDÆ, *Leach.*

(SPHINGIDES *p*, *Latreille.*)
Hawk Moths.

Genus 23: (806). SMERINTHUS, *Latreille*, *Samou.*, *Steph.*

LAÖTHOE, *Fabr.*, *Leach.*—SPECTRUM *p*, *Scop.*—AMORPHÆ, *Hüb.*

5907. 1; ocellatus*. (*Latr.* G. iv. 210.)—(*Sam.* *I.* 38.)—*Steph.* *Ill.* (*H.*) i. 112.
Sph. ocellata. *Linn.* ii. 796.—*Berk.* *S.* i. 130.—*Shaw* *N.* *M.* i. *pl.* 18.—*Don.* viii. 47. *pl.* 269.—(*Haw.* *Pr.* 5.)—*Haw.* 63. —*Stew.* ii. 135.—*Turt.* iii. 168.
Sph. Salicis. *Hüb.* *Sp.* *pl.* 15. *f.* 73.
Sph. Semipavo. *DeGeer.* G. 34. 33.
La. ocellata. (*Leach* *E.* *E.* ix. 130.)
Pap. sextæ &c. *Mouff.* 91. 6. *cum fig.*
Ph. No. 24. *List. Goëd.* (*Angl.*) 33. *f.* 24.—*Id.* (*Lat.*) 68. *f.* 24.
Ph. magna, cinereo, &c. *Raii* *I.* 149. 3.
Eyed Hawk M. *Albin.* *pl.* viii.—*Harr.* *A.* *pl.* 5. *f.* *a–c.*—*Harr.* *V.* *M.* 29.
Eyed Willow Hawk M.: Sp. ocellata. (*Wilkes.* *pl.* 24.)

5908. 2; Populi*. (*Latr.* G. iv. 210.)—(*Sam.* *I.* 38.)—(*Kirby & Sp.* *I.* *E.* iv. 39.)—*Steph.* *Ill.* (*H.*) i. 112.
Sph. Populi. *Linn.* ii. 797.—*Berk.* *S.* i. 130.—*Don.* vii. *pl.* 241. —(*Haw.* *Pr.* 5.)—*Haw.* 64.—*Stew.* ii. 135.—*Turt.* iii. 169.
La. Populi. (*Leach* *E.* *E.* ix. 130.)
Ph. No. 25. *List. Goëd.* (*Angl.*) 34. *f.* 25.—*Id.* (*Lat.*) 71. *f.* 25.
Poplar Hawk M. *Albin.* *pl.* 57. Sp. Populi. (*Wilkes.* *pl.* 25.) —*Harr.* *A.* *pl.* 33. *f.* *a–g.*—*Harr.* *V.* *M.* 29.

---

[a] 5906 † 7. Peucedani. Vide *Appendix.*
Sp. Peucedani. *Esper.* *Sp.* ii. *pl.* xxv. *Supp.* vii. *f.* 2.
Sp. Æacus. *Hüb.* *Sp.* *pl.* 4. *f.* 22. ♀.
Zyg. Filipendulæ var. *Fabr.* iii. *a.* 286.
Burnet M.: Sp. Filipendulæ. (*Wilkes.* 45. *pl.* 91. *fig. inf.* (!)

5909. 3; Tiliæ*. (*Latr. G.* iv. 210.)—(*Sam. I.* 38.)—(*Kirby & Sp. I. E.* iii. 188. & 269.)—*Steph. Ill.* (*H.*) i. 113.
Sph. Tiliæ. *Linn.* ii. 797.—*Berk. S.* i. 131.—*Barb. G. pl.* 10. *f.* 3.—*Don.* x. 3. *pl.* 325.—(*Haw. Pr.* 5.)—*Haw.* 64.—*Stew.* ii. 135.—*Turt.* iii. 169.
La. Tiliæ. (*Leach E. E.* ix. 130.)
Albin. pl. 10.
Olive Shades, or Lime Hawk M.: Sp. Tiliæ. (*Wilkes. pl.* 23.)
Lime Hawk. *Harr. A. pl.* 20. *f. a–g.*—*Harr. V. M.* 9.

Genus 24: (807). ACHERONTIA, *Ochsen.*, (*Samou.*,) *Curtis*, *Steph.*
SPECTRUM *p*, *Scopoli.*—MANDUCÆ, *Hübner.*

5910. 1: Atropos*. (*Och.* iv. 44.)—*Curtis.* iv. *pl.* 147.—*Steph. Ill.* (*H.*) i. 114.
Sph. Atropos. *Linn.* ii. 799.—*Berk. S.* i. 131.—*Don.* ix. *pl.* 289. *pl.* 290. *l & p.*—*Haw. Pr.* 5.—*Haw.* 56.—*Stew.* ii. 136.—*Turt.* iii. 172.—(*Leach E. E.* ix. 131.)—(*Sam. I.* 30.)—(*Kirby & Sp. I. E.* i. 34; ii. 240 & 390; iii. 269.)—*Wood.* ii. 8. *pl.* 44.
Ph. maxima, &c. *Mouff.* 89. *cum fig.*—*Albin. pl.* 6.
The Jasmine Hawk M: Sp. Atropos. (*Wilkes. pl.* 19.)
Death's head. *Harr. A. pl.* 37.—*Bingley.* iii. 215.
Bee tyger Hawk M. *Harr. V. M.* 29.

Genus 25: (808). SPHINX *Auctorum.*
SPECTRUM *p*, *Scopoli.*—EUMORPHÆ *p*, *Hübner.*

5911. 1, *Carolina*[a]: Tobacco Hawk M.—*Curtis l. c.*

5912. 2, *quinquemaculatus*[b]: Yellow-spotted Unicorn H. M.—*Haw. l. c.*

5913. 3; Convolvuli*. *Linn.* ii. 798.—*Berk. S.* i. 131.—*Don.* vii. *pl.* 228, imago. 229, larva et pupa.—(*Haw. Pr.* 5.)—*Haw.* 58.—*Stew.* ii. 136.—*Turt.* iii. 176.—(*Leach E. E.* ix. 131.)—(*Sam. I.* 30.)—(*Curtis l. c. infra.*)—*Steph. Ill.* (*H.*) i. 119.
Ph. quarta &c. *Mouff.* 91. 4. *cum fig.*
Unicorn or Bind-weed Hawk M.: Sp. Convolvuli. (*Wilkes. pl.* 20, larva. *pl.* 21, imago.)
Unicorn or Convolvulus Hawk M. *Harr. A. pl.* 21. *f. a–d.*—*Harr. V. M.* 29.
Ph. No. 27. *List. Goëd.* (*Angl.*) 36. *f.* 27.—*Id.* (*Lat.*) 76. *f.* 27.

---

[a] 5911. 1, Carolina*. *Linn.* ii. 798.—*Ent. Trans.* (*Haworth.*) i. 241.—*Curtis.* v. *pl.* 195. (!)—*Steph. Ill.* (*H.*) i. 118. nota.

[b] 5912. 2, quinquemaculatus*. *Haw.* 59. (!)—*Ent. Trans.* (*Haworth.*) i. 241.—(*Curtis l. c. supra.*) (!)—*Steph. Ill.* (*H.*) i. 119. nota.
Sph. Carolina. *Don.* xi. 1. *pl.* 361. (!)

5914. 4, *Druræi*[a].

5915. 5; Ligustri*. *Linn.* ii. 799.—*Berk. S.* i. 131.—*Don.* viii. 79. *pl.* 248.—(*Haw. Pr.* 5.)—*Haw.* 59.—*Stew.* ii. 136.—*Turt.* iii. 176.—(*Leach E. E.* ix. 131.)—(*Sam. I.* 30.)—(*Kirby & Sp. I. E.* iii. 188. & 266.)—(*Curtis l. c. infra.*)—*Steph. Ill.* (*H.*) i. 121.
Ph. maxima caudacuta, &c. *Raii I.* 144.—*Id.* 146. larva.
Privet Hawk M. *Albin. pl.* vii.
Privet Hawk M.: Sp. Ligustri. (*Wilkes. pl.* 22.)—*Harr. A. pl.* 2. *f. a–h.*—*Harr. V. M.* 29.
Ph. quinta &c. *Mouff.* 91. 5. *cum fig.*—*Id.* 181. larva.

5916. 6, Pinastri*: Pine Hawk M —*Don. l. c.* *Linn.* ii. 802.—*Don.* ix. 19. *pl.* 296.—(*Haw. Pr.* 5.)—*Haw.* 59.—*Stew.* ii. 138.—*Turt.* iii. 173.—(*Leach E. E.* ix. 131.)—(*Sam. I.* 30.)—(*Curtis l. c. supra.*)—*Steph. Ill.* (*H.*) i. 121.

5917. 7, *plebeia*[b].

5918. 8, *pœcila*[c].

Genus 26: (809). DEILEPHILA, *Och.*, *Curtis*, *Steph.*

SPECTRUM *p*, *Scopoli.*—EUMORPHÆ *p*, *Hübn.*—DEILOPHILA, *Samou.*

A. Vide *Steph. Ill.* (*H.*) i. 124.

5919. 1: Euphorbiæ. (*Och.* iv. 43.)—*Curtis.* i. *pl.* 3.—*Id.* 2 *edit.* i. *pl.* 3.—*Steph. Ill.* (*H.*) i. 124.
Sph. Euphorbiæ. *Linn.* ii. 802.—*Don.* iii. *pl.* 91, 92.—(*Haw. Pr.* 5.)—*Haw.* 61.—*Stew.* ii. 137.—*Turt.* iii. 173.—(*Leach E. E.* ix. 131.)—(*Sam. I.* 30.)
Spotted Elephant Hawk M. *Harr. A. pl.* 44. *f. a, c.*—*Harr. V. M.* 23.

5920. 2: Galii*: Galium H. M.—*Leach l. c.* (*Och.* iv. 43.)—(*Curtis l. c. supra.*)—*Steph. Ill.* (*H.*) i. 125. *pl.* 12. *f.* 2.—*Id.* i. *App.* 149.—*Curtis.* 2 *edit. l. c. supra.*
Sph. Galii. *Hüb. Sp. pl.* 12. *f.* 64.—*Ent. Trans.* (*Haworth.*) i. 99. *pl.* 4.—(*Leach E. E.* ix. 131.)—(*Sam. I.* 30.)
Spotted Elephant Caterpillar. *Harr. A. pl.* 44. *f. b.* (larva.)

5921. 3. Livornica*: Striped Hawk M. *Curtis.* 2 *edit. l. c. supra.*
Accipitrina Livornica perbelle striata. *Pet. Gaz. pl.* 12. *f.* 9.
Sph. Livornica. *Hüb. Sp. pl.* 12. *f.* 65.
Sph. Kochlini. *Fuesly A.* 10. *pl.* 4.

---

[a] 5914. 4, Druræi*. *Don.* xiv. 1. *pl.* 469. (!)—(*Curtis l. c. supra.*) (!)—*Steph. Ill.* (*H.*) i. 120. nota.
Sph. Convolvuli var. *Smith & Abb.* i. *pl.* 32.

[b] 5917. 7, plebeia. *Fabr.* iii. *a.* 367.—*Steph. Ill.* (*H.*) i. 122. nota.

[c] 5918 † 8, pœcila. *Steph. Ill.* (*H.*) i. 122. nota.

Sph. lineata. *Fabr. M.* ii. 96.—(*Haw. Pr.* 5.)—*Och.* ii. 214. *Leach E. E.* ix. 130.—(*Samou.* 243.)
Dei. lineata. (*Och.* iv. 43.)—(*Curtis l. c. supra.*)—*Steph. Ill.* (*H.*) i. 126. *pl.* 12. *f.* 1.

5922. 4, *lineata*[a].

B. Vide *Steph. Ill.* (*H.*) i. 128.

5923. 5, Celerio*: Sharp winged or Silver striped H. (*Och.* iv. 43.)—(*Curtis l. c. supra.*)—*Steph. Ill.* (*H.*) i. 128.—*Curtis.* 2 *edit. l. c. supra.*)
Sph. Celerio. *Linn.* ii. 800.—*Don.* vi. *pl.* 190.—*pl.* 191. *l* & *p.* —(*Haw. Pr.* 5.)—*Haw.* 61.—*Stew.* ii. 137.—*Turt.* iii. 174. —(*Sam. I.* 30.)
Ph. inquilinus. *Harr. Ex.* 93. *pl.* 28. *f.* 1.

5924. 6, *argentata*[b].

5925. 7; Elpenor*. (*Och.* iv. 43.)—(*Curtis l. c. supra.*)—*Steph. Ill.* (*H.*) i. 131.—*Curtis.* 2 *edit. l. c. supra.*
Sph. Elpenor. *Linn.* ii. 801.—*Berk. S.* i. 132.—(*Haw. Pr.* 6.) —*Haw.* 62.—*Don.* iv. *pl.* 122.—*Stew.* ii. 137.—*Turt.* iii. 175.—(*Leach E. E.* ix. 130.)—(*Sam. I.* 30.)
Ph. No. 26. *List. Goëd.* (*Angl.*) 35. *f.* 26. larva.—*Id.* (*Lat.*) 73. *f.* 26. larva.
Ph. major cauda aucta, &c. *Raii I.* 145. 2.
Elephant. *Albin. pl.* ix.—*Harr. A. pl.* 7. *f. a–h.*—*Harr. V. M.* 23.
The Ladies Bed Straw or Elephant M.: Sp. Elpenor. (*Wilkes. pl.* 26.)

5926. 8; Porcellus*. (*Och.* iv. 43.)—(*Curtis l. c. supra.*)—*Steph. Ill.* (*H.*) i. 131.—*Curtis.* 2 *edit. l. c. supra.*
Sph. Porcellus. *Linn.* ii. 801.—*Don.* ix. 55. *pl.* 314.—(*Haw. Pr.* 6.)—*Haw.* 63.—*Stew.* ii. 137.—*Turt.* iii. 175.—(*Leach E. E.* ix. 130.)—(*Sam. I.* 30.)
The small Elephant H. M.: Sp. porcellus. (*Wilkes. pl.* 16.) —*Harr. V. M.* 24.

---

[a] 5922. 4, lineata. *Curtis.* 2 *edit. l. c. supra.*
Sph. lineata. *Fabr. S. E.* 541.—*Don.* vi. *pl.* 204. *f.* 1. (!)—*Haw.* 60. (!)—*Stew.* ii. 138. (!)—*Turt.* iii. 174. (!)
Sph. Daucus. *Cramer. pl.* cxxv. *f. D.*
Dei. Daucus. *Steph. Ill.* (*H.*) i. 126. nota.

[b] 5924. 6, argentata. *Steph. Ill.* (*H.*) i. 130. nota.
Sph. argentata. *Ent. Trans.* (*Haworth.*) i. *p.* 334. (!)

Familia VII: (120). SESIIDÆ *mihi.*

(SPHINGIDES, *Latreille.*—SPHINGIDÆ B, *Leach.*)
Humming Bird Hawk Moth.

Genus 27: (810). MACROGLOSSA, *Och., Steph.*

MACROGLOSSUM, *Scop., Samou.*—SESIA *p, Fabr., Leach.*

5927. 1; Stellatarum*. (*Och.* iv. 42.)—*Steph. Ill.* (*H.*) i. 133.
Sp. Stellatarum. *Linn.* ii. 803.—*Barb.* G. *pl.* 10. *f.* 27.—*Berk. S.* i. 132.—*Don.* vii. *pl.* 155.—(*Haw. Pr.* 6.)—*Haw.* 65.—*Stew.* ii. 138.—*Turt.* iii. 179.—(*Kirby* & *Sp. I. E.* ii. 229.)
Se. Stellatarum. (*Leach E. E.* ix. 131.)
Ma. Stellatarum. (*Sam. I.* 26.)—(*Kirby* & *Sp. I. E.* iii. 557 & 571.)
Pap. med. diurnæ, &c. *Mouff.* 105. 12.
Humming-bird Sphinx. *Harr. A. pl.* 24. *f. f–b.*—*Harr. V. M.* 13.

Genus 28: (811). SESIA, *Fabricius, Leach, Samou., Curtis, Steph.*

MACROGLOSSA *p, Ochsenheimer.*—MACROGLOSSUM *p, Scopoli.*—BOMBYCIÆ, *Hübner.*

5928. 1, Fuciformis*. *Steph. Ill.* (*H.*) i. 134.
Sp. Fuciformis. *Linn.* ii. 803.
Se. Bombyliformis. *Fabr.* iii. *a.* 382.
Sp. Bombyliformis: Narrow-bordered Bee Hawk M. (*Haw. Pr.* 6?)—*Haw.* 68.—*Curtis* i. *pl.* 40.—*Turt.* iii. 180.
Se. Bombyciformis. (*Leach E. E.* ix. 131.)—(*Sam. I.* 37.)

5929. 2; Bombyliformis*: Broad-bordered Bee Hawk M. *Steph. Ill.* (*H.*) i. 135.
Sp. Bombyliformis. *Och.* ii. 189.
Sp. Fuciformis. *Fabr.* iii. *a.* 381.—*Hüb. Sp. pl.* 9. *f.* 55.—*Don.* iii. *pl.* 87.—(*Haw. Pr.* 6?)—*Haw.* 67.—*Stew.* ii. 138.—*Turt.* iii. 180.—(*Curtis l. c. supra.*)
Se. fusiformis. (*Leach E. E.* ix. 131.)—(*Sam. I.* 37.)
Pap. diurnæ minimæ primæ, &c. *Mouff.* 105. 5. *cum fig.*
Clear-winged humming bird Sp. *Harr. V. M.* 13.—*Harr. Ex.* 52. *pl.* xiii. *f.* 2.

5930. 3, *Cimbiciformis*[a].

Familia VIII: (121). ÆGERIIDÆ *mihi.*

(ZYGÆNIDES *p, Latr.*—ZYGÆNIDÆ *p, Leach.*)
Clear-winged Hawk Moths.

Genus 29: (812). TROCHILIUM, *Scopoli, Steph.*

ÆGERIA *p, Fabr., Leach, Samou.*—SESIA *p, Fabr.*

5931. 1; Apiformis*. *Steph. Ill.* (*H.*) i. 137.

---

[a] 5930. 3, Cimbiciformis. *Steph. Ill.* (*H.*) i. 135. nota.

Sp. Apiformis. *Linn.* ii. 804.—*Don.* i. 55. *pl.* 25.—*Linn. Trans.* (*Lewin.*) iii. *pl.* 3. *f.* 1–5.—(*Haw. Pr.* 6.)—*Haw.* 68. —*Stew.* ii. 139.—*Turt.* iii. 180.—(*Kirby & Sp. I. E.* i. 209.)
Sp. Crabroniformis. *Hüb. Sp. pl.* 8. *f.* 51.
Æg. Apiformis. (*Leach E. E.* ix. 131.)—(*Sam. I.* 1.)
Hornet M. *Harris Ex.* 15. *pl.* iii. *f.* 7.—*Harr. V. M.* 31.

5932. 2; Crabroniformis*: The lunar Hornet M. *Steph. Ill.* (*H.*) i. 138.
Sp. Crabroniformis. *Linn. Trans.* (*Lewin.*) iii. *pl.* 3. *f.* 6–10. —(*Haw. Pr.* 6.)—*Haw.* 69.—*Turt.* iii. 180.—*Don.* xiii. 7. *pl.* 436.
Se. Bembiciformis. *Hüb. Sp. pl.* 20. *f.* 98. ♀.
Æg. Crabroniformis. (*Leach E. E.* ix. 131.)—(*Sam. I.* 1.)

Genus 30: (813). ÆGERIA, *Fabricius, Leach, Samou., Curtis.*
TROCHILIUM *p, Scopoli.*—SESIA *p, Latr.*

5933. 1: Asiliformis*. (*Sam. I.* 1.)—(*Curtis l. c. infra.*)—*Steph. Ill.* (*H.*) i. 139.
Le. Asiliformis. *Fabr.* iii. *a.* 383.
Sp. Asiliformis: Clear underwing. (*Haw. Pr.* 6.)—*Haw.* 69. —*Don.* xi. 67. *pl.* 384.
Sp. Tabaniformis. *Bork. E. S.* ii. 45.
Se. Œstriformis. *Kirby & Sp. I. E.* i. *pl.* 3. *f.* 2.

5934 † 2, Spheciformis*. (*Sam. I.* 1.)—(*Curtis l. c. infra.*)—*Steph. Ill.* (*H.*) i. 140. *pl.* 11. *f.* 1.
Sp. Spheciformis: Black and white horned Clear-wing.—*Haw. l. c.* *Vill.* ii. 203.—*Hüb. Sp. pl.* 16. *f.* 77. ♂. *f.* 78. ♀.—*Haw.* 71.
Se. Sphegiformis. *Fabr.* iii. *a.* 383.
Sp. semizonatus. (*Haw. Pr.* 6.)
Sp. Ichneumoniformis. *Bork. E. S.* ii. 43.
In Mus. *Brit.*, D. *Ingpen, Latham, Stone et Vigors.*

5935. 3, Ichneumoniformis. *Curtis.* ii. *pl.* 53.—*Steph. Ill.* (*H.*) i. 140.
Se. Ichneumoniformis. *Fabr.* iii. *a.* 385.
Sp. Vespiformis: 6-belted Clear-wing.—*Haw. l. c.* *Hüb. Sp. pl.* vi. *f.* 39.—(*Haw. Pr.* 6.)—*Haw.* 70.—(*Sam. I.* 1.)

5936 † 4, Chrysidiformis. (*Curtis l. c. supra.*)—*Steph. Ill.* (*H.*) i. 141.
Sp. Chrysidiformis: Fiery Clear-wing.—*Haw. l. c.* *Hüb. Sp. pl.* 8. *f.* 53.—*Haw.* 71.
Sp. bicinctus. (*Haw. Pr.* 6.) In Mus. D. *Haworth.*

5937. 5; Cynipiformis*. (*Sam. I.* 1.)—*Steph. Ill.* (*H.*) i. 141. *pl.* 11. *f.* 2. ♂.
Se. Cynipiformis. *Och.* ii. 151.
Sp. Asiliformis. *Bork. E. S.* ii. 41.

Se. Vespiformis. *Fabr.* iii. *a.* 385.
Sp. Vespiformis. *Turt.* iii. 182.
Sp. chrysorrhœa: Golden-tail Hawk M. *Don.* iv. *pl.* 116.—*Stew.* ii. 139.—*Turt.* iii. 182.
Æg. Vespiformis. (*Leach E. E.* ix. 131.)—(*Curtis l. c. supra.*)
♂; Sp. Cynipiformis: Yellow-legged Clear-wing. *Haw.* 69.
Sp. Tenthrediniformis. (*Haw. Pr.* 6.)
♀; Sp. Œstriformis: Yellow-tailed Clear-wing. (*Haw. Pr.* 6.)—*Haw.* 70.
Æg. Œstriformis. (*Sam. I.* 1.)
Sp. Tipuliformis: Small Bee Moth. *Berk. S.* i. 132.

5938. 6; Tipuliformis*: Currant Hawk M. (*Leach E. E.* ix. 131.) (*Sam. I.* 1.)—(*Curtis l. c. supra.*)—*Steph. Ill.* (*H.*) i. 142.
Sp. Tipuliformis. *Linn.* ii. 804.—*Don.* ii. *pl.* 52, 53.—(*Haw. Pr.* 6.)—*Haw.* 70.—*Stew.* ii. 139.—*Turt.* iii. 181.
Clear-winged or lesser humming bird Sp. *Harris Ex.* 14. *pl.* 111. *f.* 8.—(*Harr. V. M.* 14.)

5939. 7; Mutillæformis*. (*Curtis l. c. supra.*)—*Steph. Ill.* (*H.*) i. 142.
Sp. Mutillæformis. *Lasp.* 26. *f.* 15–17.
Sp. Culiciformis. *Hüb. Sp. pl.* 7. *f.* 45. *pl.* 19. *f.* 91.—*Haw.* 71.
Sp. Myopæformis. *Bork. E. S.* ii. 169.
Sp. zonatus: red belted Sp. *Don.* vi. 35. *pl.* 195.
Sp. zonata: red bellied Hawk M. *Stew.* ii. 139.—*Turt.* iii. 181.
Æg. Culiciformis. (*Sam. I.* 1.)

5940. 8, Culiciformis*. *Steph. Ill.* (*H.*) i. 143. *pl.* 10. *f.* 3.
Sp. Culiciformis. *Linn.* ii. 804.—(*Haw. Pr.* 6?)—*Turt.* iii. 181. (!)

5941. 9; Stomoxyformis*. *Steph. Ill.* (*H*). i. 144. *pl.* 11. *f.* 4.
Sp. Stomoxyformis. *Hüb. Sp. pl.* 2. *f.* 47. ♀.
Sp. Stomoxyformis. *Ent. Trans.* (*Haworth.*) i. 334.
Sp. Culiciformis. *Scop. C.* 188.
Æg. Culiciformis. (*Curtis l. c. supra.*)

5942. 10; Formiciformis*. (*Sam. I.* 1.)—(*Curtis l. c. supra.*)—*Steph. Ill.* (*H.*) i. 144. *pl.* 11. *f.* 3.
Sp. Formiciformis: Flame tipped red-belt.—*Haw. l. c. Vill.* ii. 104.—*Lasp. S.* 24. *pl. f.* 11. ♂. *f.* 12. ♀.—*Haw.* 72.
Sp. Tenthrediniformis. *Bork. E. S.* ii. 39.
Sp. flammeus. (*Haw. Pr.* 6.)

5943 † 11. Ephemeræformis. *Steph. Ill.* (*H.*) i. 145.
Sp. Ephemeræformis: Beltless Clear-wing. *Haw.* 72.
Tinea fenestrella. (*Haw. Pr.* 35.) In Mus. D. Donovan.

Sectio III. LEP.-NOCTURNA, *Latreille.*

(Phalæna, *Linné.*)

Moths, Bustards, Millers, Soles, Owls.

Subsectio 1. LEP.-POMERIDIANA *mihi.*

(Ph.-Attacus *et* Bombyx, *et* Noctua *et* Tinea *p, Linné.*)

Feathered Full-bodied Moths.

Familia IX: (122). HEPIALIDÆ *mihi.*

(Phalæna-Noctua *p, Linné, &c.*—Bombycites *p, Latr.*—Bombycidæ *p, Leach.*)

Genus 31: (814). HEPIALUS, *Fabr., Haw., Leach, Samou., (Kirby,) Curtis, Steph.* *Swifts.*

Hepiolus, *Illiger.*—Bombyx *p, Hübn.*—Noctua *p, Esper.*

A. Antennis in utroque sexû simplicibus.

5944. 1; Hectus*: Gold Sw.—*Haw. l.* c. *Fabr.* iii. *b.* 6.—*Haw.* 144.—(*Leach E. E.* ix. 132.)—(*Sam. I.* 21.)—(*Curtis l. c. infra.*) —*Steph. Ill.* (*H.*) ii. 4.
Ph. No. Hectus. *Linn.* ii. 833.—*Stew.* ii. 206.—*Don.* viii. 60. *pl.* 274. *f.* 3. ♂.—*Turt.* iii. 340.
Bo. Hectus. *Hüb. Bo. pl.* 49. *f.* 208. ♂. *f.* 209. ♀.
He. Hectator. (*Haw. Pr.* 19.)
♀. He. Jodutta: dingy Sw.—*Haw. l.* c. *Schr. B.* ii. 303.
He. Joduttator. (*Haw. Pr. App.* 4.)
Bo. lupulina. *Hüb. B.* i. *pl.* iv. *f. T. S.*
♂. No. Nemorosa. *Esper. No. pl.* LXXXI. *f.* 5.

5945. 2; lupulinus*. *Fabr.* iii. *b.* 6.—(*Curtis l. c. infra.*)—*Steph. Ill.* (*H.*) ii. 5.
♀, He. fuscus: brown Sw. *Haw.* 141.—(*Sam. I.* 21.)
Ph. No. lupulina. *Linn.* ii. 833.
He. lupulator. (*Haw. Pr.* 19.)
He. carnator. (*Haw. Pr. App.* 4.)
Common Small Sw. *Harr. V. M.* 47.
Bo. Flina. (*Wien. V.* 61.)—*Hüb. Bo. pl.* 49. *f.* 210. ♂. *f.* 211. ♀.
He. Cora. *Schra. B.* ii. 304.
β; *Steph. Ill.* (*H.*) *l.* c.
He. angulum: tawny Sw. *Haw.* 142.—(*Sam. I.* 21.)
He. angulator. (*Haw. Pr.* 19.)
γ; *Steph. Ill.* (*H.*) *l.* c.
He. obliquus: silver Sw.—*Haw. l.* c. *Fabr.* iii. *b.* 6.—*Haw.* 142.—(*Sam. I.* 21.)
He. obliquator. (*Haw. Pr. App.* 4.)
δ; *Steph. Ill.* (*H.*) *l.* c.
He. nebulosus: spotted silver Sw. *Haw.* 143.—(*Sam. I.* 21.)

He. nebulosator. (*Haw. Pr. App.* 4)
ε; *Steph. Ill.* (*H.*) *l. c.*
He. nebulosus β. *Haw.* 143.
ζ; *Steph. Ill.* (*H.*) *l.* c.
He. nebulosus γ. *Haw.* 143.

5946. 3; Humuli*: Otter M.—*Berk. l. c. Fabr.* iii. *b.* 5.—*Haw.* 138.—(*Sam. I.* 21.)—(*Kirby & Sp. I. E.* iii. 90, 271, 306, 335.)—(*Curtis l. c. infra.*)—*Steph. Ill.* (*H.*) ii. 6.
Ph. No. Humuli. *Linn.* ii. 833.—*Berk. S.* i. 139.—*Don.* viii. 59. *pl.* 274. *f.* 1. ♂.—*f.* 2. ♀.—*Stew.* ii. 206.
He. Humulator. (*Haw. Pr.* 19.)
Ph. He. Humuli. *Turt.* iii. 340.
Ghost M. *Harr. Ex.* 17. *pl.* iv. *f. a–d.*—*Harr. V. M.* 27.

5947. 4; Velleda*: Beautiful Sw. *Haw.* 141.—(*Curtis l. c. infra.*)—*Steph. Ill.* (*H.*) ii. 6.
Bo. Velleda. *Hüb. Bo. pl.* 50. *f.* 212.
Ph. Mappa: Map-winged Sw. *Don.* x. 95. *pl.* 36. *f.* 3.
He. Mappa. (*Sam. I.* 21.)

5948. 5, carnus. *Fabr.* iii. *b.* 6.—(*Ing. Inst.* 91.)—(*Curtis l. c. infra.*)—*Steph. Ill.* (*H.*) ii. 7. *pl.* 13. *f.* 1. ♂.
♀, He. Jodutta. *Fabr.* iii. *b.* 6?—*Hüb. Bo. pl.* 50. *f.* 213.

B. Antennis pectinatis, aut serratis.

5949. 6; sylvinus*: Tawny and brown Sw.—*Curtis l. c. Och.* iii. 109.—*Curtis.* v. *pl.* 185. ♂ & ♀.—*Steph. Ill.* (*H.*) ii. 7.
Ph. No. sylvina. *Linn.* ii. 834.
He. sylvinator. (*Haw. Pr. App.* 4.)
He. cruxator. (*Haw. Pr.* 19.)
He. angulatus. *Fabr. Sp.* ii. 506.
He. Crux. *Fabr.* iii. *b.* 7.
Bo. lupulinus. *Hüb. Bo. pl.* 48. *f.* 205. ♂. *f.* 206. ♀.
He. lupulinus: orange Sw. *Haw.* 140.—(*Sam. I.* 21.)
No. flina. *Esper.* iv. *pl.* lxxxii. *f.* 5–7.
He. Hamma. *Schra. B.* ii. 302.
He. Fauna. *Schra. B.* ii. 305.
Golden Swift. *Harris Ex.* 19. *pl.* iv. *f. f.*—*Harr. V. M.* 47.
Large evening Swift. *Harris Ex. pl.* xiii. *f.* 6.
Orange or Evening Sw. *Harr. A. pl.* 22. *f. h–m.*—*Harr. V. M.* 47.

Genus 32: (815). ZEUZERA, *Latr.*, *Leach*, *Samou.*, *Steph.*

COSSUS *p*, *Fabr.*—HEPIALUS *p*, *Schra.*—BOMBYX *p*, *Haw.*—PHALÆNA-NOCTUA *p*, *Linn.*

5950. 1: Æsculi*. *Latr.* G. iv. 217.—(*Leach E. E.* ix. 132.)—(*Samou.* 246.)—*Steph. Ill.* (*H.*) ii. 8.
Ph. No. Æsculi. *Linn.* ii. 833.—*Don.* v. 27. *pl.* 152.

Ph. No. Pyrina. *Linn. F. No.* 1150.
Bo. Æsculus. (*Haw. Pr.* 7.)
Bo. Pyrinus. *Haw.* 89.
Ph. Cossus Æsculi. *Turt.* iii. 342.
Ph. Bo. Æsculi. *Stew.* ii. 157.
Phal. 14 &c. *Mouff.* 92. *fig. infra.*
Wood Leopard M. *Harris Ex.* 11. *pl.* II. *f.* 3. 4.—*Harr. V. M.* 34.

Genus 33: (816). COSSUS, *Fabr., Leach, Samou.,* (*Kirby,*) *Curtis, Steph.*

HEPIALUS *p, Schr.*—TEREDINES, *Hüb.*—BOMBYX *p, Haw.*—PHALÆNA-BOMBYX *p, Linn.*

5951. 1; ligniperda*. *Fabr. S. E.* ii. 569.—(*Leach E. E.* ix. 132.) —(*Samou.* 246.)—(*Kirby & Sp. I. E.* iii. 89, 124, 174, 201, 203, 269 & 352.—*Id.* iv. 14.)—*Curtis.* ii. *pl.* 60.—*Steph. Ill.* (*H.*) ii. 9. —*Starke E.* ii. 365.
Ph. Bo. Cossus. *Linn.* ii. 827.—*Berk. S.* i. 138.—*Don.* iv. *pl.* 114.—*Stew.* ii. 153.
Bo. Cossus. (*Haw. Pr.* 7.)—*Haw.* 89.
Ph. Cossus Cossus. *Turt.* iii. 341.
Pa. No. 39. *List.* Goëd. (*Angl.*) 49. *f.* 39.—*Id.* (*Latr.*) 105. *f.* 39.—*Albin. pl.* xxxv.
Goat M.: Ph. Cossus. *Wilkes.* 15. *pl.* 31.—*Harr. A. pl.* 23. —*Harr. V. M.* 27.—*Bingley.* iii. 221.

Familia X: (123). NOTODONTIDÆ *mihi.*

(PHALÆNA BOMBYX *p, et* PH.-NOCTUA *p, Linné.*)

(BOMBYCITES *p, et* NOCTUELITÆ *p, Latr.*—BOMBYCIDÆ *p, Leach.*)

Genus 34: (817). PYGÆRA, *Och., Samou., Steph.*

LARIA *p, Schra.*—MELALOPHÆ *p, Hübn.*

5952. 1; bucephala*. *Och.* iii. 235.—(*Sam. I.* 35.)—*Steph. Ill.* (*H.*) ii. 12.
Ph. Bo. bucephala. *Linn.* ii. 816.—*Berk. S.* i. 134.—*Don.* i. *pl.* 3. ♂.—*Stew.* 149.—*Turt.* iii. 208.
Bo. bucephalus. (*Haw. Pr.* 7.)—*Haw.* 92.
Ph. lunula. *DeGeer. G.* 38. 55.
Phal. majores 11 &c. *Mouff.* 92. *f.* 11.
Pa. No. 95. *List.* Goëd. (*Angl.*) 88. *f.* 95.—*Id.* (*Latr.*) 213. *f.* 95.—*Albin. pl.* xxiii. *f.* 33. *a–d.*
Buff-tip M.: Ph. bucephala. (*Wilkes.* 21. *pl.* 43.)—*Harr. A. pl.* 39. *f. a–c.*—*Harr. V. M.* 48.

Genus 35: (818). CLOSTERA, *Hoffmansegg*, (*Samou.*,) (*Kirby*,) *Steph.*

PYGÆRA A., *Och.*—MELALOPHÆ *p*, *Hüb.*—LARIA *p*, *Schra.*

5953 † 1. Anastomosis[a].

5954. 2; reclusa*: Small Chocolate-tip.—*Haw. l. c.* (*Sam. I.* 11.) —*Steph. Ill.* (*H.*) ii. 13.
Bo. reclusa. *Fabr.* iii. *a.* 447.—*Esper.* iii. *pl.* 51. *f.* 6. 7.— *Haw.* 131.
Ph. anastomosis: Scarce Ch.-tip.—*Don. l. c.* *Scop. C. No.* 502? —*Don.* iv. *pl.* 124.—*Id.* iv. *pl.* 129. *f.* 4.
La. reclusa. (*Leach E. E.* ix. 132.)
† β, *Steph. Ill.* (*H.*) *l. c. pl.* 16. *f.* 1. β in Mus. *D. Vigors.*

5955 † 3, anachoreta: Scarce Chocolate-tip.—*Haw. l. c.* *Steph. Ill.* (*H.*) ii. 13.
Bo. anachoreta. *Fabr.* iii. *a.* 447.
Bo. anachoretus. (*Haw. Pr.* 10.)—*Haw.* 131.
Bo. curtula. *Esper.* iii. *pl.* 51. *f.* 1–4. In Mus. *Brit.*

5956. 4; curtula*. (*Sam. I.* 11.)—*Steph. Ill.* (*H.*) ii. 14.
Ph. Bo. curtula. *Linn.* ii. 823.—*Stew.* ii. 152.—*Turt.* iii. 212.
Bo. curtulus. (*Haw. Pr.* 10.)—*Haw.* 130.
La. curtula? (*Leach E. E.* ix. 132.)
Bo. anachoreta. *Esper.* iii. *pl.* 51. *f.* 5.
Ph. alticaudata furcata. *DeGeer.* G. 39. 61.
Albin. pl. lxxxviii. *f.* a?–e.
Chocolate tip M. *Harr. V. M.* 49.

Genus 36: (819). EPISEMA, (*Och.*,) *Treitschke*, *Steph.*

GRAPHIPHORA *p*, *Hübner.*

5957. 1; cæruleocephala*. (*Och.* iv. 65.)—*Och. Tr.* v. *a.* 112. —*Steph. Ill.* (*H.*) ii. 15.
Ph. Bo. cæruleocephala. *Linn.* ii. 826.—*Berk. S.* i. 132.— *Don.* iii. 75. *pl.* 100.—*Stew.* ii. 153.—*Turt.* iii. 208.
Bo. cæruleocephalus. (*Haw. Pr.* 10.)—*Haw.* 105.
Bo. cæruleocephala. (*Kirby & Sp. I. E.* i. 197.)—(*Sam. I.* 6.)
La. cæruleocephala. (*Leach E. E.* ix. 132.)
Pa. No. 47. *List. Goëd.* (*Angl.*) 55. *f.* 47.—*Id.* (*Latr.*) 121. *f.* 47.
Albin. pl. xiii. *f.* 17. *a–e.*
Black thorn M.: Ph. cæruleocephala. (*Wilkes.* 6. *pl.* 12.)
Figure-of-8 M. *Harr. A. pl.* 30. *f. a–d.*—*Harr. V. M.* 23.

---

[a] 5953 † 1. Anastomosis. (*Kirby & Sp. I. E.* iii. 261.)—*Steph. Ill.* (*H.*) ii. 13. nota.
Ph. Bo. Anastomosis. *Linn.* ii. 824.—*Esper.* iii. *pl.* 52. *f.* 1–4.— *Stew.* ii. 152. (!)—*Turt.* iii. 212. (!)

Genus 37: (820). CERURA, *Schra.*, *Leach*, *Samou.*, (*Ingpen*,) (*Kirby*,) *Curtis*, *Steph.*

ANDRIÆ, *Hübn.*—HARPYIA A., *Och.*

5958. 1, bicuspis*. (*Ing. Inst.* 90.)—*Curtis l. c. supra.*—*Steph. Ill.* (*H.*) ii. 16. *pl.* 13. *f.* 3.
Bo. bicuspis. *Bork. E. S.* iii. 380.—*Hüb. Bo. pl.* 10. *f.* 36?
Ha. bicuspis. *Och.* iii. 26?

5959. 2, integra* *mihi.* *Steph. Ill.* (*H.*) ii. 16. *pl.* 15. *f.* 3.
Ce. bicuspis. *Fischer E. R.* i. 63. *pl.* III. *f.* 2?

5960. 3; furcula*. *Schra. B.* ii. 292?—(*Leach E. E.* ix. 132.)—(*Sam. I.* 10.)—(*Curtis l. c. infra.*)—*Steph. Ill.* (*H.*) ii. 17.
Ph. Bo. furcula. *Linn.* ii. 823.—*Don.* viii. 55. *pl.* 272.—*Stew.* ii. 151.—*Turt.* iii. 226.
Bo. furculus. (*Haw. Pr.* 10.)—*Haw.* 103.
Ph. diurna-minor. *DeGeer.* G. 38. 54.
Kitten M.: Ph. furcula. (*Wilkes.* 13. *pl.* 29. *f.* 2.)
Small Puss or Kitten M. *Harr. V. M.* 41.

5961. 4, arcuata* *mihi.* *Steph. Ill.* (*H.*) ii. 17.
Ce. forficula. *Zetter.*—*Fischer E. R.* i. 62. *pl.* III. *f.* 1?

5962. 5, latifascia*? broad-barred Kitten M. *Curtis.* iv. *pl.* 193.—*Steph. Ill.* (*H.*) i. 18.
Ce. bidens *mihi* olim.

5963. 6; fuscinula*. (*Curtis l. c. infra.*)—*Steph. Ill.* (*H.*) ii. 18. *pl.* 15. *f.* 1.
Bo. fuscinula. *Hüb. Bo. pl.* 10. *f.* 37.
Ha. bifida var. *Och.* iii. 31.

5964. [7; bifida*.] (*Sam. I.* 10.)—(*Curtis l. c. infra.*)—*Steph. Ill.* (*H.*) ii. 19. *pl.* 15. *f.* 2.
Bo. bifida. *Hüb. Bo. pl.* 10. *f.* 38.
Bo. furcula. *Esper.* iii. *pl.* 19. *f.* 6.

5965. 8; Vinula*. *Schra. B.* ii. 291.—(*Leach E. E.* ix. 132.)—(*Sam. I.* 10.)—(*Kirby & Sp. I. E.* iii. 150 & 283.—*Id.* iv. 38 & 213.)—(*Curtis l. c. infra.*)—*Steph. Ill.* (*H.*) ii. 19.
Ph. Bo. Vinula. *Linn.* ii. 815.—*Berk. S.* i. 134.—*Don.* iii. *pl.* 85.—*Turt.* iii. 203.—*Stew.* ii. 146.
Bo. Vinulus. (*Haw. Pr.* 7.)—*Haw.* 86.
Ph. diurna major. *DeGeer.* G. 36. 53.
Pa. No. 20. *List. Goëd.* (*Angl.*) 27. *f.* 20. *a–c.*—*Id.* (*Latr.*) 5. *f.* 20. *a–c.*—*Albin. pl.* XI.
Puss M.: Ph. vinula. (*Wilkes.* 13. *pl.* 29. *f.* 1.)—*Harr. A. pl.* 38. *f. a–e.*—*Harr. V. M.* 41.
β; Bo. minax. *Hüb. Bo. pl.* 56. *f.* 243. ♂?
Ce. minax. (*Sam. I.* 10.)

5966. 9; Erminea*. (*Curtis l. c. infra.*—*Steph. Ill.* (*H.*) ii. 20.

Bo. Erminea. *Hüb. Bo. pl.* 9. *f.* 35. ♀.
Ha. Erminea. *Och.* iii. 24.

Genus 38: (821). STAUROPUS, *Germar, Samou.*, (*Kirby*,) *Steph.*

HARPYIA B. *p*, *Och.*

5967. 1: Fagi*. (*Sam. I.* 39.)—(*Kirby & Sp. I. E.* iii. 193.)—*Steph. Ill.* (*H.*) ii. 21.
Ph. Bo. Fagi. *Linn.* ii. 816.—*Don.* xii. *pl.* 328. *mala.*—*Stew.* ii. 147.—*Turt.* iii. 200.
Bo. Fagus. (*Haw. Pr.* 7.)—*Haw.* 85.
Staphylinus 2. *Mouff.* 197, 198. *cum figuris.* (larvæ.)
Lobster Caterpillar. *Albin. pl.* LVIII.

Genus 39: (822). NOTODONTA *mihi.*

NOTODONTA A., *Och.*, *Samou.*—PTILODONTES *p*, *Hüb.*

5968. 1; dromedarius*. *Och.* iii. 53.—(*Sam. I.* 31.)—*Steph. Ill.* (*H.*) ii. 22.
Ph. Bo. dromedarius. *Linn.* ii. 827.—*Don.* x. 67. *pl.* 350. *f.* 1.
Bo. dromedarius: Iron Prominent. (*Haw. Pr.* 8.)—*Haw.* 100.
La. dromedarius. (*Leach E. E.* ix. 132.)
Rusty Prominent M. *Harr. V. M.* 41.
β; Bo. dromedarulus: Small Iron Prominent. (*Haw. Pr.* 8.)—*Haw.* 101.—*Ent. Trans.* (*Haworth.*) i. 335.
No. dromedarulus. (*Sam. I.* 31.)
Ph. Zebu. *Don.* xii. 1. *pl.* 397. *f.* 1.

5969. 2, perfusca. *Steph. Ill.* (*H.*) ii. 23. *pl.* 14. *f.* 2.
Bo. perfuscus. (*Haw. Pr.* 8?)—*Haw.* 100?
No. perfuscus. (*Sam. I.* 31.)
Dark Prominent. *Harris Ex.* 54. *pl.* xiii. *f.* 5? *var?*—*Harr. V. M.* 41?
No. torva. (*Ing. Inst.* 91.)

5970. 3; ziczac*. *Och.* iii. 48.—(*Sam. I.* 31.)—*Steph. Ill.* (*H.*) ii. 23.
Ph. Bo. ziczac. *Linn.* ii. 827.—*Berk. S.* i. 138.—*Don.* iv. 29. *pl.* 119.—*Stew.* ii. 153.—*Turt.* iii. 209.
Bo. ziczacus: Pebble prominent. (*Haw. Pr.* 8.)—*Haw.* 99.
Ph. ziczac trituberculata. *DeGeer.* G. 87.
La. ziczac. (*Leach E. E.* ix. 132.)
Ph. No. 21. *List. Goëd.* (*Angl.*) 30. *f.* 21.—*Id.* (*Latr.*) 63. *f.* 21.—*Albin. pl.* xiv. *f.* 20. *e–h.*
Ozier or Pebble M.: Ph. ziczac. (*Wilkes.* 12. *pl.* 28.)
Pebble M. *Harr. V. M.* 39.

Genus 40: (823). LEIOCAMPA *mihi.*

NOTODONTA C., *Och.*—PTILODONTES *p*, *Hübn.*

5971. 1; dictæa*. *Steph. Ill.* (*H.*) ii. 25.

Ph. Bo. dictæa. *Linn.* ii. 826.
Ph. Bo. Tremula. *Linn.* ii. 826.
Bo. Tremulus. (*Haw. Pr.* 8.)—*Haw.* 99.
Ph. trepida. *Don.* vii. *pl.* 239. *f.* 1.—*Stew.* ii. 154.
La. trepida. *Leach E. E.* ix. 132.
No. trepida. (*Sam. I.* 31.)
Swallow M. *Harr. V. M.* 47.

5972. 2; dictæoides*: Lesser Swallow Prominent.—*Haw. l. c.* (*Ing. Inst.* 91.)—*Steph. Ill.* (*H.*) ii. 25.
Bo. dictæoides. *Esper.* iii. *pl.* 84.—*Hüb. Bo. pl.* 6. *f.* 24, 25. —*Ent. Trans.* (*Haworth.*) i. 335.
Bo. gnoma. *Fabr.* iii. *a.* 443.

Genus 41: (824). LOPHOPTERYX *mihi.*

NOTODONTA B., *Och.*—PTILODONTES *p*, *Hüb.*—NOCTUA *p*, *Samou.*—NOTODONTA *p*, *Samou.*

5973. 1; camelina*: Coxcomb Prominent. *Steph. Ill.* (*H.*) ii. 26.
Ph. Bo. camelina. *Linn.* ii. 832.—*Don.* vi. 7. *pl.* 183.—*Stew.* ii. 156.—*Turt.* iii. 213.—(*Kirby & Sp. I. E.* iii. 205.)
Bo. camelinus. (*Haw. Pr.* 8.)—*Haw.* 96.
Bo. capuzina. *Esper.* iii. *pl.* 70. *f.* 1–5.
Not. camelina. (*Leach E. E.* ix. 134.)—(*Sam. I.* 31.)
Noc. camelina. (*Sam. I.* 29.)
Albin. pl. lxix. *f. a, c, d, e.*

5974. 2; cuculla*. *Steph. Ill.* (*H.*) ii. 27.
Bo. cucullina. (*Wien. V.* 311.)—*Hüb. Bo. pl.* 5. *f.* 20. ♀.
Bo. cucullus. *Haw.* 97.
Ph. cuculla: Maple Prominent. *Don.* x. 31. *pl.* 338. *f.* 1.
Not. cuculla. (*Sam. I.* 31.)
Albin. pl. lxix. *f. b*? larva.

5975 † 3: carmelita*. *Steph. Ill.* (*H.*) ii. 27. *pl.* 14. *f.* 3.
Bo. carmelita. *Esper. Bo.* iii. *pl.* LXXXXI. *f.* 1. ♂.
Bo. capuzina. *Hüb. Bo. pl.* 5. *f.* 21. ♂.
Bo. capuzinus. (*Haw. Pr.* 8.)
Ph. capuzina. *Turt.* iii. 213.
Lo. cucullina. (*Ing. Inst.* 91.) In Mus. *Brit. et D. Vigors.*

Genus 42: (825). PTILODONTIS *mihi.*

NOTODONTA D. *p*, *Och.*, *Samou.*—PTILODONTES *p*, *Hübn.*—NOCTUA *p*, *Leach*, *Samou.*

5976. 1; palpina*. *Steph. Ill.* (*H.*) ii. 28.
Ph. Bo. palpina. *Linn.* ii. 828.—*Turt.* iii. 213.
Bo. palpina. *Hüb. Bo. pl.* 1. *f.* 16. ♂.
Bo. palpinus. (*Haw. Pr.* 8.)—*Haw.* 96.
Noc. palpina. (*Leach E. E.* ix. 134.)—(*Sam. I.* 30.)

Not. palpina. (*Sam. I.* 31.)
Pale prominent M. *Harr. V. M.* 41.

Genus 43: (826). PTILOPHORA *mihi.*

NOTODONTA D. *p, Och.*—PTILODONTES *p, Hübn.*

5977. 1: variegata*. (*Ing. Inst.* 91.)—*Steph. Ill.* (*H.*) ii. 29. *pl.* 14. *f.* 1. ♂.
Ph. Bo. variegata. *Villers Ent.* ii. 160.
Bo. plumigera. (*Wien. V.* 61?)—*Fabr.* iii. *a.* 462.—*Hüb. Bo. pl.* 4. *f.* 13. ♂. *pl.* 58. *f.* 150. ♀.
Pt. Aceris *mihi* olim.

Genus 44: (827). CHAONIA *mihi.*

NOTODONTA D. *p, Och.*—NOCTUA *p, Fabr.*

5978. 1: Roboris*. *Steph. Ill.* (*H.*) ii. 30.
No. Roboris. *Fabr.* iii. *b.* 35.
Ph. No. Roboris. *Stew.* ii. 185.—*Turt.* iii. 307.
Bo. Roboreus: Lunar Marbled brown. (*Haw. Pr.* 10.)—*Haw.* 104.—(*Sam. I.* 7.)
Ph. Roboris. *Don.* ix. 25. *pl.* 299.
Bo. chaonia. *Hüb. Bo. pl.* 3. *f.* 10, 11.
La. Roboris. (*Leach E. E.* ix. 132.)

5979. 2; dodonea*. *Steph. Ill.* (*H.*) ii. 30.
Bo. dodonea. *Hüb. Bo. pl.* 3. *f.* 8.—(*Sam. I.* 7.)
Bo. dodoneus: Marbled brown. *Haw. Pr.* 104.
Bo. trimacula. *Esper.* iii. *pl.* 46. *f.* 1–3.
Ph. trimacula. *Don.* x. 71. *pl.* 352. *f.* 2.
La. trimacula. (*Leach E. E.* ix. 132.)
Bo. tripartita. *Bork. E. S.* iii. 424.
Bo. Ilicis. *Fabr.* iii. *a.* 434.
β, *Steph. Ill.* (*H.*) *l. c.*
Bo. Querneus: Dark marbled brown. *Haw.* 104.
Bo. Quercea. (*Sam. I.* 7.)

5980 † 3. Querna[a].

Genus 45: (828). PETASIA *mihi.*

BOMBYX *p, Haworth.*—XYLENA *p, Ochsen.*

5981. 1; Cassinea*. *Steph. Ill.* (*H.*) ii. 32.
Bo. Cassinea. (*Wien. V.* 61.)—*Fabr.* iii. *a.* 460.—(*Sam. I.* 6.)
Bo. cassineus: Sprawler. (*Haw. Pr.* 9.)—*Haw.* 106.
Ph. Bo. cassinea. *Don.* xii. 4. *pl.* 397. *f.* 2.
Bo. Sphinx. *Esper. Bo.* iii. *pl.* xlix. *f.* 1–3.

---

[a] 5980 † 3. Querna. *Steph. Ill.* (*H.*) ii. 31. nota.
Bo. Querna. *Fabr.* iii. *a.* 449.—*Hüb. Bo. pl.* 3. *f.* 9.

Genus 46: (829). PERIDEA *mihi.*

NOTODONTA *D. p, Och.*

5982. 1; serrata*. *Steph. Ill.* (*H.*) ii. 33.
Bo. trepida. *Fabr.* iii. *a.* 449.—*Turt.* iii. 213?
Bo. tremula. *Hüb. Bo. pl.* 70. *f.* 30.
Ph. tritopha. *Don.* x. 91. *pl.* 359.
Bo. tritophus: Great Prominent. (*Haw. Pr.* 8.)—*Haw.* 98.
Bo. serrata. *Thunb. I. S.* (*Akermann.*) *p.* 47.
La. tritopha. (*Leach E. E.* ix. 132.)
Not. tritopus. (*Sam. I.* 31.)

Genus 47: (830). ENDROMIS, *Och., Samou.,* (*Kirby,*) *Steph.*

DIMORPHÆ, *Hübn.*—BOMBYX *p, Schra., Haw.*—DORVILLIA, *Leach.*

5983. 1; versicolora*. *Och.* iii. 16.—(*Sam. I.* 16.)—*Steph. Ill.* (*H.*) ii. 34.
Ph. Bo. versicolor. *Linn.* ii. 817.—*Berk. S.* i. 134.—*Don.* v. 63. *pl.* 158.—*Stew.* ii. 147.—*Turt.* iii. 203.—*Ent. Trans.* (*Neale.*) i. 323. *pl.* 9. ♂. ♀. *and larvæ.*
Bo. versicolorus. (*Haw. Pr.* 7.)—*Haw.* 80.
Do. versicolor. (*Leach E. E.* ix. 132.)
Glory of Kent, A Moth: Ph. versicolora. (*Wilkes.* 44. *pl.* 89.)
Kentish Glory. *Harr. V. M.* 27.

Familia XI: (124). BOMBYCIDÆ *mihi.*

(PHALÆNA-ATTACUS *et* BOMBYX, *Linné.*—BOMBYCITES LEGITIMÆ, *Latr.*—BOMBYCIDÆ *p, Leach.*)

Genus 48: (831). AGLIA[a].

Genus 49: (832). SATURNIA, *Schrank, Leach, Samou.,* (*Kirby,*) *Steph.*

BOMBYX *p, Fabr., Haw.*—HERAEÆ, *Hübn.*—ATTACUS, *Germar.*

5985. 1; Pavonia-minor*. (*Leach E. E.* ix. 132.)—(*Sam. I.* 37.)—(*Kirby & Sp. I. E.* iii. 150, 176, 186, 221, 266, 279 & 322.—*Id.* iv. 212.)—*Steph. Ill.* (*H.*) ii. 37.

---

[a] Genus 48: (831). AGLIA, *Ochsenheimer, Steph.*

ECHIDNÆ, *Hübn.*—PH. ATTACUS *p, Linn.*—SATURNIA *p, Schrank.*—BOMBYX *p, Haworth.*

5984 ‡ 1. Tau. *Och.* iii. 12.—*Steph. Ill.* (*H.*) ii. 37. nota.
Ph. At. Tau. *Linn.* ii. 811.—*Martyn. Art. Birch.* (!)—*Turt.* iii. 198. (!)
Bo. Tau. *Hübn. Bo. pl.* 13. *f.* 51. ♂. *f.* 52. ♀.
Bo. Tanus: Tau Emperor. (*Haw. Pr.* 8.)—*Haw.* 79.

Ph. At. Pavonia-minor. *Linn.* ii. 810.—*Berk. S. pl.* 10. *f.* 7. ♂.—*Stew.* ii. 144.
Bo. Pavonia minor. *Stark. E.* ii. 366.
Ph. Bo. Pavonia. *Linn. F.* 1099.—*Don.* i. 1. *pl.* 1. ♂. viii. *pl.* 254. ♀.—*Turt.* iii. 198.
Ph. Pavonia. *Berk. S.* i. 132.
Ph. Pavionella. *Scop. C.* 192.
Bo. Pavonus. (*Haw. Pr.* 8.)—*Haw.* 78.
Ph. Pavunculus. *DeGeer. G.* 35. 39.
Bo. Carpini. *Hüb. Bo. pl.* 14. *f.* 53. ♂. *f.* 54. ♀. *Albin. pl.* xxv. *f.* 37. *a–h.*
Emperor M.: Ph. Pavonia. (*Wilkes.* 16. *pl.* 32, larva; *pl.* 33, imago.)—*Harr. A. pl.* 25. *a–i, t.*—*Harr. V. M.* 23.

Genus 50: (833). LASIOCAMPA, *Schrank., Leach, Samou.,* (*Kirby,*) *Curtis, Steph.*

GASTROPACHA C. D., *Ochsenheimer.*

5986. 1; Rubi*. *Schrank B.* ii. 153.—(*Leach E. E.* ix. 132.)—(*Sam. I.* 24.)—(*Curtis l. c. infra.*)—*Steph. Ill.* (*H.*) ii. 39.
Ph. Bo. Rubi. *Linn.* ii. 813.—*Don.* ii. 87. *pl.* 69. ♂.—*Stew.* ii. 145.—*Turt.* iii. 203.
Bo. Rubus. (*Haw. Pr.* 7.)—*Haw.* 83.
Ph. Rubi. *Berk. S.* i. 133.
Bo. Trifolii. *Schæffer Icon.* 157. *pl.* clxxviii. *f.* 3.
Fox: (Devil's Gold ring.) *Albin. pl.* lxxxi. *a–d.*—*Harr. V. M.* 25.
Fox coloured M.: Ph. Rubi. (*Wilkes.* 25. *pl.* 54.)

5987. 2; Trifolii*. *Schrank B.* ii. 154.—(*Leach E. E.* ix. 132.)—(*Sam. I.* 24.)—(*Curtis l. c. infra.*)—*Steph. Ill.* (*H.*) ii. 40.
Ph. Bo. Trifolii. *Vill.* ii. 136.—*Linn. Trans.* (*Lewin.*) iii. *pl.* 4. *f.* 1–4.—*Turt.* iii. 200.
Bo. Trifolius. (*Haw. Pr.* 7.)—*Haw.* 83.
Ph. Dumeti. *Rhen. Mag.* (*Fuesly.*) iii. 151.
Grass Egger: Ph. Trifolii. (*Wilkes.* 23. *pl.* 47. *fig. sup.* ♀.)—*Harr. V. M.* 23.

5988. [3: Medicaginis:] Medick Eggar.—*Curtis l. c.* (*Biche. Add.* 13.)—(*Ing. Inst.* 91.)—*Curtis.* iv. *pl.* 181.—*Steph. Ill.* (*H.*) ii. 41.
Bo. Medicaginis. *Rhein. M.* (*Bork.*) i. 363.
Bo. Trifolii. *Panz. F.* xix. *f.* 23.
Bo. Quercus. *Wilkes. pl.* xlvii.
Grass Egger. (*Wilkes.* 23. *pl.* 47. *fig. inf.* ♂.)

5989. 4; Quercus*. *Schrank B.* ii. 154.—*Steph. Ill.* (*H.*) ii. 40.
Ph. Bo. Quercus. *Linn.* ii. 814.

5990. [5; Roboris*.] *Schrank B.* ii. 154?—*Steph. Ill.* (*H.*) ii. 41.

Ph. Quercus. *Barbut* G. *pl.* 10. *f.* 25. ♂.—*Berk.* S. i. 133. —*Don.* iii. 83. *pl.* 104. *f.* 1. ♂. *f.* 2. ♀.—*Stew.* ii. 146.— *Turt.* iii. 201.
Bo. Quercus. *Fabr.* iii. *a.* 423.—(*Haw. Pr.* 7.)—*Haw.* 81.
La. Quercus. (*Leach E. E.* ix. 132.)—(*Sam. I.* 24.)—(*Kirby & Sp. I. E.* iii. 557 & 571.)—(*Curtis l. c. supra.*)
Pa. nona et decima. *Mouff.* 92. *fig. superiores.*
Pa. No. 88. *List.* Goëd. (*Angl.*) 85. *f.* 88.—*Id.* (*Lat.*) 203. *f.* 88.
Great Egger. *Albin. pl.* xviii.—(*Wilkes.* 22. *pl.* 46.)—*Harr. A. pl.* 29. *f. a–f.*—*Harr. V. M.* 21.
Var. Bo. Spartii. *Hüb. Bo. pl.* 39. *f.* 173. ♂. *pl.* 52. *f.* 224. ♀.
La. Spartii. (*Ing. Inst.* 91.)
5991 † 6. Dumeti[a].

Genus 51: (834). TRICHIURA *mihi.*

LASIOCAMPA *p, Schrank, Leach, Samou.*—GASTROPACHA E *a, Och.*—CLISIOCAMPA *p, Curtis.*

5992. 1; Cratægi*. *Steph. Ill.* (*H.*) ii. 43.
Ph. Bo. Cratægi. *Linn.* ii. 823.—*Don.* iv. 23. *pl.* 117.— *Stew.* ii. 151.—*Turt.* iii. 218.
La. Cratægi. (*Leach E. E.* ix. 132.)—(*Sam. I.* 24.)
Bo. Cratægus. (*Haw. Pr.* 10.)—*Haw.* 105.
Cl. Cratægi. (*Curtis. fo.* 229.)
Ph. albicaudata. *DeGeer.* G. 36. 44.
Var. Bo. Mali. *Fabr.* iii. *a.* 434.
Bo. Avellanæ. *Fabr. M.* ii. 116.
Bo. pallidus: pale Oak Egger. *Haw.* 105.
Albin. pl. xxxiv. *f.* 54. *a–d.*—*Id.* lxxxviii. *f. i.*
Oak Egger. *Harr. V. M.* 23.

Genus 52: (835). PŒCILOCAMPA *mihi.*

ERIOGASTER *p, Samou.*

5993. 1; Populi*. *Steph. Ill.* (*H.*) ii. 44.
Ph. Bo. Populi. *Linn.* ii. 818.—*Berk. S.* i. 134.—*Don.* ix. 41. *pl.* 307. ♀.—*Stew.* ii. 148.—*Turt.* iii. 203.
Bo. Populeus. (*Haw. Pr.* 10.)—*Haw.* 127.
La. Populi. (*Leach E. E.* ix. 132.)
Er. Populi. (*Sam. I.* 16.)
December M.: Ph. Populi. (*Wilkes.* 23. *pl.* 48.)—*Harr. V. M.* 21.
Albin. pl. lxxxv.

---

[a] 5991 † 1. Dumeti. *Schrank B.* ii. 154.—*Steph. Ill.* (*H.*) ii. 41. nota.
Ph. Bo. Dumeti. *Linn.* ii. 815.—*Turt.* iii. 202. (!)—*Stew.* ii. 146. (!)
Bo. Dumeti. *Hüb. Bo. pl.* 37. *f.* 164. ♂.

Genus 53: (836). ERIOGASTER, *Germar*, *Samou.*, *Steph.*

LACHNEIDES, *Hübn.*—LASIOCAMPA *p*, *Schr.*, *Leach.*—GASTROPACHA E., *Och.*

5994. 1; lanestris*. (*Sam. I.* 16.)—(*Kirby & Sp. I. E.* iii. 223, 265 & 266.)—*Steph. Ill.* (*H.*) ii. 45.
Ph. Bo. Lanestris. *Linn.* ii. 815.—*Don.* vi. 73. *pl.* 310. ♀.—*Stew.* ii. 146.—*Turt.* iii. 203.
Bo. Lanestrosus. (*Haw. Pr.* 10.)
Bo. Lanestrus. *Haw.* 124.
Ph. Lanestris. *Berk. S.* i. 134.
La. Lanestris. (*Leach E. E.* ix. 132.)
Small Egger. (*Albin. pl.* xix. *f.* 26. *a–d.*)—(*Wilkes.* 25. *pl.* 53.)—*Harr. A. pl.* 25. *f. k–o.*—*Harr. V. M.* 21.

Genus 54: (837). CNETHOCAMPA *mihi.*

GASTROPACHA *p*, *Och.*—CLISIOCAMPA *p*, *Curtis.*—LASIOCAMPA *p*, (*Kirby.*)

5995 † 1. processionea[a].
5996 † 2, Pityocampa. *Steph. Ill.* (*H.*) ii. 48.
Bo. Pityocampa. *Fabr.* iii. *a.* 431.—*Hüb. Bo. pl.* 36. *f.* 161. ♀.—(*Kirby & Sp. I. E.* i. 131.)
La. Pityocampa. (*Kirby & Sp. I. E.* iii. 254.) In Mus. *Brit.*

Genus 55: (838). CLISIOCAMPA, *Curtis*, *Steph.*

LASIOCAMPA *p*, *Schr.*, *Leach*, *Samou.*—GASTROPACHA E. γ., *Och.*—TRICHODA[b], *Hübn.*—TRICHODIA *mihi olim.*

5997. 1, castrensis. *Steph. Ill.* (*H.*) ii. 48. *pl.* 13. *f.* 2. ♂.—*Curtis.* v. *pl.* 229.
Ph. Bo. Castrensis. *Linn.* ii. 818.—*Stew.* ii. 148.—*Turt.* iii. 205.
Bo. castrensis. *Hüb. Bo. pl.* 40. *f.* 177. ♂. 178. ♀.
Bo. Castrensus: Ground Lackey. *Haw.* 128.
Bo. Castrosus. (*Haw. Pr.* 10.)
La. Castrensa. (*Sam. I.* 24.)
Ph. tesseraria pratensis. *DeGeer.* G. 36. 43.
Ph. Neustria β. *Linn. F. No.* 1102.

5998. 2; Neustria*. *Curtis l. c. supra.*—*Steph. Ill.* (*H.*) ii. 49.
Ph. Bo. Neustria. *Linn.* ii. 818.—*Berk. S.* i. 135.—*Stew.* ii. 148.—*Don.* iii. 61. *pl.* 95.—*Turt.* iii. 205.

---

[a] 5995 † 1. processionea. *Steph. Ill.* (*H.*) ii. 47. nota.
Ph. Bo. processionea. *Linn.* ii. 819.—*Martyn.* (!)—*Stew.* ii. 148. (!)—*Turt.* iii. 204. (!)
Bo. processionea: processionary M.—*K. & Sp. l. c. Hüb. Bo. pl.* 36. *f.* 159. ♂. *f.* 160. ♀.—(*Kirby & Sp. I. E.* i. 130 & 478.—*Id.* ii. 23.)
Cl. processionea. (*Curtis. folio* 229. (!)

[b] TRICHODA: Genus Infusoriam. Vide *Don.* i. *p.* 21. *&c.*

Bo. Neustrius: barred Tree Lackey. (*Haw. Pr.* 10.)—*Haw.* 129.
La. Neustria. (*Leach E. E.* ix. 132.)—(*Sam. I.* 24.)—(*Kirby & Sp. I. E.* iii. 80.)
Ph. No. 89. *List. Goëd.* (*Angl.*) 85. *f.* 89.—*Id.* (*Lat.*) 204. *f.* 89.
Lacquey. *Albin. pl.* xix. *f.* 27. *e–l.*—(*Wilkes D. pl.* 10. *f.* 8.)—(*Wilkes.* 21. *pl.* 45.)—*Harr. A. pl.* 17. *a–f.*—*Harr. V. M.* 33.
β; *Steph. Ill. l. c.* Bo. bilineatus: striped Tree Lackey. *Haw.* 130. Lackey likeness. *Harr. V. M.* 33?

Genus 56: (839). EUTRICHA *mihi.*

EUTRICHÆ, *Hübn.*—GASTROPACHA A. *p*, *Och.*—LASIOCAMPA *p*, *Schra.*—ODONESTIS *p*, *Curtis.*

5999 † 1, Pini *? *Steph. Ill.* (*H.*) ii. 50.
Ph. Bo. Pini. *Linn.* ii. 814.—*Don.* v. 103. *pl.* 177. *pl.* 178. larva.—*Stew.* ii. 145.—*Turt.* iii. 102.—(*Leach E. E.* ix. 132.)
Bo. Pinus. (*Haw. Pr.* 7.)—*Haw.* 80.
Od. Pini. *Curtis.* i. *pl.* 7.
Ph. Pini. *Berk. S.* i. 133.
Ga. Pinus. (*Sam. I.* 68.)
Wild Pine tree Lappit M.: Ph. Pini. (*Wilkes.* 29. *pl.* 61.)
Pine tree Lappet. *Harr. V. M.* 33. In Mus. *Brit. et D. Vigors.*

Genus 57: (840). ODONESTIS, *Germar*, *Steph.*

LASIOCAMPA *p*, *Schra.*—BOMBYX, *Leach.*—GASTROPACHA B., *Och.*—ODENESIS, *Samou.*, (*Kirby.*)—ODONESTIS *p*, *Curtis.*

6000. 1; Potatoria *. *Germar.*—(*Curtis. vide* i. *folio* 7.)—*Steph. Ill.* (*H.*) ii. 51.
Ph. Bo. potatoria. *Linn.* ii. 813.—*Don.* v. *pl.* 148.—*Stew.* ii. 145.—*Turt.* iii. 201.
Bo. potatorius. (*Haw. Pr.* 7.)—*Haw.* 84.
Ph. potatoria. *Berk. S.* i. 133.
Bo. potatoria. (*Leach E. E.* ix. 132.)—(*Kirby & Sp. I. E.* i. 390.)
Ode. potatoria. (*Samou.* 247. *pl.* 12. *f.* 3.)—(*Kirby & Sp. I. E.* iii. 99. 221.)
The Drinker Caterpillar: Ph. No. 82. *List. Goëd.* (*Angl.*) 82. *f.* 82.—*Id.* (*Lat.*) 195. *f.* 82.—*Albin. pl.* xvii.
Drinker M.: Ph. potatoria. (*Wilkes D. pl.* 8. *f.* 2.)—(*Wilkes.* 27. *pl.* 58.)—*Harr. A. pl.* 42. *f. a, m, n, p, & t.*—*Harr. V. M.* 21.

Genus 58: (841). GASTROPACHA, *Och.*, *Samou.*, *Curtis*, *Steph.*

LASIOCAMPA *p*, *Schra.*

6001. 1; Quercifolia *. *Och.* iii. 247.—(*Sam. I.* 18.)—(*Kirby & Sp. I. E.* iii. 36 & 99.)—*Curtis.* i. *pl.* 24.—*Steph. Ill.* (*H.*) ii. 52.
Ph. Bo. Quercifolia. *Linn.* ii. 812.—*Don.* vii. *pl.* 332.—*Stew.* ii. 144.—*Turt.* iii. 199.
Ph. Quercifolia. *Berk. S.* i. 133.

Bo. Quercifolius. (*Haw. Pr.* 7.)—*Haw.* 95.
Bo. Quercifolia. (*Leach E. E.* ix. 132.)—(*Kirby & Sp. I. E.* ii. 222.)
Caterpillar with Lappets. *Albin. pl.* xvi.
Lappit M.: Ph. Quercifolia. (*Wilkes D. pl.* 9. *f.* 1.)—(*Wilkes.* 27. *pl.* 57.)—*Harr. A. pl.* 43. *f. a–c.*—*Harr. V. M.* 33.

6002 † 2. Ilicifolia[a].

6003 † 3. Populifolia[b].

Familia XII: (125). ARCTIIDÆ *mihi.*

(PHALÆNA BOMBYX *p*, *et* PH.-NOCTUA *p*, *Linné.*—NOCTUO-BOMBYCITES, *Latr.*—ARCTIIDÆ *p*, *Leach.*)

Genus 59: (842). HYPOGYMNA, *Hübner, Steph.*, (*Kirby.*)

LARIA[c] *p*, *Schra.*—HYPOGYMNÆ *p*, *Hüb.*—LIPARIS *p*, *Och.*, *Samou.*

6004. 1: dispar*. (*Kirby & Sp. I. E.* iii. 102 & 223.)—*Steph. Ill.* (*H.*) ii. 55.
Ph. Bo. dispar. *Linn.* ii. 821.—*Berk. S.* i. 136.—*Don.* v. 67. *pl.* 163.—*Stew.* ii. 150.—*Turt.* iii. 207.
Bo. disparus. (*Haw. Pr.* 8.)—*Haw.* 88.
Bo. dispar. (*Kirby & Sp. I. E.* i. 206.)
La. dispar. (*Leach E. E.* ix. 132.)
Li. dispar. (*Sam. I.* 25.)
Gypsey M.: Ph. dispar. (*Wilkes D. pl.* 10. *f.* 2.)—(*Wilkes.* 22. *pl.* 42.)—*Harr. V. M.* 29.

Genus 60: (843). PSILURA *mihi.*

LARIA *p*, *Schra.*, *Leach.*—LIPARIS *p*, *Och.*, *Samou.*—HYPOGYMNÆ *p*, *Hüb.*

6005. 1; Monacha*. *Steph. Ill.* (*H.*) ii. 57.
Ph. Bo. Monacha. *Linn.* ii. 821.—*Berk. S.* i. 136.—*Don.* vii. *pl.* 227.—*Stew.* ii. 150.—*Turt.* iii. 211.
Bo. Monachus. (*Haw. Pr.* 7.)—*Haw.* 87.
Bo. Monacha. (*Kirby & Sp. I. E.* i. 206.)
La. Monacha. (*Leach E. E.* ix. 132.)
Li. Monacha. (*Sam. I.* 25.)
Black Arches, a Moth: Ph. Monacha. (*Wilkes D. pl.* 4. *f.* 7. ♂. ♀.)—(*Wilkes.* 19. *pl.* 39.)—*Harr. V. M.* 9.

---

[a] 6002 † 2. Ilicifolia. *Och.* iii. 240.—*Steph. Ill.* (*H.*) ii. 53. nota.
Ph. Bo. Ilicifolia. *Linn.* ii. 813.—*Turt.* iii. 199. (!)
Bo. Betutifolia. *Esper.* iii. *pl.* 7. *f.* 2.

[b] 6003 † 3. Populifolia. *Och.* iii. 245.—*Steph. Ill.* (*H.*) ii. 53. nota.
Bo. Populifolia. *Fabr.* iii. *a.* 420.—*Hüb. Bo. pl.* 43. *f.* 189. ♂.—*Martyn V. M.* (!)

[c] LARIA. Vide *pars* i. *p.* 80.—LIPARIS: Genus Piscium. Vide *Will. Ich. App.* 17.

Genus 61: (844). DASYCHIRA *mihi.*

ORGYIA *p, Och.*—DASYCHIRÆ *p, Hübn.*—LARIA *p, Schra., Leach, Samou.*

6006. 1; fascelina*. *Steph. Ill.* (*H.*) ii. 58.
Ph. Bo. fascelina. *Linn.* ii. 825.—*Berk. S.* i. 137.—*Don.* xvi. *pl.* 576.—*Stew.* ii. 152.—*Turt.* iii. 208.
Bo. fascelinus. (*Haw. Pr.* 8.)—*Haw.* 102.
Bo. Medicaginis. *Hüb. Bo. pl.* 21. *f.* 81.
Ph. porrecta cana. *DeGeer. G.* 38. 57.
La. fascelina. (*Leach E. E.* ix. 132.)—(*Sam. I.* 24.)—(*Kirby & Sp. I. E.* iii. 223.)
Pa. No. 80. *List. Goëd.* (*Angl.*) 81. *f.* 80.—*Id.* (*Lat.*) 192. *f.* 80.
Albin. pl. xxvi. *f.* 39. *e–h.*
Black Tussock M.: Ph. fascelina. (*Wilkes.* 30. *pl.* 62.)
Dark Tussock M. (*Wilkes* D. *pl.* 9. *f.* 5.)—*Harr. V. M.* 51.

6007. 2; pudibunda*. *Steph. Ill.* (*H.*) ii. 59.
Ph. Bo. pudibunda. *Linn.* ii. 824.—*Berk. S.* i. 137.—*Don.* v. 59. *pl.* 160.—*Stew.* ii. 152.—*Turt.* iii. 207.
Bo. pudibundus. (*Haw. Pr.* 8.)—*Haw.* 101.
La. pudibunda. (*Leach E. E.* ix. 132.)—(*Sam. I.* 24.)—(*Kirby & Sp. I. E.* iii. 223.)
Bo. Juglandis. *Hüb. Bo. pl.* 21. *f.* 84, 85.
Ph. porrecta alba. *DeGeer.* G. 38. 56.
Pa. No. 81. *List. Goëd.* (*Angl.*) 82. *f.* 81.—*Id.* (*Lat.*) 193. *f.* 81.
Albin. pl. xxvi. *f.* 38. *a–d.*
Yellow Tussock M.: Ph. pudibunda. (*Wilkes.* 30. *pl.* 63.)—*Harr. A. pl.* 15.
Light Tussock M. *Wilkes* D. *pl.* 10. *f.* 4.
Pale Tussock M. *Harr. V. M.* 51.

Genus 62: (845). DEMAS *mihi.*

COLOCASIA, *Och.*, (*Samou.*)—BOMBYX *p, Haworth.*

6008. 1; Coryli*. *Steph. Ill.* (*H.*) ii. 60.
Ph. Bo. Coryli. *Linn.* ii. 823.—*Berk. S.* i. 136.—*Don.* ix. 45. *pl.* 309.—*Stew.* ii. 151.—*Turt.* iii. 211.
Bo. Corylus. (*Haw. Pr.* 9.)—*Haw.* 102.
La. Coryli. (*Leach E. E.* ix. 132.)
Bo. Coryli. (*Sam. I.* 6.)
Co. Coryli. (*Och.* iv. 63.)
Albin. pl. xc. *f. e–h.*
Nut-tree Tussock M.: Ph. Coryli. (*Wilkes.* 31. *pl.* 66.)—*Harr. V. M.* 49.

Genus 63: (846). ORGYIA.

LARIA *p, Schra.*—DASYCHIRÆ *p, Hübn.*—ORGYIA *p, Och.*

6009. 1; antiqua*. *Och.* iii. 221.—*Steph. Ill.* (*H.*) ii. 61.

Ph. Bo. antiqua. *Linn.* ii. 825.—*Berk. S.* i. 137.—*Don.* i. *pl.* 16.—*Stew.* ii. 157.—*Turt.* iii. 226.
Bo. antiquus. (*Haw. Pr.* 10.)—*Haw.* 132.
Bo. antiqua. (*Sam. I.* 6.)
Ph. gonostigma. *Scop. C.* 199.
Ph. paradoxa. *DeGeer.* G. 36.
Ph. No. 79. *List. Goëd.* (*Angl.*) 80. *f.* 79. ♀.—*Id.* (*Lat.*) 191. *f.* 79. ♀.—*Albin. pl.* lxxxix. *a–e.*
Red-spot Tussock M.: Ph. antiqua. (*Wilkes.* 30. *pl.* 64.)
Common Vapourer. *Harr. A. pl.* 20. *f. h–p.*—*Harr. V. M.* 52.

6010. 2; gonostigma*. *Och.* iii. 218.—*Steph. Ill.* (*H.*) ii. 61.
Ph. Bo. gonostigma. *Linn.* ii. 826.—*Berk. S.* i. 137.—*Don.* ix. 61. *pl.* 316.—*Stew.* ii. 157.—*Turt.* ii. 226.
Bo. gonostigmatus. (*Haw. Pr.* 10.)—*Haw.* 132.
Bo. gonostigmata. (*Sam. I.* 7.)
Ph. antiqua. *Scop. C.* 198.
Pa. No. 78. *List. Goëd.* (*Angl.*) 78. *f.* 78 *a.* ♂. 79. *f.* 78 *b.* ♀. —*Id.* (*Lat.*) 185. *f.* 78 *a.* ♂. 187. *f.* 78 *b.* ♀.
Albin. pl. xc. *f. a–d.*
Orange Tussock M.: Ph. gonostigma. (*Wilkes.* 31. *pl.* 65.)
Scarce Vapourer. *Harr. A. pl.* 14. *f. a–g.*—*Harr. V. M.* 51.

Genus 64: (847). LÆLIA *mihi.*

LEUCOMÆ *p?* *Hüb.*—ARCTIA *p, Curtis.*

6011. 1: cænosa. *Steph. Ill.* (*H.*) ii. 63.
Bo. cænosa. *Hüb. Bo. pl.* 51. *f.* 218. ♂.
Ar. cænosa: Whittlesea Arctia. *Curtis.* ii. *pl.* 68. ♂. ♀.
Læ. pyrrhopus *mihi olim.*

Genus 65: (848). LEUCOMA *mihi.*

LEUCOMÆ *p, Hübn.*—LIPARIS *p, Och.*—LARIA *p, Schrank.*—ARCTIA** *p, Leach, Samou.*—ARCTIA *p, Curtis.*

6012 † 1, Vau-nigra*. *Steph. Ill.* (*H.*) ii. 64. *pl.* 16. *f.* 2. ♀.
Bo. Vau-nigrum. *Fabr.* iii. *a.* 458.
Bo. nivosa. *Hüb. No. pl.* 18. *f.* 71.
Bo. V. nigrosus. (*Haw. Pr.* 8.)
Bo. V. nigrus: the black V. *Haw.* 107.
Ar. V. nigra. (*Sam. I.* 5.)—(*Curtis l. c. fo.* 68.)
In Mus. *Brit.* D. *Haworth, Swainson, et Vigors.*

6013. 2; Salicis*. *Steph. Ill.* (*H.*) ii. 64.
Ph. Bo. Salicis. *Linn.* ii. 282.—*Berk. S.* i. 136.—*Don.* i. 65. *pl.* 30.—*Stew.* ii. 151.—*Turt.* iii. 215.
Bo. Salicinus. (*Haw. Pr.* 8.)—*Haw.* 107.
Ar. Salicis. (*Leach E. E.* ix. 133.)—(*Sam. I.* 5.)—(*Curtis l. c. supra.*)—(*Kirby & Sp. I. E.* iii. 176 & 222.)

Ph. apparens. *DeGeer. G.* 37. 46.
Pa. No. 87. *List.* Goëd. (*Angl.*) 84. *f.* 87.—*Id.* (*Lat.*) 202. *f.* 87.—*Albin. pl.* lxxxiv. *f. a–d.*
White Sattin M.: Ph. Salicis. (*Wilkes.* 21. *pl.* 41.)—*Harr. A. pl.* 5.—*Harr. V. M.* 43.

Genus 66: (849). PORTHESIA *mihi.*

LARIA *p, Schrank.*—LIPARIS *p, Och.*—LEUCOMÆ *p, Hübn.*—ARCTIA** *p, Leach, Samou.*—ARCTIA *p, Curtis.*

6014. 1; chrysorrhœa*. *Steph. Ill.* (*H.*) ii. 65.
Ph. Bo. chrysorrhœa. *Linn.* ii. 822.—*Berk. S.* i. 136.—*Don.* i. *pl.* 10.—*Stew.* ii. 150.—*Turt.* iii. 217.
Bo. chrysorrhœus. (*Haw.* 108.)
Bo. chrysorrhœa. (*Kirby & Sp. I. E.* i. 30.)
Ar. chrysorrhœa. (*Leach E. E.* ix. 133.)—(*Sam. I.* 5.)—(*Kirby & Sp. I. E.* iii. 173 & 219.)—(*Curtis l. c. fo.* 68.)
Pa. No. 90? *List.* Goëd. (*Angl.*) 86. *f.* 90?—*Id.* (*Lat.*) 206. *f.* 90?
Albin. pl. lxxxvii. *f. e–i.*
Yellow-tail M.: Ph. chrysorrhœa. (*Wilkes.* 28. *pl.* 59.)—*Harr. A. pl.* 25. *f. p–s.*—*Harr. V. M.* 47.
♂; Bo. aurifluus: spotted yellow-tail.—*Haw. l. c. Esper. Bo. pl.* xxxix. *f.* 6.—(*Haw. Pr.* 8.)—*Haw.* 108.

6015. 2; auriflua*. *Steph. Ill.* (*H.*) ii. 66.
Ph. Bo. phæorrhœa. *Curtis on Brown-tail Moth, cum figura.*—*Don.* x. 4. *pl.* 555.
Bo. phæorrhœus. (*Haw. Pr.* 8.)—*Haw.* 109.
Bo. phæorrhœa. (*Kirby & Sp. I. E.* i. 204.)
Ar. phæorrhœa. (*Leach E. E.* ix. 133.)—(*Sam. I.* 5.)—(*Curtis l. c. supra.*)
♂; Bo. auriflua. *Fabr.* iii. *a.* 458.—*Hüb. Bo. pl.* 18. *f.* 68. ♂. *f.* 69. ♀.
Brown-tail M. *Albin. pl.* lxxxvii. *f. a–d.*—*Harr. V. M.* 47.

Genus 67: (850). HYPERCOMPA, *Hübner? Steph.*

HYPERCOMPÆ *p, Hübner.*—EYPREPIA B. *p, Och.*—ARCTIA *p, Schra.*—NOCTUA *p, Esper.*—CALLIMORPHA *p, Latr., Leach, Samou.*

6016. 1; Dominula*. *Steph. Ill.* (*H.*) ii. 67.
Ph. No. Dominula. *Linn.* ii. 834.—*Berk. S.* i. 139.—*Don.* iv. *pl.* 141.—*Stew.* ii. 158.—*Turt.* iii. 225.
No. Dominulina. (*Haw. Pr.* 16.)
Bo. Domina. *Hüb. Bo. pl.* 27. *f.* 117. ♂. *f.* 118. ♀.
Cal. Dominula. (*Latr. G.* iv. 221.)—(*Leach E. E.* ix. 133.)—(*Sam. I.* 8.)
Albin. pl. xxii. *f.* 31. *a–d.*

Scarlet Tyger M.: Ph. Dominula. (*Wilkes D. pl.* 12. *f.* 3.) —(*Wilkes.* 19. *pl.* 38.)—*Harr. A. pl.* 40. *f. a–d.*—*Harr. V. M.* 51.

Genus 68: (851). EUTHEMONIA *mihi.*

EYPREPIA *p, Och., Curtis.*—HYPERCOMPÆ *p, Hüb.*—ARCTIA *p, Schrank.*

6017. 1; Russula*. *Steph. Ill.* (*H.*) ii. 68.
Ph. Bo. Russula. *Linn.* ii. 830.—*Berk. S.* i. 138.—*Don.* vi. 81. *pl.* 214.—*Stew.* ii. 156.—*Turt.* iii. 220.
Bo. Sannius. (*Haw. Pr.* 10.)—*Haw.* 133.
Ar. Russula. (*Leach E. E.* ix. 133.)—(*Sam. I.* 5.)
Ey. Russula. *Och.* iii. 309.—*Curtis.* i. *pl.* 21.
Clouded Buff: Ph. quadra. *Harr. V. M.* 15.
♂, Ph. Bo. Sannio. *Linn. F.* 1135.

Genus 69: (852). ARCTIA *mihi.*

ARCTIA *p, Schra., Latr., Leach.*—EYPREPIA *p, Och., Curtis.*—HYPERCOMPÆ *p, Hüb.*—CALLIMORPHA *p,* (*Kirby.*)

6018. 1; Caja*. *Schra. B.* ii. 152.—(*Leach E. E.* ix. 133.)—(*Sam. I.* 5.)—*Steph. Ill.* (*H.*) ii. 69.
Ph. Bo. Caja. *Linn.* ii. 819.—*Berk. S.* i. 135.—*Barb.* G. *pl.* 10. *f.* 38.—*Don.* i. *pl.* 15.—*Stew.* ii. 149.—*Turt.* iii. 223.
Bo. Cajus. (*Haw. Pr.* 7.)—*Haw.* 93.
Cal. Caja. (*Kirby & Sp. I. E.* ii. 252; iii. 89, 176 & 223.)
Ey. Caja. *Och.* iii. 335.—(*Curtis l. c. fo.* 21.)
Ph. erinacea. *DeGeer.* G. 36. 45.
Pa. No. 99. *List.* Goëd. (*Angl.*) 89. *f.* 99.—*Id.* (*Lat.*) 217. *f.* 99.—*Albin. pl.* xx.
Great Tyger M.: Ph. Caja. (*Wilkes* D. *pl.* 3. *f.* 4.)—(*Wilkes.* 18. *pl.* 36.)
Large Tyger M. *Harr. A. pl.* 12. *f.* g–m.—*Harr. V. M.* 51.
β; *Steph. l. c. supra.*—*Ernst. Pap.* iv. *pl.* cxxxix. *f.* 187 *g, fem.*
γ; *Steph. l. c. supra.*—*Ernst. Pap.* iv. *pl.* cxl. *f.* 187, *o.*
ε; *Steph. l. c. supra.*—*Ernst. Pap.* iv. *pl.* cxl. *f.* 187, *n.*
† η; *Steph. l. c. supra.*—*Ernst. Pap.* iv. *pl.* cxxxix. *f.* 187, *d.*
θ; *Steph. l. c. supra.*—*Ernst. Pap.* iv. *pl.* cxl. *f.* 187, *k.*
ι, *Steph. l. c. supra.*—*Ernst. Pap.* iv. *pl.* cxli. *f.* 187, *i.*
† κ, *Steph. l. c. supra.*—*Ernst. Pap.* iv. *pl.* cxl. *f.* 187, *l.*
η in *Mus. Brit.* κ in Mus. D. *Stone.*

6019 † 2. matronula[a].

6020. 3; Villica*. *Schr. B.* ii. 152.—(*Leach E. E.* ix. 133.)—(*Sam. I.* 5.)—(*Kirby & Sp. I. E.* iii. 176.)—*Steph. Ill.* (*H.*) ii. 71.

---

[a] 6019 † 2. matronula. *Schra. B.* ii. 152.—*Steph. Ill.* (*H.*) ii. 70. nota.
Ph. Bo. matronula. *Linn.* ii. 835.—*Turt.* iii. 222. (!)
Bo. Matrona. *Hüb. Bo. pl.* 55. *f.* 259. ♂. *pl.* 32. *f.* 138. ♀.

Ph. Bo. Villica. *Linn.* ii. 820.—*Berk. S.* i. 135.—*Don.* ii. *pl.* 71.—*Stew.* ii. 149.—*Turt.* iii. 222.
Bo. Villicus. (*Haw. Pr.* 7.)—*Haw.* 94.
Ey. Villica. *Och.* iii. 330.—(*Curtis l. c. supra.*)
Albin. pl. xxi. *f.* 29. *a–d.*
Cream-spot Tyger M.: Ph. Villica. (*Wilkes D. pl.* 11. *f.* 7.)—(*Wilkes.* 18. *pl.* 37.)—*Harr. A. pl.* 4.—*Harr. V. M.* 51.
β; *Steph. l. c. supra.*—*Ernst. Pap.* iv. *pl.* cli. *f.* 196, *o.*

6021 ‡ 4. Hebe[a].

6022 ‡ 5. purpurea[b].

6023 ‡ 6. aulica[c].

Genus 70: (853). NEMEOPHILA *mihi.*

EYPREPIA B. *p, Och.*—HYPERCOMPÆ *p, Hübn.*—CALLIMORPHA *p, Latr.?*—ARCTIA *p, Schr.*

6024. 1; Plantaginis*. *Steph. Ill.* (*H.*) ii. 73.
Ph. Bo. Plantaginis. *Linn.* ii. 820.—*Berk. S.* i. 135.—*Don.* iv. 69. *pl.* 134.—*Stew.* ii. 150.—*Turt.* iii. 221.
Bo. Plantaginis. (*Haw. Pr.* 10.)—*Haw.* 94.
Ar. Plantaginis. (*Leach E. E.* ix. 133.)—(*Sam. I.* 5.)
Wood Tyger M.: Ph. Plantaginis. (*Wilkes.* 24. *pl.* 50.)
Small Tyger M. (*Wilkes D. pl.* 1. *f.* 6.)—*Harr. A. pl.* 16. *f. g–m.*—*Harr. V. M.* 51.
Var. β; *Steph. Ill.* (*H.*) *l. c.*—*Ernst. Pap.* iv. *pl.* cxlvii. *f.* 192, *b.*
Bo. hospita. (*Wien. V.* 310.)—*Bork. E. S.* iii. 470.
Var. γ; *Steph. l. c. supra.*—*Ernst. Pap.* iv. *pl.* cxlvii. *f.* 192, *a.*
Ph. Alpicola. *Scop. C. No.* 507?
Var. δ, *Steph. l. c. supra.*—*Ernst. Pap.* iv. *pl.* cxlv. *f.* 191, *f.*
Var. ε. *Steph. l. c. supra.*—*Ernst. Pap.* iv. *pl.* cxlvi. *f.* 191, *p.*

Genus 71: (854). PHRAGMATOBIA *mihi.*

EYPREPIA D., *Och.*—ARCTIA *p, Schra.*—CALLIMORPHA *p, Samou.*

6025. 1; fuliginosa*. *Steph. Ill.* (*H.*) ii. 74.
Ph. No. fuliginosa. *Linn.* ii. 836.—*Berk. S.* i. 139.—*Don.* iii. 21. *pl.* 80.—*Stew.* ii. 159.—*Turt.* iii. 236.

---

[a] 6021 ‡ 4. Hebe. *Schra. B.* ii. 152.—*Steph. Ill.* (*H.*) ii. 71. nota.
Ph. Bo. Hebe. *Linn.* ii. 820.
Bo. Hebe. *Hüb. Bo. pl.* 30. *f.* 129. ♂.

[b] 6022 ‡ 5. purpurea. *Schra. B.* ii. 153.—*Steph. Ill.* (*H.*) ii. 71. nota.
Ph. Bo. purpurea. *Linn.* ii. 828.—*Stew.* ii. 154. (!)—*Turt.* iii. 221. (!)
Bo. purpurea. *Hüb. Bo. pl.* 33. *f.* 142. ♀. *pl.* 53. *f.* 229. ♂. *var.*

[c] 6023 ‡ 6. aulica. *Schra. B.* ii. 152.—*Steph. Ill.* (*H.*) ii. 71. nota.
Ph. Bo. aulica. *Linn.* ii. 829.—*Martyn V. M.* (!)—*Stew.* ii. 154. (!) *Turt.* iii. 213. (!)
Bo. aulica. *Hüb. Bo. pl.* 32. *f.* 139. ♂.

No. fuliginosina. (*Haw. Pr.* 16.)
Ca. fuliginosa. (*Sam. I.* 8.)
Spotted red & white underwing M.: Ph. fuliginosa. (*Wilkes.* 23. *pl.* 49.)
Ruby Tyger. *Harr. A. pl.* 27. *f. l–n.*—*Harr. Ex.* 32. *pl.* viii. *f.* 7. ♂.—*Harr. V. M.* 51.

Genus 72: (855.) SPILOSOMA *mihi, Curtis.*

ARCTIA *p, Schra.*—EYPREPIA *p, Och.*

6026. 1; Menthastri*. (*Curtis l. c. infra.*)—*Steph. Ill.* (*H.*) ii. 75.
Bo. Menthastri. *Fabr.* iii. *a.* 452.
Ph. Menthastri. *Don.* vi. 23. *pl.* 189.
Bo. Menthrastrus. (*Haw. Pr.* 9.)
Ph. Bo. lubricepeda *α*. *Gmel.* iv. 2434.
Ph. Erminea. *Linn. Trans.* (*Marsham.*) i. 70. *pl.* 1. *f.* 1.—*Stew.* ii. 154.—*Turt.* iii. 214.
Bo. Ermineus: Large Ermine. *Haw.* 111.
Ar. Erminea. (*Kirby & Sp. I. E.* iii. 173.)
Ar. Menthrastri. (*Leach E. E.* ix. 133.)—(*Sam. I.* 5.)
Ph. Bo. lubricepeda, ♀. *Linn.* ii. 829.
Ph. No. 96. *List. Goëd.* (*Angl.*) 88. *f.* 96.—*Id.* (*Lat.*) 213. *f.* 96.—*Albin. pl.* xxiv. *f. g–k*; (the dissembler.)
Great Ermine M.: Ph. lubricepeda. (*Wilkes.* 20. *pl.* 40.)—*Harr. A. pl.* 38. *f. k, l.*—*Harr. V. M.* 23.
β; *Steph. Ill. l. c. supra.*—*Ernst. Pap.* iv. *pl.* clviii. *f.* 204, *d.*
γ; *Steph. l. c. supra.*—*Ernst. Pap.* iv. *pl.* clviii. *f.* 204, *h.*
δ; *Steph. l. c. supra.*—*Albin. l. c. fig.*
ε; *Steph. Ill.* (*H.*) *l. c. pl.* 16. *f.* 3.
† *ζ*, *Steph. Ill. l. c. supra.*—Sp. Walkerii. *Curtis.* ii. *pl.* 92.
ξ in Mus. D. *Walker.*

6027. [2, Urticæ*.] (*Curtis l. c. supra.*)—*Steph. Ill.* (*H.*) ii. 76.
Bo. Urticæ: Dingy White.—*Haw. l. c.* *Hüb. Bo. pl.* 35. *f.* 134?—*Ent. Trans.* (*Haworth.*) i. 336.—(*Haw.* 144.)

6028. 3; papyratia*. *Steph. Ill.* (*H.*) ii. 76.
Ph. papyratia. *Linn. Trans.* (*Marsham.*) i. 72. *pl.* 1. *f.* 4.—*Stew.* ii. 155.—*Turt.* iii. 214.—*Don.* xvi. *pl.* 571.
Bo. papyratius. (*Haw. Pr.* 9.)—*Haw.* 112.
Ar. papyritia. (*Leach E. E.* ix. 133.)—(*Sam. I.* 5.)
Sp. papyritia. (*Curtis l. c. supra.*)
Albin. pl. xxi. *f.* 30. *e–h.*
Water Ermine. *Harr. V. M.* 25.

6029. 4; lubricepeda*. (*Curtis l. c. supra.*)—*Steph. Ill.* (*H.*) ii. 77.
Ph. Bo. lubricepeda. *Linn.* ii. 829.—*Barb. G. pl.* 10. *f.* 69.—*Berk. S.* i. 138.—*Linn. Trans.* (*Marsham.*) i. 71. *pl.* 1. *f.* 2. *Stew.* ii. 155.—*Turt.* iii. 214.—*Don.* xvi. *pl.* 568.

Bo. lubricepedatus. (*Haw. Pr.* 8.)—*Haw.* 110.
Ar. lubricepeda. (*Leach E. E.* ix. 133.)—(*Sam. I.* 5.)
Ph. Lepus. *DeGeer. G.* 37. 47.
Pa. No. 93. *List. Goëd.* (*Angl.*) 87. *f.* 93.—*Id.* (*Lat.*) 210. *f.* 93.—*Albin. pl.* xxiv. *f.* 35. *a–e.*
Spotted buff M.: Ph. lubricepeda. (*Wilkes.* 21. *pl.* 43.)
Cream dot stripe. *Harr. A. pl.* 17. *f. g–i.*—*Harr. V. M.* 45.

6030 † 5, radiata. (*Curtis l. c. supra.*)—*Steph. Ill.* (*H.*) ii. 77.
Bo. radiatus: Radiated Buff. *Ent. Trans.* (*Haworth.*) i. 336.
In Mus. D. *Haworth.*

Genus 73: (856). DIAPHORA *mihi.*

ARCTIA *p, Schra., Leach, Samou.*—EYPREPIA E., *Och.*

6031. 1; mendica*. *Steph. Ill.* (*H.*) ii. 78.
Ph. Bo. mendica. *Linn.* ii. 822.—*Linn. Trans.* (*Marsham.*) i. 72. *pl.* 1. *f.* 3.—*Stew.* ii. 153.—*Turt.* iii. 214.—*Don.* xi. 77. *pl.* 388.
Bo. mendicus. (*Haw. Pr.* 9.)—*Haw.* 112.
Ar. mendica. (*Leach E. E.* ix. 133.)—(*Sam. I.* 5.)
Spotted Muslin, or seven spot Ermine. *Harr. A. pl.* 35. *f. m.* —*Harr. V. M.* 25.

Genus 74: (857). PENTHOPHERA, *Germar, Curtis, Steph.*

LIPARIS *p, Och.*—PENTHROPHERA, (*Samou.*)

6032 † 1. Rubea[a].

6033 † 2. nigricans. *Curtis.* v. *pl.* 213.—*Steph. Ill.* (*H.*) ii. 79.
Ph. Bo. Morio. *Linn.* ii. 828? In Mus. D. Dale.

Genus 75: (858). PSYCHE, *Schrank,* (*Kirby,*) *Steph.*

NUDARIA *p, Haw.*—TINEA *p, Hüb.*

6034. 1, fusca*. (*Ing. Inst.* 91.)—*Steph. Ill.* (*H.*) ii. 80. *pl.* 18. *f.* 3. ♂. *f.* 4. ♀.
Nu. fusca: the brown Muslin. *Haw.* 157.—(*Sam. I.* 31.)
Ps. calvella. *Och.* iii. 172?
Ti. hirsutella. *Hüb. Ti. pl.* 1. *f.* 3?
Ti. fenestrella. (*Haw. Pr.* 35.)

Genus 76: (859). FUMEA, *Haw. Ent. Trans., Steph.*

PSYCHE *p, Schra.,* (*Kirby.*)—CANEPHORÆ, *Hüb.*—BOMBYX *p, Fabr.*—TINEA *p, Hüb.*—FUMARIA, *Haw. olim.,* (*Samou.*)

6035. 1; nitidella*. *Steph. Ill.* (*H.*) ii. 82.

---

[a] 6032 † 1. Rubea. *Germar.*—(*Samou.* 247.) (!)—*Steph. Ill.* (*H.*) ii. 79. nota.
Bo. Rubea. *Fabr.* iii. *a.* 445.—*Hüb. Bo. pl.* 56. *f.* 240. ♂. *pl.* 16. *f.* 60, 61. ♀.

Ti. nitidella. *Hüb. Ti. pl.* 1. *f.* 6.—(*Haw. Pr.* 35.)
Ps. Carpini. *Schrank B.* ii. *B.* 2. 90.
Bo. nana. *Bork. E. S.* iii. 283.
Fu. nitida: Shining Chimney Sweep. *Haw.* 374.

6036. 2; pulla*. *Steph. Ill.* (*H.*) ii. 82.
Bo. pulla. *Esper.* iii. *pl.* 44. *f.* 8.
Ti. plumella. (*Wien. V.* 133?)—*Hüb. Ti. pl.* 1. *f.* 7.—(*Haw. Pr.* 31.)
Fu. plumea: lesser Chimney Sweep. *Haw.* 374.
Small Chimney Sweep. *Harr. V. M.* 47?

6037. 3; muscella*. *Steph. Ill.* (*H.*) ii. 82.
Bo. muscella. *Fabr.* iii. *a.* 482.
Bo. atra. *Esper.* iii. *pl.* xliv. *f.* 7.
Bo. graminella. *Vieweg. T. V.* i. 68?
Bo. pulla. *Brahm. I. K.* 501.
Ti. muscella. *Hüb. Ti. pl.* 2. *f.* 8.
Ti. plumistrella. *Hüb. Ti. pl.* 31. *f.* 213.—(*Haw. Pr.* 34.)
Fu. plumistrea: Chimney Sweeper's Boy. *Haw.* 374.—(*Sam. I.* 17.)

6038. 4, Bombycella. *Steph. Ill.* (*H.*) ii. 83.
Ti. Bombycella. *Hüb. Ti. pl.* 1. *f.* 4?
Ti. lucidella. (*Haw. Pr.* 34.)
Fu. muscea: transparent Chimney-Sweeper. *Haw.* 373.

6039. 5, pectinea. *Steph. Ill.* (*H.*) ii. 83.
Bo. pectinella. *Fabr.* iii. *a.* 482.
Ti. pectinella. *Hüb. Ti. pl.* 1. *f.* 5.—(*Haw. Pr.* 35.)
Fu. pectinea: the light Chimney Sweeper. *Haw.* 373.

Genus 77: (860). NUDARIA, *Haworth*, (*Sam. I.*) *Steph.*

PHALÆNA-ATTACUS *p*, *Linné*.—BOMBYX *p*, *Fabr.*—LITHOSIA D., *Och.*

A. Alæ elongato-ellipticæ.

6040. 1; munda*: the Muslin. *Haw.* 156.—(*Sam. I.* 31.)—*Steph. Ill.* (*H.*) ii. 84.
Ph. At. mundana. *Linn.* ii. 812.
Bo. munda. *Fabr.* iii. *a.* 482.
Bo. nuda. *Hüb. Bo. pl.* 17. *f.* 63, 64.
Ph. mundanata. (*Haw. Pr.* 27.)

6041. [2; Hemerobia*.] *Steph. Ill.* (*H.*) ii. 84.
Bo. Hemerobia. *Hüb. Bo. pl.* 17. *f.* 65.

B. Alæ breviores, rotundatæ.

6042. 3: senex*. *Steph. Ill.* (*H.*) ii. 84. *pl.* 18. *f.* 2.
Bo. senex. *Hüb. Bo. pl.* 55. *f.* 236. ♂. *f.* 237. ♀.
Nu. rotunda: round-winged Muslin. *Haw.* 156.—(*Sam. I.* 31.)

Genus 78: (861). HETEROGENEA, *Knoch, Steph.*

HEPIALUS *p, Fabr.*

6043. 1, Asellus. *Steph. Ill.* (*H.*) ii. 85. *pl.* 17. *f.* 2.
Hep. Asellus. (*Wien. V.* 65.)—*Fabr.* iii. *b.* 7.
Het. cruciata. *Knoch. Beitr.* iii. 63. *pl.* 3.

Genus 79: (862). LIMACODES, (*Latr.*,) *Steph.*

BOMBYX *p, Fabr.*—HEPIALUS *p, Fabr.*—TORTRIX *p, Hüb.*—PHALÆNA *p, Don.*—APODA, *Haworth, Samou.*—HETEROGENEA *p, Knoch.*

6044. 1; Testudo*. *Steph. Ill.* (*H*). ii. 86.
Hep. Testudo. (*Wien. V.* 65.)—*Fabr.* iii. *b.* 7.—(*Kirby & Sp. I. E.* iii. 135, nota.)
Ap. Testudo. *Haw.* 137.—(*Sam. I.* 4.)
To. Testudinana. *Hüb. To. pl.* 26. *f.* 164. ♂. *f.* 165. ♀.
Ph. funalis: Festoon M. *Don.* iii. *pl.* 76.—*Stew.* ii. 195.
No. funalina. (*Haw. Pr.* 19.)
Albin. pl. lxviii. *f.* *e–g.*
Small Oak Egger M. (*Wilkes.* 43. *pl.* 88.)

Subsectio 2. LE.-NOCTURNA *mihi.*

(PH.-NOCTUA, *et* PH.-TINEA *p, Linné.*)

Full-bodied Moths.

Familia XIII: (126). LITHOSIIDÆ *mihi.*

Genus 80: (863). CALLIMORPHA, *Latreille, Steph.*

PHALÆNA-NOCTUA *p, Linné.*—BOMBYX *p, Fabr.*—SETINA *p, Schra.*—LITHOSIA *p, Haw.*

6045. 1; Jacobææ*. (*Leach E. E.* ix. 132.)—(*Sam. I.* 8.)—*Steph. Ill.* (*H.*) ii. 90.
Ph. No. Jacobææ. *Linn.* ii. 839.—*Berk. S.* i. 139.—*Don.* ii. *pl.* 45.
No. Jacobeina. (*Haw. Pr.* 16.)
Li: Jacobeæ. *Haw.* 150.
Ph. Bo. Jacobeæ. *Stew.* ii. 159.—*Turt.* iii. 220.
Pa. No. 54. *List. Goëd.* (*Angl.*) 61. *f.* 54.—*Id.* (*Lat.*) 135. *f.* 54.—*Albin. pl.* xxxiv. *f.* 55. *e–i.*
Pink underwing. *Harr. A. pl.* 4. *fig. sup.*—*Harr. V. M.* 55.
Cinnabar M.: Ph. Jacobææ. (*Wilkes.* 26. *pl.* 55.)

6046. 2; miniata*. *Steph. Ill.* (*H.*) ii. 90.
Ph. Ge. miniata. *Forst. C.* i. 75.
Li. miniata. *Haw.* 149.
Bo. rosea. *Fabr.* iii. *a.* 485.—*Stew.* ii. 158.—*Turt.* iii. 229.
Ph. rosea. *Don.* ii. 9. *pl.* 40. *fig. inf.*
No. rosina. (*Haw. Pr.* 13.)

Ca. rosea. (*Leach E. E.* ix. 132.)—(*Sam. I.* 8.)
Bo. rubicunda. (*Wien. V.* 68.)—*Hüb. Bo. pl.* 26. *f.* 111.
Red arches. *Harr. A.* 30. *f. p.*—*Harr. V. M.* 9.

Genus 81: (864). EULEPIA, *Curtis, Steph.*

ARCTIA *p, Schra.*—LITHOSIA *p, Latr., Samou.*—EYPREPIA A. *p, Och.*

6047. 1, grammica*. (*Curtis l. c. infra.*)—*Steph. Ill.* (*H.*) ii. 91. *pl.* 17. *f.* 3.
Ph. Bo. grammica. *Linn.* ii. 831.—*Stew.* ii. 156.—*Turt.* iii. 220.—*Don.* xiii. 37. *pl.* 450.
Bo. grammica. *Hüb. Bo. pl.* 28. *f.* 122. ♂. *f.* 123. ♀. *pl.* 56. *f.* 241. ♂. *f.* 242. ♀.
Bo. grammicus: Feathered Footman. (*Haw. Pr.* 10.)—*Haw.* 134.
Bo. striata. *Bork. E. S.* iii. 471.
Li. grammica. (*Sam. I.* 25.)

6048. 2: Cribrum. *Curtis.* ii. *pl.* 56.—*Steph. Ill.* (*H.*) ii. 92.
Ph. Bo. cribrum. *Linn.* ii. 831.

Genus 82: (865). DEIOPEIA *mihi, Curtis.*

PHALÆNA-TINEA *p, Linné.*—BOMBYX *p, Fabr.*—NOCTUA *p, Esper.*—LITHOSIA *p, Haw.*—EYPREPIA A. *p, Och.*

6049. 1, pulchella. *Steph. Ill.* (*H.*) ii. 93.
Ph. Ti. pulchella. *Linn.* ii. 884.
Li. pulchella: crimson speckled. *Haw.* 150.—(*Sam. I.* 25.)
Bo. pulchra. *Hüb. Bo. pl.* 26. *f.* 113.
Eu. pulchra. *Curtis.* iv. *pl.* 169.
Pet. Gaz. pl. 3. *f.* 3.

Genus 83: (866). LITHOSIA, *Fabr., Haw., Leach, Samou., Curtis, Steph.*

SETINA *p, Schra.*—HIPPOCRITÆ, *Hüb.*—CALLIMORPHA *p, Latr.*—BOMBYX *p, Hüb.*—NOCTUA *p, Esper.*

6050. 1; aureola*. *Och.* iii. 140.—*Steph. Ill.* (*H.*) ii. 94. *pl.* 18. *f.* 1.
Bo. aureola. *Hüb. Bo. pl.* 24. *f.* 98. ♂.
No. aurantina. (*Haw. Pr.* 13.)
Li. aurantia: orange Footman. *Haw.* 147.—(*Sam. I.* 25.)—(*Curtis l. c. infra.*)
No. cincta. *Esper.* iv. *pl.* 93. *f.* 6, 7.

6051. 2, helvola*. *Och.* iii. 133.—*Steph. Ill.* (*H.*) ii. 94.
Bo. helvola. *Hüb. Bo. pl.* 23. *f.* 95. ♀.
No. deplana. *Esper.* iv. *pl.* 93. *f.* 1, 2.
No. complana ♀. *Esper.* iv. *pl.* 92. *f.* 8.

6052. 3; flava*. *Fabr. S.* 461.—*Haw.* 147.—(*Sam. I.* 25.)—(*Curtis l. c. infra.*)—*Steph. Ill.* (*H.*) ii. 95.
No. flavina: straw-coloured Footman. (*Haw. Pr.* 13.)

6053. 4; gilveola*. *Och.* iii. 137?—*Steph. Ill.* (*H.*) ii. 95.
Bo. cinereola. *Hüb. Bo. pl.* 23. *f.* 91?

6054. 5; complana*. *Haw.* 147.—(*Leach E. E.* ix. 133.)—*Leach Z. M.* i. *pl.* 49. *fig.* 3.—(*Sam. I.* 25.)—(*Curtis l. c. infra.*)—*Steph. Ill.* (*H.*) ii. 95.
Ph. No. complana. *Linn.* ii. 840.—*Stew.* ii. 186.—*Turt.* iii. 292.
No. Complanina: common Footman. (*Haw. Pr.* 13.)
Albin. pl. lxx. *f. e–h.*
Footman M. *Harr. V. M.* 27. Inlaid. *Harr. V. M.* 33?
♂; Bo. plumbeola. *Hüb. Bo. pl.* 24. *f.* 100.
Bo. caniola. *Hüb. Bo. pl.* 81. *f.* 220.

6055. 6; depressa*. *Och.* iii. 132.—*Steph. Ill.* (*H.*) ii. 96.
No. depressa. *Esper. No. pl.* 93. *f.* 3.
Bo. ochreola. *Hüb. Bo. pl.* 23. *f.* 96. ♂.
Li. ochreola. (*Curtis l. c. infra.*)

6056. 7; griseola*. *Haw.* 147.—(*Sam. I.* 25.)—(*Curtis l. c. infra.*)—*Steph. Ill.* (*H.*) ii. 96.
Bo. griseola. *Hüb. Bo. pl.* 23. *f.* 97.
No. luteolina: dun Footman. (*Haw. Pr. App.* 3.)

6057. 8; plumbeolata* *mihi. Steph. Ill.* (*H.*) ii. 96.

6058 † 9, muscerda. *Och.* iii. 143.—*Curtis.* i. *pl.* 36.—*Steph. Ill.* (*H.*) ii. 97.
Li. perla. *Fabr. Supp.* 462.
Ti. perlella. *Fabr.* iii. *b.* 292.
Ne. cinerina. *Esper.* ii. 67.
No. Pudorina. *Esper.* iv. *pl.* cxcvi. *f.* 4, 5.
In Mus. D. *Haworth* et Sparshall.

6059. 10; quadra*. *Fabr. Sup.* 459.—*Haw.* 146.—(*Leach E. E.* ix. 133.)—(*Sam. I.* 25.)—(*Curtis l. c. supra.*)—*Steph. Ill.* (*H.*) ii. 97.
Ph. No. Quadra. *Linn.* ii. 840.—*Don.* ix. 39. *pl.* 306.—*Berk. S.* i. 139.—*Stew.* ii. 185.—*Turt.* iii. 292.
No. Quadrina: Large Footman. (*Haw. Pr.* 13.)
Bo. Quadra. *Hüb. Bo. pl.* 24. *f.* 101. ♂. *f.* 102. ♀.
Yellow July M.: Ph. quadra. (*Wilkes.* 24. *pl.* 52.)
♂, Li. deplana. *Fabr. Sup.* 459.

Genus 84: (867). GNOPHRIA *mihi.*

LITHOSIA *p, Fabr., Haw., Samou.*—BOMBYX *p, Hüb.*—PHALÆNA-NOCTUA *p, Linné.*

6060. 1; rubricollis*. *Steph. Ill.* (*H.*) ii. 98.
Ph. No. rubricollis. *Linn.* ii. 840.—*Don.* x. 68. *pl.* 350. *f.* 3.—*Turt.* iii. 230.

No. rubricollina. (*Haw. Pr.* 13.)
Li. rubricollis: the Black Footman. *Haw.* 149.—(*Leach E. E.* ix. 133.)—(*Sam. I.* 25.)
Ph. Bo. rubricollis. *Stew.* ii. 159.
Red neck M. *Harr. A. pl.* 43. *f. p.*—*Harr. V. M.* 41.

Genus 85: (868). SETINA *mihi.*

PHALÆNA-TINEA *p, Linné.*—TINEA *p, Fabr.*—BOMBYX *p, Hüb.* NOCTUA *p, Esp.*—LITHOSIA *p, Fabr., Haw., Leach, Samou.*—SETINA *p, Schra.*

6061. 1: irrorea*. *Schra. B.* ii. 166.—*Steph. Ill.* (*H.*) ii. 98.
Ph. Ti. irrorella. *Linn.* ii. 885.—*Turt.* iii. 363.
No. irrorina: Dew M. (*Haw. Pr.* 13.)
Bo. irrorea. *Hüb. Bo. pl.* 25. *f.* 105.
Li. irrorata. *Fabr. Sup.* 461.—*Haw.* 148.—(*Leach E. E.* ix. 133.)
Li. irrorea. (*Sam. I.* 25.)

6062. [2: roscida*.] *Schra. B.* ii. 166?—*Steph. Ill.* (*H.*) ii. 99.
Li. roscida. *Fabr. Sup.* 462.
Bo. roscida. *Hüb. Bo. pl.* 25. *f.* 106. ♂. *f.* 107. ♀?

6063. 3; eborina*: Four-spot, small Footman.—*Haw. l. c. Schra. B.* ii. *B.* 2. 166.—*Steph. Ill.* (*H.*) ii. 99.
Li. eborina. *Fabr. Sup.* 462.—*Hüb. Bo. pl.* 24. *f.* 104.—(*Haw. Pr.* 13.)—*Haw.* 147.—(*Sam. I.* 25.)
Ph. Ti. mesomella. *Linn.* ii. 886?—*Turt.* iii. 364.
β; No. lutarina. (*Haw. Pr.* 13.)
Li. lutarella: Four-spot, yellow Footman. *Haw.* 148.
Li. luterella. (*Sam. I.* 25.)

Familia XIV: (127). NOCTUIDÆ *mihi.*

(PH.-NOCTUA, *et* BOMBYX *p, Linné.*—NOCTUA, *Fabr., Haw., &c.*)

Genus 86: (869). TRIPHÆNA, (*Och.*,) *Treit.*

6064. 1; orbona*: Lesser Yellow underwing.—*Haw. l. c.*
No. orbona. *Fabr.* iii. *b.* 57.—*Haw.* 161.—*Don.* x. 36. *pl.* 343. *f.* 2.—(*Leach E. E.* ix. 134.)—(*Sam. I.* 29.)
No. orbonina. (*Haw. Pr.* 11.)
No. Comes. *Hüb. No. pl.* 111. *f.* 521.
No. subsequa. *Esper.* iv. *pl.* civ. *f.* 1–3.
No. Pronuba minor. *Villars E.* ii. 279.
August Yellow underwing. *Harr. V. M.* 55.
β; No. subsequa: Lunar Yellow underwing. *Haw.* 161.—(*Sam. I.* 30.)

6065 † 2. subsequa. *Och. Tr.* v. *a.* 258.
No. subsequa. (*Wien. V.* 79.)—*Hüb. No. pl.* 23. *f.* 106.
No. consequa. *Hüb. No. pl.* 23. *f.* 105? In Mus. D. Curtis?

6066. 3; Pronuba*. (*Och.* iv. 69.)—*Och. Tr.* v. *a.* 260.
Ph. No. Pronuba. *Linn.* ii. 842.—*Berk. S.* i. 141.—*Don.* ix. 49. *pl.* 311.—*Stew.* ii. 183.—*Turt.* iii. 295.
No. Pronubina. (*Haw. Pr.* 11.)
No. Pronuba. *Fabr.* iii. *b.* 56.—*Hüb. No. pl.* 22. *f.* 103.—*Haw.* 160.—(*Leach E. E.* ix. 134.)—(*Sam. I.* 30.)—(*Kirby & Sp. I. E.* iii. 189.)
Pa. No. 41. *List. Goëd.* (*Angl.*) 52. *f.* 41.—*Id.* (*Lat.*) 114. *f.* 41.
Albin. pl. lxxii. *f. a–d.*
Great Yellow underwing M.: Ph. pronuba. (*Wilkes D. pl.* 9. *f.* 7.)—(*Wilkes.* 2. *pl.* 1.)—*Harr. A. pl.* 39. *f. d–f.*—*Harr. V. M.* 53.

6067. [4; innuba*.] *Och. Tr.* v. *a.* 265.
No. Pronuba. *Esper.* iv. *pl.* cii. *f.* 2–5.

6068. 5: fimbria*. (*Och.* iv. 69.)—*Och. Tr.* v. *a.* 266.
Ph. No. fimbria. *Linn.* ii. 842.—*Don.* vi. 69. *pl.* 208.—*Stew.* ii. 183.—*Turt.* iii. 296.
No. fimbriina. (*Haw. Pr.* 11.)
No. fimbria. *Hüb. No. pl.* 22. *f.* 102.—*Haw.* 161.—(*Leach E. E.* ix. 134.)—(*Sam. I.* 29.)
No. Solani. *Fabr.* iii. *b.* 57.
Ph. Domiduca. *Berl. M.* (*Hufnagle.*) iii. 404.
Broad-bordered Yellow underwing. *Harr.* 21. *pl.* v. *f.* 2.—*Harr. V. M.* 55.

6069. 6; interjecta*. (*Och.* iv. 69.)—*Och. Tr.* v. *a.* 253.
No. interjecta: least broad-border.—*Haw. l. c. Hüb. No. pl.* 23. *f.* 107.—*Haw.* 162.—(*Sam. I.* 29.)
No. fimbriina minor. (*Haw. Pr.* 12.)
Third Yellow underwing. *Harr. V. M.* 55.

6070. 7; Janthina*. (*Och.* iv. 69.)—*Och. Tr.* v. *a.* 269.
No. Janthina: lesser broad-border.—*Haw. l. c.* (*Wien. V.* 78.)—*Fabr.* iii. *b.* 59.—*Hüb. No. pl.* 21. *f.* 100.—(*Haw. Pr.* 12.)—*Haw.* 162.—(*Leach E. E.* ix. 134.)—(*Sam. I.* 29.)
No. Janthe. *Bork. E. S.* iv. 109.
No. Fimbria minor. *Villers E.* ii. 278. *pl.* 5. *f.* 24.
Ph. Domiduca. *Fuesl. A.* iii. *pl.* 16.
No. Janthia. (*Sam. I.* 29.)
The Yellow underwing. (*Wilkes D. pl.* 11. *f.* 5.)

Genus 87: (870). CERIGO *mihi.*

POLIA D., *Och.*—POLIA C., *Treit.*

6071. 1: texta*.
No. texta. *Esper.* iv. *pl.* cviii. *f.* 6.
No. connexa. *Hüb. No. pl.* 23. *f.* 109. *pl.* 118. *f.* 548.
No. Cytherea. *Fabr.* iii. *b.* 57.—*Haw.* 161.—(*Sam. I.* 29.)
No. Cytherina. (*Haw. Pr.* 12.)

Ph. Matura. *Berl. M.* (*Hufnagle.*) iii. 414.
Straw-coloured underwing. *Harr. V. M.* 59.
♀, No. prospicua: brown-bordered Yellow underwing. *Haw.* 161: nec Hubneri.

Genus 88: (871). LYTÆA *mihi.*

APAMEA *p, Och.*

6072. 1: umbrosa*.
No. umbrosa. *Hüb. No. pl.* 97. *f.* 456, 457.—(*Sam. I.* 30.)
No. radicea var. *Esper.* iv. *pl.* cxliii. *f.* 3.
No. sextrigata: six-striped Rustic. *Haw.* 228.

Genus 89: (872). CHARÆAS *mihi.*

APAMEA *et* AGROTIS *p, Och.*—EPISEMA *p, Treit.*—AGROTIS *p, Curtis.*

6073. 1: cespitis*.
No. cespitis. (*Wien. V.* 82.)—*Fabr.* iii. *b.* 68.—*Hüb. No. pl.* 91. *f.* 428.
No. Hordei. *Schrank B.* ii. 351.
Ag. cespitis. (*Curtis l. c. fo.* 165.)

6074. [2, confinis] *mihi.*
Ag. autumnalis. (*Curtis l. c. supra?*)

6075. 3: fusca*.
Bo. fuscus: barred feathered Rustic. (*Haw. Pr.* 9.)
No. fusca. *Haw.* 119 & 204.—(*Sam. I.* 29.)
Ag. fusca. (*Curtis l. c. supra.*)
No. lutulenta. (*Wien. V.* 81?)—*Hüb. No. pl.* 33. *f.* 159?

6076. 4, consimilis* *mihi.*

6077 † 5, orthostigma* *mihi.* In Mus. D. *Stone.*

6078. 6: Æthiops*. *Steph. Ill.* (*H.*) ii. *pl.* 20. *f.* 1.
No. Æthiops. *Hüb. No. pl.* 116. *f.* 538.
No. nigra: black Rustic. *Haw.* 192.—(*Sam. I.* 29.)
Ag. nigra: (albicolon Fab.?) *Curtis l. c. supra.*

6079. 7: Graminis*.
Ph. Bo. Graminis. *Linn.* ii. 830.—*Stew.* ii. 158.—*Turt.* iii. 229.—*Don.* xiii. 53. *pl.* 458.—(*Kirby & Sp. I. E.* i. 178.)
Bo. Graminis. (*Haw. Pr.* 10.)
No. Graminis. *Hüb. No. pl.* 102. *f.* 480, 481.—*Haw.* 117 & 222.—(*Sam. I.* 29.)
No. tricuspis. *Hüb. No. pl.* 30. *f.* 143.
Antler M. *Harr. Ex.* 23. *pl.* v. *f.* 7.—(*Harr. V. M.* 9.)

Genus 90: (873). RUSINA *mihi.*

AGROTIS *p, Och.*—BOMBYX *p, Esper.*

6080. 1; ferruginea*.

Bo. ferruginea. *Esper.* iii. *pl.* xlvii. *f.* 5, 6.
No. tenebrosa. *Hüb. No. pl.* 33. *f.* 158.—*Id. pl.* 107. *f.* 503.
No. nigricans. *Hüb. No. pl.* 116. *f.* 538.
Bo. phæus. *Marsh. MSS.*—(*Haw. Pr.* 11.)
No. phæa: feathered Rustic. *Haw.* 133 & 205.—(*Sam. I.* 30.)
♀; No. obsoletissima: brown Rustic. *Haw.* 207.—(*Sam. I.* 29.)

Genus 91: (874). AGROTIS, (*Hübner,*) (*Samou.,*) *Treit., Curtis.*—
BOMBYX *p, Esper., Haworth.*

6081. 1, lunigera *mihi.* *Steph. Ill.* (*H.*) ii. *pl.* 20. *f.* 2.

6082. 2; Corticea*. (*Och.* iv. 66.)—*Och. Tr.* v. *a.* 158.
Ag. clavigera. (*Curtis l. c. infra.*)
♂; No. Corticea. *Hüb. No. pl.* 30. *f.* 145.
Bo. exclamationis. *Esper. Bo.* iii. *pl.* lxiv. *f.* 1, 2.
Bo. clavigerus. (*Haw. Pr.* 9.)
No. clavigera: Heart and Club. *Haw.* 114 & 219.—(*Sam. I.* 29.)
♀; No. sordida. *Hüb. No. pl.* 32. *f.* 154.
No. subfusca: brown Heart and Club. *Haw.* 114 & 219.—(*Sam. I.* 30.)

6083. 3, æqua*. *Och. Tr.* v. *a.* 150.—(*Curtis l. c. infra.*)
No. æqua. *Hüb. No. pl.* 122. *f.* 564.
No. margaritosa: the pearly Underwing. *Haw.* 218.—(*Sam. I.* 29.)
Ag. margaritosa. (*Ing. Inst.* 90.)
† β, No. majuscula: dark Pearl Underwing. *Haw.* 218.—(*Sam. I.* 29.) β in Mus. D. *Hatchett, Haworth et Stone.*

6084. 4; Segetum*. (*Och.* iv. 66.)—(*Curtis l. c. infra.*)
No. Segetum. (*Wien. V.* 252. *pl.* 1. *a, b. f.* 3.)—(*Sam. I.* 30.)
No. Segetis. *Fabr.* iii. *b.* 61.—*Hüb. No. pl.* 31. *f.* 146.
Ph. No. Segetis. *Stew.* ii. 183.
Bo. caliginosa. *Esper. Bo.* iii. *pl.* lxiv. *f.* 3.
Var. Bo. fuscosa. *Esper. Bo.* iii. *pl.* lxiv. *f.* 4.
No. corticea: pointed Dart. *Haw.* 116 & 218.
No. connexa: Chain-shot Dart. *Haw.* 116 & 218.—(*Sam. I.* 29.)
No. venosa: broad-veined Dart. *Haw.* 116 & 218.—(*Samou.* 402.)
No. spinula: brindled Dart. *Haw.* 115 & 218.—(*Sam. I.* 30.)
No. nigricornutus: black Dart. (*Haw. Pr.* 9.)—*Haw.* 117 & 219.—(*Sam. I.* 29.)
No. monilea: Necklace Dart. *Haw.* 115 & 219.—(*Sam. I.* 29.)
No. subatrata: dark Dart. (*Haw. Pr.* 9.)—*Haw.* 116 & 219.—(*Sam. I.* 30.)
No. pectinata: pectinated Dart. *Haw.* 115 & 219.—(*Sam. I.* 30.)

No. catænata: brindled Heart and Club. *Haw.* 114 & 219.—(*Sam. I.* 29.)
Ag. affinis *mihi.*—(*Curtis l. c. infra.*)
Ag. monostigma. (*Curtis l. c. infra?*)
Heart and Dart M. *Harr. V. M.* 31.

6085. 5; suffusa*. (*Och.* iv. 66.)—(*Curtis l. c. infra.*)
No. suffusa. (*Wien. V.* 80.)—*Fabr.* iii. *b.* 71.—*Haw.* 217.—(*Sam. I.* 30.)
Bo. spinula. *Esper.* iii. *pl.* lxiii. *f.* 6, 7.—*Don.* x. 52. *pl.* 345. *f.* 2, 3.
No. Upsilon. *Berl. M.* (*Hufnagle.*) iii. 416.
No. Stigmaticus: dark Sword-grass. (*Haw. Pr.* 9.)
β; Ph. Bo. spinifera. *Villers E.* ii. 74.
Bo. spinifera: small Sword-grass. (*Haw. Pr.* 9.)
No. spinifera. *Haw.* 114 & 217.—(*Sam. I.* 30.)

6086 † 6, annexa*. *Och. Tr.* v. *a.* 154.—*Steph. Ill.* (*H.*) ii. *pl.* 22. *f.* 2.
No. subterranea: tawny Shoulder. *Haw.* 171, non Fabricii.
Ag. subterranea. (*Ing. Inst.* 90.)—(*Curtis l. c. infra.*)
In Mus. D. *Curtis et Stone.*

6087. 7; valligera. (*Och.* iv. 67.)—*Och. Tr.* v. *a.* 163.
No. valligera. (*Wien. V.* 80.)—*Fabr.* iii. *b.* 72.
Bo. clavis. *Esper.* iii. *pl.* lxiii. *f.* 5.—*Don.* x. 36. *pl.* 340. *f.* 2.
Ph. No. clavifera. *Villers E.* ii. 174.
No. sagittifera: Archer's Dart. (*Haw. Pr.* 9.)—*Haw.* 118 & 224.)—(*Sam. I.* 30.)
Ag. sagittifera. (*Curtis l. c. infra.*)
♀. Bo. trigonalis. *Esper.* iii. *pl.* lxxxv. *f.* 6.
Var. Ag. hibernica. *Haworth MSS.*—(*Curtis l. c. infra.*)

6088. 8: radia*. (*Curtis l. c. infra.*)
Bo. radius. (*Haw. Pr.* 10.)
No. radia: Shuttle-shaped Dart. *Haw.* 119 & 223.—(*Sam. I.* 30.)

6089. 9: radiola*. (*Curtis l. c. infra.*)—*Steph. Ill.* (*H.*) ii. *pl.* 20. *f.* 3.
No. radiola. *Haw. MSS.*
Ag. radiatus. (*Ing. Inst.* 90.)

6090. 10, sagittifera*. (*Och.* iv. 67.)—*Och. Tr.* v. *a.* 172.
No. sagittifera. *Hüb. No. pl.* 114. *f.* 532. ♀.

6091. 11; lineolata*. (*Curtis l. c. infra.*)
No. lineolata: lineolated Dart. *Haw.* 223.—(*Sam. I.* 29.)
No. vitta. *Hüb. No. pl.* 115. *f.* 533. ♂; *f.* 534. ♀?

6092. 12, pupillata*. (*Curtis l. c. infra.*)
No. pupillata: pupilled Dart. *Haw.* 118 & 223.—(*Sam. I.* 30.)

6093. 13, Aquilina*. (*Och.* iv. 66.)—*Och. Tr.* v. *a.* 134.

No. Aquilina. (*Wien. V.* 80.)—*Hüb. No. pl.* 29. *f.* 135. ♂; *pl.* 115. *f.* 535. ♀.
No. nigrofusca. *Esper.* iv. *pl.* cxxvii. *f.* 6.
No. domestica. *Fabr.* iii. *b.* 23.

6094. 14, Tritici*. (*Och.* iv. 66?)—*Och. Tr.* v. *a.* 137?
Ph. No. Tritici. *Linn.* ii. 855?
No. fictilis. *Hüb. No. pl.* 101. *f.* 479. ♂?
No. eruta. *Hüb. No. pl.* 136. *f.* 623. ♀?
No. pratincola. *Hüb. No. pl.* 123. *f.* 567. ♂. *var.*
No. recussa. *Hüb. No. pl.* 138. *f.* 630. ♂. *var.*
No. albilinea: White-line Dart. *Haw.* 223.—(*Sam. I.* 28.)
Ag. albilinea. (*Curtis l. c. infra.*)

6095. 15; ocellina*. (*Och.* iv. 66.)—*Och. Tr.* v. *a.* 129.
No. ocellina. (*Wien. V.* 313.)—*Hüb. No. pl.* 27. *f.* 129. ♀; *pl.* 131. *f.* 599. ♂.
Bo. Phytheuma. *Esper.* iv. *pl.* cxcii. *f.* 3, 4.

6096. 16. valligera*. (*Curtis l. c. infra?*)
No. valligera: Wedge-barred Dart. *Haw.* 222.—(*Sam. I.* 30.)

6097. 17, venosa *mihi.*

6098. 18, Hortorum* *mihi.*

6099. 19; nigricans*. (*Curtis l. c. infra.*)
Ph. No. nigricans. *Linn.* ii. 855.
No. nigricans: Garden Dart. *Haw.* 221.—(*Sam. I.* 29.)
No. fuliginea. *Hüb. No. pl.* 131. *f.* 602.
β; No. concolorina. (*Haw. Pr.* 19.)
No. concolor: plain Quaker. *Haw.* 243.
Orthosia! concolor. (*Curtis l. c. fo.* 237!)
γ, No. chrysostigmatus. (*Haw. Pr.* 9.)
No. carbonea. *Hüb. No. pl.* 143. *f.* 700, 701.

6100. [20; fumosa*]: dark Rustic.—*Haw. l. c.* (*Och.* iv. 66.)—*Och. Tr.* v. *a.* 140.
No. fumosa. (*Wien. V.* 81.)—*Fabr.* iii. *b.* 115.—*Hüb. No. pl.* 32. *f.* 153.—*Haw.* 221.—(*Sam. I.* 29.)
No. rubricans. *Esper.* iv. *pl.* cxxx. *f.* 23.

6101. 21; obeliscata*. (*Curtis l. c. infra?*)
No. obeliscata: Square-spot Dart. *Haw.* 222.—(*Sam. I.* 29.)
No. obelisca. *Hüb. No. pl.* 26. *f.* 123?

6102. 22; ruris*: rufous Dart. (*Och.* iv. 66.)—*Och. Tr.* v. *a.* 246.—(*Curtis l. c. infra.*)
No. ruris. *Hüb. No. pl.* 89. *f.* 416.—*Haw.* 222.—(*Sam. I.* 30.)
♂; No. sordida: striped Square Spot. *Haw.* 222.—(*Sam. I.* 30.)

6103. [23, dubia.]
No. dubia: White line. *Haw.* 222.

6104. 24, subgothica*. (*Ing. Inst.* 90.)—(*Curtis l. c. infra.*)—*Steph. Ill.* (*H.*) ii. *pl.* 22. *f.* 3.
No. subgothica: Gothic Dart. *Haw.* 224.

6105. 25; exclamationis*. (*Och.* iv. 67.)—*Och. Tr.* v. *a.* 160.—(*Curtis l. c. infra.*)
Ph. No. exclamationis. *Linn.* ii. 850.—*Stew.* ii. 184.
No. exclamationis: Heart and Dart. (*Haw. Pr.* 17.)
No. exclamationis. *Hüb. No. pl.* 51. *f.* 149.—*Haw.* 219.—(*Sam. I.* 29.)
Ph. Clavus. *Berl. M.* (*Hufnagle.*) iii. 298.
♀. Var. No. picea: pitchy Dart. *Haw.* 220.—(*Sam. I.* 30.)
Ag. picea. (*Curtis l. c. infra.*)

6106 † 26, nebulosa *mihi.* *Steph. Ill.* (*H.*) ii. *pl.* 22. *f.* 1.
In Mus. D. *Raddon.*

6107. 27: cinerea*. (*Och.* iv. 67.)—(*Och. Tr.* v. *a.* 178.)—(*Curtis.* iv. *pl.* 165.)
No. cinerea. *Hüb. No. pl.* 33. *f.* 155, 156.
Bo. denticulatus. (*Haw. Pr.* 11.)
No. denticulata: light feathered Rustic. *Haw.* 133 & 205.—(*Sam. I.* 29.)
† ♀, No. obscura. *Hüb. No. pl.* 33. *f.* 157; *pl.* 104. *f.* 490.

Genus 92: (875). GRAPHIPHORA, (*Hüb.*,) (*Och.*,) (*Samou.*)

NOCTUA, *Treitschke.*—AGROTIS *p*, *Och.*

6108. 1, rhomboidea. *Och. Tr.* v. *a.* 231.—*Steph. Ill.* (*H.*) ii. *pl.* 19. *f.* 1. ♂.
No. rhomboidea. *Esper. No.* iv. *pl.* civ. *f.* 3.
No. stigmatica. *Hüb. No. pl.* 100. *f.* 470.

6109. 2, renigera.
No. renigera. *Hüb. No. pl.* 82. *f.* 384. ♂.
Ag. renigera. *Och. Tr.* v. *a.* 197.—(*Ing. Inst.* 90.)

6110. 3, latens.
No. latens. *Hüb. No. pl.* 89. *f.* 419?
Ag. latens. *Och. Tr.* v. *a.* 204?

6111. 4, pyrophila.
No. pyrophila. (*Wien. V.* 71.)—*Fabr.* iii. *b.* 98.
Ag. pyrophila. *Och. Tr.* v. *a.* 204.—(*Ing. Inst.* 90.)
Bo. radicea. *Esper.* iv. *pl.* cxliii. *f.* 1, 2.
No. tristis. *Fabr.* iii. *b.* 118.
No. simulans. *Fabr.* iii. *b.* 177.

6112 † 5, lunulina.
No. lunulina: Crescent Striped. (*Haw. Pr.* 19.)—*Haw.* 192.
In Mus. D. *Haworth.*

6113. 6, crassa*: stout Dart.—*Haw. l. c.*

No. crassa. *Hüb. No. pl.* 32. *f.* 152?—*Haw.* 220.—(*Sam. I.* 30.)
Bo. segetum. *Esper.* iii. *pl.* lx. *f.* 5?
Ag. crassa. *Och. Tr.* v. *a.* 166?

6114. 7; augur*. (*Och.* iv. 68.)—*Och. Tr.* v. *a.* 210.
No. augur. *Fabr.* iii. *b.* 61.—*Hüb. No. pl.* 31. *f.* 145.—*Haw.* 220.—(*Sam. I.* 28.)
No. cincta. *Marsham MSS.*
No. cinctina: double Dart. (*Haw. Pr.* 17.)
No. Omega. *Esper.* iv. *pl.* cxxxi. *f.* 2.
No. assimulans. *Bork. E. S.* iv. 209.

6115. 8; brunnea*: purple Clay.—*Haw. l. c. Och. Tr.* v. *a.* 219.
No. brunnea. (*Wien. V.* 83.)—*Hüb. No. pl.* 26. *f.* 121.—*Haw.* 223.—(*Sam. I.* 29.)
No. lucifera. *Esper.* iv. *pl.* cxlii. *f.* 6.
No. Fragrariæ. *Bork. E. S.* iv. 497.

6116. 9; candilesequa*. (*Och.* iv. 68.)—*Och. Tr.* v. *a.* 217.
No. candilesequa. (*Wien. V.* 72.)—*Hüb. No. pl.* 85. *f.* 397.

6117. 10; tristigma*. (*Och.* iv. 68.)—*Och. Tr.* v. *a.* 243.
No. ditrapezium. *Bork. E. S.* iv. 515.—*Hüb. No. pl.* 24. *f.* 115.

6118. 11; triangulum*. (*Och.* iv. 68.)—*Och. Tr.* v. *a.* 240.
Ph. triangulum. *Berl. M.* (*Hufnagle.*) iii. 306.
Ph. ditrapezium. (*Wien. V.* 312.)
No. sigma. *Hüb. No. pl.* 106. *f.* 497.—*Haw.* 225.—(*Sam. I.* 30.)
No. sigmina: double Square-spot. (*Haw. Pr.* 18.)
Ph. No. sigma. *Don.* xvi. *pl.* 562.

6119 † 12. depuncta[a].

6120. 13; baja*: dotted Clay.—*Haw. l. c. Och. Tr.* v. *a.* 215.
No. baja. (*Wien. V.* 77.)—*Hüb. No. pl.* 25. *f.* 119.—*Haw.* 224.—(*Sam. I.* 28.)
No. tricomma. *Esper.* iv. *pl.* clxvii. *f.* 6.

6121, 14; erythrocephala*: barred Chesnut.—*Haw. l. c.*
No. erythrocephala. (*Wien. V.* 77?)—*Hüb. No. pl.* 37. *f.* 176?—*Haw.* 227.—(*Sam. I.* 29.)
No. silene. *Bork. E. S.* iv. 741.
No. Vaccinii var. *Esper.* iv. *pl.* clxii. *f.* 1, 2.
Cr. erythrocephala. *Och. Tr.* v. *b.* 405.

---

[a] 6119 † 12. depuncta. (*Och.* iv. 68.)—*Och. Tr.* v. *a.* 229.
Ph. No. depuncta. *Linn.* ii. 858.
No. mendosa. *Hüb. No. pl.* 26. *f.* 120. ♂. *pl.* 107. 502. ♀.
No. depunctina. (*Haw. Pr. App.* 5.) (!)

6122. 15; festiva*: ingrailed Clay.—*Haw. l. c.* (*Och.* iv. 68.)—*Och. Tr.* v. *a.* 224.
No. festiva. (*Wien. V.* 77.)—*Hüb. No. pl.* 99. *f.* 467, 468, 469.—*Haw.* 226.—(*Sam. I.* 29.)
No. Primulæ. *Esper.* iv. *pl.* cxxxvi. *f.* 5.
No. mendica. *Fabr.* iii. *b.* 93.
β; No. subrufa: rufous Clay. *Haw.* 227.—(*Sam. I.* 30.)

6123. 16; C. nigrum*. (*Och.* iv. 68.)—*Och. Tr.* v. *a.* 237.
Ph. No. C. nigrum. *Linn.* ii. 852.
No. C. nigrina: setaceous Hebrew Character. (*Haw. Pr.* 18.)
No. C. nigrum. *Hüb. No. pl.* 24. *f.* 111.—*Haw.* 226.—(*Sam. I.* 29.)
Bo. Gothica var. *Esper.* iii. *pl.* lxxvi. *f.* 3.
No. Nun-atrum. *Bork. E. S.* iv. 495.
No. Triangulum. *Naturf.* ix. 126.

6124 † 17, musiva. (*Och.* iv. 69.)—*Och. Tr.* v. *a.* 247.
No. musiva. *Hüb. No. pl.* 25. *f.* 118. In Mus. *Brit.*

6125. 18, albimacula. *Steph. Ill.* (*H.*) ii. *pl.* 19. *f.* 3.
Gr. ravida. (*Ing. Inst.* 91.)

6126. 19; plecta*. (*Och.* iv. 69.)—*Och. Tr.* v. *a.* 248.
Ph. No. plecta. *Linn.* ii. 851.—*Turt.* iii. 299.
No. plectina: Flame Shoulder. (*Haw. Pr.* 19.)
No. plecta. *Hüb. No. pl.* 25. *f.* 117.—*Haw.* 226.—(*Sam. I.* 30.)

6127. 20; punicea*: small Square-spot.—*Haw. l. c.*
No. punicea. *Hüb. No. pl.* 25. *f.* 115.—*Haw.* 228.—(*Sam. I.* 30.)
No. Dahlii. *Och. Tr.* v. *a.* 222?
No. radica. *Esper.*?

Genus 93: (876). SEMIOPHORA *mihi.*

EPISEMA *p, Och.*—HETEROMORPHÆ *p, Hüb.*—NOCTUA *p, Treit.*—BOMBYX *p, Villers.*

6128. 1; gothica*.
Ph. No. gothica. *Linn.* ii. 851.—*Stew.* ii. 190.—*Turt.* iii. 320.
Bo. gothicus. (*Haw. Pr.* 11.)
No. gothica. *Haw.* 119 & 226.—(*Sam. I.* 29.)
No. Nun-atrum. (*Wien. V.* 78.)—*Hüb. No. pl.* 24. *f.* 112. ♀.
Bo. C. nigrum. *Villers E.* ii. 183.
Hebrew Character. *Harr. V. M.* 19.

Genus 94: (877). ORTHOSIA, (*Och.*,) *Treit.*, *Curtis.*

A. Alis anticis apicibus subrotundatis. (Antennis in masculis plus minusve pectinatis.)

6129. 1; instabilis*. (*Och.* iv. 79.)—*Och. Tr.* v. *b.* 204.—(*Curtis l. c. infra.*)

No. instabilis. (*Wien. V.* 76.)—*Fabr.* iii. *b.* 44.—*Hüb. No. pl.* 35. *f.* 165.
No. contracta. *Esper.* iv. *pl.* cxlvii. *f.* 4.
No. trigutta. *Esper.* iv. *pl.* cli. *f.* 2.
Bo. subsetaceus. (*Haw. Pr.* 9.)
No. subsetacea: clouded Drab-M. *Haw.* 120 & 241.—(*Sam. I.* 30.)
No. nebulosa: scarce Clouded Drab-M. *Haw.* 120 & 241.—(*Sam. I.* 29.)
No. fuscata: dark Drab-M. *Haw.* 122 & 241.—(*Sam. I.* 29.)
No. angusta: narrow-winged Drab-M. *Haw.* 122 & 241.—(*Sam. I.* 28.)
Albin. pl. lxxvi. *f. a–d.*

6130. 2; intermedia* *mihi.*

6131. 3; gracilis*. (*Och.* iv. 79.)—*Och. Tr.* v. *b.* 217.
No. gracilis. (*Wien. V.* 76.)—*Fabr.* iii. *b.* 48.—*Hüb. No. pl.* 35. *f.* 168.
No. gracillina. (*Haw. Pr. App.* 3?)
No. lepida. *Bork. E. S.* iv. 600.
No. collinita. *Esper.* iv. *pl.* clii. *f.* 6.
No. subplumbea: Lead-coloured Drab-M. *Haw.* 121 & 242.—(*Sam. I.* 30.)
Or. subplumbea. (*Curtis l. c. infra.*)

6132. 4; munda*. (*Och.* iv. 79.)—*Och. Tr.* v. *b.* 208.—(*Curtis l. c. infra.*)
No. munda. (*Wien. V.* 16.)—*Fabr.* iii. *b.* 48.
No. gemina. *Esper.* iii. *pl.* lii. *f.* 5, 6.—(*Haw. Pr.* 19.)
No. geminoidina. (*Haw. Pr.* 19.)
No. geminata: Twin-spotted Drab-M. *Haw.* 121 & 242.—(*Sam. I.* 29.)
β, No. bimaculata: ferruginous Drab-M. *Haw.* 121 & 242.—(*Sam. I.* 28.)
No. Lota: (munda.) *Hüb. No. pl.* 35. *f.* 166.
Or. bimaculata. (*Curtis l. c. infra.*)

6133. 5; sparsa*. (*Curtis l. c. infra.*)
Bo. sparsus. (*Haw. Pr.* 9.)
No. sparsa: powdered Quaker-M. *Haw.* 122 & 242.—(*Sam. I.* 30.)

6134. 6; pallida*. (*Curtis l. c. infra.*)
No. pallida: pale Quaker-M. *Haw.* 242.—(*Sam. I.* 30.)

6135. 7; stabilis*. (*Och.* iv. 79.)—*Och. Tr.* v. *b.* 223.—(*Curtis l. c. infra.*)
No. stabilis. (*Wien. V.* 76.)—*Hüb. No. pl.* 36. *f.* 171.
No. Cerasi. *Fabr.* iii. *b.* 44.—*Haw.* 123 & 243.—(*Sam. I.* 29.)
Bo. Cerasus: common Quaker-M. (*Haw. Pr.* 9.)

Ph. No. Cerasi. *Stew.* ii. 185.—*Turt.* iii. 311.
Albin. pl. lxxv. *f.* a–e.
Quaker M. *Harr. V. M.* 41.
β; Bo. junctus. (*Haw. Pr.* 9.)
No. juncta: conjoined Quaker-M. *Haw.* 123 & 243.—(*Sam. I.* 29.)
Or. juncta. (*Curtis l. c. infra.*)
Albin. pl. lxxvi. *f.* e–h.
γ; No. rufannulata: red-ringed Quaker-M. *Haw.* 243.
Or. rufannulata. (*Curtis l. c. infra.*)

6136. 8; miniosa*. (*Och.* iv. 79.)—*Och. Tr.* v. *b.* 228.—(*Curtis l. c. infra.*)
No. miniosa. (*Wien. V.* 88.)—*Fabr.* iii. *b.* 43.—*Hüb. No. pl.* 36. *f.* 174.—*Haw.* 241.—(*Sam. I.* 29.)
Bo. rubricosa. *Esper.* iv. *pl.* lxxv. *f.* 3, 4.
Ph. No. serrata. *Marsh. MSS.*
No. serratina. (*Haw. Pr.* 15.)
Blossom underwing. *Harr. V. M.* 55.

6137. 9; cruda*. *Och. Tr.* v. *b.* 230.—(*Curtis l. c. infra.*)
No. cruda. (*Wien. V.* 77.)
No. ambigua. *Hüb. No. pl.* 36. *f.* 173.
No. pulverulenta. *Esper.* iv. *pl.* lxxvi. *f.* 5, 6.
Bo. nanus. (*Haw. Pr.* 9.)
Bo. nana: small Quaker-M. *Haw.* 123 & 244.—(*Sam. I.* 29.)
Albin. pl. lxxiv. *f.* a–e.

6138. [10; pusilla*.] (*Curtis l. c. infra.*)
No. pusilla: dwarf Quaker-M. *Haw.* 124 & 244.—(*Sam. I.* 30.)

B. Alis anticis apicibus acutis; margine postico subrepando. (Antennis in masculis sæpissimè ciliatis.)

6139. 11; litura*. (*Och.* iv. 80.)—*Och. Tr.* v. *b.* 242.—(*Curtis l. c. infra.*)
Ph. No. litura. *Linn.* ii. 853.—*Stew.* ii. 194.—*Turt.* iii. 331.
No. litura. *Hüb. No. pl.* 27. *f.* 127.—*Haw.* 232.—(*Sam. I.* 29.)
No. liturina: brown-spot Pinion. (*Haw. Pr. App.* 3.)
No. Polluta. *Esper.* iv. *pl.* cxxvii. *f.* 5.
No. costina. (*Haw. Pr.* 15.)

6140. 12; Pistacina*. (*Och.* iv. 80.)—*Och. Tr.* v. *b.* 239.—(*Curtis l. c. infra.*)
No. Pistacina. *Fabr.* ii. 109.—*Hüb. No. pl.* 28. *f.* 121.—*Haw.* 231.—(*Sam. I.* 30.)
No. sexpunctata. *Marsham MSS.*
No. sexpunctina: the pale beaded Chesnut. (*Haw. Pr.* 15.)
No. serina. *Esper.* iv. *pl.* clvi. *f.* 1, 2.
β; No. Lychnidis. *Fabr.* ii. 106.

No. canaria. *Esper. pl.* clvi. *f.* 5, 6.
No. lineola: dark-beaded Chesnut. *Haw.* 231.—(*Sam. I.* 29.)
Ph. No. lineola. *Don.* x. 94. *pl.* 360. *f.* 2.
Or. lineola. (*Curtis l. c. infra.*)
γ; No. Schænobæna. *Esper.* iv. *pl.* clvii. *f.* 2, 3.
No. sphærulatina: beaded Chesnut. (*Haw. Pr.* 13.)—*Haw.* 230.—(*Sam. I.* 30.)
Ph. No. sphærulatoria. *Marsham MSS.*
Or. sphærulatina. (*Curtis l. c. infra.*)
δ; No. Rubetra. *Esper.* iv. *pl.* clvi. *f.* 5, 6.
No. ferrea: Iron Chesnut. *Haw.* 231.—(*Sam. I.* 29.)
Or. ferrea. (*Curtis l. c. infra.*)
ε; No. venosa: veiny Chesnut. *Haw.* 232.—(*Samou.* 440.)
Or. venosa. (*Curtis l. c. infra.*)

6141. 13; lunosa*. (*Curtis.* v. *pl.* 237.)
No. lunosa: Lunar Underwing M. *Haw.* 230.—(*Sam. I.* 29.)

6142. 14; Lota*. (*Och.* iv. 79.)—*Och. Tr.* v. *b.* 212.—(*Curtis l. c. infra.*)
Ph. Bo. Lota. *Linn.* ii. 830.
Bo. Lotus. (*Haw. Pr.* 9.)
No. Lota: red-line Quaker-M. *Haw.* 122 & 242.—(*Sam. I.* 29.)
No. munda: (Lota.) *Hüb. No. pl.* 35. *f.* 167.

6143. 15; flavilinea*. (*Curtis l. c. infra.*)—*Steph. Ill.* (*H.*) ii. *pl.* 19. *f.* 2.
No. flavilinea: yellow-line Quaker-M. *Haw.* 243.—(*Sam. I.* 29.)
Ph. Bo. Hippophaës. *Vill. E.* ii. 178?

6144. 16; macilenta*. (*Och.* iv. 79?)—*Och. Tr.* v. *b.* 215?—(*Curtis l. c. infra.*)
No. macilenta. *Hüb. No. pl.* 89. *f.* 418?—*Haw.* 239.—(*Sam. I.* 29.)
No. unimaculina: Brick M. (*Haw. Pr.* 15.)

6145. 17; Upsilon*. (*Och.* iv. 79.)—*Och. Tr.* v. *b.* 210.—(*Curtis l. c. supra.*)
No. Ypsilon. (*Wien. V.* 78.)—*Hüb. No. pl.* 29. *f.* 136.—(*Sam. I.* 30.)
No. corticea. *Esper.* iv. *pl.* cxlv. *f.* 2, 3.
No. fissipuncta: dingy Shears M. *Haw.* 197.
Dismal. *Harr. V. M.* 21.

Genus 95: (878). MYTHIMNA.

MYTHIMNA B., (*Och.*,) (*Samou.*)

6146. 1; turca*. (*Och.* iv. 78.)—*Och. Tr.* v. *b.* 181.
Ph. No. turca. *Linn.* ii. 847.
No. turcina: double line. (*Haw. Pr.* 15.)

No. turca. *Hüb. No. pl.* 45. *f.* 218.—*Haw.* 250.—(*Sam. I.* 30.)
Ph. volupia. *Naturf.* (——) ix. 123.

6147. 2; grisea*: bright-eyed Clay.—*Haw. l. c.*
No. grisea. *Fabr.* iii. *b.* 69.—*Haw.* 229.—(*Sam. I.* 29.)
No. unipunctina. *Marsh. MSS.*—(*Haw. Pr.* 15.)

6148. 3, Lithargyria. (*Och.* iv. 78.)—*Och. Tr.* v. *b.* 183.
No. Lithargyria. *Bork. E. S.* iv. 696.—*Hüb. No. pl.* 46. *f.* 225.
No. Ferrago. *Fabr.* iii. *b.* 76.
No. Punctum-album. *Villers E.* ii. 282.
No. ferruginea. *Scriba. B.* ii. 149. *pl.* x. *f.* 5.

6149. 4; conigera*. (*Och.* iv. 78.)—*Och. Tr.* v. *b.* 190.
No. conigera. (*Wien. V.* 84.)—*Fabr.* iii. *b.* 113.—*Hüb. No. pl.* 46. *f.* 222.—*Haw.* 239.—(*Sam. I.* 29.)
No. conigerina. (*Haw. Pr.* 14.)
No. floccida. *Esper.* iv. *pl.* cxxiii. *f.* 5.
Brown-line bright-eye. *Harr. V. M.* 35.

Genus 96: (879). SEGETIA *mihi.*
MYTHIMNA C., *Och.*, *Treit.*

6150. 1; xanthographa*.
No. xanthographa. (*Wien. V.* 83.)—*Fabr. M.* ii. 171.—*Hüb. No. pl.* 29. *f.* 138.—*Turt.* iii. 328.
No. tetragona. *Marsh. MSS.*—*Haw.* 205.—(*Sam. I.* 30.)
No. tetragonina: the square spot Rustic. (*Haw. Pr.* 18.)

6151: 2: neglecta*: neglected Rustic.—*Haw. l. c.*—*Steph. Ill.* (*H.*) ii. *pl.* 21. *f.* 1.
No. neglecta. *Hüb. No. pl.* 34. *f.* 160. ♀.—*Haw.* 205.
My. neglecta. *Och. Tr.* v. *b.* 199.

Genus 97: (880). CARADRINA, (*Och.*,) (*Samou.*,) *Treit.*

A. Alis ferè absque stigmatibus: abdomine crasso.

6152. 1; trilinea*. (*Och.* iv. 80.)—*Och. Tr.* v. *b.* 272.
No. trilinea. (*Wien. V.* 84.)—*Hüb. No. pl.* 45. *f.* 216.—*Haw.* 249.—(*Sam. I.* 30.)
No. trilinearina: equal Treble-lines. (*Haw. Pr.* 19.)
No. trigrammica. *Esper.* vi. *pl.* cxxiii. *f.* 6.
No. Quercus. *Fabr.* iii. *b.* 22.
No. evidens. *Thunb. I. S.* 2. *pl. et fig. adj.*
Triple-lines. *Harr. V. M.* 55.
β; No. approximans: inequal Treble-lines. *Haw.* 249.—(*Sam. I.* 28.)
γ; No. semifuscans: clouded Treble-lines. *Haw.* 249.—(*Sam. I.* 30.)

6153. 2; bilinea*: dark Treble-lines.—*Haw. l. c.* (*Och.* iv. 80.) *Och. Tr.* v. *b.* 275.
No. bilinea. *Hüb. No. pl.* 45. *f.* 217.—*Haw.* 249.—(*Sam. I.* 28.)

B. Alis stigmatibus ordinariis distinctis. a. *Abdomine crasso.*

6154. 3; ambigua*. (*Och.* iv. 30?)—*Och. Tr.* v. *b.* 262?
No. ambigua. *Fabr.* iii. *b.* 48?
No. ambiguina. (*Haw. Pr. App.* 3.)
No. Plantaginis. *Hüb. No. pl.* 125. *f.* 576.
No. xanthographa: dotted Rustic. *Haw.* 206.—(*Sam. I.* 30.)

6155. [4; redacta*.]
No. redacta: lesser dotted Rustic. *Haw.* 206.—(*Sam. I.* 30.)
No. depunctina. (*Haw. Pr. App.* 3.)

6156. 5; Alsines*. (*Och.* iv. 80.)—*Och. Tr.* v. *b.* 266.
No. Alsines. *Bork. E. S.* iv. 608.—*Hüb. No. pl.* 125. *f.* 577.
No. egens: garden Rustic. *Haw.* 206.—(*Sam. I.* 29.)

6157. [6; implexa*.]
No. implexa. *Hüb. No. pl.* 88. *f.* 414.

6158. [7; lævis*.]
No. lævis: grey Rustic. *Haw.* 207.—(*Sam. I.* 29.)

6159 † [8. sordida*.]
Ph. No. sordida. *Marsh. MSS.*
No. sordida: sordid Rustic. *Haw.* 207. In Mus. ——?

b. *Abdomine graciliore; alis posticis albidis.*

6160. 9; Morpheus*. (*Och.* iv. 80.)—(*Sam. I.* 29.)—*Och. Tr.* v. *b.* 249.

6161. [10; Sepii*]: mottled Rustic.—*Haw. l.* c.
No. Sepii. *Hüb. No. pl.* 34. *f.* 161.—*Haw.* 206.—(*Sam. I.* 30.)
No. radica. *Esper.* iv. *pl.* cli. *f.* 4?
Ph. No. pulla. *Linn. Trans.* (*Beckwith.*) ii. 5. *pl.* 1. *f.* 7–9.—*Stew.* ii. 196.—*Turt.* iii. 324.
No. pullina. (*Haw. Pr.* 19.)

6162. 11; cubicularis*: pale mottled Willow.—*Haw. l. c.* (*Och.* iv. 80.)—*Och. Tr.* v. *b.* 251.
No. cubicularis. (*Wien. V.* 72?)—*Hüb. No. pl.* 89. *f.* 417.—*Haw.* 208.—(*Sam. I.* 29.)
Ph. clavipalpis. *Scop. C.* 213.
No. segetum. *Esper.* iv. *pl.* cl. *f.* 4, 5.
No. 4-punctata. *Fabr.* iii. *b.* 22.—*Stew.* ii. 182.

6163. 12; superstes: powdered Rustic.—*Haw. l.c.* (*Och.* iv. 80.)—*Och. Tr.* v. *b.* 260.
No. blanda. *Hüb. No. pl.* 34. *f.* 162.—*Haw.* 208.—(*Sam. I.* 29.)

6164. 13: glareosa*. (*Och.* iv. 80.)—*Och. Tr.* v. *b.* 247.—(*Ing. Inst.* 90.)—*Steph. Ill.* (*H.*) ii. *pl.* 21. *f.* 2.
No. glareosa. *Esper.* iv. *pl.* cxxviii. *f.* 3.
No. margaritacea. *Bork. E. S.* iv. 215.

No. I-intactum. *Hüb. No. pl.* 28. *f.* 130.
No. I-geminum. *Godart. Dup.* vi. 75. *pl.* 77. *f.* 6.
Orthosia Hebræica. (*Curtis. folio* 237.)

Genus 98: (881). GLÆA, *Hübner.*

CERASTIS[a], *Och.*

A. Corpore haud depresso.

6165. 1: rubricosa*.
No. rubricosa. (*Wien. V.* 77.)—*Fabr.* iii. *b.* 110.—*Hüb. No. pl.* 91. *f.* 430.
No. mucida. *Esper.* iv. *pl.* cxlviii. *f.* 4.
No. pilicornis. *Bork. E. S.* iv. 536.
No. rufa: red Chesnut. *Ent. Trans.* (*Hatchett.*) i. 244. *pl.* 5. *fig. sup.*—*Haw.* 232.—(*Sam. I.* 30.)
Var. No. mista. *Hüb. No. pl.* 109. *f.* 509.

6166 † 2, rubiginea*? dotted Chesnut.—*Haw. l. c.*
No. rubiginea. (*Wien. V.* 86.)—*Fabr.* iii. *b.* 31.—*Hüb. No. pl.* 38. *f.* 183.—*Haw.* 230.
No. tigrina. *Esper.* iv. *pl.* cxxiii. *f.* 3, 4.
No. sulphurago. *Bork. E. S.* iv. 679. In Mus. *Brit.*

B. Corpore depresso.

6167. 3; Satellitia*.
Ph. No. Satellitia. *Linn.* ii. 855.—*Don.* v. 81. *pl.* 168.—*Stew.* ii. 193.—*Turt.* iii. 321.
No. Satellitina: Satellites. (*Haw. Pr.* 15.)
No. Satellitia. *Hüb. No. pl.* 38. *f.* 182.—*Haw.* 229.—(*Sam. I.* 30.)
No. transversa. *Berl. M.* (*Hufn.*) iv. 418.

6168. 4; Vaccinii*.
Ph. No. Vaccinii. *Linn.* ii. 852.
No. Vaccina: the Chesnut. (*Haw. Pr.* 15.)
No. Vaccinii. *Hüb. No. pl.* 37. *f.* 177.—*Haw.* 233.—(*Sam. I.* 30.)
Albin. pl. xxiii. *f.* 34. *e–h.*

6169. [5; spadicea*]: dark Chesnut.—*Haw. l. c.*
No. spadicea. *Hüb. No. pl.* 37. *f.* 179.—*Haw.* 233.—(*Sam. I.* 30.)
No. castanea. *Marsh. MSS.*
No. castanina. (*Haw. Pr.* 15.)

6170. [6; subnigra*].
No. subnigrina. (*Haw. Pr.* 15.)
No. subnigra: black Chesnut M. *Haw.* 234.—(*Sam. I.* 30.)

---

[a] CERASTIS: Genus Serpentorum. Vide *Gen. Zool.*

6171. 7; polita*: netted Chesnut.—*Haw. l. c.*
No. polita. (*Wien. V.* 85.)—*Hüb. No. pl.* 37. *f.* 178.—*Haw.* 233.—(*Sam. I.* 30.)
No. Ligula. *Esper.* iv. *pl.* clxvi. *f.* 3.

Genus 99: (882). AMPHIPYRA, (*Och.*,) (*Samou.*,) *Treit.*

PYROPHILÆ *p*, *Hüb.*

6172. 1; Pyramidea*. (*Och.* iv. 70.)—*Och. Tr.* v. *a.* 285.
Ph. No. Pyramidea. *Linn.* ii. 856.—*Don.* vi. 31. *pl.* 193.—*Stew.* ii. 193.—*Turt.* iii. 327.
No. Pyramidina. (*Haw. Pr.* 12.)
No. Pyramidea. *Hüb. No. pl.* 8. *f.* 36.—*Haw.* 163.—(*Sam. I.* 30.)
Copper underwing. *Harr. V. M.* 55.

Genus 100: (883). PYROPHILA *mihi.*

PYROPHILÆ *p*, *Hüb.*—AMPHIPYRA *p*, *Treit.*

6173. 1; Tragopogonis*: Mouse.—*Haw. l. c.*
Ph. No. Tragopogonis. *Linn.* ii. 855.—*Stew.* ii. 193.—*Turt.* iii. 333.
No. Tragopogina. (*Haw. Pr.* 18.)
No. Tragopoginis. *Hüb. No. pl.* 8. *f.* 40.—*Haw.* 164.
No. Tragopoginus. (*Sam. I.* 30.)

6174. 2; tetra*: Mahogany.—*Haw. l. c.*
No. tetra. *Fabr.* iii. *b.* 21.—*Hüb. No. pl.* 8. *f.* 39.—*Haw.* 164.—(*Sam. I.* 30.)
Ph. tragopoginis. *Don.* vii. *pl.* 223. *f.* 2?
Albin. pl. xxvii. *f.* 41. *h–k.*

Genus 101: (884). DYPTERYGIA *mihi.*

XYLENA D., *Och.*

6175. 1; Pinastri*: O, G. Moth.
Ph. No. Pinastri. *Linn.* ii. 851.—*Don.* x. 58. *pl.* 347. *f.* 2.—*Stew.* ii. 190.—*Turt.* iii. 328.
No. Pinastrina: Bird's wing. (*Haw. Pr.* 18.)
No. Pinastri. *Hüb. No. pl.* 51. *f.* 246.—*Haw.* 172.—(*Sam. I.* 30.)
No. Dypterygia. *Berl. M.* (*Hufnagle.*) iii. 300.

Genus 102: (885). NÆNIA *mihi.*

LEMURES *p*, *Hüb.*—MORMO *p*, *Och.*—MANIA *p*, *Treit.*

6176. 1; typica*.
Ph. No. typica. *Linn.* ii. 857.
No. typicina. (*Haw. Pr.* 17.)
No. typica. *Haw.* 195.—(*Sam. I.* 30.)

No. venosa. *Hüb. No. pl.* 13. *f.* 61.
No. excusa. *Esper.* iv. *pl.* cxcvii. *f.* 1–3.
Albin. pl. xv. *f.* 21. *a–d.*
Dark or common Gothic M. *Harr. A. pl.* 22. *f. d–g.*—*Harr. V. M.* 25.

Genus 103: (886). XYLINA, *Treitschke.*

XYLENÆ *p, Hüb.*—XYLENA *p, Och.*, (*Samou.*)

6177. 1; Lambda*.
No. Lambda. *Fabr.* iii. *b.* 106?—*Haw.* 131.—(*Sam. I.* 29.)
No. Lambdina: grey Shoulder Knot. (*Haw. Pr.* 17.)
No. rhizobitha. *Hüb. No. pl.* 50. *f.* 242.

6178 † 2, pulla*. (*Och.* iv. 86.)—(*Ing. Inst.* 91.)
No. pulla. (*Wien. V.* 76.)—*Hüb. No. pl.* 49. *f.* 238.
In Mus. *Brit.*

6179. 3, semibrunnea*: tawny Pinion.—*Haw. l. c. Steph. Ill.* (*H.*) ii. *pl.* 21. *f.* 3.
No. semibrunnea. *Haw.* 171.—(*Sam. I.* 30.)
No. brunnina. (*Haw. Pr.* 12.)

6180. 4, petrificata*. (*Och.* iv. 86.)—*Och. Tr.* v. c. 23.—*Steph. Ill.* (*H.*) ii. *pl.* 23. *f.* 1.
No. petrificata. (*Wien. V.* 75.)—*Fabr.* iii. *b.* 123.
No. petrificosa. *Hüb. No. pl.* 49. *f.* 239.
No. umbrosa. *Esper.* iv. *pl.* cxxxiii. *f.* 5, 6.
Ph. socia. *Berl. M.* (*Hufn.*) iii. 418.

6181. 5, conspicillaris*. (*Och.* iv. 86.)—(*Ing. Inst.* 91.)—*Och. Tr.* v. c. 26.
Ph. No. conspicillaris. *Linn.* ii. 849.
No. conspicillina: Silver Cloud. (*Haw. Pr.* 19.)
No. conspicillaris. *Hüb. No. pl.* 49. *f.* 236, 237.—*Haw.* 171.—(*Sam. I.* 29.)
Ph. No. leuconota. *Don.* xiii. 44. *pl.* 453. *f.* 3.

6182. [6; perspicillaris*.] (*Och.* iv. 87?)—(*Ing. Inst.* 91.)
Ph. No. perspicillaris. *Linn.* ii. 849?—*Turt.* iii. 323?
No. perspicillina. (*Haw. Pr.* 12?)
La. perspicillaire. *Pap. d'Eur.* vi. 161. *pl.* ccliii. *f.* 383.

6183. 7; putris*. (*Och.* iv. 86.)—*Och. Tr.* v. c. 29.
Ph. No. putris. *Linn.* ii. 850.—*Turt.* iii. 337.
No. putrina. (*Haw. Pr.* 19.)
No. putris: the Flame M. *Haw.* 172.—(*Sam. I.* 30.)
No. lignosa. *Hüb. No. pl.* 50. *f.* 245.
Ph. subcorticalis. *Berl. M.* (*Hufn.*) iii. 308.
Albin. pl. lxxix. *f. a–d.*
Small Sword-grass likeness. *Harr. V. M.* 35.

Genus 104: (887). CALOCAMPA *mihi.*

XYLENA *p, Och.*—XYLINA *p, Treit.*

6184. 1; exoleta*.
Ph. No. exoleta. *Linn.* ii. 849.—*Berk. S.* i. 141.—*Don.* vi. 19. *pl.* 187.—*Stew.* ii. 189.—*Turt.* iii. 336.
No. exoletina. (*Haw. Pr.* 12.)
No. exoleta. *Hüb. No. pl.* 50. *f.* 244.—*Haw.* 168.—(*Sam. I.* 29.)
Sword-grass M.: Ph. exoleta. (*Wilkes.* 8. *pl.* 18.)—*Harr. V. M.* 29.

6185. 2: vetusta*. *Steph. Ill.* (*H.*) *pl.* 23. *f.* 1.
No. vetusta. *Hüb. No. pl.* 97. *f.* 459.

Genus 105: (888). XYLOPHASIA *mihi.*

XYLENA B., *Och.*

6186. 1; lithoxylea*: light Arches.—*Haw. l. c.*
No. lithoxylea. *Fabr.* iii. *b.* 123.—*Haw.* 169.—(*Sam. I.* 29.)
No. sublustris var. *Esper.* iv. *pl.* cxxxiii. *f.* 2.
No. musicalis. *Esper.* iv. *pl.* cxxxxvii. *f.* 5.

6187. 2, sublustris.
No. sublustris. *Esper.* iv. *pl.* cxxxiii. *f.* 1.
No. lithoxylea. *Hüb. No. pl.* 49. *f.* 240.

6188. 3; polyodon*.
Ph. No. polyodon. *Linn.* ii. 853.—*Stew.* ii. 191.
No. polyodina. (*Haw. Pr.* 18.)
No. polyodon. *Haw.* 186.—(*Sam. I.* 30.)
No. radicea. (*Wien. V.* 81.)—*Fabr.* iii. *b.* 125.—*Hüb. No. pl.* 17. *f.* 82.
No. occulta. *Esper.* iv. *pl.* cxxxii. *f.* 3, 4.
No. monoglypha. *Berl. M.* (*Hufn.*) iii. 308.
Dark Arches. *Harr. V. M.* 9.

6189. 4; hepatica*: clouded-bordered Brindle.—*Haw. l. c.*
Ph. No. hepatica. *Linn.* ii. 853.
No. hepaticina. (*Haw. Pr.* 18.)
No. hepatica. *Haw.* 169.—(*Sam. I.* 29.)
No. putris. *Hüb. No. pl.* 50. *f.* 241.
Ph. No. epomis. *Marsh. MSS.*
No. epomina. (*Haw. Pr.* 18.)

6190 † 5, hirticornis*: the hairy-horned Brindle.—*Haw. l. c.*
No. hirticornis. *Ent. Trans.* (*Haw.*) i. 336.
Olim in Mus. D. *Hatchett.*

6191. 6; epomidion*: clouded Brindle.—*Haw. l. c.*
No. epomidion. *Haw.* 170.—(*Sam. I.* 29.)
No. epomidina. (*Haw. Pr.* 18.)
No. characterea. *Hüb. No. pl.* 18. *f.* 133?

6192. 7; rurea*: dark tawny.—*Haw. l. c.*
No. rurea. *Fabr.* iii. *b.* 125. (!)
Ph. No. rurea. *Stew.* ii. 189.—*Turt.* iii. 338.
No. combusta. *Hüb. No. pl.* 79. *f.* 366. ♂.—*Haw.* 170.—(*Sam. I.* 29.)
No. Alopecurus. *Esper.* iv. *pl.* cxlvii. *f.* 3.
No. luculenta. *Esper.* iv. *pl.* cxxxiii. *f.* 3, 4.

6193. 8, scolopacina*: slender clouded Brindle.—*Haw. l. c.*
No. scolopacina. *Esper.* iv. *pl.* cxx. *f.* 1.—(*Sam. I.* 30.)
No. abbrevina. (*Haw. Pr.* 18.)
No. abbreviata. *Haw.* 170.

6194 † 9, rectilinea.
No. rectilinea. *Hüb. No. pl.* 51. *f.* 248.
La Saxone. *Pap. d'Eur.* vi. 163. *pl.* ccliv. *f.* 385.
In Mus. D. *Marshall.*

Genus 106: (889). HADENA, *Schrank,* (*Samou.,*) *Treit.*

A. Alis anticis margine postico subangulato.

6195. 1, amica. *Och. Tr.* v. *a.* 332?—*Steph. Ill.* (*H.*) ii. *pl.* 23. *f.* 2.
No. satura: barred Arches. *Haw.* 187.—(*Sam. I.* 30.)

6196. 2; adusta*. (*Och.* iv. 72.)—*Och. Tr.* v. *a.* 339.
No. adusta. *Esper. No. pl.* cxlix. *f.* 1–2.
No. aquilina. *Bork. E. S.* iv. 381.
No. valida. *Hüb. No. pl.* 133. *f.* 606. ♂. *f.* 607, 608. ♀.
No. porphyrea. *Scriba B.* ii. *pl.* 10. *f.* 4.
Ph. No. duplex. *Marsh. MSS.*
No. duplexina. (*Haw. Pr.* 18.)
No. duplex: dark Brocade. *Haw.* 190.—(*Sam. I.* 29.)

6197. 3: satura. *Och. Tr.* v. *a.* 333.
No. satura. (*Wien. V.* 83.)—*Hüb. No. pl.* 16. *f.* 75.
No. porphyrea. *Esper. No.* iv. *pl.* cxlv. *f.* 5.

6198. 4; remissa*: gothic Brocade.—*Haw. l. c.* (*Och.* iv. 72.)
No. remissa. *Hüb. No. pl.* 90. *f.* 423.—*Haw.* 189.
No. gemina. *Hüb. No. pl.* 102. *f.* 482.
No. satura. *Bork. E. S.* iv. 377.
Ha. gemina. *Och. Tr.* v. *a.* 345.

6199. 5, oblonga.
No. oblonga: brown-pinioned Brocade. *Haw.* 188.

6200. 6; Thalassina*. *Och. Tr.* v. *a.* 342.
No. Thalassina. *Bork. E. S.* iv. 383.
No. Achates. *Hüb. No. pl.* 106. *f.* 498. ♂. *pl.* 133. *f.* 610.—(*Sam. I.* 28.)
No. gemina. *Hüb. No. pl.* 102. *f.* 483. ♀.
No. humeralis: pale shouldered Brocade. *Haw.* 190.

6201. 7; Genistæ*. (*Och.* iv. 72.)—*Och. Tr.* v. *a.* 349.
No. Genistæ. *Bork. E. S.* iv. 355.—*Hüb. No. pl.* 134. *f.* 611. ♂. *f.* 612. ♀.
No. W. Latinum. *Bork. E. S.* iv. 378.—*Esper.* iv. *pl.* cxxxvi. *f.* 1, 2.
No. rectilinea: light Brocade. *Haw.* 189.—(*Sam. I.* 30.)
No. leucoples. *Marsham MSS.*
No. leucoplina. (*Haw. Pr.* 18.)
Ph. No. dives. *Don.* x. 71. *pl.* 352. *f.* 1.
Gothic M. (*Wilkes D. pl.* 6. *f.* 2.)—*Harr. V. M.* 29.

6202. 8; contigua*. (*Och.* iv. 72.)—*Och. Tr.* v. *a.* 352.
No. contigua. (*Wien. V.* 82.)—*Hüb. No. pl.* 18. *f.* 85. ♀. *pl.* 133. *f.* 609. ♂.
No. Spartii. *Bork. E. S.* iv. 352.
No. Ariæ. *Esper.* iv. *pl.* clx. *f.* 8.
No. dives. *Marsham MSS.*—*Haw.* 189.—(*Sam. I.* 29.)
No. divesina: the beautiful Brocade. (*Haw. Pr.* 18.)
Var. No. pulchellina. (*Haw. Pr.* 18.)

6203 † 9, obscura.
No. obscura: the dingy Brocade. *Haw.* 189.
In Mus. D. *Haworth.*

B. Alæ anticæ margine postico subrotundato.

a. Vide *Haw.* 196.—(Noct.-fissonotatæ.)

6204. 10, glauca*: glaucous Sheers.—*Haw. l. c.* (*Och.* iv. 71.) —*Och. Tr.* v. *a.* 322.
No. glauca. *Hüb. No. pl.* 87. *f.* 410.—*Haw.* 197.—(*Sam. I.* 29.)

6205. 11; plebeia*: the Sheers.—*Haw. l.* c.
Ph. No. plebeia. *Linn.* ii. 853?—*Mus. Linn. teste Haworthi.* —*Haw.* 198.—(*Sam. I.* 29.)
No. plebeina. (*Haw. Pr.* 18.)
Var. No. dentina: yellow-striped Sheers.—*Haw. l. c.* (*Wien. V.* 82.) —*Fabr.* iii. *b.* 69.—*Hüb. No.* 87. *f.* 408.—*Haw.* 198.— (*Sam. I.* 29.)
No. nana. *Esper.* iv. *pl.* cxxvii. *f.* 2, 3.

6206. [12; leucostigma*.]
No. leucostigma: the pale Sheers. *Haw.* 198.—(*Sam. I.* 29.)

6207. 13, ochracea*. *Steph. Ill.* (*H.*) ii. *pl.* 23. *f.* 3.
No. ochracea: the tawny Sheers. *Haw.* 199.—(*Sam. I.* 29.)
Ha. peregrina. *Och. Tr.* v. *a.* 330?
Polia ochracea. (*Curtis.* vi. *folio* 248.)

6208. 14; Lithorhiza*: the early Grey.—*Haw. l.* c.
No. Lithorhiza. *Bork. E. S.* iv. 339.
No. areola. *Esper.* iv. *pl.* cxli. *f.* 4.

No. operosa. *Hüb. No. pl.* 88. *f.* 398.—*Haw.* 185.—(*Sam. I.* 29.)

b. Vide *Haw.* 195.—(Noct.-clathratæ.)—HELIOPHOBUS *p, Boisd.*

6209. 15; Cucubali*: the Campion.—*Haw. l.* c. (*Och.* iv. 71.)—*Och. Tr.* v. *a.* 311.

No. Cucubali. (*Wien. V.* 84.)—*Hüb. No. pl.* 12. *f.* 56.—*Haw.* 196.—(*Sam. I.* 29.)

No. rivularis. *Fabr.* iii. *b.* 101.

No. triangularis. *Thunb. I. S.* 3. *fig. sup. dext.*

6210. 16; Capsincola*: the Lychnis.—*Haw. l.* c. (*Och.* iv. 72.) *Och. Tr.* v. *a.* 308.

No. Capsincola. (*Wien. V.* 84.)—*Hüb. No. pl.* 12. *f.* 57.—*Haw.* 196.—(*Sam. I.* 29.)

No. Capsincolina. (*Haw. Pr. App.* 3.)

No. bicruris. *Berl. M.* (*Hufnagle.*) iii. 302.

6211. 17; Saponariæ*. *Och. Tr.* v. *a.* 303.

No. Saponariæ. *Bork. E. S.* iv. 370.

No. typica: (Saponariæ.) *Hüb. No. pl.* 12. *f.* 58.

No. calcatrippe. *Vieweg. T. V.* ii. 71.

Ph. No. reticulata. *Villers E.* ii. 254.

No. marginosa: the bordered Gothic. *Haw.* 195.—(*Sam. I.* 29.)

No. typicoidina. (*Haw. Pr.* 17.)

Genus 107: (890). HELIOPHOBUS, *Boisduval.*

HADENA *p, Schrank.*—BOMBYX *p, Fabr.*—DAHNEA *mihi* (*olim.*)

6212. 1; Popularis*.

Bo. Popularis. *Fabr.* iii. *a.* 484.

Bo. Popularius. (*Haw. Pr.* 9.)

No. Popularis: feathered Gothic M. *Haw.* 117 & 226.—(*Sam. I.* 30.)

No. Graminis. (*Wien. V.* 82.)—*Hüb. No. pl.* 12. *f.* 59.

No. Lolii. *Esper.* iii. *pl.* xlviii. *f.* 1–5.

Ph. typicoides. *Don.* xv. i. *pl.* 505.

6213. 2, leucophæus. *Steph. Ill.* (*H.*) ii. *pl.* 24. *f.* 1.

No. leucophæa. (*Wien. V.* 82.)—*Hüb. No. pl.* 17. *f.* 80.

Bo. fulminea. *Fabr.* iii. *a.* 484.

† ♂. Bo. vestigialis. *Esper.* iii. *pl.* liii. *f.* 4, 5.

♀, Bo. ravida. *Esper.* iv. *pl.* cxlv. *f.* 1.

Genus 108: (891). MAMESTRA, (*Och.,*) *Treit.*

MAMESTRIA, (*Samou.*)

6214. 1; furva*. (*Och.* iv. 77.)—*Och. Tr.* v. *b.* 154.

No. furvina. (*Haw. Pr.* 19.)

No. furva. (*Wien. V.* 81.)—*Hüb. No. pl.* 87. *f.* 407.

6215. 2; Pisi*. (*Och.* iv. 76.)—*Och. Tr.* v. *b.* 128.
Ph. No. Pisi. *Linn.* ii. 854.—*Berk. S.* i. 143.—*Don.* ii. *pl.* 52. —*Stew.* ii. 192.—*Turt.* iii. 323.
No. Pisina. (*Haw. Pr.* 15.)
No. Pisi. *Hüb. No. pl.* 91. *f.* 429.—*Haw.* 193.—(*Leach E. E.* ix. 134.)—(*Sam. I.* 30.)
The Broom M.: Ph. Pisi. (*Wilkes.* 4. *pl.* vii.)
Albin. pl. xxxii. *f.* 51. *c–f.*
Favourite or Broom M. *Harr. V. M.* 25.

6216. 3, splendens. (*Och.* iv. 76?)—*Och. Tr.* v. *b.* 131?
No. splendens. *Hüb. No. pl.* 85. *f.* 400. ♀?

6217. 4; oleracea*. (*Och.* iv. 77.)—*Och. Tr.* v. *b.* 132.
Ph. No. oleracea. *Linn.* ii. 853.—*Berk. S.* i. 142.—*Stew.* ii. 191.—*Turt.* iii. 328.
No. oleracina: Bright-line Brown-eye. (*Haw. Pr.* 15.)
No. oleracea. *Hüb. No. pl.* 18. *f.* 87.—*Haw.* 193.—(*Sam. I.* 29.)
Ph. No. monstrosa. *Villers E.* ii. 248.
Albin. pl. xxvii. *f.* 40. *a–d.*
White-line brown Eye. *Harr. V. M.* 21.

6218. 5: Suasa*. (*Och.* iv. 77.)
No. Suasa. (*Wien. V.* 83.)—*Hüb. No. pl.* 91. *f.* 426. ♀.
No. dissimilis. *Vieweg. T. V.* ii. 65.
No. W. Latinum var. *Esper.* iv. *pl.* cxxxvi. *f.* 3.
No. Leucographa. *Esper.* iv. *pl.* cl. *f.* 3.
No. dens-canis: the Dog's tooth. *Haw.* 190.—(*Sam. I.* 29.)

6219. 6: nigricans*. *Steph. Ill.* (*H.*) ii. *pl.* 24. *f.* 2.
No. nigricans. *Vieweg. T. V.* ii. 66?
No. abjecta. *Hüb. No. pl.* 116. *f.* 539?

6220. 7; Brassicæ*. (*Och.* iv. 77.)—*Och. Tr.* v. *b.* 150.
Ph. No. Brassicæ. *Linn.* ii. 852.—*Stew.* ii. 184.—*Turt.* iii. 300.
No. Brassicina: Cabbage M. (*Haw. Pr.* 18.)
No. Brassicæ. *Hüb. No. pl.* 18. *f.* 88.—*Haw.* 191.—(*Sam. I.* 29.)—(*Kirby & Sp. I. E.* i. 29.)
Albin. pl. xxviii. *f.* 42, 43. *a–e.*—*Id. pl.* xxix. *f.* 45. *e–h.*—*Id. pl.* lxvii. *f. a–d.*—*Id.* lxxviii. *f. a–d.*
Old Gentlewoman. *Harr. V. M.* 27.

6221 † [8, albidilinea*.]
No. albidilinea: White-line Black. *Haw.* 191.
In Mus. D. *Hatchett.*

6222. 9, albicolon. *Och. Tr.* v. *b.* 147.—*Steph. Ill.* (*H.*) ii. *pl.* 24. *f.* 3.
No. albicolon. *Hüb. No. pl.* 117. *f.* 542, 543.

6223. 10; Aliena*. (*Och.* iv. 77.)—*Och. Tr.* v. *b.* 139.

No. Aliena. *Hüb. No. pl.* 94.*f.* 441.
No. contigua: large Nutmeg. *Haw.* 192.—(*Sam. I.* 29.)

6224. 11; Chenopodii*. (*Och.* iv. 77.)—*Och. Tr.* v. *b.* 144.
No. Chenopodii. (*Wien. V.* 82.)—*Fabr.* iii. *b.* 68.—*Hüb. No. pl.* 18.*f.* 86.—*Haw.* 192.—(*Sam. I.* 29.)
No. Chenopodina: Nutmeg. (*Haw. Pr. App.* 3.)
No. infraina. (*Haw. Pr.* 18.)
Ph. No. Chenopodii. *Stew.* ii. 184.
Albin. pl. xxix.*f.* 44. *e–d.*
Small old Gentlewoman. *Harr. V. M.* 27.

6225. 12; Persicariæ*. (*Och.* iv. 77.)—*Och. Tr.* v. *b.* 156.
Ph. No. Persicariæ. *Linn.* ii. 847.—*Don.* ix. 63. *pl.* 317.—*Turt.* iii. 333.
No. Persicarina. (*Haw. Pr.* 17.)
No. Persicariæ. *Hüb. No. pl.* 13. *f.* 64.—*Haw.* 191.—(*Sam. I.* 30.)
Ph. Sambuci. *Berl. M.* (*Hufn.*) iii. 308.
Albin. pl. lxxvii.*f. a–d.*
Dot M. *Harr. A. pl.* 24.*f. a–e.*—*Harr. V. M.* 21.

Genus 109: (892). EUPLEXIA *mihi.*

HADENA D., *Och.*—PHLOGOPHORA B., *Treit.*

6226. 1; lucipara*.
Ph. No. lucipara. *Linn.* ii. 857.—*Don.* vii. *pl.* 230. *f.* 2.—*Stew.* ii. 193.—*Turt.* iii. 328.
No. luciparina. (*Haw. Pr.* 19.)
No. lucipara. *Hüb. No. pl.* ii. *f.* 55.—*Haw.* 210.—(*Sam. I.* 29.)
Bo. flavomacula. *Fabr. M.* ii. 117.
Ph. dubia. *Berl. M.* (*Hufnagle.*) iii. 404.
Small angleshades or Chevalier. *Harr. V. M.* 43.

Genus 110: (893). HAMA *mihi.*

APAMEA C. *p*, (*Och.*)—APAMEA *p*, (*Samou.*)—LUPERNIA, *Boisd.*

6227. 1; basilinea*.
No. basilinea. (*Wien. V.* 78.)—*Fabr.* iii. *b.* 125.—*Hüb. No. pl.* 91.*f.* 427.—*Haw.* 194.—(*Sam. I.* 28.)
No. basilineina: rustic Shoulder Knot. (*Haw. Pr.* 18.)
No. nebulosa. *Vieweg. T. V. pl.* 1.*f.* 6.

6228. 2; testacea*.
No. testacea. *Hüb. No. pl.* 29.*f.* 139. ♂.
No. lunatostrigata: lesser flounced Rustic. *Haw.* 194.—(*Sam. I.* 29.)
Var. No. unca: flounced Rustic. *Haw.* 194.—(*Sam. I.* 30.)
Var. No. X. notata: tawny X. *Haw.* 194.—(*Sam. I.* 30.)
6229 † 3, connexa.

Ap. connexa. (*Och.* iv. 76?)
No. connexa. *Bork. E. S.* iv. 360.
No. Elota. *Hüb. No. pl.* 98. *f.* 462.
No. pabulitricola. *Scriba. B.* iii. 259. *pl.* xviii. *f.* 3.
In Mus. D. *Haworth?*

Genus 111: (894). APAMEA, (*Och.*,) *Treit.*

GORTYNA *p*, *Treit.*

6230. 1: fibrosa*. (*Och.* iv. 75.)
No. fibrosa. *Hüb. No. pl.* 82. *f.* 385.
No. lunina: the Crescent. (*Haw. Pr.* 19.)—*Haw.* 209.—(*Sam. I.* 19.)
Go. leucostigma. *Och. Tr.* v. *b.* 331.
β, No. leucostigma. *Hüb. No. pl.* 80. *f.* 375.

6231. 2; nictitans*. (*Och.* iv. 75.)—*Och. Tr.* v. *b.* 82.
Ph. No. nictitans. *Linn.* ii. 847.
No. nictitina: Golden ear. (*Haw. Pr.* 18.)
No. chrysographa. (*Wien. V.* 313.)—*Hüb. No. pl.* 46. *f.* 221.
No. auricula. *Haw.* 240.—(*Sam. I.* 28.)
Ph. auricula. *Don.* xii. 5. *pl.* 397. *f.* 3.
No. cinerago. *Fabr. Sup.* 445.
Ear M. *Harr. V. M.* 21.
β; No. erythrostigma: red Dot. *Haw.* 240.—(*Sam. I.* 29.)

6232. 3; didyma*. (*Och.* iv. 75.)—*Och. Tr.* v. *b.* 86.
No. nictitans. *Esper.* iv. *pl.* cxxvi. *f.* 6.
No. lugensina. (*Haw. Pr.* 18.)
No. lugens: rustic Mourner. *Marsh. MSS.*—*Haw.* 212.—(*Sam. I.* 29.)

6233. 4; secalina*: Small clouded Brindle.—*Haw. l. c.*
No. secalina. *Hüb. No. pl.* 89. *f.* 420.—*Haw.* 210.—(*Sam. I.* 30.)
No. leucostigma. *Esper.* iv. *pl.* clix. *f.* 7.
No. Lambda. *Vieweg. T. V.* ii. 82.

6234 † 5. unanimis. (*Och.* iv. 75.)
No. unanimis. *Hüb. No. pl.* 120. *f.* 556? In Mus. D. Dale.

6235. 6; oculea*.
Ph. No. oculea. *Linn. F. No.* 1215.
Ph. No. nictitans. *Villers E.* ii. 220.
No. Phœbe. *Marsh. MSS.*
No. Phœbina. (*Haw. Pr.* 18.)
No. oculea: common Rustic. *Haw.* 211.—(*Sam. I.* 29.)
Ap. didyma var. *Och. Tr.* v. *b.* 86.

6236. 7; I-niger*.
No. I-niger: Letter I. *Haw.* 211.—(*Sam. I.* 29.)
Ap. didyma var. *Och. Tr.* v. *b.* 86.

6237. 8; rava*.
No. rava: the Russet. *Haw.* 209.—(*Sam. I.* 30.)
No. didyma. *Esper.* iv. *pl.* cxxvi. *f.* 7.
Ap. didyma var. *Och. Tr.* v. *b.* 86.
β; No. furcina: the Flame Furbelow. (*Haw. Pr.* 19.)
No. furca. *Marsh. MSS.*—*Haw.* 209.—(*Sam. I.* 29.)

6238. 9; ophiogramma*. (*Och.* iv. 75.)—*Och. Tr.* v. *b.* 91.
No. ophiogramma. *Esper.* iv. *pl.* clxxxii. *f.* 2.—*Hüb. No. pl.* 76. *f.* 355.
No. bilobina: double lobed. (*Haw. Pr.* 18.)
No. biloba. *Marsh. MSS.*—*Haw.* 209.—(*Sam. I.* 28.)

Genus 112: (895). MIANA *mihi.*

APAMEA B., (*Och.*)

6239. 1; literosa*. *Steph. Ill.* (*H.*) iii. *pl.* 25. *f.* 1.
No. literosa: rosy Minor. *Haw.* 213.—(*Sam. I.* 29.)

6240. 2; strigilis*: Marbled Minor.—*Haw. l. c.*
Ph. No. strigilis. *Linn.* ii. 851.
No. strigalina. (*Haw. Pr.* 19.)
No. præduncula. (*Wien. V.* 89.)—*Hüb. No. pl.* 20. *f.* 95.—*Haw.* 213.—(*Sam. I.* 30.)
Minor beauty. *Harr. V. M.* 9.
β; No. strigilis. *Haw.* 214.—(*Sam. I.* 30.)

6241. 3; latruncula*: tawny Marbled Minor.—*Haw. l. c.*
No. latruncula. *Hüb. pl.* 20. *f.* 94.—*Haw.* 214.—(*Sam. I.* 29.)
No. ærata. *Esper.* iv. *pl.* cxlvi. *f.* 4.

6242. [4; Æthiops*.]
No. Æthiops: the Blackamoor. *Haw.* 215.—(*Sam. I.* 28.)
No. ærata var. *Esper.* iv. *pl.* cxlvi. *f.* 6.

6243. 5; humeralis*.
No. humeralis: the cloaked Minor. *Haw.* 215.—(*Sam. I.* 29.)
No. furuncula. *Hüb. No. pl.* 117. *f.* 545?
No. bicoloria. *Bork. E. S.* iv. 190?

6244. [6; terminalis*.]
No. terminalis: the flounced Minor. *Haw.* 215.—(*Sam. I.* 30.)
No. ærata var. *Esper.* iv. *pl.* cxlvi *f.* 5.
Ap. suffuruncula. (*Och.* iv. 76?)

6245. 7, rufuncula.
No. rufuncula: plain red Minor. *Haw.* 216.—(*Sam. I.* 30.)

6246. 8; fasciuncula*.
No. fasciuncula: middle-barred Minor. *Haw.* 215.—(*Sam. I.* 29.)

6247 † 9, minima*.
No. minima: least Minor. (*Haw. Pr.* 19.)—*Haw.* 216.—(*Sam. I.* 29.) In Mus. *D. Bentley, Chant et Haworth.*

Genus 113: (896). CELÆNA *mihi.*

APAMEA *p, Och.?*

6248. 1, renigera *mihi.* *Steph. Ill.* (*H.*) iii. *pl.* 25. *f.* 2.

6249. 2: leucographa. *Steph. Ill.* (*H.*) iii. *pl.* 25. *f.* 3.
No. leucographa. (*Wien. V.* 83?)—*Hüb. No. pl.* 88. *f.* 411? *pl.* 124. *f.* 572?

6250. 3: Lancea?
No. Lancea. *Esper.* iv. *pl.* clxvi. *f.* 7?

6251. [4, hibernica.] *Haworth MSS.*

Genus 114: (897). SCOTOPHILA *mihi.*

TRACHEA C. *p,* (*Och.,*) *Treit.*—ACHATIÆ *p,* (*Hübn.*)—ILARUS *p, Boisd.*

6252. 1; porphyrea*.
No. porphyrea. (*Wien. V.* 83.)—*Hüb. No. pl.* 19. *f.* 93. ♀. *pl.* 100. *f.* 473. ♂.
No. picta. *Fabr.* iii. *b.* 91.
No. concinna. *Esper.* iv. *pl.* clii. *f.* 1.
No. lepida. *Esper.* iv. *pl.* clii. *f.* 2.
No. birivia. *Bork. E. S.* iv. 206.
Ph. No. varia. *Villers E.* ii. 276.
No. Ericæ. *Marsh. MSS.*—*Don.* x. 93. *pl.* 360. *f.* 1.—*Haw.* 224.—(*Sam. I.* 29.)
No. Ericina. (*Haw. Pr.* 18.)
True lover's knot. *Harr. Ex.* 22. *pl.* v. *f.* 5.—*Harr. V. M.* 33.

Genus 115: (898). ACHATIA.

TRACHEA C. *p, Treit.*—ACHATIÆ *p, Hübn.*—ACHATEA, *Curtis.* —ILARUS *p, Boisd.*

6253. 1; piniperda*.
No. piniperda. *Naturf.* (*Kob.*) xxi. 27. *pl.* 2.—*Esper.* iv. *pl.* cxxv. *f.* 1–6.
Bo. spreta. *Fabr.* iii. *a.* 455.
Ac. spreta. *Curtis.* iii. *pl.* 117.
Ph. No. Pini. *Villers E.* ii. 278.
No. flammea. *Fabr.* iii. *b.* 248.—*Hüb. No. pl.* 101. *f.* 476.
β, No. ochroleuca. *Hüb. No. pl.* 39. *f.* 91.

Genus 116: (899). ACTEBIA *mihi.*

TRACHEA B., (*Och.,*) (*Samou.,*) *Treit.*

6254. 1, præcox*?
Ph. No. præcox. *Linn.* ii. 854.—*Don.* vi. 79. *pl.* 213.—*Stew.* ii. 192.—*Turt.* iii. 327.
No. præcocina. (*Haw. Pr.* 15.)
No. præcox: Portland M. *Haw.* 201.—(*Sam. I.* 30.)
No. præceps. *Hüb. No. pl.* 15. *f.* 70.

Genus 117: (900). TRACHEA, (*Och.*,) (*Samou.*,) *Treit.*

6255. 1, Atriplicis. *Och. Tr.* v. *b.* 66.
Ph. No. Atriplicis. *Linn.* ii. 854.—*Berk. S.* i. 143.—*Don.* viii. 25. *pl.* 262. *f.* 1.—*Stew.* ii. 192.—*Turt.* iii. 325.
No. Atriplicina. (*Haw. Pr.* 14.)
No. Atriplicis. *Hüb. No. pl.* 17. *f.* 83.—*Haw.* 196.—(*Sam. I.* 28.)
Wild Arrach M.: Ph. Atriplicis. (*Wilkes.* 3. *pl.* 2.)—*Harr. V. M.* 9.

Genus 118: (901). VALERIA, (*Germar.*)

MISELIA *p*, *Hübner.*—BOMBYX *p*, *Fabr.*—POLIA *p*, *Boisd.*

6256. 1, oleagina*. (*Ing. Inst.* 91.)
Bo. oleagina. (*Wien. V.* 59.)—*Fabr.* iii. *a.* 440.
Bo. oleaginus: the Green-brindled Dot. (*Haw. Pr.* 9.)
No. oleagina. *Hüb. No. pl.* 7. *f.* 33.—*Haw.* 120 & 202.
Ph. oleagina. *Sowerb. B. M.* i. *pl.* 37.—*Don.* xiii. 13. *pl.* 439.

Genus 119: (902). MISELIA, (*Och.*,) *Treit.*, *Curtis.*

MISELIÆ *p*, *Hübn.*—POLIA *p*, *Boisd.*

6257 † 1, bimaculosa. (*Och.* iv. 73.)—*Och. Tr.* v. *a.* 408.—(*Ing. Inst.* 91.)—*Curtis.* iv. *pl.* 177.
Ph. No. bimaculosa. *Linn.* ii. 856.
No. bimaculosa. *Hüb. No. pl.* 7. *f.* 32. ♂.
No. bimaculosa Italica. *Esper.* iv. *pl.* clvii. *f.* 5. In Mus. *Brit.*

6258. 2; Oxyacanthæ*. (*Och.* iv. 73.)—*Och. Tr.* v. *a.* 405.—(*Curtis l. c. supra.*)
Ph. No. Oxyacanthæ. *Linn.* ii. 852.—*Berk. S.* i. 142.—*Don.* v. 75. *pl.* 165.—*Stew.* ii. 191.—*Turt.* iii. 324.
No. Oxyacanthina: Green-brindled Crescent. (*Haw. Pr.* 16.)
No. Oxyacanthæ. *Hüb. No. pl.* 7. *f.* 31.—*Haw.* 201.—(*Sam. I.* 29.)
Ealing's Glory: Ph. Oxyacanthæ. (*Wilkes.* 12. *pl.* 27.)—*Harr. A. pl.* 43. *f.* d–f.—*Harr. V. M.* 27.
Pa. No. 23. *List. Goëd.* (*Angl.*) 32. *f.* 23.—*Id.* (*Lat.*) 67. *f.* 23. *Albin. pl.* xiv. *f.* 19. *a–d.*

6259. 3; Aprilina*. (*Och.* iv. 73.)—*Och. Tr.* v. *a.* 411.—(*Curtis l. c. supra.*)
Ph. No. Aprilina. *Linn.* ii. 847.—*Stew.* ii. 188.—*Turt.* iii. 329.
No. Aprilina major. *Berl. M.* (*Hufnagle.*) iii. 422.
No. Aprilina. (*Haw. Pr.* 15.)—*Haw.* 200.—(*Sam. I.* 28.)
No. runica. (*Wien. V.* 70.)—*Fabr.* iii. *b.* 102.—*Hüb. No. pl.* 15. *f.* 71.
Ph. runica. *Don.* x. 75. *pl.* 354 *f.* 1.
Marvel du Jour. *Harr. V. M.* 19.

6260. 4; compta*. (*Och.* iv. 72.)—(*Curtis l. c. supra.*)
No. compta. (*Wien. V.* 70.)—*Fabr.* iii. *b.* 95.—*Hüb. No. pl.* 11. *f.* 53.—*Haw.* 179.—(*Sam. I.* 29.)
No. variegatina: the Marbled Coronet. (*Haw. Pr.* 17.)
Ph. No. transversalis. *Villers E.* ii. 281.
Mi. comta. *Och. Tr.* v. *a.* 389.
β; Ph. No. X-scriptum. *Sowerb. B. M.* i. *pl.* 55.

6261 † 5, albimacula*. *Och. Tr.* v. *a.* 391.
No. albimacula. *Bork. E. S.* iv. 149.
No. concinna: (conserta). *Hüb. No. pl.* 11. *f.* 51.
No. compta. *Esper. No.* iv. *pl.* cxvii. *A. f.*
Mi. concinna. (*Och.* iv. 72.)—(*Curtis l. c. supra.*)
In Mus. *Brit.*

Genus 120: (903). POLIA, (*Hübn.*,) (*Samou.*,) *Treit.*, *Curtis.*
AGROTIS *p*, *Och.*

6262. 1; advena*. (*Och.* iv. 74.)—*Och. Tr.* v. *b.* 39.—(*Curtis l. c. infra.*)
No. advena. (*Wien. V.* 77.)—*Hüb. No. pl.* 17. *f.* 81.—*Haw.* 187.—(*Sam. I.* 28.)
No. celata. *Marsh. MSS.*
No. celatina: pale-shining brown. (*Haw. Pr.* 17.)

6263. [2, nitens*.] (*Curtis l. c. infra.*)
No. nitensina. (*Haw. Pr.* 17.)
No. nitensina: tawny and silver M. *Haw.* 188.

6264. 3; bimaculosa*.
No. bimaculosa. *Esper.* iv. *pl.* cxxxii. *f.* 1, 2.
No. Thapsi. *Bork. E. S.* iv. 567.
No. plebeia: (nebulosa.) *Hüb. No. pl.* 16. *f.* 78.
Ph. No. grandis: grey Arches. *Don.* x. 52. *pl.* 341. *f.* 1.
No. grandina. (*Haw. Pr.* 17.)
No. grandis. *Haw.* 185.—(*Sam. I.* 29.)
Po. nebulosa. (*Och.* iv. 74.)—*Och. Tr.* v. *b.* 48.—(*Curtis l. c. infra.*)
No. polyodon. *Fabr.* iii. *b.* 114?
Grey Arches. (*Harr. V. M.* 9.)

6265. 4, occulta*. (*Och.* iv. 74.)—*Och. Tr.* v. *b.* 52.—*Curtis.* vi. *pl.* 248.
Ph. No. occulta. *Linn.* ii. 849.
No. occultina: great Brocade. (*Haw. Pr. App.* 3.)
No. occulta. *Hüb. No. pl.* 17. *f.* 79.—*Haw.* 186.
No. occulata. (*Sam. I.* 29.)

6266. 5; tincta*. (*Och.* iv. 74.)—*Och. Tr.* v. *b.* 43.—(*Curtis l. c. supra.*)
No. tincta. *Brahm. I. K.* ii. 395.
No. trimaculosa. *Esper.* iv. *pl.* cxxxi. *f.* 5.

No. advena. *Vieweg. T. V.* ii. 55.
No. hepatica: (tincta). *Hüb. No. pl.* 16.*f.* 77.
No. argentina: silvery Arches M. *Haw.* 186.—(*Sam. I.* 28.)
No. occulta. *Fabr.* iii. *b.* 88?

6267. 6; Herbida*. (*Och.* iv. 74.)—*Och. Tr.* v. *b.* 56.—(*Curtis l. c. supra.*)
No. Herbida. (*Wien. V.* 313.)—*Hüb. No. pl.* 16.*f.* 76.
No. mixtina. (*Haw. Pr.* 17.)
No. mixta: green Arches M. *Haw.* 187.
No. Ligustrina. (*Haw. Pr.* 15.)
No. Prasina. (*Wien. V.* 82.)—*Fabr.* iii. *b.* 95.
No. Egregia. *Esper.* iv. *pl.* cxix. *f.* 7.
No. Jàspidea. *Bork. E. S.* iv. 438.

6268 † 7, Templi. *Och. Tr.* v. *b.* 23.—*Steph. Ill.* (*H.*) iii. *pl.* 26. *f.* 1.
No. Templi. *Thunb. I. S.* iv. 56. *pl.* 4. *fig. sup. dext.*
Ag. Templi. (*Och.* iv. 67.)—(*Ing. Inst.* 90.)
In Mus. D. *Curtis, Stone* et Wilmore.

6269 † 8, Polymita. (*Och.* iv. 73.)—*Och. Tr.* v. *b.* 24.—*Steph. Ill.* (*H.*) iii. *pl.* 26. *f.* 2.
Ph. No. Polymita. *Linn.* ii. 855.—*Stew.* ii. 184.
No. ridens: (Polymita.) *Hüb. No. pl.* 4. *f.* 20.
No. seladonia. *Fabr.* iii. *b.* 103? In Mus. D. *Vigors.*

6270. 9; flavocincta*. (*Och.* iv. 74.)—*Och. Tr.* v. *b.* 27.—(*Curtis l. c. supra.*)
No. flavocincta. (*Wien. V.* 72.)—*Fabr.* iii. *b.* 114.—*Don.* x. 23. *pl.* 334.—*Haw.* 182.—(*Sam. I.* 29.)
No. flavocinctina: large Ranunculus M. (*Haw. Pr.* 17.)
Ph. No. tricolor. *Villers E.* ii. 282.
No. flavicincta major. *Esper.* iv. *pl.* cliii. *f.* 1–3.
No. dysodea. *Bork. E. S.* iv. 262.
Ranunculus M. (*Wilkes.* 7. *pl.* 14. *fig. sup.*)—*Harr. V. M.* 41.

6271. 10; dysodea*. (*Och.* iv. 73.)—*Och. Tr.* v. *b.* 16.—(*Curtis l. c. supra.*)
No. dysodea. (*Wien. V.* 72.)—*Hüb. No. pl.* 10. *f.* 47. mala.
No. ranunculina. (*Haw. Pr.* 17.)—*Haw.* 183.—(*Sam. I.* 30.)
No. chrysozona. *Bork. E. S.* iv. 264.
No. flavicincta minor. *Esper.* iv. *pl.* cliii. *f.* 6, 7.
No. Spinaciæ. *Vieweg. T. V.* ii. 70.
Ph. No. ornata. *Villers E.* ii. 280.
Ranunculus M. (*Wilkes.* 7. *pl.* 14. *fig. inf.*)

6272. 11; serena*. (*Och.* iv. 73.)—*Och. Tr.* v. *b.* 12.—(*Curtis l. c. supra.*)
No. serena. (*Wien. V.* 84.)—*Fabr.* iii. *b.* 101.—*Hüb. No. pl.* 11. *f.* 54.—*Haw.* 184.—(*Sam. I.* 30.)
No. bicolor. *Naturforscher.* ix. 137.

No. bicolorata. *Berl. M.* (*Hufn.*) iii. 410.
Ph. No. Par. *Don.* x. 32. *pl.* 338. *f.* 3.
No. parina: broad-barred White. (*Haw. Pr.* 18.)
Harris Ex. 19. *pl.* iv. *f.* 8.

6273. 12; seladonia*.
No. seladonia: brindled Green. *Haw.* 199.—(*Sam. I.* 30.)
No. virensina. (*Haw. Pr.* 15.)
Ph. virescens. *DeGeer.* ii. *pl.* 16. *f.* 24?

6274. 13, Chi. (*Och.* iv. 74.)—*Och. Tr.* v. *b.* 9.
Ph. No. Chi. *Linn.* ii. 846.—*Stew.* ii. 188.—*Turt.* iii. 331.
No. Chiina: the July Chi. (*Haw. Pr.* 17.)
No. Chi. *Hüb. No. pl.* 10. *f.* 49.—*Don.* xii. 35. *pl.* 406.—*Haw.* 183.—*Ent. Trans.* (*Haworth.*) i. 242.—(*Sam. I.* 29.)
Albin. pl. lxxxiii. *f.* C, D.

Genus 121: (904.) APATELA.

APATELÆ *p*, *Hübner*.—ACRONICTA *p*, *Och*.—ACRONYCTA *p*, *Curtis*, *fo.* 136.—BOMBYX *p*, *Fuesly*.

6275. 1; Leporina*: the Miller.—*Haw. l. c.*
Ph. No. Leporina. *Linn.* ii. 838.—*Don.* x. 7. *pl.* 327. *f.* 1.
No. Leporina. *Hüb. No. pl.* 3. *f.* 15.—(*Haw. Pr.* 17.)—*Haw.* 182.—(*Sam. I.* 29.)
No. Bradyporina. *Hüb. No. pl.* 124. *f.* 570. ♂.; *f.* 571. ♀.
Ac. Leporina. (*Curtis l. c. supra.*)

6276. 2; Bradyporina*. *Steph. Ill.* (*H.*) iii. *pl.* 26. *f.* 3.
Ac. Bradyporina. *Och. Tr.* v. *a.* 9.—(*Curtis l. c. supra.*)
Bo. Leporina. *Fuesly. M.* ii. 16. *pl.* 1.—*Turt.* iii. 215.
No. Leporina. *Hüb. No. pl.* 4. *f.* 16.

6277. 3; Aceris*.
Ph. No. Aceris. *Linn.* ii. 846.—*Berk. S.* i. 141.—*Don.* x. 15. *pl.* 330.—*Stew.* ii. 188.—*Turt.* iii. 331.
No. Acerina. (*Haw. Pr.* 17.)
No. Aceris. *Hüb. No. pl.* 3. *f.* 13.—*Haw.* 176.—(*Sam. I.* 23.)—(*Kirby & Sp. I. E.* iii. 173.)
Ac. Aceris. (*Och.* iv. 62.)—*Id.* v. *a.* 11.—(*Curtis l. c. supra.*)
Albin. pl. lxxxiii. *f.* *a–d.*
Sycamore Tussock M.: Ph. Aceris. (*Wilkes.* 32. *pl.* 67.)
Sychamore M. *Harr. V. M.* 47.
β; No. infuscata: dark Sycamore. *Haw.* 177.—(*Sam. I.* 29.)
No. Aceris. *Hüb. No. pl.* 3. *f.* 14.

Genus 122: (905). ACRONYCTA, *Och.*, *Tr.*, *Curtis.*

APATELA *p*, *Hübner*.—ACRONICTA, (*Och.*)

6278. 1; megacephala*: Poplar Grey.—*Haw. l. c.* (*Och.* iv. 62.)—*Och. Tr.* v. *a.* 11.—(*Curtis l. c. infra.*)
No. megacephala. (*Wien. V.* 67.)—*Hüb. No. pl.* 2. *f.* 10. *pl.* 3. *f.* 11.—*Haw.* 177.—(*Sam. I.* 29.)

No. macrocephala. *Villers E.* ii. 175.
No. areaina. (*Haw. Pr.* 17?)
Sychamore Likeness. *Harr. V. M.* 47.

6279. 2: Ligustri*. *Och. Tr.* v. *a.* 20.—(*Curtis l. c. infra.*)
No. Ligustri. (*Wien. V.* 70.)—*Fabr.* iii. *b.* 102.—*Hüb. No. pl.* 5. *f.* 21.—*Haw.* 178.—(*Leach E. E.* ix. 134.)—(*Sam. I.* 29.)
No. Coronina: the Coronet. (*Haw. Pr.* 17.)
The Crown M. (*Harr. V. M.* 19?)
β, No. Coronula: the dark Coronet. *Haw.* 179.—(*Sam. I.* 29.)
No. Coronulina. (*Haw. Pr.* 17.)

6280. 3, Alni. (*Och.* iv. 62.)—*Och. Tr.* v. *a.* 16.—(*Curtis l. c. infra.*)
Ph. No. Alni. *Linn.* ii. 845.—*Turt.* iii. 322.—*Don.* x. 8. *pl.* 327. *f.* 2.
No. Alnina: the Alder. (*Haw. Pr.* 17.)
No. Alni. *Hüb. No. pl.* 1. *f.* 3.—*Haw.* 180.—(*Sam. I.* 28.)
No. degener. (*Wien. V.* 70.)

6281 † 4, strigosa. *Och. Tr.* v. *a.* 23.
No. strigosa. (*Wien. V.* 88.)—*Fabr.* iii. *b.* 32.
No. favillacea: (strigosa.) *Hüb. No. pl.* 1. *f.* 2.
In Mus. D. *Haworth.*

6282. 5; Psi*. (*Och.* iv. 62.)—*Och. Tr.* v. *a.* 30.—(*Curtis l. c. infra.*)
Ph. No. Psi. *Linn.* ii. 846.—*Berk. S.* i. 141.—*Don.* iv. *pl.* 133.—*Stew.* ii. 187.—*Turt.* iii. 330.
No. Psiina. (*Haw. Pr.* 17.)
No. Psi. *Haw.* 181.—(*Sam. I.* 30.)—(*Kirby & Sp. I. E.* iii. 263.)
No. tridens. *Hüb. No. pl.* 1. *f.* 4.
Albin. pl. lxxxvi. *f.*
Dagger M.: Ph. Psi. (*Wilkes.* 28. *pl.* 60.)
Grey Dagger. *Harr. A. pl.* 15.—*Harr. V. M.* 19.

6283. 6; tridens*. (*Och.* iv. 62.)—*Id.* v. *a.* 26.—(*Curtis l. c. infra.*)
No. tridens. (*Wien. V.* 67.)—*Fabr.* iii. *b.* 105.—*Haw.* 181.—(*Sam. I.* 30.)
No. Psi. *Hüb. No. pl.* 1. *f.* 5.
Albin. pl. lxxxvi. *f.*
Dark Dagger. *Harr. V. M.* 19.

6284 † 7, cuspis?* *Och. Tr.* v. *a.* 32?
No. cuspis. *Hüb. No. pl.* 108. *f.* 504? In Mus. D. *Hatchett.*

6285. 8; auricoma*: the Scarce Dagger.—*Haw. l. c.* (*Och.* iv. 62.)—*Och. Tr.* v. *a.* 35.—(*Curtis l. c. infra.*)

No. auricoma. (*Wien. V.* 67.)—*Fabr.* iii. *b.* 106.—*Hüb. No. pl.* 2. *f.* 8.—*Haw.* 180.—(*Sam. I.* 28.)
No. Pepli. *Hüb. No. pl.* 134. *f.* 614. ♂. teste Och.

6286. [9, similis*.] (*Curtis l. c. infra.*)
No. similis: the scarce Knot-grass. *Haw.* 180.—(*Sam. I.* 30.)
β, No. Menyanthedis: the light Knot-grass. *Haw.* 180.—(*Sam. I.* 29.)

6287. 10, Menyanthedis. (*Och.* iv. 62.)—*Och. Tr.* v. *a.* 34.—(*Curtis l. c. infra.*)—*Steph. Ill.* (*H.*) iii. *pl.* 27. *f.* 1.
No. Menyanthedis. *Hüb. No. pl.* 2. *f.* 6, 7.
β? Ac. Salicis. *Curtis.* iii. *pl.* 136.

6288. 11; Rumicis*. (*Och.* iv. 62.)—*Och. Tr.* v. *a.* 38.—(*Curtis l. c. supra.*)
Ph. No. Rumicis. *Linn.* ii. 852.—*Berk. S.* i. 142.—*Don.* iv. *pl.* 126.—*Stew.* ii. 190.—*Turt.* iii. 335.
No. Rumicina. (*Haw. Pr.* 17.)
No. Rumicis. *Hüb. No. pl.* 2. *f.* 9.—*Haw.* 178.—(*Sam. I.* 30.)
Albin. pl. xxii. *f.* 32. *e–i.*
Bramble M.: Ph. Rumicis. (*Wilkes.* 26. *pl.* 56. imago.)
Knot-grass. *Harr. V. M.* 29.

6289. 12; Euphorbiæ*: the Spurge.—*Haw. l. c.* (*Och.* iv. 62?) *Och. Tr.* v. *a.* 40.
No. Euphorbiæ. (*Wien. V.* 67?)—*Fabr. E. S.* iii. 103?—*Hüb. No. pl.* 3. *f.* 12.—*Haw.* 178.—(*Sam. I.* 29.)
No. Cyparissiæ. *Hüb. No. pl.* 134. *f.* 615. ♀. teste Och.

6290. 13, Euphrasiæ*. (*Och.* iv. 62.)—*Och. Tr.* v. *a.* 43.
No. Euphorbiæ. *Esper. pl.* cxvii. *f.* 1–3.
No. Esulæ. *Hüb. No. pl.* 134. *f.* 613.
Albin. pl. lxxxviii. *fig. f.* larva.

Genus 123: (906). BRYOPHILA, *Ochsenheimer.*

JASPIDIÆ, *Hübner.*—PŒCILIA, *Schrank,* (*Och.,*) (*Samou.*)

6291. 1; glandifera*. *Och.* v. *a.* 58.
No. glandifera. (*Wien. V.* 70.)—*Hüb. No. pl.* 5. *f.* 24.
No. Lichenes. *Fabr.* iii. *b.* 104.—*Haw.* 203.—(*Sam. I.* 29.)
Ph. Lichenes. *Don.* vii. *pl.* 223. *f.* 3.
Ph. No. Lichenis. *Stew.* ii. 188.—*Turt.* iii. 330.
No. Lichenina: marbled green. (*Haw. Pr.* 16.)
Bishop. *Harr. V. M.* 13.

6292. 2; perla*: marbled Beauty.—*Haw. l. c.* *Och.* v. *a.* 61.
No. perla. (*Wien. V.* 70.)—*Hüb. No. pl.* 5. *f.* 25.—*Haw.* 203.—(*Sam. I.* 30.)
No. glandifera. *Bork. E. S.* iv. 128.

Genus 124: (907.) DIPHTHERA, (*Hüb.*,) (*Och.*,) *Treit.*

6293 † 1. ludifica[a].

6294. 2: runica*. (*Och.* iv. 63.)
Ph. No. runica. *Gmel.* iii. 2561.
No. runica. *Haw.* 200.—(*Sam. I.* 30.)
No. Aprilina. *Fabr.* iii. *b.* 103.—*Hüb. No. pl.* 5. *f.* 22.
Ph. No. Aprilina. *Don.* x. 57. *pl.* 347. *f.* 1.
No. Ludificiana. (*Haw. Pr.* 15.)
Scarce Marvel du Jour. *Harr. A. pl.* 42. *f.e.*—*Harr. V. M.* 19.

6295 [† 3, Orion.] (*Och.* iv. 63.)
No. Orion. *Esper. No. pl.* cxvii. *f.* 4–7. In Mus. *Brit.*

Genus 125: (908). THYATIRA, (*Och.*,) (*Samou.*,) *Treit.*, *Curtis.*

6296. 1; derasa*. (*Och.* iv. 77.)—*Och. Tr.* v. *b.* 165.—(*Curtis l. c. infra.*)
Ph. No. derasa. *Linn.* ii. 851.—*Don.* vii. 17. *pl.* 223. *f.* 1.—*Stew.* ii. 190.—*Turt.* iii. 320.
No. derasina. (*Haw. Pr.* 14.)
No. derasa. *Hüb. No. pl.* 14. *f.* 66.—*Haw.* 244.—(*Sam. I.* 29.)
No. Pyritoides. *Berl. Mag.* (*Hufn.*) iii. 400.
Buff Arches. (*Wilkes D. pl.* 6. *f.* 3.)—*Harr. V. M.* 9.

6297. 2; batis*. (*Och.* iv. 77.)—*Och. Tr.* v. *b.* 162.—*Curtis.* ii. *pl.* 72.
Ph. No. batis. *Linn.* ii. 836.—*Don.* i. *pl.* 33. *f.* 1.
No. batisina. (*Haw. Pr.* 16.)
No. batis. *Haw.* 245.—(*Leach E. E.* ix. 134.)—(*Sam. I.* 28.)
Peach blossom. (*Wilkes D. pl.* 10. *f.* 6.)—*Harr. Ex.* 53. *pl.* xiii. *f.* 3.—*Harr. V. M.* 13.

Genus 126: (909). ERIOPUS[b], *Treit.*

---

[a] 6293 † 1. ludifica. (*Och.* iv. 68.)—*Och. Tr.* v. *a.* 50.
Ph. No. ludifica. *Linn.* ii. 848.—*Turt.* iii. 329. (!)
No. ludifica. *Hüb. No. pl.* 5. *f.* 23.

[b] Genus 126: (909). ERIOPUS, *Treit.*

LAGOPUS*, *Latr.* ?—HADENA *p*, *Och.*, (*Samou.*)

6298 † 1. Pteridis. *Och. Tr.* v. *a.* 366.
No. Pteridis. *Fabr.* iii. *b.* 90.—*Hüb. No. pl.* 13. *f.* 65.—(*Sam. I.* 30.) (!)
No. manicata. *Villers E.* ii. 275. *pl.* 5. *f.* 22.
No. formosa. *Bork. E. S.* iv. 362.
No. Juventina. *Cramer.* iv. *pl.* 400. *f. N. N.*

---

* LAGOPUS: Genus Avium. Vide *Gen. Zool.* xi. 286.

Genus 127: (910). CALYPTRA, (*Ochsen.*)

PH. BOMBYX, *Linné, &c.*—CALPE *p, Treit.*—GONOPTERA, *Latr.*

6299. 1; Libatrix*. (*Och.* iv. 78.)
Ph. Bo. Libatrix. *Linn.* ii. 831.—*Berk. S.* i. 138.—*Don.* vi. 85. *pl.* 216.—*Stew.* ii. 158.—*Turt.* iii. 299.
Bo. Libatrus: Furbelow M. (*Haw. Pr.* 8.)
No. Libatrix. *Hüb. No. pl.* 93. *f.* 436.—*Haw.* 844.—(*Sam. I.* 29.)
Pa. No. 30. *List.* Goëd. (*Angl.*) 40. *f.* 30.—*Id.* (*Lat.*) 85. *f.* 30. *Albin. pl.* xxxii. *f.* 50. *a–c.*
Herald M. *Harr. Ex.* 10. *pl.* 1. *f. c–f.*—*Harr. V. M.* 31.

Genus 128: (911). CEROPACHA *mihi.*

TETHEA B., (*Och.*,) (*Samou.*)—CYMATOPHORA B. *p, Treitschke.*

A. Alæ vix elongatæ, latæ; posticæ haud abbreviatæ.

a. *Corpore gracile.*

6300. 1; fluctuosa*.
No. fluctuosa. *Hüb. No. pl.* 44. *f.* 212.—*Haw.* 252.—(*Sam. I.* 29.)
No. similina: Satin Carpet. (*Haw. Pr.* 14.)

6301. 2; duplaris*.
Tinea duplaris. *Linn. F. No.* 1357.
No. duplaris: Lesser Satin Carpet. *Haw.* 253.—(*Sam. I.* 29.)
No. bicolor. *Esper.* iv. *pl.* cxcvii. *f.* 4.
No. bipuncta. *Bork. E. S.* iv. 627.
No. undosa. *Hüb. No. pl.* 44. *f.* 211.
No. dissimilina. (*Haw. Pr.* 14.)

b. *Corpore subcrasso.*

6302. 3; diluta*.
No. diluta. (*Wien. V.* 87.)—*Fabr.* iii. *b.* 86.—*Hüb. No. pl.* 43. *f.* 206.—*Haw.* 251.—(*Sam. I.* 29.)
No. dilutina: lesser Lutestring. (*Haw. Pr. App.* 3.)
No. octogena. *Esper.* iv. *pl.* cxxviii. *f.* 6.
Bo. undata. *Fabr.* iii. *a.* 451.
Bo. fasciculosa. *Bork. E. S.* iii. 326.
Half-mourner. *Harr. A. pl.* 35. *f. f–g.* *Harr. V. M.* 31.

6303. 4; Or*.
No. Or. (*Wien. V.* 87.)—*Fabr.* iii. *b.* 86.—*Hüb. No. pl.* 43. *f.* 210.
No. flavicornis: Poplar Lutestring. *Haw.* 252.—(*Sam. I.* 29.)
No. octogena. *Esper.* iv. *pl.* cxxviii. *f.* 5.
No. consobrina. *Bork. E. S.* iv. 622.
No. gemina. *Linn. Trans.* (*Beckwith.*) ii. 4. *pl.* 1. *f.* 4–6.—*Don.* x. 59. *pl.* 347. *f.* 3.—*Stew.* ii. 196.—*Turt.* iii. 324.
No. 8 græcum. *Villers E.* ii. 251.

6304. 5, octogesima. *Steph. Ill.* (*H.*) iii. *pl.* 27. *f.* 2.
No. octogesima. *Hüb. No. pl.* 43. *f.* 209?
No. octogena. *Esper.* iv. *pl.* cxxviii. *f.* 4.
No. Or. *Bork. E. S.* iv. 620.

B. Alæ anticæ elongatæ, angustæ; posticæ breves.

6305. 6; flavicornis*.
Ph. No. flavicornis. *Linn.* ii. 856.—*Don.* x. 72. *pl.* 352. *f.* 3.
No. flavicornina. (*Haw. Pr.* 19.)
No. luteicornis: yellow horned. *Haw.* 252.
No. lenticornis. (*Sam. I.* 29.)

6306. 7; ridens*.
No. ridens. *Fabr.* iii. *b.* 119.—*Haw.* 202.—(*Sam. I.* 30.)
No. ridensina: Frosted-green. (*Haw. Pr.* 15.)
No. xanthoceros. *Hüb. No. pl.* 43. *f.* 205.
No. chrysoceras. *Linn. Trans.* (*Beckwith.*) ii. 3. *pl.* 1. *f.* 1–3. —*Stew.* ii. 195.
No. erythrocephala. *Esper.* iv. *pl.* cxxi. *f.* 1–3.
No. flavicornis. (*Wien. V.* 72.)
No. putris. *Naturforscher.* viii. 108.

Genus 129: (912). TETHEA, (*Och.*,) (*Samou.*)
CYMATOPHORA A., *Treit.*

6307. 1; subtusa*. (*Och.* iv. 64.)
No. subtusa. (*Wien. V.* 88.)—*Hüb. No. pl.* 44. *f.* 213.—*Haw.* 250.—(*Sam. I.* 30.)
No. tenuina: the Olive. (*Haw. Pr.* 14?)

6308. 2; retusa*. (*Och.* iv. 64.)
Ph. No. retusa. *Linn.* ii. 858.
No. retusa. *Hüb. No. pl.* 44. *f.* 214.—*Haw.* 251.—(*Sam. I.* 30.)
No. Capreæ. *Fabr.* iii. *b.* 58?
Ph. No. chrysoglossa. *Linn. Trans.* (*Beckwith.*) ii. 6. *pl.* 1. *f.* 10–11.—*Don.* x. 67. *pl.* 350. *f.* 2.—*Stew.* ii. 196.—*Turt.* iii. 325.
No. chrysoglossina: Double Kidney. (*Haw. Pr.* 14.)

6309. [3; gracilis*.]
No. gracillina. (*Haw. Pr.* 14.)
No. chrysoglossidina. (*Haw. Pr. App.* 3.)
No. gracilis: slender bodied. *Haw.* 251.—(*Sam. I.* 29.)

Genus 130: (913). BOMBYCIA, *Hübner?*
TETHEA B. *p*, *Och.*—BOMBYX *p*, *Esper.*—CYMATOPHORA B. *p*, *Treit.*—BOMBYCIÆ *p*, *Hübner.*

6310. 1; Viminalis*: Minor Shoulder Knot.—*Haw. l. c.*
No. Viminalis. *Fabr.* iii. *b.* 72.
No. Saliceti. *Bork. E. S.* iv. 632.

Bo. stricta. *Esper.* iii. *pl.* lxxxiv. *f.* 5.
No. scripta. *Hüb. No. pl.* 10. *f.* 50.—*Haw.* 213.—(*Sam. I.* 30.)
Bolton's Beauty. *Harr. V. M.* 17.

Genus 131: (914). CYMATOPHORA.

CYMATOPHORA A, b, *Treitschke.*

6311. 1; Oo*.
Ph. Bo. Oo. *Linn.* ii. 832.—*Berk. S.* i. 139.—*Don.* v. 107. *pl.* 179.—*Stew.* ii. 156.
No. Ooina. (*Haw. Pr.* 14.)
No. ferruginago. *Hüb. No. pl.* 41. *f.* 95.—*Haw.* 238.—(*Sam. I.* 29.)
Scallop-winged Oak M.: Ph. Oo. (*Wilkes.* 6. *pl.* 11.)
Heart M. *Harr. V. M.* 31.
β; No. renata. *Fabr.* iii. *b.* 85.
No. renago: dark Heart M. *Haw.* 238.—(*Sam. I.* 30.)

Genus 132: (915). COSMIA, (*Och.*,) (*Samou.*,) *Treit.*

COSMIÆ, (*Hüb.*)

6312. 1; diffinis*. (*Och.* iv. 84.)—*Och. Tr.* v. *b.* 386.
Ph. No. diffinis. *Linn.* ii. 848.—*Stew.* ii. 189.—*Turt.* iii. 321.
No. diffina. (*Haw. Pr.* 16.)
No. diffinis. *Hüb. No. pl.* 42. *f.* 202.—*Haw.* 247.—(*Sam. I.* 29.)
White-spotted Pinion. *Harr. Ex.* 24. *pl.* v. *f.* 8.—*Harr. V. M.* 39.

6313. 2; affinis*. (*Och.* iv. 84.)—*Och. Tr.* v. *b.* 389.
Ph. No. affinis. *Linn.* ii. 848.
No. affina: lesser-spotted Pinion. (*Haw. Pr.* 16.)
No. affinis. *Hüb. No. pl.* 42. *f.* 201.—*Haw.* 247.—(*Sam. I.* 28.)
Albin. pl. xxxi. *f.* 49. *f.*–*i.*
Double-spotted Pinion. *Harr. V. M.* 29.

6314. 3: Pyralina*. (*Och.* iv. 84.)—*Och. Tr.* v. *b.* 392.
No. Pyralina. (*Wien. V.* 88.)—*Hüb. No. pl.* 42. *f.* 203.—lunar-spotted Pinion.—*Haw.* 247.—*Ent. Trans.* (*Hatchett.*) i. 327. *pl.* 9. *f.* 1.—(*Sam. I.* 30.)
No. corusca. *Esper.* iv. *pl.* cxxxv. *f.* 4, 5.

6315. 4; trapetzina*. (*Och.* iv. 84.)—*Och. Tr.* v. *b.* 383.
Ph. No. trapetzina. *Linn.* ii. 836.—*Turt.* iii. 310.
No. trapezina. (*Haw. Pr.* 15.)
No. trapetzina. *Hüb. No. pl.* 42. *f.* 200.—*Haw.* 246.—(*Sam. I.* 30.)
Ph. rhombica. *Berl. M.* (*Hufn.*) iii. 296.
Dun-bar. *Harr. A. pl.* x. *f. n*–*r.*—*Harr. V. M.* 11.

6316. 5: fulvago*. (*Och.* iv. 84.)—*Och. Tr.* v. *b.* 389.
No. fulvago. (*Wien. V.* 86.)—*Hüb. No. pl.* 41. *f.* 198, 199.

No. paleacea. *Esper.* iv. *pl.* cxxii. *f.* 3, 4.
No. gilvago. *Bork. E. S.* iv. 683.
No. angulago: angle-striped Sallow. *Ent. Trans.* (*Hatchett.*) i. 244. *pl. fig. sup.*—*Haw.* 239.—(*Sam. I.* 28.)

Genus 133: (916). XANTHIA, (*Hübner,*) (*Samou.,*) *Treit., Curtis.*

6317. 1; flavago*. (*Curtis l. c. infra.*)
No. flavago: pink-barred Sallow.—*Haw. l. c.* *Fabr.* iii. *b.* 76. —(*Haw. Pr.* 14.)—*Haw.* 236.—(*Sam. I.* 29.)
No. togata. *Esper.* iv. *pl.* cxxiv. *f.* 1.
No. ochreago. *Bork. E. S.* iv. 671.
No. silago. *Hüb. No. pl.* 40. *f.* 1.
Albin. pl. lxviii. *f. a–d.*

6318. 2; fulvago*. (*Curtis l. c. infra.*)
Ph. No. fulvago. *Linn.* ii. 858.—*Stew.* ii. 194.—*Turt.* iii. 315.
No. fulvago. *Fabr.* iii. *b.* 73.—(*Haw. Pr.* 14.)—*Haw.* 236.—(*Sam. I.* 29.)
Ph. No. crocea. *Villers E.* ii. 279.
No. flavescens. *Esper.* iv. *pl.* cxxii. *f.* 2.
No. cerago. *Hüb. No. pl.* 94. *f.* 444. *pl.* 40. *f.* 190.
Ph. rubago. *Don.* x. 32. *pl.* 338. *f.* 2.
Ph. No. cerago. *Stew.* ii. 194?
Albin. pl. xxxiii. *f.* 52. *a–d.*
Sallow M.: Ph. citrago. (*Wilkes.* 5. *pl.* 8. *imago.*)—*Harr. V. M.* 41.

6319. [3; gilvago*]: lemon Sallow.—*Haw. l. c.*
No. gilvago. *Fabr.* iii. *b.* 76?—(*Haw. Pr.* 14.)—*Haw.* 237.—(*Sam. I.* 29.)
No. cerago. *Hüb. No. pl.* 94. *f.* 445.

6320. 4; aurago*: barred S.—*Haw. l. c.* (*Och.* iv. 83.)—(*Curtis l. c. infra.*)
No. aurago. (*Wien. V.* 86.)—*Fabr.* iii. *b.* 74.—*Hüb. No. pl.* 41. *f.* 196, 197.—(*Haw. Pr.* 14.)—*Haw.* 236.—(*Sam. I.* 28.)
No. prætexta. *Esper.* iv. *pl.* cxxiv. *f.* 2.
No. fucata. *Esper.* iv. *pl.* cxxiv. *f.* 3, 4.
No. rutilago. *Bork. E. S.* iv. 674.

6321. 5, centrago. *Curtis.* ii. *pl.* 84.
No. centrago: centre-barred Sallow. *Haw.* 236.—(*Sam. I.* 29.)
No. xerampelina. *Hüb. No. pl.* 90. *f.* 421. ♀.

6322. 6; citrago*: orange S.—*Haw. l. c.* (*Och.* iv. 83.)—(*Curtis l. c. supra.*)
Ph. No. citrago. *Linn.* ii. 857.—*Berk. S.* i. 143.—*Stew.* ii. 194.—*Turt.* iii. 316.
No. citrago. *Hüb. No. pl.* 39. *f.* 188.—(*Haw. Pr.* 14.)—*Haw.* 238.—(*Sam. I.* 29.)

No. ochreago. *Hüb. B.* ii. 1. *pl.* 1. *f. D.*
Sallow M. (*Wilkes.* 5. *pl.* 8. *larva.*)

6323. 7, fimbriago *mihi.*

6324. 8; croceago*. (*Och.* iv. 83.)—*Och. Tr.* v. *b.* 360.
No. croceago. (*Wien. V.* 86.)—*Fabr.* iii. *b.* 73.—(*Haw. Pr.* 14.)—*Hüb. No. pl.* 40. *f.* 189.—*Haw.* 238.—(*Sam. I.* 29.)
No. fulvago. *Esper.* iv. *pl.* clxxxvi. *f.* 3, 4.
No. aurantiago. *Don.* v. 15. *pl.* 150. *f.* 2, 3.—*Stew.* ii. 195.
Ph. ferruginata. *Turt.* iii. 254?
Albin. pl. xv. *f.* 22. *e–h.*
Orange Upperwing. *Harr. V. M.* 57.

6325. 9; rufina*. (*Och.* iv. 83.)—*Och. Tr.* v. *b.* 347.
Ph. Bo. rufina. *Linn.* ii. 830.
No. rufina. *Fabr.* iii. *b.* 33.—*Hüb. No. pl.* 38. *f.* 184.
No. catænata. *Esper.* iv. *pl.* cxxiii. *f.* 1, 2.
No. punica. *Bork. E. S.* iv. 300.
No. ferrugina. *Marsh. MSS.*—(*Haw. Pr.* 15.)
Ph. No. helvola. *Linn. F. No.* 1142.
No. helvola: flounced Rustic. *Haw.* 229.—(*Sam. I.* 29.)
Xa. helvola. (*Curtis l. c. supra.*)

Genus 134: (917). GORTYNA, (*Och.*,) (*Samou.*,) *Treit.*, *Curtis.*
XANTHIA *p*, (*Samou.*)

6326. 1; micacea*: Rosy Rustic.—*Haw. l. c.* (*Och.* iv. 83.)—*Curtis.* vi. *pl.* 252.
No. micacea. *Esper. No.* vi. *pl.* cxlv. *f.* 6.
No. cypriaca. *Hüb. No. pl.* 46. *f.* 224.—*Haw.* 227.—(*Sam. I.* 29.)

6327. 2; flavago*. (*Och.* iv. 83.)—*Och. Tr.* v. *b.* 335.—(*Curtis l. c. supra.*)
No. flavago. (*Wien. V.* 86.)—*Hüb. No. pl.* 39. *f.* 186, 187.
No. ochracea. *Hüb. B.* i. 1, 2. *M?*
Ph. No. Lappæ. *Don.* x. 35. *pl.* 340.
No. Lappina. (*Haw. Pr.* 14.)
No. ochraceago: frosted Orange. *Haw.* 234.—(*Sam. I.* 29.)
No. rutilago. *Fabr.* iii. *b.* 75.—(*Samou.* 252.)
Mottled Orange. *Harr. A. pl.* 35. *f. a–e.*—*Harr. V. M.* 39.

6328 † 3. Luteago[a].

---

[a] 6328 † 3. Luteago. *Och. Tr.* v. *b.* 338.
No. Luteago. (*Wien. V.* 86.)—*Hüb. No. pl.* 39. *f.* 184? ♂.
Xa. Luteago. (*Samou.* 252.) (!)
No. brunneago. *Esper. No.* iv. *pl.* cxcvi. *f.* 2, 3.
No. lutea. *Bork. E. S.* iv. 684.

Genus 135: (918). NONAGRIA, (*Och.*,) (*Samou.*,) *Treit.*

6329. 1; Typhæ*. (*Och.* iv. 82.)—*Och. Tr.* v. *b.* 327.
No. Typhæ. *Marsh. MSS.*—*Hüb. No. pl.* 88. *f.* 415.—*Haw.* 173.—(*Sam. I.* 30.)
No. Arundinis. *Fabr.* iii. *b.* 30.
No. Typhina: Bullrush M. (*Haw. Pr.* 12.)
Var. No. nervosa. *Esper.* iv. *pl.* cxl. *f.* 1?
No. fraterna. *Bork. E. S.* iv. 724?

6330 † 2, pilicornis*.
No. pilicornis: hairy-horned Wainscot. *Ent. Trans.* (*Haworth.*) i. 336. In Mus. D. *Hatchett.*

6331. 3, crassicornis.
No. crassicornis: large Wainscot. *Haw.* 173.—(*Sam. I.* 29.)

6332 † 4, Cannæ. (*Och.* iv. 82.)—*Och. Tr.* v. *b.* 325.
No. Algæ. *Esper.* iv. *pl.* cxl. *f.* 1, 2.
No. Arundinis. *Hüb. No. pl.* 83. *f.* 386.—(*Sam. I.* 28.)
In Mus. D. *Haworth et Stone.*

Genus 136: (919). LEUCANIA, (*Och.*,) (*Samou.*,) *Treit.*, *Curtis.*
HELIOPHILÆ, (*Hübner.*)

6333. 1: comma*. (*Och.* iv. 81.)—*Och. Tr.* v. *b.* 302.—(*Curtis l. c. infra.*)
Ph. No. Comma. *Linn.* ii. 850.—*Turt.* iii. 331.
No. Commaina: Shoulder-stripe Wainscot. (*Haw. Pr.* 12.)
No. Comma. *Haw.* 174.—(*Sam. I.* 29.)
No. pallens. *Esper.* iv. *pl.* cx. *f.* 2.
No. turbida. *Hüb. No. pl.* 47. *f.* 228.
♀; No. congrua. *Hüb. No. pl.* 135. *f.* 616.
No. ramilinea. *Haworth MSS.*

6334. 2, littoralis.
Le. litoralis. *Curtis.* iv. *pl.* 157.

6335 † 3, obsoleta*. (*Och.* iv. 81.)—*Och. Tr.* v. *b.* 301.
No. obsoleta. *Hüb. No. pl.* 48. *f.* 233.
In Mus. *Brit. et* D. *Chant.*

6336. 4; impura*. (*Och.* iv. 81.)—*Och. Tr.* v. *b.* 294.
No. impura. *Hüb. No. pl.* 85. *f.* 396.
No. fuligosina: Smoky Wainscot. (*Haw. Pr.* 12.)—*Haw.* 174.
No. fuliginosa. (*Sam. I.* 29.)
Le. obsoleta. (*Curtis l. c. supra.*)
Albin. pl. xxxiii. *f.* 53. *e–h.*
Wainscot M. *Harr. V. M.* 57.

6337. [5; punctina*]: dotted-bordered Wainscot.—*Haw. l. c.* (*Curtis l. c. supra.*)

No. punctina. (*Haw. Pr.* 13.)—*Haw.* 174.—(*Sam. I.* 30.)
No. Ectypa. *Hüb. No. pl.* 48. *f.* 231?

6338. 6; arcuata* *mihi.*

6339. 7; pallens*. (*Och.* iv. 81.)—*Och. Tr.* v. *b.* 290.—(*Curtis l. c. supra.*)
Ph. No. pallens. *Linn.* ii. 838.—*Stew.* ii. 182.—*Turt.* iii. 291.
No. pallensina: Common Wainscot. (*Haw. Pr.* 13.)
No. pallens. *Hüb. No. pl.* 48. *f.* 234.—*Haw.* 175.—(*Sam. I.* 29.)

6340. 8; rufescens*. (*Curtis l. c. supra.*)
No. rufescens: red Wainscot. *Haw.* 175.—(*Sam. I.* 30.)
No. rufescensina. (*Haw. Pr.* 13.)
Leu. straminea. (*Och.* iv. 81?)—*Och. Tr.* v. *b.* 297?

6341. 9; suffusa* *mihi.*

6342. 10; ochracea* *mihi.*

6343. 11: fluxa*.
No. fluxa. *Hüb. No. pl.* 88. *f.* 413.
♀: No. fulva. *Hüb. No. pl.* 106. *f.* 496.
Le. fulva. (*Curtis l. c. supra.*)

6344. 12; Phragmatidis*.
No. Phragmatidis. *Hüb. No. pl.* 47. *f.* 230.
No. semicana. *Esper.* iv. *pl.* clxxxix. *f.* 5?
No. pygmina: Small Wainscot. (*Haw. Pr.* 13.)—*Haw.* 176.—(*Sam. I.* 30.)
Le. pygmina. *Curtis l. c. supra.*

6345. [13, pallida*] *mihi.*

6346. 14, neurica*.
Noc. neurica. *Hüb. No. pl.* 82. *f.* 381. ♂. *pl.* 144. *f.* 659, 660. ♂. *f.* 661. ♀.

6347 † 15, geminipuncta*. (*Curtis l. c. supra.*)
Noc. geminipuncta: Twin-spotted Wainscot. *Haw.* 176.—*Ent. Trans.* (*Hatchett.*) i. 327. *pl.* 9. *f.* 2.—(*Sam. I.* 29.)
Noc. extrema. *Hüb. No. pl.* 88. *f.* 412?
In Mus. *Brit.* D. *Hatchett et Milne.*

6348. 16: pudorina. (*Och.* iv. 81.)—*Och. Tr.* v. *b.* 299.—(*Curtis l. c. supra.*)
No. pudorina. (*Wien. V.* 85.)—*Hüb. No. pl.* 86. *f.* 401.
Leu. impudens. (*Ing. Inst.* 91.)
♀, No. impudens. *Hüb. No. pl.* 47. *f.* 229.

6349 † 17, unipuncta.
No. unipuncta: White-speck. *Haw.* 174.
Simyra punctosa. *Och. Tr.* v. *b.* 287? In Mus. D. *Haworth.*

Genus 137: (920). SIMYRA, (*Och.*,) *Treit.*

6350. 1, musculosa. (*Och.* iv. 81.)—*Och. Tr.* v. *b.* 286.—(*Ing. Inst.* 91.)
No. musculosa. *Hüb. No. pl.* 78. *f.* 363.
No. pudorina. *Bork. E. S.* iv. 720.

6351. 2: venosa. (*Och.* iv. 81.)—*Och. Tr.* v. *b.* 281.
No. venosa. *Bork. E. S.* iv. 716.
No. degener. *Hüb. No. pl.* 81. *f.* 380.
No. atomina: powdered Wainscot. (*Haw. Pr.* 13.)—*Haw.* 175.—(*Sam. I.* 28.)

6352. 3, nervosa*. (*Och.* iv. 81.)—*Och. Tr.* v. *b.* 283.
No. nervosa: tawny veined W.—*Haw. l. c.* (*Wien. V.* 85.)—*Fabr.* iii. *b.* 23.—*Hüb. No. pl.* 47. *f.* 326.—*Haw.* 176.—(*Sam. I.* 29.)
No. oxyptera. *Esper.* iv. *pl.* cxxx. *f.* 6.

Genus 138: (921). PHLOGOPHORA.

PHLOGOPHORA A., *Treit.*—HADENA C., *Och.*

6353. 1; meticulosa*. *Och. Tr.* v. *a.* 373.
Ph. No. meticulosa. *Linn.* ii. 845.—*Berk. S.* i. 140.—*Don.* iv. 83. *pl.* 139.—*Stew.* ii. 187.—*Turt.* iii. 319.
No. meticulosina. (*Haw. Pr.* 16.)
No. meticulosa. *Hüb. No. pl.* 14. *f.* 67.—*Haw.* 244.—(*Leach E. E.* ix. 134.)—(*Sam. I.* 29.)
Pa. Nos. 44 & 56. *List. Goëd.* (*Angl.*) 54. *f.* 44. 62. *f.* 56.—*Id.* (*Lat.*) 118. *f.* 44. 137. *f.* 56.
Albin. pl. xxx. *f.* 46–47. *a–e.*
Angleshades M.: Ph. meticulosa. (*Wilkes.* 3. *pl.* 3.)—*Harr. A. pl.* 41. *f.* c–e.—*Harr. V. M.* 43.

Genus 139: (922). CUCULLIA, *Schrank*, (*Samou.*,) *Curtis*, (*Kirby.*)

TRIBONOPHORÆ, *Hüb.*—CENROTÆPTERYX, *Leach MSS.*

A. Alis denticulatis.

6354. 1; Scrophulariæ*. (*Och.* iv. 88.)—(*Curtis l. c. infra.*)
No. Scrophulariæ. *Hüb. No. pl.* 55. *f.* 267.—*Haw.* 167.—(*Sam. I.* 30.)
Ph. No. Verbasci. *Don.* viii. 13. *pl.* 257?
Water Betony M.: Ph. Verbasci. (*Wilkes.* 7. *pl.* 15.)
Water Betony likeness. *Harr. V. M.* 17.

6355. 2; Verbasci*. (*Och.* iv. 88.)—(*Curtis l. c. infra.*)
Ph. No. Verbasci. *Linn.* ii. 850.—*Berk. S.* i. 142.—*Stew.* ii. 189.—*Turt.* iii. 336.
No. Verbascina: Mullein M. (*Haw. Pr.* 12.)
No. Verbasci. *Hüb. No. pl.* 55. *f.* 266.—*Haw.* 167.—(*Leach E. E.* ix. 134.)—(*Sam. I.* 30.)

Albin. pl. xiii. *f.* 18. *f–k.*
Water Betony Sword-grass M. *Harr. A. pl.* viii. *f. a–e.*—*Harr. V. M.* 17.

B. Alis integris.

6356. 3: Asteris*: Starwort.—*Haw. l. c.* (*Och.* iv. 88.)—*Curtis.* i. *pl.* 45.
No. Asteris. (*Wien. V.* 312.)—*Fabr.* iii. *b.* 121.—*Haw.* 168.—(*Sam. I.* 28.)
No. Asterina. (*Haw. Pr.* 12.)

6357 † 4: Thapsiphaga*. *Och. Tr.* v. c. 120? In Mus. *Brit.*

6358. 5; Umbratica*. (*Och.* iv. 88.)—(*Curtis l. c. supra.*)
Ph. No. Umbratica. *Linn.* ii. 849.—*Don.* viii. 26. *pl.* 262. *f.* 2.—*Stew.* ii. 189.—*Turt.* iii. 337.
No. Umbraticina. (*Haw. Pr.* 12.)
No. Umbratica: large pale Shark.—*Haw. l. c.* *Hüb. No. pl.* 54. *f.* 263.—*Haw.* 165.—(*Sam. I.* 30.)
Shark. *Harr. V. M.* 43.

6359. 6; Lactucæ*: Lettuce S.—*Haw. l. c.* (*Och.* iv. 88.)—(*Curtis l. c. supra.*)
No. Lactucæ. (*Wien. V.* 76.)—*Fabr.* iii. *b.* 367.—*Haw.* 166.—(*Sam. I.* 29.)
No. lucifuga. *Hüb. No. pl.* 54. *f.* 262.

6360. 7; Tanaceti*: Tansy S.—*Haw. l. c.* (*Och.* iv. 88.)—(*Curtis l. c. supra.*)
No. Tanaceti. (*Wien. V.* 73.)—*Fabr.* iii. *b.* 121.—*Hüb. No. pl.* 54. *f.* 265.—*Haw.* 165.—(*Leach E. E.* ix. 134.)—(*Sam. I.* 30.)

6361. 8; lucifuga*: large dark S.—*Haw. l. c.* (*Och.* iv. 88.)—(*Curtis l. c. supra?*)
No. lucifuga. (*Wien. V.* 312.)—*Hüb. No. pl.* 74. *f.* 262.—*Haw.* 166.—(*Sam. I.* 29.)

6362. 9, Chamomillæ*: Chamomile S.—*Haw. l. c.* (*Och.* iv. 88.)—(*Curtis l. c. supra.*)
No. Chamomillæ. (*Wien. V.* 73.)—*Fabr. E. S.* iii. *b.* 121.—*Hüb. No. pl.* 54. *f.* 264.—*Haw.* 165.—(*Sam. I.* 29.)

6363. [10; fissina*]: twin-tailed S.—*Haw. l. c.* (*Curtis l. c. supra.*)
No. fissina. (*Haw. Pr.* 12.)—*Haw.* 166.—(*Sam. I.* 29.)

6364. 11. Gnaphalii[a].

6365. 12, Absinthii*. (*Och.* iv. 87.)—(*Curtis l. c. supra.*)
Ph. No. Absinthii. *Linn.* ii. 845.—*Don.* ix. 35. *pl.* 304.—*Stew.* ii. 187.—*Turt.* iii. 321.

---

[a] 6364 † 11. Gnaphalii. (*Och.* iv. 87.)—(*Curtis l. c. supra.*) (!)
No. Gnaphalii. *Hüb. No. pl.* 126. *f.* 582.

No. Absinthina: Wormwood M. (*Haw. Pr.* 12.)
No. Absinthii. *Hüb. No. pl.* 55. *f.* 258.—*Haw.* 168—(*Sam. I.* 28.)
Ph. punctigera. *Berl. M.* (*Hufn.*) iii. 416.

6366 ‡ 13, Artemisiæ. (*Och.* iv. 87.)
No. Artemisiæ. (*Wien. V.* 312.)—*Fabr.* iii. *b.* 78.—*Hüb. No. pl.* 53. *f.* 259.—(*Sam. I.* 28.)
No. Argentea. *Fuesly A.* i. *pl.* 5. *f.* 1–7. In Mus. *Brit.*

Genus 140: (923). CALOPHASIA *mihi.*

XYLENA *p, Och.*—XYLINA *p, Treit.*

6367. 1, Linariæ*.
No. Linariæ. (*Wien. V.* 73)—*Fabr.* iii. *b.* 92.—*Hüb. No. pl.* 52. *f.* 252.—(*Kirby & Sp. I. E.* iii. 255.)
No. stictica. *Fabr. M.* ii. 173.
Ph. lunula. *Berl. M.* (*Hufn.*) iii. 394.

Genus 141: (924). CHARICLEA *mihi, Curtis,* (*Kirby.*)

XYLENA D. *γ, Och.*—XYLINA *p, Treit.*

6368. 1, Delphinii*. *Curtis.* ii. *pl.* 76.
Ph. No. Delphinii. *Linn.* ii. 857.—*Berk. S.* i. 143.—*Don.* x. 17. *pl.* 331.—*Stew.* ii. 193.—*Turt.* iii. 322.
No. Delphina. (*Haw. Pr.* 16.)
No. Delphinii. *Haw.* 248.—(*Sam. I.* 29.)
Pease-blossom M.: Ph. Delphinii. (*Wilkes.* 3. *f.* 4.)—*Harr. V. M.* 13.

Genus 142: (925). EREMOBIA *mihi.*

XANTHIA A. *p, Och.*—XANTHIA *p, Curtis, fo.* 84.—PHYTOMETRA *p, Haw.*

6369. 1: ochroleuca*.
No. ochroleuca. (*Wien. V.* 87.)—*Esper.* iv. *pl.* cxxvi. *f.* 1, 4.
No. flammea: (ochroleuca.) *Hüb. No. pl.* 19. *f.* 92.
No. citrina. (*Haw. Pr.* 14.)—*Haw.* 237.—(*Sam. I.* 29.)
Ph. citrina: Dusky Sallow M. *Don.* x. 36. *pl.* 340. *f.* 2.
Xa. citrina. (*Curtis l. c. supra.*)
Var. Phy. bifasciosa: double-barred Brown. *Haw.* 266.

Genus 143: (926). ABROSTOLA, *Och.,* (*Samou.*)

PLUSIÆ *p, Hüb.*

6370. 1; triplasia*. (*Och.* iv. 88.)
Ph. No. triplasia. *Linn.* ii. 854.—*Stew.* ii. 192.—*Turt.* iii. 335.
No. triplacea: dark Spectacle M. *Haw.* 245.—(*Sam. I.* 30.)
No. Asclepiades. *Hüb. No. pl.* 55. *f.* 268.

6371. 2; Asclepiadis*. (*Och.* iv. 88.)
No. Asclepiadis. (*Wien. V.* 91.)—*Fabr.* iii. *b.* 117.—*Haw.* 246.—(*Sam. I.* 28.)

No. triplasia. *Hüb. No. pl.* 55.*f.* 269.
Ph. No. triplacia. *Don.* ix. 23. *pl.* 298.
Ph. tripartita. *Berl. M.* (*Hufn.*) iii. 414.
Spectacle. *Harr. V. M.* 45.

6372 † 3. Urticæ?
No. Urticæ. *Hüb. No. pl.* 137.*f.* 625? In Mus. D. —— ?

6373. 4, illustris. (*Och.* iv. 88.)
No. illustris: Purple shades.—*Haw. l. c. Fabr.* iii. *b.* 84.—*Hüb. No. pl.* 56.*f.* 274.—*Haw.* 258.—(*Sam. I.* 29.)
♀, No. cuprea. *Esper.* iv. *pl.* cx.*f.* 4.

Genus 144: (927). CELSIA *mihi*[a].

Genus 145: (928). PLUSIA, *Och.*, (*Kirby.*)

Plusiæ *p*, *Hübn.*—Phytometra 1, *Haw.*—Campæa *p*, *Lamarck.*

6375. 1; Iota*. (*Och.* iv. 90.)—*Och. Tr.* v. c. 181.
Ph. No. Iota. *Linn.* ii. 844.—*Turt.* iii. 318.
No. Iotina. (*Haw. Pr.* 16.)
No. Iota. *Hüb. No. pl.* 52.*f.* 282.—(*Sam. I.* 29.)
Phy. Iota γ. *Haw.* 256.
No. pulchrina. (*Haw. Pr.* 16.)
No. interrogationis. *Esper.* iv. *pl.* cxiii.*f.* 1, 2.
Golden Y. Moth. *Harr. V. M.* 59.
β, No. Gamma-aurina. (*Haw. Pr.* 16.)

6376. 2; percontationis*. (*Och.* iv. 90?)
No. Iota. *Esper.* iv. *pl.* cxiii.*f.* 3, 4.
Ph. No. Iota. *Don.* viii. 33. *pl.* 256.*f.* 1.
Phy. Iota α. *Haw.* 256.

6377. 3, inscripta?
No. inscripta. *Esper. No. pl.* cxiii.*f.* 5?

6378. 4, interrogationis. (*Och.* iv. 90.)—*Och. Tr.* v. c. 190.
Ph. No. interrogationis. *Linn.* ii. 844.—*Stew.* ii. 86.
No. subpurpurina. (*Haw. Pr.* 16.)
Ph. aurosignata. *Don.* xiii. 43. *pl.* 453.
Phy. interrogationis: Yorkshire Y. *Haw.* 257.
No. interrogationis. (*Sam. I.* 29.)
No. Æmula. *Bork. E. S.* iv. 789.
Harr. Ex. 21. *pl.* v.*f.* 3.
Scarce Silver Y. Moth. *Harr. V. M.* 59.

---

[a] Genus 144: (927). CELSIA *mihi.*

Polia C., *Och.*—Plusia C., *Treitschke.*—Phalæna-Bombyx *p*, *Linné.*

6374 † 1. viridis.
Ph. Bo. Celsia. *Linn.* ii. 831.—*Martyn V. M.* (!)
No. Celsia. *Hüb. No. pl.* 15.*f.* 72. ♂.*f.* 73. ♀.

6379. 5; Gamma*. (*Och.* iv. 90.)—*Och. Tr.* v. c. 185.
Ph. No. Gamma. *Linn.* ii. 843.—*Berk. S.* i. 141.—*Don.* viii. 34. *pl.* 265. *f.* 2.—*Stew.* ii. 186.
No. Gammina. (*Haw. Pr.* 16.)
No. Gamma. *Hüb. No. pl.* 58. *f.* 283.—(*Sam. I.* 29.)—(*Kirby & Sp. I. E.* iii. 255.)
Phy. Gamma. *Haw.* 256.
Pa. No. 14. *List. Goëd.* (*Angl.*) 20. *f.* 14.—*Id.* (*Lat.*) 41. *f.* 14.
Y. Moth. *Albin. pl.* lxxix. *f. e–h.*—*Harr. V. M.* 59.
Silver Y. Moth: Ph. Gamma. (*Wilkes.* 34. *pl.* 69.)
Golden Y. M. (*Wilkes D. pl.* 11. *f.* 3.)

6380 † 6, circumflexa*? (*Och.* iv. 90.)—*Och. Tr.* v. c. 179.
Ph. No. circumflexa. *Linn.* ii. 844.
No. circumflexina. (*Haw. Pr.* 16.)
Phy. circumflexa: Essex Y. *Haw.* 257.
No. circumflexa. *Hüb. No. pl.* 58. *f.* 285.—(*Sam. I.* 29.)
Ph. flexuosa: Yorkshire Y. *Don.* xii. 53. *pl.* 412.
In Mus. D. Donovan, *et Swainson.*

6381 † 7, biloba *mihi.* In Mus. D. *Swainson.*

6382 † 8, aurifera. (*Och.* iv. 89.)—*Och. Tr.* v. c. 168.—(*Ing. Inst.* 91.)
No. aurifera. *Hüb. No. pl.* 98. *f.* 463.
In Mus. *Brit. et* D. *Ingpen.*

6383. 9; chrysitis*. (*Och.* iv. 89.)—*Och. Tr.* v. c. 169.
Ph. No. chrysitis. *Linn.* ii. 843.—*Berk. S.* i. 140.—*Don.* iv. 77. *pl.* 137.—*Stew.* ii. 186.—*Turt.* iii. 316.
No. chrysitis. *Hüb. No. pl.* 56. *f.* 272.—(*Leach E. E.* ix. 134.)—(*Sam. I.* 29.)
No. chrysitina. (*Haw. Pr.* 16.)
Phy. chrysitis. *Haw.* 255.
Albin. pl. lxxi. *f. a–d.*
The Burnished M. (*Wilkes D. pl.* 11. *f.* 8.)
Burnished brass. *Harr. A. pl.* 22. *f. a–c.*—*Harr. V. M.* 15.

6384. 10; orichalcea*. (*Och.* iv. 89.)—*Och. Tr.* v. c. 173.
No. orichalcea. *Fabr.* iii. *b.* 77.—*Hüb. No. pl.* 57. *f.* 278.—(*Sam. I.* 29.)
Phy. orichalcea. *Haw.* 255.
No. Chrysa. *Esper.* iv. *pl.* cxli. *f.* 2.
Ph. No. ærifera. *Sowerb. B. M.* i. *pl.* 29.
Scarce burnished Brass. *Harr. Ex.* 25. *pl.* vi. *f.* 4.
Burnished Brass likeness. *Harr. V. M.* 15.

6385. 11, bractea. (*Och.* iv. 89.)—*Och. Tr.* v. c. 176.
No. bractea. (*Wien. V.* 314.)—*Fabr.* iii. *b.* 78.—*Hüb. No. pl.* 57. *f.* 279.—(*Sam. I.* 29.)
No. bracteina. (*Haw. Pr.* 16.)

Phy. bractea: Gold spangle. *Haw.* 255.
Ph. No. bractea. *Stew.* ii. 186.—*Sowerb. B. M.* i. *pl.* 28.
Ph. No. securis. *Villers E.* ii. 271. *pl.* v. *f.* 10.

6386. 12; Festucæ*. (*Och.* iv. 89.)—*Och. Tr.* v. *c.* 165.
Ph. No. Festucæ. *Linn.* ii. 845.—*Berk. S.* i. 141.—*Don.* ii. *pl.* 46.—*Stew.* ii. 187.—*Turt.* iii. 317.
No. Festucina. (*Haw. Pr.* 16.)
Phy. Festucæ. *Haw.* 254.
No. *Festucæ*. (*Leach E. E.* ix. 134.)—(*Sam. I.* 29.)
Albin. pl. lxxxiv. *f. e–h.*
Gold spot M.: Ph. *Festucæ*. (*Wilkes.* 8. *pl.* 17.)—*Harr. V. M.* 43.

Genus 146: (929). HELIOTHIS, *Och.*, (*Samou.*)
HELIOTHENTES, *Hüb.*

6387. 1; marginata*. (*Och.* iv. 91.)—*Och. Tr.* v. c. 232.
No. marginata. *Fabr.* iii. *b.* 88.—*Don.* v. *pl.* 150. *f.* 1.—*Turt.* iii. 321.
No. marginago: bordered Sallow. (*Haw. Pr.* 14.)—*Haw.* 235.—(*Sam. I.* 29.)
Ph. No. limbata. *Stew.* ii. 188.
β; No. rutila. (*Wien. V.* 86.)—*Hüb. No. pl.* 39. *f.* 155.—(*Haw. Pr.* 14.)
Ph. No. rutilago. *Turt.* iii. 316.
No. umbra. *Bork. E. S.* iv. 672.
No. umbrago. *Esper.* iv. *pl.* clxxxv. *f.* 7, 8.

6388. 2: peltigera*. (*Och.* iv. 91.)—*Och. Tr.* v. c. 227.
No. peltigera. (*Wien. V.* 89.)—*Hüb. No. pl.* 63. *f.* 310.
No. scutigera. *Bork. E. S.* iv. 93.
No. florentina. *Esper.* iv. *pl.* cxxxv. *f.* 2.
No. barbara. *Fabr.* iii. *b.* 111.
Ph. straminea. *Don.* ii. 65. *pl.* 61.—*Stew.* ii. 194.—*Turt.* iii. 297.
No. strammina. (*Haw. Pr.* 16.)—(*Sam. I.* 30.)
Phy. straminea: bordered Straw. *Haw.* 263.

6389. 3; dipsacea*. (*Och.* iv. 91.)—*Och. Tr.* v. *c.* 220.
Ph. No. dipsacea. *Linn.* ii. 856.—*Don.* x. 9. *pl.* 327. *f.* 3.
No. dipsacina. (*Haw. Pr.* 12.)
No. dipsacea. *Hüb. No. pl.* 63. *f.* 311.—(*Sam. I.* 29.)
Phy. dipsacea: Marbled Clover. *Haw.* 263.
Ph. viriplaca. *Berl. M.* (*Hufn.*) iii. 406.
Teazle. *Harr. V. M.* 47?

Genus 147: (930). ANARTA, *Och.*, (*Samou.*,) *Curtis.*
PHYTOMETRA *p*, *Haw.*
A. Corpore subcrasso.

6390. 1, Myrtilli*. (*Och.* iv. 90.)—*Curtis.* iii. *pl.* 145.

Ph. No. Myrtilli. *Linn.* ii. 852.—*Don.* vii. *pl.* 231.—*Stew.* ii. 191.—*Turt.* iii. 339.
No. Myrtillina. (*Haw. Pr.* 12.)
No. Myrtilli. *Haw.* 162.—(*Sam. I.* 29.)
Ph. Ericæ. *Berl. M.* (*Hufn.*) iii. 292.
Beautiful Yellow underwing. *Harr. Ex.* 18. *pl.* iv. *f. e.*—*Harr. V. M.* 53.
Var. No. albirena: Small Yellow underwing. *Haw.* 163.—(*Sam. I.* 28.)

6391 † 2. cordigera [a].

B. Corpore gracili.

6392. 3; Heliaca*. (*Och.* iv. 91.)—*Och. Tr.* v. c. 212.
No. Heliaca. (*Wien. V.* 94.)—*Hüb. No. pl.* 64. *f.* 316.
No. Arbuti. *Fabr.* iii. *b.* 126. (!)—(*Sam. I.* 28.)
No. Arbutina. (*Haw. Pr.* 12.)
Ph. No. Arbuti. *Stew.* ii. 191.—*Turt.* iii. 339.—*Don.* x. 52. *pl.* 343. *f.* 3.
Phy. Arbuti: Minute Yellow underwing. *Haw.* 265.
An. Arbuti. *Curtis l. c. supra.*
No. fasciola. *Esper.* iv. *pl.* clxiii. *f.* 1.
Ph. domestica. *Berl. M.* (*Hufn.*) iii. 406.
Small Yellow underwing. (*Wilkes* D. *pl.* 9. *f.* 2.)—*Harr. A. pl.* 53. *f. o.*—*Harr. V. M.* 55.

Genus 148: (931). ACONTIA, *Och.*

PHYTOMETRA *p*, *Haw.*

6393. 1; luctuosa*. (*Och.* iv. 92.)—*Och. Tr.* v. c. 247.
No. luctuosa. *Hüb. No. pl.* 62. *f.* 305, 306.—(*Sam. I.* 29.)
Phy. luctuosa: four-spotted. *Haw.* 264.
No. Italica. *Fabr.* iii. *b.* 37?

6394 † 2. aprica. (*Och.* iv. 92?)—*Och. Tr.* v. c. 239?
No. aprica. *Hüb. No. pl.* 80. *f.* 371?
No. atro-albina. (*Haw. Pr.* 17.)
No. albo-ater: the Nun. *Haw.* 184. In Mus. D. ——?

6395. 3, Solaris. (*Och.* iv. 92.)—(*Ing. Inst.* 90.)—*Och. Tr.* v. c. 244.
No. Solaris. (*Wien. V.* 90.)—*Hüb. No. pl.* 62. *f.* 307. ♂: *f.* 308. ♀.
No. albicollis. *Fabr.* iii. *b.* 36.
No. rupicola. *Bork. E. S.* iv. 82.
Ph. lucida. *Berl. M.* (*Hufn.*) iii. 302.

---

[a] 6391 † 2. cordigera. (*Och.* iv. 90.)—(*Curtis l. c. supra.*) (!)
No. cordigera. *Thunb. Mus.* vi. *f.* 4.
No. albirena. *Hub. No. pl.* 21. *f.* 99.—*nec* Haworthi 163. *vide supra.*

6396 † 4, Caloris. (*Och.* iv. 92.)—*Och. Tr.* v. c. 241.
No. Caloris. *Hüb. No. pl.* 80. *f.* 372. In Mus. *D. Curtis.*

6397 † 5, nigrirena.
Phy. nigrirena: black Kidney. *Haw.* 266.
In Mus. *D. Swainson.*

Genus 149: (932). DESMOPHORA *mihi.*

6398 † 1, elegans*?
Ph. catæna: Brixton Beauty. *Sowerb. B. M.* i. *pl.* 14.
No. catæna. *Haw.* 184.—(*Sam. I.* 29.) In Mus. D. *Curtis.*

Genus 150: (933). ERASTRIA, (*Och.*,) (*Samou.*,) *Curtis.*

EROTYLÆ *p*, *Hüb.*—PHYTOMETRA 2. *p*, *Haw.*—PHALÆNA-TORTRIX *p*, *Linné.*—BOMBYX *p*, *et* PYRALIS *p*, *Fabr.*

6399. 1: sulphurea*. (*Och.* iv. 92.)—(*Curtis l. c. infra.*)
Ph. Py. sulphuralis. *Linn.* ii. 881.—*Don.* x. 33. *pl.* 339. *f.* 1. —*Stew.* ii. 180.
No. sulphurea. *Hüb. No. pl.* 60. *f.* 291.—(*Sam. I.* 30.)
Phy. sulphurea: spotted Sulphur. *Haw.* 262.
Py. sulphuralis. (*Haw. Pr.* 29.)—*Panz. F.* viii. *f.* 18.
Bo. lugubris. *Fabr.* iii. *a.* 467.
No. Arabica. *Bork. E. S.* iv. 807.
Pyr. trabealis. *Villers E.* ii. 445.

6400. 2: uncana. (*Och.* iv. 92.)—*Curtis l. c. infra.*
Ph. Ge. uncana. *Linn.* ii. 875.—*Turt.* iii. 343.
Phy. unca: Silver hook. *Haw.* 262.—(*Samou.* 403.)
Py. unca. (*Sam. I.* 35.)
No. unca. *Hüb. No. pl.* 60. *f.* 293.

6401. 3, Bankiana.
Py. Bankiana. *Fabr.* iii. *b.* 242. (!)—*Stew.* ii. 172.—*Turt.* iii. 342.
Py. Banksianalis. (*Haw. Pr.* 31.)
Phy. Bankiana: Silver barred. *Haw.* 261.
No. Olivea. *Hüb. No. pl.* 60. *f.* 292.
Er. argentula. (*Och.* iv. 92.)—*Curtis l. c. infra.*

6402. 4, venustula. (*Och.* iv. 92.)—*Curtis l. c. infra.*
No. venustula. *Hüb. No. pl.* 60. *f.* 294.
Phy. venustula: Rosy Marbled. *Haw.* 265.

6403. 5, minuta. (*Och.* iv. 92.)—*Och. Tr.* v. c. 266.
No. minuta. *Hüb. No. pl.* 95. *f.* 451.
Phy. minuta: small Marble. *Haw.* 265.

6404 † 6, ostrina. (*Och.* iv. 92.)—*Curtis.* iii. *pl.* 140.
No. ostrina. *Hüb. No. pl.* 85. *f.* 399. In Mus. D. *Curtis.*

6405 † 7, apicosa.
Phy. apicosa: blossom tip. *Haw.* 261.
In Mus. D. *Haworth et Stone.*

6406. 8; fuscula*. (*Och.* iv. 92.)—*Och. Tr.* v. c. 257.
No. fuscula. (*Wien. V.* 89.)—*Hüb. No. pl.* 60. *f.* 297.
No. polygramma. *Esper.* iv. *pl.* cxlvi. *f.* 7.
No. præduncula. *Bork. E. S.* iv. 175.
Phy. fusca: Marbled White spot. *Haw.* 261.
No. fusca. (*Samou.* 403.)

6407. [9, albidilinea*.]
Phy. albidilinea: Marbled White line. *Haw.* 261.—(*Sam. I.* 403.)

Genus 151: (934). PHYTOMETRA.

ANTHOPHILA, *Och.*—PHYTOMETRA 3 *p*, *Haworth.*

6408. 1; ænea*: small Purple barred. *Haw.* 266.
No. ænea. (*Wien. V.* 85.)—*Hüb. No. pl.* 75. *f.* 350.—(*Sam. I.* 28.)
No. latruncula. *Esper.* iv. *pl.* clxiii. *f.* 2.
No. olivacea. *Vieweg. T. V.* ii. 85.
Ph. Laccata. *Scop. C. No.* 563.

Genus 152: (935). ACOSMETIA *mihi.*

ANTHOPHILA *p*, *Och.*

6409. 1: lutescens.
Phy. lutescens: reddish Buff. *Haw.* 260.

6410. 2: caliginosa.
No. caliginosa. *Hüb. No. pl.* 100. *f.* 474.
An. infida. (*Och.* iv. 93.)
An. caliginosa. *Och. Tr.* v. c. 286.

6411. 3; rufula.
Phy. rufa: small rufous. *Haw.* 260.
No. ruferculina. (*Haw. Pr.* 13?)

6412. 4, lineola *mihi.*

6413. 5; arcuosa*.
Phy. arcuosa: small dotted Buff. *Haw.* 260.—(*Sam. I.* 28.)
No. arcuina. (*Haw. Pr.* 13.)

Genus 153: (936). SCOPELOPUS[a].

Genus 154: (937). STILBIA *mihi.*

PHALÆNA *p*, *Haw.*—GEOMETRA *p*, *Hüb.*—OPHIUSA? *Ing.*

6415. 1, anomalata*.

---

[a] Genus 153: (936). SCOPELOPUS *mihi.*

PHYTOMETRA 2 *p*, *Haw.*

6414 † 1. inops *mihi.*
Phy. scopulæpes; brush-Foot. *Haw.* 260. (!) In Mus. Brit.?

Ph. anomalata: the Anomalous. *Ent. Trans.* (*Haworth.*) i. 336.
Op.? anomala. (*Ing. Inst.* 91.)
Geo. hybridata. *Hüb. Geo.*

Genus 155: (938). OPHIUSA, *Och.*

ASCALAPHÆ *p*, *Hübn.*—BOMBYX *p*, *Esper.*

6416. 1; lusoria*. (*Och.* iv. 93.)—*Och. Tr.* v. c. 289.
Ph. Bo. lusoria. *Linn.* ii. 831.—*Don.* x. 76. *pl.* 354. *f.* 2.
No. lusina. (*Haw. Pr.* 13?)
No. lusoria. *Hüb. No. pl.* 65. *f.* 318.—(*Sam. I.* 29.)
Phy. lusoria: Black-neck. *Haw.* 259.

6417 † 2, ludicra. (*Och.* iv. 93.)—*Och. Tr.* v. *c.* 292.
No. ludicra. *Hüb. No. pl.* 65. *f.* 319.
Phy. ludicra: Scarce Black-neck. *Haw.* 259.
In Mus. *D. Swainson.*

6418. 3, crassiuscula.
Phy. crassiuscula: double-barred. *Haw.* 259.

6419 † 4. *grandirena*[a].

Genus 156: (939). CATEPHIA, *Och.*

PHYTOMETRA 3 *p*, *Haw.*

6420 † 1, leucomelas.
Ph. No. leucomelas. *Linn.* ii. 856.
No. leucomelina. (*Haw. Pr.* 12.)
Phy. leucomelas: large 4-spotted. *Haw.* 264.
No. Alchemista. *Hüb. No. pl.* 62. *f.* 303.
No. convergens. *Fabr.* iii. *b.* 100. In Mus. D. *Haworth.*

6421. 2, trifasciata *mihi.*

Genus 157: (940). MORMO, (*Samou.*)

MORMO *p*, *Och.*—HEMIGEOMETRA *p*, *Haworth.*—LEMURES *p*, *Hübner.*—MANIA *p*, *Treit.*

6422. 1; maura*. (*Och.* iv. 70.)
Ph. No. maura. *Linn.* ii. 843.—*Don.* vii. *pl.* 30. *f.* 1.—*Stew.* ii. 183.—*Turt.* iii. 298.
No. maurina: great Brown-bar. (*Haw. Pr.* 11.)
No. maura. *Hüb. No. pl.* 67. *f.* 326.—*Haw.* 269.—(*Leach E. E.* ix. 134.)—(*Sam. I.* 29.)
No. Lemur. *Naturf.* (*Hufnagle.*) vi. 112. *pl.* 5. *f.* 1.
Old Lady M. *Harr. E.* 9. *pl.* 1. *f. a, b.*—*Harr. V. M.* 33.

---

[a] 6419 † 4. grandirena: great Kidney. *Haw.* 264. (!)
Olim in Mus. D. Leach ex Museo Portlandico.—*Haw. l. c.*

Genus 158: (941). CATOCALA, *Schrank*, *Curtis*, (*Kirby*.)

BLEPHARA, *Hüb*.—HEMIGEOMETRA *p*, *Haw*.

6423. 1: Fraxini*. (*Och*. iv. 95.)—*Curtis l. c. infra*.
Ph. No. Fraxini. *Linn*. ii. 843.—*Berk*. *S*. i. 140.—*Don*. v. 89. *pl*. 171. *pl*. 172. *larva*.—*Stew*. ii. 184.—*Turt*. iii. 294.
No. Fraxina. (*Haw*. *Pr*. 11.)
No. Fraxini. *Hüb*. *No*. *pl*. 68. *f*. 327.—(*Leach E*. *E*. ix. 134.) —(*Sam*. *I*. 29.)
He. Fraxini. *Haw*. 267.
Clifden Nonpareil, a Moth: Ph. Fraxini. (*Wilkes*. 44. *pl*. 90.) —*Harr*. *A*. *pl*. 31. *f*. *a–e*.—*Harr*. *V*. *M*. 37.

6424 † 2. Elocata[a].

6425. 3; Nupta*. (*Och*. iv. 95.)—*Och*. *Tr*. v. c. 337.—(*Curtis l. c. infra*.)
Ph. No. Nupta. *Linn*. ii. 841.—*Berk*. *S*. i. 140.—*Don*. vii. *pl*. 224.—*Stew*. ii. 183.—*Turt*. iii. 294.
No. Nuptina. (*Haw*. *Pr*. 11.)
No. Nupta. *Hüb*. *No*. *pl*. 69. *f*. 330.—(*Leach E*. *E*. ix. 134.) —(*Sam*. *I*. 29.)
He. Nupta. *Haw*. 268.
No. concubina. *Bork*. *E*. *S*. iv. 21.
Red underwing. *Albin*. *pl*. lxxx. *f*. *a–d*.—*Harr*. *A*. *pl*. 18. *f*. *g–m*.—*Harr*. *V*. *M*. 55.
The great Red underwing. (*Wilkes* D. *pl*. 6. *f*. 1.)—*Id*. *pl*. 10. *f*. 7.
Willow Red underwing: Ph. Pacta. (*Wilkes*. 17. *pl*. 35.)

6426 † 4. Pacta[b].

6427. 5; sponsa*. (*Och*. iv. 95.)—*Och*. *Tr*. v. c. 343.—(*Curtis l. c. infra*.)
Ph. No. sponsa. *Linn*. ii. 841.—*Don*. ix. 77. *pl*. 324.—*Stew*. ii. 182.—*Turt*. iii. 293.
No. sponsina. (*Haw*. *Pr*. 11.)
No. sponsa. *Hüb*. *No*. *pl*. 71. *f*. 333.—(*Sam*. *I*. 30.)—(*Kirby & Sp*. *I*. *E*. iii. 260.)
He. sponsa: dark Crimson underwing. *Haw*. 268.

6428. 6; promissa*. (*Och*. iv. 95.)—(*Ing*. *Inst*. 90.)—(*Curtis l. c. infra*.)

---

[a] 6424 † 2. Elocata. (*Och*. iv. 95.)—*Curtis*. v. *pl*. 217. (!)
No. Elocata. *Esper*. iv. *pl*. xcix. *f*. 1, 2.
No. Nupta. *Fabr*. iii. *b*. 53.
No. Marita. *Hüb*. *No*. *pl*. 105. *f*. 494. ♀.
No. Uxor. *Hüb*. *No*. *pl*. 69. *f*. 328. ♀. In Mus. D. Curtis.

[b] 6426 † 4. Pacta. (*Och*. iv. 95.)—*Och*. *Tr*. v. c. 352.
Ph. No. Pacta. *Linn*. ii. 841.—*Berk*. *S*. i. 140. (!)—*Turt*. iii. 294. (!) —(*Leach E*. *E*. ix. 134.)—(*Kirby & Sp*. *I*. *E*. iii. 260.)

No. promissa. (*Wien. V.* 90.)—*Fabr.* iii. *b.* 54.—*Hüb. No. pl.* 71. *f.* 334.—(*Leach E. E.* ix. 134.)—(*Sam. I.* 30.)
He. promissa: light Crimson underwing. *Haw.* 268.
No. Mneste. *Hüb. No. pl.* 123. *f.* 569.
Crimson underwing: Ph. Nupta. (*Wilkes.* 33. *pl.* 68.)—*Harr. A. pl.* 19. *f.* g–l.—*Harr. V. M.* 55.

6429. 7: conjuncta*. (*Och.* iv. 95.)—*Och. Tr.* v. c. 347.—(*Curtis l. c. supra.*)
No. conjuncta. *Esper.* iv. *pl.* c. *f.* 1, 2.
No. conjuga. *Hüb. No. pl.* 71. *f.* 335.—(*Sam. I.* 29.)
He. conjuga: lesser Crimson underwing. *Haw.* 269.

Genus 159: (942). BREPHA, *Hüb.*, (*Samou.*,) *Curtis.*

BREPHOS, (*Och.*)—HEMIGEOMETRA *b*, *Haw.*

A. Antennis haud submoniliformibus; (in masculis pectinatis.)

6430. 1; Parthenias*. (*Och.* iv. 96.)—(*Curtis l. c. infra.*)
Ph. No. Parthenias. *Linn.* ii. 835.—(*Sam. I.* 30.)
No. notha. *Hüb. No. pl.* 74. *f.* 343, 344.
Bo. parthenias. (*Haw. Pr.* 10.)
Ph. Bo. parthenias. *Turt.* iii. 222.—*Stew.* ii. 156.—*Don.* vii. 80. *pl.* 246. *f.* 1.
Bo. vidua. (*Wien. V.* 91.)—*Fabr. E. S.* i. *a.* 468.
Orange underwing. (*Wilkes D. pl.* 11. *f.* 2.)—*Harr. A. pl.* 35. *f.* i.—*Harr. V. M.* 55.

6431 † 2, puella. (*Och.* iv. 96.)—*Och. Tr.* v. c. 385.
No. puella. *Esper.* iv. *pl.* cvi. *f.* 2, 3.
No. spuria. *Hüb. No. pl.* 74. *f.* 345. In Mus. *Brit.*

B. Antennis submoniliformibus; (in masculis haud pectinatis.)

6432. 3; notha*. (*Och.* iv. 96.)—*Curtis.* iii. *pl.* 121.
No. Parthenias. *Hüb. No. pl.* 74. *f.* 341, 342.
Ne. notha: light Orange underwing. *Haw.* 269.
No. notha. (*Sam. I.* 29.)
Orange underwing companion. *Harr. V. M.* 55.

Genus 160: (943). EUCLIDIA, (*Och.*,) (*Samou.*,) *Treit.*

EUCLIDIÆ, *Hübn.*—PHYTOMETRA *p*, *Haw.*—CAMPÆA *p*, *Lamarck.*

6433. 1; glyphica*. (*Och.* iv. 96.)—*Och. Tr.* v. c. 390.
Ph. No. glyphica. *Linn.* ii. 838.—*Turt.* iii. 306.
No. glyphicina. (*Haw. Pr.* 12.)
No. glyphica. *Hüb. No. pl.* 75. *f.* 347.—(*Sam. I.* 29.)
Phy. glyphica: the Burnet. *Haw.* 265.
Burnet companion. *Harr. V. M.* 15.

6434 † 2. triquetra[a].

6435. 3; Mi*. (*Och.* iv. 96.)—*Och. Tr.* v. c. 395.
Ph. No. Mi. *Linn.* ii. 838.
No. Miina. (*Haw. Pr.* 12.)
No. Mi. *Hüb. No. pl.* 75. *f.* 346.—(*Sam. I.* 29.)
Phy. Mi. *Haw.* 265.
Ph. Graminis. *Berl. M.* (*Hufnagle.*) iii. 412.
Mask M. *Harr. A. pl.* 41. *f.* 5.—Shipton M. *Harr. V. M.* 43.

Subsectio 3. LE.-SEMIDIURNA *mihi.*

Familia XV: (128). GEOMETRIDÆ.

(PHALÆNA-GEOMETRA *p, Linné, &c.*—PHALÆNA *p, Fabr.*)

Slender-bodied Moths. Larvæ: *Loopers.*

Genus 161: (944). PSYCHOPHORA, *Kirby*: (*vide Bombyx Sabini. Parry Voyage: App.*)

PSODOS (PSOÏDOS) *p, Treit.*

6436 † 1, trepidaria: Black Mountain M.—*Haw. l. c.*
Ge. trepidaria. *Hüb. Ge. pl.* 66. *f.* 343?—*Haw.* 281.—(*Sam. I.* 19.) In Mus. D. Dale, *Haworth et Vigors.*

6437 † 2, alpinata*: the Gold Four-spot.—*Haw. l. c.*
Ge. alpinata. (*Wien. V.* 115.)—*Hüb. Ge. pl.* 38. *f.* 197.
Ph. equestrata. *Fabr.* iii. *b.* 179.—(*Haw. Pr.* 26.)—*Haw.* 345.
Ge. equestraria. *Esper.* v. *pl.* l. *f.* 1.
In Mus. D. *Curtis, Dale et Haworth.*

Genus 162: (945). SPERANZA, *Curtis.*

FIDONIA *p, Treit.*

6438. 1; limbaria*. (*Curtis l. c. infra.*)
Ph. limbaria. *Fabr.* iii. *b.* 141. (!)—*Stew.* ii. 163.—*Turt.* iii. 237.
Ge. limbaria. (*Haw. Pr.* 23.)—*Haw.* 286.—(*Sam. I.* 18.)
Ge. conspicuata. (*Wien. V.* 116.)
Ge. conspicuaria. *Hüb. Ge. pl.* 22. *f.* 117, 118.
Frosted Yellow. *Harr. Ex.* 22. *pl.* v. *f.* 4.—*Harr. V. M.* 59.

6439. 3, sylvaria. (*Ing. Inst.* 91.)—*Curtis.* v. *pl.* 225.

6434 † 2. triquetra. (*Och.* iv. 96.)—*Och. Tr.* v. c. 393.
No. triquetra. (*Wien. V.* 94.)—*Fabr.* iii. *b.* 34.—*Hüb. No. pl.* 75. *f.* 348.
No. fortificata. *Fabr.* iii. *b.* 52.
Pyr. fascialis. *Villers E.* ii. 450. *pl.* vi. *f.* 28.
No. triquetra. (*Sam. I.* 30.) (!)

Genus 163: (946). FIDONIA.

FIDONIA *p, Treit.*—BUPALUS? *p, Curtis folio* 33.

6440. 1: fuliginaria*.
Ph. Ge. fuliginaria. *Linn. F.* 327.
Ge. fuliginaria: waved Black. (*Haw. Pr.* 24.)—*Haw.* 281.—(*Sam. I.* 18.)
Bu.? fuliginarius. (*Curtis l. c. supra.*)
Ph. carbonaria. *Fabr.* iii. *b.* 154.
Ge. carbonaria. *Hüb. Ge. pl.* 28. *f.* 151.
† ♀, Ph. lunulata. *Fabr.* iii. *b.* 194.

6441. 2; atomaria*: common Heath.—*Haw. l. c. Och. Tr.* vi. *a.* 286.
Ph. Ge. atomaria. *Linn.* ii. 862.—*Don.* vii. 85. *pl.* 248. *f.* 1, 2.—*Stew.* ii. 165.—*Turt.* iii. 238.
Ge. atomaria. *Hüb. Ge. pl.* 25. *f.* 136.—(*Haw. Pr.* 21.)—*Haw.* 280.—(*Sam. I.* 18.)
Bu.? atomarius. (*Curtis l. c. supra.*)
Ph. Aceraria. *Berl. M.* (*Hufnagle.*) iv. 520.
Ph. pennata. *Scop. C.* 228. ♂.
Ph. Isoscelata. *Scop. C.* 225. ♀.
The Heath M. (*Wilkes D. pl.* 1. *f.* 3.)
Brown heath. *Harr. V. M.* 31.
β; Ge. glarearia: yellow Heath.—*Haw. l. c.* (*Wien. V.* 106)?—*Hüb. Ge. pl.* 25. *f.* 131?—*Haw.* 280.—(*Sam. I.* 18.)
Pale Heath. *Harr. V. M.* 31.
γ; Ge. roseidaria: light Heath.—*Haw. l. c. Hüb. Ge. pl.* 24. *f.* 128.—*Haw.* 280.—(*Sam. I.* 19.)

6442. [3, carbonaria.]
Ge. carbonaria: black Heath. *Haw.* 281.—(*Sam. I.* 18.)

6443. 4: ericetaria*.
Ph. Ge. ericetaria. *Vill. E.* ii. 329. *pl.* vi. *f.* 9. ♀?
Ge. ericetaria: bordered Grey. (*Haw. Pr.* 21.)—*Haw.* 278.—(*Sam. I.* 18.)
Bu. ericetarius. (*Curtis l. c. supra.*)
Ge. plumistraria. *Hüb. Ge. pl.* 24. *f.* 127. ♂. *pl.* 81. *f.* 417, 418. ♂. *pl.* 419, 420.
♀, Ge. subfimbriata: the four-barred. *Ent. Trans.* (*Haworth.*) i. 336.

Genus 164: (947). BUPALUS, *Leach, Samou., Curtis.*

FIDONIA *p, Treit.*

6444. 1; Piniarius*. *Leach E. E.* ix. 134.—(*Samou.* 253.)—(*Curtis l. c. infra.*)
Ph. Ge. Piniaria. *Linn.* ii. 861.—*Don.* x. 27. *pl.* 336.—*Stew.* ii. 163.—*Turt.* iii. 237.

Ge. Piniaria: bordered White.—*Haw. l. c.* *Hüb. Ge. pl.* 22. *f.* 119.—(*Haw. Pr.* 21.)—*Haw.* 278.—(*Sam. I.* 19.)
Ph. Ge. Tiliaria. *Linn. F. No.* 1234. ♀.

6445. 2: favillacearius. *Curtis.* i. *pl.* 33.
Ge. favillacearia. *Hüb. Ge. pl.* 26. *f.* 139.—*Haw.* 278.—(*Sam. I.* 18.)
Ph. Ge. mediopunctaria. *Don.* xiii. 59. *pl.* 461. *f.* 1.
The Grey Scolloped bar. *Harr. A. pl.* 33. *f. m.*—*Harr. V. M.* 25.
β? Ge. Belgiaria. *Fabr. S.* 451.—*Hüb. Ge. pl.* 26. *f.* 140. ♂.

Genus 165: (948). ANISOPTERYX *mihi.*

FIDONIA *p, Treit.*

6446. 1; leucophearia*.
Ge. leucophearia. (*Wien. V.* 101.)—*Hüb. Ge. pl.* 37. *f.* 195. —*Haw.* 279.—(*Sam. I.* 18.)
Spring Usher. *Harr. A. pl.* 43. *f. m, n, o, q.*—*Harr. V. M.* 57.
β; Ge. nigricaria: dark-bordered Usher.—*Haw. l. c.* *Hüb. Ge. pl.* 35. *f.* 181.—*Haw.* 279.—(*Sam. I.* 18.)
Ge. marmorinaria. *Esper.* v. *pl.* xxxvii. *f.* 1.
♀ β, Ge. luctuaria: mourning Widow. *Haw.* 279.—(*Sam. I.* 18.) *Albin. pl.* cxvii. *f. a–d?*

6447. 2; Æscularia*.
Ge. Æscularia. (*Wien. V.* 101.)—*Hüb. Ge. pl.* 36. *f.* 189.—*Haw.* 306.—(*Sam. I.* 18.)
Ge. murinaria. *Esper.* v. *pl.* xxxii. *f.* 5, 6.
Ge. cineraria. *Marsh. MSS.*—(*Haw. Pr.* 23.)

Genus 166: (949). LAMPETIA *mihi.*

FIDONIA *p, Treit.*

6448. 1; stictaria*.
Ge. stictaria: dotted border. *Marsh. MSS.*—(*Haw. Pr.* 21.) —*Haw.* 286.
Ge. progemmaria. *Hüb. Ge. pl.* 35. *f.* 183.
Ge. capreolaria. *Esper.* v. *pl.* xxxvi. *f.* 8, 9.
Ge. strictaria. (*Sam. I.* 19.)
Cross wing. *Harr. V. M.* 59.
Albin. pl. xliv. *f.* 72. *a–d.*

6449. [2, connectaria*.]
Ge. connectaria: connecting Umbre.—*Haw.* 285.—(*Sam. I.* 18.)

6450. 3; prosapiaria*.
Ph. Ge. prosapiaria. *Linn.* ii. 864.—*Turt.* iii. 238.
Ge. prosapiaria: scarce Umbre. (*Haw. Pr.* 21.)—*Haw.* 285. —(*Sam. I.* 19.)

Ge. aurantiaria. *Hüb. Ge. pl.* 35. *f.* 184.
β, Ph. Ge. testacearia. *Vill. E.* ii. 321.—(*Haw. Pr. App.* 4.)

6451. 4; defoliaria*.
Ph. Ge. defoliaria. *Linn.* ii. 225.—*Stew.* ii. 167.—*Turt.* iii. 240.
Ge. defoliaria. *Hüb. Ge. pl.* 35. *f.* 182.—(*Haw. Pr.* 21.)—*Haw.* 284.—(*Sam. I.* 18.)
Ph. pulveraria. *Berl. M.* (*Hufnagle.*) iv. 518.
Mallow M. or Mottled Umbre. *Harr. A. pl.* 14. *f. n–r.*
Mottled Umbre. (*Wilkes D. pl.* 8. *f.* 6.)—Ph. defoliata. (*Wilkes.* 36. *pl.* 72.)—*Harr. V. M.* 53.
Albin. pl. c. *f. e–h.*
β; Ph. diversaria. *Fabr.* iii. *b.* 157. (!)?—*Turt.* iii. 244.

Genus 167: (950). AMPHIDASIS *mihi.*

AMPHIDASIS *p, Treit.*—BISTON *p, Samou.*

6452. 1; pilosaria*. (*Och. Tr.* v. *b.* 434.)—*Id.* vi. *a.* 240.
Ge. pilosaria. (*Wien. V.* 100.)—*Hüb. Ge. pl.* 34. *f.* 176.
Ge. plumaria. *Esper.* v. *pl.* xxxv. *f.* 1, 2.
Ph. pedaria. *Fabr. E. S.* iii. *b.* 148.
Ge. pedaria. (*Haw. Pr.* 20.)—*Haw.* 274.
Bi. pedaria. (*Sam. I.* 6.)
Ge. hyemaria. *Bork. E. S.* v. 193.
Albin. pl. xci. *f. e–i.*
Brindle M. (*Wilkes.* 42. *pl.* 87.)
Pale brindled Beauty. (*Harr. V. M.* 15.)—(*Id.* 17.)

6453. 2; hispidaria*. (*Och. Tr.* v. *b.* 434.)—*Id.* vi. *a.* 247.
Ge. hispidaria. (*Wien. V.* 99.)—(*Haw. Pr.* 20.)—*Haw.* 274.
Bi. hispidaria. (*Sam. I.* 6.)
Ph. Ursularia. *Don.* xiii. 31. *pl.* 447.
Small brindled Beauty. *Harr. V. M.* 15.

Genus 168: (951). BISTON, *Leach, Samou.*, (*Kirby.*)

AMPHIDASIS *p, Treit.*

6454. 1; prodromarius*. (*Leach E. E.* ix. 134.)—(*Samou.* 253.)
Ph. prodromaria. (*Wien. V.* 99.)—*Fabr.* iii. *b.* 159.—*Don.* vii. 7. *pl.* 219.—*Stew.* ii. 165.—*Turt.* iii. 245.
Ge. prodromaria. (*Haw. Pr.* 20.)—*Haw.* 272.
Ge. marmoraria. *Esper.* v. *pl.* xxviii. *f.* 1–8.
Ph. strataria. *Berl. M.* (*Hufnagle.*) iv. 514.
Oak Beauty. (*Wilkes D. pl.* 5. *f.* 5.)—*Harr. Ex.* 53. *pl.* xiii. *f.* 4.—*Harr. V. M.* 11.

6455. 2; Betularius*. (*Leach E. E.* ix. 134.)—(*Samou.* 253.)
Ph. Ge. Betularia. *Linn.* ii. 862.—*Berk. S.* i. 144.—*Don.* vii. 55. *pl.* 237.—*Stew.* ii. 165.—*Turt.* iii. 245.
Ge. Betularia. (*Haw. Pr.* 20.)—*Haw.* 272.

Ge. Ulmaria. *Bork. E. S.* v. 181.
Albin. pl. xl. *f.* 64. *a–d.* ♂.—*pl.* xli. *f.* 66. *a–d.* ♀.—*Id. pl.* xci. *f. a–d.*—*Id. pl.* xcii. *f. a–d.*
Spotted Elm M.: Ph. Betularia. (*Wilkes.* 38. *pl.* 77.)
Peppered M. (*Wilkes D. pl.* 5. *f.* 1.)—*Harr. A. pl.* 18. *f. a–f.* —*Harr. V. M.* 39.

6456. 3; hirtarius*. (*Leach E. E.* ix. 134.)—(*Samou.* 253.)
Ph. Ge. hirtaria. *Linn. F. No.* 1236.—*Turt.* iii. 241.
Ge. hirtaria. *Hüb. Ge. pl.* 33. *f.* 175.—(*Haw. Pr.* 20.)—*Haw.* 273.
Ph. atomaria. *Berl. M.* (*Hufnagle.*) iv. 510.
Albin. pl. xxxix. *f.* 62. *a–d.*
Brindled Beauty. (*Wilkes D. pl.* 11. *f.* 11.)—(*Wilkes* 35. *pl.* 70.)—*Harr. A. pl.* 9. *f. a–f.*—*Harr. V. M.* 11.—*Id.* 15.
β; Ge. congeneraria: forked striped Brindle. *Hüb. Ge. pl.* 33. *f.* 174?—*Haw.* 273.—(*Sam. I.* 18.)
Ge. contiguaria. *Bork. E. S.* v. 187.
γ, Ge. fumaria: dark Brindle. *Haw.* 273.—(*Sam. I.* 18.)

Genus 169: (952). METRA *mihi.*

CROCALLIS *p, Treit.*

6457. 1; pennaria*.
Ph. Ge. pennaria. *Linn.* ii. 861.—*Don.* viii. 86. *pl.* 287. *f.* 2. —*Stew.* ii. 163.—*Turt.* iii. 233.
Ge. pennaria. *Hüb. Ge. pl.* 3. *f.* 14.—(*Haw. Pr.* 22.)—*Haw.* 290.—(*Sam. I.* 19.)
October M.: Ph. penriaria. (*Wilkes.* 38. *pl.* 79.)—*Harr. V. M.* 39?
November. *Harr. A. pl.* 43. *f. g–i.*—*Harr. V. M.* 37.
† β, Ge. bifidaria: the feathered Thorn. *Haw.* 291 *et* 301.

Genus 170: (953). CROCALLIS *mihi.*

CROCALLIS *p, Treit.*

6458. 1; elinguaria*. (*Och. Tr.* v. *b.* 431.)—*Id.* vi. *a.* 153.
Ph. Ge. elinguaria. *Linn.* ii. 862.
Ge. elinguaria. *Hüb. Ge. pl.* 4. *f.* 20.—(*Haw. Pr.* 22.)—*Haw.* 291.—(*Sam. I.* 18.)
Albin. pl. xxxix. *f.* 63. *e–h.*—*Id. pl.* xlii. *f.* 67. *e–h.*
Scolloped Oak. (*Harr. V. M.* 43.)

Genus 171: (954). ——

ENNOMOS D. *p, Treit.*

6459. 1; bidentata*.
Ph. Ge. bidentata. *Linn. F. No.* 1255.
Ge. bidentaria: scoloped Hazel. (*Haw. Pr.* 22.)—*Haw.* 291. —*Turt.* iii. 233.—(*Sam. I.* 18.)

Ge. dentaria. *Hüb. Ge. pl.* 3. *f.* 12. ♀.—*Turt.* iii. 235.
Ph. Ge. dentaria. *Fabr.* iii. *b.* 137. (!)—*Stew.* ii. 161.
Ge. obscurata. (*Wien. V.* 104?)
Albin. pl. xcvi. *f. a–c.*
Scolloped winged broad-bar. *Harr. A. pl.* 10. *f. s–z.*—*Harr. V. M.* 11.

Genus 172: (955). GEOMETRA, *Leach? Samou.*, (*Kirby.*)
ENNOMOS D. *p, Treit.*

6460. 1; Alniaria*.
Ph. Ge. Alniaria. *Linn.* ii. 860.—*Stew.* ii. 161.—*Turt.* iii. 234.
Ge. Alniaria: Canary-shouldered Thorn. (*Haw. Pr.* 22.)—*Haw.* 293.—(*Sam. I.* 18.)

6461. 2; Canaria*.
Ph. Ge. Canaria. *Hüb. B.* ii. *pl.* 4. *f. Y.* 1, 2.
Ge. Tiliaria. *Hüb. Ge. pl.* 5. *f.* 23.—(*Haw. Pr.* 22.)
Ge. Alniaria β. *Haw.* 294.

6462. 3; ochraria* *mihi.*

6463. 4; erosaria*: September Thorn.—*Haw. l. c.* (*Haw. Pr.* 22.)—*Haw.* 293.—(*Sam. I.* 18.)
Ge. erosaria. (*Wien. V.* 103.)—*Hüb. Ge. pl.* 5. *f.* 25.
Ph. crassaria. *Fabr.* iii. *b.* 135?
Ge. Tiliaria. *Esper.* v. *pl.* x. *f.* 3–7?

6464. 5; Quercinaria*: plain August Thorn.—*Haw. l. c. Hüb. Ge. pl.* 5. *f.* 24.—(*Haw. Pr.* 22.)—*Haw.* 294.—(*Sam. I.* 19.)
♀. Ge. Alniaria. *Hüb. Ge. pl.* 5. *f.* 26.
Ph. Ge. Alniaria. *Don.* viii. 61. *pl.* 275. *f.* 1.
Ge. Tiliaria: freckled August Thorn. *Haw.* 294.—(*Sam. I.* 19.)

6465. 6; Quercaria*. *Hüb. Ge. pl.* 80. *f.* 411, 412?
Ge. angularia: clouded August Thorn. (*Haw. Pr.* 22.)—*Haw.* 294.—(*Sam. I.* 18.)

6466. 7; angularia*. (*Wien. V.* 103.)—*Hüb. Ge. pl.* 5. *f.* 22.

6467. [8; Carpiniaria*]: flounced Thorn.—*Haw. l. c. Hüb. Ge. pl.* 5. *f.* 27.—(*Haw. Pr.* 22.)—(*Sam. I.* 18.)
Ph. erosaria. *Esper.* v. *pl.* xi. *f.* 1, 2.
Ge. Carpiniaria β. *Haw.* 295.
Ph. dentaria. *Villers E.* ii. 301.

6468. 9; fuscantaria*. (*Haw. Pr.* 22.)
Ge. Carpiniaria α. *Haw.* 295.

6469. 10; illunaria*: early Thorn.—*Haw. l. c. Hüb. Ge. pl.* 7. *f.* 36, 37.—(*Haw. Pr.* 22.)—*Haw.* 292.—(*Sam. I.* 18.)
Ge. lunaria var. 2, 3. (*Wien. V.* 282.)
Ge. bilunaria. *Esper.* v. *pl.* xiii. *f.* 1, 2.
Ge. unilunaria. *Esper.* v. *pl.* xiv. *f.* 6.

Ph. Ge. ustularia. *Don.* iii. 25. *pl.* 82?—*Stew.* ii. 162?—
Turt. iii. 233.
Scollop-winged M. (*Wilkes.* 39. *f.* 31.)

6470. [11; juliaria*]: July Thorn. *Haw.* 293.—(*Sam. I.* 18.)
Ge. lunaria var. 4. (*Wien. V.* 282.)
Ge. fulvolunaria. *Esper.* v. *pl.* xiv. *f.* 9, 10?
Albin. pl. xlii. *f.* 68. *a–d.*

6471. 12; lunaria*: lunar Thorn.—*Haw. l. c.* (*Wien. V.* 103.)
—*Hüb. Ge. pl.* 7. *f.* 33.—(*Haw. Pr.* 22.)—*Haw.* 292.—(*Leach E. E.* ix. 134.)—(*Sam. I.* 18.)
Ph. lunaria. *Fabr.* iii. *b.* 136.—*Don.* iv. 63. *pl.* 132.—*Stew.* ii. 161.—*Turt.* iii. 235.
Ge. lunularia. *Hüb. B.* i. 27. *pl.* 3. *f. T.* 1, 2.
Albin. pl. xv. *f. a–d.*
Hawthorn M. (*Wilkes.* 40. *pl.* 83.)

6472. 13: delunaria*. *Hüb. Ge. pl.* 7. *f.* 34.—(*Ing. Inst.* 90.)

6473. 14, sublunaria *mihi.*

6474. 15; illustraria*: purple Thorn.—*Haw. l. c. Hüb. Ge. pl.* 7. *f.* 35.—(*Haw. Pr.* 22.)—*Haw.* 291.—(*Sam. I.* 18.)
Ge. lunaria var. 5. (*Wien. V.* 283.)
Ge. quadrilunaria. *Esper.* v. *pl.* xii. *f.* 5, 6.
Ph. Phœbaria. *Schrank B.* ii. 13.
Ph. tetralunaria. *Berl. M.* (*Hufnagle.*) iv. 506.
Albin. pl. xlii. *f.* 69. *e–h.*

Genus 173: (956). PERICALLIA *mihi,* (*Kirby.*)

ENNOMOS D. *p, Treit.*

6475. 1; Syringaria*: Lilac Beauty.—*Haw. l. c.*
Ph. Ge. Syringaria. *Linn.* ii. 860.—*Berk. S.* i. 144.—*Stew.* ii. 161.—*Turt.* iii. 234.—*Don.* vi. 3. *pl.* 181.
Ge. Syringaria. *Hüb. Ge. pl.* 6. *f.* 29.—(*Haw. Pr.* 22.)—*Haw.* 293.—(*Sam. I.* 19.)
Richmond Beauty: Ph. Syringaria. (*Wilkes D. pl.* 10. *f.* 9.) —(*Wilkes.* 39. *pl.* 82.)—*Harr. V. M.* 13.

Genus 174: (957). ——

ENNOMOS D. *p, Treit.*—HIPPARCHUS *p, Leach, Samou.*

6476. 1; Prunaria*.
Ph. Ge. Prunaria. *Linn.* ii. 861.—*Berk. S.* i. 144.—*Don.* i. 47. *pl.* 27.—*Stew.* ii. 162.—*Turt.* iii. 237.
Ge. Prunaria. *Hüb. Ge. pl.* 23. *f.* 123. ♂. *f.* 122. ♀.—(*Haw. Pr.* 22.)—*Haw.* 283.—(*Sam. I.* 19.)
Hi. Prunaria. (*Leach E. E.* ix. 134.)—(*Samou.* 253.)
Hi. Prunatus. (*Sam. I.* 22.)
Ph. fulvularia. *Berl. M.* (*Hufn.*) iv. 518.

Ph. corticalis. *Scop. C.* 216.
Albin. pl. xliii. *f.* 70. *a–c.*—*Id. pl.* c. *f. a–d.*
Orange M.: Ph. Prunaria. (*Wilkes.* 40. *pl.* 84.)—*Harr. V. M.* 37.
β, Ge. Corylaria. *Esper.* v. *pl.* xviii. *f.* 1, 3.
Ph. Prunaria var. *Don.* ix. 13. *pl.* 293. *f.* 3.
γ, Ge. dimidiata. *Fabr.* iii. *b.* 169. (!)—*Stew.* ii. 162.—*Turt.* ii. 250.
Ph. sordiata. *Schrank B.* ii. 19.

Genus 175: (958). ——

ENNOMOS D. *p*, *Treit.*, (*Kirby.*)

6477. 1; Cratægata*.
Ph. Ge. Cratægata. *Linn.* ii. 868.—*Berk. S.* i. 145.—*Stew.* ii. 168.
Ge. Cratægaria. *Hüb. Ge. pl.* 6. *f.* 32.—*Turt.* iii. 254.—(*Haw. Pr.* 26.)—*Haw.* 298.—(*Sam. I.* 18.)—(*Kirby & Sp. I. E.* iii. 97.)
Albin. pl. xl. *f.* 65. *e–h.*—*Id. pl.* xlvi. *f.* 76. *f.* 77. *a–e.*—*Id. pl.* xcv. *f. e–h.*
Brimstone M.: Ph. Cratægata. (*Wilkes.* 39. *pl.* 80.)—*Harr. A. pl.* 29. *f. g, l, m.*
Yellow or Brimstone. *Harr. V. M.* 59.

Genus 176: (959). OURAPTERYX, *Leach, Samou.*

ACÆNA, *Treit.*—URAPTERYX, (*Kirby.*)

6478. 1; Sambucaria*. (*Leach E. E.* ix. 134.)—(*Leach Z. M.* i. 80. *pl.* 35. *f.* 2.—(*Samou.* 253.)
Ph. Ge. Sambucaria. *Linn.* ii. 860.—*Berk. S.* i. 144.—*Stew.* ii. 160.—*Turt.* iii. 233.—*Don.* v. 85. *pl.* 170.
Ge. Sambucaria. *Hüb. Ge. pl.* 6. *f.* 28.—(*Haw. Pr.* 21.)—*Haw.* 297.—(*Sam. I.* 19.)
Ph. No. 10. *List. Goëd.* (*Lat.*) 26. *f.* 10.—*Id.* (*Angl.*) 14. *f.* 10.
Albin. pl. xciv. *f. a–d.*
Swallow-tail M.: Ph. Sambucaria. (*Wilkes.* 38. *pl.* 78.)—*Harr. V. M.* 47.

Genus 177: (960). PHALÆNA, *Latreille, Leach, Samou.*

ELLOPIA A. *p*, *Treit.*—BOMBYX *p*, *Esper.*—CAMPÆA *p*, *Lamarck.*

6479. 1; margaritaria*. *Fabr.* iii. *b.* 131.—(*Leach E. E.* ix. 134.)—(*Samou.* 252.)
Ph. Ge. margaritata. *Linn.* ii. 568. (865.)—*Don.* xvi. 5. *pl.* 543.
Ge. margaritaria: the light Emerald.—*Haw. l. c.* *Hüb. Ge. pl.* 3. *f.* 13.—(*Haw. Pr.* 21.)—*Haw.* 299.—(*Sam. I.* 18.)

Noctua! margaritaria. (*Sam. I.* 29.)
Bo. sesquitriàta. *Bork. E. S.* iii. 454.
Bo. sesquitriataria. *Esper.* v. iii. *pl.* lii. *f.* 1, 2.
Ge. bupleuraria. *Villers E.* iv. 495.
Ge. vernaria. *Berl. M.* (*Hufman.*) iv. 506.
Cross-barred Housewife. *Harr. V. M.* 31.

Genus 178: (961). ELLOPIA *mihi.*

ELLOPIA B, *Treit.*—CAMPÆA *p, Lamarck.*

6480. 1; fasciaria*. (*Och. Tr.* v. *b.* 430.)—*Id.* vi. *a.* 97.
Ph. Ge. fasciaria. *Linn.* ii. 862.—*Turt.* iii. 244.
Ge. fasciaria: barred Red.—*Haw. l. c.* *Hüb. Ge. pl.* 1. *f.* 5. *pl.* 87. *f.* 446, 447.—(*Haw. Pr.* 23.)—*Haw.* 301.—(*Sam. I.* 18.)
Ph. Neustriaria. *Berl. M.* (*Hufnagle.*) iv. 520.

Genus 179: (962). HIPPARCHUS, *Leach, Samou.*

GEOMETRA, *Treit.*—PHALÆNA *p, Latr.*

6481. 1; Papilionarius*. (*Leach E. E.* ix. 134.)—(*Samou.* 253.)
Ph. Ge. Papilionaria. *Linn.* ii. 864.—*Berk. S.* i. 144.—*Don.* viii. 85. *pl.* 287. *f.* 1.—*Stew.* ii. 164.—*Turt.* iii. 236.
Ge. Papilionaria. *Hüb. Ge. pl.* 2. *f.* 6.—(*Haw. Pr.* 21.)—*Haw.* 298.—(*Sam. I.* 19.)
The large Emerald. *Harr. Ex.* 52. *pl.* xiii. *f.* 1.—*Harr. V. M.* 23.

6482. 2; vernarius*.
Ph. Ge. vernaria. *Linn.* ii. 858.—*Stew.* ii. 159.—*Turt.* iii. 231.
Ge. vernaria. *Hüb. Ge. pl.* 2. *f.* 7.
Ge. chrysoprasaria. *Esper.* v. *pl.* v. *f.* 1–4.
Ge. Æruginaria. *Bork. E. S.* v. 43.
Ph. Ge. volutata. *Villers E.* ii. 373?—*Turt.* iii. 258?
Ge. volutaria. *Haw.* 298.—(*Sam. I.* 19.)
Ph. Ge. lucidata. *Don.* iii. 67. *pl.* 97.—*Stew.* ii. 163.
Ge. lucidata. (*Haw. Pr.* 21.)
The small Emerald M. *Harr. Ex.* 32. *pl.* 8. *f.* 8.—*Harr. V. M.* 23.

6483. 3, viridatus*.
Ph. Ge. viridata. *Linn.* ii. 865.—*Berk. S.* i. 145.—*Stew.* ii. 163.
Ge. viridaria. *Hüb. Ge. pl.* 2. *f.* 11.
Ge. vernaria: the small Grass Emerald. *Haw.* 300.—(*Sam. I.* 19.)
Albin. pl. xlviii. *f.* 80. *a–c.*
Small green Housewife. *Harr. V. M.* 31.

6484. 4, clorarius?
Ge. clorarius. *Hüb. Ge. pl.* 68. *f.* 352?
Ph. Ge. lactearia. *Linn.* ii. 858?

6485. 5; Cythisaria*.
Ge. Cythisaria. (*Wien. V.* 97.)—*Hüb. Ge. pl.* 1.*f.* 2.
Ph. prasinaria. *Fabr.* iii. *b.* 151.
Ge. prasinaria: Grass Emerald. (*Haw. Pr.* 21.)—*Haw.* 299. —(*Sam. I.* 19.)
Ph. pruinata. *Berl. M.* (*Hufnagle.*) iv. 520.
Ph. Genistaria. *Villers E.* ii. 328.
Green brown M.: Ph. Papilionaria? (*Wilkes.* 37. *pl.* 75.)
Green Carpet. *Harr. Ex.* 25. *pl.* vi. *f.* 5.

Genus 180: (963). CLEORA.

CLEORA *p, Curtis.*—GEOMETRA *p, et* BOARMIA *p, Treit.*

A. GEOMETRA B. *p, Treit.*

6486. 1; bajularia*.
Ge. bajularia. (*Wien. V.* 97.)—*Hüb. Ge. pl.* 1.*f.* 3.
Ph. ditaria. *Fabr.* iii. *b.* 152.—*Don.* vi. 51. *pl.* 202. *f.* 1.—*Stew.* ii. 166.—*Turt.* iii. 242.
Ge. ditaria: blotch'd Emerald. (*Haw. Pr.* 21.)—*Haw.* 299.—(*Sam. I.* 18.)
Ph. pustulata. *Berl. M.* (*Hufnagle.*) iv. 520.
Maid of Honor M. *Harr. A. pl.* 41.*f. r.*—*Harr. V. M.* 31.

B. BOARMIA *p, Treit.*

6487. 2; Lichenaria*.
Ge. Lichenaria. (*Wien. V.* 100.)—*Hüb. Ge. pl.* 31.*f.* 164.—(*Haw. Pr.* 20.)—*Haw.* 280.—*Don.* x. 41. *pl.* 342. *f.* 1.—(*Sam. I.* 18.)
Ge. cineraria. *Bork. E. S.* v. 165.
Lime-Moss M. (*Wilkes.* 37. *pl.* 76.)
Brussels' lace. *Harr. V. M.* 33.

6488. 3: viduaria. (*Ing. Inst.* 90.)
Ge. viduaria. (*Wien. V.* 101.)—*Hüb. Ge. pl.* 31. *f.* 165. *pl.* 70. *f.* 364.
Ph. angularia. *Thunb. I. S.* iv. 59. *fig. adj.*

6489. 4, teneraria. (*Ing. Inst.* 90.)
Ge. teneraria. *Hüb. Ge. pl.* 67. *f.* 348.
Cl. tæneraria. (*Curtis l. c. infra.*)
Ge. glabraria. *Hüb. Ge. pl.* 65. 339. ♂?—*Id. pl.* 31. *f.* 162. ♀?

6490. 5: cinctaria. *Curtis.* ii. *pl.* 88.—(*Ing. Inst.* 90.)
Ge. cinctaria. (*Wien. V.* 101.)—*Hüb. Ge. pl.* 31. *f.* 166.
Ge. pascuaria. *Esper. Ge. pl.* xliii. *f.* 1, 2.

6491. 6: pictaria*.
Ge. pictaria. *Thunb. I. S.* i. 6. *fig. adj.*

Genus 181: (964). ALCIS, *Curtis.*

BOARMIA *p, Treit.*, (*Kirby.*)—HEMEROPHILA *p, mihi*: (*olim.*)

6492. 1; repandaria*. (*Curtis l. c. infra.*)
Ph. Ge. repandata. *Linn.* ii. 866.—*Stew.* ii. 164.—*Don.* x. 21. *pl.* 333. *f.* 1.
Ge. repandaria. (*Wien. V.* 101.)—*Hüb. Ge. pl.* 30. *f.* 161.—(*Haw. Pr.* 20.)—*Haw.* 275.—(*Sam. I.* 19.)
Mottled Beauty. *Harr. V. M.* 11.

6493. [2, destrigaria*.] (*Curtis l. c. infra.*)
Ge. destrigaria: light mottled Beauty. *Haw.* 275.—(*Sam. I.* 18.)
Ge. conspersaria. (*Haw. Pr.* 20.)

6494. 3, muraria. *Curtis l. c. infra.*

6495. 4: conversaria*. (*Curtis l. c. infra.*)
Ge. conversaria: large Carpet.—*Haw. l. c.* *Hüb. Ge. pl.* 62. *f.* 321. *pl.* 76. *f.* 393.—*Haw.* 302.—*Don.* xv. *pl.* 514.—*Ent. Trans.* (*Hatchett.*) i. 328. *pl.* 9. *f.* 3.—(*Sam. I.* 13.)

6496. 5: sericearia. *Curtis.* iii. *pl.* 113.—(*Ing. Inst.* 90.)

6497. 6; consortaria*. (*Curtis l. c. supra.*)
Ph. consortaria. *Fabr.* iii. *b.* 137.—*Don.* x. 22. *pl.* 333. *f.* 2.
Ge. consortaria. *Hüb. Ge. pl.* 32. *f.* 168.—(*Haw. Pr.* 20.)—*Haw.* 275.—(*Sam. I.* 18.)
Pale Oak Beauty. *Harr. V. M.* 11.

6498. 7: Roboraria*. (*Curtis l. c. supra.*)
Ph. Roboraria. *Fabr.* iii. *b.* 137.—*Don.* xv. *pl.* 527.
Ge. Roboraria. *Hüb. Ge. pl.* 32. *f.* 169.—*Haw.* 275.—(*Sam. I.* 19.)
Ph. leucophearia. *Villers E.* ii. 323.
Ge. grandaria: the great Oak Beauty. (*Haw. Pr.* 20.)
Albin. pl. xciii. *f. a–d.*

6499 [† 8, piperitaria.]
Ge. piperitaria. *Haw. MSS.* In Mus. D. *Haworth.*

6500. 9; rhomboidaria*. (*Curtis l. c. supra.*)
Ge. rhomboidaria. (*Wien. V.* 101.)—*Hüb. Ge. pl.* 29. *f.* 154. *pl.* 32. *f.* 170. *pl.* 95. *f.* 488.—*Haw.* 276.—(*Sam. I.* 19.)
Ge. gemmaria. *Bork. E. S.* v. 156.
Noctua! rhomboidaria. (*Sam. I.* 30.)
Willow Beauty. *Harr. V. M.* 13.

6501. 10, consobrinaria: tawny Beauty.—*Haw. l. c.*
Ge. consobrinaria. *Hüb. Ge. pl.* 29. *f.* 152?—*Haw.* 276.—(*Sam. I.* 18.)
Al. australaria. *Curtis l. c. supra.*
Ge. Devoniaria. *Haworth MSS.*

6502. 11, fimbriaria *mihi;* (nec *Hübneri:* vide *Curtis l. c. supra.*)

Genus 182: (965). HEMEROPHILA *mihi.*

ACIDALIA *p, Treit.*

6503. 1; abruptaria*.
Ph. abruptaria. *Thunb. I. S.* iii. 59. *pl.* 4. *fig. adj.*
Ph. Ge. suberaria. *Don.* vii. 91. *pl.* 251. *f.* 1.—*Stew.* ii. 162. —*Turt.* iii. 236.
Ge. petrificata. *Hüb. Ge. pl.* 52. *f.* 267.
Ge. suberaria. (*Haw. Pr.* 21.)—*Haw.* 284.—(*Sam. I.* 19.)
Waved Umbre. (*Wilkes D. pl.* 11. *f.* 10.)—*Harr. V. M.* 53.

Genus 183: (966). BOARMIA *mihi.*

BOARMIA *p, Treit.*—CLEORA *p, Curtis folio* 88.

6504. 1; tetragonaria*.
Ge. tetragonaria. *Haworth. MSS.*
Cl. tetragonaria. (*Curtis l. c. supra.*)
Bo. trigonaria *mihi olim.*

6505. 2; Abietaria*. (*Och. Tr.* v. *b.* 433.)—*Id.* vi. *a.* 204.
Ge. Abietaria. (*Wien. V.* 101.)—*Hüb. Ge. pl.* 30. *f.* 160.—*Haw.* 276.—(*Sam. I.* 18.)
Cl. Abietaria. (*Curtis l. c. supra.*)
Ge. Gemmaria Abietis. *Esper.* v. *pl.* lii. *f.* 5.
Ingrailed. *Harr. V. M.* 33.

6506. 3; crepuscularia*. (*Och. Tr.* v. *b.* 433.)—*Id.* vi. *a.* 190.
Ge. crepuscularia: small Ingrailed.—*Haw. l. c.* (*Wien. V.* 101.)—*Hüb. Ge. pl.* 30. *f.* 158.—*Haw.* 277.—(*Sam. I.* 18.)
Cl. crepuscularia. (*Curtis l. c. supra.*)
Ge. biundularia. *Esper.* v. *pl.* xl. *f.* 3, 4.
Ph. similaria. *Berl. M.* (*Hufnagle.*) iv. 512.

6507. 4, consonaria*. (*Och. Tr.* v. *b.* 433.)
Ge. consonaria: brindled Grey.—*Haw. l. c.* *Hüb. Ge. pl.* 30. *f.* 157.—*Haw.* 277.—(*Sam. I.* 18.)
Cl. consonaria. (*Curtis l. c. supra.*)

6508. 5; strigularia* *mihi.*

6509. 6; extersaria*. (*Och. Tr.* v. *b.* 433.)—*Id.* vi. *a.* 216.
Ge. extersaria: brindled White-spot.—*Haw. l. c.* *Hüb. Ge. pl.* 30. *f.* 159.—*Haw.* 277.—(*Sam. I.* 18.)
Cl. extersaria. (*Curtis l. c. supra.*)

6510. 7; punctularia*: Grey Birch.—*Haw. l. c.*
Ge. punctularia. *Hüb. Ge. pl.* 61. *pl.* 317.—(*Haw. Pr. App.* 4.)—*Haw.* 277.—(*Sam. I.* 18.)
Cl. punctularia. (*Curtis l. c. supra.*)
Ge. punctulata. (*Wien. V.* 101.)—*Bork. E. S.* v.

Genus 184: (967). GRAMMATOPHORA *mihi.*

FIDONIA *p, Treit.*

6511. 1; Vauaria*.
Ph. Ge. Vauaria. *Linn.* ii. 863.—*Berk. S.* i. 144.—*Don.* vi. 37. *pl.* 196.—*Stew.* ii. 166.—*Turt.* iii. 241.
Ge. Wauaria: the common V. Moth.—*Haw. l. c.* *Hüb. Ge. pl.* 11. *f.* 55.—*Haw.* 283.
Ge. Vauaria. (*Haw. Pr.* 21.)—(*Sam. I.* 19.)
Ge. Viduaria. *Müll. F. F.* 49.
Pa. No. 12. *List. Goëd.* (*Angl.*) 19. *f.* 12.—*Id.* (*Lat.*) 19. *f.* 12.
Albin. pl. xlvii. *f.* 78. *a–d.*
Gooseberry M.: Ph. Wauaria. (*Wilkes.* 41. *pl.* 86.)
L. Moth. *Harr. A. pl.* 34. *f. g–i.*—*Harr. V. M.* 35.

6512 † [2, Vau nigraria*]: Sooty V.—*Haw. l. c.*
Ge. Vau nigraria: the slender-bodied Black V. *Ent. Trans.* (*Hatchett.*) i. 244. *pl.* 7. *f.* 3.—*Haw.* 282.—(*Sam. I.* 19.)

Genus 185: (968). AZINEPHORA *mihi.*

FIDONIA *p, Treit.*

6513. 1; pulveraria*: barred Umbre.—*Haw. l. c.*
Ph. Ge. pulveraria. *Linn.* ii. 862.—*Turt.* iii. 244.
Ge. pulveraria. *Hüb. Ge. pl.* 39. *f.* 203.—(*Haw. Pr.* 23.)—*Haw.* 301.—(*Sam. I.* 19.)
Albin. pl. xcvi. *f. d–f.*
Freckled broad bar. *Harr. A. pl.* 42. *f. o.*

Genus 186: (969). CABERA *mihi.*

CABERA *p, Treit.*

6514. 1; pusaria*. (*Och. Tr.* v. *b.* 437.)—*Och. Tr.* vi. *a.* 344.
Ph. Ge. pusaria. *Linn.* ii. 864.—*Stew.* ii. 167.—*Turt.* iii. 239.
Ge. pusaria: common white Wave.—*Haw. l. c.* *Hüb. Ge. pl.* 17. *f.* 87.—(*Haw. Pr.* 23.)—*Haw.* 290.—(*Sam. I.* 19.)
Ph. strigata. *Scop. C.* 218.
Albin. pl. cxviii. *f. e–h.*
Pale waved. *Harr. A. pl.* 44. *f. h.*—*Harr. V. M.* 57.

6515. 2; rotundaria*.
Ge. rotundaria: round winged Wave. *Haw.* 289.—(*Sam. I.* 19.)

6516. 3; exanthemata*.
Ph. exanthemata. *Scop. C.* 218.
Ge. exanthemaria. *Esper.* v. *pl.* xxxiii. *f.* 3, 4.
Ge. striaria: common Wave.—*Haw. l. c.* *Hüb. Ge. pl.* 17. *f.* 88.—(*Haw. Pr.* 23.)—*Haw.* 289.—(*Sam. I.* 19.)
Albin. pl. xcii. *f. e–h.*—*Id. pl.* xcix. *f. a–d.*
β; Ge. arenosaria: sandy Wave. *Haw.* 289.—(*Sam. I.* 18.)
γ? Ge. approximaria: twin-striped Wave. (*Haw. Pr.* 23.)—*Haw.* 289.

Genus 187: (970). CYCLOPHORA *mihi*, (*Kirby*.)

CABERA *p*, *Treit*.

6517. 1; omicronaria*: Mocha.—*Haw. l.* c.
Ge. omicronaria. (*Wien. V.* 107.)—*Hüb. Ge. pl.* 13. *f.* 65.—(*Haw. Pr.* 24.)—*Haw.* 312.—(*Sam. I.* 18.)
Ph. annularia. *Fabr. E. S.* iii. *b.* 147.

6518. 2; ocellaria*: false Mocha.—*Haw. l.* c.
Ge. ocellaria. *Hüb. Ge. pl.* 13. *f.* 64.—(*Haw. Pr.* 24.)—*Haw.* 312.—(*Sam. I.* 18.)
Ge. albiocellaria. *Esper.* v. *pl.* xliii. *f.* 7?
Ge. Hübneraria. (*Haw. Pr.* 24.)
Mocha stone. *Harr. V. M.* 45.
Albin. pl. l. *f.* 85. *a–d.*
β? Ge. punctaria. *Hüb. Ge. pl.* 13. *f.* 67.
Ph. Ge. porata. *Linn.* ii. 866?—*Stew.* ii. 164?

6519. 3; pendularia*: Birch Mocha.—*Haw. l. c.*
Ph. Ge. pendularia. *Linn. F. No.* 1244.
Ge. pendularia. *Hüb. Ge. pl.* 13. *f.* 66.—(*Haw. Pr.* 24.)—*Haw.* 311.—(*Sam. I.* 19.)
Pale Mocha Stone. *Harr. V. M.* 45.
Albin. pl. xlix. *f.* 82. *f. a–d.*
β; Ph. circularia. *Fabr. Sup.* 451.
Ph. albipunctata. *Berl. M.* (*Hufnagle.*) iv. 526.
Buff Argus M.: Ph. amataria. (*Wilkes.* 37. *pl.* 74. *fig. sup.*)—*Harr. V. M.* 9.

6520 † 4, albicinctata*.
Ph. albicinctata: the white-lined Black. *Haw.* 344.
In Mus. D. *Hatchett.*

6521. 5; orbicularia*: dingy Mocha.—*Haw. l. c.*
Ge. orbicularia. *Hüb. Ge. pl.* 13. *f.* 60.—(*Haw. Pr.* 24.)—*Haw.* 311.—(*Sam. I.* 19.)

6522. 6; punctaria*.
Ph. Ge. punctaria. *Linn.* ii. 859.—*Stew.* ii. 160.—*Turt.* iii. 232.
Ge. punctaria. (*Haw. Pr.* 23.)—*Haw.* 312.—(*Sam. I.* 19.)
Ge. fultaria. *Villers E.* ii. 330?
Ph. Ge. communifasciata. *Don.* xiii. 47. *pl.* 456.
Buff Argus M.: Ph. amataria. (*Wilkes.* 37. *pl.* 74. *fig. inf.*)
Maiden's Blush. *Harr. V. M.* 13.
β? Ge. subangularia: scarce Angled. *Haw.* 313.

6523. 7: trilinearia*: Clay Triple-lines.—*Haw. l. c.*
Ge. trilinearia. *Bork. E. S.* v. 502.
Ge. linearia. *Hüb. Ge. pl.* 13. *f.* 68.—(*Haw. Pr.* 23.)—*Haw.* 314.—(*Sam. I.* 18.)

Genus 188: (971). BRADYEPETES *mihi.*

ENNOMOS B. *p, Treit.*, (*Kirby.*)

6524. 1; amataria*.
Ph. Ge. amataria. *Linn.* ii. 859.—*Berk. S.* i. 143.—*Stew.* ii. 160.—*Turt.* iii. 232.—(*Kirby & Sp. I. E.* iii. 151. *nota.*)
Ph. Ge. amatorio: Buff Argus. *Don.* ii. 74. *pl.* 33. *f.* 2.
Ge. amataria. *Hüb. Ge. pl.* 10. *f.* 52.—(*Haw. Pr.* 22.)—*Haw.* 296.—(*Sam. I.* 18.)
Ph. vibicaria. *Berl. M.* (*Hufnagle.*) iv. 514.
Blood vein. *Harr. V. M.* 53.

6525. 2; advenaria*: little Thorn.—*Haw. l. c.*
Ge. advenaria. *Hüb Ge. pl.* 9. *f.* 45.—(*Haw. Pr.* 23.)—*Haw.* 296.—(*Sam. I.* 18.)

6526 † 3. dilectaria[a].

6527. 4; apiciaria*: bordered Beauty.—*Haw. l.* c.
Ge. apiciaria. (*Wien. V.* 104.)—*Hüb. Ge. pl.* 9. *f.* 47.—(*Haw. Pr.* 23.)—*Haw.* 295.
Ph. Ge. vespertaria. *Don.* vii. 47. *pl.* 233. *f.* 3.—*Stew.* ii. 167.—*Turt.* iii. 241.
Ge. apicaria. *Esper.* v. *pl.* xv. *f.* 3.
Ge. apriciaria. (*Sam. I.* 18.)
Ge. marginaria. *Villers E.* ii. 303. *pl.* vi. *f.* 3.
Essex broad border. (*Harr. V. M.* 15.)

6528. 5, vespertaria.
Ge. vespertaria. *Linn.* ii. 864.—*Thunb. I. S.* 5. *fig. adj.*
Ge. parallelaria. (*Wien. V.* 104.)—*Hüb. Ge. pl.* 9. *f.* 43, 44.
Ge. affiniaria. *Bork. E. S.* v. 136.
Ph. repandaria. *Berl. M.* (*Hufn.*) iv. 508.

6529. 6; dolabraria*.
Ph. Ge. dolabraria. *Linn.* ii. 861.—*Don.* x. 63. *pl.* 349. *f.* 1.—*Stew.* ii. 161.—*Turt.* iii. 235.
Ge. dolabraria. *Hüb. Ge. pl.* 8. *f.* 42.—(*Haw. Pr.* 23.)—*Haw.* 295.—(*Leach E. E.* ix. 134.)—(*Sam. I.* 18.)
Ph. Ustulataria. *Berl. M.* (*Hufn.*) iv. 516.
Scorched Wing. (*Wilkes D. pl.* 3. *f.* 1.)—*Harr. V. M.* 57.

Genus 189: (972). ASPILATES, *Treitschke.*

CABERA *p, Treit.*—ASPITATES, *Treit. olim.*

6530 † 1, purpuraria. (*Och. Tr.* v. *b.* 431.)—*Id.* vi. *a.* 127.
Ph. Ge. purpuraria. *Linn.* ii. 864.—*Stew.* ii. 167. (!)—*Turt.* iii. 247. (!)

---

[a] 6526 † 3. dilectaria: Small Thorn.—*Haw. l. c.*
Ge. dilectaria. *Hüb. Ge. pl.* 8. *f.* 38–39?—*Haw.* 296. (!)

Ge. purpuraria: purple-barred Yellow.—*Haw. l. c. Hüb. Ge. pl.* 38. *f.* 198, 199.—*Haw.* 310.
Ph. cruentaria. *Berl. M.* (*Hufnagle.*) iv. 516.
In Mus. D. *Swainson.*

6531. 2; citraria*. (*Och. Tr.* v. *b.* 431.)—*Id.* vi. *a.* 139.
Ge. citraria: yellow Belle.—*Haw. l. c. Hüb. Ge. pl.* 40. *f.* 212. —*Haw.* 288.—(*Sam. I.* 18.)
Ge. trifoliaria. (*Haw. Pr.* 23.)

6532. 3; gilvaria. (*Och. Tr.* v. *b.* 431.)—*Id.* vi. *a.* 132.
Ge. gilvaria: Straw Belle.—*Haw. l. c.* (*Wien. V.* 102.)—*Hüb. Ge. pl.* 38. *f.* 201.—*Haw.* 287.—(*Sam. I.* 18.)

6533. 4; plumbaria*.
Ph. plumbaria. *Fabr.* iii. *b.* 160. (!)—*Stew.* ii. 166.—*Turt.* iii. 246.
Ge. palumbaria. (*Wien. V.* 202.)—*Hüb. Ge. pl.* 42. *f.* 221.
Ge. plumbaria. (*Haw. Pr.* 23.)—*Haw.* 287.—(*Sam. I.* 19.)
Ph. mucronata. *Scop. C.* 222.
Bell or Belle. *Harr. V. M.* 13.

6534. 5; respersaria*: Grass Wave.—*Haw. l. c.*
Ge. respersaria: lesser Grass Wave.—*Haw. l. c. Hüb. Ge. pl.* 23. *f.* 125.—*Haw.* 289.—(*Sam. I.* 19.)
β; Ge. strigillaria: the ingrailed Wave.—*Haw. l. c. Hüb. Beyt.* i. 2. *K.*—*Haw.* 288.
γ; Ge. inæquaria: the larger Grass Wave. *Haw.* 288.—(*Sam. I.* 18.)
Ge. majuscularia. (*Haw. Pr.* 23.)

Genus 190: (973). LARENTIA *mihi.*

LARENTIA *p, Treit.*

6535. 1; cervinaria*. (*Och. Tr.* v. *b.* 440.)—*Och. Tr.* vi. *b.* 76.
Ge. cervinata. (*Wien. V.* 111.)—*Hüb. Ge. pl.* 62. *f.* 318.
Ph. Ge. clavaria. *Marsh. MSS.*
Ge. clavaria: Mallow M. (*Haw. Pr.* 22.)—*Haw.* 302.—(*Sam. I.* 18.)

6536. 2; Chenopodiata*: Aurelian's plague: *olim.*
Ph. Ge. Chenopodiata. *Linn.* ii. 868.—*Stew.* ii. 169.
Ge. Chenopodiaria: small Mallow M. (*Haw. Pr.* 22.)—*Haw.* 302.—(*Sam. I.* 18.)
Ge. mensuraria. (*Wien. V.* 111.)—*Hüb. Ge. pl.* 37. *f.* 193.
Ph. limitata. *Scop. C.* 228.
Ph. Ge. Chenipodiata. *Turt.* iii. 260.
Shaded broad bar. *Harr. A. pl.* 33. *f. n.*—*Harr. V. M.* 9.

6537. 3; bipunctaria*. (*Och. Tr.* v. *b.* 441.)—*Och. Tr.* vi. *b.* 87.
Ph. bipunctaria. *Fabr.* iii. *b.* 145.
Ge. bipunctaria: Chalk Carpet. *Haw.* 303.—(*Sam. I.* 18.)

Ge. bipunctata. *Hüb. Ge. pl.* 53. *f.* 276.
Ph. Undulata. *Scop. C.* 223?
Chalk-pit M. (*Harr. V. M.* 39.)

6538. 4; multistrigaria*.
Ge. multistrigaria: mottled Grey. *Haw.* 306.
Ge. multistrigata. (*Sam. I.* 18.)

Genus 191: (974). CIDARIA *mihi.*

CIDARIA *et* XERENE *p, Treit.*

6539. 1; didymata*.
Ph. Ge. didymata. *Linn.* ii. 872.
Ge. didymaria: twin-spot Carpet. (*Haw. Pr.* 23.)—*Haw.* 306. —(*Sam. I.* 18.)
Ge. scabrata. *Hüb. Ge. pl.* 44. *f.* 229?

6540. 2, munitata*.
Ge. munitata. *Hüb. Ge. pl.* 66. *f.* 346.—(*Sam. I.* 18.)
Ph. munitata: the rufous Carpet. *Haw.* 328?
Ph. Ge. tristrigaria. *Don.* xiii. 60. *pl.* 461. *f.* 2.

6541. 3; unidentaria*.
Ge. unidentaria: dark-barred Twin-spot. (*Haw. Pr.* 23.)—*Haw.* 308.—(*Sam. I.* 19.)
Ge. quadrifasciaria. *Hüb. Ge. pl.* 55. *f.* 284.
Albin. pl. xcvii. *f. e–h.*

6542. 4; quadrifasciaria*. (*Och. Tr.* v. *b.* 442?)—*Och. Tr.* vi. *b.* 146?
Ph. Ge. quadrifasciaria. *Linn. F. No.* 1253.
Ge. quadrifasciaria: large Twin-spot. (*Haw. Pr. App.* 4.)—*Haw.* 307.—(*Sam. I.* 19.)
Ge. badiata. *Bork. E. S.* v. 344.
β, Ge. Ligustrata. *Hüb. Ge. pl.* 55. *f.* 282.
γ, Ph. Ge. undulataria. *Villers E.* ii. 322. *pl.* vi. *f.* 6.
Ge. undulataria. (*Haw. Pr. App.* 4.)

6543. 5; ferrugaria*. (*Och. Tr.* v. *b.* 442.)—*Och. Tr.* vi. *b.* 148.
Ph. Ge. ferrugata. *Linn. F. No.* 1292.—*Turt.* iii. 263.
Ge. ferrugata. *Hüb. Ge. pl.* 55. *f.* 285.
Ge. ferrugaria: red Twin-spot. (*Haw. Pr.* 23.)—*Haw.* 308. —(*Sam. I.* 18.)
Ge. Alchemillaria. *Esper.* v. *pl.* xl. *f.* 5, 6.
Ge. corculata. *Berl. M.* (*Hufnagle.*) iv. 616.
Var. Ge. spadicearia. (*Wien. V.* 113.)
Ge. Linariaria. *Bork. E. S.* v. 381.

6544. [6, Salicaria*.]
Ge. Salicaria: striped Twin-spot. *Haw.* 309.—(*Sam. I.* 19.)

6545. 7, latenaria.
Ge. latenaria. (*Curtis Cat. No.* 45.)

6546. 8; miaria*. (*Och. Tr.* v. *b.* 443.)—*Och. Tr.* vi. *b.* 159.
Ge. miaria. (*Wien. V.* 113.)
Ge. miata. *Hüb. Ge. pl.* 57. *f.* 292.
Ge. miataria. *Esper.* v. *pl.* xlv. *f.* 2, 3.
Ph. viridata. *Fabr.* iii. *b.* 152. (!)—*Berk. S.* i. 145?—(*Haw. Pr.* 28?)
Ge. viridaria. *Villers E.* ii. 320.—*Stew.* ii. 166.—*Haw.* 304.—(*Sam. I.* 19.)
Ph. rectangulata. *Berl. M.* (*Hufnagle.*) iv. 606.
Ge. pectinitaria. *Fuesly M.* i. 218.—(*Haw. Pr.* 23.
Green Carpet. *Harr. V. M.* 17.

6547. 9; olivaria*. (*Och. Tr.* v. *b.* 442.)—*Och. Tr.* vi. *b.* 157.
Ge. olivata. (*Wien. V.* 112.)—*Hüb. Ge. pl.* 59. *f.* 307.
Ge. olivaria: Beech Green Carpet. *Haw.* 304.—(*Sam. I.* 18.)
♀, Ge. alboviridata. (*Haw. Pr.* 26.)
♂, Ge. alboviridaria. (*Haw. Pr.* 23.)
Ph. pectinitaria. *Don.* xiv. 33. *pl.* 479. *f.* 1.

6548. 10; implicaria*.
Ph. Ge. implicata. *Villers E.* ii. 386.
Ge. implicaria. (*Haw. Pr. App.* 4.)—*Haw.* 303.—(*Sam. I.* 18.)
Ge. montanata. (*Wien. V.* 113.)—*Hüb. Ge. pl.* 48. *f.* 248.
Cid. montanaria. (*Och. Tr.* v. *b.* 444.)—*Och. Tr.* vi. *b.* 201.
Silver ground. *Harr. V. M.* 29.

6549. 11; fluctuata*.
Ph. Ge. fluctuata. *Linn.* ii. 871.—*Stew.* ii. 171.—*Turt.* iii. 257.
Ge. fluctuata. *Hüb. Ge. pl.* 48. *f.* 249.—(*Sam. I.* 18.)
Ph. fluctuata: the Garden Carpet. (*Haw. Pr.* 25.)—*Haw.* 333.
Albin. pl. xcix. *f. e–h.*

6550. [12; costovata*.]
Ph. costovata: the short-barred Carpet. (*Haw. Pr.* 25.)—*Haw.* 334.
Ge. costovata. (*Sam. I.* 18.)

6551. 13; propugnata*: the flame Carpet.—*Haw. l. c.*
Ge. propugnata. (*Wien. V.* 112.)—*Hüb. Ge. pl.* 55. *f.* 286.—(*Sam. I.* 19.)
Ph. propugnata. *Fabr.* iii. *b.* 188.—(*Haw. Pr. App.* 5.)—*Haw.* 354.
Ci. propugnaria. (*Och. Tr.* v. *b.* 442.)—*Och. Tr.* vi. *b.* 141.
Ph. designata. *Berl. M.* (*Hufnagle.*) iv. 612.
Ph. lynceata. *Don.* x. 65. *pl.* 349. *f.* 3.

Genus 192: (975). HARPALYCE *mihi.*

CIDARIA *p, Treit.*

6552. 1; fulvata*.
Ge. fulvata. (*Wien. V.* 112.)—*Forst. C.* i. 76.—*Hüb. Ge. pl.* 57. *f.* 297.—(*Sam. I.* 18.)

Ph. fulvata: the barred Yellow. (*Haw. Pr.* 27.)—*Haw.* 328.
Ph. sociata. *Fabr.* iii. *b.* 198.—*Stew.* ii. 173.—*Turt.* iii. 264. —(*Haw. Pr.* 26.)
Small clouded Yellow. (*Wilkes D. pl.* 1. *f.* 2.)
Clouded Yellow. *Harr. A. pl.* 35. *f.* 1.—*Harr. V. M.* 59.

6553. 2; ocellata*.
Ph. Ge. ocellata. *Linn.* ii. 870.
Ge. ocellata. (*Sam. I.* 18.)—*Hüb. Ge. pl.* 48. *f.* 252.
Ph. ocellata. (*Haw. Pr.* 25.)—*Haw.* 331.
Ph. tridentata. *Berl. M.* (*Hufnagle.*) iv. 618.
Ph. lynceata. *Fabr.* iii. *b.* 189. (!)—(*Haw. Pr. App.* 5.)—*Stew.* ii. 171.—*Turt.* iii. 259.
Ge. fasciata. *Scop. C.* 221.—*Villers E.* ii. 364.
Purple bar. *Harr. V. M.* 11.
(*Wilkes D. pl.* 10. *f.* 3.)

6554. 3; Galiata.
Ge. Galiata. (*Wien. V.* 113.)—*Hüb. Ge. pl.* 53. *f.* 272.—(*Sam. I.* 18.)
Ph. Galiata: the Galium Carpet. *Haw.* 332.

6555 † [4, unilobata.]
Ph. unilobata: the blunt-angled Carpet. *Haw.* 331.
Ge. unilobata. (*Sam. I.* 19.) In Mus. D. *Haworth.*

6556 † [5, quadriannulata.]
Ph. quadriannulata: the Necklace Carpet. *Haw.* 331.
In Mus. D. *Haworth.*

6557. 6: sinuata*.
Ge. sinuata. (*Wien. V.* 114.)—*Hüb. Ge. pl.* 56. *f.* 288.—(*Sam. I.* 19.)
Ph. sinuata: the royal Mantle. *Haw.* 326.
Ph. regalata. (*Haw. Pr.* 25.)

6558. 7; derivata*.
Ge. derivata. (*Wien. V.* 113.)—*Hüb. Ge. pl.* 56. *f.* 289.—(*Sam. I.* 18.)
Ph. derivata. *Haw.* 326.
Ge. violacea nigrostriata. *Villers E.* ii. 378.
Ph. fimbriata. *Fabr.* iii. *b.* 187?
The Streamer. (*Wilkes D. pl.* 11. *f.* 9.)—*Harr. Ex.* 26. *pl.* vi. *f.* 7.—*Harr. V. M.* 45.

6559. 8; rubidata*: the Flame.—*Haw. l. c.*
Ge. rubidata. (*Wien. V.* 188.)—*Hüb. Ge. pl.* 56. *f.* 290.
Ph. rubidata. *Fabr.* iii. *b.* 180.—*Haw.* 325.
Ge. rubiadata. (*Sam. I.* 19.)

6560. 9: Berberata*.
Ge. Berberata. (*Wien. V.* 113.)—*Hüb. Ge. pl.* 56. *f.* 287.—(*Sam. I.* 18.)

6561. 10; biangulata*: the cloaked Carpet.—*Haw. l. c.*
Ph. biangulata. *Marsh. MSS.*—(*Haw. Pr.* 26.)—*Haw.* 326.
Ge. picata. *Hüb. Ge. pl.* 84. *f.* 435.
Ge. biangulata. (*Sam. I.* 18.)
A Carpet M. (*Wilkes* D. *pl.* 11. *f.* 4.)
Short Cloak Carpet. *Harr. Ex.* 31. *pl.* viii. *f.* 4.—*Harr. V. M.* 17.

6562. 11; unangulata*.
Ph. unangulata: the sharp-angled Carpet. *Marsh. MSS.*—(*Haw. Pr.* 25.)—*Haw.* 332.
Ge. unangulata. (*Sam. I.* 19.)
The White Stripe M. (*Wilkes* D. *pl.* 5. *f.* 6.)

6563. 12; sylvaticata*.
Ph. sylvaticata: the Wood Carpet. *Haw.* 332.
Ge. aquata. *Hüb. Ge. pl.* 75. *f.* 410?
Ge. sylvaticata. (*Sam. I.* 19.)

6564. 13; subtristata*.
Ph. subtristata: the common Carpet. (*Haw. Pr.* 25.)—*Haw.* 332.
Ph. contristata. *Don.* xv. 12. *pl.* 510. *f.* 2.
Ge. subtristata. (*Sam. I.* 19.)
A White stripe M. (*Wilkes* D. *pl.* 8. *f.* 5.)
Carpet M. *Harr. V. M.* 17.
β, Ph. degenerata: the degenerate Carpet. *Haw.* 333.
Ge. degenerata. (*Sam. I.* 18.)

6565. 14, tristata: the small Argent and Sable.—*Haw. l. c.*
Ph. Ge. tristata. *Linn.* ii. 869.—*Turt.* iii. 256.
Ge. tristata. (*Wien. V.* 113.)—*Hüb. Ge. pl.* 49. *f.* 254.
Ph. tristata. *Fabr.* iii. *b.* 183.—*Haw.* 333.
Ph. contristata. (*Haw. Pr.* 25.)

6566. 15; silaceata*.
Ge. silaceata. (*Wien. V.* 113.)—*Hüb. Ge. pl.* 59. *f.* 303. ♂. *pl.* 93. *f.* 477, 478. ♀.
♀; Ph. silaceata: the small Phœnix. *Haw.* 323.
♂; Ph. insulata: the insulated Carpet. *Marsh. MSS.*—(*Haw. Pr.* 26.)—*Haw.* 330.
Ge. insulata. (*Sam. I.* 18.)
Ph. cuneata. *Don.* xiv. 52. *pl.* 487. *f.* 2.

6567. 16; ruptata*.
Ge. ruptata. *Hüb. Ge. pl.* 57. *f.* 295.—(*Sam. I.* 19.)
Ph. ruptata: the broken-barr'd Carpet. *Haw.* 327.—*Don.* xiv. 34. *pl.* 479. *f.* 2.
Ph. interruptata. (*Haw. Pr.* 26.)
Albin. pl. xcviii. *f. a–d.*

6568. 17; miata*: the autumn Green Carpet.

Ph. Ge. miata. *Linn.* ii. 869.—*Turt.* iii. 255.
Ph. miata. *Fabr.* iii. *b.* 180.—(*Haw. Pr.* 26.)—*Haw.* 328.—*Don.* xiv. 35. *pl.* 479. *f.* 3.
Ph. luctuata. *Fabr.* iii. *b.* 195.
Ge. miata. (*Sam. I.* 18.)
Albin. pl. 1. *f.* 86. *e–h.*

6569. 18; psittacata*: the red-green Carpet.
Ge. psittacata. (*Wien. V.* 112.)—*Hüb. Ge. pl.* 43. *f.* 227.—(*Sam. I.* 19.)
Ph. psittacata. *Fabr.* iii. *b.* 195.—*Haw.* 329.
Ph. siterata. *Berl. M.* (*Hufnagle.*) iv. 522.
Ph. rubroviridata. *Marsh. MSS.*—(*Haw. Pr.* 26.)—*Don.* xiv. 48. *pl.* 485. *f.* 3.

6570. 19; immanata*.
Ph. immanata: the dark marbled Carpet. *Marsh. MSS.*—(*Haw. Pr.* 26.)—*Haw.* 323.
Ge. immanata. (*Sam. I.* 18.)

6571. [20, amœnata] *mihi.*

6572. 21; marmorata*: the marbled Carpet.—*Haw. l. c.*
Ph. marmorata. *Fabr.* iii. *b.* 192.—*Haw.* 324.
Ge. marmorata. (*Sam. I.* 18.)
Ph. punctum-notata. *Marsh. MSS.*—(*Haw. Pr.* 26.)
Ph. omicrata. *Don.* xv. 11. *pl.* 510. *f.* 1?

6573. [22, concinnata] *mihi.*

6574. 23; comma-notata*: yellow marbled Carpet.—*Haw. l. c.*
Ph. comma-notata. *Marsh. MSS.*—(*Haw. Pr.* 26.)—*Haw.* 325.
Ge. comma-notata. (*Sam. I.* 18.)

6575. 24; centum-notata*: common marbled Carpet.—*Haw. l. c.*
Ph. centum-notata. *Fabr.* iii. *b.* 191.—(*Haw. Pr.* 25.)—*Haw.* 324.
Ge. centum-notata. (*Sam. I.* 18.)
Ge. rupata. (*Wien. V.* 113.)—*Hüb. Ge. pl.* 59. *f.* 305. *pl.* 86. *f.* 445?
Ph. truncata. *Berl. M.* (*Hufnagle.*) iv. 602?
Hornsey Beauty. *Harr. V. M.* 11.

6576. [25; saturata*] *mihi.*

6577. [26; perfuscata*.]
Ge. perfuscata: the brown marbled Carpet. *Haw.* 325.—(*Sam. I.* 19.)

6578. 27; suffumata*.
Ge. suffumata. (*Wien. V.* 316.)—*Hüb. Ge. pl.* 59. *f.* 306.—(*Sam. I.* 19.)
Ph. suffumata: the Water Carpet. *Haw.* 323.
Ph. insititiata. *Marsh. MSS.*—(*Haw. Pr.* 25.)

Genus 193: (976). STEGANOLOPHIA *mihi.*

CIDARIA *p, Treit.*

6579. 1; Prunata*.
Ph. Ge. Prunata. *Linn.* ii. 869.—*Stew.* ii. 170.—*Turt.* iii. 254.
Ph. Prunaria. *Don.* vii. 45. *pl.* 233. *f.* 1.
Ph. Prunata. *Fabr.* iii. *b.* 178.—(*Haw. Pr.* 25.)—*Haw.* 322.
Ge. Prunata. (*Sam. I.* 19.)
Cliefden Carpet. (*Wilkes* D. *pl.* 2. *f.* 5?)
Clouded Carpet. *Harr. Ex.* 30. *pl.* viii. *f.* 1.—*Harr. V. M.* 17.
Phœnix. (*Harr. V. M.* 39.)

Genus 194: (977). ELECTRA *mihi.*

CIDARIA *p, Treit.*

6580. 1; Spinachiata*.
Ph. Spinachiata: the Spinach. (*Haw. Pr.* 26.)—*Haw.* 341.
Ge. marmorata. *Hüb. Ge. pl.* 54. *f.* 279.
Ge. Spinaciata. (*Sam. I.* 19.)

6581. 2; comitata*: dark Spinach.— *Haw. l. c.*
Ph. Ge. comitata. *Linn.* ii. 868.—*Stew.* ii. 169.—*Turt.* iii. 253.
Ph. comitata. *Fabr.* iii. *b.* 176.—(*Haw. Pr.* 26.)—*Haw.* 342.
Ge. comitata. (*Sam. I.* 18.)
Ge. chenopodiata. (*Wien. V.* 112.)—*Hüb. Ge. pl.* 58. *f.* 299.
Albin. pl. xlvii. *f.* 79. *f–i.*
July Arrach M.: Ph. chenopodiata. (*Wilkes.* 36. *pl.* 71.)—*Harr. V. M.* 9.
β. Ph. Ge. dotata. *Mus. Linn.*

6582. 3, Populata.
Ph. Ge. Populata. *Linn.* ii. 868.—*Turt.* iii. 253.
Ge. Populata. *Hüb. Ge. pl.* 58. *f.* 300.—*Ent. Trans.* (*Haworth.*) i. 337.—(*Sam. I.* 19.)
Ph. Ge. cuspidata. *Stew.* ii. 164.

6583. 4; testata*.
Ph. Ge. testata. *Linn. F. No.* 1262.
Ph. testata: the Chevron. *Haw.* 342.
Ge. testaceata. (*Haw. Pr.* 26.)
Ge. testata. (*Sam. I.* 19.)
Broad Chivern'd. *Harr. V. M.* 17?

6584. [5, achatinata.]
Ge. achatinata. *Hüb. Ge. pl.* 58. *f.* 301.

6585. 6; Pyraliata*.
Ge. Pyraliata. (*Wien. V.* 117.)—*Hüb. Ge. pl.* 58. 302.—(*Sam. I.* 19.)
Ph. Populata: barred Straw. (*Haw. Pr.* 26.)—*Haw.* 341.
Albin. pl. xlv. *f.* 75. *d–g.*

6586. 7, imbutata.
Ge. imbutata. *Hüb. Ge. pl.* 78.*f.* 403.

Genus 195: (978). IDAËA.

IDAËA A, *Treit.*

6587. 1, dealbata*. (*Och. Tr.* v. *b.* 446.)—*Och. Tr.* vi. *b.* 259.
Ph. Ge. dealbata. *Linn.* ii. 870.
Ge. dealbata. (*Sam. I.* 18.)
Ph. dealbata. (*Haw. Pr.* 27.)—*Haw.* 317.
Ge. dealbaria. *Hüb. Ge. pl.* 41.*f.* 214.
Ph. lineata. *Scop. C.* 218.
Black veined. *Harr. V. M.* 53.

Genus 196: (979). ABRAXAS, *Leach, Samou.*, (*Kirby.*)

ZERENE *p*, *Treit.*

6588. 1, pantaria.
Ge. pantaria. *Linn.* ii. 863.—*Stew.* ii. 166.—(*Sam. I.* 19.)
Ph. pantaria. *Haw.* 317.
Ge. pantaria. (*Wien. V.* 115.)—*Hüb. Ge. pl.* 16. *f.* 84.—(*Haw. Pr.* 21.)
Ph. Ulmata. *Fabr.* iii. *b.* 176.
The Panther. (*Harr. V. M.* 39.)

6589. 2: Ulmata*. (*Leach E. E.* ix. 134.)—(*Samou.* 253.)
Ph. Ulmata. *Fabr.* iii. *b.* 176.—*Don.* ix. 11. *pl.* 293. *f.* 1.—*Turt.* iii. 254.
Ge. Ulmaria. *Hüb. Ge. pl.* 76. *f.* 391, 392.—(*Haw. Pr.* 21.)
Ph. Ulmata: scarce Magpie. *Haw.* 317.
Ph. Ge. sylvata. *Villers E.* ii. 363.
Ge. pantherata. *Bork. E. S.* v. 470.
Clouded Magpie. *Harr. V. M.* 35.

6590. 3; Grossulariata*. (*Leach E. E.* ix. 134.)—(*Samou.* 253.)
Ph. Ge. Grossulariata. *Linn.* ii. 867.—*Berk. S.* i. 145.—*Stew.* ii. 168.—*Don.* i. *pl.* 4.—*Turt.* iii. 253.
Ge. Grossulariaria. (*Wien. V.* 115.)—*Hüb. Ge. pl.* 16. *f.* 81, 82.—(*Haw. Pr.* 21.)
Ph. Grossulariata: the Magpie. *Haw.* 316.
Pa. No. 9. *List. Goëd.* (*Angl.*) 13. *f.* 9.—*Id.* (*Lat.*) 24. *f.* 9.
Albin. pl. xliii. *f.* 71. *d–g.*
Gooseberry M.: Ph. Grossulariata. (*Wilkes.* 41. *pl.* 85.)
Currant M. *Harr. A. pl.* 12. *f. f–h.*
Great or Large Magpie M. (*Wilkes. D. pl.* 11. *f.* 6.)—*Harr. V. M.* 35.

Genus 197: (980). ——[a]

Genus 198: (981). XERENE, *Treitschke.*

6592. 1; hastata*.
Ph. Ge. hastata. *Linn.* ii. 870.—*Stew.* ii. 170.—*Don.* iv. 55. *pl.* 129. *f.* 123.—*Turt.* iii. 256.
Ph. hastata. (*Haw. Pr.* 25.)—*Haw.* 336.
Ge. hastata. (*Sam. I.* 18.)
The Mottled Beauty, a Moth. (*Wilkes* D. *pl.* 4. *f.* 6.)
Argent and Sable. *Harr. A. pl.* 15.—*Harr. V. M.* 9.

6593. 2; albicillata*. (*Och. Tr.* v. *b.* 444.)—*Och. Tr.* vi. *b.* 228.
Ph. Ge. albicillata. *Linn.* ii. 870.—*Stew.* ii. 170.—*Don.* vi. 52. *pl.* 202. *f.* 2.—*Turt.* iii. 256.
Ph. albicillata: the beautiful Carpet. (*Haw. Pr.* 24.)—*Haw.* 337.
Ge. albicillata. (*Sam. I.* 18.)
Ph. vestalis. *Naturforscher.* xiii. 30. *pl.* 3. *f.* 7. *a, b.*
Clifden Beauty. *Harr. V. M.* 13?

6594. 3; procellata*. (*Och. Tr.* v. *b.* 444.)—*Och. Tr.* vi. *b.* 213.
Ge. procellata. (*Wien. V.* 114.)—*Hüb. Ge. pl.* 48. *f.* 251.—*Turt.* iii. 257.—(*Sam. I.* 19.)
Ph. procellata: the chalk Carpet. (*Haw. Pr.* 24.)
Ph. porcellata. *Don.* vi. 53. *pl.* 502. *f.* 3.
Cliefden Beauty. (*Wilkes* D. *pl.* 2. *f.* 2.)
Clifden Beauty. *Harr. Ex.* 22. *pl.* v. *f.* 6.

6595. 4; adustata*. (*Och. Tr.* v. *b.* 444.)—*Och. Tr.* vi. *b.* 225.
Ge. adustata. (*Wien. V.* 114.)—*Hüb. Ge. pl.* 15. *f.* 75.—(*Sam. I.* 18.)
Ph. adustata: the scorched Carpet. (*Haw. Pr.* 25.)—*Haw.* 337.
The large Blue-bordered. *Harr. Ex.* 20. *pl.* v. *f.* 1.
Clifden Carpet. *Harr. V. M.* 17.

6596. 5; rubiginata*. (*Och. Tr.* v. *b.* 444.)—*Och. Tr.* vi. *b.* 223.
Ge. rubiginata. (*Wien. V.* 114.)—*Hüb. Ge. pl.* 48. *f.* 250.—(*Sam. I.* 19.)
Ph. rubiginata: the blue-bordered Carpet. *Haw.* 338.
Ph. abbreviata. *Marsh. MSS.*—(*Haw. Pr.* 25.)
Ge. albaria. *Villers E.* ii. 321.
Ge. bicolorata. *Berl. M.* (*Hufnagle.*) iv. 608.
Blue-border. (*Wilkes* D. *pl.* 4. *f.* 5.)—*Harr. V. M.* 13.

---

[a] Genus 197: (980). ——

XERENE *p*, *Treit.*

6591 ‡ 1. melanaria.
Ph. Ge. melanaria. *Linn.* ii. 862.
Ge. melanaria. *Hub. Ge. pl.* 16. *f.* 86.
Ph. melanata: the yellow underwinged Magpie. *Haw.* 316. (¹)

6597. [6, contaminata?]
Ge. contaminata. *Berl. M.* (*Hufnagle.*) iv. 614?

Genus 199: (982). YPSIPETES *mihi.*

ACIDALIA *p, Treit.*

6598. 1; impluviata*.
Ge. impluviata. (*Wien. V.* 109.)—*Hüb. Ge. pl.* 43. *f.* 223.—(*Sam. I.* 18.)
Ph. impluviata: May Highflyer. *Haw.* 321.
Ge. trifasciata. *Bork. E. S.* v. 308.
Ph. literata. *Don.* xiv. 81. *pl.* 499. *f.* 2.
July highflyer likeness. *Harr. V. M.* 25.

6599. 2; elutata*.
Ge. elutata. (*Wien. V.* 109.)—*Hüb. Ge. pl.* 45. *f.* 224. *pl.* 74. *f.* 381–385.—(*Sam. I.* 18.)
Ph. elatata. *Haw.* 321.
Ph. Ge. virgata. *Villers E.* ii. 379?
Albin. pl. xliv. *f.* 73. *e–h.*
July Sallow M.: Ph. fluctuata? (*Wilkes.* 26. *pl.* 73.)
July high Flyer. *Harr. V. M.* 25.
β, Ph. fuscoundata. *Don.* xi. 73. *pl.* 386. *f.* 3.—or yellow-striped Highflyer. *Haw.* 321.
Ge. fuscoundata. (*Sam. I.* 18.)

6600. 3, horridaria?
Ge. horridaria. (*Haw. Pr.* 24?)

Genus 200: (983). PHIBALAPTERYX *mihi.*

ACIDALIA *p, et* ASPILATES *p, Treit.*

6601. 1; tersata*.
Ge. tersata. (*Wien. V.* 109.)—*Hüb. Ge. pl.* 52. *f.* 268.—(*Sam. I.* 19.)
Ph. tersata. *Haw.* 339.
The Fern. *Harr. V. M.* 25.

6602. 2; vitalbata*.
Ge. vitalbata. (*Wien. V.* 109.)—*Hüb. Ge. pl.* 22. *f.* 269.—(*Sam. I.* 19.)
Ph. vitalbata: the small waved-Umber. *Haw.* 340.

6603. 3; lignata*.
Ge. lignata. *Hüb. Ge. pl.* 52. *f.* 270.—(*Sam. I.* 18.)
Ph. lignata: the oblique Carpet. *Haw.* 340.
Ge. areolaria. (*Haw. Pr.* 23.)
Ph. lineataria. *Don.* xiv. 47. *pl.* 485. *f.* 1, 2.

6604 † 4, cognata* *mihi.* In Mus. D. *Curtis.*

6605. 5, lineolata.
Ge. lineolata. (*Wien. V.* 102.)—*Hüb. Ge. pl.* 60. *f.* 311.—(*Sam. I.* 18.)

Ph. lineolata: the Oblique striped. *Haw.* 340.
Ph. virgata. *Berl. M.* (*Hufnagle.*) iv. 608.

6606 † 6. polygrammata?
Ge. polygrammata. *Hüb. Ge. pl.* 54. *f.* 277?
In Mus. D. —— ?

6607 † 7, angustata*.
Ph. angustata: the narrow-barred Carpet. *Haw.* 340.
Ge. angustata. (*Sam. I.* 18.)
In Mus. D. *Haworth, J. Latham? et Milne.*

6608. 8; badiata*.
Ge. badiata. (*Wien. V.* 111.)—*Hüb. Ge. pl.* 56. *f.* 291.—(*Sam. I.* 18.)
Ph. badiata. *Haw.* 325.
Ph. Berberata. *Don.* xiv. 65. *pl.* 493. *f.* 1.
Shoulder stripe. (*Harr. V. M.* 45.)

Genus 201: (984). SCOTOSIA *mihi.*

ACIDALIA *p, Treitschke.*

6609. 1; Rhamnata*.
Ge. Rhamnata. (*Wien. V.* 109.)—*Hüb. Ge. pl.* 52. *f.* 271.—(*Sam. I.* 19.)
Ph. Rhamnata: the dark Umber. (*Haw. Pr.* 25.)—*Haw.* 339.
Ph. transversata. *Berl. M.* (*Hufnagle.*) iv. 600.

6610. 2; vetulata*.
Ge. vetulata. (*Wien. V.* 109.)—*Hüb. Ge. pl.* 51. *f.* 263.—(*Sam. I.* 19.)
Ph. vetulata: the brown Scolop. *Haw.* 320.
Ph. fuscata. (*Haw. Pr.* 27.)

6611. 3, Sparsaria.
Ge. Sparsaria. *Hüb. Ge. pl.* 22. *f.* 116.
Ge. roraria. *Esper.* v. *pl.* xxiv. *f.* 4.

Genus 202: (985). TRIPHOSA *mihi.*

ACIDALIA *p, Treitschke.*

6612. 1; dubitata*.
Ph. Ge. dubitata. *Linn.* ii. 866.—*Don.* vii. 82. *pl.* 246. *f.* 2.—*Stew.* ii. 163.—*Turt.* iii. 249.
Ph. dubitata. (*Haw. Pr.* 25.)—*Haw.* 318.
Ge. dubitata. (*Sam. I.* 18.)
Ph. fuliginata. *Berl. M.* (*Hufnagle.*) iv. 610.
Tissue. *Harr. V. M.* 51.

6613. 2; cinereata* *mihi.*

6614. 3; cervinata*.
Ge. cervinata. *Hüb. Ge. pl.* 51. *f.* 266.—(*Sam. I.* 18.)
Ph. cervinata: the scarce Tissue. *Haw.* 318.

Ac. ancipitata. (*Och. Tr.* v. *b.* 440.)
Ac. certata. *Hübner Verz: teste Treit. Och. Tr.* vi. *b.* 72.

Genus 203: (986). ——

ACIDALIA *p, Treitschke.*

6615. 1; bilineata*.
Ph. Ge. bilineata. *Linn.* ii. 868.—*Stew.* ii. 169.—*Don.* viii. 86. *pl.* 287. *f.* 3.—*Turt.* iii. 258.
Ph. bilineata: the yellow Shell. (*Haw. Pr.* 26.)—*Haw.* 343.
Ge. bilineata. (*Sam. I.* 18.)
Yellow shoulder stripe. *Harr. V. M.* 45.

Genus 204: (987). ——

ACIDALIA *p, Treitschke,* (*Kirby.*)

6616. 1; undulata*.
Ph. Ge. undulata. *Linn.* ii. 867.—*Stew.* ii. 168.—*Don.* x. 43. *pl.* 342. *f.* 3.—*Turt.* iii. 253.
Ph. undulata. *Haw. Pr.* 27.—*Haw.* 320.
Ge. undulata. (*Sam. I.* 19.)
Ph. serrata. *Berl. M.* (*Hufnagle.*) iv. 522.
Scollop Shell. (*Wilkes D. pl.* 11. *f.* 1.)—*Harr. Ex.* 12. *pl.* 11. *f.* 5, 6.—*Harr. V. M.* 43.

Genus 205: (988). CHARISSA, *Curtis.*

GNOPHOS *p, Treitschke.*

6617. 1: obscuraria: Dark Annulet.—*Haw. l.* c. (*Curtis l.* c. *infra.*)
Ge. obscuraria. (*Wien. V.* 108.)—*Hüb. Ge. pl.* 27. *f.* 146.—*Haw.* 314.
Ph. lividata. *Fabr.* iii. *b.* 170.
Ge. Anthracinaria. *Esper.* v. *pl.* xxv. *f.* 3–5.

6618. 2; dilucidaria.
Ge. dilucidaria. (*Wien. V.* 315.)—*Hüb. Ge. pl.* 2. *f.* 143.

6619. 3; pullata: brown Annulet.—*Haw. l.* c. (*Curtis l.* c. *infra.*)
Ge. pullata. (*Wien. V.* 108.)
Ge. pullaria. *Hüb. Ge. pl.* 27. *f.* 145.—*Haw.* 314.—(*Sam. I.* 19.)
Ph. Ge. quadripustulata. *Don.* xiii. 63. *pl.* 463.
Ge. dilucidaria. *Esper.* v. *pl.* xlix. *f.* 3.

6620. [4, serotinaria.] (*Curtis l.* c. *infra.*)
Ge. serotinaria: large Mocha. *Hüb. Ge. pl.* 28. *f.* 147?—*Haw.* 311.

6621 † 5, operaria. *Curtis.* iii. *pl.* 105.
Ge. operaria *Hüb. Ge. pl.* 69. *f.* 359.
In Mus. D. Curtis, Dale, *Haworth et Stone.*

Genus 206: (989). PACHYCNEMIA *mihi.*

FIDONIA *p, Treitschke.*

6622. 1; Hippocastanaria*.
Ge. Hippocastanaria. *Hüb. Ge. pl.* 36. *f.* 186.
Ph? anomalata. *Ent. Trans.* (*Haworth.*) i. 337.

Genus 207: (990). LOZOGRAMMA *mihi.*

ASPILATES *p, Treitschke.*

6623. 1; petraria*.
Ge. petraria. *Hüb. Ge. pl.* 21. *f.* 113.
Ph. petrata. *Haw.* 344.
Ge. petrata. (*Sam. I.* 19.)
Ph. chlorosata. *Scop. C.* 222.
Ge. virgaria. *Bork. E. S.* v. 61.
Brown Silver lines. *Harr. V. M.* 35.

Genus 208: (991). APLOCERA *mihi.*

LARENTIA *p, Treitschke.*

6624. 1; plagiata*.
Ph. Ge. plagiata. *Linn.* ii. 869.—*Stew.* ii. 169.
Ph. plagiata: the slender Treble-bar. *Haw. Pr.* 25.—*Haw.* 318.
Ph. duplicata. *Fabr.* iii. *b.* 193.—*Don.* vii. 46. *pl.* 233. *f.* 2. —(*Haw. Pr. App.* 5.)—*Turt.* iii. 261.
Ge. duplicata. (*Sam. I.* 18.)
Ge. plagiata. (*Sam. I.* 19.)
The Treble-barred Moth. (*Wilkes* D. *pl.* 6. *f.* 4.)
Treble or triple bars. *Harr. A. pl.* 28. *f. k.*—*Harr. V. M.* 11.

6625. 2, cæsiata*?
Ge. cæsiata. (*Wien. V.* 112.)—*Hüb. Ge. pl.* 53. *f.* 275.—(*Sam. I.* 18.)
Ph. cæsiata: the February Carpet. *Haw.* 330.
β, Ph. infrequentata: the scarce Carpet. *Haw.* 330.
Ge. cyanata. *Hüb. Ge. pl.* 62. *f.* 319?

6626. 3, flavicinctata.
Ge. flavicinctata. *Hüb. Ge. pl.* 68. *f.* 354.

Genus 209: (992). CHESIAS, *Treitschke.*

6627. 1; Spartiata*. (*Och. Tr.* v. *b.* 437.)—*Id.* vi. *a.* 331.
Ph. Spartiata. *Fabr.* iii. *b.* 181.—*Don.* x. 42. *pl.* 342. *f.* 2.—*Haw. Pr.* 27.—*Haw.* 339.
Ge. Spartiata. (*Sam. I.* 19.)
Ph. Ge. Spartiata. *Fuesly Ar. pl.* 11.
Streak. (*Harr. V. M.* 45.)

6628. 2; obliquaria*. (*Och. Tr.* v. *b.* 437.)—*Id.* vi. *a.* 340.
Ge. obliquaria. (*Wien. V.* 102.)

Ph. Bombycata: the Chevron.—*Don. l. c.* *Hüb. B.* i. 2. *K.* —(*Haw. Pr. App.* 5.)—*Don.* xi. 74. *pl.* 386. *f.* 4.
Ph. rufata: the Broom-tip.—*Haw. l. c.* *Fabr.* iii. *b.* 181?—*Haw.* 322.—*Stew.* ii. 170.
Ge. rufata. (*Sam. I.* 19.)
Broom buft tip. *Harr. V. M.* 49.

Genus 210: (993). ——

CHESIAS *p, Treitschke.*

A. Antennis pectinatis.

6629. 1; simulata*.
Ge. simulata. *Hüb. Ge. pl.* 66. *f.* 345.—(*Sam. I.* 19.)

B. Antennis haud pectinatis.

6630. 2; variata*.
Ge. variata. (*Wien. V.* 110.)—(*Sam. I.* 19.)
Ph. Juniperata. (*Haw. Pr.* 26.)
Ph. variata: the grey Carpet. *Ent. Trans.* (*Hatchett.*) i. 245. *pl.* 7. *f.* 1.—*Haw.* 327.

6631. 3, fulvata.
Ph. fulvata. *Fabr.* iii. *b.* 188?
Ge. pinetata. *Bork. E. S.* v. 373?
Ge. obeliscata. *Hüb. Ge. pl.* 57. *f.* 296?

6632. 4: Juniperata*.
Ph. Ge. Juniperata. *Linn.* ii. 871.
Ge. Juniperata. *Hüb. Ge. pl.* 57. *f.* 294.—(*Sam. I.* 18.)

Genus 211: (994). ——

ACIDALIA *p, Treitschke.*

6633. 1; dilutata*.
Ge. dilutata. (*Wien. V.* 109.)—*Hüb. Ge. pl.* 36. *f.* 188. ♂.—(*Samou.* 443.)
Ph. dilutata: the November. *Haw.* 319.
Ge. 4-fasciata. *Bork. E. S.* v. 294.
Ge. affiniata. *Bork. E. S.* v. 294.
Ph. omicrata. *Haw. Pr.* 25.
Ph. inscriptata. *Don.* xv. 25. *pl.* 517.
Albin. pl. xlv. *f.* 74. *a–c.*
The grey waved. *Harr. Ex.* 31. *pl.* viii. *f.* 3.
β; Ph. fimbriata: the bordered November. *Haw.* 320.
Ge. fimbriata. (*Sam. I.* 18.)

Genus 212: (995). CHEIMATOBIA *mihi,* (*Kirby.*)

ACIDALIA *p, et* FIDONIA *p, Treitschke.*—HYBERNIA *p, Latr.?*

6634. 1; vulgaris*: Winter M.—*Haw. l. c.*
Ph. Ge. brumata. *Linn.* ii. 874.—*Stew.* ii. 172.—*Turt.* iii. 260.

Ge. brumaria. *Hüb. Ge. pl.* 37. *f.* 191. *pl.* 80. *f.* 415. *pl.* 99. *f.* 509. ♀.—*Haw. Pr.* 23.—*Haw.* 305.—(*Sam. I.* 18.)
Ph. hyemata. *Berl. M.* (*Hufnagle.*) iv. 612.
Noctua brumata. (*Kirby & Sp. I. E.* i. 206.)

6635. 2; rupicapraria*: Early M.—*Haw. l. c.*
Ge. rupicapraria. (*Wien. V.* 105.)—*Hüb. Ge. pl.* 42. *f.* 222.
Ph. Ge. primaria. *Marsh. MSS.*—*Haw. Pr.* 23.—*Haw.* 305.
Ge. primaria. (*Sam. I.* 19.)
Albin. pl. xlix. *f.* 83. *e–g?*

Genus 213: (996). LOBOPHORA *mihi, Curtis,* (*Kirby.*)

CHESIAS *p, Treitschke.*—ACIDALIA *p, Treitschke.*—PTERAPHERAPTERYX, *Leach olim.*

A. Tibiis posticis bicalcaratis.

6636. 1: polycommata*. *Curtis.* ii. *pl.* 81.
Ge. polycommata. (*Wien. V.* 109.)—*Hüb. Ge. pl.* 36. *f.* 190. ♀.

6637. 2, costæstrigata*. (*Curtis l. c. supra.*)
Ph. costæstrigata: the twin-striped Pinion. *Haw.* 319.
Ge. rupestrata. *Hüb. Ge. pl.* 37. *f.* 192?
Ge. costastrigata. (*Sam. I.* 18.)

6638. 3; dentistrigata*.
Ph. dentistrigata: the early Tooth-striped. *Haw.* 320.
Ge. dentistrigata. (*Sam. I.* 18.)
Ge. lobulata. *Hüb. Ge. pl.* 70. *f.* 362.
Lo. lobulata. (*Curtis l. c. supra.*)

6639. 4; viretata*. (*Curtis l. c. supra.*)—(*Ing. Inst.* 91.)
Ge. viretata. *Hüb. Ge. pl.* 44. *f.* 230.—(*Sam. I.* 19.)
Ph. viretata: the brindle-barred Yellow. *Haw.* 329.
Ph. trinotata. *Don.* xiv. 79. *pl.* 499. *f.* 1.

B. Tibiis posticis quadricalcaratis.

6640. 5; hexapterata*. (*Curtis l. c. supra.*)
Ph. hexapterata. *Fabr.* iii. *b.* 193.—*Don.* vi. 29. *pl.* 192.—(*Haw. Pr.* 28.)—*Haw.* 356.—*Stew.* ii. 169.—*Turt.* iii. 261.—(*Kirby & Sp. I. E.* ii. 352.)
Ge. hexapterata. (*Sam. I.* 18.)
Ph. halterata. *Berl. M.* (*Hufnagle.*) iv. 608.
Var. ♀. Ph. zonata. *Thunb. I. S.* iv. 60. *fig. adj.?*
Prominent M. *Harr. V. M.* 41.

6641. 6; sexalisata*. (*Curtis l. c. supra.*)
Ge. sexalisata. *Hüb. Ge. pl.* 44. *f.* 228.—(*Sam. I.* 19.)
Ph. sexalisata: the small Seraphim. *Haw.* 356.
Ge. sex-alata. *Bork. E. S.* v. 304.
Seraphim. (*Harr. V. M.* 43.)

Genus 214: (997). EUPITHECIA, *Curtis.*

LAURENTIA *p, Treitschke.*

6642. 1; Linariata*. *Curtis.* ii. *pl.* 64.
Ph. Linariata: the beautiful Pug.—*Haw. l. c.* (*Wien. V.* 113.) —*Fabr.* iii. *b.* 190.—(*Haw. Pr. App.* 5.)—*Haw.* 364.
Ge. linariata. (*Sam. I.* 18.)

6643. 2; pulchellata* *mihi.*

6644. 3; rectangulata*.
Ph. Ge. rectangulata. *Linn.* ii. 872.
Ph. rectangulata: the green Pug. (*Haw. Pr.* 28.)—*Haw.* 363.
Ge. rectangulata. *Hüb. Ge. pl.* 45. *f.* 235.—(*Sam. I.* 19.)
Ph. viridulata. *Berl. M.* (*Hufnagle.*) iv. 524.
The Olive M. *Harr. V. M.* 37?—*Harr. Ex.* 15. *pl.* iii. *f.* 7?

6645. 4, subærata.
Ge. subærata. *Hüb. Ge. pl.* 90. *f.* 463?

6646. 5, nigrosericeata*.
Ph. nigrosericeata: the black-silk Pug. *Haw.* 363.

6647 † 6, sericeata*.
Ph. sericeata: the Satin Pug. *Haw.* 363.
In Mus. D. J. *Latham.*

6648. 7; V ata*.
Ph. V ata: the V Pug. *Haw.* 364.
Geo. V ata. (*Sam. I.* 19.)
Ge. coronata. *Hüb. Ge. pl.* 72. *f.* 372, 373?

6649. 8; Strobilata*.
Ge. Strobilata. *Bork. E. S.* v. 352.—*Hüb. Ge. pl.* 87. *f.* 449, 450.
Ph. bistrigata: the double-striped Pug. *Haw.* 361.

6650. 9; rufifasciata*.
Ph. rufifasciata: the red-barred Pug. *Haw.* 361.
Ge. rufifasciata. (*Sam. I.* 19.)

6651. 10; lævigata*: Juniper P.—*Haw. l. c.*
Ph. lævigata. *Fabr.* iii. *b.* 197.—(*Haw. Pr.* 28.)—*Haw.* 362.
Ge. lævigata. (*Sam. I.* 18.)

6652. [11; exiguata*.] (*Curtis l. c. supra.*)
Ge. exiguata. *Hüb. Ge. pl.* 73. *f.* 379.

6653. 12; trimaculata*.
Ph. trimaculata: the mottled Pug. *Haw.* 362.
Ge. trimaculata. (*Sam. I.* 19.)

6654. [13; ochreata*] *mihi.*

6655. 14; abbreviata* *mihi.* (*Curtis l. c. supra.*)
Ge. abbreviata. (*Sam. I.* 18.)

6656. 15, singulariata*.
Ph. Ge. singulariata. *Villers E.* ii. 383. *pl.* 6. *f.* 20.
Ph. singulariata: the grey P. *Haw.* 360.—(*Sam. I.* 19.)

6657. 16; nebulata*.
Ph. nebulata: the brindled P. *Haw.* 360.

6658. 17; albipunctata*.
Ph. albipunctata: the speckled P. *Haw.* 360.

6659. 18, subumbrata*.
Ge. subumbrata. (*Wien. V.* 110.)—*Hüb. Ge. pl.* 45. *f.* 233.
Ph. subumbrata: the small brindled P. *Haw.* 361.

6660. 19, subfasciata* *mihi.*

6661. 20; minutata*.
Ge. minutata. (*Wien. V.* 110.)—*Hüb. Ge. pl.* 45. *f.* 237. *pl.* 88. *f.* 454.
Ph. vulgata: the common P. *Haw.* 359.
Ge. vulgata. (*Sam. I.* 19.)

6662. 21, subfuscata*.
Ph. subfuscata: the brown-grey P. *Haw.* 360.
Ge. subfuscata. (*Sam. I.* 19.)
Ph. denotata?

6663. 22, innotata*.
Ph. innotata. *Bork. E. S.* v. 332.—*Hüb. Ge. pl.* 86. *f.* 441, 2.

6664. 23; Absinthiata*.
Ph. Ge. Absinthiata. *Linn. F. No.* 296.
Ph. Absinthiata: the Wormwood P. (*Haw. Pr.* 28.)—*Haw.* 359.
Ge. Absinthiata. *Hüb. Ge. pl.* 88. *f.* 453?—(*Sam. I.* 18.)

6665. 24, notata*.

6666. 25, elongata*.
Ph. elongata: the long-winged P. *Haw.* 358.
Ge. elongata. (*Sam. I.* 18.)
Ge. Pimpinellata. *Hüb. Ge. pl.* 86. *f.* 443, 4. ♀.

6667. 26; simpliciata*.
Ph. simpliciata: the plain P. *Haw.* 359.

6668. 27; pusillata*.
Ge. pusillata. (*Wien. V.* 110.)—*Fabr.* iii. *b.* 204?—*Turt.* iii. 266.—*Hüb. Ge. pl.* 73. *f.* 378.—(*Sam. I.* 19.)
Ph. pusillata: the small grey P. (*Haw. Pr. App.* 5.)—*Haw.* 359.
Albin. pl. xciii. *f. e–h?*

6669. 28, plumbeolata*.
Ph. plumbeolata: the lead-coloured P. *Haw.* 360.
Ge. plumbeolata. (*Sam. I.* 19.)

670. 29; subfulvata*.

Ph. subfulvata: the Tawny speck. *Haw.* 357.
Ge. subfulvata. (*Sam. I.* 19.)
Ph. fulvata. (*Haw. Pr.* 27.)
Lar. oxydata. *Och. Tr.* vi. *b.* 114?

6671. 30, cognata *mihi.*

6672. 31; succenturiata*.
Ph. Ge. succenturiata. *Linn.* ii. 872.—*Turt.* iii. 262.—*Hüb. Ge. pl.* 89. *f.* 459.
Ph. succenturiata: the bordered Lime Speck. (*Haw. Pr.* 25.)—*Haw.* 358.
Ge. succenturiata. (*Sam. I.* 19.)

6673. 32; Centaureata*.
Ge. Centaureata. (*Wien. V.* 114.)—*Hüb. Ge. pl.* 46. *f.* 240. *pl.* 88. *f.* 452.—(*Sam. I.* 18.)
Ph. Centaureata: the Lime Speck. (*Haw. Pr. App.* 5.)—*Haw.* 358.
Ph. succenturiata. *Berl. M.* (*Hufnagle.*) iv. 329.
Ph. oblongata. *Thunb. I. S.* i. 14. 33. *pl. adj.*
Ph. signata. *Scop. C.* 231.
Lime Speck. *Harr. V. M.* 44.

6674. 33, piperitata* *mihi.*

6675. 34; angustata*.
Ph. angustata: the narrow-winged P. *Haw.* 362.

6676. 35; variegata*.
Ph. variegata: the marbled P. *Haw.* 362.
Ph. insigniata. *Hüb. B.* ii. 4. *pl.* 3. *f. U. S.?*
Eu. insignata. (*Curtis l. c. supra.*)

6677. 36, consignata*.
Ge. consignata. *Bork. E. S.* v. 315.—*Hüb. Ge. pl.* 47. *f.* 245.—(*Sam. I.* 18.)
Ph. consignata: the pinion spotted P. *Haw.* 357.

6678. 37; venosata*: the netted P. *Haw. l. c.*
Ph. venosata. *Fabr.* iii. *b.* 197.—*Haw.* 357.—*Turt.* iii. 263.
Ge. venosata. (*Sam. I.* 19.)
Ph. decussata: the pretty Widow M. *Don.* viii. 38. *pl.* 266. *f.* 3.—(*Haw. Pr.* 28.)—*Stew.* ii. 173.—*Turt.* iii. 256.
Ph. diversata. *Fabr.* iii. *b.* 183?

Genus 215: (998). MINOA.

MINOA A, *Treitschke.*

6679. 1; Chærophyllata*. (*Och. Tr.* vi. 445.)—*Och. Tr.* vi. *b.* 251.—the looping Chimney Sweeper.—*Haw. l. c.*
Ph. Chærophyllata. *Linn.* ii. 866.—*Berk. S.* i. 145.—*Don.* vii. 48. *pl.* 253. *f.* 4.—*Stew.* ii. 167.—*Turt.* iii. 257.—(*Haw. Pr.* 27.)—*Haw.* 345.

Ph. atrata. *Linn. F. No.* 1274.
Ge. Chærophyllata. (*Sam. I.* 18.)
Chimney Sweeper. *Harr. A. pl.* 30. *f. o.*—*Harr. V. M.* 45.

6680. 2; Euphorbiata*. (*Och. Tr.* v. *b.* 445.)—*Och. Tr.* vi. *b.* 249.
Ph. Euphorbiata. (*Wien. V.* 116.)—*Fabr.* iii. *b.* 197.—*Don.* v. 31. *pl.* 153. *f.* 1.—(*Haw. Pr.* 27.)—*Stew.* ii. 171.—*Turt.* iii. 263.—*Haw.* 345.
Ge. Euphorbiaria. *Hüb. Ge. pl.* 15. *f.* 78.—(*Sam. I.* 18.)
Ge. murinata. *Scop. C.* 229.—*Villers E.* ii. 368.
Ph. fuscata. *Berl. M.* (*Hufnagle.*) iv. 524.
Ph. sordiata. (*Haw. Pr.* 26.)

6681 † 3. tinctaria. (*Och. Tr.* v. *b.* 445.)
Ge. tinctaria. *Hüb. Ge. pl.* 23. *f.* 121?
Ge. lutearia. *Fabr.* iii. *b.* 143.—*Esper.* v. *pl.* xxiv. *f.* 1.
An hujus generis? In Mus. D. ——?

6682 † 4. niveata[a].

Genus 216: (999). BAPTA *mihi.*

XERENE *p*, *Treitschke.*—MACARIA *p*, *Curtis.*

6683. 1; bimaculata*.
Ph. Ge. bimaculata. *Villers E.* ii. 373. *pl.* vi. *f.* 17.
Ph. bimaculata: the white Pinion-spotted. (*Haw. Pr.* 27.)—*Haw.* 356.
Ge. bimaculata. (*Sam. I.* 18.)
Ge. taminata. (*Wien. V.* 116.)
Ge. taminaria. *Hüb. Ge. pl.* 17. *f.* 90.
Ma. bimaculata. (*Curtis fo.* 132.)

6684. 2; punctata*.
Ph. punctata. *Fabr.* iii. *b.* 197. (!)—*Stew.* ii. 171.—*Turt.* iii. 263.
Ge. punctata. (*Sam. I.* 19.)
Ph. nubeculata: the clouded Silver. *Marsh. MSS.*—(*Haw. Pr.* 25.)—*Haw.* 355.
Ge. temerata. (*Wien. V.* 116.)
Ge. temeraria. *Hüb. Ge. pl.* 17. *f.* 91.
Ma. punctata. (*Curtis fo.* 132.)

Genus 217: (1000). EMMELESIA *mihi.*

A. ACIDALIA *et* CIDARIA *p*, *Treitschke.*

6685. 1; decolorata*.

---

[a] 6682 † 4. niveata. (*Och. Tr.* v. 446.)—*Och. Tr.* vi. *b.* 254.
Ph. niveata. *Scop. C.* 539.
Ge. nivearia. (*Wien. V.* 116.)—*Hüb. Ge. pl.* 41. *f.* 217.
Ph. nivearia. *Fabr.* iii. *b.* 129. (!)—*Stew.* ii. 59. (!)—*Turt.* iii. 231. (!)
Ge. farinata. *Bork. E. S.* v. 489.

Ge. decolorata. *Hüb. Ge. pl.* 47. *f.* 243.—(*Sam. I.* 18.)
Ph. decolorata: the sandy Carpet. *Haw.* 328.

6686. 2; Alchemillata*.
Ph. Ge. Alchemillata. *Linn.* ii. 869?—*Stew.* ii. 170?—*Turt.* iii. 262.
Ph. Alchemillata: the Rivulet. *Haw. Pr.* 26.—*Haw.* 335.
Ge. Alchemillata. *Hüb. Ge. pl.* 50. *f.* 261.—(*Sam. I.* 18.)

6687. 3; affinitata* *mihi.*

6688. 4; rivulata*.
Ge. rivulata. (*Wien. V.* 109.)—*Hüb. Ge. pl.* 50. *f.* 259.—(*Sam. I.* 19.)
Ph. rivulata: the middle Rivulet. *Haw.* 335.

6689. [5; nassata*]: the small Rivulet.
Ph. nassata. *Fabr.* iii. *b.* 204.—(*Haw. Pr. App.* 5.)—*Haw.* 335.
Ge. nassata. (*Sam. I.* 18.)

6690. 6, ericetata.
Ge. ericetata. *Dale? MSS.*

6691. 7; albulata*: the grass Rivulet.
Ph. albulata. (*Wien. V.* 109.)—*Fabr.* iii. *b.* 204.—*Haw.* 336.
Ge. albulata. *Hüb. Ge. pl.* 50. *f.* 257.—(*Sam. I.* 18.)

6692. 8, trigonata*.
Ph. trigonata: the small blue-border. *Haw. Pr.* 25.—*Haw.* 338.
Ge. trigonata. (*Sam. I.* 19.)

6693. 9, blandiata.
Ge. blandiata. (*Wien. V.* 109?)—*Hüb. Ge. pl.* 50. *f.* 258?

6694. 10; unifasciata*.
Ph. unifasciata: the single-barred Rivulet. *Haw.* 335.
Ge. unifasciata. (*Sam. I.* 19.)
Ge. Salicata. *Hüb. Ge. pl.* 53. *f.* 273?
Ac. Salicaria. (*Och. Tr.* v. *b.* 440.)—*Och. Tr.* vi. *b.* 64.

6695. 11; bifasciata*.
Ph. bifasciata: the double-barred Rivulet. *Haw.* 334.

6696. 12; rusticata*: the least Carpet.—*Haw. l.* c.
Ph. rusticata. (*Wien. V.* 113.)—*Fabr.* iii. *b.* 189.—(*Haw. Pr. App.* 5.)—*Haw.* 364.
Ge. rusticata. *Hüb. Ge. pl.* 46. *f.* 241.—(*Sam. I.* 19.)

6697 † 13, rubricata.
Ge. rubricata. (*Wien. V.* 110.)—*Fabr.* iii. *b.* 201.
Ge. rubricaria. *Hüb. Ge. pl.* 111. *f.* 487.
In Mus. D. *J. Standish.*

6698 † 14. purpurata[a].

B. ACIDALIA *p, Treit.*, (*Kirby.*)

6699. 15; sylvata*: the waved Carpet.—*Haw. l. c.*
Ph. sylvata. (*Wien. V.* 109.)—(*Haw. Pr. App.* 5.)—*Haw.* 329.
Ge. sylvata. *Hüb. Ge. pl.* 44. *f.* 231.—(*Sam. I.* 19.)
Ph. testaceata. *Don.* xiv. 51. *pl.* 487. *f.* 1.

6700. 16; candidata*: the small white Wave.—*Haw. l. c.*
Ph. candidata. (*Wien. V.* 110.)—*Haw.* 352.
Ge. candidaria. *Hüb. Ge. pl.* 19. *f.* 101.
Ph. immutata. *Fabr.* iii. *b.* 203.
Ph. albulata. *Berl. M.* (*Hufnagle.*) iv. 616.
Ge. candidulata. (*Sam. I.* 18.)

6701. 17; luteata*: the small yellow Wave.—*Haw. l. c.*
Ph. luteata. (*Wien. V.* 110.)—*Fabr.* iii. *b.* 199.—(*Haw. Pr. App.* 5.)—*Haw.* 352.
Ge. lutearia. *Hüb. Ge. pl.* 19. *f.* 103.
Ph. flavostrigata. *Don.* xi. 73. *pl.* 386. *f.* 1, 2.
Ge. luteata. (*Sam. I.* 18.)

C. FIDONIA *p, Treit.*—MACARIA *p, Curtis.*

6702. 18; heparata*.
Ge. heparata. (*Wien. V.* 116.)—(*Sam. I.* 18.)
Ge. hepararia. *Hüb. Ge. pl.* 11. *f.* 58.
Ph. heparata: dingy Shell. *Haw.* 343.
Ma. heparata. (*Curtis fo.* 132.)

Genus 218: (1001). HERCYNA *mihi.*

FIDONIA *p, et* XERÈNE *p, Treitschke.*—MACARIA *p, Curtis,* (*Kirby.*)

6703. 1; clathrata*.
Ph. Ge. clathrata. *Linn.* ii. 867.—*Stew.* ii. 168.—*Don.* vii. 86. *pl.* 248. *f.* 2.—*Turt.* iii. 256.
Ge. clathraria. *Hüb. Ge. pl.* 25. *f.* 112.
Ph. clathrata: the latticed Heath. (*Haw. Pr.* 27.)—*Haw.* 348.
Ge. clathrata. (*Sam. I.* 18.)
Ph. retialis. *Scop. C.* 217.
Scarce Heath. *Harr. V. M.* 31.
Var. Ph. radiata: the radiated Heath. *Haw.* 348.
Ph. retata: the netted Heath. *Haw.* 348.
Ge. retata. (*Sam. I.* 19.)
Ma. clathrata. (*Curtis folio* 132.)

---

[a] 6698 † 14. purpurata.
Ph. Ge. purpurata. *Linn. F. No.* 1302.—*Schæff. Ic.* 205. *f.* 4, 5?—*Turt.* iii. 265. (!)
An hujus generis?

6704. 2; macularia*.
Ph. Ge. macularia. *Linn.* ii. 862.—*Stew.* ii. 171.—*Turt.* iii. 263.
Ge. maculata. (*Wien. V.* 115.)—*Hüb. Ge. pl.* 25. *f.* 135.—(*Sam. I.* 18.)
Ph. maculata. *Don.* vii. 93. *pl.* 251. *f.* 3.—(*Haw. Pr.* 26.)—*Haw.* 343.
Ma. maculata. (*Curtis fo.* 132.)
Speckled Yellow. (*Wilkes D. pl.* 1. *f.* 5.)—*Harr. A. pl.* 28. *f.* l.—*Harr. V. M.* 59.

6705. 3: quadrimaculata*.
Ph. quadrimaculata: Pinion-spotted Yellow. *Ent. Trans.* (*Hatchett.*) i. 245. *pl.* 7. *f.* 3.—*Haw.* 343.
Ge. quadrimaculata. (*Sam. I.* 19.)
Ma. quadrimaculata. (*Curtis fo.* 132.)

Genus 219: (1002). HYRIA *mihi.*

FIDONIA *p, Treitschke.*—PYRALIS *p, Wien. V.*

6706. 1: auroraria*: Purple bordered Gold.—*Haw. l. c.*
Ge. auroraria. *Hüb. Ge. pl.* 12. *f.* 63.—*Haw.* 310.
Py. auroralis. (*Wien. V.* 124.)
Ge. variegata. *Fabr.* iii. *b.* 205.
Ge. sanguinaria. *Hüb. B.* i. *pl.* 3. *f. S. S.* 21.
Ph. muricata. *Berl. M.* (*Hufnagle.*) iv. 606.

Genus 220: (1003). PTYCHOPODA *mihi.*—(*Curtis fo.* 132.)

A. Tibiis posticis in masculis haud calcaratis, in fœminis bicalcaratis.

a. ACIDALIA *p, Treit.*

6707. 1; osseata*.
Ph. osseata. (*Wien. V.* 110.)—(*Sam. I.* 19.)
Ge. ossearia. *Hüb. Ge. pl.* 19. *f.* 102.
Ph. osseata: the dwarf Cream-wave. *Haw.* 353.

6708. 2; subochreata* *mihi.*

6709 † 3, marginepunctata *mihi.* In Mus. D. *Dale, Haworth et Stone.*

b. IDAËA *p, Treit.*

6710. 4; lividata*.
Ph. Ge. lividata. *Linn. F. No.* 1285.
Ge. lividata. (*Sam. I.* 18.)
Ph. lividata: the small Dotted-wave. *Haw.* 355.
Ph. scutata. *Fabr.* iii. *b.* 202.
Ge. scutularia. *Hüb. Ge. pl.* 14. *f.* 72.
Ph. scutulata. (*Haw. Pr.* 27?)
Ph. dimidiata. *Bork. E. S.* v. 522.

6711. 5; trigeminata*.

Ph. trigeminata: the treble Twin-spot. *Haw.* 354.
Ge. scutularia. *Hüb. Ge. pl.* 14. *f.* 73.
Ge. trigeminata. (*Sam. I.* 19.)

c. IDAËA *p, Treit.*

6712. 6; decoraria*.
Ge. decoraria. (*Wien. V.* 117.)—*Hüb. Ge. pl.* 14. *f.* 71?
Ph. cinerata. *Fabr.* iii. *b.* 200?

6713. 7; dilutaria*.
Ge. dilutaria. *Hüb. Ge. pl.* 19. *f.* 100.
Ph. dilutata: the small fan-footed Wave. *Haw.* 353.
Ge. dilutata. (*Sam. I.* 18.)
Ph. decorata. (*Haw. Pr.* 28.)

6714. 8; fimbriolata* *mihi.*

6715. 9; virgulata*.
Ge. virgularia. *Hüb. Ge. pl.* 19. *f.* 104?
Ge. virgulata. (*Sam. I.* 19.)
Ph. virgulata: the small Dusty-wave. *Haw.* 354.
Ph. virgata. (*Haw. Pr. App.* 5?)

6716. 10; subsericeata*.
Ph. subsericeata: the satiny Wave. *Haw.* 352.
Ge. subsericeata. (*Sam. I.* 19.)

6717. 11; inornata*.
Ph. inornata: the plain Wave. *Haw.* 349.
Ge. inornata. (*Sam. I.* 18.)
Ph. murinata. *Fabr.* iii. *b.* 198. (!)?—*Turt.* iii. 263?
Ida. suffusata. *Och. Tr.* vi. *b.* 272?

6718. 12; remutata*.
Ph: Ge. remutata. *Linn.* ii. 872.
Ge. remutata. (*Sam. I.* 19.)
Ph. remutata: the false ribband Wave. (*Haw. Pr.* 27.)—*Haw.* 349.
Ge. remutaria. *Hüb. Ge. pl.* 18. *f.* 98?
Ph. trilineata. *Berl. M.* (*Hufnagle.*) iv. 606?

6719. 13; aversata*.
Ph. Ge. aversata. *Linn.* ii. 869.
Ge. aversata. (*Sam. I.* 18.)
Ph. aversata: the ribband Wave. (*Haw. Pr.* 27.)—*Haw.* 349.
Ge. aversaria. *Hüb. Ge. pl.* 11. *f.* 56. ♂.—*Id. pl.* 75. *f.* 388. ♀.

6720. 14, fuliginata.
Ph. fuliginata: the dingy Wave. *Haw.* 350.

6721. 15, fumata. *Dale? MSS.*
Ph. pallidata. (*Wien. V.* 109?)
Ge. pallidaria. *Hüb. Ge. pl.* 18. *f.* 96?

B. Tibiis posticis in masculis haud calcaratis, in fœminis 4-calcaratis.

a. IDAEA p, *Treit.*

6722. 16; incanata.
Ph. Ge. incanata. *Linn.* ii. 871.
Ph. incanata: the Mullein Wave. *Haw.* 350.
Ge. immutaria. *Hüb. Ge. pl.* 20. *f.* 108.
Ge. incanata. (*Sam. I.* 18.)

6723. [17, contiguaria.]
Ge. contiguaria. *Hüb. Ge. pl.* 20. *f.* 105.
Ph. contiguata: the tooth-striped Wave. *Haw.* 350.

6724. 18; lactata*.
Ph. lactata: the pale cream Wave. *Haw.* 351.
Ph. nemorata. *Haw. Pr.* 27.
Ge. sericeata. *Hüb. Ge. pl.* 78. *f.* 404?
Ge. lactata. (*Sam. I.* 18.)

6725. [19; sublactata*.]
Ph. sublactata: the broad-striped cream-Wave. *Haw.* 351.
Ge. sublactata. (*Sam. I.* 19.)

6726. 20; centrata*.
Ge. centrata. *Fabr.* iii. *b.* 200?
Ge. remutaria. *Hüb. Ge. pl.* 18. *f.* 98?
Ph. floslactata: the cream-Wave. *Haw.* 351.
Ge. floslactata. (*Sam. I.* 18.)

b. IDAEA p, *Treit.*

6727. 21; ornata*.
Ge. ornata. (*Wien. V.* 117.)
Ge. ornataria. *Hüb. Ge. pl.* 14. *f.* 70.
Ph. ornata: the Lace border. *Haw.* 355.—*Turt.* iii. 265.
Ph. institata. *Berl. M.* (*Hufnagle.*) iv. 526.
Ge. paludata. (*Sam. I.* 19.)
Laced border. *Harr. V. M.* 15.

6728 † 22, limboundata[a].

6729. 23, immutata.
Ph. Ge. immutata. *Linn.* ii. 871.—*Stew.* ii. 171.—*Turt.* iii. 266.
Ph. immutata: the lesser cream-Wave. *Haw. Pr.* 27.—*Haw.* 352.
Ge. sylvestraria. *Hüb. Ge. pl.* 18. *f.* 97.
Ac. sylvestrata. (*Och. Tr.* v. *b.* 439.)
Ge. immulata. (*Sam. I.* 18.)

6730 † 24, subroseata.

---

[a] 6728 † 22, ? limboundata.
Ph. limboundata: the large Lace-border. *Haw.* 355. (')

Ph. subroseata: the rosy Wave. *Haw.* 351.
Ge. subroseata. (*Sam. I.* 19.) In Mus. D. *Haworth et Stone.*

C. Tibiis posticis in utroque sexû 4-calcaratis.

a. XERENE *p, Treit.*

6731. 25; marginata*.
Ph. Ge. marginata. *Linn.* ii. 870.—*Stew.* ii. 171.—*Don.* ix. 12. *pl.* 293. *f.* 2.—*Turt.* iii. 255.
Ge. marginata. (*Sam. I.* 18.)
Ph. Staphyleata. *Scop. C.* 221.
Ge. marginaria. *Hüb. Ge. pl.* 15. *f.* 79.
Ph. marginata. (*Haw. Pr.* 25.)—*Haw.* 338.
Clouded border. *Harr. Ex.* 26. *pl.* vi. *f.* 6.—*Harr. V. M.* 15.
Var. β; Ge. nævaria. *Hüb. Ge. pl.* 15. *f.* 80.
Ph. navata. (*Haw. Pr.* 25.)
Var. γ; Ge. pollutaria. *Hüb. Ge. pl.* 15. *f.* 77.
Ph. pollulata. (*Haw. Pr.* 25.)

b. GEOMETRA *p, Treit.*

6732. 26; putataria*.
Ph. Ge. putataria. *Linn.* ii. 859.—*Stew.* ii. 160.—*Turt.* iii. 232.
Ge. putataria. *Hüb. Ge. pl.* 2. *f.* 10.—*Haw.* 300.—(*Sam. I.* 19.)—*Scop. C.* 211.
Ge. lactearia? little Emerald. (*Haw. Pr.* 21.)
Pale Green Housewife. *Harr. V. M.* 31.

D. Tibiis posticis in masculis bicalcaratis, in fœminis 4-calcaratis.

GEOMETRA *p, Treit.*—MACARIA *p, Curtis.*

6733. 27; Thymiaria*.
Ph. Ge. Thymiaria. *Linn.* ii. 859?
Ge. Thymiaria: common Emerald. (*Haw. Pr.* 21.)—*Haw.* 300.—(*Sam. I.* 19.)
Ph. Ge. strigata. *Villers E.* ii. 378.
Ph. fimbriata. *Berl. M.* (*Hufnagle.*) iv. 604.
Ge. Æstivaria. *Bork. E. S.* v. 32.—*Hüb. Ge. pl.* 2. *f.* 9.
Ph. vernaria. *Fabr.* iii. *b.* 129.—*Don.* ix. 47. *pl.* 310.
Ma. Thymiaria. (*Curtis fo.* 132.)
Albin. pl. xlviii. *f.* 81. *d–g.*
Small green Housewife. *Harr. A. pl.* 3.
Large green Housewife. *Harr. V. M.* 31.

6734. [28, vernaria*.]
Ge. vernaria. *Esper.* v. *pl.* 1. *f.* 6–9?

6735 † 29, vibicaria.
Ph. Ge. vibicaria. *Linn.* ii. 859.—*Stew.* ii. 66.—*Haw.* 310.
Ge. vibicaria. (*Wien. V.* 117.)—*Hüb. Ge. pl.* 10. *f.* 50.
Ph. rubrofasciata. *Berl. M.* (*Hufnagle.*) iv. 612.
Ph. cruentata. *Scop. C.* 226.
An hujus generis? In Mus. D. *Swainson.*

6736 † 30. Lactearia[a].

Genus 221: (1004). MACARIA, *Curtis*.

ENNOMOS *p*, *Treitschke*, (*Kirby*.)

A. Tibiis posticis in masculis haud calcaratis, in fœminis 4-calcaratis.

6737. 1; imitaria*.
Ge. imitaria*: Small Blood vein.—*Haw. l.c. Hüb. Ge. pl.* 10. *f.* 51.—(*Haw. Pr.* 22.)—*Haw.* 297.—(*Sam. I.* 18.)
Ma. imitata. (*Curtis l. c. infra.*)

6738. 2; variegata*.
Ph. variegata. *Scop. C.* 222.
Ge. strigillaria. *Hüb. Ge. pl.* 20. *f.* 109.
Ph. strigillata: the subangled Wave. *Haw.* 350.
Ph. strigillaria. (*Ing. Inst.* 91.)
Ge. strigilata. (*Sam. I.* 19.)
Ph. mediata. *Fabr.* iii. *b.* 198?

6739 † 3, emutaria. (*Ing. Inst.* 91.)
Ge. emutaria. *Hüb. Ge. pl.* 63. *f.* 323. In Mus. D. *Stone*.

B. Tibiis posticis in utroque sexû 4-calcaratis.

6740. 4; notata*. (*Curtis l. c. infra.*)
Ph. Ge. notata. *Linn.* ii. 866.—*Turt.* iii. 251.
Ge. notata: the Peacock M. (*Sam. I.* 18.)
Ph. notata. (*Haw. Pr.* 27.)—*Haw.* 346.
Ge. notataria. *Hüb. Ge. pl.* 11. *f.* 53.—*Id. pl.* 61. *f.* 316.
Ph. exustata. *Berl. M.* (*Hufnagle.*) iv. 600.

6741. 5; alternata*.
Ge. alternata. (*Wien. V.* 106.)
Ge. lituraria: (alternaria.) *Hüb. Ge. pl.* 61. *f.* 315.
Ph. prænotata: the sharp-angled Peacock. *Haw.* 346.
Ge. prænotata. (*Sam. I.* 19.)
Ma. prænotata. (*Curtis l. c. infra.*)

6742 † 6, subrufata.
Ph. subrufata: the slender-striped Rufous. *Haw.* 347.
In Mus. D. *Haworth*.

6743. 7, liturata*. *Curtis.* iii. *pl.* 132.
Ph. Ge. liturata. *Linn. F. No.* 1273.
Ph. liturata: the tawny-barred Angle. *Haw.* 346.
Ge. liturata. (*Sam. I.* 18.)
Ge. lituraria. *Hüb. Ge. pl.* 11. *f.* 54.
Ge. alternaria: (lituraria.) *Hüb. Ge. pl.* 61. *f.* 314.

---

[a] 6736 † 30. Lactearia.
Ph. Ge. Lactearia. *Linn.* ii. 858.—*Turt.* iii. 231. (!)
An hujus generis?—MINOA *p*, *Treit.*, (*Kirby.*)

Ph. obscurata. (*Haw. Pr.* 27.)
Ph. igneata. *Berl. M.* (*Hufnagle.*) iv. 612.

6744. 8, præatomata. (*Curtis l. c. supra.*)
Ph. præatomata: the dingy Angled. *Haw.* 345.

6745. 9, unipunctata.
Ph. unipunctata: the white Spot. *Haw.* 345.

C. Tibiis posticis in masculis haud calcaratis, in fœminis bicalcaratis.

6746. 10, limbata.
Ph. limbata: the bordered Checquer. *Haw.* 346.

6747. 11; emarginata*. (*Curtis l. c. supra.*)
Ph. Ge. emarginata. *Linn.* ii. 866.—*Turt.* iii. 252.
Ge. emarginataria. *Hüb. Ge. pl.* 20. *f.* 107.
Ph. emarginata: the scolop'd Double line. *Haw.* 347.
Ge. emarginata. (*Sam. I.* 18.)
Ph. demandata. *Fabr.* iii. *b.* 167.—(*Haw. Pr.* 27.)
Ph. quadripunctata. *Don.* xiv. *pl.* 493. *f.* 3.
Ph. erosata. *Berl. M.* (*Hufnagle.*) iv. 526.
♀; Ph. dimidiata: the small Scolop. (*Haw. Pr.* 27.)—*Haw.* 347.
Ma. dimidiata. (*Curtis l. c. supra.*)
Ge. dimidiata. (*Sam. I.* 18.)
Ph. rumigerata. *Don.* xiv. *pl.* 493. *f.* 2.

Genus 222: (1005). ENNOMOS, *Treitschke.*

BOMBYX *p, Borkhausen.*—FALCARIA *p, Haworth.*—PYRALIS *p, Haworth Pr.*—PLATYPTERYX *p, Samou.*—LASPERIA? *Germar.*

6748. 1; flexula*.
Ph. flexula. *Fabr.* iii. *b.* 166.
Fa. flexula: the beautiful Hook-tip. *Haw.* 154.
Ge. flexularia. *Hüb. Ge. pl.* 4. *f.* 19.
Py. flexulalis. (*Haw. Pr.* 28.)
Ph. Ge. sinuata. *Villers E.* ii. 334.
Ph. flexula. (*Sam. I.* 34.)

Familia XVI: (129). PLATYPTERICIDÆ.

(PH.-GEOMETRA *p, et* PH.-BOMBYX *p, Linné.*—PHALÆNA *p, Fabr.*) *Hook-tips.*

Genus 223: (1006.) PLATYPTERYX, *Laspeyres, Leach, Samou.,* (*Kirby.*)

DREPANA *p, Schrank,* (*Kirby.*)—PYRALIS *p, Haw. Pr.*—FALCARIA *p, Haworth.*

6749. 1; lacertula*.
Ge. lacertula. (*Wien. V.* 64.)—*Hüb. Bo. pl.* 12. *f.* 49. ♀.
Ph. Ge. lacertinaria. *Linn.* ii. 860.—*Berk. S.* i. 144.—*Don.* vii. 92. *pl.* 251. *f.* 2.—*Stew.* ii. 160.—*Turt.* iii. 233.

Ph. Scincula. *Hüb. Ge. pl.* 12. *f.* 50. ♂.
Ge. dentaria. *Thunb. Diss.* i. 5.
Py. lacertinalis. (*Haw. Pr.* 28.)
Fa. Lacertula. *Haw.* 153.
Ph. lacertanaria. (*Leach E. E.* ix. 134.)—(*Sam. I.* 254.)
Scallap Hook-tip. (*Harr. V. M.* 49.)
β; Fa. curvula: the bordered Hook-tip. *Haw.* 154.
Ph. curvula. (*Sam. I.* 34.)

6750. [2: cultraria*.]
Ph. cultraria. (*Leach E. E.* ix. 134.)—(*Kirby & Sp. I. E.* iii. 260.)
Fa. Lacertula β. *Haw.* 154.

Genus 224: (1007). DREPANA, *Lespeyres, Leach, Samou.*

DREPANA *p, Schrank.*—PYRALIS *p, Haw. Pr.*—FALCARIA *p, Haworth.*—BOMBYX *p, Hübner.*—PLATYPTERYX *p, Och.*, (*Kirby.*)

6751. 1; falcataria*. (*Leach E. E.* ix. 134.)—(*Samou.* 254.)
Ph. Ge. falcataria. *Linn.* ii. 859.—*Turt.* iii. 233.
Py. falcatarialis. (*Haw. Pr.* 28.)
Bo. falcula. (*Wien. V.* 64.)—*Hüb. Bo. pl.* 11. *f.* 44.
Fa. falcula. *Haw.* 152.
Pebble Hook-tip. *Harr. V. M.* 49.

6752. 2; hamula*. (*Sam. I.* 34.)
Bo. hamula. (*Wien. V.* 64.)—*Hüb. Bo. pl.* 12. *f.* 46, 47.
Fa. hamula: Oak Hook-tip. *Haw.* 153.
Ph. falcata. *Fabr.* iii. *b.* 165. (!)—*Stew.* ii. 162.—*Turt.* iii. 243.
Py. falcatalis. (*Haw. Pr.* 28.)
Ph. binaria. *Berl. Mag.* (*Hufnagle.*) iv. 516.
Albin. pl. lxv. *f. a–d.*
Wild rose M.: Ph. Lacertinaria? (*Wilkes.* 14. *pl.* 30.)
Brown Hook-tip. *Harr. A. pl.* 41. *f. a–b.*—*Harr. V. M.* 49.

6753. [3, uncula*.]
Bo. uncula. *Hüb. Bo. pl.* 12. *f.* 45.
Bo. Uncinula. *Bork. E. S.* iii. 461.
Ph. hamula var. (*Och.* iv. 97.)

6754. 4: unguicula*.
Bo. unguicula. *Hüb. Bo. pl.* 12. *f.* 48.
Fa. unguicula: the barred Hook-tip. *Haw.* 153.
Py. cultrarialis. (*Haw. Pr.* 28.)
Bo. sicula. *Esper.* iii. *pl.* lxxiv. *f.* 4–7.
Ph. cultraria. *Fabr.* iii. *b.* 133?

6755 † 5, fasciata *mihi.*
Fa. uncula: the scarce Hook-tip. *Haw.* 153.
In Mus. *D. Swainson.*

Genus 225: (1008). CILIX, *Leach, Samou.*

PHALÆNA-ATTACUS *p, Linné?*—BOMBYX *p, Fabricius.*—PLATYPTERYX *p, Latr.*—FALCARIA *p, Haworth.*

6756. 1; compressa*. (*Leach E. E.* ix. 134.)—(*Samou.* 254.)
Bo. compressa. *Fabr.* iii. 455.
Bo. compressus. (*Haw. Pr.* 10.)—*Haw.* 110.
Ph. Bo. compressa. *Don.* vii. *pl.* 239. *f.* 2.—*Stew.* ii. 155.—*Turt.* iii. 216.
Bo. spinula. (*Wien. V.* 64.)—*Hüb. Bo. pl.* 11. *f.* 40.
Fa. spinula: the Chinese character. *Haw.* 154.
Ph. At. ruffa. *Linn.* ii. 1068?
Ge. modesta. *Naturforscher.* xiii. 27. *pl.* 3. *f.* 4.
Ph. glaucata. *Scop. C. No.* 221. *pl.* xxxii. *f.* 549.
Albin. pl. lxv. *f. e–h.*
Goose Egg M. *Harr. V. M.* 21.

Familia XVII: (130). PYRALIDÆ, *Leach.*

(PHALÆNA-PYRALIS, *Linné, &c.*)—CRAMBITES *p, et* TORTRICES *p, Latr.*)

Genus 226: (1009). HYPENA, *Schrank.*

CRAMBUS *p, Fabr., Haworth.*—PYRALIS *p, Hübner, Haw. Pr.*—HERMINIA, *Latr., Leach, Samou.,* (*Kirby.*)

6757. 1; proboscidalis*. (*Och. Tr.* vi. *b.* 311.)
Ph. Py. proboscidalis. *Linn.* ii. 881.—*Stew.* ii. 180.
Py. proboscidalis. *Hüb. Py. pl.* 2. *f.* 7.—(*Haw. Pr.* 28.)
Cr. proboscideus. *Fabr. Sup.* 465.
Cr. proboscidatus: the Snout. *Haw.* 365.
He. proboscidalis. (*Leach E. E.* ix. 134.)—(*Samou.* 253.)
Ph. rostralis. *Berk. S.* i. 146.
Ph. Ge. proboscidalis. *Turt.* iii. 274.
Ph. ensalis. *Fabr.* iii. *b.* 221.
Snout Eggar likeness. *Harr. A. pl.* 31. *f. f–h.*—*Harr. V. M.* 23.

6758 † 2, obesalis*? (*Och. Tr.* vi. *b.* 312.)
Py. crassalis. *Hüb. Py. pl.* 2. *f.* 8.—(*Haw. Pr.* 29.)
Cr. crassatus: the pinion Snout. *Haw.* 366.
He. crassalis. (*Sam. I.* 21.) In Mus. D. *Haworth.*

6759. 3; rostralis*. (*Och. Tr.* vi. *b.* 312.)
Ph. Py. rostralis. *Linn.* ii. 881.—*Stew.* ii. 180.
Py. rostralis. *Hüb. Py. pl.* 2. *f.* 10.—(*Haw. Pr.* 29.)
Ph. Ge. rostralis. *Turt.* iii. 274.
He. rostralis. (*Sam. I.* 21.)
Cr. rostratus: the buttoned Snout. *Haw.* 366.
Eggar likeness. *Harr. V. M.* 23?
β; Cr. palpatus: the dark Snout. *Haw.* 366.

Ph. palpalis. *Fabr.* iii. *b.* 223. (!)
Py. palpalis. *Hüb. Py. pl.* 2. *f.* 9?—*Stew.* ii. 180.—*Turt.* iii. 275.

γ; Cy. vittatus: the cream edged Snout. *Haw.* 367.
Py. vittalis. (*Haw. Pr.* 29.)
He. vittalis. (*Sam. I.* 21.)
Py. radiatalis. *Hüb. Py. pl.* 20. *f.* 134?

6760. 4, crassalis*. (*Och. Tr.* vi. *b.* 311.)
Cra. crassalis. *Fabr. S.* 466.
Py. achatalis. *Hüb. Py. pl.* 2. *f.* 12. ♀. *pl.* 27. *f.* 172. ♂.—(*Haw. Pr.* 29.)
Cr. achatatus: the beautiful Snout. *Haw.* 367.
He. achatalis. (*Sam. I.* 21.)

Genus 227: (1010). POLYPOGON, *Schrank.*

HERMINIA *p*, *Latreille, Samou.*—CRAMBUS *p*, *Fabr.*, *Haworth.*—PYRALIS *p*, *Hüb.*, *Haw. Pr.*—HERMINIA, *Treitschke.*

6761. 1, cribralis.
Py. cribralis. *Hüb. Py. pl.* 1. *f.* 2.

6762. 2; barbalis*.
Ph. Ge. barbalis. *Linn.* ii. 881.—*Turt.* iii. 273.
Cr. barbatus. *Fabr. Sup.* 464.—*Haw.* 368.
Py. barbalis: the common Fan-foot. (*Haw. Pr.* 29.)
He. barbalis. (*Sam. I.* 21.)
Py. pectitalis. *Hüb. Py. pl.* 19. *f.* 122.
Ph. Ge. strigillata. *Linn.* ii. 872.—*teste Haworthi.*
Albin. pl. lxxiv. *f. e–h.*
Snout. *Harr. A. pl.* 43. *f. l, t.*—*Harr. V. M.* 43.
The Fan-footed. *Harr. Ex.* 25. *pl.* vi. *f.* 2.—*Harr. V. M.* 25.

6763. 3: derivalis*.
Py. derivalis. *Hüb. Py. pl.* 3. *f.* 19.
Cr. derivatus: the Clay Fan-foot. *Haw.* 369.
He. derivalis. (*Sam. I.* 21.)

6764. 4; tarsicrinalis*.
Py. tarsicrinalis. *Hüb. Py. pl.* 1. *f.* 5.
Cr. tarsicrinatus: the Fan-foot. *Haw.* 369.
He. tarsicrinalis. (*Sam. I.* 21.)
Py. tentaculalis. (*Haw. Pr.* 29.)—*Turt.* iii. 274?

6765. 5; nemoralis*.
Ph. nemoralis. *Fabr.* iii. *b.* 222. (!)—*Stew.* ii. 180.—*Turt.* iii. 275.
Py. grisealis. (*Wien. V.* 120.)—*Hüb. Py. pl.* 1. *f.* 4.—(*Haw. Pr.* 30.)
He. nemoratus: the small Fan-foot. *Haw.* 370.
He. nemoralis. (*Sam. I.* 21.)

6766. 6, emortualis.

Py. emortualis. *Hüb. Py. pl.* 1. *f.* 1.—(*Haw. Pr.* 30.)
Cr. emortuatus: the olive Crescent. *Haw.* 369.
Py. glaucinalis. (*Wien. V.* 120.)
Ge. olivaria. *Bork. E. S.* v.

Genus 228: (1011). MADOPA *mihi.*

HYPENA *p, Treitschke.*

6767. 1, Salicalis*.
Ph. Salicalis. *Fabr.* iii. *b.* 227.—*Turt.* iii. 277.
Py. Salicalis. (*Wien. V.* 285. *pl.* 1. *a. f.* 5.)—*Hüb. Py. pl.* 1. *f.* 3.—(*Haw. Pr.* 30.)
Cr. Salicatus: the lesser Bell. *Haw.* 370.
Her. Salicalis. (*Sam. I.* 21.)

Genus 229: (1012). CLEDEOBIA *mihi.*

6768. 1; angustalis*.
Py. angustalis. *Hüb. Py. pl.* 4. *f.* 21.—(*Haw. Pr.* 31.)
Cr. angustatus: the small Snout. *Haw.* 368.
He. angustalis. (*Sam. I.* 21.)
Py. curtalis. (*Wien. V.* 120.)
♀; Py. Bombycalis. *Hüb. Py. pl.* 19. *f.* 123.
Cr. Bombycatus: the long-tailed Snout. *Haw.* 368.
He. Bombycalis. (*Sam. I.* 21.)
Py. rufescalis. (*Haw. Pr.* 30?)
Py. erigalis. *Fabr.* iii. *b.* 237.

6769. 2; albistrigalis*.
Cr. albistrigatus: the white-line Snout. *Haw.* 368.
He. albistrigalis. (*Sam. I.* 21.)

6770. 3, costæstrigalis.
Cr. costæstrigalis. *Haworth? MSS.*

Genus 230: (1013). ——

6771. 1, undulalis.
Py. undulalis. *Haworth MSS.*

6772. 2, bistrigalis *mihi.*

Genus 231: (1014). AGLOSSA, *Latreille,* (*Kirby.*)

CRAMBUS *p, Fabr., Haworth.*—PYRALIS, *Schrank.*

6773. 1; pinguinalis*. *Latr. H.* xiv. 229.—(*Kirby & Sp. I. E.* iii. 156. *nota.*)
Ph. Py. pinguinalis. *Linn.* ii. 882.—(*Haw. Pr.* 29.)—*Stew.* ii. 181.—*Turt.* iii. 279.
Cr. pinguinalis. (*Sam. I.* 13.)
Py. pinguinalis. *Hüb. Py. pl.* 4. *f.* 24.—(*Leach E. E.* ix. 135.) —(*Sam. I.* 35.)
Cr. pinguis. *Fabr. Sup.* 468.

Cr. pinguinatus: the large Tabby. *Haw.* 371.
Cr. pinguinalis. (*Kirby & Sp. I. E.* i. 135.)
He. pinguinalis. (*Sam. I.* 21.)
Tabby. *Harr. V. M.* 47.
β, Cr. pinguiculatus: the Tabby. *Haw.* 372.
Py. pinguiculalis. (*Haw. Pr.* 29.)

6774. 2; capreolatus*.
Py. capreolata. *Hüb. Py. pl.* 23. *f.* 153.
Cr. capreolatus: the small Tabby. *Haw.* 372.
Py. capreolatus. (*Sam. I.* 35.)

6775. 3: dimidiatus*[a].

Genus 232: (1015). PYRALIS *mihi, Samou.*

CRAMBUS *p, Fabr., Haworth.*—BOTYS *p, Latreille.*—AGROTERA *p, Schrank.*—ASOPIA, *Treitschke.*

6776. 1; farinalis*. *Hüb. Py. pl.* 15. *f.* 95.—(*Haw. Pr.* 31.)—(*Sam. I.* 35.)
Ph. Py. farinalis. *Linn.* ii. 881.—*Turt.* iii. 273.
Cr. farinatus: the Meal Moth. *Haw.* 374.

6777 ‡ 2, marginatus.
Cr. marginatus: the scarce Meal Moth. *Haw.* 374.
In Mus. D. *Raddon et Swainson.*

6778. 3; glaucinalis*. (*Haw. Pr.* 31.)—(*Sam. I.* 35.)
Ph. Py. glaucinalis. *Linn.* ii. 881.—*Turt.* iii. 273.
Cr. glaucinatus: the double-striped.—*Haw. l. c.* *Fabr. Sup.* 464.—*Haw.* 374.
Py. nitidalis. *Hüb. Py. pl.* 15. *f.* 98.

Genus 233: (1016). AGROTERA, *Schrank?*

ASOPIA A. *p, Treitschke.*

6779. 1; costalis*.
Ph. costalis. *Fabr.* iii. *b.* 240. (!)—*Stew.* ii. 181.—*Turt.* iii. 284.
Cr. costalis. *Haw.* 375.
Py. costalis. (*Haw. Pr.* 29.)—(*Sam. I.* 35.)
Py. fimbrialis. *Hüb. Py. pl.* 15. *f.* 77.
Gold Fringe. *Harr. V. M.* 27.

Genus 234: (1017). ——

ASOPIA B. *p, Treitschke.*

6780. 1; flamealis*.

---

[a] 6775. 3: dimidiatus*.
Py. dimidialis. (*Haw. Pr.* 29.) (!)
Cr. dimidiatus: the Tea Tabby. *Haw.* 372. (!)
He. dimidiatus. (*Sam. I.* 21.) (!)

Py. flamealis. *Hüb. Py. pl.* 15. *f.* 99.—(*Haw. Pr.* 29.)
Cr. flameatus: the rosy Flounced. *Haw.* 375.
He. flamealis. (*Sam. I.* 21.)

Genus 235: (1018). SIMAËTHIS, *Leach, Samou.*

ANTHOPHILA, *Haworth.*—XYLOPODA, *Latr.*—ASOPIA B. *p*, *Treitschke.* *Nettle-taps.*

6781. 1; Fabriciana*.
Ph. To. Fabriciana. *Linn.* ii. 880.
An. Fabricii: Fabricius' Nettle-tap. *Haw.* 471.
Ph. To. Urticana. (*Wien. V.* 132.)—*Villers E.* ii. 418.
Ph. Ti. Oxyacanthella. *Linn.* ii. 886.
An. Oxyacanthæ: autumn Nettle-tap. *Haw.* 471.
Si. dentana. (*Leach E. E.* ix. 135.)—(*Samou.* 254.)
Si. dentata. (*Sam. I.* 38.)
Aso. alternalis. (*Och. Tr.* vi. *b.* 317.)
Nettle-tap. *Harr. V. M.* 51.
To. dentana. *Hüb. To. pl.* 1. *f.* 4, 5.—(*Sam. I.* 41.)

6782. 2; pariana*.
Ph. To. pariana. *Linn.* ii. 880.—*Hüb. To. pl.* 1. *f.* 1, 2.—*Turt.* iii. 357.
An. par: double-barred Nettle-tap. *Haw.* 471.
As. parialis. (*Och. Tr.* vi. *b.* 317.)

6783. 3; lutosa*.
An. lutosa: early Nettle-tap. *Haw.* 472.
To. lutosa. (*Sam. I.* 42.)

6784. 4, Myllerana*.
Py. Myllerana. *Fabr.* iii. *b.* 277.
An. Mylleri: Myller's Nettle-tap. *Haw.* 472.
To. Mylleri. (*Sam. I.* 42.)

6785. 5, punctosa.
An. punctosa: narrow Silver-dotted. *Haw.* 472.

Genus 236: (1019). ENNYCHIA, *Treitschke.*

PYRAUSTA *p*, *Curtis.*

6786. 1; octomaculata*.
Ge. octomaculata. *Linn. M.* 540.
Ph. atralis. *Fabr.* iii. *b.* 241.—*Don.* viii. 39. *pl.* 266. *f.* 4.—*Stew.* ii. 181.—*Turt.* iii. 284.
Py. atralis. (*Haw. Pr.* 29.)—*Haw.* 388.
Bo. atralis. (*Sam. I.* 7.)
Py. atralis. (*Curtis l. c. infra.*)
Py. guttalis. (*Wien. V.* 124.)—*Hüb. Py. pl.* 12. *f.* 75.
No. trigutta. *Esper.* iv. *pl.* clxiii. *f.* 6.

White-spot. (*Wilkes D. pl.* 9. *f.* 3.)—*Harr. A.* 27. *f. p.*—*Harr. V. M.* 43.

6787. 2; anguinalis*. (*Och. Tr.* vi. *b.* 319.)
Py. anguinalis*: the wavy-barred Sable.—*Haw. l. c.*
Py. anguinalis. *Hüb. Py. pl.* 5. *f.* 32.—*Haw.* 389.
Bo. anguinalis. (*Sam. I.* 7.)
Pyr. anguinalis. (*Curtis l. c. infra.*)

6788. 3, cingulata: the silver-barred Sable.—*Haw. l. c.*
Ph. Ge. cingulata. *Linn.* ii. 874.—*Turt.* iii. 267.
Py. cingulalis. *Hüb. Ge. pl.* 5. *f.* 30.—(*Haw. Pr.* 29.)—*Haw.* 389.
Bo. cingulatus. (*Sam. I.* 7.)
Pyr. cingulalis. *Curtis.* iii. *pl.* 128.

6789 † 4, fascialis*?
Py. fascialis: the Flounced. *Haw.* 390. In Mus. D. *Haworth.*

Genus 237: (1020). PYRAUSTA, *Schrank*, (*Curtis.*)

PHALÆNA-GEOMETRA *p*, *Linné.*—CRAMBUS *p*, *Fabricius.*—BOTYS *p*, *Latr.*, *Leach*, *Samou.*

6791. 1; purpuralis*. (*Curtis folio* 128.)
Ph. Py. purpuralis. *Linn.* ii. 883.—*Don.* x. 34. *pl.* 339. *f.* 2.—*Turt.* iii. 283.
Py. purpuralis. (*Haw. Pr.* 29.)—*Haw.* 388.
Bo. purpuralis. (*Sam. I.* 7.)
Py. coccinalis. *Hüb. Py. pl.* 6. *f.* 37.
Crimson and Gold. (*Wilkes D. pl.* 5. *f.* 2.)—*Harr. A. pl.* 28. *f. m.*—*Harr. V. M.* 19.

6792. 2; punicealis*. (*Curtis l. c. supra.*)
Py. punicealis: the Purple and Gold.—*Haw. l. c.* *Fabr.* iii. *b.* 239.—(*Haw. Pr.* 29.)—*Hüb. Py. pl.* 6. *f.* 34?—*Haw.* 239.
Bo. punicealis. (*Sam. I.* 7.)

6793. [3; ostrinalis*.] (*Curtis l. c. supra.*)
Py. ostrinalis: the scarce Purple and Gold.—*Haw. l. c.* *Hüb. Py. pl.* 17. *f.* 113.—*Haw.* 389.
Bo. ostrinalis. (*Sam. I.* 7.)

6794. 4; Porphyrialis*.
Py. Porphyrialis. *Fabr.* iii. *b.* 239.
Py. Porphyrialis. *Hüb. Py. pl.* 6. *f.* 36?—(*Haw. Pr.* 29.)
Py. porphyralis: the Porphyry. *Haw.* 389.
Pyr. porphyralis. (*Curtis l. c. supra.*)
Bo. Porphyrialis. (*Sam. I.* 7.)

6795. 5; cespitalis*. (*Curtis l. c. supra.*)
Ph. cespitalis: the Straw-barred. *Fabr.* iii. *b.* 238.
Py. cespitalis. *Hüb. Py. pl.* 6. *f.* 39.—*Haw.* 390.
Bo. cespitalis. (*Sam. I.* 7.)

6796. [6; sordidalis*.] (*Curtis l. c. supra.*)
Py. sordidalis: the dingy Straw-barred.—*Haw. l. c. Hüb. Py. pl.* 7. *f.* 40.—*Haw.* 391.
Bo. sordidalis. (*Sam. I.* 7.)

Genus 238: (1021). HYDROCAMPA, *Latreille*, (*Kirby.*)
BOTYS *et* NYMPHULA *p*, *Schrank.*

6797. 1; literalis*: the lettered China-mark.—*Haw. l. c.*
Py. literalis. (*Wien. V.* 121.)—*Hüb. Py. pl.* 13. *f.* 86.—(*Haw. Pr.* 30.)—*Haw.* 384.
Bo. literalis. (*Sam. I.* 7.)
Ph. argentalis. *Fabr.* iii. *b.* 240?

6798. 2; Sambucata*: the garden China-mark.—*Haw. l. c.*
Ph. Sambucata. *Fabr.* iii. *b.* 215.
Py. Sambucalis. (*Wien. V.* 124.)—*Hüb. Py. pl.* 13. *f.* 81.—(*Haw. Pr.* 30.)—*Haw.* 383.
Bo. Sambucata. (*Sam. I.* 7.)
Albin. pl. xxxvii. *f.* 59. *a–d.*
Brown China Mark. *Harr. V. M.* 37.

6799. 3; Potamogata*.
Ph. Ge. Potamogata. *Linn.* ii. 873.—*Don.* xi. 9. *pl.* 363. *f.* 1.
Py. Potamogalis. (*Haw. Pr.* 30.)—*Haw.* 382.
Bo. Potamogata. (*Leach E. E.* ix. 135.)
Bo. Potamogeta. (*Sam. I.* 7.)
Py. Nymphealis. *Hüb. Py. pl.* 13. *f.* 85.
Brown China Mark. *Harr. A. pl.* 41. *f. o–q.*
Large China Mark. *Harr. Ex.* 30. *pl.* viii. *f.* 2.

6800. 4; Nymphæata*.
Ph. Ge. Nymphæata. *Linn.* ii. 873.—*Stew.* ii. 172.—*Turt.* iii. 271.
Py. Nymphealis: the beautiful China-mark. (*Haw. Pr.* 30.)—*Haw.* 383.
Py. Potamogalis. *Hüb. Py. pl.* 13. *f.* 82.
Ph. stagnata. *Don.* xi. 10. *pl.* 363. *f.* 2.
Bo. Nymphæata. (*Sam. I.* 7.)

6801. 5; Lemnata*: the small China-mark.—*Haw. l. c.*
Ph. Ge. Lemnata. *Linn.* ii. 874.—*Don.* viii. 37. *pl.* 266. *f.* 1, 2.—*Stew.* ii. 172.—*Turt.* iii. 271.
Py. Lemnalis. *Hüb. Py. pl.* 13. *f.* 83. ♂. *f.* 84. ♀.—(*Haw. Pr.* 30.)—*Haw.* 384.
Bo. Lemnata. (*Leach E. E.* ix. 135.)—(*Sam. I.* 7.)
Small white China Mark. *Harr. A. pl.* 7. *f. k–m.*—*Harr. V. M.* 37.
♀; Ph. uliginata. *Fabr.* iii. *b.* 214.

6802. 6; Stratiotata*: the ringed China-mark.—*Haw. l. c.*
Ph. Ge. Stratiotata. *Linn.* ii. 873.—*Turt.* iii. 270.

Py. Stratiotalis. *Hüb. Py. pl.* 13. *f.* 87. ♂.—(*Haw. Pr.* 30.) —*Haw.* 383.
Bo. Stratiotalis. (*Sam. I.* 7.)—(*Kirby & Sp. I. E.* iv. 56 & 74.)
Ph. Ge. paludata. *Fabr.* iii. *b.* 213. (!)—(*Haw. Pr.* 25.)—*Turt.* iii. 270.
Rush China Mark. *Harr. V. M.* 35.

Genus 239: (1022). ——

6803. 1; hybridalis*: the rush Veneer.—*Haw. l. c.*
Py. hybridalis. *Hüb. Py. pl.* 17. *f.* 114.—*Haw.* 386.
Bo. hybridalis. (*Sam. I.* 7.)
Tin. Noctuella. (*Wien. V.* 136.)
Rush Vanear. *Harr. V. M.* 53.

Genus 240: (1023). DIAPHANIA *mihi.*

6804. 1, lucernalis*? the transparent China-mark.—*Haw. l. c.*
Py. lucernalis. *Hüb. Py. pl.* 17. *f.* 108.—(*Haw. Pr.* 30.)—*Haw.* 384.

Genus 241: (1024). ——

Botys *p*, *Latreille.*—Nymphula *p*, *Schrank.*

6805. 1; forficalis*: the garden Pebble.—*Haw. l. c.*
Ph. Py. forficalis. *Linn.* ii. 882.—*Turt.* iii. 275.
Py. forficalis. *Hüb. Py. pl.* 9. *f.* 58.—(*Haw. Pr.* 30.)—*Haw.* 377.
Cr. forficatus. *Fabr. Sup.* 467.
Bo. forficalis. (*Sam. I.* 7.)

Genus 242: (1025). BOTYS.

Botys *p*, *Latreille.*—Nymphula *p*, *Schr.*—Margaritia *p*, (*Kirby.*)

6806. 1; Urticata*.
Ph. Ge. Urticata. *Linn.* ii. 873.—*Berk. S.* i. 145.—*Don.* x. 64. *pl.* 349. *f.* 2.—*Stew.* ii. 172.—*Turt.* iii. 269.
Py. Urticalis. *Hüb. Py. pl.* 12. *f.* 78.—(*Haw. Pr.* 30.)—*Haw.* 382.
Bo. Urticata. (*Sam. I.* 7.)
Pa. No. 61. *List. Goëd.* (*Angl.*) 66. *f.* 61.—*Id.* (*Lat.*) 150. *f.* 61.
Magpie M. *Albin. pl.* xxxvii. *f.* 60. *i–m.*
Small Magpie M. *Harr. A. pl.* 6.—*Harr. V. M.* 35.

Genus 243: (1026). MARGARITIA *mihi*, (*Kirby.*)

Botys *p*, *Latreille, Leach, Samou.*—Nymphula *p*, *Schrank.*

6807 † 1, diversalis.
Py. diversalis. *Hüb. Py. pl.* 16. *f.* 102. In Mus. *Brit.*

6808. 2, longalis*.
Py. longalis: the long-winged Pearl. *Haw.* 379.

Py. elongalis. (*Haw. Pr.* 30.)
Bo. longalis. (*Sam. I.* 7.)
Py. glabralis. *Hüb. Py. pl.* 18. *f.* 117?.
Py. lancealis. (*Wien. V.* 121?)—*nec* Fabricii.

6809. 3; verticalis*.
Ph. Py. verticalis. *Linn.* ii. 882.—*Berk. S.* i. 146.—*Don.* xvi. *pl.* 556.—*Stew.* ii. 181.—*Turt.* iii. 277.
Py. verticalis. *Hüb. Py. pl.* 9. *f.* 57.—(*Haw. Pr.* 30.)—*Haw.* 376.
Bo. verticalis. (*Sam. I.* 7.)
Albin. pl. lxxiii. *f. a–d.*
Mother of Pearl M.: Ph. verticalis. (*Wilkes.* 24. *pl.* 51.)
Small Magpie likeness. *Harr. A. pl.* 33. *f. h–l.*—*Harr. V. M.* 35.

6810. 4, palealis: the Sulphur.—*Haw. l. c.*
Ph. palealis. *Fabr.* iii. *b.* 231.
Py. palealis. (*Wien. V.* 123.)—*Hüb. Py. pl.* 11. *f.* 70.—*Haw.* 378.
Bo. pulealis. (*Sam. I.* 7.)

6811. 5, centrostrigalis *mihi*.

6812. 6; limbalis*: the lesser Pearl.—*Haw. l. c.*
Py. limbalis. *Hüb. Py. pl.* 11. *f.* 72, 73.—*Haw* 378.
Bo. limbalis. (*Sam. I.* 7.)
Bo. cinctalis. (*Och. Tr.* vi. *b.* 314.)

6813. 7; hyalinalis*: the scarce Pearl.—*Haw. l. c.*
Py. hyalinalis. *Hüb. Py. pl.* 11. *f.* 74.—*Haw.* 377.
Bo. hyalinalis. (*Sam. I.* 7.)

6814. 8, glabralis: the dingy Pearl.—*Haw. l. c.*
Ph. glabralis. *Fabr. E. S.* iii. *b.* 277.
Py. glabralis. *Hüb. Py. pl.* 10. *f.* 65.—*Haw.* 380.
Bo. glabralis. (*Sam. I.* 7.)

6815. 9; angustalis*.
Py. angustalis: the narrow-winged Pearl. *Haw.* 379.
Bo. angustulus. (*Sam. I.* 7.)

6816. 10; terminalis*.
Py. terminalis: the bordered Pearl. *Haw.* 379.
Bo. terminalis. (*Sam. I.* 7.)

6817 † 11, pallidalis.
Py. pallidalis: the delicate Pearl. *Haw.* 379.
In Mus. D. *Haworth.*

6818. 12; Thapsalis*. (*Och. Tr.* vi. *b.* 314.)—the straw China-mark.—*Haw. l. c.*
Ph. Verbascalis. *Fabr.* iii. *b.* 212?

Py. Verbascalis. *Hüb. Py. pl.* 9. *f.* 59.—(*Haw. Pr.* 30.)—*Haw.* 381.
Bo. Verbascalis. (*Sam. I.* 7.)

6819. [13; ochrealis*]: the small straw China-mark.—*Haw. l.* c.
Py. ochrealis. *Fabr.* iii. *b.* 231.
Py. ochrealis. *Hüb. Py. pl.* 22. *f.* 146.—*Haw.* 381.
Bo. ochrealis. (*Sam. I.* 7.)

6820. 14, longipedalis.
Py. longipedalis. *Dale? MSS.*

6821. 15; Verbascalis*: the rusty China-mark.—*Haw. l.* c.
Py. Verbascalis. (*Wien. V.* 121.)
Py. arcualis. *Hüb. Py. pl.* 12. *f.* 80.—(*Haw. Pr.* 30.)—*Haw.* 381.
Bo. arcualis. (*Sam. I.* 7.)
Large golden China Mark. *Harr. V. M.* 37.

6822. 16; flavalis*: the gold China-Mark.—*Haw. l.* c.
Py. flavalis. *Hüb. Py. pl.* 11. *f.* 69.—*Haw.* 382.
Bo. flavalis. (*Sam. I.* 7.)
Small golden China Mark. *Harr. V. M.* 37.

6823. 17; ferrugalis*: the rusty Dot.—*Haw. l. c.*
Py. ferrugalis. *Hüb. Py. pl.* 9. *f.* 54.—(*Haw. Pr.* 30.)—*Haw.* 382.
Bo. ferrugalis. (*Sam. I.* 7.)

6824. 18; pulveralis*.
Py. pulveralis. *Hüb. Py. pl.* 17. *f.* 109.

6825. 19; cineralis*: the cinereous Pearl.—*Haw. l.* c.
Ph. cineralis. *Fabr.* iii. *b.* 230.
Py. cineralis. *Hüb. Py. pl.* 10. *f.* 66.—(*Haw. Pr.* 30.)—*Haw.* 380.
Bo. cineralis. (*Sam. I.* 7.)

6826. 20; fimbrialis* *mihi.*

6827. 21, uliginosalis.
Py. uliginosalis. *Curtis? MSS.*

6828. 22; lutealis*: the pale Straw.—*Haw. l. c.*
Py. lutealis. *Hüb. Py. pl.* 22. *f.* 145.—*Haw.* 380.
Bo. lutealis. (*Sam. I.* 7.)
Py. secalis. (*Haw. Pr.* 31.)—*Turt.* iii. 279.—(*Kirby & Sp. I. E.* i. 172.)

6829. 23; tetragonalis*.
Py. tetragonalis: the Diamond-spot. *Haw.* 385.
Bo. tetragonalis. (*Sam. I.* 7.)

6830 † 24, cilialis.
Py. cilialis. *Hüb. Py. pl.* 18. *f.* 119. In Mus. *D. Curtis.*

6831. 25: sericealis*: the straw Dot.—*Haw. l. c.*
Py. sericealis. (*Wien. V.* 122.)—*Hüb. Py. pl.* 9. *f.* 56.—*Haw.* 381.
Bo. sericealis. (*Sam. I.* 7.)
Py. Leeana. *Fabr.* iii. *b.* 242. (!)—*Don.* x. 87. *pl.* 357. *f.* 1.—*Stew.* ii. 173.—*Turt.* iii. 342.
Py. Leeanalis. (*Haw. Pr.* 31.)
Ph. Ge. limbata. *Linn.* ii. 873?—*Turt.* iii. 269?

6832. 26, margaritalis.
Ph. margaritalis. (*Wien. V.* 123.)—*Fabr.* iii. *b.* 226.
Ph. extimalis. *Scop. C. No.* 614?

6833. 27; elutalis*: the chequered Straw.—*Haw. l. c.*
Py. elutalis. *Hüb. Py. pl.* 10. *f.* 62.—(*Haw. Pr.* 30.)—*Haw.* 378.
Bo. elutalis. (*Sam. I.* 7.)

Genus 244: (1027). SCOPULA, *Schrank.*

6834. 1, Prunalis. (*Och. Tr.* vi. *b.* 313.)
Py. Prunalis. (*Wien. V.* 121.)
Py. leucophealis. *Hüb. Py. pl.* 12. *f.* 77.
♀? Py. elutalis. (*Wien. V.* 121.)
Py. albidalis. *Hüb. Py. pl.* 18. *f.* 118.

6835. 2; nebulalis*: the dusky Brindled.—*Haw. l. c.* (*Och. Tr.* vi. *b.* 313.)
Py. nebulalis. *Hüb. Py. pl.* 8. *f.* 51.—(*Haw. Pr.* 31.)—*Haw.* 386.
Bo. nebulalis. (*Sam. I.* 7.)

6836. 3; nivealis*.
Ph. nivealis. *Fabr.* iii. *b.* 232.
Py. nivealis: the white Brindled. (*Haw. Pr.* 31.)—*Haw.* 385.
Bo. nivealis. (*Sam. I.* 7.)
Py. umbralis. *Hüb. Py. pl.* 8. *f.* 52.
China Mark Likeness. *Harr. A. pl.* 29. *f. p–s.*—*Harr. V. M.* 37.

6837 † 4. sticticalis*.

6838. 5, dentalis*? the starry Brindle.
Py. dentalis. (*Wien. V.* 120.)—*Hüb. Py. pl.* 4. *f.* 25.—*Haw.* 385.
No. fulminans. *Fabr.* iii. *b.* 104.
Ph. ramalis. *Fabr.* iii. *b.* 230.
No. radiata. *Esper.* iv. *pl.* cxxvi. *f.* 2, 3.
An hujus generis?

---

* 6837 † 4. sticticalis.
Ph. Py. sticticalis. *Linn.* ii. 883.—*Turt.* iii. 281. (!)

Genus 245: (1028). NOLA, *Leach, Samou.*

HERCYNA *p, Treitschke.*—PH.-TINEA *p, Linné.*—BOTYS *p, Samou.* —BRIOPHILA? *p,* (*Kirby.*)

6839. 1; Monachalis*.
Py. Monachalis: the small Black-arches. *Haw.* 386.

6840. 2; strigulalis*: the least Black-arches.—*Haw. l. c.*
Py. strigulalis. *Hüb. Py. pl.* 3. *f.* 16.—(*Haw. Pr.* 31.)—*Haw.* 387.—(*Kirby & Sp. I. E.* iii. 232.)
Bo. strigulalis. (*Sam. I.* 7.)

6841. 3; cucullatella*.
Ph. Ti. cucullatella. *Linn.* ii. 889.
Py. cucullatalis: the short-cloaked. (*Haw. Pr.* 29.)—*Haw.* 387.
Bo. cucullatalis. (*Sam. I.* 7.)
Py. palliolalis. *Hüb. Py. pl.* 23. *f.* 149.
No. palliolalis. (*Leach E. E.* ix. 135.)—(*Sam. I.* 30.)
To. palliolatis. (*Sam. I.* 42.)

6842. [4; fuliginalis*] *mihi.*

Subsectio 4. LE.-VESPERTINA *mihi.*

(PH.-TORTRIX, -TINEA *p, et* -ALUCITA, *Linné.*)

Familia XVIII: (131). TORTRICIDÆ.

(PHALÆNA-TORTRIX, *Linné.*—PYRALIS, *Fabr.*—TORTRIX, *Hüb.*, *Haw.*—PYRALITES *p, Latr.*) *Twisters.*

Genus 246: (1029). CHLOEPHORA *mihi.*

NOCTUA *p, Haw. Pr.*

6843. 1; Fagana*.
Py. Fagana. *Fabr.* iii. *b.* 243.—*Stew.* ii. 174.—*Turt.* iii. 343.
No. Fagina. (*Haw. Pr.* 15.)
To. Fagana. *Haw.* 395.—(*Leach E. E.* ix. 135.)—(*Sam. I.* 41.)
Ph. Fagana. *Don.* viii. 75. *pl.* 280.
To. prasinana. (*Wien. V.* 125.)—*Hüb. To. pl.* 25. *f.* 158.—*Berk. S.* i. 146.
Green M. with Silver lines: Ph. prasinana. (*Wilkes.* 6. *pl.* 13.)
Green Silver lines. *Harr. A. pl.* 10. *f.* i–m.—*Harr. V. M.* 35. *Albin. pl.* 31.
♂; Py. sylvana. *Fabr.* iii. *b.* 244.

6844. 2; prasinana*.
Ph. To. prasinana. *Linn.* ii. 875.—*Stew.* ii. 173.—*Turt.* iii. 342.
No. prasinina. (*Haw. Pr.* 15.)
Py. prasinaria. *Fabr.* iii. *b.* 243.
Ph. prasinana? *Don.* ii. *pl.* 40. *fig. sup.*
To. Quercana. (*Wien. V.* 125.)—*Hüb. To pl.* 25. *f.* 159.

To. prasinana. *Haw.* 395.—(*Kirby & Sp. I. E.* iii. 278.)
Scarce Silver lines. *Harr. A. pl.* 30. *f.* e–h.—*Harr. V. M.* 35.

Genus 247: (1030). TORTRIX *Auctorum.*

6845. 1; clorana*. *Hüb. To. pl.* 25. *f.* 160.—*Haw. Pr.* 32.—*Haw.* 397.—(*Leach E. E.* ix. 135.)—(*Sam. I.* 41.)
Ph. To. clorana. *Linn.* ii. 876.—*Stew.* ii. 174.—*Turt.* iii. 343.—(*Kirby & Sp. I. E.* iii. 279.)
Small green Oak M.: Ph. viridana. (*Wilkes.* 4. *pl.* 5.)
Cream-bordered Green Pea. *Harr. V. M.* 29.

6846. 2; viridana*. *Haw. Pr.* 32.
Ph. To. viridana. *Linn.* ii. 875.—*Berk. S.* i. 146.—*Don.* iv. 95. *pl.* 144.—*Stew.* ii. 174.—*Turt.* iii. 343.—(*Sam. I.* 42.)
To. viridana. *Hüb. To. pl.* 25. *f.* 156.—(*Haw. Pr.* 32.)
Albin. pl. lxxii. *f. e–h.*
Pea Green M. *Harr. A. pl.* 10.—1, 2, *y*, *z*.—*Harr. V. M.* 27.

6847. 3; flavana*: the plain Yellow. *Haw.* 397.
To. flavana. *Hüb. To. pl.* 25. *f.* 157.

6848. 4; unitana. *Hüb. To. pl.* 19. *f.* 123.
Ph. Viburnana. (*Wien. V.* 128.)

6849 † 5, Pillerana. (*Wien. V.* 126?)
Py. Pillerana. *Fabr.* iii. *b.* 251?
To. luteolana. *Hüb. To. pl.* 21. *f.* 136?
To. Galiana *Museorum.* In Mus. D. *Bentley, Chant, et Stone.*

Genus 248: (1031). LOZOTÆNIA *mihi.*

To.-OBLIQUANÆ *p*, *Haworth.*

6850. 1; Forsterana*.
Py. Forsterana. *Fabr.* iii. *b.* 251. (!)—*Stew.* ii. 177.—*Turt.* iii. 346.
To. Forsterana: the Forsterian. (*Haw. Pr.* 32.)—*Haw.* 421.—(*Sam. I.* 41.)
Albin. pl. lxii. *f. a–d.*

6851. 2; Avellana*.
Ph. To. Avellana. *Linn.* ii. 877.—*Stew.* ii. 176.—*Turt.* iii. 350.
To. Avellana: the Hazel Tortrix. (*Haw. Pr.* 33.)—*Haw.* 421.—(*Sam. I.* 41.)
To. Sorbiana. *Hüb. To. pl.* 18. *f.* 113.

6852. 3; Carpiniana*.
To. Carpiniana: the dark Oblique-bar.—*Haw. l. c. Hüb. To. pl.* 18. *f.* 116.—*Haw.* 422.—(*Sam. I.* 41.)
Ph. To. Heperana. *Stew.* ii. 176?

6853. 4; Ribeana*.
To. Ribeana: the common Oblique-bar.—*Haw. l. c. Hüb. To. pl.* 18. *f.* 114.—*Haw.* 423.—(*Sam. I.* 42.)

6854. [5; Grossulariana*] *mihi.*

6855. 6; Cerasana*.
To. Cerasana: the hollow Oblique-bar.—*Haw. l. c. Hüb. To. pl.* 19. *f.* 119.—*Haw.* 423.

6856. 7; lævigana*.
To. lævigana. (*Wien. V.* 129?)—*Fabr.* iii. *b.* 253?

6857. 8; Corylana*.
Py. Corylana. *Fabr.* iii. *b.* 260.
To. Corylana: the great Checquered. *Haw.* 422.
To. textana. *Hüb. To. pl.* 18. *f.* 115.

6858. 9; Rosana*.
Ph. To. Rosana. *Linn.* ii. 876.—*Stew.* ii. 176.—*Turt.* iii. 345.
To. Rosana: the Rose. (*Haw. Pr.* 32.)—*Haw.* 424.—(*Sam. I.* 42.)
Ph. Ameriana. *Villers E.* iii. 393.—(*Haw. Pr.* 34.)

6859. 10; Oxyacanthana*.
To. Oxyacanthana: the White-thorn.—*Haw. l. c. Hüb. To. pl.* 18. *f.* 117.—(*Haw. Pr.* 33.)—*Haw.* 425.—(*Sam. I.* 42.)

6860. 11; Viburnana*.
Py. Viburnana. *Fabr.* iii. *b.* 257?
To. Viburnana: the Viburnian. *Haw.* 425.

6861. 12, subocellana *mihi.*

6862. 13; fuscana*.
Ph. To. fuscana. *Fabr.* iii. *b.* 251.—*Turt.* iii. 346.
To. fuscana: the great Brown. (*Haw. Pr.* 33.)—*Haw.* 424.

6863. [14; Branderiana*.]
Ph. To. Branderiana. *Linn.* ii. 877?
To. Branderiana: the Branderian. (*Haw. Pr.* 34.)—*Haw.* 424.
Py. Pasquayana. *Fabr.* iii. *b.* 248?
β? To. Cratægana. *Hüb. To. pl.* 17. *f.* 107.

6864. 15; oporana*.
Ph. To. oporana. *Linn.* ii. 876.—*Berk. S.* i. 146.—*Stew.* ii. 175.—*Turt.* iii. 352.
To. oporana: the great Hook-tipped.—*Haw. l. c. Hüb. To. pl.* 18. *f.* 112.—(*Haw. Pr.* 32.)—*Haw.* 427.—(*Sam. I.* 42.)
To. Hermanniana. (*Wien. V.* 317.)
Plumb-tree M.: Ph. oporana. (*Wilkes.* 17. *pl.* 34.)
Scorched Hook-tip. *Harr. V. M.* 49.
Albin. pl. xxxvi. *f.* 58. *f–l.*

6865. 16; fulvana*.
To. fulvana. (*Wien. V.* 128.)
To. Pyrastrana. *Hüb. To. pl.* 20. *f.* 124.
Py. Gerningana: the Gerningean.—*Haw. l. c. Haw.* 428.

6866. 17; Xylosteana*.
Ph. To. Xylosteana. *Linn.* ii. 876?—*Turt.* iii. 446?
To. Xylosteana: the forked Red-bar. (*Haw. Pr.* 32.)—*Haw.* 428.—(*Sam. I.* 42.)
To. obliquana. *Fabr.* iii. *b.* 257?
To. characterana. *Hüb. To. pl.* 20. *f.* 125?

6867. 18; Roborana*.
To. Roborana. *Hüb. To. pl.* 20. *f.* 126.

6868. 19; obliquana*.
Py. obliquana. *Fabr.* iii. *b.* 257? (!)—*Stew.* ii. 175?—*Turt.* iii. 349.
To. obliquana: the oblique-barred. (*Haw. Pr.* 34.)—*Haw.* 421.
Barred Hook-tip. *Harr. A. pl.* 41.

6869. 20; costana*.
Py. costana. *Fabr.* iii. *b.* 252.
To. costana: the straw Oblique-bar. *Haw.* 423.—(*Sam. I.* 41.)
To. Gnomana. *Hüb. To. pl.* 21. *f.* 131.
Ph. To. Betulana. *Don.* xi. 26. *pl.* 369. *f.* 2.

6870. 21, biustulana* *mihi.*

6871. 22; Modeeriana*.
Ph. To. Modeeriana. *Linn.* ii. 880?—*Turt.* iii. 351.
To. Modeeriana: the Modeerian. *Haw.* 423.

6872. 23; Acerana*.
To. Acerana: the Maple.—*Haw. l. c. Hüb. To. pl.* 19. *f.* 118.—*Haw.* 425.—(*Sam. I.* 41.)

6873. 24; trifasciana*.
Py. trifasciana. *Fabr.* iii. *b.* 248.
To. trifasciana: the Afternoon. *Haw.* 426.

6874. 25; Grotiana*.
Py. Grotiana. *Fabr.* iii. *b.* 272.
To. Grotiana: the Grotian. (*Haw. Pr.* 34.)—*Haw.* 426.
To. ochreana. *Hüb. To. pl.* 21. *f.* 134.
Var? To. flavana. *Hüb. To. pl.* 21. *f.* 133?

6875. 26; croceana*.
To. croceana: the Saffron.—*Haw. l. c. Hüb.* 19. *f.* 120.—*Haw.* 424.

6876. 27; cruciana*.
Ph. To. cruciana. *Linn.* ii. 880.
To. cruciana: the red Cross. *Haw.* 426.

6877. [28; cinerana*.]
Py. cinerana. *Fabr.* iii. *b.* 257?
To. cruciana β. *Haw.* 427.

6878. 29; Holmiana*.

Ph. To. Holmiana. *Linn.* ii. 878.—*Turt.* iii. 352.
To. Holmiana: the Holmian.—*Haw. l. c.* *Hüb. To. pl.* 7. *f.* 39.—(*Haw. Pr.* 33.)—*Haw.* 427.—(*Sam. I.* 41.)

6879. 30, Schreberiana?
To. Schreberiana. *Panz. F.* vii. *f.* 19?—*Turt.* iii. 352?

Genus 249: (1032). AMPHISA, *Curtis.*

6880. 1, Gerningiana.
To. Gerningiana. (*Wien. V.* 318?)
Py. Gerningiana. *Fabr.* iii. 265?
To. pectinana. *Hüb. To. pl.* 17. *f.* 108.
Am. pectinana. (*Curtis l. c. infra.*)

6881 † 2. Walkeri. *Curtis.* v. *pl.* 209. In Mus. D. Curtis et Walker.

Genus 250: (1033). DITULA *mihi.*

6882. 1; angustiorana*.
To. angustiorana: the narrow-winged Red-bar. *Haw.* 429.

6883. [2; rotundana*.]
To. rotundana: the round-tipped Red-bar. *Haw.* 429.

6884. 3; porphyriana*.
To. porphyriana: bright Oblique-barred.—*Haw. l. c.* *Hüb. To. pl.* 5. *f.* 26.—*Haw.* 461.

6885. [4; nebulana*.]
Ph. To. nebulana. *Don.* xi. 13. *pl.* 364. *f.* 3.
To. nebulana: the clouded iron. *Haw.* 461.—(*Sam. I.* 42.)
Ph. To. piceana. *Linn.* ii. 877?

6886. 5; sylvana*.
To. sylvana: red Blotch-back.—*Haw. l. c.* *Hüb. To. pl.* 20. *f.* 128.—*Haw.* 462.

6887. 6; Asseclana*.
To. Asseclana: barred Blotch-back.—*Haw. l. c.* *Hüb. To. pl.* 4. *f.* 19?—*Haw.* 462.

6888. [7; Æthiopiana*.]
To. Æthiopiana: the Negro. *Haw.* 462.

6889. 8; scriptana*.
To. scriptana: short-barred White.—*Haw. l. c.* *Hüb. To. pl.* 17. *f.* 110.—*Haw.* 431.

6890. 9, semifasciana*.
To. semifasciana: short-barred Grey. *Haw.* 431.—(*Sam. I.* 43.)

Genus 251: (1034). ANTITHESIA *mihi.*

To.-MARMORANÆ *p*, *Haworth.*

6891. 1; corticana*.
To. corticana: the marbled Long-cloak.—*Haw. l. c.* *Hüb. To. pl.* 3. *f.* 13.—*Haw.* 432.—(*Sam. I.* 41.)

6892. 2; Betuletana*.
To. Betuletana: the Birch Long-cloak. *Haw.*432.—(*Sam. I.*41.)

6893. 3; tripunctana*.
Py. tripunctana. *Fabr.* iii. *b.* 283.
To. tripunctana: the common Long-cloak. *Haw.* 432.
To. tripunctata. (*Haw. Pr.* 34?)
To. variegana. *Hüb. To. pl.* 3. *f.* 14.
Ph. Cynosbana. *Don.* x. 79. *pl.* 355. *f.* 3.

6894. 4; Pruniana*.
To. Pruniana: the Lesser Long-cloak.—*Haw. l. c. Hüb. To. pl.* 3. *f.* 15.—*Haw.* 433.—(*Sam. I.* 42.)

6895. 5, pullana*.
To. pullana: the Dingy-marbled. *Haw.* 434.

6896. 6; marginana*.
To. marginana: the bordered Long-cloak. *Haw.* 433.

6897. 7: oblongana*.
To. oblongana: the narrow Long-cloak. *Haw.* 433.

6898. 8: Gentianæana*.
To. Gentianæana: the Gentian.—*Haw. l. c. Hüb. To. pl.* 3. *f.* 12.—*Haw.* 433.

6899. 9; Salicella*.
Ph. Ti. Salicella. *Linn.* ii. 887.—*Berk. S.* i. 147.—*Stew.* ii. 198.—*Turt.* iii. 367.
Ti. Salicella. (*Haw. Pr.* 39.)
To. Salicana: the White-backed. *Haw.* 430.—*Hüb. To. pl.* 3. *f.* 11.—(*Sam. I.* 42.)
Provence Rose M.: Ph. Salicella. (*Wilkes.* 5. *pl.* 10.)—*Harr. V. M.* 41.

Genus 252: (1035). SPILONOTA *mihi.*

6900 † 1, nubiferana*.
To. nubiferana: the cloudy White. *Haw.* 431.—(*Sam. I.* 42.)
Ph. To. Wahlboomina. *Linn.* ii. 879?
In Mus. D. *Curtis, Hatchett, Haworth et Stone.*

6901. 2; Cynosbatella*.
Ph. Ti· Cynosbatella. *Linn.* ii. 887.
To. Cynosbana: the Black-cloaked.—*Haw. l. c.* (*Haw. Pr.* 33.)—*Stew.* ii. 178.—*Turt.* iii. 359.
To. ocellana. *Hüb. To. pl.* 4. *f.* 18.
Py. luxana. *Fabr.* iii. *b.* 255?

6902. 3; aquana*.
To. aquana: the Brown-cloaked.—*Haw. l. c. Hüb. To. pl.* 4. *f.* 17.—*Haw.* 430.

6903. 4; trimaculana*.
To. trimaculana: the triple Blotched. *Haw.* 442.

6904. 5: fœnella*.
Ph. Ti. fœnella. *Linn.* ii. 888.—*Turt.* iii. 368.
To. fœnana: the White-foot. *Haw.* 439.
Py. Scopoliana. (*Wien. V.* 129.)—*Fabr.* iii. *b.* 281.
Ph. To. interrogationana. *Don.* ii. 75. *pl.* 65. *f.* 1.—*Stew.* ii. 179.—*Turt.* iii. 351.
To. interrogatiana. (*Haw. Pr.* 32.)
To. tibialana. *Hüb. To. pl.* 7. *f.* 4.

6905. 6; rusticana*.
Py. rusticana. *Fabr.* iii. *b.* 254?
To. rusticana: the tawny Blotch-back. (*Haw. Pr.* 33.)—*Haw.* 442.—(*Sam. I.* 42.)

6906. 7, Pflugiana*.
Py. Pflugiana. *Fabr.* iii. *b.* 253.
To. Pflugiana: the Pflugian. *Haw.* 441.

6907. 8; Stræmiana*.
Py. Stræmiana. *Fabr.* iii. *b.* 255.
To. Stræmiana: the Stræmian. (*Haw. Pr.* 33.)—*Haw.* 441.
To. similana. *Hüb. To. pl.* 7. *f.* 41.
Ph. To. bimaculana. *Don.* xiii. 55. *pl.* 459.

6908. 9; trigeminana* *mihi.*

6909. 10; sticticana*.
Py. sticticana. *Fabr.* iii. *b.* 270. (!)—*Turt.* iii. 354.
To. sticticana. (*Sam. I.* 42.)
To. profundana. *Hüb. To. pl.* 4. *f.* 21.
To. stictana: the brown Blotch-back. (*Haw. Pr.* 34.)—*Haw.* 442.

6910. 11; costipunctana*.
To. costipunctana: the lesser Blotch-back. *Haw.* 443.

6911. 12; tetragonana* *mihi.*

6912. 13; nigricostana*.
To. nigricostana: the black-edged Marble. *Haw.* 438.

6913. 14; ustulana*.
To. ustulana: the scorched Bluntwing. *Haw.* 467.

6914. 15; comitana*.
To. comitana: the Cream Short-cloak.—*Haw. l. c. Hüb. To. pl.* 3. *f.* 16.—*Haw.* 434.—(*Sam. I.* 41.)

Genus 253: (1036). ——

6915. 1; dorsana*.
To. dorsana. *Hüb. To. pl.* 7. *f.* 36?
To. fimbriana: the brown Bordered. *Haw.* 446.—(*Sam. I.* 41.)

Genus 254: (1037). PSEUDOTOMIA *mihi.*

6916. 1, obscurana* *mihi.*

6917. 2; fraternana*.
To. fraternana: the cinereous-silver Barred. *Haw.* 449.—(*Sam. I.* 41.)
To. Strobilana. *Hüb. To. pl.* 12. *f.* 70.

6918. 3; atromargana*.
To. atromargana: the black Bordered. *Haw.* 446.—(*Sam. I.* 41.)

6919. 4; Strobilella*.
Ph. Ti. Strobilella. *Linn.* ii. 802.—*Stew.* ii. 199.—*Turt.* iii. 371.
To. Strobilana: the light Silver-striped. *Haw.* 448.—(*Sam. I.* 42.)
To. Argyrana. *Hüb. To. pl.* 8. *f.* 46.

6920. 5, sequana*: the silver Blotch-back.—*Haw. l. c.*
To. sequana. *Hüb. To. pl.* 8. *f.* 44.—*Haw.* 446.—(*Sam. I.* 42.)

6921. 6, Petiverella*.
Ph. Ti. Petiverella. *Linn.* ii. 895.
To. Petiverana: the Petiverian. *Haw.* 445.
To. montana. (*Wien. V.* 127.)

6922 † 7, concinnana *mihi.* In Mus. D. *Rudd.*

6923. 8; simpliciana*.
To. simpliciana: the plain Silver-fringed. *Haw.* 444.
To. lunulana. (*Wien. V.* 127.)—*Hüb. To. pl.* 7. *f.* 35?

6924. 9; Jacquiniana*.
Py. Jacquiniana. *Fabr.* iii. *b.* 258.
To. Jacquiniana: the Jacquinian. *Haw.* 444.
Var. To. inquinitana. *Hüb. To. pl.* 8. *f.* 43.

6925. 10; strigana*.
Py. strigana. *Fabr.* iii. *b.* 282?
To. strigana: the plain Gold-fringed. *Haw.* 444.

6926. 11; atropurpurana*.
To. atropurpurana: the purple Black. *Haw.* 467.

6927. 12; nigricana*.
Py. nigricana. *Fabr.* iii. *b.* 276. (!)—*Turt.* iii. 357.
To. nigricana: the black Striped-edge. *Haw.* 458.—(*Sam. I.* 42.)

6928. 13, proximana*.
To. proximana: pale brown Striped-edge. *Haw.* 458.

6929. 14; puncticostana* *mihi.*

6930. 15; Trauniana*.
Py. Trauniana. *Fabr.* iii. *b.* 259.
To. Trauniana: the Traunian.—*Haw. l. c.* *Hüb. To. pl.* 7. *f.* 38.—*Haw.* 444.

6931. 16; Populana*.
Py. Populana. *Fabr.* iii. *b.* 258.
To. Populana: the pygmy Y. *Haw.* 447.—(*Sam. I.* 42.)

6932. 17; trigonana* *mihi.*

6933. 18; Lediana*.
Ph. To. Lediana. *Linn.* ii. 879.—*Clerk. P. pl.* x. *f.* 12.
To. Lediana: the Ledian.—*Haw.* 459.

6934. 19; comitana*.
To. comitana. (*Wien. V.* 131.)
To. piciæana. *Hüb. To. pl.* 12. *f.* 72.
To. Abietana *mihi olim.*

6935. 20; Gundiana*: the Gundian.—*Haw. l. c.*
To. Gundiana. *Hüb. To. pl.* 8. *f.* 42.—*Haw.* 448.

6936. 21; compositella*.
Ti. compositella. *Fabr.* iii. *b.* 316. (!)—*Stew.* ii. 200.—*Turt.* iii. 374.
Py. composana. *Fabr. S.* 480.
To. composana: the triple-striped Blotch-back. *Haw.* 447.—(*Sam. I.* 41.)
To. compositana. (*Haw. Pr.* 33.)

6937. 22; dorsana*.
Py. dorsana. *Fabr.* iii. *b.* 282?
To. dorsana: the single-striped Blotch-back. (*Haw. Pr.* 33.)—*Haw.* 447.

6938. 23; aurana*.
Py. aurana. *Fabr.* iii. *b.* 279. (!)—*Don.* ii. 53. *pl.* 57. *f.* 2.—*Stew.* ii. 178.—*Turt.* iii. 358.
To. aurana: the double Orange-spot. (*Haw. Pr.* 33.)—*Haw.* 445.—(*Sam. I.* 41.)
Py. mediana. *Fabr.* iii. *b.* 284.
To. mediana. *Hüb. To. pl.* 28. *f.* 179.
Py. fulvana. *Fabr. S.* 477.
Ti. fulvella. *Fabr.* iii. *b.* 317?

6939. 24; nitidana*.
Py. nitidana. *Fabr.* iii. *b.* 276. (!)—*Turt.* iii. 357.
To. nitidana: dark Silver-striped. (*Haw. Pr.* 34.)—*Haw.* 448
To. nitida. (*Sam. I.* 42.)

Genus 255: (1038). STEGANOPTYCHA *mihi.*

To.-DORSANÆ *p, Haworth.*

6940. 1; tetraquetrana*.
To. tetraquetrana: the square-barred Single-dot. *Haw.* 454.—(*Sam. I.* 42.)

6941. 2; unipunctana*.

To. unipunctana: the marbled Single-dot. *Haw.* 454.—(*Sam.* *I.* 42.)

6942. 3, triquetrana*.
To. triquetrana: the angle-barred Single-dot. *Haw.* 454.

6943. 4, angulana*.
To. angulana: the angle-striped Single-dot. *Haw.* 455.

6944. 5; Bœberana*.
Py. Bœberana. *Fabr.* iii. *b.* 285.
To. Bœberana: the Bœberian. *Haw.* 450.
Ph. Ti. nisella. *Linn.* ii. 888?

6945. 6; Rubiana*.
Ph. Rubiana. *Villers E.* ii. 409?
To. Rubiana: the blotch-backed Grey. (*Haw. Pr.* 34.)—*Haw.* 450.—(*Sam. I.* 42.)
Ph. Pavonana. *Don.* i. 61. *pl.* 58. *f.* 3. *pl.* 59. *f.* 1. *aucta.*—*Stew.* ii. 179.

6946. 7; cuspidana*.
To. cuspidana: the pointed Bar. *Haw.* 451.

6947. 8; stictana*.
To. stictana: the spotted Red. *Haw.* 451.

6948. 9; rhombifasciana*.
To. rhombifasciana: the square Bar. *Haw.* 451.

6949. 10; cinerana*.
To. cinerana: the mottled Grey. (*Haw. Pr.* 33.)—*Haw.* 451.—(*Sam. I.* 41.)

Genus 256: (1039). ANCHYLOPERA *mihi.*

6950. 1; retusana*.
To. retusana: the variable Red. *Haw.* 453.

6951. 2; subuncana*.
To. subuncana: the red Hook-tip. *Haw.* 453.
Py. Mitterbachiana. *Panz. F.* 81. *f.* 23?
To. festivana. *Hüb. To. pl.* 9. *f.* 52?

6952. 3, obtusana*.
To. obtusana: the blunt-winged Blotch-back. *Haw.* 453.

6953. 4; unculana*.
To. unculana: the hook-tipped Blotch-back. *Haw.* 453.

6954. 5; Lundana*.
Py. Lundana. *Fabr.* iii. *b.* 282.
To. Lundiana: the Lundian. *Don.* xi. 39. *pl.* 374. *f.* 1?—*Haw.* 452.—(*Sam. I.* 42.)
To. Corylana. *Hüb. To. pl.* 9. *f.* 53.
To. badiana. (*Wien. V.* 126?)

6955. 6; fractifasciana*.
To. fractifasciana: the Broken-barred. *Haw.* 466.

6956. 7; siculana*.
To. siculana: hook-tipped Streak.—*Haw. l. c* *Hüb. To. pl.* 13. *f.* 79.—*Haw.* 452.

6957. 8, diminutana*.
To. diminutana: Festoon Tortrix. *Haw.* 452.

6958 † 9, funalana* *mihi.* In Mus. D. *Stone.*

6959. 10; uncana*.
To. uncana: the Bridge.—*Haw. l. c.* *Hüb. No. pl.* 13. *f.* 76. —*Haw.* 451.
Ph. To. geminana. *Don.* xi. 29. *pl.* 370. *f.* 1.

6960 † 11, biarcuana* *mihi.* In Mus. D. *Chant, et Stone.*

Genus 257: (1040). ——

6961. 1; harpana*.
To. harpana: the hooked Marble.—*Haw. l. c.* *Hüb. To. pl.* 13. *f.* 77.—*Haw.* 437.—(*Sam. I.* 41.)
Py. lætana. *Fabr.* iii. *b.* 258.

6962: 2, nigromaculana*.
To. nigromaculana: the beautiful Marble. *Haw.* 436.

6963. 3; nævana*.
To. nævana. *Hüb. To. pl.* 41. *f.* 261.

6964. 4; sociana*.
To. sociana: the white Short-cloak. *Haw.* 434.

6965. 5; incarnana*.
To. incarnana: the marbled Short-cloak. *Haw.* 435.—*Hüb. To. pl.* 30. *f.* 191?—(*Sam. I.* 41.)

6966 † 6, Paykulliana.
Py. Paykulliana: the Paykullian.—*Haw. l. c.* *Fabr.* iii. *b.* 272.
To. Paykulliana. (*Haw. Pr.* 33.)—*Haw.* 435.
In Mus. D. *Haworth.*

6967. 7; fimbriana*.
To. fimbriana. *Thunb. I. S.* 44. *pl.* 2. *f.* 3.
To. sesquilunana: the double Crescent. *Haw.* 435.

6968. 8; subocellana*.
Ph. To. subocellana: the retuse Marble.—*Haw. l. c.* *Don.* xii. 59. *pl.* 380. *f.* 1.—*Haw.* 437.—(*Sam. I.* 42.)

6969. 9, Asseclana.
To. Asseclana. *Hüb. To. pl.* 30. *f.* 194.
To. decorana: the obtuse Marble. *Haw.* 437.

6970. 10; Mitterbacheriana*.
To. Mitterbacheriana. (*Wien. V.* 129.)—*Hüb. To. pl.* 30. *f.* 192.

To. Mitterbachiana: the Mitterbachian. *Haw.* 463.
Ph. To. trimaculana. *Don.* xi. 25. *pl.* 369. *f.* 1.
To. Mitterbachina. (*Sam. I.* 42.)

Genus 258: (1041). SEMASIA *mihi*, (*Kirby.*)

To.-SIGNANÆ *p*, *Haw.*—ERMINEA *p*, (*Kirby.*)

6971. 1; Pomonella*.
Ph. Ti. Pomonella. *Linn.* ii. 892.—*Berk.* i. 147.—*Stew.* ii. 179.
Er. Pomonella. (*Kirby & Sp. I. E.* iii. 123.)
Py. Pomana. *Fabr.* iii. *b.* 279.—(*Haw. Pr.* 32.)—*Turt.* iii. 358.
To. Pomonana. *Haw.* 457.
To. Pomona. (*Sam. I.* 42.)—(*Leach E. E.* ix. 135.)
The Codling M.: Ph. Pomonella. (*Wilkes.* 5. *pl.* 9.)—*Harr. V. M.* 17.
Apple-tree M.: Ph. Pomonilla. *Harr. V. M.* 9.

6972. 2; splendana*.
To. splendana. *Hüb. To. pl.* 6. *f.* 31.

6973. 3; grossana*.
To. grossana: the smoky Marble. *Haw.* 438.

6974. 4; Wœberana*.
Py. Wœberana. *Fabr.* iii. *b.* 259.
To. Wœberana: the Wœberian. (*Wien. V.* 126.)—*Haw.* 457.
To. Wœberiana. (*Sam. I.* 42.)
To. unguicana. (*Haw. Pr.* 33?)—*Turt.* iii. 359?
To. ornatana. *Hüb. To. pl.* 6. *f.* 32.
Ph. Ti. unguicella. *Linn.* ii. 887?

6975. 5; Rheediella*: the Rheedian.—*Haw. l. c.*
Ph. Ti. Rheediella. *Linn.* ii. 898.—*Don.* xii. 49. *pl.* 377. *f.* 1.—*Turt.* iii. 376.
To. Rheediana. *Haw.* 405.
To. aurana. *Hüb. To. pl.* 4. *f.* 22.

6976. 6; lanceolana*.
To. lanceolana. *Hüb. To. pl.* 13. *f.* 80.
To. Ulicetana: the light Striped-edge. *Haw.* 458.

6977. 7; Hypericana*.
To. Hypericana: the yellow Striped-edge.—*Haw. l. c.* *Hüb. To. pl.* 4. *f.* 23.—*Haw.* 458.

6978. 8, perlepidana*.
To. perlepidana: the beautiful Crescent. *Haw.* 458.—(*Sam. I.* 42.)
Affinis Tiniæ Jungiellæ. *Linn. F. No.* 1410.—*Haw. l. c.*

6979. 9, pupillana.
Ph. To. pupillana. *Linn.* ii. 880.—*Turt.* iii. 348.
Py. pupillana. *Fabr.* iii. *b.* 255. (!)

To. pupillana. (*Haw. Pr.* 32.)
To. Absinthiana: Wormwood Tortrix.—*Haw. l. c.* *Hüb. To. pl.* 6. *f.* 34.—*Haw.* 457.—(*Sam. I.* 41.)

6980. 10; fulvana* *mihi.*
To. pupillana: the fulvous Sealed.—*Haw. l. c.* *Hüb. To. pl.* 4. *f.* 20.—*Haw.* 455.

6981. 11; cana*.
To. cana: the hoary Sealed. *Haw.* 456.—(*Sam. I.* 41.)

6982. 12; Scopoliana*.
To. Scopoliana: Scopolian. *Haw.* 456.

6983. 13, rufana *mihi.*

Genus 259: (1042). ——

6984. 1; arcuana*: the Arched.—*Haw. l. c.*
Ph. To. arcuana. *Linn.* ii. 877.—*Don.* xi. 11. *pl.* 364. *f.* 1.—*Turt.* iii. 350.
To. arcuana. *Hüb. To. pl.* 6. *f.* 33.—(*Haw. Pr.* 33.)—*Haw.* 403.
Ph. To. Lambergiana. *Scop. C.* 234.

Genus 260: (1043). APHELIA *mihi.*

To.-INOPIANÆ *p*, *Haworth.*

6985. 1; egenana*.
To. egenana: the dusty Drab. *Haw.* 469.

6986. 2; pauperana*.
To. pauperana: the spotted Drab. *Haw.* 469.—(*Sam. I.* 42.)

6987. 3; egestana*.
To. egestana: the lesser Drab. *Haw.* 470.—(*Sam. I.* 41.)

6988. 4; plagana*.
To. plagana: the broad-streaked Drab. *Haw.* 470.

6989. 5, expallidana*.
To. expallidana: the pale Drab. *Haw.* 469.

Genus 261: (1044). ——

CNEPHASIA *p*, *Curtis.*

6990. 1; quadripunctana*. (*Curtis l. c. supra.*)
To. quadripunctana: dotted Drab. *Haw.* 468.
To. Pratana. *Hüb. To. pl.* 36. *f.* 227, 228.

6991. [2; Cantiana.] *Curtis l. c. supra.*

Genus 262: (1045). CNEPHASIA, *Curtis.*

6992. 1, Penziana?
To. Penziana. *Thunb. I. S.* 43. *pl.* 2. *f.* 1?
Cn. bellana. *Curtis.* iii. *pl.* 100.

6993. [2, octomaculana.] *Haw. MSS.*—(*Curtis l. c. supra.*)

6994. 3; longana*. (*Curtis l. c. supra.*)
To. longana: Long-winged. *Haw.* 463.

6995. 4; ictericana*.
To. ictericana: the jaundiced Drab. *Haw.* 469.

6996. 5, sinuana*.
To. sinuana. (*Wien. V.* 131.)—*Hüb. To. pl.* 33. *f.* 212.

6997. 6; obsoletana* *mihi.*

6998. 7; assinana*: the large grey Elm.—*Haw. l. c.* (*Curtis l. c. supra.*)
To. assinana. *Hüb. To. pl.* 16. *f.* 101.—*Haw.* 464.
Py. cretana. *Fabr.* iii. *b.* 250.

6999. 8; interjectana*. (*Curtis l. c. supra.*)
To. interjectana: lesser Grey Elm. *Haw.* 464.
To. Masculana. *Hüb. To. pl.* 16. *f.* 98?

7000. 9; Logiana*. (*Curtis l. c. supra.*)
Ph. To. Logiana. *Linn.* ii. 879?
To. Logiana: the Logian. *Haw.* 464.—(*Sam. I.* 41.)
To. pascuiana. *Hüb. To. pl.* 16. *f.* 99.

7001. 10; rectifasciana*. (*Curtis l. c. supra.*)
To. rectifasciana: straight-barred Elm. *Haw.* 465.
Ph. To. trifasciana. *Don.* xi. 30. 370. *f.* 2.
To. hybridana. *Hüb. To. pl.* 38. *f.* 238?

7002. 11; aurifasciana* *mihi.*

7003. 12; Resinella*.
Ph. Ti. Resinella. *Linn.* ii. 892.
To. Turionana. *Hüb. To. pl.* 35. *f.* 220, 221.

Genus 263: (1046). ——

7004. 1; nubilana*.
To. nubilana: the smoky Grey.—*Haw. l. c. Hüb. To. pl.* 17. *f.* 111.—*Haw.* 467.—(*Sam. I.* 42.)

7005. 2; perfuscana*.
To. perfuscana: the deep Brown. *Haw.* 467.

Genus 264: (1047). ORTHOTÆNIA *mihi.*
To.-Fascianæ *p, Haworth.*

7006. 1; quadrimaculana*.
To. quadrimaculana: blotched Drab. *Haw.* 468.
To. antiquana. *Hüb. To. pl.* 34. *f.* 213, 214.

7007. 2; fasciana*.
Py. fasciana. *Fabr.* iii. *b.* 261.
To. fasciana: straight Barred. (*Haw. Pr.* 32.)—*Haw.* 460.—(*Sam. I.* 41.)
To. rusticana. *Hüb. To. pl.* 11. *f.* 66.
Ph. To. heperana. *Stew.* ii. 176?
♀; To. fasciolana.
Ph. To. biliturana. *Don.* xi. 32. *pl.* 371. *f.* 2. ♀.

7008. 3; Urticana*.

To. Urticana: barred Nettle.—*Haw. l. c.* *Hüb. No. pl.* 11. *f.* 65.—*Haw.* 460.—(*Sam. I.* 42.)
To. lacunana. (*Wien. V.* 318?)

7009. 4; micana*.
To. micana: Silver dotted. *Haw.* 460.—*Hüb. To. pl.* 5. *f.* 28?

7010. 5; undulana*.
To. undulana. (*Wien. V.* 131.)
Py. rivellana. *Fabr.* iii. *b.* 280.
To. conchana: Silver-striped.—*Haw. l. c.* *Hüb. To. pl.* 17. *f.* 106.—*Haw.* 460.

7011. 6; marmorana*: the marbled Dog's tooth.—*Haw. l. c.*
To. marmorana. *Hüb. To. pl.* 5. *f.* 25.—*Haw.* 450.
To. achatana. (*Wien. V.* 131?)

7012. 7; obsoletana* *mihi.*

7013. 8, bistrigana* *mihi.*

7014. 9, Pinetana.
To. Pinetana: Silver-marbled.—*Haw. l. c.* *Hüb. To. pl.* 10. *f.* 57?—*Haw.* 461.

7015. 10; politana*.
To. politana: the red-barred Grey. *Haw.* 465.

7016. 11, fuligana*: the dark-barred Grey.—*Haw. l. c.*
To. fuligana. *Hüb. To. pl.* 17. *f.* 109.—*Haw.* 465.

7017. 12, pulchellana*.
To. pulchellana: the dark-barred Grey. *Haw.* 429.

7018. 13; bifasciana*.
To. bifasciana: the double-barred Orange. *Haw.* 468.

7019. 14; aurofasciana*.
To. aurofasciana: gold Barred. *Haw.* 468.

7020. 15; furfurana*.
To. furfurana: mottled Bran. *Haw.* 466.

7021. 16, subsequana*.
To. subsequana: the faint Silver-striped. *Haw.* 448.—(*Sam. I.* 42.)

7022. 17, Hastiana.
Ph. To. Hastiana. *Linn.* ii. 878.—*Turt.* iii. 350.
To. Hastiana: the Hastian.—*Haw. l. c.* *Hüb. To. pl.* 29. *f.* 186.—*Haw.* 462.

Genus 265: (1048). ——

7023. 1, communana*.
Py. communana. *Fabr.* iii. *b.* 259.
To. communana: the Cock's head. *Haw.* 443.
Var.? Py. Cuivana. *Fabr. S.* 478?
Py. marmorana. *Fabr. S.* 477?

Genus 266: (1049). PŒCILOCHROMA *mihi.*

7024. 1; Udmanniana*.
Ph. To. Udmanniana. *Linn.* ii. 880.—*Don.* v. 33. *pl.* 153 *f.* 1–3.—*Stew.* ii. 178.—*Turt.* iii. 347.
To. Udmanniana: the Udmannian. *Haw.* 449.—(*Sam. I.* 42.)
Py. Solandrana. *Fabr.* iii. *b.* 254.
To. Achatana. *Hüb. To. pl.* 9. *f.* 49.
To. Udmaddiana. (*Haw. Pr.* 32.)

7025. 2; Sparmanniana*.
Py. Sparmanniana. *Fabr.* iii. *b.* 285.
Ph. To. Sparmanniana. *Gmel.* iv. 2518.—*Stew.* ii. 179?
To. Sparmanniana: the Sparmannian. *Haw.* 440.
β, Py. profundana. *Fabr.* iii. *b.* 258.
To. profundana: the marbled Diamond-back. (*Haw. Pr.* 33.) —*Haw.* 440.
γ, Py. Brunnichana. *Fabr.* iii. *b.* 258?—*Turt.* iii. 349.
To. Brunnichana: the Brunnichian. (*Haw. Pr.* 33.)—*Haw.* 440.
δ, Py. trapezana. *Fabr.* iii. *b.* 255.
To. trapezana: the testaceous Diamond-back. *Haw.* 441.—(*Sam. I.* 42.)

7026. 3; Solandriana*.
Ph. To. Solandriana. *Linn.* ii. 878.—*Turt.* iii. 347.
To. Solandriana: the Solandrian. (*Haw. Pr.* 32.)—*Haw.* 449.—(*Sam. I.* 42.)
Py. Udmanniana. *Fabr.* iii. *b.* 254.
To. semimaculana. *Hüb. To. pl.* 9. *f.* 48.

7027. 4; maculana*.
Py. maculana. *Fabr.* iii. *b.* 251.
To. maculana: the black Double-blotched. *Haw.* 440.—(*Sam. I.* 42.)

7028. 5, semifuscana*.
To. semifuscana. *Haworth MSS.*

7029. 6: piceana*.
To. piceana: the shining Pitch. *Haw.* 440.—(*Sam. I.* 42.)

7030. 7, maurana*: the great Double-bar.—*Haw. l. c.*
To. maurana. *Hüb. To. pl.* 19. *f.* 122.—*Haw.* 459.

Genus 267: (1050). PTYCHOLOMA *mihi.*

7031. 1; Lecheana*: the Lechean.—*Haw. l. c.*
Ph. To. Lecheana. *Linn.* ii. 877.—*Turt.* iii. 350.
To. Lecheana. *Hüb. To. pl.* 11. *f.* 67.—(*Haw. Pr.* 32.)—*Haw.* 403.—(*Sam. I.* 41.)

Genus 268: (1051). EUCHROMIA *mihi.*

7032. 1, purpurana*: the Purple.—*Haw. l. c.*
To. purpurana. *Haw.* 400.

7033. 2, fulvipunctana: the Tawney-dotted.—*Haw. l. c.*
To. fulvipunctana. *Haw.* 400.
Py. sponsana. *Fabr.* iii. *b.* 246?

Genus 269: (1052). LOPHODERUS *mihi*.

7034. 1; ministranus*: Yellow-barred Iron.—*Haw. l. c.*
Ph. To. ministrana. *Linn.* ii. 877.—*Turt.* iii. 346.—*Don.* xii. 60. *pl.* 380. *f.* 2.
To. ministrana. (*Haw. Pr.* 33.)—*Haw.* 398.
To. ferrugana. *Hüb. To. pl.* 10. *f.* 56.

7035. 2, subfasciana *mihi*.

Genus 270: (1053). SARROTHRIPUS, *Curtis*.

To.-PALPANÆ, *Haw.*

7036. 1; degeneranus*: large Marbled.—*Haw. l.* c. (*Curtis l.* c. *infra.*)
To. degenerana. *Hüb. To. pl.* 2. *f.* 8?—*Haw.* 406.—(*Sam. I.* 41.)
Ph. To. bifasciana. *Don.* x. 85. *pl.* 357. *f.* 3?

7037. 2; dilutanus*: large Brown.—*Haw. l. c.* (*Curtis l. c. infra.*)
To. dilutana. *Hüb. To. pl.* 2. *f.* 6.—*Haw.* 406.
Py. rivagana. *Fabr.* iii. *b.* 266?

7038. 3; Afzelianus*: Afzelian.—*Haw. l.* c. (*Curtis l.* c. *infra.*)
Ph. To. Afzeliana. *Gmel.* iv. 2518?—*Stew.* ii. 179.
To. Afzeliana. (*Haw. Pr.* 33.)—*Haw.* 407.—(*Sam. I.* 41.)

7039. 4; Lathamianus*: Lathamian.—*Haw. l. c.* (*Curtis l. c. infra.*)
Ph. To. Lathamiana. *Gmel.* iv. 2518?—*Stew.* ii. 179.
To. Lathamiana. *Haw.* 407.
Ph. To. Ilicana. *Don.* x. 85. *pl.* 357. *f.* 2.

7040. 5; Ilicanus*: large Holly.—*Haw. l. c.*
Py. Ilicana. *Fabr.* iii. *b.* 266. (!)—*Stew.* ii. 175.—*Turt.* iii. 352.
To. Ilicana. (*Haw. Pr.* 32.)—*Haw.* 407.—(*Sam. I.* 41.)
To. punctulana. *Hüb. To. pl.* 2. *f.* 9.
Sa. punctulanus. (*Curtis l.* c. *infra.*)

7041. 6; ramosanus*. *Curtis.* i. *pl.* 29.
To. ramosana. *Hüb. To. pl.* 2. *f.* 10.

7042 † 7, Stonanus: *Curtis MSS.*
Sa. ramulana *mihi*. In Mus. *D. Stone.*

Genus 271: (1054). PERONEA, *Curtis*.

To.-ASPERANÆ, *Haworth.*

A. Alis anticis in medio fasciculo elevato squamarum.

7043. 1: profanana*. (*Curtis l. c. infra.*)

Py. profanana. *Fabr.* iii. *b.* 268. (!)—*Turt.* iii. 353.
To. profanana: the rusty Button. (*Haw. Pr.* 34.)—*Haw.* 412.
—*Don.* xii. 51. *pl.* 377. *f.* 3.

7044. 2; striana*. (*Curtis l. c. infra.*)
To. striana: the brown Button. *Haw.* 413.

7045. 3, substriana* *mihi.* (*Curtis l. c. infra.*)

7046. 4: brunneana* *mihi.* (*Curtis l. c. infra.*)

7047. 5: vittana* *mihi.* (*Curtis l. c. infra.*)

7048. 6, spadiceana*. (*Curtis l. c. infra.*)
To. spadiceana: the bay-shouldered Button. *Haw.* 412.—
(*Sam. I.* 42.)

7049. 7; consimilana* *mihi.* (*Curtis l. c. infra.*)

7050. 8; Desfontainiana*.
Py. Desfontainiana. *Fabr.* iii. *b.* 268. (!)—*Turt.* iii. 353.
To. Desfontainiana: the Desfontainian. (*Haw. Pr.* 34.)—
Haw. 413.—(*Sam. I.* 41.)
To. sericeana. *Hüb. To. pl.* 14. *f.* 83.
Pe. sericeana. (*Curtis l. c. infra.*)

7051. 9, fulvocristana *mihi.* (*Curtis l. c. infra.*)

7052. 10, albovittana *mihi.* (*Curtis l. c. infra.*)

7053. 11: fulvovittana *mihi.* (*Curtis l. c. infra.*)

7054. 12; cristalana*. (*Curtis l. c. infra.*)
Ph. To. cristalana. *Don.* iii. 13. *pl.* 77. *f.* 1, 2.—*Stew.* ii. 180.
—*Turt.* iii. 347.
To. christallana. (*Haw. Pr.* 32.)

7055. 13, subvittana *mihi.* (*Curtis l. c. infra.*)

7056. 14, cristana.
Py. cristana. *Fabr.* iii. *b.* 267.
To. cristana: the white Button.—*Haw. l. c.* *Hüb. To. pl.* 28.
f. 176.

7057. 15, albipunctana. *Haworth MSS.*
Py. ephippiana. *Fabr. S.* 479?

B. Alis anticis haud cristatis.

7058. 16; umbrana*. (*Curtis l. c. infra.*)
To. umbrana: the dark-streaked Button.—*Haw. l. c.* *Hüb.*
To. pl. 10. *f.* 59.—*Haw.* 411.—(*Sam. I.* 42.)

7059. 17: divisana*. (*Curtis l. c. infra.*)
To. divisana. *Hüb. To. pl.* 31. *f.* 198.

7060 † 18, strigana *mihi.* (*Curtis l. c. infra.*) In Mus. D. *Stone.*

7061. 19: radiana*. (*Curtis l. c. infra.*)
To. radiana: the Buff-edged.—*Haw. l. c.* *Hüb. To. pl.* 28.
f. 177.—*Haw.* 412.

7062. 20; centrovittana*. *Haworth MSS.*—(*Curtis l. c. infra.*)

7063. 21, ramostriana *mihi.* (*Curtis l. c. infra.*)

7064. 22: combustana. (*Curtis l. c. infra.*)
To. combustana. *Hüb. To: pl.* 37. *f.* 234.

7065. 23; albistriana*. (*Curtis l. c. infra.*)
To. albistriana: the Grey-streak. *Haw.* 412.

7066. 24, autumnana. (*Curtis l. c. infra.*)
To. autumnana. *Hüb. To. pl.* 39. *f.* 247.

7067. 25, subcristana *mihi.* (*Curtis l. c. infra.*)

7068. 26, coronana*.
To. coronana. *Thunb. I. S.* 18. *pl.* 2. *fig. adj.*
To. eximiana: the marbled Chesnut. *Haw.* 413.
Pe. eximiana. (*Curtis l. c. infra.*)
To. examiana. (*Sam. I.* 41.)

7069. 27; Byringerana*. (*Curtis l. c. infra.*)
To. Byringerana. *Hüb. To. pl.* 10. *f.* 61.

7070. 28; obsoletana* *mihi.*

7071. 29; favillaceana*.
To. favillaceana: the Ash-coloured.—*Haw. l. c. Hüb. To. pl.* 11. *f.* 62.—*Haw.* 409.

7072. 30; tristana*.
To. tristana: the lesser Ash-coloured.—*Haw. l. c. Hüb. To. pl.* 9. *f.* 50.—*Haw.* 410.

7073. 31; reticulana*.
To. reticulana: the checquered Grey. *Haw.* 409.

7074. 32: ruficostana. *Curtis.* i. *pl.* 16.

7075 † 33: bistriana*. (*Curtis l. c. supra.*)
To. bistriana: the double Bay-streak. *Haw.* 399.
In Mus. D. *Curtis, Haworth, et Stone*

7076. 34; albicostana *mihi.* (*Curtis l. c. supra.*)

7077. 35: similana* *mihi.* (*Curtis l. c. supra.*)

C. An Novum Genus?

7078. 36, latifasciana*. (*Curtis l. c. supra.*)
To. latifasciana: the broad Barred. *Haw.* 414.—(*Sam. I.* 41.)

7079. 37; plumbosana*. (*Curtis l. c. supra.*)
Py. plumbana. *Fabr.* iii. *b.* 269?
To. scabrana. *Hüb. To. pl.* 10. *f.* 58?
To. plumbosana: the Lead-coloured. *Haw.* 415.

7080. 38; Boscana*. (*Curtis l. c. supra.*)
Py. Boscana. *Fabr.* iii. *b.* 269?
To. Boscana: the Boscan. *Haw.* 415.

7081. 39; trigonana* *mihi.*

7082. 40; Schalleriana.
Ph. To. Schalleriana. *Linn.* ii. 879.
To. Schalleriana: the Schallerian. (*Haw. Pr.* 33.)—*Haw.* 416.—(*Sam. I.* 42.)

7083. 41; rufana.
Py. rufana. *Fabr.* iii. *b.* 263.
To. rufana: the red Triangle.—*Haw. l. c.* *Hüb. To. pl.* 20. *f.* 127.—*Haw.* 417.—(*Sam. I.* 42.)

7084. 42; costimaculana *mihi.*

7085 † 43, Logiana*.
To. Logiana. (*Wien. V.* 130.)—*Hüb. To. pl.* 11. *f.* 64.
Ph. To. Logiana. *Turt.* iii. 353? In Mus. D. *Stone.*

7086. 44; asperana*.
Py. asperana. *Fabr.* iii. *b.* 269.—*Stew.* ii. 177.
To. asperana: the White-shouldered. (*Haw. Pr.* 32.)—*Haw.* 414.—(*Sam. I.* 41.)

7087. 45; variegana*.
Py. variegana. *Fabr.* iii. *b.* 254.
To. variegana: the common Rough-wing. *Haw.* 414.
To. cristana. *Hüb. To. pl.* 10. *f.* 55.

7088. 46; borana*.
Py. borana. *Fabr.* iii. *b.* 270?
To. borana: the crested Buff. *Haw.* 415.—(*Sam. I.* 41.)

Genus 272: (1055). PARAMESIA *mihi.*

7089. 1; subtripunctulana* *mihi.*

7090. 2; gnomana*.
Ph. To. gnomana. *Linn.* ii. 876.—*Turt.* iii. 352.
To. gnomana: the Dial. (*Haw. Pr.* 32.)—*Haw.* 417.—(*Sam. I.* 41.)
To. Steneiriana. *Hüb. To. pl.* 27. *f.* 170.
Var.? Ph. To. notana. *Don.* xi. 27. *pl.* 369. *f.* 3, 3?

7091. 3; bifidana*.
To. bifidana: the Fork-striped. *Haw.* 418.—(*Sam. I.* 41.)

7092. 4; tripunctulana*.
To. tripunctulana: the rusty Treble-spot. *Haw.* 417.
To. tripunctana. *Hüb. To. pl.* 20. *f.* 129.—(*Sam. I.* 42.)

7093. 5; cerusana*.
To. cerusana: the white Treble-spot.—*Haw. l. c.* *Hüb. To. pl.* 11. *f.* 63.—*Haw.* 416.—(*Sam. I.* 41.)

Genus 273: (1056). LEPTOGRAMMA, *Curtis MSS.*

7094. 1; literana*.

Ph. To. literana. *Linn.* ii. 876.
To. literana: the black-sprigged green.—*Haw. l. c. Hüb. To. pl.* 15. *f.* 89.—(*Haw. Pr.* 32.)—*Don.* x. 78. *pl.* 355. *f.* 2.—*Haw.* 411.—(*Sam. I.* 41.)

7095. 2; squamana*: Green tufted M.—*Haw. l.* c.
Py. squamana. *Fabr.* iii. *b.* 270.—*Don.* v. 51. *pl.* 157. *f.* 7.—*Stew.* ii. 178.—*Turt.* iii. 354.
To. squamana: the scaly Green. (*Haw. Pr.* 34.)—*Haw.* 410.—(*Sam. I.* 42.)
To. squamulana. *Hüb. To. pl.* 15. *f.* 93 & 94.

7096. 3; tricolorana*.
To. tricolorana: the tricoloured Green. *Haw.* 411.—(*Sam. I.* 42.)
To. irrorana var. *Hüb. To. pl.* 15. *f.* 95.

7097. 4; irrorana*.
To. irrorana. *Hüb. To. pl.* 15. *f.* 96.

7098. 5; fulvomixtana* *mihi.*

7099. 6; scabrana*.
Py. scabrana. *Fabr.* iii. *b.* 271. (!)—*Stew.* ii. 174.—*Turt.* iii. 355.
To. scabrana: the grey Rough-wing. (*Haw. Pr.* 32.)—*Haw.* 410.
To. irrorana var. *Hüb. To. pl.* 15. *f.* 97.

Genus 274: (1057). GLYPHISIA *mihi.*

TORTRICES-EMARGANÆ, *Haw.*

7100. 1; emargana*.
Py. emargana. *Fabr.* iii. *b.* 271. (!)—*Stew.* ii. 174.—*Turt.* iii. 355.
To. emargana: the checquered Notch-wing. (*Haw. Pr.* 32.)—*Haw.* 403.—(*Sam. I.* 41.)

7101. 2; excavana*.
To. excavana: the Iron Notch-wing. *Haw.* 403.—(*Sam. I.* 41.)
Ph. To. emargana var. *Don.* iii. 91. *pl.* 106. *f.* 5.

7102. 3, effractana*.
To. effractana: the common Notch-wing.—*Haw. l. c. Hüb. To. pl.* 28. *f.* 175.—*Haw.* 403.
Ph. To. emargana. *Don.* iii. 91. *pl.* 106. *f.* 1.
To. affractana. (*Sam. I.* 41.)

7103. 4; caudana*.
Py. caudana. *Fabr.* iii. *b.* 271. (!)—*Stew.* ii. 174.—*Turt.* iii. 355.
To. caudana: the shallow Notch-wing. (*Haw. Pr.* 32.)—*Haw.* 404.—(*Sam. I.* 41.)

7104 † [5, ochracea*] *mihi.* In Mus. D. *Stone.*

Genus 275: (1058). DICTYOPTERYX *mihi.*

7105. 1; contaminana*.
To. contaminana: the checquered Pebble.—*Haw. l. c. Hüb. To. pl.* 22.*f.* 142?—(*Sam. I.* 41.)
Ph. Py. xylosteana. (*Haw. Pr.* 32.)
Harr. Ex. 94. *pl.* xxviii.*f.* 2, 3.

7106. 2; ciliana*.
To. ciliana: the White-fringed.—*Haw. l.* c. *Hüb. To. pl.* 27. *f.* 171.—*Haw.* 419.—(*Sam. I.* 41.)
Ph. To. obscurana. *Don.* xii. 40. *pl.* 374.*f.* 2.
Orange Hook-tip. *Harr. V. M.* 49?

7107. 3; rhombana*.
To. rhombana: the Dark-checquered.—*Haw. l. c. Hüb. To. pl.* 27.*f.* 173.—*Haw.* 418.—(*Sam. I.* 42.)

7108. 4; plumbana*.
To. plumbana. *Hüb. To. pl.* 9.*f.* 54.
To. plumbeolana: the clouded Straw. *Haw.* 420.—(*Sam. I.* 42.)

7109. 5; Lœflingiana*.
Ph. To. Lœflingiana. *Linn.* ii. 878.—*Don.* iii. 49. *pl.* 90.—*Stew.* ii. 177.—*Turt.* iii. 356.
To. Lœflingiana: the Lœflingian. (*Haw. Pr.* 33.)—*Haw.* 420.—(*Sam. I.* 41.)
To. Ectypana. *Hüb. To. pl.* 30.*f.* 190.

7110. 6; Forskåleana*.
Ph. To. Forskåleana. *Linn.* ii. 878.—*Turt.* iii. 355.
To. Forskåliana: the *Forskålian.*—*Haw. l. c. Hüb. To. pl.* 22. *f.* 143.—(*Haw. Pr.* 33.)—*Haw.* 420.—(*Sam. I.* 41.)

Genus 276: (1059). ——

7111. 1, ochraceana* *mihi.*

Genus 277: (1060). CHEIMATOPHILA *mihi.*

7112. 1; castaneana*.
To. castaneana: the chesnut Tortrix. *Haw.* 410.
To. hyemana: the Winter Tortrix. *Haw.* 413.
To. hyemalis. (*Sam. I.* 41.)
To. mixtana. *Hüb. To. pl.* 34.*f.* 215.
β? To. nigripunctana (*mihi olim.*)

Genus 278: (1061). ARGYROTOZA *mihi.*

7113. 1; Bergmanniana*.
Ph. To. Bergmanniana. *Linn.* ii. 878.—*Don.* v. 49. *pl.* 157. *f.* 1–6.—*Stew.* ii. 177.—*Turt.* iii. 356.
To. Bergmanniana: the Bergmannian. (*Haw. Pr.* 33.)—*Haw.* 404.—(*Sam. I.* 41.)
To. Rosana. *Hüb. To. pl.* 22.*f.* 137.

7114. 2; Conwayana*: Conwayian. (*Haw. Pr.* 33.)—*Haw.* 405.
Py. Conwayana. *Fabr.* iii. *b.* 277. (!)—*Stew.* ii. 178.—*Turt.* iii. 357.
To. Lediana. *Hüb. To. pl.* 24. *f.* 151?

7115. 3; subaurantiana* *mihi.*

7116. 4; Hoffmanseggiana*: Hoffmanseggian.—*Haw. l. c.*
To. Hoffmanseggiana. *Hüb. No. pl.* 24. *f.* 150?—*Haw.* 405.

7117. 5; permixtana*: Orange and Black.—*Haw. l. c.*
To. permixtana. *Hüb. To. pl.* 12. *f.* 75.—*Haw.* 406.

7118. 6; Daldorfiana*: Daldorfian. *Haw.* 405.
Py. Daldorfiana. *Fabr.* iii. *b.* 246.
Ph. To. dimidiana. *Don.* xi. 12. *pl.* 364. *f.* 2.

Genus 279: (1062). ARGYROLEPIA *mihi.*

7119 † 1. Lathoniana: the Silver-spotted.—*Haw. l. c. Hüb. To. pl.* 30. *f.* 189.—*Haw.* 402. In Mus. D. ——?

7120. 2, Bentleyana: Bentleyan. *Haw.* 403.
Ph. To. Bentleyana. *Don.* x. 85. *pl.* 357. *f.* 1.

7121. 3; Turionana*: orange-spotted. *Haw.* 399.—(*Ing. Inst.* 91.)
Ph. Ti. Turionella. *Linn.* ii. 802.—*Stew.* ii. 200.
Py. sponsana. *Fabr.* iii. *b.* 246?

7122. 4; gemmana*. *Hüb. To. pl.* 43. *f.* 269.

7123. 5, æneana*: Silvery Broad Bar.—*Haw. l. c. Hüb. To. pl.* 30. *f.* 188.—*Haw.* 404.

7124. 6; tesserana*: the Tessellated.—*Haw. l. c.*
To. tesserana. (*Wien. V.* 126.)
To. tesserana. *Hüb. To. pl.* 23. *f.* 144.—*Haw.* 427.
Py. Heisana. *Fabr.* iii. *b.* 262.

7125. [7; decimana*.]
To. decimana. (*Wien. V.* 317?)—*Hüb. To. pl.* 23. *f.* 145?
Py. decimana. *Fabr.* iii. *b.* 261?

7126. 8; Baumanniana*: Baumannian. (*Haw. Pr.* 34.)—*Haw.* 404.—(*Sam. I.* 41.)
Py. Baumanniana. *Fabr.* iii. *b.* 262.
Var. To. Hartmanniana. *Hüb. To. pl.* 24. *f.* 146.

Genus 280: (1063). EUPŒCILIA *mihi.*

7127. 1; maculosana*.
To. maculosana: the small Black-spotted. *Haw.* 438.

7128. 2; angustana*: the barred Marble.—*Haw. l. c.*
To. angustana. *Hüb. To. pl.* 12. *f.* 74.—*Haw.* 438.—(*Sam. I.* 41.)
Ph. fasciella. *Don.* xiii. 41. *pl.* 452.

7129. 3; pygmeana*.
To. pygmeana: the dingy Dwarf. *Haw*. 439.

7130. 4; nana*.
To. nana: the barred Dwarf. *Haw*. 439.—(*Sam*. *I*. 42.)

7131. 5, luteolana* *mihi*.

7132. 6, dubitana*.
To. dubitana. *Hüb*. *To*. *pl*. 12. *f*. 71.

7133. 7; Sodaliana*.
To. Sodaliana: the brindled Marble. *Haw*. 436.

7134. 8; bilunana*.
To. bilunana: the hoary Double-crescent. *Haw*. 436.

7135. 9; albana*.
To. albana: the light Marbled. *Haw*. 436.

Genus 281: (1064). ——

7136. 1; roseana*.
To. roseana: the Rosy. *Haw*. 401.

7137. 2, rubroseana*.
To. rubroseana: the dingy Rosy. *Haw*. 402.

7138. 3; ruficiliana*
To. ruficiliana: the red Fringe. *Haw*. 402.—(*Sam*. *I*. 42.)
Ti. ciliella. *Hüb*. *To*. *pl*. 26. *f*. 180.—(*Haw*. *Pr*. 38.)

7139. 4; griseana*.
To. griseana: oblique-barred Grey. *Haw*. 402.

7140. 5, marginana* *mihi*.

7141. 6, margaritana.
To. margaritana: the orange-barred Pearl. *Haw*. 401.

7142. 7. Smeathmanniana*: the Smeathmannian.—*Haw*. *l*. *c*.
Py. Smeathmanniana. *Fabr*. iii. *b*. 249. (!)—*Stew*. ii. 176.—*Turt*. iii. 345.
To. Smeathmanniana. (*Haw*. *Pr*. 33.)—*Haw*. 400.—(*Sam*. *I*. 42.)
To. badiana. *Hüb*. *To*. *pl*. 23. *f*. 147.

Genus 282: (1065). PHTHEOCHROA *mihi*.

7143. 1; rugosana*: the Rough-wing.—*Haw*. *l*. *c*.
To. rugosana. *Hüb*. *To*. *pl*. 14. *f*. 82.—*Haw*. 431.—(*Sam*. *I*. 42.)
To. strigillana. (*Haw*. *Pr*. 32.)
Ph. V albana. *Don*. xi. 31. *pl*. 371. *f*. 1.

Genus 283: (1066). LOZOPERA *mihi*.

7144. 1; alternana *mihi*.
To. Stephensiana. *Blunt MSS*.

7145. 2; Straminea*.
To. Straminea: short-barred Straw. *Haw.* 401.—(*Sam. I.* 43.)

7146. 3; Fabriciana*: the Fabrician.—*Haw. l. c.*
To. Fabriciana. *Hüb. To. pl.* 23. *f.* 149.—*Haw.* 401.

7147. 4, Francillana*: the Francillonian.—*Haw. l. c.*
Py. Francillana. *Fabr.* iii. *b.* 264. (!)—*Turt.* iii. 352.—*Don.* x. 77. *pl.* 355. *f.* 1.
To. Francillana. (*Haw. Pr.* 34.)—*Haw.* 401.
To. Baumanniana. *Hüb. To. pl.* 23. *f.* 148.

Genus 284: (1067). XANTHOSETIA *mihi.*

7148. 1; Zœgana*: Zœgian.—*Haw. l. c. Hüb. To. pl.* 55. *f.* 256. —(*Haw. Pr.* 32.)—*Haw.* 398.—(*Sam. I.* 42.)
Ph. To. Zœgana. *Linn.* ii. 876.—*Don.* iii. 92. *pl.* 106. *f.* 2.—*Stew.* ii. 175.—*Turt.* iii. 348.
Clouded Straw. *Harr. V. M.* 45.—*Harr. Ex.* 11. *pl.* 11. *f.* 1, 2.

7149. 2; ferrugana*: clouded Iron. *Haw.* 398.

7150. 3; hamana*: hook-marked Straw.—*Haw. l. c. Hüb. To. pl.* 22. *f.* 140.—(*Haw. Pr.* 32.)—*Haw.* 397.—(*Sam. I.* 41.)
Ph. To. Hamana. *Linn.* ii. 876.—*Stew.* ii. 175.—*Turt.* iii. 348.
Py. Zœgana β. *Fabr.* iii. *b.* 256.

7151. 4; diversana*: crossed Straw.—*Haw. l. c. Hüb. To. pl.* 22. *f.* 139.—*Haw.* 397.—(*Sam. I.* 41.)

7152. 5; inopiana*: plain Drab. *Haw.* 469.
Ti. tetricella. *Fabr.* iii. *b.* 303?

Genus 285: (1068). PHIBALOCERA *mihi.*

7153. 1; Quercana*.
Py. Quercana. *Fabr.* iii. *b.* 271.—*Don.* iii. 93. *pl.* 106. *f.* 3.—*Stew.* ii. 175.—*Turt.* iii. 355.—(*Sam. I.* 42.)
To. Quercana: the Long-horned. (*Haw. Pr.* 33.)—*Haw.* 399.
To. Fagana. *Hüb. To. pl.* 24. *f.* 153.

Genus 286: (1069). HYPERCALLIA *mihi.*

7154. 1; Christiernana*.
Ph. To. Christiernana. *Linn.* ii. 877.—*Don.* ii. 45. *pl.* 20. *f.* 1. —*Stew.* ii. 176.—*Turt.* iii. 345.—(*Sam. I.* 41.)
To. Chrystiernana. (*Haw. Pr.* 33.)
To. Christiernana: the Christiernian. *Haw.* 399.

Genus 287: (1070). ORTHOTELIA *mihi.*

7155. 1, venosa*. *Haworth MSS?*

Familia XIX: (132). YPONOMEUTIDÆ *mihi.*

(PHALÆNA-TORTRIX *p, et* PH.-TINEA, *Linné, &c.*—PYRALITES *p, Latr.*) TINEA *p, Fabricius.*

Genus 288: (1071). DEPRESSARIA, *Haworth, Curtis.*

VOLUCRA, *Latr.* *Flat-body M.*

7156. 1; Heracleana*.
Ph. To. Heracleana. *Linn.* ii. 880?—*Stew.* ii. 178.
To. Heracleana. (*Haw. Pr.* 31.)
De. Heraclei: the Cow-Parsnip. *Haw.* 505.—(*Curtis l. c. infra.*)
Albin. pl. xxxviii. *f.* 61. *a–d.*

7157. 2; Umbellana*.
Py. Umbellana. *Fabr.* iii. *b.* 286.
Ti. Umbellella. *Fabr. S.* 484.
De. Umbellarum: the large-streaked. *Haw.* 506.—(*Curtis l. c. infra.*)

7158. 3; putridella*.
Ti. putridella. *Hüb. Ti. pl.* 33. *f.* 244.
De. putrida: the Brown-veined. *Haw.* 509.—(*Curtis l. c. infra.*)

7159. 4; venosa*: the Wainscot Flat-body. *Haw.* 506.—(*Curtis l. c. infra.*)

7160. 5; apicella*?
Ti. apicella. *Hüb. Ti. pl.* 14. *f.* 94.
De. apiosa: the brindled Flat-body. *Haw.* 509.—(*Curtis l. c. infra.*)
To. apiana. (*Haw. Pr.* 31?)

7161. 6, nervosa*: the coarse Wainscot. *Haw.* 506.—(*Curtis l. c. infra.*)

7162. 7; badiella.
Ti. badiella. *Hüb. Ti. pl.* 14. *f.* 92.
De. badia: the brown Brindled. *Haw.* 509.—(*Curtis l. c. infra.*)

7163. 8; characterosa*: the lesser Flat-Body. *Haw.* 511.—(*Curtis l. c. infra.*)

7164. 9: liturella*.
Ti. liturella. *Hüb. Ti. pl.* 12. *f.* 83.
De. liturosa: the purple Flat Body. *Haw.* 508.—(*Curtis l. c. infra.*)

7165 † 10. Bluntii. *Curtis.* v. *pl.* 221.
In Mus. D. Curtis, Dale, et Parsons?

7166. 11; applana*: the common Flat-Body. *Haw.* 510.—(*Curtis l. c. supra.*)
Py. applana. *Fabr.* iii. *b.* 285.
Ti. applanella. *Fabr. S.* 484.

Ti. applana. (*Sam. I.* 41.)
Ti. Cicutella. *Hüb. Ti. pl.* 12. *f.* 79.
To. Cicutana. (*Haw. Pr.* 31.)

7167. 12; curvipunctosa*: the Curve Dotted. *Haw.* 511.—(*Curtis l. c. supra.*)
Ti. curvipunctosa. (*Sam. I.* 41.)

7168. 13; albipunctella*.
Ti. albipunctella. *Hüb. Ti. pl.* 22. *f.* 149.
De. albipuncta: the rufous Brindled. *Haw.* 510.—(*Curtis l. c. supra.*)

7169. 14; ocellana*.
Py. ocellana. *Fabr.* iii. 272. (!)—*Stew.* ii. 175.—*Turt.* iii. 355.
Ti. characterella. (*Wien. V.* 137.)
Ti. signella. *Hüb. Ti. pl.* 12. *f.* 80.
De. signosa: the red Letter. *Haw.* 508.—(*Curtis l. c. supra.*)
To. signana. (*Haw. Pr.* 31.)
Ti. signosa. (*Sam. I.* 41.)

7170. 15; gilvella*.
Ti. gilvella. *Hüb. Ti. pl.* 14. *f.* 96.
De. gilvosa: the brindled Straw. *Haw.* 507.—(*Curtis l. c. supra.*)
To. gilvana. (*Haw. Pr.* 34.)

7171. 16; costosa*: the dingy Straw. *Haw.* 508.—(*Curtis l. c. supra.*)

7172. 17; irrorella* *mihi.*

7173. 18; Carduella.
Ti. Carduella. *Hüb. Ti. pl.* 66. *f.* 439.

7174. 19; atomella*.
Ti. atomella. *Hüb. Ti. pl.* 35. *f.* 240.
De. atomosa: the Powdered. *Haw.* 507.—(*Curtis l. c. supra.*)

7175. 20; Sparmanniana*.
Py. Sparmanniana. *Fabr.* iii. *b.* 285. (!)—*Turt.* iii. 360.
Ti. flavella. *Hüb. Ti. pl.* 14. *f.* 97.
De. flavosa: the Straw-coloured. *Haw.* 507.—(*Curtis l. c. supra.*)

7176. 21, immaculana *mihi.*

7177. 22; Yatesana*.
Py. Yatesana. *Fabr.* iii. *b.* 274. (!)
Ph. To. Yeatiana. *Stew.* ii. 177.—*Turt.* iii. 356.
Ph. albidana. *Don.* xi. 50. *pl.* 377. *f.* 2.
De. Yeatsii: Yates's. *Haw.* 509.—(*Curtis l. c. supra.*)

7178. 23; Alstræmeriana*.
Ph. To. Alstræmeriana. *Linn.* ii. 879.—(*Haw. Pr.* 31.)—*Turt.* iii. 357.

De. Alstræmeria: Alstræmer's. *Haw.* 508.—(*Curtis l. c. supra.*)
Ti. Alstræmeri. (*Sam. I.* 41.)
Ti. Puella. *Hüb. Ti. pl.* 12. *f.* 82.

7179. 24; purpurea*: the Lesser purple. *Haw.* 511.—(*Curtis l. c. supra.*)
Ti. purpurea. (*Sam. I.* 41.)

Genus 289: (1072). ANACAMPSIS, *Curtis.*

RECURVARIA, *Haworth*, (*Kirby.*)—VOLUCRA *p? Latr.?*

7180. 1; Juniperella*.
Ph. Ti. Juniperella. *Linn.* ii. 893.
An. Juniperi. (*Curtis l. c. infra.*)
Re. Juniperi: the Juniper. *Haw.* 548.

7181. 2; Populella*.
Ph. Ti. Populella. *Linn.* ii. 892.—(*Haw. Pr.* 39.)
An. Populi. (*Curtis l. c. infra.*)
Re. Populi: the Poplar. *Haw.* 548.

7182. 3; rusticella*.
Ti. rusticella. *Hüb. Ti. pl.* 3. *f.* 17.
An. rustica. (*Curtis l. c. infra.*)
Re. rustica: the dark Brown. *Haw.* 548.

7183. 4, longicornis. *Curtis.* iv. *pl.* 189.

7184. 5; Listerella*.
Ph. Ti. Listerella. *Linn.* ii. 896?—*Turt.* iii. 374?
An. Listeri. (*Curtis l. c. supra.*)
Re. Listeri: Lister's. *Haw.* 548.

7185. 6; nebulea*. (*Curtis l. c. supra.*)
Re. nebulea: the dotted Brown. *Haw.* 549.
Ti. Populella. *Hüb. Ti. pl.* 3. *f.* 21.

7186. 7; Betulea*. (*Curtis l. c. supra.*)
Re. Betulea: the Birch. *Haw.* 549.
Ti. Betulinella. *Hüb. Ti. pl.* 3. *f.* 20?
Ti. Turpella. (*Wien. V.* 139.)

7187. 8; rhombella*.
Ti. rhombella. *Hüb. Ti. pl.* 40. *f.* 277.
Re. rhombea: the black-speck'd Grey. *Haw.* 549.
An. rhombea. (*Curtis l. c. supra.*)

7188. 9; cinerella*.
Ph. Ti. cinerella. *Linn.* ii. 891.—*Turt.* iii. 373.
Re. cinerella: the Cinereous. *Haw.* 547.
An. cinerea. (*Curtis l. c. supra.*)

7189. 10; subcinerea*. (*Curtis l. c. supra.*)
Re. subcinerea: the Subcinereous. *Haw.* 548.
Ti. terrella. *Hüb. Ti. pl.* 25. *f.* 170?

7190. 11, lutarea*. (*Curtis l. c. supra.*)
Re. lutarea: the Clay-coloured. *Haw.* 549.
Ti. Verbascella. *Hüb. Ti. pl.* 14. *f.* 98?

7191. 12; dodecella*.
Ph. Ti. dodecella. *Linn.* ii. 802.—*Stew.* ii. 200.—*Turt.* iii. 372.
Re. dodecea: the Small black-speck'd Grey. *Haw.* 549.
An. dodecea. (*Curtis l. c. supra.*)

7192. 13; aspera*. (*Curtis l. c. supra.*)
Re. aspera: the Rough. *Haw.* 550.
Ti. Schellenbergella. *Fabr. S.* 501?

7193. 14; Mouffettella*.
Ph. Ti. Mouffettella. *Linn.* ii. 896?—*Turt.* iii. 375?
Ti. pedisequella. *Hüb. Ti. pl.* 14. *f.* 95?
To. pedisequana. (*Haw. Pr.* 31.)
Re. punctifera: the dotted Grey. *Haw.* 551.
An. punctifera. (*Curtis l. c. supra.*)

7194. 15, nigra*. (*Curtis l. c. supra.*)
Re. nigra: the dusted Black. *Haw.* 550.

7195. 16; sarcitella*.
Ph. Ti. sarcitella. *Linn.* ii. 888.—*Turt.* iii. 369.
Re. sarcitella: the White-shouldered Woollen M. *Haw.* 550.
Ti. sarcitella. (*Haw. Pr.* 38.)—(*Kirby & Sp. I. E.* i. 230.)
Ph. fascitella. *Stew.* ii. 198.
An. sarcitea. (*Curtis l. c. supra.*)

7196. 17; lacteella*.
Ti. lacteella. (*Wien. V.* 139?)

7197. 18, Hübneri*. (*Curtis l. c. supra.*)
Re. Hübneri: Hübner's. *Haw.* 551.
Ti. Granella. *Hüb. Ti. pl.* 24. *f.* 165.

7198. 19, domestica*. (*Curtis l. c. supra.*)
Re. domestica: the Domestic. *Haw.* 551.

7199. 20, affinis*. (*Curtis l. c. supra.*)
Re. affinis: the brindled Brown. *Haw.* 551.

7200. 21; diffinis*. (*Curtis l. c. supra.*)
Re. diffinis: the greater Brindled-brown. *Haw.* 551.

7201. 22, contigua*. (*Curtis l. c. supra.*)
Re. contigua: the light Brindled-brown. *Haw.* 552.

7202. 23, sequax*. (*Curtis l. c. supra.*)
Re. sequax: the Brindled Brown. *Haw.* 552.
Ti. sexpunctella. *Fabr.* iii. *b.* 313?

7203. [24, proxima*.] (*Curtis l. c. supra.*)
Re. proxima: the beautiful Brindled-brown. *Haw.* 552.

7204. 25; tricolorella*.
Ti. tricolorella: the tricoloured. *Ent. Trans.* (*Haworth.*) i. 338.

7205. 26, luctuella*.
Ti. luctuella. *Hüb. Ti. pl.* 21. *f.* 144?

7206. 27, decorella*.
Ti. decorella: the neat. *Ent. Trans.* (*Haworth.*) i. 338.

7207. 28, subrosea*. (*Curtis l. c. supra.*)
Re. subrosea: the little Blossom. *Haw.* 553.

7208. 29, marmorea*. (*Curtis l. c. supra.*)
Re. marmorea: the beautiful Marble. *Haw.* 553.

7209. 30, guttifera*. (*Curtis l. c. supra.*)
Re. guttifera: the white-dotted Black. *Haw.* 553.

7210. 31, atra. (*Curtis l. c. supra.*)
Re. atra: the little Black. *Haw.* 553.
Ti. exiguella. *Fabr.* iii. *b.* 328?—*Hüb. Ti. pl.* 39. *f.* 266?

7211. 32, maculella*.
Ti. maculella. *Fabr.* iii. *b.* 306.—*Turt.* iii. 369.
Re. maculea: the short-barred White. *Haw.* 552.
An. maculea. (*Curtis l. c. supra.*)

7212. 33; Tremella*.
Ti. Tremella. (*Wien. V.* 139.)
Ti. Blattariella. *Hüb. Ti. pl.* 22. *f.* 148.
Re. Blattariæ: the Horse-shoe. *Haw.* 553.
An. Blattariæ. (*Curtis l. c. supra.*)

7213. 34; nivella*.
Ti. nivella. *Fabr.* iii. *b.* 335. (!)—*Stew.* ii. 203.—*Turt.* iii. 380.
Re. nivella: the black-dotted White. *Haw.* 554.
An. nivea. (*Curtis l. c. supra.*)

7214. 35; aleella*.
Ti. aleella. *Fabr.* iii. *b.* 317. (!)—*Turt.* iii. 374.
Ti. nanella. *Hüb. Ti. pl.* 39. *f.* 267.
Re. nana: the Black-clouded. *Haw.* 554.
An. nana. (*Curtis l. c. supra.*)

7215. 36, interruptella*.
Ti. interruptella. *Hüb. Ti. pl.* 17. *f.* 116?
Re. interrupta: the Brown-streak. *Haw.* 554.
An. interrupta. (*Curtis l. c. supra.*)

7216. 37, fulvescens. (*Curtis l. c. supra.*)
Re. fulvescens: the Tawny. *Haw.* 554.

7217. 38, fuscescens. (*Curtis l. c. supra.*)
Re. fuscescens: the faint-dotted Brown. *Haw.* 555.

7218. 39; alternella*.
Ti. alternella. *Hüb. Ti. pl.* 22. *f.* 151.—(*Haw. Pr.* 35.)
Re. alterna: the black-spotted White. *Haw.* 556.

7219. 40, angustella*.

Ti. angustella. *Hüb. Ti. pl.* 26. *f.* 177.—(*Haw. Pr.* 37.)
Re. angusta: the black-spotted Brimstone. *Haw.* 557.

7220. [41, albimaculea*.]
Re. albimaculea: the white-spotted Black. *Haw.* 557.

7221. 42, quadripuncta*.
Re. quadripuncta: the four-spotted. *Haw.* 557.
Ti. Scopollella. *Hüb. Ti. pl.* 36. *f.* 246.—*Turt.* iii. 376.

Genus 290: (1073). LOPHONOTUS *mihi.*

7222. 1, fasciculellus *mihi.*

Genus 291: (1074). ——

7223 † 1. emarginella.
Ph. Ti. emarginella. *Don.* xi. 90. *pl.* 392. *f.* 3.
In Mus. D. Donovan.

Genus 292: (1075). CHELARIA, *Haworth.*

7224. 1; rhomboidella*.
Ph. Ti. rhomboidella. *Linn.* ii. 891.—*Turt.* iii. 373.
Ti. conscriptella. *Hüb. Ti. pl.* 41. *f.* 283.
Ch. conscripta: the Lobster-clawed. *Haw.* 526.
Ph. Hübnerella. *Don.* xi. 63. *pl.* 382. *f.* 2.

Genus 293: (1076). ——

RECURVARIA*** *p, Haworth.*

7225. 1; Silacella*.
Ti. Silacella. *Hüb. Ti. pl.* 17. *f.* 117.
Re. silacea: the dingy Straw. *Haw.* 555.

7226. 2; rufescens*.
Re. rufescens: the dwarf Wainscot. *Haw.* 555.

7227. 3; nebulella* *mihi.*

7228. 4; ochroleucella* *mihi.*

7229. 5; lucidella* *mihi.*

7230. 6, falciformis*.
Re. falciformis: the Hook-tipped. *Haw.* 555.

Genus 294: (1077). ——

7231. 1; tinctella*.
Ti. tinctella. *Hüb. Ti. pl.* 31. *f.* 214.

7232. 2; cinerella*.
Ti. cinerella. *Hüb. Ti. pl.* 25. *f.* 173.

7233. 3; unitella*.
Ti. unitella. *Hüb. Ti. pl.* 22. *f.* 147.
Ti. fusco-aurella: the Brown-gold. *Haw.* 569?

Genus 295: (1078). MACROCHILA *mihi.*

YPSOLOPHUS *p*, *Haw.*—CRAMBUS *p*, *Fabr.*

7234. 1; fasciella*.
Ti. fasciella. *Hüb. Ti. pl.* 16. *f.* 111.
Yp. fasciellus: the Long-winged. *Haw.* 545.

7235. 2; parenthesella*.
Ph. Ti. parenthesella. *Linn.* ii. 890.—*Turt.* iii. 370.
Yp. parenthesellus: the Parenthesis. *Haw.* 540.
Ti. rostrella. *Hüb. Ti. pl.* 17. *f.* 113.—(*Haw. Pr.* 37.)

7236. 3; marginella*.
At. marginella. *Fabr.* iii. *b.* 333. (!)
Ph. Ti. marginella. *Don.* ii. 58. *pl.* 58. *f.* 2.—*Turt.* iii. 370, 380.
Li. marginata. *Fabr. S.* 461.
Ti. striatella. *Hüb. Ti. pl.* 23. *f.* 154?—(*Haw. Pr.* 38.)
Yp. marginellus: the white Bordered. *Haw.* 540.
Ph. marginatella. *Stew.* ii. 203.

7237. 4; bicostella*.
Ph. Ti. bicostella. *Linn.* ii. 890.—*Turt.* iii. 370.
Ti. bicostella. *Hüb. Ti. pl.* 17. *f.* 115.—(*Haw. Pr.* 38.)
Yp. bicostellus: the light Streak. *Haw.* 539.

7238 † 5. aristella[a].

7239. 6; palpella*.
Yp. palpellus: the dingy Streak. *Haw.* 545.

Genus 296: (1079). ENICOSTOMA *mihi.*

RECURVARIA *p*, *Haw.*—ŒCOPHORA *p*, *Latr.*

7240. 1; Thunbergana*.
Py. Thunbergana. *Fabr.* iii. *b.* 268.
To. Thunbergana. (*Haw. Pr.* 32.)
Re. Thunbergii: Thunberg's. *Haw.* 547.
Ti. lobella. *Hüb. Ti. pl.* 35. *f.* 238.

7241. 2; Geoffroyella*.
Ph. Ti. Geoffroyella. *Linn.* ii. 896.
Ti. Geoffroyella. *Hüb. Ti. pl.* 18. *f.* 123.—(*Haw. Pr.* 36.)
Re. Geoffroyi: Geoffroy's. *Haw.* 556.

Genus 297: (1080). DASYCERA.

DASYCERUS[b], *Haworth.*—ŒCOPHORA *p*, *Latr.*—ÆCOPHORA *p*, *Samou.*

7242. 1; Oliviella*.

---

[a] 7238 † 5. aristella.
Ph. Ti. aristella. *Linn.* ii. 894.—*Turt.* iii. 381. (!)
An hujus generis?

[b] DASYCERUS: Genus Coleopterorum. Vide *Latr. G.* iii. 19.

Ti. Oliviella. *Fabr.* iii. *b.* 316.
Da. Olivieri: Olivier's. *Haw.* 524.
Ti. Æmulella. *Hüb. Ti. pl.* 32. *f.* 222.
Ti. fulgidella. *Marsh. MSS.*

7243. 2; sulphurella*.
Ti. sulphurella. *Fabr.* iii. *b.* 345. (!)—(*Haw. Pr.* 38.)—*Turt.* iii. 383.
Da. sulphureus: the yellow Underwinged. *Haw.* 524.
Ph. Ti. sulphuratella. *Stew.* ii. 204.
♀; Al. flavella. *Fabr.* iii. *b.* 332.—(*Sam. I.* 1.)
Ti. flavella. (*Haw. Pr.* 38?)
Ti. cornutella. *Fabr. S.* 492.

Genus 298: (1081). ADELA, *Latr., Leach, Samou.*

NEMOPHORA, (*Hoffmansegg.*)—NEMAPOGON *p, Schrank.*—ALUCITA *p, Fabr.*—CAPILLARIA, *Haworth.*

7244. 1; Latreillella*. (*Latr.* G. iv. 224.)
Ti. Latreillella. *Hüb. Ti. pl.* 52. *f.* 355, 356?

7245. 2; Sulzella*. (*Latr.* G. iv. 224.)
Ph. Ti. Sulzella. *Linn.* ii. 896.—*Turt.* iii. 382.
Ti. Sulzella. *Hüb. Ti. pl.* 18. *f.* 121.—(*Haw. Pr.* 37.)
Ca. Sultzii: Sultz's. *Haw.* 520.
Var.? Ph. Ti. Podaëlla. *Linn.* ii. 896.—*Don.* viii. 42. *pl.* 267. *f.* 3.—*Stew.* ii. 205.—*Turt.* iii. 382.

7246. 3; DeGeerella*. (*Latr.* G. iv. 224.)—(*Leach E. E.* ix. 133.)—(*Sam. I.* 1.)
Ph. Ti. DeGeerella. *Linn.* ii. 895.—*Don.* viii. 41. *pl.* 267. *f.* 1, 2.—*Stew.* ii. 205.—*Turt.* iii. 382.
Ti. Geerella. *Hüb. Ti. pl.* 19. *f.* 130.—(*Haw. Pr.* 37.)
Ca. DeGeerii: DeGeer's. *Haw.* 519.
β? Al. striatella. *Fabr. M.* ii. 257.—(*Haw. Pr.* 37.)
Ph. striella. *Stew.* ii. 204.

7247. 4; fasciella*.
Al. fasciella. *Fabr.* iii. *b.* 342. (!)—*Turt.* iii. 383.—*Stew.* ii. 204.
Ca. fascia: the copper Japan. *Haw.* 520.
Ti. Schiffermyllerella. *Hüb. Ti. pl.* 19. *f.* 132.—(*Haw. Pr.* 37.)

7248. 5, cuprella*.
Al. cuprella. *Fabr.* iii. *b.* 342.
Ca. cuprella: the Scabious Long-horn. *Haw.* 521.
Ti. viridella. *Hüb. Ti. pl.* 19. *f.* 128?—(*Haw. Pr.* 37?)

7249. 6; viridella*.
Al. viridella. *Fabr.* iii. *b.* 341.
Ca. viridis: the green Long-horn. *Haw.* 520.
Ti. Sphingiella. *Hüb. Ti. pl.* 19. *f.* 129.—(*Haw. Pr.* 37.)
Japan Long-horn. *Harr. V. M.* 31.

7250. [7, Reaumerella*.] (*Latr. G.* iv. 224.)
Ph. Reaumerella. (*Wien. V.* 143.)—*Turt.* iii. 383.
Ph. Ti. Reaumerella. *Linn.* ii. 895?—(*Haw. Pr.* 39?)

7251. 8, Frischella*.
Ph. Ti. Frischella. *Linn.* ii. 896.—*Turt.* iii. 382.
Ca. Frischii: the Frischian. *Haw.* 521.

7252. 9; Swammerdammella*.
Ph. Ti. Swammerdammella. *Linn.* ii. 895.—*Turt.* iii. 382.
Ti. Swammerdammella. *Hüb. Ti. pl.* 62. *f.* 410 & 411.—(*Haw. Pr.* 37.)
Ca. Swammerdami: Swammerdam's. *Haw.* 522.
Golden Long-horn. *Harr. V. M.* 31?

7253. 10; Panzerella*.
Al. Panzerella. *Fabr.* iii. *b.* 339.
Ca. Panzeri: Panzer's. *Haw.* 522.
Ti. Swammerdammella. *Hüb. Ti. pl.* 19. *f.* 127.
Ti. Panzerella. (*Haw. Pr.* 37.)—*Hüb. Ti. pl.* 62. *f.* 412.

7254. 11; Ti. Robertella*.
Ph. Ti. Robertella. *Linn.* ii. 896.—*Turt.* iii. 382.
Al. Pilella. *Fabr.* iii. *b.* 339.
Ti. pillella. *Hüb. Ti. pl.* 34. *f.* 235.—(*Haw. Pr.* 37.)
Ca. pilea: the pale-brown Long-horn. *Haw.* 522.

Genus 299: (1082). ——

DIURNEA 3 *p*, *Haw.*

7255. 1; Tortricella*.
Ti. Tortricella. *Hüb. Ti. pl.* 2. *f.* 11.
Di. Tortricea: the clouded Lead. *Haw.* 503.
Ti. Tortricea. (*Sam. I.* 41.)

7256. 2; nubilea*.
Di. nubilea: the clouded Brown. *Haw.* 503.
Ti. nubilea. (*Sam. I.* 41.)

Genus 300: (1083). ——

CAPILLARIA 2 *p*, *Haw.*

7257. 1; Tesserella*.
Ca. tessellea: the pale chequered Brown. *Haw.* 522.

7258. 2; pubicornis*.
Ca. pubicornis: the pale downy-horned. *Haw.* 523.

Genus 301: (1084). ——

DIURNEA 3 *p*, *Haw.*—GEOMETRA *p*, *Haw.*

7259. 1; Salicella*.
Ti. Salicella. *Hüb. Ti. pl.* 2. *f.* 9.
Di. Salicis: the rosy Day Moth. *Haw.* 504.

Ti. Salicis. (*Sam. I.* 41.)
♀; Geo. incompletaria: the incomplete. *Haw.* 305.—(*Sam. I.* 18.)

7260. 2; gelatella*.
Ph. Py. gelatella. *Linn.* ii. 883.—*Stew.* ii. 196.—*Turt.* iii. 362.
Di. gelata: the autumnal Dagger. *Haw.* 502.
Ti. gelatella. (*Sam. I.* 41.)
To. gelatana. *Hüb. To. pl.* 42. *f.* 266. ♂–♀.
Ph. congelatella. *Clerck Ic. pl.* 8. *f.* 5.
♀. Ph. paradoxa. *Sulzer. pl.* 23. *f.* 21, 22.

Genus 302: (1085). ——

DIURNEA 2, *Haw.*—GEOMETRA *p*, *Haw.*

7261. 1; Phryganella*.
Ti. Phryganella. *Hüb. Ti. pl.* 2. *f.* 10.—(*Haw. Pr.* 35.)
Di. Phryganea: the drab Day M. *Haw.* 503.
Ti. Phryganea. (*Sam. I.* 41.)
♀; Geo. apteraria: the Apterous. *Haw.* 306.
Geo. apteria. (*Sam. I.* 18.)
Ph. Ti. Lichenella. *Linn.* ii. 899?—*Turt.* iii. 378?
Ti. Lichenum. (*Kirby & Sp. I. E.* i. 463?)

Genus 303: (1086). DIURNEA, (*Kirby.*)

DIURNEA 1, *Haw.*

7262. 1; Fagella*.
Ti. Fagella. *Fabr.* iii. *b.* 300.—(*Haw. Pr.* 36.)
Di. Fagi. *Haw.* 501.
Ti. Fagi. (*Sam. I.* 41.)
March Dagger. *Harr. V. M.* 19.
Albin. pl. xxxvi. *f.* 57. *a–e.*

7263. 2; Novembris*: the November Dagger. *Haw.* 502.
Ti. Novembris. (*Sam. I.* 41.)—(*Kirby & Sp. I. E.* iv. 510.)

Genus 304: (1087). EPIGRAPHIA *mihi.*

DIURNEA 4, *Haworth.*

7264. 1; Avellanella*.
Ti. Avellanella. *Hüb. Ti. pl.* 4. *f.* 27.

7265. 2; Steinkelnerana*.
Py. Steinkelnerana. *Fabr.* iii. *b.* 267.
Di. Steinkelneri: Steinkelner's. *Haw.* 504.
Ti. Characterella. *Hüb. Ti. pl.* 4. *f.* 26.—(*Haw. Pr.* 36.)

Genus 305: (1088). MELANOLEUCA *mihi.*

YPONOMENTA *p*, *Fabricius.*—ERMINEA *p*, *Haw.*

7266. 1, Echiella.
Ti. Echiella. (*Wien. V.* 140.)—*Hub. Ti. pl.* 15. *f.* 105.

Al. bipunctella. *Fabr.* iii. *b.* 384.
Yp. Echiella. (*Latr.* G. iv. 222.)—(*Sam. I.* 44.)

7267. 2: pusiella*.
Ph. Ti. pusiella. *Linn.* ii. 884.—*Turt.* iii. 368.
Ti. Lithospermella. *Hüb. Ti. pl.* 15. *f.* 104.
Ti. sequella. (*Wien. V.* 140.)

7268. 3, dodecea*.
Er. dodecea: the Scarce Ermine. *Haw.* 514.
Ti. decorella. *Hüb. Ti. pl.* 44. *f.* 303.

7269 † 4, funerella?*
Ti. funerella. *Fabr.* iii. *b.* 309?—*Hüb. Ti. pl.* 15. *f.* 85?
Er. funerea: the Funereal. *Haw.* 515. In Mus. D. *Curtis.*

Genus 306: (1089). YPONOMEUTA, *Fabr., Leach, Samou.*
ERMINEA, *Haworth.*

A. Alis anticis nigro punctatis.

7270. 1; Evonymella*. (*Latr.* G. iv. 222.)—(*Leach E. E.* ix. 133.)—(*Sam. I.* 44.)
Ph. Ti. Evonymella. *Linn.* ii. 885.—*Stew.* ii. 197.—*Don.* xi. 80. *pl.* 355. *f.* 4.—*Turt.* iii. 362.
Ti. Evonymella. *Fabr.* iii. *b.* 289.—(*Haw. Pr.* 35.)—*Hüb. Ti. pl.* 13. *f.* 88.
Er. Evonymi: the Full-spotted E. *Haw.* 512.
Distaff Ermine. *Harr. V. M.* 25.

7271. 2; irrorella*. (*Sam. I.* 44.)
Ti. irrorella. *Hüb. Ti. pl.* 14. *f.* 93.—(*Haw. Pr.* 35.)
Er. irrorea: the Surrey Ermine. *Haw.* 512.

7272. 3; rorella*.
Ti. rorella. *Hüb. Ti. pl.* 34. *f,* 234.—(*Haw. Pr.* 35.)
Er. rorea: the few-spotted E. *Haw.* 513.
Ermine. *Albin. pl.* lxx. *f. a–d.*

7273. 4; padella*. (*Latr. G.* iv. 222.)—(*Sam. I.* 44.)
Ph. Ti. padella. *Linn.* ii. 885.—*Stew.* ii. 197.—*Turt.* iii. 362.
Ti. padella. (*Haw. Pr.* 35.)
Ph. Evonymella. *Berk. S.* i. 146.—*Don.* i. 23. *pl.* 9.
Er. Padi: the common Er. *Haw.* 513.
Small Ermine M.: Ph. Evonymella. (*Wilkes.* 4. *pl.* 5.)—*Harr. A. pl.* 3. *f. k–m.*—*Harr. V. M.* 23.

7274. 5; plumbella*. (*Sam. I.* 44.)
Ti. plumbella. *Fabr.* iii. 290.—(*Haw. Pr.* 35.)
Er. plumbea: the Kent Er. *Haw.* 513.

7275. 6; Curtisella*.
Ph. Ti. Curtisella. *Don.* ii. 77. *pl.* 65. *f.* 4.
Ti. Curtisella. (*Haw. Pr.* 38.)
Er. Curtisella: Curtis's. *Haw.* 514.
Ti. cænobitella. *Hüb. Ti. pl.* 45. *f.* 309.

B. Alis anticis haud punctatis.

7276. 7, Cratægella.
Ph. Ti. Cratægella. *Linn.* ii. 885.—*Turt.* iii. 368.

7277. 8: ambiguella*.
Ti. ambiguella. *Hüb. Ti. pl.* 22. *f.* 153.
Er. ambigua: the small Brown-bar. *Haw.* 514.

7278. 9; leucatella*.
Ph. Ti. leucatella. *Linn.* ii. 891.—*Turt.* iii. 371.
Ti. leucatella. *Hüb. Ti. pl.* 21. *f.* 146.—(*Haw. Pr.* 37.)
Er. leucatea: the small White-bar. *Haw.* 514.

7279. 10, bifasciella?*
Ti. bifasciella. *Fabr.* iii. *b.* 309?

7280. 11; comptella*.
Ti. comptella. *Hüb. Ti. pl.* 13. *f.* 89.
Er. compta: the Peacock's feather. *Haw.* 515.
Ph. Ti. apiella. *Don.* ii. 54. *pl.* 57. *f.* 3.

7281. 12; lutarella?*
Ti. lutarella. *Hüb. Ti. pl.* 25. *f.* 168?
Er. lutarea: the muddy Ermine. *Haw.* 515.

7282. 13; subfasciella* *mihi.*

7283. 14; Cæsiella*.
Ti. Cæsiella. *Hüb. Ti. pl.* 25. *f.* 172?
Er. cæsia: the Purple-edged. *Haw.* 516.

7284. 15; semifusca*.
Er. semifusca: the long-winged White-back. *Haw.* 517.
Ph. Ti. Pruniella. *Don.* ii. 57. *pl.* 58.—61. *pl.* 59. *f.* 2. *aucta.*
Ph. To. Pruniella. *Turt.* iii. 349.

7285. 16; mendicella*.
Ti. mendicella. *Hüb. Ti. pl.* 26. *f.* 179.
Er. mendica: the purple White-back. *Haw.* 517.

7286. 17; albistria*.
Er. albistria: the purple White-back. *Haw.* 517.

7287. 18; Pruniella*.
Ph. Ti. Pruniella. *Linn. F. No.* 1386?—*Stew.* ii. 202.—*Turt.* iii. 362.
Ti. Pruniella. *Hüb. Ti. pl.* 26. *f.* 175.—(*Haw.* 38.)
Er. Pruni: the White-back. *Haw.* 516.
Ti. semialbella. (*Haw. Pr.* 38.)
Al. Ephippella. *Fabr.* iii. *b.* 330.

7288. 19; tetrapodella*.
Ph. Ti. tetrapodella. *Mus. Marsham.*
Ph. Ti. tetrapodella. *Linn.* ii. 890?

7289. 20, ocellea* *mihi.*

7290. 21, subocellea* *mihi.*

7291. 22; ossea*.

Er. ossea: the Cream-coloured. *Haw.* 517.

7292. 23; curvella*.

Ph. Ti. curvella. *Linn. F. No.* 1387.

Er. curva: the brindled Ermine. *Haw.* 516.

7293. 24; Clematella*.

Ti. Clematella. *Fabr.* iii. *b.* 306. (!)—*Stew.* ii. 199.—*Turt.* iii. 370.

Rec. Clematea: the barred White. *Haw.* 552.

Ti. arcella. *Fabr.* iii. *b.* 305?

Ti. repandella. *Hüb. Ti. pl.* 37. *f.* 256.

An. Clematea. (*Curtis fo.* 189.)

Genus 307: (1090). ARGYROSETIA *mihi.*

TI.-METALLICÆ *p, Haworth.*

7294. 1; Gœdartella*: Gœdart's.—*Haw. l. c.*

Ph. Ti. Gœdartella. *Linn.* ii. 897.—*Stew.* ii. 201.

Ti. Gœdartella. *Hüb. Ti. pl.* 20. *f.* 133.—(*Haw. Pr.* 37.)—*Haw.* 571.

Ph. semiargentella. *Don.* ii. 76. *pl.* 65. *f.* 2, 3.—*Turt.* iii. 363.

Ti. semiargentella. (*Haw. Pr.* 39.)

7295. 2; semifasciella*.

Ti. semifasciella: the short Gold-bar. *Haw.* 570.

7296. 3; I W-ella*.

Ti. I W-ella: the Gold I W. *Haw.* 569.

7297. 4; I V-ella*.

Ti. I V-ella: the Gold I V. *Haw.* 570.

7298. 5; literella*.

Ti. literella: the Greek-lettered Gold. *Haw.* 570.

7299. 6; aurivittella*.

Ti. aurivittella: the golden Ribband. *Haw.* 570.

7300. 7, aurifasciella* *mihi.*

Genus 308: (1091). ARGYROMIS? (*Curtis MSS?*)

TI.-METALLICÆ *p, Haworth.*

7301. 1; Blancardella*: Blancard's.—*Haw. l. c.*

Ti. Blancardella. *Fabr.* iii. *b.* 327. (!)—*Stew.* ii. 202.—*Turt.* iii. 378.—*Haw.* 575.

7302. 2; Schreberella*: Schreber's.—*Haw. l. c.*

Ti. Schreberella. *Fabr.* iii. *b.* 327.—*Stew.* ii. 202.—*Turt.* iii. 377.—*Haw.* 575.

7303. 3; Cydoniella*: the bright speck'd Gold.—*Haw. l. c.*

Ti. Cydoniella. *Fabr.* iii. *b.* 323.—*Hüb. Ti. pl.* 39. *f.* 271?—*Haw.* 575.

7304. 4; Klemanella*: Kleman's.—*Haw. l. c.*
Ti. Klemanella. *Fabr.* iii. *b.* 326.—*Hüb. Ti. pl.* 29. *f.* 201.—(*Haw. Pr.* 38.)—*Haw.* 576.

7305. 5; Mespilella*: the silver-spotted Gold.—*Haw. l. c.*
Ti. Mespilella. *Hüb. Ti. pl.* 39. *f.* 272.—*Haw.* 576.

7306. 6, Rayella*: Ray's.—*Haw. l. c.*
Ph. Ti. Rayella. *Linn.* ii. 898.—*Turt.* iii. 377.
Ti. Rayella. *Fabr.* iii. *b.* 167?—*Hüb. Ti. pl.* 29. *f.* 200.—(*Haw. Pr.* 38.)—*Haw.* 577.

7307. 7; tristrigella*.
Ti. tristrigella: the treble Gold stripe. *Haw.* 576.

7308. 8, trifasciella*: the tawny Treble-bar.—*Haw. l. c.*
Ph. Ti. Lyonetella. *Linn.* ii. 897?—*Turt.* iii. 376?

7309 † 9. Myllerella[a].

7310. 10; Harrisella*: Harris's.—*Haw. l. c.*
Ph. Ti. Harrisella. *Linn.* ii. 899.—*Fabr.* iii. *b.* 327. (!)—*Stew.* ii. 202.—*Turt.* iii. 377.
Ti. Harrisella. *Fabr.* iii. *b.* 327.—*Haw.* 577.

7311. 11; Cramerella*.
Ti. Cramerella: Cramer's.—*Haw. l. c. Fabr.* iii. *b.* 327. (!)—*Stew.* ii. 202.—*Turt.* iii. 378.—*Haw.* 578.
Ti. Prunifoliella. *Hüb. Ti. pl.* 28. *f.* 191.
Ph. Ti. Bonnetella. *Linn.* ii. 897?—*Turt.* iii. 376?

7312. 12; hortella*: the Porcelain.—*Haw. l. c.*
Ti. hortella. *Fabr.* iii. *b.* 327.—*Haw.* 579.
Ph. Cramerella. *Don.* xi. 89. *pl.* 392. *f.* 1?

7313. 13; sylvella*: the dark Porcelain.—*Haw. l. c.*
Ti. sylvella. *Haw.* 579.
Ph. Blancardella. *Don.* xi. 90. *pl.* 392. *f.* 2?

7314. 14; cuculipenella*: the Cuckoo's Feather.—*Haw. l. c.*
Ti. cuculipenella. *Hüb. Ti. pl.* 28. *f.* 192?—*Haw.* 579.

7315. 15; Corylifoliella*: the Hazel Red.—*Haw. l. c.*
Ti. Corylifoliella. *Hüb. Ti. pl.* 28. *f.* 194.—(*Haw. Pr.* 37.)—*Haw.* 580.

7316. 16; Alnifoliella*.
Ti. Alnifoliella. *Hüb. Ti. pl.* 28. *f.* 193.—(*Haw. Pr.* 37.)

7317. 17; obscurella* *mihi.*

7318. 18; rufipunctella*.
Ti. rufipunctella: the red-and-white Barr'd. *Haw.* 580.

---

[a] 7309 † 9. Myllerella.
Ti. Myllerella. *Fabr.* iii. *b.* 325.—*Turt.* iii. 377. (!)

Genus 309: (1092). HERIBEÏA *mihi.*

TI.-METALLICÆ *p, Haw.*—GRACILLARIA? *p,* (*Kirby.*)

7319. 1, Haworthella *mihi.*

7320. 2; humerella*.
Ti. humerella. *Hüb. Ti. pl.* 42. *f.* 292.

7321. 3; Forsterella*: *Forster's.*—*Haw. l. c.*
Ti. Forsterella. *Fabr.* iii. *b.* 328.—(*Haw. Pr.* 39.)—*Haw.* 577.

7322. [4; simpliciella]* *mihi.*

7323. 5, cognatella *mihi.*

7324. 6; Clerckella*.
Ph. Ti. Clerckella. *Linn.* ii. 899.
Ti. Clerckella: Clerck's.—*Haw. l. c. Fabr.* iii. *b.* 329.—(*Haw. Pr.* 39.)—*Haw.* 578.

7325. 7; nivella* *mihi.*
Ti. cerasifoliella. *Hüb. Ti. pl.* 28. *f.* 190?

7326. 8; punctaurella*.
Ti. punctaurella: the golden Dot. *Haw.* 578.

7327. 9; unipunctella* *mihi.*

7328. 10; semiaurella* *mihi.*
Ti. malifoliella. *Hüb. Ti. pl.* 28. *f.* 195?

Genus 310: (1093). MICROSETIA *mihi.*

TI.-METALLICÆ *p, Haworth.*

7329. 1; Subbistrigella*.
Ti. subbistrigella: the double Silver-bar. *Haw.* 581.

7330. 2; obsoletella* *mihi.*

7331. 3; exiguella*.
Ti. exiguella. *Fabr.* iii. *b.* 328?

7332. 4; cinereo-punctella*.
Ti. cinereo-punctella: the grey dotted Brown. *Haw.* 381.

7333. 5; Stipella*: the triple Gold-spotted.—*Haw. l. c.*
Ti. Stipella. *Hüb. Ti. pl.* 20. *f.* 138.—*Haw.* 568.

7334. 6; guttella*: the white-spotted Sable.—*Haw. l. c.*
Ti. guttella. *Hüb. Ti. pl.* 26. *f.* 176.—*Haw.* 582.

7335. 7; quadrella*: the silver-spotted Sable.—*Haw. l. c.*
Ti. quadrella. *Fabr.* iii. *b.* 298?—*Hüb. Ti. pl.* 42. *f.* 293?—*Haw.* 582.

7336. 8; sequella*.
Ti. sequella: the silver-blotched. *Haw.* 583.

7337. 9; pulchella*.
Ti. pulchella: the small Argent and Sable. *Haw.* 582.

7338. 10; nigrella*: the small double Silver-bar.—*Haw. l. c.*
Ti. nigrella. *Hüb. Ti. pl.* 41. *f.* 285.—*Haw.* 583.

7339. 11; trimaculella*.
Ti. trimaculella: the cream-spotted Sable. *Haw.* 583.

7340. 12; subbimaculella*.
Ti. subbimaculella: the twin-spot Sable. *Haw.* 585.

7341. 13; nigrociliella* *mihi.*

7342 † 14, unifasciella*.
Ti. unifasciella: the silver-barred Brown. *Haw.* 584.
In Mus. D. *Haworth.*

7343. 15; mediofasciella*.
Ti. mediofasciella: the central Silver-bar. *Haw.* 584.

7344. 16; aurella*: the diamond-barred Pygmy.—*Haw. l. c.*
Ti. aurella. *Fabr.* iii. *b.* 329.—*Haw.* 584.

7345. 17; posticella*.
Ti. posticella: the pygmy Silver-bar. *Haw.* 584.
Ti. Hübnerella. *Hüb. Ti. pl.* 34. *f.* 236.

7346 † 18, violaceella*.
Ti. violaceella: the violet Pygmy. *Haw.* 585.
In Mus. D. *Haworth.*

7347. 19; floslactella*.
Ti. floslactella: the cream Pygmy. *Haw.* 585.

7348. 20; atri-capitella*.
Ti. atri-capitella: the black-headed Pygmy. *Haw.* 585.

7349. 21; ruficapitella*.
Ti. ruficapitella: the red-headed Pygmy. *Haw.* 586.

7350. 22; pygmæella*.
Ti. pygmæella: the least Pygmy. *Haw.* 586.

B. An hujus generis?

7351. 23; sericiella*.
Ti. sericiella: the Satin Pygmy. *Haw.* 585.

7352. 24; aurofasciella* *mihi.*

7353. 25; Gleichella*: Gleiche's.—*Haw. l. c.*
Ti. Gleichella. *Fabr.* iii. *b.* 323. (!)—*Turt.* iii. 376.—*Haw.* 582.
Ph. Gleichenella. *Stew.* ii. 201.

7354. 26; quadriguttella*: the four-spotted Gold.—*Haw. l. c.*
Ti. 4-guttella. *Haw.* 574.

Genus 311: (1094). ŒCOPHORA, *Latr.*

GLYPHYPTERYX, *Curtis.*—TI.-METALLICÆ *p*, et GRACILLARIA *p*, *Haworth.*—ÆCOPHORA, *Samou.*

7355. 1: Linneella*. (*Latr.* G. iv. 223.)—(*Leach E. E.* ix. 133.)
—(*Sam. I.* 1.)

Ph. Ti. Linneella. *Linn.* ii. 898.—*Turt.* iii. 377.
Gl. Linneella. *Curtis.* iv. *pl.* 152.

7356. 2; Rœsella*. (*Latr. G.* iv. 223.)—(*Leach E. E.* ix. 133.) —(*Sam. I.* 1.)
Ph. Ti. Rœsella. *Linn.* ii. 898.—*Stew.* ii. 202.—*Turt.* iii. 377.
Ti. Rœsella: Rœsel's.—*Haw. l. c. Hüb. Ti. pl.* 20. *f.* 135.—*Haw.* 573.
Gl. Rœsella. (*Curtis l. c. supra.*)

7357. 3; Schæfferella*.
Ph. Ti. Schæfferella. *Linn.* ii. 898.—*Don.* v. 99. *pl.* 175.—*Stew.* ii. 201.
Ti. Schæfferella: Schæffer's.—*Haw. l. c. Hüb. Ti. pl.* 20. *f.* 136. (*aucta.*)—(*Haw. Pr.* 37.)—*Turt.* iii. 376.—*Haw.* 574.
Gl. Schæfferella. (*Curtis l. c. supra.*)

7358 † 4. bimaculella*? the orange-blotched Black.—*Haw. l. c.*
Ti. bimaculella. *Haw.* 574. In Mus. D. I. Latham.

7359. 5; eximia*.
Gr. eximia: the Nonpareil. *Haw.* 532.

7360. 6; metallella*.
Ph. Ti. metallella. (*Wien. V.* 144?)

Genus 312: (1095). PANCALIA *mihi.*

ŒCOPHORA *p*, *Latreille.*—GRACILLARIA *p*, *Haworth.*

7361. 1; Leuwenhoekella*.
Ph. Ti. Leuwenhoekella. *Linn.* ii. 897.—*Turt.* iii. 377.
Ti. Leuwenhoekella: Leuwenhoek's. (*Haw. Pr.* 37.)—*Haw.* 574.

7362. 2; fusco-ænea*.
Po. fusco-ænea: the Brown-brassy. *Haw.* 537.

7363. 3; fusco-cuprea*.
Po. fusco-cuprea: the Brown-copper. *Haw.* 537.

7364. 4: Merianella*.
Ph. Ti. Merianella. *Linn.* ii. 897.—*Stew.* ii. 201.—*Turt.* iii. 375.
Te. Merianella. (*Haw. Pr.* 37.)
Gr. Merianæ: Merian's. *Haw.* 531.

Genus 313: (1096). ——

GRACILLARIA *p*, *et* TINEA *p*, *Haworth.*

7365. 1; guttea*.
Gr. guttea: the white-spotted Brown. *Haw.* 531.
Ti. Merianella. *Hüb. Ti. pl.* 20. *f.* 134.

7366. 2; Fyeslella*.
Ti. Fyeslella. *Fabr.* iii. *b.* 318.—(*Haw. Pr.* 39.)

Ph. triguttella. *Don.* xi. 63. *pl.* 382. *f.* 1.
Gr. Fueslii: Fuesly's. *Haw.* 531.

7367. 3; Erxlebella?*
Ti. Erxlebella. *Fabr.* iii. *b.* 340?
Ti. fusco-cuprella: the Brown-copper. *Haw.* 569.

7368. 4; fusco-viridella*.
Ti. fusco-viridella: the Brown-green. *Haw.* 569.

Genus 314: (1097). ——

TINEA-METALLICÆ *p, Haworth.*

7369. 1; cinctella*.
Ph. Ti. cinctella. *Linn.* ii. 891.—*Turt.* iii. 371.
Ti. cinctella: the Silver barred Sable. *Hüb. Ti. pl.* 21. *f.* 142. —(*Haw. Pr.* 37.)—*Haw.* 581.

7370. 2, albistrigella* *mihi.*

Genus 315: (1098). ——

PORRECTARIA *p, et* GRACILLARIA *p, Haworth.*

7371. 1, grandipennis*.
Po. grandipennis: the great Raven-Feather. *Haw.* 536.

7372. 2, Picæpennis*.
Po. Picæpennis: the Pye-feather. *Haw.* 536.

7373. 3; cylindrella*.
Ti. cylindrella. *Fabr.* iii. *b.* 308.—(*Haw. Pr.* 39.)
Gr. cylindrea: the buff-blotched Slender. *Haw.* 527.
Ti. tristella. *Hüb. Ti. pl.* 32. *f.* 218.

7374. 4; serratella*.
Ph. Ti. serratella. *Linn.* ii. 802.—(*Kirby & Sp. I. E.* i. 462.)
Po. Coracipennis β. *Haw.* 536.

7375. 5; Coracipennella*.
Ti. Coracipenella. *Hüb. Ti. pl.* 30. *f.* 209.
Po. Coracipennis: the small Raven-feather. *Haw.* 536.

7376. 6; obscurella*.
Ti. obscurella. *Fabr.* iii. *b.* 312?
Po. obscurea: the Brown-feather. *Haw.* 536.

7377. 7, Gryphipennella*.
Ti. Gryphipenella. *Hüb. Ti. pl.* 30. *f.* 206.
Po. Gryphipennis: the Vulture-feather. *Haw.* 537.

7378. 8; lutarea*.
Po. lutarea: the shining Clay. *Haw.* 537.

7379. 9; ochroleucella* *mihi.*

7380. 10; nigricella* *mihi.*

7381. 11; flavicaput*.
Po. flavicaput: the yellow-headed Black. *Haw.* 536.

Genus 316: (1099). ——

TI.-METALLICÆ *p*, *Haworth.*

7382. 1; scissella*.
Ti. scissella: the oblong Gold-head.—*Haw. l. c.* *Hüb. Ti. pl.* 39. *f.* 270.—*Haw.* 580.
Ti. angustipennella.——?

Genus 317: (1100). ——

PORRECTARIA *p*, *Haworth.*

7383. 1; spissicornis*.
Po. spissicornis: the thick-horned Green. *Haw.* 537.

7384. 2; Trifolii* *mihi.*

Genus 318: (1101). PORRECTARIA, *Haworth.*

7385. 1; Anatipennella*.
Ti. Anatipenella. *Hüb. Ti. pl.* 29. *f.* 186.
Po. Anatipennis: the Goose-feather. *Haw.* 534.

7386. 2, ornatipennella*.
Ti. ornatipenella. *Hüb. To. pl.* 29. *f.* 199.
Po. ornatipennis: the silver-streaked Hook-tip. *Haw.* 534.

7387. 3; ochrea*: the silver-streaked. *Haw.* 533.
Pter. ochrodactylus. *Fabr.* iii. *b.* 345?
The Spear Moth. *Harr. V. M.* 47?—*Harr. Ex.* 14. *pl.* iii. *f.* 2–5?

7388. 4; Gallipennella*.
Ti. Gallipenella. *Hüb. Ti. pl.* 29. *f.* 200.
Po. Gallipennis: the Cock's-feather. *Haw.* 534.

7389. 5, lineolea*: the Red-speck'd. *Haw.* 534.

7390. 6, albicosta*: the White-edged. *Haw.* 535.

7391. 7; leucapennella*.
Ti. leucapenella. *Hüb. Ti. pl.* 30. *f.* 205.
Po. leucapennis: the Lead-coloured. *Haw.* 535.

7392. [8; argentula*.]
Ph. Ti. argentula. *Marsham. MSS.*

Genus 319: (1102). ——

PORRECTARIA *p*, *et* RECURVARIA *p*, *Haworth.*

7393. 1, auritella*.
Ti. auritella. *Hübner.*

7394. 2, Cygnipennella*.

Ti. Cygnipennella. *Hüb. Ti. pl.* 30. *f.* 207.
Po. Cygnipennis: the Swan's Feather. *Haw.* 536.

7395. 3, semialbella* *mihi.*

7396 † 4, triatomea.
Po. triatomea: the treble-atom'd White. *Haw.* 535.
In Mus. D. *Haworth.*

7397. 5; floslactis*.
Po. floslactis: the Cream-colour'd. *Haw.* 535.

7398. 6, rufo-cinerea*.
Po. rufo-cinerea: the Red-brindled. *Haw.* 535.

7399. 7; lucidella* *mihi.*

7400. 8; Oleella*.
Ti. Oleella. *Fabr.* iii. *b.* 308?
Po. Oleæ: the small Shining Brown. *Haw.* 535.

7401. 9; rufipennella*.
Ti. rufipennella. *Hüb. Ti. pl.* 30. *f.* 204.
Po. rufipennis: the Red-feather. *Haw.* 537.

7402. 10; fulvescens*.
Po. fulvescens. *Haworth? MSS?*

7403. 11; marginea*.
Re. marginea: the bordered Straw. *Haw.* 556.

Genus 320: (1103). ——

RECURVARIA*** *p, Haworth.*

7404. 1; Lambdella*.
Ph. Ti. Lambdella. *Don.* ii. 53. *pl.* 57. *f.* 2.—*Turt.* iii. 363.
Ti. Lambdella. (*Haw. Pr.* 38.)
Re. Lambda: the tawny Crescent. *Haw.* 556.

7405. 2; lunaris*.
Re. lunaris: the lesser tawny Crescent. *Haw.* 556.

7406. 3; lutarella*.
Ti. lutarella. *Hüb. Ti. pl.* 25. *f.* 168?

7407. 4; saturatella* *mihi.*

7408. 5, Panzerella*.
Ph. Ti. Panzerella. *Don.* iii. 94. *pl.* 106. *f.* 4?—*Turt.* iii. 362?
Ti. Donovanella. (*Haw. Pr.* 38.)

7409. 6; flavifrontella*: the yellow-head.—*Haw. l. c.*
Ti. flavifrontella. *Fabr. S.* 305.—*Hüb. Ti. pl.* 18. *f.* 126.—(*Haw. Pr.* 38.)—(*Haw.* 393.)—*Ent. Trans.* (*Haworth.*) i. 338.

Familia XX: (133). TINEIDÆ *mihi.*

(PHALÆNA-TINEA *p, Linné.*—NOCTUO-BOMBYCITES *p, et* PYRALITES *p, Latr.*—CRAMBIDÆ *et* TINEIDÆ *p, Leach.*)

(TINEA *p, et* ALUCITA *p, Fabricius.*)

Genus 321: (1104). GALLERIA, *Fabricius, Haworth, Leach, Samou.*, (*Kirby.*)

A. Alis anticis posticè rotundatis.

7410. 1, alvearia*. *Fabr. S.* 463.—(*Leach E. E.* ix. 135.)
Ga. alvea: the honey Moth. *Haw.* 392.—(*Sam. I.* 17.)
To. grisella. *Fabr.* iii. *b.* 289?
Bo. cinereola. *Hüb. Bo. pl.* 23. *f.* 91.

B. Alis anticis posticè emarginatis.

7411. 2; cercana*. *Fabr. S.* 462.
Ph. Ge. cereana. *Linn.* ii. 874.
Ti. cerella. *Fabr.* iii. *b.* 287.—*Hüb. Ti. pl.* 4. *f.* 25.
Ga. cerea: the Honey-comb Moth. *Haw.* 392.—(*Sam. I.* 17.)
Ph. Ti. mellonella. *Linn.* ii. 888.—*Don.* viii. 79. *pl.* 283.—*Stew.* ii. 199.—*Turt.* iii. 369.
Ti. mellonella. (*Kirby & Sp. I. E.* i. 31. 230.: ii. 266.)
No. mellonina. (*Haw. Pr.* 13.)

Genus 322: (1105). ILYTHIA, (*Latreille.*)

CRAMBUS *p, Fabr., Haw.*—HERMINIA *p, Sam.*

7412. 1; colonella*.
Ph. Ti. colonella. *Linn.* ii. 883.—*Don.* viii. 28. *pl.* 263. *f.* 2. —*Stew.* ii. 196.—*Turt.* iii. 361.
Ti. colonella. *Hüb. Ti. pl.* 4. *f.* 23.
Cr. colonatus: the green-shaded. *Haw.* 374.
Cr. colonum. *Fabr. Sup.* 469.
No. colonina. (*Haw. Pr.* 13.)
He. colonalis. (*Sam. I.* 21.)

Genus 323: (1106). MELIA, *Curtis.*

LITHOSIA *p, Fabricius, Haworth.*

7413. 1; socia*. (*Curtis folio* 201.)
Ph. Ti. sociella. *Linn.* ii. 883.
Li. socia: the pale shoulder. *Haw.* 151.
Ti. tribunella. (*Wien. V.* 319.)—*Hüb. Ti. pl.* 4. *f.* 22.
No. sociina. (*Haw. Pr.* 13.)
He. socia. (*Sam. I.* 21.)

7414 † 2, bipunctana*. (*Curtis l. c. supra.*)
To. bipunctata. *Ent. Trans.* (*Haworth.*) i. 337.
Ti. sociella. *Hüb. Ti. pl.* 4. *f.* 24. In Mus. D. *Hatchett.*

Genus 324: (1107). ——

MELIA p, *Curtis.*

7415 † 1, flammea*?
Me. flammea. *Curtis.* v. *pl.* 201. In Mus. *D. Dale.*

7416 † 2, sericea*.
Me. sericea. *Curtis l. c. supra.* In Mus. D. *Curtis et Dale?*

Genus 325: (1108). EUDOREA, *Curtis,* (*Kirby.*)

SCOPARIA[a], *Haw.*—SCOPEA, *Haw.*—CHILO p, *Zinck. Sommers.*

7417. 1, pallida. *Curtis l. c. infra.*

7418. 2; Pyralella*.
Ti. Pyralella. *Hüb. Ti. pl.* 24. *f.* 167.—(*Haw. Pr.* 36.)
Sc. Pyralea: the yellow-stigmaed Grey. *Haw.* 499.
Ti. Pyralea. (*Sam. I.* 41.)
Eu. Pyralea. (*Curtis l. c. supra.*)

7419. 3; tristrigella* *mihi.*

7420. 4; dubitalis*.
Py. dubitalis. *Hüb. Py. pl.* 8. *f.* 49.
Sc. dubita: the hoary Grey. *Haw.* 499.
Eu. dubita. (*Curtis l. c. infra.*)

7421. 5; cembrella*.
Ph. Ti. cembrella. *Linn.* ii. 892?—*Stew.* ii. 200.—*Turt.* iii. 372.
Sc. Cembræ: the large Grey. *Haw.* 498.
Eu. Cembræ. (*Curtis l. c. infra.*)

7422. 6; subfusca*. (*Curtis l. c. infra.*)
Sc. subfusca: the drab Grey. *Haw.* 498.

7423. 7, murana. *Curtis.* iv. *pl.* 170.

7424. 8; lineola*. *Curtis l. c. supra.*

7425. 9; Resinea*.
Sc. Resinea: the Resin Grey. *Haw.* 499.
Eu. Resinea. *Curtis l. c. supra.*

7426. 10; Mercurella*.
Ph. Ti. Mercurella. *Linn.* ii. 892.—(*Haw. Pr.* 39.)
Sc. Mercurea: the Small Grey. *Haw.* 499.
Eu. Mercurea. (*Curtis l. c. supra.*)
Ti. Cratægella. *Hüb. Ti. pl.* 34. *f.* 231.

7427. 11; angustea*. *Curtis l. c. supra.*

Genus 326: (1109). PHYCITA, *Curtis.*

PHYCIS[b] p, *Fabr., Haworth, Leach, Samou.*

7428. 1: nebulella*.

---

[a] SCOPARIA: Genus Plantarum. Vide *Linn.* iii. 123.
[b] PHYCIS: Piscis Antiquorum.

Ti. nebulella. *Hüb. Ti. pl.* 23. *f.* 157.—(*Haw. Pr.* 38.)
Ph. nebulea: the ermine Knot-horn. *Haw.* 494.—(*Curtis l. c. infra.*)

7429. 2; diluta*: the powdered Knot-horn. *Haw.* 495.—(*Curtis l. c. infra.*)
Ti. dilutella. *Hüb. Ti. pl.* 10. *f.* 69?
Ti. canella. (*Wien. V.* 135?)

7430. 3; elutella*.
Ti. elutella. *Hüb. Ti. pl.* 24. *f.* 163.—(*Haw.* 38.)
Ph. elutea: the cinereous Knot-horn. *Haw.* 496.—(*Curtis l. c. infra.*)

7431. 4: rufa*: the rufous Knot-horn. *Haw.* 497.—(*Curtis l. c. infra.*)

7432. 5; angustella*.
Ti. angustella. *Hüb. Ti. pl.* 10. *f.* 68?
Ph. angustea: the small Ermine Knot-horn. *Haw.* 497.—(*Curtis l. c. infra.*)

7433. 6; semirufa*: the red-streaked Knot-horn. *Haw.* 496.—(*Curtis l. c. infra.*)

7434. 7; bistriga*: the double-striped Red Knot-horn. *Haw.* 496.—(*Curtis l. c. infra.*)
Ti. bistriga. (*Sam. I.* 41.)

7435. 8: Gemina*: the twin-barred Knot-horn. *Haw.* 497.

7436. 9; tumidella*. *Germ. M.* (*Zincken g. Sommer.*) iii. 136.
To. tumidana. (*Wien. V.* 130?)
Ti. verrucella. *Hüb. Ti. pl.* 11. *f.* 73.—(*Haw. Pr.* 36.)
Ph. verrucea: the warted Knot-horn. *Haw.* 494.—(*Curtis l. c. infra.*)

7437. 10; consociella*. *Germ. M.* (*Zincken g. Sommer.*) iii. 138.—(*Curtis l. c. infra.*)
Ti. consociella. *Hüb. Ti. pl.* 48. *f.* 328.

7438. 11; fascia*: the broad-barred Knot-horn. *Haw.* 496.—(*Curtis l. c. infra.*)
Ph. suavella. *Germ. M.* (*Zincken g. Sommer.*) iii. 140.

7439. 12: advenella*. *Germ. M.* (*Zincken g. Sommer.*) iii. 141.

7440. 13; marmorea*: the marbled Knot-horn. *Haw.* 495.—(*Curtis l. c. infra.*)

7441. 14, Porphyrea*. (*Curtis l. c. infra.*)

7442. 15; Rhenella*. *Germ. M.* (*Zincken g. Sommer.*) iii. 166.
Tin. Rhenella. *Schiffermüller.*—*Charp. & Z. S.* 178.
Ti. palumbella. *Hüb. Ti. pl.* 10. *f.* 70.

Ph. palumbea: the dove-coloured Knot-horn. *Haw.* 494.—(*Curtis l. c. infra.*)

7443. 16: pinguis*: the tabby Knot-horn. *Haw.* 493.—*Curtis.* v. *pl.* 233.

7444. 17; formosa*: the beautiful Knot-horn. *Haw.* 494.—(*Curtis l. c. supra.*)

7445. 18, obtusa*: the blunt-winged Knot-horn. *Haw.* 495.—(*Curtis l. c. supra.*)
Ti. obtusella. *Hüb. Ti. pl.* 31. *f.* 215?

7446. 19; ornatella*. *Germ. M.* (*Zincken g. Sommer.*) iii. 154.
Ti. ornatella. (*Wien. V.* 319.)
Ti. criptella. *Hüb. Ti. pl.* 11. *f.* 77.—(*Haw. Pr.* 36.)
Ph. cryptea: the speckled Knot-horn. *Haw.* 493.—(*Curtis l. c. supra.*)

7447 † 20, Abietella*. *Germ. M.* (*Zincken g. Sommer.*) iii. 160.
Ti. Abietella. (*Wien. V.* 138.)—*Fabr. M.* ii. 245.
Ti. Decuriella. *Hüb. Ti. pl.* 11. *f.* 74.
Ph. Decuriella. (*Curtis l. c. supra.*)
In Mus. D. *Bentley*, Dale *et Stone.*

7448. 21; Roborella*. *Germ. M.* (*Zincken g. Sommer.*) iii. 147.
Ti. Roborella. (*Wien. V.* 138.)
Ti. spissicella. *Fabr. Sp. I.* ii. 289.—*Hüb. Ti. pl.* 11. *f.* 75.—(*Haw. Pr.* 36.)
Ti. spissicornis. *Fabr. S.* 463.—(*Sam. I.* 41.)
Ph. spissicornis: the dotted Knot-horn. *Haw.* 492.—(*Curtis l. c. supra.*)

7449. 22; cristella*. *Germ. M.* (*Zincken g. Sommer.*) iii. 152?
Ti. cristella. *Hüb. Ti. pl.* 11. *f.* 76.—(*Haw. Pr.* 36.)
Ph. cristea: the purplish Knot-horn. *Haw.* 492.—(*Curtis l. c. supra.*)

7450. 23; legatella*. *Germ. M.* (*Zincken g. Sommer.*) iii. 149.
Ti. legatella. *Hüb. Ti. pl.* 11. *f.* 71?—(*Haw. Pr.* 36.)
Ph. legatea: the plain Knot-horn. *Haw.* 492.—(*Curtis l. c. supra.*)

7451. 24; fusca*: the brown Knot-horn. *Haw.* 493.—(*Curtis l. c. supra.*)

7452. 25; palumbella*. *Germ. M.* (*Zincken g. Sommer.*) iii. 151.
Ti. palumbella. (*Wien. V.* 138.)—*Fabr.* iii. *b.* 302.
Ti. contubernella. *Hüb. Ti. pl.* 11. *f.* 72.—(*Haw. Pr.* 36.)
Ph. contubernea: the mealy Knot-horn. *Haw.* 493.—(*Curtis l. c. supra.*)
Ti. contubernia. (*Sam. I.* 41.)

Genus 327: (1110). ONCOCERA *mihi.*

CRAMBUS *p, Fabricius.*—PALPARIA *p, Haw.*—PHYCIS *p, Zincken g. Sommer.*—PHYCITA *p, Curtis fo.* 233.—YPONOMEUTA *p, Samou.*

A. Palpis erectis.

7453. 1; Cardui*.
Pa. Cardui: the Thistle Ermine. *Haw.* 484.
Cr. Cardui. (*Sam. I.* 13.)
Ph. Cardui. (*Curtis l. c. supra.*)
Ti. Cribella. *Hüb. Ti. pl.* 10. *f.* 67?—(*Haw. Pr.* 35.)
Ypo. Cribella. (*Sam. I.* 44.)
Thistle Ermine. *Harr. V. M.* 25.

7454. 2; carnella*.
Ph. Ti. carnella. *Linn.* ii. 887.—(*Haw. Pr.* 36.)
Ti. carnella. *Hüb. Ti. pl.* 10. *f.* 66.
Pa. carnea: the rosy Veneer. *Haw.* 484.
Cr. carnea. (*Sam. I.* 13.)
Ph. carnea. (*Curtis l. c. supra.*)
Purple Vanear. *Harr. V. M.* 51?

7455. [3; sanguinella*.]
Ti. sanguinella. *Hüb. Ti. pl.* 10. *f.* 65.—(*Haw. Pr.* 36.)
Pa. sanguinea: the buff-edged rosy Veneer. *Haw.* 484.
Cr. sanguinea. (*Sam. I.* 13.)
Ph. sanguinea. (*Curtis l. c. supra.*)
Ph. Ti. carnella. *Don.* v. 35. *pl.* 153. *f.* 5.—*Turt.* iii. 365.

B. Palpis porrectis.

7456. 4, miniosella.
Phy. miniosella. *Germ. Mag.* (*Zincken g. Sommer.*) iii. 126.

7457. [5, Lotella]?
Ti. Lotella. *Hüb. Ti. pl.* 48. *f.* 334?

7458 † 6, ocellea*.
Pa. ocellea: the Necklace Veneer. *Haw.* 436.
Cr. ocellea. (*Sam. I.* 13.) In Mus. D. *Haworth.*

7459. 7; ahenella*.
Ti. ahenella. (*Wien. V.* 135?)—*Hüb. Ti. pl.* 9. *f.* 58, 9?
Ti. æneella. *Hüb. Ti. pl.* 6. *f.* 41. ♂?
Cr. obscuratus: the dingy Snout. *Haw.* 367.
Her. obscuralis. (*Sam. I.* 21.)
Pa. tetrix: the mouse-coloured Veneer. *Haw.* 486.
Cr. tetrix. (*Curtis l. c. folio* 109.)

Genus 328: (1111). CRAMBUS, *Fabricius, Leach, Samou., Curtis.*

PALPARIA *p, Haworth.*—TRICHOSTOMA, *Zinck. S.* (*olim.*)—CHILO *p, Zinck. S.*

A.

7460. 1; Lythargyrellus*.
Ti. Lythargyrella. *Hüb. Ti. pl.* 33. *f.* 227.

7461. 2; argyreus*. *Fabr. S.* 471.
Pa. argyrea: the streaked-Satin Veneer. *Haw.* 486.
Striped Vanear. *Harr. V. M.* 53.

7462. 3; Arbustorum*. *Fabr. S.* 472.
Pa. arbustea: the yellow-satin Veneer. *Haw.* 486.
Cr. Arborum. (*Sam. I.* 13.)

7463. 4; argentellus*.
Ph. Ti. argentellus. *Linn.* ii. 895?—*Stew.* ii. 202?—*Turt.* iii. 365?
Pa. argentea: the white-satin Veneer. *Haw.* 486.
Ti. perlella. (*Wien. V.* 134.)—*Hüb. Ti. pl.* 6. *f.* 40.
Silver Vanear. *Harr. V. M.* 53.

7464. [5; dealbellus*.]
Ti. dealbella. *Thunb. Diss.* (*Wenner.*) 84.

B.

7465. 6, hamellus*.
Ti. hamella. *Thunb. Diss.* (*Wenner.*) 84. *pl. adj. f.* 3.
Ti. ensigerella. *Hüb. Ti. pl.* 54. *f.* 367. (667.)
Pa. baccæstria: the pearl-streak Veneer. *Haw.* 488.

7466 † 7, tentaculellus*. (*Curtis l. c. supra.*)
Ti. tentaculella. *Hüb. Ti. pl.* 83. *f.* 230.
Pa. tentaculea: the Portland Veneer. *Haw.* 487.
In Mus. D. *Stone.*

7467. 8; pascuellus*.
Ph. Ti. pascuella. *Linn.* ii. 886.—*Stew.* ii. 197.—*Turt.* iii. 365.
Ti. pascuella. *Hüb. Ti. pl.* 5. *f.* 31.—(*Haw. Pr.* 36.)
Cr. pascuum. *Fabr. S.* 471.
Cr. pascuorum. (*Leach E. E.* ix. 135.)—(*Sam. I.* 13.)
Pa. pascuea: the inlaid Veneer. *Haw.* 488.
† β, Cr. pascella. *Ahrens F.* iii. *f.* 16. β in Mus. D. *Stone.*

7468. 9: Dumetellus*.
Ti. Dumetella. *Hüb. Ti. pl.* 58. *f.* 389.
Ph. Ti. pascuella. *Scop. C. No.* 621.

7469. 10, ericellus*.
Ti. ericella. *Hüb. Ti. pl.* 54. *f.* 371. (271.)

7470. 11; pratellus*.
Ph. Ti. pratella. *Linn.* ii. 886.—*Stew.* ii. 197.—*Turt.* iii. 364.
Ti. pratella. *Hüb. Ti. pl.* 5. *f.* 29.—*pl.* 60. *f.* 401.—(*Haw. Pr.* 36.)
Cr. pratorum. *Fabr. S.* 471.—(*Leach E. E.* ix. 135.)—(*Sam. I.* 13.)
Pa. pratea: the dark inlaid Veneer. *Haw.* 488.

7471. [12; angustellus*] *mihi.*

7472. 13; hortuellus*.
Ti. hortuella. *Hüb. Ti. pl.* 7. *f.* 46.

Pa. hortuéa: the garden Veneer. *Haw.* 490.
Cr. hortorum. (*Sam. I.* 13.)

7473. [14; cespitellus*.]
Ti. cespitella. *Hüb. Ti. pl.* 7. *f.* 45.
Pa. cespitea: the straw-coloured Veneer. *Haw.* 490.
Cr. cespitis. (*Sam. I.* 13.)
Ti. strigella. *Fabr.* iii. *b.* 297.
Cr. strigatus. *Fabr. S.* 472.—(*Sam. I.* 13.)
Ti. chrysonuchella. (*Wien. V.* 134.)
Chi. hortuellus var. *Germ. M.* (*Zincken g. Sommer.*) ii. 62.

7474. 15, montanellus *mihi.*

7475. 16, marginellus *mihi.*

7476. 17; tristis*.
Pa. tristis: the dingy Veneer. *Haw.* 486.
Te. tristella. *Fabr.* iii. *b.* 296?
Ti. lignella. *Hüb. Ti. pl.* 9. *f.* 57?

7477. 18; culmellus*.
Ti. culmella. *Linn.* ii. 886.
Ti. straminella. (*Wien. V.* 134.)—*Hüb. Ti. pl.* 7. *f.* 49.—(*Haw. Pr.* 36.)
Pa. striga: the small straw-coloured Veneer. *Haw.* 490.
Grass close-wing. *Harr. V. M.* 57?

7478. 19, auriferellus*.
♂, Ti. auriferella. *Hüb. Ti. pl.* 9. *f.* 62.
Pa. aurifera: the dark dwarf Veneer. *Haw.* 491.
Cr. auriferus. (*Curtis l. c. supra.*)
♀, Ti. barbella. *Hüb. Ti. pl.* 9. *f.* 61.
Pa. barbella: the dwarf Veneer. *Haw.* 490.
Cr. barbus. (*Curtis l. c. supra.*)

7479. 20, pygmæus.
Pa. pygmæa. *Haworth MSS.*

C.

7480. 21, radiellus. *Curtis.* iii. *pl.* 109.
Ti. radiella. *Hüb. Ti. pl.* 47. *f.* 325.

7481. 22, margaritellus.
Ti. margaritella. (*Wien. V.* 134.)—*Fabr.* iii. *b.* 295.—*Hüb. Ti. pl.* 6. *f.* 39.
Cr. margaritaceus. *Fabr. S.* 470.—(*Curtis l. c. supra.*)

7482. 23, latistrius*. (*Curtis l. c. supra.*)
Pa. latistria: the broad-streak Veneer. *Haw.* 485.
Ch. Leachellus. *Germ. M.* (*Zincken g. Sommer.*) iii. 114.

7483. 24; Pinetellus*.
Ph. Ti. pinetella. *Linn.* ii. 886.—*Don.* viii. 27. *pl.* 263. *f.* 1.—*Stew.* ii. 197.—*Turt.* iii. 364.

Cr. Pineti. *Fabr. S.* 470.—(*Leach E. E.* ix. 135.)—(*Sam. I.* 13.)
Pa. Pineti: the Pearl Veneer. *Haw.* 487.
Ti. conchella. *Hüb. Ti. pl.* 6. *f.* 38.—(*Haw. Pr.* 36.)
Ph. virginella. *Scop. C. No.* 629.

7484. 25; chrysonuchellus*.
Ti. chrysonuchella. *Scop. C. No.* 628.
Ti. Gramella. *Fabr.* iii. *b.* 300. (!)—*Stew.* ii. 198.—*Turt.* iii. 367.
Ti. campella. *Hüb. Ti. pl.* 7. *f.* 44.—(*Haw. Pr.* 36.)
Pa. campea: the powdered Veneer. *Haw.* 489.

7485 † 26, rorellus*.
Ti. rorella. *Linn.* ii. 886.
Ti. Craterella. *Scop. C. No.* 627.
Ti. linetella. *Fabr.* iii. *b.* 292.
Cr. lineatus. *Fabr. S.* 470.
Ti. chrysonuchella. *Hüb. Ti. pl.* 7. *f.* 43.—(*Curtis l.* c. *supra.*)
In Mus. D. *Stone.*

7486. 27; falsellus*.
Ti. falsella. (*Wien. V.* 134.)—*Fabr.* iii. *b.* 295.—*Hüb. Ti. pl.* 5. *f.* 50.—(*Haw. Pr.* 36.)
Pa. falsa: the chequered Veneer. *Haw.* 488.
Ti. abruptella. *Thunb. Diss.* (*Wenner.*) 83. *pl. f.* 2.
Cr. falsa. (*Sam. I.* 13.)

7487. 28; luteellus*.
Ti. luteella. (*Wien. V.* 134.)
Ti. ochrella. *Hüb. Ti. pl.* 8. *f.* 55. ♂.
Ti. exsoletella. *Hüb. Ti. pl.* 7. *f.* 48. ♀.—(*Haw. Pr.* 36.)
Pa. rorea: the barred Veneer. *Haw.* 488.
Cr. Rosea. (*Sam. I.* 13.)

7488. 29; inquinatellus*.
Ti. inquinatella. (*Wien. V.* 134.)—*Hüb. Ti. pl.* 8. *f.* 54.—(*Haw. Pr.* 36.)
Pa. geniculea: the elbow-striped Veneer. *Haw.* 488.
Cr. geniculea. (*Sam. I.* 13.)

7489 † 30: aridellus*. (*Curtis l.* c. *supra.*)
Ti. aridella. *Thunb. Diss.* (*Wenner.*) 83. *pl. f.* 1.
In Mus. D. *Stone.*

7490. 31; selasellus*.
Ti. selasella. *Hüb. Ti. pl.* 60. *f.* 405. ♂. 406. ♀.

7491. 32; fuscelinellus.
Ti. fuscelinella. *Schrank B.* ii. 100.

7492. 33; obtusellus* *mihi.*

7493. 34; petrificellus*.
Ti. petrificella. *Hub. Ti. pl.* 7. *f.* 47.

Pa. petrificea: the common Veneer. *Haw.* 485.
Cr. petrificea. (*Sam. I.* 13.)
Vanear. *Harr. V. M.* 53.

7494. 35; nigristriellus* *mihi.*
Ti. deliella. *Hüb. Ti. pl.* 60. *f.* 402, 403?

7495. 36; aquilellus.
Ti. aquilella. *Hüb. Ti. pl.* 8. *f.* 52.

7496. 37; paleellus*.
Ti. paleella. *Hüb. Ti. pl.* 8. *f.* 51.—(*Haw. Pr.* 36.)
Pa. paleea: the large yellow Veneer. *Haw.* 485.
Ti. exoletella. (*Wien. V.* 134.)
Yellow Vanear. *Harr. V. M.* 53.

7497. 38; culmorum*. *Fabr. S.* 471.
Ti. culmella. *Hüb. Ti. pl.* 8. *f.* 50.—(*Haw. Pr.* 36.)
Pa. culmea: the large brown-edged Veneer. *Haw.* 485.
Ph. culmella. *Berk. S.* i. 146?—*Stew.* ii. 198?—*Turt.* iii. 365?
Cr. culmorum. (*Sam. I.* 13.)

Genus 329: (1112). CHILO, *Zinck. Sommer.*

CRAMBUS *p, et* LITHOSIA *p, Fabr.*—PALPARIA *p, Haworth.*

7498. 1; forficellus*. *Charp. & Z. S.* 123.
Ti. forficella. *Thunb. Diss.* (*Wenner.*) 85. *pl. adj. f.* 4.
Pa. consorta: the aquatic Veneer. *Haw.* 483.
Cr. consorta. (*Sam. I.* 13.)
Ch. mucronellus. *Germ. M.* (*Zincken g. Sommer.*) ii. 39.
♂; Ti. consortella. *Hüb. Ti. pl.* 32. *f.* 220.—(*Haw. Pr.* 36.)
♀; Ti. lanceolella. *Hüb. Ti. pl.* 43. *f.* 296.
Var. ♂, Pa. hirta: the hairy-horned Veneer. *Haw.* 483.

7499. 2, fumeus.
Pa. fumea: the smoky Veneer. *Haw.* 483.

7500. 3, punctigerellus *mihi.*

7501. 4: Phragmitellus. *Germ. M.* (*Zincken g. Sommer.*) ii. 36.
Ti. Phragmitella. *Hüb. Ti. pl.* 43. *f.* 297. ♂.—*f.* 298. ♀.
Pa. rhombea: the wainscot Veneer. *Haw.* 483.

7502. 5: gigantellus*. *Germ. M.* (*Zincken g. Sommer.*) ii. 38.
Ti. gigantella. (*Wien. V.* 135.)—*Hüb. Ti. pl.* 8. *f.* 53.
Pa. gigantella: the gigantic Veneer. *Haw.* 482.
Cr. gigantea. (*Sam. I.* 13.)
Lith. convoluta. *Fabr. S.* 460.

7503. 6; caudellus*.
Ph. Ti. caudella. *Linn.* ii. 894.—(*Haw. Pr.* 35.)
Pa. caudea: the Hook-tip Veneer. *Haw.* 482.
Cr. caudea. (*Sam. I.* 13.)
Ti. mucronella. (*Wien. V.* 136.)—*Fabr.* iii. *b.* 298.

Cr. mucronatus. *Fabr. S.* 473.
Ti. acuminella. *Hüb. Ti. pl.* 41. *f.* 284. ♀.

Genus 330: (1113). PLUTELLA, *Schrank*, (*Kirby.*)

PALPARIA *p*, *Haw.*—YPSOLOPHUS *p*, *Fabr.*, *Haw.*

7504. 1; Acinacidella*.
Ti. Acinacidella. *Hüb. Ti. pl.* 34. *f.* 237.
Pa. acinacidea: the Narrow-winged Veneer. *Haw.* 482.
Cr. acinacidea. (*Sam. I.* 13.)

7505. 2; nemorella*.
Ph. Ti. nemorella. *Linn.* ii. 887.—*Stew.* ii. 198.
Ti. cultrella. *Hüb. Ti. pl.* 41. *f.* 282.—(*Haw. Pr.* 35.)
Pa. cultrea: the pale Hook-tip Veneer. *Haw.* 482.
Cr. cultrea. (*Sam. I.* 13.)

7506. 3; dentella*: the tooth-streaked Hook-tip.—*Haw. l. c.*
Yp. dentellus. *Fabr. S.* 508.—*Haw.* 538.
Ti. harpella. (*Wien. V.* 136.)—*Hüb. Ti. pl.* 16. *f.* 110.—(*Haw. Pr.* 35.)
Albin. pl. lxxiii. *f. e–h.*

7507. 4: scabrella*.
Ph. Ti. scabrella. *Linn.* ii. 891.—*Turt.* iii. 381.
Ti. pterodactylella. *Hüb. Ti. pl.* 15. *f.* 102.
Yp. pterodactylellus: the wainscot Hook-tip. *Haw.* 539.
Ti. bifissella. (*Wien. V.* 319.)

7508. 5: subfalcatella.
Ti. falcella. *Hüb. Ti. pl.* 16. *f.* 112.—(*Haw. Pr.* 35?)

7509. 6, asperella.
Ph. Ti. asperella. *Linn.* ii. 891.—*Stew.* ii. 204.—*Turt.* iii. 381.
To. asperellana: the rough White. *Haw.* 416.
Ti. asperella. *Hüb. Ti. pl.* 15. *f.* 101.
Yp. asperellus: the checquered Hook-tip. *Haw.* 539.
Ph. Ti. falcatella. *Don.* x. 81. *pl.* 355. *f.* 5.

Genus 331: (1114). YPSOLOPHUS, *Fabricius.*

YPONOMEUTA *p*, *Samou.*—CEROSTOMA, *Latr.*—ALUCITA *p*, *Fabr.*, (*olim.*)

7510. 1; mucronellus*: the netted Hook-tip. *Haw.* 539.
Ti. mucronella. *Hüb. Ti. pl.* 15. *f.* 99.

7511. 2; Persicellus*.
Yps. bifasciatus: the double-barred. *Haw.* 544.
Yps. Nemorum. *Fabr. S.* 508?—(*Leach E. E.* ix. 135.)
Ti. Nemorum. (*Sam. I.* 41.)

7512. 3; costellus*: the White-shouldered.—*Haw. l. c. Fabr. S.* 509. (!)—*Haw.* 542.—*Turt.* iii. 381.

Ti. costella. *Hüb. Ti. pl.* 16. *f.* 107.
Ph. costatella. *Stew.* ii. 203.

7513. [4; ochroleucus*]: the buff White-shoulder. *Haw.* 542.

7514. [5; ermineus*]: the ermined White-shoulder. *Haw.* 542.

7515. [6; ustulatus*]: the streaked White-shoulder. *Haw.* 542.

7516. 7; variellus*.
Ti. variella. *Hüb. Ti. pl.* 16. *f.* 106.
Yps. variellus: the variable Autumn. *Haw.* 543.

7517. [8; quinquepunctatus*]: the five-spotted. *Haw.* 544.

7518. [9; lutosus*]: the dotted Drab. *Haw.* 545.

7519. [10; flaviciliatus*]: the yellow-fringed White. *Haw.* 545.

7520. 11; rufimitrellus*.
Ti. rufimitrella. *Hüb. Ti. pl.* 18. *f.* 124.

7521. 12; fissellus*.
Ti. fissella. *Hüb. Ti. pl.* 16. *f.* 108.
Yps. fissus: the Broad Streak. *Haw.* 543.

7522. 13; radiatellus*.
Ph. Ti. radiatella. *Don.* iii. 14. *pl.* 77. *f.* 3, 4.—*Turt.* iii. 370.
Ti. radiatella. (*Haw. Pr.* 39.)
Yps. radiatus: the Radiated. *Haw.* 543.

7523. 14; maurellus*.
Ti. maurella. *Hüb. Ti. pl.* 18. *f.* 122.

7524. 15; vittellus*: the black Back. *Haw.* 541.
Ph. Ti. vittella. *Linn.* ii. 890.
Ti. vittella. *Hüb. Ti. pl.* 24. *f.* 164.—(*Haw. Pr.* 38.)
Ti. vittatus. (*Leach E. E.* ix. 135.)
Ce. dorsatus. *Latr. G.* iv. 233. *pl.* 16. *f.* 6?

7525. 16; sequellus*: the small Marvel du Jour. *Haw.* 541.
Ph. Ti. sequella. *Linn.* ii. 885.—*Turt.* iii. 363.
Ti. sequella. *Hüb. Ti. pl.* 15. *f.* 103.—(*Haw. Pr.* 35.)
Yps. sequellus. (*Sam. I.* 44.)

7526. 17; Hesperidellus*.
Ti. Hesperidella. *Hüb. Ti. pl.* 25. *f.* 169.—(*Haw. Pr.* 38.)
Yps. vittatus. *Fabr. S.* 506.
Yps. Hesperidis: the gray Streak. *Haw.* 541.
Ph. Ti. porrectella. *Linn.* ii. 894?

7527. 18; Xylostella*: the Honeysuckle.—*Haw. l. c.*
Ph. Ti. Xylostella. *Linn.* ii. 890.—*Stew.* ii. 203.
Yp. Xylostei. *Fabr. S.* 503.—*Haw.* 540.
Ti. Xylostella. *Hüb. Ti. pl.* 17. *f.* 119.—(*Haw. Pr.* 38.)

Genus 332: (1115). EUPLOCAMUS, *Latreille, Leach, Samou.*
PHYCIS *p, Och.*—PYRALIS *p, Scopoli.*

7528 † 1. Anthracinellus[a].

7529 † 2. tessellus[b].

7530. 3: mediellus*. (*Ing. Inst.* 90.)
Ti. mediella. *Hüb. Ti. pl.* 3. *f.* 19.

Genus 333: (1116). TINEA *Auctorum. Clothes-Moth.*

7531. 1; tapetzella*: black-cloaked Woollen-M.—*Haw. l. c. Fabr.* iii. 303.—*Hüb. Ti. pl.* 13. *f.* 91.—(*Haw. Pr.* 35.)—*Haw.* 562.—(*Kirby & Sp. I. E.* i. 230.)
Ph. Ti. tapetzella. *Linn.* ii. 888.—*Stew.* ii. 198.—*Turt.* iii. 368.
Pha. tapetzella. *Berk. S.* i. 147.
Py. tapezana. *Fabr. S.* 480.

7532. 2; semifulvella*: the Fulvous tip. *Haw.* 562.

7533. 3; vestianella*?
Ph. Ti. vestianella. *Linn.* ii. 888.—*Turt.* iii. 369.
Ti. vestianella. (*Haw. Pr.* 37.)
Ti. saturella: the dark brindled Woollen-M. *Haw.* 562.

7534. 4; fuscipunctella*: the brown-dotted Woollen-M. *Haw.* 562.

7535. 5; pellionella*: the single spotted Woollen-M.—*Haw. l. c. Fabr.* iii. 304.—*Hüb. Ti. pl.* 3. *f.* 15.—(*Haw. Pr.* 37.)—(*Leach E. E.* ix. 133.)—*Haw.* 563.—(*Kirby & Sp. I. E.* i. 230.)—*Stark E.* ii. 369.
Ph. Ti. pellionella. *Linn.* ii. 888.—*Stew.* ii. 198.
Physis pellionella. (*Sam. I.* 33.)

7536. 6; nigripunctella*: the many-spotted Yellow. *Haw.* 564.

7537. 7, albipunctella*: the white-speckled Black. *Haw.* 564.

7538. 8, flavescentella: the triple-spotted Buff. *Haw.* 564.

7539. 9; Lappella*: the triple-spotted Yellow.—*Haw. l. c. Hüb. Ti. pl.* 37. *f.* 252.—*Haw.* 564.
Ph. Ti. Lappella. *Linn.* ii. 889?—*Stew.* ii. 203?
Ti. tripunctella. *Fabr.* iii. *b.* 312.—*Don.* xi. 64. *pl.* 382. *f.* 3.

---

[a] 7528 † 1. Anthracinellus.
Ti. anthracinella. (*Wien. V.* 319.)—*Hüb. Ti. pl.* 33. *f.* 224.
Ti. guttella. *Fabr.* iii. *b.* 293.—*Turt.* iii. 364. (!)
Eu. guttellus. (*Latr. G.* iv. 223.)—(*Leach E. E.* ix. 133.)—(*Samou.* 249. (!)
Ph. Anthracina. (*Och.* iii. 122.)—*Id.* iv. 51.

[b] 7529 † 2. tessellus.
Ph. Ti. tessella. *Linn.* ii. 889?—*Turt.* iii. 370. (!)—An hujus generis?

7540. 10, sulphurella*: the Sulphur. *Haw.* 564.

7541. 11, Destructor *mihi.* *Zool. J.* (*Stephens.*) i. 453 nota.
Ti. flavifrontella. *Latr.* G. iv. 224?

7542. 12; ustella*: the white-backed Black. *Haw.* 565.
Ph. Ti. ustella. *Linn.* ii. 890.
Ph. Ti. flavella. *Mus. Marsham.*
Ti. ferruginella. *Hüb. Ti. pl.* 51. *f.* 348.

7543. 13; cloacella*: the dark-mottled Woollen. *Haw.* 563.
Ph. Ti. fuscella. *Linn.* ii. 893?—*Turt.* iii. 372?

7544. 14; granella*: the mottled Woollen. (*Haw. Pr.* 38.)—*Haw.* 563.
Ph. Ti. granella. *Linn.* ii. 889.—*Stew.* ii. 203.—*Turt.* iii. 380.

7545. 15; parasitella*: the light-brindled Woollen.—*Haw. l. c.* *Hüb. Ti. pl.* 3. *f.* 16.—(*Haw. Pr.* 37.)—*Haw.* 563.

Genus 334: (1117). LEPIDOCERA *mihi.*

YPSOLOPHUS *p, Haworth.*

7546. 1; Taurella*.
Ti. Taurella. *Hüb. Ti. pl.* 27. *f.* 188.
Yp. Taurellus: the little Bull. *Haw.* 546.

7547. 2; setella*.
Ph. Ti. setella. *Marsham MSS.*

7548. 3; mediopectinella*.
Ti. mediopectinella. (*Haw. Pr.* 35.)
Yp. mediopectinellus: the middle Feathered. *Haw.* 545.

7549. 4, Chenopodiella.
Ti. Chenopodiella. *Hüb. Ti. pl.* 46. *f.* 320.

Genus 335: (1118). INCURVARIA, *Haworth.*

RECURVARIA *p, Haworth.*

7550. 1; masculella*.
Ti. masculella. (*Wien. V.* 143.)—*Hüb. Ti. pl.* 18. *f.* 125. ♀. —(*Haw. Pr.* 35.)
Ti. muscalella. *Fabr.* iii. *b.* 314.
In. muscula: the feathered Diamond-Back. *Haw.* 559.
Var. Ti. bipunctella. (*Haw. Pr.* 35.)

7551. 2, pectinella*: the feathered Twin-spot. *Haw.* 559.
Ti. pectinella. *Fabr.* iii. *b.* 310?
Ph. Ti. trigonella. *Linn.* ii. 891?—*Turt.* iii. 373?

7552. 3; Oehlmanniella*.
Ti. Oehlmanniella. *Hüb. Ti. pl.* 27. *f.* 184.—(*Haw. Pr.* 37.)
In. Oehlmanni: Oehlman's. *Haw.* 560.

7553. [4, spuria*]: the Treble-spotted. *Haw.* 560.

7554. 5; tripunctella*.
Rec. tripuncta: the Treble-spotted. *Haw.* 557.

Genus 336: (1119). ——

TINEA-MACULATÆ *p*, *Haworth*.

7555. 1; oppositella*: the two-spotted Brown.—*Haw. l. c.*
Ti. oppositella. *Fabr. S.* 486. (!)—*Hüb. Ti. pl.* 21. *f.* 141.—(*Haw. Pr.* 37.)—*Stew.* ii. 204.—*Turt.* iii. 383.—*Haw.* 567.

7556. 2; 4-punctella*: the double-spotted Brown.—*Haw. l. c.*
Ti. 4-punctella. *Fabr.* iii. *b.* 311.—*Haw.* 567.

7557. 3; minutella*: the double Gold-spotted.—*Haw. l. c.*
Ti. minutella. *Linn.* ii. 893.—*Turt.* iii. 374.—(*Haw. Pr.* 39.)—*Haw.* 568.
Ti. similella. *Hüb. Ti. pl.* 27. *f.* 182?

7558. 4; atrella: the two-spotted Black.—*Haw. l. c.*
Ti. atrella. *Hüb. Ti. pl.* 40. *f.* 278.—*Haw.* 567.

7559 † 5, miscella.
Ti. miscella: the Yellow-dotted. *Haw.* 580.
In Mus. D. *Haworth*.

7560 † 6, Knockella*: Knock's.—*Haw. l. c.*
Ti. Knockella. *Fabr.* iii. *b.* 318.—*Haw.* 568.
In Mus. D. *Haworth*.

7561. 7, cerusella*: the triple-spotted White.—*Haw. l. c.*
Ti. cerusella. *Hüb. Ti. pl.* 27. *f.* 183.—(*Haw. Pr.* 39.)—*Haw.* 567.

7562 † 8. formosella*.
Ti. formosella: the Gold striped Sable. *Ent. Trans.* (*Haworth.*) i. 337. In Mus. D. ——?

7563. 9; Albinella*: Albin's.—*Haw. l. c.*
Ph. Ti. Albinella. *Linn.* ii. 897.—*Turt.* iii. 375.
Ti. Albinella. *Fabr.* iii. *b.* 319.—*Haw.* 581.

7564. 10, Megerlella.
Ti. Megerlella. *Hüb. Ti. pl.* 44. *f.* 307.

Genus 337: (1120). LAMPRONIA *mihi*.

TI.-METALLICÆ *p*, *Haworth*.

7565. 1; capitella*: the triple-spotted Black.—*Haw. l. c.*
Ph. Ti. capitella. *Linn.* ii. 894?—*Stew.* ii. 200.—(*Haw. Pr.* 39.)—*Turt.* iii. 373.
Ti. capitella. *Fabr.* iii. *b.* 315. (!)—*Haw.* 565.

7566. 2; prælatella*: the spotted Violet.—*Haw. l. c.*
Ti. prælatella. *Fabr.* iii. *b.* 315.—*Hüb. Ti. pl.* 26. *f.* 251.—*Haw.* 566.

7567. 3: rupella*? the four-spotted Black.—*Haw. l. c.*
Ti. rupella. *Fabr.* iii. *b.* 315.—*Hüb. Ti. pl.* 36. *f.* 250.—*Haw.* 565.

7568. 4; flavipunctella*.
Ti. flavipunctella: the four-spotted Brown. *Haw.* 566.

7569. 5; marginepunctella* *mihi.*

7570. 6; melanella*.
Ti. melanella: the white-speckled Black. *Haw.* 266.

7571. 7; corticella*.
Ti. corticella: the golden-speckled Black. (*Kirby & Sp. I. E.* i. 198.)—*Haw.* 266.
Ph. Ti. corticella. *Linn.* ii. 893?—*Stew.* ii. 200?

7572. 8; atrella* *mihi.*

7573. 9; subpurpurella*.
Ti. subpurpurella: the purple Underwing. *Haw.* 571.
Ti. Knockella. *Hüb. Ti. pl.* 38. *f.* 260?

7574. 10; purpurella*.
Ti. purpurella: the purple Upperwing. *Haw.* 571.
Ti. Goldeggella. *Hüb. Ti. pl.* 37. *f.* 258?

7575. 11; auropurpurella*.
Ti. auropurpurella: the Gold-brindled Purple. *Haw.* 572.
Ti. Sparmannella. *Fabr.* iii. *b.* 324?

7576. 12; rubroaurella*.
Ti. rubroaurella: the red-gold. *Haw.* 572.
Ti. fibulella. *Fabr.* iii. *b.* 326?

7577. 13; Helwigella*.
Ti. Helwigella. *Hüb. Ti. pl.* 38. *f.* 263.
Ti. rubrifasciella: the red-barred Gold. *Haw.* 572.

7578. 14, sanguinella*.
Ti. sanguinella: the scarlet-barred Gold. *Haw.* 572.

7579. 15; Calthella*.
Ph. Ti. Calthella. *Linn.* ii. 895.—*Turt.* iii. 382.
Ti. Calthella: the small Gold. (*Haw. Pr.* 38.)—*Haw.* 573.
Ti. pusillella. *Hüb. Ti. pl.* 50. *f.* 341.

7580. 16, concinnella* *mihi.*

7581. 17; Seppella*: Sepp's.—*Haw. l. c.*
Ti. Seppella. *Fabr.* iii. *b.* 320? (!)—(*Haw. Pr.* 38.)—*Haw.* 573.—*Turt.* iii. 375.

7582. 18; amœnella*.
Ti. amœnella. *Hüb. Ti. pl.* 57. *f.* 388.

7583. 19, bistrigella.
Ti. bistrigella: the silver-striped Gold. *Haw.* 573.

Genus 338: (1121). ——

7584. 1; auroguttella* *mihi.*

Genus 339: (1122). GRACILLARIA, *Haworth.*

ALUCITA *p, Fabr.*

7585. 1; nebulea*: the nebulous Slender. *Haw.* 532.

7586. 2; Meleagripennella*.
Ti. Meleagripennella: the Turkey's Feather.—*Haw. l. c. Hüb. Ti. pl.* 28. *f.* 189.—(*Haw. Pr.* 38.)—*Haw.* 578?

7587. 3; anastomosis*: the confluent Barr'd. *Haw.* 530.
Ti. Syringella. *Fabr.* iii. *b.* 328?

7588. 4; cinerea*: the double-barr'd Slender. *Haw.* 530.

7589. 5; V flava*: the yellow V. *Haw.* 530.
Ph. Ti. padella. *Linn.* ii. 894?

7590. 6; versicolor*: the Changeable. *Haw.* 531.
Ti. ustulatella. *Fabr.* iii. *b.* 307?

7591. 7; substriga*: the obscure striped. *Haw.* 532.

7592. 8; semifascia*: the semi White-bar. *Haw.* 528.

7593. 9; purpurea*: the triangle-marked Purple. *Haw.* 528.

7594. 10; stigmatella*.
Ti. stigmatella. *Fabr.* iii. 304. (!)—(*Haw. Pr.* 39.)—*Stew.* ii. 199.—*Turt.* iii. 369.
Ti. triangulella. *Panz. F.* xviii. *f.* 23.
Gra. trigona: the triangle-marked Red. *Haw.* 529.
Ti. Upupæpenella. *Hüb. Ti. pl.* 30. *f.* 203.

7595. 11; ochracea*: the triangle-marked Ochre. *Haw.* 528.

7596. 12; Thunbergella*.
Ti. Thunbergella. *Fabr.* iii. *b.* 326.
Gr. Thunbergii: Thunberg's. *Haw.* 529.

7597. 13; hemidactylella*.
Ti. hemidactylella. (*Wien. V.* 144.)—*Hüb. Ti. pl.* 40. *f.* 276.
Gr. hemidactyla: the mottled Red. *Haw.* 527.
Ph. Ti. punctella. *Linn.* ii. 890.

7598. 14; rufipennella*.
Ti. rufipennella. *Hüb. Ti. pl.* 30. *f.* 204.

7599. 15, elongella*.
Ph. Ti. elongella. *Linn.* ii. 890?—*Turt.* iii. 370?
Ti. signipenella. *Hüb. Ti. pl.* 29. *f.* 196.—(*Haw. Pr.* 38.)
Gr. signipennis: the plain Red. *Haw.* 527.
Ti. ciliella. (*Wien. V.* 136?)

7600. 16; violacea*: the violaceous Slender. *Haw.* 528.

7601. 17; roscipennella*.

Ti. roscipenella. *Hüb. Ti. pl.* 29. *f.* 198.
Gr. roscipennis: the livid Slender. *Haw.* 528.

7602. 18: leucapennella.
Ti. leucapennella. *Hüb. Ti. pl.* 30. *f.* 205.

7603. 19; præangusta: the Poplar Slender. *Haw.* 530.

Familia XXI: (134). ALUCITIDÆ, *Leach.*

(PTEROPHORITES, *Latreille.*—PHALÆNA-ALUCITA, *Linné.*)
Plumes.

Genus 340: (1123). PTEROPHORUS, *Geoffroy, Haworth, Leach, Samou., Curtis.*

ALUCITA *p, Schrank,* (*Kirby.*)

A.

7604. 1; pentadactylus*. *Fabr.* iii. *b.* 348.—(*Leach E. E.* ix. 135.)—(*Sam. I.* 35.)—(*Curtis l. c. infra.*)
Ph. Al. pentadactyla. *Linn.* ii. 900.—*Berk. S.* i. 147.—*Don.* iv. 5. *pl.* 110.—*Stew.* ii. 205.—*Turt.* iii. 385.
Al. pentadactyla: the large White Plume. (*Haw. Pr.* 39.)—*Haw.* 475.
Plumed M. *Harr. A. pl.* 1. *f. o–q.*
White plumed M. *Harr. V. M.* 41.

7605 † 2, spilodactylus. *Curtis.* iv. *pl.* 161.
In Mus. D. *Curtis,* et Sparshall.

7606. 3; bipunctidactylus. (*Sam. I.* 35.)—(*Curtis l. c. supra.*)
Ph. Al. bipunctidactylus. *Villers E.* ii. 535.
Al. bipunctidactyla: the grey Wood Plume. (*Haw. Pr. App.* 6.)—*Haw.* 476.

7607. 4; fuscodactylus*. (*Sam. I.* 35.)—(*Curtis l. c. supra.*)
Ph. Al. fuscodactylus. *Villers E.* ii. 535.
Al. fuscodactyla: the brown Wood Plume. (*Haw. Pr.* 39.)—*Haw.* 476.

7608. 5; pterodactylus*. *Fabr.* iii. *b.* 347.—(*Sam. I.* 35.)—(*Curtis l. c. supra.*)
Ph. Al. pterodactyla. *Linn.* ii. 900.—*Turt.* iii. 384.
Al. pterodactyla: the common Plume.—*Haw. l. c.* *Hüb. Al. pl.* 1. *f.* 4.—(*Haw. Pr. App.* 6.)—*Haw.* 475.
Ph. didactyla: Brown Feathered M. *Berk. S.* i. 147.—*Stew.* ii. 205.—*Turt.* iii. 384.
Brown plumed M. *Harr. A. pl.* 30. *f. i–l.*—*Harr. V. M.* 39.

7609. 6; monodactylus*. (*Sam. I.* 35.)—(*Curtis l. c. supra.*)
Ph. Al. monodactyla. *Linn.* ii. 899?—*Stew.* ii. 383.
Al. monodactyla: the hoary Plume. (*Haw. Pr. App.* 5.)—*Haw.* 476.

7610. 7, tephradactylus*. (*Curtis l. c. supra.*)
Al. tephradactyla. *Hüb. Al. pl.* 4. *f.* 17.

7611. 8; tridactylus*. *Fabr.* iii. *b.* 346.—(*Sam. I.* 35.)—(*Curtis l. c. supra.*)
Al. tridactyla. (*Haw. Pr. App.* 5.)—*Haw.* 477.
Ph. Al. tridactyla. *Linn.* ii. 899.—*Stew.* ii. 205.

7612. 9, niveidactylus* *mihi.*

7613. 10; tetradactylus*. (*Sam. I.* 35.)—(*Curtis l. c. supra.*)
Ph. Al. tetradactyla. *Linn.* ii. 900.
Ph. Al. didactyla. *Scop. C. No.* 265.
Ph. Al. tridactyla. *Villers E.* ii. 532.
Al. tetradactyla: the white-shafted Plume. (*Haw. Pr. App.* 6.)—*Haw.* 477.

7614. 11; citridactylus. *Haworth MSS.*—(*Curtis l. c. supra.*)

7615. 12; ochrodactylus*. (*Curtis l. c. supra.*)
Al. ochrodactyla. *Fabr.* iii. *b.* 345?—(*Haw. Pr. App.* 6.)

7616. 13, galactodactylus*. (*Sam. I.* 35.)—(*Curtis l. c. supra.*)
Al. galactodactyla: the spotted White Plume.—*Haw. l. c. Hüb. Al. pl.* 1. *f.* 2.—(*Haw. Pr. App.* 6.)—*Haw.* 475.
Al. albodactylus. *Fabr.* iii. *b.* 348?

7617. 14; leucodactylus*. (*Sam. I.* 35.)—(*Curtis l. c. supra.*)
Al. leucodactyla. *Hüb. Al. pl.* 1. *f.* 5.
Al. leucodactyla: the lemon Plume. (*Haw. Pr. App.* 6.)—*Haw.* 477.

7618. 15; lunædactylus*. (*Sam. I.* 35.)—(*Curtis l. c. supra.*)
Al. lunædactyla: the crescent Plume. *Haw.* 477.
Al. phæodactyla. *Hüb. Al. pl.* 3. *f.* 14, 15.

7619. 16; pallidactylus*. (*Sam. I.* 35.)—(*Curtis l. c. supra.*)
Al. pallidactyla: the pale Plume. *Haw.* 478.
Al. ochrodactyla. *Hüb. Al. pl.* 3. *f.* 12, 13?

7620. 17; migadactylus*. *Fabr.* iii. *b.* 348?—(*Curtis l. c. supra.*)
Al. migadactyla: the Chalk-pit Plume. (*Haw. Pr. App.* 6.)—*Haw.* 478.
Pt. megadactylus. (*Sam. I.* 35.)

7621. 18, phæodactylus* *mihi.* (*Curtis l. c. supra.*)

B.

7622. 19; trigonodactylus*. (*Sam. I.* 35.)—(*Curtis l. c. supra.*)
Al. trigonodactyla: the triangle Plume. *Haw.* 478.

7623. 20; rhododactylus*. *Fabr.* iii. *b.* 347.—(*Sam. I.* 35.)—(*Curtis l. c. supra.*)
Al. rhododactyla: the Rose Plume.—*Haw. l. c. Hüb. Al. pl.* 2. *f.* 8.—(*Haw. Pr.* 39.)—*Haw.* 478.

7624. 21; calodactylus*. *Fabr.* iii. *b.* 346.—(*Sam. I.* 35.)—(*Curtis l. c. supra.*)
Al. calodactyla: the beautiful Plume.—*Haw. l. c.* *Hüb. Al. pl.* 2. *f.* 7.—(*Haw. Pr.* 39.)—*Haw.* 478.

7625. 22; tesseradactylus*. *Fabr.* iii. *b.* 347.—(*Sam. I.* 35.)—(*Curtis l. c. supra.*)
Ph. Al. tesseradactyla. *Linn.* ii. 900.—*Turt.* iii. 384.
Al. tesseradactyla: the marbled Plume. (*Haw. Pr. App.* 6.)—*Haw.* 479.
Gray Plumed. *Harr. V. M.* 41.

7626. 23, punctidactylus*. (*Sam. I.* 35.)—(*Curtis l. c. supra.*)
Al. punctidactylus: the brindled Plume. *Haw.* 479
Al. acanthadactyla. *Hüb. Al. pl.* 5. *f.* 23, 24?

7627. 24; didactylus*. (*Leach E. E.* ix. 135.)—(*Sam. I.* 35.)—(*Curtis l. c. supra.*)
Ph. Al. didactyla. *Linn.* ii. 899.—*Don.* ix. 65. *pl.* 318.
Al. didactyla: the spotted rusty Plume. (*Haw. Pr.* 39.)—*Haw.* 479.
Al. trichodactyla. *Hüb. Al. pl.* 2. *f.* 9.

7628. [25; heterodactylus*.] (*Sam. I.* 35.)
Ph. Al. heterodactyla. *Villers E.* ii. 535.
Al. heterodactyla: the spotted black Plume. (*Haw. Pr.* 39.)—*Haw.* 479.

7629. 26, microdactylus*. (*Curtis l. c. supra.*)
Al. parvidactyla: the small Plume. *Haw.* 480.
Pt. macrodactylus. (*Sam. I.* 35.)

Genus 341: (1124). ALUCITA, *Scopoli, Leach, Samou.*

PTEROPHORUS *p, Fabr., Haw.*—ORNEODES, *Latr.*

7630. 1; hexadactyla*. (*Haw. Pr.* 39.)—(*Leach E. E.* ix. 135.)—(*Sam. I.* 2.)
Ph. Al. hexadactyla. *Linn.* ii. 900.—*Berk. S.* i. 148.—*Don.* iv. 75. *pl.* 136.—*Stew.* ii. 205.—*Turt.* iii. 385.
Pt. hexadactylus: the six-cleft Plume.—*Haw. l. c.* *Fabr.* iii. *b.* 349.—*Haw.* 480.
Twenty-plumed. *Harr. V. M.* 41.—*Harr. Ex.* 13. *pl.* ii. *f.* 7. *pl.* iii. *f.* 1, *aucta.*
Many-feathered Moth.—*Berk. l. c.*

7631. 2, polydactyla. *Hüb. Al. pl.* 6. *f.* 28.

7632 † 3, pæcilodactyla *mihi.*

In Mus. D. *Haworth.*

Ordo II: (9). DIPTERA.

(Antliata, *Fabricius*.—Halteriptera, *Clairville*.)

Familia I: (135). CULICIDÆ, *Latreille*.

(Culex, *Linné*, *&c.*—Tipulidæ *p*, *Leach*.—Culicides, *Latr.*, *olim.*)

Genus 1: (1125). CULEX *Auctorum*. *Gnat*: Larvæ; *Loggerheads*.

7633. 1; annulatus*. *Fabr. A.* 35.—*Turt.* iii. 663.—*Zool. Journ.* (*Steph.*) i. 452.
Cu. pipiens. (*Samou.* 71. *pl.* 9. *f.* 5. ♀.)

7634. 2; affinis*. *Zool. Journ.* (*Steph.*) i. 452.

7635. 3; Calopus*. *Meig. Zw.* i. 3?—*Zool. Journ.* (*Steph.*) i. 452.

7636. 4; cantans*. *Meig. Zw.* i. 6. *pl.* 1. *f.* 9. ♂.—*Zool. Journ.* (*Steph.*) i. 453.

7637. 5; fumipennis*. *Zool. Journ.* (*Steph.*) i. 453.

7638. 6; ornatus*. *Meig. Zw.* i. 5.—*Zool. Journ.* (*Steph.*) i. 454.
Cu. equinus. *Meig. Kl.* i. 3.

7639. 7; sylvaticus*. *Meig. Zw.* i. 6.—*Zool. Journ.* (*Steph.*) i. 454.
Cu. fasciatus. *Meig. Kl.* i. 4.

7640. 8, maculatus. *Meig. Zw.* i. 6.—*Zool. Journ.* (*Steph.*) i. 454.

7641. 9; lateralis*. *Meig. Zw.* i. 5.

7642. 10; nemorosus*. *Meig. Zw.* i. 4.—*Zool. Journ.* (*Steph.*) i. 454.
Cu. reptans. *Meig. Kl.* i. 3.

7643. 11; domesticus?* *Meig. Zw.* i. 8?—*Zool. Journ.* (*Steph.*) i. 455.

7644. 12; pipiens*. *Linn.* ii. 1002.—*Berk.* i. 166.—*Stew.* ii. 268.—*Turt.* iii. 663.—*Shaw.* G. Z. vi. 387.—(*Kirby & Sp. I. E.* i. 113.—*Id.* ii. 361.)—(*Bingley.* iii. 327.)—*Zool. Journ.* (*Steph.*) i. 455.—*Wood.* ii. 97. *pl.* 67.—*Stark. E.* ii. 376.

7645. 13; bicolor*. *Meig. Zw.* i. 9.—*Zool. Journ.* (*Steph.*) i. 456.

7646. 14; marginalis*. *Zool. Journ.* (*Steph.*) i. 455.

7647. 15; lutescens*. *Fabr. A.* 35.—*Turt.* iii. 664.—*Zool. Journ.* (*Steph.*) i. 456.

7648. 16, punctatus. *Meig. Zw.* i. 9.—*Zool. Journ.* (*Steph.*) i. 456.

7649. 17; rufus*. *Meig. Zw.* i. 7.—*Zool. Journ.* (*Steph.*) i. 456.

7650. 18, concinnus *mihi*.

Genus 2: (1126). ANOPHELES, *Meigen, Steph., Curtis.*

CULEX, *Linné, &c.*

7651. 1; bifurcatus*. *Meig. Zw.* i. 11.—*Zool. Journ.* (*Steph.*) i. 456.—*Curtis.* v. *pl.* 210.
Cu. bifurcatus. *Linn.* ii. 1002.—*Berk.* i. 166.—*Stew.* ii. 269.
Cu. trifurcatus. *Fabr. A.* 35.—*Turt.* iii. 664.
Cu. claviger. *Meig. Kl.* i. 4. *pl.* i. *f.* 8.

7652. 2; grisescens* *mihi.* *Zool. Journ.* (*Steph.*) iii. 160.

7653. 3; maculipennis*. *Meig. Zw.* i. 11. *pl.* 1. *f.* 17. ♀.—*Zool. Journ.* (*Steph.*) i. 457.—*Curtis l. c. supra.*

7654. 4, plumbeus* *mihi.* *Zool. Journ.* (*Haliday & Steph.*) iii. 160.

Familia II: (136). TIPULIDÆ, *Leach.*

(TIPULA, *Linné, &c.*)

Genus 3: (1127). CORETHRA, *Meigen, Samouelle.*

TIPULA *p, DeGeer.*—CHIRONOMUS *p, Fabr.*

7655. 1, plumicornis*. *Meig. Zw.* i. 15. *pl.* 1. *f.* 22. ♂.
Ch. plumicornis. *Fabr. A.* 42.
Ti. cristallina. *DeGeer.* vi. 149.
Ti. hafniensis. *Gmel.* v. 2826.
Co. lateralis. *Meig. Kl.* i. 8.—*Panz. F.* cix. *f.* 16.
Straw coloured plumed Gnat. *Gor. & Pr.* i. *pl.* 2. *f.* 1, 2? larva and pupa.

7656. 2; culiciformis*. *Meig. Zw.* i. 16.—(*Samou. I.* 12.)
Ti. culiciformis. *DeGeer.* vi. 144. *pl.* 23. *f.* 3–12.

Genus 4: (1128). CHIRONOMUS, *Meigen, Samou., Curtis.*

A. Alis nudis. a. *Halteribus pallidis.*

7657. 1; plumosus*. *Fabr. A.* 37.—(*Kirby & Sp. I. E.* iii. 288.—*Id.* iv. 54.)
Ti. plumosa. *Linn.* ii. 974.—*Berk. S.* i. 161.—*Don.* i. *pl.* 49?—*Stew.* ii. 252.—*Turt.* iii. 590.

7658. 2; annularius*. *Meig. Zw.* i. 21.
Ti. annularia. *DeGeer.* vi. 146. *pl.* 19. *f.* 14, 15.

7659. 3; grandis*. *Megerle.*—*Meig. Zw.* i. 21.

7660. 4; prasinus*. *Meig. Zw.* i. 22.

7661. 5, pallens. *Meig. Zw.* i. 22.

7662. 6; flaveolus*. *Meg.*—*Meig. Zw.* i. 23.

7663. 7; palustris* *mihi.*

7664. 8; tentans*. *Fabr. A.* 38.
Ch. vernalis. *Meig. Kl.* i. 13.

7665. 9; flavicollis*. *Meig. Zw.* i. 24.

7666. 10; dorsalis*. *Meig. Zw.* i. 25.

7667. 11; uliginosus* *mihi.*

7668. 12; fasciatus* *mihi.*

7669. 13, pilipes. *Meg.—Meig. Zw.* i. 26.

7670. 14; littorellus*. *Meig. Zw.* i. 26.

7671. 15; flavicornis* *mihi.*

7672. 16; riparius*. *Meig. Zw.* i. 23.

7673. 17; nigricans* *mihi.*

7674. 18; fuscescens* *mihi.*

7675. 19; pallipes* *mihi.*

7676. 20; brunnipes* *mihi.*

7677. 21; rufipes* *mihi.*

7678. 22; tendens*. *Fabr. A.* 39.

7679. 23; fuscipennis*. *Meig. Zw.* i. 35.

7680. 24; tarsalis* *mihi.*

7681. 25; pedellus*. *Meig. Zw.* i. 28.
Ti. littoralis β. *Linn. F. No.* 1759.
Ti. pedella. *DeGeer.* vi. 146. *pl.* 19. *f.* 12, 13.
Ch. cantans. *Fabr. A.* 45.
Ti. littoralis. *Schrank A. No.* 874.—*Berk. S.* i. 162.

7682. 26: gibbus*. *Fabr. A.* 41.—*Panz. F.* cix. *f.* 20. ♂. *f.* 21. ♀.

7683. 27; scutellatus*. *Meig. Zw.* i. 33.

7684. 28; terminalis*. *Meig. Zw.* i. 34.

7685. 29; flexilis*. *Fabr. A.* 40.
Ti. flexilis. *Linn.* ii. 975.—*Turt.* iii. 591.

7686. 30; viridulus*. *Fabr. A.* 33.
Ti. viridula. *Linn.* ii. 975.—*Turt.* iii. 593.
Chi. littoralis. *Meig. Kl.* i. 14.

7687. 31; virescens*. *Meig. Zw.* i. 31.
Ti. littoralis. *Stew.* ii. 252?—*Turt.* iii. 592?

7688. 32; annulimanus* *mihi.*

7689. 33; autumnalis* *mihi.*

7690. 34; pusillus*. *Fabr. A.* 45.
Ti. pusilla. *Linn.* ii. 975.—*Stew.* ii. 252.—*Turt.* iii. 593.

7691. 35; vitripennis*. *Meig. Zw.* i. 32.

7692. [36; assimilis*] *mihi.*
Ti. macrocephala. *Turt.* iii. 593?

7693. 37; sticticus*. *Meig. Zw.* i. 37.
Ti. stictica. *Fabr. S. I.* ii. 407.

Ti. stricta. *Fabr.* iv. 245.
Ch. Histrio. *Meig. Kl.* i. 19.

7694. 38; Histrio*. *Fabr. A.* 41.

7695. 39; maculipennis*. *Meig. Zw.* i. 38.

7696. 40; nubeculosus*. *Meig. Zw.* i. 37.

7697. 41; humeralis* *mihi.*

7698. 42; lætus*. *Meig. Zw.* i. 38.

7699. 43; hyalipennis* *mihi.*

7700. 44; nigricornis* *mihi.*

7701. 45; nitidus*. *Meig. Zw.* i. 35.

7702. 46; ater* *mihi.*

7703. 47; carbonarius*. *Meig. Zw.* i. 40.

7704. 48; plebeius*. *Meg.—Meig. Zw.* i. 40.

7705. 49; tibialis*. *Meig. Zw.* i. 40.

7706. 50; albimanus*. *Meig. Zw.* i. 40.
Ch. annularis. *Meig. Kl.* i. 17.—*Panz. F.* cix. *f.* 19.

7707. 51; bicinctus*. *Meg.—Meig. Zw.* i. 41.

7708. 52; trifasciatus*. *Meig. Zw.* i. 42.—*Panz. F.* 109. 18.

7709. 53; sylvestris*. *Fabr. A.* 47.
Ch. vibratorius. *Meig. Kl.* i. 16.

7710. 54; tricinctus*. *Meig. Zw.* i. 41.

7711. 55; ornatus*. *Meig. Zw.* i. 43.

7712. 56; motitator*. *Fabr. A.* 38.
Ti. motitatrix. *Linn.* ii. 974.—*Berk. S.* i. 162.—*Stew.* ii. 252.
—*Turt.* iii. 591.

7713. 57; tremulus*. *Fabr. A.* 40.
Ti. tremula. *Linn.* ii. 975.

7714. 58; annulipes*. *Meig. Zw.* i. 42.

7715. 59; apicalis* *mihi.*

7716. 60; rubicundus*. *Meig. Zw.* i. 35.

7717. 61; pygmæus*. *Meig. Zw.* i. 36?

7718. 62; melaleucus*. *Meig. Zw.* i. 39.

7719. 63; trimaculatus* *mihi.*

b. *Halteribus atris.*

7720. 64, plumipes *mihi.*
Ch. obscurus. *Fabr. A.* 40?

7721. 65, tristis. *Wied.—Meig. Zw.* i. 48.

7722. 66; nitidicollis* *mihi.*

7723. 67; stercorarius*. *Meig. Zw.* i. 46.
Tipula stercoraria. *DeGeer.* vi. 149. *pl.* 22. *f.* 14–20. *pl.* 23. *f.* 1.
Ch. chiopterus. *Meig. Kl.* i. 17.

7724. 68; aterrimus*. *Meig. Zw.* i. 47.

7725. 69; byssinus*. *Meig. Zw.* i. 46.
Ti. byssina. *Schr. B.* iii. *No.* 2330.

7726. 70; minimus*. *Meig. Zw.* i. 47.

7727. 71; nigerrimus* *mihi.*

B. Alis pilosis. a. *Halteribus atris.*

7728. 72; fuscus*. *Meig. Zw.* i. 52.

7729. 73; picipes*. *Meig. Zw.* i. 52.

b. *Halteribus pallidis.*

7730. 74, flavipes. *Meig. Zw.* i. 50.

7731. 75; vernus*. *Meig. Zw.* i. 49.

7732. 76; luteicornis* *mihi.*

7733. 77; luteipennis* *mihi.*

7734. 78; Junci*. *Meig. Zw.* i. 50.

7735. 79; cingulatus* *mihi.*

7736. 80; fuscipes*. *Meig. Zw.* i. 49.

7737. 81; pallicornis* *mihi.*

7738. 82; affinis*. *Wied. Z. M.* i. 66.—*Meig. Zw.* i. 51.
Ch. viridulus. *Meig. Kl.* i. 14.

7739. 83; flavescens* *mihi.*

7740. 84; albidus*. *Wied.*—*Meig. Zw.* i. 51.

7741. 85; nanus*. *Meig. Zw.* i. 50.

7742. 86; flabellatus*. *Meig. Zw.* i. 51.

Genus 5: (1129). ——

CHIRONOMUS *p*, *Meg.*—CHIRONOMUS, *Curtis.*

7743 † 1, æstivus.
Ch. æstivus. *Curtis.* ii. *pl.* 90. In Mus. D. *Bentley.*

7744. 2; fulvus* *mihi.*

7745. 3; ferruginatus* *mihi.*

Genus 6: (1130). SPHÆROMIAS *mihi.*

7746. 1; annulitarsis* *mihi.*

7747. 2; varipes *mihi.*

7748 † 3, albomarginatus* *mihi.* In Mus. D. *Cooper.*

Genus 7: (1131). TANYPUS, *Meigen, Samou.*

CHIRONOMUS *p*, *Fabr.*

7749. 1; varius*. *Meig. Zw.* i. 56. *pl.* 2. *f.* 12. ♂.

Ch. varius. *Fabr. A.* 41.
Ta. punctatus. *Meig. Kl.* i. 21.

7750. 2; variegatus* *mihi.*

7751. 3; zonatus*. *Meig. Zw.* i. 59.
Chir. zonatus. *Fabr. A.* 44.
Tip. zonata. *Stew.* ii. 252.

7752. 4; pallidulus*. *Meg.—Meig. Zw.* i. 65.

7753. 5: carneus. *Meig. Zw.* i. 67.
Ch. carneus. *Fabr. A.* 41.

7754. 6; notatus*. *Meig. Zw.* i. 58.

7755. 7; melanurus*. *Meig. Zw.* i. 59.

7756. 8; ornatus* *mihi.*

7757. 9; elegans* *mihi.*

7758. 10; monilis*. *Meig. Zw.* i. 60.
Ti. monilis. *Linn.* ii. 975.—*Berk. S.* i. 162.—*Stew.* ii. 252.
Ti. maculata. *DeGeer.* vi. 151. *pl.* 27. *f.* 15–19.

7759. 11; pictipennis*. *Meig. Zw.* i. 61.
Ta. cinctus. *Meig. Kl.* i. 22.—*Panz. F.* cv. *f.* 6.—(*Samou. I.* 40.)

7760. 12; concinnus* *mihi.*

7761. 13; nebulosus*. *Meig. Zw.* i. 57.
Ta. littoralis. *Meig. Kl.* i. 22.

7762. 14; punctatus*. *Meig. Zw.* i. 58.
Ch. punctatus. *Fabr. A.* 43.
Ta. nebulosus. *Meig. Kl.* i. 23.

7763. 15; choreus*. *Meig. Zw.* i. 62.
♂. Ta. fasciatus. *Meig. Kl.* i. 21.
♀. Ta. sylvaticus. *Meig. Kl.* i. 24.

7764. 16; culiciformis*. *Meig. Zw.* i. 63.
Tip. culiciformis. *Linn.* ii. 978.—*Stew.* ii. 253.

7765. 17; nervosus*. *Hoffmansegg.—Meig. Zw.* i. 64.

7766. 18; pallens* *mihi.*

7767. 19; ferruginicollis*. *Meg.—Meig. Zw.* i. 64.

7768. 20; Melanops*. *Wied.—Meig. Zw.* i. 65.

7769. 21; Arundineti*. *Meig. Zw.* i. 66.
Ti. Arundineti. *Linn.* ii. 974.—*Turt.* iii. 592.

7770. 22; præcox*. *Meig. Zw.* i. 62.

Genus 8: (1132). PRIONOMYIA *mihi.*

CRATOPOGON C, *Meigen.*—CHIRONOMUS *p*, *Fabr.*—CULEX *p*, *Fabr.*—SERROMYIA? *Megerle.*

7771. 1; femorata*.

Chi. femoratus. *Fabr. A.* 45.
Cer. femoratus. *Meig. Kl.* i. 28. *pl.* 2. *f.* 4. mas.

7772. 2; morio*.
Cul. morio. *Fabr. A.* 36.—*Stew.* ii. 269.—*Turt.* iii. 664.

7773. 3; atra*.
Cer. ater. *Meg.*—*Meig. Zw.* i. 84.

7774. 4; rufitarsis*.
Cer. rufitarsis. *Meig. Zw.* i. 83.

7775. 5; pusilla* *mihi.*

7776. 6, armata.
Cer. armatus. *Meig. Zw.* i. 83.

Genus 9: (1133). PALPOMYIA, *Megerle?*
CERATOPOGON B, *Meig.*—CHIRONOMUS *p*, *Fabr.*

7777. 1, succincta.
Cer. succinctus. *Hoffm.*—*Meig. Zw.* i. 85.

7778. 2; basalis* *mihi.*

7779. 3; scutellata* *mihi.*

7780. 4; hortulana*.
Cer. hortulanus. *Meig. Zw.* i. 81.

7781. 5; rufipes*.
Cer. rufipes. *Meig. Zw.* i. 81.

7782. 6; serripes*.
Cer. serripes. *Meig. Zw.* i. 82.

7783. 7. spinipes.
Cer. spinipes. *Meig. Zw.* i. 81.—*Panz. F.* ciii. *f.* 14.

Genus 10: (1134). CERATOPOGON, *Meigen.*
CHIRONOMUS *p*, *Fabr.*

7784. 1; nigerrimus* *mihi.*

7785. 2; holosericeus*. *Meig. Zw.* i. 70.

7786. 3; floralis*. *Meig. Zw.* i. 70.

7787. 4; palustris*. *Meig. Zw.* i. 71.
Tipula. *Schll. G. M. pl.* 37. *f.* 2?

7788. 5; brunnipes*. *Meig. Zw.* i. 71.

7789. 6; albicornis*. *Meig. Zw.* i. 74.

7790. 7; communis*. *Meig. Zw.* i. 70.
Chi. communis. *Fabr. A.* 44.

7791. 8; pallipes*. *Meig. Zw.* i. 74.

7792. 9; albipennis. *Meg.*—*Meig. Zw.* i. 73.
Cer. ambiguus. *Meig. Kl.* i. 32.

7793. 10; leucopeza. *Meig. Zw.* i. 72.

7794. 11; exigua* *mihi.*

7795. 12; albitarsis. *Weid. M.* i. 67?

7796. 13; Stigma*. *Meig. Zw.* i. 73. *pl.* 2. *f.* 18.

7797. 14; obsoletus*. *Meig. Zw.* i. 76.

7798. 15; bimaculatus* *mihi.*

7799. 16; angustatus* *mihi.*

7800. 17; lutescens* *mihi.*

7801. 18; assimilis* *mihi.*

7802. 19; niveipennis*. *Meig. Zw.* i. 73.

7803. 20; lucorum*. *Meig. Zw.* i. 72.

Genus 11: (1135). CULICOIDES, *Latreille.* *Midge.*
CULEX *p, Linné.*—CERATOPOGON *p, Meigen.*

7804. 1; punctata*. *Latr.* G. iv. 252.
Cul. pulicaris. *Linn.* ii. 1003.—*Berk. S.* i. 166.—*Stew.* ii. 269. —*Turt.* iii. 664.
Cer. pulicaris. *Meig. Zw.* i. 75. *pl.* 2. *f.* 17.

7805. 2; variegata* *mihi.*

Genus 12: (1136). LABIDOMYIA *mihi.*
FORCIPOMYIA, *Megerle.*

7806. 1; bipunctata*.
Tipula bipunctata. *Linn.* ii. 978.
Cer. bipunctata. *Meig. Zw.* i. 74.
Cer. trichopterus. *Meig. Kl.* i. 31.
For. pictipennis. *Megerle; teste Meigen.*

7807. 2; nemorosa*.
Cer. nemorosus. *Meig. Zw.* i. 75.
Tan. nemoralis. *Meig. Kl.* i. 24.

7808. 3; albipuncta* *mihi.*

7809. 4; brunnipes* *mihi.*

7810. 5; costalis* *mihi.*

Genus 13: (1137). LASIOPTERYX *mihi.*
LASIOPTERA B, *Meigen.*

7811. 1; obfuscata*.
Las. obfuscata. *Hoffm.*—*Meig. Zw.* i. 90.

7812. 2; pusilla?*
Las. pusilla. *Wied.*—*Meig. Zw.* i. 91?

7813. 3, confinis* *mihi.*

7814. 4; elegantissima* *mihi.*

7815. 5; inops* *mihi.*

Genus 14: (1138). DIOMYZA? *Megerle.*

LASIOPTERA A? *Meig.*

7816. 1, fuliginosa * *mihi.*

7817. 2, rubra * *mihi.*
An Tip. Berberina var. *Schr. A. No.* 885?

7818. 3; gigantea * *mihi.*

Genus 15: (1139). CECIDOMYIA, *Latreille, Curtis,* (*Kirby.*)

OLIGOTROPHUS, *Latr.*—CHIRONOMUS *p, Fabr.*

7819. 1; Oxyacanthæ * mihi.
Ce. grandis. *Meig. Zw.* i. 94?

7820. 2; Leacheana * *mihi.*

7821. 3; verna *. *Curtis.* iv. 178.

7822. 4; Klugii *. *Meig. Zw.* i. 95.

7823 † 5, Kirbii *. *Westwood MSS.* In Mus. D. *Westwood.*

7824. 6; nigra *. *Meig. Zw.* i. 95. *pl.* 3. *f.* 11. ♀.—(*Curtis l. c. supra.*)

7825. 7; iridescens * *mihi.*

7826. 8; Ribesii *. *Meg.*—*Meig. Zw.* i. 98.

7827. 9; lateralis *. *Meig. Zw.* i. 96.—(*Curtis l. c. supra.*)

7828. 10; palustris *. *Meig. Zw.* i. 96.—(*Curtis l. c. supra.*)
Tipula palustris. *Linn.* ii. 978.—*Stew.* ii. 254.—*Turt.* iii. 595.

7829. 11; Tritici *. (*Kirby & Sp. I. E.* iv. 503.)—(*Curtis l. c. supra.*)
Tipula Tritici. *Linn. Trans.* (*Kirby.*) iv. 230.—v. 106. *pl.* 4. *f.* 1–3.—*Stew.* ii. 254.—*Turt.* iii. 593.—(*Kirby & Sp. I. E.* i. 28.)—*Loudon M.* (*Kirby.*) i. 227. *f.* 91. *a.*
The Wheat Fly. *Bingley.* iii. 311.

7830. 12; nigricollis *. *Meig. Zw.* i. 97.—(*Curtis l. c. supra.*)

7831. 13; griseola *. *Meig. Zw.* i. 97.

7832. 14; griseicollis *. *Meig. Zw.* i. 97.

7833. 15; fuscicollis *. *Meig. Zw.* i. 97.

7834. 16; longicornis *.
Tipula longicornis. *Linn.* ii. 97.—*Turt.* iii. 595.

7835. 17; flavicans * *mihi.*

7836. 18; carnea *. *Meig. Zw.* i. 98.

7837. 19; rosea * *mihi.*

7838. 20; melanocephala * *mihi.*
Tip. minutissima. *Stew.* ii. 254?

7839. 21; bicolor *. *Meig. Zw.* i. 98.

7840. 22; lutea*. *Meig. Kl.* i. 40. *pl.* 2. *f.* 10, 11.—(*Samou. I.* 9.)—(*Curtis l. c. supra.*)

7841. 23, flava*. *Meig. Zw.* i. 99.—(*Curtis l. c. supra.*)

7842. 24; maculipennis* *mihi.*

7843. 25; fuscipennis*. *Meig. Zw.* i. 98.

7844. 26; cucullata*. *Meig. Zw.* i. 96.

Genus 16: (1140). CAMPYLOMYZA, *Wiedemann.*

7845. 1; atra*. *Meig. Zw.* i. 102.
Cecid. atra. *Meig. Kl.* i. 40.

7846. 2; Aceris*. *Meig. Zw.* i. 102.

7847. 3; bicolor*. *Wied.*—*Meig. Zw.* i. 102.

7848. 4; flavipes*. *Meig. Zw.* i. 102. *pl.* 3. *f.* 6. ♀.

Genus 17: (1141). PSYCHODA, *Latreille, Samou.*

BIBIO, *Geoffroy.*—TINEARIA, *Schellenburg.*—TRICHOPTERA, *Meig. Kl.*

7849. 1; nigrofusca* *mihi.*
Ps. obscura. *Macquart?*—*Boir. M.* ii. 360?

7850. 2; aterrima* *mihi.*

7851. 3; palustris*. *Meig. Zw.* i. 105. *pl.* 2. *f.* 18.

7852. 4; ocellaris*. *Meig. Zw.* i. 105. *pl.* 2. *f.* 14, 17.
Tip. hirta. *Stew.* ii. 253?—*Turt.* iii. 595?

7853. 5; concinna* *mihi.*
Ps. variegata. *Macquart?*—*Boir. M.* ii. 360?

7854. 6; trifasciata*. *Meig. Zw.* i. 105.
Trich. trifasciata. *Meig. Kl.* i. 44. *pl.* 2. *f.* 20.

7855. 7; nubila*. *Meg.*—*Meig. Zw.* i. 107.

7856. 8; fuliginosa*. *Meig. Zw.* i. 107.

7857. 9; humeralis*. *Hoffm.*—*Meig. Zw.* i. 106.

7858. 10; phalænoides*. *Latr. Gen.* iv. 251.
Tipula phalænoides. *Linn.* ii. 977.—*DeGeer.* vi. 158. *pl.* 27. *f.* 6–9.—*Berk. S.* i. 162.—*Stew.* ii. 253.—*Turt.* iii. 595.—(*Samou. I.* 35.)
Ps. muraria. *Latr. H.* xiv. 293.

7859. 11; canescens*. *Meig. Zw.* i. 106.

7860. 12; nervosa*. *Meig. Zw.* i. 106.
Tipula nervosa. *Schr. B.* iii. *sp.* 2350.

7861. 13, nana* *mihi.*

Genus 18: (1142). ERIOPTERYX *mihi.*

ERIOPTERA, *Meigen.*—POLYRAPHIA, *Megerle.*

A. Vide *Meigen Zw.* i. 109.

7862. 1; maculata*.
Eri. maculata. *Meig. Zw.* i. 109.

7863. 2; flavescens*.
Tipula flavescens. *Linn.* ii. 973.
Eri. flavescens. *Meig. Zw.* i. *pl.* 4. *f.* 9. ala.

7864. 3; lutea*.
Eri. lutea. *Meig. Zw.* i. 110.

7865. 4, montana.
Eri. montana. *Meig. Zw.* i. 110.

7866. 5, tænionota.
Eri. tænionota. *Wied.*—*Meig. Zw.* i. 111.
Poly. pallidipennis. *Megerle: (teste Meig. l. c.)*

7867. 6; fuscipennis*.
Eri. fuscipennis. *Meig. Zw.* i. 111.

7868. 7, fusca *mihi.*

7869. 8; trivialis*.
Eri. trivialis. *Hoffm.*—*Meig. Zw.* i. 112.

7870. 9; lineata*.
Eri. lineata. *Meig. Zw.* i. 111.

7871. 10; Stigma* *mihi.*

7872. 11; geniculata* *mihi.*

7873. 12; analis* *mihi.*

7874. 13; varia*.
Eri. varia. *Hoffm.* ?—*Meig. Zw.* i. 115?

B. Vide *Meigen Zw.* i. 112. a. *Alis densè villosis.*

7875. 14, obscura.
Eri. obscura. *Meig. Zw.* i. 113.

7876. 15, grisea.
Eri. grisea. *Meig. Zw.* i. 112.

7877. 16, atra.
Eri. atra. *Fabr. A.* 33.—*Meig. Kl.* i. *pl.* 3. *f.* 8. mas. 9. fœm.

7878. 17; murina*.
Eri. murina. *Meig. Zw.* i. 113.

7879. 18; pallida* *mihi.*

7880. 19; tincta* *mihi.*

7881. 20, ochracea.
Eri. ochracea. *Hoffm.*—*Meig. Zw.* i. 114.

b. *Alis ferè nudis.*

7882. 21; virescens* *mihi.*

7883. 22; hyalina* *mihi.*

7884. 23; pallidipennis* *mihi.*

7885. 24; halterata* *mihi.*

7886. 25; pallipes* *mihi.*

7887. 26; pallidula* *mihi.*

C. Vide *Meigen Zw.* i. 114.

7888. 27; imbuta*. *Wied.—Meig. Zw.* i. 114. *pl.* 4. *f.* 8.

7889. 28, cinerascens*. *Meig. Zw.* i. 114. *pl.* 4. *f.* 6. antenna.

7890. 29; nodicornis* *mihi.*

Genus 19: (1143). LEPTORHINA *mihi.*

LIMNOBIA *p, Meig.*

7891. 1; flava* *mihi.*

7892. 2; bicolor* *mihi.*
Lim. longirostris. *Wied.—Meig. Zw.* i. 146. *pl.* 5. *f.* 1. caput. —(*Curtis folio* 50.)

Genus 20: (1144). ——

7893. 1; thoracicus* *mihi.*

7894. 2; cognatus* *mihi.*

Genus 21: (1145). DICRANOMYIA *mihi.*

LIMNOBIA *p, Meig.*—FURCOMYIA *p, Meger.*—UNOMYIA *p, Meger.*

A. Vide *Meig. Zw.* i. 146. *pl.* 6. *f.* 7. ala.—GONOMYIA, *Megerle.*

7895. 1; sulphurea* *mihi.*

7896. 2, tenella.
Lim. tenella. *Hoffm.—Meig. Zw.* i. 146.
Lim. limbata. *Wiedemann*: (*teste Meig. l. c.*)
Go. tricolor. *Megerle*: (*teste Meig. l. c.*)

B. Vide *Meig. Zw.* i. 133. *pl.* 6. *f.* 5. ala.

7897. 3; nigripes* *mihi.*

7898. 4; lutea*.
Lim. lutea. *Meig. Zw.* i. 133.—(*Curtis folio* 50.)
Furcomyia pallida. *Megerle*: (*teste Meig. l. c.*)

7899. 5; inusta*.
Lim. inusta. *Meig. Zw.* i. 135.

7900. 6; modesta*.
Lim. modesta. *Wied.—Meig. Zw.* i. 134.

7901. 7; iridescens* *mihi.*

7902. 8; chorea*.
Lim. chorea. *Wied.—Meig. Zw.* i. 134.
Unomyia nubila. *Megerle:* (*teste Meig. l. c.*)

7903. 9; Didyma*.
Lim. Didyma. *Meig. Zw.* i. 135.

7904. 10; Dumetorum*.
Lim. Dumetorum. *Meig. Zw.* i. 136.

C. Vide *Meig. Zw.* i. 141. *pl.* 4. *f.* 18. ala.

7905. 11; glabrata*?
Lim. glabrata. *Wied.—Meig. Zw.* i. 142?
Ti. versipellis. *Harr. Ex.* 160. *pl.* xlviii. *f.* 8?

Genus 22: (1146). LIMNOBIA.

LIMNOBIA *p, Meig.*—UNOMYIA *p, Megerle.*—LIMONIA *p, Latreille.*

A. Vide *Meig. Zw.* i. 133. *pl.* 4. *f.* 19. ala.

7906. 1, fusca. *Meig. Zw.* i. 133.—(*Curtis l. c. infra.*)

B. Vide *Meig. Zw.* i. 125. *pl.* 6. *f.* 2. ala.

7907. 2; fuscipennis*. *Meig. Zw.* i. 125.—(*Curtis l. c. infra.*)

7908. 3, lucorum*. *Meig. Zw.* i. 125.

7909. 4, discicollis. *Meg.—Meig. Zw.* i. 125.

7910. 5; xanthura* *mihi.*
Ti. lentus. *Harr. Ex.* 160. *pl.* xlviii. *f.* 7?

7911. 6; cingulata* *mihi.*

C. Vide *Meig. Zw.* i. 126. *pl.* 6. *f.* 3. ala.

7912. 7, nemoralis. *Meig. Zw.* i. 126.—(*Curtis l. c. infra.*)
Lim. xanthopyga. *Wiedemann:* (*teste Meig. l. c.*)

7913. 8; leucophæa*. *Hoffm.—Meig. Zw.* i. 127.

7914. 9, plebeia. *Meig. Zw.* i. 127.

D.

7915. 10; aprica* *mihi.*

E. Vide *Meig. Zw.* i. 137. *pl.* 4. *f.* 13. ala.

7916. 11, Stigma. *Meig. Zw.* i. 138.

7917. 12; tripunctata*. *Meig. Zw.* i. 138. *pl.* 4. *f.* 13.—(*Curtis l. c. infra.*)
Tipula tripunctata. *Fabr. Sp.* ii. 405.
Tipula Phragmatidis. *Schra. No.* 860.

7918. 13; sexpunctata*. *Meig. Zw.* i. 139.
Tipula sexpunctata. *Fabr. Sp.* ii. 405.
Limonia sexpunctata. *Meig. Kl.* i. 59. *pl.* 3. *f.* 15.

7919. 14; analis*. *Meig. Zw.* i. 141.
Limonia flavipes. *Meig. Kl.* i. 59.

7920. 15, pabulina. *Meig. Zw.* i. 140.

7921. 16; flavipes*. *Meig. Zw.* i. 150.
Tipula flavipes. *Fabr. A.* 30.

7922. 17; nubeculosa*. *Meig. Zw.* i. 140.
Tipula punctata. *Stew.* ii. 251?
Tipula 4-maculata. *Turt.* iii. 585?

F. Vide *Meig. Zw.* i. 127. *pl.* 4. *f.* 20. ala.

7923. 18; albifrons?* *Meig. Zw.* i. 137.

7924. 19; ferruginea*. *Meig. Zw.* i. 128.—(*Curtis l. c. infra.*)
Limonia flavescens. *Meig. Kl.* i. 56.
Tipula flavescens. *Schell.* G. *M. pl.* 37. *f.* 1. mas.

7925. 20; punctum*. *Meig. Zw.* i. 128.

7926. 21; dispar*. *Meger.*—*Meig. Zw.* i. 129.
Lim. fuscipes. *Wiedemann*: (*teste Meig. l. c.*)

7927. 22; fulvescens*. *Hoffm.*—*Meig. Zw.* i. 127.

7928. 23; lineola. *Meig. Zw.* i. 128.

7929. 24, ochracea. *Meig. Zw.* i. 129.

G. Vide *Meig. Zw.* i. 141. *pl.* 4. *f.* 18. ala.

7930. 25; xanthoptera*. *Meig. Zw.* i. 141.—(*Curtis l. c. infra.*)

7931 † 26. replicata[a].

7932. 27; quadrinotata*. *Meig. Zw.* i. 144.
Tipula quadrinotata. *Fabr. Mus.*: (*teste Meig. l. c.*)

H. Vide *Meig. Zw.* i. 131. *pl.* 6. *f.* 4. ala.

7933. 28, littoralis. *Meig. Zw.* i. 131.—(*Curtis l. c. infra.*)

7934. 29, geniculata. *Hoffm.*—*Meig. Zw.* i. 124.

I. Vide *Meig. Zw.* i. 122. *pl.* 4. *f.* 15. ala.

7935. 30; transversa*. *Meig. Zw.* i. 123.

7936. 31; ocellaris*. *Meig. Zw.* i. 152.—*Curtis.* i. *pl.* 50.
Tipula ocellaris. *Linn.* ii. 973.—*Turt.* iii. 587.

7937. 32; picta*. *Meig. Zw.* i. 123.—(*Curtis l. c. supra.*)
Tipula picta. *Fabr. A.* 29.
Tipula —— *Schell.* G. *M. pl.* 3. *f.* 1.

7938. 33; punctata*. *Meig. Zw.* i. 122. *pl.* 4. *f.* 15. ala.
Tipula ocellaris. *Fabr. Mus.*: (*teste Meig. l. c.*)
Tipula annulata. *Turt.* iii. 587?—(*Kirby & Sp. I. E.* iv. 54?)

---

[a] 7931 † 26. replicata. *Meig. Zw.* i. 142.
Tipula replicata. *Linn.* ii. 973.—*DeGeer.* vi. 138. *pl.* 20. *f.* 1–16.—*Turt.* iii. 588. (!)—(*Kirby & Sp. I. E.* iii. 116.)

K. Vide *Meig. Zw.* i. 121. *pl.* 4. *f.* 16. ala.

7939. 34; marmorata*. *Hoffm.*—*Meig. Zw.* i. 121.
Limónia maculata. *Meig. Kl.* i. 61.

7940. 35, fasciata. *Meig. Zw.* i. 121. *pl.* 4. *f.* 16. ala.
Ti. fasciata. *Fabr. A.* 30.

L. Vide *Meig. Zw.* i. 119. *pl.* 6. *f.* 1. ala.

7941. 36, pictipennis. *Meig. Zw.* i. 119.

M. Vide *Meig. Zw.* i. 148. *pl.* 5. *f.* 8. ala.

7942. 37; immaculata*. *Meig. Zw.* i. 148.—(*Curtis l. c. supra.*)

N. Vide *Meig. Zw.* i. 147. *pl.* 5. *f.* 7. ala.

7943. 38; punctipennis*. *Meig. Zw.* i. 147. *pl.* 5. *f.* 2. caput. *f.* 3. os. *f.* 7. ala.
Limonia hybrida. *Meig. Kl.* i. 57.

7944. 39; stictica*. *Meig. Zw.* i. 148.

O. Vide *Meig. Zw.* i. 132. *pl.* 5. *f.* 6. ala.

7945. 40; fimbriata*. *Meig. Zw.* i. 132.—(*Curtis l. c. supra.*)

7946. 41, pilipes. *Meig. Zw.* i. 150.—(*Curtis l. c. supra.*)
Tipula pilipes. *Fabr. A.* 30.

Genus 23: (1147). RHIPIDIA, *Meigen.*

7947. 1, maculata. *Meig. Zw.* i. 153. *pl.* 11. ♂.

Genus 24: (1148). PEDICIA, *Latreille, Samou.*

LIMNOBIA *p, Meig.*

7948. 1, rivosa*. *Latr.* G. iv. 255.
Tipula rivosa. *Linn.* ii. 971.—*Stew.* ii. 250.—*Meig. Kl.* i. 23. *pl.* iii. *f.* 14.—(*Samou.* 291.)—(*Kirby & Sp. I. E.* i. *pl.* 4. *f.* 4.)
Tipula triangularis. *Fabr. A.* 27.—*Stew.* ii. 250.—*Turt.* iii. 585.

7949. [2, crassipes] *mihi.*

Genus 25: (1149). CTENOPHORA, *Meigen, Samou., Curtis.*

TANIPTERA, *Latr.*

A. Vide *Meig. Zw.* i. 160. *pl.* 5. *f.* 16. antennæ.

7950 † 1, ornata. *Megerle.*—*Meig. Zw.* i. 166.—*Curtis.* i. *pl.* 5.
♂ in Mus. D. *Dale.*

7951 † 2, flaveolata. *Fabr. A.* 18.—*Meig. Kl.* i. 87. *pl.* 4. *f.* 18. fœm.—(*Curtis 2 edit. l. c. supra.*)
Reaum. Ins. v. *pl.* 1. *f.* 14–16. mas.
Tip. crocata. *Schr. A.* 854. In Mus. D. *Beck, et Haworth.*

7952. 3; pectinicornis*. *Meig. Zw.* i. 160.—(*Curtis l. c. supra.*)—*Curtis 2 edit. l. c. supra.*

Tip. pectinicornis. *Linn.* ii. 970.—*Stew.* ii. 250.—*Turt.* iii. 584.—*Schæf. Ic.* cvi. *f.* 5, 6.—*Wood.* ii. 85. *pl.* 64.
Tip. nigro-crocea. *DeGeer.* vi. 152.
Tip. variegata. *Fabr. Sp.* ii. 402.—*Stew.* ii. 251.—*Turt.* iii. 585.
Tip. splendor. *Harr. Ex.* 56. *pl.* xiv. *f.* 3.

7953. 4; consobrina* *mihi.*

B. Vide *Meig. Zw.* i. 158. *pl.* 5. *f.* 15. antenna.

7954. 5; nigricornis*. *Meig. Zw.* i. 159.—(*Curtis* 2 *edit. l. c. supra.*)
Ct. atrata. *Meig. Kl.* i. 85.

7955. 6, atrata*. *Meig. Zw.* i. 158.—(*Samou.* 291.)—(*Curtis l. c. supra.*)—(*Curtis* 2 *edit. l. c. supra.*)
Tip. atrata. *Linn.* ii. 972.
Tip. ichneumonea. *DeGeer.* vi. 138. *pl.* 19. *f.* 10.

C. Vide *Meig. Zw.* i. 156. *pl.* 5. *f.* 14. antennæ.

7956. 7; bimaculata*. *Meig. Zw.* i. 156.—(*Curtis l. c. supra.*)—*Curtis* 2 *edit. l. c. supra.*
Tip. bimaculata. *Linn.* ii. 972.—*Stew.* ii. 251.—*Turt.* iii. 587.—*Schæff. Ic.* cxi. *f.* 5, 6.

7957. 8; paludosa?* *Fabr. A.* 19?—*Curtis* 2 *edit. l. c. supra.*

Genus 26: (1150). NEPHROTOMA, *Meigen.*

7958. 1; dorsalis*. *Meig. Zw.* i. 202. *pl.* 5. *f.* 19–22. ♂.
Tip. dorsalis. *Fabr. A.* 28.

Genus 27: (1151). TIPULA *Auctorum.*

LIMONIA *p. Meig. Kl.* *Daddy Long-legs*: "*Harry Long-legs or Taylor Fly.*" Albin.

A. Vide *Meig. Zw.* i. 192. *pl.* 6. *f.* 9. ala.

7959. 1, Scurra. *Hoffmansegg.*—*Meig. Zw.* i. 198.

7960. 2; interrupta* *mihi.*

7961. 3; Histrio*. *Fabr. A.* 28.
Tip. flavo-maculata. *DeGeer.* vi. 137. *pl.* 19. *f.* 2, 3.
Tip. cornicina. *Meig. Kl.* i. 71.

7962. 4; pallida* *mihi.*

7963. 5; maculosa*. *Hoffm.*—*Meig. Zw.* i. 197.
Tip. maculata. *Meig. Kl.* i. 71.

7964. 6; quadrifaria*. *Meig. Zw.* i. 199.

7965. 7; cornicina*. *Linn.* ii. 972.—*Berk. S.* i. 161.—*Stew.* ii. 251.—*Turt.* iii. 586.

7966. 8; pratensis*. *Linn.* ii. 972.—*Stew.* ii. 251.—*Turt.* iii. 586.—*Schæff. Ic.* xv. *f.* 5.

7967. 9, imperialis*? *Meger.*—*Meig. Zw.* i. 196. *pl.* 6. *f.* 9. fœm.

7968. 10; crocata*. *Linn.* ii. 971.—*Don.* ii. 29. *pl.* 48. *f.* 1.—*Berk. S.* i. 161.—*Stew.* ii. 250.—*Turt.* iii. 584.—(*Kirby & Sp. I. E.* ii. 359 & 382.)
Tip. flavo-fasciata. *DeGeer.* vi. 137.
Tip. perpulcher. *Harr. Ex.* 159. *pl.* xlviii. *f.* 6.
B. Vide *Meig. Zw.* i. 170. *pl.* 6. *f.* 8. ala.

7969. 11, nigra. *Linn.* ii. 971.—(*Ing. Inst.* 92.)
Ti. verticillata. *Fabr.* iv. 237.
Ti. cœnosus. *Harr. Ex.* 161. *pl.* xlviii. *f.* 9?

7970. 12, Stigma *mihi.*

7971. 13; vernalis*. *Meig. Zw.* i. 182.
Tip. divagor. *Harr. Ex.* 161. *pl.* xlviii. *f.* 10?

7972. 14; Lineola*. *Megerle.*—*Meig. Zw.* i. 181.
Tip. pendens. *Harr. Ex.* 56. *pl.* xiv. *f.* 4.

7973. 15; hortulana*. *Meig. Zw.* i. 177.

7974. 16; pabulina*. *Meig. Zw.* i. 180.

7975. 17, lateralis. *Meig. Zw.* i. 174.

7976. 18; flavolineata*. *Meig. Zw.* i. 185.

7977. 19; vittata*. *Meig. Zw.* i. 171.
Tip. oleracea. *Samou.* 71. *pl.* 9. *f.* 2.

7978. 20; nubeculosa*. *Meig. Zw.* i. 174.
Tip. Hortorum. *Fabr. A.* 24.—*Stew.* ii. 250.—*Turt.* iii. 585.
Tip. unca. *Hoffm. Mus.*: (*teste Meig. l.* c.)

7979. 21; Hortorum*. *Linn.* ii. 971.—*Berk. S.* i. 161?

7980. 22; varipennis*. *Hoffm.*—*Meig. Zw.* i. 183.
Tip. rivosa. *Mus. Fabr*: (*teste Meig. l.* c.)

7981. 23; hortensis*. *Hoffm.*—*Meig. Zw.* i. 178.
Tip. Hortorum. *Meig. Kl.* i. 69.

7982. 24; marmorata*. *Meig. Zw.* i. 179.

7983. 25; rufina*. *Meig. Zw.* i. 176.

7984. 26, suffusa *mihi.*

7985. 27; pagana*. *Meig. Zw.* i. 184.
Tip. plicata. *Meig. Kl.* i. 73.

7986. 28; pruinosa*. *Hoffm.*—*Meig. Zw.* i. 191.

7987. 29; bimaculosa* *mihi.*
Tip. selenitica. *Hoffm.?*—*Meig. Zw.* i. 187?

7988. 30; lunata*. *Linn.* ii. 972.—*Berk. S.* i. 161.—*Stew.* ii. 251.—*Turt.* iii. 585.—*Schæff. Ic.* 162. *f.* 5. mas. 6. fœm.

7989. 31; fascipennis*. *Hoffm.*—*Meig. Zw.* i. 187.

7990. 32; ochracea*. *Meig. Zw.* i. 186.
Tip. lunata. *Fabr. Sp.* ii. 402?
Albin. pl. lxi. *f. e–g.*

7991. 33, flavipennis *mihi.*

7992. 34, flavilinea *mihi.*

7993. 35; oleracea*. *Linn.* ii. 971.—*Berk. S.* i. 161.—*Stew.* ii. 250.—*Turt.* iii. 585.—*DeGeer.* vi. 134. *f.* 18. *f.* 12, 13.—*Bingley.* iii. 310.—(*Kirby & Sp. I. E.* i. 181 nota.—ii. 360 nota, 361.—*Id.* iii. 295, 551, 552.)
Tip. terrestris. *Harr. Ex.* 56. *pl.* xiv. *f.* 2.

7994. 36; brachypteryx* *mihi.*

7995. 37; gigantea*. *Schr. A. No.* 845.—*Meig. Zw.* i. 170.
Ti. sinuata. *Fabr. A.* 23.—(*Kirby & Sp. I. E.* iii. 37.)
Tip. rivosa. *Don.* ii. 30. *pl.* 48. *f.* 2.—*Berk. S.* i. 160.—*Stew.* ii. 250.—*Turt.* iii. 584.
Tip. nubilosus. *Harr. Ex.* 55. *pl.* xiv. *f.* 1.

7996. 38; lutescens*. *Fabr. A.* 24.
Tip. fulvipennis. *DeGeer.* vi. 135.
Tip. discimacula. *Megerle.*

7997. 39; glabricollis* *mihi.*

7998. 40; fimbriata*. *Meig. Zw.* i. 190.

Genus 28: (1152). DOLICHOPEZA, *Curtis.*

7999. 1, Sylvicola. *Curtis.* ii. *pl.* 62.

Genus 29: (1153). PTYCHOPTERYX, *Leach.*
PTYCHOPTERA, *Meigen.*

8000. 1; albimana*.
Pty. albimana. *Fabr. A.* 21.—*Meig. Zw.* i. 207. *pl.* 6. *f.* 17.

8001. 2; contaminata.
Tip. contaminata. *Linn.* ii. 972.—*Berk. S.* i. 161.—*Stew.* ii. 251.—*Turt.* iii. 585.—*Schæff. Ic.* 196. *f.* 3. male. 48. *f.* 7. fœm.
Tip. fuscipes. *Gmel.* v. 2814.

8002. 3; paludosa*.
Pty. paludosa. *Meig. Zw.* i. 207.

8003. 4; affinis* *mihi.*

8004. 5; subnebulosa* *mihi.*

Genus 30: (1154). TRICHOCERA, *Meigen.* *Tell-tales,* Harris.
NEMATOCERA *p, Megerle.*—LIMONIA *p, Latreille.*

8005. 1; fuscata*. *Meig. Zw.* i. 212.

8006. 2; parva*. *Meig. Zw.* i. 213.

8007. 3; hiemalis*. *Meig. Zw.* i. 213.—(*Kirby & Sp. I. E.* ii. 444 nota.)
Tip. hiemalis. *DeGeer.* vi. 141. *pl.* 21. *f.* 1, 2.
Tri. perennis. *Mus. Hoffm.*: (*teste Meig. l.* c.)
Tip. saltator. *Harr. Ex.* 57. *pl.* xiv. *f.* 5.

8008. 4; regelationis*. *Meig. Zw.* i. 214. *pl.* 7. *f.* 9. fœm.
Tip. regelationis. *Linn.* ii. 973.—*Stew.* ii. 251.

8009. 5; maculipennis*. *Meig. Zw.* i. 214.

8010. 6; annulata*. *Meig. Zw.* i. 215.

Genus 31: (1155). ANISOMERA, *Hoffmansegg.*

8011 † 1. obscura. *Hoffm.*—*Meig. Zw.* i. 210. *pl.* 7. *f.* 5–8?—(*Curtis Cat. No.* 45.) In Mus. D. Curtis.

Genus 32: (1156). DIXA, *Meigen.*

8012. 1, aprilina. *Meig. Zw.* i. 218. *pl.* 7. *f.* 12. mas.

8013. 2, maculata. *Meig. Zw.* i. 219.

8014. 3; variegata* *mihi.*

Genus 33: (1157). BOLITOPHILA, *Meigen.*

LEPTOCERA, *Megerle.*—MACROCERA *p*, *Meig. Kl.*

8015 † 1, fusca. *Meig. Zw.* i. 212. *pl.* 8. *f.* 3, 4.
Mac. hybrida. *Meig. Kl.* i. 47. In Mus. *Brit.*

Genus 34: (1158). MACROCERA, *Meigen.*

8016. 1; fasciata*. *Meig. Zw.* i. 223. *pl.* 8. *f.* 5. mas.

8017. 2; lutea*. *Meig. Zw.* i. 223.—*Panz. F.* cv. *f.* 7.

8018. 3; lineola* *mihi.*

8019. 4; centralis*. *Meig. Zw.* i. 225.

8020. 5; angulata*. *Meig. Zw.* i. 224.

8021. 6; phalerata*. *Meig. Zw.* i. 223.

Genus 35: (1159). MYCETOBIA, *Meigen.*

8022. 1; pallipes*. *Megerle.*—*Meig. Zw.* i. 280. *pl.* 8. *f.* 10.

Genus 36: (1160). PLATYURA, *Meigen, Curtis.*

CEROPLATUS *p*, *Bosc.*?—SCIARA *p*, *Fabr.*—ASINDULUM, *Latreille.*

A. Vide *Meig. Zw.* i. 232. *pl.* 8. *f.* 14. ala.

8023. 1; atrata*. *Meig. Zw.* i. 233.—(*Curtis l. c. infra.*)
Ce. atratus. *Fabr. A.* 16.

B. Vide *Meig. Zw.* i. 235. *pl.* 8. *f.* 19. ala.

8024. 2; lineata*. *Meig. Zw.* i. 234. (!)—(*Curtis l. c. infra.*)

Sc. lineata. *Fabr. A.* 57. (!)
Musca lineata. *Turt.* iii. 633.
Musca striata. *Stew.* ii. 264.

8025. 3; laticornis*. *Meig. Zw.* i. 238. *pl.* 8. *f.* 19–21.—(*Curtis l. c. infra.*)

8026. 4; signata*. *Meig. Zw.* i. 238.

8027. 5; rufipes*. *Hoffm.*—*Meig. Zw.* i. 241.—(*Curtis l. c. infra.*)

8028. 6; pallipes* *mihi.*

8029. 7; semirufa*. *Meig. Zw.* i. 237.—(*Curtis l. c. infra.*)

8030. 8; nemoralis*. *Meig. Zw.* i. 236.—(*Curtis l. c. infra.*)

8031. 9; flavipes*. *Meig. Zw.* i. 237.—*Curtis.* iii. *pl.* 134.

8032. 10; cingulata* *mihi.*—(*Curtis l. c. supra.*)

8033. 11; ochracea*. *Meig. Zw.* i. 240.—(*Curtis l. c. supra.*)

8034. 12; discoloria*. *Meig. Zw.* i. 239.—(*Curtis l. c. supra.*)

8035. 13; abdominalis* *mihi.*

8036. 14; interrupta* *mihi.*

Genus 37: (1161). ——

8037. 1; zonatus* *mihi.*

Genus 38: (1162). SCIOPHILA, *Hoffm.*, *Meigen.*
PLATYURA *p*, *Fabr.*

A. Vide *Meig. Zw.* i. 247. *pl.* 9. *f.* 6. ala.

8038. 1; marginata*. *Meig. Zw.* i. 249.

8039. 2; maculata*. *Meig. Zw.* i. 248.
Pl. maculata. *Fabr. A.* 33.

8040. 3, annulata*. *Meig. Zw.* i. 247.

8041. 4; ferruginea*. *Meig. Zw.* i. 249.

B. Vide *Meig. Zw.* i. 251. *pl.* 9. *f.* 8. ala.

8042. 5; vitripennis*. *Meig. Zw.* i. 251.

8043. 6; ochracea* *mihi.*

C. Vide *Meig. Zw.* i. 251. *pl.* 9. *f.* 7. ala.

8044. 7; hirta*. *Hoffm.*—*Meig. Zw.* i. 251.

8045. 8; pallipes* *mihi.*

Genus 39: (1163). LEIA, *Meigen.*

A. Vide *Meig. Zw.* i. 254. *pl.* 9. *f.* 13. caput.

8046. 1, nitidicollis. *Meig. Zw.* i. 255.

B. Vide *Meig. Zw.* i. 255. *pl.* 9. *f.* 12. ala.

8047. 2; flavicornis*. *Meig. Zw.* i. 255. *pl.* 9. *f.* 11. fœm.

8048. 3; fascipennis*. *Megerle.—Meig. Zw.* i. 255.

8049. 4; fasciola*. *Meig. Zw.* i. 256.

C. Vide *Meig. Zw.* i. 257. *pl.* 9. *f.* 18. ala.

8050. 5; analis*. *Meig. Zw.* i. 257.
Mycetophila dubia. *Meig. Kl.* i. 92.

Genus 40: (1164). MYCETOPHILA, *Meigen.*

SCIARA *p, Fabr.*—MUSCA *p, Villers.*

A. Vide *Meig. Zw.* i. 260. *pl.* 9. *f.* 15. ala. a. *Alis maculatis.*

8051. 1, ornata* *mihi.*

8052. 2, lutea. *Meig. Zw.* i. 263.

8053. 3; arcuata*. *Meig. Zw.* i. 261.

8054. 4; lunata*. *Meig. Zw.* i. 260.
Sc. lunata. *Fabr. A.* 58.
My. lunata. *Meig. Kl.* i. 90. *pl.* 5. *f.* 2, 3.

8055. 5; fuscicornis. *Meig. Zw.* i. 261?

8056. 6; nebulosa* *mihi.*

8057. 7, lineola*. *Meig. Zw.* i. 262. *pl.* 9. *f.* 15.

8058. 8; ruficollis*. *Meg.—Meig. Zw.* i. 262.

b. *Alis immaculatis.*

8059. 9; luteipennis* *mihi.*

8060. 10; flavipennis* *mihi.*

B. Vide *Meig. Zw.* i. 267. *pl.* 9. *f.* 21. ala.

8061. 11; maculosa*. *Meig. Zw.* i. 268.
Odontophila maculosa. *Megerle*: (*teste Meig. l.* c.)

8062. 12; discoidea*. *Meig. Zw.* i. 268.

C. Vide *Meig. Zw.* i. 266. *pl.* 9. *f.* 20. ala.

8063. 13; lateralis*. *Meig. Zw.* i. 266.

8064. 14; fusca*. *Meig. Zw.* i. 266.
Ti. Fungorum. *DeGeer.* vi. 142.

8065. 15; cingulata* *mihi.*

8066. 16; semifusca*. *Meig. Zw.* i. 267.

D. Vide *Meig. Zw.* i. 269. *pl.* 9. *f.* 13. ala.

8067. 17; ornaticollis*. *Meig. Zw.* i. 269.

8068. 18; cucullata* *mihi.*

8069. 19; lugens*. *Wiedemann.—Meig. Zw.* i. 269.

8070. 20; analis*. *Megerle.—Meig. Zw.* i. 269.

8071. 21; zonata* *mihi.*

8072. 22; flaviceps*. *Meig. Zw.* i. 270.
8073. 23; nigra*. *Meig. Zw.* i. 270.

Genus 41: (1165). CORDYLA, *Meigen.*
ODONTOPHILA *p*, *Megerle.*

8074. 1; crassicornis*. *Meig. Zw.* i. 275.

Genus 42: (1166). MOLOBRUS, *Latreille.*
HIRTEA *p*, *Fabr.*—SCIARA *p*, *Fabr.*

A. Halteribus fuscis.

8075. 1; Thomæ*.
Tip. Thomæ. *Linn.* ii. 976.
Sci. Thomæ. *Meig. Zw.* i. 278. *pl.* 4. *f.* 3.

8076. 2; Morio*.
♀; Sc. Morio. *Fabr. A.* 57.
♂; Hi. forcipata. *Fabr. A.* 55.—*Stew.* ii. 253.—*Turt.* iii. 594.
Sc. florilega. *Meig. Kl.* i. 98.

8077. 3; fuliginosa* *mihi.*

8078. 4; fuscipes*.
Sc. fuscipes. *Meig. Zw.* i. 280.

8079. 5; obscura* *mihi.*

8080. 6; fucata*.
Sc. fucata. *Meg.*—*Meig. Zw.* i. 281.

8081. 7; præcox*.
Sc. præcox. *Meig. Zw.* i. 279.

8082. 8; ruficauda*.
Sc. ruficauda. *Meig. Zw.* i. 280.

8083. 9; xanthopus* *mihi.*

8084. 10; vitripennis*.
Sc. vitripennis. *Hoffm.*—*Meig. Zw.* i. 281.

8085. 11; fenestrata*.
Sc. fenestrata. *Meig. Zw.* i. 281.

8086. 12; fuscipennis*.
Sc. fuscipennis. *Meig. Zw.* i. 282.

8087. 13; scatopsoides*.
Sc. scatopsoides. *Meig. Zw.* i. 282.

8088. 14; sylvatica*.
Sc. silvatica. *Meig. Zw.* i. 283.

8089. 15; nervosa*.
Sc. nervosa. *Meig. Zw.* i. 283.

B. Halteribus flavis aut pallidis.

8090. 16; flavipes*.
Sc. flavipes. *Meig. Zw.* i. 283.—*Panz. F.* ciii. *f.* 15.

8091. 17; phæopus* *mihi.*

8092. 18; bicolor*.
Sc. bicolor. *Meg.*—*Meig. Zw.* i. 284.

8093. 19; pallipes*.
Chironomus pallipes. *Fabr. A.* 45.
Sc. pallipes. *Meig. Zw.* i. 284.

8094. 20; hyalipennis*.
Sc. hyalipennis. *Meig. Zw.* i. 285.

8095. 21; brunnipes*.
Sc. brunnipes. *Meig. Zw.* i. 286.

8096. 22; pusillus*.
Ti. pusilla. *Meig. Zw.* i. 286.

8097. 23; nemoralis*.
Sc. nemoralis. *Meig. Zw.* i. 287.

8098. 24; longipes*.
Sc. longipes. *Meig. Zw.* i. 286.

8099. 25; hirticornis*.
Sc. hirticornis. *Meig. Zw.* i. 287.

Genus 43: (1167). SIMULIUM, *Latreille.*

SIMULIA, *Meigen.*—CULEX *p, Linné.*—BIBIO *p, Olivier.*—SCATOPSE *p, Fabr.*—HIRTEA *p, Schellen.*—ATRACTOCERA, *Meig. Kl.*

8100. 1; ornata*. *Meig. Zw.* i. 290.
Atr. regelationis. *Meig. Kl.* i. 94.

8101. 2; reptans*. *Meig. Zw.* i. 291.—(*Kirby & Sp. I. E.* i. 128.)
Cu. reptans. *Linn.* ii. 1003.—*Turt.* iii. 664.
Tip. erythrocephala. *DeGeer.* vi. 161. *pl.* 28. *f.* 5, 6. mas.
Atr. varipes. *Megerle*: (*teste Meig. l. c.*)

8102. 3; varia*. *Meig. Zw.* i. 292.

8103 † 4. lineata. *Meig. Zw.* i. 293. In Mus. *Brit.*?

8104. 5; variegata*. *Hoffm.*—*Meig. Zw.* i. 292.
Hir. livida. *Schell.* G. *M. pl.* 38. *f.* 3.

8105. 6; affinis* *mihi.*

8106. 7; luteicornis* *mihi.*

8107. 8; picipes* *mihi.*

8108. 9; flavipes* *mihi.*

8109. 10; elegans*. *Meig. Zw.* i. 296.

8110. 11; auricoma*. *Meig. Zw.* i. 296.

8111. 12; latipes*. *Meig. Zw.* i. 297.

8112. 13; nigra*. *Meig. Zw.* i. 297.

8113. 14. equinus[a].

Genus 44: (1168). THYRIDOPHILA *mihi.*

8114. 1; pyrrhopa* *mihi.*

Genus 45: (1169). SCATOPSE, *Geoffroy.*

SCIARA *p, Fabr.*—CERIA, *Scopoli.*

8115. 1; notata*. *Meig. Zw.* i. 300. *pl.* 10. *f.* 13.
Tip. notata. *Linn.* ii. 977.
Tip. latrinarum. *DeGeer.* vi. 160. *pl.* 28. *f.* 1–4.
Sc. albipennis. *Fabr. A.* 55.
Ce. decemnodia. *Scop. C. No.* 950.

8116. 2; nigra*. *Meig. Zw.* i. 300.

8117. 3; punctata*. *Meig. Zw.* i. 301.

8118. [4, Nectarea.] *Marsham MSS.*

8119. 5; picipes* *mihi.*

Genus 46: (1170). DILOPHUS, *Meigen.*

HIRTEA *p, Fabr.*—BIBIO *p, Olivier.*—PULLATA *p, Harr.*

8120. 1; vulgaris*. *Meig. Zw.* i. 306. *pl.* 11. *f.* 1. mas. *f.* 8. ala fœm.
Tip. febrilis. *Linn.* ii. 976.—*Berk. S.* i. 162.—*Stew.* ii. 253.—*Turt.* iii. 594.
Tip. forcipata. *Schran. B.* iii. 2339.
Pu. parvus. *Harr. Ex.* 77. *pl.* xxii. *f.* 7, 8.
♂; Di. senilis. *Megerle:* (*teste Meig. l. c.*)
♀; Di. costalis. *Megerle:* (*teste Meig. l. c.*)

8121. 2; marginatus*. *Meig. Zw.* i. 307.
Bi. nigrita. *Olivier:* (*teste Meig. l. c.*)

8122. 3; femoratus*. *Meig. Zw.* i. 307.

8123. 4; tenuis*. *Hoffm.*—*Meig. Zw.* i. 308.

Genus 47: (1171). BIBIO, *Geoffroy, Curtis,* (*Kirby.*)

HIRTEA *p, Fabr.*—PULLATA *p, Harr.*

8124. 1; Pomonæ*. *Meig. Zw.* i. 312.—(*Curtis l. c. infra.*)
Hir. Pomonæ. *Fabr. A.* 53.
Tip. Marci fulvipes. *DeGeer.* vi. 160.
Tip. Pomonæ. *Herbst. Gem. Nat.* viii. *pl.* 338. *f.* 5.—*Don.* ix. 27. *pl.* 300.—*Stew.* ii. 252.—*Turt.* iii. 594.
Pu. Funestus. *Harr. Ex.* 77. *pl.* xxii. *f.* 3, 4.

---

[a] 8113 † 14. equinus.
Culex equinus. *Linn.* ii. 1003.—*Turt.* iii. 664. (!)

8125. 2; Marci*. *Meig. Zw.* i. 311.—(*Curtis l. c. infra.*)
♀; Tip. Marci. *Linn.* ii. 976.—*Stew.* ii. 252.—*Turt.* iii. 593.—*Panz. F.* xcv. *f.* 20.
Hi. Marci. (*Kirby & Sp. I. E.* ii. 361.)
♂; Tip. brevicornis. *Linn.* ii. 976.—*Turt.* iii. 593.
Tip. febrilis. *Schr. A.* 878.
Empis! pennipes. *Samou.* 72. *pl.* 9. *f.* 6!
Pu. Funerosus. *Harr. Ex.* 76. *pl.* xxii. *f.* 1, 2.

8126. 3; hortulanus*. *Meig. Zw.* i. 310.—(*Kirby & Sp. I. E.* i. 192.)—(*Curtis l. c. infra.*)
Tip. hortulana. *Linn.* ii. 977.—*Berk. S.* i. 162.—*Stew.* ii. 353.—*Turt.* iii. 593.
Hir. hortulana. *Schell. G. M. pl.* 39. *f.* 1. fœm. *f.* 2. mas.
Pu. Citrius. *Harr. Ex.* 77. *pl.* xxii. *f.* 5, 6.

8127. 4; ferruginatus*. *Meig. Zw.* i. 316.—(*Curtis l. c. infra.*)
Tip. ferruginata. *Linn.* ii. 976.
Tip. flavicaudis. *DeGeer.* vi. 160.

8128. 5; villosus*. *Meig. Zw.* i. 313.

8129. 6; Johannis*. *Meig. Zw.* i. 314.—(*Curtis l. c. infra.*)
Tip. Johannis. *Linn.* ii. 976.—*DeGeer.* vi. 159. *pl.* 27. *f.* 12–20.
Tip. Pyri. *Fabr. E. S.* iv. 249.
Hir. præcox. *Fabr. A.* 31.
Hir. hyalina. *Meig. Kl.* i. 110.

8130. 7; nigripes*. *Meig. Zw.* i. 315.—(*Curtis l. c. infra.*)
Pu. Minimus. *Harr. Ex.* 77. *pl.* xxii. *f.* 9?

8131. 8; venosus*. *Meig. Zw.* i. 315.—*Curtis.* iii. *pl.* 138.

8132. 9; vernalis*. *Meig. Zw.* i. 315.—(*Curtis l. c. supra.*)
Pu. Miniusculus. *Harr. Ex.* 77. *pl.* xxii. *f.* 10?

8133. 10; geniculatus* *mihi.*

8134. 11; flavipennis* *mihi.*

8135. 12; clavipes*. *Meig. Zw.* i. 317.—(*Curtis l. c. supra.*)
Hir. Johannis. *Fabr. A.* 52.

8136. 13, lanigerus. *Hoffm.*—*Meig. Zw.* i. 317.—(*Curtis l. c. supra.*)

8137 † 14. dorsalis. *Megerle.*—*Meig. Zw.* i. 318.—(*Curtis l. c. supra.!*) In Mus. D. ——

Genus 48: (1172). ASPISTES, *Hoffmansegg, Meigen,* (*Kirby.*)

8138 † 1, Chelseaensis* *mihi.* In Mus. D. *Haworth.*

8139. 2, obscurus *mihi.*

Genus 49: (1173). RHYPHUS, *Latreille, Curtis.*

SCIARA *p, Fabr.*—RHAGIO *p, Fabr.*—ANISOPUS, *Meig. Kl.*—SYLVICOLÆ *p, Harr.*

8140. 1; fuscatus*. *Meig. Zw.* i. 321. *pl.* 11. *f.* 18. ♀.—(*Curtis l. c. infra.*)
Sc. fuscata. *Fabr. A.* 58.
An. fuscus. *Meig. Kl.* i. 103.

8141. 2; punctatus*. *Meig. Zw.* i. 322.—(*Curtis l. c. infra.*)
Sc. punctata. *Fabr. A.* 59.
Musca bilineata. *Gmel.* v. 2866.
An. nebulosus ♂.—*Meig. Kl.* i. 103. *pl.* 6. *f.* 4.
Sy. Cælebs. *Harr. Ex.* 102. *pl.* xxxi. *f.* 8?

8142. 3; fenestralis*. *Meig. Zw.* i. 323.—*Curtis.* iii. *pl.* 102.
Tip. fenestralis. *Scop. C.* 322.
Sc. cincta. *Fabr. A.* 60.
Musca succincta. *Gmel.* v. 2866.
An. nebulosus ♀. *Meig. Kl.* i. 103.
Sy. brevis. *Harr. Ex.* 104. *pl.* xxxi. § 3.

8143. 4; variegatus* *mihi.*

Familia III: (137). ASILIDÆ, *Leach.*

(ASILUS, *Linné, &c.*—ASILICI, *Latr.*)

Genus 50: (1174). ASILUS *Auctorum.*

DASYPOGON *p, Fabr.*—ERAX *p, Scopoli.*

8144. 1; forcipatus*. *Linn.* ii. 1008.—*Berk. S.* i. 167.—*Stew.* ii. 273.—*Turt.* iii. 676.—(*Curtis l. c. infra.*)
As. cinereus. *DeGeer.* vi. *pl.* 14. *f.* 5–9.
As. Tipuloides. *Harr. Ex.* 64. *pl.* xvii. *f.* 3.

8145. 2, obscurus*. *Meig. Zw.* ii. 315.

8146. 3; albiceps*. *Meig. Zw.* ii. 312?
As. albipes. (*Curtis l. c. infra.*)

8147. 4, cristatus*. *Hoffm.*—*Meig. Zw.* ii. 322?
As. delecta. *Harr. Ex.* 64. *pl.* xvii. *f.* 4 & 5.

8148. 5; trigonus*. *Meig. Zw.* ii. 322.

8149. 6, cingulatus*. *Fabr. A.* 172.

8150. 7; varipes*. *Meig. Zw.* ii. 328?
As. maculosus. *Harr. Ex.* 64. *pl.* xvii. *f.* 6.

8151. 8; calceatus*. *Meg.*—*Meig. Zw.* ii. 316?

8152. 9; atricapillus*. *Fall. D. S.* (*Asili.*) 105.

8153. 10; opacus*. *Gürtler.*—*Meig. Zw.* ii. 315.—(*Curtis l. c. infra.*)

8154. 11; æstivus*. *Schra. A. No.* 996.—(*Curtis l. c. infra.*)
As. niger. *DeGeer.* vi. *pl.* 14. *f.* 12.
As. tibialis. *Fall. D. S.* (*Asil.*) 9.
As. leucopogon. *Mus. Brit.*

8155. 12, cognatus* *mihi.*

8156. 13, germanicus. *Linn.* ii. 1008.—*Curtis.* i. *pl.* 46.
♀, As. tibialis. *Fabr. E. S.* iv. 383.

8157. 14; crabroniformis*: Hornet Fly.—*Berk. l. c. Linn.* ii. 1007.—*Harr. Ex.* 65. *pl.* xvii. *f.* 1, 2.—*Berk. S.* i. 167.—*Don.* v. 109. *pl.* 180.—*Stew.* ii. 272.—*Turt.* iii. 674.—(*Kirby & Sp. I. E.* ii. 361.)—(*Samou.* 72. *pl.* 9. *f.* 9.)—(*Curtis l. c. supra.*)

Genus 51: (1175). DASYPOGON, *Meig.*, *Leach*, *Samou.*, *Curtis.*

ERAX *p*, *Scop.*

8158. 1, punctatus. *Fabr. A.* 165.—*Panz. F.* xlv. *f.* 24.—(*Sam. I.* 14.)—(*Curtis folio* 153.)
♂, Da. diadema. *Fabr. A.* 164.—*Panz. F.* xlv. *f.* 23.

Genus 52: (1176). LEPTARTHRUS *mihi.*

DASYPOGON *p*, *Meig.*, *Curtis.*

8159. 1; brevirostris*.
Da. brevirostris. *Meig. Zw.* ii. 273.—*Curtis.* iv. *pl.* 153.
♂. Da. longitarsis. *Fallen D. S.* (*Asil.*) 13.
♀. Da. armillatus. *Fallen D. S.* (*Asil.*) 12.

Genus 53: (1177). LAPHRIA, *Fabr.*, *Leach*, *Samou.*, *Curtis.*

ERAX *p*, *Scopoli.*

8160. 1; nigra*. *Meig. Zw.* ii. 293.—*Ahrens. F.* ii. *f.* 24.—*Curtis.* ii. *pl.* 94.

8161 † 2. gilva[a].

8162 † 3. gibbosa[b].

8163 † 4. atra[c].

8164 † 5. flava[d].

---

[a] 8161 † 2. gilva.
As. gilvus. *Linn.* ii. 1007.—*Stew.* ii. 272. (!)—*Turt.* iii. 676. (!)
As. rufus. *DeGeer.* vi. 97. *pl.* 13. *f.* 15.

[b] 8162 † 3. gibbosa.
As. gibbosus. *Linn.* ii. 1007.—*Stew.* ii. 272. (!)—*Turt.* iii. 674. (!)
As. bombylius. *DeGeer.* vi. 96. *pl.* 13. *f.* 6.

[c] 8163 † 4. atra. *Linn.* ii. 1007.—*Stew.* ii. 272. (!)—*Turt.* iii. 675. (!)
As. atra. *Meig. Zw.* ii. 302. *pl.* 20. *f.* 24.
La. violacea. *Meig. Kl.* i. 262.

[d] 8164 † 5. flava. *Meig. Zw.* ii. 288.
As. flavus. *Linn.* ii. 1007.—*Stew.* ii. 272. (!)—*Panz. F.* xxxix. *f.* 23, 24.—*Turt.* iii. 675. (!)

8165 † 6. marginata[a].

Genus 54: (1178). ——

DIOCTRIA *p, Meigen.*

8166. 1; atricapilla*.
Di. atricapilla. *Fallen.* D. S. (*Asil.*) 7.

8167. 2; semihyalina*.
Di. semihyalina. *Hoffmansegg.—Meig. Zw.* ii. 254.

Genus 55: (1179). DIOCTRIA, *Meigen, Leach, Samou.*, (*Kirby.*)
SYLVICOLÆ II., *Harris.*—ERAX *p, Scop.*

8168. 1; œlandica*. *Fabr. A.* 149.—(*Sam. I.* 15.)
As. œlandicus. *Linn.* ii. 1008.—*Stew.* ii. 273.—*Herbst. Nat.* viii. 119. *pl.* 346. *f.* 5.—*Turt.* iii. 679.
As. Morio. *Berk. S.* i. 167?
Sy. lugubris. *Harr. Ex.* 159. *pl.* xlviii. *f.* 1.

8169. 2; rufipes*. *Meig. Zw.* ii. 242.
As. rufipes. *DeGeer.* vi. 97. *pl.* 17.
Di. frontalis. *Meig. Kl.* i. 257.
Di. flavipes var. 1. *Fallen.* D. S. (*Asil.*) 7.
Sy. Cursor. *Harr. Ex.* 159. *pl.* xlviii. *f.* 2.

8170. 3; varipes*. *Meig. Zw.* ii. 245?

8171. 4; lateralis*. *Meig. Zw.* ii. 249?
Sy. informis. *Harr. Ex.* 159. *pl.* xlviii. *f.* 3?

8172. 5; annulata*. *Meig. Zw.* ii. 251?

8173. 6, podagricus. *Mus. Marsham.*
Di. flavipes. *Fall.* D. S. (*Asili.*) 7. 2?

8174. 7; nigripes*. *Meig. Zw.* ii. 246.
Di. fuscipennis. *Fall.* D. S. (*Asili.*) 7. 3.

Genus 56: (1180). GONIPES, *Latreille, Leach, Samou.*
LEPTOGASTER, *Meigen.*—DASYPOGON *p, Fabr.*

8175. 1; cylindricus*. *Latr. G.* iv. 301.
As. cylindrica. *DeGeer.* vi. 99. *pl.* 14. *f.* 13.
As. tipuloides. *Fabr. E. S.* iv. 172.—*Berk. S.* i. 167?—*Stew.* ii. 273.—*Turt.* iii. 678.
Go. tipuloides. (*Sam. I.* 19.)

8176. 2, fuscus.
Sc. fuscus. *Meig. Zw.* ii. 344.

---

[a] 8165 † 6. marginata. *Meig. Zw.* ii. 291.
As. marginatus. *Linn.* ii. 1008.—*DeGeer.* vi. 97. *pl.* 14. *f.* 1.—*Stew.* ii. 273. (!)

B. HYBOTINÆ, *Meigen*.—EMPIS *p*, *Fabr*.—EMPIDES *p*, *Latr*.

Genus 57: (1181). HYBOS, *Fabr*.

DASYPOGON *p*, *Fabr*.—ACROMYIA, *Bonelli*.

8177. 1; funebris*. *Fabr. A.* 145.
As. culiciformis. *Fabr.* iv. 389.—*Stew.* ii. 273.—*Turt.* iii. 680.
Em. clavipes. *Fabr.* iv. 403.

8178. 2; vitripennis*. *Meig. Zw.* ii. 348.

8179. 3; pilipes*. *Meig. Zw.* ii. 349.

8180. 4; fumipennis*. *Meig. Zw.* ii. 349.

8181. 5, Leachianus *mihi*.

8182. 6; flavipes. *Fabr. A.* 145.—*Meig. Zw.* ii. 348. *pl.* 21. *f.* 20.
As. clavipes. *Mus. Marsham.*

8183. 7; Marshamanus *mihi*.
As. culiciformis. *Mus. Marsham.*

8184. 8; brunnipes* *mihi*.

8185. 9; nervosus* *mihi*.

Genus 58: (1182). OCYDROMIA, *Hoffmansegg*.

8186. 1; scutellata*. *Meig. Zw.* ii. 354.

8187. 2; rufipes*. *Meig. Zw.* ii. 353.

8188. 3; flavipes*. *Megerle.*—*Meig. Zw.* ii. 353.

8189. 4, ruficollis. *Meig. Zw.* ii. 353. *pl.* 21. *f.* 24.
Em. flavicollis. *Mus. Marsham.*

8190. 5, glabricula. *Fallen.* D. S. (*Empid.*) 33.—*Meig. Zw.* ii. 352. *pl.* 21. *f.* 23.

8191. 6; dorsalis* *mihi*.

Genus 59: (1183). OEDALEA, *Meigen*.

8192. 1; minuta*. *Meig. Zw.* ii. 356.
Em. minuta. *Fall. D. S.* (*Empid.*) 32.

8193. 2; hybotina*. *Meig. Zw.* ii. 356. *pl.* 21. *f.* 27.
Em. hybotina. *Fall. D. S.* (*Empid.*) 31.

Familia IV: (138). EMPIDÆ, *Leach*.

(EMPIS, *Linné*, &c.—EMPIDES, *Latr.*)

A. EMPIDIÆ, *Meig.*

Genus 60: (1184). BRACHYSTOMA, *Meigen*.

BACCHA *p*, *Fabr*.

8194. 1; vesiculosa*. *Meig. Zw.* iii. 13. *pl.* 22. *f.* 8, 9.
Ba. vesiculosa. *Fabr. A.* 200.

8195. 2; longicornis*. *Meig. Zw.* iii. 12. *pl.* 22. *f.* 6, 7.

Genus 61: (1185). HILARA, *Meig.*, *Curtis*, (*Kirby.*)

TACHYDROMIA *p*, *Fabr.*—BIBIO *p*, *Panzer.*

A. Vide *Meig. Zw.* iii. 3.

8196. 1, cilipes*. *Meig. Zw.* iii. 3. *pl.* 22. *f.* 3.
Em. clavipes. *Harr. Ex.* 150. *pl.* xliv. *f.* 3. ♂.

8197. 2; globulipes*. *Hoffm.*—*Meig. Zw.* iii. 3.—(*Curtis l. c. supra.*)
Em. maura. *Fabr. A.* 139.—*Meig. Kl.* i. 222. *pl.* 2. *f.* 28.—(*Kirby & Sp. I. E.* ii. 7.)
Bibio senilis. *Panz. F.* liv. *f.* 3.

8198. 3; chorica*. *Meig. Zw.* iii. 4.—(*Curtis l. c. supra.*)
Em. chorica. *Fall. D. S.* (*Empid.*) 24.
Em. affinis. *Mus. Marsham.*

8199. 4; nigrina*. *Meig. Zw.* iii. 4.—(*Curtis l. c. supra.*)
Em. nigrina. *Fall. D. S.* (*Empid.*) 24.

8200. 5; clypeata*. *Meig. Zw.* iii. 4.—(*Curtis l. c. supra.*)

8201. 6; manicata*. *Meig. Zw.* iii. 5.—(*Curtis l. c. supra.*)
Em. Corynodis. *Mus. Marsham.*

8202. 7; modesta*. *Meig. Zw.* iii. 10.—(*Curtis l. c. supra.*)

8203. 8; interstincta*. *Meig. Zw.* iii. 6.—(*Curtis l. c. supra.*)
Em. interstincta. *Fall. D. S.* (*Empid.*) 24.

8204. 9; fuscipes*. *Meig. Zw.* iii. 6.—(*Curtis l. c. supra.*)
Ta. fuscipes. *Fabr. A.* 144.
Ta. plumbea. *Fabr. A.* 144.
Em. intermedia. *Fall. D. S.* (*Empid.*) 23.
Em. albida. *Meig. Kl.* i. 227.

8205. 10; quadrivittata*. *Wied.*—*Meig. Zw.* iii. 7.—(*Curtis l. c. supra.*)

8206. 11; pruinosa*. *Megerle.*—*Meig. Zw.* iii. 7.—(*Curtis l. c. supra.*)

8207. 12; univittata*. *Meig. Zw.* iii. 9?

8208. 13; lugubris*. *Meig. Zw.* iii. 10.

8209. 14, litorea. *Meig. Zw.* iii. 8.—(*Curtis l. c. supra.*)
Em. litorea. *Fall. D. S.* (*Empid.*) 24.

8210. 15; lurida*. *Meig. Zw.* iii. 8.—(*Curtis l. c. supra.*)
Em. lurida. *Fall. D. S.* (*Empid.*) 22.

8211. 16; tenella*. *Meig. Zw.* iii. 9.
Em. tenella. *Fall. D. S.* (*Empid.*) 25.

B. Vide *Meig. Zw.* iii. 11.

8212. 17; flavipes*. *Meig. Zw.* iii. 11.—(*Curtis l. c. supra.*)
Em. acephala. *Panz. F.* liv. *f.* 24.

8213. 18, obscura. *Meig. Zw.* iii. 11.—(*Curtis l. c. supra.*)
C. Vide *Meig. Zw.* iii. 11.

8214. 19, fasciata. *Meig. Zw.* iii. 11.—(*Curtis l. c. supra.*)

Genus 62: (1186). GLOMA, *Meigen.*

8215. 1, fuscipennis. *Meig. Zw.* iii. 14. *pl.* 22. *f.* 10–12.

Genus 63: (1187). ——

8216. 1, bipunctata.
Em. bipunctata. *Mus. Marsham.*

8217. 2, Evanida.
Em. Evanida. *Mus. Marsham.*

Genus 64: (1188). PACHYMERIA *mihi.*
EMPIS B. *p*, *Meigen.*

8218. 1; ruralis*.
Em. ruralis. *Meig. Zw.* iii. 40.
Em. confidens. *Harr. Ex.* 151. *pl.* xliv. *f.* 7.

8219. 2; aprica* *mihi.*

Genus 65: (1189). EMPIS *Auctorum.*

8220. 1; pennipes*. *Linn.* ii. 1003.—*Panz. F.* lxxiv. *f.* 18.—*Stew.* ii. 269.—*Turt.* iii. 665.—(*Sam. I.* 16.)—*Wood.* ii. 100. *pl.* 68.
Em. ciliata. *Fall. D. S.* (*Empid.*) 20.
Em. pennata. *Schrank A. No.* 987.
Em. longirostris. *Meig. Kl.* ii. 223.

8221. 2; umbrina*. *Hoffm.*—*Meig. Zw.* iii. 41.

8222. 3; decora*. *Meig. Zw.* iii. 24. *pl.* 22. *f.* 18.

8223. 4; pennaria*. *Fallen. D. S.* (*Empid.*) 20.

8224. 5; vitripennis*. *Meig. Zw.* iii. 25.

8225. 6; lepidopus*. *Meig. Zw.* iii. 23.

8226. 7; leucoptera*. *Meig. Zw.* iii. 27.

8227. 8; vernalis*. *Meig. Zw.* iii. 27.

8228. 9; chioptera*. *Fallen. D. S.* (*Empid.*) 21.—*Meig. Zw.* iii. 27. *pl.* 22. *f.* 19.
Em. crassipes. *Schrank A. No.* 988.

8229. 10; Bistortæ*. *Meig. Zw.* iii. 29.

8230. 11; pilipes*. *Meig. Zw.* iii. 31.

8231. 12; punctata*. *Fabr. A.* 142.

8232. 13; testacea*. *Fabr. A.* 141.

8233. 14; stercorea*. *Linn.* ii. 1004.—*Stew.* ii. 270.—*Turt.* iii. 667.
Em. pertinax. *Harr. Ex.* 150. *pl.* xliv. *f.* 5.

8234. [15; forcipata*.] *Mus. Marsham.*

8235. 16; lutea*. *Meig. Zw.* iii. 37.
Em. certus (oertus). *Harr. Ex.* 150. *pl.* xliv. *f.* 6.

8236. 17; tessellata*. *Fabr. A.* 140.
Em. livida. *Fabr. A.* 139.—*Turt.* iii. 666.
Em. sordida. *Mus. Marsham.*

8237. 18, affinis*? *mihi.*

8238. 19; livida*. *Linn.* ii. 1003.—*DeGeer.* vi. 101. *pl.* 14. *f.* 14.
—*Harr. Ex.* 149. *pl.* xliv. *f.* 1.—*Stew.* ii. 270.
Em. lineata. *Fabr. A.* 141.
♀, Em. fugeo. *Harr. Ex.* 150. *pl.* xliv. *f.* 4.
Var.? Em. melanopa. *Mus. Marsham.*
Em. constans. *Harr. Ex.* 150. *pl.* xliv. *f.* 2?

8239. 20, cognata *mihi.*

Genus 66: (1190). PLATYPTERYGIA *mihi.*

EMPIS *p, Linné, Curtis.*—PLATYPTERA, *Meig.* (*olim.*)

8240. 1, borealis.
Em. borealis. *Linn.* ii. 1003.—*Turt.* iii. 665.—(*Sam. I.* 16.)
—*Curtis.* i. *pl.* 18.

Genus 67: (1191). RHAMPHOMYIA, *Hoffmansegg.*

PLATYPTERA *p, Meig.?*

8241 † 1. cinerea[a].

8242. 2; sulcata*. *Meig. Zw.* iii. 46.
Em. fixus: *Harr. Ex.* 151. *pl.* xliv. *f.* 8.

8243. 3; plumipes*. *Meig. Zw.* iii. 47.

8244. 4; nigripes*. *Meig. Zw.* iii. 48.
Em. nigripes. *Fabr. A.* 141.
Em. crassirostris. *Fall. D. S.* (*Empid.*) 31.
Em. vicanus. *Harr. Ex.* 151. *pl.* xliv. *f.* 9.

8245. 5; atra*. *Meig. Zw.* iii. 45.

8246. 6, rugicollis. *Meig. Zw.* iii. 46.

8247. 7, variabilis. *Meig. Zw.* iii. 51.
Em. variabilis. *Fall.* D. S. (*Empid.*) 29.

8248. 8; culicina*. *Meig. Zw.* iii. 52.

---

[a] 8241 † 1. cinerea. *Meig. Zw.* iii. 43.
Em. cinerea. *Fabr. A.* 141.—*Turt.* iii. 666. (!)
Em. tipularia. *Fall. D. S.* (*Empid.*) 27.

Em. culicina. *Fall. D. S.* (*Empid.*) 28.
Em. avidus. *Harr. Ex.* 151. *pl.* xliv. *f.* 10.

8249. 9, tenuirostris. *Meig. Zw.* iii. 52.
Em. tenuirostris. *Fall. D. S.* (*Empid.*) 29.

8250. 10, flava. *Meig. Zw.* iii. 59.
Em. flava. *Fall. D. S.* (*Empid.*) 30.

8251. 11; flavipes* *mihi.*

8252. 12; longipes*. *Meig. Zw.* iii. 55. *pl.* 22. *f.* 3. mas.

8253. 13; obscura*.

8254. 14; albipennis*. *Meig. Zw.* iii. 59.
Em. albipennis. *Fall. D. S.* (*Empid.*) 30.

8255. 15; umbripennis*. *Meig. Zw.* iii. 54.
Em. umbripennis. *Fall. D. S.* (*Empid.*) 30. 34.

8256. 16; gibba*. *Meig. Zw.* iii. 58.
Em. gibba. *Fall. D. S.* (*Empid.*) 32. 41.

8257. 17; holosericea*. *Meig. Kl.* i. 231.

Genus 68: (1192). ENICOPTERYX *mihi.*

RHAMPHOMYIA *p, Meig.*

8258. 1; fusca*.
Em. fusca. *Mus. Marsham.*

8259. 2; infuscata*.
Rh. infuscata. *Meig. Zw.* iii. 53. *pl.* 23. *f.* 4. ala.

8260. 3; hyalipennis* *mihi.*
Rh. anomalipennis. *Meig. Zw.* iii. 55.

B. TACHYDROMIÆ, *Meig.*—SICUS, *Latr.*, (*Kirby.*)

Genus 69: (1193). HEMERODROMIA, *Hoffmansegg.*

8261. 1; monostigma*. *Hoffm.*—*Meig. Zw.* iii. 62. *pl.* 23. *f.* 6.

8262. 2; Meigeniana* mihi.

8263. 3, Mantispa. *Meig. Zw.* iii. 64. *pl.* 23. *f.* 9. ala.
Ta. Mantispa. *Panz. F.* ciii. *f.* 16.
Em. melanocephala. *Fabr.* iv. 407.
Si. raptor. *Latr. G.* iv. 104. *pl.* 16. *f.* 11, 12.

Genus 70: (1194). TACHYDROMIA, *Meig.*

MUSCA *p, Linn.*—CALOBATA *p, Fabr.*—PLATYPALPUS, *Marc.*

8264. 1; arrogans*. *Meig. Zw.* iii. 68.
Mu. arrogans. *Linn.* ii. 995.
Ta. cimicoides. *Fabr. A.* 144.

8265. 2; annulimana*. *Meig. Zw.* iii. 69.

8266. 3; connexa*. *Meig. Zw.* iii. 70.
Ta. cimicoides. *Meig. Kl.* i. 239.

8267. 4; fuscipennis. *Fall. D. S.* (*Empid.*) 14. 19.

8268. 5; nervosa*. *Meig. Zw.* iii. 72.

8269. 6; venosa* *mihi.*

8270. 7; annulipes*. *Meig. Zw.* iii. 77.

8271. 8; annulata*. *Fallen D. S.* (*Empid.*) 7.—*Meig. Zw.* iii. 77.

8272. 9; fascipes*. *Meig. Zw.* iii. 78.

8273. 10; flavipes*. *Fabr. A.* 142.
Ta. vulgaris. *Meig. Kl.* i. 237.
Si. flavipes. (*Kirby & Sp. I. E.* iii. 669.)

8274. 11; bicolor*. *Fabr. A.* 143.

8275. 12; pallidiventris*. *Meig. Zw.* iii. 82.
Ta. flavipes. *Fall. D. S.* (*Empid.*) 6.

8276. 13; candicans. *Fall. D. S.* (*Empid.*) 10.

8277. 14, grossipes.
Em. grossipes. *Mus. Marsham.*

8278. 15; dissimilis*. *Fall. D. S.* (*Empid.*) 9.

8279. 16; fasciata*. *Meig. Zw.* iii. 86. *pl.* 23. *f.* 22.

8280. 17, flavipalpis*. *Meig. Zw.* iii. 74.

8281. 18; fulvipes*. *Megerle.*—*Meig. Zw.* iii. 78.

8282. 19; ventralis*. *Megerle.*—*Meig. Zw.* iii. 85.

8283. 20, calceata. *Meig. Zw.* iii. 87.

8284. 21; pallipes*. *Fall. D. S.* (*Empid.*) 8.

8285. 22; nigritarsis*. *Fall. D. S.* (*Empid.*) 34.

8286. 23; albiseta*. *Panz. F.* ciii. *f.* 17.
Ta. assimilis. *Fall. D. S.* (*Empid.*) 8.

8287. 24; nigra*. *Meig. Zw.* iii. 75.

8288. 25; nigrina*. *Meig. Zw.* iii. 76.

8289. 26; minuta*. *Meig. Zw.* iii. 76.

8290. 27; exigua*. *Meig. Zw.* iii. 81.

8291. 28; castanipes*. *Meig. Zw.* iii. 79.

8292. 29; agilis*. *Meig. Zw.* iii. 80.

8293. 30; tibialis* *mihi.*

8294. 31; apicalis* *mihi.*

8295. 32; consorta* *mihi.*

8296. 33; lutea*. *Meig. Zw.* iii. 89.

8296*. 34; cursitans*. *Fabr. A.* 144. (!)
Mu. cursitans. *Stew.* ii. 261.—*Turt.* iii. 612.

Familia V: (139). DOLICHOPIDÆ, *Leach.*

DOLYCHOPODES, *Latr.*—DOLYCHOPUS, *Fabr.*

A. PLATYPEZINÆ, *Meig.*

Genus 71: (1195). PLATYPEZA, *Meigen.*

8297. 1; boletina*. *Fallen D. S.* (*Platypt.*) 4. 1.

8298. 2; atra*. *Fall. D. S.* (*Platypt.*) 6. 3.

Genus 72: (1196). CALLOMYIA, *Meigen.*

8299. 1; aterrima* *mihi.*
Cal. antennata. *Fall.*—*Meig. Zw.* iv. 15?

B. MEGACEPHALI, *Meig.*—SYRPHIÆ *p*, *Latr.*

Genus 73: (1197). PIPUNCULUS, *Latr.*, *Leach*, *Samou.*
CEPHALOPS, *Fallen.*

8300. 1; campestris*. *Latr. G.* iv. 332.—(*Leach E. E.* x. 130.) —(*Sam. I.* 34.)

8301. 2; geniculatus*. *Meig. Zw.* iv. 20.

8302. 3; sylvaticus*. *Meig. Zw.* iv. 20.

8303. 4; Dubrensis *mihi.*
Pi. flavipes. *Meig. Zw.* iv. 21?

8304. 5, ruralis*. *Meig. Zw.* iv. 22.

C. DOLICHOPODES, *Meig.*

Genus 74: (1198). RHAPHIUM, *Meigen.*

8305. 1; longicorne*. *Meig. Zw.* iv. 28.

8306 † 2, macrocerum*. *Wied.*—*Meig. Zw.* iv. 29.
In Mus. D. *Westwood.*

Genus 75: (1199). DIAPHORUS, *Meigen.*

8307. 1; flavocinctus*. *Hoffmansegg.*—*Meig. Zw.* iv. 33. *pl.* 34. *f.* 8.

8308. 2; Winthemi*. *Meig. Zw.* iv. 34.

Genus 76: (1200). CHRYSOTUS, *Meigen.*

8309. 1; neglectus*. *Meig. Zw.* iv. 41. *pl.* 35. *f.* 10.
Do. neglectus. *Wied. Z. M.* i. 74.

8310. 2; nigripes*. *Meig. Zw.* iv. 42.
Do. nigripes. *Fabr. A.* 269.

8311. 3; læsus*. *Meig. Zw.* iv. 43.
Do. læsus. *Wied. Z. M.* i. 75.

8312. 4; femoralis*. *Meg.*—*Meig. Zw.* iv. 42.

8313. 5; cilipes*. *Meig. Zw.* iv. 41.

8314. 6; tibialis* *mihi.*

8315. 7; copiosus*. *Meig. Zw.* iv. 41.

8316. 8; lætus*. *Wied.—Meig. Zw.* iv. 43.

Genus 77: (1201). PORPHYROPS, *Meigen.*

8317. 1; diaphanus*. *Meig. Zw.* iv. 46. *pl.* 35. *f.* 5.
Do. diaphanus. *Fabr. A.* 270.
Mu. ludeus. *Harr. Ex.* 157. *pl.* xlvii. *f.* 3?

8318. 2; splendidus*.
Po. auricollis. *Meig. Zw.* iv. 47?
Mu. semiargentata. *Turt.* iii. 606?—*Stew.* ii. 261?

8319. 3, argyrius*. *Meig. Zw.* iv. 46.

8320. 4, argentinus. *Meig. Zw.* iv. 47.

8321. 5; leucocephalus*. *Meig. Zw.* iv. 49.

8322. 6, thoracicus*. *Meig. Zw.* iv. 53.

8323. 7, consobrinus* *mihi.*

8324. 8; aulicus*. *Meig. Zw.* iv. 48.

8325. 9; quadrifasciatus*. *Meig. Zw.* iv. 48.
Do. quadrifasciatus. *Fabr. A.* 269.

8326. 10, tarsalis* *mihi.*

Genus 78: (1202). PSILOPUS, *Meigen.*

8327. 1; platypterus*. *Meig. Zw.* iv. 36.
Do. platypterus. *Fabr. A.* 270.

8328. 2; consobrinus* *mihi.*

8329. 3; contristans*. *Meig. Zw.* iv. 37.
Do. contristans. *Wied. Z. M.* i. 72.

8330. 4; lobipes*. *Meig. Zw.* iv. 38.

8331. 5; lugens*. *Meig. Zw.* iv. 38.

8332. 6; obscurus*. *Meig. Zw.* iv. 39. (!)

8333. 7; nervosus*. *Meig. Zw.* iv. 36.
Do. nervosus. *Lehmann* D. 40.

8334. 8; regalis*. *Meig. Zw.* iv. 35.

Genus 79: (1203). SYBISTROMA, *Meigen.*

8335. 1; discipes*. *Meig. Zw.* iv. 71.
Do. discipes. *Ahr. F.* iv. *f.* 24.
Do. gracilipes. *Leach? MSS.*

8336. 2, patellipes. *Meig. Zw.* iv. 72. (!)
Do. frontalis. *Leach? MSS.*

Genus 80: (1204). MEDETERUS, *Fischer, Curtis.*
HYDROPHORUS, *Fallen.*

8337. 1, regius*. *Meig. Zw.* iv. 60.—(*Curtis l. c. infra.*

Do. regius. *Fabr. A.* 267.
Mu. virens. *Panz. F.* liv. *f.* 16?

8338. 2; viridis*. *Meig. Zw.* iv. 60.—(*Curtis l. c. infra.*)

8339. 3; rostratus*. *Meig. Zw.* iv. 61.
Do. rostratus. *Fabr. A.* 269.

8340. 4, notatus*. *Meig. Zw.* iv. 62.—*Curtis.* iv. *pl.* 162.
Do. notatus. *Fabr. A.* 269. (!)
Mu. notata. *Stew.* ii. 261.—*Turt.* iii. 613.

8341. 5, abbreviatus.
Mu. abbreviatus. *Mus. Marsham.*

8342. 6, xanthopterus *mihi.*

8343. 7; flavipes*. *Meig. Zw.* iv. 61.

8344. 8; præcox*. *Meig. Zw.* iv. 64.

8345. 9; prodromus*. *Meig. Zw.* iv. 64.

8346. 10; truncorum*. *Winthem.*—*Meig. Zw.* iv. 67.

8347. 11; nebulosus*. *Meig. Zw.* iv. 68.
Hy. nebulosus. *Fall.* D. S. (*Dolich.*) 3, 4.

8348. 12; tenellus*. *Wied. Z. M.* i. 73.—*Meig. Zw.* iv. 69.

Genus 81: (1205). DOLICHOPUS, *Latr.*, *Leach*, *Samou.*, (*Kirby.*)
RHAGIO *p*, *Schrank.*—SATYRA, *Meig.* (*olim.*)

8349. 1; nobilitatus*. *Fabr. A.* 268.—(*Sam. I.* 15.)
Mu. nobilitatus. *Linn.* ii. 995.—*Turt.* iii. 613.
Mu. joco. *Harr. Ex.* 157. *pl.* xlvii. *f.* 1.

8350. 2; plumicornis*. *Meig. Zw.* iv. 83. (!)

8351. 3; atratus*. *Hoffm.*—*Meig. Zw.* iv. 76.

8352. 4; picipes*. *Winthem.*—*Meig. Zw.* iv. 76.

8353. 5; nigricornis*. *Megerle.*—*Meig. Zw.* iv. 82.

8354. 6; chalybeus*. *Wied. Z. M.* i. 72.—*Meig. Zw.* iv. 79.

8355. 7; metallicus* *mihi.*

8356. 8; æneovirens* *mihi.*

8357. 9; nigroæneus* *mihi.*

8358. 10; tarsalis* *mihi.*

8359. 11; angulatus*.
Do. nitidus. *Meig. Zw.* iv. 80. (!)

8360. 12; nitidus*. *Fall. D. S.* (*Dolich.*) 12. 9.

8361. 13; subflexuosus* *mihi.*

8362. 14; ungulatus*. *Fabr. A.* 266.—*Meig. Zw.* iv. 80.
Mu. ungulata. *Linn.* ii. 995.—*Panz. F.* xx. *f.* 21.—*Stew.* ii. 261.—*Turt.* iii. 613.

Nemotelus æneus. *DeGeer.* vi. 78. *pl.* xi. *f.* 14–22.
Mu. ludicrus. *Harr. Ex.* 157. *pl.* xlvii. *f.* 2?

8363. 15, nigritarsis *.
Do. ruralis. *Meig. Zw.* iv. 94?

8364. 16; simplex *. *Meig. Zw.* iv. 85.

8365. 17, pennitarsis. *Fall. D. S.* (*Dolich.*) ii. 6.—*Meig. Zw.* iv. 91.

8366. 18, plumitarsis. *Fall. D. S.* (*Dolich.*) 10. 4.—*Meig. Zw.* iv. 89.

8367. 19, popularis. *Hoffm.*—*Fall. D. S.* (*Dolich.*) 11. 9.—*Meig. Zw.* iv. 91.

8368. 20; acuticornis *. *Wied. Z. M.* i. 74.—*Meig. Zw.* iv. 94.

8369. 21; plebeius *. *Meig. Zw.* iv. 99. (!)

8370. 22; xanthogaster *. *Meig. Zw.* iv. 99. (!)

8371. 23; viridis *. *Wied.*—*Meig. Zw.* iv. 100.

8372. 24; gratiosus *. *Meig. Zw.* iv. 100?

8373. 25; relictus *. *Meig. Zw.* iv. 77.

8374. 26; æratus * *mihi.*

8375. 27; nigripennis *. *Fall. D. S.* (*Dolich.*) 15, 16.—*Meig. Zw.* iv. 102. *pl.* 35. *f.* 20. caput. An hujus generis?

Familia VI: (140). RHAGIONIDÆ, *Leach.*

RHAGIONIDES, *Latr.*—LEPTIDES, *Meig.*

Genus 82: (1206). LEPTIS, *Fabr.*

ATHERIX *p*, *Fabr.*—SYLVICOLÆ *p*, *Harr.*

8376. 1; aurata *. *Meig. Zw.* ii. 99.
Rh. atratus. *Fabr. E. S.* iv. 276.
Rh. tomentosus. *Fabr.* iv. 275.
Mu. cingulata. *Don.* xiii. 67. *pl.* 465.
♂, Sy. secretus. *Harr. Ex.* 101. *pl.* xxxi. *f.* 6.
♀, Sy. solitaneous. *Harr. Ex.* 102. *pl.* xxxi. *f.* 10.

8377. 2, helveola *. *Meig. Zw.* ii. 100.

8378. 3, flaveola *. *Meig. Zw.* ii. 100.
Anthrax genius. *Fanz. F.* liv. *f.* 4.

8379. 4, Diadema *. *Meig. Zw.* ii. 101.
Rh. aureus. *Meig. Kl.* i. 302.

8380. 5, affinis * *mihi.*

Genus 83: (1207). RHAGIO, *Olivier, Leach, Samou.*

LEPTIS, *Fabr., Meig. Zw.*—NEMOTELUS *p*, *DeGeer.*—ATHERIX *p*, *Fabr.*—DY.-SYLVICOLÆ, *Harr.*

8381. 1; lineola *. *Fabr.* iv. 275.

Rh. albifrons. *Meig. Kl.* i. 300.
Sy. Monachus. *Harr. Ex.* 102. *pl.* xxxi. *f.* 9.

8382. 2; strigosus. *Meig. Kl.* i. 299.
Sy. derelictus. *Harr. Ex.* 102. *pl.* xxxi. *f.* 7.

8383. 3; punctatus* *mihi.*

8384. 4; Tringarius*. *Fabr.* iv. 372.
Rh. vanellus. *Fabr.* iv. 272.
Musca tringaria. *Linn.* ii. 982.
Nem. scolopaceus var. *DeGeer.* vi. 69. *pl.* 9. *f.* 10.
Sy. solivagus. *Harr. Ex.* 101. *pl.* xxxi. *f.* 4.

8385. 5; immaculatus*. *Meig. Kl.* i. 301.
Sy. reconditus. *Harr. Ex.* 101. *pl.* xxxi. *f.* 3.

8386. 6; annulatus*.
Nem. annulatus. *DeGeer.* vi. 69.

8387. 7; Scolopaceus*. *Fabr.* iv. 271.—*Panz. F.* xiv. *f.* 19.—(*Sam. I.* 36.)
Musca Scolopacea. *Linn.* ii. 982.—*Stew.* ii. 263.—*Turt.* iii. 632.
Sy. solitarius. *Harr. Ex.* 100. *pl.* xxxi. *f.* 1, 2.
Var.? Sy. monotropus. *Harr. Ex.* 101. *pl.* xxxi. *f.* 5.

Genus 84: (1208). ATHERIX, *Meig.*, *Leach*, *Samou.*, *Curtis.*

ANTHRAX *p*, BIBIO *p*, *et* LEPTIS *p*, *Fabr.*

8388. 1, Ibis*. *Meig. Zw.* ii. 105.—*Curtis.* i. *pl.* 26.—(*Ing. Inst.* 91.)
♂, Lep. Ibis. *Fabr. A.* 70?
Sy. melancholia. *Harr. Ex.* 103. *pl.* xxxi. § 2. *f.* 1.
♀, Ant. Titanus. *Fabr. A.* 126.
At. maculatus. *Latr. G.* iv. 289.—(*Sam. I.* 5.)
Nodutis maculata. *Megerle.* (*teste Meig.*)

8389. 2, marginata*? *Meig. Zw.* ii. 106.—(*Curtis l. c. supra.*)
Musca marginata. *Gmel.* v. 2829.—*Stew.* ii. 255.
Musca Atherix. *Don.* xvi. 17. *pl.* 549.

8390. 3; melæna*. *Hoffm.*—*Meig. Zw.* ii. 109.

8391. 4, crassicornis. *Panz. F.* cv. *f.* 10.
Lep. griseola. *Fall. D. S.* (*Ath.*) 12. 7.

Familia VII: (141). MYDASIDÆ, *Leach.*

MYDASII, *Latreille.*—XYLOTOMÆ, *Meig.*

Genus 85: (1209). THEREVA, *Latr.*, *Leach*, *Samou.*

BIBIO *p*, *Fabr.*—NEMOTELUS *p*, *DeGeer.*—DI.-SYLVICOLÆ *p*, *Harr.*

8392. 1; plebeia*. *Meig. Zw.* ii. 117.—(*Leach E. E.* x. 130.)—(*Sam. I.* 41.)

Musca plebeia. *Linn.* ii. 979.—*Stew.* ii. 255.—*Turt.* iii. 635.
Bi. strigata. *Fabr. A.* 67.
Bi. rustica. *Fall. D. S.* (*Anthr.*) 4. 2.
Ne. fasciatus. *DeGeer.* vi. 76. *pl.* 11. *f.* 1.
Sy. Monos. *Harr. Ex.* 103. *pl.* xxxi. § 2. *f.* 3.

8393. 2; cincta*. *Meig. Zw.* ii. 117?

8394. 3; tæniata. *Meig. Zw.* ii. 120.

8395. 4, funebris. *Meig. Zw.* ii. 121.
Bi. lugubris. *Meig. Kl.* i. 214.

8396. 5, bipunctata. *Meig. Zw.* ii. 121.

8397. 6; fulva*. *Meig. Zw.* ii. 123.

8398. 7; flavilabris*. *Meig. Zw.* ii. 122.
Ly. Unicus. *Harr. Ex.* 103. *pl.* xxxi. § 2. *f.* 2.

8399. 8; anilis*. *Meig. Zw.* ii. 125.
Musca anilis. *Linn.* ii. 982.
Bi. flavipes. *Fabr. A.* 67.
Bi. anilis. *Panz. F.* v. *f.* 23.
Bi. sordida. *Panz. F.* xcviii. *f.* 19.

8400. 9, albipennis*. *Meig. Zw.* ii. 119?

8401. 10; annulata*. *Meig. Zw.* ii. 126.
Bi. anilis. *Fabr. A.* 68.
Bi. annulata. *Fabr. A.* 68.

Familia VIII: (142). TABANIDÆ, *Leach.*

TABANUS, *Linné, &c.*—TABANII, *Latreille.*

Genus 86: (1210). TABANUS *Auctorum. Horse-Fly, Breeze.*

8402. 1; micans*. *Meig. Zw.* ii. 34. *pl.* 13. *f.* 20. *caput.*—(*Curtis l. c. infra.*)
Ta. austriacus. *Fabr. A.* 96?
Ta. niger. *Don.* xvi. 47. *pl.* 564.

8403. 2, signatus? *Meig. Zw.* ii. 34.—(*Curtis l. c. infra.*)

8404. 3; autumnalis*. *Linn.* ii. 1000.—*Stew.* ii. 267.—*Turt.* iii. 657.—(*Sam. I.* 40.)—(*Curtis l. c. infra.*)
Ta. bovinus. *Harr. Ex.* 27. *pl.* vii. *f.* 1.

8405. 4; bovinus*. *Linn.* ii. 1000.—*Berk. S.* i. 166.—*Panz. F.* ii. *f.* 20.—*Stew.* ii. 267.—*Turt.* iii. 657?—(*Sam. I.* 40.)—*Wood.* ii. 92. *pl.* 66.—(*Curtis l. c. infra.*)
Ta. tropicus. *Harr. Ex.* 28. *pl.* vii. *f.* 2?

8406. 5, tarandinus. *Linn. F.* 1884.

8407. 6; paganus. *Fabr. A.* 99. (!)—*Stew.* ii. 267.—*Turt.* iii. 659.—(*Curtis l. c. infra.*)—(*Sam. I.* 40.)

8408. 7; bromius*. *Linn.* ii. 1001.
Ta. maculatus. *DeGeer.* vi. 89.
Ta. autumnalis. *Harr. Ex.* 28. *pl.* vii. *f.* 4.

8409. 8; montanus*. *Meig. Zw.* ii. 55.—(*Curtis l. c. infra.*)

8410. 9; luridus*. *Fallen.*—*Meig. Zw.* ii. 55.—(*Curtis l. c. infra.*)

8411. 10, solstitialis*. *Meig. Zw.* ii. 56.—(*Curtis l. c. infra.*)

8412. 11; tropicus*. *Linn.* ii. 1001.—*Panz. F.* xiii. *f.* 22.—*Stew.* ii. 267.—*Turt.* iii. 659.—(*Samou.* 71. *pl.* 9. *f.* 4.)—(*Curtis l. c. infra.*)

8413. 12; rusticus*. *Fabr. A.* 99.—*Panz. F.* xiii. *f.* 21.—*Turt.* iii. 659.—(*Curtis l. c. infra.*)

8414. 13, fulvus. *Meig. Zw.* ii. 61.
Ta. sanguisorba. *Harr. Ex.* 23. *pl.* vii. *f.* 3.
Ta. alpinus. *Schrank B.* iii. 2534.—*Curtis.* ii. *pl.* 78.

8415. 14, bimaculatus. ——?

Genus 87: (1211). HÆMATOPOTA, *Meig.*, *Leach*, *Samou.*, (*Kirby.*)

8416. 1, ocellata. *Megerle?*
Ta. pluvialis. *Harr. Ex.* 29. *pl.* vii. *f.* 8?

8417. 2; pluvialis*. *Meig. Zw.* ii. 78.—(*Sam. I.* 20.)
Tab. pluvialis. *Linn.* ii. 1001.—*Berk. S.* i. 166.—*Don.* v. 21. *pl.* 151. *f.* 3.—*Stew.* ii. 267.—*Turt.* iii. 660.
Ta. cæcutiens. *Harr. Ex. pl.* vii. *f.* 8.
♂; Tab. hyetomantis. *Schrank B.* iii. 2536.

8418. 3, italica. *Meig. Kl.* i. 163.
Ta. sanguisuga. *Harr. Ex.* 29. *pl.* vii. *f.* 6.

8419. 4; equorum*. *Meig. Zw.* i. 80.

Genus 88: (1212). CHRYSOPS, *Meig.*, *Leach*, *Samou.*

8420. 1; cæcutiens*. *Fabr. A.* 110.—(*Sam. I.* 11.)
Ta. cæcutiens. *Berk. S.* i. 166.—*Don.* iv. 61. *pl.* 131.—*Stew.* ii. 268.—*Turt.* iii. 662.
♂. Ch. lugubris. *Fabr. A.* 113.
Ch. viduatus. *Fabr. A.* 113.

8421. 2; relictus*. *Meig. Zw.* ii. 69.
Ch. viduatus. *Meig. Kl.* i. 158. *pl.* 9. *f.* 12.
Ta. cæcutiens. *Panz. F.* xiii. *f.* 24.
Ta. nubilosus. *Harr. Ex.* 28. *pl.* vii. *f.* 5.

8422. 3, consimilis *mihi.*

Familia IX: (143). BOMBYLIDÆ, *Leach.*

(Bombylius, *Linné.*—Bombyliarii, *Latreille.*)

Genus 89: (1213). BOMBYLIUS *Auctorum.* *Humble-bee Fly.*

8423. 1, posticus. *Fabr. A.* 131.
Bo. micans. *Meig. Kl.* i. 183.

8424. 2; major*. *Linn.* ii. 1009.—*Berk. S.* i. 167.—*Don.* ii. 79. *pl.* 66.—*Stew.* ii. 273.—*Turt.* iii. 681.—(*Sam. I.* 7.)
Bo. variegatus. *DeGeer.* vi. 107. *pl.* 15. *f.* 10.
Bo. sinuatus. *Mikan B.* 35. *pl.* 2. *f.* 4.
Bombylius. *Harr. Ex. pl.* xlvii. *f.* 2.

8425. 3; medius*. *Linn.* ii. 1009.—*Berk. S.* i. 168.—*Don.* v. 5. *pl.* 146. *f.* 1.—*Stew.* ii. 274.—*Turt.* iii. 681.—(*Sam. I.* 7.)
Bo. major. (*Samou.* 295. *pl.* 9. *f.* 10.)—*Wood.* ii. 106. *pl.* 71.
Bo. discolor. *Mikan B.* 27. *pl.* 2. *f.* 1.
Bombylius. *Harr. Ex. pl.* xlvii. *f.* 1.

8426. 4, pictus. *Panz. F.* xxiv. *f.* 24.
Bo. planicornis. *Fabr. A.* 129.

8427. 5, cinerascens. *Mikan B.* 50. *pl.* 3. *f.* 10.

8428. 6, minor*. *Linn.* ii. 1009.—*Berk. S.* i. 168.—*Stew.* ii. 274. —*Don.* xv. 65. *pl.* 536.—*Turt.* iii. 682.—(*Sam. I.* 7.)
Bo. venosus. *Mikan B.* 42.

8429. 7, ctenopterus. *Mikan B.* 45. *pl.* 3. *f.* 8.—(*Ing. Inst.* 91.)

Genus 90: (1214). PLOAS, *Latr.*

Conophorus, *Meig. olim.*

8430 † 1, virescens? *Fabr. A.* 136?—*Meig. Zw.* ii. 231. *pl.* 19. *f.* 6?
Co. maurus. *Meig. Kl.* i. 191. *pl.* 10. *f.* 17?
Pl. hirticornis. (*Latr.* G. iv. 312. *pl.* 15. *f.* 7?)
In Mus. *Brit.*, *et* D. *Curtis?*

Familia X: (144). ANTHRACIDÆ, *Leach.*

Anthrax, *Fabricius.*—Anthracii, *Latreille.*—Bombyliarii *p, Meigen.*

Genus 91: (1215). LOMATIA, *Meig.*

Stygia, *Meig. olim.*

8431 † 1. Lo. lateralis.
St. lateralis. *Meig. Zw.* ii. 140.
An. Belzebul. *Panz. F.* xlv. *f.* 16.

8432 ‡ 2, Lo. Belzebub.
An. Belzebub. *Fabr. A.* 124. In Mus. *Brit.*

Genus 92: (1216). ANTHRAX, *Fabricius, Leach, Samou.*, (*Kirby,*) *Curtis.*

Bibio *p, Fabr.*—Nemotelus *p, DeGeer.*

8433. 1, flava. *Hoffm.*—*Meig. Zw.* ii. 143.—(*Curtis l. c. infra.*)

An. hottentottus. *Latr.* G. iv. 310.—(*Sam.* I. 3.)
Mu. hottentotta. *Don.* xiv. 69. *pl.* 494.

8434. 2, circumdata. *Hoffm.*—*Meig. Zw.* ii. 143.—(*Curtis l. c. infra.*)
An. hottentotta. *Linn.* ii. 981.
Ne. hottentottus. *DeGeer.* vi. 77. *pl.* 11. *f.* 7.
An. fasciatus. *Meig. Kl.* i. 200.
Bo. Hottentottus. *Turt.* iii. 686.

8435. 3, cingulata. *Meig. Zw.* ii. 145. *pl.* 17. *f.* 9. ♂.

8436 † 4. Abaddon[a].

8437. 5, ornata. *Hoffm.*—*Curtis.* i. *pl.* 9.

8438. 6, Pandora. *Fabr. A.* 121.—*Meig. Zw.* ii. 170. *pl.* 17. *f.* 12. —(*Ing. Inst.* 91.)

8439 † 7. Maura[b].

8440 † 8. semiatra[c].

8441 † 9. sinuata[d].

Familia XI: (145). ACROCERIDÆ, *Leach.*

INFLATA, *Latreille.*—MUSCA *p*, *Linné.*

Genus 93: (1217). HENOPS, *Illiger, Curtis.*

OGCODES, *Latreille, Leach, Samou.*—SYRPHUS *p*, *Panzer.*—NEMOTELUS *p*, *Schæff.*

8442. 1; gibbosus*. *Fabr. A.* 333.—(*Curtis l. c. infra.*)
Mu. gibbosa. *Linn.* ii. 987.
He. leucomelas. *Meig. Kl.* i. 151. *pl.* 8. *f.* 30.
Og. gibbosus. (*Sam.* I. 32.)

8443. 2; marginatus*. *Meig. Zw.* iii. 100. *pl.* 24. *f.* 12.—(*Curtis.* iii. *pl.* 110.)
Sy. gibbosus. *Panz. F.* xliv. *f.* 21.

---

[a] 8936 † 4. Abaddon. *Fabr. E. S.* iv. 262.—(*Sam.* I. 3.) (!)

[b] 8439 † 7. Maura. *Fabr. E. S.* iv. 258.—*Panz. F.* xxxii. *f.* 19.
Mu. Maura. *Linn.* ii. 981.—*Stew.* ii. 256. (!)
Bo. Maurus. *Turt.* iii. 685. (!)
An. bifasciata. *Meig. Zw.* ii. 156. *pl.* 17. *f.* 15. ala.
An. dæmon. *Panz. F.* xlv. *f.* 16.

[c] 8440 † 8. semiatra. *Hoffm.*—*Meig. Zw.* ii. 157. *pl.* 17. *f.* 14. ala.
Mu. Morio. *Linn.* ii. 981.—*Berk. S.* i. 163. (!)—*Panz. F.* xxxii. *f.* 18.—*Stew.* ii. 256. (!)

[d] 8441 † 9. sinuata. *Fall. D. S.* (*Ant.*) 6. 1.—*Meig. Zw.* ii. 159. *pl.* 17. *f.* 18. ala.
An. Morio. *Fabr. E. S.* iv. 257.
As. Morio. *Stew.* ii. 257. (!)
Bo. Morio. *Turt.* iii. 685. (!)
Ne. Morio. *DeGeer.* vi. 78. *pl.* 11. *f.* 13.
Mu. Anthrax. *Schra. A. No.* 893.

Genus 94: (1218). ACROCERA, *Meigen, Leach, Samou.*
SYRPHUS *p, Panzer.*—NEMOTELUS *p, Schæff.*

8444. 1: globula*. *Meig. Zw.* iii. 95.
Sy. Globulus. *Panz. F.* lxxxvi. *f.* 20.
Ac. gibbosa. (*Sam. I.* 1.)

8445. 2; albipes*. *Meig. Zw.* iii. 96.
Ne. globulus var. *Fall. D. S.* (*Strat.*) 4.

Familia XII: (146). STRATIOMYDÆ, *Latreille.*

Genus 95: (1219). PACHYGASTER, *Meigen, Curtis.*
VAPPO, *Latr., Leach, Samou.*—NEMOTELUS *p, Panz.*—SARGUS *p, Fallen.*

8446. 1; ater*. *Meig. Zw.* iii. 102.—(*Curtis l. c. infra.*)
Ne. ater. *Panz. F.* liv. *f.* 5.
Va. ater. *Fabr. A.* 254.—(*Sam. I.* 43.)
Sa. pachygaster. *Fall. D. S.* (*Strat.*) 13

8447. 2; Leachii* *mihi.*—*Curtis.* i. *pl.* 42.

Genus 96: (1220). NEMOTELUS *Auctorum.*

8448. 1; uliginosus*. *Fabr.* iv. 269.—*Meig. Zw.* iii. 114. *pl.* 25. *f.* 19. fœm.—(*Sam. I.* 28.)
Mu. uliginosa. *Linn.* ii. 982.—*Turt.* iii. 655.—*Don.* xv. 31. *pl.* 519.
Str. mutica. *Fabr. S. I.* ii. 419.

8449. 2; Pantherinus*. *Meig. Zw.* iii. 115. *pl.* 25. *f.* 20. mas.
Mu. Pantherina. *Linn.* ii. 980.
Ne. marginatus. *Fabr. A.* 83. (!)
Mu. marginella. *Gmel.* v. 2836.—*Stew.* ii. 256.
♂; Ne. uliginosus. *Panz. F.* xlvi. *f.* 21.
♀; Ne. marginatus. *Panz. F.* xlvi. *f.* 22.
Str. mutica. *Schr. B.* iii. 2389.
Mu. marginata. *Turt.* iii. 655.

8450. 3; nigrinus*. *Fall. D. S.* (*Strat.*) 6.
Ne. nigritus. *Panz. F.* cvii. *f.* 17.

8451. 4; brevirostris*. *Meig. Zw.* iii. 117.

Genus 97: (1221). SARGUS *Auctorum.*
NEMOTELUS *p, DeGeer.*
A. Vide *Meig. Zw.* iii. 110.

8452. 1; politus*. *Fabr. A.* 257.
Mu. polita. *Linn.* ii. 894.—*Stew.* ii. 261.—*Turt.* iii. 611.
Nem. auratus. *DeGeer.* vi. 81.
Sar. splendens. *Meig. Kl.* i. 144.
Mu. vitreus. *Harr. Ex.* 48. *pl.* xi. *f.* 9, 10. ♂, ♀.
♀; Sar. cyaneus. *Fabr. A.* 258. var.

8453. 2; flavicornis. *Meig. Zw.* iii. 112.
Mu. parvulus. *Harr. Ex.* 48. *pl.* xi. *f.* 11?

8454. 3; formosus*. *Meig. Zw.* iii. 110.
Mu. aurata. *Fabr.* iv. 335.—*Don.* iv. 91. *pl.* 142. *f.* 1.—*Stew.* ii. 261.—*Turt.* iii. 610.
♂; Sar. auratus. *Fabr. A.* 257.
♀; Sar. xanthopterus. *Fabr. A.* 255.—*Meig. Kl. pl.* 8. *f.* 16–18.
Nem. flavogeniculatus. *DeGeer.* vi. 81.
Mu. cicur. *Harr. Ex.* 47. *pl.* xi. *f.* 88. ♂. ♀.

B. Vide *Meig. Zw.* iii. 106.

8455. 4; flavipes*. *Meig. Zw.* iii. 108.

8456. 5; nitidus*. *Meig. Zw.* iii. 108.

8457. 6; infuscatus*. *Meig. Zw.* iii. 107.
Sar. auratus. *Meig. Kl.* i. 143.
Sar. cuprarius ♀. *Fall. D. S.* (*Strat.*) 15.

8458. 7; cuprarius*. *Meig. Zw.* iii. 106.
Mu. cupraria. *Linn.* ii. 994.—*Stew.* ii. 261.—*Turt.* iii. 610.
Sar. cupreus. (*Sam. I.* 37.)
Mu. indicus. *Harr. Ex.* 47. *pl.* xi. *f.* 77. ♂, ♀?

8459. 8; Reaumuri*. *Meig. Zw.* iii. 109.
Reaumur Ins. iv. 22. *pl.* 5–8.

Genus 98: (1222). ODONTOMYIA, *Meig. Kl.*, *Leach*, *Samou.*

STRATIOMYS *p*, *Meig. Zw.*

8460. 1; tigrina*. *Meig. Kl.* i. 130.—(*Sam. I.* 31.)
Str. tigrina. *Panz. F.* lviii. *f.* 20.
β, Mu. geniculata. *Marsh. MSS.*

8461 † 2, argentata*. *Meig. Kl.* i. 131.
Str. argentata. *Panz. F.* lxxi. *f.* 20. ♂.—cviii. *f.* 10. ♀.
In Mus. *Brit.*, et D. *Kirby.*

8462 † 3. Microleon[a].

8463. 4; Hydroleon*. *Meig. Kl.* i. 131.—(*Sam. I.* 31.)
Mu. Hydroleon. *Linn.* ii. 980.—*Berk. S.* i. 165.—*Stew.* ii. 256.—*Turt.* iii. 630.
Oxycera Hydroleon. (*Sam. I.* 32.)
♀; Str. angulata. *Panz. F.* lviii. *f.* 19.
♂; Str. Hydroleon. *Panz. F.* vii. *f.* 21.

---

[a] 8462 † 3. Microleon.
Ox. Microleon. (*Sam. I.* 31.) (!)
Mu. Microleon. *Linn.* ii. 980.—*DeGeer.* vi. 64. *pl.* 9. *f.* 1, 2.—*Turt.* iii. 629. (!)

8464. 5; viridula*.
Str. viridula. *Fabr. A.* 84.—*Panz. F.* lviii.*f.* 18.
Str. marginata. *Fabr. A.* 84.
Str. canina. *Panz. F.* lviii.*f.* 23.
Od. dentata. *Meig. Kl.* i. 130.

8465. 6; felina*.
Str. felina. *Panz. F.* lviii.*f.* 22.—*Ent. Trans.*(*Haworth.*)i. 254.
Str. vulpina. *Panz. F.* lviii.*f.* 24.
Odo. vulpina. (*Sam. I.* 31.)
Str. flavissima. *Fabr. MSS.*
Mu. Mycroleon. *Harr. Ex.* 46. *pl.* xi.*f.* 5.

8466 † 7, trimaculata. *Mus. Brit.*
Str. Hydrodromia. *Meig. Zw.* iii. 146? In Mus. *Brit.*

8467. 8; Hydropota*.
Str. Hydropota. *Meig. Zw.* iii. 147.

8468. 9: ornata*.
Str. ornata. *Meig. Zw.* iii. 144.
Od. furcata. *Meig. Kl.* i. 129. *pl.* vii.*f.* 22.

Genus 99: (1223). STRATIOMYS, *Geoffroy.*

ODONTOMYIA *p, Samou.*—HIRTEA *p, Scop.*

8469. 1; Potamida*. *Meig. Zw.* iii. 137.
Str. Chamæleon, ♂. *Meig. Kl.* i. 126. *pl.* vii.*f.* 18.
Mu. Chamæleon. *Harr. Ex.* 44. *pl.* xi.*f.* 1.

8470. 2; Chamæleon. *Meig. Zw.* iii. 134.—(*Sam. I.* 39.)—(*Kirby & Sp. I. E.* ii. 230 & 285.—*Id.* iii. 99.—*Id.* iv. 11 & 54.)
Mu. Chamæleon. *Linn. F. No.* 1780.—*Berk. S.* i. 162.—*Don.* i. 67. *pl.* 31. *f.* 1. *pl.* 35. larva.—*Stew.* ii. 256.—*Turt.* iii. 628.—*Shaw* G. *Z.* vi. 379. *pl.* 105. *fig. sup.*
♀, Str. nigrodentata. *Meig. Kl.* i. 126. *pl.* vii.*f.* 19.

8471. 3; furcata*. *Fabr. A.* 78.
Stra. panthaleon. *Fall.* D. *S.* (*Strat.*) 7.
Odo. furcata. (*Sam. I.* 31.)
Mu. singularius. *Harr. Ex.* 45. *pl.* xi.*f.* 2.

8472. 4, riparia*. *Meig. Zw.* iii. 138.
Stra. strigata. *Meig. Kl.* i. 124.
Mu. tenebricus. *Harr. Ex.* 45. *pl.* xi.*f.* 3?

8473. 5; strigata*. *Fabr. A.* 80.—*Panz. F.* xii.*f.* 20.
Hirtæa longicornis. *Scop. Carn. No.* 999.
♀; Stra. thoracica. *Fabr. A.* 79.
♂; Stra. villosa. *Meig. Kl.* i. 125.
♀; Stra. nubeculosa. *Meig. Kl.* i. 125.

8474. 6; triangulata* *mihi.*

Genus 100: (1224). CLITELLARIA, *Meig.*, *Leach*, *Samou.*
EPHIPPIUM, *Latr.*

8475. 1, ephippium*. *Meig. Zw.* iii. 122.—(*Sam. I.* 11.)
Mu. ephippium. *Don.* xvi. 37. *pl.* 559.—(*Mill. B. E. pl.* 1. *f.* 6.)
Eph. thoracicum. *Latr. Gen.* iv. 276.
Mu. Inda. *Schrank A.* 891.

Genus 101: (1225). OXYCERA, *Meig.*, *Leach*, *Samou.*

8476. 1; pulchella*. *Meig. Zw.* iii. 125.
Oxy. hypoleon. *Meig. Kl.* i. 137. *f.* 8. *f.* 3. ♂.
Mu. hypoleon. *Don.* v. 6. *pl.* 146. *f.* 3.—*Stew.* ii. 256.—*Turt.* iii. 631.
Mu. tardigradus. *Harr. Ex.* 46. *pl.* xi. *f.* 6.

8477. 2; trilineata*. *Meig. Zw.* iii. 126.—(*Sam. I.* 32.)
Str. trilineata. *Fabr.*—*Panz. F.* i. *f.* 13.
Mu. trilineata. *Linn.* ii. 980.—*Don.* v. 25. *pl.* 151. *f.* 5.—*Stew.* ii. 256.—*Turt.* iii. 630.
Mu. Hydroleon. *Harr. Ex.* 46. *pl.* xi. *f.* 4.

8478. 3: muscaria*. *Meig. Zw.* iii. 126.
Str. muscaria. *Fabr.*—*Panz. F.* cviii. *f.* 15. ♂. *f.* 16. ♀.

Familia XIII: (147). XYLOPHAGIDÆ.

(XYLOPHAGI, *Meigen.*—STRATIOMYDÆ *p*, *Latr.*)

Genus 102: (1226). BERIS, *Latr.*, *Leach*, *Samou.*, (*Kirby.*)
STRATIOMYS *p*, *Geoff.*—ACTINA, *Meig.* (*olim.*)

A. Scutello octo-spinoso.

8479. 1; abdominalis* *mihi.*

8480. 2; lucida* *mihi.*
Mu. nitens. *Mus. Marsham.*

8481. 3; fuscipes*. *Meig. Zw.* ii. 8.

B. Scutello sex-spinoso.

8482. 4; femoralis*. *Meig. Zw.* ii. 6.

8483. 5; nigra*. *Meig. Zw.* ii. 7.

8484. 6; similis*.
Mu. similis. *Forst. C.* i. 97.—*Stew.* ii. 257.—*Turt.* iii. 632.
Be. nigripes. *Meig. Zw.* ii. 7?

8485. 7; chalybeata*. *Meig. Zw.* ii. 4.
Mu. chalybeata. *Forst. C.* i. 95.—*Stew.* ii. 257.—*Turt.* iii. 632.
Actina atra. *Meig. Kl.* i. 118.
Mu. crassipes. *Mus. Marsh.*
♀; Mu. sexdentata. *Fabr. A.* 87. (!)—*Stew.* ii. 256.

8486. 8; clavipes*. *Meig. Zw.* ii. 5.—(*Sam. I.* 6.)

Mu. clavipes. *Linn.* ii. 981.
Stra. clavipes. *Panz. F.* ix. *f.* 19.
8487. 9; vallata*. *Meig. Zw.* ii. 5.
Stra. clavipes. *Fabr. A.* 86.
Mu. vallata. *Forst. C.* i. 96.—*Stew.* ii. 257.—*Turt.* iii. 632.
Be. nigritarsis. *Latr. G.* iv. 273.—(*Sam. I.* 6.)
♀; Mu. sexdentata. *Turt.* iii. 631?
C. Scutello quadri-spinoso.
8488. 10; tibialis*. *Meig. Zw.* ii. 3.

Genus 103: (1227). XYLOPHAGUS, *Meig.*, *Leach*, (*Kirby.*)
NEMOTELUS *p*, *DeGeer.*—ACTINA *p*, *Meig.*—BERIS *p*, *Latr.*
8489 † 1, ater*. *Fabr. A.* 64.
♂. Empis subulata. *Panz. F.* liv. *f.* 23. In Mus. *Brit. et* D. *Stone?*

Familia XIV: (148). SYRPHIDÆ, *Leach.*

(MUSCA *p*, *Linné.*—SYRPHUS, *Fabr.*—SYRPHIÆ, *Latr.*—SYRPHII, *Meig.*)

Genus 104: (1228). MICRODON, *Illiger*, *Curtis.*
MULIO *p*, *Fabr.*—STRATIOMYS *p*, *Panz.*—APHRITIS, *Latr.*, *Leach*, *Samou.*
8490. 1; Apiformis. *Meig. Zw.* iii. 163.—*Curtis.* i. *pl.* 70.
Mus. Apiformis. *DeGeer.* vi. 56. *pl.* vii. *f.* 18–20.
Mus. mutabilis. *Linn. F. No.* 1807. (nec descriptio.)
Stra. pigra. *Schrank B.* iii. 2387.
Mus. nova. *Schra. A. No.* 889.
Mul. apiarius. *Fabr. A.* 185.
Ap. auropubescens. (*Latr.* G. iv. 330. *pl.* 16. *f.* 7, 8.)—(*Sam. I.* 4.)
8491. 2, mutabilis. *Meig. Zw.* iii. 164.
Mus. mutabilis. *Linn.* ii. 985.
Mus. plebeia. *Schra. A. No.* 890.
Stra. conica. *Panz. F.* xii. *f.* 21.

Genus 105: (1229). CHRYSOTOXUM, *Meig.*, *Leach*, *Samou.*
MULIO *p*, *et* MILESIA, *Fabr.*
8492. 1; bicinctum*. *Latr. G.* iv. 329.
Mus. bicincta. *Linn.* ii. 985.—*Don.* x. 54. *pl.* 346. *f.* 2?
Syr. bicinctus. *Panz. F.* xlv. *f.* 18.
Mus. callosus. *Harr. Ex.* 61. *pl.* xv. *f.* 18.
8492*. 2; hortensis. *Meig. Zw.* iii. 173.
8493. 3; arcuatum*. *Latr.* G. iv. 327.—(*Sam. I.* 11.)
Mus. arcuata. *Linn.* ii. 985.—*Stew.* ii. 265.—*Turt.* iii. 645.—*Schæff. Ic. pl.* 73. *f.* 8.
Mus. imbellis. *Harr. Ex.* 60. *pl.* xv. *f.* 16.

8494. 4; intermedium*. *Meig. Zw.* iii. 169.
Mul. arcuatus var. ♀. *Fall. D. S.* (*Syrph.*) 5.

8495. 5; marginatum*. *Meig. Zw.* iii. 171.
Mul. fasciolatus. *Fall. D. S.* (*Syrph.*) 5.
Mus. cautus. *Harr. Ex.* 60. *pl.* xv. *f.* 15.

8496. 6; fasciolatum*.
Mus. fasciolata. *DeGeer.* vi. 55. *pl.* 7. *f.* 13.
Mil. Vespiformis. *Fabr. A.* 188.
Syr. arcuatus. *Panz. F.* ii. *f.* 10.

Genus 106: (1230). CERIA, *Fabr., Leach, Samou., Curtis.*
CONOPS *p, Schrank.*

8497 ‡ 1, Conopsoides. *Meig. Zw.* iii. 160.—*Curtis.* iv. *pl.* 186.
Mus. Conopsoides. *Linn.* ii. 982.
Cer. clavicornis. *Fabr. A.* 173.
Syr. conopseus. *Panz. F.* xliv. *f.* 20.
Con. vaginicornis. *Schrank B.* iii. *No.* 2561.
In Mus. *Brit.* et D. Heysham?

Genus 107: (1231). DOROS, *Meig., Leach, Samou.*
MILESIA *p, Fabr. A.*—MULIO *p, Fabr. S.*—SCÆRA *p, Fallen.*

8498. 1; conopseus*. (*Sam. I.* 15.)
Syr. conopseus. *Fabr. S. I.* ii. 429.
Mus. conopsea. *Turt.* iii. 645.
Syr. coarctatus. *Panz. F.* xlv. *f.* 22.
Mus. profuges. *Harr. Ex.* 81. *pl.* xxiv. *f.* 25.

Genus 108: (1232). ASCIA, *Megerle.*
MERODON *p, Fabr.*—MILESIA *p, Fallen.*

8499. 1; omicron*.
Mus. omicron. *Mus. Marsham.*
♀. Mus. femorata. *Mus. Marsham.*

8500. 2; cingulata* *mihi.*

8501. 3; podagrica*. *Meig. Zw.* iii. 186.
Mer. podagrica. *Fabr. A.* 198.
Syr. podagricus. *Panz. F.* lix. *f.* 16.
Mus. ichneumonia. *Schra. A. No.* 911.
Mus. elongata. *Schrank B.* iii. *No.* 2414.
♂; Mus. molio. *Harr. Ex.* 111. *pl.* xxxiii. *f.* 54.
♀; Mus. tenur. *Harr. Ex.* 112. *pl.* xxxiii. *f.* 59.

8502. 4, lanceolata. *Meig. Zw.* iii. 187?

8503. 5; floralis*. *Meig. Zw.* iii. 188.

8504. 6; dispar*. *Meig. Zw.* iii. 188. *pl.* 27. *f.* 27. ♂. *f.* 28. ♀.
Mus. dispar. *Schrank B.* iii. *No.* 2415.

8505. 7; bicincta* *mihi.*

8506 † 8, geniculata. *Meig. Zw.* iii. 192. (!) In Mus. *Brit.*

Genus 109: (1233). SPHEGINA, *Meig.*
MILESIA *p*, *Fallen.*

8507. 1; clunipes*. *Meig. Zw.* iii. 194. *pl.* 28. *f.* 5.
Mil. clunipes. *Fallen D. S.* (*Syrph.*) 12.

8508 † 2, nigra. *Meig. Zw.* iii. 195. In Mus. *Brit. et* D. *Hope.*

Genus 110: (1234). BACCHA, *Fabr.*

8509. 1; elongata*. *Fabr. A.* 200.—*Meig. Zw.* iii. 197. *pl.* 28. *f.* 13.
8510. 2; nigripennis*. *Meig. Zw.* iii. 200.
8511. 3; tabida*. *Meig. Zw.* iii. 199.
8512. 4; splegina*. *Meig. Zw.* iii. 198.
8513. 5, scutellata*. *Meig. Zw.* iii. 197.
8514. 6; obscuripennis*. *Meig. Zw.* iii. 199.
Mus. perexilis. *Harr. Ex.* 81. *pl.* xxiv. *f.* 24.

Genus 111: (1235). PSARUS, *Fabr.*, *Leach.*

8515 † 1, abdominalis. *Fabr. A.* 211.—*Meig. Zw.* iii. 174. *pl.* 27. *f.* 12. In Mus. *Brit.*

Genus 112: (1236). PARAGUS, *Latr.*
SCÆVA *p*, *et* MULIO *p*, *Fabr.*

8516 † 1, bicolor. *Latr.* G. iv. 326.—*Coqueb. Ic.* 26. *f.* 9.
Mul. bicolor. *Fabr. A.* 186.
Mus. melanochrysa. *Gmel.* v. 2879. In Mus. *Brit.?*
8517 † 2, arcuatus. *Meig. Zw.* iii. 179. *pl.* 27. *f.* 20, 21. In Mus. *Brit.?*
8518. 3; obscurus*. *Megerle.*—*Meig. Zw.* iii. 183.
8519. 4; femoratus*. *Megerle.*—*Meig. Zw.* iii. 184.

Genus 113: (1237). EUMERUS, *Meigen.*
PIPIZA *p*, *Fallen.*

8520. 1; ruficornis*. *Meig. Zw.* iii. 206.
8521. 2; ornatus*. *Meig. Zw.* iii. 204.
8522. 3; strigatus*. *Meig. Zw.* iii. 208.
Pip. strigata. *Fallen D. S.* (*Syrph.*) 61.
Syr. annulatus. *Panz. F.* lx. *f.* 11.
8523. 4; funeralis*. *Megerle.*—*Meig. Zw.* iii. 208.
8524. 5; Selene*. *Meig. Zw.* iii. 210.

Genus 114: (1238). PSILOTA, *Meigen.*

8525. 1; Anthracina*. *Meig. Zw.* iii. 256. *pl.* 29. *f.* 18. 20.
Mu. diræ. *Harr. Ex.* 112. *pl.* xxxiii. *f.* 57.
8526. [2; nubila*] *mihi.*

Genus 115: (1239). CHRYSOGASTER, *Meigen.*

ERISTALIS *p, Fabr.*—MILESIA *p, Latr.*

A. Antennæ capitulo ovato.

8527. 1; fumipennis* *mihi.*

8528. 2; cæmiteriorum*. *Meig. Zw.* iii. 268.
Er. cæmiteriorum. *Fabr. A.* 246.
Sy. cæmiteriorum. *Panz. F.* lxxxii. *f.* 17.
Er. solstitialis. *Fall. D. S.* (*Syr.*) 56.
Mu. cæmiteriorum. *Don.* xii. 20. *pl.* 401. *f.* 2?

8529. 3; ænea*. *Meig. Zw.* iii. 270.

8530. 4; chalybeata*. *Meig. Zw.* iii. 267.

8531. 5; nigroænea* *mihi.*

8532. 6; viduata*. *Meig. Zw.* iii. 269.
Mu. viduata. *Linn.* ii. 994.—*Turt.* iii. 611.

8533. 7; metallica*. *Meig. Zw.* iii. 267.
Er. metallicus. *Fabr. A.* 246.

8534. 8; violacea*. *Meig. Zw.* iii. 267.

8535. 9; splendens*. *Meig. Zw.* iii. 266.

8536. 10; splendida*. *Megerle.*—*Meig. Zw.* iii. 271.

8537. 11; discicornis*. *Megerle.*—*Meig. Zw.* iii. 270.

8538. 12; grandicornis*. *Meig. Zw.* iii. 270.
Mu. politus. *Harr. Ex.* 81. *pl.* xxiv. *f.* 26?

B. Antennæ capitulo elongato.

8539. 13; nobilis*. *Meig. Zw.* iii. 272.
Er. nobilis. *Fall. D. S.* (*Syr.*) 57.

8540. 14; elegans*. *Wied.*—*Meig. Zw.* iii. 272. *pl.* 30. *f.* 9. antenna.

Genus 116: (1240). PIPIZA, *Meigen.*

ERISTALIS *p, Fabricius.*—MILESIA *p, Latr.*

A. Vide *Meig. Zw.* iii. 250.

8541. 1; melancholica*. *Meig. Zw.* iii. 251.

8542. 2; vitripennis*. *Meig. Zw.* iii. 254.

8543. 3; carbonaria*. *Meig. Zw.* iii. 251.

8544. 4; chalybeata*. *Meig. Zw.* iii. 252.

8545. 5; funebris*. *Meig. Zw.* iii. 250.

8546. 6; lugubris*. *Meig. Zw.* iii. 250.
Er. lugubris. *Fabr. A.* 246.
Mu. mœsta. *Gmel.* v. 2874.

B. Vide *Meig. Zw.* iii. 242.

8547. 7; noctiluca*. *Meig. Zw.* iii. 244.

Mu. noctiluca. *Linn.* ii. 986.—*Don.* x. 55. *pl.* 346. *f.* 4.
Sy. rosarum. *Panz. F.* xcv. *f.* 21.
Mu. tristor. *Harr. Ex.* 107. *pl.* xxxii. *f.* 42.

8548. 8, signata*. *Meig. Zw.* iii. 246.

8549. 9; Artemis*. *Meig. Zw.* iii. 244.

8550. 10, lucida. *Meig. Zw.* iii. 247.
Mu. rosarum. *Marsham MSS.*

8551. 11; geniculata*. *Meig. Zw.* iii. 245.

8552. 12; bimaculata*. *Meig. Zw.* iii. 246.

8553. 13; guttata*. *Meig. Zw.* iii. 247.

8554. 14; notata*. *Meig. Zw.* iii. 246.

8555. 15, hyalipennis *mihi.*

Genus 117: (1241). CHEILOSIA, *Hoffmansegg.*
SCÆVA *p, Fabricius.*

8556. 1; granditarsa*.
Mu. granditarsa. *Forst. C.* i. 99.—*Stew.* ii. 263.—*Turt.* iii. 626.
Sy. lobatus. *Meig. Zw.* iii. 336.

8557. 2; Ocymi*.
Sc. Ocymi. *Fabr. A.* 252.
Sy. Ocymi. *Panz. F.* lxxxii. *f.* 18.

8558. 3; Rosarum*.
Sc. Rosarum. *Fabr. A.* 251.—*Panz. F.* cviii. *f.* 14.

8559. 4; albimanus*.
Sc. albimanus. *Fabr. A.* 253.
Mu. albimana. *Stew.* ii. 266.—*Turt.* iii. 651.

8560. 5; submaculata* *mihi.*

8561. 6; scutata*.
Sy. scutatus. *Meig. Zw.* iii. 333.
Sc. albimana. *Fall. D. S.* (*Syr.*) 46.
Mu. albimana ♀. *Turt.* iii. 651.

8562. 7; manicata*.
Sy. manicatus. *Meig. Zw.* iii. 336.

8563. 8; irregularis* *mihi.*

8564. 9; clypeata*.
Sy. clypeatus. *Meig. Zw.* iii. 335.

8565. 10; peltata*.
Sy. peltatus. *Meig. Zw.* iii. 334. *pl.* 30. *f.* 31, 32.
Mu. navus. *Harr. Ex.* 109. *pl.* xxxiii. *f.* 50?

8566. 11; stictica*.
Sy. sticticus. *Meig. Zw.* iii. 332.
Mu. dexter. *Harr. Ex.* 104. *pl.* xxxii. *f.* 46?

8567. 12; maculosa*.

Sy. maculosus. *Meig. Zw.* iii. 330.
Mu. facultas. *Harr. Ex.* 109. *pl.* xxxiii. *f.* 49?

8568. 13; gracilis*.
Sy. gracilis. *Meig. Zw.* iii. 328.

8569. 14; scalaris*.
Sy. scalaris. *Fabr. E. S.* iv. 308.—*Panz. F.* xlv. *f.* 20.
Sc. mellina. *Fall. D. S.* (*Syr.*) 46.

8570. 15; melliturga*.
Sy. melliturgus. *Meig. Zw.* iii. 329.

8571. 16, mellaria*.
Sy. mellarius. *Meig. Zw.* iii. 328.

8572. 17, mellina.
Mu. mellina. *Linn.* ii. 988.—*Turt.* iii. 650.
Sy. noctilucus. *Panz. F.* lxxii. *f.* 24.
Sc. Rosarum. *Fall. D. S.* (*Syr.*) 47.

Genus 118: (1242). ——

SCÆVA *p, Fallen.*

8573. 1; tæniatus*.
Sy. tæniatus. *Meig. Zw.* iii. 325. *pl.* 30. *f.* 35, 36. ♂.

8574. 2; scriptus*.
Mu. scripta. *Linn.* ii. 987.—*Stew.* ii. 266.—*Turt.* iii. 651.—*Schell. D. pl.* 10. *f.* 2.
Sc. menthastri. *Fall. D. S.* (*Syr.*) 48.
Mu. invisito. *Harr. Ex.* 83. *pl.* xxiv. *f.* 31.

8575. 3; Menthastri*.
Mu. Menthastri. *Linn.* ii. 987.—*Berk. S.* i. 163.—*Stew.* ii. 266.—*Turt.* iii. 651.

8576. 4; Melissæ*.
Sy. Melissæ. *Meig. Zw.* iii. 326.

8577. 5; pictus*.
Sy. pictus. *Meig. Zw.* iii. 326.

8578. 6; Philanthus*.
Sy. Philanthus. *Meig. Zw.* iii. 327?

Genus 119: (1243). SCÆVA.

SCÆVA *p, Fabr.*

A. Abdomine sublineari, depresso.

8579. 1; Iris*.
Sy. Iris. *Meig. Zw.* iii. 320.

8580. 2; decora*.
Sy. decorus. *Meig. Zw.* iii. 319.
Am. Ribesii. *Mus. Marsham.*
Mu. timeo. *Harr. Ex.* 107. *pl.* xxxii. *f.* 44?

8581. 3; cincta*. *Fall. D. S.* (*Syr.*) 45.

8582. 4; auricollis*.
Sy. auricollis. *Meig. Zw.* iii. 318.

8583. 5; umbellatarum*. *Fabr. A.* 250.
Mu. agilitas. *Harr. Ex.* 108. *pl.* xxxii. *f.* 45?

8584. 6, balteata*.
Mu. balteata. *DeGeer.* vi. 52.
Sy. nectareus. *Fabr.* iv. 309.—*Panz. F.* lxxxii. *f.* 19.
Mu. alternans. *Mus. Marsham.*
Mu. alternata. *Schr. A. No.* 908.
Mu. canabina. *Gmel.* v. 2864.
♂; Mu. scitule. *Harr. Ex.* 111. *pl.* xxxiii. *f.* 55.
♀; Mu. scitulus. *Harr. Ex.* 105. *pl.* xxxii. *f.* 33.

8585. 7, hyalinata. *Fall. D. S.* (*Syr.*) 43.
Mu. comtus. *Harr. Ex.* 108. *pl.* xxxii. *f.* 47.

B. Abdomine subovato, paulò convexo.

8586. 8, subflexuosa *mihi.*

8587. 9; Grossulariæ*.
Sy. Grossulariæ. *Meig. Zw.* iii. 306.
Mu. formosus. *Harr. Ex.* 107. *pl.* xxxii. *f.* 43.

8588. 10; Ribesii.
Mu. Ribesii. *Linn.* ii. 987.—*DeGeer.* vi. 47. *pl.* 6. *f.* 3–13.—*Don.* xii. 21. *pl.* 401. *f.* 3.—*Stew.* ii. 266.—*Turt.* iii. 649.
Mu. blandus. *Harr. Ex.* 106. *pl.* xxxii. *f.* 38.

8589. 11; undulata* *mihi.*

8590. 12; vitripennis*. *Megerle?*
Sy. vitripennis. *Meig. Zw.* iii. 308.

8591. 13; tricincta*. *Fall. D. S.* (*Syr.*) 41.
Sc. bifasciata. *Mus. Fabr.*: (*teste Meig.* iii. 309.)

8592. 14; nitidicollis*. *Megerle?*
Sy. nitidicollis. *Meig. Zw.* iii. 308.

8593. 15; bifasciata*. *Fabr. A.* 248.
Mu. interrupta. *Gmel.* v. 2879.
Mu. eligans. *Harr. Ex.* 105. *pl.* xxxii. *f.* 34.

8594. 16; elegans* *mihi.*

8595. [17, Corollæ*.] *Fabr. A.* 250.
Sc. olitoria. *Fall. D. S.* (*Syr.*) 43.
My. pyrorum. *Schrank B.* iii. 2430.

8596. 18; topiaria*.
Sy. topiarius. *Meig. Zw.* iii. 305.

8597. 19; lata* *mihi.*

8598. 20; luniger*.
Sy. luniger. *Meig. Zw.* iii. 300.

8599. 21; transfuga*.
Mu. transfuga. *Mus. Marsham.*
8600. 22; arcuata*. *Fall. D. S.* (*Syr.*) 42.
Sc. Pyrastri. *Mus. Fabr.*: (*teste Meigen* iii. 301.)
8601. 23; Pyrastri*.
Mu. Pyrastri. *Linn.* ii. 987.—*Berk. S.* i. 163.—*Don.* xii. 19. *pl.* 401. *f.* 1.—*Stew.* ii. 266.—*Turt.* iii. 649.
Mu. Rosæ. *DeGeer.* vi. 49. *pl.* 6. *f.* 14–21.
Sc. transfuga. *Mus. Fabr.*: (*teste Meig.* iii. 302.)
Sy. Pyrastri. (*Sam. I.* 41.)
Albin. pl. lxvi. *a–d.*
8602. 24; selenitica*.
Sy. seleniticus. *Meig. Zw.* iii. 304. *pl.* 30. *f.* 21.
Mu. mellina. *Harr. Ex.* 80. *pl.* xxiv. *f.* 23.
8603. 25; nigrocærulea* *mihi.*
8604. 26; albostriata*. *Fall. D. S.* (*Syr.*) 42.
8605. 27; lunulata*.
Sy. lunulatus. *Meig. Zw.* iii. 299.
8606. 28; venustus*.
Sy. venustus. *Meig. Zw.* iii. 299.

C. ERISTALIS *p, Fabr.*—ELOPHILUS *p, Latr.*

8607. 29; glaucia*. *Panz. F.* civ. *f.* 16.
Mu. glaucia. *Linn.* ii. 986.
8608. 30; nobilis*.
Sy. nobilis. *Meig. Zw.* iii. 316.

Genus 120: (1244). ——

ERISTALIS *p, Fabr.*—SCÆVA *p, Fullen.*

8609. 1; ornatus*.
Sy. ornatus. *Meig. Zw.* iii. 298.
Mu. pedissequus. *Harr. Ex.* 61. *pl.* xv. *f.* 19.
8610. 2; festivus*.
Mu. festiva. *Linn.* ii. 986.—*Stew.* ii. 265.—*Turt.* iii. 647.
Mu. citrofasciata. *DeGeer.* vi. 53.
Mu. arcuata. *Don.* xii. 84. *pl.* 424. *f.* 2.
Mu. anteambulo. *Harr. Ex.* 60. *pl.* xv. *f.* 17.

Genus 121: (1245). SYRPHUS.

ERISTALIS *p, Fall., Fabr. A.*

8611. 1; lucorum*. *Meig. Zw.* iii. 313. *pl.* 30. *f.* 27.
Mu. lucorum. *Linn.* ii. 985.—*Turt.* iii. 643.
Mu. pellucens. *Harr. Ex.* 82. *pl.* xxiv. *f.* 28.
8612. 2; maculipennis* *mihi.*
8613. 3, œstraceus*.
Mu. œstracea. *Linn.* ii. 985.—*Turt.* iii. 642.

Sy. rupestris. *Panz. F.* lix. *f.* 13.
Mu. illustratus. *Harr. Ex.* 104. *pl.* xxxii. *f.* 32.

8614. 4; assimilis* *mihi.*

8615. 5; ruficornis*. *Meig. Zw.* iii. 278.—*Panz. F.* lxxvii. *f.* 20.
Er. ruficornis. *Fabr. A.* 243.
Mu. rutilo. *Harr. Ex.* 80. *pl.* xxiv. *f.* 21.

8616. 6, grossus*? *Meig. Zw.* iii. 281.
Er. grossa. *Fall. D. S.* (*Syr.*) 53.
Mu. Corydon. *Harr. Ex.* 106. *pl.* xxxii. *f.* 36.

8617. 7; flavipes*. *Panz. F.* liv. *f.* 10.

8618. 8; fulvicornis*. *Meig. Zw.* iii. 288.

8619. 9; variabilis*. *Panz. F.* lx. *f.* 10.
Er. ater. *Fabr. A.* 246.
Er. nigrita. *Fabr. A.* 244.

8620. 10; nigrinus*. *Meig. Zw.* iii. 282.
Mu. funebres. *Harr. Ex.* 106. *pl.* xxxii. *f.* 37.

8621. 11; scutellatus*. *Meig. Zw.* iii. 284. *pl.* 30. *f.* 29, 30. (fœm.)
Er. scutellata. *Fall. D. S.* (*Syr.*) 55.
Mu. inundata. *Mus. Marsham.*
Mu. tarditas. *Harr. Ex.* 106. *pl.* xxxii. *f.* 40.

8622. 12, flavicornis. *Meig. Zw.* iii. 285.
Er. flavicornis. *Fabr. A.* 244.

8623. 13; vulpinus*. *Meig. Zw.* iii. 292.

8624. 14; nigripes*. *Meig. Zw.* iii. 282.

8625. 15; chalybeatus*. *Megerle.*—*Meig. Zw.* iii. 294.

8626. 16, funeralis. *Meig. Zw.* iii. 292.

8627. 17, mutabilis. *Meig. Zw.* iii. 283.
Er. mutabilis. *Fall. D. S.* (*Syr.*) 54.

8628. 18; Chlorus*. *Meig. Zw.* iii. 284.

8629. 19; urbanus*. *Meig. Zw.* iii. 287.

8630. 20; pusillus* *mihi.*

8631. 21, means. *Fabr. Sup.* 562.

8632. 22; viduus*. *Meig. Zw.* iii. 282.
Mu. viduata. *Fabr.* iv. 336.
Mu. bardus. *Harr. Ex.* 106. *pl.* xxxii. *f.* 39?

8633. 23; albitarsis*. *Meig. Zw.* iii. 290.

8634. 24; curialis*. *Meig. Zw.* iii. 287.

8635. 25; antiquus*. *Meig. Zw.* iii. 291.

8636. 26; subfasciatus* *mihi.*

8637. 27; picipes* *mihi.*

8638. 28; caliginosus* *mihi.*
Mu. fastuosa. *Stew.* ii. 265?

8639. 29; costalis* *mihi.*

8640 † 30. devius[a].

Genus 122: (1246). ERISTALIS, *Fabricius*, (*Kirby*.)
HELOPHILUS, *Leach*, *Samou.*—ELOPHILUS, *Latr.*

A. Setâ antennarum nudâ. a. *Tibiis posticis simplicibus.*

8641. 1; sepulchralis*. *Fabr. A.* 245.
Mu. sepulchralis. *Linn.* ii. 991.
Mu. interpunctus. *Harr. Ex.* 59. *pl.* xv. *f.* 13.
♀; Er. tristis. *Fabr. A.* 303.
Sy. tristis. *Panz. F.* lxxxii. *f.* 16.
Mu. ater. *Harr. Ex.* 58. *pl.* xv. *f.* 11.

8642. 2; æneus*. *Fabr. A.* 244.
Sy. æneus. *Panz. F.* lxxx. *f.* 15.
Mu. melanius. *Harr. Ex.* 59. *pl.* xv. *f.* 12?

8643. 3: sericeus *mihi.*

8644. 4; floreus*. *Fabr. A.* 233.
Mu. florea. *Linn.* ii. 984.—*Stew.* ii. 264.—*Turt.* iii. 639.
Sy. florcus. *Fanz. F.* xli. *f.* 21.
Mu. Atropos. *Schr. A.* 904.
Mu. ablectus. *Harr. Ex.* 41. *pl.* x. § 3. *f.* 2.

8645. 5, cryptarum. *Fabr. A.* 235.

b. *Tibiis posticis compressis.*

8646. 6; apiformis*. *Meig. Zw.* iii. 390.
Sy. apiformis. *Fall. D. S.* (*Syr.*) 28.
Albin. pl. lxiii. *f. e–g?*

8647. 7; Hortorum*. *Meig. Zw.* iii. 387.

8648. 8; vulpinus*. *Megerle.*—*Meig. Zw.* iii. 388.

8649. 9; tenax*. *Fabr. A.* 238.
Mu. tenax. *Linn.* ii. 984.—*Harr. Ex.* 41. *pl.* x. 111. *f.* 1.—*Berk. S.* i. 163.—*Stew.* ii. 265.—*Turt.* iii. 641.—*Don.* xvi. 66. *pl.* 574.—*Shaw. G. Z.* vi. 380. *pl.* 106.
Mu. porcina. *DeGeer.* vi. 45.
Sy. tenax. *Panz. F.* xiv. *f.* 23, 24.
He. tenax. (*Sam. I.* 21.)
Mu. Arbustorum. *Schra. A.* 902.

8650. 10, fumipennis *mihi.*

B. Setâ antennarum plumosâ.

8651. 11; similis*. *Meig. Zw.* iii. 392.
Sy. similis. *Fall. D. S.* (*Syr.*) 25.

8652. 12; Fossarum*. *Megerle.*—*Meig. Zw.* iii. 393.

---

[a] 8640 † 30. devius. *Fabr. A.* 185. (!)
Mu. devia. *Linn.* ii. 985.—*Stew.* ii. 265. (!)—*Turt.* iii. 644. (!)

8653. 13; Pratorum*. *Megerle.*—*Meig. Zw.* iii. 393.

8654. 14; Arbustorum*. *Fabr. A.* 236.
Mu. Arbustorum. *Linn.* ii. 984.
Mu. horticola var. *DeGeer.* vi. 60.
Sy. arbustorum. *Panz. F.* xiv. *f.* 22.
Mu. cinctus. *Harr. Ex.* 43. *pl.* x. *O.* iii. *f.* 6.

8655. 15; Nemorum*. *Fabr. A.* 234.
Mu. Nemorum. *Linn.* ii. 984.—*Berk. S.* i. 163.—*Stew.* ii. 264.—*Turt.* iii. 640.
Mu. Lyra. *Harr. Ex.* 42. *pl.* x. *O.* iii. *f.* 5.

8656. 16; Rupium*. *Fabr. A.* 24.
Sy. cryptarum. *Panz. F.* xc. *f.* 18.
Mu. paralleli. *Harr. Ex.* 43. *pl.* x. *O.* iii. *f.* 7.
Var.? Mu. lineolæ. *Harr. Ex.* 58. *pl.* xv. *f.* 10.

8657. 17; Horticola*. *Meig. Zw.* iii. 396.
Mu. Horticola. *DeGeer.* vi. 60. *pl.* 8. *f.* 12.
Mu. lineatus. *Harr. Ex.* 42. *pl.* x. *O.* iii. *f.* 4?
Sy. flavicinctus. *Fall.* D. *S.* (*Syr.*) 24.

8658 † 18, elegans* *mihi.* In Mus. *D. Ingpen.*

8659. 19; intricarius*. *Fabr. A.* 232.
Mu. intricaria. *Linn.* ii. 985.—*Stew.* ii. 265.—*Turt.* iii. 638.
Mu. fuscus. *Harr. Ex.* 42. *pl.* x. *O.* iii. *f.* 3.

8660. 20, caliginosus* *mihi.*

Genus 123: (1247). HELOPHILUS, *Meigen.*

ERISTALIS *p, Fabr.*—RHINGIA *p, Fabr.*

8661. 1, trivittatus. *Meig. Zw.* iii. 373.
Er. trivittatus. *Fabr. A.* 235.
Mu. parallelus. *Harr. Ex.* 57. *pl.* xv. *f.* 8.

8662. 2; pendulus*. *Meig. Zw.* iii. 373.—(*Sam. I.* 21.)
Mu. pendula. *Linn.* ii. 984.—*Don.* i. 69. *pl.* 31. *f.* 2.—*Stew.* ii. 264.—*Turt.* iii. 638.—*Shaw* G. *Z.* vi. 382. *pl.* 105. *fig. inf.*
Mu. trilineata: (trilenva). *Harr. Ex.* 58. *pl.* xv. *f.* 9.

8663. 3; lineatus*. *Meig. Zw.* iii. 369. *pl.* 32. *f.* 7. ♂.
Rh. lineata. *Fabr. A.* 223.
Rh. muscaria. *Fabr. A.* 223.—*Panz. F.* xx. *f.* 24.

8664 † 4. Frutetorum[a].

8665. 5; lunulatus*. *Meig. Zw.* iii. 370. *pl.* 32. *f.* 9. (*abdomen.*)
Er. lunatus. (*Ing. Inst.* 91.)

---

[a] 8664 † 4. Frutetorum. *Meig. Zw.* iii. 375.
Er. Frutetorum. *Fabr. A.* 236. (!)
Sy. versicolor. *Fabr.* iv. 285.—*Panz. F.* lxxxii. *f.* 14.
Sy. femoralis. *Fall. D. S.* (*Syrph.*) 31. 27.
Mu. Frutetorum. *Turt.* iii. 641. (!)—*Stew.* ii. 264. (!)

8666. 6; transfugus*. *Meig. Zw.* iii. 371. *pl.* 32. *f.* 8. ♂.
Mu. transfuga. *Linn.* ii. 987.

Genus 124: (1248). TROPIDIA, *Meigen.*

ERISTALIS *p, Fallen.*

8667. 1, milesiformis. *Meig. Zw.* iii. 346. *pl.* 31. *f.* 14. (ala.)—(*Ing. Inst.* 92.)—(*Curtis Cat. No.* 45.)
Er. milesiformis. *Fall. D. S.* (*Syr.*) 52.
Mu. scitus. *Harr. Ex.* 107. *pl.* xxxii. *f.* 41.

Genus 125: (1249). XYLOTA.

MILESIA *p, Fabr.*—XYLOTA A., *Meigen,* (*Kirby.*)

8668. 1; pipiens*.
Mu. pipiens. *Linn.* ii. 988.—*Harr. Ex.* 109. *pl.* xxxii. *f.* 48.—*Berk. S.* i. 164.—*Panz. F.* xxxii. *f.* 20.—*Stew.* ii. 266.—*Turt.* iii. 652.
Mi. pipiens. (*Sam. I.* 27.)

Genus 126: (1250). ——

XYLOTA B. *p, Meigen.*—MILESIA *p, Latr.*

8669. 1; ignava*. *Meig. Zw.* iii. 221.
Sy. ignavus. *Panz. F.* lx. *f.* 4.

8670. 2; segnis*. *Meig. Zw.* iii. 220.
Mu. segnis. *Linn.* ii. 988.—*Stew.* ii. 266.—*Turt.* iii. 644.
Sy. segnis. *Panz. F.* lx. *f.* 3.
Mu. Brassicariæ. *Don.* v. 17. *pl.* 151. *f.* 1.
Mu. longisco. *Harr. Ex.* 83. *pl.* xxiv. *f.* 30.

8671. 3; lenta*. *Meig. Zw.* iii. 222.
Sy. piger. *Panz. F.* lx. *f.* 5.

8672. 4; Sylvarum*. *Meig. Zw.* iii. 223.
Mu. sylvarum. *Linn.* ii. 985.
Mu. fucatus. *Harr. Ex.* 83. *pl.* xxiv. *f.* 29.
Sy. impiger. *Panz. F.* xlv. *f.* 21.

8673. [5; ærata*] *mihi.*

Genus 127: (1251). SPILOMYIA, *Meigen?*

XYLOTA B. *p, Meig.*—MILESIA *p, Fallen.*

8674. 1, femorata.
Mu. femorata. *Linn.* ii. 988?—*Turt.* iii. 645.

Genus 128: (1252). MILESIA, *Fabricius, Curtis.*

MILESIA D. b. *p, Meigen.*

8675. 1: speciosa. *Meig. Zw.* iii. 234.—*Curtis.* i. *pl.* 34.
Mi. speciosa. *Fabr. A.* 188.

8676 † 2. vespiformis[a].

Genus 129: (1253). MERODON, *Fabricius, Curtis.*
MILESIA *p, Latreille, Leach, Samou.*

8677 † 1, clavipes. *Fabr. A.* 195.—*Meig. Zw.* iii. 351. *pl.* 31. *f.* 22. ♀. —*Curtis.* ii. *pl.* 98. ♂.
Mu. curvipes. *Gmel.* iv. 2871. In Mus. *Brit.*

8678 † 2. annulata[b].

Genus 130: (1254). CRIORHINA, *Hoffmansegg.*
MILESIA B. *Meigen.*—ERISTALIS, *Latr., Leach, Samou.*

8679. 1; Ranunculi*. (*Ing. Inst.* 91.)
Sy. Ranunculi. *Panz. F.* xci. *f.* 21.
Mu. personatus. *Harr. Ex.* 79. *pl.* xxiv. *f.* 20? ♀.

8680. 2; asilica*. (*Ing. Inst.* 91.)
Sy. asilicus. *Fall. D. S.* (*Syr.*) 22.

8681. 3, floccosa. (*Ing. Inst.* 91.)
Mi. floccosa. *Meig. Zw.* iii. 238.
Mu. Alopece. *Mus. Marsham.*

§ HELIOPHILUS *p, Meigen.*

8682. 4; *Oxyacanthæ**. (*Ing. Inst.* 91.)
Mi. *Oxyacanthæ.* *Meig. Zw.* iii. 237.
Er. Narcissi. (*Sam. I.* 16.)

8683. 5; berberina*. (*Ing. Inst.* 91.)
Er. berberinus. *Fabr. A.* 240.
Mi. biberina. *Meig. Zw.* iii. 237. *pl.* 29. *f.* 9.

Genus 131: (1255). SERICOMYIA, *Latr., Meigen, Leach, Samou.*

8684. 1; mussitans*. *Meig. Zw.* iii. 345.—(*Ing. Inst.* 91.)
Sy. mussitans. *Fabr. A.* 225.—*Panz. F.* i. *f.* 15.
Mu. fulvus. *Harr. Ex.* 80. *pl.* xxiv. *f.* 22.

8685. 2; borealis*. *Meig. Zw.* iii. 343. *pl.* 31. *f.* 9.—(*Ing. Inst.* 91.)
Sy. borealis. *Fall. D. S.* (*Syr.*) 20.
Mu. lappona. *DeGeer.* vi. 61. *pl.* 8. *f.* 14.
Mu. silentis. *Harr. Ex.* 59. *pl.* xv. *f.* 14.

8686. 3; lappona*. *Meig. Zw.* iii. 344.
Mu. lappona. *Linn.* ii. 983.
Se. Lapponum. (*Sam. I.* 37.)—(*Kirby & Sp. I. E.* iv. *pl.* 4. *f.* 5.)

---

[a] 8676 † 2. vespiformis. *Meig. Zw.* iii. 232.
Mu. vespiformis. *Linn.* ii. 986.—*Turt.* iii. 647. (!)
Sy. vespiformis. *Panz. F.* xc. *f.* 19.
Sy. apiformis. *Fabr.* iv. 300.

[b] 8678 † 2. annulatus. (*Samou.* 298.) (!)
Me. annulata. *Fabr.* iv. 296.

Genus 132: (1256). VOLUCCELLA, *Geoffroy*, *Leach*, *Samou.*, (*Kirby.*)

PTEROCERA *p*, *Meigen*, (*olim.*)—CONOPS *p*, *Scop.*

8687. 1; inanis*. *Latr. G.* iv. 322.—(*Sam. I.* 43.)
Mu. inanis. *Linn. F. No.* 1825.—*Berk. S.* i. 164?—*Don.* xiv. 59. *pl.* 490. *f.* 1.—*Stew.* ii. 257.—*Turt.* iii. 636.—*Samou.* 71. *pl.* 9. *f.* 3.
Mu. apivora. *DeGeer.* vi. 28. *pl.* 3. *f.* 4.
Sy. micans. *Fabr. A.* 224.
Mu. annulatus. *Harr. Ex.* 40. *pl.* x. *f.* 4.

8688. 2; inflata*. *Meig. Zw.* iii. 405. *pl.* 32. *f.* 28.—(*Ing. Inst.* 91.)
Sy. inflatus. *Fabr. A.* 226.
Vo. dryophila. *Schrank B.* iii. 2476.

8689. 3; pellucens*. *Meig. Zw.* iii. 404.—(*Sam. I.* 43.)
Mu. pellucens. *Linn. F. No.* 1826.—*Berk. S.* i. 164.—*Stew.* ii. 257.—*Turt.* iii. 637.—*Shaw* G. *Z.* vi. 385.—*Anon. Rem. L.* G. 31. *pl.* 6.
Sy. pellucens. *Panz. F.* i. *f.* 17.
Sy. putescens. *Schell. G. M. pl.* 8. *f.* 2.
Mu. fera. *Harr. Ex.* 39. *pl.* x. *f.* 2.

8690. 4; plumata*. *Meig. Zw.* iii. 403.
Mu. plumata. *DeGeer.* vi. 58. *pl.* 8. *f.* 4–9.
Sy. mystaceus. *Fabr. A.* 224.—*Panz. F.* viii. *f.* 22.
Mu. mystacea. *Harr. Ex.* 39. *pl.* x. *f.* 1.—*Don.* xiv. 15. *pl.* 471. *f.* 1.—*Stew.* ii. 264.—*Turt.* iii. 637.
Vo. mystaceus. (*Sam. I.* 43.)
Mu. bombylans var. *Linn.* ii. 983.
Vo. apiaria. *Schrank B.* iii. 2475.

8691. 5; bombylans*. *Latr.* G. iv. 322.—(*Sam. I.* 43.)
Mu. bombylans. *Harr. Ex.* 40. *pl.* x. *f.* 3.—*Don.* v. 23. *pl.* 151. *f.* 4.—*Stew.* ii. 264.—*Turt.* iii. 637.
Sy. bombylans. *Panz. F.* viii. *f.* 21.
Mu. plumosa. *Gmel.* v. 2867.
Var.? Mu. melanopyrrha. *Forst. C.* i. 98?—*Stew.* ii. 258.—*Turt.* iii. 604.

Genus 133: (1257). BRACHYOPA, *Hoffm.*, *Meigen.*

RHINGIA *p*, *Fallen.*

8692. 1; conica*. *Meig. Zw.* iii. 261.
Mu. conica. *Panz. F.* lx. *f.* 20.
Rh. testacea. *Fall. D. S.* (*Syr.*) 34. 4.

8693. 2, pallida *mihi.*

Genus 134: (1258). RHINGIA, *Scop.*, *Fabr.*, *Leach*, *Samou.*, *Curtis*, (*Kirby.*)

CONOPS *p*, *Linné.*—STOMOXYS *p*, *Gmel.*—VOLUCELLA *p*, *Geoff.*

8694. 1; rostrata*. *Fabr. A.* 222.—(*Sam. I.* 36.)—(*Curtis l. c. infra.*)
Co. rostrata. *Linn.* ii. 1004.—*Stew.* ii. 271.—*Turt.* iii. 670.—*Panz. F.* lxxxvii. *f.* 22.—*Shaw* G. Z. vi. 396. *pl.* 111. *fig. inf.*
St. rostrata. *Stew.* ii. 271.
Var. Mu. nasatus. *Harr. Ex.* 105. *pl.* xxxii. *f.* 35.

8695. 2; campestris*. *Meig. Zw.* iii. 259.—(*Ing. Inst.* 91.)—(*Curtis.* iv. *pl.* 182.)
Rh. rostrata var. *Fall. D. S.* (*Syr.*) 33.
Rhingia. *Schell. G. M. pl.* viii.
Mu. macrocephala. *Harr. Ex.* 82. *pl.* xxiv. *f.* 27.

Familia XV: (149). STOMOXYDÆ, *Meigen.*

Genus 135: (1259). STOMOXYS *Auctorum.*

CONOPS *p*, *Linné.*—EMPIS *p*, *Scop.*—SIPHONA *p*, *Meig.*

8696. 1, sibirita. *Fabr. A.* 280.—*Meig. Zw.* iv. 160. *pl.* 38. *f.* 1–4.
St. irritans. *Panz. F.* v. *f.* 24.
♂, St. longipes. *Gmel.* v. 2892.
♀, St. grisea. *Fabr. A.* 279.

8697. 2; irritans*. *Meig. Zw.* iv. 162.
Co. irritans. *Linn.* ii. 1004.—*Shaw G. Z.* vi. 395. *pl.* 111. *f. m.*
St. pugens. *Fabr. A.* 282.

8698. 3; calcitrans*. *Fabr. A.* 280.—*Stew.* ii. 270.—*Turt.* iii. 668.—(*Samou.* 298. *pl.* 9. *f.* 7.)—(*Kirby & Sp. I. E.* i. 48, 110 & 145.)
Co. calcitrans. *Linn.* ii. 1004.—*Berk. S.* i. 167.—*Shaw* G. Z. vi. 395. *pl.* 111. *fig. sup.?*
Mu. calcitrans. *Harr. Ex.* 141. *pl.* xli. *f.* 41.
♀, St. tessellata. *Fabr. A.* 281.
Mu. pungens. *DeGeer.* vi. 39. *pl.* 4. *f.* 12–18.

8699. 4, stimulans? *Meig. Zw.* iv. 161. *pl.* 38. *f.* 8–10.
St. irritans. *Fabr. A.* 281.—*Stew.* ii. 271.—*Turt.* iii. 669.
Mu. irritans. *Harr. Ex.* 148. *pl.* xliii. *f.* 74?

Genus 136: (1260). BUCENTES, *Latr.*, *Leach*, *Samou.*

SIPHONA *p*, *Meig.*—STOMOXYS *p*, *Fabr.*

8700. 1; geniculata*.
Mu. geniculata. *DeGeer.* vi. 20. *pl.* 2. *f.* 19–23.
Si. geniculata. *Meig. Zw.* iv. 155.
St. minuta. *Fabr. A.* 282.
Bu. cinereus. *Latr. G.* iv. 339.—(*Sam. I.* 7.)

8701. 2; cinereus.
Si. cinereus. *Meig. Zw.* iv. 156.

8702. 3, tachinaria.
Si. tachinaria. *Meig. Zw.* iv. 156.
St. cristata. *Fabr. A.* 281?
St. geniculata. *Fall. D. S.* (*Hæmatomyzides.*) 5. 2?
Mu. urbanus. *Harr. Ex.* 153. *pl.* xlv. *f.* 85?

8703. 4; nigrovittata.
Si. nigrovittata. *Meig. Zw.* iv. 157.

Familia XVI: (150). CONOPIDÆ, *Leach.*

Genus 137: (1261). ZODION, *Latr., Leach, Samou.*

MYOPA *p, Fabr.*

8704. 1; cinereum*. *Meig. Zw.* iv. 138. *pl.* 37. *f.* 6, 7.
My. cinerea. *Fabr. A.* 181.
My. tibialis. *Fabr. A.* 182. (*teste Meig.*)
Zo. conopsoides. (*Latr.* G. iv. 337. *pl.* 15. *f.* 8.)—(*Sam. I.* 44.)

Genus 138: (1262). MYOPA *Auctorum.*

STOMOXOIDES, *Schæff.*—STOMOXYS *p, Fabr.*—ASILUS *p, Geoff.* —SICUS, *Scop.*

8705. 1; atra*. *Fabr. A.* 179.—*Panz. F.* xii. *f.* 24.
♂; My. cinerascens. *Meig. Kl.* i. 287.
♀; My. maculata. *Meig. Kl.* i. 287.
My. annulata. *Fabr. A.* 181.
Var. My. micans. *Meig. Kl.* i. 289.
My. femorata. *Fabr. A.* 181.

8706. [2; pusilla*.] *Meig. Zw.* iv. 150?

8707. 3; fasciata*. *Meig. Kl.* i. 286.
My. ephippium. *Fabr. A.* 180.
Co. fusca. *Harr. Ex.* 72. *pl.* xx. *f.* 6, 7?

8708. 4; ferruginea*. *Fabr. A.* 178.
Co. ferruginea. *Linn.* ii. 1005.—*Stew.* ii. 271.—*Turt.* iii. 672.
Co. buccæ. *Harr. Ex.* 71. *pl.* xx. *f.* 5, 9.

8709. 5; dorsalis*. *Fabr. A.* 178.—(*Sam. I.* 28.)
My. ferruginea. *Panz. F.* xxii. *f.* 24.
My. testacea. *Fabr. S.* ii. 468.
My. grandis. *Meig. Kl.* i. 284.
Co. cessans. *Harr. Ex.* 71. *pl.* xx. *f.* 4.

8710. 6; testacea*. *Fabr. A.* 179.
Co. testaceus. *Linn.* ii. 1006.
Co. buccata. *Gmel.* v. 2895.

8711. 7; buccata*. *Fabr. A.* 179.
Co. buccata. *Linn.* ii. 1006.—*Turt.* iii. 672.
Harr. Ex. pl. xx. *f.* 8?

8712. 8; picta*. *Panz. F.* liv. *f.* 22.—(*Sam. I.* 28.)

Genus 139: (1263). CONOPS *Auctorum.*

ASILUS *p*, *Geoff.*—EMPIS *p*, *Scop.*

A. Abdominis basi coarctato.

8713. 1; rufipes*. *Fabr. A.* 176.
Co. petiolata. *Don.* xiii. 39. *pl.* 451.

B. Abdominis basi haud coarctato.

8714. 2; vesicularis*. *Linn.* ii. 1005.—*Stew.* ii. 271.—*Turt.* iii. 670.
Co. macrocephala. *Berk. S.* i. 167?
♂, Co. cylindrica. *Meig. Kl.* i. 275.

8715 † 3. macrocephala[a].

8716. 4; flavipes*. *Linn.* ii. 1005.—*Stew.* ii. 271.—*Panz. F.* lxx. *f.* 21, 22.
Co. macrocephala. (*Samou.* 72. *pl.* 9. *f.* 8.)
Co. vesicularis. *Harr. Ex.* 70. *pl.* xx. *f.* 1.
♀; Co. trifasciata. *DeGeer.* vi. 104.
Var. Co. melanocephala. *Meig. Kl.* i. 278.

8717. 5; aculeata*. *Linn.* ii. 1005.—(*Sam. I.* 12.)
Co. scutellata. *Meig. Kl.* i. 276.
Co. macrocephala. *Harr. Ex.* 71. *pl.* xx. *f.* 2, 3?

8718. 6; ceriæformis*. *Meg.*—*Meig. Zw.* iv. 132. *pl.* 36. *f.* 26?

8719. 7; quadrifasciata*. *DeGeer.* vi. 104. *pl.* 15. *f.* 1.
Co. aculeata. *Fabr. A.* 174.—*Turt.* iii. 670.

Familia XVII: (151). ŒSTRIDÆ, *Leach.*

MUSCIDES I., *Latr.*—ASTOMATA, *Dumeril.*

Genus 140: (1264). ŒSTRUS *Auctorum.*

8720 † 1, pictus. *Megerle.*—*Meig. Zw.* iv. 172.—*Curtis.* iii. *pl.* 106.
Œ. Leachii. *Samouelle MSS.*

8721. 2; Ovis*: Grey Fly.—*Berk. l. c. Linn.* ii. 970.—*Berk. S.* i. 160.—*Linn. Trans.* (*Clark.*) iii. 313. *pl.* 23. *f.* 14–17.—*Stew.* ii. 249.—*Turt.* iii. 582.—*Don.* xvi. 19. *pl.* 550.—*Clark E. pl.* 2. *f.* 16–20.—*Shaw G. Z.* vi. 364. *pl.* 102. *fig. inf.*—(*Leach A.* 1.)—*Bingley.* iii. 309.—(*Samou.* 301.)—(*Kirby & Sp. I. E.* i. 157.)—(*Curtis l. c. supra.*)

8722. 3, Bovis*. (*Linn.* ii. 969. Gasterophilo Equo confunditur.)—*Linn. Trans.* (*Clark.*) iii. 291. *pl.* 25. *f.* 1–6.—*Stew.* ii. 248.—*Turt.* iii. 581.—*Clark E. pl.* 2. *f.* 1–10.—*Shaw* G. *Z.* vi. 358. *pl.* 102. *fig. sup.*—(*Leach A.* 1.)—(*Sam. I.* 31.)—(*Kirby & Sp. I. E.* i. 149.)—(*Curtis l. c. supra.*)—*Bingley.* iii. 304.
Larvæ; *Wornils, Wormals, Warbles.*—*Bingley l. c.*

---

[a] 8715 † 3. macrocephala. *Linn.* ii. 1005.—*Stew.* ii. 271. (!)—*Turt.* iii. 671. (!)—*Meig. Zw.* iv. 125. *pl.* 36. *f.* 27.

8723. ‡4; Ericetorum*. *Leach A.* 2.—(*Curtis l. c. supra.*)
Œs. hæmorrhoidalis β. *Linn. Trans.* (*Clark.*) iii. 294.

Genus 141: (1265). GASTEROPHILUS, *Leach, Samou., Curtis,* (*Kirby.*)

GASTRUS, *Meigen.*—ŒSTRUS *p, Linné.*

8724. 1; hæmorrhoidalis*. (*Leach A.* 2.)—(*Sam. I.* 18.)—(*Curtis l. c. supra.*)
Œs. hæmorrhoidalis. *Linn.* ii. 970.—*Linn. Trans.* (*Clark.*) iii. 308. *pl.* 23. *f.* 10–13.—*Stew.* ii. 248.—*Turt.* iii. 582.—*Clark E. pl.* 1. *f.* 17–23.—*Shaw G. Z.* vi. 366.—(*Kirby & Sp. I. E.* i. 146.)
Œs. hæmoridalis. *Berk. S.* i. 160.
Œs. Equi var. *Fabr.* iv. 232.

8725. 2, nasalis. (*Curtis l. c. supra.*)
Œs. nasalis. *Linn.* ii. 969.—*Stew.* ii. 249.—(*Kirby & Sp. I. E.* i. 159. *et nota.*)
Œs. veterinus. *Fabr. A.* 230.—*Linn. Trans.* (*Clark.*) iii. 312. *pl.* 23. *f.* 18, 19.—*Turt.* iii. 582.—*Clark E. pl.* 1. *f.* 24–27.—*Shaw G. Z.* vi. 367.
Ga. veterinus. (*Sam. I.* 18.)
Œs. Equi. *Fabr.* iv. 232.

8726. 3, Clarkii. *Leach A.* 2.—*Meig. Zw.* iv. 180. (!)—(*Curtis l. c. supra.*)

8727 † 4, Salutiferus*. *Curtis.* iii. *pl.* 146.
Œs. Salutiferus. *Clark E. Sup. pl.* 1. *f.* 33, 34.
Œs. Salutaris. *Id. pl.* 1. *f.* 35, 36. In Mus. D. Clark, et Curtis.

8728. 5; Equi*. *Leach A.* 2.—(*Sam. I.* 18.)—(*Curtis l. c. infra.*)
Œs. Equi. *Linn. Trans.* (*Clark.*) iii. 298. *pl.* 23. *f.* 7–9.—*Stew.* ii. 248.—*Turt.* iii. 581.—*Clark E. pl.* 1. *f.* 1–16.—*Shaw* G. *Z.* vi. 359. *pl.* 102. *fig. med.*—*Bingley.* iii. 306.—(*Kirby & Sp. I. E.* i. 146.)—*Wood.* ii. 81. *pl.* 63.—*Stark E.* ii. 388.
Œs. Bovis. *Linn.* ii. 969. *desc. sola.*—*Berk. S.* i. 160.—(*Samou.* 70. *pl.* 9. *f.* 1.)
Breeze or Gad Fly.—*Bot-Fly.*—Larvæ; *Bots.*

Familia XVIII: (152). MUSCIDÆ, *Leach.*

Genus 142: (1266). PHASIA, *Latr., Leach, Samou.,* (*Kirby.*)

CONOPS *p, Linné.*—SYRPHUS *p, et* THEREVA *p, Fabr.*

8729. 1, hemiptera. *Meig. Zw.* iv. 191.
Ph. variabilis. (*Linn. Trans.* (*Leach.*) xi. 421.)—(*Sam. I.* 33.)
♂, Sy. hemipterus. *Fabr.* iv. 284.
Mu. hemiptera. *Turt.* iii. 639.—*Don.* xii. 95. *f.* 429.
Th. coleoptrata. *Panz. F.* lxxiv. *f.* 13.
♀, Sy. affinis. *Fabr.* iv. 284.

Th. affinis. *Panz. F.* lxxiv.*f.* 16.
Var. Sy. coleoptratus. *Fabr. S. I.* ii. 423.
Th. subcoleoptrata. *Panz. F.* lxxiv.*f.* 14.
Mu. subcoleoptrata. *Stew.* ii. 264.

8730. 2; subcoleoptrata*. *Meig. Zw.* iv. 190. *pl.* 39. *f.* 13. ♀.
Co. subcoleoptrata. *Linn.* ii. 1006.—*Wood.* ii. 102. *pl.* 69.

8731. 3; semicinerea*. *Meig. Zw.* iv. 199. *pl.* 39. *f.* 14.

Genus 143: (1267). GYMNOSOMA, *Meigen, Leach, Samou.*
TACHINA *p, et* SYRPHUS, *Fabr.*—OCYPTERA *p, Latr.*

8732. 1; rotundata*. *Meig. Zw.* iv. 204.—(*Sam. I.* 20.)
Mu. rotundata. *Linn.* ii. 991.—*Panz. F.* xx.*f.* 19.—*Don.* xii. 83. *pl.* 424. *f.* 1.—*Stew.* ii. 259.—*Turt.* iii. 605.
Mu. cerinus. *Harr. Ex.* 36. *pl.* ix.*f.* 6.

8733 † 2. globosa[a].

Genus 144: (1268). PHANIA, *Meigen.*

8734. 1; thoracica*. *Meig. Zw.* iv. 220.

8735. 2; curvicauda*. *Meig. Zw.* iv. 221. *pl.* 40. *f.* 10.
Ta. curvicauda. *Fall. D. S.* (*Musc.*) 17. 33.

Genus 145: (1269). MILTOGRAMMA, *Meigen.*

8736. 1, œstracea. *Meig. Zw.* iv. 229. *pl.* 40. *f.* 25?
Ta. œstracea. *Fall. D. S.* (*Musc.*) 10. 17.

Genus 146: (1270). GONIA, *Meigen.*

8737. 1; auriceps*. *Meig. Zw.* v. 5.

8738. 2; nervosa*. *Winthem.*—*Meig. Zw.* v. 4.

Genus 147: (1271). TRIXA, *Meigen.*

8739. 1, dorsalis. *Meig. Zw.* iv. 225.

8740. 2; variegata*. *Meig. Zw.* iv. 225. (!)

Genus 148: (1272). TACHINA, *Illiger, Leach, Samou.,* (*Kirby.*)
ECHINOMYIA *p, Latr.*

8741. 1; ferox*. *Panz. F.* civ.*f.* 20.
Mu. rotundata. *Harr. Ex.* 35. *pl.* ix.*f.* 2.

8742. 2; fera*. *Fabr. A.* 308.—(*Sam. I.* 40.)
Mu. fera. *Linn.* ii. 991.—*Stew.* ii. 258.—*Turt.* iii. 604.
Mu. reccumbo. *Harr. Ex.* 35. *pl.* ix.*f.* 4.

8743. 3; Virgo*. *Meig. Zw.* iv. 243.

---

[a] 8733 † 2. globosa. *Meig. Zw.* iv. 206. *pl.* 40. *f.* 22. ♀.
Sy. globosa. *Fabr. S. I.* ii. 432.
Mu. globosa. *Stew.* ii. 265. (!)—*Turt.* iii. 609. (!)
Gym. dispar. *Fall. D. S.* (*Rhyzomiz.*) 9. 2.

8744. 4; lurida*. *Fabr. A.* 310.

8745. 5; leucocoma*. *Meig. Zw.* iv. 244.

8746. 6; echinata*. *Meig. Zw.* iv. 245.
Mu. repens. *Harr. Ex.* 37. *pl.* ix. *f.* 10?

8747. 7; ursina*. *Meig. Zw.* iv. 245.

Genus 149: (1273). ECHINOMYIA, *Dumeril, Leach, Samou.*, (*Kirby.*)
TACHINA *p*, *Fabr.*

8748. 1; grossa*. (*Latr.* G. iv. 343.)—(*Sam. I.* 16.)—(*Kirby & Sp. I. E.* iii. 37.—iv. 149, 150.)
Mu. grossa. *Linn.* ii. 991.—*Harr. Ex.* 35. *pl.* ix. *f.* 1.—*Don.* x. 53. *pl.* 346. *f.* 1.—*Stew.* ii. 259.—*Turt.* iii. 605.—*Shaw G. Z.* vi. 385.

Genus 150: (1274). ——
TACHINA *p*, *Fallen.*
A. Oculis pubescentibus.

8749. 1; cæsia*.
Ta. cæsia. *Fall. D. S.* (*Musc.*) 27. 55.—*Meig. Zw.* iv. 247.
Mu. orior. *Harr. Ex.* 143. *pl.* xlii. *f.* 48.

8750. 2; consobrina*.
Ta. consobrina. *Meig. Zw.* iv. 248.

8751. 3; Radicum*.
Mu. Radicum. *Fabr. A.* 300.—*Turt.* iii. 606.
Ta. lurida. *Fall. D. S.* (*Musc.*) 26. 54.

8752. 4, cognata* *mihi.*

8753. 5; Histrio*.
Ta. Histrio. *Meig. Zw.* iv. 250. (!)

8754. 6; plumbea* *mihi.*

8755. 7; apicalis* *mihi.*

8756. 8, puparum*.
Mu. puparum. *Fabr. A.* 299.
Ta. tricincta. *Fall. D. S.* (*Musc.*) 26. 53.

8757. 9; quadripustulata*.
Ta. quadripustulata. *Fabr. A.* 309.
Ta. æstuans. *Fall. D. S.* (*Musc.*) 30. 61.

8758. 10; variegata*.
Ta. variegata. *Meig. Zw.* iv. 256.

8759. 11; bimaculata* *mihi.*

8760. 12; hortulana*.
Ta. hortulana. *Meig. Zw.* iv. 330.

8761. 13; arvensis*.
Ta. arvensis. *Meig. Zw.* iv. 337.

8762. 14; aratoria*.

Ta. aratoria. *Meig. Zw.* iv. 338.
Mu. convolo. *Harr. Ex.* 143. *pl.* xlii. *f.* 49?

8763. 15; nigrolineata* *mihi.*

8764. 16; schistacea*.
Ta. schistacea. *Meig. Zw.* iv. 414.

8765. 17; quadricincta* *mihi.*

8766. 18; vulpina*.
Ta. vulpina. *Fall. D. S.* (*Musc.*) 23. 47.
Mu. investigator. *Harr. Ex.* 35. *pl.* ix. *f.* 3?

8767. 19; testaceipes* *mihi.*

8768 † 20. Histrix[a].

8769. 21; ænea*.
Ta. ænea. *Meig. Zw.* iv. 273.
Mu. Mano. *Harr. Ex.* 142. *pl.* xlii. *f.* 45?

8770. 22; viridis*.
Ta. viridis. *Fall. D. S.* (*Musc.*) 25. 51.—*Meig. Zw.* iv. 258.

8771. 23; Tremula*.
Mu. Tremula. *Linn.* ii. 991.—*Stew.* ii. 259.
Mu. obsidianus. *Harr. Ex.* 37. *pl.* ix. *f.* 8.

8772. 24; chalybeata*.
Ta. chalybeata. *Meig. Zw.* iv. 271.
Mu. promans. *Harr. Ex.* 143. *pl.* xlii. *f.* 47?

B. Oculis nudis.

8773. 25; sylvatica*.
Ta. sylvatica. *Fall. D. S.* (*Musc.*) 12. 20.

8774. 26, festinans*.
Ta. festinans. *Fallen.*—*Meig. Zw.* iv. 384.

8775. 27; læta*.
Ta. læta. *Wied.*—*Meig. Zw.* iv. 381.

8776. 28; morosa*.
Ta. morosa. *Meig. Zw.* iv. 314.

8777. 29; continua*.
Mu. continua. *Panz. F.* lx. *f.* 19.

8778. 30; melanocephala.
Ta. melanocephala. *Meig. Zw.* iv. 281.

8779. 31; dubia* *mihi.*

8780. 32; nana* *mihi.*

Genus 151: (1275). MELANOPHORA, *Meig. olim.*

TACHINA *p*, *Meig.*—OCYPTERA *p*, *Fallen.*—TEPHRITIS *p*, *Fabr.*

8781. 1; roralis*.

---

[a] 8768 † 20. Histrix. *Fabr. A.* 310?
Mu. hirsuta. *Don.* xiv. 60. *pl.* 490. *f.* 2. (¹) In Mus. *Brit.*

Mu. roralis. *Linn.* ii. 993.—*Turt.* iii. 608.
Mu. grossificationis. *Linn.* ii. 996.—(*Haw.* xviii.)—*Turt.* iii. 618.—*Don.* xii. 71. *pl.* 419.
Te. grossificationis. (*Sam. I.* 40.)
Mu. interventum. *Harr. Ex.* 144. *pl.* xlii. *f.* 57.

Genus 152: (1276). LEUCOSTOMA, *Meig. olim.*

TACHINA *p, Meig.*

8782. 1; gagatina*.
Ta. gagatina. *Meig. Zw.* iv. 287.

8783. 2; venosa*.
Ta. venosa. *Meig. Zw.* iv. 288.

8784. 3; lepida*.
Ta. lepida. *Meig. Zw.* iv. 289.

8785. 4; simplex*.
Oc. simplex. *Fall. D. S.* (*Rhyomyz.*) 8. 10.
Ta. simplex. *Meig. Zw.* iv. 289.

8786. 5; floralis*.
Ta. floralis. *Fall. D. S.* (*Musc.*) 36. 74.

8787. 6; terminalis*.
Ta. terminalis. *Meig. Zw.* iv. 323.

8788. 7, exigua*.
Ta. exigua. *Meig. Zw.* iv. 567. (367.)

8789. 8; carbonaria*.
Mu. carbonaria. *Panz. F.* liv. *f.* 15.

8790. 9; venosa* *mihi.*

Genus 153: (1277). METOPIA, *Meig. olim.*

TACHINA *p, Meig.*

8791. 1, leucocephala*?
Mu. leucocephala. *Panz. F.* liv. *f.* 14.
Mu. labiata. *Fabr. A.* 304.
Mu. argyrostoma. *Mus. Marsham.*
Mu. argentata. *Stew.* ii. 258?

Genus 154: (1278). TRIARTHRIA *mihi.*

TACHINA *p, Meig.*

8792. 1; bicolor*.
Ta. bicolor. *Meig. Zw.* iv. 354.

8793. 2; spinipennis*.
Ta. spinipennis. *Meig. Zw.* iv. 350.
Mu. confluo. *Harr. Ex.* 143. *pl.* xlii. *f.* 50?

8794. 3; albicollis*.
Ta. albicollis. *Meig. Zw.* iv. 350.

Genus 155: (1279). EXORISTA, *Meig.*

8795. 1; larvarum*. (*Illig. M.* (*Meig.*) ii. 280.)
Mu. larvarum. *Linn.* ii. 992.—*Stew.* ii. 259.—*Turt.* iii. 606. —(*Kirby & Sp. I. E.* i. 188. *nota.* 267. iv. 224.)
Ta. rustica. *Fall. D. S.* (*Musc.*) 5. 5.

8796. 2; simulans*.
Ta. simulans. *Meig. Zw.* iv. 306.

8797. 3; nitidula*.
Ta. nitidula. *Meig. Zw.* iv. 297.

8798. 4; cincta*.
Ta. cincta. *Meig. Zw.* iv. 297.

8799. 5; devia*.
Ta. devia. *Meig. Zw.* iv. 301.

8800. 6; agrestis*.
Ta. agrestis. *Fall. D. S.* (*Musc.*) 6. 7.

8801. 7; Vidua*.
Ta. Vidua. *Meig. Zw.* iv. 315.

8802. 8; dorsalis*.
Ta. dorsalis. *Meig. Zw.* iv. 325.

8803. 9; longipes*.
Ta. longipes. *Meig. Zw.* iv. 341.
Mu. dimano. *Harr. Ex.* 142. *pl.* xlii. *f.* 46?

8804. 10; inanis*.
Ta. inanis. *Meig. Zw.* iv. 342.

8805. 11, pacifica*.
Ta. pacifica. *Meig. Zw.* iv. 342.

8806. 12; spreta*.
Ta. spreta. *Meig. Zw.* iv. 343.

Genus 156: (1280). ERIOTHRIX, *Meigen.*

8807. 1; lateralis*. (*Illig. M.* (*Meig.*) ii. 279.)
Mu. lateralis. *Fabr. A.* 314.—*Don.* i. 69. *pl.* 31. *f.* 3.—*Stew.* ii. 259.—*Turt.* iii. 607.
Mu. restituo. *Harr. Ex.* 36. *pl.* ix. *f.* 5.

Genus 157: (1281). OCYPTERYX, *Leach, Samou.*

OCYPTERA, *Meig.*—SYRPHUS *p*, *Panz.*

8808. 1; brassicaria*. *Fabr. A.* 312.—*Meig. Zw.* iv. 211. *pl.* 39. *f.* 29.—*Stew.* ii. 259.—*Turt.* iii. 606.—(*Sam. I.* 31.)
Mu. cylindrica. *DeGeer.* vi. 16. *pl.* 1. *f.* 12–14.
Sy. segnis. *Panz. F.* xxii. *f.* 22.

8809. 2; interrupta*. *Meig. Zw.* iv. 213.

Genus 158: (1282). DEXIA, *Meigen.*

OCYPTERA *p, Fabr.*—VOLUCCELLA *p, Schr.*

A. "Articulis primo secundoque abdominis æqualibus." *Meig.*

8810. 1, compressa*. *Meig. Zw.* v. 41.
Mu. compressa. *Fabr.* iv. 327.
Oc. rufa. *Fabr. A.* 314.
Mu. rutillus. *Harr. Ex.* 86. *pl.* xxv. *f.* 20.
Mu. rufiventris. *Fall. D. S.* (*Musc.*) 41. 8.
Vol. lurida. *Schrank B.* iii. 2483.
Mu. lateralis. *Panz. F.* vii. *f.* 22.

8811. 2, Maura. *Meig. Zw.* v. 39. *pl.* 43. *f.* 20. ♂.
Mu. Maura. *Fabr. A.* 302.
Mu. halterata. *Panz. F.* liv. *f.* 13.

8812. 3; caminaria*. *Meig. Zw.* v. 39.

8813. 4; melania*. *Meig. Zw.* v. 40.
Mu. putris. *Mus. Marsham.*

8814. 5; volvulus*. *Meig. Zw.* v. 35. *pl.* 43. *f.* 19.
Oc. volvulus. *Fabr. A.* 314.
Mu. cylindrica. *Fall. D. S.* (*Musc.*) 43. 13.
Mu. agilis. *Harr. Ex.* 86. *pl.* xxv. *f.* 19.

8815. 6; Anthracina*. *Meig. Zw.* v. 36.

8816. 7; melanoptera*. *Meig. Zw.* v. 36.
Mu. melanoptera. *Fall. D. S.* (*Musc.*) 52. 34.
Vo. roralis. *Schrank B.* iii. 2502?
Mu. atratus. *Harr. Ex.* 37. *pl.* ix. *f.* 7?

8817. 8; nana*. *Meig. Zw.* v. 37.

8818. 9; leucozona*. *Meig. Zw.* v. 37.
Mu. leucozona. *Panz. F.* civ. *f.* 19.
Mu. nigripes. *Fall. D. S.* (*Musc.*) 42. 12.
Mu. solivagus. *Harr. Ex.* 85. *pl.* xxv. *f.* 15?

8819. 10; nigripes*. *Meig. Zw.* v. 38.
Mu. nigripes. *Fabr. A.* 293.—*Panz. F.* civ. *f.* 18.
Mu. lateralis. *Fall. D. S.* (*Musc.*) 42. 11.

8820. 11; albifrons*.
Mu. albifrons. *Mus. Marsham.*

B. "Articulo primo abdominis secundo breviori." *Meig.*

8821. 12, picta. *Wied.*—*Meig. Zw.* v. 44.
Mu. procedo. *Harr. Ex.* 141. *pl.* xli. *f.* 42?

8822. 13; rustica*. *Meig. Zw.* v. 46.
Mu. rustica. *Fabr. A.* 296.
Mu. provenio. *Harr. Ex.* 142. *pl.* xli. *f.* 43.
Mu. tachinoides. *Fall. D. S.* (*Musc.*) 44. 17.
Mu. longipes. *Mus. Marsham.*

8823. 14; cinerea* *mihi.*

8824. 15, grisescens*. *Meig. Zw.* v. 45.
Mu. grisescens. *Fall. D. S.* (*Musc.*) 44. 16.

8825. 16; canina*. *Meig. Zw.* v. 47.
Mu. canina. *Fabr. A.* 296.—*Stew.* ii. 258.—*Turt.* iii. 601.
Mu. volets. *Harr. Ex.* 84. *pl.* xxv. *f.* 12.
Mu. chrysostoma. *Mus. Marsham.*

Genus 159: (1283). MESEMBRINA, *Meig.*, (*Kirby.*)

8826. 1; meridiana*. *Meig. Zw.* v. 11. *pl.* 42. *f.* 25.
Mu. meridiana. *Linn.* ii. 989.—*Harr. Ex.* 37. *pl.* ix. *f.* 9.—*Stew.* ii. 257.—*Turt.* iii. 597.—*Don.* xiv. 16. *pl.* 471. *f.* 2.—(*Sam. I.* 28.)—(*Kirby & Sp. I. E.* ii. 89.)—*Wood.* ii. 90. *pl.* 65.

Genus 160: (1284). SARCOPHAGA, *Meigen*, (*Kirby.*)

8827. 1; Mortuorum*. *Meig. Zw.* v. 16.
Mu. Mortuorum. *Linn.* ii. 989.—*Turt.* iii. 600.—*Don.* xv. 5. *pl.* 507.
Oc. Mortuorum. (*Sam. I.* 31.)
Mu. vomitoria. *Fabr. A.* 290.—*Stew.* ii. 258.—*Turt.* iii. 600.
Mu. chrysocephala. *DeGeer.* vi. 30.

8828. 2; carnaria*. *Meig. Zw.* v. 18.
Mu. carnaria. *Linn.* ii. 990.—*Stew.* ii. 258.—*Turt.* iii. 598.—*Shaw* G. Z. vi. 385. *pl.* 107. *fig. sup.*—(*Kirby & Sp. I. E.* i. 254.—ii. 202.)
Mu. vivipara major. *DeGeer.* vi. 31. *pl.* 3. *f.* 5–18.

8829. 3; striata*. *Meig. Zw.* v. 21.
Mu. striata. *Fabr. A.* 288.
Mu. pernix. *Harr. Ex.* 84. *pl.* xxv. *f.* 13?

8830. 4; albiceps*. *Meig. Zw.* v. 22.

8831. 5; sinuata*. *Meig. Zw.* v. 24.

8832. 6; pumila*. *Meig. Zw.* v. 24.

8833. 7; hæmorrhoidalis*. *Meig. Zw.* v. 28.
Mu. hæmorrhoidalis. *Fall. D. S.* (*Musc.*) 39. 2.
Mu. carnaria. *Berk.* i. 164.—*Harr. Ex.* 84. *pl.* xxv. *f.* 14.

8834. 8; cruenta*. *Meig. Zw.* v. 28.

8835. 9, hæmorrhoa*. *Meig. Zw.* v. 29.

Genus 161: (1285). MUSCA *Auctorum. Fly.*

A. Vide *Meig.* v. 51.

8836. 1; Cæsar*. *Linn.* ii. 989.—*Berk. S.* i. 164.—*Harr. Ex.* 87. *pl.* xxv. *f.* 23.—*Stew.* ii. 257.—*Turt.* iii. 599.—(*Sam. I.* 27.)—(*Kirby & Sp. I. E.* i. 254.)
Var.? Mu. fulges. *Harr. Ex.* 87. *pl.* xxv. *f.* 21?
Var.? Mu. ingredior. *Harr. Ex.* 140. *pl.* xli. *f.* 33.
Mu. redeo. *Harr. Ex.* 140. *pl.* xli. *f.* 34.

8837. 2; cornicina*. *Fabr. A.* 289.

8838. 3; regina*. *Meig. Zw.* v. 58.

8839. 4; Cæsarion. *Hoffm.—Meig. Zw.* v. 57.

8840. 5; illustris*. *Meig. Zw.* v. 54.

8841. 6; albipennis*. *Meig. Zw.* v. 58.

8842. 7; serena*. *Meig. Zw* v. 59.

8843. 8, cadaverina*? *Linn.* ii. 989.—*Berk. S.* i. 164.—*Stew.* ii. 258.—*Turt.* iii. 600.

B. Vide *Meig. Zw.* v. 60.

8844. 9; vomitoria*: Blue Flesh Fly.—*Berk. l. c. Linn.* ii. 989. —*Harr. Ex.* 86. *pl.* xxv. *f.* 18.—*Panz. F.* x. *f.* 19.—*Berk.* i. 164. —*Shaw* G. Z. vi. 383. *pl.* 107. *fig. inf.*—(*Sam. I.* 28.)—(*Kirby & Sp. I. E.* i. 254.—iv. 229.)
Mu. carnivora. *Fabr.* iv. 313.
Mu. carnaria cærulea. *DeGeer.* vi. 29.

8845. 10; erythrocephala*. *Meig. Zw.* v. 62.
Mu. vomitoria minimus. *Harr. Ex.* 87. *pl.* xxv. *f.* 22?

8846. 11; nana* *mihi.*

8847. 12; cærulea*. *Meig. Zw.* v. 63.

8848. 13; equestris*. *Meig. Zw.* v. 57?

8849. 14, semicærulea* *mihi.*

8850. 15; azurea*. *Fall. D. S.* (*Musc.*) 46. 19.—*Meig. Zw.* v. 63.
Mu. pervenio. *Harr. Ex.* 139. *pl.* xli. *f.* 29?

8851. [16; carnarida*.] *Mus. Marsham.*
Mu. ventito. *Harr. Ex.* 139. *pl.* xli. *f.* 30?

8852. 17; Lanio*. *Fabr. A.* 287.—*Panz. F.* liv. *f.* 11.
Mu. reviso. *Harr. Ex.* 141. *pl.* xli. *f.* 39.

8853. 18; nigromarginata* *mihi.*
Mu. ausus. *Harr. Ex.* 85. *pl.* xxv. *f.* 16?

8854. 19; Vespillo*. *Fabr. A.* 292.
Vol. cervina. *Schrank B.* iii. 2496?
Mu. recurro. *Harr. Ex.* 85. *pl.* xxv. *f.* 17?

8855. 20; atramentaria*. *Meig. Zw.* v. 65.
Mu. reverto. *Harr. Ex.* 140. *pl.* xli. *f.* 35?

8856. 21; rudis*. *Fabr. A.* 287.
Mu. remigro. *Harr. Ex.* 113. *pl.* xli. *f.* 37?

8857. 22; domestica*: House Fly. *Linn.* ii. 990.—*Harr. Ex.* 142. *pl.* xli. *f.* 44.—*Berk.* i. 165.—*Stew.* ii. 258.—*Schell. G. M. pl.* 1. —*Turt.* iii. 598.—(*Sam. I.* 27.)—(*Kirby & Sp. I. E.* i. 48.—ii. 362, 472 *nota.*—iv. 104.)

8858. 23; vitripennis*. *Meig. Zw.* v. 73. *pl.* 43. *f.* 34.

8859. 24; corvina*. *Fabr. S. I.* ii. 440.

Mu. Tau. *Schrank A. No.* 931.
Mu. autumnalis. *DeGeer*. vi. 41.
♂; Mu. nigripes. *Panz. F.* lx.*f.* 13.
♀; Mu. ludifica. *Fabr.* iv. 320.—*Panz. F.* cv. *f.* 13.

8860. 25; hortorum*. *Fall. D. S.* (*Musc.*) 52. 33.—*Meig. Zw.* v. 73.

8861. 26; pabulorum*. *Fall. D. S.* (*Musc.*) 51. 31.—*Meig. Zw.* v. 75.
Mu. prolapsa. *Harr. Ex.* 139. *pl.* xli. *f.* 32?

8862. 27; stabulans*. *Fall. D. S.* (*Musc.*) 52. 32.—*Meig. Zw.* v. 75. *pl.* 43. *f.* 35. ala.
Mu. cinerascens. *Wied. Z. M.* i. 79.
Mu. prodeo. *Harr. Ex.* 141. *pl.* xli. *f.* 41.

8863. 28; cæsia*. *Meig. Zw.* v. 76.

8864. 29; cyanella*. *Meig. Zw.* v. 77.

8865. 30; versicolor*. *Meig. Zw.* v. 77.

8866. 31; maculata*. *Meig. Zw.* v. 78.
♂; Mu. vulpina. *Fabr. A.* 292.
Mu. compunctus. *Harr. Ex.* 113. *pl.* xxxiii. *f.* 24.
♀; Mu. maculata. *Fabr. A.* 287.—*Harr. Ex.* 140. *pl.* xli. *f.* 38? —*Panz. F.* xliv. *f.* 22.—*Turt.* iii. 598.—*Don.* xiii. 25. *pl.* 445. *f.* 1.

8867. 32; meditabunda*. *Fabr. A.* 297.—*Panz. F.* xliv. *f.* 23.
Mu. serpo. *Harr. Ex.* 144. *pl.* xlii. *f.* 54.

Genus 162: (1286). ANTHOMYIA, *Meigen, Leach, Samou.*, (*Kirby.*)

A. "Seta antennarum plus minusve plumosa. a. *Pedibus nigris.*
* Oculis pubescentibus." *Meigen.*

8868. 1; lardaria*. *Meig. Zw.* v. 83.
Mu. lardaria. *Fabr. A.* 285.
Mu. levidus. *Harr. Ex.* 124. *pl.* xxxvi. *f.* 52.

8869. 2, albolineata. *Meig. Zw.* v. 83.
Mu. albolineata. *Fall. D. S.* (*Musc.*) 54. 38.
Mu. lardaria. *Mus. Marsham.*

8870. 3, incana. *Wied.*—*Meig. Zw.* v. 84.
Mu. nemorum. *Fall. D. S.* (*Musc.*) 55. 39.
Mu. vicanus. *Harr. Ex.* 152. *pl.* xlv. *f.* 78?

8871. 4, lucorum*. *Meig. Zw.* v. 85.
Mu. lucorum. *Fall. D. S.* (*Musc.*) 55. 40.
Mu. celsus. *Harr. Ex.* 125. *pl.* xxxvi. *f.* 53.

8872. 5, lugubris*. *Meig. Zw.* v. 87.
Mu. erectus. *Harr. Ex.* 125. *pl.* xxxvi. *f.* 55.

8873. 6; umbratica*. *Meig. Zw.* v. 88.

8874. 7; obscurata*. *Meig. Zw.* v. 89.

8875. 8; hirticeps*. *Meig. Zw.* v. 197.

** "Oculis nudis." *Meigen.*

8876. 9; quadrimaculata*. *Meig. Zw.* v. 92.
Mu. quadrimaculata. *Fall. D. S.* (*Musc.*) 63. 61.
Mu. cornuta. *Fabr. A.* 298.

8877. 10; quadrum*. *Meig. Zw.* v. 93.
Mu. quadrum. *Fabr. A.* 297.
Mu. subpuncta. *Fall. D. S.* (*Musc.*) 80. 97.
Mu. meditabunda. *Mus. Marsham.*
Mu. reversio. *Harr. Ex.* 146. *pl.* xliii. *f.* 62.

8878. 11, hilaris. *Meig. Zw.* v. 94. *pl.* 44. *f.* 9.
Mu. bilaris. *Fall. D. S.* (*Musc.*) 57. 44.

8879. 12; conica*. *Wied. Z. M.* i. 79.—*Meig. Zw.* v. 98.

8880. 13; sociata*. *Meig. Zw.* v. 98.

8881. 14; æqualis*. *Meig. Zw.* v. 99.

8882. 15, cinerella*. *Meig. Zw.* v. 100.
Mu. cinerella. *Fall. D. S.* (*Musc.*) 77. 91.
Mu. reditus. *Harr. Ex.* 146. *pl.* xliii. *f.* 61?

8883. 16; Cardui*. *Meig. Zw.* v. 104.

8884. 17; ancilla*. *Meig. Zw.* v. 105?

8885. 18; obtusipennis*. *Meig. Zw.* v. 193.
Mu. obtusipennis. *Fall. D. S.* (*Musc.*) 57. 46.

8886. 19; crassirostris*. *Meig. Zw.* v. 107.
Mu. flavipennis. *Fall. D. S.* (*Musc.*) 59. 52.

8887. 20; semicinerea*. *Wied. Z. M.* i. 84.—*Meig. Zw.* v. 108.
Mu. hyalinata. *Fall. D. S.* (*Musc.*) 64. 62.

8888. 21; fumosa*. *Meig. Zw.* v. 109.

8889. 22; denigrata*. *Meig. Zw.* v. 110.

8890. 23; asella*. *Meig. Zw.* v. 110.

8891. 24; nigrita*. *Meig. Zw.* v. 110.
Mu. nigrita. *Fall. D. S.* (*Musc.*) 60. 53.

b. "*Pedibus plus minusve fulvis.* * Oculis pubescentibus." *Meig.*

8892. 25; erratica*. *Meig. Zw.* v. 111.
Mu. erratica. *Fall. D. S.* (*Musc.*) 77. 92.
Mu. validus. *Harr. Ex.* 124. *pl.* xxxvi. *f.* 50.

8893. 26; errans*. *Meig. Zw.* v. 112.
Mu. revolo. *Harr. Ex.* 124. *pl.* xxxvi. *f.* 51?

8894. 27; vagans*. *Meig. Zw.* v. 112.
Mu. vagans. *Fall. D. S.* (*Musc.*) 78. 93.

8895. 28; signata*. *Meig. Zw.* v. 113.
Mu. extranea. *Mus. Marsham.*

8896. 29, variegata*. *Meig. Zw.* v. 114.

8897. 30; Populi*. *Meig. Zw.* v. 115.
Mu. pallida. *Fabr. A.* 295?
Mu. testacea. *DeGeer.* vi. 42.
** " Oculis nudis." *Meig.*

8898. 31, testacea. *Meig. Zw.* v. 116.
Mu. testacea. *Fabr.* iv. 320.—*Panz. F.* lx. *f.* 14.
Mu. scutellaris. *Fall.* D. *S.* (*Musc.*) 90. 123.
Mu. subventus. *Harr. Ex.* 145. *pl.* xliii. *f.* 60.

8899. 32, pagana*. *Meig. Zw.* v. 116.
Mu. pagana. *Fabr. A.* 288.
Mu. princeps. *Harr. Ex.* 151. *pl.* xlv. *f.* 76?

8900. 33; Angelicæ*. *Meig. Zw.* v. 117.
Mu. Angelicæ. *Scop. C. No.* 880.
Mu. deceptoria. *Gmel.* v. 2844.

8901. 34; urbana*. *Meig. Zw.* v. 118.
Mu. lenis. *Harr. Ex.* 147. *pl.* xliii. *f.* 66.

8902. 35; impuncta*. *Meig. Zw.* v. 118.
Mu. impuncta. *Fall.* D. *S.* (*Musc.*) 79. 96.
Mu. virescens. *Mus. Marsham.*

8903. 36; Marshami* *mihi.*
Mu. cinerea. *Mus. Marsham.*

8904. 37; separata*. *Meig. Zw.* v. 119.

8905. 38; modesta*. *Wied.*—*Meig. Zw.* v. 119.

8906. 39; pertusa*. *Meig. Zw.* v. 119.

8907. 40; tetrastigma*. *Meig. Zw.* v. 120.

8908. 41; uliginosa*. *Meig. Zw.* v. 121.
Mu. uliginosa. *Fall.* D. *S.* (*Musc.*) 81. 100.
Mu. deduco. *Harr. Ex.* 125. *pl.* xxxvi. *f.* 54.

8909. 42. coarctata*. *Meig. Zw.* v. 130.
Mu. coarctata. *Fall.* D. *S.* (*Musc.*) 84. 108.

8910. 43; strigosa*. *Meig. Zw.* v. 131.
Mu. strigosa. *Fabr. A.* 296.
Mu. conica. *Fall.* D. *S.* (*Musc.*) 83. 107.
Mu. vagans. *Panz. F.* lix. *f.* 18.

8911. 44; nigrimana*. *Meig. Zw.* v. 132.
Mu. Lancifer. *Harr. Ex. pl.* xxxvi. *f.* 59.

8912. 45; femorata* *mihi.*

8913. 46; nigrifrons* *mihi.*

8914. 47; tibialis* *mihi.*

B. " Seta antennarum plerumque nuda. a. *Pedibus fulvis. Oculis nudis.*" Meig.

8915. 48; diaphana*. *Wied. Z. M.* i. 81.—*Meig. Zw.* v. 189.

8916. 49; inanis*. *Meig. Zw.* v. 189.
Mu. inanis. *Fall.* D. *S.* (*Musc.*) 91. 127.

8917. 50, posticata. *Meig. Zw.* v. 190?

8918. 51; ornata*. *Meig. Zw.* v. 191.

8919. 52; varia*. *Meig. Zw.* v. 187.

8920. 53; fulgens*. *Meig. Zw.* v. 183.

8921. 54; mitis*. *Meig. Zw.* v. 183.
Mu. lustrator. *Harr. Ex.* 148. *pl.* xliii. *f.* 72.

8922. 55, bicolor. *Hoffm.*—*Meig. Zw.* v. 185.

8923. 56; solennis*. *Meig. Zw.* v. 187.

8924. 57; Winthemi*. *Meig. Zw.* v. 186.

b. "*Pedibus nigris: oculis nudis.*" Meig.

8925. 58; irritans*. *Meig. Zw.* v. 134.
Mu. irritans. *Fall. D. S.* (*Musc.*) 62. 58.

8926. 59; bidens*. *Meig. Zw.* v. 135.

8927. 60; palæstica. *Meig. Zw.* v. 135.

8928. 61; militaris*. *Meig. Zw.* v. 136.

8929. 62; meteorica*. *Meig. Zw.* v. 137.
Mu. meteorica. *Linn.* ii. 993.—*Fanz. F.* i. *f.* 19.—*Stew.* ii. 260.
—*Turt.* iii. 609.—(*Kirby & Sp. I. E.* i. 147.)
Mu. vaccarum. *DeGeer.* vi. 41. *pl.* 5. *f.* 1.
Mu. Mantos. *Harr. Ex.* 148. *pl.* xliii. *f.* 73. ♀.

8930. 63; armipes*. *Meig. Zw.* v. 138.
Mu. armipes. *Fall. D. S.* (*Musc.*) 75. 86.

8931. 64; armata*. *Meig. Zw.* v. 139.

8932. 65, Lepida*. *Wied. Z. M.* i. 82.—*Meig. Zw.* v. 140.

8933. 66; scalaris*. *Meig. Zw.* v. 141.
Mu. scalaris. *Fabr. A.* 305.

8934. 67; canicularis*. *Meig. Zw.* v. 143.
Mu. canicularis. *Linn.* ii. 992.—*Stew.* ii. 259.—*Turt.* iii. 607.
Mu. domestica minor. *DeGeer.* vi. 14.
Mu. socio. *Harr. Ex.* 147. xliii. *pl.* 69. ♂.
Mu. constans. *Harr. Ex.* 149. *pl.* xliii. *f.* 75. ♀?
Var.? Mu. socio minor. *Harr. Ex.* 153. *pl.* xlv. *f.* 83.

8935. 68; dentipes*. *Meig. Zw.* v. 144.
Mu. dentipes. *Fabr. A.* 303.
Mu. domitor. *Harr. Ex.* 148. *pl.* xliii. *f.* 71.

8936. 69; hortorum* *mihi.*

8937. 70; floricola*. *Meig. Zw.* v. 145.

8938. 71; triangula*. *Meig. Zw.* v. 148.
Mu. triangula. *Fall. D. S.* (*Musc.*) 74. 82.

8939. 72; pacifica*. *Meig. Zw.* v. 149.
Mu. allevo. *Harr. Ex.* 147. *pl.* xliii. *f.* 97.

8940. 73; Diadema*. *Meig. Zw.* v. 151.

8941. 74; pusilla*. *Meig. Zw.* v. 151.

8942. 75; gibbera*. *Meig. Zw.* v. 152.

8943. 76; sepia*. *Meig. Zw.* v. 152.

8944. 77; tristis*. *Meig. Zw.* v. 153.

8945. 78; carbonaria*. *Meig. Zw.* v. 154?

8946. 79; fumigata*. *Wied.—Meig. Zw.* v. 154.

8947. 80; glabricula*. *Meig. Zw.* v. 155.
Mu. glabricula. *Fall. D. S.* (*Musc.*) 75. 87.
Mu. structus. *Harr. Ex.* 147. *pl.* xliii. *f.* 68?

8948. 81; Stygia*. *Meig. Zw.* v. 155.

8949. 82; luctuosa*. *Meig. Zw.* v. 156.

8950. 83; nigella*. *Meig. Zw.* v. 156.

8951. 84; aterrima*. *Meig. Zw.* v. 157.

8952. 85; ærea*. *Meig. Zw.* v. 157.
Mu. ærea. *Fall. D. S.* (*Musc.*) 76. 89.

8953. 86; pratensis*. *Meig. Zw.* v. 158?
Mu. solor. *Harr. Ex.* 146. *pl.* xliii. *f.* 65.

8954. 87; transversa* *mihi.*

8955. 88; leucostoma*. *Meig. Zw.* v. 160.
Mu. leucostoma. *Fall. D. S.* (*Musc.*) 61. 56.
Mu. diabolus. *Harr. Ex.* 126. *pl.* xxxvi. *f.* 58?

8956. 89; Anthrax*. *Meig. Zw.* v. 161.

8957. 90; tarsalis* *mihi.*

8958. 91; dispar* *mihi.*

8959. 92; triquetra*. *Meig. Zw.* v. 162.
Mu. triquetra. *Fall. D. S.* (*Musc.*) 73. 80.
Mu. ambulans. *Schrank B.* iii. 2469.
Mu. ornate. *Harr. Ex.* 146. *pl.* xliii. *f.* 64.

8960. 93; pratincola (pratinicola.)* *Meig. Zw.* v. 162.
Mu. pratincola. *Panz. F.* cvii. *f.* 12.

8961. 94; pluvialis*. *Meig. Zw.* v. 163.—(*Sam. I.* 3.)
Mu. pluvialis. *Linn.* ii. 992.—*Stew.* ii. 260.—*Turt.* iii. 607.—*Don.* xi. 33. *pl.* 372.
Mu. litus. *Harr. Ex.* 119. *pl.* xxxv. *f.* 32.

8962. 95; floralis*. *Meig. Zw.* v. 165.
Mu. floralis. *Fall. D. S.* (*Musc.*) 71. 76.

8963. 96; promissa*. *Meig. Zw.* v. 166.

8964. 97; stigmatica*. *Meig. Zw.* v. 167?

8965. 98; radicum*. *Meig. Zw.* v. 168.
Mu. radicum. *Linn.* ii. 992.
Mu. decore. *Harr. Ex.* 146. *pl.* xliii. *f.* 63. ♀.

An. Brassicæ. *Wied. Z. M.* i. 78.
Scat. ceparum. (*Kirby & Sp. I. E.* i. 190?)
8966. 99; æstiva*. *Meig. Zw.* v. 169.
8967. 100; muscaria*. *Meig. Zw.* v. 170.
Stomoxys muscaria. *Fabr. A.* 282.
8968. 101; spreta*. *Meig. Zw.* v. 171.
8969. 102; platura*. *Meig. Zw.* v. 171.
8970. 103; melanura*. *Meig. Zw.* v. 172.
8971. 104; striolata*. *Meig. Zw.* v. 173.
Mu. striolata. *Fall.* D. *S.* (*Musc.*) 71. 77.
8972. 105; fugax*. *Meig. Zw.* v. 174.
8973. 106; fuscula*. *Meig. Zw.* v. 174.
Mu. fuscula. *Fall.* D. *S.* (*Musc.*) 86. 113.
8974. 107; nigricornis* *mihi.*
8975. 108; ruficeps*. *Meig. Zw.* v. 177.
8976. 109; picipes*. *Meig. Zw.* v. 178.

Genus 163: (1287). DRYMEIA, *Meigen?*
8977. 1; hispida* *mihi.*

Genus 164: (1288). ERIPHIA, *Meigen?*
8978. 1; aterrima* *mihi.*

Genus 165: (1289). CŒNOSIA, *Meigen.*
8979. 1; Fungorum*. *Meig. Zw.* v. 211.
Mu. Fungorum. *DeGeer.* vi. 42. *pl.* 5. *f.* 1–7.
8980. 2; tigrina*. *Meig. Zw.* v. 212.
Mu. tigrina. *Fabr. A.* 297. (!)—*Turt.* iii. 602.
Mu. rapax. *Stew.* ii. 260.
Mu. quadrum. *Fall.* D. *S.* (*Musc.*) 80. 99.
8981. 3; nemoralis*. *Meig. Zw.* v. 212. *pl.* 45. *f.* 9.
8982. 4; rufina*. *Meig. Zw.* v. 213.
Mu. rufina. *Fall.* D. *S.* (*Musc.*) 92. 132.
8983. 5; sexnotata*. *Meig. Zw.* v. 213.
8984. 6; intermedia*. *Meig. Zw.* v. 214.
Mu. intermedia. *Fall.* D. *S.* (*Musc.*) 87. 115.
8985. 7; murina*. *Meig. Zw.* v. 215.
8986. 8; means*. *Meig. Zw.* v. 216.
8987. 9; nigra*. *Meig. Zw.* v. 216.
8988. 10; minima*. *Meig. Zw.* v. 217.
8989. 11, assimilis* *mihi.*
8990. 12; basalis* *mihi.*
8991. 13; semicinerea* *mihi.*
8992. 14, pallida *mihi.*

Genus 166: (1290). LISPE, *Meigen.*

8993. 1, tentaculata*. *Meig. Zw.* v. 227. *pl.* 45. *f.* 15.
Mu. tentaculata. *DeGeer.* vi. 42.

Genus 167: (1291). CORDYLURA, *Fallen.*
ACOLASTE, *Meig. olim.*

8994. 1; pubera*. *Meig. Zw.* v. 230. *pl.* 45. *f.* 22. ♀.
Mu. pubera. *Linn.* ii. 994.
Mu. asiliformis. *Mus. Marsham.*

8995. 2, ciliata*. *Meig. Zw.* v. 231.
Mu. umbrosa. *Mus. Marsham.*

8996. 3; rufipes. *Meig. Zw.* v. 232. (!)

8997. 4, albipes. *Fall. D. S.* (*Scatom.*) 9. 8.—*Meig. Zw.* v. 233.
Mu. ochroleuca. *Mus. Marsham.*

8998. 5, nervosa. *Meig. Zw.* v. 234. *pl.* 45. *f.* 21. ♂.

8999. 6; armipes*. *Meig. Zw.* v. 234.
Mu. semiflava. *Panz. F.* lix. 19?

9000. 7; striolata. *Meig. Zw.* v. 235. (!)

9001. 8; spinimana*. *Fall. D. S.* (*Scatom.*) 7. 3.

9002. 9; obscura*. *Fall. D. S.* (*Scatom.*) 9. 6.

Genus 168: (1292). SCATOPHAGA, *Meigen.*
SCATOMYZA, *Fallen.*—VOLUCELLA *p, Schr.*

9003. 1; fucorum. *Fall. D. S.* (*Scatom.*) 5. 5.—*Meig. Zw.* v. 253.

9004. 2; litorea. *Fall. D. S.* (*Scatom.*) 4. 4.—*Meig. Zw.* v. 254.

9005 † 3, rufipes. *Meig. Zw.* v. 253. (!) In Mus. *Brit.*

9006. 4; merdaria*. *Meig. Zw.* v. 249.
Mu. merdaria. *Fabr. A.* 306.

9007. 5; cineraria*. *Meig. Zw.* v. 251.

9008. 6; spurca*. *Meig. Zw.* v. 250.

9009. 7; lutaria*. *Meig. Zw.* v. 250.
Mu. lutaria. *Fabr. A.* 306.
Mu. scybalaria. *Mus. Marsham.*—*Don.* x. 56. *pl.* 345. *f.* 5?
Mu. leucophæus. *Harr. Ex.* 73. *pl.* xxi. *f.* 3.

9010. 8; inquinata*. *Meig. Zw.* v. 250.

9011. 9; analis*. *Meg.*—*Meig. Zw.* v. 251.

9012. 10; stercoraria*. *Meig. Zw.* v. 248.
Mu. stercoraria. *Linn.* ii. 996.—*Berk. S.* i. 165.—*Stew.* ii. 262.—*Turt.* iii. 615.
Vo. scybalaria. *Schrank B.* iii. 2499.
Mu. putris. *Harr. Ex.* 73. *pl.* xxi. *f.* 1, 2.

9013. 11; scybalaria*. *Meig. Zw.* v. 247.
Mu. scybalaria. *Linn.* ii. 996.—*Stew.* ii. 261.—*Turt.* iii. 615.

Genus 169: (1293). DRYOMYZA, *Fallen.*

9014. 1, cingulata *mihi.*

9015. 2; flaveola*. *Meig. Zw.* v. 256.
Mu. flaveola. *Fabr. A.* 306.
Dr. vetula. *Fall. D. S.* (*Sciomyz.*) 16. 1.
Mu. concolor. *Mus. Marsham.*
Mu. mellinus. *Harr. Ex.* 74. *pl.* xxi. *f.* 4.

9016. 3, Fungorum.
Mu. Fungorum. *Mus. Marsham.*

9017. 4; anilis*. *Fall. D. S.* (*Sciomyz.*) 16. 2.

9018. 5; præusta*. *Meig. Zw.* v. 257.

Genus 170: (1294). SAPROMYZA, *Fallen.*
TEPHRITIS *p, Fabr.*

9019. 1; rorida*. *Fall. D. S.* (*Ortal.*) 32. 7.—*Meig. Zw.* v. 259.
Mu. mulsus. *Harr. Ex.* 116. *pl.* xxxiv. *f.* 20?

9020. 2; obsoleta*. *Fall. D. S.* (*Ortal.*) 31. 6.—*Meig. Zw.* v. 260.
Mu. flava. *Fabr.* iv. 355.—*Stew.* ii. 262.—*Turt.* iii. 620.

9021. 3; pallida*. *Fall. D. S.* (*Ortal.*) 32. 8.—*Meig. Zw.* v. 260.

9022. 4; flava*. *Meig. Zw.* v. 260.
Mu. flava. *Linn.* ii. 997.—*Berk. S.* i. 165.—*Shaw* G. Z. vi. 385.

9023. 5; quadripunctata*. *Meig. Zw.* v. 262.
Mu. quadripunctata. *Fabr.* iv. 356.—*Turt.* iii. 620.

9024. 6; sexpunctata*. *Meig. Zw.* v. 262.

9025. 7; præusta*. *Meig. Zw.* v. 264.

Genus 171: (1295). PALLOPTERA, *Fallen.*
SAPROMYZA *p, Meigen.*—DACUS *p, Fabr.*—TRUPANEA *p, Schr.*

9026. 1; arcuata*. *Fall. D. S.* (*Ortal.*) 25. 3.
Sa. arcuata. *Meig. Zw.* v. 269.

9027. 2; trimacula*.
Sa. trimacula. *Meig. Zw.* v. 267.
Mu. minutus. *Harr. Ex.* 74. *pl.* xxi. *f.* 7.

9028. 3; decempunctata*. *Fall. D. S.* (*Ortal.*) 30. 1.
Sa. decempunctata. *Meig. Zw.* v. 270. *pl.* 46. *f.* 12.

9029. 4, inusta.
Sa. inusta. *Meig. Zw.* v. 267.
Mu. Dianæ. *Mus. Marsham.*

9030. 5; Umbellatarum*.
Da. Umbellatarum. *Fabr. A.* 277. (!)—*Turt.* iii. 619.
Mu. gangrenosa. *Panz. F.* lix. *f.* 22.

9031. 6; ustulata*. *Fall. D. S.* (*Ortal.*) 24. 2.

9032 † 7, unicolor*.

Mu. unicolor. *Fabr. A.* 307.
Pa. marginella. *Fall. D. S.* (*Ortal.*) 25. 4. In Mus. *D. Cooper.*
9033. 8; notata*. *Fall. D. S.* (*Ortal.*) 30. 3.

Genus 172: (1296). ORTALIS, *Fallen.*
SCATOPHAGA *p, et* TEPHRITIS *p, Fabr.*—TRUPANEA *p, Schr.*
9034. 1; oscillans*. *Meig. Zw.* v. 281.
9035. 2; lugubris*. *Meig. Zw.* v. 279.
9036. 3; mœrens*. *Meig. Zw.* v. 280. (!)
9037. 4; lacustris*. *Meig. Zw.* v. 280.
Mu. germinationis. *Linn.* ii. 998?—*Stew.* ii. 263?—*Turt.* iii. 620?
9038. 5; nigrina*. *Wied.*—*Meig. Zw.* v. 279.
9039. 6; Cerasi*. *Meig. Zw.* v. 282. *pl.* 46. *f.* 26. ala.
Mu. Cerasi. *Linn.* ii. 998.—*Stew.* ii. 263.—*Turt.* iii. 621.
Te. Cerasi. (*Kirby & Sp. I. E.* i. 196.)
Or. uliginosæ. *Fall. D. S.* (*Ortal.*) 19. 4.
Var.? Te. Mali. *Fabr. A.* 320.
Te. morio. *Fabr. A.* 322.
Mu. frondescentiæ. *Linn.* ii. 999?
9040. 7; Syngenesiæ*. *Meig. Zw.* v. 283.
Te. Syngenesiæ. *Fabr. A.* 321.
Or. Juncorum. *Fall. D. S.* (*Ortal.*) 19. 5.
Mu. Urticæ. *Schrank A.* 969.
9041. 8, ornata. *Meig. Zw.* v. 277.
9042. 9, picta*. *Meig. Zw.* v. 276. *pl.* 46. *f.* 28. ala. (!)
9043. 10; Urticæ*. *Fall. D. S.* (*Ortal.*) 17. 1.—*Meig. Zw.* 275. *pl.* 46. *f.* 19.
Mu. Urticæ. *Linn.* ii. 998.—*Turt.* iii. 620.
9044. 11; crassipennis*. *Fall. D. S.* (*Ortal.*) 18. 2.—*Meig.* v. 273. *pl.* 46. *f.* 23. ala.
Sc. crassipennis. *Fabr. A.* 209.
9045. 12; marmorea*. *Meig. Zw.* v. 274. *pl.* 46. *f.* 21. ala.
Sc. marmorea. *Fabr. A.* 209.
Mu. hyalinata. *Panz. F.* lx. *f.* 24.

Genus 173: (1297). SEIOPTERA, (*Kirby.*)
TEPHRITIS *p, Latr.*—ORTALIS *p, Meigen.*—TYROPHAGA *p, Curtis.*
9046. 1; vibrans*. (*Kirby & Sp. I. E.* ii. 305 & 359.)
Mu. vibrans. *Linn.* ii. 996.—*DeGeer.* vi. 17. *pl.* 1. *f.* 19, 20. —*Harr. Ex.* 121. *pl.* xxxv. *f.* 39.—*Berk. S.* i. 165.—*Stew.* ii. 262.—*Turt.* iii. 618.—*Don.* x. 54. *pl.* 346. *f.* 3.—xiii. 71. *pl.* 467. *f.* 1.
Te. vibrans. (*Sam. I.* 40.)
Ty. vibrans. (*Curtis folio* 126.)

Genus 174: (1298). TEPHRITIS, *Latr.*, *Leach*, *Samou.*, *Curtis*.

DACUS *p*, DICTYA *p*, *et* SCATOPHAGA *p*, *Fabr.*—TRUPANEA *p*, *Schra.*—TRYPETA, *Meig.*

9047. 1; cornuta*. *Curtis*. vi. *pl.* 241.
Mu. cornuta. *Fabr.* iv. 357.

9048. 2; Tussilaginis*.
Mu. Tussilaginis. *Fabr.* iv. 359.
Try. Tussilaginis. *Meig. Zw.* v. 319. *pl.* 48. *f.* 27: (*nec* 26.)
Tru. Acanthi. *Schrank B.* iii. 2509.

9049. 3; Lappæ*.
Try. Lappæ. *Meig.* v. 318.
Mu. Arctii. *Fanz. F.* xxii. *f.* 23.

9050. 4, Arctii. (*Curtis l. c. supra.*)
Mu. Arctii. *DeGeer*. vi. 21. *pl.* 2. *f.* 6–14.
Te. solstitialis. *Panz. F.* ciii. *f.* 22.

9051. 5; marginata*. (*Curtis l. c. supra.*)
Te. marginata. *Fall.* D. *S.* (*Ortalidæ.*) 7. 8.
Try. marginata. *Meig. Zw.* v. 322. *pl.* 49. *f.* 15. *ala.*

9052. 6; Wiedemanni*. (*Curtis l. c. supra.*)
Try. Wiedemanni. *Meig. Zw.* v. 320. *pl.* 49. *f.* 2. (!)

9053. 7; Winthemi*.
Try. Winthemi. *Meig. Zw.* v. 320. *pl.* 48. *f.* 25. (*nec* 27.) *ala.*

9054. 8; Abrotani*. (*Curtis l. c. supra.*)
Try. Abrotani. *Meig. Zw.* v. 314. *pl.* 48. *f.* 21. *ala.*

9055. 9; Artemisiæ. *Fabr. A.* 317.
Try. Artemisiæ. *Meig. Zw.* v. 314. *pl.* 48. *f.* 20.
Tep. interrupta. *Fall.* D. *S.* (*Ortal.*) 5. 4.
Mu. fasciata. *Mus. Marsham.*
Mu. perelegand. *Harr. Ex.* 74. *pl.* xxi. *f.* 5?

9056. 10; Zoë*. (*Curtis l. c. supra.*)
Try. Zoë. *Wiedemann.*—*Meig. Zw.* v. 315. *pl.* 48. *f.* 14. ♂. *f.* 15. ♀.

9057. 11; cognata*.
Try. cognata. *Wiedemann.*—*Meig. Zw.* v. 315. *pl.* 48. *f.* 19.

9058. 12; Onopordinis*. *Fall.* D. *S.* (*Ortal.*) 15. 25.—(*Sam. I.* 40.)—(*Curtis l. c. supra.*)
Mu. Onopordinis. *Fabr. S. I.* ii. 455.—*Stew.* ii. 263.
Try. Onopordinis. *Meig. Zw.* v. 316. *pl.* 48. *f.* 24. *ala.*

9059. 13; Centaureæ*. *Fabr. A.* 322.
Try. Centaureæ. *Meig. Zw.* v. 324. *pl.* 49. *f.* 8. *ala.*

9060. 14; discoidea*. (*Curtis l. c. supra.*)
Di. discoidea. *Fabr. A.* 326.
Te. Centaureæ. *Fall.* D. *S.* (*Ortal.*) 16. 26.
Mu. cæsio. *Harr. Ex.* 75. *pl.* 21. *f.* 8.

9061. 15; rivosa*.
Mu. rivosa. *Mus. Marsham.*

9062. 16; Cardui*. (*Sam. I.* 40.)—(*Kirby & Sp. I. E.* iii. 274.) —(*Curtis l. c. supra.*)
Mu. Cardui. *Linn.* ii. 998.—*Stew.* ii. 263.—*Turt.* iii. 621
Try. Cardui. *Meig. Zw.* v. 326. *pl.* 49. *f.* 9. *ala.*

9063. 17; cuspidata*. (*Curtis l. c. supra.*)
Try. cuspidata. *Meig. Zw.* v. 328. *pl.* 49. *f.* 5.

9064. 18; Solstitialis*. (*Curtis l. c. supra.*)
Mu. Solstitialis. *Linn.* ii. 999.—*Berk. S.* i. 165?—*Stew.* ii. 263. —*Turt.* iii. 621.—*Don.* ix. 15. *pl.* 294?
Try. Solstitialis. *Meig. Zw.* v. 329. *pl.* 49. *f.* 10. *ala.*
Tru. Leucacanthi. *Schrank B.* iii. 2507.
Da. Dauci. *Fabr. A.* 277.
♀; Da. hastatus. *Fabr. A.* 276.

9065. 19; flavipes* *mihi.*

9066. 20; stylata*. *Fabr. A.* 275. (!)—(*Curtis l. c. supra.*)
Mu. stylata. *Stew.* ii. 262.—*Turt.* iii. 619.
Mu. Jacobeæ. *Panz. F.* xcvii. *f.* 22.
Tru. Circii. *Schrank B.* iii. 2505.

9067. 21; pugionata*.
Try. pugionata. *Meig. Zw.* v. 330. *pl.* 49. *f.* 11. *ala.*

9068. 22; Arnicæ*. *Fabr. A.* 316.—(*Curtis l. c. supra.*)
Mu. Arnicæ. *Linn.* ii. 997.
Mu. arcuata. *Panz. F.* xcviii. *f.* 22.
Mu. Onopordinis? *Don.* ii. 67. *pl.* 62?—*Turt.* iii. 622?
Tru. sphærocephalas. *Schrank B.* iii. 2515.
Mu. miliaria. *Schrank B.* iii. 968.

9069. 23; corniculata*. *Fall. D. S.* (*Ortal.*) 8. 11.
Mu. biarcuata. *Mus. Marsham.*
Tep. biarcuata. (*Curtis l. c. supra.*)

9070. 24; laticauda*.
Try. laticauda. *Meig. Zw.* v. 339. *pl.* 50. *f.* 11. *ala.*
Mu. Achillææ. *Mus. Marsham.*

9071. 25; parietina*. (*Curtis l. c. supra.*)
Mu. parietina. *Linn.* ii. 996.
Try. parietina. *Meig. Zw.* v. 334. *pl.* 50. *f.* 7. *ala.*
Tep. pantherina. *Fall. D. S.* (*Ortal.*) 10. 14.

9072. 26; guttularis*. (*Curtis l. c. supra.*)
Try. guttularis. *Meig. Zw.* v. 341. (!)

9073. 27; flavicauda*.
Try. flavicauda. *Meig. Zw.* v. 336. *pl.* 50. *f.* 11. *ala.*

9074. 28; Absinthii*. *Fabr. A.* 322.—(*Curtis l. c. supra.*)
Try. Absinthii. *Meig. Zw.* v. 340. *pl.* 50. *f.* 12. *ala.*

9075. 29; confusa*. (*Curtis l. c. supra.*)
Try. confusa. *Meig. Zw.* v. 337. *pl.* 50. *f.* 9. *ala.*

9076. 30, Heraclei. *Fabr. A.* 277.
Mu. Heraclei. *Linn.* ii. 998.
Try. Heraclei. *Meig. Zw.* v. 338. *pl.* 50. *f.* 1.

9077 † 31. Hyoscyami. *Fall. D. S.* (*Ortal.*) 9. 12.—(*Curtis l. c. supra.*)
Mu. Hyoscyami. *Linn.* ii. 998.—*Stew.* ii. 262.
Try. Hyoscyami. *Meig. Zw.* v. 337. *pl.* 50. *f.* 2. In Mus. Brit.?

9078. 32; Leontodontis*. *Fall. D. S.* (*Ortal.*) 9. 13.—(*Curtis l. c. supra.*)
Mu. Leontodontis. *DeGeer.* vi. 24. *pl.* 2. *f.* 15–18.
Mu. stellata. *Fanz. F.* xx. *f.* 23.
Da. Scabiosæ. *Fabr. A.* 278.
Tep. parietina. *Fabr. A.* 319?
Mu. cinereus. *Harr. Ex.* 75. *pl.* xxi. *f.* 11.

9079. 33; radiata*. *Fabr. A.* 319.—(*Curtis l. c. supra.*)
Try. radiata. *Meig. Zw.* 343. *pl.* 50. *f.* 3.

9080. 34; eluta*. *Meig. Zw.* v. 346. *pl.* 50. *f.* 13? *ala.*

9081. 35; obsoleta*. *Wied.*—*Meig. Zw.* iv. 349.

9082. 36; Sonchi*. *Fall. D. S.* (*Ortal.*) 14. 23.
Mu. Sonchi. *Linn.* ii. 998?

9083. 37; Colon*. (*Curtis l. c. supra.*)
Try. Colon. *Meig. Zw.* v. 346.

9084. 38, Serratulæ. *Fall. D. S.* (*Ortal.*) 14. 22.
Mu. Serratulæ. *Linn.* ii. 997.—*Turt.* iii. 620.

9085. 39; pallens*. (*Curtis l. c. supra.*)
Try. pallens. *Meig. Zw.* v. 347. *pl.* 50. *f.* 5.

9086. 40; marginepunctata* *mihi.*

9087. 41; innotata* *mihi.*

9088 † 42. basalis* *mihi.*
Mu. permundus. *Harr. Ex.* 74. *pl.* xxi. *f.* 6.

Genus 175: (1299). LAUXANIA, *Meigen.*
SARGUS *p*, *Fabr.*

9089. 1; cylindricornis*. *Fabr. A.* 212.—*Panz. F.* cv. *f.* 11.
Mu. chrysoptera. *Schrank B.* iii. 2470.
La. rufitarsis. *Latr. H.* xiv. 390.

9090. [2; basalis*] *mihi.*

9091. 3, ænea*. *Fall. D. S.* (*Ortal.*) 28. 3.

9092. 4; geniculata*. *Meig. Zw.* v. 298.
Sa. geniculatus. *Fabr. A.* 257.

9093. 5; atrimana*. *Meg.*—*Meig. Zw.* v. 299.

9094. 6; lupulina*. *Meig. Zw.* v. 301.
Mu. lupulina. *Fabr. A.* 298.

Genus 176: (1300). LONCHÆA, *Fallen.*

9095. 1; chorea*. *Fall. D. S.* (*Ortal.*) 26. 1.—*Meig. Zw.* v. 304. *pl.* 47. *f.* 31.
Mu. chorea. *Fabr. A.* 304.

9096. 2; nigrimana*. *Meig. Zw.* v. 306.

9097. 3; vaginalis*. *Fall. D. S.* (*Ortal.*) 26. 2.

9098. 4; pusilla*. *Meig. Zw.* v. 305.

9099. 5; tarsata*. *Fall. D. S.* (*Ortal.*) 26. 3.

9100. 6, nigra. *Meig. Zw.* v. 305.

9101. 7, ænea*. *Meig. Zw.* v. 306.

9102. 8; latifrons*. *Meig. Zw.* v. 308.

Genus 177: (1301). ——

9103. 1. splendida *mihi.*
Mu. stagnorum. *Mus. Marsham.*

Genus 178: (1302). PSILA, *Meigen.*

SCATOPHAGA *p, Fabr.*—VOLUCELLA *p, Schrank.*

9104. 1; fimetaria*. *Meig. Zw.* v. 356.
Mu. fimetaria. *Linn.* ii. 996.—*Turt.* iii. 615.
Mu. flava. *Fanz. F.* xx. *f.* 22.
Mu. rufescens. *Mus. Marsham.*
Mu. luteus. *Harr. Ex.* 116. *pl.* xxxiv. *f.* 16.

9105. 2; pallida*. *Meig. Zw.* v. 357.
Sca. pallida. *Fall. D. S.* (*Opomyz.*) 9. 2.

9106. 3, bicolor. *Meig. Zw.* v. 358.

9107. 4; pectoralis. *Fall.?*—*Meig. Zw.* v. 358?

9108. 5; nigricornis*. *Meig. Zw.* v. 359.

9109. 6; Rosæ*. *Meig. Zw.* v. 358.
Te. Rosæ. *Fabr. A.* 319.

9110. 7; nigra*. *Meig. Zw.* v. 359.
Sc. nigra. *Fall. D. S.* (*Opomyz.*) 9. 4.

Genus 179: (1303). LOXOCERA, *Meig.*, *Leach*, *Samou.*

MULIO *p, et* SYRPHUS *p, Fabr.*—NEMOTELUS *p, Schr.*

9111. 1; ichneumonea*. *Meig. Zw.* v. 363.—(*Sam. I.* 26.)
Mu. ichneumonea. *Linn.* ii. 986.—*Turt.* iii. 646.
Mu. aristata. *Panz. F.* lxxiii. *f.* 24.
Ne. albisetus. *Schrank B.* iii. 2406.
Mu. longerro. *Harr. Ex.* 152. *pl.* xlv. *f.* 79.

9112. 2; elongata*. *Meig. Zw.* v. 364.

9113. 3; sylvatica*. *Meig. Zw.* v. 365. *pl.* 51. *f.* 22.
Mu. cruenta. *Mus. Marsham.*

Genus 180: (1304). CHYLIZA, *Fallen.*

SARGUS *p*, *Fabricius.*

9114. 1; leptogaster*. *Fall. D. S.* (*Opomyz.*) 7. 2.
Mu. leptogaster. *Panz. F.* liv. *f.* 19.
Sa. scutellatus. *Fabr. A.* 257.

9115. 2; atriseta*. *Meig. Zw.* v. 369. *pl.* 51. *f.* 26?

Genus 181: (1305). ——

LISSA[a], *Meig.*—CHYLIZA *p*, *Fallen.*—OCYPTERA *p*, *Fabr.*

9116. 1; dolium*.
Oc. dolium. *Fabr. A.* 315.
Ch. loxocerina. *Fall. D. S.* (*Opomyz.*) 6. 1.
Li. loxocerina. *Meig. Zw.* v. 370. *pl.* 52. *f.* 1–4.

Genus 182: (1306). CALOBATA, *Meigen.*

MOSILLUS, *Latr.*—MOCILLUS, *Samou.*

9117. 1; petronella*. *Meig. Zw.* v. 377.
Mu. petronella. *Linn.* ii. 994.—*Turt.* iii. 611.
Ca. corrigiolata. *Fall. D. S.* (*Opomyz.*) 2. 3.
Schellen. G. *M. pl.* 6. *f.* 2.

9118. 2; cibaria*. *Meig. Zw.* v. 378.
Mu. cibaria. *Linn.* ii. 995.
Ca. cothurnata. *Fall. D. S.* (*Opomyz.*) 2. 2.

9119. 3; cothurnata*. *Meig. Zw.* v. 379. *pl.* 52. *f.* 18, 19.
Mu. cothurnata. *Panz. F.* liv. *f.* 20.

9120. 4, sellata. *Meig. Zw.* v. 380. (!)

9121. 5; ephippium*. *Meig. Zw.* v. 380.
Mu. ephippium. *Fabr.* iv. 338.—*Panz. F.* xxvi. *f.* 21.

Genus 183: (1307). MICROPEZA, *Meigen.*

CALOBATA *p*, *Fabr.*—RHAGIO *p*, *Schrank.*

9122. 1; lateralis*. *Meig. Zw.* v. 383. *pl.* 53. *f.* 5.

9123. 2; corrigiolata*. *Meig. Zw.* v. 384. *pl.* 53. *f.* 6.
Mu. corrigiolata. *Linn.* ii. 995.
Ca. filiformis. *Fabr. A.* 263.—(*Sam. I.* 8.)
Mu. pedo. *Harr. Ex.* 115. *pl.* xxxiv. *f.* 16.

Genus 184: (1308). SEPSIS, *Fallen, Curtis.*

MICROPEZA *p*, *Latreille.*—TEPHRITIS *et* CALOBATA *p*, *Fabr.*—
TYROPHAGA *p*, *Curtis.*

A. Alis immaculatis.

9124. 1; cylindrica*. *Meig. Zw.* v. 290.—(*Curtis l. c. fo.* 245.)
Ca. cylindrica. *Fabr. A.* 263.
Le. nitidula. *Fall. D. S.* (*Ortal.*) 21.

---

[a] LISSA: Genus Crustaceorum. Vide *Leach Z. M.* ii. 69.

9125. 2; Leachi*. *Meig. Zw.* v. 291.—(*Curtis l. c. fo.* 245.)

9126. 3; putris*. *Fall. D. S.* (*Ortal.*) 21.—(*Curtis l. c. fo.* 245.)
Mu. putris α. *Linn.* ii. 993.—*Berk. S.* i. 165.
Tyr. putris. (*Curtis folio* 126.)

9127. 4; nigricornis*. *Meig. Zw.* v. 291.

9128. 5; annulipes*. *Meig. Zw.* v. 292.—*Curtis.* vi. *pl.* 245.

9129. 6; flavipes* *mihi.*

B. Alis apice puncto nigro.

9130. 7; ruficornis*. *Meig. Zw.* v. 288.

9131. 8; nigripes*. *Meig. Zw.* v. 289.—(*Curtis l. c. supra?*)

9132. 9; flavimana*. *Meig. Zw.* v. 288.

9133. 10; hilaris*. *Meig. Zw.* v. 288.—(*Curtis l. c. supra.*)

9134. 11; Cynipsea*. *Fall. D. S.* (*Ortal.*) 23.—(*Curtis l. c. supra.*)
Mu. Cynipsea. *Linn.* ii. 997.—*Stew.* ii. 262.—*Turt.* iii. 618.
The lesser Mu. vibrans. *Harr. Ex.* 122. *pl.* xxxv. *f.* 43.

9135. 12; fulgens*. *Hoffmansegg.*—*Meig. Zw.* v. 287.

9136. 13; motatoria.
Mu. motatoria. *Mus. Marsham.*

9137. 14; violacea*. *Meig. Zw.* v. 289.

9138. 15; Punctum*. *Fall. D. S.* (*Ortal.*) 22.—(*Curtis l. c. supra.*)
Te. Punctum. *Fabr. A.* 351.
Mu. Stigma. *Panz. F.* lx. *f.* 21.

9139. 16; ornata*. *Meig. Zw.* v. 290.

Genus 185: (1309). PIOPHILA, *Fallen.*

TEPHRITIS *p, Fabr.*—TYROPHAGA, (*Kirby,*) *Curtis.*

9140. 1; Casei*. *Fall. D. S.* (*Heteromyza.*) 9. 1.
Mu. Casei. *Linn. F. No.* 1850. β.
Mu. putris var. β. *Linn.* ii. 993.—*Berk. S.* i. 165.—*Stew.* ii. 260.—*Turt.* iii. 610.
Try. Casei: (putris). (*Kirby & Sp. I. E.* i. 226.—iv. 9.)—*Curtis.* iii. *pl.* 126.
The Cheese Fly. *Bingley.* iii. 318.

9141. 2; atrata*. *Meig. Zw.* v. 396.
Mu. atrata. *Fabr.* iv. 333.
Pi. vulgaris. *Fall. D. S.* (*Heteromyz.*) 9. 2.

9142. 3; nigrimana*. *Meig. Zw.* v. 396. *pl.* 54. *f.* 5.

9143. 4, nigricornis*. *Meig. Zw.* v. 397.

9144. 5; scutellaris*. *Fall. D. S.* (*Heteromyz.*) 10. 3.

Genus 186: (1310). HOMALURA, *Meigen.*

PLANURIA, (*Megerle.*)

9145. 1; tarsata*. *Megerle.*—*Meig. Zw.* v. 399. *pl.* 54. *f.* 6–9.
Mu. leucopa. *Mus. Marsham.*

Genus 187: (1311). PLATYSTOMA, *Meig.*

DICTYA *p*, *Fabr.*—TRUPANEA *p*, *Schrank.*

9146. 1; seminationis*. *Latr.* G. iv. 354.
Mu. seminationis. *Fabr.* iv. 355.
Mu. nævosus. *Harr. Ex.* 75. *pl.* xxi. *f.* 10?
Mu. pulverulenta. *Mus. Marsham.*

9147 † 2. umbrarum[a].

Genus 188: (1312). ULIDIA, *Meigen.*

CHRYSOMYZA, *Fallen.*—TEPHRITIS *p*, *Fabr.*

9148. 1; demandata*. *Meig. Zw.* v. 386. *pl.* 53. *f.* 12.
Mu. demandata. *Fabr. S.* 564.
Chr. splendida. *Fall. D. S.* (*Scenopinii.*) 4. 1.

9149. 2; erythrophthalma*. *Meg.*—*Meig. Zw.* v. 387.

Genus 189: (1313). ACTORA, *Meigen.*

9150. 1; æstuum*. *Meig. Zw.* v. 403. *pl.* 54. *f.* 16–20.

Genus 190: (1314). HELEOMYZA, *Meig.*

HELCOMYZA, *Curtis.*—SCATOPHAGA *p*? *Fabr.*

9151. 1; ustulata. *Meigen MSS.*—*Curtis.* ii. *pl.* 66.

9152. 2, centropunctata. *Meigen MSS.*

9153 † 3, lacustris. *Meigen MSS.* In Mus. *Brit.*

Genus 191: (1315). MYCETOMYZA? *Fallen.*

9154. 1; obscuripennis*. *Meig. MSS.*

9155. 2; fuscicornis* *mihi.*

9156. 3; apicalis *mihi.*

9157. [4; ruficornis] *mihi.*

9158. 5, pallida *mihi.*

9159. 6; marginepunctata*. *Leach MSS.*

9160. 7, nigrofasciata* *mihi.*

9161. 8; tripunctata* *mihi.*

9162. 9; rufo-testacea* *mihi.*

9163. 10, nebulosa *mihi.*

9164. 11; subvittata* *mihi.*

Genus 192: (1316). TETANOCERA, *Dumeril.*

SCATOPHAGA *p*, *et* DICTYA *p*, *Fabr.*—OTITIS *p*, *Latr.*

9165. 1; marginata*.

---

[a] 9147 † 2. umbrarum. *Meig. Zw.* v. 392.
Mu. umbrarum. *Fabr.* iv. 350.—*Stew.* ii. 262. (!)—*Turt.* iii. 617. (!)
Mu. fulviventris. *Schrank A. No.* 953.

Mu. marginata. *Fabr.* iv. 345.—*Panz. F.* xxxii. *f.* 22.
Mu. tristis. *Harr. Ex.* 115. *pl.* xxxiv. *f.* 13.

9166. 2, varipes*? *mihi.*
Te. variegata. *Meigen MSS?*

9167. 3; Umbrarum*.
Di. Umbrarum. *Fabr. A.* 325.
Os. Argus. *Fabr. A.* 216?

9168. 4; obsoleta* *mihi.*

9169. 5, consobrina* *mihi.*

9170. 6, ustulata*. *Leach MSS.*
Mu. crocus. *Harr. Ex.* 118. *pl.* xxxiv. *f.* 27.

9171. 7; testacea*.
Mu. testacea. *Mus. Marsham.*
Mu. varicus. *Harr. Ex.* 114. *pl.* xxxiv. *f.* 12?

9172. 8; rufa*.
Mu. rufa. *Mus. Marsham.*

9173. 9, rufifrons*.
Mu. rufifrons. *Panz. F.* xliv. *f.* 24.

9174. 10; Graminum*.
Sc. Graminum. *Fabr. A.* 205.
Di. fulvifrons. *Latr. H.* xiv. 384.
Mu. buccata. *Mus: Marsham.*

9175. 11; flavifrons*.
Mu. flavifrons. *Panz. F.* lx. *f.* 22.
Mu. contigua. *Mus. Marsham.*

9176. 12; obliterata?*
Sc. obliterata. *Fabr. A.* 205?
Mu. varicus. *Harr. Ex.* 115. *pl.* xxxiv. *f.* 14.

9177. 13; bimaculata* *mihi.*
Mu. vagus. *Harr. Ex.* 153. *pl.* xlv. *f.* 82.

Genus 193: (1317). SEPEDON, *Latr.*, *Leach*, *Samou.*

BACCHA *p*, *et* SCATOPHAGA *p*, *Fabr.*—SYRPHUS *p*, *Rossi.*—MULIO *p*, *Schell.*

9178. 1; palustris*. *Latr. H.* xiv. 386.—(*Sam. I.* 37.)
Ba. sphegea. *Fabr. A.* 199. (!)
Mus. sphegea. *Stew.* ii. 265.—*Turt.* iii. 646.
Mus. rufipes. *Panz. F.* lx. *f.* 23.
Mul. dentipes. *Schell.* G. *M. pl.* 16.
Mus. simulator. *Harr. Ex.* 152. *pl.* xlv. *f.* 77.

Genus 194: (1318). OSCINIS, *Fabr.*

SCATOPHAGA *p*, *Fabr.*—TETANOCERA *p*, *Latr.*

9179. 1; planifrons*.
Mu. planifrons. *Fabr. S.* 565.

9180. 2, melanoleuca* *mihi.*

Genus 195: (1319). ——

9181. 1; nervosa* *mihi.*

9182. 2; decempunctata* *mihi.*

Genus 196: (1320). GEOMYZA, *Fallen.*

9183. 1; combinata*.
Mu. combinata. *Linn.* ii. 997.

9184. 2; abdominalis* *mihi.*

9185. 3; minatrix*.
Mu. minatrix. *Mus. Marsham.*

9186. 4, quaterna*. *Leach? MSS.*

9187. 5, trimaculata *mihi.*

9188. 6; pallipes* *mihi.*
Tephritis maculata. *Ahr. F.* iii. *f.* 22?

9189. 7, rufescens* *mihi.*

9190. 8; flava*.
Mu. flava. *Mus. Marsham.*

9191. 9; pallida* *mihi.*

9192. 10; arcuata* *mihi.*

9193. 11; ventralis* *mihi.*

9194. 12; dubia* *mihi.*

Genus 197: (1321). ——
DACUS *p, Fabr.*—TEPHRITIS *p, Samou.*

9195. 1; pulchella*.
Mu. pulchella. *Fabr.* iv. 352.—*Don.* ii. 19. *pl.* 366.
Te. pulchella. (*Sam. I.* 40.)
Mu. muliebris. *Harr. Ex.* 75. *pl.* xxi. *f.* 9.

Genus 198: (1322). ——
SCATOPHAGA *p, Fabr.*—MADIZA, *Fallen?*

9196. 1; cucularia*.
Mu. cucularia. *Linn.* ii. 995.—*Turt.* iii. 615.

9197. 2; maculicornis* *mihi.*

9198. 3; longicornis* *mihi.*

9199. 4; acuta* *mihi.*

9200. 5; xanthoptera* *mihi.*

9201. 6; hyalinata* *mihi.*

9202. 7, flavescens.
Mu. flavescens. *Mus. Marsh.*

9203. 8; geniculata* *mihi.*

Genus 199: (1323). MEROMYZA? *Meigen.*

9204. 1; pusilla*.
Mu. pusilla. *Mus. Marsham.*

9205. 2, bipunctata*.
Mu. bipunctata. *Mus. Marsham.*

9206. 3; incrassata*.
Mu. incrassata. *Mus. Marsham.*

Genus 200: (1324). CHLOROPS, *Meigen.*

OSCINIS *p*, *Latr.*—TEPHRITIS *p*, *Fabr.*

9207. 1; lineatus*.
Mu. lineata. *Fabr. A.* 215.—*Stew.* ii. 262.

9208. 2; pratensis. *Meigen MSS.*

9209. 3; pumilionis*.
Mu. pumilionis. *Act. H.* (*Bierkand.*) 1778.—*Linn. Trans.* (*Markwick.*) ii. 78. *pl.* 15.—*Stew.* ii. 260.—*Turt.* iii. 609.—(*Kirby & Sp. I. E.* i. 172.)
Mu. lepidus. *Harr. Ex.* 121. *pl.* xxxv. *f.* 40.
The Hessian Fly? *Bingley.* iii. 316.

9210. 4, quadristrigatus*.
Mu. 4-strigata. *Don.* xiii. 72. *pl.* 467. *f.* 2?

9211. 5; puncticornis. *Meigen MSS.*

9212. 6; Frit?*
Mu. *Frit.* *Linn.* ii. 994.—*Stew.* ii. 260.—*Turt.* iii. 609.

9213. 7; Delta*. *Meigen MSS.*

9214. 8; geniculatus. *Meigen MSS.*

9215. 9; scutellatus*. *Meig.*—*Panz. F.* civ. *f.* 21.

9216. 10, nitidulus*.
Mu. nitidula. *Mus. Marsham.*

9217. 11; lucidus* *mihi.*

9218. 12; ventralis* *mihi.*

9219. 13; caliginosus* *mihi.*

9220. 14; tarsalis* *mihi.*

9221. 15; denticornis*. *Meig.*—*Panz. F.* civ. *f.* 22.

9222. 16; lateralis* *mihi.*

9223. 17; minutissimus* *mihi.*

9224. 18; nigricornis* *mihi.*

9225. 19, femoralis* *mihi.*

9226. 20, obscurus* *mihi.*

9227. 21; pusillus* *mihi.*

9228. 22, luteicornis* *mihi.*

Genus 201: (1325). ——

9229. 1; fulvifrons* *mihi.*

Genus 202: (1326). SCENOPINUS, *Latr.*, *Leach*, *Samou.*
NEMOTELUS *p*, *DeGeer.*—ATRICHIA, *Schra.*—CONA, *Schell.*
9230. 1; fenestralis*. *Fabr. A.* 535.
Mu: fenestralis. *Linn.* ii. 981.—*Turt.* iii. 608.
At. spoliata. *Schrank B.* iii. 2404.
Schell. G. M. 13. *f.* 1, 2.
9231. 2; sulcicollis*. *Meg.*—*Meig. Zw.* iv. 114.
9232. 3, vitripennis. *Meig. Zw.* iv. 115.
9233. 4, senilis. *Fabr. A.* 336.
9234. 5, domesticus. *Meig. Zw.* iv. 116.
Sc. fenestralis var. *Fall. D. S.* (*Scenop.*) 5. 1.
9235. 6, niger*. *Meig. Zw.* iv. 116.—(*Sam. I.* 37.)
He. niger. *DeGeer.* vi. 76. *pl.* 9. *f.* 5.
Ne. tarsatus. *Panz. F.* xcviii. *f.* 20.
Sc. ater. *Fall. D. S.* (*Scenop.*) 5. 2.
9236. 7; rugosus*. *Fabr. A.* 336.
Mu. funestalus. *Fabr.* iv. 330.
9237. 8, nigripes*. *Meg.*—*Meig. Zw.* v. 117.

Genus 203: (1327). OINOPOTA, *Kirby.*
MOSILLUS *p*, *Latr.*—MOCILLUS, *Samou.*
9238. 1, cellaris.
Mu. cellaris. *Linn.* ii. 993.—*Berk. S.* i. 165.—*Stew.* ii. 260. —*Turt.* iii. 609.
Moc. cellarius. (*Samou.* 299.)
Mu. erythrophthalma. *Hellw.*—*Panz. F.* xvii. *f.* 24.

Genus 204: (1328). EPHYDRA, *Fallen.*
9239. 1; littoralis*. *Fall.?*
9240. 2; fumipennis* *mihi.*

Genus 205: (1329). NOTIPHILA, *Fallen.*
9241. 1, riparia. *Meig.?*
9242. 2, cinerea. *Meig.?*

Genus 206: (1330). ——
9243. 1, obsoleta*.
Mu. obsoleta. *Mus. Marsham.*
9244. 2, semiobscura.
Mu. semiobscura. *Mus. Marsham.*
9245. 3; sedecimmaculata* *mihi.*
9246. 4; punctipennis* *mihi.*
9247. 5; tarsalis* *mihi.*

Genus 207: (1331). OCHTHERA, *Latreille*, *Leach*, *Samou.*
MACROCHIRA, *Meig.*—TEPHRITIS *p*, *Fabr.*
9248 † 1, Mantis. *Latr. H.* xiv. 391.—*Latr.* G. *pl.* 15. *f.* 10.—(*Sam. I.* 31.)

Mu. Mantis. *DeGeer*. vi. 143. *pl*. 8. *f*. 15, 16.
Te. manicata. *Fabr. A*. 323. In Mus. *Brit*.

Genus 208: (1332). BORBORUS, *Meigen*.
SPHÆROCERA, *Latr*.

9249. 1; grossipes*.
Mu. grossipes. *Linn*. ii. 988.

9250. [2, curvipes*.]
Mu. curvipes. *Latr. H*. xiv. 394.

9251. 3; nitidus* *mihi*.

9252. 4; pallipes* *mihi*.

9253. 5; fuscipes* *mihi*.

9254. 6, obsoletus* *mihi*.

9255. 7; gonymelas *mihi*.

9256. 8; subsultans*.
Mu. subsultans. *Linn*. ii. 993.—*Turt*. iii. 608.

9257. 9; nervosus* *mihi*.

9258. 10; frontalis* *mihi*.

9259. 11; subæneus* *mihi*.

9260. 12; geniculatus* *mihi*.

9261. 13; longicornis* *mihi*.

9262. 14, atronitens* *mihi*.

9263. 15; longipes* *mihi*.

9264. 16; ramosus* *mihi*.

9265. 17; aterrimus* *mihi*.

9266. 18, opacus* *mihi*.

9267. 19, brunneus* *mihi*.

9268. 20, phæopterus* *mihi*.

9269. 21, lucidus* *mihi*.

9270. 22; pallidus *mihi*.

9271. 23, platycephalus* *mihi*.

Genus 209: (1333). ——

9272. 1, hirtipes* *mihi*.
Hippobosca marina. *Montagu MSS*.

Genus 210: (1334). ——

9273. 1; dilatatus* *mihi*.

Genus 211: (1335). PHORA, *Latr*.
TRINEURA, *Meig*.—TEPHRITIS *p*, *Fabr*.—NODA, *Schell*.

9274. 1; convexa*.
Mu. convexa. *Mus. Marsham*.

9275. 2; rufipes*.
Tr. rufipes. *Meig. Kl.* i, *b.* 313. *pl.* 15. *f.* 23.
Ph. pallipes. (*Latr.* G. iv. 360.)

9276. 3, consorta* *mihi.*

9277. 4, ochropa* *mihi.*

9278. 5, serrata?*
Mu. serrata. *Linn.* ii. 993?—*Turt.* iii. 608?

9279. 6; longipes* *mihi.*

9280. 7, rufescens *mihi.*

9281. 8, annulata.
Mu. annulata. *Mus. Marsham.*

9282. 9, pusilla* *mihi.*

9283. 10, cursitans.
Mu. cursitans. *Mus. Marsham.*

9284. 11, aterrima*. *Latr. H.* xiv. 394.
Te. aterrima. *Fabr. A.* 323.
Tr. atra. *Meig. Kl.* i, *b.* 313. *pl.* 15. *f.* 22.
Mu. semiserrata. *Mus. Marsham.*

Genus 212: (1336). ——

9285. 1; culicoides* *mihi.*

9286. 2; cognata* *mihi.*

9287. 3, lonchopteroides* *mihi.*

9288. 4, assimilis* *mihi.*

§ B. OXYPTERÆ, *Meigen.*

Genus 213: (1337). LONCHOPTERYX.

LONCHOPTERA, *Meigen.*—DIPSA, *Fallen.*

9289. 1; lutea*. *Fanz. F.* cviii. *f.* 20, 21.
Di. furcata. *Fall. D. S.* (*Phytomyzides.*)

9290. 2; lacustris*. *Meig. Zw.* iv. 107.

9291. 3, riparia*. *Meig. Zw.* iv. 108.

9292. 4, palustris. *Meig. Zw.* iv. 109.

9293. 5, punctum. *Meig. Zw.* iv. 110.

9294. 6, Leachii *mihi.*

9295. 7, tristis. *Meig. Zw.* iv. 110. *pl.* 36. *f.* 11. (!)

Ordo III: (10). HOMALOPTERA, *MacLeay*.

(Diptera *p*, *Linné*, *&c.*—Antliata *p*, *Fabr.*—Omaloptera, *Leach.*)

Familia I: (153). HIPPOBOSCIDÆ, *Leach*.

Genus 1: (1338). HIPPOBOSCA *Auctorum*. *Forest or Horse-Fly*.

Nirmomyia, *Nitzsch*.

9296. 1; equina*. *Linn.* ii. 1010.—*Berk. S.* i. 168.—*Don.* viii. 21. *pl.* 261. *f.* 1.—*Stew.* ii. 274.—*Turt.* iii. 687.—*Shaw G. Z.* vi. 401. *pl.* 114.—(*Leach E. E.* ix. 130.)—*Leach E. I.* 9. *pl.* xxvi. *f.* 4–7.—(*Samou.* 302. *pl.* 9. *f.* 11.)—(*Kirby & Sp. I. E.* i. 227.)

Genus 2: (1339). ORNITHOMYIA, *Latreille, Leach, Samou.*

9297. 1; avicularia*. *Leach E. I.* 15. *pl.* xxv. *f.* 4, 5.—(*Sam. I.* 32.)—(*Kirby & Sp. I. E.* i. 111.)
Hi. avicularia. *Linn.* ii. 1010.—*Stew.* ii. 274.—*Turt.* iii. 687?—*Shaw G. Z.* vi. 402. *pl.* 115?

9298. 2; viridis*. *Latr. H.* xiv. 402.—*Leach E. I.* 14. *pl.* xxv. *f.* 1–3.—(*Sam. I.* 32.)
Hi. viridis. *DeGeer.* vi. 285. *pl.* 16. *f.* 21, 22.
Hi. Corvi. *Scop. C. No.* 1026.
Hi. avicularia. *Don.* viii. 22. *pl.* 261. *f.* 2.

Genus 3: (1340). CRATERINA, *Olfers, Samou., Curtis.*

Stenepteryx, *Leach.*—Ornithomyia *p*, *Latreille.*

9299. 1; Hirundinis*. *Olfers.*—(*Sam. I.* 13.)—*Curtis* iii. *pl.* 122.
Hi. Hirundinis. *Linn.* ii. 1010.—*Berk. S.* i. 168.—*Don.* viii. 45. *pl.* 268. *f.* 1.—*Stew.* ii. 274.—*Turt.* iii. 687.—*Shaw G. Z.* vi. 403.—*Wood.* ii. 109. *pl.* 72.
St. Hirundinis. *Leach E. I.* 16. *pl.* xxv. *f.* 9–11.
Or. Hirundinis. (*Kirby & Sp. I. E.* i. 111.)

Genus 4: (1341). OXYPTERUM, *Kirby, Leach, Samou.*

Ornithomyia *p*, *Olivier.*

9300. 1; pallidum*. *Leach E. I.* 17. *pl.* xxv. *f.* 12–14.
Or. pallida. *Oliv. E. M.* viii. 544.

9301. 2; Kirbyanum*. *Leach E. I.* 17. *pl.* xxv. *f.* 15–16.—(*Sam. I.* 32.)

Genus 5: (1342). HÆMOBORA, *Curtis.*

9302 † 1, pallipes. *Curtis.* i. *pl.* 14. In Mus. D. *Samouelle.*

Genus 6: (1343). MELOPHAGUS, *Latreille, Leach, Samou., Curtis.*

Melophila, *Nitzsch.*

9303. 1; Ovinus*. *Latr. H.* xiv. 403.—*Leach E. I.* 18. *pl.* xxvi. *f.* 14, 15.—(*Sam. I.* 27.)—*Curtis.* iii. *pl.* 142.

Hi. Ovina. *Linn.* ii. 1011.—*Don.* viii. 46. *pl.* 268. *f.* 2.—*Stew.* ii. 274.—*Turt.* iii. 687.—*Shaw* G. *Z.* vi. 403.

Familia II: (154). NYCTERIBIDÆ, *Leach.*

Genus 7: (1344). NYCTERIBIA, *Latreille, Samou.*

PHTHIRIDIUM, *Hermann, Leach.*—HIPPOBOSCA *p, Schrank.*—CELERIPES, *Montagu.*—PEDICULUS *p, Linné.*

9304. 1: Hermanni*. (*Sam. I.* 31.)
Hi. Vespertilionis. *Schrank B. No.* 2587.
Ce. Vespertilionis. *Linn. Trans.* (*Mont.*) xi. 11. *pl.* 3. *f.* 5.
Ph. biarticulatum. *Herm. M. A.* 124. *pl.* 6. *f.* 1.
Ph. Hermanni. *Leach E. B. Sup.* i. 446. *pl.* 23.—*Leach Z. M.* iii. 55. *pl.* 144. *fig. sup.* ♀. *fig. med.* ♂.—*fig. inf. pes auctus.*

9305. 2: Latreillii*.
Pe. Vespertilionis. *Linn. F. No.* 1941.
Ny. pedicularia. *Latr. H.* ix. 403?
Ph. Latreillii. *Leach Z. M.* iii. 56.

ORDO IV: (11). APHANIPTERA, *Kirby.*

(SUCTORIA, *Latr.*—SIPHONAPTERA, *Latr.*—APTERA *p, Linné.*—RHYNGOTA *p, Fabr.*—ROPHOTEIRA, *Clairv.*

Familia I: (155). PULICIDÆ.

Genus 1: (1345). PULEX *Auctorum.* *Flea.*

9306. 1; irritans*. *Linn.* ii. 1021.—*Barbut G. pl.* 18.—(*Berk. S.* i. 170.)—*Stew.* ii. 232.—*Shaw N. M.* v. *pl.* 178.—*Shaw* G. *Z.* vi. 456. *pl.* 122.—*Turt.* iii. 701.—*Leach E. E.* ix. 126.—*Samou.* 234.—*Bingley.* iii. 347.—*Wood.* ii. *pl.* 77.
Alb. Sp. pl. 41.—*Baker M. pl.* 13. *f.* 6.

9307. 2; Canis* *mihi.* (*Sam. I.* 35.)

9308 † 3, Bovis. *Leach MSS.* In Mus. *Brit.*

9309. 4, Talpæ*. (*Sam. I.* 35.)—*Curtis.* iii. *pl.* 114.

9310 † 5, Meles. *Leach MSS.* In Mus. *Brit.*

9311 † 6, Sciurorum*. *Olfers.* 98.
Pe. Sciuri. (*Sam. I.* 35.) In Mus. *Brit.*

9312 † 7, Erinacei. *Leach MSS.* In Mus. *Brit.*

9313 † 8, Leporis. *Leach MSS.* In Mus. *Brit.*

9314. 9; Vespertilionis* *mihi.*

9315. 10; Columbæ* *mihi.*

9316. 11; Hirundinis* *mihi.* (*Sam. I.* 35.)

9317. 12; fasciatus*. *Latr. H.* xiv. 412.

Ordo V: (12). APTERA.

(Aptera *p*, *Linné.*—Antliata *p*, *Fabr.*—Arachnida-Parasita, *Latreille.*—Pododunera, *Clairv.*—Anoplura, *Leach.*)

Familia I: (156). PEDICULIDÆ, *Leach.*

(Pediculus *p*, *Linné.*—Hemiptera-epizoica, *Nitzsch.*)

Genus 1: (1346). PEDICULUS *Auctorum.* *Louse:* ova *Nits.*

9318. 1; humanus*: Body-louse, or Tailor's Louse.—*Leach l. c.* *Linn.* ii. 1016.—*Barbut G. pl.* 18. *gen.* 4.—(*Berk. S.* i. 170.)—*Stew.* ii. 278.—*Shaw G. Z.* vi. 450.—*Turt.* iii. 695.—*Leach E. E.* ix. 77.—*Leach Z. M.* iii. 66.—*Samou.* 142.—*Bingley.* iii. 345.
Pe. capitis. *DeGeer:* (*teste Nitzsch.*)
Alb. Sp. pl. 42.—*Swam. B. N. pl.* 1. *f.* 3–6.

9319. 2; cervicalis*: Head-louse.—*Leach l.* c. *Latr. G.* i. 168. —*Leach E. E.* ix. 77.—*Leach Z. M.* iii. 66.—*Samou.* 143.
Pe. vestimenti. *DeGeer:* (*teste Nitzsch.*)

Genus 2: (1347). PHTHIRUS, *Leach*, *Samou.*, (*Kirby.*)

9320. 1, inguinalis*. *Leach E. E.* ix. 77.—*Leach Z. M.* iii. 65.—*Samou.* 142.
Pe. inguinalis. *Redi Ex. pl.* xix. *fig. sup.*
Pe. Pubis: Crab L.—*Berk. l.* c. *Linn.* ii. 1017.—*Berk. S.* i. 170.—*Stew.* ii. 279.—*Turt.* iii. 695.
Albin. Sp. pl. 49. *f. sinist.*

Genus 3: (1348). HÆMATOPINUS, *Leach*, *Samou.*, (*Kirby.*)

9321. 1; Suis*. *Leach E. E.* ix. 77.—(*Leach E. B. Sup.* i. *pl.* 24.) —*Leach Z. M.* iii. 65. *pl.* 146.—(*Samou.* 143.)
Pe. Suis. (*Linn.* ii. 1017.)—(*Stew.* ii. 279.)—*Wood.* ii. 123. *pl.* 76.
Pe. Urius. *Germ. M.* (*Nitzsch.*) iii. 305.

9322. 2, sphærocephalus.
Pe. sphærocephalus. (*Germ. M.* (*Nitz.*) iii. 305.)

9323 † 3, Cervi.
Pe. Cervi. *Redi Ex. pl.* xxiii. *fig. sup.*—(*Linn.* ii. 1017.)—(*Stew.* ii. 279.)
Pe. crassicornis. (*Germ. M.* (*Nitz.*) iii. 305.)
The Louse of the Deer. *Alb. Sp. pl.* lxiii. *fig. dext.*

9324. 4, Vituli*.
Pe. Vituli. (*Linn.* ii. 1018.)—*Berk. S.* i. 170.—*Stew.* ii. 279. —*Turt.* iii. 696.

9325. 5, eurysternus. (*Germ. M.* (*Nitz.*) iii. 305?)

9326. 6, Caballi*. *Leach MSS.*

9327. 7; Pe. Asini. *Redi Ex. pl.* xxi.—(*Linn.* ii. 1018.)—*Turt.* iii. 696.

Familia II: (157). NIRMIDÆ, *Leach.*

(PEDICULUS *p*, *Linné.*—ORTHOPTERA-EPIZOICA, *Nitzsch.*)
(RICINUS, *DeGeer.*—NIRMUS, *Hermann.*)

Genus 4: (1349). TRICHODECTES, *Nitzsch.*

9328. 1, crassus. (*Germ. M.* (*Nitzsch.*) iii. 295.)
Pe. Melis. *Fabr. A.* 341.

9329. 2, latus*. (*Germ. M.* (*Nitzsch.*) iii. 296.)
Ri. Canis. *DeGeer.* vii. *pl.* 4. *f.* 16.
Pe. setosus. *Olfers.* 84.

9330. 3, subrostratus*. (*Germ. M.* (*Nitzsch.*) iii. 296.)
Pe. Canis. *O. Fabr. F. p.* 215?

9331. 4, sphærocephalus. (*Germ. M.* (*Nitzsch.*) iii. 296.)
Pe. Ovis. (*Linn.* ii. 1017.)—*Schra. A. p.* 502. *pl.* 1. *f.* 8–9.—(*Turt.* iii. 696.)

9332 † 5. Climax. (*Germ. M.* (*Nitzsch.*) iii. 296.)

9333. 6, scalaris*. (*Germ. M.* (*Nitzsch.*) iii. 296.)
Pe. Bovis. *Linn.* ii. 1017.—*Berk. S.* i. 170.—*Stew.* ii. 279.—*Turt.* iii. 696.

9334 † 7. longicornis. (*Germ. M.* (*Nitzsch.*) iii. 296.)
Pe. Cervi. *Redi Ex. pl.* xxiii. *fig. inf.*
The Louse of the Stag. *Alb. Sp. pl.* lxiii. *fig. inf.*

9335. 8, Equi*.
Pe. Equi. (*Linn.* ii. 1018.)—(*Stew.* ii. 279.)—(*Turt.* iii. 696.)

9336 † 9. exilis. (*Germ. M.* (*Nitzsch.*) iii. 296.)

9337. 10, retusus. (*Germ. M.* (*Nitzsch.*) iii. 296.)

Genus 5: (1350). DOCOPHORUS, (PHILOPEDON) *Nitzsch.*

9338. 1, ocellatus. (*Germ. M.* (*Nitzsch.*) iii. 920.)
Pe. ocellatus. *Scop. C. No.* 1038.
Pe. Corvinis. *Fabr. A.* 344.—*Stew.* ii. 279.

9339. 2, atratus. (*Germ. M.* (*Nitzsch.*) iii. 920.)
Ni. adustus. *Olfers.* 87.
Pulex Corvi. *Redi Ex. pl.* xvi.
Pe. Corvi. (*Linn.* ii. 1018.)—*Stew.* ii. 279.—*Turt.* iii. 696.
The Louse of the Crow. (*Alb. Sp. pl.* xlvii. *fig. sup.*)—*Shaw G. Z.* vi. *pl.* 119. *fig. sup.*)

9340. 3, communis. (*Germ. M.* (*Nitzsch.*) iii. 920.)
Pe. Emberizæ. *Fabr. A.* 349.—*Stew.* ii. 279.—*Turt.* iii. 699.
Ri. Emberizæ. *DeGeer.* vii. *pl.* iv. *f.* 9.
Pe. Curvirostræ. *Panz. F.* li. *f.* 23.—(*Stew.* ii. 281.)
Pe. Pyrrhulæ. (*Schr. A.* 506.)—(*Stew.* ii. 281.)
Pe. Chloridis. (*Schr. A.* 506.)—(*Stew.* ii. 281.)
Pe. Citrinellæ. (*Schr. A.* 507.)—(*Stew.* ii. 281.)

Pe. Rubeculæ. (*Schr. A.* 507.)—(*Stew.* ii. 281.)
Ni. globifer. *Olfers.* 91.

9341. 4, Leontodon. (*Germ. M.* (*Nitzsch.*) iii. 290.)
Pe. Sturni. *Schra. Beit. pl.* v. *f.* 11.—*Stew.* ii. 281.

9342. 5, platyrhynchus. (*Germ. M.* (*Nitzsch.*) iii. 290.)
Pe. hæmatopus. *Scop. C. No.* 1035.
Pe. Strigis. *Fabr. A.* 343.—*Stew.* ii. 280.—*Turt.* iii. 279.

9343. 6, excisus. (*Germ. M.* (*Nitzsch.*) iii. 290.)
Pe. Hirundinis. *Linn.* ii. 1020.—*Stew.* ii. 281.—*Turt.* iii. 699.

9344 † 7, pertusus. (*Germ. M.* (*Nitzsch.*) iii. 290.)
Pe. Fulicæ. *Stew.* ii. 280.

9345. 8, icterodes. (*Germ. M.* (*Nitzsch.*) iii. 290.)
Pulex Cygni secundi generis. *Redi Ex. pl.* ix. *fig. inf.*
Another Swan Louse. (*Albin. Sp. pl.* 48. *fig. sinist.*)

9346. 9, melanocephalus. (*Germ. M.* (*Nitzsch.*) iii. 290.)

9347. 10, auratus. (*Germ. M.* (*Nitzsch.*) iii. 290.)

9348. 11, latifrons. (*Germ. M.* (*Nitzsch.*) iii. 290.)

9349. 12, tricolor. (*Germ. M.* (*Nitzsch.*) iii. 290.)

9350. 13, incompletus. (*Germ. M.* (*Nitzsch.*) iii. 290.)

Genus 6: (1351). NIRMUS *Auctorum.*
NIRMUS (PHILOPEDON), *Nitzsch.*

9351. 1, discocephalus. (*Germ. M.* (*Nitzsch.*) iii. 291.)

9352. 2, cameratus. (*Germ. M.* (*Nitzsch.*) iii. 291.)
Pe. Tetraonis. *Linn.* ii. 1020?—*Stew.* ii. 281?
Pe. Lagopodis. (*Linn.* ii. 1020?)—*Stew.* ii. 281?

9353. 3, fenestratus. (*Germ. M.* (*Nitzsch.*) iii. 291.)

9354. 4, uncinosus. (*Germ. M.* (*Nitzsch.*) iii. 291.)

9355. 5, Argulus. (*Germ. M.* (*Nitzsch.*) iii. 291.)
Pe. Picæ. (*Linn.* ii. 1018?)—*Stew.* ii. 279?—*Turt.* iii. 697.
Pu. Picæ. *Redi Ex. pl.* v.
The Louse of the Magpie. (*Alb. pl.* 45. *fig. sup.*)

9356. 6, gracilis. (*Germ. M.* (*Nitzsch.*) iii. 291.)

9357. 7, decipiens. (*Germ. M.* (*Nitzsch.*) iii. 291.)
Pe. Recurvirostræ. (*Linn.* ii. 1019.)—*Stew.* ii. 280.—*Turt.* iii. 698.

9358. 8, piceus. (*Germ. M.* (*Nitzsch.*) iii. 291.)

9359. 9, attenuatus. (*Germ. M.* (*Nitzsch.*) iii. 291.)
Pe. Ortygometræ. *Schr. A.* 503?—*Stew.* ii. 281?

9360. 10, fissus. (*Germ. M.* (*Nitzsch.*) iii. 291.)
Pe. Charadrii. (*Linn.* ii. 1019.)—(*Stew.* ii. 280?)—(*Turt.* iii. 698?)

9361. 11, punctatus. (*Germ. M.* (*Nitzsch.*) iii. 291.)

9362. 12, eugrammicus. (*Germ. M.* (*Nitzsch.*) iii. 291.)

9363. 13, minutus. (*Germ. M.* (*Nitzsch.*) iii. 291.)
Pulex Fulicæ. *Redi Ex. pl.* iv. *f.* iii.
The Louse of the More-Hen. *Albin Sp. pl.* xlv. *fig. sinist.*

9364. 14, glaucus.
Pe. hæmatopodis. (*Linn.* ii. 1019.)—*Stew.* ii. 280.—*Turt.* iii. 698.

Genus 7: (1352). LIPEURUS, (PHILOPEDON) *Nitzsch.*

9365. 1, versicolor. (*Germ. M.* (*Nitzsch.*) iii. 292.)
Pe. Ciconiæ. (*Linn.* ii. 1019.)—*Frisch. I.* viii. *pl.* vi.—*Stew.* ii. 280.

9366. 2, luridus *. (*Germ. M.* (*Nitzsch.*) iii. 292.)
Pulex Fulicæ. *Redi Ex. pl.* iv. *f.* ii.
The Louse of the More-Hen. (*Albin Sp. pl.* xlv. *fig. dext.*)—(*Shaw* G. Z. vi. *pl.* 120. *fig. med.*)

9367. 3; squalidus. (*Germ. M.* (*Nitzsch.*) iii. 292.)
Pe. Anatis. *Fabr. A.* 345.—*Stew.* ii. 281.

9368. 4, temporalis. (*Germ. M.* (*Nitzsch.*) iii. 292.)
Ri. Mergi. *DeGeer.* vii. *pl.* iv. *f.* 13.
Pe. Mergi. *Fabr. A.* 345.—*Stew.* ii. 280.

9369. 5, jejunus. (*Germ. M.* (*Nitzsch.*) iii. 292.)
Pe. Anseris. *Linn.* ii. 1018.—*Schra. A.* 503.—*Stew.* ii. 180.—*Turt.* iii. 699.
Ni. crassicornis. *Olfers.* 88.
Pulex Anseris sylvestris. *Redi Ex. pl.* x. *fig. dext.*
The Louse of the wild Goose. (*Alb. Sp. pl.* 48. *fig. dext.*)—(*Shaw* G. Z. vi. *pl.* 120. *fig. sin. sup.*)

9370. 6, variabilis [a].

9371. 7, ebræus. (*Germ. M.* (*Nitzsch.*) iii. 293.)
Pe. Gruis. (*Linn.* ii. 1019.)—*Stew.* ii. 280.
Pulex Gruis. *Redi Ex. pl.* iii.
The Louse of the Crane. *Albin. Sp. pl.* 44. *fig. sup.*

9372. 8, quadripustulatus. (*Germ. M.* (*Nitzsch.*) iii. 293.)

9373. 9; Baculus *. (*Germ. M.* (*Nitzsch.*) iii. 293.)
Pe. Columbæ. (*Linn.* ii. 1020.)—*Stew.* ii. 281.—*Turt.* iii. 699.—*Don.* ix. 7. *pl.* 291.
Ni. filiformis. *Olfers.* 90.
Pu. Columbæ majoris. *Redi Ex. pl.* ii. *fig. sup.*
The Louse of the Pigeon. *Albin Sp. pl.* 43. *fig. inf.*

9374 † 10, obtusus *.
Pe. Ardeæ. (*Linn.* ii. 1019.)—*Stew.* ii. 280.

---

[a] 9370. 6, variabilis? (*Germ. M.* (*Nitzsch.*) iii. 293.)

Pu. Ardeæ. *Redi Ex. pl.* vi.
The Louse of the Heron. (*Albin Sp. pl.* 45. *fig. inf.*)—(*Shaw G. Z.* vi. *pl.* 121. *fig. med.*)

9375 † 11, bilineatus.
Pe. Vagelli. *Fabr. A.* 346.—*Stew.* ii. 280.

Genus 8: (1353). GONIODES, (PHILOPEDON) *Nitzsch.*

9376. 1, falcicornis [a].

9377. 2, chelicornis. (*Germ. M.* (*Nitzsch.*) iii. 293?)

9378. 3, dissimilis [b].

9379. 4, dispar. (*Germ. M.* (*Nitzsch.*) iii. 294.)

9380. 5, Stylifer [c].

9381. 6, paradoxus. (*Germ. M.* (*Nitzsch.*) iii. 294.)
Pediculus of the Quail. (*Shaw* G. *Z.* vi. *pl.* 121. *fig. sinist.*)

9382. 7, hologaster [d].

9383. 8, compar. (*Germ. M.* (*Nitzsch.*) iii. 294.)

9384. 9, microthorax. (*Germ. M.* (*Nitzsch.*) iii. 294.)
Pediculus of the Partridge. (*Shaw G. Z.* vi. *pl.* 121. *fig. dext.*)

9385. 10, rectangulus [e].

Genus 9: (1354). COLPOCEPHALUM, (LIOTHEUM) *Nitzsch.*

9386. 1. Zebra. (*Germ. M.* (*Nitzsch.*) iii. 298.)

9387. 2, flavescens. (*Germ. M.* (*Nitzsch.*) iii. 298.)

9388. 3, subæquale. (*Germ. M.* (*Nitzsch.*) iii. 298.)

9389. 4, ochraceum. (*Germ. M.* (*Nitzsch.*) iii. 299.)
Pulex Avis Pluvialis. *Redi Ex. pl.* xi. *fig. sup.*
The Louse of the Cormorant. (*Albin Sp. pl.* xlvi. *fig. sup.*)—(*Shaw* G. *Z.* vi. *pl.* 120. *fig. dext. sup.*)

---

[a] 9376. 1, falcicornis. (*Germ. M.* (*Nitzsch.*) iii. 293.)
Pe. Pavonis. (*Linn.* ii. 1019.)—*Panz. F.* li. *f.* 19. ♀.—*Stew.* ii. 280. (!)—*Turt.* iii. 698. (!)
Ni. tetragonocephalus. *Olfers.* 90.
Pulex Pavonis. *Redi Ex. pl.* xiv. ♂.
The Louse of the Peacock. *Albin Sp. pl.* l. *fig. inf.*—(*Shaw G. Z.* vi. *pl.* 119. *fig. inf.*)

[b] 9378. 3, dissimilis. (*Germ. M.* (*Nitzsch.*) iii. 294.)

[c] 9380. 5, Stylifer. (*Germ. M.* (*Nitzsch.*) iii. 294.)
Pe. Meleagris. *Schr. A.* 504. *pl.* 1. *f.* 4.

[d] 9382. 7, hologaster. (*Germ. M.* (*Nitzsch.*) iii. 294.)
Ri. Gallinæ. *DeGeer.* vii. *pl.* iv. *f.* 15.

[e] 9385. 10, rectangulus. (*Germ. M.* (*Nitzsch.*) iii. 294.)

Genus 10: (1355). MENOPON, (LIOTHEUM) *Nitzsch.*

9390. 1, pallidum[a].

9391. 2, stramineum[b].

9392. 3, cuculare. (*Germ. M.* (*Nitzsch.*) iii. 300.)
Pulex Sturni candidi. *Redi Ex. pl.* xvii. ♂.
The Louse of the Starling. *Albin Sp. pl.* liii. *fig. sup.*

9393. 4, mesoleucum. (*Germ. M.* (*Nitzsch.*) iii. 300.
Ri. Cornicis. *DeGeer.* vii. *pl.* iv. *f.* 11. *pupa.*
Ni. Cornicis. *Latr.* G. i. 169?—*Samou.* 143?

9394. 5, minutum. (*Germ. M.* (*Nitzsch.*) iii. 300.)
Pe. Currucæ. *Schrank Beit. pl.* v. *f.* 1.

9395. 6, phanerostigmaton. (*Germ. M.* (*Nitzsch.*) iii. 300.)
Pe. fasciatus. *Scop. C. No.* 1040.
Pe. Cuculi. *Fabr. A.* 345.—*Stew.* ii. 280.—*Turt.* iii. 697.

Genus 11: (1356). TRINOTON, (LIOTHEUM) *Nitzsch.*

9396. 1, conspurcatum. (*Germ. M.* (*Nitzsch.*) iii. 300.)
Pe. Cygni. (*Linn.* ii. 1018.)—*Stew.* ii. 280?
Pe. Anseris. *Roem. G. pl.* xxix. *f.* 4.
Pul. Cygni. *Redi Ex. pl.* 8.
The Louse of the Swan. (*Albin. Sp. pl.* xlviii. *fig. inf.*)

9397. 2, luridum. (*Germ. M.* (*Nitzsch.*) iii. 300.)
Pul. Anseris sylvestris. *Redi Ex. pl.* x. *fig. inf.*
The Louse of the wild Goose. (*Albin. Sp. pl.* 47. *fig. sinist.*)—(*Shaw G. Z.* vi. *pl.* 120. *fig. dext. inf.*)

9398. 3, lituratum. (*Germ. M.* (*Nitzsch.*) iii. 300.)
Ri. Lari. *DeGeer.* vii. *pl.* 4. *f.* 12?

9399. 4, obscurum.
Pe. Sternæ. *Linn.* ii. 1019.—*Stew.* ii. 280.—*Turt.* iii. 698.
Ni. fornicatus. *Olfers.* 89.

Genus 12: (1357). LÆMOBOTHRION, (LIOTHEUM) *Nitzsch.*

9400 † 1, giganteum. (*Germ. M.* (*Nitzsch.*) iii. 301.)
Pe. maximus. *Scop. C. No.* 1036.
Pe. Buteonis. (*Fabr. A.* 347.)—*Stew.* ii. 279.—*Turt.* iii. 696.

9401. 2, hasticeps*. (*Germ. M.* (*Nitzsch.*) iii. 302.)
Ni. hasticeps. *Olfers.* 87.

---

[a] 9390. 1, pallidum. (*Germ. M.* (*Nitzsch.*) iii. 299.)
Pe. Caponis. (*Linn.* ii. 1020.)—*Stew.* ii. 281. (!)—*Turt.* iii. 699. (!)
Pe. Gallinæ. *Panz. F.* li. *f.* 21.—*Stew.* ii. 280. (!)—*Turt.* iii. 698. (!)
Pulex Capi. *Redi Ex. pl.* xvi?
Ni. trigonocephalus. *Olfers.* 90.

[b] 9391. 2, stramineum. (*Germ. M.* (*Nitzsch.*) iii. 300.)
Pe. Meleagris. (*Linn.* ii. 1020.)—*Stew.* ii. 280. (!)—*Turt.* iii. 698. (!)

Pe. Tinnunculi. (*Linn.* ii. 1018.)—*Stew.* ii. 279.—*Turt.* iii. 696.
Pulex Tinnunculi. *Redi Ex. pl.* xiii.
The Louse of the Kestril. (*Albin Sp. pl.* 1. *fig. sup.*)
9402 † 3, atrum. (*Germ. M.* (*Nitzsch.*) iii. 302.)
Pulex Fulicæ. *Redi Ex. pl.* iv. *f.* 1.
The Louse of a Coot. (*Albin. Sp. pl.* 44. *fig. inf.*)

Genus 13: (1358). EUREUM, *Nitzsch.*

9403. 1, cimicoides. (*Germ. M.* (*Nitzsch.*) iii. 301.)
Ni. truncatus. *Olfers.* 91?
9404 † 2. Malleus. (*Germ. M.* (*Nitzsch.*) iii. 301.)

Genus 14: (1359). PHYSOSTOMUM, *Nitzsch.*

9405. 1, irascens. (*Germ. M.* (*Nitzsch.*) iii. 302.)
Pe. Motacillæ. *Fabr. A.* 349?—*Stew.* ii. 281?
9406. 2, nitidissimum. (*Germ. M.* (*Nitzsch.*) iii. 302.)
Ni. pterocephalus. *Olfers.* 91.
Ri. Fringillæ. *DeGeer.* vii. *pl.* 4. *f.* 6.

Ordo VI: (13). HEMIPTERA, *Linné.*

(Rhyngota *p*, *Fabr.*—Hemimeroptera *p*, *Clairv.*—Hemiptera-Heteroptera, *Latr.*)

Sectio I. TERRESTRIA. (Cimex, *Linné.*)

Familia I: (158). CIMICIDÆ, *Leach.*

(He.-membranaceæ, *Latr.*—Acanthia, *Fabr.*)

Genus 1: (1360). CIMEX *Auctorum.* *Bug.*

9407. 1; lectularius*. *Linn.* ii. 715.—*Berk. S.* i. 115.—*Barbut Gen.*—*Stew.* ii. 101.—*Turt.* iii. 608.—*Bingley.* iii. 181.—*Shaw G. Z.* vi. 161.—*Leach E. E.* ix. 122.—*Samou.* 223.—(*Kirby & Sp. I. E.* i. 106.)

Genus 2: (1361). ANEURUS, *Curtis.*

Acanthia *p*, *Fabr. olim.*—Aradus *p*, *Fabr.*

9408. 1; lævis*. *Curtis.* ii. *pl.* 86.
Ar. lævis. *Fabr. R.* 119.
Ci. lævis. *Stew.* ii. 102.—*Turt.* iii. 611.

Genus 3: (1362). ARADUS, *Fabr.*, *Curtis*, (*Kirby.*)

Coreus *p*, *Schellenberg.*

9409. 1; depressus*. *Fabr. R.* 119.—*Wolff.* 129. *pl.* 13. *f.* 123. (*Kirby & Sp. I. E.* iii. 615.)—(*Curtis l. c. infra.*)

9410. 2, Betulæ*. *Fabr. R.* 119.—(*Kirby & Sp. I. E.* iii. 330.) —*Curtis l. c. infra.*
Ci. Betulæ. *Linn.* ii. 718.—*DeGeer.* iii. 305. *pl.* 15. *f.* 16, 17. —*Berk. S.* i. 115.—*Stew.* ii. 102.—*Turt.* iii. 611.

9411 † 3, corticalis. *Fabr. R.* 119.—*Curtis.* v. *pl.* 230.
Ci. corticalis. *Linn.* ii. 718.—*Berk. S.* i. 115.—*Stew.* ii. 102. —*Turt.* iii. 611. In Mus. D. Dale, et Kirby.

Genus 4: (1363). AGRAMMA, *Westwood MSS.*

TINGIS *p*, *Fall.*

9412. 1; læta*. (*Ing. Inst.* 90.)
Ti. læta. *Fall. C. S.* 40.

Genus 5: (1364). TINGIS *Auctorum.*

A. "Thorace mutico." *Fallen.*

9413. 1; capitata*. *Wolff.* 131. *pl.* 13. *f.* 124.—*Fall. C. S.* 40.

9414. 2; antica* mihi.
Ti. capitata β. *Fall. C. S.* 40.

B. "Thorace carinato." a. "*Antennis nudis.*" Fall.

9415. 3; costata*. *Fabr. R.* 125.—*Panz. F.* xxiii. *f.* 23.

9416. 4; Cardui*. *Fabr. R.* 125.—*Leach E. E.* ix. 123.—*Samou.* 223.
Ci. Cardui. *Linn.* ii. 718.—*DeGeer.* iii. 309. *pl.* 16. *f.* 1.—*Turt.* ii. 614.

9417. 5; concinna* *mihi.*

9418. 6; nigrina*. *Fall. C. S.* 37.

9419. 7; Humuli*. *Fabr. P.* 126.

9420. 8; dilatata*.
Ci. dilatatus. *Mus. Marsham.*

9421. 9; parvula*. *Fall. C. S.* 37.

9422. 10; reticulata*.
Ci. reticulata. *Mus. Marsham.*

9423. 11, erythrophthalma.
Ci. erythrophthalmus. *Mus. Marsham.*

9424. 12; Cassidea*. *Fall. C. S.* 37.

9425. 13; carinata*. *Panz. F.* xcix. *f.* 20.

9426. 14; pallida* *mihi.*

9427. 15; pusilla*. *Fall. C. S.* 38.

b. "*Antennis pilosis.*" Fall.

9428. 16, obscura* *mihi.*
Ac. clavicornis. *Fabr.* iv. 70.—*Panz. F.* xxiii. *f.* 23.
Ci. clavicornis. *Linn.* ii. 717.—*Berk. S.* i. 115.—*Stew.* ii. 102. —*Turt.* iii. 608.

Genus 6: (1365). DICTYONOTA, *Curtis*, (*Kirby*.)
TINGIS *p*, *Fabr.*

9429. 1; crassicornis*. *Curtis*. iv. *pl.* 154.
Ti. crassicornis. *Fall. C. S.* 38.
Ci. marginatus. *Mus. Marsham.*

9430. 2; Eryngii*. (*Curtis l. c. supra.*)
Ti. Eryngii. *Latr. H.* xii. 253.

9431. 3, spinifrons.
Ti. spinifrons. *Fall. C. S.* 38.

9432. 4, cristata.
Ti. cristata. *Panz. F.* xcix. *f.* 19.

9433. 5; foliacea*.
Ti. foliacea. *Fall. C. S.* 39.

9434. 6, Pyri.
Ti. Pyri. *Fabr. R.* 126.—*Turt.* ii. 614.

9435. 7; rufescens* *mihi.*

Familia II: (159.) PENTATOMIDÆ, *Leach.*

(HE.-LONGILABRES *p*, *Latr.*)

Genus 7: (1366). TETYRA, *Leach*, *Samou.*, (*Kirby.*)
TETYRA *p*, *Fabricius.*—SCUTELLERA *p*, *Latreille.*

9436. 1: obliqua*. *Leach MSS?*—(*Ing. Inst.* 90.)

9437. 2: maura*. *Fabr. R.* 136.—(*Sam. I.* 40.)
Ci. maurus. *Linn.* ii. 716.

9438 ‡ 3. lineata[a].

9439. 4, fuliginosa. *Fabr. R.* 139?
Ci. fuliginosus. *Linn.* ii. 716.—*Schæff. Ic.* 11. *f.* 10–12.—*Turt.* ii. 619. (!)
Te. trilineata *mihi.*

9440. 5, fulvicornis.
Sc. fulvicornis. *Faun. F. cum icone.*

9441. 6; inuncta*. *Fabr. R.* 139.—(*Sam. I.* 40.)
Ci. inunctus. *Panz. F.* xxxvi. *f.* 24.—*Stew.* ii. 103.—*Turt.* ii. 619.

9442. 7; Scarabæoides*. *Fabr. R.* 143.
Ci. Scarabæoides. *Linn.* ii. 716.—*Wolff. Ci.* i. 4. *pl.* 1. *f.* 4.—*Berk. S.* i. 115.—*Stew.* ii. 102.—*Turt.* ii. 619.

---

[a] 9438 ‡ 3. lineata. *Leach E. E.* ix. 121.
Ci. lineatus. *Linn.* ii. 716.
Te. nigrolineata. *Fabr. R.* 135.
Ci. nigrolineatus. *Don.* xiv. 19. *pl.* 473. (!)

Genus 8: (1367). THYREOCORIS, *Schrank, Leach.*
SCUTELLERA *p, Latreille.*—TETYRA *p, Fabricius.*
9443. 1, globus. *Leach E. E.* ix. 121.
Ci. globus. *Wolff. Ci.* i. 3. *pl.* 1. *f.* 3.

Genus 9: (1368). ——
CYDNUS *p? Fall.*—PENTATOMA *p, Latr.*
9444. 1, umbrinus.
Ci. umbrinus. *Wolff.*—*Panz. F.* xciii. *f.* 15.

Genus 10: (1369). CYDNUS, *Fabricius, Leach, Samou., Curtis, (Kirby.)*
PENTATOMA *p, Latreille.*
9445. 1; bicolor*. (*Curtis l. c. infra.*)
Ci. bicolor. *Linn.* ii. 722.—*Panz. F.* xxxii. *f.* 11.—*Berk. S.* i. 116.—*Don.* ix. 21. *pl.* 297.—*Stew.* ii. 104.—*Turt.* ii. 649.—*Wood.* i. 107. *pl.* 38.
Ci. nubilosa. *Harr. Ex.* 90. *pl.* xxvi. *f.* 8.
9446. 2; tristis*. *Fabr. R.* 185?
Ci. spinipes. *Schrank A.* 527?
Ci. morio. *Panz. F.* xxxii. *f.* 15.
9447. 3, Morio*. *Fabr. R.* 184.—(*Curtis l. c. infra.*)
Ci. Morio. *Linn.* ii. 722.—*Panz. F.* xxxii. *f.* 16.—*Turt.* ii. 650.
9448. 4, dubius*. *Curtis.* ii. *pl.* 74.
Ci. dubius. *Scop. C.* 121.
Ci. albo-marginatus. *Schra. A.* 531.
Cy. marginatus *mihi.* (*Curtis l. c. supra.*)
9449. 5; biguttatus*. (*Curtis l. c. supra.*)
Ci. biguttatus. *Linn.* ii. 722.—*Turt.* ii. 649.
9450. 6; albomarginatus*.
Ci. albomarginatus. *Fabr. R.* 179.—*Panz. F.* xxxiii. *f.* 20.

Genus 11: (1370). PENTATOMA, *Olivier, Leach, Samou., Curtis, (Kirby.)*
CYDNUS *p, Fabr., Leach, Samou.*
9451. 1; oleracea*. (*Curtis l. c. infra.*)—(*Kirby & Sp. I. E.* iv. 166.)
Ci. oleraceus. *Linn.* ii. 722.—*Panz. F.* xxxii. *f.* 12.—*Berk. S.* i. 116.—*Stew.* ii. 104.—*Turt.* ii. 649.
Cy. oleraceus. *Leach E. E.* ix. 121.—(*Sam. I.* 14.)
9452. 2; cærulea*. *Curtis.* i. *pl.* 20.
Ci. cæruleus. *Linn.* ii. 722.—*Berk. S.* i. 116.—*Stew.* ii. 104.—*Turt.* ii. 649.
β, Pe. cyanea. *Mus. Marsham.*
9453. 3; melanocephala*. (*Curtis l. c. supra.*)

Cy. melanocephalus. *Fabr. R.* 187. (!)
Ci. melanocephalus. *Panz. F.* xxvi. *f.* 24.—*Stew.* ii. 104.—*Don.* xii. 99. *pl.* 431.—*Turt.* ii. 651.
Ælia melanocephala. (*Sam. I.* 1.)

9454. 4; perlata*. (*Curtis l. c. supra.*)
Cy. perlatus. *Fabr. R.* 187.
Ci. perlatus. *Panz. F.* xxxiii. *f.* 22.

9455 † 5. ornata[a].

9456. 6; festiva*. (*Curtis l. c. supra.*)
Ci. festivus. *Linn.* ii. 723.—*Panz. F.* vi. *f.* 19.—*Stew.* ii. 104.—*Turt.* ii. 648.
Ci. dominulus. *Scop. C.* 362.

9457 † 7, concinna *mihi.*
Pe. picta. *Leach? MSS.*—(*Curtis l. c. supra.*)
In Mus. D. *Vigors.*

9458 † 8. dumosa[b].

9459. 9; Juniperina*. (*Curtis l. c. supra.*)
Ci. Juniperinus. *Linn.* ii. 722.—*Panz. F.* xxxiii. *f.* 14.—*Berk. S.* i. 116.—*Stew.* ii. 103.—*Turt.* ii. 643.

9460. 10; prasina*. *Leach E. E.* ix. 121.—(*Sam. I.* 33. *pl.* 5. *f.* 6.)—(*Curtis l. c. supra.*)
Ci. prasinus. *Linn.* ii. 722.—*Don.* iv. *pl.* 123.—*Stew.* ii. 103.—*Turt.* ii. 643.
Ci. viridis. *Harr. Ex.* 88. *pl.* xxvi. *f.* 1.

9461. 11; dissimilis*. (*Curtis l. c. supra.*)
Ci. dissimilis. *Fabr. R.* 167.—*Panz. F.* xxxiii. *f.* 13.

9462. 12; Baccarum*. (*Curtis l. c. supra.*)
Ci. Baccarum. *Linn.* ii. 721.—*Panz. F.* xxxiii. *f.* 18.—*Berk. S.* i. 115.—*Stew.* ii. 103.—*Turt.* ii. 647.
Ci. subater. *Harr. Ex.* 90. *pl.* xxvi. *f.* 7 ?

9463. 13; Lynx*. (*Curtis l. c. supra.*)
Ci. Lynx. *Fabr. R.* 168.—*Panz. F.* xxxiii. *f.* 15.
β, Ci. grisea. (*Curtis l. c. supra.*)

9464 † 14. punctata[c].

9465. 15; lurida*. (*Curtis l. c. supra.*)
Ci. luridus. *Fabr. R.* 157. (!)—*Don.* iii. *pl.* 98.—*Stew.* ii. 103.—*Turt.* ii. 623.

9466. 16; rufipes*. (*Curtis l. c. supra.*)

---

[a] 9455 † 5. ornata.
Ci. ornatus. *Linn.* ii. 728.—*Panz. F.* xxxiii. *f.* 19.—*Turt.* ii. 647. (!)

[b] 9458 † 8. dumosa.
Ci. dumosus. *Linn.* ii. 721.—*Panz. F.* xxxiii. *f.* 16.—*Turt.* ii. 644. (!)

[c] 9464 † 14. punctata.
Ci. punctatus. *Linn.* ii. 720.—*Turt.* ii. 624. (!)

Ci. rufipes. *Linn.* ii. 719.—*Wolff. Ci.* 1. 9. *pl.* 1. *f.* 9.—*Harr. Ex.* 89. *pl.* xxvi. *f.* 6.—*Stew.* ii. 103. *pl.* viii. *f.* 5.—*Turt.* ii. 623. —(*Kirby & Sp. I. E. pl.* ii. *f.* 5.)

9467. 17; Custos*. (*Curtis l. c. supra.*)
Ci. Custos. *Fabr. R.* 157.

9468. 18; bidens*. *Leach E. E.* ix. 121.—(*Sam. I.* 33.)—(*Curtis l. c. supra.*)
Ci. bidens. *Linn.* ii. 718.—*Panz. F.* xxvi. *f.* 22.—*Turt.* ii. 623.

Genus 12: (1371). ACANTHOSOMA, *Curtis.*

PENTATOMA *p, Olivier, Curtis.*

9469. 1, grisea.
Ci. griseus. *Linn.* ii. 721.—*Panz. F.* xxxiii. *f.* 17.—*Stew.* ii. 103.—*Turt.* ii. 646.—(*Kirby & Sp. I. E.* i. 358.)
Pe. grisea. (*Curtis l. c. fo.* 20.)

9470. 2; hæmorrhoidalis*. *Curtis.* i. *pl.* 28.
Ci. hæmorrhoidalis. *Linn.* ii. 720.—*Don.* vii. 6. *pl.* 218. *f.* 2. —*Stew.* ii. 103.—*Turt.* ii. 626.—(*Kirby & Sp. I. E.* iii. 341.)
Ci. pabulinus. *Harr. Ex.* 88. *pl.* xxvi. *f.* 2.

9471. 3; liturata*. (*Curtis l. c. supra.*)
Ci. lituratus. *Fabr. R.* 170.—*Panz. F.* xl. *f.* 19.

9472. 4; agathina*. (*Curtis l. c. supra.*)
Ci. agathinus. *Fabr. R.* 170.—*Wolff. Ci.* ii. 58. *pl.* 6. *f.* 55.

Genus 13: (1372). ÆLIA, *Fabricius, Leach, Samou.*

9473. 1; acuminata*. *Fabr. R.* 189.—*Leach E. E.* ix. 121.—*Samou.* 221.
Ci. acuminatus. *Panz. F.* xxxii. *f.* 17.—*Berk. S.* i. 116.—*Don.* iv. 28. *pl.* 118. *f.* 2.—*Stew.* ii. 104.—*Turt.* ii. 651.

Familia III: (160). COREIDÆ, *Leach.*

(HE.-LONGILABRES *p, Latr.*—CORISIÆ, *Latr.*)

Genus 14: (1373). COREUS, *Fabricius, Leach, Samou., Curtis,* (*Kirby.*)

9474. 1; marginatus*. *Fabr. R.* 192.—*Leach E. E.* ix. 121.—*Samou.* 222. *pl.* 5. *f.* 7.—(*Kirby & Sp. I. E.* ii. 378.—*Id.* iii. 109.) *Curtis l. c. infra.*
Ci. marginatus. *Linn.* ii. 719.—*Berk. S.* i. 115.—*Stew.* ii. 103. —*Turt.* ii. 630.
Ci. Tipularious. *Harr. Ex.* 89. *pl.* xxvi. *f.* 3.

9475. 2, scapha*? *Fabr. R.* 193.—*Wolff. Ci.* ii. 69. *pl.* 7. *f.* 66. —*Curtis.* iv. *pl.* 174.

9476. 3; quadratus*. *Fabr. R.* 199.—*Wolff. Ci.* ii. 70. *pl.* 7. *f.* 67. —(*Curtis l. c. supra.*)
Co. rhomboideus. (*Sam. I.* 12.)
Co. venator. *Don.* xi. 41. *pl.* 375.

9477. 4, scabricollis *mihi.*
Co. scabricornis. *Panz. F.* xcix. *f.* 21?—(*Curtis l. c. supra?*)
9478 † 5, V. album. *Kirby MSS.* In Mus. *D. Kirby, et Waterhouse.*
9479. 6; denticulatus*. *Scop. C. No.* 365.—*Wolff. Ci.* ii. 71. *pl.* 7. *f.* 68.—(*Curtis l. c. supra.*)
Co. hirticornis. *Fabr. R.* 198.—(*Sam. I.* 12.)

Genus 15: (1374). ——
LYGÆUS *p, Burrell.*
9480. 1; micropterus*.
Ly. micropterus. *Ent. Trans.* (*Burrell.*) i. 73. *pl.* 3.—(*Sam. I.* 26.)

Genus 16: (1375). ——
COREUS II., *Latreille.*—LYGÆUS *p, Sam.*
9481. 1; nugax*.
Co. nugax. *Fabr. R.* 200.—*Wolff. Ci.* i. 50. *pl.* 3. *f.* 30.
Ci. gonymelas. *Don.* vii. 5. *pl.* 218. *f.* 1.—*Stew.* ii. 107.—*Turt.* ii. 668.
Ly. nugax. (*Sam. I.* 26.)

Genus 17: (1376). ALYDUS, *Fabricius,* (*Kirby.*)
COREUS III., *Latreille.*
9482. 1, calcaratus*. *Fabr. R.* 251. (!)—(*Ing. Inst.* 90.)—(*Kirby & Sp. I. E.* iii. 615.)
Ci. calcaratus. *Linn.* ii. 732.—*DeGeer.* iii. 280. *pl.* 14. *f.* 23, 24.
♀, Ci. niger. *Marsh.? MSS.*
9483. 2, æneus*.
Ci. æneus. *Marsh.? MSS.*

Genus 18: (1377). PYRRHOCORIS, *Fallen,* (*Kirby.*)
LYGÆUS *p, Fabricius, Leach, Samou.*
9484. 1, apterus.
Ci. apterus. *Linn.* ii. 727.—*Berk. S.* i. 117.—*Stew.* ii. 105.—*Turt.* ii. 663.
Ly. apterus. *Fabr. R.* 227.—(*Leach E. E.* ix. 122.)—(*Sam. I.* 26.)

Genus 19: (1378). CORIZUS, *Fallen.*
LYGÆUS *p, Fabr.*
9485. 1, Hyoscyami.
Ci. Hyoscyami. *Linn.* ii. 726.—*Panz. F.* lxxix. *f.* 21.—*Berk. S.* i. 117.—*Stew.* ii. 105.—*Turt.* ii. 657.
Ly. Hyoscyami. *Fabr. R.* 218.—(*Sam. I.* 26.)
Ci. apterus. *Harr. Ex.* 91. *pl.* xxvi. *f.* 12.—(*Samou. pl.* 5. *f.* 8.)
9486 † 2, Equestris.

Ci. equestris. *Linn.* ii. 726.—*Panz. F.* lxxix. *f.* 19.—*Berk. S.* i. 117.—*Stew.* ii. 105.—*Turt.* ii. 656.

In Mus. *Brit.*, *et D. Hope.*

Genus 20: (1379). KLEIDOCERYS, *Westwood MSS.*

LYGÆUS *p*, *Fallen.*

9487. 1; capitatus*.
Co. capitatus. *Fabr. R.* 201.—*Wolff. Ci.* ii. 75. *pl.* 8. *f.* 72.

9488. 2; magnicornis*.
Co. magnicornis. *Fabr. R.* 200.

9489. 3; crassicornis*.
Ci. crassicornis. *Fabr. R.* 201.—*Schæff. Ic. pl.* 13. *f.* 10.—*Turt.* ii. 666. (!)

9490. 4; Caricis*.
Ci. Caricis. *Mus. Marsham.*
Ly. claviculus. *Fallen C. S.* 64?

9491. 5; Resedæ*.
Ci. Resedæ. *Panz. F.* xl. *f.* 20.

9492. 6; centralis*. *Westwood MSS.*

9493. 7; immunis *mihi.*

9494. 8; costalis* *mihi.*

Genus 21: (1380). PACHYMERUS, *St. Fargeau?*

LYGÆUS *p*, *Fabricius.*

A. "Capite dilatato, thorace immarginato." *Fall.*

9495. 1; Urticæ*.
Ly. Urticæ. *Fabr. R.* 231. (!)
Ci. Urticæ. *Stew.* ii. 106.—*Turt.* ii. 665.
Ci. viror. *Harr. Ex.* 91. *pl.* xxvi. *f.* 13?

9496. 2; sanguineus* *mihi.*

9497. 3; annularis* *mihi.*

9498. 4, Thymi.
Ly. Thymi. *Wolff.*—*Fall. C. S.* 63.

B. "Capite angustato, fronte brevi, subacutâ; thorace sæpiùs marginato." *Fall.*

9499. 5; Pini*.
Ci. Pini. *Linn.* ii. 729.—*Wolff. Ci.* ii. 74. *pl.* 8. *f.* 71.—*Stew.* ii. 106.—*Turt.* ii. 664.

9500. [6; lynceus*.]
Ly. lynceus. *Fabr. R.* 231. (!)
Ci. lynceus. *Stew.* ii. 106.—*Turt.* ii. 665.

9501. 7; nubilus*.
Ly. nubilus. *Fall. C. S.* 65.
Ci. arenanis. *Linn.* ii. 729?—*Stew.* ii. 106?—*Turt.* ii. 668?

9502. 8; agrestis*.
Ly. agrestis. *Fall. C. S.* 66.

9503. 9; erraticus.
Ly. erraticus. *Fabr. R.* 232.

9504. 10, plebeius*.
Ly. plebeius. *Fall. C. S.* 67.

9505. 11; chiragra*.
Ly. chiragra. *Fabr. R.* 233.

9506. 12, Echii*.
Ly. Echii. *Fabr. R.* 235.

9507. 13; Rolandri*.
Ci. Rolandri. *Linn.* ii. 729.—*Schæff. Ic.* 87. *f.* 7.—*Stew.* ii. 106.—*Turt.* ii. 664.

9508. 14; apicalis*.
Ci. apicalis. *Mus. Marsham.*

9509. 15; sylvaticus*.
Ly. sylvaticus. *Fabr. R.* 229.

9510. 16; transversus*.
Ci. transversus. *Mus. Marsham.*

9511. 17; podagricus*.
Ly. podagricus. *Fabr. R.* 232. (!)
Ci. podagricus. *Turt.* ii. 666.

9512. 18; quadratus*.
Ly. quadratus. *Fabr. R.* 232.—*Coqueb. Ic.* 9. *f.* 12.

9513. 19; brachypterus* *mihi.*

9514. 20; pedestris*.
Ly. pedestris. *Fall. C. S.* 71.—*Panz. F.* xcii. *f.* 14?

9515. 21; rusticus*.
Ly. rusticus. *Fall. C. S.* 70.

9516 ad 9522: 7 sp. adhuc examinandæ. (28 sp.)

Genus 22: (1381). PLINTHISUS, *Westwood MSS.*
Lygæus *p*, *Latr.*

9523. 1; brevipennis*.
Ly. brevipennis. *Latr.* G. iii. 123.

Genus 23: (1382) ——
Lygæus *p*, *Latr.*—Miris *p*, *Panz.*

9524. 1; Abietis*.
Ci. Abietis. *Linn.* ii. 726.—*Berk. S.* i. 118.—*Stew.* ii. 107.—*Turt.* ii. 694.
Mi. Abietis. *Panz. F.* xcii. *f.* 22.
Var.? Ci. ferrugineus. *Linn.* ii. 730.

Genus 24: (1383). HYLOPHILA, (*Kirby*.)
SALDA *p*, *Fabr.*

9525. 1; Nemorum*.
Ci. Nemorum. *Linn. F.* 953.—(*Kirby & Sp. I. E.* i. 108.)

9526. 2; bipunctata*.
Ci. bipunctata. *Mus. Marsham.*
Sa. pratensis. *Fabr. R.* 116?

9527. 3; Serratulæ*.
Sa. Serratulæ. *Fabr. R.* 115. (!)
Ci. Serratulæ. *Stew.* ii. 102.—*Turt.* ii. 610.

9528. [4, campestris*.]
Sa. campestris. *Fabr. R.* 116.

9529. 5, sylvestris*.
Sa. sylvestris. *Fabr. R.* 116.—*Turt.* ii. 613.

9530. 6, affinis* *mihi.*

9531. 7; minuta*.
Ly. minutus. *Fall. C. S.* 73.

9532. 8; pygmæa*.
Ly. pygmæus. *Fall. C. S.* 73.

9533. 9; exilis*.
Ly. exilis. *Fall. C. S.* 73.

9534. 10; cursitans*.
Ly. cursitans. *Fall. C. S.* 74.

Genus 25: (1378). ——

9535. 1; leucoptera* *mihi.*
Salda coleoptrata. *Fall. C. S.* 31?

9536. [2; ruficeps*] *mihi.*

Genus 26: (1384). ORTHONOTUS, *Westwood.*
CAPSUS *p*, *Fall.*

9537. 1, pallicornis.
Sa. pallicornis. *Fabr. R.* 115.
Cicada aptera. *Linn. F. No.* 894.

9538. 2; rufifrons*.
Ca. rufifrons. *Fall. C. S.* 105.

9539. [3, leucocephalus*.] (*Ing. Inst.* 90.)
Ci. leucocephalus. *Mus. Marsham.*

9540. 4; albipes* *mihi.*

9541. 5; pilosus*.
Ci. pilosus. *Mus. Marsham.*

9542. 6, nigripes *mihi.*

Genus 27: (1385). HETEROTOMUS, *Latr.*
CAPSUS *p*, *Fabr.*

9543. 1; spissicornis*.
Ly. spissicornis. *Fabr.* iv. 181.—*Panz. F.* ii. *f.* 16.

Ci. spissicornis. *Don.* iv. 71. *pl.* 135.—*Stew.* ii. 104.—*Turt.* ii. 674.
Ca. spissicornis. (*Sam. I.* 8.)

9544. 2; magnicornis*.
Ly. magnicornis. *Fall. C. S.* 99.

9545. 3, atricornis* *mihi.*

Genus 28: (1386). CAPSUS, *Fabr.*, *Leach*, *Samou.*, (*Kirby.*)

9546. 1; ater*. *Fabr. R.* 241.—*Leach E. E.* ix. 122.—*Samou.* 222.
Ci. ater. *Linn.* ii. 725.—*Turt.* ii. 671.

9547. [2; tyrannus*.] *Fabr. R.* 242.

9548. 3, semiflavus*.
Ci. semiflavus. *Linn.* ii. 725.
Ca. flavicollis. *Fabr. R.* 243. (!)
Ci. flavicollis. *Stew.* ii. 104.—*Turt.* iii. 672.
Ca. ruficollis. (*Sam. I.* 8.)

9549. 4; tricolor*. *Fabr. R.* 246.
Ly. danicus. *Fall. C. S.* 93?
Ci. rubens. *Harr. Ex.* 90. *pl.* xxvi. *f.* 10?
Var.? Ca. seticornis. *Fabr. R.* 244?
Ca. capilaris. *Fabr. R.* 244?
Ci. melinus. *Harr. Ex.* 90. *pl.* xxvi. *f.* 11?

9550. 5; aterrimus* *mihi.*

9551. 6; albomarginatus*. *Fabr. R.* 245.
Ci. flavomarginatus. *Don.* vii. 79. *pl.* 245.—*Stew.* ii. 106.—*Turt.* ii. 680.

9552. [7; gothicus*.] *Fabr. R.* 244.
Ci. gothicus. *Linn.* ii. 726.—*Turt.* ii. 673.

9553. [8; concinnus*] *mihi.*
Ca. scutellaris. *Fabr. R.* 245?
Ci. superciliosus. *Linn.* ii. 728?

9554. 9; mutabilis*. *Fall. C. S.* 98.

9555. 10; fuscipes* *mihi.*

9556. 11, unifasciatus. *Fabr. R.* 243.
Mi. semiflavus. *Wolff. Ci.* iv. 154. *pl.* 15. *f.* 48.

9557. 12; flavomaculatus*. *Fabr. R.* 247.

9558. 13, umbratilis. *Fabr. R.* 246.
Ci. umbratilis. *Turt.* ii. 672.

9559. 14; triguttatus*.
Ci. triguttatus. *Linn.* ii. 729.

9560 ad 9566: 7 sp. adhuc examinandæ.

9567 † 22. trifasciatus[a].

---

[a] 9567 † 22. trifasciatus. *Fabr. R.* 244.
Ci. trifasciatus. *Linn.* ii. 725.—*Turt.* ii. 672. (!)

Genus 29: (1387). ——

CAPSUS *p, Fabr.?*

9568. 1; pulcherrimus*.
Ca. bifasciatus. *Fabr. R.* 242?
Ci. clavatus. *Linn.* ii. 729?

Genus 30: (1388). ——

LYGÆUS *p, Fabr.*

9569. 1; tunicatus*.
Ly. tunicatus. *Fabr. R.* 233.

Genus 31: (1389). PŒCILOSOMA *mihi.*

LYGÆUS *p, et* CAPSUS *p, Fabr.*

9570. 1; pratensis*.
Ly. pratensis. *Fabr. R.* 234.
Ci. pratensis. *Linn.* ii. 728?—*Berk. S.* i. 117.—*Stew.* ii. 106.
—*Turt.* ii. 669.
Ci. prasinus. *Mus. Marsham.*

9571. 2; campestris*.
Ci. campestris. *Linn.* ii. 728.

9572. 3, rubicunda*.
Ly. rubicundus. *Fall. C. S.* 84.

9573. 4, rufipennis*.
Ly. rufipennis. *Fall. C. S.* 84.

9574. 5, maculipennis* *mihi.*

9575. 6; Kalmii*.
Ci. Kalmii. *Linn.* ii. 728?

9576. 7; cordigera* *mihi.*

9577. 8, flavovaria*.
Ca. flavovarius. *Fabr. R.* 243.
Ci. umbratilis. *Berk. S.* i. 117?

9578. 9, immunis* *mihi.*

9579. 10; pistacinæ*.
Ly. pistacinæ. *Fall. C. S.* 86.

9580. 11; varipes* *mihi.*

9581. 12, punctulata*.
Ly. punctulatus. *Fall. C. S.* 87.

9582. 13; concinna* *mihi.*

9583. 14, tripustulata*.
Ly. tripustulatus. *Fabr. R.* 239.

9584. 15; Gyllenhalii*.
Ly. Gyllenhalii. *Fall. C. S.* 88.

9585. 16; variabilis*.
Ly. variabilis. *Fall. C. S.* 88.

9586. 17, Umbellatarum.
Ly. Umbellatarum. *Fanz. F.* xciii. *f.* 19.

9587. 18, mutabilis*.
Ci. mutabilis. *Linn.* ii. 731.

9588. 19; ambigua*.
Ly. ambiguus. *Fall. C. S.* 89.

9589. 20; rubricata*.
Ly. rubricatus. *Fall. C. S.* 91.

9590. 21; rosea*.
Ly. roseus. *Fabr. R.* 238.

9591. 22; sanguinea*.
Ly. sanguineus. *Fabr. R.* 238.

9592. 23; hæmoptera.
Ci. hæmopterus. *Mus. Marsham.*

9593 ad 9609: 17 sp. adhuc examinandæ. (40 sp.)

Cimices incertæ sedes: 20 sp.—LYGÆUS *p, et* CAPSUS *p, Fabr.*

9610. 1, Filicis*.
Ci. Filicis. *Linn.* ii. 718.—*Berk. S.* i. 115.—*Stew.* ii. 102.—*Turt.* ii. 613.
Ac. Filicis. *Wolff. Ci.* ii. 46. *pl.* 5. *f.* 43.

9611. 2; tenellus*.
Ly. tenellus. *Fall. C. S.* 90.
Ci. chlorocephalus. *Mus. Marsham.*

9612 † 3. leucocephalus[a].

9613. 4; pulicarius*.
Ly. pulicarius. *Fall. C. S.* 95.

9614 ad 9630: 17 sp. adhuc examinandæ.

Genus 32: (1390). AZINECERA *mihi.*

9631. 1; dispar* *mihi.*
Ca. dolabricornis. *Kirby MSS?*

Genus 33: (1391). ——

LYGÆUS *p, Fabr.*

9632. 1; Stigma* *mihi.*
Ci. lævigatus. *Mus. Marsham.*

9633. 2; rufus* *mihi.*
Ci. danicus. *Mus. Marsham.*

9634. 3; ferrugatus*.
Ly. ferrugatus. *Fabr. R.* 236.
Ci. roseo-maculatus. *De Geer.*

---

[a] 9612 † 3. leucocephalus.
Ci. leucocephalus. *Linn.* ii. 723.—*Turt.* ii. 670. (¹)

9635. 4; Chenopodii*.
Ly. Chenopodii. *Fall. C. S.* 74.
Var.? Ly. quadripunctatus. *Fabr. R.* 235.
Ci. quadrimaculatus. *Stew.* ii. 105.
9636. 5; bipunctatus*.
Ly. bipunctatus. *Fabr. R.* 235.
9637. 6; binotatus*.
Ly. binotatus. *Fabr. R.* 235.
9638. 7; lævigatus*.
Mi. lævigatus. *Panz. F.* xciii. *f.* 21.
9639. 8; pabulinus*.
Ci. pabulinus. *Linn.* ii. 727.—*Turt.* ii. 693.
9640. 9; contaminatus*.
Ly. contaminatus. *Fall. C. S.* 76.
9641. 10; nassatus*.
Ly. nassatus. *Fabr. R.* 236.
9642. 11; rugicollis*.
Ly. rugicollis. *Fall. C. S.* 76.
9643. 12; Tanaceti*.
Ly. Tanaceti. *Fall. C. S.* 77?
9644. 13; angulatus*.
Ly. angulatus. *Fall. C. S.* 76.
9645. 14; vernans* *mihi.*
9646. 15; seladonius*.
Ly. seladonius. *Fall. C. S.* 77.
9647. 16; molliculus*.
Ly. molliculus. *Fall. C. S.* 77?
9648. 17; chlorizans*.
Ly. chlorizans. *Panz. F.* xviii. *f.* 21.
9649. 18; Coryli*.
Ci. Coryli. *Linn.* ii. 733.—*Stew.* ii. 106.—*Turt.* ii. 668.
9655. 24; Arbustorum*.
Ly. Arbustorum. *Fabr. R.* 238.
9657. 26; viridulus*.
Ly. viridulus. *Fall. C. S.* 90.
9658. 27; Paykullii*.
Ly. Paykullii. *Fall. C. S.* 92?

Genus 34: (1392). ——
Capsus *p, Fall.*

9659. 1; vittatus*.
Ci. vittatus. *Don.* vii. 95. *pl.* 252. *f.* 1.—*Stew.* ii. 105.—*Turt.* ii. 677.
9660. 2; Caricis*.
Ca. Caricis. *Fall. C. S.* 102.

Genus 35: (1393). ——

CAPSUS *p*, *Fall.*

9661. 1; atricapillus*.

Ci. atricapillus. *Mus. Marsham.*

9662. 2; collaris*.

Ca. collaris. *Fall. C. S.* 103?

9663. 3; pallidus* *mihi.*

9664. 4; tripunctatus* *mihi.*

Genus 36: (1394). PHYTOCORIS, *Fallen?*

LYGÆUS *p*, *et* CAPSUS *p*, *Fabr.*

9665. 1; striatus*.

Ci. striatus. *Linn.* ii. 730.—*Panz. F.* xciii. *f.* 22.—*Stew.* ii. 107.—*Turt.* ii. 694.

Ci. striatus: Fine streaked Bugkin. *Berk. S.* i. 118.

9666. 2; striatellus*.

Ly. striatellus. *Fabr. R.* 236.

Ci. quadripunctatus. *Don.* iii. 77. *pl.* 101. *f.* 1–3.—*Turt.* ii. 680.

Ci. pulligo. *Harr. Ex.* 89. *pl.* xxvi. *f.* 4.

9667. 3, scriptus.

Ci. scriptus. *Fabr. R.* 234.

9668. 4, campestris.

Ci. campestris. *Stew.* ii. 106.—*Turt.* ii. 669.

9670. 6, sexguttatus*.

Ly. sexguttatus. *Fabr. R.* 237.

9672. 8; elongatus*.

Ci. elongatus. *Mus. Marsham.*

9674. 10; Populi*.

Ci. Populi. *Linn.* ii. 731.—*Berk. S.* i. 118.—*Don.* vii. 95. *pl.* 252. *f.* 2.—*Turt.* ii. 670.

9675. [11, umbratilis*.]

Ci. umbratilis. *Linn.* ii. 728.—*Stew.* ii. 104.

9676 † 12. Ulmi[a].

Genus 37: (1395). MIRIS, *Fabr.*, *Leach*, *Samou.*, (*Kirby.*)

9677. 1; dolabratus*. *Fall. C. S.* 107.

Ci. dolabratus. *Linn.* ii. 730.—*Turt.* ii. 692.

Mi. lateralis. *Fabr. R.* 254.—*Wolff. Ci.* iii. 115. *pl.* 11. *f.* 109. ♂?

♀. Mi. abbreviatus. *Wolff. Ci.* iii. 116. *pl.* 11. *f.* 110.

Ci. subapterus. *Mus. Marsham.*

9678. 2; ferrugatus*. *Fall. C. S.* 107.

Mi. dolabratus. *Fabr. R.* 253?

---

[a] 9676 † 12. Ulmi.

Ci. Ulmi. *Linn.* ii. 731.—*Berk. S.* i. 118. (!)—*Stew.* ii. 107. (!)—*Turt.* ii. 694. (!)

9679. 3; lævigatus*. *Fabr. R.* 253.
Ci. lævigatus. *Linn.* ii. 730.—*Turt.* ii. 692.
Ci. pallidus. *Harr. Ex.* 90. *pl.* xxvi. *f.* 9.
Ci. pallescens. *Mus. Marsham.*

9680. 4; calcaratus*. *Fall. C. S.* 110.
β; Ci. virens. *Linn.* ii. 730?

9681. 5; nudicornis* *mihi.*

9682. 6; erraticus*. *Fall. C. S.* 111.
Ci. erraticus. *Linn.* ii. 731.
Mi. hortorum. *Wolff. Ci.* iv. 160. *pl.* 16. *f.* 154.

9683. 7; longicornis*. *Fall. C. S.* 108.

9684. 8; ruficornis*. *Fall. C. S.* 112.

9685. 9; holsatus*. *Fabr. R.* 254.

9686. 10; Marshami*.
Ci. Marshami. *Gmel.?*—*Turt.* ii. 692.
Ci. pallescens. *Don.* iii. 78. *pl.* 101. *f.* 5, 6.—*Stew.* ii. 107.

9687. 11, ferus. *Fabr. R.* 255?
Ci. ferus. *Linn.* ii. 731?—*Turt.* ii. 693.
β; Mi. vagans. *Fabr. R.* 255?—*Leach E. E.* ix. 122.—*Samou.* 222.

9688. 12; dilatatus* *mihi.*

9689. 13; xanthoceras* *mihi.*

Genus 38: (1396). MYODOCHA[a].

Genus 39: (1397). NEIDES, *Latr.*, *Curtis.*
BERYTUS, *Fabr.*, *Leach*, *Samou.*—GERRIS *p*, *Fabr. olim.*

9691. 1; tipularia*. *Latr.* G. iii. 120.—(*Curtis l. c. infra.*)
Ci. tipularius. *Linn.* ii. 733.—*Turt.* ii. 692.
By. tipularius. *Leach E. E.* ix. 122.—*Samou.* 222.

9692. 2; clavipes*. (*Curtis l. c. infra.*)
By. clavipes. *Fabr. R.* 265.

9693. 3; tripunctata* *mihi.*

9694. 4, elegans. *Curtis.* iv. *pl.* 150.

Familia IV: (161). REDUVIIDÆ.

(HE.-NUDICOLLES, *Latr.*)

Genus 40: (1398). REDUVIUS, *Fabr.*, *Leach*, *Samou.*, (*Kirby.*)

9695. 1; personatus*. *Fabr. R.* 267.—(*Leach E. E.* ix. 122.)—(*Sam. I.* 36.)—(*Kirby & Sp. I. E.* ii. 297.)

---

[a] Genus 38: (1396). MYODOCHA, *Latr.*, *Leach*, *Samou.*
9690 † 1. Tipuloides. (*Latr. G.* iii. 126.)—(*Leach E. E.* ix. 122.)—(*Samou.* 223.) (!)
Ci. Tipuloides. *DeGeer.* v. 354. *pl.* 35. *f.* 18.

Ci. personatus. *Linn.* ii. 724.—*Berk. S.* i. 117.—*Stew.* ii. 108. —*Turt.* ii. 694.
Ci. anulata. *Harr. Ex.* 89. *pl.* xxvi. *f.* 5.

Genus 41: (1399). NABIS? *Latr.*, (*Kirby?*)
REDUVIUS *p*, *Fabr.*

9696. 1, annulatus?*
Re. annulatus. *Fabr. R.* 271?
Ci. annulatus. *Turt.* ii. 696?

9697. 2; subapterus*.
Ci. subapterus. *DeGeer.* iii. 287. *pl.* 15. *f.* 10.—(*Kirby & Sp. I. E.* ii. 391.)
Re. apterus. *Fabr. R.* 281.

9698. 3; annulipes* *mihi.*

9699. 4; cingulatus* *mihi.*

9700. 5; sublineatus* *mihi.*

9701. 6; obsoletus* *mihi.*

9702. 7; tripunctatus*.
Ci. tripunctatus. *Mus. Marsham.*

9703. 8; hirtipes *mihi.*

9704. 9; creticola*.
Ci. creticola. *Mus. Marsham.*

Genus 42: (1400). PLOIARIA, *Scop.*, *Leach*, *Samou.*, (*Kirby.*)
GERRIS *p*, *Fabr.*

9705. 1; vagabunda*. *Latr. H.* xii. 162.—(*Leach E. E.* ix. 122.) —(*Sam. I.* 34.)
Ci. vagabundus. *Linn.* ii. 732.—*DeGeer.* iii. 332. *pl.* 17. *f.* 1, 2. —*Berk. S.* i. 119.—*Stew.* ii. 107.—*Turt.* ii. 691.

9706. 2; erratica*.
Ge. erratica. *Fall. C. S.* 117.

9707. 3; obscura* *mihi.*

Familia V: (162). ACANTHIIDÆ, *Leach.*

(HE.-OCULATÆ, *Latr.*—CIMICIDES II., *Latr. olim.*)

Genus 43: (1401). ACANTHIA, *Latr.*, *Leach*, *Samou.*, (*Kirby.*)
ACANTHIA *p*, *et* SALDA *p*, *Fabr.*

9708. 1; littoralis.
Ci. littoralis. *Linn.* ii. 717.—*DeGeer.* iii. 278. *pl.* 14. *f.* 17.—*Stew.* ii. 102.—*Turt.* ii. 610.
Ac. Zosteræ. *Fabr.* iv. 68.
Var. Sa. flavipes. *Fabr. R.* 114.

9709. 2; saltatoria*.
Ci. saltatorius. *Linn.* ii. 729.—*Wolff. Ci.* ii. 77. *pl.* 8. *f.* 74.—*Stew.* ii. 106.—*Turt.* ii. 671.

Ac. maculata. *Latr. H.* xii. 243.—*Leach E. E.* ix. 123.—*Samou.* 225.

9710. [3, pallipes*.] *Fabr.* iv. 71?

9711. 4, agilis*.
Ci. agilis. *Mus. Marsham.*
Sa. striata. *Fabr. R.* 114?

9712. 5, pilosa.
Sa. pilosa. *Fall. C. S.* 29.

9713. 6, marginalis.
Sa. marginalis. *Fall. C. S.* 30.

9714. 7; elegantula*.
Sa. elegantula. *Fall. C. S.* 30.

9715. 8; lateralis*.
Sa. lateralis. *Fall. C. S.* 30.

Familia VI: (163). HYDROMETRIDÆ, *Leach.*

(HE.-PLOTERES, *Latr.*—CIMICIDES I., 2, *Latr. olim.*)

Genus 44: (1402). HYDROMETRA, *Latr., Leach, Samou., Curtis,* (*Kirby.*)

AQUARIUS *p, Schellenb.*—EMESSA, *Fall.*

9716. 1; stagnorum*. *Fabr. R.* 258. (!)—*Leach E. E.* ix. 123. —*Samou.* 224.—*Curtis.* i. *pl.* 32.
Ci. stagnorum. *Linn.* ii. 732.—*Berk. S.* i. 118.—*Don.* ii. 5. *pl.* 38.—*Stew.* ii. 107.—*Turt.* ii. 689.—*Shaw* G. Z. vi. 167. *pl.* 57. *fig. med.*
Aq. paludum. *Schellenberg H.*

Genus 45: (1403). VELIA, *Latr., Leach, Samou., Curtis,* (*Kirby.*)

HYDROMETRA *p, Fabr.*—GERRIS *p, Fall.*

9717. 1; rivulorum*. *Latr. H.* xii. 270.—*Ent. Trans.* (*Haworth.*) i. 252.—*Leach E. E.* ix. 123.—*Samou.* 224.—*Curtis.* i. *pl.* 2.
Ge. rivulorum. *Fabr.* iv. 189.

9718. [2; currens*.] *Fabr. R.* 259.—*Leach E. E.* ix. 123.—*Curtis 2 edit.* 1. *fo.* 2.

Genus 46: (1404). GERRIS, *Latr., Leach, Samou.,* (*Kirby.*)

HYDROMETRA *p, Fabr.*—AQUARIUS *p, Schell.*

9719. 1; paludum*. *Fabr.* iv. 188.—*Stoll. Ci. pl.* 9. *f.* 63.—*Latr.* G. iii. 133.—*Leach E. E.* ix. 123.—*Samou.* 224. *pl.* 5. *f.* 5?

9720. [2; brachypteryx*] *mihi.*

9721. 3; rufo-scutellata*. *Latr.* G. iii. 134.—*Stoll. Ci. pl.* 15. *f.* 108.

9722. 4, thoracica* *mihi.*

9723. 5; lacustris*. *Fabr.* iv. 187.—*Latr. G.* iii. 134.

Ci. lacustris. *Linn.* ii. 732.—*Berk. S.* i. 118.—*Don.* iv. 27. *pl.* 118?—*Stew.* ii. 107.—*Turt.* ii. 689.—*Shaw* G. Z. vi. 167.

9724. 6, phæoptera* *mihi.*

9725. 7; nana* *mihi.*

Ge. lacustris var. C. *Latr.* G. iii. 135.

9726. 8; aterrima* *mihi.*

Sectio II. AQUATICA, *Leach.*

(Hydrocorisiæ, *Latreille.*)

Familia VII: (164). NEPIDÆ, *Leach.*

(Nepides, *Latreille.*—Nepa, *Linné, &c.*)

Genus 47: (1405). RANATRA, *Fabr., Leach, Samou.,* (*Kirby.*)

Hepa *p, Geoffroy.*

9727. 1; linearis*. *Fabr. R.* 109.—*Don.* iii. 87. *pl.* 105.—*Leach E. E.* ix. 124.—*Samou.* 225.

Ne. linearis. *Linn.* ii. 714.—*Berk. S.* i. 114.—*Stew.* ii. 100. —*Turt.* ii. 607.—*Shaw* G. Z. vi. 159. *pl.* 56. *fig. sup.*—*Wood.* i. 103. *pl.* 37.

Genus 48: (1406). NEPA *Auctorum. Water Scorpion.*

Hepa *p, Geoffroy.*

9728. 1; cinerea*. *Linn.* ii. 714.—*Berk. S.* i. 114.—*Don.* i. 41. *pl.* 18.—*Stew.* ii. 100.—*Turt.* ii. 607.—*Shaw G. Z.* vi. 157. *pl.* 55. *fig. inf., pl.* 56. *fig. inf. dext. et sinist.*—*Leach E. E.* ix. 124.—*Samou.* 225. *pl.* 5. *f.* 4.—(*Kirby & Sp. I. E.* iii. 616.)

Genus 49: (1407). NAUCORIS, *Geoffroy, Leach, Samou.,* (*Kirby.*)

9729. 1; cimicoides*. *Fabr. R.* 110.—*Don.* xi. 61. *pl.* 381.—*Leach E. E.* ix. 124.—(*Sam. I.* 28.)

Ne. cimicoides. *Linn.* ii. 714.—*Berk. S.* i. 114.—*Stew.* ii. 100. —*Turt.* ii. 608.—*Shaw G. Z.* vi. 158. *pl.* 56. *fig. inf. med.*

Familia VIII: (165). NOTONECTIDÆ, *Leach.*

(Notonectides, *Latr.*—Notonecta, *Linné, &c.*)

Genus 50: (1408). NOTONECTA *Auctorum. Boat-fly.*

9730. 1; furcata*. *Fabr. R.* 102.—*Ent. Trans.* (*Haworth.*) i. 98. —*Don.* xvi. *pl.* 560. *f.* 2.—*Leach E. E.* ix. 124.—*Linn. Trans.* (*Leach.*) xii. 12.—*Samou.* 226.—(*Curtis l. c. infra.*)

9731. 2; maculata*. *Oliv. E. M.* viii. 388.—*Don.* xvi. *pl.* 560. *f.* 1.—*Linn. Trans.* (*Leach.*) xii. 12.—*Samou.* 226.—(*Millard. pl.* 1. *f.* 2.)—*Curtis.* i. *pl.* 10.

No. glauca α. *Latr.* G. iii. 150.

9732. 3; glauca*. *Linn.* ii. 712.—*Berk. S.* i. 113.—*Don.* iii. 7. *pl.* 73.—*Stew.* ii. 99.—*Shaw* G. Z. vi. 155. *pl.* 54. *fig. sup.*—

Turt. ii. 606.—*Leach E. E.* ix. 124.—*Linn. Trans.* (*Leach.*) xii. 13.—(*Kirby & Sp. I. E.* i. 108.)—*Samou.* 227. *pl.* 5. *f.* 3.—*Wood.* i. 101. *pl.* 36.—(*Curtis l. c. supra.*)

Genus 51: (1409). PLOA, (PLEÄ) *Leach, Samou.*, (*Kirby.*)

9733. 1; minutissima*. *Linn. Trans.* (*Leach.*) xii. 14.—*Samou.* 227.
No. minutissima. *Fourc. P.* i. 220.—*Berk. S.* i. 114.—*Stew.* ii. 99.—*Turt.* ii. 605.—*Shaw* G. *Z.* vi. 156. *pl.* 54. *fig. med.*

Genus 52: (1410). SIGARA, *Leach, Samou.*

9734. 1; minutissima*. *Linn. Trans.* (*Leach.*) xii. 14.—*Samou.* 227.
No. minutissima. *Linn.* ii. 713.

Genus 53: (1411). CORIXA, *Geoffroy, Leach, Samou.*

SIGARA *p, Fabricius.*

A. "Elytris ad apicem subgradatim acuminatis." *Leach.*

9735. 1; coleoptrata*. *Linn. Trans.* (*Leach.*) xii. 16.—*Samou.* 228.
Si. coleoptrata. *Fabr. R.* 105?—*Ent. Trans.* (*Haworth.*) i. 98.

B. "Elytris ad apicem rotundatis." *Leach.*

a. "*Elytris thoraceque rugulosis.*" Leach.

9736. 2; rivalis* *mihi.*

9737. 3; lacustris* *mihi.*

9738. 4; fossarum*. *Linn. Trans.* (*Leach.*) xii. 17.—*Samou.* 228.

9739. 5; striata*. *Linn. Trans.* (*Leach.*) xii. 16.—*Samou.* 228.
No. striata. *Linn.* ii. 712.—*Don.* v. 101. *pl.* 176. *f.* 1?

9740. 6; stagnalis*. *Linn. Trans.* (*Leach.*) xii. 16.—*Samou.* 228.

9741. 7; lateralis*. *Linn. Trans.* (*Leach.*) xii. 17.—*Samou.* 228.

9742. 8; dorsalis*. *Linn. Trans.* (*Leach.*) xii. 17.—*Samou.* 229.

b. "*Elytris thoraceque glaberrimis, lævibus.*" Leach.

9743. 9; Geoffroyi*. *Linn. Trans.* (*Leach.*) xii. 17.—*Samou.* 229.
La. Corise. *Geoffroy H.* i. 478. *pl.* 9. *f.* 7.
Si. striata. *Panz. F.* 1. *f.* 7.
No. striata. *Berk. S.* i. 114.—*Don.* v. 101. *pl.* 176. *f.* 2, 3.—*Stew.* ii. 99.—*Turt.* ii. 606.—*Shaw* G. *Z.* vi. 155. *pl.* 54. *fig. inf.*

9744. 10; affinis*. *Linn. Trans.* (*Leach.*) xii. 18.—*Samou.* 229.

ORDO VII: (14). HOMOPTERA, *Leach.*

(RHYNGOTA *p, Fabr.*—HEMIPTERA *p, Linné.*—HEMIPTERA-HOMOPTERA, *Latr.*)

Familia I: (166). CICADIIDÆ, *Leach.*

Genus 1: (1412). CICADA, *Geoff., Leach, Samou.*

TETTIGONIA, *Fabr.*

9745 ‡ 1. *Orni*[a].

9746. 2: hæmatodes. *Linn.* ii. 707.—*Leach E. E.* ix. 125.
Ti. sanguinea. *Fabr. R.* 39.
Ci. anglica. (*Samou.* 447. *pl.* 5. *f.* 2.)

Familia II: (167). FULGORIDÆ.

(FULGORELLÆ, *Latr.*—FULGORA *et* CICADA *p, Linné.*

Genus 2: (1413). FULGORA[b].

Genus 3: (1414). ——

FLATA, *Fabr., Leach, Samou.,* (*Kirby.*)—CIXIUS *p, Samou.*

9748. 1; reticulata*.
Fl. reticulata. (*Samou.* 230.)

9749. 2; venosa*.
Ci. venosa. *Mus. Marsham.*

9750. 3; membranacea*.
Ci. membranacea. *Mus. Marsham.*

9751. 4; nebulosa* *mihi.*

9752. 5; nervosa*.
Ci. nervosa. *Linn.* ii. 709.—*DeGeer.* iii. *pl.* 12. *f.* 1, 2.—*Turt.* ii. 596.—(*Sam. I.* 11.)

9753. 6; bipunctata* *mihi.*

9754. 7; punctifrons* *mihi.*

9755. 8; X-notata* *mihi.*

9756. 9; rufescens* *mihi.*

9757. 10; obscura* *mihi.*

9758. 11; varipes* *mihi.*

9759. 12; lineola* *mihi.*

---

[a] 9745 ‡ 1. Orni. *Linn.* 707.—*Stew.* 2 *edit.* ii. 120. (!)
Te. punctata. *Fabr. S.* 516.

[b] Genus 2: (1413). FULGORA, *Linn., Don., Stew.*
9747 † 1. Europæa. *Linn.* ii. 704.—*Panz. F.* xx. *f.* 16.—*Don.* vi. 65. *pl.* 203. (!)

Genus 4: (1415). ISSUS, *Fabr.*, *Leach*, *Samou.*

9760. 1; coleoptratus*. *Fabr. R.* 99.—(*Sam. I.* 23.)
Ce. coleoptrata. *Panz. F.* ii. *f.* 11.
Ci. dilatata. *Don.* 81. *pl.* 138. *f.* 5, 6.—*Stew.* ii. 99.

Genus 5: (1416). CIXIUS, *Latr.*, *Leach*, *Samou.*
FLATA *p*, *Fabr.*

9761. 1; Cynosbati*.
Fl. Cynosbati. *Fabr. R.* 54.
Ce. Dionysii. *Panz. F.* xxxiv. *f.* 24.

9762. 2; trifasciatus* *mihi.*

9763. 3; cunicularius*.
Ci. cunicularia. *Linn.* ii. 711.

9764. 4; pilosus*. *Latr. H.* xii. 311.

9765. 5; phæopterus* *mihi.*

9766. 6; fuliginosus* *mihi.*

9767. 7; dorsalis* *mihi.*
Fl. albicincta. *Germ. M.* (*Germar.*) iii. 199?

9768. 8; stigmaticus*.
Fl. stigmatica. *Germ. M.* (*Germar.*) iii. 199.

9769. 9; leporinus*.
Ci. leporina. *Linn.* ii. 711.—*Panz. F.* lxi. *f.* 19.

9770. 10; obliquus* *mihi.*

9771. 11; contaminatus*.
Fl. contaminata. *Germ. M.* (*Germar.*) iii. 196.

Genus 6: (1417). ASIRACA, *Latr.*, *Leach*, *Samou.*
DELPHAX *p*, *Fabr.*

9764*. 1; clavicornis*. *Latr. H.* xii. 316.—(*Sam. I.* 5.)
De. clavicornis. *Fabr. R.* 83.—*Coqueb.* i. *pl.* 8. *f.* 7.—(*Sam. I.* 14.)

9765*. 2; apicalis* *mihi.*

9766*. 3; dubia*.
Ci. dubia. *Panz. F.* xxxv. *f.* 20.
As. grisea. *Latr. H.* xii. 317.

Genus 7: (1418). DELPHAX, *Fabr.*—(*Kirby.*)

9767*. 1; striata*. *Fabr. R.* 84.

9768*. 2; marginata*. *Fabr. R.* 84.—*Coqueb.* iii. *pl.* 21. *f.* 4.

9769*. 3; flavescens*. *Fabr. R.* 84.

9770*. 4; pellucida*. *Fabr. R.* 84.

Genus 8: (1419). ——
DELPHAX *p*, *Germ.*

9771*. 1; hemiptera*.
De. hemiptera. *Germ. M.* (*Germar.*) iii. 217.

9772. 2; aterrima* *mihi.*

Familia III: (168). CERCOPIDÆ, *Leach.*

(MEMBRACIDES *et* CICADELLÆ, *Latreille.*—CICADA *p, Linné.*)

Genus 9: (1420). CENTROTUS, *Fabr.*

9773. 1; Genistæ*. *Fabr. R.* 21. (!)—*Panz. F.* l.*f.* 20.
Ci. Genistæ. *Stew.* ii. 96.—*Turt.* ii. 578.
Me. Genistæ. (*Sam. I.* 27.)

Genus 10: (1421). MEMBRACIS, *Fabr., Leach, Samou.*
CENTROTUS, (*Kirby.*)

9774. 1; cornuta*. *Fabr.* iv. 14.—(*Leach E. E.* ix. 125.)—(*Sam. I.* 231.)
Ci. cornuta. *Linn.* ii. 705.—*Berk. S.* i. 112.—*Don.* ii. 27. *pl.* 83.—*Stew.* ii. 97.—*Turt.* ii. 578.

Genus 11: (1422). LEDRA, *Fabr., Leach, Samou.,* (*Kirby.*)
MEMBRACIS *p, Oliv.*

9775. 1; aurita*. *Fabr. R.* 24.—(*Kirby & Sp. I. E. pl.* ii. *f.* 4.)—(*Sam. I.* 24.)
Le. aurata. (*Samou.* 231.)
Ci. aurita. *Linn.* ii. 706.

Genus 12: (1423). CERCOPIS, *Fabr., Leach, Samou.,* (*Kirby.*)

9776. 1; sanguinolenta*. *Fabr. R.* 92.—*Leach E. E.* ix. 125.—*Samou.* 231. *pl.* 5.*f.* 1.
Ci. sanguinolenta. *Linn.* ii. 708.—*Don.* ii. 45. *pl.* 54.*f.* 1.—*Stew.* ii. 97.—*Turt.* ii. 584.

Genus 13: (1424). TETTIGONIA, *Oliv., Leach, Samou.*

9777. 1; bifasciata*.
Ci. bifasciata. *Linn.* ii. 706.—*Panz. F.* vii.*f.* 20.
Cer. bifasciata. (*Kirby & Sp. I. E.* iv. 511.)

9778. 2; impressa* *mihi.*

9779. 3; spumaria*.
Ci. spumaria. *Linn.* ii. 708.—*Berk. S.* i. 112.—*Don.* ii. 45. *pl.* 54.*f.* 2.—*Stew.* ii. 97.—*Turt.* ii. 585.
Var. Ci. lateralis. *Linn.* ii. 709.—*Panz. F.* vi. *f.* 24.—*Berk. S.* i. 113.—*Stew.* ii. 98.—*Turt.* ii. 593.
Ci. vittata. *Fabr. R.* 96.
Ci. lineata. *Fabr.* iv. 52.—*Turt.* ii. 593.
Ci. leucocephalus. *Linn.* ii. 709.—*Berk. S.* i. 112.—*Stew.* ii. 98.—*Turt.* ii. 585.
Ci. capitata. *Fabr.* iv. 56.
Ci. flavicollis. *Mus. Marsham.*
Ci. albiceps. *Mus. Marsham.*
Ci. vertumnus. *Mus. Marsham.*
Ci. carnaria. *Mus. Marsham.*
Ci. rufescens. *Mus. Marsham.*

9780. [4; leucophthalma*.]
Ci. leucophthalma. *Linn.* ii. 709.—*Turt.* ii. 585.
9781. 5, pusilla* *mihi.*

Genus 14: (1425). ——
TETTIGONIA I. 2, *Latr.*

9782. 1, sanguinicollis.
Ce. sanguinicollis. *Fabr. R.* 94.
Ci. thoracica. *Panz. F.* lxi. *f.* 18.
† ♀? Ci. hæmorrhoa. *Panz. F.* lxi. *f.* 16.

Genus 15: (1426). IASSUS, *Fabr., Leach, Samou.*
TETTIGONIA *p, Latr.*—CERCOPIS *p, Fabr.*

9783. 1; lanio*. *Fabr. R.* 86.—(*Sam. I.* 23.)
Ci. lanio. *Linn.* ii. 710.—*Panz. F.* iv. *f.* 23; xxii. *f.* 10.
Ci. viridis. *Don.* ii. 48. *pl.* 54. *f.* 3.

9784. 2; incarnatus* *mihi.*

9785. 3, griseus*.
Ci. grisea. *Fabr. R.* 96?

9786. 4, os-album*.
Ci. os-album. *Mus. Marsham.*

9787. 5; Populi*.
Ci. Populi. *Mus. Marsham.*
Ia. brunneus. *Fabr. R.* 87?

9788. 6; binotatus* *mihi.*

9789. 7; geminatus* *mihi.*

9790. 8; ornatus*.
Ci. ornata. *Mus. Marsham.*

9791. 9, variegatus* *mihi.*

Genus 16: (1427). ——

9792. 1; humeralis* *mihi.*

9793. 2; flavovaria* *mihi.*

9794. 3; unicolor* *mihi.*

9795. 4, rustica.
Ci. rustica. *Fabr.* iv. 54.—*Turt.* ii. 587.

9798. 7, flava*.
Ci. flava. *Mus. Marsham.*

9802. 11; canaliculata.
Ci. canaliculata. *Mus. Marsham.*

9809. 18, glauca.
Ci. glauca. *Mus. Marsham.*

Genus 17: (1428). ——
IASSUS *p, Fabr.*

9810. 1; viridis*.

Ci. viridis. *Linn.* ii. 711.—*Berk. S.* i. 113.—*Stew.* ii. 98.—*Turt.* ii. 594.—*Wood.* i. 99. *pl.* 35.
Ia. viridis. (*Sam. I.* 23.)

9811. 2; interruptus*.
Ci. interrupta. *Linn.* ii. 710.—*Berk. S.* i. 113.—*Stew.* ii. 98.—*Turt.* ii. 593.
Ia. interruptus. (*Sam. I.* 23.)

9812. 3; acuminatus*.
Ci. acuminata. *Fabr. R.* 76.

9813. 4; nebulosus* *mihi.*

9814. 5, brachypteryx*. *Kirby? MSS.*

Genus 18: (1429). ——
IASSUS *p, et* CERCOPIS *p, Fabr.*

A. Maculati.

9815. 1; obscura* *mihi.*

9816. 2; excavata* *mihi.*

9817. 3; irrorata* *mihi.*

9818. 4; atra* *mihi.*

9821. 7, variegata.
Ci. variegata. *Fabr.* iv. 55.

9822. 8, cuspidata.
Ci. cuspidata. *Fabr. R.* 79. (!)—*Stew.* ii. 98.—*Turt.* ii. 598.

B. Lineati.

9824. 10; lineata.
Ia. lineatus. *Fabr. R.* 87.

9825. 11; concinna* *mihi.*

9826. 12; pulchella* *mihi.*

9827. 13; striata*.
Ci. striatus. *Fabr. R.* iv.—*Berk. S.* i. 113.—*Stew.* ii. 98?—*Turt.* ii. 586.

9828. 14; flavostrigata.
Ci. flavostrigata. *Don.* viii. 87. *pl.* 288. *f.* 2.

C. Fasciati.

9831. 17; trifasciata.
Ce. trifasciata. *Fabr. R.* 98.—*Coqueb. Ill. Ic.* 8. *f.* 10.

9832. 18; bifasciata.
Ci. bifasciata. *Don.* xi. 75. *pl.* 387.

9835. 21; Serratulæ.
Ci. serratulæ. *Fabr.* iv. 41. (!)—*Stew.* ii. 98.—*Turt.* ii. 596.

9836. 22; nitidula*.
Ci. nitidula. *Don.* viii. 87. *pl.* 288. *f.* 1.

Genus 19: (1430). ULOPA, *Fallen.*

9837. 1; Ericæ*.
Ce. Ericæ. *Ahr. F.* iii. *f.* 24.

9840. 4; obliqua* *mihi*.

9846. 10; reticulata*.
Ce. reticulata. *Fabr. R.* 98?

Genus 20: (1451). ——
IASSUS *p*, CERCOPIS *p*, *et* CICADA *p*, *Fabr.*

9847. 1, acephala.
Ci. acephala. *Mus. Marsham.*

9850. 4, virescens*.
Ci. virescens. *Fabr. R.* 79.

9851. 5, Rosæ*?
Ci. Rosæ. *Linn.* ii. 712.—*Berk. S.* i. 113.—*Stew.* ii. 98.—*Turt.* ii. 598.

9852. 6; suturalis.
Ci. suturalis. *Mus. Marsham.*

9853. 7, sanguinea* *mihi*.

9854. 8; Urticæ.
Ci. Urticæ. *Fabr. R.* 77.

9855. 9; centromacula* *mihi*.
Ci. vitrea. *Fabr. R.* 79?

9856. 10; flavescens*.
Ci. flavescens. *Fabr. R.* 79.

9857. 11; Ulmi*.
Ci. Ulmi. *Linn.* ii. 711.—*Berk. S.* i. 113.—*Stew.* ii. 98.—*Turt.* ii. 598.

9858. 12; aurata*.
Ci. aurata. *Linn.* ii. 711.—*Turt.* ii. 598.

9859. 13; splendidula*.
Ci. splendidula. *Fabr. R.* 79.
Ci. nitidula. *Turt.* ii. 598.

9860. 14; quadrinotata.
Ci. quadrinotata. *Fabr. R.* 78.

9862. 16; apicalis* *mihi*.

9863. 17; curvatula* *mihi*.

9864. 18; picta*.
Ci. picta. *Fabr. R.* 77.

9865. 19; varia* *mihi*.

9866. 20; punctata.
Ci. punctata. *Fabr. R.* 78.—*Turt.* ii. 597.

9869. 23; fulgida.
Ia. fulgidus. *Fabr. R.* 87.—*Stew.* ii. 98.—*Turt.* ii. 597.
Ci. vittata. *Mus. Marsham.*

9870. 24; Quercus.
Ci. Quercus. *Fabr. R.* 79.

9872. 26; flammigera.
Ci. flammigera. *Mus. Marsham.*

9873. 27; hæmatica* *mihi.*

9874. 28; pulchella* *mihi.*

9875. 29; obscura* *mihi.*

9876. 30; pallida* *mihi.*

Familia IV: (169). PSYLLIDÆ[a], *Latreille.*

Genus 21: (1432). PSYLLA, *Geoff., Leach, Samou.*
CHERMES, *Linné,* (*Kirby.*)

9877. 1; Alni*. *Latr.* G. iii. 169.—*Leach E. E.* ix. 125.—*Samou.* 231.
Ch. Betulæ Alni. (*Linn.* ii. 738.)—*DeGeer.* iii. 148. *pl.* 10. *f.* 8.—*Berk. S.* i. 121.—(*Stew.* ii. 111.)—*Turt.* ii. 711.—*Shaw G. Z.* vi. 186.

9878. 2. Graminis.
Ch. Graminis. (*Linn.* ii. 737.)—*Berk. S.* . 121.—(*Stew.* ii. 111.)—(*Turt.* ii. 710.)

9879. 3, Ulmi.
Ch. Ulmi. (*Linn.* ii. 737.)—(*Stew.* ii. 111.)—(*Turt.* ii. 710.)

9880. 4. Cerastii.
Ch. Cerastii. (*Linn.* ii. 737.)—(*Stew.* ii. 111.)—*Turt.* ii. 710.

9881. 5, Pyri.
Ch. Pyri. (*Linn.* ii. 737.)—*DeGeer.* iii. 141. *pl.* 9. *f.* 2.—*Berk. S.* i. 121.—(*Stew.* ii. 111.)—*Turt.* ii. 710.—*Shaw G. Z.* vi. 187. *pl.* 59. *fig. sup.*

9882. 6; Sorbi.
Ch. Sorbi. *Linn.* ii. 738.—*Berk. S.* i. 121.—(*Stew.* ii. 111.)—*Turt.* ii. 710.

9883. 7. Calthæ.
Ch. Calthæ. (*Linn.* ii. 738.)—(*Stew.* ii. 111.)—*Turt.* ii. 710.

9884. 8; Buxi.
Ch. Buxi. (*Linn.* ii. 738.)—(*Stew.* ii. 112.)—*Shaw G. Z.* vi. 187. *pl.* 59. *fig. inf.*—*Wood.* i. 114. *pl.* 40.
Reaum. I. iii. 29. *pl.* 1–14.

9885. 9, Urticæ.

---

[a] In this and the three remaining Families I have done little more than bring the various references of English writers together; the vastness of the science of Entomology, as may be readily conceived by the foregoing pages, and a constant residence in the metropolis, having prevented me from devoting that attention to these sadly neglected, though highly interesting, groups, which is requisite in order to obtain a correct knowledge either of the species or their affinities.

Ch. Urticæ. (*Linn.* ii. 738.)—*DeGeer.* iii. 134. *pl.* 9. *f.* 17–19. —*Berk. S.* i. 121.—(*Stew.* ii. 112.)—*Turt.* ii. 711.

9886. 10; Betulæ.
Ch. Betulæ. (*Linn.* ii. 738.)—(*Stew.* ii. 112.)—*Turt.* ii. 711. —(*Sam. I.* 10.)

9887. 11; Quercus.
Ch. Quercus. *Linn.* ii. 738.—*Berk. S.* i. 121.—(*Stew.* ii. 112.)

9888. 12; *Fagi.*
Ch. *Fagi.* (*Linn.* ii. 738.)—(*Stew.* ii. 112.)—*Turt.* ii. 710. *Reaum. I.* iii. *pl.* 26. *f.* 1–6.

9889. 13; Abietis.
Ch. Abietis. (*Linn.* ii. 738.)—*DeGeer.* iii. 99. *pl.* 8. *f.* 1–3.—*Berk. S.* i. 121.—(*Stew.* ii. 112.)—(*Turt.* ii. 711.)

9890. 14, Salicis.
Ch. Salicis. (*Linn.* ii. 739.)—(*Stew.* ii. 112.)—*Turt.* ii. 711.

9891. 15; Fraxini.
Ch. Fraxini. (*Linn.* ii. 739.)—*Berk. S.* i. 122.—(*Stew.* ii. 112.) —(*Turt.* ii. 711.)—(*Kirby & Sp. I. E. pl.* xxviii. *f.* 18.)

9892. 16; Aceris.
Ch. Aceris. (*Linn.* ii. 739.)—(*Stew.* ii. 112.)—*Turt.* ii. 711.

9893. 17; Pini.
Ch. Pini. G*melin*?—(*Stew.* ii. 112.)

9894. 18, Pruni.
Ch. Pruni. G*melin*?—(*Stew.* ii. 112.)

9895. 19, Cratægi.
Ch. Cratægi. G*melin*?—(*Stew.* ii. 112.)

9896. 20. Euonymi.
Ch. Euonymi. G*melin*?—(*Stew.* ii. 112.)

9897. 21. Senecionis.
Ch. Senecionis. G*melin*?—(*Stew.* ii. 112.)

9898. 22. Lichenis.
Ch. Lichenis. G*melin*?—*Stew.* ii. 112.

9899. 23; castanea.
Ch. castanea. G*melin*?—*Stew.* ii. 112.

9900. 24, rubra.
Ch. rubra. G*melin*?—*Stew.* ii. 112.

9901 † 25. Amygdali Persicæ [a].

9902 † 26. *Ficus* [b].

---

[a] 9901 † 25. Amygdali Persicæ. (*Fabr. R.* 304.)
Ch. Amygdali Persicæ. *Stew.* ii. 111. (!)
Reaum. I. iv. *pl.* 1. *f.* 1, 2.

[b] 9902 † 26. *Ficus.*
Ch. Ficus. *Linn.* ii. 739.—*Stew.* ii. 112. (!)
Reaum. I. iii. *pl.* 29. *f.* 17–24.

Genus 22: (1433). LIVIA, *Latr.*, *Leach*, *Samou.*, (*Kirby.*)
DIRAPHIA, *Illiger.*

9903. 1; Juncorum*. (*Leach E. E.* ix. 125.)—(*Samou.* 232. *pl.* 5. *f.* 11.)
Ps. Juncorum. *Latr. F.* 322. *pl.* xii. *f.* 3.—*Ent. Trans.* (*Haworth.*) i. 252.

Familia V: (170). THRIPIDÆ.
(PHYSAPI, *Latreille.*)

Genus 23: (1434). THRIPS *Auctorum.*

9904. 1; aculeata*. *Fabr. R.* 312.

9905. 2; Ulmi*. *Fabr. R.* 313.—*Turt.* ii. 716.
Th. corticis. *DeGeer.* iii. 11. *pl.* 1. *f.* 8–13.

9906. 3; Urticæ*. *Fabr. R.* 313.—*Schrank Beyt. pl.* 1. *f.* 25, 26. —*Turt.* ii. 716.

9907. 4; Physapus*. *Linn.* ii. 743.—*DeGeer.* iii. 6. *pl.* 1. *f.* 1. —*Berk. S.* i. 122.—*Stew.* ii. 114.—*Turt.* ii. 716.—*Shaw G. Z.* vi. 199. *pl.* 63.—*Leach E. E.* ix. 126.—*Samou.* 232. *pl.* 5. *f.* 12.—(*Kirby & Sp. I. E.* i. 128: ii. 331.)—*Wood.* i. 118. *pl.* 42.

9908. 5; Juniperina*. *Linn.* ii. 743.—*DeGeer.* iii. 10. *pl.* 1. *f.* 5. —*Berk. S.* i. 122.—*Stew.* ii. 114.—*Turt.* ii. 716.—(*Sam. I.* 41.)

9909. 6; fasciata*. *Linn.* ii. 743.—*Sulzer. pl.* 7. *f.* 48.—*Berk. S.* i. 133.—*Stew.* ii. 114.—*Turt.* ii. 717.—(*Sam. I.* 41.)

9910. 7; variegata. *Gmel.*—*Turt.* ii. 717.

9911. 8; obscura*. *Müll. Z. D. Pr. No.* 1084.

9912. 9; minutissima*. *Linn.* ii. 743.—*Stew.* ii. 114.—*Turt.* ii. 717.—(*Sam. I.* 41.)

Familia VI: (171). APHIDÆ, *Leach.*
(APHIDII, *Latr.*—APHIS, *Linné.*) *Plant-lice.*

Genus 24: (1435). APHIS, *Linné*, *Leach*, *Samou.*, (*Kirby.*)

A. "Abdomen bicorniculatum. (Antennæ setaceæ, elongatæ)." *Latr.*

9913. 1; Pruni. (*Fabr. R.* 296.)—*DeGeer.* iii. 49. *pl.* 2. *f.* 1–8. —(*Stew.* ii. 110.)—*Turt.* ii. 704.—(*Sam. I.* 4.)

9914. 2, Pomi. (*Latr. G.* iii. 173.)
Ap. Mali. (*Fabr. R.* 298.)—*DeGeer.* iii. 53. *pl.* 3. *f.* 29.—(*Stew.* ii. 111.)—*Turt.* ii. 706.—*Shaw G. Z.* vi. *pl.* 58. *fig. sup. sinist.*

9915. 3, Juniperi*. (*Fabr. R.* 300.)—*DeGeer.* iii. 56. *pl.* 4. *f.* 7, 8. —(*Stew.* ii. 110.)—*Turt.* ii. 707.—(*Sam. I.* 4.)

9916. 4, Viciæ. (*Linn.* ii. 735.)—*DeGeer.* iii. 58. *pl.* 2. *f.* 14, 15. —*Turt.* ii. 708.

9917. 5; Millefolii*. *Fabr. R.* 296.—*DeGeer.* iii. 60. *pl.* 4. *f.* 1–5.

—(*Stew.* ii. 110.)—*Shaw* G. *Z.* vi. 170. *pl.* 58. *fig. sup. dext.*—(*Sam. I.* 4.)

9918. 6; Rosæ*. (*Linn.* ii. 734.)—*Harr. Ex.* 66. *pl.* xviii. *f.* 1–3. —*Berk. S.* i. 119.—(*Stew.* ii. 110.)—*Turt.* ii. 706?—*Shaw G. Z.* vi. 171. *pl.* 58. *fig. med.*—(*Sam. I.* 4.)—*Bingley.* iii. 186.—*Wood.* i. 112. *pl.* 39.

9919. 7; Salicis*. (*Linn.* ii. 736.)—*Berk. S.* i. 120.—(*Stew.* ii. 111.)—*Linn. Trans.* (*Curtis.*) vi.—*Turt.* ii. 707.—*Shaw* G. *Z.* vi. 170.—(*Sam. I.* 4.)
Reaum. I. iii. *pl.* 22. *f.* 2.

9920. 8; Ribis. (*Linn.* ii. 733.)—*Berk. S.* i. 119.—(*Stew.* ii. 110.) —*Turt.* ii. 703.—(*Sam. I.* 4.)
Reaum. I. iii. *pl.* 22. *f.* 7–10.

9921. 9, Cardui. (*Linn.* ii. 735.)—*Berk. S.* i. 120.—(*Stew.* ii. 110.) —*Turt.* ii. 703. (704.)

9922. 10, Absinthii. (*Linn.* ii. 735.)—*Berk. S.* i. 120.—(*Stew.* ii. 110.)—*Turt.* ii. 705.—(*Sam. I.* 4.)

9923. 11, Euonymi. (*Fabr. R.* 297.)—*Turt.* ii. 705.

9924. 12, Avenæ. (*Fabr. R.* 297.)—*Turt.* ii. 705.
Ap. avenæ sativæ. (*Stew.* ii. 110.)—(*Sam. I.* 4.)

9925. 13, Viburni. *Scop. C. No.* 396.—*Stew.* ii. 111.—(*Sam. I.* 4.)

9926. 14, Tanaceti. (*Linn.* ii. 733.)—*Berk. S.* i. 120.—(*Stew.* ii. 110.)—*Turt.* ii. 706.—(*Sam. I.* 4.)

9927. 15, Papaveris. *Fabr. R.* 299.—*Turt.* ii. 707.

9928. 16, Sonchi. (*Linn.* ii. 735.)—*Berk. S.* i. 120.—(*Stew.* ii. 110.)—(*Turt.* ii. 708.)—(*Sam. I.* 4.)

9929. 17, Brassicæ. (*Linn.* ii. 734.)—*Harr. Ex.* 66. *pl.* xviii. *f.* 3–6. —*Berk. S.* i. 120.—(*Stew.* ii. 110.)—*Turt.* ii. 707.—(*Sam. I.* 4.)

9930. 18. Althæa. *Harr. Ex.* 66. *pl.* xviii. *f.* 7–9.

9931. 19, Quercus.
Ap. Quercus. *Linn.* ii. 735.—*Berk. S.* i. 120.—*Stew.* ii. 111. —(*Sam. I.* 4.)

B. "Abdomen bituberculatum." (Antennæ sæpe filiformes.) *Latr.*

9932. 20; Pini*. (*Linn.* ii. 736.)—*DeGeer.* iii. 27. *pl.* 6. *f.* 9–16. —(*Stew.* ii. 110.)—*Turt.* ii. 707.—(*Sam. I.* 4.)

9933. 21; Pineti. (*Fabr. R.* 300.)—*DeGeer.* iii. 39. *pl.* 6. *f.* 27–33.—*Turt.* ii. 707.

9934. 22; Alni. (*Fabr. R.* 298.)—*DeGeer.* iii. 47. *pl.* 3. *f.* 15–17. —*Berk. S.* i. 121.—(*Stew.* ii. 110.)—(*Sam. I.* 4.)

9935. 23, Aceris. (*Linn.* ii. 736.)—*Berk. S.* i. 121.—*Turt.* ii. 703. (704.)
Ap. Aceris platanoides. (*Stew.* ii. 111?)—(*Sam. I.* 4?)
Reaum. I. iii. *pl.* 22. *f.* 6–10.

9936. 24, Rumicis. (*Linn.* ii. 734.)—*Berk. S.* i. 119.—*Turt.* ii. 703. (704.)
Ap. Rumicis Lapathi. (*Stew.* ii. 110.)—(*Sam. I.* 4.)

9937. 25, Betulæ. (*Linn.* ii. 735.)—*Berk. S.* i. 120.—*Stew.* ii. 110. —*Turt.* ii. 705.—(*Sam. I.* 4.)
Reaum. I. iii. *pl.* 22. *f.* 2.

9938. 26, Atriplicis. (*Fabr. R.* 298.)—*Berk. S.* i. 121?—(*Stew.* ii. 111.)—*Turt.* ii. 706.—(*Sam. I.* 4.)

9939. 27, Tiliæ. (*Linn.* ii. 734.)—*DeGeer.* iii. *pl.* 5. *f.* 1–5.—*Berk. S.* i. 119.—(*Stew.* ii. 110.) (tibiæ.)—*Turt.* ii. 707.—*Shaw G. Z.* vi. 171.—(*Sam. I.* 4.)

9940. 28, Roboris. *Linn.* ii. 735.—*Turt.* ii. 707.

9941. 29, Ulmi.
Ap. Ulmi. *Linn.* ii. 733.—*DeGeer.* iii. 81. *pl.* 5. *f.* 7–18.—*Berk. S.* i. 119?—(*Stew.* ii. 110.)—*Turt.* ii. 706.—(*Shaw G. Z.* vi. *pl.* 58. *fig. inf.*)—(*Sam. I.* 4.)

C. Incertæ sedes.

9942. 30, Lychnidis. (*Linn.* ii. 294.)—*Berk. S.* i. 119.—(*Stew.* ii. 110.)—*Turt.* ii. 703.—(*Sam. I.* 4.)

9943. 31, Capreæ. (*Fabr. R.* 294.)—(*Stew.* ii. 110.)—*Turt.* ii. 703.—(*Sam. I.* 4.)

9944. 32, Sambuci. (*Linn.* ii. 734.)—*Berk. S.* i. 119.—(*Stew.* ii. 110.)—*Turt.* ii. 703.—(*Sam. I.* 4.)

9945. 33, Cerasi. (*Fabr. R.* 295.)—*Turt.* ii. 703.
Ap. Pruni Cerasi. (*Stew.* ii. 110.)—(*Sam. I.* 4.)

9946. 34, Pistinacæ. (*Linn.* ii. 734.)—(*Turt.* ii. 703. (704.)

9947. 35, Nymphææ. (*Linn.* ii. 734.)—(*Turt.* ii. 703. (704.)

9948. 36. Ægopodii. *Scop. C. No.* 399.
Ap. Ægopodii podagrariæ. (*Stew.* ii. 110.)

9949. 37. Dauci. (*Fabr. R.* 299.)—*Stew.* ii. 110.—*Turt.* ii. 706. —(*Sam. I.* 4.)

9950. 38, Ligustici. (*Fabr. R.* 301.)
Ap. Ligustici Scotici. (*Stew.* ii. 110.)—(*Sam. I.* 4.)

9951. 39. Acetosæ. (*Linn.* ii. 734.)—*Turt.* ii. 708.—(*Sam. I.* 4.)

9952. 40. Padi. (*Linn.* ii. 734.)—(*Stew.* ii. 110.)—(*Turt.* ii. 708.) —(*Sam. I.* 4.)
Reaum. I. iii. *pl.* 23. *f.* 9, 10.

9953. 41. Lactucæ. *Linn.* ii. 735.—(*Stew.* ii. 110.)—(*Turt.* ii. 708.)—(*Sam. I.* 4.)
Reaum. I. iii. *pl.* 22. *f.* 3–5.

9954. 42. Cirsii. *Linn. F. No.* 987.—(*Stew.* ii. 110.)

9955. 43. Jaceæ. *Linn. F. No.* 991.—*Berk. S.* i. 120.—(*Stew.* ii. 110.)—*Turt.* ii. 708.—(*Sam. I.* 4.)

9956. 44. Plantaginis. *Bonnet.*—(*Stew.* ii. 111.)—(*Sam. I.* 4.)

9957. 45. Leucanthemi. *Scop. C. No.* 404.—(*Stew.* ii. 111.)—(*Sam. I.* 4.)

9958. 46. Scabiosæ. *Scop. C. No.* 405.—(*Stew.* ii. 111.)—(*Sam. I.* 4.)

9959. 47. Fabæ. *Scop. C. No.* 408.—(*Stew.* ii. 111.)—(*Turt.* ii. 710.)—(*Sam. I.* 4.)—*Bingley.* iii. 189.

9960. 48. Craccæ. *Scop. C. No.* 407.—*Berk. S.* i. 120.—(*Stew.* ii. 110.)—(*Sam. I.* 4.)

9961. 49. Vitis[a].

Genus 25: (1436). ERIOSOMA, *Leach MSS.*, *Samou.*
MYZOXYLA, *Latr.*, (*Kirby.*)—MYZOXYLUS, *Stark.*

A.

9962. 1; Fraxini*.
Ap. Fraxini. (*Fabr. R.* 297?)—(*Stew.* ii. 110?)—*Turt.* ii. 705? —(*Sam. I.* 4.)

9963. 2; Abietis*.
Ap. Pineti. *DeGeer.* iii. *pl.* 6. *f.* 19–26?

9964. 3; Cratægi* *mihi.*

9965. 4, Urticæ.
Ap. Urticæ. *Linn.* ii. 736?—*Frisch. I.* 8. 34. *pl.* 17.—*Turt.* ii. 706.
Ap. Urticata. (*Stew.* ii. 110.)—(*Sam. I.* 4.)

9966. 5; Mali*. *Leach MSS.*—(*Sam. I.* 16.)
Ap. lanigera. *Illiger.*—*Hort. Trans.* (*Sir J. Banks.*) ii. 162. *pl.* 11.
Co. Mali. *Bingley.* iii. 200.

B.

9967. 6; Quercus* *mihi.*

9968. 7; Piceæ*.
Ap. Piciæ. *Panz. F.* lxxviii. *f.* 22.

9969. 8. Tremulæ.
Ap. Tremulæ. *Linn.* ii. 736.—(*Stew.* ii. 111.)—(*Sam. I.* 4.)

9970. 9; Xylostei.
Ap. Xylostei. *DeGeer.* iii. 96. *pl.* 7. *f.* 8.—*Turt.* ii. 709.

9971. 10, Bursaria.
Ap. Bursaria. *Linn.* ii. 756.—(*Stew.* ii. 111.)—(*Sam. I.* 4.)
Ap. Bursaria. *Turt.* ii. 703. (704.)
Swam. B. N. pl. 45. *f.* xxii–xxv.

9972. 11, Fagi.
Ap. Fagi. *Linn.* ii. 735.—*Berk. S.* i. 120.—(*Stew.* ii. 110.)—*Turt.* ii. 705.—(*Sam. I.* 4.)
Reaum. I. iii. *pl.* 22. *f.* 1.

---

[a] 9961. 49. Vitis. *Scop. C. No.* 398.—*Turt.* ii. 708. (!)

9973. 12. Populi. (*Linn.* ii. 736.)—(*Stew.* ii. 111.)—*Turt.* ii. 705. —(*Sam. I.* 4.)

Genus 26: (1437). ——

9974. 1; Pinicola* *mihi.*

Genus 27: (1438). ALEYRODES, *Latr., Leach, Samou.,* (*Kirby.*)
PHALÆNA-TINEA *p, Linné.*—PHALÆNA *p, Geoff.*

9975. 1; Chelidonii*. *Latr. G.* iii. 174.—*Leach E. E.* ix. 126. —*Samou.* 233.—(*Kirby & Sp. I. E.* iii. 89. 261.)
Ph. Ti. proletella. *Linn.* ii. 889.—*Roem. G. I. pl.* 23. *f.* 18.—*Stew.* ii. 199.

9976. 2; immaculata* *mihi.*

9977. 3; bifasciata* *mihi.*

9978. 4; gigantea* *mihi.*

9979. 5; dubia* *mihi.*

Familia VII: (172). COCCIDÆ, *Leach.*

(GALLINSECTA, *Latreille.*)

Genus 28: (1439). DORTHESIA, *Bosc, Leach,* (*Kirby.*)
CIONOPS, *Leach.*

9980. 1, Characias. *Bosc.*—(*Latr.* G. iii. 175.)—(*Leach E. E.* ix. 126.)—(*Kirby & Sp. I. E.* iii. 183.)
Co. Characias. *Dorthes.*—*Fabr. R.* 311.

9981. [2, dubia.] *Kirby & Sp. I. E.* iii. 183.
Co. dubia. *Panz. F.* xxxv. *f.* 21.

9982. 3, floccosa. (*Kirby & Sp. I. E.* iii. 183.)
Co. floccosa. *DeGeer.* vii. 604. *pl.* xliv. *f.* 26.

9983. 4, Melampyri.
(Vide *Kirby & Sp. I. E.* iii. 184.)

9984. 5, cimiciformis.
Ci. cimiciformis. *Leach MSS.*

9985. 6, cataphractus.
Co. cataphractus. *Shaw N. M.* v. *pl.* 182.—*Shaw G. Z.* vi. 194. *pl.* 62.—*Stew.* ii. 114.—*Turt.* ii. 714.—(*Sam. I.* 12.)

Genus 29: (1440). COCCUS, *Linn., Leach, Haw., Samou.,* (*Kirby.*)
Scale-insect.

9986. 1, Phalaridis. *Linn.* ii. 742.—*Berk. S.* i. 122.—(*Stew.* ii. 113.)—*Turt.* iii. 714.—(*Sam. I.* 12.)
Geoff. H. i. 512. *pl.* 10. *f.* 5.

9987 † 2. Cacti[a].

---

[a] 9987 † 2. Cacti: Cochineal Insect. *Linn.* ii. 742.—*DeGeer.* vi. *pl.* 80. *f.* 12–14.—*Shaw G. Z.* vi. 191. *pl.* 61.—(*Samou.* 233.) (!)

9988. 3. Vitis[a].

9989 † 4. Hesperidum[b].

9990. 5, Quercus. (*Linn.* ii. 740.)—(*Stew.* ii. 113.)—(*Turt.* ii. 712.)—(*Sam. I.* 12.)
Reaum. I. iv. *pl.* 6. *f.* 1–4. 8–10.

9991. 6, Betulæ. (*Linn.* ii. 740.)—*Berk. S.* i. 122.—(*Stew.* ii. 113.)—*Turt.* ii. 713.

9992. 7, Carpini. (*Linn.* ii. 740.)—(*Stew.* ii. 113.)—(*Sam. I.* 12.)

9993. 8. Ulmi. (*Linn.* ii. 740.)—*DeGeer.* v. 28. *f.* 7.—(*Stew.* ii. 113.)—*Turt.* ii. 713.—(*Sam. I.* 12.)

9994. 9. Coryli. (*Linn.* ii. 741.)—(*Stew.* ii. 113.)—*Turt.* ii. 713. —(*Sam. I.* 12.)
Reaum. I. v. *pl.* 3. *f.* 4–10.

9995. 10. Tiliæ. (*Linn.* ii. 741.)—(*Stew.* ii. 113.)—*Turt.* ii. 713. —(*Sam. I.* 12.)
Reaum. I. v. *pl.* 3. *f.* 1–3.

9996. 11. Capræææ. *Linn.* ii. 741.—*DeGeer.* vi. *pl.* 28. *f.* 13.—(*Stew.* ii. 113.)—*Turt.* ii. 713.—(*Sam. I.* 12.)

9997. 12. Salicis. *Linn.* ii. 741.—(*Stew.* ii. 113.)—(*Sam. I.* 12.)

9998. 13. polonicus. (*Linn.* ii. 741.)—(*Stew.* ii. 113.)—*Shaw* G. Z. vi. 194.—(*Sam. I.* 12.)

9999. 14. Fragariæ. G*mel.*?—(*Stew.* ii. 113.)—*Turt.* ii. 715.—(*Sam. I.* 12.)
Phil. Trans. 1765. 91. *pl.* 10.

10000. 15. Pilosellæ. (*Linn.* ii. 742.)—*Act. H.* 1742. *pl.* 2.—(*Stew.* ii. 113.)—(*Sam. I.* 12.)

10001. 16. Uva ursi. (*Linn.* ii. 742.)—(*Stew.* ii. 113.)—(*Sam. I.* 12.)
Co. Arbuti. (*Fabr. R.* 310.)

10002 † 17. Adonidum[c].

10003. 18. Oxyacanthæ. (*Linn.* ii. 742.)—(*Stew.* ii. 113.)—(*Sam. I.* 12.)
Reaum. I. iv. *pl.* 6. *f.* 11, 12.
Co. Cratægi. (*Fabr. R.* 310.)

10004. 19. Serratulæ. (*Fabr. R.* 310.) (!)—(*Stew.* ii. 113.)—*Turt.* ii. 714.—(*Sam. I.* 12.)

---

[a] 9988. 3. Vitis. *Linn.* ii. 741.—*Ent. Trans.* (*Haworth.*) i. 307. (!)
Reaum. I. iv. *pl.* 6. *f.* 5–7.

[b] 9989 † 4. Hesperidum. *Linn.* ii. 739.—*Berk. S.* i. 122. (!)—(*Stew.* ii. 113!)—(*Turt.* ii. 712!)—*Shaw G. Z.* vi. 190. *pl.* 60. *fig. sup.*—*Ent. Trans.* (*Haworth.*) i. 307. (!)

[c] 10002 † 17. Adonidum. *Linn.* ii. 740.—*Shaw G. Z.* vi. 190.—*Ent. Trans.* (*Haworth.*) i. 308. (!)

10005. 20. Persicæ[a].

10006. 21. Abietis. *Latr. H.* xii. 389.—(*Stew.* ii. 113.)—(*Sam. I.* 12.)

10007. 22. Mespili. *Gmelin?*—(*Stew.* ii. 113.)—(*Sam. I.* 12.)

10008. 23; Alni. *Gmelin?*—(*Stew.* ii. 113.)—*Shaw.* vi. *pl.* 60. *fig. med.*—(*Sam. I.* 12.)

10009. 24. fuscus. *Gmelin?*—*Stew.* ii. 113.—(*Sam. I.* 12.)

10010. 25. variegatus. *Gmelin?*—*Stew.* ii. 113.—(*Sam. I.* 12.) *Reaum. I.* iv. *pl.* 5. *f.* 3. *a.*

10011. 26, conchiformis. *Gmelin?*—*Stew.* ii. 113.—(*Sam. I.* 12.) *Reaum. I.* iv. 5. *f.* 7.

10012. 27. Aceris. (*Fabr. R.* 308.)—(*Sam. I.* 12.)

---

[a] 10005. 20. Persicæ. (*Fabr. R.* 307.)—(*Stew.* ii. 113.)—(*Sam. I.* 12.) —*Bingley.* iii. 197.

Co. Persicorum. *Roem. G. pl.* 11. *f.* 9.—*Shaw G. Z.* vi. *pl.* 60. *fig. inf.*

APPENDIX

AD PARTEM SECUNDAM.

Page. No.

1\. 5794: Papilio Machaon *: Swallow-tail B. (*Wilkes D. pl.* 2. *f.* 1.)

3\. 5800; Colias Edusa*. (*Curtis. folio* 242.)—Clouded Yellow B. (*Wilkes* D. *pl.* 9. *f.* 8.)

4\. 5800, Co. —— Hyale. *Curtis.* vi. *pl.* 242.

5\. 5810; Mancipium Cardamines*. (*Steph. Nom.* 38.)—Orange-tip B. (*Wilkes* D. *pl.* 12. *f.* 4.)

9\. 5823; Melitæa Selene*: The Pearl-border *F.* (*Wilkes* D. *pl.* 7. *f.* 4.)

10\. 5827; Argynnis Adyppe*: The High-brown *F.* (*Wilkes* D. *pl.* 2. *f.* 4.)

— 5828; Ar. —— Aglaia*: The Darkened Green *F.* (*Wilkes* D. *pl.* 3. *f.* 6.)

11\. 5829; Ar. —— Paphia*: The Great *F.* (*Wilkes* D. *pl.* 7. *f.* 6.)

— 5830; Vanessa C. album*: The Comma B. (*Wilkes* D. *pl.* 3. *f.* 7.)

12\. 5831; Va. —— Polychloros*: The Great Tortoise-shell B. (*Wilkes* D. *pl.* 2. *f.* 6.)

— 5832; Va. —— Urticæ*: The Small Tortoise-shell B. (*Wilkes* D. *pl.* 9. *f.* 9.)

— 5833; Va. —— Io*: The Peacock B. (*Wilkes* D. *pl.* 1. *f.* 4.)

13\. 5834; Va. —— Atalanta*: The Admirable B. (*Wilkes* D. *pl.* 2. *f.* 3: *pl.* 7. *f.* 5.)

14\. 5836; Cynthia Cardui*: The Painted Lady B. (*Wilkes* D. *pl.* 4. *f.* 2: *pl.* 12. *f.* 6.)

— 5839; Apatura Iris*: The Purple Emperor B. (*Wilkes* D. *pl.* 6. *f.* 6. ♂: *pl.* 10. *f.* 1. ♀.)

— 5841; Limenitis Camilla*: The White Admirable. (*Wilkes* D. *pl.* 1. *f.* 7: *pl.* 8. *f.* 1.)

16\. 5848; Hipparchia Semele*: The Rock Underwing B. (*Wilkes* D. *pl.* 5. *f.* 4.)

— 5849; Hi. —— Galathea*: The Marbled B. (*Wilkes* D. *pl.* 9. *f.* 5.)

17\. 5850; Hi. —— Tithonus*: The Orange Field B. (*Wilkes* D. *pl.* 12. *f.* 5.)

Page. No.
20. 5860, Thecla Pruni. *Curtis.* vi. *pl.* 264.
— 5865; Th. —— Quercus*: The Purple Hair Streak B. (*Wilkes* D. *pl.* 3. *f.* 3.)
22. 5871, Lycæna Hippothoë. (*Loudon M.* i. 55. *f.* 24.)
24. 5880, Polyommatus Icarius?—Ly. Agathon. G*odart*?
— 5881; Po. —— Alexis*: The Ultra-marine B. (*Wilkes* D. *pl.* 3. *f.* 2.)
28. linea 1. *Pro* B. *lege* b.
29.†5901. Anthrocera Scabiosæ.—Zy. Minos. *Schrank.*
30. 5905; An. —— Filipendulæ*: The Burnet M. (*Wilkes* D. *pl.* 7. *f.* 2.)
— 5907; Smerinthus ocellatus*: The Eyed Hawk M. (*Wilkes* D. *pl.* 8. *f.* 4.)
— 5908; Sm. —— Populi*: The Poplar Hawk M. (*Wilkes* *pl.* 4. *f.* 4.)
— 5909; Sm. —— Tiliæ*: The Lime Hawk. (*Wilkes* *pl.* 8. *f.* 3.)
31. —— ACHERONTIA.—BRACHYGLOSSA, *Boisduval.*
— 5910; Ac. —— Atropos*: The Bee Tyger M. (*Wilkes* D. *pl.* 12. *f.* 1.)
32. 5915; Sphinx Ligustri*: The Privet Hawk M. (*Wilkes* D. *pl.* 5. *f.* 3.)
33. 5925; Deilephila Elpenor*: The Elephant M. (*Wilkes* D. *pl.* 9. *f.* 4.)
— 5926; De. —— Porcellus*: The Small Elephant M. (*Wilkes* D. *pl.* 10. *f.* 5.)
39. 5950; Zeuzera Æsculi*: The Wood Leopard M. (*Wilkes* D. *pl.* 5. *f.* 7.)—(*Loudon M.* ii. 67. fig. 16.)
— 5951; Cossus ligniperda*: The Goat M. (*Wilkes* D. *pl.* 4. *f.* 3.)
— —— PYGÆRA:—SERICARIA *p, Latreille.*
— 5952; Py. —— bucephala*: The Buff-tip M. (*Wilkes* D. *pl.* 6. *f.* 7.
40. —— CLOSTERA.—SERICARIA *p, Latreille.*
41. —— CERURA.—FURCULA, *Lamarck.*—DICRANURA, *Latr.*
— 5960; Ce. —— furcula*: Furcula Salicis. *Lamarck.* iii. 562.
— 5965; Ce. —— Vinula*: The Puss M. (*Wilkes* D. *pl.* 12. *f.* 2.)
42. —— STAUROPUS.—FURCULA *p, Lamarck.*
— 5970; Notodonta ziczac*: The Pebble M. (*Wilkes* D. *pl.* 6. *f.* 5.)
— 5971; Leiocampa dictæa*: The Swallow M. (*Wilkes* D. *pl.* 9. *f.* 6.)
43. 5975. Dele †.
— —— PTILODONTIS.—ORTHORINIA, *Boisduval.*
44. —— PETASIA.—ASTEROSCOPUS, *Boisduval.*

Page. No.

46. 5985; Saturnia Pavonia minor*: The Emperor M. (*Wilkes D. pl.* 4. *f.* 1. ♀: *pl.* 7. *f.* 1. ♂.)—(*Loudon M.* ii. 68. *fig.* 17.)

— 5990; Lasiocampa Roboris*: The Great Eggar. (*Wilkes D. pl.* 1. *f.* 1.)

49. —— Genus 55, b. BOMBYX[a].

53. 6015; Porthesia auriflua*.—Arctia phæorrhœa. (*Loudon M.* (*Bree.*) ii. 66. *fig.* 14.)

54. —— EUTHEMONIA.—CHELONIA *p*, *Godart.*

— —— ARCTIA.—CHELONIA *p*, *Godart.*

55. —— NEMEOPHILA.—CHELONIA *p*, *Godart.*

60. —— EULEPIA.—EMYDIA, *Boisduval.*

— —— DEIOPEIA.—EUCHELIA, *Boisduval.*

62. 6061, Setina irrorella*. *Steph. Ill.* (*H.*) ii. 99. *pl.* 17. *f.* 1.

— 6062, Se. —— roscida*. *Steph. Ill.* (*H.*) ii. 100.

— 6063; Se. —— eborina*. *Steph. Ill.* (*H.*) ii. 100.

63. 6066. Dele *Albin. pl.* lxxii. *f. a–d.*

— 6067. Adde *Albin. pl.* lxxii. *f. a–d.*

64. 6072. LYTÆA *mihi.*

— 6072*. Ly. —— leucographa. (*Steph. Nom.* 40.)—*Steph. Ill.* (*H.*) ii. 199.—Noctua leucographa. (*Wien. V.* 83?)—*Hüb. No. pl.* 88. *f.* 411.

— 6072**. Ly. —— albimacula. (*Steph. Nom.* 40.)—Graphiphora albimacula. *Supra. No.* 6125.—(*Steph. Ill.* (*H.*) ii. 137. *pl.* 19. *f.* 3.)

64. 6078: Charæas nigra*. *Steph. Ill.* (*H.*) ii. 110. *pl.* 20. *f.* 2. (nec *f.* 1.)—Ch. Æthiops. *Supra. No.* 6078.

65. 6081, Agrotis lunigera. *Steph. Ill.* (*H.*) ii. 113. *pl.* 20. *f.* 3. (nec *f.* 2.)

66. 6089; Ag. —— radiola*. *Steph. Ill.* (*H.*) ii. 119. *pl.* 20. *f.* 1. (nec *f.* 3.)

67. 6096; Ag. —— cuneigera* *nec* valligera. Vide *Steph. Ill.* (*H.*) ii. 123.

68. 6108, Graphiphora subrosea. *Steph. Ill.* (*H.*) ii. 200. *pl.* 19. *f.* 1. ♂.

70. 6125, Dele: Vide *supra. No.* 6072**.

74. —— CARADRINA § A. GRAMMESIA *mihi.*

— 6152; Grammesia trilinea*. *Steph. Ill.* (*H.*) ii. 152.

— 6153; Gr. —— bilinea*. *Steph. Ill.* (*H.*) ii. 152.

78. 6182: Xylina conspicillaris β*. Vide *Steph. Ill.* (*H.*) ii. 169. —Dele Ph. No. perspicillaris. *Linné?* &c.

---

[a] Genus 55, b. BOMBYX *Auctorum.*

49. 5998*. 1; Mori*.

Ph. Bo. Mori.—*Linn.* ii. 817.—*Stew.* ii. 142.—(*Kirby & Sp. I. E.* i. 334.)

The Silkworm M. *Albin. pl.* 12. *f.* 16.—*Harr. A. pl.* 13. *f. a–f.*—*Bingley* iii. 217.

Page. No.

79. —— CALOCAMPA.—XYLINA *p*, *Curtis*.

— 6184; Ca. —— exoleta*.—Ph. No. exoleta. *Don*. vi. 19. *pl*. 187. larva.—Xy. exoleta. *Curtis*. vi. *pl*. 256.

— 6185; Ca. vetusta*. Dele *Steph. Ill.* (*H.*) *pl*. 23. *f*. 1.—Adde *Don*. vi. 19. *pl*. 187. imago.

— 6189; Xylophasia rurea*. *Steph. Ill.* (*H.*) ii. 176.—No. rurea. *Fabr*. iii. *b*. 125. (!)—Ph. No. rurea. *Stew*. ii. 189. (!)—*Turt*. iii. 338. (!)—Ph. No. hepatica. *Linn.*?

80. 6192; Xy. —— combusta*. *Steph. Ill.* (*H.*) ii. 177.—Dele No. rurea. *Fabr.*, *Stew*, *et Turt*.

87. 6249: Celæna leucographa.—Apamea Haworthi. *Curtis*. vi. *pl*. 260?—Dele No. leucographa. (*Wien. V.*? *&* *Hüb.*?)

112. †6424, Catocala Elocata[a].

118. ‡6457*. Metra? monilis[b].

126. †6512. Grammatophora Vau-nigraria*. In Mus. D. *Hatchett*.

146. 6673; Eupithecia Centaureata*: The Lime Speck. *Harr. A. pl*. 19. *f*. *a–f*.

148. 6697, Dele †.

149. —— *Pro* HERCYNA *lege* ARTE.

172. †6881. Amphisa Walkerana. (*Loudon M.* i. 54. *fig*. 21.)

183. 7025; Pœcilochroma profundana*. (*Steph. Nom.* 47.)—Dele Pœ. Sparmanniana, et Py. Sparmanniana. *Fabr.*, *Gmel.*, *et Stew*.

209. 7357; Œcophora Schæfferella*.—Gly. Schæfferella. (*Ing. Inst*. 90.)

216. 7447, Dele †.

DIPTERA.

233. 7651; Anopheles bifurcatus*. Anapheles bifurcatus. (*Loudon M.* i. 54. *fig*. 22.)

236. 7723; Chironomus stercorarius*.—Tip. stercoraria. (*Kirby & Sp. I. E.* iii. 143.)

268. 8340, Medeterus notatus*. (*Ing. Inst.* 91.)

281. 8514; Baccha obscuripennis*. (*Ing. Inst.* 91.)

285. 8598; Scæva lunigera* *nec* luniger.

286. 8608; Sc. —— venusta* *nec* venustus.

295. †8720, Œstrus pictus. (*Ing. Inst.* 91.)

300. 8790; Leucostoma nervosa* *nec* venosa.

HOMALOPTERA.

327. †9302, Hæmabora pallipes. (*Ing. Inst.* 92.)

---

[a] 112 † 6424, Catocala Elocata. *Loudon M.* i. 272. *fig*. 136.

[b] 118 ‡ 6457*. Metra? monilis.
Noctua monilis. *Fab*. iii. b. 46. (!)—*Stew*. ii. 85. (!)—Phytometra monilis. *Haw*. 258..

INSECTA DUBIA.

1. Nitidula (Helophorus) fusca. *Turt.* ii. 112.
2. Ni. —— reticularia. *Turt.* ii. 111.
3. Ni. —— psyllia. *Marsh.* i. 138.
4. Silpha (Ips) 6-pustulata. *Turt.* ii. 104.
5. Dermestes ochropus. *Marsh.* i. 64.
6. Hister 4-guttatus. *Marsh.* i. 95.
7. Buprestis fuliginosa. *Forst. C.* i. 51.—*Stew.* ii. 75.—*Turt.* ii. 418.—An. Dasytes?
8. Bostrichus melanocephalus. *Turt.* ii. 88.
9. Curculio badensis. *Stew.* ii. 55.
10. Cu. —— 5-maculatus. *Berk. S.* i. 98.—*Stew.* ii. 56.—*Turt.* ii. 221.—An. Rh. Scanicus. *Gyll.* iii. 118?
11. Cu. —— Cerasi. *Stew.* ii. 56.
12. Cu. —— violaceus. *Turt.* ii. 236.—An. Magdalis?
13. Chrysomela aterrima. *Marsh.* i. 183.—An. Phalacrus?
14. Ch. —— chalcea. *Marsh.* i. 183.—An. Phalacrus?
15. Cryptocephalus glabratus. *Stew.* ii. 52.
16. Cr. —— 4-fasciatus. *Stew.* ii. 51.—*Turt.* ii. 186.
17. Hispa cornigera. *Stew.* ii. 52.
18. Crioceris nigrina. *Marsh.* i. 223.
19. *Lytta nitidula.* *Stew.* ii. 85.
20. Phalæna (Bombyx) Cerasi. *Stew.* ii. 145.
21. Ph. —— (Geometra) oblongata. *Stew.* ii. 165.
22. Ph. —— (Geometra) bidentata. *Stew.* ii. 194.
23. Ph. —— (Geometra) strigata. *Fabr.* iii. *b.* 171.—*Turt.* iii. 51.
24. Ph. —— (Tinea) Arbutella. *Fabr.* iii. *b.* 306.—*Turt.* iii. 370.
25. Ph. —— (Tinea) viridella. *Fabr.* iii. *b.* 312.—*Turt.* iii. 372.
26. Ph. —— (Tinea) tædella. *Stew.* ii. 200.
27. Ichneumon lineator. *Fabr.* ii. 168. (!)
28. Culex? ciliaris. *Stew.* ii. 269.
29. Tipula vernans. *Stew.* ii. 253.
30. Empis forcipata. *Stew.* ii. 270.
31. Musca cristata. *Stew.* ii. 261.
32. Mu. —— Hieracii. *Fabr.* iv. 201.—*Stew.* ii. 263.—*Turt.* iii. 622.
33. Mu. —— æstuans. *Turt.* iii. 620.
34. Mu. —— sericea. *Don.* xiii. 26. *pl.* 445. *f.* 2.
35. Mu. —— festiva. *Don.* xv. *pl.* 511.—An. Dolichopus?
36. Mu. seminationis. *Don.* iv. 45. *pl.* 125.

EX GENERUM.

FINIS.

LONDON:

PRINTED BY RICHARD TAYLOR,

RED LION COURT, FLEET-STREET.

Lightning Source UK Ltd.
Milton Keynes UK
UKHW050152241118
332685UK00031B/1942/P